R. Gupta's®

POPULAR MASTER GUIDE

SNAP

SYMBIOSIS NATIONAL APTITUDE TEST

For

Admission to MBA Programmes in Symbiosis Colleges

by

RPH Editorial Board

2027
EDITION

RAMESH PUBLISHING HOUSE, NEW DELHI

Published by
O.P. Gupta *for* Ramesh Publishing House

Admin. Office
12-H, New Daryaganj Road, Opp. Officers' Mess,
New Delhi-110002 ✆ 23275224, 23245124

E-mail: info@rameshpublishinghouse.com
For Online Shopping: www.rameshpublishinghouse.com

Showroom
- Balaji Market, Nai Sarak, Delhi-110006 ✆ 23282525 📱 9354373464
- 4457, Nai Sarak, Delhi-110006

Book Code: R-1651

ISBN: 978-93-5012-437-6

Price: ₹ 720

Printed at: Deepak Offset, Delhi

CONTENTS

STUDY MATERIAL

Previous Years' Paper

Symbiosis National Aptitude Test (SNAP), 2025

(Exam held in December 2025)*

1. Daniel Wheeler, at the head of his own table, was a charming host, and his ________ entirely disappeared as the talk ran along on subjects grave or gay, but of no personal ________.

A. gloom, potent B. cheer, intent
C. melancholy, import D. joy, consequence

2. Select the most appropriate antonym of the given word.
PREPOSTEROUS

A. Absurd B. Fantastic
C. Insane D. Logical

3. Which of the following should fill the blank?
There is no such movie ________ you mention.

A. which B. that
C. as D. who

4. Choose the part of the sentence that contains a grammatical error. (Ignore punctuations)
(a) Even though every country has/(b) diverse sets of experience, I believe/(c) the global community can reach similar targets/(d) and common policies that deepen universal connection

A. (a) B. (b)
C. (c) D. (d)

5. Fill in the blanks with appropriate words given in the options:
China's ultra-rich are ______ to make high-profile philanthropic donations to shore ______ their reputations and avoid tax inspections, not least as China's equivalent of HMRC has powers to arrest suspected evaders and hold them incommunicado for months.

A. enthusiastic, in B. competing, with
C. working, off D. scrambling, up

6. Identify the figure of speech in the following sentence.
It was claimed to be the most hi-tech train the country has ever seen but it got derailed in its first run.

A. Hyperbole B. Irony
C. Metaphor D. Paradox

7. The word "best" has been used in different parts of speech given in column I, Match it to the sentences given in column II.

Column I	Column II
1. Noun	(*i*) You need to find out which program works **best** on your computer.
2. Adverb	(*ii*) At the wedding he'll be wearing his **best** suit.
3. Adjective	(*iii*) She can afford the **best** of everything.
4. Verb	(*iv*) My goal is to **best** the competition.

A. 1-(*ii*) 2-(*i*) 3-(*iii*) 4-(*iv*)
B. 1-(*iii*) 2-(*i*) 3-(*ii*) 4-(*iv*)
C. 1-(*iii*) 2-(*iv*) 3-(*ii*) 4-(*i*)
D. 1-(*i*) 2-(*iii*) 3-(*ii*) 4-(*iv*)

8. Select the most appropriate meaning of the given idiom/phrase.
Golden handshake

A. Shaking hands with golden gloves
B. Good amount of money on early retirement
C. Getting paid for doing nothing
D. Cracking a good deal.

9. Which of the following is a compound word?

A. Hopscotch B. Frightened
C. Acronym D. Trimester

10. In which of the following sentences has the phrasal verb been incorrectly used?

A. The end product turned out okay in spite of the many production issues
B. The judge issued an arrest warrant against the official who failed to turn up despite judicial summons
C. You need to turn in the assignment by Monday
D. I am going to turn myself in so that I can get up early. Good night!

1. C	**2.** D	**3.** C	**4.** D	**5.** D	**6.** B	**7.** B	**8.** B	**9.** A	**10.** D

* **Based on memory.**

11. Fill in the blanks:

At a certain season of our life we are __________ to consider every spot as the possible site of a house. I have thus __________ the country on every side within a dozen miles of where I live.

A. accustomed, observed
B. habitual, surveyed
C. adapted, travelled
D. accustomed, surveyed

12. In the following question, a sentence has been given in Active/Passive Voice. Out of the four alternatives suggested, select the one which best expresses the same sentence in Passive/Active Voice.

The mechanic remodelled the bike to help it sell at a high price.

1. The bike has been remodelled by the mechanic to help it sell at a high price.
2. The bike is remodelled by the mechanic to help it sell at a high price.
3. The bike was sold at a high price by the mechanic after remodelling.
4. The bike was remodelled by the mechanic to help it sell at a high price.

A. 1 B. 2
C. 3 D. 4

13. Identify the correct parts of speech for the underlined words:

The <u>principal</u> of the school emphasized that the <u>principal</u> reason for the policy change was to improve student safety.

A. adjective, adverb B. noun, adverb
C. adjective, noun D. noun, adjective

14. Find the most appropriate word to be fit in the blank.

Last week, the central bank __________ all banks to provide a three-month moratorium on repayment of term loans

A. asked B. order
C. provided D. directed

15. Select the word with the correct spelling.

A. innosence B. indictment
C. disastrus D. embarass

16. Choose the missing number in the following grid.

8	24	12
7	?	24
9	45	32

A. 42 B. 36
C. 37 D. 32

17. Which of the following numbers will replace the question mark (?) in the following series?

17, 24, 32, 38, ?

A. 46 B. 62
C. 70 D. 68

18. Choose the option which should replace '?'.

B, C, E, G, K, M, ?

A. P B. O
C. Q D. S

19. Arrange the words given below in meaningful order

1. Write
2. Read
3. Choosing topic
4. Publish
5. Format

A. 3, 5, 4, 1, 2 B. 3, 5, 1, 4, 2
C. 3, 4, 1, 5, 2 D. 3, 1, 5, 4, 2

20. In the following question, four groups of three numbers are given. In each group the third number is related to the first and second numbers by a Logic/Rule/ Relation. Three are similar on basis of same Logic/ Rule/Relation.

Select the odd one out from the given alternatives.

A. (16,24,42) B. (34,81,61)
C. (14,23,25) D. (18,72,81)

21. 22 boys are standing in front of Kunal in a Queue 12 boys are standing to the back of Rohit in the same queue. If total number of boys is 30 then the number of boys standing in between Kunal and Rohit is

A. 3 B. 4
C. 6 D. 5

22. Santa has an old computer with a defective keyboard. When he types A, the letters ABCA appear on the screen. When he types B, the letters BACB appear on the screen. When he types C, the letters ABCB appear on the screen. He typed the following word having 48 alphabets:

ABCCBAABCCBA....CBA

What will be the 100th alphabet from the left in the word that appears on the screen?

A. A B. B
C. C D. Cannot be determined

23. Pointing to a photograph, Ram said, "He is the only son of only daughter of my grandfather." How is Ram related to the man in the photograph?

A. Brother B. Cousin
C. Brother-in-law D. Can't be determined

11. D	12. D	13. D	14. D	15. B	16. A	17. B
18. C	19. D	20. B	21. B	22. A	23. D	

24. Sheena said that Kawan's grandson is the only son of her sister. How is Kawan related to Sheena if Kawan has a wife?

A. Father
B. Father-in-law
C. Uncle
D. Cannot be determined

25. If 1st Jan 2000 was a Saturday, on which of the following dates does the first Monday fall in August 2005?

A. 1st B. 4th
C. 3rd D. 6th

26. At what time between 7 and 8 o'clock will the hands of a clock be in the same straight line but not together?

A. $5\frac{5}{11}$ past 7 B. $4\frac{5}{11}$ past 7
C. $5\frac{4}{11}$ past 7 D. $5\frac{3}{11}$ past 7

27. If it was Sunday on 15 June 2007, what is the day of the week on 14 July 2009?

A. Tuesday
B. Friday
C. Monday
D. Thursday

28. In the question, three statements are given, followed by two conclusions, I and II. You have to consider the statements to be true even if it seems to be at variance from commonly known facts. You have to decide which of the given conclusions, if any, follows from the given statements.

Statement 1: All phones are desktops.
Statement 2: All computers are televisions.
Statement 3: No desktop is a computer.

Conclusion I: Some televisions can be phones.
Conclusion II: No computer is a phone.

A. Only Conclusion I follows
B. Only Conclusion II follows
C. No Conclusion follows
D. Both Conclusion I and Conclusion II follows

29. Each of three friends knows whether the other two have passed or failed in an examination, but does not know his own result. The teacher comes and says, "At least one has failed". If none of the 3 persons knew their own results even after the teacher's statement, then which of the following is true?

A. One student has failed.
B. Two students have failed.
C. Two or more students have failed.
D. All three have failed.

30. Brazil wants to avoid the collapse of its economy. They can do this by increasing their GDP by 40% of its present value. Once the GDP increases by 40%, it is not dicult to increase the GDP by 50%.

If the above statements are true, then which of the following statements are also true?

A. Brazil's economy will fail if it is not able to increase its GDP by 50%
B. Brazil's GDP will have a 50% increase if its economy fails
C. Brazil's economy will not fail if it can increase its economy by 50%
D. The 10% increase in GDP is not possible even though Brazil increases its GDP by 40%

31. Ravindra Jadeja is obviously a bad batsman. During the last Australia tour earlier this year, he was the least scorer, in comparison to all other batsmen in the Indian team.

Which of the following, if true, does most to weaken the argument given above?

A. Even though Jadeja scored the least number of runs he was exceptional in bowling and fielding.
B. Jadeja started his career as a spin bowler and then he progressed to become an all-rounder.
C. During the Australia tour, at the order of the captain, Jadeja played with an experimental heavy bat.
D. In the Australia tour, two years back, Jadeja was one of the top run getters of the Indian team.

32. A foreign national entered a prohibited island and was killed by the indigenous tribe residing in the island. The tribe has not had any contact with the outside world for thousands of years. The host country is under pressure from the foreign nation to arrest the tribal people who killed the person. Which of the following is the best course of action that the country should take?

A. Issue a public apology for the mishap and ensure that no such incidents happen in the future.
B. Arrest the tribal people who killed the person since doing so will help to mitigate the stress.
C. Explain that it was the foreign national who was at fault for trespassing into a prohibited area.
D. Conduct awareness camps for the tribal people so that they do not kill any other person in the future.

24. A	**25.** A	**26.** A	**27.** D	**28.** D	**29.** C	**30.** C	**31.** D	**32.** C

33. Every year, many people become sick because of tick infestation in their homes. After someone becomes sick, experts are often hired to eradicate the ticks. These experts look in damp areas of the house since ticks are almost always found in places where there is substantial moisture. If one wishes to avoid tick infestation, then, one should make sure to keep the plumbing of the house in good condition to prevent leakage that could serve as a breeding ground for ticks. Which of the following options is an assumption on which the argument depends?

A. Ticks cannot be filtered from the air
B. House-owners have a good knowledge about plumbing to gauge the health of the pipes.
C. Ticks cannot grow in dry areas.
D. Ticks themselves do not cause moisture

34. A shoe company produces 200 shoes per week. The manager wants to ensure that quality standards remain constant from week to week. The manager, therefore, claims that out of 200 shoes produced per week, 50 shoes will be rejected.

The main flaw in the manager's plan is that the plan assumes that:

A. Shoes manufacturers cannot accept all shoes that are produced
B. The overall quality of the shoes would not be improved if the total number of shoes produced were reduced
C. Each shoe that is reviewed is worthy of being reviewed
D. The 150 shoes that are accepted will be of the same quality from week to week

35. *Read the following passage and answer the questions based on it.*

With reduced human activities and improvement of water quality of the Beas during Covid-19 lockdown, Indus river dolphins — one of the world's rarest mammals — are being sighted more regularly at Beas Conservation Reserve and have also started venturing out to other areas, as far as about 50 km away from its hotspots in the Beas. Two of the four freshwater dolphin species in the world are found in India. These are the Ganges and Indus river dolphins. Beas Conservation Reserve in Punjab is home to the only population of Indus river dolphins in the country.

Which of the following can be concluded from the above passage?

A. If there will be no human movement then there will be more chances of improving water quality.
B. If water quality is good then there are more chances of frequent spotting of fishes or mammals in the river.
C. Dolphins are the rarest among mammals and they can be seen more frequently if the water is clean.
D. All of the above

36. Given below in a statement followed by some conclusions. You have to take the given statement to be true even if it seems to be at variance with the commonly known facts and then decide which of the given conclusions logically follow(s) from the given statement.

Statement: Children without proper care and nourishment are likely to be a victim of malnutrition.

Conclusions:

I. The number of child deaths due to malnutrition has been on rise since 2000.
II. All children who are suffering from malnutrition weren't taken care of well in their formative years.

A. Only conclusion I follows
B. Only conclusion II follows
C. Both I and II follow
D. Neither of them follows

37. **Statements:** African countries will face a rise in population and at the same time depletion in the resources in the coming decade.

Conclusions:

I. Developing countries are facing a similar problem today.
II. Governments of African countries will have problems in providing excellent quality of life to its citizens.

A. Only conclusion I follows
B. Only conclusion II follows
C. Neither conclusion I nor II follow
D. Both the conclusions follow

38. The average weight of 10 students in a class is 30 kg. If the weight of the teacher A is included, the average becomes 33 kg. If the weight of teacher B is also included, the average weight of the class becomes 34 kg. The weight of the teacher B is

A. 45 kg B. 63 kg
C. 70 kg D. 75 kg

39. If $\frac{3}{4}$th of A = $\frac{1}{3}$rd of B = 30% of C, then find the ratio between A, B and C.

A. 4 : 8 : 9 B. 6 : 9 : 10
C. 4 : 9 : 10 D. 4 : 9 : 12

33. D	**34.** D	**35.** B	**36.** D	**37.** B	**38.** A	**39.** C

40. P sold a watch to Q at ₹ y. Q sold the same watch to R at 10% loss. R sold the same watch to T at 33.33% profit. If the watch was purchased by T at ₹ 1176, then find out what is $\frac{11}{7}$ times ₹ y.

A. ₹ 1420　　B. ₹ 1860
C. ₹ 1540　　D. ₹ 1280

41. Find out the smallest number which when individually divided by 16, 17, and 20 leave remainders 13, 14 and 17 respectively.

A. 1357　　B. 1460
C. 1268　　D. 1456

42. If $\sin\theta = \frac{4}{5}$, then find out the value of $\frac{\tan\theta+1}{\tan\theta-1}$?

(Assume that $0 \le \theta \le 90$)

A. 5　　B. 3
C. 1　　D. 7

43. Suman is standing on the bank of a river. A tree is standing on the opposite bank of the river. Suman observes the top of the tree at an angle of 60 degrees. If the width of the river is 40 m then what is the height of the tree?

A. $\frac{20}{\sqrt{3}}$　　B. $40\sqrt{3}$
C. $10\sqrt{7}$　　D. $20\sqrt{5}$

44. If the amount received at the end of 2nd and 3rd year at Compound Interest on a certain Principal is ₹ 2640 and ₹ 2772 respectively, what is the rate of interest?

A. 6%　　B. 7%
C. 5%　　D. 4%

45. In a university student election, five students, namely P, Q, R, S, and T, participated for the post of general secretary. The division of the votes is represented in a bar graph given below.

What is the ratio of the number of votes received by the two highest polling students to the number of votes received by the two lowest polling students?

A. 5:11　　B. 11:5
C. 5:2　　D. 2:5

46. If the value of $\log_n 75 = p$ and the value of $\log_n 45 = q$, then the value of $\log_n 15$ is ?

A. $\frac{p+q}{3}$　　B. $\frac{p+q}{2}$
C. $\frac{2p+2q}{3}$　　D. $\frac{3p+3q}{2}$

47. Find the value of log 2 + log 4 + log 8 + log 16 + ... + log 2^n, if log 2 = x

A. $\frac{n(n+1)x}{2}$　　B. $\frac{n(n+1)(2n+1)x}{6}$
C. nx　　D. $n(n+1)x$

48. A class has 60 students who appeared for two different tests English and Computers. It is known that 40 students passed in English while 45 students passed in Computers. It is known that 5 students failed in both the subjects. What is the number of students who passed in both the subjects?

A. 22　　B. 30
C. 24　　D. 25

49. If a string is chosen at random from all the possible permutations of the letters of the word 'COMPUTER' what is the probability that the vowel will appear together?

A. $\frac{3}{168}$　　B. $\frac{5}{84}$
C. $\frac{3}{84}$　　D. $\frac{3}{28}$

50. In a 400 metre race, A beats B by 40 metres and C by 80 metres. If B and C run a 200 metre race at the same speeds as before, by how much does B beat C?

A. 20 metres　　B. 22.22 metres
C. 18.88 metres　　D. 23.33 metres

51. Three friends A, B and C can complete a project in n days. Together, they take 1/5th of the time that A alone would have taken to complete the project or 2/3rd of the time B alone would have taken to complete the project. Find the time that C alone would take to complete the project.

A. $10n$ days　　B. $\frac{5n}{2}$ days
C. $8n$ days　　D. $\frac{15n}{2}$ days

40. C	**41.** A	**42.** D	**43.** B	**44.** C	**45.** B
46. A	**47.** A	**48.** B	**49.** D	**50.** B	**51.** D

52. What is the sum to infinity of the below sum?

$$\frac{6}{2\times3\times4}+\frac{7}{3\times4\times5}+\frac{8}{4\times5\times6}+...$$

A. $\frac{1}{2}$ B. $\frac{4}{5}$
C. $\frac{2}{3}$ D. $\frac{3}{4}$

53. In a factory, 40% of the bulbs produced are blue in colour, 20% of the bulbs are green in colour and the rest are red in colour. 30% of the red bulbs are defective while the total percentage of defective bulbs is 24%. 80% of the blue bulbs are not defective. If the number of defective blue bulbs is 120, what is the number of non-defective green bulbs?

A. 60 B. 120
C. 240 D. 360

54. A shopkeeper bought 23 pens, 60 pencils and 28 erasers and paid his dealer ₹ 2530. The next day, he bought 17 pens, 44 pencils and 21 erasers and paid his dealer ₹ 1890. One day, he bought only 4 pens and 16 pencils. How much did he pay the dealer that day?

A. 80
B. 120
C. 160
D. Cannot be determined

55. The given system of linear equations has no solution:

$9ax + 2y = 3$

$4x + (2 - a)y = 2$

Then which of the following is true?

A. $3a - 4 = 0$ B. $3a - 2 = 0$
C. $3a + 2 = 0$ D. $3a + 4 = 0$

56. What is the probability that a three digit palindrome number is divisible by 9?

A. $\frac{1}{10}$ B. $\frac{1}{9}$
C. $\frac{1}{8}$ D. $\frac{1}{7}$

57. 2 dice are rolled simultaneously, what is the probability that the sum of the outcomes is a prime number and also, at least one of the 2 individual outcomes is a prime number?

A. $\frac{5}{18}$ B. $\frac{1}{3}$
C. $\frac{11}{36}$ D. $\frac{1}{4}$

58. In the following question, select the related letters from the given alternatives.

Praise : Criticism : : Urge : ?

A. Force B. Request
C. Deter D. Increase

Instructions (Qs. No. 59 and 60): *6 friends A, B, C, D, E and F are living in 6 consecutive floors numbered from 1 to 6, 1 being the lowermost floor and 6 being the topmost floor. A and E live on even numbered floors while C lives on a floor above both of them. D lives on a floor above C. F does not live on the bottom most floor.*

59. Who lives on floor number 3?

A. B B. C
C. D D. F

60. Who lives on floor number 5?

A. D B. C
C. F D. B

EXPLANATORY ANSWERS

1. "Melancholy" means sadness, gloom, or a depressed state of mind. The sentence says that Daniel Wheeler was a charming host and that this feeling completely disappeared as the conversation continued on various subjects. The phrase "of no personal import" is a standard English expression meaning "of no personal importance or significance." Thus, both words fit the meaning and tone of the sentence perfectly. The other options either create an unnatural meaning or do not suit the context as accurately.

2. The word "preposterous" refers to something absurd, ridiculous, unreasonable, or completely lacking common sense. Therefore, its antonym should describe something sensible, rational, and based on sound reasoning. "Logical" exactly conveys this opposite meaning because a logical idea follows reason and makes sense. The options "Absurd" and "Insane" are close in meaning to preposterous, while "Fantastic" generally means extraordinary or imaginative and is not its opposite. Hence, "Logical" is the correct antonym.

3. The sentence requires the idiomatic structure "no such + noun + as." Therefore, "There is no such movie as you mention" is the grammatically correct expression. Here, "as" acts as a relative word referring to the kind of movie being mentioned. The sentence means that no movie of the type or description mentioned by the speaker exists. The alternatives "which," "that," and "who" do not fit this established grammatical pattern, making "as" the only correct choice.

52. C	**53.** C	**54.** B	**55.** A	**56.** B	**57.** A	**58.** C	**59.** D	**60.** B

4. The error lies in part (d). The phrase "reach similar targets and common policies" is grammatically incorrect because the verb "reach" can appropriately take "targets" as its object, but it cannot logically take "policies" as an object in the same structure. One can reach targets, but one generally adopts, develops, or formulates policies. Thus, the sentence lacks parallelism and proper verb usage in this part. The other parts are grammatically acceptable in the given context.

5. The expression "scrambling to" means making hurried or desperate efforts to achieve something, which fits the context of ultra-rich individuals trying to protect their reputations. The phrasal verb "shore up" means to strengthen, support, or improve something that is weak or threatened. Therefore, "scrambling to make high-profile philanthropic donations to shore up their reputations" conveys the intended meaning perfectly. The remaining options either do not form correct idiomatic expressions or fail to fit the context logically.

6. Irony occurs when there is a contrast between what is expected and what actually happens. In the sentence, the train was described as the most advanced and high-tech train the country had ever seen, creating an expectation of superior performance and reliability. However, it derailed on its very first run, which is the exact opposite of what one would expect from such a technologically advanced train. This sharp contrast between expectation and reality makes the figure of speech irony.

7. **1. Noun → *(iii)* She can afford the best of everything:** Here, "best" functions as a noun because it refers to the finest or highest-quality things. It is the object of the preposition "of" and names a thing rather than describing one.

2. Adverb → *(i)* You need to find out which program works best on your computer: In this sentence, "best" modifies the verb "works" and tells how well the program functions. Therefore, it is used as an adverb.

3. Adjective → *(ii)* At the wedding he'll be wearing his best suit: Here, "best" describes the noun "suit" by indicating its quality. Since it modifies a noun, it functions as an adjective.

4. Verb → *(iv)* My goal is to best the competition: In this sentence, "best" means "to defeat" or "to outperform." It expresses an action and is therefore used as a verb.

Thus, the correct matching is 1-(*iii*), 2-(*i*), 3-(*ii*), 4-(*iv*), making Option B the correct answer.

8. A "golden handshake" refers to a large financial payment or attractive compensation package given to an employee when they retire early, resign, or leave an organization. Such payments are often offered to senior employees as an incentive for voluntary retirement or termination of service. The expression has nothing to do with literally shaking hands, being paid for doing nothing, or simply making a profitable deal. Therefore, option B correctly conveys the meaning of the idiom.

9. A compound word is formed by combining two separate words to create a new word with its own meaning. "Hopscotch" is made up of the words "hop" and "scotch," making it a true compound word. "Frightened" is a derived word formed by adding a suffix, "Acronym" is a single lexical word referring to an abbreviation, and "Trimester" is formed through prefixation rather than the joining of two independent words. Hence, "Hopscotch" is the correct answer.

10. The phrasal verb "turn oneself in" means to surrender oneself to the police or authorities after committing or being accused of a crime. In the given sentence, it is incorrectly used to suggest going to bed early so as to wake up early, which is not its meaning. In the other sentences, "turned out" means resulted, "turn up" means appear or arrive, and "turn in" means submit, all of which are correctly used. Therefore, option D contains the incorrect use of a phrasal verb.

11. The phrase "accustomed to" is a standard English expression meaning "used to" or "inclined to." Thus, "we are accustomed to consider every spot as the possible site of a house" fits the context naturally. In the second blank, "surveyed" means examined or inspected carefully, which matches the idea of looking over the surrounding countryside in search of possible house sites. "Observed" is too limited in meaning, while the other options do not fit grammatically or contextually. Therefore, "accustomed, surveyed" is the most appropriate pair.

12. The given sentence is in the active voice and uses the simple past tense: "The mechanic remodelled the bike." To convert it into the passive voice, the object "the bike" becomes the subject, and the verb changes to "was remodelled." The agent is introduced with "by the mechanic," while the purpose clause remains unchanged. Thus, the correct passive form is "The bike was remodelled by the mechanic to help it sell at a high price." Hence, option 4 is the correct answer.

13. In the phrase "The principal of the school," the word "principal" refers to the head of the school and functions as a noun. In the phrase "the principal reason," the word "principal" means main, chief, or most important and describes the noun "reason." Since it modifies a noun, it functions as an adjective in the second instance. Therefore, the correct identification of the parts of speech is noun followed by adjective, making option D the correct answer.

14. The sentence refers to an official instruction issued by the central bank to all banks regarding a three-month moratorium on loan repayments. The verb "directed" is most appropriate because it means officially ordered or instructed an authority or institution to take a particular action. "Asked" is too weak for a regulatory directive, "order" is grammatically incorrect in the sentence as given, and "provided" does not fit the meaning. Therefore, "directed" is the correct choice.

15. "Indictment" is the correctly spelled word and refers to a formal accusation or charge of a serious crime. The other options contain spelling errors: "innosense" should be "innocence," "disastrus" should be "disastrous," and "embarass" should be "embarrass." Since only "indictment" is correctly spelled, option B is the correct answer.

16. $8 \times (1 + 2) = 8 \times 3 = 24$

$7 \times (2 + 4) = 7 \times 6 = 42$

$9 \times (3 + 2) = 9 \times 5 = 45$

So, the missing number is 42.

17. 17 + (1 × 7) = 17 + 7 = 24
24 + (2 × 4) = 24 + 8 = 32
32 + (3 × 2) = 32 + 6 = 38
38 + (3 × 8) = 38 + 24 = 62
So, the missing number is 62.

18. B = 2, C = 3, E = 5, G = 7, K = 11, M = 13.
These are letters placed at prime-number positions in the alphabet.
The next prime number after 13 is 17.
The 17th letter of the alphabet is Q, so the missing letter is Q.

19. First, a topic is chosen before any writing starts.
After choosing the topic, the content is written.
Then the written matter is formatted properly.
After formatting, it is published, and finally it is read by people.
So, the meaningful order is Choosing topic → Write → Format → Publish → Read.

20. (16,24): (1 + 6) × (2 + 4) = 7 × 6 = 42
(14,23): (1 + 4) × (2 + 3) = 5 × 5 = 25
(18,72): (1 + 8) × (7 + 2) = 9 × 9 = 81
But in (34,81), (3 + 4) × (8 + 1) = 7 × 9 = 63, not 61.
So, (34,81,61) is the odd one out.

21. 22 boys are standing in front of Kunal, so Kunal's position from the front = 22 + 1 = 23rd.
12 boys are standing behind Rohit, and total boys are 30, so Rohit's position from the front = 30 − 12 = 18th.
The boys between Rohit and Kunal are at positions 19, 20, 21, and 22.
Therefore, the number of boys standing between them is 4.

22. Each typed alphabet produces 4 alphabets on the screen.
So, the 100th screen alphabet will come from the 100 ÷ 4 = 25th typed alphabet.
The typed word follows the repeating pattern "ABCCBA," which has 6 letters.
25 ÷ 6 leaves remainder 1, so the 25th typed alphabet is A.
When A is typed, "ABCA" appears, and its 4th alphabet is A.
Therefore, the 100th alphabet from the left is A.

23. "The only daughter of my grandfather" may refer to Ram's mother or Ram's aunt, depending on whether the grandfather is maternal or paternal.
If she is Ram's mother, then her only son would be Ram himself.
If she is Ram's aunt, then her only son would be Ram's cousin.
Since "himself" is not given in the options and the exact side of the grandfather is not specified, the relationship cannot be determined with certainty.
Therefore, the best answer is Can't be determined.

24. Kawan's grandson is the only son of Sheena's sister.
This means Sheena's sister's son is Kawan's grandson.
So, Sheena's sister is Kawan's daughter, and since Sheena is her sister, Sheena is also Kawan's daughter.
The statement also says Kawan has a wife, confirming Kawan is male.
Therefore, Kawan is Sheena's father.

25. 1st January 2000 = Saturday.
Counting the years and leap-year adjustment, 1st January 2005 falls on Saturday.
In 2005, August 1 falls on Monday.
Therefore, the first Monday in August 2005 is on 1st August.

26. For the hands of a clock to be in the same straight line but not together, they must be diametrically opposite to each other. This means the angle between the hour hand and the minute hand must be exactly 180°. Since the time lies between 7 and 8 o'clock, let the required time be m minutes past 7.
The hour hand at 7 o'clock is at:
$$7 \times 30 = 210°$$
After m minutes, the hour hand moves:
$$\frac{m°}{2}$$
So, the position of the hour hand is:
$$210 + \frac{m}{2}$$
The minute hand moves 6° per minute, so after m minutes its position is:
$$6\,m$$
Since the hands are in a straight line but not together:
$$\left|\left(210 + \frac{m}{2}\right) - 6\,m\right| = 180$$
Taking the valid case for a time between 7 and 8:
$$210 + \frac{m}{2} - 6\,m = 180$$
$$210 - \frac{11m}{2} = 180$$
$$\frac{11m}{2} = 30$$
$$11\,m = 60$$
$$m = \frac{60}{11} = 5\frac{5}{11}$$
Therefore, the hands of the clock will be in the same straight line but opposite each other at $5\frac{5}{11}$ minutes past 7. This satisfies the condition that they are on a straight line while not coinciding.
Hence, Option A is the correct answer.

27. 15 June 2007 = Sunday.
15 June 2007 to 15 June 2008 = 366 days = 2 odd days, so 15 June 2008 = Tuesday.
15 June 2008 to 15 June 2009 = 365 days = 1 odd day, so 15 June 2009 = Wednesday.
15 June 2009 to 14 July 2009 = 29 days = 1 odd day.
Therefore, 14 July 2009 = Thursday.

28. Statement 1 says all phones are desktops, so every phone must lie inside the desktop group. Statement 3 says no desktop is a computer, which means desktops and computers are completely separate groups. Since all phones are desktops, phones also cannot be computers; therefore, "No computer is a phone" definitely follows. Statement 2 says all computers are televisions, so computers are inside televisions, but it does not say that all televisions are computers. Hence, televisions may still include things other than computers, and phones can possibly be among those televisions. So, "Some televisions can be phones" is a possibility conclusion, while "No computer is a phone" is a definite conclusion. Therefore, both conclusions follow.

29. If only one student had failed, that failed student would see that the other two had passed. After the teacher said "At least one has failed," he would immediately know that he himself had failed. But the question says none of the three knew their own result even after the statement. Therefore, exactly one failure is impossible. This means there must be two or more students who failed.

30. The statement says that Brazil can avoid economic collapse by increasing its GDP by 40%. Therefore, if Brazil manages to increase its GDP by 50%, it has already exceeded the required 40% increase needed to prevent collapse. Hence, a 50% increase guarantees that the economy will not fail. The other options either assume facts not given in the statements or contradict the information provided. Therefore, Option C is the correct answer.

31. The argument concludes that Jadeja is a bad batsman based only on his poor performance in one particular Australia tour. Option D weakens this conclusion by showing that on an earlier tour to the same country, he was among the highest run-scorers for India. This demonstrates that a single poor series is insufficient to judge his overall batting ability. Therefore, the statement directly challenges the argument's generalization and is the strongest weakening option.

32. The island is a prohibited area, and the foreign national entered it despite the restrictions. The tribe has remained isolated for thousands of years and cannot reasonably be expected to understand or follow modern legal systems. Arresting them or conducting awareness campaigns would violate their protected isolation and could create further problems. Therefore, the most appropriate response is to explain that the incident occurred because the foreign national trespassed into a restricted area, making Option C the best course of action.

33. The argument states that ticks are generally found in damp areas and concludes that preventing plumbing leaks will help avoid tick infestation. For this conclusion to be valid, it must be assumed that moisture is a cause of tick infestation rather than a result of it. If ticks themselves could create moisture, then even a house with perfectly maintained plumbing could develop damp conditions after ticks arrived, making leak prevention ineffective. Thus, the argument depends on the assumption that ticks do not cause moisture. Therefore, Option D is the correct answer.

34. The manager wants to maintain constant quality standards and therefore plans to reject exactly 50 out of 200 shoes every week. The flaw is that rejecting a fixed number of shoes does not guarantee that the quality of the accepted shoes will remain constant. Some weeks the production quality may be high, while in others it may be low. The plan assumes that accepting 150 shoes each week will automatically result in the same quality level from week to week, which need not be true. Therefore, Option D identifies the main flaw in the manager's reasoning.

35. The passage states that during the Covid-19 lockdown, reduced human activities led to improved water quality in the Beas River, and as a result, Indus river dolphins were sighted more frequently and even moved beyond their usual habitats. This supports the conclusion that better water quality increases the likelihood of spotting aquatic animals such as mammals and possibly other river species. Option A introduces a broader cause-and-effect relationship not directly established by the passage, while Option C incorrectly states that dolphins are the rarest among all mammals and overgeneralizes the conclusion. Therefore, Option B is the most appropriate conclusion based on the information given in the passage.

36. The statement states that children who do not receive proper care and nourishment are likely to become victims of malnutrition. This establishes a possible relationship between lack of care, poor nutrition, and malnutrition. However, Conclusion I talks about an increase in child deaths due to malnutrition since the year 2000, whereas the statement provides no information about death rates, trends, or any time period. Therefore, Conclusion I does not follow. Conclusion II states that all children suffering from malnutrition were not cared for well in their formative years. The statement does not say that poor care is the only cause of malnutrition; there may be other causes as well. Hence, Conclusion II is also not justified. Therefore, neither conclusion follows.

37. The statement says that African countries will experience a rise in population along with depletion of resources in the coming decade. A larger population combined with fewer available resources is likely to put pressure on governments and make it difficult to provide a high standard of living, healthcare, education, employment, and other essential services. Therefore, Conclusion II logically follows. However, Conclusion I refers to developing countries in general facing a similar problem today. The statement provides information only about African countries in the future and does not mention the present situation of other developing countries. Hence, Conclusion I does not follow. Therefore, only Conclusion II follows.

38. Average weight of 10 students = 30 kg

Total weight of 10 students = $10 \times 30 = 300$ kg

Including teacher A, total persons = 11 and average = 33 kg

Total weight = $11 \times 33 = 363$ kg

Including teacher B, total persons = 12 and average = 34 kg

Total weight = $12 \times 34 = 408$ kg

Weight of teacher B = $408 - 363 = 45$ kg

39. Given that $\frac{3}{4}$ of A, $\frac{1}{3}$ of B, and 30% of C are equal, let their common value be x.

$$\frac{3}{4}\text{A} = x \Rightarrow \text{A} = \frac{4x}{3}$$

$$\frac{1}{3}\text{B} = x \Rightarrow \text{B} = 3x$$

$$30\% \text{ of C} = x \Rightarrow \frac{3}{10}\text{C} = x \Rightarrow \text{C} = \frac{10x}{3}$$

Therefore,

$$\text{A : B : C} = \frac{4x}{3} : 3x : \frac{10x}{3}$$

Multiplying all terms by 3 to remove the fractions:

$$\text{A : B : C} = 4x : 9x : 10x$$

$$\text{A : B : C} = 4 : 9 : 10$$

Hence, the ratio between A, B and C is 4 : 9 : 10, so option C is the correct answer.

40. P sold the watch to Q at ₹ y.

Q sold it to R at 10% loss:

$$\text{R's CP} = 90\% \text{ of } y = \frac{9y}{10}$$

R sold it to T at $33.33\% = \frac{1}{3}$ profit:

$$\text{T's price} = \frac{9y}{10} \times \frac{4}{3}$$

$$1176 = \frac{6y}{5}$$

$$y = 980$$

Now, $\frac{11}{7} \times y = \frac{11}{7} \times 980 = 1540$

Therefore, the required value is ₹ 1540.

41. The number leaves remainders 13, 14, and 17 when divided by 16, 17, and 20 respectively.

Each remainder is 3 less than the divisor, so the required number + 3 must be exactly divisible by 16, 17, and 20.

LCM of 16, 17, and 20 = 1360.

Therefore, required number = 1360 – 3 = 1357.

42. Given, $\sin\theta = \frac{4}{5}$

Using the identity,

$$\sin^2\theta + \cos^2\theta = 1$$

$$\cos\theta = \frac{3}{5}$$

Therefore, $$\tan\theta = \frac{\sin\theta}{\cos\theta} = \frac{\frac{4}{5}}{\frac{3}{5}} = \frac{4}{3}$$

Now, $$\frac{\tan\theta + 1}{\tan\theta - 1} = \frac{\frac{4}{3}+1}{\frac{4}{3}-1} = \frac{\frac{4}{5}}{\frac{1}{3}} = 7$$

Hence, the required value is 7. Therefore, Option D is the correct answer.

43. The width of the river is 40 m and the angle of elevation of the top of the tree from Suman's position is 60°. Considering the river width as the base and the tree height as the perpendicular side of a right-angled triangle, we have:

$$\tan 60° = \frac{\text{Height of tree}}{\text{Width of river}}$$

$$\sqrt{3} = \frac{h}{40}$$

$$h = 40\sqrt{3}$$

Therefore, the height of the tree is $40\sqrt{3}$ m, making Option B the correct answer.

44. The amount at the end of the 2nd year is ₹ 2640 and at the end of the 3rd year is ₹ 2772. The increase from the 2nd year to the 3rd year represents one year's compound interest on ₹ 2640.

Interest for 1 year = 2772 – 2640 = 132

$$\text{Rate} = \frac{132}{2640} \times 100 = 5\%$$

Therefore, the rate of compound interest is 5%, so Option C is the correct answer.

45. From the bar graph:

P = 250 votes
Q = 150 votes
R = 100 votes
S = 300 votes
T = 200 votes

The two highest polling students are S and P.

300 + 250 = 550

The two lowest polling students are R and Q.

100 + 150 = 250

Therefore, the required ratio is:

550 : 250 = 11 : 5

Hence, the ratio of the votes received by the two highest polling students to the votes received by the two lowest polling students is 11 : 5. Therefore, Option B is the correct answer.

46. Given,

$$\log_n 75 = p, \log_n 45 = q$$

Adding both,

$$p + q = \log_n 75 + \log_n 45$$

$$p + q = \log_n (75 \times 45)$$

$$75 \times 45 = 3375 = 153$$

$$p + q = \log_n 15^3 = 3 \log_n 15$$

$$\log_n 15 = \frac{p+q}{3}$$

Therefore, Option A is correct.

47. Given,

$$\log 2 = x$$

$$\log 2 + \log 4 + \log 8 + ... + \log 2^n$$

$$= \log 2^1 + \log 2^2 + \log 2^3 + ... + \log 2^n$$

$$= x + 2x + 3x + ... + nx$$

$= x(1 + 2 + 3 + ... + n)$

$= \frac{n(n+1)x}{2}$

Therefore, Option A is correct.

48. Out of 60 students, 5 failed in both subjects. Therefore, the number of students who passed in at least one of the two subjects is:

$$60 - 5 = 55$$

Let the number of students who passed in both English and Computers be x. Using the formula for two sets:

$$n(E \cup C) = n(E) + n(C) - n(E \cap C)$$

Substituting the given values:

$$55 = 40 + 45 - x$$

$$55 = 85 - x$$

$$x = 30$$

Thus, 30 students passed in both English and Computers, making Option B the correct answer.

49. The word COMPUTER contains 8 distinct letters, of which O, U, and E are vowels. Therefore, the total number of arrangements of the letters is: 8!

To find the arrangements in which all the vowels appear together, treat the three vowels as a single unit. Then the units to be arranged are:

(C, M, P, T, R, OUE)

Thus, there are 6 units, which can be arranged in: 6! ways. The three vowels within the group can themselves be arranged in: 3! ways.

Hence, the number of favourable arrangements is:

$$6! \times 3!$$

Therefore, the required probability is:

$$\frac{6! \times 3!}{8!} = \frac{720 \times 6}{40320} = \frac{4320}{40320} = \frac{3}{28}$$

Thus, the probability that all the vowels appear together is $\frac{3}{28}$. Hence, Option D is the correct answer.

50. When A runs 400 metres, B runs 360 metres.

$$A : B = 400 : 360 = 10 : 9$$

Similarly, when A runs 400 metres, C runs 320 metres.

$$A : C = 400 : 320 = 5 : 4$$

Therefore,

$$B : C = \frac{9}{10} : \frac{4}{5} = \frac{9}{10} : \frac{8}{0} = 9 : 8$$

In a 200 metre race, when B completes 200 metres, C covers:

$$200 \times \frac{8}{9} = \frac{1600}{9} = 177.78 \text{ metres}$$

Hence, B beats C by:

$$200 - 177.78 = 22.22 \text{ metres}$$

Therefore, B beats C by 22.22 metres, so Option B is the correct answer.

51. Together A, B and C complete the project in n days. So, A alone takes $5n$ days.

Also, $n = \frac{2}{3}$ of B's time, so B alone takes $\frac{3n}{2}$ days.

$$\frac{1}{n} = \frac{1}{5n} + \frac{2}{3n} + \frac{1}{C}$$

$$\frac{1}{C} = \frac{1}{n} - \frac{13}{15n} = \frac{2}{15n}$$

$$C = \frac{15n}{2}$$

Therefore, Option D is correct.

52. The series is:

$$\frac{6}{2 \times 3 \times 4} + \frac{7}{3 \times 4 \times 5} + \frac{8}{4 \times 5 \times 6} + ...$$

General Term:

$$\frac{n+4}{n(n+1)(n+2)}$$

This can be written as:

$$\frac{2}{n} - \frac{3}{n+1} + \frac{1}{n+2}$$

On adding the terms, most middle terms cancel out. The remaining sum becomes:

$$1 - \frac{1}{3} = \frac{2}{3}$$

Therefore, Option C is correct.

53. The number of defective blue bulbs is 120, and 80% of the blue bulbs are non-defective. Therefore, 20% of the blue bulbs are defective. Since blue bulbs constitute 40% of the total production, defective blue bulbs represent 20% of 40%, i.e., 8% of the total bulbs.

8% of Total = 120

$$\text{Total bulbs} = \frac{120 \times 100}{8} = 1500$$

Green bulbs:

20% of 1500 = 300

Total defective bulbs:

24% of 1500 = 360

Red bulbs:

$$1500 - 600 - 300 = 600$$

Defective red bulbs:

30% of 600 = 180

Defective green bulbs:

$$360 - 120 - 180 = 60$$

Non-defective green bulbs:

$$300 - 60 = 240$$

Therefore, the number of non-defective green bulbs is 240, so Option C is the correct answer.

54. Let the cost of a pen, a pencil and an eraser be x, y and z respectively.

$$23x + 60y + 28z = 2530$$

$$17x + 44y + 21z = 1890$$

Multiply the first equation by 3 and the second equation by 4:

$$69x + 180y + 84z = 7590$$
$$68x + 176y + 84z = 7560$$

Subtracting the second equation from the first:

$$x + 4y = 30$$

The required amount for 4 pens and 16 pencils is:

$$4x + 16y = 4(x + 4y)$$
$$= 4 \times 30 = 120$$

Therefore, the amount paid to the dealer was ₹ 120, so Option B is the correct answer.

55. For a pair of linear equations to have no solution, the lines must be parallel. Therefore, the condition is:

$$\frac{a_1}{a_2} = \frac{b_1}{b_2} = \frac{c_1}{c_2}$$

Given equations:

$$9ax + 2y = 3$$
$$4x + (2 - a)y = 2$$

Comparing with $a_1x + b_1y + c_1 = 0$ and $a_2x + b_2y + c_2 = 0$:

$$a_1 = 9a$$
$$b_1 = 2$$
$$c_1 = -3$$
$$a_2 = 4$$
$$b_2 = 2 - a$$
$$c_2 = -2$$

Since there is no solution,

$$\frac{9a}{4} = \frac{2}{2-a}$$

$$9a(2 - a) = 8$$
$$18a - 9a^2 = 8$$
$$9a^2 - 18a + 8 = 0$$
$$(3a - 2)(3a - 4) = 0$$
$$a = \frac{2}{3}$$

or $$a = \frac{4}{3}$$

Now check the second condition:

For $a = \frac{2}{3}$,

$$\frac{9a}{4} = \frac{2}{2-a} = \frac{3}{2}$$

and $$\frac{-3}{-2} = \frac{3}{2}$$

This gives infinitely many solutions, not no solution.

For $a = \frac{4}{3}$,

$$\frac{9a}{4} = \frac{2}{2-a} = 3$$

but

$$\frac{-3}{-2} = \frac{3}{2}$$

The ratios are not all equal, so the system has no solution. Hence,

$$a = \frac{4}{3}$$

$$3a - 4 = 0$$

Therefore, Option A is the correct answer.

56. A three-digit palindrome is of the form ABA, where A can take values from 1 to 9 and B can take values from 0 to 9. Therefore, the total number of three-digit palindromes is:

$$9 \times 10 = 90$$

For a palindrome ABA to be divisible by 9, the sum of its digits must be a multiple of 9.

$$A + B + A = 2A + B$$

The possible multiples of 9 within the range of 2A + B are 9, 18, and 27.

For 2A + B = 9, the valid palindromes are:

171, 252, 333, 414

Total = 4.

For 2A + B = 18, the valid palindromes are:

585, 666, 747, 828, 909

Total = 5.

For 2A + B = 27, the valid palindrome is: 999

Total = 1.

Hence, the number of favourable palindromes is:

$$4 + 5 + 1 = 10$$

Therefore,

$$\text{Probability} = \frac{10}{90} = \frac{1}{9}$$

Thus, the probability that a three-digit palindrome is divisible by 9 is $\frac{1}{9}$. Therefore, Option B is the correct answer.

57. When two dice are rolled simultaneously, the total number of possible outcomes is:

$$6 \times 6 = 36$$

The sum of the outcomes must be a prime number, and at least one of the two numbers must itself be a prime number (2, 3, 5).

For sum 3:

(1, 2), (2, 1)

Favourable outcomes = 2.

For sum 5:

(2, 3), (3, 2)

Favourable outcomes = 2.

For sum 7:

(2, 5), (5, 2), (3, 4), (4, 3)

Favourable outcomes = 4.

For sum 11:

(5, 6), (6, 5)

Favourable outcomes = 2.

Thus, total favourable outcomes are:

$$2 + 2 + 4 + 2 = 10$$

Therefore,

$$\text{Probability} = \frac{10}{36} = \frac{5}{18}$$

Hence, the required probability is $\frac{5}{18}$, so Option A is the correct answer.

58. "Praise" and "Criticism" are opposites in meaning. Praise means expressing approval, while criticism means expressing disapproval. Similarly, "Urge" means to encourage or persuade someone to do something, whereas "Deter" means to discourage or prevent someone from doing it. Therefore, the relationship in both pairs is that of antonyms. Hence, Option C (Deter) is the correct answer.

59. A and E live on even-numbered floors, so they must occupy floors 2 and 4 in some order, because C has to live above both of them. Therefore, C must live on floor 5 and D must live above C, so D lives on floor 6. F does not live on the bottommost floor, so F cannot live on floor 1. The only remaining suitable floor for F is floor 3. Hence, Option D (F) is correct.

60. Since A and E occupy floors 2 and 4 in some order, C must live above both of them. D lives above C, so the only possible arrangement is C on floor 5 and D on floor 6. Therefore, the person living on floor number 5 is C. Hence, Option B (C) is correct.

Previous Years' Paper

Symbiosis National Aptitude Test (SNAP), 2024

(Exam held in December 2024)*

1. The following sentence has been split into four segments. Identify the segment that contains an inappropriately used preposition.

The hiker stumbled upon a beautiful waterfall / at the end of the trail / and marvelled on its majestic beauty / for several minutes.

A. The hiker stumbled upon a beautiful waterfall
B. at the end of the trail
C. and marvelled on its majestic beauty
D. for several minutes

2. Select the grammatically incorrect sentences

(A) The thing the Time Traveller held in his hand was a glittering metallic framework, scarcely larger than a small clock, and very delicately made.
(B) There was ivory in it, and some transparent crystalline substance.
(C) And now I must be explicit, for this that follows — unless his explanation is to be accepted — an absolutely unaccountable thing.
(D) Taking one of the small octagonal tables that were scattered about the room, and setting it in front of the fire, with two legs on the hearthrug.
(E) On this table he placed the mechanism.

A. AC B. BD
C. BE D. CD

3. Select the option that represents the given sentence in the passive voice:

The reforms could have affected a large number of marginal workers.

A. A large number of marginal workers were affected by the reforms.
B. A large number of marginal workers could have been affected by the reforms.
C. A large number of marginal workers could have affected the reforms.
D. A large number of marginal workers can be affected by the reforms.

Instructions (Qs. No. 4): *The sentence given with blanks is to be filled with an appropriate word(s). Four alternatives are suggested for the question. Choose the correct alternative and click the button corresponding to it.*

4. Pakistan has approached the United Nations after India ________ to revoke the Indus water treaty.

A. told B. call
C. threatened D. talked

Instructions (Qs. No. 5): *Choose the option that would fit both the given blanks.*

5. The jury failed to grasp the full _____ of her testimony.
You have to pay duties on goods that you _____ into the country.

A. extent B. meaning
C. import D. bring

6. Identify the figure of speech in the below-given sentence.

I am feeling awfully good today.

A. Oxymoron B. Synecdoche
C. Anti-Climax D. Euphemism

7. Select the antonym of
Diffidence

A. Timorousness B. Meekness
C. Timidity D. Confidence

8. Select the option which best captures the meaning of the compound word given below:
Thoroughfare

A. A festivity or exhibition
B. A valuable deal
C. A main road
D. A fight

9. Identify the correct phrasal verb from the list given below in the given sentence:

The team continued to slack off on their responsibilities, causing the manager to _________ on them.

A. cut into B. look up
C. go off D. pick out

1. C	2. D	3. B	4. C	5. C	6. A	7. D	8. C	9. C

* Based on memory.

10. Select the word with the correct spelling.

A. idiosyncrasy B. posession
C. wierd D. jist

11. Identify the origin/source of the given word below: Check

A. German B. Persian
C. Italian D. Polish

12. Which of the following statements are grammatically correct?

(A) Looking around the class, only a few hands were raised in response
(B) I was under the impression that the firm was being managed by the engineer from China
(C) Being the first born, she has always had a lot of responsibility on her shoulders

A. A only B. C only
C. A and B only D. B and C only

13. Choose the sentence in which the word 'fall' is incorrectly used.

FALL

A. The team fell apart as soon as differences started cropping up
B. The boy stumbled on the brick and fell down on the road
C. After the disastrous event, Sam had a big fall out with his family
D. His hair started falling off as soon as he turned fifty

14. The *MBA* (1) is hardly a prerequisite for success, but it (2) certainly *helps* (3), and it has been getting more *important* (4) in recent years. *Most* (5) MBA programs equip their graduates to understand *how* (6) to deal with **many** of the important questions that their organizations will need to *tackle* (7) over time, *and* (8) that they will face in their careers. The above italicized numbered words will be correctly represented by the following parts of speech:

A. 1-Noun, 2-Pronoun, 3-Adverb, 4-Adjective, 5-Adjective, 8-Conjunction
B. 1-Noun, 3-Adjective, 4-Adjective, 6-Adverb, 7-Adverb, 8-Article
C. 1-Noun, 2-Pronoun, 4-Adjective, 7-Adverb, 8-Conjunction
D. 2-Pronoun, 3-Verb, 4-Adjective, 7-Verb, 8-Conjunction

15. Four sentences are given below. These sentences, when rearranged in proper order, form a logical and meaningful paragraph. Rearrange the sentences and choose the correct order as the answer.

(A) One of the most prominent and romantic characters in the history of the Oriental world is Cyrus the Great.
(B) The territory over which Cyrus bore rule extended nearly three thousand miles from east to west, and fifteen hundred miles from north to south.
(C) He was the founder and organizer of the greatest empire the world has seen.
(D) Even with such a big empire, it is a matter of regret that so little is really known of the Persian hero, both in the matter of events and also of exact dates.

A. ACBD B. CABD
C. ABDC D. CBAD

Instructions (Qs. No. 16): *The following question has a series of numbers which follow a definite pattern except for one number. Identify the incorrect term of the series.*

16. 4, 18, 48, 100, 180, 295

A. 18 B. 295
C. 180 D. 48

Instructions (Qs. No. 17): *What should come in place of question mark(?) in the following number series.*

17. 5, 8, 10, 16, 20, 32, ?

A. 38 B. 36
C. 44 D. 40

18. In each of the following question select the related numbers/words/letters from the given options.

78 : 225 : : 89 : ?

A. 305 B. 289
C. 324 D. 296

Instructions (Qs. No. 19 and 20): *In each question below are given two or three statements followed by two conclusions numbered I and II. You have to take the given statements to be true even if they seem to be at variance with commonly known facts and then, decide which of the given conclusions logically follow(s) from the given statements, disregarding commonly known facts.*

19. Statements:

1. Some letters are pins.
2. Some keys are eyes.
3. All letters are keys.

Conclusions:

1. Some pins are not letters.
2. All pins are letters.

A. Only 1 follows B. Only 2 follows
C. Either 1 or 2 follows D. Both 1 and 2 follows

10. A	**11.** B	**12.** D	**13.** D	**14.** D	**15.** A	**16.** B	**17.** D	**18.** B	**19.** C

20. Statements:

Some pins are boards.

Some boards are white.

No pin is a box.

Conclusion I: Some boxes are not boards

Conclusion II: No pins are white

A. If only conclusion I follows

B. If only conclusion II follows

C. If either conclusion I or II follows

D. If neither conclusion I or II follows

21. Select the odd word/letters/number/number pair from the given alternatives.

A. 1 B. 243

C. 256 D. 32

Instructions (Qs. No. 22-24): *An organization decides to distribute seven types of laptops on week days starting from Monday to Sunday. The seven laptops to be distributed are of Sony, Acer, Asus, Dell, HP, Lenovo and HCL but not necessarily in the same order. Dell laptops are to be given on Saturday. Lenovo laptops are given just after Asus laptops and Asus laptops are not given on Wednesday. HP laptops are given three days after the distribution of Asus laptops. Acer laptops are distributed just after Sony laptop. HCL laptops are distributed before Acer laptops. HP laptops are distributed after Dell laptops.*

22. If HCL is related with Tuesday, Sony is related with Wednesday, similarly Lenovo is related with which day?

A. Thursday B. Friday

C. Saturday D. Sunday

23. On which day of the week Lenovo laptops are distributed?

A. Tuesday B. Wednesday

C. Friday D. Sunday

24. Which laptops are distributed just between the days when Lenovo and HP are distributed?

A. Asus B. Dell

C. Sony D. HCL

25. If 'DANGER' is written as 'RDEAGN' in a certain code, then how would 'PARALLEL' be written in that code?

A. LAPAELLR B. LPALEALR

C. LPEALRLA D. LPLEARAL

26. Rohit is the son of Rajesh's father's brother's. How is Rohit related to Rajesh?

A. Cousin B. Son-in-law

C. Nephew D. Grandfather

27. Pointing to a photograph, Mukesh said, 'This man's father is the father-in-law of my sister's husband's mother-in-law. How is the man in the photograph related to Mukesh?

A. Father B. Grandfather

C. Brother D. Uncle

28. At what time, between 4 and 5 o'clock, will the minute and hour hands be together?

A. 24 and 5/7 minutes past 4 o'clock.

B. 26 and 9/11 minutes past 4 o'clock.

C. 21 and 9/11 minutes past 4 o'clock.

D. 20 and 7/9 minutes past 4 o'clock.

29. If 1st July, 1977 was a Friday then 1st July 1970 was a

A. Wednesday B. Thursday

C. Sunday D. Tuesday

30. A watch gains 7 seconds in 4 minutes and is correctly set at 10:00 AM. What time will it show at 11:00 PM on the same day?

A. 11:21:05 PM B. 11:22:45 PM

C. 11:21:25 PM D. 11:21:55 PM

Instructions (Qs. No. 31 and 32): *In each of the questions given below which one of the five answer figures on the right should come after the problem figures on the left, if the sequence were continued.*

31. Problem figures

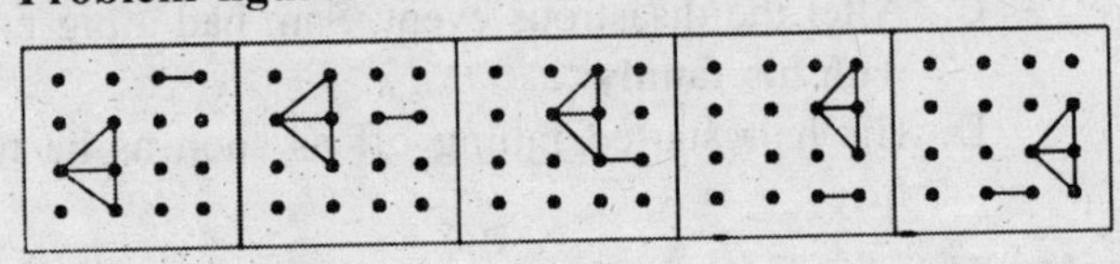

Answer figures

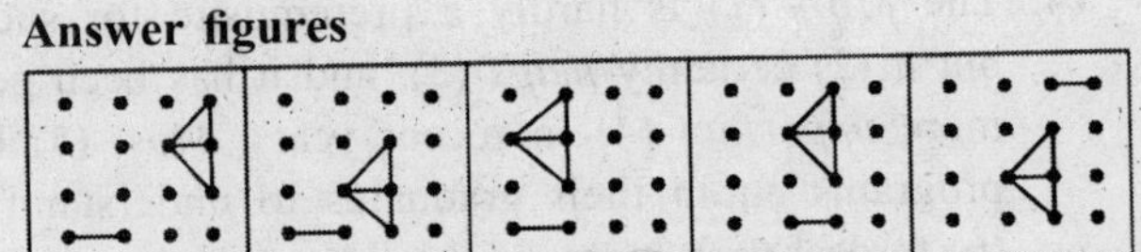

(1) (2) (3) (4) (5)

A. (1) B. (2)

C. (3) D. (4)

32. Problem figures

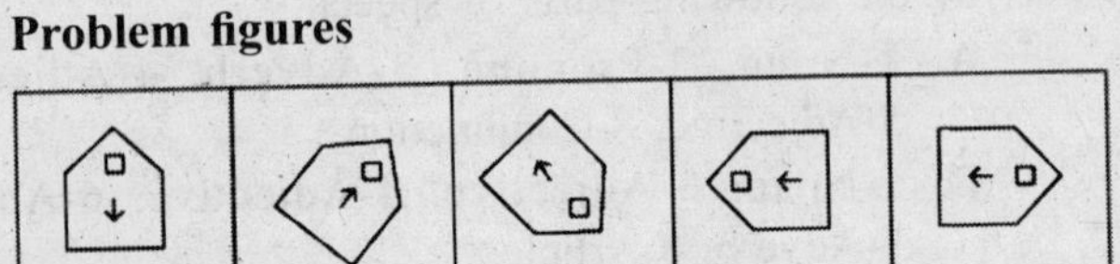

Answer figures

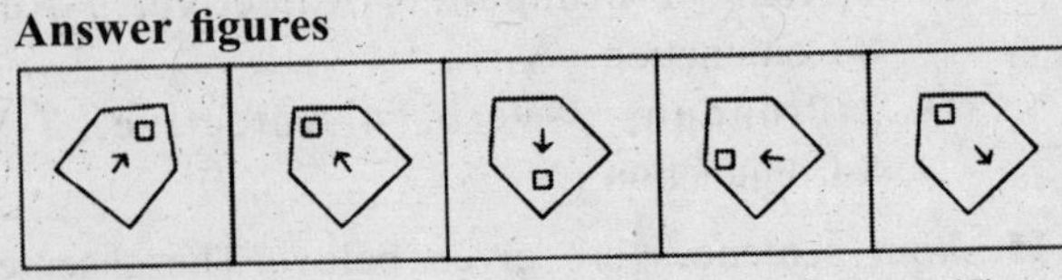

(1) (2) (3) (4) (5)

A. (1) B. (2)

C. (3) D. (4)

20. D **21.** C **22.** C **23.** C **24.** B **25.** C **26.** A **27.** A **28.** C **29.** A **30.** B **31.** B **32.** B

33. If \$ is an operator such that 2 \$ 5 = 32, what is the value of $\frac{4\$3}{2\$5}$?

A. 0 B. 1
C. 2 D. 3

Instructions (Qs. No. 34): *Study the following information carefully and answer the given question:*

In a certain code language
'Make your trip interesting' is written as 'de la pa ni'
'your trip be customized' is written as 'ha ni ta de'
'customized trip always good' is written as 'ta fa ka ni'
And 'good make me happy' is written as 'ka ro pa na'

34. Which of the following represents 'good trip always interesting'?

A. de ka ni la B. fa ni ka ro
C. na fa de ha D. ni fa la ka

Instructions (Qs. No. 35): *A statement is given followed by two assumptions numbered I and II. Consider the statement to be true even if it is at variance with commonly known facts. You have to decide which of the assumptions, if any, follow from the given statement.*

35. Statement: After the Reserve Bank of India's adoption of a flexible inflation-targeting framework from August 2016, the rules of monetary policy have changed in India, with the central bank becoming even more focused on anchoring inflation and inflation expectations than ever before.

Assumptions:

I. There was a perceived need or rationale for the RBI to shift its monetary policy framework in 2016, suggesting that the previous rules may have been inadequate.

II. Monetary policy can be adjusted and adapted to changing economic conditions and goals, such as inflation targeting.

A. Only I B. Only II
C. Both I and II D. Either I or II

36. Effect: At least twenty school children were seriously injured while going for a school picnic during the weekend.

Which of the following can be a probable cause of the above effect?

A. The teacher accompanying the school children fell ill during the journey.
B. The bus in which the children were travelling met with an accident while taking turn on the main highway.
C. The driver of the bus in which the children were travelling did not report after the break at the halting place on their journey.
D. The school authority banned all school picnics for the next six months with immediate effect

Instructions (Qs. No. 37): *In the given question below, a statement is given followed by two courses of action numbered I and II. On the basis of the information given, you have to assume everything in the statement to be true, and then decide which of the suggested courses of action logically follow(s) for pursuing.*

37. Statement : A large number of people from city X are admitted in hospitals and are diagnosed to be suffering from malaria.

Courses of action :

I. The municipal authority of the city X should take immediate steps to carry out extensive fumigation.
II. City X should be vacated immediately.

A. Only I B. Either I or II
C. Neither I nor II D. Both I and II

Instructions (Qs. No. 38): *In the question below, a statement is given followed by two arguments. Choose the most appropriate option.*

38. Statement: Companies serious about more harmonious relationships with the contractual workforce should offer them a pathway to up-skill and be employed as permanent workers over time.

Argument:

I. Building a good relationship with the contractual workforce will help both the companies and the employees to progress.
II. The policy of offering contractual workers permanent status is likely to face opposition from existing permanent staff.

A. I weakens the statement, whereas II strengthens it.
B. Both I and II weaken the statement.
C. I strengthens the statement, whereas II weakens it.
D. Both I and II strengthen the statement.

Instructions (Qs. No. 39): *In the following question, two statements numbered I and II are given. There may be cause and effect relationship between the two statements. These two statements may be the effect of the same cause or independent causes. These statements may be independent causes without having any relationship.*

39. Statements:

I. The country has seen no major breakthrough in medical research in the past few years.
II. The government, two years back, had banned patents and royalties on medicines.

A. Statement I is the cause and Statement II is its effect.
B. Statement II is the cause and Statement I is its effect.
C. Both the statements I and II are independent causes
D. Both the statements I and II are effects of independent causes.

33. C	**34.** D	**35.** C	**36.** B	**37.** A	**38.** C	**39.** B

Instructions (Qs. No. 40): *The question given below consists of some statements. You need to decide which of the statements is/are the cause of the situation or effect of the situation and accordingly choose your answer from the alternative given below each question.*

40. Statements:

I. A new covid variant named 'Omicron' is spreading rapidly in the country. Lots of new cases are being reported in a short period of time.

II. The central government is in a dilemma on whether to proceed with a lockdown or not.

A. Statement I and Statement II are effects of some common cause

B. Statement I is the cause and Statement II is its effect

C. Statement I and Statement II are independent causes

D. Statement II is the cause and Statement I is its effect

41. If N = $a*b$ where a, b are prime numbers, then what are the number of factors of $\frac{N^3}{b^2}$

A. 4 B. 6
C. 8 D. 9

42. X is a single digit number such that the LCM of the two digit number X6 and 54 is 108. What is the value of X?

A. 3 B. 4
C. 2 D. 1

43. If one of the roots of $ax^2 + bx + c = 0$ is half of the other root, then which of the following conditions must be true?

A. $b^2 - a^2 = 9ac$ B. $4b^2 = 9ac$
C. $b^2 = 18ac$ D. $2b^2 = 9ac$

44. In a two-digit number, the difference between the digits is 4, and 10 times the number is equal to 14 times the sum of the number obtained by adding the digits of the original and the number formed by reversing the digits. Find the number.

A. 84 B. 26
C. 62 D. 51

45. A large cube of volume 64 cubic metre is broken down into small cubes of volume 8 cubic metre each. The percentage increase in the surface area is

A. 50% B. 100%
C. 200% D. 75%

46. A, B and C are employed to do a piece of work for ₹ 700. A and B are supposed to finish $\frac{17}{28}$th of work together. Then, Amount that is paid to C is?

A. ₹ 250 B. ₹ 275
C. ₹ 150 D. ₹ 200

47. A tank is fitted with 3 pipes A, B and C. A takes twice as much time to fill the tank as B takes to empty the tank. B takes twice as much time to empty the tank as C takes to fill the tank. If it is known that C takes 10 hours to fill the tank, in how much time three quarters of the tank will be filled provided that all the taps were opened simultaneously?

A. 7.5 hours B. 10 hours
C. 5 hours D. 15 hours

48. Sam and Ram run a 100 m race. Sam beats Ram by 20 m. Sam gives Ram a head start such that both Ram and Sam will be meeting each other at the 75 m mark. What will be the result of the race?

A. Ram loses by 5 m.
B. Ram loses by 10 m.
C. Ram loses by 15 m.
D. The race ends in a dead heat.

49. If cosec(30°) + cot(60°) = y. What is the value of y?

A. $\frac{2\sqrt{3}+1}{\sqrt{3}}$ B. $\frac{\sqrt{3}+1}{\sqrt{3}}$
C. $\frac{3\sqrt{2}+1}{\sqrt{2}}$ D. $\frac{2\sqrt{2}+1}{\sqrt{2}}$

50. If $a^2 = b^3 = c^4$, $b^2 = d^5$, then find the value of $\log_a b \times \log_c d$.

A. $\frac{8}{45}$ B. $\frac{4}{15}$
C. $\frac{8}{15}$ D. $\frac{16}{45}$

51. Krishna's piggy bank contains coins in the denominations of ₹ 1, ₹ 2, ₹ 5 in the ratio of 3 : 4 : 5 respectively. If the total value of all the coins combined is ₹ 1296, what is the number of ₹ 2 coins in the piggy bank?

A. 108 B. 432
C. 216 D. 144

52. A person travels one fourth of his distance with 20 km/hr, half of the distance with 40 km/hr and the rest of the distance with 10 km/hr. What is his average speed throughout the journey ?

A. 20 km/hr B. 25 km/hr
C. 30 km/hr D. 40 km/hr

53. An arithmetic progression is such that the 12th term is 42. What should the common difference be, such that the product of the first, second and 33rd terms is maximum?

A. 3/2 B. 2
C. 0 D. –2

40. B	41. C	42. A	43. D	44. A	45. B	46. B	47. B	48. A	49. A	50. D	51. D	52. A	53. C

54. In a carom tournament in school all the kids played against each other exactly once. The number of games involving a girl and a boy is 150 and the number of games involving two girls is 60 less than the number of games involving two boys, find the total strength of the school?

A. 23 B. 24
C. 25 D. 35

55. A bag contains 4 red balls, 5 green balls and 6 yellow balls. Two balls are drawn at random from this bag. What is the probability that exactly one of the balls is yellow in colour?

A. $\frac{21}{54}$ B. $\frac{18}{35}$
C. $\frac{5}{9}$ D. $\frac{2}{5}$

56. A reduction of 20% in the price of rice enables a customer to purchase 4 kg more for ₹ 120. Then find original price of rice per kg

A. ₹ 20 B. ₹ 7.50
C. ₹ 8.25 D. ₹ 12

57. The time required for a sum of money to become 31 times of itself at 15% Simple Interest per annum is?

A. 150 years B. 200 years
C. 205 years D. 175 years

58. A trader mixes 20 kg of rice at ₹ 15 per kg with 25 kg of rice at ₹ 20 per kg and sells the mixture at ₹ 24 per kg. Then find his profit/loss percent.

A. 33.33% profit B. 35% loss
C. 25% loss D. 35% profit

Instructions (Qs. No. 59 and 60): *Study the given below Pie chart and answer the following questions. The total no. of chocolates are 350.*

Percentage of chocolates with each

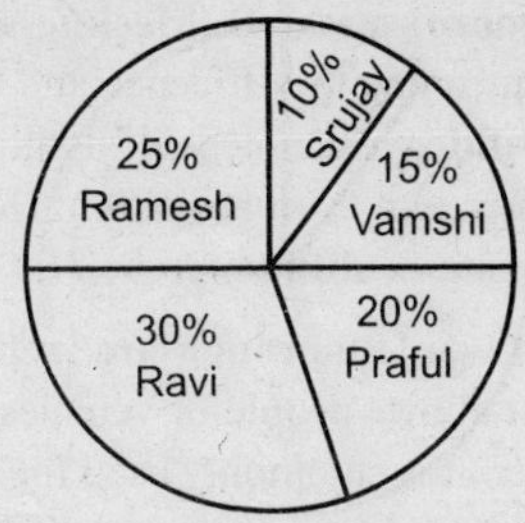

59. What is the difference between the person receiving highest and lowest number of chocolates?

A. 105 B. 80
C. 90 D. 70

60. What is the average number of chocolates received by Ravi, Praful and Srujay?

A. 80 B. 70
C. 50 D. 60

EXPLANATORY ANSWERS

1. The error lies in the preposition **"on"** used after the verb **"marvelled."** The correct idiomatic usage is **"marvelled at"** when expressing wonder or admiration for something. Thus, the phrase should read **"and marvelled at its majestic beauty."** The other segments are grammatically correct and fit naturally within the sentence. Therefore, segment C contains the inappropriately used preposition.

2. Sentence **(C)** is grammatically incorrect because the expression is incomplete and does not form a proper sentence structure. The phrase requires additional words to convey a complete meaning. Sentence **(D)** is also incorrect because it is merely a participial phrase beginning with **"Taking"** and lacks an independent clause with a finite verb. In contrast, sentences **(A), (B), and (E)** are grammatically acceptable and convey complete ideas in the context. Hence, the incorrect sentences are **(C)** and **(D).**

3. To convert the sentence into the passive voice, the object **"a large number of marginal workers"** becomes the subject of the new sentence. The modal construction **"could have affected"** changes to **"could have been affected"** in the passive form. The phrase **"by the reforms"** is retained to indicate the agent performing the action. Option B preserves both the meaning and tense of the original sentence accurately, making it the correct answer.

4. The sentence is **"Pakistan has approached the United Nations after India _______ to revoke the Indus water treaty."** The verb **"threatened"** correctly combines with the infinitive **"to revoke"** and conveys the meaning that India warned of a possible action. Option A (**told**) requires an object, Option B (**call**) is grammatically incorrect in this context, and Option D (**talked**) does not fit the meaning of the sentence. Therefore, **"threatened"** is the most suitable and grammatically correct choice.

5. The same word must fit both blanks. In the first sentence, **"import"** as a noun means **importance, significance, or meaning**, so **"The jury failed to grasp the full import of her testimony"** is correct. In the second sentence, **"import"** as a verb means **to bring goods into a country from abroad**, making **"You have to pay duties on goods that you import into the country"** perfectly correct. Since the word fits grammatically and contextually in both sentences with different meanings, **"import"** is the correct answer.

54. C	**55.** B	**56.** B	**57.** B	**58.** D	**59.** D	**60.** B

6. The sentence **"I am feeling awfully good today"** contains an apparent contradiction in expression. The word **"awfully"** is originally associated with something unpleasant or bad, while **"good"** conveys a positive meaning. When words with contrasting meanings are used together for emphasis, the figure of speech is called an **oxymoron**. The expression is intended to intensify the feeling of being good rather than convey a literal contradiction.

7. **Diffidence** means lack of self-confidence, shyness, hesitation, or modesty arising from uncertainty about one's abilities. Options A (**Timorousness**), B (**Meekness**), and C (**Timidity**) are similar in meaning to diffidence and therefore cannot be antonyms. **Confidence** denotes self-assurance, belief in one's abilities, and freedom from hesitation. Hence, **confidence** is the exact opposite of **diffidence**.

8. The compound word **"thoroughfare"** refers to a road, street, or passage that allows people or vehicles to travel from one place to another. It is commonly used for an important public road with regular traffic movement. The other options do not relate to the meaning of the word. Therefore, **"a main road"** best captures the meaning of **thoroughfare**.

9. The phrasal verb **"go off on someone"** is a common idiomatic expression meaning **to become very angry with someone and criticize or scold them strongly**. In the sentence, the team's continued neglect of their responsibilities causes the manager to react angrily toward them. Therefore, **"go off on them"** fits both the meaning and the grammatical structure perfectly. The other options do not naturally combine with **"on them"** in this context. Hence, **Option C** is the correct answer.

10. The word **"idiosyncrasy"** is correctly spelled and refers to a distinctive habit, mannerism, or characteristic peculiar to an individual. Option B is incorrect because the correct spelling is **"possession."** Option C is incorrect because the correct spelling is **"weird."** Option D is incorrect because the correct spelling is **"gist."** Hence, **"idiosyncrasy"** is the only correctly spelled word among the given options.

11. The word **"check"** traces its origin to the Persian word **"shah,"** meaning **"king."** The term entered the game of chess and was used in expressions indicating that the king was under threat. Through linguistic development and transmission into other languages, it evolved into the English word **"check."** Therefore, the source or origin of the word **"check"** is **Persian**.

12. Sentence **(B)** is grammatically correct. The passive construction **"was being managed by the engineer from China"** is properly formed, and the sentence expresses a complete and clear idea without any grammatical error.

    Sentence **(C)** is also grammatically correct. The introductory phrase **"Being the first born"** correctly refers to **"she,"** and the sentence logically conveys that she has had greater responsibilities because of her position in the family.

    Sentence **(A)** is incorrect because **"Looking around the class"** is a dangling modifier. The phrase should describe the person who was looking around, but it incorrectly appears to modify **"only a few hands."** Therefore, only **(B)** and **(C)** are grammatically correct.

13. The expression **"fall off"** is generally used when something becomes detached from a surface or object, such as leaves falling off a tree or a button falling off a shirt. Hair normally **"falls out,"** not **"falls off."** Therefore, the correct expression would be **"His hair started falling out as soon as he turned fifty."** The uses of **"fell apart," "fell down,"** and **"fall out"** in the other sentences are grammatically and contextually correct. Hence, option **D** contains the incorrect usage of the word **"fall."**

14. The numbering in the passage refers to the words immediately preceding the numbers. Thus, **(2) "it"** is a **pronoun** because it stands in place of **"The MBA." (3) "helps"** is a **verb** as it expresses the action performed by the subject **"it." (4) "important"** is an **adjective** because it describes the significance of the MBA. **(7) "tackle"** is a **verb** since it denotes the action of dealing with or handling important questions. **(8) "and"** is a **conjunction** because it joins two parallel clauses: **"that their organizations will need to tackle over time"** and **"that they will face in their careers."** Therefore, Option **D** correctly represents all the identified parts of speech.

15. Sentence **(A)** introduces the subject, **Cyrus the Great**, making it the natural opening sentence. Sentence **(C)** logically follows by explaining who he was and highlighting his achievement as the founder of a great empire. Sentence **(B)** then elaborates on the vast extent of that empire, providing supporting details. Finally, sentence **(D)** concludes the paragraph by expressing regret that so little is known about such an important historical figure despite the size of his empire. Therefore, the most logical and coherent order is **A → C → B → D (ACBD).**

16. The series is:

    4, 18, 48, 100, 180, 295

    Checking the pattern:

    $1 \times 2^2 = 4$

    $2 \times 3^2 = 18$

    $3 \times 4^2 = 48$

    $4 \times 5^2 = 100$

    $5 \times 6^2 = 180$

    The next term should be:

    $6 \times 7^2 = 6 \times 49 = 294$

    But the given term is 295.

    Therefore, 295 is the incorrect term in the series.

17. The series is: 5, 8, 10, 16, 20, 32, ?

    Odd-position terms: 5, 10, 20, ?

    $5 \times 2 = 10$

    $10 \times 2 = 20$

    $20 \times 2 = 40$

    Even-position terms: 8, 16, 32

    $8 \times 2 = 16$

    $16 \times 2 = 32$

    Thus, the missing term is: 40

    Therefore, the correct answer is 40.

18. Given: 78 : 225 :: 89 :?

For 78: $7 + 8 = 15$

$$15^2 = 225$$

Applying the same pattern to 89:

$$8 + 9 = 17$$

$$17^2 = 289$$

Therefore,

$$89 : 289$$

Hence, the correct answer is 289.

19. From the statements, **all letters are keys** and **some letters are pins**. This means that there is definitely an overlap between **letters** and **pins**, but no definite information is given about all pins. Therefore, Conclusion **2 (All pins are letters)** does not directly follow from the statements.

At the same time, Conclusion **1 (Some pins are not letters)** also does not directly follow because the statements do not provide any information about pins that lie outside the group of letters.

However, Conclusions **1** and **2** form a complementary pair: **either all pins are letters, or some pins are not letters**. Both cannot be true simultaneously, but one of them must be true. Hence, **either Conclusion 1 or Conclusion 2 follows**. Therefore, the correct answer is **C (Either 1 or 2 follows)**.

20. Conclusion I does not follow because the statements say **some pins are boards** and **no pin is a box**, but they do not give any definite relation between **boxes** and **boards**. It is possible that some boxes are boards, all boxes are boards, or no boxes are boards. Conclusion II also does not follow because **some boards are white** and **some pins are boards**, but it is not definite that pins and white things are completely separate. Some pins may be white or may not be white. Therefore, neither conclusion follows.

21. The numbers are: 1, 243, 256, 32

Now check them as fifth powers:

$$1 = 1^5$$

$$32 = 2^5$$

$$243 = 3^5$$

But,

$$256 = n^5$$

Also,

$$256 = 2^8 = 4^4 = 16^2$$

So, 256 is not a perfect fifth power, while the other three numbers are perfect fifth powers.

Therefore, 256 is the odd one out.

22. Dell is fixed on **Saturday** and HP is after Dell, so HP must be on **Sunday**. Since HP is three days after Asus, Asus must be on **Thursday**, and Lenovo, being just after Asus, comes on **Friday**. The remaining order becomes **Monday-HCL, Tuesday-Sony, Wednesday-Acer, Thursday-Asus, Friday-Lenovo, Saturday-Dell, Sunday-HP**. Since HCL on Monday is related to Tuesday and Sony on Tuesday is related to Wednesday, Lenovo on Friday is related to **Saturday**.

23. HP is distributed after Dell, and Dell is already fixed on **Saturday**, so HP can only be on **Sunday**. HP is given three days after Asus, so Asus must be on **Thursday**. Lenovo is given just after Asus, so Lenovo must be distributed on **Friday**. Hence, Lenovo laptops are distributed on Friday.

24. From the correct arrangement, Lenovo laptops are distributed on **Friday**, while HP laptops are distributed on **Sunday**. The only day between Friday and Sunday is **Saturday**. Dell laptops are fixed on Saturday as per the given condition. Therefore, the laptop distributed just between Lenovo and HP is **Dell**.

25. In the code, **DANGER** is written as **RDEAGN** by arranging letters alternately from the end and the beginning. For **DANGER**, the order becomes: last letter **R**, first letter **D**, second last **E**, second letter **A**, third last **G**, and third letter **N**. Applying the same pattern to **PARALLEL**, we take **L, P, E, A, L, R, L, A**. Therefore, **PARALLEL** is written as **LPEALRLA**.

26. Rajesh's father's brother means Rajesh's **uncle**. Rohit is the son of Rajesh's father's brother, so Rohit is the son of Rajesh's uncle. The son of one's uncle is one's **cousin**. Therefore, Rohit is related to Rajesh as his **cousin**.

27. Mukesh's sister's husband's mother-in-law is Mukesh's **mother**, because the mother-in-law of Mukesh's sister's husband is his wife's mother. The father-in-law of Mukesh's mother is Mukesh's **paternal grandfather**. The man in the photograph has this paternal grandfather as his father. Therefore, the man in the photograph is Mukesh's **father**.

28. At 4:00, the hour hand is at:

$$4 \times 30° = 120°$$

The minute hand is at: 0°

Relative speed of minute hand with respect to hour hand:

$$6° - 0.5° = 5.5° \text{ per minute}$$

Time required to gain 120°:

$$\frac{120}{5.5} = \frac{120}{11/2} = \frac{240}{11}$$

$$= 21\frac{9}{11} \text{ minutes}$$

Therefore, the minute and hour hands will coincide at 21 and 9/11 minutes past 4 o'clock.

29. The period from 1st July 1970 to 1st July 1977 is:

$$1977 - 1970 = 7 \text{ years}$$

Leap years within this period = 1972, 1976 = 2 leap years

Ordinary years = 5

$$\text{Total Odd Days} = (2 \times 2) + (5 \times 1)$$

$$= 4 + 5 = 9 \text{ odd days}$$

$$9 \text{ (mod 7)} = 2 \text{ odd days}$$

Given that 1st July 1977 was Friday, moving 2 days backward:

Friday − 2 = Wednesday

Therefore, 1st July 1970 was Wednesday.

30. Time elapsed from 10:00 AM to 11:00 PM: 13 hours

$$13 \times 60 = 780 \text{ minutes}$$

The watch gains 7 seconds in 4 minutes.

Gain in 780 minutes:

$$\frac{780}{4} \times 7 = 195 \times 7 = 1365 \text{ seconds}$$

Converting 1365 seconds into minutes and seconds:

$$1365 = 22 \times 60 + 45$$
$$= 22 \text{ minutes } 45 \text{ seconds}$$

Therefore, when the actual time is 11:00:00 PM, the watch will show:

$$11 : 00 : 00 + 00 : 22 : 45$$
$$= 11 : 22 : 45 \text{ PM}$$

Hence, the correct answer is 11:22:45 PM.

31. The triangle moves in a fixed positional sequence from one box to the next. From Box 1 to Box 2, it moves upward; from Box 2 to Box 3, it moves right; from Box 3 to Box 4, it again moves right; and from Box 4 to Box 5, it moves downward. Continuing this movement pattern, the triangle should next move left by one position. This placement matches **Answer Figure (2)**, while the small horizontal line remains at the bottom-left position in the relevant options. Therefore, the correct answer is **B: (2)**.

32. Observe three simultaneous patterns:

1. Outer figure rotates 45° clockwise in every step.
2. Arrow rotates 45° anticlockwise in every step.
3. Small square moves one position clockwise along the boundary.

From the 5th figure:

- Rotate the outer figure 45° clockwise.
- Rotate the arrow ← → ↖.
- Move the square from right → upper-left.

Only Option (2) satisfies all three conditions.

33. From the given condition: 2 $ 5 = 32

Since, $2^5 = 32$

the operator $ represents exponentiation:

$$a \,\$\, b = a^b$$

Now,

$$\frac{4\$3}{2\$5} = \frac{4^3}{2^5}$$

Calculating the numerator:

$$43 = 4 \times 4 \times 4 = 64$$

Calculating the denominator:

$$2^5 = 32$$

Therefore,

$$\frac{64}{32} = 2$$

Hence, the value of the expression is 2. Therefore, the correct answer is C (2).

34. From the given codes:

Make your trip interesting = de la pa ni

your trip be customized = ha ni ta de

Common words your and trip give common codes de and ni.

Now,

customized trip always good = ta fa ka ni

Here, trip = ni, customized = ta, so remaining codes fa and ka represent always and good.

Also,

good make me happy = ka ro pa na

So, good = ka and make = pa.

Therefore:

trip = ni, always = fa, good = ka, interesting = la

Hence,

good trip always interesting = ka ni fa la

The same set of codes is:

ni fa la ka

Therefore, the correct answer is D: ni fa la ka.

35. The statement says that after adopting a **flexible inflation-targeting framework in August 2016**, the RBI changed the rules of monetary policy and became more focused on controlling inflation and inflation expectations. This implies that there was a reason or necessity for introducing a new framework, suggesting that the earlier approach was not considered fully adequate for the desired objectives. Hence, **Assumption I** follows.

The statement also indicates that the RBI was able to modify its monetary policy framework to place greater emphasis on inflation targeting. This shows that monetary policy is not fixed and can be adjusted according to changing economic requirements and policy goals. Therefore, **Assumption II** also follows.

Since both assumptions are logically implied by the statement, the correct answer is **C: Both I and II**.

36. The effect states that **at least twenty school children were seriously injured while going for a school picnic**. For such a large number of children to be injured simultaneously, a major incident such as a **bus accident** is the most probable cause. Option A may create inconvenience but does not directly lead to injuries. Option C relates to the driver's absence and does not explain the injuries. Option D is a consequence or reaction after the incident, not a cause. Therefore, the most probable cause is that the bus met with an accident.

37. The statement mentions that a large number of people in City X have been diagnosed with **malaria**, which is a mosquito-borne disease. Therefore, taking immediate preventive and control measures such as **extensive fumigation** is a logical and practical course of action to reduce mosquito breeding and prevent further spread of the disease. Hence, Course of Action I follows.

Course of Action II, which suggests that the entire city should be vacated immediately, is an extreme and impractical measure. Malaria outbreaks are generally controlled through medical treatment, sanitation, mosquito control, and public health measures rather than evacuating an entire city. Therefore, Course of Action II does not follow. Hence, only Course of Action I follows.

38. The statement recommends that companies should help contractual workers acquire new skills and eventually become permanent employees. Argument **I** supports this view because better relations between companies and contractual workers can improve productivity, trust, job satisfaction, and long-term growth for both parties. Thus, it provides a positive reason in favour of the statement and strengthens it.

Argument II points out a possible obstacle, namely resistance from existing permanent employees. Such opposition could make the implementation of the proposed policy more difficult and less effective. Therefore, instead of supporting the statement, it highlights a drawback and weakens it. Hence, Argument I strengthens the statement, whereas Argument II weakens it.

39. Statement II mentions that the government banned patents and royalties on medicines two years earlier. Patents and royalties often encourage pharmaceutical companies and researchers to invest time, money, and resources in developing new medicines and technologies. Removing such incentives may reduce investment in research activities.

Statement I says that there has been no major breakthrough in medical research during the past few years. A reduction in incentives for innovation could logically contribute to fewer significant research achievements. Therefore, Statement II can reasonably be viewed as the cause, and Statement I as the effect. Hence, the correct answer is B.

40. Statement I describes a situation in which a new COVID variant, **Omicron**, is spreading rapidly and a large number of new cases are being reported within a short period. Such a sudden rise in infections creates concern about public health, hospital capacity, and disease control measures.

Statement II states that the central government is in a dilemma about whether to impose a lockdown. This uncertainty arises because the rapid spread of the virus may require stricter restrictions, while a lockdown can also have economic and social consequences. Thus, the government's dilemma is a direct result of the worsening COVID situation described in Statement I.

Therefore, the rapid spread of Omicron and the increase in cases (**Statement I**) act as the **cause**, while the government's uncertainty regarding a lockdown (**Statement II**) is the **effect**. Hence, the correct answer is B.

41. Given:

$$N = a \times b$$

where a and b are prime numbers.

$$\frac{N^3}{b^2} = \frac{(ab)^3}{b^2} = \frac{a^3b^3}{b^2} = a^3b$$

The prime factorization of the number is:

$$a^3 \times b^1$$

For a number of the form:

$$p^m q^n$$

Number of factors:

$$(m + 1)(n + 1)$$

Therefore,

$$(3 + 1)(1 + 1) = 4 \times 2 = 8$$

Hence, the number of factors of $\frac{N^3}{b^2}$ is 8.

Therefore, the correct answer is C (8).

42. Given: LCM (X6, 54) = 108

Prime factorization of 54:

$$54 = 2 \times 3^3$$

Prime factorization of 108:

$$108 = 2^2 \times 3^3$$

Since the LCM is 108, the two-digit number X6 must be a factor of 108 and should not introduce any additional prime factor.

Checking the options:

For X = 3,

$$X6 = 36$$

Prime factorization of 36:

$$36 = 2^2 \times 3^2$$

Now,

$$\text{LCM}(36, 54) = 2^2 \times 3^3 = 108$$

For X = 4,

$$X6 = 46$$

46 is not a factor of 108, so it cannot give LCM 108.

For X = 2,

$$X6 = 26$$

26 contains the prime factor 13, so the LCM will be greater than 108.

For X = 1,

$$X6 = 16$$

$$\text{LCM}(16, 54) = 432$$

$$432 \neq 108$$

Therefore, $X = 3$

Hence, the correct answer is A(3).

43. Let the roots of

$$ax^2 + bx + c = 0$$

be r and $2r$.

Using the relations between roots and coefficients:

$$r + 2r = -\frac{b}{a}$$

$$3r = -\frac{b}{a}$$

$$r = -\frac{b}{3a}$$

Also, $$r(2r) = \frac{c}{a}$$

$$2r^2 = \frac{c}{a}$$

Substituting $$r = -\frac{b}{3a}$$

Gives $$2\left(\frac{b^2}{9a^2}\right) = \frac{c}{a}$$

Multiplying both sides by $9a^2$,

$$2b^2 = 9ac$$

Therefore, the required condition is:

$$2b^2 = 9ac$$

Hence, the correct answer is D ($2b^2 = 9ac$).

44. Let the two-digit number be:

$$10x + y$$

where x is the tens digit and y is the units digit.

Given: $$x - y = 4$$

The reversed number is: $10y + x$

The sum of the digits is: $x + y$

According to the question:

$$10(10x + y) = 14[(x + y) + (10y + x)]$$
$$100x + 10y = 14(2x + 11y)$$
$$100x + 10y = 28x + 154y$$
$$72x = 144y$$
$$x = 2y$$

Using:

$$x - y = 4$$
$$2y - y = 4$$
$$y = 4$$
$$x = 8$$

Therefore, the number is:

Hence, the correct answer is A (84).

45. Given:

$$V_1 = 64 \text{ m}^3$$

Side of the large cube:

$$a_1 = \sqrt[3]{64} = 4 \text{ m}$$

Surface area of the large cube:

$$A_1 = 6a_1^2$$
$$= 6 \times 4^2$$
$$= 6 \times 16 = 96 \text{ m}^2$$

Volume of each small cube:

$$V_2 = 8 \text{ m}^3$$

Side of each small cube:

$$a_2 = \sqrt[3]{8} = 2 \text{ m}$$

Number of small cubes:

$$N = \frac{64}{8} = 8$$

Surface area of one small cube:

$$6a_2^2 = 6 \times 2^2$$
$$= 6 \times 4 = 24 \text{ m}^2$$

Total surface area of all small cubes:

$$A_2 = 8 \times 24 = 192 \text{ m}^2$$

Increase in surface area:

$$A_2 - A_1 = 192 - 96 = 96 \text{ m}^2$$

Percentage increase:

$$\frac{96}{96} \times 100 = 100\%$$

Therefore, the percentage increase in the surface area is 100%.

Hence, the correct answer is B (100%).

46. Total work = 1

A and B together complete: $\frac{17}{28}$

Therefore, C completes the remaining work:

$$1 - \frac{17}{28} = \frac{28 - 17}{28} = \frac{11}{28}$$

Total payment: ₹ 700

C's share:

$$700 \times \frac{11}{28} = 25 \times 11 = ₹\ 275$$

Therefore, the amount paid to C is ₹ 275.

Hence, the correct answer is B (₹ 275).

47. Pipe C fills the tank in: 10 hours

Therefore,

$$\text{Rate of C} = \frac{1}{10}$$

Pipe B takes twice as much time as C to empty the tank:

$$2 \times 10 = 20 \text{ hours}$$

Therefore,

$$\text{Rate of B} = -\frac{1}{20}$$

Pipe A takes twice as much time as B to fill the tank:

$$2 \times 20 = 40 \text{ hours}$$

Therefore,

$$\text{Rate of A} = \frac{1}{40}$$

Net rate when all three pipes are opened:

$$\frac{1}{40} - \frac{1}{20} + \frac{1}{10}$$

Taking LCM 40,

$$= \frac{1 - 2 + 4}{40} = \frac{3}{40}$$

Thus, $\frac{3}{40}$ of the tank is filled in one hour.

Time required to fill $\frac{3}{4}$ of the tank:

$$\frac{\frac{3}{4}}{\frac{3}{40}} = \frac{3}{4} \times \frac{40}{3} = 10 \text{ hours}$$

Therefore, three-quarters of the tank will be filled in 10 hours.

Hence, the correct answer is B (10 hours).

48. Sam beats Ram by 20 m in a 100 m race.

So, when Sam runs: 100 m

Ram runs: 80 m

Therefore, speed ratio:

$$\text{Sam} : \text{Ram} = 100 : 80 = 5 : 4$$

They meet at the 75 m mark.

When Sam covers: 75 m

$$\text{Ram covers:} \quad 75 \times \frac{4}{5} = 60 \text{ m}$$

So Ram must have received a head start of:

$$75 - 60 = 15 \text{ m}$$

Now, from the 75 m mark to the finish, both have:

$$100 - 75 = 25 \text{ m}$$

When Sam covers the remaining 25 m, Ram covers:

$$25 \times \frac{4}{5} = 20$$

So Ram reaches:

$$75 + 20 = 95 \text{ m}$$

Therefore, Ram loses by:

$$100 - 95 = 5 \text{ m}$$

Hence, the correct answer is A (Ram loses by 5 m).

49. Given:

$$\text{cosec } 30° + \cot 60° = y$$

We know:

$$\text{cosec } 30° = \frac{1}{\sin 30°} = \frac{1}{1/2} = 2$$

and $\cot 60° = \frac{1}{\sqrt{3}}$

Therefore, $y = 2 + \frac{1}{\sqrt{3}}$

Taking LCM $\sqrt{3}$, $y = \frac{2\sqrt{3}+1}{\sqrt{3}}$

Hence, the correct answer is A

50. Given:

$$a^2 = b^3 = c^4$$

Let $a^2 = b^3 = c^4 = k$

Then, $a = k^{1/2},\ b = k^{1/3},\ c = k^{1/4}$

Also,

$$b^2 = d^5$$
$$(k^{1/3})^2 = d^5$$
$$k^{2/3} = d^5$$
$$d = k^{2/15}$$

Now, $\log_a b = \frac{\log k^{1/3}}{\log k^{1/2}} = \frac{1/3}{1/2} = \frac{2}{3}$

and $\log_c d = \frac{\log k^{2/15}}{\log k^{1/4}} = \frac{2/15}{1/4} = \frac{8}{15}$

Therefore,

$$\log_a b \times \log_c d = \frac{2}{3} \times \frac{8}{15} = \frac{16}{45}$$

Hence, the correct answer is D $\left(\frac{16}{45}\right)$.

51. Let the number of ₹ 1, ₹ 2 and ₹ 5 coins be:

$$3x,\ 4x,\ 5x$$

respectively.

Value of ₹ 1 coins:

$$1 \times 3x = 3x$$

Value of ₹ 2 coins:

$$2 \times 4x = 8x$$

Value of ₹ 5 coins:

$$5 \times 5x = 25x$$

Total value of all coins:

$$3x + 8x + 25x = 1296$$
$$36x = 1296$$
$$x = \frac{1296}{36} = 36$$

Number of ₹ 2 coins:

$$4x = 4 \times 36 = 144$$

Therefore, the number of ₹ 2 coins in the piggy bank is 144.

Hence, the correct answer is D (144).

52. Let the total distance be: 4D

Distance covered at km/hr:

$$\frac{1}{4} \times 4\text{D} = \text{D}$$

Time taken: $\frac{\text{D}}{20}$

Distance covered at 40 km/hr:

$$\frac{1}{2} \times 4\text{D} = 2\text{D}$$

Time taken: $\frac{2\text{D}}{40} = \frac{\text{D}}{20}$

Remaining distance:

$$4\text{D} - \text{D} - 2\text{D} = \text{D}$$

Distance covered at 10 km/hr: D

Time taken: $\frac{\text{D}}{10}$

Total distance: 4D

Total time:

$$\frac{\text{D}}{20} + \frac{\text{D}}{20} + \frac{\text{D}}{10} = \frac{\text{D} + \text{D} + 2D}{20}$$
$$= \frac{4\text{D}}{20} = \frac{\text{D}}{5}$$

Average speed:

$$\frac{\text{Total Distance}}{\text{Total Time}} = \frac{4\text{D}}{\text{D}/5} = 20 \text{ km/hr}$$

Therefore, the average speed throughout the journey is 20 km/hr.

Hence, the correct answer is A (20 km/hr).

53. Given:

$$a_{12} = 42$$

For an AP:

$$a_n = a + (n - 1)d$$

So, $a + 11d = 42$

$$a = 42 - 11d$$

Now,

$$a_1 = 42 - 11d$$
$$a_2 = 42 - 10d$$
$$a_{33} = 42 + 21d$$

Product:

$$\text{P}(d) = (42 - 11d)(42 - 10d)(42 + 21d)$$

Let $k = 42$:

$$\text{P}(d) = (k - 11d)(k - 10d)(k + 21d)$$
$$(k - 11d)(k - 10d) = k^2 - 21kd + 110d^2$$
$$\text{P}(d) = (k^2 - 21kd + 110d^2)(k + 21d)$$
$$\text{P}(d) = k^3 - 331kd^2 + 2310d^3$$

Differentiating:

$$P'(d) = -662kd + 6930d^2$$

$$P'(d) = d(-662k + 6930d)$$

So, $d = 0$

At $d = 0$, $P''(d) = -662k + 13860d$

$$P''(0) = -662k$$

Since $k = 42$,

$$P''(0) < 0$$

Therefore, $d = 0$ gives the maximum product.

Hence, the correct answer is C (0).

54. Let the number of girls be g and boys be b.

Games involving one girl and one boy:

$$gb = 150$$

Games involving two girls: $\frac{g(g-1)}{2}$

Games involving two boys: $\frac{b(b-1)}{2}$

Given: $\frac{g(g-1)}{2} = \frac{b(b-1)}{2} - 60$

$$b(b-1) - g(g-1) = 120$$

$$b^2 - b - g^2 + g = 120$$

$$(b-g)(b+g-1) = 120$$

Now, from $gb = 150$, the suitable pair is:

$$g = 10,\ b = 15$$

Checking:

$$gb = 10 \times 15 = 150$$

$$(b-g)(b+g-1) = (15-10)(15+10-1) = 5 \times 24 = 120$$

Total strength:

$$g + b = 10 + 15 = 25$$

Hence, the correct answer is C (25).

55. Red balls: 4

Green balls: 5

Yellow balls: 6

Total balls: $4 + 5 + 6 = 15$

Total number of ways to draw 2 balls:

$$^{15}C_2 = \frac{15\times14}{2} = 105$$

For exactly one yellow ball, we must select:

1 yellow ball from 6

and

1 non-yellow ball from $(4 + 5) = 9$

Favourable outcomes:

$$^6C_1 \times {}^6C_1 = 6 \times 9 = 54$$

Required probability: $\frac{54}{105}$

Dividing numerator and denominator by 3,

$$\frac{54}{105} = \frac{18}{35}$$

Therefore, the probability that exactly one of the two balls drawn is yellow is: $\frac{18}{35}$

Hence, the correct answer is B $\left(\frac{18}{35}\right)$.

56. Let the original price of rice be:

P per kg

Quantity purchased for ₹ 120 before reduction: $\frac{120}{P}$

After a 20% reduction, the new price becomes:

$$P - \frac{20P}{100} = 0.8P$$

Quantity purchased for ₹ 120 after reduction:

$$\frac{120}{0.8P} = \frac{150}{P}$$

Given that the customer can purchase 4 kg more:

$$\frac{150}{P} - \frac{120}{P} = 4$$

$$\frac{30}{P} = 4$$

$$P = \frac{30}{4}$$

$$P = 7.50$$

Thus, the original price of rice was: ₹ 7.50 per kg

Hence, the correct answer is B (₹ 7.50).

57. Let the principal be: P

Amount becomes 31 times the principal:

$$A = 31P$$

Simple Interest: $SI = A - P = 31P - P = 30P$

Using the Simple Interest formula:

$$SI = \frac{PRT}{100}$$

$$30P = \frac{P\times15\times T}{100}$$

Cancelling : $30 = \frac{15T}{100}$

$$T = \frac{30\times100}{15} = 200$$

Therefore, the time required is: 200 years

Hence, the correct answer is B (200 years).

58. Cost price of 20 kg rice at ₹ 15 per kg:

$$20 \times 15 = 300$$

Cost price of 20 kg rice at ₹ 20 per kg:

$$25 \times 20 = 500$$

Total cost price:

$$300 + 500 = 800$$

Total quantity:

$$20 + 25 = 45 \text{ kg}$$

Selling price of the mixture:

$$45 \times 24 = 1080$$

Profit:

$$1080 - 800 = 280$$

Profit percentage:

$$\frac{280}{800} \times 100 = 35\%$$

Therefore, the trader earns a 35% profit.

Hence, the correct answer is D (35% profit).

59. Total chocolates: 350

Highest percentage = Ravi = 30%

$$30\% \text{ of } 350 = \frac{30 \times 350}{100} = 105$$

Lowest percentage = Srujay = 10%

$$10\% \text{ of } 350 = \frac{10 \times 350}{100} = 35$$

Difference: $105 - 35 = 70$

Therefore, the difference between the highest and lowest number of chocolates is 70.

Hence, the correct answer is D (70).

60. Ravi receives:

$$30\% \text{ of } 350 = 105$$

Praful receives:

$$20\% \text{ of } 350 = 70$$

Srujay receives:

$$10\% \text{ of } 350 = 35$$

Average number of chocolates:

$$\frac{105 + 70 + 35}{3} = \frac{210}{3} = 70$$

Therefore, the average number of chocolates received by Ravi, Praful and Srujay is 70.

Hence, the correct answer is B (70).

YOUR SPACE

Previous Years' Paper

Symbiosis National Aptitude Test (SNAP), 2023*

(Exam held on 10-12-2023)

1. The proverb "A rolling stone gathers no moss" emphasizes that:

A. One should be content with what they have and avoid change.

B. Patience and perseverance are key to achieving success.

C. Frequent movement leads to a lack of progress.

D. New experiences bring valuable knowledge and growth.

2. In the following question, a sentence is given with a phrasal verb in bold. Read each sentence carefully to find out whether there is any error in the phrasal verb, if there is any error, choose the correct phrasal verb from given options. If the phrasal verb given in bold is correct, mark (D) as your answer.

She **narrowed** the exam **down** in the first attempt.

A. opted for   B. nailed down

C. passed out   D. No Error

3. The question below consists of a set of labelled sentences. Out of the four options given, select the most logical order of sentences to form a coherent paragraph.

P : When the game of life is finally over there is no second chance to correct our errors.

Q : Time is the greater equalizer of all mankind.

R : Time offers opportunity but demands a sense of regard.

S : It has taken away the best and the worst of us without regard of either.

A. QSRP   B. RSQP

C. QRSP   D. RSPQ

4. In the following question, the sentences have been given in Active/Passive Voice. From the given alternatives, choose the one which best expresses the given sentence in Passive/Active Voice as your answer.

You can play with these kittens quite safely.

A. These kittens can played with quite safely.

B. These kittens can played with you quite safely.

C. These kittens can be played with you quite safely.

D. These kittens can be played with quite safely.

5. The speaker describes their overwhelming workload as a ………. drowning them in tasks.

A. Rushing river

B. Sparkling fountain

C. Gentle breeze

D. Heavy stone

6. In the following questions, find out which part of the sentence has an error. The letter corresponding to that part is your answer. If there is no mistake, the answer is (D) 'No error'.

If I was you, (A)/I would not allow them (B)/ such leverage in the business. (C)/ No error (D).

7. In the following questions, choose the word opposite in meaning of the highlighted word given in the sentence.

The journey was becoming **treacherous**, but they continued on.

A. perilous   B. mutinous

C. seditious   D. secure

8. In the following question, out of the four alternatives, choose the one which best expresses the meaning of the highlighted word given in the sentence.

The soldier showed an **exemplary** courage.

A. flawed   B. faulty

C. ideal   D. boisterous

1. C	2. B	3. A	4. D	5. A	6. A	7. D	8. C

* **Based on memory.**

9. Fill in the blank with an appropriate preposition:

We must mail the letter .... Friday.

A. by B. since
C. until D. for

10. Choose the correct question tag to complete each sentence.

I guess he is in a hurry, ...?

A. doesn't he B. is he
C. isn't he D. does he

11. The librarian asked everyone to be more .... as she needed silence to concentrate on filing the new book.

A. respectful B. enthusiastic
C. confused D. talkative

12. Having .... the report all night, Sarah was exhausted but satisfied with the final product.

A. written B. write
C. to write D. writes

13. Choose the correct spelling of the word given in bold.

The rich youth cynically declared that the **penacea** for all speeding tickets was a big enough bribe.

A. panacea B. penecea
C. panacia D. panacya

14. Choose the correct conjunction for each sentence.

.... my best friend Hasan is five years older than me, he looks very young.

A. Until B. When
C. Although D. Because

15. In the question below has two blanks, each blank indicating that something has been omitted. Choose the set of words for each blank that best fits the meaning of the sentence as a whole.

After a .... three months of stock market activity, the price of bonds .... last week.

A. tumultuous, stabilised
B. hectic, fell
C. variety, surged
D. continuous, soared

16. Roy bought articles at ₹ 20 per item. He sells them for ₹ 2, ₹ 4 and ₹ 6 and so on. He intends to make at least a profit of 40%. What will be the minimum number of articles he should sell?

A. 25 B. 27
C. 28 D. None of these

17. A and B have bags full of one rupee coins. Total value of the amount with them is ₹ 45. There was a hole in both the bags due to which 5 one rupee coins were dropped from each bag. Product of the remaining number of coins in the bags is now 124. Find the product of the number of coins in the bags initially.

A. 350 B. 324
C. 200 D. None of these

18. Imran sold a mobile phone for ₹ 17600 after giving 12% discount on its marked price. He earned a profit of 10%. Find the profit percentage earned by him when no discount is given.

A. 12% B. 22%
C. 25% D. None of these

19. During an event, 160 persons participated. The participants were grouped into male and female. There were 1240 apples to be distributed among them. If each male gets 7 apples and each female gets 12 apples. Find the number of females present at the event.

A. 24 B. 26
C. 28 D. 30

20. A container has a mixture containing three fourth of milk and remaining is water. After diluting the mixture with water, final mixture contains half milk and half water. How much of the liquid will be extracted and replaced with water?

A. 1/2 B. 1/4
C. 1/3 D. None of these

21. Shopkeeper mixes two types of sugar. If the mixes them in the ratio 3 : 2, his profit is 10% and if he mixes them in the ratio 2 : 3, his profit is 5%. If the selling price of the mixture is ₹ 40, then find the ratio of cost prices of both types of sugar.

A. 2 : 3 B. 19 : 24
C. 17 : 20 D. None of these

22. 4 different positive numbers are written in ascending order. 1/4th of the average of all the 4 numbers is 24 less than the greatest of these numbers. If the average of the first 3 numbers is 32, the greatest number among the given number is:

A. 32 B. 34
C. 36 D. None of these

9. A	**10.** C	**11.** A	**12.** A	**13.** A	**14.** C	**15.** B
16. B	**17.** B	**18.** C	**19.** A	**20.** C	**21.** B	**22.** A

23. A and B complete a certain work together in 2 days working 5 hours a day. B and C complete the same work in 4 days working 5 hours a day. C and A complete the same in 2 days working 6 hours a day. In how many days can A finish the work alone if he works for 5 hours a day?

A. 3 B. 5
C. 12 D. 15

24. Three friends Sonu, Monu and Romu play cricket. They scored 342 runs in total. If the ratio of runs scored by Sonu and Monu is 3 : 2 and the ratio of runs scored by Monu and Romu is also 3 : 2, then find the number of runs scored by Romu.

A. 96 B. 64
C. 72 D. None of these

25. Letters of the word MINUTE be arranged such that the vowels occupy only odd positions.

A. 32 B. 36
C. 48 D. None of these

26. In how many ways 7 different chocolates can be distributed to A, B and C such that at least one of them gets exactly one chocolate?

A. 1050 B. 1176
C. 1218 D. None of these

27. To select a group out of three, an unblased die is rolled. If the dice shows 1, then group X is chosen. If it shows 2 or 3, then group Y is chosen and if it shows 4, 5 or 6, then group Z is chosen. Group X has 3 girls, 2 boys, Group Y has 3 girls and 4 boys. Group Z has 4 girls and 5 boys. The dice is rolled and the group is chosen, if it is known that girl is chosen, then find probability that she is chosen from group Y.

A. 7/90 B. 63/293
C. 90/293 D. None of these

28. Mohini and Rohini are classmates. Mohini speaks truth 6 out of 9 times and Rohini speaks truth 3 out of 6 times. Find the probability that they contradict each other on certain fact.

A. 1/2 B. 2/3
C. 4/5 D. None of these

29. x, y and z are three consecutive numbers and it is given that log $(1 + xz) = 2A$, then find A.

A. log x B. log y
C. log z D. None of these

30. If in a decreasing AP, sum of all the terms except the first term is –75 while the sum of all the terms is 0. If the difference between the 9th term and the 13th term is 20, find the first term of the AP?

A. 25 B. 30
C. 35 D. None of these

31. From each vertex of an equilateral triangle with area of 240 square units, corners are cut to form a regular hexagon. Area of hexagon will be:

A. 120 B. 160
C. 180 D. None of these

32. A flower bed has an area of 125, 375 and 250 square meters. What is the maximum rows of flowers a farmer can grow if the width of a row is 3 square meters?

A. 249 B. 246
C. 250 D. None of these

33. A person observes the top of a building from a point, the angle of elevation is 30 degrees. He moves 500 m towards the building and angle of elevation of the top of the building changes to 75 degrees. Find the height of the building.

A. $125(\sqrt{3}+1)$ B. $250(\sqrt{3}+1)$
C. $500(\sqrt{3}+1)$ D. None of these

State	% population below Education index	Proportion of Males and Females	
		Below Education Index	Above Education Index
Gujarat	15	3 : 2	2 : 1
Delhi	20	1 : 3	3 : 1
Rajasthan	18	3 : 4	1 : 4
MP	12	2 : 5	2 : 3
Punjab	11	3 : 5	4 : 5
UP	16	2 : 3	5 : 1
Kerala	10	1 : 5	3 : 2

34. If male population above Education Index for UP is 1.6 lakhs, then find the total population of UP in lakhs (approximately).

A. 2.28 B. 2.96
C. 3.67 D. 3.69

35. 3 + 2 = 7, 51 + ? = 76, 32 – 8 = –32, 12 + 3 = 21

Find the value of ?.

A. 7 B. 25
C. 5 D. None of these

23. A	**24.** C	**25.** B	**26.** C	**27.** C	**28.** A	**29.** B
30. C	**31.** B	**32.** A	**33.** A	**34.** A	**35.** C	

36. Find the next term of the series:

1, 3, 7, 13, 21, 31, ...

A. 67 B. 43
C. 31 D. 52

37. Choose the correct option Tue, Wed, Fri, Mon, ?, Wed.

A. Wed B. Fri
C. Thu D. Sat

38. If 8055 & 238945 are spelled as BOSS and ZEBRAS then how LEETS is spelled?

A. 23375 B. 13357
C. 13572 D. None of these

39. Suresh Bhai who is father-in-law of wife of Mahesh Bhai has only one son. The only Brother-in-law of Mahesh is Paresh & Suresh is Ramesh's Grandfather. If Mahesh has no Sister-in-law, what is Paresh to Ramesh?

A. Father
B. Brother
C. Maternal Uncle
D. None of these

40. Find logically correct sequence:

1. Pune
2. Aga Khan Palace
3. World
4. India
5. Maharashtra

A. 21534 B. 21354
C. 21543 D. 15425

41. In a 24 hours time clock, if angle (measured in clockwise directions) between long hand & short hand of a clock is represented by some variable P, then find [(Value of P at 6:00) + (Value of P at 18:00)]/(Value of P at 12:00) = ?

A. 3 B. 2
C. 4 D. None of these

42. Mohini and Rohini are classmates. Mohini speaks truth 6 out of 9 and Rohini speaks truth 3 out of 6. Probability that they contradict each other on certain fact?

A. 1/2
B. 4/5
C. 2/3
D. None of these

43. Find the number that will complete the given series;

15, 21, 27, 33, 35, 39, ..., 51, 55

A. 47 B. 49
C. 45 D. 50

44. Arrange from bottom to top:

1. Knees
2. Elbow
3. Head
4. Tongue
5. Neck
6. Foot

A. 345216
B. 345261
C. 345621
D. None of these

45. Find the correct Calendar Repetition of 2003:

A. 2032 B. 2034
C. 2025 D. 2028

Directions (Qs. 46-47): *In each of the questions below are given some statements followed by two conclusions. You have to take the given statements to be true even if they seem to be at variance with commonly known facts. Read all the conclusions and then decide which of the given conclusions logically follows from the given statements disregarding commonly known fact.*

Given answer:

A. If only conclusion I follows
B. If only conclusion II follows
C. If either conclusion I or II follows
D. If neither conclusion I nor II follows

46. Statements:

Only a few Chocolate are table
Only a few Tables are Plants
Only a few Plants are Fan

Conclusions:

I. Some fan is Chocolate is a possibility.
II. All fan are Tables.

47. Statements:

Only a few Ant are Cow
No Cow is Joker
Only a few joker are Deer

Conclusions:

I. Some Deer are Ant
II. Some Ant are Joker

36. B	**37.** B	**38.** D	**39.** C	**40.** C	**41.** B
42. A	**43.** A	**44.** D	**45.** B	**46.** A	**47.** D

48. Give a statement followed by two assumptions numbered I and II. Consider the statement and the following assumptions and decide which of the assumptions is/are implicit in the statement.

Statements:

All the species of plants and animals are part of biodiversity and ecosystems and play major role in the overall bealth of the environment.

Assumptions:

I. Preserve or create a backyard habitat. Save as many native plants as you can when building or landscaping.

II. One of the major goals of sustainability is to preserve biodiversity. All life on Earth is connected through the flow of energy (planetary food web), and each time a species becomes endangered or lost to extinction, one more part of that energy flow is lost.

A. Neither assumption I nor II is implicit
B. Both assumption I and II are implicit
C. Only assumption I is implicit
D. Only assumption II is implicit

49. Statement: The Indian Council of Medical Research issued a warning two years ago, based on studies conducted in hospitals that resistance to antibiotics was found in 50% of patients. A large number of infants were dying due to infections that did not respond to treatment.

Courses of Action:

I. The revision of antibiotics classes by the World Health Organisation in its list of essential medicines is a welcome step in the global initiative to push back against antimicrobial resistance, the phenomenon of bacteria becoming resistant even to the most potent drugs.

II. Close scrutiny of these by national stewardship programmes such as those initiated by the ICMR is needed.

Which of the following will be the course of action?

A. Only I B. Both I and II
C. Either I or II D. Only II

Directions (Qs. 50-51): *Read the following information carefully and mark the correct answer to the questions given below.*

Sampada Apartment is a housing society formed by a group of professors of a University. It has six flats on a floor in two rows facing North and South which are allotted to prof. Purohit, Prof. Qureshi, Prof. Rathor, prof. Sawant, prof. Tripathy and Prof. Usman. Prof. Qureshi gets a North facing flat and is not next to Prof. Sawant's flat. Prof. Sawant and Prof. Usman get their flats which are diagonally opposite to each other. Prof. Rathor gets a south facing flat which is next to Prof. Usman's flat, Prof. Tripathy's flat is North facing.

50. Which of the following professors get South facing flats?

A. Prof. Qureshi, Prof. Tripathy and prof. Sawant
B. Prof. Usman, Prof. Tripathy and Prof. Purohit
C. Prof. Usman, Prof. Rathor and Prof. Purohit
D. None of these

51. If the flats of Prof. Tripathy and Prof. Purohit are interchanged, whose flat will be next to that of Prof. Usman?

A. Prof. Rathor B. Prof. Tripathy
C. Prof. Usman D. None of these

52. There has been a considerable drop in sales of four wheelers during the past six months when compared to the number of four wheelers sold during this period last year.

Which of the following can the probable cause of the above phenomenon?

(*a*) The govt. has imposed higher excise duty on four wheelers at the beginning of this year.
(*b*) The petrol prices have risen considerably during the past eight months.
(*c*) The rate of interest on home and car loans have been rising for the past seven months.

A. All (*a*), (*b*) and (*c*)
B. (*a*) and (*c*) only
C. (*b*) and (*c*) only
D. (*b*) only

53. 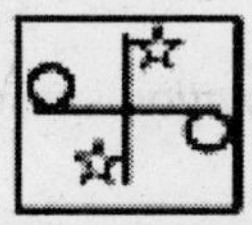

The water image of the above figure is:

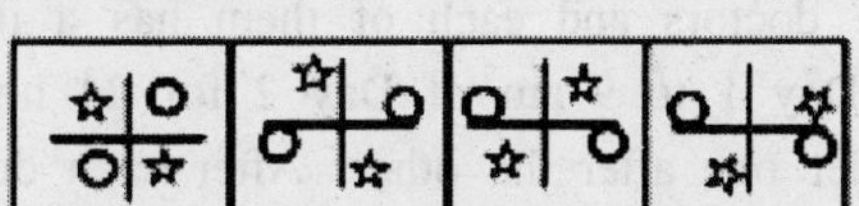

(*a*) (*b*) (*c*) (*d*)

A. (*d*) B. (*c*)
C. (*a*) D. (*b*)

48. D	49. B	50. C	51. A	52. A	53. D

54. Statement: The government has decided to construct an eight-lane super highway across the state to facilitate fast movement of vehicles.

Which of the following can be an **assumption** which is implicit in the above statement?

A. The government has adequate resources to construct the proposed super highway

B. The people of the state may protest against the Government's decision as their farm land will be taken over the Government for constructing the highway

C. The Government may find it difficult to enroll a suitable contractor for constructing the highway

D. There is no other highway which can be used for transportation of goods across the state

55. Statement: Thousands of farmers is drought-hit western Madhya Pradesh have been protesting for over a week demanding that authorities increase the minimum price for their crops and waive their bank loans. Madhya Pradesh is one of several states that have suffered droughts and crop failures in recent years. Which of the following courses of action will help to improve the current scenario?

(*i*) Government should punish those involved in the protests.

(*ii*) Cloud seeding should be used in such situations.

(*iii*) Government should constitute a team to assess the severity of drought and convince that appropriate compensation will be paid.

A. Only (*i*)

B. Both (*i*) and (*ii*)

C. Only (*iii*)

D. Both (*ii*) and (*iii*)

Directions (Qs. 56-57): *Based on the information, answer the questions which follow.*

In a department of a hospital, Amir, Sahil, Vaibhav and Gaurav are doctors and each of them has a duty from 9 am of Day 1 to 9 am of Day 2 for 24 hours in the same order one after the other. After each duty, the doctor gets an off for 3 days, including the day he is relieved from duty. For example, if Amir has a duty on Monday, his next duty will now fall on Friday and this rotation goes on.

56. If 10th July 2020, Gaurav was on duty, who will be on duty on 26th September 2020?

A. Amir

B. Sahil

C. Vaibhav

D. Gaurav

57. If on 14th February 2020, Sahil was on duty, then who did the maximum number of duties in that month?

A. Amir

B. Sahil

C. Vaibhav

D. Gaurav

58. Effect: The temple at the religious site wears a deserted look with the number of devotees trickling down.

Which of the following can be a possible causes of the above effect?

A. A structural engineer had visited the temple a month back and had declared the structure unsafe.

B. The temple is facing drastic depletion of its funds which had accumulated over the years due to offerings made by devotees.

C. The local corporation decided to donate a huge amount money to the temple for its renovation.

D. The village housing the religious site has qualified priests to perform religious ceremonies.

Directions (Qs. 59-60): *Answer the question with the help of the information given below.*

One word can be formed with the letters of the English alphabet in order with position numbered 5th, 24th and 16th and AND as a suffix.

59. Find the suffix of the word whose first three alphabets are 19th, 8th and 18th position in the English alphabet and have opposite meanings of the word formed earlier.

A. ACT	B. AND
C. INK	D. NAL

60. Find the suffix of the word which can be formed using the alphabets in order with position numbered 3rd, 15th, 14th, 20th and 18th and have the same meaning as the word formed in the above question:

A. ACT	B. AND
C. INK	D. NAL

54. A	**55.** C	**56.** B	**57.** A	**58.** A	**59.** C	**60.** A

EXPLANATORY ANSWERS

1. **(C):** The word 'moss' symbolizes inactivity and lack of progress. A rolling stone keeps moving, avoiding such stagnation.

2. **(B):** The phrasal verb 'narrowed down' means remove less important options to make it easier to choose. 'Opted for' means 'to choose', 'nailed down' means 'succeed of achieve something', 'passed out' means 'distribute' and 'passed away' means 'die'. Hence, option (B) is the correct answer.

3. **(A):** Q will be the first sentence as it gives an introduction to the paragraph that talks about 'Time'. Only option (A) starts with Q. So, it will be the answer.

4. **(D):** In passive voice the focus shifts to object being acted upon. The construction of the sentence emphasizes the safety aspect and removes the agent (year) performing the action.

5. **(A):** 'rushing river' completes the metaphor. It creates a vivid image of being overwhelmed by a constant flow of work, similar to being swept away by a strong current.

6. **(A):** "If I were you" is the correct formation as per the conditional rules.

7. **(D):** The word **treacherous** means 'presenting hidden or unpredictable dangers'. **Perilous** means 'full of danger or risk', **mutinous** means 'wilful or disobedient', **seditious** means 'inciting or causing people to rebel against the authority of a state monarch' and **secure** means 'certain to remain safe and unthreatened'. Thus, secure is the antonym of treacherous.

8. **(C):** Exemplary means commendable, ideal and admirable etc. thus, option (C) i.e., ideal is the correct answer.

9. **(A):** When we talk about a deadline for a future event, we use 'by'. We use 'until' when we talk about a continuous process that will stop in the future.

11. **(A):** 'Respectful' is the most appropriate choice because libraries are places where quiet is expected. It indicates consideration for others.

12. **(A):** The past participle 'written' is needed because the action of writing happened before Sarah's exhaustion and satisfaction.

14. **(C):** 'Although' is a conjunction used to introduce a contrasting clause.

16. **(B):** Let Roy bought N number of articles.

$\therefore$ Cost price $= 20$ N

Total S.P. to get 40% profit $= 20\text{N}\times\dfrac{140}{100} = 28$ N

Now, $2 + 4 + 6 + \ldots\ldots$ Nth term $= 28\text{N}$

$\Rightarrow 2(1 + 2 + 3 + \ldots\ldots\text{N}) = 28\text{N}$

$\Rightarrow 2\left[\dfrac{\text{N}(\text{N}+1)}{2}\right] = 28\text{N}$

$\Rightarrow \text{N}(\text{N} + 1)^2 = 28\text{N}$

$\Rightarrow \text{N} + 1 = 28$

$\Rightarrow \text{N} = 27$

$\therefore$ Minimum number of articles $= 27$.

17. **(B):** Let two bags A and B have x and $(45 - x)$ number of coins.

According to the question,

$$(x - 5)(45 - x - 5) = 124$$
$$(x - 5)(40 - x) = 124$$
$$\Rightarrow \quad 45x - x^2 - 200 = 124$$
$$\Rightarrow \quad x^2 - 45x + 324 = 0$$
$$\Rightarrow \quad (x - 9)(x - 36) = 0$$

$\therefore x = 9$ and 36

Product $= 9 \times 36 = 324$

18. **(C):** Imran's mobile phone selling price $=$ ₹ 17600

Discount $= 12\%$

$\therefore \quad$ Marked price $= 17600\dfrac{100}{(100-12)}$

$= 20000$

Profit $= 12\%$

$\therefore \quad$ Cost price $= 17600+\dfrac{100}{(100+10)}$

$= 16000$

$\therefore \quad$ Required profit % $= \dfrac{(20000-16000)}{16000}\times 100$

$= 25\%$

19. **(A):** Here, $\text{m} + \text{f} = 160$ ...(*i*)

and $7\text{m} + 12\text{f} = 1240$ ...(*ii*)

From (*i*) and (*ii*), we get that

$(12 - 7)\text{m} = 1920 - 1240 = 680$

$\therefore \text{m} = \dfrac{680}{5} = 136$

$\therefore$ Number of female present at the event

$= 160 - 136 = 24$.

20. **(C):** Milk : Water $= \dfrac{3}{4}:\dfrac{1}{4}$

When x ml mixture is replaced by x and water then,

$\therefore$ Milk water $= \left(\frac{3}{4}-\frac{3x}{4}\right):\left(\frac{1}{4}-\frac{x}{4}+x\right)$

$= 50 : 50$

$(3 - 3x) : (1 + 3x) = 1 : 1$

$\therefore 6x = 2$ or $x = \frac{1}{3}$.

21. (B): Let cost price of two types of sugar are y and z ₹/kg.

Here, profit% = 10%

$\therefore (3xy + 2xz) \times \frac{110}{100} =$ S.P. ...(*i*)

Similarly, $(2xy + 3xz) \times \frac{105}{100} =$ S.P. ...(*ii*)

From equation (*i*) and (*ii*), we get

$$(3y + 2z) \times \frac{110x}{100} = (2y + 3z) \times \frac{105x}{100}$$

$$(3y + 2z) \times 22 = (2y + 3z) \times 21$$

$$66y - 42y = 63z - 44z$$

$$24y = 19z$$

$\therefore \frac{y}{z} = \frac{19}{24}$.

22. (A): Average of first 3 number = 32

$\therefore$ Sum of first 3 numbers = 32 × 3 = 96

Let 4th number is d.

Then, $\left(\frac{96+d}{4}\right)$ gives the average of those 4 numbers.

According to the question,

$$d - \left(\frac{96+d}{4}\right) \times \frac{1}{4} = 24$$

$$16d - d = 24 \times 16 + 96$$

$$d = 32$$

23. (A): (A + B) total working hours = 2 × 5 = 10h

(B + C) total working hours = 4 × 5 = 20h

(A + C) total working hours = 2 × 6 = 12h

Now, L.C.M. of 10, 20 and 12 = 60

$\therefore$ A + B efficiency = 6

B + C efficiency = 3

C + A efficiency = 5

From this,

A − B = 5 − 3 = 2

$\therefore$ A = 4

Hence, A will finish work in $\frac{60}{4}$ = 15 hours

If he work 5 hours a day, then, will finish

$= \frac{15}{5}$

= 3 days.

24. (C): Sonu : Monu : Romu

3 : 2 : ×3

×2 3 : 2

9 : 6 : 4

Total run scored = 342.

$(9 + 6 + 4)x = 342$

$\Rightarrow x = \frac{342}{19} = 18$

$\therefore$ Number of run scored by Romu = 4 × 18 = 72.

25. (B):

V	C	V	C	V	C
1st		3rd		5th	

Vowel → E, I, U

Consonant → M, N, T

Arrange = Three vowels occupy three odd positions in 3! ways and remaining 3 consonant will occupy three positions in 3! ways.

$\therefore$ Total number of ways = 3! × 3! = 36.

26. (C):

A	B	C	
6	0	1	$\Rightarrow \frac{7!}{6!} \times 3! = 42$
5	1	1	$\Rightarrow \frac{7!}{5!} \times \frac{3!}{2!} = 126$
4	2	1	$\Rightarrow \frac{7!}{2! \times 4!} \times 3! = 630$
3	3	1	$\Rightarrow \frac{7!}{3! \times 3!} \times \frac{3!}{2!} = 420$

$\therefore$ Total number of ways distribution

= 42 + 126 + 630 + 420

= 1218.

27. (C): Probability of choosing a group $x = \frac{1}{6}$

Probability of choosing a group $y = \frac{2}{6} = \frac{1}{3}$

Probability of choosing a group $z = \frac{3}{6} = \frac{1}{2}$

Now, Probability of choosing a girl from group $x = \frac{3}{5}$

Probability of choosing a girl from group $y = \frac{3}{7}$

Probability of choosing a girl from group $z = \frac{4}{9}$

$\therefore$ Probability of choosing a girl

$$P(G) = \frac{3}{5}\times\frac{1}{6}+\frac{3}{7}\times\frac{1}{3}+\frac{4}{9}\times\frac{1}{2}$$

$$= \frac{1}{10}+\frac{1}{7}+\frac{2}{9}=\frac{63+90+140}{630}$$

$$= \frac{293}{630}$$

$\therefore$ Probability that she is chosen from group y

$$P\left(\frac{Y}{G}\right) = P\left(\frac{Y \cap G}{P(G)}\right)$$

$$= \frac{\frac{1}{7}}{\frac{293}{630}}$$

$$= \frac{90}{293}$$

28. (A): Probability for Mohini speaks truth P(M)

$$= \frac{6}{9}=\frac{2}{3}$$

Probability for Rohini speaks truth $P(R) = \frac{3}{6}=\frac{1}{2}$

$\therefore$ Probability that they contradict each other on certain facts $= P(M) . P(R') + P(M') . P(R)$

$$= \frac{2}{3}\times\left(1-\frac{1}{2}\right)+\left(1-\frac{2}{3}\right)\times\frac{1}{2}$$

$$= \frac{2}{3}\times\frac{1}{2}+\frac{1}{3}\times\frac{1}{2}=\frac{1}{2}$$

29. (B): $y = (x + 1)$ and $z = (x + 2)$

$\log(1 + xz) = 2A$

$\log(1 + x(x + 2)) = 2A$

$\log(1 + x^2 + 2x) = 2A$

$\log(1 + x)^2 = 2A$

$2 \log(1 + x) = 2A$

$\therefore A = \log(1 + x) = \log(y)$

30. (C): Let first term $= a$ and last term $= l$ and common difference $= d$.

Difference between 9th and 13th term = 20

$a + (9 - 1) . d - \{a + (13 - 1) . d\} = 20$

$- 4d = 20$

$\Rightarrow d = - 5$

Now for a series of n terms (says)

Now, first term $(a) - 75$ = Sum (s)

$\Rightarrow s = a - 75$

and sum = 0 + last term (1)

$\Rightarrow$ sum = 1

$\therefore a - 75 = 1 = a + (n - 1)(-5)$

$\therefore n - 1 = 15$

$\Rightarrow n = 16$

Sum of the series except last term = 0

$$\left(\frac{n-1}{2}\right)[2a+(n-1-1)d]=s$$

$$\left(\frac{16-1}{2}\right)[2\times a+(16-2(-5)]=0$$

$a = 35.$

31. (B):

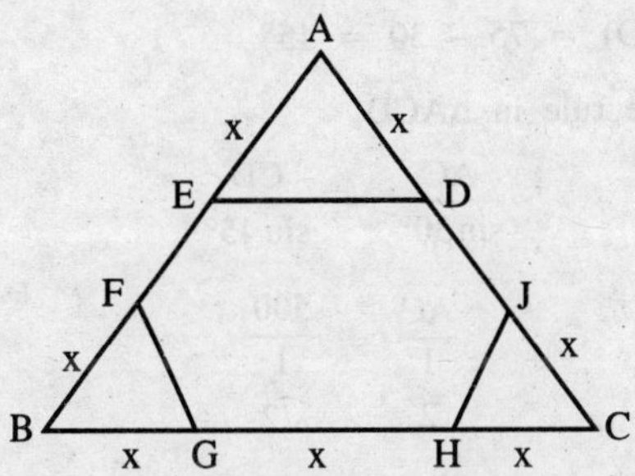

Let the side of the equilateral Δ is $3x$ units then, Area of equilateral $\Delta = \frac{\sqrt{3}}{4} \times (\text{side})^2$

$$240 = \frac{\sqrt{3}}{4} \times (\text{side})^2$$

$$= \frac{\sqrt{3}}{4} \times (3x)^2$$

Again, when this triangle converted into regular hexagon of each side x unit.

Then, In ΔADE, AE = AD = DE = x unit

$\therefore$ By symmetry, ΔADE = ΔBFG = ΔCJH are equilateral triangle of same area,

Area of ΔADE $= \frac{\sqrt{3}}{4}(x^2)$

$\therefore$ Area of hexagonal $= \frac{9\sqrt{3}}{4}(x^2)-\frac{3\sqrt{3}}{4}(x)^2$

$$6\times\frac{\sqrt{3}}{4}(x)^2=\frac{6\times 240}{9}$$

= 160 unit.

32. (A): Maximum number of flower $= \left[\frac{125}{3}\right] = 41$

Maximum number of flower $= \left[\frac{375}{3}\right] = 125$

Maximum number of flower $= \left[\frac{250}{3}\right] = 83$

$\therefore$ Maximum rows of flower = 41 + 125 + 83 = 249.

33. (A): Let height of the building AB = h m and BC = x m.

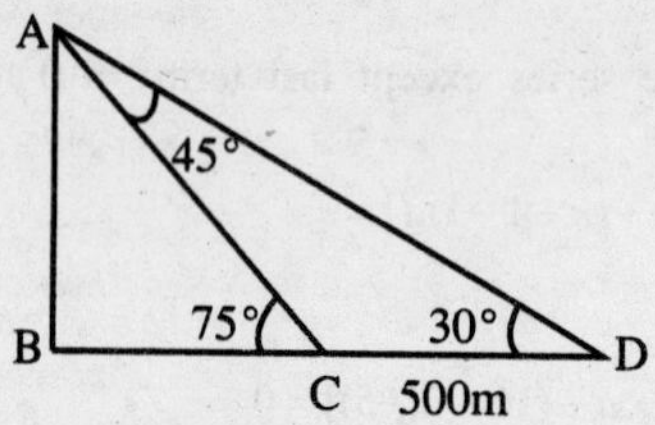

From ΔACD, ∠ACB = ∠CDA + ∠CAD

75° = (30°) + (∠CAD)

$\therefore$ (∠CAD) = 75 − 30 = 45°

From sine rule in ΔACD,

$$\frac{AC}{\sin 30°} = \frac{CD}{\sin 45°}$$

$$\Rightarrow \quad \frac{AC}{\frac{1}{2}} = \frac{500}{\frac{1}{\sqrt{2}}}$$

$$\therefore \quad AC = 250\sqrt{2}$$

From ΔABC,

$$\sin 75° = \frac{AB}{AC} = \frac{h}{250\sqrt{2}}$$

$$\sin(45° + 30°) = \frac{h}{250\sqrt{2}}$$

sin 45°, cos 30° + cos 45°

$$\sin 30° = \frac{h}{250\sqrt{2}}$$

$$\frac{1}{\sqrt{2}}.\frac{\sqrt{3}}{2} + \frac{1}{\sqrt{2}}.\frac{1}{2} = \frac{h}{250\sqrt{2}}$$

$$\frac{\sqrt{3}+1}{2\sqrt{2}} = \frac{h}{250\sqrt{2}}$$

$$\therefore h = 125\left(\sqrt{3}+1\right).$$

34. (A): Let total population of UP = N

Then, population above education index $= N \times \left(\frac{100-16}{100}\right)$

= 0.84 N

Male population above education index

$$= \frac{5}{6} \times 0.84N = 0.7N$$

$$\therefore \quad N\frac{1.6}{0.7} \approx 2.28 \text{ lakhs}$$

35. (C): Pattern is: $3 + 2 = 3 + (2)^2 = 7$

$32 - 8 = 32 - (8)^2 = -32$

$12 + 3 = 12 + (3)^2 = 12 + 9 = 21$

$51 + (?)^2 = 76$

$\Rightarrow \sqrt{76-51} = ?$

$\therefore$? = 5.

36. (B): 1 → 3 → 7 → 13 → 21 → 31 → [43]

+2, +4, +6, +8, +10, +12

Here, the next term = 43.

37. (B): Tue → Wed → Fri → Mon → Fri → Wed

+1 days, +2 days, +3 days, +4 days, +5 days

Here, the missing day = Friday.

38. (D):

Letter	B	O	S	Z	E	R	A
Code	8	0	5	2	3	9	4

$\therefore$ Code of LEETS = 13375.

39. (C):

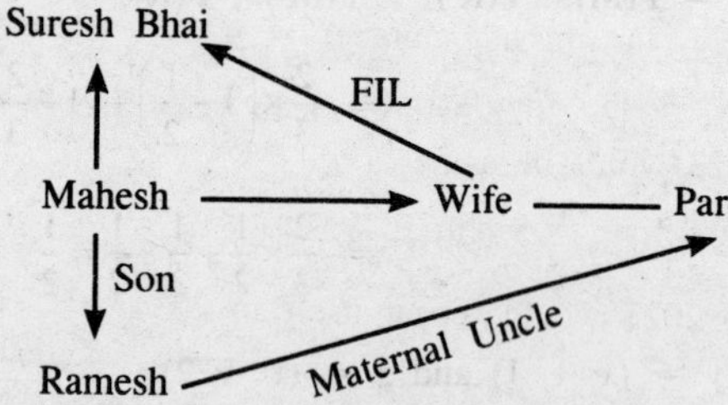

$\therefore$ Paresh is Ramesh maternal uncle.

40. (C): Logical order:

2. Aga Khan Palace
1. Pune
5. Maharashtra
4. India
3. World.

41. (B): Value of P at 6:00 = 180°

Value of P at 18:00 = 540°

Value of P at 12:00 = 360°

$$\therefore \frac{180+540}{360} = 2.$$

42. (A): Probability of Mohini speaking truth P(M)

$$=\frac{6}{9}=\frac{2}{3}$$

Probability of Mohini not speaking truth P(M)

$$=1-\frac{2}{3}=\frac{1}{3}$$

Probability of Rohini speaking truth P(R)

$$=\frac{3}{6}=\frac{1}{2}$$

Probability of Rohini not speaking truth = P(R°)

$$=\frac{1}{2}$$

Required probability = P(M) . P(R′) + P(R) R′(M)

$$=\frac{2}{3}\times\frac{1}{2}+\frac{1}{2}\times\frac{1}{3}$$

$$=\frac{1}{3}+\frac{1}{6}$$

$$=\frac{1}{2}.$$

43. (A): 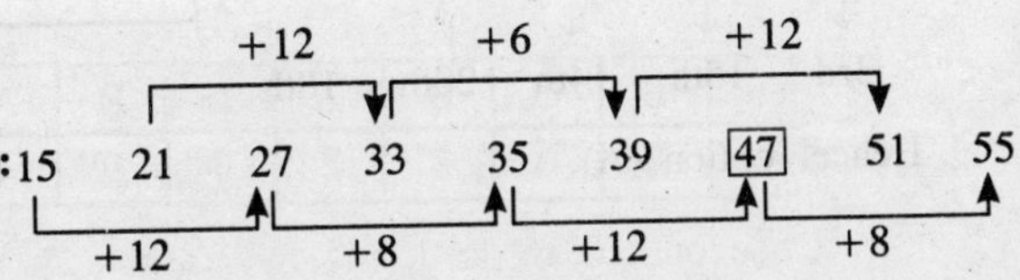

Here, the missing number = 47.

45. (B): 2032, 2028 will be leap years, so the calendar of 2023 won't be same

Also 2022 = 2020 (leap year) + 3

So 2023 would repeat itself in 2034.

46. (A): 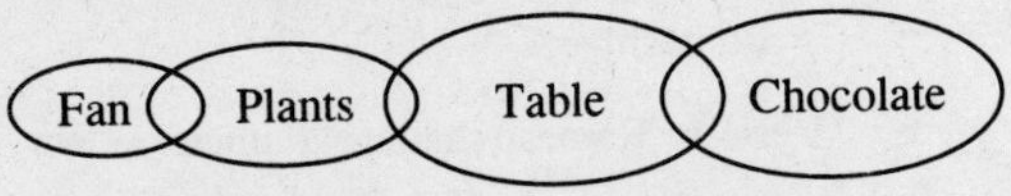

47. (D): 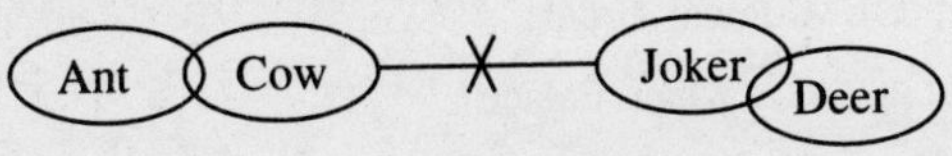

48. (D): With reference to the given statement Only assumption II is implicit i.e. One of the major goals of sustainability is to preserve biodiversity. All life on Earth is connected through the flow of energy (planetary food web), and each time a species becomes endangered or lost to extinction, one more part of that energy flow is lost ,

49. (B): Course of action I follows because with a graded approach to the use of antibiotics, under which some medicines are reserved for the most resistant microbes, the WHO list can stop their misuse as broad-spectrum treatments. Course of action II also follows because a close inspection is also needed of the national supervisory programmes to help people to get rid out of it

For Q.No. 50-51:

Let the six professors be denoted as P, Q, R, S, T and U.

The first thing we should focus on doing is to identify which professors are in the North facing flats and which professors are in the South facing flats.

From the last two statements, viz: "R gets a south facing flat which is next to U's flat" and "T's flat is North Facing".

At this point we also know that "S and U get flats which are diagonally opposite to each other" and since we know that U faces South, S would naturally face North; and we can also deduce that both S and U must be occupying corner flats.

Further from the statement: "Q gets a North facing flat and it is not next to S/s flat" we realise two things:

T, S, Q are in the three North facing flats and since S and Q are not adjacent to each other, they must be both occupying corner flats while T would be occupying the middle flat in the North facing flats.

Consequently P would be in a South facing flat and the three people in South facing flats would be P, R, U. Also since R is next to U and we have already deduced that U is in the corner flat, we know that R would be in a middle flat.

The figure changes to the following:

Flat Direction	Corner Flat	Middle Flat	Corner Flat	Professors facing Sooth and North
South Facing Flats				R (corner flat), R (middle flat), U (corner flat)
North Facing Flats				T (in middle flat), S & Q (corner flat)

Consequently we also realise that there are essentially the following 2 ways of arranging the 6 people:

Possibility 1:

Flat Direction	Corner Flat	Middle Flat	Corner Flat	Professors facing Sooth and North
South Facing Flats	P	R	U	R (corner flat), R (middle flat), U (corner flat)
North Facing Flats	S	T	Q	T (in middle flat), S & Q (corner flat)

Possibility 2:

Flat Direction	Corner Flat	Middle Flat	Corner Flat	Professors facing Sooth and North
South Facing Flats	U	R	P	P (corner flat), R (middle flat), U (corner flat)
North Facing Flats	Q	T	S	T (in middle flat), S & Q (corner flat)

Based on these reactions and deductions we can solve the questions.

50. (C): P, R and U get South facing flats as deduced above and hence, option (C) is the correct answer.

51. (A): From the final conclusions above, it is quite clear that there is no change in U's neighbour if P and T interchanged their flats. Option (A) is correct.

52. (A): All the three causes will effect the sales of four wheelers.

54. (A): Government decision to construct super highway shows that Government has sufficient resources to construct it.

55. (C): (i) is not an appropriate course of action as it not known from the statement weather the protest are violent or peaceful (ii) is preventive measure not a corrective action. As it is mentioned in the statement that the given state has faced several drought like situations over the recent years, so (iii) is an appropriate course of action.

56. (B): Each person get his next duty after the gap of 3 days.

As on 10th July 2020, Gaurav was on duty. So, his next duty was on 14th, 18th, 22nd .... July

Now, number of days between 10th July to 26th Sept.

$= 21 + 31 + 26$

$= 78$ days.

$= (4 \times 19 + 2)$ days.

$\therefore$ On 24th September Gaurav was on duty and on 26th September Sahil was on duty.

57. (A): As February 2020 is 29 days (Leap year)

Then, person having duty on 1st February, must have their duty on 29th February 2020.

Now, number of days between 1st and 14th = 13 days.

$= 4 \times 3 + 1$ odd day.

So, Amir was on duty since on 1st February and 29th February

Hence, Amir did the maximum number of duties.

59. (C): As per the question, we can easily from the word

E	X	P	A	N	D
↓	↓	↓			
5th	24th	16th			

First three letters are:

S H R [I N K]
↑ as suffix

Hence, option (C).

60. (A): "SHRINK" is the word was the answer formed in the previous question.

According to the question,

C	O	N	T	R	[A C T]
↓	↓	↓			
3rd	15th	14th	20th	18th	

Hence, option (A).

Previous Paper (Solved)

Symbiosis National Aptitude Test (SNAP), 2022*

Directions (Qs. No. 1-5): *Read the following passage and answer the questions given below:*

Wayanad, Congress President Rahul Gandhi's choice as his second constituency, is arguably the safest seat for the Congress in south India and among the safest in the entire country. Mr. Gandhi's main rival will be P.P. Suneer of the CPI. The BJP has a marginal presence in the constituency. The Congress chief had been undecided over contesting from Wayanad, considering the multiple battles of perception that it brings. First, while he has been pointedly trying to define the 2019 election as a fight to save India from advancing Hindutva, he will end up facing off with the Left, an adversary in Kerala but an ally nationally. Second, he will be accused of "running away" from Amethi. Third, the more fundamental question: whether contesting two seats will be seen as a sign of weakness. After weighing the pros and cons, on which he received **divergent** opinions, the decision was taken. Congress sources close to Mr. Gandhi are **emphatic** that they sense no challenge in Amethi. But the volatility in Uttar Pradesh, despite the formation of the SP-BSP alliance and its declaration of support for Mr. Gandhi in Amethi, is palpable. The battle lines in the critical State have not been defined yet, and Prime Minister Narendra Modi's campaign blitz has only begun. They said, the reason for national leaders contesting a second seat is less because of any insecurity in one seat and more for expanding their personal imprint to critical territories, as Mr. Modi himself did in 2014 by contesting from Vadodara in Gujarat and Varanasi in U.P.

1. Which seat according to the passage, is the safest seat for Congress in South India?

A. Kannur
B. Wayanad
C. Kozhikode
D. Pathanamthitta

2. Which word is closest to the meaning of the word 'Emphatic' in the passage given?

A. Hesitant
B. Tentative
C. Ambiguous
D. Decisive

3. What is the reason for national leaders less contesting for a second seat?

A. Over insecurity in one state
B. Expanding their personal imprint to critical territories
C. Both (A) and (B)
D. Because of election lines not being defined properly

4. Why is Congress President undecided in contesting from Wayanad?

A. Battle with the left
B. Accused of running away from Amethi
C. Sign of weakness
D. All the three

5. Which word is closest to the meaning of the word 'divergent' in the passage given?

A. Correspondent
B. Homogenous
C. Similar
D. Diverse

6. The sentences given with blanks are to be filled with an appropriate word(s). Four alternatives are suggested for each question. For each question, choose the correct alternative.

Do your best and we'll back you ______

A. on
B. in
C. out
D. up

7. The sentences given with blanks are to be filled with an appropriate word(s). Four alternatives are suggested for each question. For each question, choose the correct alternative.

Sanjay ______ his mother in the morning every day.

A. calls in
B. calls up
C. calls off
D. calls down

8. In each of the questions, four alternatives are given for the Idiom/Phrase. Choose the alternative which best expresses the meaning of the Idiom/Phrase.

An axe to grind

A. Attack aggressively
B. Suffer a lot
C. Betray somebody
D. Have a selfish interest

9. A sentence/a part of the sentence is underlined. Four alternatives are given to the underlined part which will improve the sentence. Choose the correct alternative. In case no improvement is needed, select the option "No improvement".

I'm staying with some friends who <u>are owning</u> a farm.

A. will be owning
B. own
C. have been owning
D. No improvement

* Based on memory

10. The sentences given with blanks are to be filled with an appropriate word(s). Four alternatives are suggested for each question. For each question, choose the correct alternative.

Most children remain ______ school between the ages of seven and ten.

A. in B. under
C. at D. inside

11. In the following question, out of the four alternatives, select the alternative which will improve the bracketed part of the sentence. In case no improvement is needed, select "No improvement".

All (was surprising) to find that he was not with them.

A. was surprised B. were surprised
C. is surprised D. No improvement

12. In the following question, the sentence given with blank to be filled in with an appropriate word. Select the correct alternative out of the four and indicate it by selecting the appropriate option.

The diamond necklace was _______ too extravagant for a simple dinner party.

A. distance B. long
C. very much D. far

13. Select the antonym of the given word.

TENDER

A. soft B. warm
C. gentle D. rough

14. Select the synonym of the given word.

GARRULOUS

A. guttural B. throaty
C. concise D. talkative

15. Select the most appropriate option to fill in the blank.

She got a lucrative job of a translator because she was ______ in French.

A. Deficient B. Sufficient
C. Proficient D. Efficient

16. Raja is the son of Varun's father's sister. Sameer is son of Ganga, who is mother of Jai and grandmother of Varun. Nitin is father of Varsha and grandfather of Raja. Ganga is the wife of Nitin. How is Raja related to Ganga?

A. Son B. Nephew
C. Grandson D. Data inadequate

17. While going to his office Tejveer meets Rakesh who is related to Tejveer because Tejveer is Deepak's father who is married to Garima. Garima is the daughter of Rakesh. Garima has a daughter named Isha. How is Tejveer related to Isha?

A. Uncle B. Grandfather
C. Father-in-law D. Grandmother

18. Pointing to a lady in a photograph, Dev said, "This woman is my sister's father's son-in-law's wife." How is woman related to the Dev.

A. Wife
B. Mother
C. Sister
D. Can't say

19. Five friends namely L, M, N, R and S are standing in a row facing south, but not necessarily in the same order.

- Only M is between L and S.
- N is to the immediate right of S.
- R is to the immediate left of L.

Who is occupying the middle position in the row?

A. L B. S
C. M D. R

20. Hemant walks from his house and moves 4 metres towards North. He turns left at 90° to his right and moves 3 metres. He takes another 90° turn to the right and walks 1 metre. From here Hemant turns 90° to the left and moves for another 1 metre. He finally turns 90° to the right and moves for another 3 metres, his final destination. What is the distance between the starting and destination points?

A. 4 metres B. 5 metres
C. 7 metres D. 8 metres

21. Five persons are standing in a line in a ceremony. One of the two persons standing at the extreme ends is a General Manager and the other one is a Senior Manager. The Deputy General Manager is standing to the right of Assistant General Manager. A Deputy Manager is to the left of the Senior Manager. The Assistant General Manager is standing between the General Manager and the Deputy General Manager.

Counting from the left, what is the position of the Deputy General Manager?

A. First B. Second
C. Third D. Fifth

Directions (Qs. No. 22 & 23): *Complete the series by replacing the '?' mark.*

22. XZ 16 BD, FH 64 JL, NP 144 RT, VX 256 ZB, '?'

A. DF 400 HJ
B. DF 515 HJ
C. CE 400 HJ
D. DF 324 HJ

23. 15 ABC 28, 18 DEF 32, 21 HI 36, '?', 27 MNO 44

A. 23 JKL 38 B. 24 JKL 40
C. 25 JKL 42 D. 26 JKL 43

24. What does the area marked 1 in the figure given below represent?

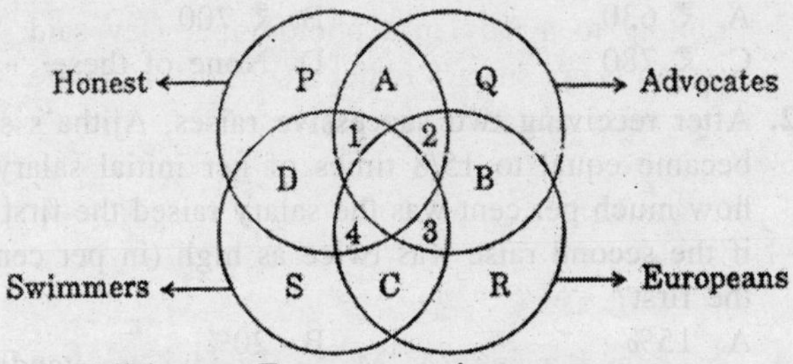

A. All honest European swimmers
B. All honest advocates who are swimmers
C. All non-European advocates who are honest swimmers
D. All non-Europeans who are honest swimmers

25. Non-Americans who are unsuccessful and dishonest are represented by which letter in the figure given below?

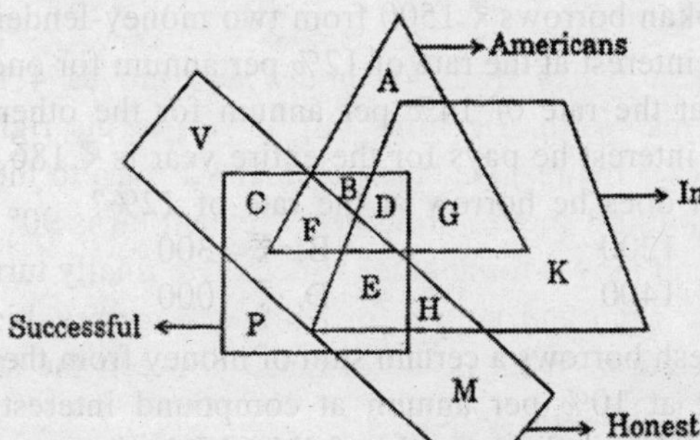

A. A B. K
C. M D. C

26. P, Q, R, S, T, U and V are sitting in a circle facing the centre. S, who is second to the right hand side of R, is not to the immediate right of V. U is not between V and T. P is between R and Q. If Q and R interchange places so as to be like T and V, then which of the following is true?
A. S is third to the right of R
B. T is second to the left of R
C. Q is fourth to the right of T
D. None of these

27. Five boys are standing in a line facing the wall. Each boy is either wearing red, green, yellow, white or blue dress. The boy dressed in yellow is not standing at any end of the line. The boy dressed in red is not standing at any end of the line. What colour dress has been worn by the boy who is standing to the right side of the boy dressed in yellow?
A. White B. Green
C. Blue D. Cannot be determined

28. Five shirts are placed on a shelf in a heap. Red shirt is above the Blue shirt and Green shirt is below the Orange shirt. Blue shirt is above the Orange shirt and White shirt is below the Green shirt. The colour of the shirt between Red and Orange shirts is
A. White B. Green
C. Blue D. None of these

29. M, N, O and P are all distinct individuals. M is the daughter of N. N is the son of O and O is the father of P. If B is the son of N and B has one brother, D, then which of the following statements is/are true?
I. M is the sister of D.
II. D and B are brothers.
III. O is the grandfather of D.
A. I only B. I and III only
C. III only D. All these

30. Pointing out to the man on the stage, Madhu said that he is the brother of the daughter of the wife of my husband. How is the man on the stage related to Madhu?
A. Son B. Husband
C. Cousin D. Nephew

31. X introduces Y saying, "He is the husband of the grand daughter of the father of my father." How is Y related to X?
A. Brother B. Son
C. Brother-in-law D. Son-in-law

Directions (Qs. No. 32 & 33): *Complete the series by replacing the '?' mark.*

32. 4A, 12D, 28G, 60J, 124M, '?'
A. 248Q B. 252P
C. 256T D. 246S

33. P3C, R5F, T8I, V12L, '?'
A. Y17O B. X17M
C. X17O D. X16O

34. A cube, on whose sides letters have been written, is shown below in different positions as can be seen from different directions. Which letter will replace '?' mark?

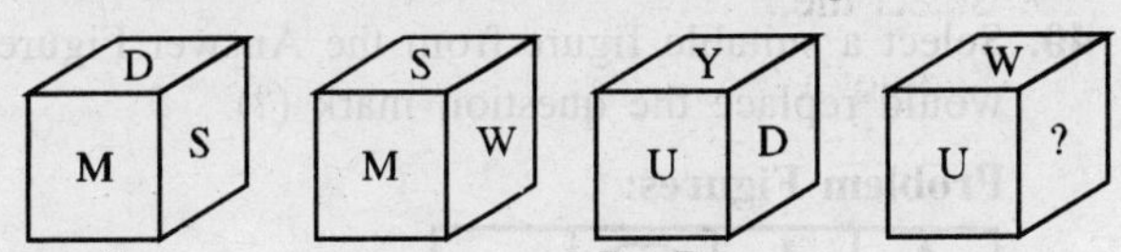

A. S B. D
C. Y D. W

35. Six sides of a cube are coloured in the following manner.

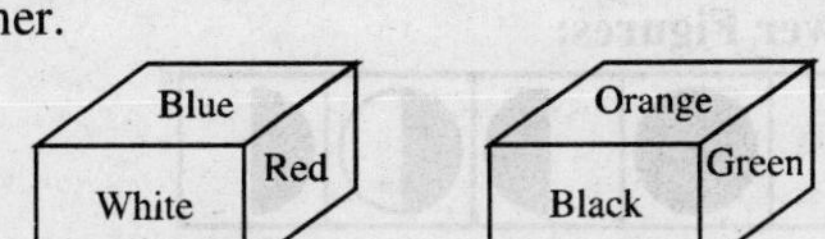

If blue and orange are opposite and red is on the top, which colour will be at the bottom?
A. Orange B. White
C. Black D. Green

36. The sides of a cube are painted in different colours. Black side is opposite to red. White side is between black and red. Green side is adjacent to grey and blue side is adjacent to green. What colour will be on the side opposite to the white side of the cube?

A. Blue B. Green
C. Grey D. Data insufficient

37. A large cube is painted on all its six faces. It is then cut into a certain number of smaller identical cubes. It was found that among the smaller cubes, there were eight cubes which didn't have any face painted. How many small cubes have exactly three faces painted?

A. 6 B. 8
C. 27 D. None, of these

38. Choose from the given diagrams the one that illustrates the relationship among three classes
China, India, Asia

A. 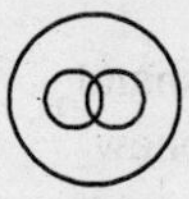 B. 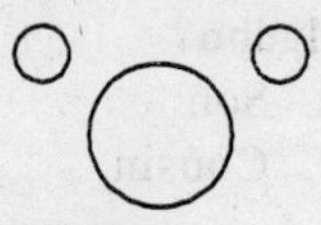

C.  D. 

39. Select a suitable figure from the Answer Figures that would replace the question mark (?).

Problem Figures:

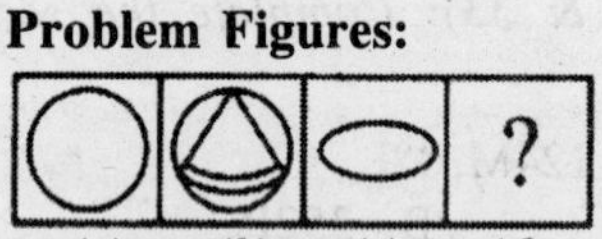

(a) (b) (c) (d)

Answer Figures:

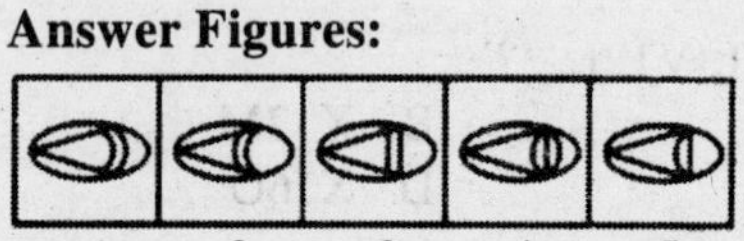
1 2 3 4 5

A. 1 B. 2
C. 3 D. 4

40. Select a suitable figure from the Answer Figures that would replace the question mark (?).

Problem Figures:

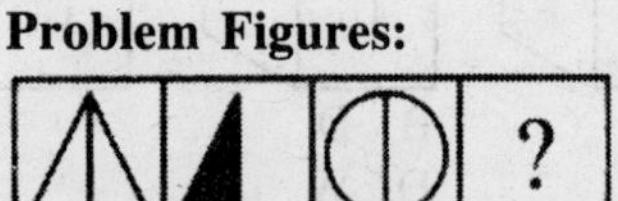

(a) (b) (c) (d)

Answer Figures:

1 2 3 4 5

A. 1 B. 2
C. 3 D. 5

41. A shopkeeper allows a discount of 10% on the marked price of an item but charges a sales tax of 8% on the discounted price. If the customer pays ₹ 680.40 as the price including the sales tax, then what is the marked price of the item?

A. ₹ 630 B. ₹ 700
C. ₹ 780 D. None of these

42. After receiving two successive raises, Ajitha's salary became equal to 15/8 times of her initial salary. By how much per cent was the salary raised the first time if the second raise was twice as high (in per cent) as the first?

A. 15% B. 20%
C. 25% D. 30%

43. Find the principal if compound interest is charged on the principal at the rate of $16\frac{2}{3}$ % per annum for two years and the sum becomes ₹ 196.

A. ₹ 140 B. ₹ 154
C. ₹ 150 D. None of these

44. Ashokan borrows ₹ 1500 from two money-lenders. He pays interest at the rate of 12% per annum for one loan and at the rate of 14% per annum for the other. The total interest he pays for the entire year is ₹ 186. How much does he borrow at the rate of 12%?

A. ₹ 1200 B. ₹ 1300
C. ₹ 1400 D. ₹ 1000

45. Mukesh borrows a certain sum of money from the ABC Bank at 10% per annum at compound interest. The entire debt is discharged in full by Mukesh on payment of two equal amounts of ₹ 1,000 each, one at the end of the first year and the other at the end of the second year. What is the approximate value of the amount borrowed by him?

A. ₹ 1852 B. ₹ 1694
C. ₹ 1736 D. ₹ 1792

46. To fill 12 vacancies there are 25 candidates of which 5 are from scheduled castes. If 3 of the vacancies are reserved for scheduled caste candidates while the rest are open to all, what is the number of ways in which the selection can be made?

A. 5456371 B. 4974400
C. 4823156 D. None of these

47. A TV tower stands vertically on a bank of a canal. From a point on the other bank directly opposite the tower, the angle of elevation of the top of the tower is 60°. From another point 20 m away from this point on the line joining this point to the foot of the tower, the angle of elevation of the top of the tower is 30°. The width of the canal is

A. 10 m B. $10\sqrt{3}$ m
C. $(10\sqrt{3})/3$ m D. None of these

48. A cylindrical box of radius 5 cm contains 10 solid spherical balls each of radius 5 cm. If the topmost ball touches the upper cover of the box, then the volume of the empty space in the box is

A. $\frac{5000}{3}\pi\ cm^3$ B. $2500\ \pi\ cm^3$

C. $\frac{2500}{3}\pi\ cm^3$ D. $5000\ \pi\ cm^3$

49. B is a sleeping partner and A working. A puts in ₹ 5,000 and B puts in ₹ 6,000. A received 12.5% of profit for managing the business and rest is divided in proportion to their capitals. A's share of profit in a profit of ₹ 880 is

A. ₹ 350 B. ₹ 400

C. ₹ 420 D. ₹ 460

50. A tin of oil was 4/5 full. When six bottles of oil were taken out and four bottles of oil were poured into it, it was 3/4 full. How many bottles of oil can the tin contain?

A. 10 B. 20

C. 30 D. 40

Directions (Qs. No. 51 to 55): *Study the following graph to answer these questions.*

The bar graph shows the performance of 3 political parties A, B and C in terms of votes acquired in the elections in a constituency during 2008–2012.

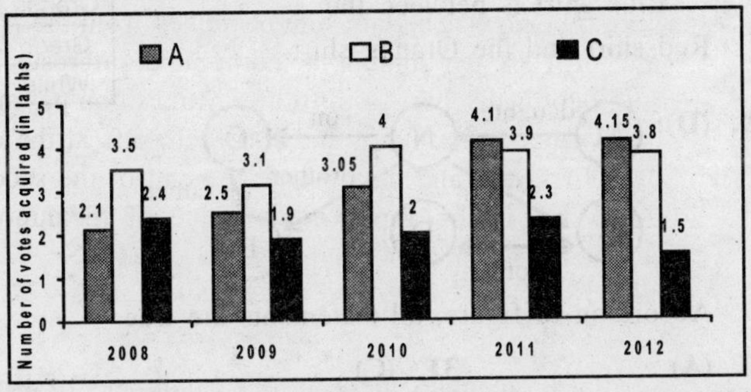

51. By what per cent is the average number of votes acquired by B more/less than that of A during 2008–2012?

A. 13% more B. 13% less

C. 15% less D. 15% more

52. How many more votes are required by C in 2013 than in 2012 to make the ratio of votes with A, 3 : 2, keeping the votes of A same in 2012?

A. 12.45 lakh B. 4.725 lakh

C. 6.225 lakh D. 1.27 lakh

53. By what per cent should votes of B increase/decrease in 2012 to equal its average during the previous 4 years?

A. Increase by 4·6% B. Increase by 5·8%

C. Decrease by 5·8% D. Decrease by 4·6%

54. In 2011, by what per cent should the votes of C increase to exceed the votes of B in the same year by 12%?

A. 51% B. 33.9%

C. 89.9% D. 53%

55. What per cent of votes were acquired by C with respect to the total votes acquired by all the three parties in 2009?

A. 25.33%

B. 41.33%

C. 33.33%

D. 61.2%

Directions (Qs. No. 56 to 60): *Study the graph to answer these questions.*

The graph below shows the number of graduates (in '000) of different streams, in six different states of a country.

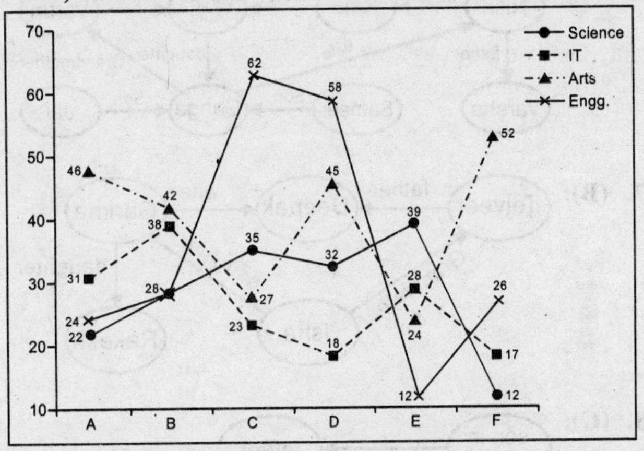

56. The total number of graduates of which two streams from all the states taken together are in the ratio 4 : 5?

A. Science and Engg.

B. Science and Arts

C. Arts and IT

D. Engg. and IT

57. In which two states is the number of IT graduates, as a percentage of the total number of graduates in that state, most nearly the same?

A. C and F B. E and B

C. A and E D. D and F

58. By what percent is the sum total of Engg. graduates in B, C and F taken together greater than the sum total of Arts graduates in D and E taken together?

A. 56 B. 68

C. 65 D. 86

59. The number of Science graduates in all the states combined is what percent of the number of graduates of all other streams taken together?

A. 28.5 B. 27.9

C. 29.7 D. 21.4

60. In which states is the difference between the number of Engg. and IT graduates higher than the difference between the number of Science and Arts graduates?

A. A, C and E

B. C, F and B

C. F, B and D

D. C, D and E

ANSWERS

1. (B) **2. (D)** **3. (C)**

4. (D) **5. (D)** **6. (D)**

7. (B) **8. (D)** **9. (B)**

10. (C) **11. (A)** **12. (D)**

13. (D) **14. (D)** **15. (C)**

16. (C):

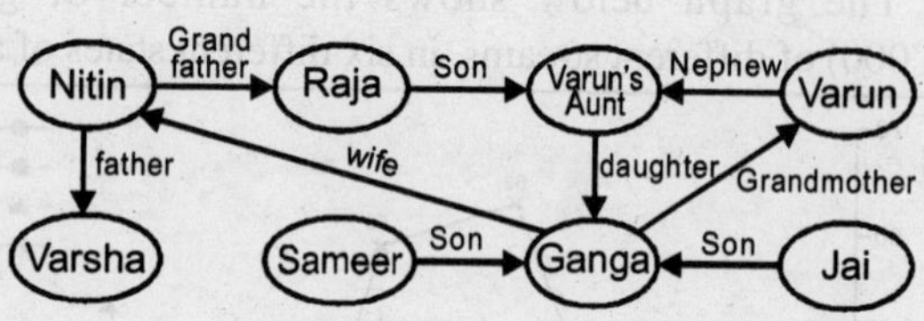

17. (B):

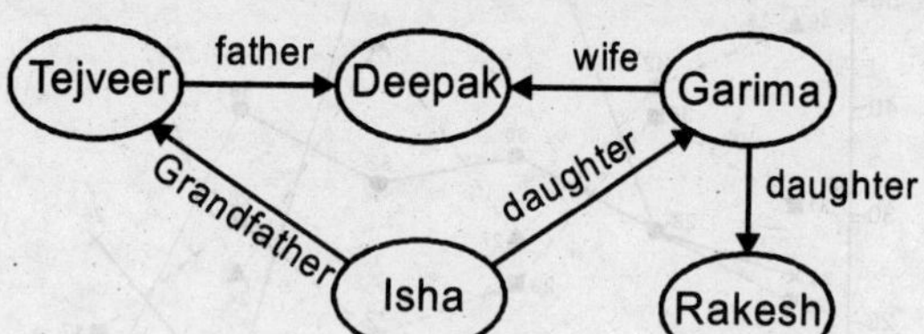

18. (C):

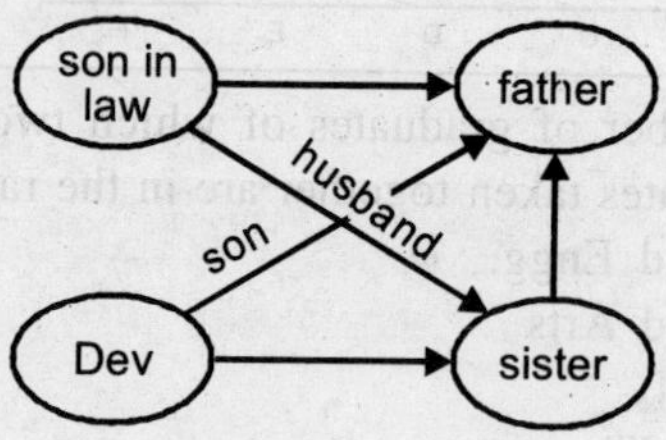

19. (C): Based on given, arrangement of five friends facing south are given below

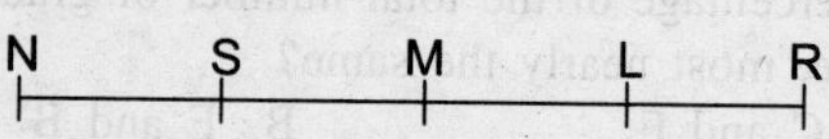

Hence, M is at the middle position.

20. (A):

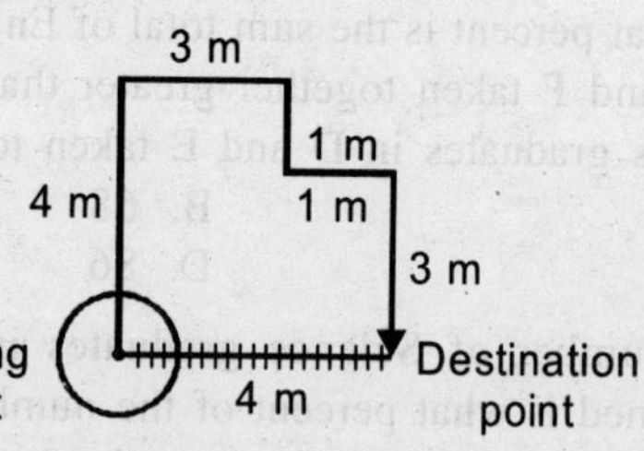

21. (C):

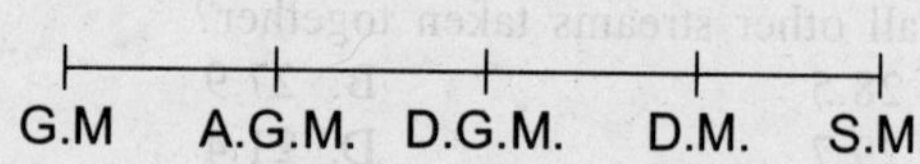

Hence, Deputy General Manager (DGM) is in third position from both ends.

22. (A):

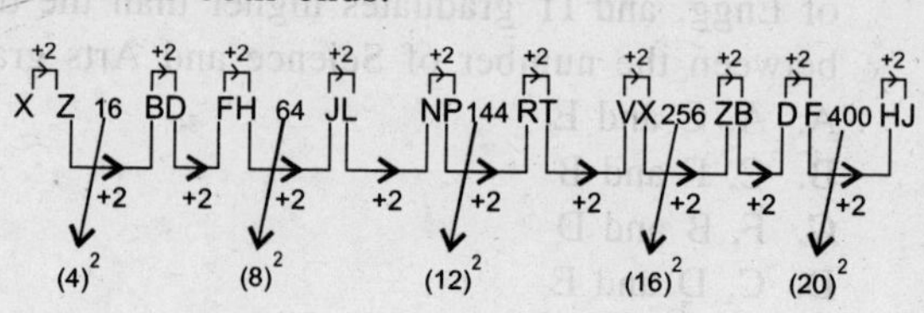

23. (B):

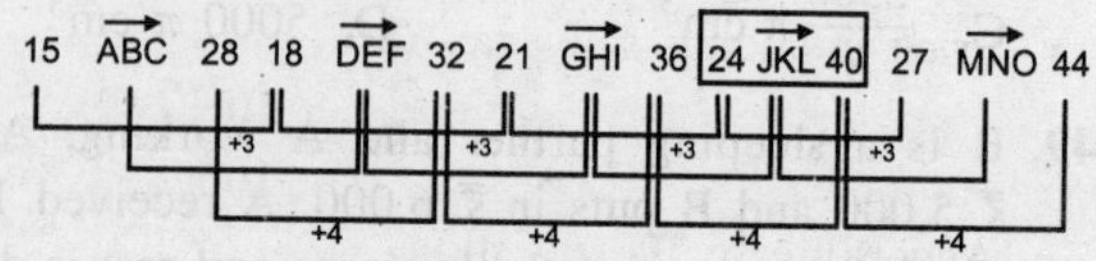

24. (B) **25. (B)**

26. (B):

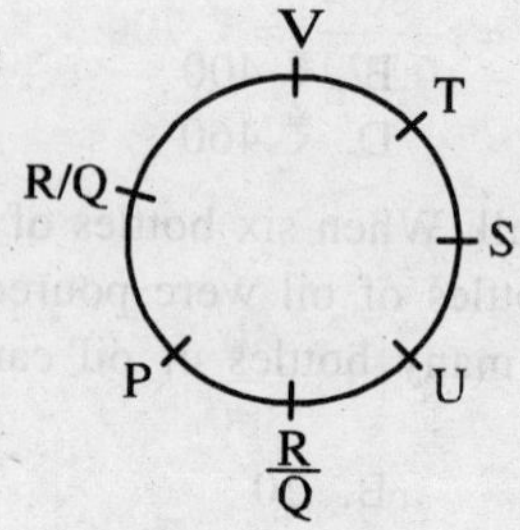

According to figure,

T is second to the left of R.

27. (D)

28. (C): The shirts are arranged in this order

Red
Blue
Orange
Green
White

∴ Blue shirt is between the Red shirt and the Orange shirt.

29. (D):

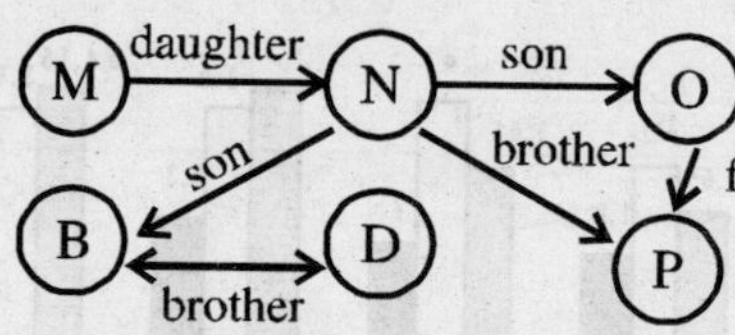

According to figure, all statements are true.

30. (A) **31. (C)**

32. (B):

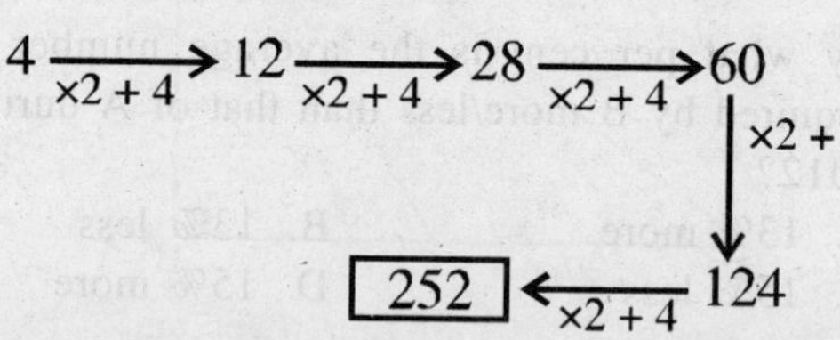

33. (C):

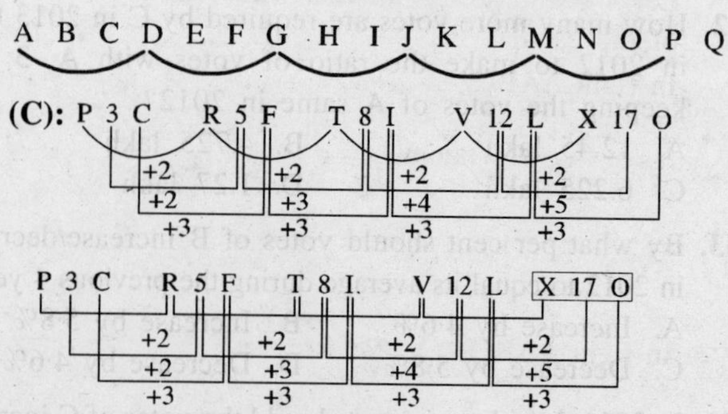

34. (C) **35. (C)** **36. (B)**

37. (B): There will be only 8 small cubes of which exactly three faces have painted in any cases.

38. (D) **39.** (A) **40.** (D)

41. (B): Let marked price = ₹ x

Discounted price = $0.9x$. (10% discount)

Price including salex tax (8%)

$$= 0.9x \times \frac{108}{100} = 0.972x$$

According to the question,

$$0.972x = 680.4$$

$$x = \frac{680.4}{0.972} = ₹\ 700.$$

42. (C)

43. (D): Let the principle is P

$$196 = \text{P}\left(1+\frac{50}{300}\right)^2$$

$$\text{P} = \frac{196 \times 36}{49} = ₹\ 144.$$

44. (A): Let he borrow ₹ x at the rate of 12%

Then he borrow ₹ $(1500 - x)$ at the rate of 14%

From question,

$$\frac{x \times 12}{100} + (1500 - x) \times \frac{14}{100} = 186$$

$$0.12x - 0.14x + 210 = 186$$

$$0.02x = 24$$

$$x = 1200.$$

45. (C) **46.** (D)

47. (A):

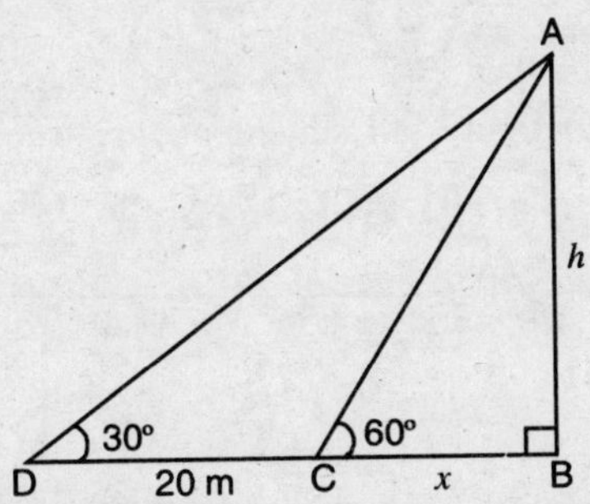

Let h be the height of the tower and x be the width of the canal.

In right ΔABC,

$$\tan 60° = \frac{h}{x}$$

$$\Rightarrow \quad \sqrt{3} = \frac{h}{x}$$

$$\Rightarrow \quad h = \sqrt{3}x$$

In right ΔABD,

$$\tan 30° = \frac{h}{x+20}$$

$$\Rightarrow \quad \frac{1}{\sqrt{3}} = \frac{h}{x+20}$$

$$\Rightarrow \quad \sqrt{3}h = x + 20$$

$$\Rightarrow \quad \sqrt{3}\left(\sqrt{3}x\right) = x + 20 \qquad \left[\because h = \sqrt{3}x\right]$$

$$3x - x = 20$$

$$\Rightarrow \quad x = 10$$

∴ The width of the canal is 10 m.

48. (C): Volume of 10 spherical balls

$$= 10 \times \frac{4}{3}\pi r^3$$

$$= 10 \times \frac{4}{3}\pi \times (5)^3$$

$$= \frac{5000}{3}\pi \text{ cm}^3$$

Height of cylinder = 10 × diameter of sphere

= 10 × 10

= 100 cm

Volume of cylinder = $\pi r^2 h$

$= \pi(5)^2 \times 100$

$= 2500\pi \text{ cm}^3$

Volume of the empty space in the box

= Volume of cylinder

– Volume of sphere

$$= \frac{2500\pi}{1} - \frac{5000\pi}{3}$$

$$= \left(\frac{7500 - 5000}{3}\right)^{\pi}$$

$$= \frac{2500}{3}\pi \text{ cm}^3.$$

49. (D): The two partners invest their capitals for the same period.

∴ Profit will be divided in the ratio of their investment.

Separate profit of A for managing the business

= 12.5% of ₹ 880

$$= \frac{25}{200} \times 880 = ₹\ 110$$

Remaining profit = ₹ 880 – 110 = ₹ 770

Ratio of the profit of A to that of B

= 5000 : 6000 = 5 : 6

A's share in the profit = $\frac{5}{11} \times 770$ = ₹ 350

Total profit of A = ₹ 350 + 110 = ₹ 460

∴ A's total share = ₹ 460.

50. (D) **51.** (D)

52. (D): 2012 A = 4.15 lakhs

C = 1.5 lakh

$$\frac{4.15}{1.5 + x} = \frac{3}{2}$$

$$4.5 + 3x = 8.30$$
$$3x = 8.30 - 4.5 = 3.8$$
$$x = \frac{3.8}{3} = 1.2666....$$
$$\cong 1.27 \text{ lakhs.}$$

53. (D): 2012

$$B = 3.8 \text{ lakh}$$
$$\text{Average} = \frac{3.9+4+3.1+3.5}{4} = \frac{14.5}{4}$$
$$= 3.625 \text{ lakh}$$
$$3.8 - 3.625 = .175$$
$$\% = \frac{.175}{3.8} \times 100$$
$$= \frac{175000}{38000} = \frac{175}{38} = 4.6\%.$$

54. (C): 2011 for C, 2.3 lakhs

for B, 3.9 lakhs

$$12\% \text{ of } 3.9 = 46800$$
$$390000 + 46800 = 436800$$

According to the question,

$$2.3 + x\% \text{ of } 2.3 = 436800$$
$$x\% \text{ of } 2.3 = 436800 - 230000$$
$$= 206800$$
$$\frac{x}{100} \times 230000 = 206800$$
$$x = \frac{206800}{2300} = \frac{2068}{23}$$
$$= 89.9\%.$$

55. (A): In 2009 for C = 1.9 lakhs

Total votes acquired by all the three parties in 2009

$$= 2.5 + 3.1 + 1.9 = 7.5 \text{ lakhs}$$

% of votes were acquired by C

$$= \frac{1.9}{7.5} \times 100$$
$$= \frac{19}{75} \times 100$$
$$= \frac{19 \times 4}{3} = \frac{76}{3}$$
$$= 25.33\%.$$

56. (A) **57. (A)** **58. (C)**

59. (B) **60. (C)**

Previous Paper (Solved)

Symbiosis National Aptitude Test (SNAP), 2021*

Directions (Qs. No. 1-7): *Read the following passage carefully and answer the questions given below it.*

India's tourism industry is experiencing a strong period of growth, driven by the burgeoning Indian middle class, growth in high spending foreign tourists, and coordinated government campaigns to promote 'Incredible India'. The tourism industry in India is substantial and vibrant, and the country is fast becoming a major global destination. India's travel and tourism industry is one of the most profitable industries in the country, and also credited with contributing a substantial amount of foreign exchange. This is illustrated by the fact that during 2006, four million tourists visited India and spent US $8.9 billion. Several reasons are cited for the growth and prosperity of India's travel and tourism industry.

Economic growth has added millions annually to the ranks of India's middle class, a group that is driving domestic tourism growth. Disposable income in India has grown by 10.11% annually from 2001-2006, and much of that is being spent on travel. Thanks in part to its booming IT and outsourcing industry a growing number of business trips are made by foreigners to India, who will often add a weekend break or longer holiday to their trip. Foreign tourists spend more in India than almost any other country worldwide. Tourist arrivals are projected to increase by over 22% per year till 2010, with a 33% increase in foreign exchange earnings recorded in 2004.

The Tourism Ministry has also played an important role in the development of the industry, initiating advertising campaigns such as the 'Incredible India' campaign, which promoted India's culture and tourist attractions in a fresh and memorable way. The campaign helped to create a colorful image of India in the minds of consumers all over the world, and has directly led to an increase in the interest among tourists. The tourism industry has helped growth in other sectors as diverse as horticulture, handicrafts, agriculture, construction and even poultry. Both directly and indirectly, increased tourism in India has created jobs in a variety of related sectors. The numbers tell the story: almost 20 million people are now working in the India's tourism industry. India's governmental bodies have also made a significant impact in tourism by requiring that each and every state of India have a corporation to administer support issues related to tourism. A new growth sector is medical tourism. It is currently growing at around 30% per annum. Medical tourist arrivals are expected to reach one million soon.

Medical tourism in Asia has grown rapidly. Medical tourism is approaching fever pitch at the tune of $4 billion US, fuelled largely by the cosmetic surgery market. One of the problems India has, despite having some world-class hospitals, is sanitation. The tourism industry of India is based on certain core nationalistic ideals and standards which are: Swaagat or welcome, Sahyog or cooperation, Soochanaa or information, Sanrachanaa or infrastructure, Suvidha or facilitation, Safaai or cleanliness and Surakshaa or security.

1. Choose the word which is MOST OPPOSITE in meaning of the word burgeoning as used in the passage?
A. thriving
B. sprouting
C. incorporating
D. decreasing

2. Find the correct statement on the basis of the given passage.
(*a*) The tourism industry in our country manages to attract the world tourist.
(*b*) India has continued to represent a place much sought after for its diversity.
(*c*) India has potential to emerge as one of the world's tourist hot spots.
A. Only (*a*) B. Only (*c*)
C. Both (*a*) and (*b*) D. All (*a*), (*b*) and (*c*)

3. Choose the word which is MOST SIMILAR in meaning to the word "cited" as used in the passage?
A. pointed out B. withheld
C. inserted D. imputed

4. Which of the following statements correctly represent(s) the reason behind the expansion of India's travel and tourism Industry?
A. India is a vast country of great beauty and diversity.
B. India's vast cultural heritage
C. The successful launch of very penetrative ad campaign
D. All of the above

5. What is the meaning of the phrase "fever pitch" as used in the passage?

* Based on memory

A. A high degree of excitement
B. A state of suffering
C. Beyond the expectations
D. Meeting the target

6. Which of the following is false in respect of India's medical tourism?
A. India's medical tourism sector is expected to experience an annual growth rate of 30%.
B. An estimated one million tourists will travel to India for healthcare procedures every year.
C. The advantage of medical treatment in India includes reduced costs not the availability of latest medical technologies.
D. All (A), (B) and (C)

7. Give a suitable title to the given passage.
A. Medical tourism in India
B. Tourism Industry–A special focus of India
C. Foreign tourist arrival and forex earning in India
D. Domestic Tourism in India

Directions (Qs. No. 8 & 9): *In each of the following sentences, there is a blank space. Below each sentence, there are four words out of which one can be used to fill the blank to make the sentence grammatically and coherently correct. Find the most appropriate word that fit into the blank contextually.*

8. The Indian corporate debt market has been ________ by several woeful issues over the years.
A. abetted
B. plagued
C. extraneous
D. stirring

9. Today, the world seems to be sitting on a volcano ready to ________.
A. incorrupt
B. interrupt
C. abrupt
D. erupt

Direction (Qs. No. 10-12): *In the following question, four sentences are given which may be grammatically and contextually incorrect. You need to find the one which has no error and mark that as your answer.*

10. A. Short people are 50 per cent like than tall people to die prematurely of heart disease, researchers reported today in a major review of three million people.
B. Brihanmumbai Municipal Corporation (BMC) will makes a short film featuring master blaster Sachin Tendulkar to spread the message of cleanliness.
C. The tunnel-like structure built by the British connecting the Delhi legislative assembly with the Red Fort will now be renovated and opens for tourists.
D. All 612 districts in India are vulnerable to climate change, but 100 districts, mostly in the eastern part of the country, are most vulnerable, a study said.

11. A. Umesh Yadav became the six Indian pace bowler to get to 150 Test wickets.
B. The government on Thursday extended for September 30, visas of all foreign nationals stuck in India due to COVID-19 pandemic, an official spokesperson said.
C. Zomato India has officially filed for dissolution and shut down of its subsidiaries in the UK and Singapore.
D. According to a report prepared by Hurun India, India added three 'unicorns' par month in 2021.

12. A. Renowned for scripting records in the field, Indian skipper Virat Kohli has been racking numbers off-field too.
B. The tech giant company- Samsung has announced the world's first 200-megapixel image sensor intended for smartphone cameras.
C. India's biggest carmaker, Maruti Suzuki, is reportedly recalling after 180,000 cars for a 'possible defect'.
D. Javelin thrower Neeraj Chopra's gold medal-win performance has been rated as one of the 10 magical moments of the Tokyo Olympic Games by World Athletics (WA).

Direction (Qs. No. 13-15): *In each of the questions given below four words are given in bold. These four words may or may not be in their correct position. The sentence is then followed by options with the correct combination of words that should replace each other in order to make the sentence grammatically and contextually correct. Find the correct combination of the words that replace each other.*

13. Replacing the office's **technology** (*a*) computer **outdated** (*b*) with new **system** (*c*) would **allow** (*d*) the workers to get a lot more done.
A. (*a*)(*c*)(*d*)(*b*)
B. (*b*)(*c*)(*a*)(*d*)
C. (*d*)(*b*)(*c*)(*a*)
D. (*c*)(*a*)(*d*)(*b*)

14. The **allure** (*a*) of the **fresh** (*b*) herbs caused the chef to **intended** (*c*) more money than he **spend** (*d*) at the market.
A. (*b*)(*a*)(*c*)(*d*)
B. (*d*)(*c*)(*a*)(*b*)
C. (*a*)(*b*)(*d*)(*c*)
D. (*c*)(*d*)(*a*)(*b*)

15. There is a **horizon** (*a*) of **outskirt** (*b*) buildings on the **industrial** (*c*) of our small town, but we tend to ignore them since they ruin the **cluster** (*d*).
A. (*d*)(*c*)(*b*)(*a*)
B. (*d*)(*c*)(*a*)(*b*)
C. (*a*)(*b*)(*d*)(*c*)
D. (*c*)(*d*)(*a*)(*b*)

Directions (Qs. No. 16-18): *Study the information given below carefully and answer the questions that follow.*

There are 10 members in the family who are sitting at some distance from each other in a park. N is 4 m in the west of M, who is 5 m in the north of O. Q is 12 m in the west of R. T is 3 m in the west of S, which is 12 m in the south of R. O is 6 m in the east of V, which is 2 m in the north of P, which is 8 m in the west of U. T is 1 m in the south of U.

16. In which direction is U with respect to M?
A. North
B. East
C. South-East
D. North-West

17. What is the shortest distance between N and Q?
A. 6 m B. 3 m
C. 7 m D. 5 m

18. Suppose an imaginary line is drawn from N to line VO such that it makes a perpendicular. The imaginary line meets VO at point Y. Then what is the perimeter of the quadrilateral NYOM?
A. 14 m B. 18 m
C. 20 m D. 24 m

Directions (Qs. No. 19-23): *Study the following information carefully and answer the questions given below.*

A certain number of persons are sitting in a linear row facing the north. Three persons are sitting between X and T, who is sitting adjacent to Q. X is on the left of Q. H is sitting 2nd to the right of Q. Two persons are sitting between H and S who is sitting at one of the ends. S is not sitting adjacent to Q, who is sitting 2nd to the right of L. L is sitting exactly in the middle of the row. J is sitting 6th from the left end. X is not sitting to the right of J. Two persons are sitting between J and R. Two persons are sitting between X and K.

19. How many persons are sitting between X and S?
A. Four B. Nine
C. Eleven D. One

20. How many persons are sitting in the row?
A. 18 B. 10
C. 17 D. 15

21. How many persons are sitting to the left of Q?
A. Nine B. Three
C. Ten D. Five

22. What is the position of H with respect to T?
A. 5th to the Left
B. 3rd to the Right
C. 5th to the Right
D. 2nd to the Left

23. Who sits exactly in the middle of K and L?
A. R B. J
C. X D. T

24. In the following question assuming the given statements to be true, find which of the conclusion(s) among given conclusions is/are definitely true and then give your answers accordingly.

Statement: B > C = D; E < F; B ≤ A; D ≤ E

Conclusions: **I.** A > C
II. B < D

A. Only conclusion I follows
B. Only conclusion II follows
C. Either conclusion I or conclusion II follow
D. Both conclusions I and II follow

Directions (Qs. No. 25-29): *Study the following information carefully and answer the given questions.*

Six persons P, Q, R, S, T, and U are living in a three-storey building such that ground floor is numbered as 1, above it is floor 2 then top floor is numbered as 3. Each of the floor has 2 flats in it as flat-X and flat-Y. Flat-X of floor-2 is immediately above flat-X of floor-1 and immediately below flat-X of floor-3. In the same way flat-Y of floor-2 is immediately above flat-Y of floor-1 and immediately below flat-Y of floor-3. Flat-X is in west of flat-Y. They all are from different cities.

The one who is from Kolkata lives immediately above U. S lives to the west of U. P and the one who is from Kolkata do not live on the same floor. Neither S nor Q is from Bangalore. U is either from Pune or from Agra. There is one floor between Q and the one who is from Delhi who lives above Q. T does not live in flat Y. The one who is from Mumbai lives on the same floor on which the one who is from Agra lives. P lives on an odd numbered floor in flat X.

25. Which of the following does not belong to the group?
A. T B. P
C. R D. S

26. Who among the following lives to the east of T?
A. The one who is from Agra
B. U
C. R
D. The one who is from Pune

27. Who lives on the bottom most floor?
A. The one who is from Pune
B. U
C. P
D. Both (A) and (C)

28. Which of the following combination is the correct?
A. T – Mumbai
B. P – Delhi
C. Q – Banglore
D. S – Mumbai

29. Who among the following lives in Flat Y?
A. P B. T
C. S D. Q

30. If 2 is added to each even digit and 1 is subtracted from each odd digit in the number 621754, then what will be the sum of number/numbers not repeated in the new number?

A. 5 B. 8
C. 6 D. 4

31. In the word 'LAVISLY', replace each vowel with the next letter in the alphabetical series and each consonant with the previous letter in the Alphabetical Series. How many letters occur more than once in this newly formed word?

A. One B. Two
C. Three D. Four

32. In the question below are given three statements followed by three conclusions I, II, and III. You have to take the given statements to be true even if they seem to be at variance from commonly known facts. Read all the given conclusions and then decide which of the following conclusions logically follows from the given statements disregarding commonly known facts.

Statements: Some Glass are Cup.
Only a few Cup are Plate.
Only Plate is Bottle.

Conclusions:

I. All Glass are Plate.
II. Some Cup are not Bottles.
III. Some Cup are not Plates.

A. Only conclusion I follows
B. Only conclusion II follows
C. Only conclusion III follows
D. Both conclusion II and III follow

33. From a point, Lokesh starts walking towards south and after walking 30 metres he turns to his right and walks 20 metres, then he turns right again and walks 30 metres. He finally turns to his left and walks 40 metres. In which direction is he with reference to the starting point?

A. North-West B. East
C. West D. South

34. In a certain code language, "RAIN" is written as "OHBQ". How is "SUMMER" written in that code language?

A. QFLNTT B. QDLLTR
C. SFNNVT D. SDNLVR

35. Select the figure that can replace the question mark (?) in the following series.

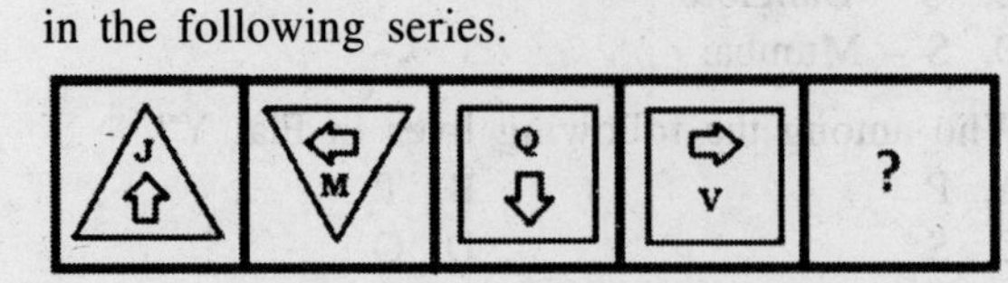

A. 
B. 
C. 
D. 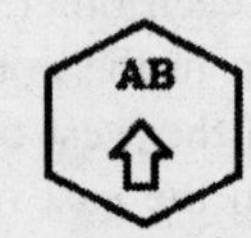

36. Select the correct mirror image of the given figure when a mirror is placed on the right of the following figure:

A. 
B. 
C. 
D. 

37. Select a suitable figure from the Answer Figures that would replace the question mark (?).

Problem Figures:

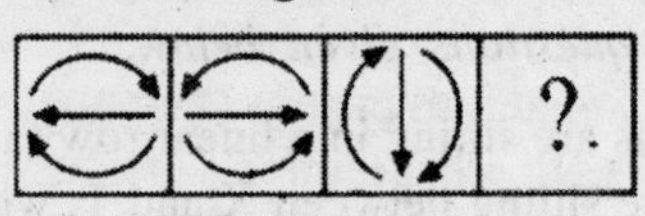

(a) (b) (c) (d)

Answer Figures:

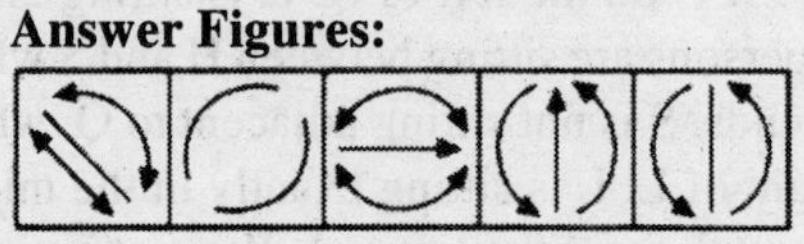

1 2 3 4 5

A. 2 B. 1
C. 3 D. 4

38. A series is given, with one term missing. Choose the correct alternative from the given ones that will complete the series.

YY, YOY, YYYY, YYOYY, YYYYYY, ?

A. YYYOYYYY
B. YYYOYYY
C. YYYYYYY
D. YYYOYYYO

39. Introducing Reeta, Monica said, "She is the only daughter of my father's only daughter." How is Monica related to Reeta?

A. Aunt B. Niece
C. Cousin D. Mother

40. If S = 19, SUN = 54 and CAKE = 20, then MISTAKE = ?

A. 78 B. 68
C. 59 D. 48

41. Butter oil worth ₹ 420 per kg and ₹ 465 kg are mixed with a third variety in the ratio 2 : 3 : 4. If the mixture is worth ₹ 440 per kg, then find the price of third variety per kg.

A. ₹ 431.25 kg
B. ₹ 445.50 kg
C. ₹ 462.5 kg
D. ₹ 224.25 kg

42. Length of a rectangle is 40 cm and breadth is 30 cm. If the length of rectangle is increased by 25% and breadth decreased by 25%, find the percentage increase or decrease in the area of rectangle.

A. Increased by 12.5%
B. Increased by 6.25%
C. Decreased by 6.25%
D. Decreased by 12.5%

43. A pipe can fill a tank in 10 hours, while another pipe can empty it in 6 hours. If both pipes are opened simultaneously, then how much time will be taken to empty the tank?

A. 12 hours B. 15 hours
C. 8 hours D. 18 hours

44. Three taps are fitted in a cistern. The empty cistern is filled by the first and the second taps in 4 hours and 5 hours respectively. The full cistern is emptied by the third tap in 3 hours. If all the three taps are opened simultaneously, then empty cistern will be fillen up in

A. $8\frac{4}{7}$ hours B. $2\frac{8}{11}$ hours
C. 9 hours 40 min. D. 10 hours 26 min

45. ₹ 4000 becomes ₹ 5200 in 3 years at simple interest. The amount that will become ₹ 14400 in 2 years at the same rate of interest is ______.

A. ₹ 11120 B. ₹ 12000
C. ₹ 14500 D. ₹ 12200

46. The length and breadth of a rectangular field are in ratio 7 : 4. If the area of the field is 157500 m^2, find the cost of fencing the field at 24 paise per cm.

A. ₹ 45450 B. ₹ 39600
C. ₹ 32080 D. ₹ 42128

47. An amount of ₹ 7330 is divided among X, Y and Z such that if their shares be reduced by ₹ 12, ₹ 47 and ₹ 71 respectively, the remainders shall be in the ratio of 2 : 3 : 7. Find Z's share.

A. ₹ 4000 B. ₹ 4200
C. ₹ 4220 D. ₹ 4350

48. If 5A = 3B = 2C, then find $\frac{A}{2}:\frac{B}{5}:\frac{3C}{5}$ is

A. 3 : 10 : 13 B. 5 : 3 : 12
C. 3 : 2 : 9 D. 5 : 8 : 9

49. Find the value of $\sqrt{13\sqrt{13\sqrt{13\sqrt{13 \ldots \infty}}}}$

A. $\sqrt{3}$ B. 13
C. $13\sqrt{2}$ D. $\sqrt{13}+13$

50. The average of n observation is ω. If the sum of $(n + 5)$ observations is μ, then what is the average of new five observations?

A. $-\left(\frac{n\omega-\mu}{5}\right)$

B. $\frac{n\omega-n\mu}{5}+\mu$

C. $\frac{n(\mu-\omega)}{5}+\mu$

D. $\frac{(\mu-\omega)}{5}+\mu$

Directions (Qs. No. 51-55) : *Study the following tables to answer these questions.*

Number of males and females staying in 6 refugee camps after a cyclone

Refugee Camp	Males	Females
I	250	350
II	400	150
III	300	275
IV	280	300
V	180	250
VI	325	300

Percentage of Children (males and females in the refugee camps)

Refugee Camp	Children %	Male children %	Female children %
I	25	40	60
II	40	75	25
III	16	25	75
IV	25	80	20
V	40	50	50
VI	24	46	54

51. What is the respective ratio of the number of adult females to the total number of female children staying in all the refugee camps together?

A. 283 : 82 B. 112 : 71
C. 82 : 243 D. 71 : 112

52. What is the total number of female children staying in all the refugee camps together?

A. 314 B. 433
C. 410 D. 343

53. What is the respective ratio of the total number of adult males in Camps I and II together to the total number of adult males in Camps V and VI together?

A. 75 : 79 B. 14 : 17
C. 79 : 75 D. 17 : 14

54. What is the total number of all refugees staying in all camps together?

A. 3520 B. 3360
C. 4100 D. None of these

55. What is the difference between the number of male children in Camp II and those is Camp VI?

A. 84 B. 14
C. 96 D. 26

Directions (Qs. No. 56-60) : *Study the following table and pie chart giving break-up of night shift employees of Raheja Group across various centres and industries to answer these questions.*

Industry Category	% age of female employees in the industry category working in night shift in Raheja Group	% age of total employees of Raheja Group working in night shift
Animation	20	18
BPO	45	32
IT	20	12
Telemarketing	60	8
Telebanking	40	14
Other	15	16

Number of employees of Raheja Group Working in night shift across all centres in India

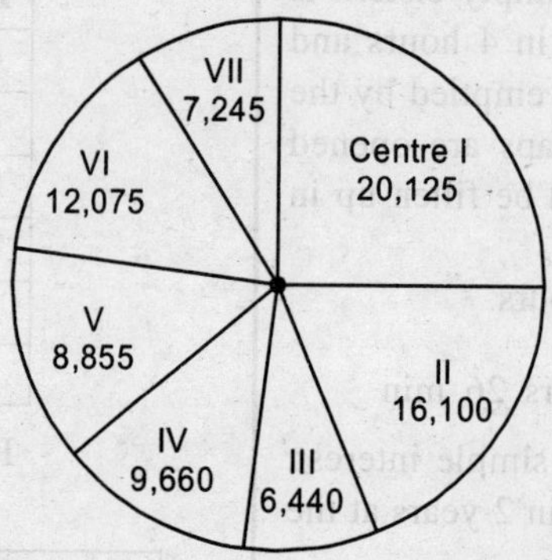

56. What is the ratio of males to females working in night shift in BPOs?

A. 11 : 9 B. 7 : 5
C. 8 : 13 D. 11 : 17

57. What is the average number of females working in night shifts from all industries together?

A. 4454 B. 8881
C. 6326 D. 5623

58. What is the total number of males working in night shifts from all industries together?

A. 48297 B. 35788
C. 26887 D. 53774

59. The number of females working in Animation is what per cent of the total number of employees working in night shift from all industries together?

A. 5.6 B. 3.6
C. 3.2 D. 4.4

60. What is the difference between total number of males and females from all industries together?

A. 13524 B. 27084
C. 27048 D. None of these

ANSWERS

1. (D)	2. (D)	3. (A)	10. (D)	11. (C)	12. (B)
4. (D)	5. (A)	6. (C)	13. (B)	14. (C)	15. (A)
7. (B)	8. (B)	9. (D)	16. (C)	17. (D)	18. (B)

For Q.No. 19-23

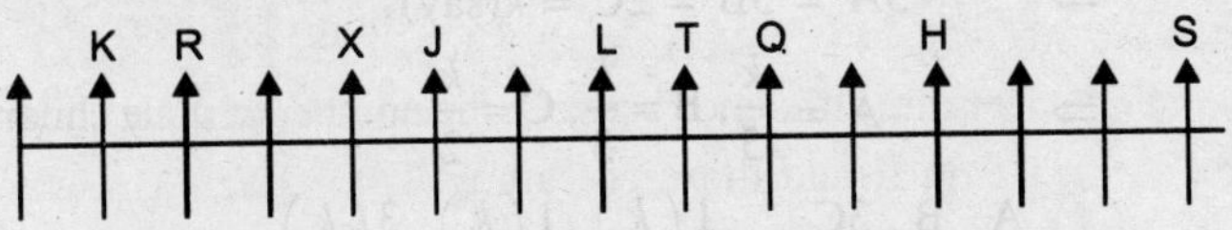

19. (B): There are 9 persons are sitting between X and S.

20. (D): There are 15 persons are sitting in the row.

21. (A): There are 9 persons are sitting to the left of Q.

22. (B): The position of H with respect to T is 3rd to the right.

23. (C): X is the middle of K and L.

24. (A) **25. (D)** **26. (C)**

27. (D) **28. (D)** **29. (D)**

30. (B):

	6	2	1	7	5	4
	↓	↓	↓	↓	↓	↓
	−1	+2	−1	+2	−1	+2
=	5	4	0	9	4	6

Sum of the digits = 5 + 4 + 0 + 9 + 4 + 6 = 28
Last digit = 8
Hence, correct option is (B).

31. (A):

L	A	V	I	S	L	Y
↓	↓	↓	↓	↓	↓	↓
K	B	U	F	R	K	X

Only K letter occurs more than once in this newly formed word.

32. (D) **33. (C)**

34. (D): Given,

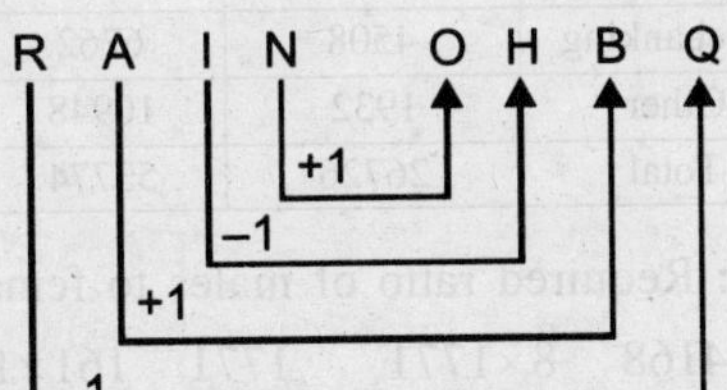

Similarly,

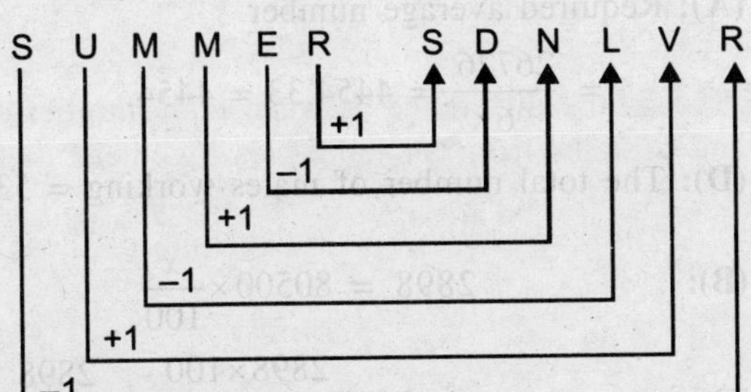

Hence, SUMMER will be written as SDNLVR.

35. (A) **36. (D)** **37. (D)**

38. (B)

39. (D):

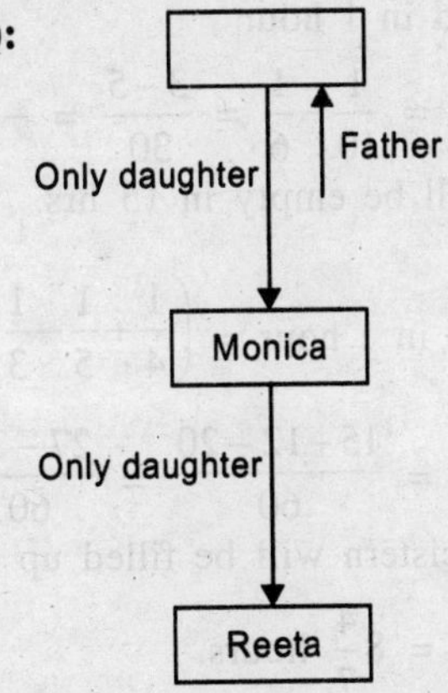

Hence, Monica is the mother of Reeta.

40. (A): Given, S = 19, SUN = 54 and CAKE = 20

$\because$ SUN = 19 + 21 + 14 = 54

CAKE = 3 + 1 + 11 + 5 = 20

Similarly, MISTAKE = 13 + 9 + 19 + 20 + 1 + 11 + 5 = 78

$\Rightarrow$ MISTAKE = 78.

41. (A): Let the price of the third variety per kg = ₹ x

Then, ₹ 420 × 2 + ₹ 465 × 3 + ₹ x × 4 = ₹ 440 × (2 + 3 + 4)

$\Rightarrow$ ₹ 840 + ₹ 1395 + ₹ $4x$ = ₹ 440 × 9

$\Rightarrow$ ₹ 2235 + ₹ $4x$ = ₹ 3960

$\Rightarrow$ ₹ $4x$ = ₹ 3960 − ₹ 2235

$\Rightarrow$ ₹ $4x$ = ₹ 1725

$\Rightarrow$ x = ₹ $\frac{1725}{4}$ = ₹ 431.25

Hence, the price of the third variety per kg = ₹ 431.25

42. (C): We have, for a rectangle length = 40 cm and breadth = 30 cm

$\therefore$ area = $l \times b = 40 \times 30\text{ cm}^2 = 1200\text{ cm}^2$

Again, increasing length = $40 \times \frac{125}{100}$ cm

$= 40 \times \frac{5}{4} = 50$ cm

and decreasing breadth

$= 30 \times \frac{75}{100}\text{ cm} = 30 \times \frac{3}{4}\text{ cm}$

$= \frac{45}{2}\text{ cm} = 22.5\text{ cm}$

$\therefore$ area = $l \times b = 50 \times 22.5\text{ cm}^2 = 1125\text{ cm}^2$

Decreasing area = $1200 - 1125 = 75\text{ cm}^2$

Hence, decrease in the area of rectangle

$= \frac{75}{1200} \times 1000\% = 6.25\%$

43. (B): Part filled in 1 hour = $\frac{1}{10}$

and, part emptied in 1 hour = $\frac{1}{6}$

∴ Net part emptied in 1 hour

$$= \frac{1}{10} - \frac{1}{6} = \frac{3-5}{30} = -\frac{2}{30} = -\frac{1}{15}$$

Hence, the tank will be empty in 15 hrs.

44. (A): Net part filled in 1 hour $= \left(\frac{1}{4} + \frac{1}{5} - \frac{1}{3}\right)$

$$= \frac{15+12-20}{60} = \frac{27-20}{60} = \frac{7}{60}$$

Hence, the empty cistern will be filled up in

$$\frac{60}{7} \text{ hours} = 8\frac{4}{7} \text{ hours.}$$

45. (B): Here, P = ₹ 4000, A = ₹ 5200, t = 3 years

∴ Simple Interest = A – P

= ₹ 5200 – ₹ 4000 = ₹ 1200

$$\therefore \quad r = \frac{\text{S.I.} \times 100}{\text{P} \times t} = \frac{1200 \times 100}{4000 \times 3} = 10\%$$

Again, A = ₹ 14400, t = 2 years,

Let principal = P

$$\text{S.I.} = \text{A} - \text{P} = ₹\ 14400 - \text{P} = \frac{\text{P} \times t \times 10}{100}$$

$$\Rightarrow ₹\ 14400 - \text{P} = \frac{\text{P}}{5}$$

$$\Rightarrow \quad ₹\ 14400 = \text{P} + \frac{\text{P}}{5} = \frac{6}{5}\text{P}$$

$$\Rightarrow \quad \text{P} = ₹\ 14400 \times \frac{5}{6}$$

$$\Rightarrow \quad \text{P} = ₹\ 12000.$$

46. (B): Let the length and breadth of a rectangular field are $7x$ and $4x$

Then, area = length × breadth

$\Rightarrow$ 157500 $m^2 = 7x \times 4x$

$\Rightarrow$ $28x^2 = 157500\ m^2$

$$\Rightarrow \quad x^2 = \frac{157500}{28} m^2$$

$\Rightarrow$ $x^2 = 5625\ m^2$

$\Rightarrow$ $x = \sqrt{5625}\ m^2$

$\Rightarrow$ x = 75 m

Perimeter of a rectangular field = $2(l + b)$

= $2(7x + 4x)$

= $2 \times 11x = 22x$

= 22 × 75 m = 1650 m

= 1650 × 100 cm

Hence, the cost of fencing the field at 24 paise per cm = 24 × 1650 × 100 paise

= 3960000 paise

$$= ₹\ \frac{3960000}{100} = ₹\ 39{,}600.$$

47. (B)

48. (C): Given that 5A = 3B = 2C

$\Rightarrow$ 5A = 3B = 2C = k(say)

$$\Rightarrow \quad \text{A} = \frac{k}{5}, \text{B} = \frac{k}{3}, \text{C} = \frac{k}{2}$$

$$\therefore \frac{\text{A}}{2} : \frac{\text{B}}{5} : \frac{3\text{C}}{5} = \frac{1}{2}\left(\frac{k}{5}\right) : \frac{1}{5}\left(\frac{k}{3}\right) : \frac{3}{5}\left(\frac{k}{2}\right)$$

$$= \frac{1}{10} : \frac{1}{15} : \frac{3}{10} = \frac{1}{10} \times 30 : \frac{1}{15} \times 30 : \frac{3}{10} \times 30 = 3 : 2 : 9.$$

49. (B): Let $x = \sqrt{13\sqrt{13\sqrt{13\sqrt{13 \ldots \infty}}}}$

Then, $x = \sqrt{13x}$

Squaring both sides, we get

$x^2 = 13x \Rightarrow x = 13.$

50. (A): The sum of n observations = $n\omega$

and the sum of $(n + 15)$ observations = μ

∴ new five observations = $\mu - n\omega$

∴ The average of new five observation

$$= \frac{\mu - n\omega}{5} = \frac{-(n\omega - \mu)}{5}$$

51. (A) **52. (C)** **53. (D)**

54. (D) **55. (C)**

For Qs. No. 56–60:

According to the given table and pie-chart:

Industry Category	Total Female Employee	Total Male Employee	Total Employees
Animation	2898	11592	14490
BPO	11922	14168	25760
IT	1932	7728	9660
Telemarketing	3864	2576	6440
Telebanking	4508	6762	11270
Other	1932	10948	12880
Total	26726	53774	80500

56. (A): Required ratio of males to females in BPOs

$$= \frac{14168}{11922} = \frac{8 \times 1771}{8 \times 1449} = \frac{1771}{1449} = \frac{161 \times 11}{161 \times 9} = \frac{11}{9} = 11 : 9.$$

57. (A): Required average number

$$= \frac{26726}{6} = 4454.33 = 4454.$$

58. (D): The total number of males working = 53774.

59. (B): $2898 = 80500 \times \frac{x}{100}$

$$\Rightarrow \quad x = \frac{2898 \times 100}{80500} = \frac{2898}{805} = 3.6$$

∴ Required % = 3.6.

60. (C): The difference between total number of males and females = 53774 – 26726 = 27048.

Previous Paper (Solved)

Symbiosis National Aptitude Test (SNAP), 2019*

SECTION-I

Directions (Qs. 1-5): *Read the following passage and answer the questions as directed.*

Two years after the World Health Organization labelled air pollution a global "public health emergency", and the House of Commons environment committee used the same phrase to describe the situation in the UK, new evidence shows that breathing unsafe air causes a loss of intelligence, particulary in the over-64s. The research, carried out in China, showed that test scores declined when people breathed toxins including nitrogen dioxide and particulate, with language ability more affected than maths. This news, though alarming, is unlikely to change anything straight away. China has been engaged in a "war against pollution" for five years, while governments and policy makers around the world already have **ample** evidence that pollution is extremely harmful. Top of the list of dangers is the way it worsens heart and lung diseases including asthma and emphysema, while one study last year suggested a link to dementia.

The problem is both global and national: urbanisation and increasing car use mean that pollution is on the rise internationally, while the UK government is under huge pressure to clean up air that has broken EU legal limits for the past eight years. Western capitals do not feature in the WHO's lists of the planet's most polluted cities, and levels of most pollutants in the UK have fallen (though not ammonia, a byproduct of farming). But increased scientific understanding of the damage of health caused by gases such as nitrogen oxides makes inexcusable the complacency of which we have seen so much, both under the current Conservative government and its coalition predecessor. London mayor Sadiq Khan's "ultra low-emission zone", which comes into force next year and will charge the most polluting vehicles to enter central London, looks set to be a game-changer. Government analysis shows clean air zones to be by far the most effective measure in reducing nitrogen oxides. But critics think it is still too ..., while leaders of other cities are calling for government funding to implement their own clean air zones, as well as a new clean air act to provide a national framework. Campaigners struggle to understand why the public outery is not loud or angry enough to force the government to act, when air pollution is thought to be a factor in shortening the lives of 40,000 people in the UK every year. So do all those who worry about air pollution, among them parents anxious about the impact on growing lungs. The usually answer is cars, and the fact that even if people would like their cities to be cleaner, they don't want restrictions on their freedom to drive.

1. As per the passage, which of the following is the result of breathing unsafe air?

A. People, particularly below 64 of years of age can lose their intelligence.
B. People, irrespective of age can lose their intelligence and sight.
C. People, particularly above 64 of years of age can lose their intelligence.
D. People tend to loose their power to think and speak.

2. Which of the following statements is/are true in context of the passage above?

(*a*) Urbanisation and increasing car use mean that pollution is on the rise internationally.
(*b*) The research, carried out in China, showed that test scores declined when people breathed toxins including nitrogen dioxide and particulate, with language ability more affected than maths.

A. Only (*a*) B. Only (*b*)
C. Both (*a*) and (*b*) D. None of these

3. Which of the following can replace the word given in bold in the passage?

A. Insufficient B. Pathetic
C. Meagre D. Plentiful

4. In the passage above, a line is given in bold. Which of the following best describes the meaning of the line given in bold?

A. Even though people would like their cities to be cleaner, they also want restrictions on their freedom to drive.
B. People want their cities to be cleaner and so they do not want to drive cars without restrictions.
C. People would like their cities to be cleaner but they don't want restrictions on their freedom.
D. Despite the fact that people want their cities to be cleaner, they do not want to compromise on their freedom to drive.

*Based on memory

5. Which of the following words can fill in the blank to make it meaningful?

A. Bold B. Timid
C. Forthcoming D. Brazen

Directions (Qs. 6-8): *Read the following passage and answer the questions as directed.*

Paragraph 1– One of the easiest ways to establish a savings habit is to participate in your employer's 401(k) plan. Funds are withheld from each paycheck and deposited into your account. If your employer matches part of your contribution – and many do! - you will accumulate yet more. A second way to consistently save is with an automatic saving tranfer program with your financial institution. You decide how much and when you want funds transferred from your checking account into a savings account. You can also use a payroll deduction plan from your employer and get the same results.

Paragraph 2– Along with how much and how often you save, what you earn on your funds will determine how fast your money grows. You cannot control what happens with interest rates or the stock market, but you can consider different types of savings vehicles that provide different returns. The simplest savings vehicle to consider is buying certificates of deposit (CDs) instead of leaving funds in a savings account. CDs usually offer higher interest rates, but they are time deposits and have ..., for early withdrawl. If you can accept not having immediate access to your funds, CDs can be an attractive savings vehicle.

6. Which of the following can be inferred as the theme of the passage?

A. Certificate of deposits is the best way to establish savings habit.
B. Establishing a consistent saving habit and also smart saving with CDs.
C. Employer's 40(k) plan provides a convenient way for consistent and smart savings.
D. Your savings decide how well you flourish and grow.

7. What is the tone of writing in the passage?

A. Satirical B. Critical
C. Didactic D. Sarcastic

8. Which of the following words can fill in the blanks to make it meaningful?

A. gifts B. lerative
C. causalities D. penalties

9. The sentences given in a question, which properly sequenced, form a coherent paragraph. Each sentence is labelled with a letter. Choose the most logical order of sentences from among the given choices to construct a coherent paragraph.

(*a*) Never in the wildest of his dreams had he ever imagined that the new addition to the business family would take the venture to such great heights. This franchise agent subsequently purchased the restaurant chain from him and gave his business an altogether different global approach.
(*b*) It was only when his business grew further that he started heading a complete fast-food restaurant chain. The business was at its peak, when a gentleman approached this man and joined the chain as a franchise agent.
(*c*) At a very young age, he had become the proud owner of a large chain of fast food restaurants that satiated the burger-pangs of many. Today, the old man sits and narrates his intriguing business journey.
(*d*) Looking back, he tells how he had started his business with just one barbecue restaurant several decades ago and how this small business eventually took its present shape.

A. *cdba* B. *abcd*
C. *bcda* D. *dcba*

10. In the question below, there are two sentences containing underlined homonyms, which may either be miss - spelt or inappropriately used in the context of the sentence. Select the appropriate and from the option given below :

(*a*) A vote of censure was passed against the chairman.
(*b*) Before release, every film is passed by the censor Board.

A. Only sentence (*a*) is incorrect
B. Only sentence (*a*) is correct
C. Both sentence (*a*) and (*b*) are correct.
D. Both sentence (*a*) and (*b*) are incorrect

11. In the following question, there are sentences that form a paragraph. Identify the sentence(s) or part(s) of sentences(s) that is/are correct in terms of grammar and usage (including spelling, punctuation and logical consistency). Then, choose the most appropriate option.

(*a*) I did not know what to make of you.
(*b*) Because you'd lived in India, I associated you more with my parents than with me.
(*c*) And yet you were unlike my cousins in Calcutta, who seem so innocent and obedient when I visited them.
(*d*) You were not curious about me in the least.
(*e*) Although you did make effort to meet me.

A. Only (*a*) B. Only (*a*) and (*b*)
C. Only (*a*) and (*e*) D. Only (*a*) and (*d*)

12. Fill in the blanks with the appropriate conjunction.

You must start at once; ______ you will be late.

A. whereas B. otherwise
C. as long as D. but

13. Choose the word closet in meaning to the given word.

Abnegation

A. Renunciation B. Excess
C. Saction D. Yielding

14. Choose the word. Which is most opposite in meaning to the given words.

Abjure

A. Renounde B. Relinquish
C. Abnegate D. Acquire

15. Choose the word closet in meaning to the given word.

lingua franca

A. Common B. Different
C. Mixture D. Share

16. Choose the word opposite in meaning to the given word.

Petulant

A. disagreeable B. dyspeptic
C. ill-humored D. amiable

17. Find the word that is the odd one out.

A. Exacerbate B. Alleviate
C. Mitigate D. Assuage

18. Change the voice of the given sentence :

I didn't realise that somebody was watching me.

A. I didn't realise that I was being watched.
B. I didn't realise that I was being watching.
C. I didn't realised that I was being watched.
D. I didn't realise that I was been watched.

19. Complete the meaning of the given sentence.

If you behaved well, your peers ______ you.

A. will respect B. would respect
C. respect D. shall respect

20. Choose the correct spelling from the given option :

The Stars were ______ in the sky.

A. scintillating B. scintilating
C. scintiellating D. scientillating

21. Choose the correct spelling from the given option :

The defence lawyer ______ there was insufficient evidence to convict his client.

A. rieterated B. reiterated
C. reiteirated D. reitiierated

22. Choose the word NOT having a prefix.

A. Distemper B. Dislike
C. Dishonest D. Disargee

23. Identify the correct figure of speech.

I must have called out to you a thousand times.

A. Personification B. Metaphor
C. Oxymoron D. Hyperbole

24. Choose the word opposite in meaning to the given word.

Servile

A. Imperious B. Humorous
C. Helpful D. Conspiratorial

25. Find a correct match of grammatical function with the usage of the word DOWN.

(*a*) Noun	(*i*)	Some people can down a whole chicken in a meal.
(*b*) Verb	(*ii*)	There is a down trend in the real state market.
(*c*) Adjective	(*iii*)	He was sent down to work in the branch office.
(*d*) Adverb	(*iv*)	In southern England the downs are so pretty.

	(*a*)	(*b*)	(*c*)	(*d*)
A.	(*iv*)	(*ii*)	(*iii*)	(*i*)
B.	(*iv*)	(*i*)	(*ii*)	(*iii*)
C.	(*i*)	(*ii*)	(*iii*)	(*iv*)
D.	(*iv*)	(*iii*)	(*ii*)	(*i*)

26. Choose the word closet in meaning to the given word.

Diaphanous

A. Transparent B. cloudy
C. foggy D. drab

27. Fill in the blanks with correct modal verb.

A. Might B. Would
C. Should D. Do

28. Change the voice of the given sentence.

Windowpanes are washed by cleaners.

A. Cleaners wash windowpanes.
B. Cleaners are washed by windowpanes.
C. Cleaners washed windowpanes.
D. Cleaners was washed windowpanes.

29. Identify the correct figure of speech.

Neeta needed new notebooks.

A. Personification
B. Metaphor
C. Alliteration
D. Hyperbole

30. Which part of speech is the given (underlined) word?

This wood will make a good <u>hiding</u> place.

A. Adverb B. Adjective
C. Noun D. Pronoun

31. Choose the correct option of the following incorrect sentence :

No matter what that I do, I can't make her happy.

A. No matter what should I do, I can't make her happy.
B. No matter what will I do, I can't make her happy.
C. No matter what I do, I can't make her happy.
D. No matter what I did, I can't make her happy.

32. Fill in the blank with proper conditional :

Suppose your car broke down in the middle of nowhere. What ______ do?

A. will you B. would you
C. may you D. did you

33. Fill in the blank with appropriate option :

I ______ anything from her in a long time.

A. haven't heard B. haven't hear
C. hasn't heard D. hasn't hear

34. Complete the meaning of the given sentence.

If you behaved well, your peers ______ you.

A. would respect B. will respect
C. would respected D. None of these

SECTION-II

35. Find the value of the given expression :

$$\sqrt{\frac{\left(3\frac{1}{4}\right)^4-\left(4\frac{1}{3}\right)^4}{\left(3\frac{1}{4}\right)^2-\left(4\frac{1}{3}\right)^2}}=?$$

A. 5 B. $5\frac{5}{12}$
C. $5\frac{7}{12}$ D. $5\frac{11}{12}$

36. Three circles with radii 1 cm each are drawn touching each other. Find the area of the shaded portion (in cm^2)?

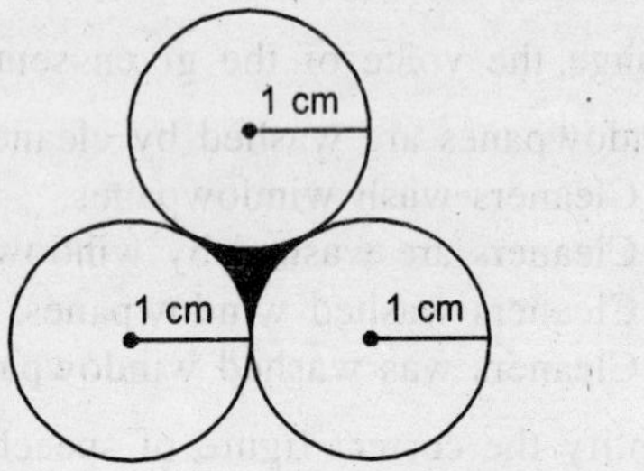

A. $\frac{(\sqrt{3}-\pi)}{2}$ B. $\frac{(2\sqrt{3}-\pi)}{2}$
C. $\frac{(\pi-\sqrt{3})}{2}$ D. $\frac{(\pi-2\sqrt{3})}{2}$

37. Two trains of length 150 m each pass each other in 20 sec., when moving in opposite direction. When they move in the same direction, they take 40 sec. to pass each other completely. Find the speed of the faster train.

A. 10.25 m/sec B. 11.25 m/sec
C. 11.75 m/sec D. 12 m/sec

38. Brigadier Rastogi travels from point A to B at 40 km/hr on bike. He travels from point B to C at 10 km/hr on cycle. The distance from A to B is equal to that of B to C. Then he travels from C to A via B at 24 km/hr by autorickshaw. Then find his average speed.

A. 12.2 km/hr
B. 12.4 km/hr
C. 19.2 km/hr
D. 19.4 km/hr

39. Rohan and Rahul are 144 km apart on point (A) and point (B), respectively. Rohan travels constantly at 8 km/hr. Rahul travels 4 km in first hour, 5 km in second hour, 6 km in third hour and so on. Find the point where they will meet.

A. 64 km from point A
B. 64 km from point B
C. Midway of A and B
D. None of the above

40. Find the unit's place digit of $(1!)^{1!} + (2!)^{2!} + (3!)^{3!} + ... + (100!)^{100!}$.

A. 4 B. 5
C. 6 D. 7

41. In how many ways can 10 books on Mechanics and 8 books on quantum physics be placed in a row such that two books on quantum physics may not be together.

A. 165 B. 176
C. 187 D. 198

42. In an institute, MBA Exam is conducted and sectional cutoff has been introduced into it. Any candidate appearing for the exam cannot qualify unless he clears the sectional cutoff. If there are 4 sections in the paper, then what is the number of ways an applicant may fail in the exam?

A. 15 B. 16
C. 32 D. 64

43. In how many ways letters of word POTICA can be arranged such that vowel occupy at odd positions?

A. 24 B. 30
C. 36 D. 42

44. Square, Circle, Hexagon and Octagon have equal perimeter. Which of the following has maximum area?

A. Square B. Circle
C. Hexagon D. Octagon

45. In how many ways can one wrap 3 kitkat, 2 fivestar and 3 bar one, if atleast one kitkat has to be there in the gift pack and the gift pack has three chocolates.

A. 3 B. 10
C. 46 D. 56

46. The difference between CI and SI for a loan is ₹ 114 when invested for 2 years. The rate of interest is 6% per annum. Find the loan amount.

A. ₹ 31,667 B. ₹ 41,667
C. ₹ 51,667 D. None of the above

47. The sum of series of n terms :

$$\log a+\log\left[\frac{a^2}{b}\right]+\log\left[\frac{a^3}{b^2}\right]+\ldots=?$$

A. $\log\left(\frac{a^{n+1}}{b^{n-1}}\right)^{\frac{n}{2}}$ B. $\log\left(\frac{a^{n-1}}{b^{n+1}}\right)^{\frac{n}{2}}$

C. $\log\left(\frac{a^{n+1}}{b^{n-1}}\right)^{n}$ D. $\log\left(\frac{a^{n-1}}{b^{n+1}}\right)^{n}$

48. What is the coefficient of Z^3 in $-7xy^2z^3a^2b^2$?

A. $-7xy^2$ B. 7
C. $-7xy^2a^2b^2$ D. -1

49. In a regular hexagon field, ropes are tied to connect all vertices (diagonal and sides). What is the number of intersection points ropes?

A. 16 B. 17
C. 19 D. 20

50. In the given diagram, $\angle CAB = 60°$ and $BC = a$, $AC = b$ and $AB = c$.

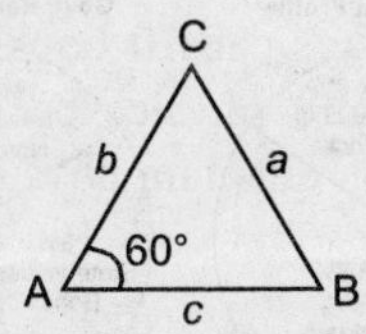

Then which of the following connect?

A. $a^2 = b^2 + c^2 - bc$ B. $a^2 = b^2 + c^2 - 2bc$
C. $a^2 = b^2 + c^2 + bc$ D. $a^2 = b^2 + c^2 + 2bc$

51. In a closed wooden box, length = 20 cm, breadth = 14 cm and height = 10 cm and thickness = 5 mm. If weight of empty box is 3.462 kg, then what is the weight of 1 cm^3 of wood?

A. 4 grams B. 5 grams
C. 6 grams D. 7 grams

52. Find the number of zeros at the end of $(5!)^{5!} + (10!)^{10!} + (50!)^{50!} + (100!)^{100!}$.

A. 11 B. 12
C. 110 D. 120

53. Ritesh is twice as good as Mitesh, Ritesh takes 30 days less than Mitesh to finish a task. How long will Ritesh and Mitesh take to complete the task together.

A. 10 days B. 20 days
C. 30 days D. 60 days

54. Mohan has a 200 litres container x litre of milk is kept in container. Mohan removes 6 litres milk and adds 5 litres water. He again replaces 6 litres of solution with water. Now milk and water are in the ratio 9 : 16. What is the quantity of mixture?

A. 6 litres B. 9 litres
C. 15 litres D. 16 litres

55. India express travels at 60 km/hr and halts for fixed time every hour. Due to halts, the average speed becomes 50 km/hr. Find the time of halt.

A. 8 minutes B. 10 minutes
C. 12 minutes D. 14 minutes

Directions (Qs. 56-59): *Answer the questions given below :*

In 2015, 100 aspirants appeared for an exam.
They had to answer four sections Maths, DI, LR and English.
The number of students who qualified in Maths = 55.
The number of students who qualified in LR = 38.
The number of students who qualified in (Maths + English) = 30.
The number of students who qualified in (LR + English) = 15.
The number of students who qualified in (Maths + LR) = 20.
The number of students who qualified in (Math + LR + English) = 5.
The number of students who qualified in DI = 22.
The number of students who qualified in (DI + LR) = 5.
The number of students who qualified in (DI + Maths) = 5.
The number of students who qualified in (DI + Math + LR = 5.
The number of students who qualified in English = 50.
Those who qualified in English could not qualify in DI section.

56. How many qualified at least two sections?

A. 35 B. 45
C. 55 D. 65

57. How many qualified in both Maths and LR but not any other subjects?

A. 5 B. 10
C. 15 D. 20

58. How many did not qualify in any section?

A. 3 B. 5
C. 12 D. None of these

59. How many qualify only in DI section?

A. 12 B. 15
C. 17 D. 21

Directions (Qs. 60-62): *Go through the table given below which contains data about production of rice in various years and change percentage compared to previous data.*

Years	Quantity of Rice (Town)	% change over previous year
20 – 21	134350	+6.25
30 – 31	1097172	+12.5
40 – 41	264280	+11.11
50 – 51	127890	–09.09
60 – 61	201924	+20.00
70 – 71	112325	–16.66
80 – 81	213465	–25.00
90 – 91	169368	+33.33
00 – 01	100956	+50.00
10 – 11	23800	–80.33

60. What is the approx. production of rice in year 1949-50?
A. 1278890 B. 264280
C. 116263 D. 140679

61. What is the difference in the production of rice in 1969-70 and 1979-80?
A. 284620 B. 134790
C. 149830 D. 23800

62. What is the approx production in 1959-60?
A. 33654 B. 168270
C. 16827 D. 201924

63. Valve A fills a bathtub in 10 hours and valve B fills the same bathtub in 15 hours. A and B are opened together, and later B was closed after some time. Total time taken to fill the batchtub is 8 hours. For how long B was open?
A. 1 hour B. 2 hours
C. 3 hours D. 4 hours

64. Average stipend of a group of students is ₹ 50 per day. The difference between maximum and minimum stipend is ₹ 45. If these students are excluded, the average decreases by ₹ 1. Minimum earning for any student lie between ₹ 42 to ₹ 47 and the number of students is a prime number where both the digit is also prime. What is the number of students initially?
A. 33 B. 35
C. 37 D. 39

65. In a smart phone factory, Robot A, B and C manufacture 25%, 35% and 40% circuits board respectively. For each of the robot faulty circuit board is 5%, 4% and 2% respectively. What is the probability that the printed circuit board is defective if 1 circuit board is picked at random?
A. 0.034 B. 0.34
C. 0.66 D. 0.066

66. A robot is 4 meters in length and placed at corner of 16 meters × 30 meters field. The robot is facing diagonally opposite corner and reaches the diagonally opposite corner in 15 seconds. What is the speed of robot?
A. 1 m/s B. 2 m/s
C. 3 m/s D. 4 m/s

67. What is the sum of integers from 113 to 113113, which are divisible by 7?
A. 16143 B. 113113
C. 16159 D. 913952088

68. Given $\dfrac{\left(\sqrt{x+4}+\sqrt{x-10}\right)^2}{(x+4)-(x-10)}=\dfrac{5}{2}$, find the value of x.
A. $\dfrac{263}{20}$ B. $\dfrac{265}{20}$
C. $\dfrac{263}{40}$ D. $\dfrac{263}{19}$

69. In a class of 50 students, 23 speak English, 15 speak Hindi and 18 speak Punjabi, 3 speak only English and Hindi, 6 speak only Hindi and Punjabi and 6 speak only English and Punjabi. If 9 can speak only English, then how many students speak all the three languages?
A. 3 B. 2
C. 5 D. 1

70. If $x=\dfrac{4ab}{a+b}$, then the value of $\dfrac{x+2a}{x-2a}+\dfrac{x+2b}{x-2b}$ is equal to.
A. 2 B. 3
C. 1 D. 4

71. Out of 80 students in a class, 25 are studying commerce, 15 mathematics and 13 physics, 3 are studying commerce and mathematics, 4 are studying mathematics and physics and 2 are studying commerce and physics, 1 student is studying all the three subjects together. How many students are not studying any of the three subject?
A. 35 B. 45
C. 40 D. 30

Directions (Qs. 72-74): *Study the diagrams given below and then answer the questions that follow :*

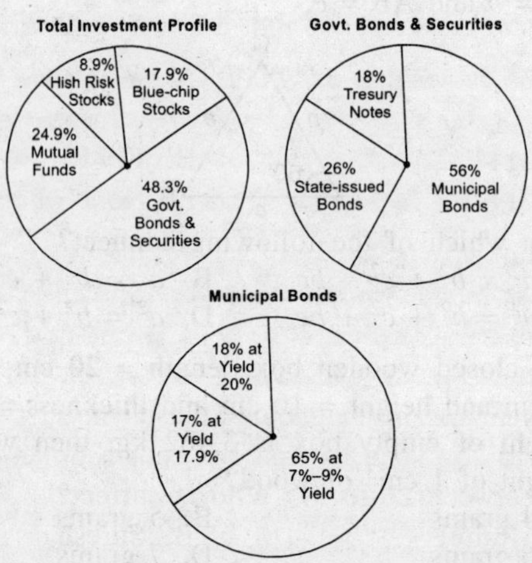

72. Approximately, how much money belonging to the investment protfolio was invested in state-issued bonds from the total investment of 11.5 crore?
A. ₹ 13876590 B. ₹ 13879065
C. ₹ 13876580 D. ₹ 13878065

73. Which of the following earned the least amount of money for the investment portfolio?
A. Municipal Bonds
B. State issued Bonds
C. Government bonds & Securities
D. Cannot be determined from the given information.

74. Which of the following was the greatest?
A. The amount of money invested in Municipal Bonds which yielded between 7% and 9%.
B. The amount of money invested in state-issued bonds.
C. The amount of money invested in High Risk Stocks.
D. The amount of money invested in Municipal Bonds which yielded over 9%.

SECTION-III

75. Royal Bengal Tiger : India :: Snow Leopard :
A. Srilanka B. Pakistan
C. Afghanistan D. Bangladesh

76. A new species lays exactly 120 out of which 50% are male and 50% are female. The female insect hatch and grow in a span of 20 days to lay eggs by themselves. On 1st April 2018, an insect laid 120 eggs. Find how many eggs will be hatched (approx.) by the end of May 2018?
A. 12960 B. 1269000
C. 12690000 D. None of these

77. A + B means A is sister of B. A/B means A is son of B. A = B means A is brother of B, A @ B means A is father of B. Which of the following shows M is grandson of P?
A. P = B @ M + N @ *s* B. L @ M @ N = K/P
C. M/T + J = L @ P D. P @ B = S @ M = N

78. India is written as 95491, then Japan is written as ______.
A. 11175 B. 11715
C. 11705 D. None of these

79. Rahul asked Shyam to find N such that N! > 10^N and must be the smallest integer. Shyam calculates and finds N is between 10 to 15, but he then asks his friends Sohan who says it is between 16 to 20. Then Shyam asks his neighbour Suresh who says it is between 21 to 25. Then he asks his cousin Sonal who says it is between 26 to 31. Who is correct?
A. Rahul B. Suresh
C. Shyam D. None of these

80. If 1st June 2013 is Saturday then 1st June 1981 is ______.
A. Monday
B. Saturday
C. Sunday
D. Thursday

81. The following figure is folded to form a block. Which symbol will appear on the opposite of ▲?

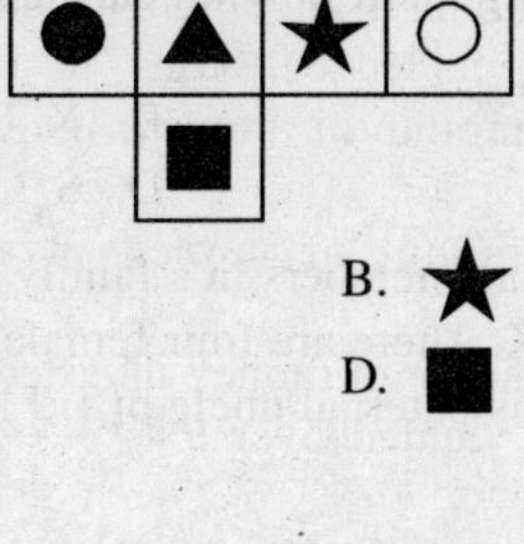

A. ● B. ★
C. ○ D. ■

82. Find the missing number in the given number table.

6	2	5	4
4	1	3	2
152	7	98	?

A. 42 B. 60
C. 56 D. None of these

83. Vijay's grandfather has an old Cuckoo clock. It takes 5 seconds for the "Cuckoo clock" to chime 5 Cuckoos. How long will it take to chime 10 Cuckoos?
A. 10 B. 11.25
C. 15 D. None of these

84. A person wants a house such that all sides of the house face North. He should build the house ______.
A. On the South Pole
B. On the equator
C. On the North Pole
D. On the Cancer line

85. Find the missing number from the below option.

19	78	20
25	144	47
16	?	13

A. 96 B. 76
C. 58 D. 29

86. Statement: Since 2018, the bulk of India's population has comprised of young working people–much more than the dependent population – children below 5 years of age and old people above 65 years of age. This trend will continue for the next 55 years.

Courses of Action:
(*a*) There will be a huge increase in the GDP of the country.
(*b*) According to a report by UNFPA, this population will be able to contribute effectively if good health facilities, education and proper infrastructure are provided to the whole population.

A. Only (*a*) follow
B. Only (*b*) follow
C. Both (*a*) and (*b*) follow
D. Neither (*a*) nor (*b*) follow

87. A clock gains 10 minutes a day. The clock was corrected at 6:00 AM. What will be the correct time when the clock shows 11:00 AM following day?
A. 10:50 AM B. 10:45 AM
C. 10:48 AM D. 10:25 AM

88. A doctor gives Vishal 3 pills to take with a gap of 30 minutes. What is the minimum time by which Vishal will get rid of his pain?

A. 1 hr 30 min B. 30 min
C. 45 min D. 60 min

89. Mobile manufacturing company 6 staff members packed 6 mobiles in 6 minutes. The management wants 60 mobiles to be packed in 60 minutes. How many staff members in total are required?
A. 60 B. 10
C. 6 D. None of these

90. Ornithologist : Bird :: Herpetologist :
A. Reptiles B. Mammals
C. Fish D. None of these

91. Amar consumed 100 laddoos from Monday to Friday. Each day, he consumed 6 more laddoos than the previous day. How many laddoos did he consume on Wednesday?
A. 15 B. 20
C. 25 D. None of these

92. Find the correct term of the following series.
0, 1, 2, 5, 20, 25, ?, 157
A. 150 B. 125
C. 130 D. None of these

93. Which answer figure will complete the pattern in the question figure?

Question figure:

Answer figure:

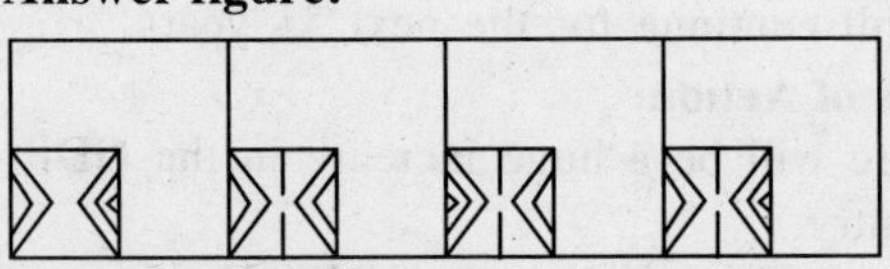

A B C D

94. Find the correct choice.

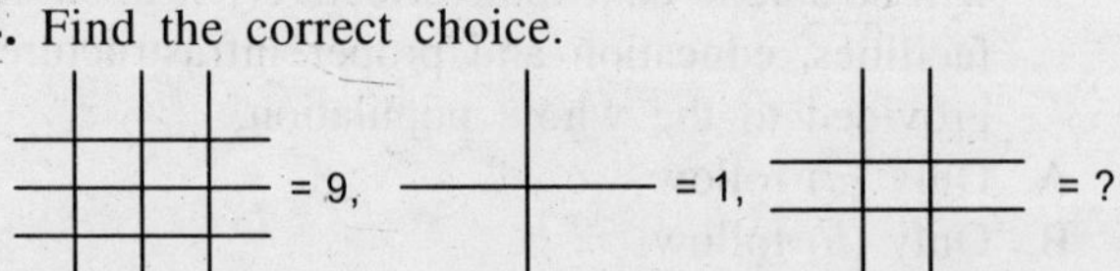

A. 5 B. 8
C. 4 D. None of these

95. Kishore says, "that man's father is my father's son". How is the man related to Kishore?
A. Brother B. Son
C. Father D. Grand Father

96. How many times does the letter 'A' appear from 0 to 100?
A. 79 B. 87
C. 68 D. None of these

97. Population of Timbaktoo (2 years ago) is 125000. Due to natural calamities people started migrating. So, population decreased at the rate of 4% per annum. How many migrated from his town in past 2 years?
A. 98000 B. 96000
C. 9600 D. 9800

Directions (Qs. 98-100): *Five movies- Do Bigha jameen, Sholay, 3 idiots, Chak de, Aanand – screening on Monday to Friday in any order. Movie screened on Friday remains till Sunday. Screening of Do Bigha Jameen and Chak de should not be first and last day. Chak de should be followed by Aanand, Sholay is immediately after Do Bigha Jameen. There is a one movie between Sholay and 3 idiots.*

98. Which movies was screened on Friday?
A. Chak de B. Sholay
C. Aanand D. 3 idiots

99. Sholay was screened on which day?
A. Tuesday B. Wednesday
C. Friday D. Monday

100. 3 idiot was screened on which day?
A. Friday B. Tuesday
C. Thursday D. Monday

Directions (Qs. 101-103): *Study the following information carefully and answer the given questions :*

Seven cars P, Q, R, S, T, Y and X are parked in a linear row facing north in such a way that no two cars parked with each other of according to alphabetical order (for ex-P is not parked with Q, Q is not parked with P and R and so on).
Some cars either of Petrol or some are Diesel variant. Y is third to the left of P. More than three cars are parked between the petrol cars. T is second to the right of Q. X is a diesel car and parked at one of the extreme end. R is a diesel car and parked forth to the right of S. All the cars are arranged in ascending order according to the distance covered by them from left to right. Car Q covers 37 km and Car R cover 50 km. X is to the right of T. Y is of Diesel variant car and no petrol variant car parked next to it.

101. Which among the following are petrol cars?
A. S and T B. P, Y, Q
C. P and S D. P, Y, S, Q

102. What can be the distance covered by car *y*?
A. 55 km B. 73 km
C. 27 km D. 41 km

103. Which among the following cars are parked at extreme ends?
A. S, T B. X, Q
C. X, S D. S, Q

104. There are nine members in a family i.e., M, N, O, X, Y, Z, I, J and K. There are four females and three married couples. M is paternal uncle of I, J has only 2 children,

K is daughter-in-law of N. O is married to N. X and Y are sons of O. Y is not married Z is married to J. N is son-in-law of J. Z is a male member.

How N's sister in law is related to J's brother in law?

A. Niece B. Nephew
C. Son D. Daughter

Directions (Qs. 105-107): *Study the following information carefully and answer the given questions :*

There are six boys i.e., Rahul, Mohit, Vikas, Gopi, Ayush and Aditya who all are of different weight. No two persons have same weight. Only two persons are lighter than Rahul. Mohit is heavier than Rahul but lighter than Vikas and Gopu. Aditya is heavier than Ayush but lighter than Gopi. Gopi is not the heaviest. The weight of 2nd heaviest person is 230 kg and the weight of lightest is 128 kg.

105. How many persons are heavier than Rahul?

A. One B. Two
C. Three D. Four

106. If the sum of weight of Ayush and Rahul is 262 and the sum of weight of Gopi and Mohit is 426, then what is the sum of weight of Rahul and Mohit?

A. 343 B. 340
C. 330 D. 300

107. Which among the following person is the 3rd heaviest?

A. Rahul B. Mohit
C. Vikas D. Gopi

Directions (Qs. 108-110): *Study the following information carefully and answer the given questions :*

Nine persons are sitting in a row. Some of them are facing North and some are facing South. C sits 2nd from one of the extreme ends. Two persons sit between C and D. M sits 3rd to the left of D. N sits 2nd to the right of M. Immediate neighbours of M faces opposite to M. O sits 2nd to the right of N. P is an immediate neighbour of O. Persons sitting in an extreme end are facing opposite direction to each other C sits 2nd to the right of O. P and C doesn't face North. Q sits 2nd to the left of B. A doesn't face south.

108. How many persons are sitting between C and M?

A. Two B. Three
C. Five D. Four

109. Who among the following pair of persons are sitting at extreme end?

A. M-B B. B-A
C. C-P D. C-A

110. How many persons are facing South?

A. Two B. Three
C. Four D. Five

SECTION-IV

111. Human Resource Development (HRD) Ministry launched Operation Digital Board to provide better digital education in school with the help of digital facilities for teaching by the which of the following year?

A. 2026
B. 2022
C. 2025
D. 2030

112. On 30th December 2018, in which of the following place, Prime Minister Narendra Modi laid the foundation stone of 50 MW LNG-based power project at Hope Town, which will be established by NTPC Vidyut Vyapar Nigam at a cost of ₹ 387.80 crore?

A. Andaman & Nicobar Islands
B. Odisha
C. Kerala
D. Tamil Nadu

113. On 31st December 2018, ______ and Israel officially quite the United Nations Educational, Scientific and Cultural Organization (UNESCO) amid concerns that UNESCO fosters anti-Israel bias?

A. Palestine
B. United States of America (USA)
C. Iran
D. Egypt

114. On 3rd January 2019, which bill passed by the Parliament that seeks to do away with the no-detention policy in schools?

A. Educational Institutions Regulation (Amendment) Bill, 2019
B. Right to Free Education for Everyone (Amendment) Bill, 2019
C. Right to Education and Employment (Amendment) Bill, 2019
D. Right of Children to Free and Compulsory Education (Amendment) Bill, 2018

115. Name the bank which has joined hands with Multilateral Investment Guarantee Agency (MIGA) to expand Development Finance in Asia and Pacific to increase the flow of private sector investment.

A. Asian Development Bank (ADB)
B. European Investment Bank (EIB)
C. International Bank for Reconstruction and Development (IBRD)
D. Eurasian Development Bank (EPB)

116. Which financial service company launched 'Sabse Khaas Loan' scheme for two-wheeler customers?
A. Mahindra & Mahindra-Financial Service Ltd.,
B. TVS credit service Ltd.,
C. Bajaj Finserve Ltd.,
D. L&T Finance

117. The Reserve Bank of India (RBI) observed Financial Literacy (FL) week on which date?
A. June 2-6, 2019 B. June 3-7, 2019
C. June 4-8, 2019 D. June 1-5, 2019

118. What C stands for in CICO which was in news recently?
A. Consumer B. Commodity
C. Currency D. Cash

119. Which think tank has indicated that India lost $30 billion in revenue collections due to trade mis invoicing in 2016?
A. NITI Aayog
B. Centre for International Policy
C. Global Financial Integrity (GFI)
D. Task Force on Financial Integrity and Economic Development

120. Name the organization which is set to create the foreign exchange trading platform for retail participants.
A. Reserve Bank of India (RBI)
B. EXIM Bank
C. Securities and Exchange Board of India (SEBI)
D. National Stock Exchange (NSE)

121. Name the country which topped in the e-Money transactions for the year 2017.
A. Singapore B. USA
C. India D. Japan

122. Name state whose Autonomous Development Councils (ADCs) has met with 15th Finance Commission headed by the chairman Shri N.K. Singh.
A. West Bengal
B. Manipur
C. Madhya Pradesh
D. Meghalaya

123. Which bank is the first to introduce repo-linked home loan product from July 1, 2019?
A. State Bank of India (SBI)
B. Bank of India
C. ICICI Bank
D. Axis Bank

124. Name the department that has relaxed income tax rules in order to facilitate fund raising by startups which is a part of the 'Startup India Vision 2024'.
A. Department of Heavy Industry
B. Department of Commerce
C. Department for Promotion of Industry and Internal Trade (DPIIT)
D. Department of Export Promotion Council

125. What is the name of the state that has become the first state in India to list Masala Bond in the overseas market?
A. Kerala B. Karnataka
C. Maharashtra D. Madhya Pradesh

126. Which of the following Indian regulatory body allows mutual funds to invest in Exchange-traded commodity derivatives (ETCD)?
A. Insurance Regulatory and Development Authority (IRDAI)
B. Reserve Bank of India (RBI)
C. Securities and Exchange Board of India (SEBI)
D. National Securities Depository Limited (NSDL)

127. Name the bank which becomes the 3rd largest bank after it merged with Dena Bank & Vijaya Bank?
A. State Bank of India
B. Bank of Baroda
C. Union Bank of India
D. Indian Bank

128. Name the state whose Health Department was conferred with the World Health Organization's (WHO) Tobacco Control Award for 2019 during World No Tobacco Day (WNTD).
A. Tamil Nadu B. Bihar
C. Andhra Pradesh D. Rajasthan

129. Name the country where the Under-20 Eurasian Athletics Championships was held.
A. Mongolia B. Tajikistan
C. Kazhakstan D. Kyrgyztan

130. The theme of Elephanta festival 2019 was–
A. "Shadja" B. "Swarang"
C. "Sarang" D. "Matanga"

ANSWERS

1	2	3	4	5	6	7	8	9	10
C	C	D	D	B	B	C	D	A	B
11	**12**	**13**	**14**	**15**	**16**	**17**	**18**	**19**	**20**
D	B	A	D	A	D	A	A	B	A
21	**22**	**23**	**24**	**25**	**26**	**27**	**28**	**29**	**30**
B	A	D	A	B	A	B	A	C	B

31	32	33	34	35	36	37	38	39	40
C	B	A	A	B	B	B	C	C	D
41	**42**	**43**	**44**	**45**	**46**	**47**	**48**	**49**	**50**
A	A	C	B	C	A	A	C	C	A
51	**52**	**53**	**54**	**55**	**56**	**57**	**58**	**59**	**60**
C	D	B	C	B	C	B	D	C	D
61	**62**	**63**	**64**	**65**	**66**	**67**	**68**	**69**	**70**
C	B	C	C	A	B	D	A	C	A
71	**72**	**73**	**74**	**75**	**76**	**77**	**78**	**79**	**80**
A	A	B	A	C	D	D	B	B	A
81	**82**	**83**	**84**	**85**	**86**	**87**	**88**	**89**	**90**
C	C	B	A	C	B	C	D	C	A
91	**92**	**93**	**94**	**95**	**96**	**97**	**98**	**99**	**100**
B	A	D	C	B	D	D	C	B	D
101	**102**	**103**	**104**	**105**	**106**	**107**	**108**	**109**	**110**
C	D	C	A	C	C	B	C	B	C
111	**112**	**113**	**114**	**115**	**116**	**117**	**118**	**119**	**120**
B	A	B	D	A	D	B	D	C	A
121	**122**	**123**	**124**	**125**	**126**	**127**	**128**	**129**	**130**
D	D	A	C	A	C	B	D	C	B

EXPLANATORY ANSWERS

35. $$\sqrt{\frac{\left(3\frac{1}{4}\right)^4-\left(4\frac{1}{3}\right)^4}{\left(3\frac{1}{4}\right)^2-\left(4\frac{1}{3}\right)^2}} = \sqrt{\frac{\left(\frac{13}{4}\right)^4-\left(\frac{13}{3}\right)^4}{\left(\frac{13}{4}\right)^2-\left(\frac{13}{3}\right)^2}}$$

$$= \sqrt{\frac{\left[\left(\frac{13}{4}\right)^2-\left(\frac{13}{3}\right)^2\right]\left[\left(\frac{13}{4}\right)^2+\left(\frac{13}{3}\right)^2\right]}{\left(\frac{13}{4}\right)^2-\left(\frac{13}{3}\right)^2}}$$

$$= \sqrt{\left(\frac{13}{4}\right)^2+\left(\frac{13}{3}\right)^2}$$

$$= 13\sqrt{\frac{1}{16}+\frac{1}{9}} = 13\sqrt{\frac{9+16}{144}}$$

$$= \frac{13}{12}\sqrt{25} = \frac{13\times 5}{12} = \frac{65}{12} = 5\frac{5}{12}$$

36. Let A, B and C are center of three circles.

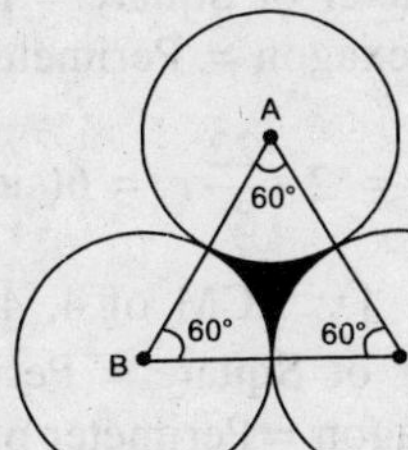

Given, Three circles with radius 1 cm each are drawn touching each other.

$\because$ AB = BC = CA = 2 cm

$\therefore$ ABC is an equilateral triangle.

Now, Area of shaded portion

= Area of ΔABC – 3 × Area of Sector A

$$= \frac{\sqrt{3}}{4}\times(2)^2 - 3\times\frac{60}{360}\times\pi(1)^2$$

$$= \sqrt{3}-\frac{1}{2}\pi = \left(\frac{2\sqrt{3}-\pi}{2}\right)\text{cm}^2$$

37. Let the speed of two trains are x m/sec and y m/sec and $x > y$

When moving in opposite direction

then, $(x + y) \times 20 = 150 + 150$

$$\Rightarrow \quad x + y = \frac{300}{20} = 15$$

$$\Rightarrow \quad x + y = 15 \quad \text{...}(i)$$

When moving is same direction

the, $(x - y) \times 40 = 150 + 150$

$$\Rightarrow \quad x - y = \frac{300}{40} = \frac{15}{2} = 7.5$$

$$\Rightarrow \quad x - y = 7.5 \quad \text{...}(ii)$$

From (i) and (ii), we get

$$2x = 22.5 \Rightarrow x = 11.25 \text{ cm/sec}$$

Hence, the speed of the faster train = 11.25 cm/sec.

38. Let distance from point A to B = distance from point B to C = x km

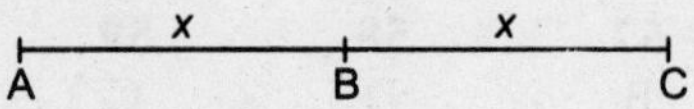

Time (from A to B) = $\frac{x}{40}$ hr

Time (from B to C) = $\frac{x}{10}$ hr

and time (from C to A) = $\frac{2x}{24} = \frac{x}{12}$ hr

$\therefore$ Average speed = $\frac{\text{All distance}}{\text{All time}}$

$$= \frac{x+x+2x}{\frac{x}{40}+\frac{x}{10}+\frac{x}{12}} = \frac{4x}{\frac{3x+12x+10x}{120}}$$

$$= \frac{4x \times 120}{25x} = \frac{480}{25} \text{ km/hr}$$

$$= 19.2 \text{ km/hr.}$$

39. Let Rohan and Rahul meets at point C after t hours.

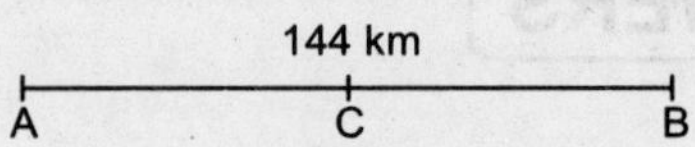

Distance travelled by Rohan in t hours = $8t$ km

and Distance travelled by Rahul in first hour

$= (3 + 1)$ km

Distance travelled by Rahul is second hour

$= (3 + 2)$ km

Distance travelled by Rahul in third hour = $(3 + 3)$ km

Distance travelled by Rahul in t hour = $(3 + t)$ km

Hence, distance travelled by Rahul

$$= (3 + 1) + (3 + 2) + (3 + 3) + \ldots + (3 + t)$$

$$= (3 + 3 + 3 + \ldots + t) + (1 + 2 + 3 + \ldots + t)$$

$$= \left\{3t + \frac{t(t+1)}{2}\right\} \text{ km}$$

Distance travelled by Rohan and Rahul = 144

$$8t + 3t + \frac{t(t+1)}{2} = 144$$

$$\Rightarrow \quad 11t + \frac{t(t+1)}{2} = 144$$

$$\Rightarrow \quad \frac{22t + t^2 + t}{2} = 144$$

$$\Rightarrow \quad t^2 + 23t = 288 \Rightarrow t^2 + 23t - 288 = 0$$

$$\Rightarrow t^2 + 32t - 9t - 288 = 0$$

$$\Rightarrow t(t + 32) - 9(t + 32) = 0$$

$$\Rightarrow (t + 32)(t - 9) = 0$$

$\Rightarrow$ $t = -32$ or 9 ($t = -32$ neglecting)

$\therefore$ $t = 9$ hours

$\therefore$ Distance travelled by Rohan in 9 hours

$= 8 \times 9 = 72$ km

$\therefore$ Distance AC = 72 km

Hence, Rohan and Rahul will meet at midway of A and B.

40. $(1!)^{1!} + (2!)^{2!} + (3!)^{3!} + (4!)^{4!} + (100!)^{100!}$

$= (1)^1 + (2 \times 1)^{2\times1} + (3 \times 2 \times 1)^{3\times2\times1} + \ldots + (100!)^{100!}$

$= 1 + 2^2 + 6^6 + (24)^{24} + \ldots (100!)^{100!}$

Sum of units place digit

$= 1 + 4 + 6 + 6 + 0 + \ldots + 0 = 17$

Hence, unit's place digit = 7.

41. Here, 10 books on Mechanics can be arrange in this ways

$PM_1PM_2PM_3PM_4PM_5PM_6PM_7PM_8PM_9PM_{10}$

Given, two books on quantum physics may not be together

Hence, the number of way = ${}^{11}C_8 = \frac{11!}{8!(11-8!)}$

$$= \frac{11!}{8!3!} = \frac{11 \times 10 \times 9(8!)}{8!3!}$$

$$= \frac{11 \times 10 \times 9}{3 \times 2} = 11 \times 5 \times 3 = 165.$$

42. Here, 4 sections in the exam

Each sections have certain cutoff marks to qualify

As, each section have two choices pass or fail

So, Total number of choice = $2 \times 2 \times 2 \times 2 = 16$

and the numbers of choice (Passed in each section)

$= 1 \times 1 \times 1 \times 1 = 1$

Hence, the number of ways in which students get fail

$= 16 - 1 = 15.$

43. Given, word "POTICA"

Here, the number of letters = 6

The number of vowels = 3 *i.e.*, (A, I, O)

$\because$ Vowel occupy odd position *i.e.*, 1st, third and fifth position remaining letters occupy even position be second, fourth and sixth position.

$\therefore$ Total number of arrangement = $3! \times 3!$

$= 3 \times 2 \times 1 \times 3 \times 2 \times 1$

$= 6 \times 6 = 36.$

44. Given that, Perimeter of Square = Perimeter of Circle = Perimeter of Hexagon = Perimeter of Octagon

$$\Rightarrow \quad 4(\text{side}) = 2 \times \frac{22}{7} r = 6(\text{side}) = 8(\text{side})$$

[$\because$ LCM of 4, 44, 6 and 8 = 264]

$\Rightarrow$ Left Perimeter of Square = Perimeter of Circle = Perimeter of Hexagon = Perimeter of Octagon = 264 m

The side of square $= \frac{264}{4} = 66$ m

$\therefore$ Area of square $= (66)^2$ m^2 $= 4356$ m^2

Radius of circle $= 264 \times \frac{7}{44} = 6 \times 7 = 42$ m

$\therefore$ Area of circle $= \pi r^2 = \frac{22}{7} \times (42)^2$

$= \frac{22}{7} \times 42 \times 42 = 22 \times 6 \times 42$

$= 5544$ m^2

Side of Hexagon $= \frac{264}{6} = 44$ m

Area of Hexagon $= 6 \times \frac{\sqrt{3}}{4} \times (44)^2$ m^2

$= 6 \times \frac{\sqrt{3}}{4} \times 44 \times 44$

$= 66 \times 44 \times 1.732$ m^2

$= 5029.73$ m^2

Side of octagon $= \frac{264}{8} = 33$ m

$\therefore$ Area of Octagon $= 2(\text{side})^2 (1+\sqrt{2}) = 2(33)^2(1+\sqrt{2})$

$= 2178 \times 2.414 = 2557.7$ m^2

Hence,

A(Circle) > A(Octagon) > A(Hexagon) > A(Square)

45. Here, Kitkat = 3, Five star = 2, Barone = 3

The total number of chocolates = 3 + 2 + 3 = 8

Number of way of seating 3 chocolates out of 8

$= {}^8C_3 = \frac{8!}{3!(8-3)!} = \frac{8!}{3!5!} = \frac{8 \times 7 \times 6}{3!}$

$= \frac{8 \times 7 \times 6}{6} = 56$

Number of way of selecting 3 chocolate out of

$5 = {}^5C_3 = \frac{5!}{3!(5-3)!} = \frac{5!}{3!2!} = \frac{5.4}{2} = 10$

Hence, number of ways of selecting 3 chocolates Such that atleast one of them is Kitkat

$= 56 - 10 = 46.$

46. Compound Interest $= P\left(1+\frac{r}{100}\right)^n - P$

$= P\left(1+\frac{6}{100}\right)^2 - P = P\left(\frac{106}{100}\right)^2 - P$

$= P\left(\frac{53}{50}\right)^2 - P = P\left[\left(\frac{53}{50}\right)^2 - 1\right]$

$= P\left[\frac{2809}{2500} - 1\right] = \frac{309}{2500}P$

Simple Interest $= \frac{P \times r \times t}{100}$

$= \frac{P \times 6 \times 2}{100} = \frac{12P}{100}$

Given, C.I. – S.I. = ₹ 114

$\frac{309}{2500}P - \frac{12P}{100} = ₹\ 114$

$\Rightarrow \frac{309P - 300P}{2500} = ₹\ 114$

$\Rightarrow 9P = ₹\ 114 \times 2500$

$\Rightarrow P = ₹\ \frac{114 \times 2500}{9}$

$= ₹\ 31666.6666$

$= ₹\ 31667$

Hence, the loan amount P = ₹ 31667.

47. $\log a + \log\left[\frac{a^2}{b}\right] + \log\left[\frac{a^3}{b^2}\right] + \ldots + \text{to } n \text{ terms}$

$= \log a + \log a^2 - \log b + \log a^3 - \log b^2 + \ldots + \text{to } n \text{ terms}$

$= (\log a + \log a^2 + \log a^3 + \ldots + \log a^n) - (\log b + \log b^2 + \ldots + \log b^{n-1})$

$= (\log a + 2\log a + 3\log a + \ldots + n\log a) - (\log b + 2\log b + 3\log b + \ldots + (n-1)\log b)$

$= (1 + 2 + 3 + \ldots + n)\log a - [1 + 2 + 3 + \ldots (n-1)]\log b$

$= \frac{n(n+1)}{2}\log a - \frac{(n-1)(n-1+1)}{2}\log b$

$= \frac{n(n+1)}{2}\log a - \frac{n(n-1)}{2}\log b$

$= \frac{n}{2}[(n+1)\log a - (n-1)\log b]$

$= \frac{n}{2}[\log a^{n+1} - \log b^{n-1}]$

$= \frac{n}{2}\left[\log\frac{a^{n+1}}{b^{n-1}}\right] = \log\left[\frac{a^{n+1}}{b^{n-1}}\right]^{\frac{n}{2}}$

48. We have, Expression $= -7xy^2z^3a^2b^2$

Co-efficint of $z^3 = -7xy^2a^2b^2$

49. In a regular hexagon

Number of sides = 6

Number of diagonals = 9

$\because$ Ropes are tied to connect all vertices (diagonal and sides)

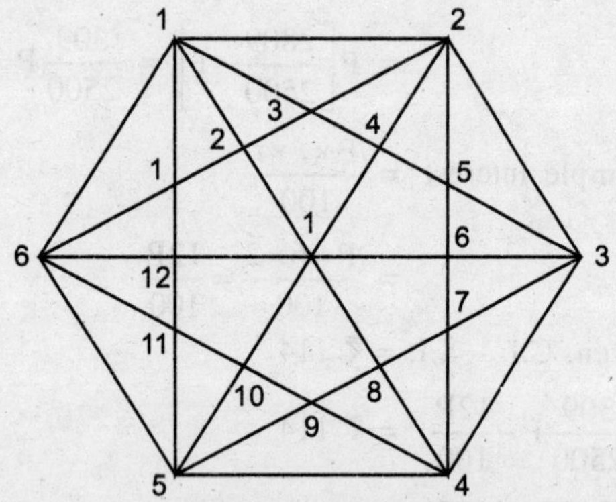

$\therefore$ The number of intersection points rope

$$= 6 + 12 + 1 = 19.$$

50. From cosine formula

$$\cos(\angle BAC) = \frac{AB^2 + AC^2 - BC^2}{2AB\cdot AC}$$

$$\Rightarrow \quad \cos 60^\circ = \frac{c^2 + b^2 - a^2}{2bc}$$

$$\Rightarrow \quad \frac{1}{2} = \frac{c^2 + b^2 - a^2}{2bc}$$

$$\Rightarrow c^2 + b^2 - a^2 = bc$$

$$\Rightarrow \quad a^2 = b^2 + c^2 - bc$$

51. Given, In a closed wooden box.

$$l = 20 \text{ cm}, b = 14 \text{ cm}, h = 10 \text{ cm}$$

and thickness = 5 mm = 0.5 cm

Volume of wood used in the box

$$= 20 \times 14 \times 10 - (20 - 2 \times 0.5)(14 - 2 \times 0.5)(10 - 2 \times 0.5)$$

$$= 2800 - (20 - 1)(14 - 1)(10 - 1)$$

$$= 2800 - 19 \times 13 \times 9$$

$$= 2800 - 2223 = 577 \text{ cm}^3$$

According to question

Weight of 577 cm^3 of wood = 3.462 kg

$$\therefore \text{Weight of 1 cm}^3 \text{ of wood} = \frac{3.462}{577} \text{ kg} = 0.006 \text{ kg}$$

$$= 0.006 \times 1000 \text{ g} = 6 \text{ gram.}$$

52. $(5!)^{5!} + (10!)^{10!} + (50!)^{50!} + (100!)^{100!}$

$= (1 \times 2 \times 3 \times 4 \times 5)^{1\times2\times3\times4\times5} + (10 \times 9!)^{10!} + (5 \times 10 \times 49!)^{50!} + (100 \times 99)^{100!}$

$= (12 \times 10)^{120} + (10 \times 9!)10! + (10 \times 5 \times 49!)^{50!} + (100 \times 99!)^{100!}$

$\because (10)^{5!} = (10)^{120}$ is a number that have 120 zeroes and $(10)^{10!}$ is a number that have more than 120 zeroes of the end

and $(10)^{50!}$ and $(100)^{100!}$ is a number that have more than 120 zeroes of the end.

Hence, number of zeroes at the end of the given expression = 120.

53. Let Mitesh finish a work in n days then Ritesh finish the same work in $(n - 30)$ days

According to question

$$2 \times \frac{1}{n} = \frac{1}{n-30}$$

$$\Rightarrow \quad 2(n - 30) = n$$

$$\Rightarrow \quad n = 60$$

Ritesh and Mitesh take to complete the task together

$$= \frac{1}{n} + \frac{1}{n-30}$$

$$= \frac{1}{60} + \frac{1}{60-30}$$

$$= \frac{1}{60} + \frac{1}{30} = \frac{1+2}{60} = \frac{3}{60} = \frac{1}{20}$$

Hence, time taken to finish the work, while working together = 20 days.

54. After removing 6 litres of milk by water then container contains

Ratio Milk : Water = $(x - 6) : 6$

Again removing 6 litres of milk by water

$\therefore$ Quantity of milk remains

$$= (x-6) - 6 \times \frac{(x-6)}{x}$$

$$= \frac{x(x-6) - 6(x-6)}{x}$$

$$= \frac{(x-6)(x-6)}{x} = \frac{(x-6)^2}{x}$$

and, quantity of water

$$= 6 + \left(6 - 6 \times \frac{6}{x}\right) = 6 + 6 - \frac{36}{x}$$

$$= 12 - \frac{36}{x} = \frac{12}{x}(x-3)$$

According to question,

$$\frac{(x-6)^2}{x} : \frac{12}{x}(x-3) = 9 : 16$$

$$\Rightarrow \frac{(x-6)^2}{12(x-3)} = \frac{9}{16} \Rightarrow 4(x-6)^2 = 27(x-3)$$

$$\Rightarrow 4(x^2 - 12x + 36) = 27(x - 3)$$

$$\Rightarrow 4x^2 - 75x + 225 = 0$$

$$\Rightarrow 4x^2 - 60x - 15x + 225 = 0$$

$$\Rightarrow 4x(x - 15) - 15(x - 15) = 0$$

$$\Rightarrow (4x - 15)(x - 15) = 0 \Rightarrow x = 15 \text{ litres.}$$

55. Speed of India express = 60 km/hr

Speed of India express with halt = 50 km/hr

LCM of 60 and 50 = 300

$\therefore$ Let distance = 300 km

then, difference in time travel

$$= \frac{300}{50} - \frac{300}{60} = 6 - 5 = 1 \text{ hr}$$

In 6 hours express halt 6 equal time interval, So each halt time

$$= \frac{1 \times 60}{6} = 10 \text{ minutes}.$$

For Q.No. 56–59:

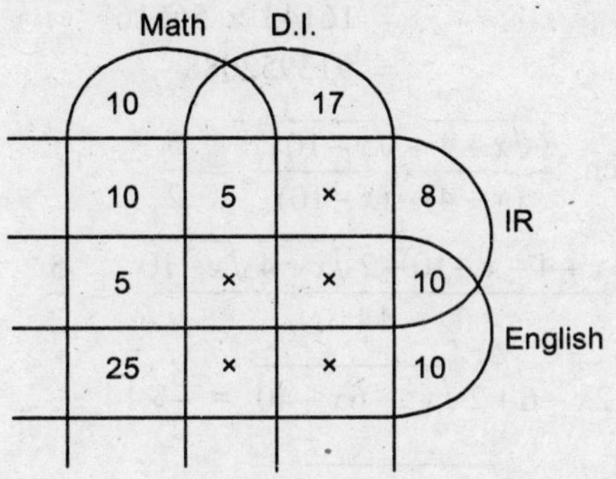

56. The number of students passed in at least two sections

$= 10 + 10 + 5 + 25 + 5 = 55.$

57. The number of student qualify in both Maths and I.R. but not any other subject = 10.

58. Number of students did not qualify in any sections

$= 100 - (10 + 17 + 8 + 10 + 10 + 10 + 25 + 5 + 5)$

$= 100 - 100 = 0.$

59. Number of student qualify in DI = 17.

60. The approx. production of rice in year 1949 – 50

$$= 127890 \times \frac{100}{100 - 09.09}$$

$$= 12789\not{0} \times \frac{11}{1\not{0}} = 140679.$$

61. Production of rice in 1969 – 70

$$= 112325 \times \frac{100}{100 - 16.66}$$

$$= 112325 \times \frac{100}{83.34}$$

$$= 112325 \times \frac{6}{5} = 134790$$

and the production of rice in 1979 – 80

$$= 213465 \times \frac{100}{100 - 25}$$

$$= 213465 \times \frac{100}{75}$$

$$= 213465 \times \frac{4}{3} = 284620$$

$\therefore$ Difference in production

$= 284620 - 134790 = 149830.$

62. Approx. production in 1959 – 60

$$= 201924 \times \frac{100}{100 + 20}$$

$$= 201924 \times \frac{100}{120}$$

$$= 201924 \times \frac{5}{6} = 168270.$$

63. Valve A fills a bathtub in 10 hours and Valve B fills bathtub in 15 hours

LCM of 10 and 15 = 30

Work done by Valve A in 1 hour = $\frac{30}{10} = 3$ units

Work done by Valve B in 1 hour = $\frac{30}{15} = 2$ units

Let Valve B was closed after t hours

Then, $3 \times 8 + 2 \times t = 30$

$\Rightarrow \quad 24 + 2t = 30$

$\Rightarrow \quad 2t = 30 - 24 = 6$

$\Rightarrow \quad t = 3$ hours

Hence Valve B was open for 3 hours.

64. Let the number of students initially = n

where, n is a two digits prime number whose both digits are also prime.

Let minimum earning of student is x

then, Maximum earning of the student is $(x + 45)$

Total stipend of n student = $50n$

Total stipend of minimum and maximum earning student

$= x + x + 45 = (2x + 45)$

Total earning of remaining students

$= 50n - (2x + 45)$

According to question,

Average stipend of remaining students

$= 50 - 1 = 49$

Total earning remaining students = $49(n - 2)$

$\therefore \; 50n - (2x + 45) = 49(n - 2)$

$\Rightarrow 50n - 2x - 45 - 49n + 98 = 0$

$\Rightarrow n - 2x + 53 = 0 \Rightarrow n = 2x - 53$

As $42 \le x \le 47 \; \therefore \; x = 45$

$n = 2x - 53 = 2 \times 45 - 53$

$= 90 - 53 = 37.$

Here, '37' is a prime number, also its two digits 3 and 7 are prime. So, number of students = 37

65. Let the total number of circuit board manufactured by A, B and C = 100

Then,

The number of circuit board manufactured by A = 25

The number of circuit board manufactured by B = 35

The number of circuit board manufactured by C = 40

Now,

The number of faulty circuit board manufactured by A

$$= 25 \times \frac{5}{100} = \frac{5}{4} = 1.25$$

The number of faulty circuit board manufactured by B

$$= 35 \times \frac{4}{100} = \frac{35}{25} = \frac{7}{5} = 1.40$$

The number of faulty circuit board manufactured by C

$$= 40 \times \frac{2}{100} = 2 \times \frac{2}{5} = \frac{4}{5} = 0.8$$

Total faulty circuit board = 1.25 + 1.4 + 0.8 = 3.45

Hence, probability for selecting a faulty circuit board

$$= \frac{3.45}{100} = 0.0345.$$

66. Let ABCD be a rectangular field of length and breadth are 30 m and 16 m

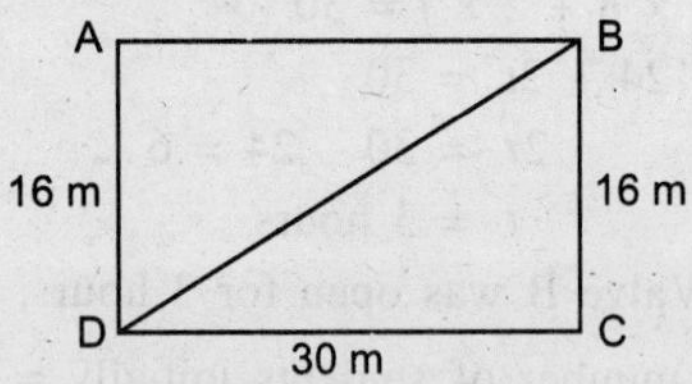

and a robot of length 4 m is placed at corner D facing towards the diagonally opposite corner B

$$\because \quad DC = \sqrt{DC^2 + CB^2}$$

$$= \sqrt{30^2 + 16^2} = \sqrt{900 + 256}$$

$$= \sqrt{1156} = 34 \text{ m}$$

Length to be covered by robot in 15 sec

= 34 – 4 = 30 m

Hence, speed of the robot $= \frac{30}{15} = 2$ m/sec.

67. Integers between 113 and 113113 which are divisible by 7 are 119, 126, 133, ..., 113113

$\because$ 119, 126, 133, 140, ..., 113113 are in A.P.

$\therefore$ First term a = 119, common difference d = 7

and t_n = 113113

$a + (n - 1)d$ = 113113

$\Rightarrow$ 119 + $(n - 1)$7 = 113113

$\Rightarrow$ $7(n - 1)$ = 113113 – 119 = 112994

$$\Rightarrow \quad n - 1 = \frac{112994}{7} = 16142$$

$\Rightarrow$ n = 16142 + 1 = 16143

$\therefore$ The number of term = 16143

$$\therefore \quad \text{Sum} = S_n = \frac{n}{2}[2a + (n-1)d]$$

$$\therefore \quad S_{16143} = \frac{16143}{2}[2 \times 119 + (16143 - 1)7]$$

$$= \frac{16143}{2}[238 + 16142 \times 7]$$

$$= \frac{16143}{2}[238 + 112994]$$

$$= \frac{16143}{2}(113232)$$

= 16143 × 56616

= 913952088.

68. Given, $\frac{(\sqrt{x+4} + \sqrt{x-10})^2}{(x+4) - (x-10)} = \frac{5}{2}$

or, $\frac{x + 4 + x - 10 + 2\sqrt{x+4}\sqrt{x-10}}{14} = \frac{5}{2}$

or, $2x - 6 + 2\sqrt{x^2 - 6x - 40} = \frac{5}{2} \times 14$

or, $2x + 2\sqrt{x^2 - 6x - 40} = 35 + 6$

or, $2x + 2\sqrt{x^2 - 6x - 40} = 41$

or, $(2x - 41) = -2\sqrt{x^2 - 6x - 40}$

or, $(2x - 41)^2 = 4(x^2 - 6x - 40)$

or, $4x^2 + 1681 - 164x = 4x^2 - 24x - 160$

or, $1681 + 160 = 164x - 24x$

or, $1841 = 140x$

or, $x = \frac{1841}{140} = \frac{263}{20}$

Hence, $x = \frac{263}{20}$

70. Given, $x = \frac{4ab}{a+b}$ ...(*i*)

$$\Rightarrow \quad \frac{x}{2a} = \frac{2b}{a+b}$$

By componendo and dividendo, we get

$$\frac{x+2a}{x-2a} = \frac{2b+a+b}{2b-a-b}$$

$$\Rightarrow \quad \frac{x+2a}{x-2a} = \frac{a+3b}{b-a} \quad ...(ii)$$

Also, by (*i*)

$$\frac{x}{2b} = \frac{2a}{a+b}$$

By componendo and dividendo, we get

$$\frac{x+2b}{x-2b} = \frac{2a+a+b}{2a-a-b}$$

$$\Rightarrow \quad \frac{x+2b}{x-2b} = \frac{3a+b}{a-b} \quad ...(iii)$$

Adding (*ii*) and (*iii*), we get

$$\frac{x+2a}{x-2a}+\frac{x+2b}{x-2b} = \frac{a+3b}{b-a}+\frac{3a+b}{a-b}$$

$$= \frac{-a-3b+3a+b}{a-b}$$

$$= \frac{2a-2b}{a-b} = 2.$$

71. Given, Total students = 80

n(Commerce students) = 25

n(Mathemtics students) = 15

n(Physics students) = 13

Venn diagram

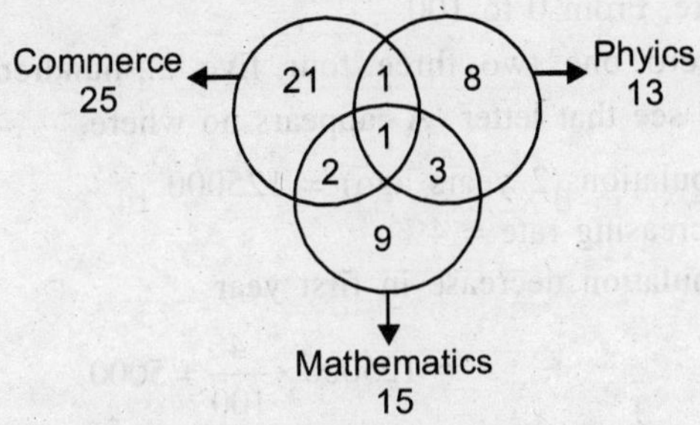

Here, Commerce and Mathematics = 3

Mathematics and Physics = 4

Commerce and Physics = 2

Students is studying all the three subject together = 1

Total students studying at least one subject

= 21 + 1 + 2 + 1 + 9 + 3 + 8 = 45

Hence, the number of students who are not studying any of the three subjects

= 80 – 45 = 35.

72. Investment Profile

Investment in Govt., Bond and Security

= 11.05 × 48.3%

$= 11.05 \times \frac{48.3}{100}$ crore

$= ₹\ 11.05 \times 48.3 \times 10000000 \times \frac{1}{100}$

$= ₹\ 533.715 \times 10000000 \times \frac{1}{100}$

$= ₹\ 5337150000 \times \frac{1}{100}$

= ₹ 53371500

Investment in state-issued Bonds

= 53371500 × 26%

$= 53371500 \times \frac{26}{100}$

= 533715 × 26

= ₹ 13876590.

76. On 1st April 2018,

Species hatch 120 eggs

There are 60 male and 60 female eggs

These 60 female eggs again hatch 60 × 120

= 7200 eggs

7200 eggs = 3600 male eggs + 3600 female eggs

Again, on 1st may 2018

3600 female eggs hatch 3600 × 60 = 21600

∵ 216000 female eggs and equal number of male eggs on 30th may 2018,

216000 female eggs hatch

= 216000 × 60 = 1296000 eggs.

77. After checking options, correction option D.

P @ B = S @ M = N

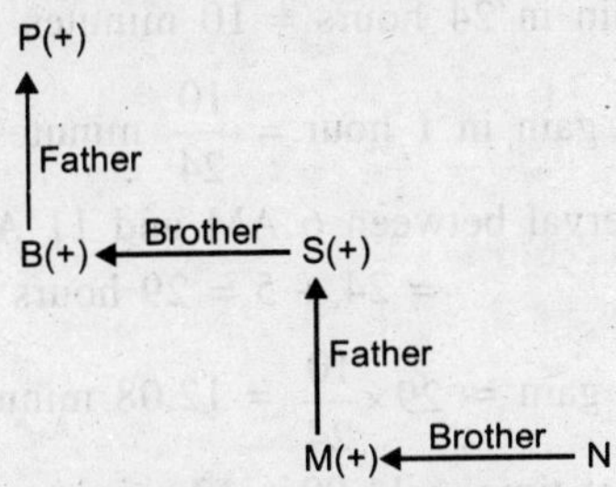

Here, M is grandson of P.

78.

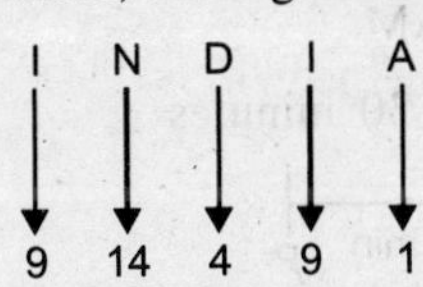

Code = 95491 ∵ 1 + 4 = 5

Similary,

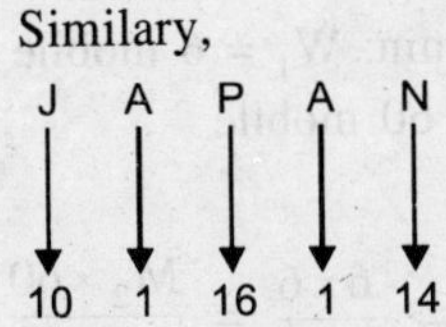

Code = 11715 ∵ 1 + 0 = 1, 1 + 6 = 7

79. ∵ $N! > 10^N$

10! = 10 × 9 × 8 × 7 × 6 × 5 × 4 × 3 × 2 × 1

= 3628800

$10! = 3628800 < 10^{10}$

For Shyam $10 < N < 15$, $N! < 10^N$

For Sohan, $16 < N < 20$, $N! < 10^N$

For Suresh, $21 < N < 25$, $N! > 10^N$

Hence, Suresh is correct.

80. We know that, same calender have interval of +6 years, +11 years, +11 years

So, 1981 have same calender year as

1987, 1998, 2009, 2015

Now, number of days between first June 2015 and first June 2013 = 365 + 365

∴ Number of odd day = 2

∵ Non-leap year = 1 odd day

Hence, 1st June 2015 = 1st June 1981 = Monday

82.

6	2	5	4
4	1	3	2
152	7	98	?

$\because \quad 6^3 - 4^3 = 216 - 64 = 152,$

$2^3 - 1^3 = 8 - 1 = 7,$

$5^3 - 3^3 = 125 - 27 = 98,$

$4^3 - 2^3 = 64 - 8 = 56$

84. A house on the South pole faces towards North will all sides.

85.

$19 \quad 78 \quad 20 \quad \rightarrow (19+20)\times 2 = 39\times 2 = 78$

$25 \quad 144 \quad 47 \quad \rightarrow (25+47)\times 2 = 72\times 2 = 144$

$16 \quad ? \quad 13 \quad \rightarrow (16+13)\times 2 = 29\times 2 = 58$

87. Clock gain in 24 hours = 10 minutes

$\therefore$ Clock gain in 1 hour = $\frac{10}{24}$ minutes

Time interval between 6 AM and 11 AM

= 24 + 5 = 29 hours

$\therefore$ Clock gain = $29\times\frac{10}{24}$ = 12.08 minutes

So, correct time = 11.00 – 12 minutes

= 10 : 48 AM

88. 3 Pills to take with a gap of 30 minutes

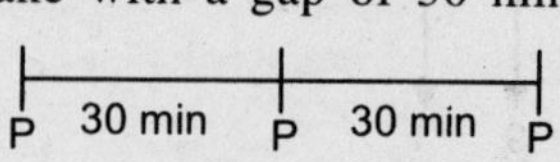

= 30 + 30 = 60 minutes

89. Here, M_1 = 6 Staff, H_1 = 6 min, W_1 = 6 mobile

M_2 = ?, H_2 = 60 min, W_2 = 60 mobile

By trick

$$\frac{M_1H_1}{W_2} = \frac{M_2H_2}{W_2} \Rightarrow \frac{6\times 6}{6} = \frac{M_2\times 60}{60}$$

$\Rightarrow \quad M_2 = 6$

Hence, the number of staff required = 6.

91. Let Amar consumed x laddoos on monday

Then, consumed on tuesday = $(x + 6)$ laddoos

on wednesday = $(x + 12)$ laddoos

Sum of all laddoos from monday to friday

$\therefore x + (x + 6) + (x + 12) + (x + 18) + (x + 24) = 100$

$\Rightarrow \quad 5x + 60 = 100$

$\Rightarrow \quad 5x = 100 - 60 = 40$

$\Rightarrow \quad x = 8$

Hence, the number of laddoos consumed on Wednesday

$= x + 12 = 8 + 12 = 20.$

92.

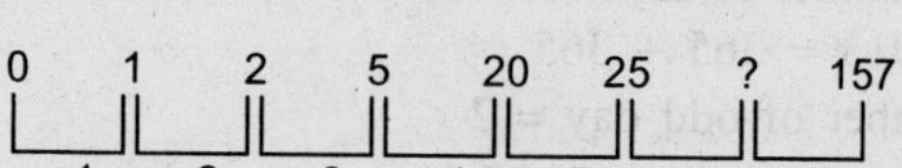

Here, $25 \times 6 = 150$

and $150 + 7 = 157$

Correct term = Missing term = 150.

94. Number of intersection = 4.

95.

Here, Man is son of Kishore.

96. Here, From 0 to 100

= zero, one, two, three, four, five, ..., hundred

We see that letter 'A' appears no where.

97. Population (2 years ago) = 125000

Decreasing rate = 4%

Population decrease in first year

$$= 125000\times\frac{4}{100} = 5000$$

Population in first year = 125000 – 5000 = 120000

Population decrease in second year

$$= 120000\times\frac{4}{100} = 4800$$

$\therefore$ Total Migration = 5000 + 4800 = 9800.

For Q.No. 98–100:

According to the given information :

Day	Movies
Moday	3 idiots
Tuesday	Do Bigha Jameen
Wednesday	Sholay
Thursday	Chak de
Friday	Aanand

For Q.No. 101–103:

According to the given information :

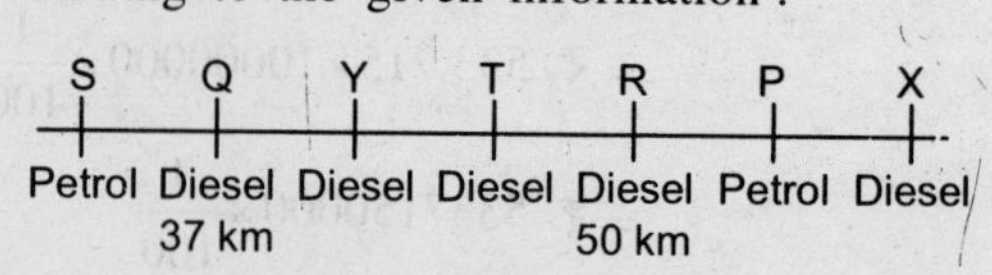

104.

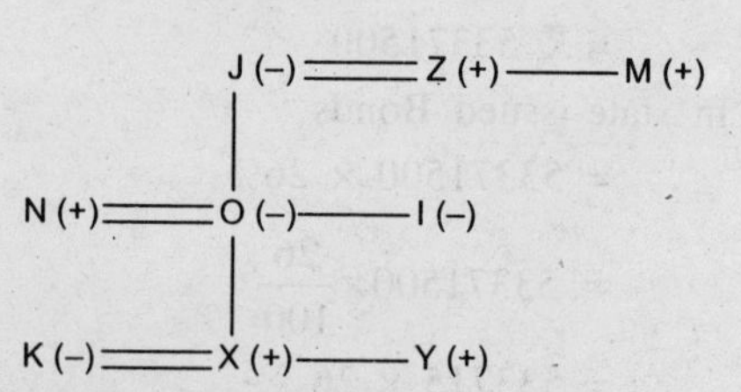

Here N's sister-in-law is niece of J's brother-in-law.

For Q.No. 105–107:

According to the given information :

Vikas > Gopi > Mohit > Rahul > Aditya > Ayush

Gopi ↓ 230 kg; Ayush ↓ 128 kg

106. Ayush + Rahul = 262 kg Ayush = 128 kg

∴ Rahul = 262 − 128 = 134 kg

and Gopi + Mohit = 426, Gopi = 230 kg

∴ Mohit = 426 − 230 = 196 kg

Hence, Rahul + Mohit = 134 + 196 = 330 kg.

For Q.No. 108–110:

According to the given information :

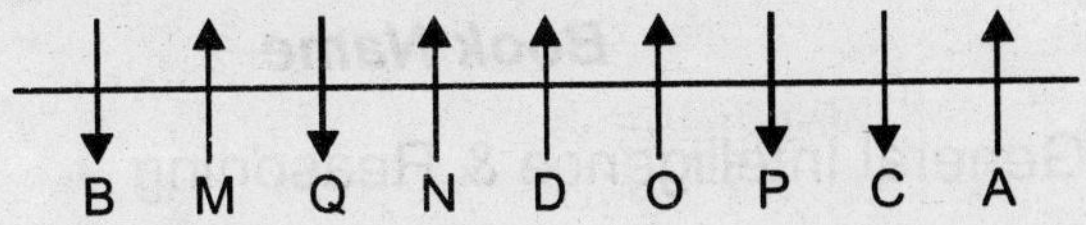

Previous Paper (Solved)

Symbiosis National Aptitude Test (SNAP), 2018*

SECTION-I

GENERAL ENGLISH

1. Choose the correct option.

Study hard ______ you will not pass the exam.

A. then B. else

C. still D. for

2. Look at the underlined part of the sentence below. The sentence is given three possible substitutions for the underlined part. If one of them is better than the underlined part, mark that as the answer. If none of the substitutions improve the sentence, mark "no improvement" as your answer.

The members of his family are coming <u>in</u> this train and I am sure that it will be a memorable trip for all of them.

A. by B. on

C. with D. No improvement

3. The options given below combine to form a meaningful sentence. Identify the erroneous statement from among the options.

1. He made me to wait
2. for him, while he
3. changed his coat
4. as it was cold.

A. 1 is incorrect B. 4 is incorrect

C. 3 is incorrect D. 2 is incorrect

4. Choose the antonym of : Ostentatious

A. Flamboyant B. Pretentious

C. Obtrustive D. Modest

5. Choose the appropriate words from the options given to fill in the blanks in the given sentence.

I wish ___________ that Harish was ill, I would have gone to see him.

A. I know B. I had known

C. I would have known D. I was knowing

6. Find the odd one out.

A. bewilder B. bemuse

C. reuse D. confuse

7. Choose the synonym of: Flagrant

A. Meek B. Flagship

C. Sweet smelling D. Blatant

8. Choose the appropriate words from the options given to fill in the blanks in the given sentence.

Tne young man said that he had no _____ riches.

A. desire of

B. desirous of

C. desire for

D. desire on

9. Find a correct match grammatical function with the usage of the word **STILL**.

Function	***Usage***
(*a*) Noun	1. Still waters run deep.
(*b*) Adjective	2. He wanted some still for a magazine.
(*c*) Verb	3. The good news makes me still my fears.
(*d*) Adverb	4. Will you still love me when I am old and grey?

	(*a*)	(*b*)	(*c*)	(*d*)
A.	3	1	4	2
B.	1	3	2	4
C.	4	3	2	1
D.	2	1	3	4

10. Choose the synonym of: Usurp

A. Capitulate B. Adjudge

C. Encroach D. Abdicate

11. The options given below combine to form a meaningful sentence. Identify the erroneous statement from among the options.

1. I am fully aware that
2. my wife is one of those
3. who is totally committed
4. to the family and its wellbeing.

A. 1 is incorrect

B. 3 is incorrect

C. 4 is incorrect

D. 2 is incorrect

12. Choose the appropriate word from the options given to fill in the blanks in the given sentence.

She kept all her ______ in the bank locker.

* Online exam held on 16/12/2018.

A. Jewellery B. Jewelries
C. Jewelery D. Jwellery

13. Find the odd man out.
A. epicurean B. hideous
C. gastronomist D. gourmet

14. Choose the correct conjunction:
Do not eat the cake, wait _____ your friends come.
A. as B. till
C. for D. when

15. Change the speech : "Wait there till I get back please." She told the boys.
A. She asked the boys to wait till I come back.
B. She requested the boys to wait there till she would return.
C. She requested the boys to wait there till she got back.
D. She requested the boys to wait there till she comes back.

16. Change the speech: "I have won the race!"
A. He said that he had won the race.
B. He said that he has won the race.
C. He said that the race has been won.
D. He said he have won the race.

17. Find a correct match of grammatical function with the usage of the word **WELL**.

Function	*Usage*
(*a*) Verb	1. You never miss the water till the well runs dry.
(*b*) Adjective	2. Tears well up in his eyes when he reads the letter.
(*c*) Noun	3. The children behaved well.
(*d*) Adverb	4. They played hard. It was a well-deserved victory.

	(*a*)	(*b*)	(*c*)	(*d*)
A.	1	3	4	2
B.	2	1	3	4
C.	1	3	2	4
D.	2	4	1	3

18. Choose the correct meaning of the following idiom: Mea Culpa
A. I'm capable B. I'm to blame
C. I love you D. I will do it

19. Look at the underlined part of the sentence below. The sentence is given three possible substitutions for the underlined part. If one of them is better than the underlined part, mark that as the answer. If none of the substitutions improve the sentence, mark "no improvement" as your answer.
The winter was such severe that even water in the taps was frozen and this created a lot of inconvenience to everyone in the family.
A. No improvement
B. so severe
C. severe such
D. severe so much

20. Select the pair that best expresses the relationship similar to that expressed in this pair.
Turmoil : Confusion
A. caustic : sarcastic
B. chaos : disorder
C. metamorphosis : transformation
D. assassination : murder

21. Some words are given below. In which order should they be arranged to give an affirmative sentence, so that the part of the day comes last?
AT(1) / EVENING (2) / AGAIN (3) / LAST (4) / MET (5) / WE (6) / HER (7) / HOSPITAL (8) / THE (9).
A. 673519248 B. 635719824
C. 657319842 D. None of these

22. In a class, Roll No. 101, states that the plural of the word CRISIS is CRISES. Roll No. 102 states that the plural is CRISISES, and Roll No. 103 states that the plural is CRISIS itself.
The Roll No. of the student which gave the correct answer is _________.
A. 103 B. 101
C. 100 D. None of these

23. There are 4 boards in a business school which have displayed the meaning of the term that means "A process in which an independent person makes an official decision that ends a legal disagreement without the need for it to be solved in court."
Board 1 : ARBITRATION
Board 2 : ARBITERATION
Board 3 : ARBETRATION
Board 4 : ARBITATION
The board that has displayed the correct spelling of the term is Board No. _______
A. 1 B. 3
C. 2 D. None of these

24. In an English Test, Column A gives a word, Column B gives its past participle and Column C gives the code of the pair as below :

Word	*Past Participle*	*Code 1*
WEAR	WEARING	1
WEAR	WEAR	2
WEAR	WORE	3
WEAR	WEARED	4

Which code should Shamu choose, if he is asked to choose the correct pair of word and its corresponding past participle?

A. 2 B. 1
C. 3 D. None of these

25. Four friends 1, 2, 3 and 4 are arguing on who is right in stating an English proverb.

1. states that the correct proverb is "A swarm in May is worth a load of hay, a swarm in June is worth a full moon; but a swarm in July is not worth a fly."
2. states that the proverb is "A swarm in May is worth a load a hay; a swarm in June is worth a looney toon; but a swarm in July is not worth a fly."
3. states that the proverb as "A swarm in May is worth a load of hay; a swarm in June is worth a Silver spoon; but a swarm in July is not worth a fly."
4. states that the proverb actually is "A swarm in May is worth a load of hay, a swarm in June is worth a golden boon; but a swarm in July is not worth a fly."

Who is correct in stating the proverb?

A. 1 B. 2
C. 3 D. 4

Read the following passage and answer the questions (Qs. Nos. 26-30) that follow.

ASSOCIATED PRESS

San Francisco, August 13, 2018

Google wants to know where you go so badly that it records your movements even when you explicitly tell it not to. An Associated Press investigation found that many Google Services on Android devices and iPhones store your location data even if you've used privacy settings that say they will prevent it from doing so. Computer Science researchers at Princeton confirmed these findings at the AP's request.

For the most part, Google is upfront about asking permission to use your location information. An app like Google Maps will remind you to allow access to location if you use it for navigating. If you agree to let it record your location over time, Google Maps will display that history for you in a "timeline" that maps out your daily movements.

Storing your minute-by-minute travels carries privacy risks and has been used by police to determine the location of suspects—such as a warrant that police in North Carolina, served on Google last year to find devices near a murder scene. So the company will let you "pause" a setting called Location History.

Google says that will prevent the company from remembering where you've been. Google's support page on the subject states, "You can turn off Location History at any time. With Location History off, the places you go are no longer stored."

That isn't true. Even with Location History paused, some Google apps automatically store time-stamped location data without asking.

For example, Google stores a snapshot of where you are when you merely open its Maps app. Automatic daily weather updates on Android phones pinpoint roughly where you are. And some searches that have nothing to do with location, like "chocolate chip cookies," or "kids science kits," pinpoint your precise latitude and longitude—accurate to the square foot—and save it to your Google account.

The privacy issue affects some two billion users of devices that run Google's Android operating software and hundreds of millions of worldwide iPhone users who rely on Google for maps or search.

Storing location data in violation of a user's preferences is wrong," said Jonathan Mayer, a Princeton computer scientist.

A researcher from Mayer's lab confirmed the AP's findings; the AP conducted its own tests on several iPhones that found the same behaviour.

"If you're going to allow users to turn off something called 'Location History,' then all the places where you maintain location history should be turned off," Mayer said. "That seems like a pretty straightforward position to have." Google says it is being perfectly clear.

"There are a number of different ways that Google may use location to improve people's experience, including: Location History, Web and App Activity, and through device-level Location Services," a Google spokesperson said. "We provide clear descriptions of these tools, and robust controls so people can turn them on or off, and delete their histories at any time." To stop Google from saving these location markers, the company says, users can turn off another setting, one that does not specifically reference location information, called "Web and App Activity" and enabled by default, that setting stores a variety of information from Google apps and websites to your Google account.

When paused, it will prevent activity on any device from being saved to your account. But leaving "Web and App Activity" on and turning "Location History" off only prevents Google from adding your movements to the "timeline", its visualization of your daily travels. It does not stop Google's collection of other location markers.

26. For the most part, Google is upfront about asking permission to use your location information. This sentence means:

A. Google is tracking your movements even when if you don't want to it.
B. Google is quite candid about asking you to allow it to show you where you are located.
C. Google is not secretive about recording your movements.
D. None of the options is correct.

27. An appropriate title to the passage would be _____.
   A. Smart phones are a boon
   B. Android, iPhones and Google
   C. Google tracks you, with or without your permission
   D. Google snapshot

28. How many users does the privacy issue affect?
   A. Everyone who owns a phone
   B. Data not available
   C. Two billion users and hundred of millions of iPhone users
   D. Half the world population

29. When the author says: 'That isn't true' to Google's claim that "you can turn off Location History at any itme.... and the places you go are no longer stored." .... He means
   A. Although Google declares that the places are no longer stored, they automatically store time stamped location data without asking.
   B. Google tracks your movement and tells you about it.
   C. The author is paranoid.
   D. Google asks permission every time it stores data.

30. What does the author mean when he says: 'It does not stop Google's collection of other location markers?
   A. If Location History is off, Google does not collect the location markers.
   B. Google only informs you of your location
   C. None of the options is correct.
   D. If you leave "Web and App Activity" on and turn "Location History" off, Google can still collect the location markers.

Read the following passage and answer the questions (Qs. Nos. 31 to 35) that follow.

FREDERIC BASTIAT, who was that rarest of creatures, a French free-market economist, wrote to this newspaper in 1846 to express a noble and romantic hope: "May all the nations soon throw down the barriers which separate them." Those words were echoed 125 years later by the call of John Lennon, who was not an economist but a rather successful global capitalist, to "imagine there's no countries". As he said in his 1971 song, it isn't hard to do. But despite the spectacular rise in living standards that has occurred as barriers between nations have fallen, and despite the resulting escape from poverty by hundreds of millions of people in those places that have joined the world economy, it is still hard to convince publics and politicians of the merits of openness. Now, once again, a queue is forming to denounce openness—*i.e.*, globalisation. It is putting at risk the next big advance in trade liberalisation and the next big reduction in poverty in the developing countries.

In Washington, DC, home of a fabled "consensus" about poor countries' economic policies, a bill before Congress devised by one of New York's senators, Charles Schumer, threatens a 27.5% tariff on imports from China if that country does not revalue its currency by an equivalent amount. In Mr Schumer's view, presumably, far too many Chinese peasants are escaping poverty. On November 4th, George Bush will escape the febrile atmosphere along Pennsylvania Avenue by visiting Argentina to attend the 34-country Summit of the Americas. There he will be greeted by a rally against "imperialism", by which is meant him personally, the Iraq war and the Free Trade Area of the Americas which he espouses. Among the hoped-for 50,000 demonstrators will be Diego Maradona, who as a footballer became rich through the game's global market and as a cocaine-addict was dependent on barrier-busting international trade; and naturally his fellow-summiteer, Hugo Chávez, who is using trade in high-priced oil to finance his "21st-century socialism" in Venezuela.

All, perhaps, the normal fun of a Latin American visit. Last week's Latinobandmetro opinion revealed that whatever the protesters may say, a clear majority in all the region's countries favour a market economy rather than a closed, state-directed one—even in Venezuela. This is, however, a difficult moment for the market economy and for relations between rich countries and poorer ones, for the Doha round of trade-liberalisation talks under the World Trade Organisation are in trouble. When it began in 2001, the round was billed as a big effort to boost growth in poor countries, and the lowering of barriers to food trade was placed at its centre. In the past few weeks, however, a fairly bold American proposal for reducing its farm protection has been greeted by a much weaker response from the European Union and none at all from Japan. And ministers from Bastiat's own country, France, have vied with one another to denounce all talk of further reform to the EU's common agricultural policy. Europe must, they say, remain an "agricultural power" even at the expense of the taxpayer and the poor, and, according to President Jacques Chirac, must fight back "liberalism". Whatever happened to Liberté, Egalité, Fraternité?

The world will find out, to some extent, next month when ministers from the 148 countries in the WTO meet in Hong Kong. The last time they gathered for such a crucial meeting was in September 2003 in Cancún, and the result was a shambles. There was a bitter row between rich countries and poor ones, and the meeting broke up in acrimony. At that stage, however, there was still plenty of time to repair the damage. For in effect the deadline for the Doha round comes in June 2007, when the trade-negotiating authority granted by Congress to President Bush expires. But, although that leaves more than a year and a half after Hong Kong, the complexity of a negotiation involving 148 countries and scores of highly technical issues means that the deal really needs to be done during 2006, with the political framework for it set early on—which essentially means in Hong Kong.

The case for Selfish Generosity

Trade-liberalisation rounds are arcane affairs about which free-traders are often thought to cry wolf. The previous talks, known as the Uruguay round, went through lots of brinkmanship and delays before they were completed. The result was still disappointing in many ways, especially to developing countries, and yet since the round's completion in 1993 the world economy has grown lustily and the biggest developing countries, China, India and Brazil, have all burst on to the global trading scene. Would the world really be hurt if the EU merely refuses to expose its farmers to more competition?

The likeliest outcome both from the Hong Kong meeting and the eventual Doha agreement is a compromise—as always. The European position is feeble but not risible, for it has offered an overall average cut in its farm tariffs of 39%, up from 25% only a month ago, though with rather a lot of loopholes that could severely limit the benefits. France, and other European farm protectionists, may prove more flexible than they currently imply: this is hardly the first time they have promised to man the barricades shortly before striking a deal. Yet though some sort of fudge in Hong Kong must be likely, with the Americans lowering their ambition and the Europeans raising theirs a little, such an outcome would still represent both a missed opportunity and a risk.

The missed opportunity is that Doha has offered the first proper chance to involve developing countries in trade negotiations—they now make up two-thirds of the WTO members—but also thereby to use a full exchange of agricultural, industrial and service liberalisations to make a big advance in free trade that could benefit a wide range of countries. Some of that progress may still be made, even in a fudged deal: Brazil, for example, stands to benefit hugely from freer trade in agriculture, so it should be willing to promote other concessions in return. India is reluctant to cut its own farm tariffs but has a big interest in liberalising trade in services, wanting more freedom in everything from finance to health care to entertainment. But if the rich world could gird itself to be more ambitious on agriculture the gains would be even greater: help for the poorest countries, making the rich look generous; better access to the biggest and richest developing countries for western companies; and a rise in global income in a decade's time of $300 billion a year (says the World Bank), which would thus help everyone.

The risk is that failure to agree on a new wave of openness during a period (the past two years) in which the world economy has been growing at its fastest for three decades, with more countries sharing in that growth than ever before, will set a sour political note for what may well be tougher times ahead. A turn away from trade liberalisation just ahead of an American recession, say, or a Chinese economic slowdown, could open up a chance not just for a slowdown in progress but for a rollback. Currently, for example, the Schumer bill to put a penal tariff on Chinese goods looks unlikely to pass. If American unemployment were rising and world trade talks had turned acrimonious, that might change. So might the political wind in many developing countries.

If so, that would be a tragedy for the whole world. Although the case for reducing poverty by sending more aid to the poorest countries has some merit, the experience of China, South Korea, Chile and India shows that the much better and more powerful way to deal with poverty is to use the solution that worked in the past in America, western Europe and Japan: open, trading economies, exploiting the full infrastructure of capitalism (including financial services—see our survey on microfinance) amid a rule of law provided by government. In other words, globalisation. To paraphrase Samuel Johnson, anyone who is tired of that is tired of life.

Source : www.economist.com

31. Who was John Lennon?

A. A singer

B. An economist

C. A writer

D. An industrialist

32. According to the article, why is Frederic Bastiat called the "Rarest of creaures"?

A. Because he wrote to newspaper

B. Because he was a friend of John Lennon

C. Because he was a German industrialist

D. Because he was a French who promoted a free market economy

33. According to the article, the better way to deal with poverty is:

A. By reducing interest rates

B. By moving away from democracy

C. By giving more aid to poor countries

D. By exploiting the full infrastructure of capitalism

34. "Trade-liberalization rounds are arcane affairs." The adjective arcane means:

A. Incomprehensible B. International

C. Well known D. Transparent

35. As per the article, India's position with respect to the talks is:

A. Higher tariff for agriculture and liberalization for manufacuring sector.

B. Higher tariff for agriculture and liberalization for services sector.

C. Reduced taxes for agriculture and liberalization for services sector.

D. Reduced taxes for agriculture and liberalisation for manufacturing sector.

SECTION-II

QUANTITATIVE, DATA INTERPRETATION

36. In the following diagram there are four semi-circular arcs and a shaded region. The diameter of largest semi-circle is 28 cm and of the smallest is 7 cm. The area of shaded region is ______.

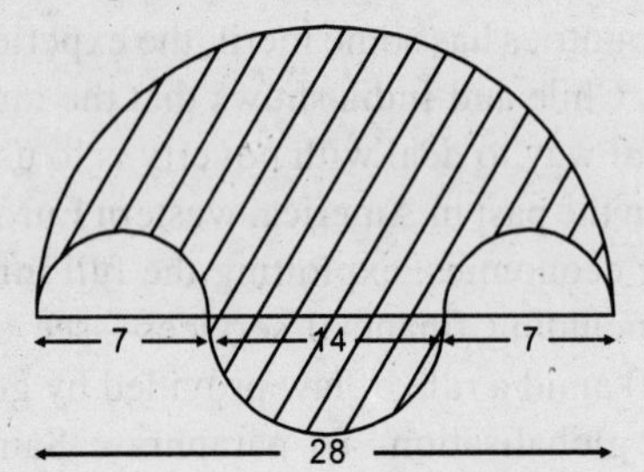

A. 98.75π B. 120.5π
C. 110.25π D. 105.5π

37. Big Bang Theory scast wishes to find out the number of ways in which the word "ASTRONAUT" can be scrambled. They find that the number of ways in which it can be put to an unscrambling puzzle are ______.
A. 60480 B. 181440
C. 362880 D. 90720

38. In the figure given below $L_1 \parallel L_2$ then x = ____.

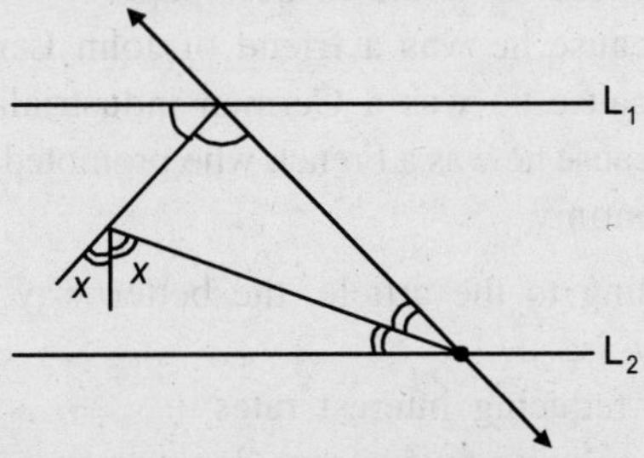

A. 45° B. 30°
C. 40° D. 60°

39. A man purchased a TV and if the price of TV is 150% of price of fridge, then the price of fridge is what percentage of the total cost of TV and fridge?
A. 45 B. 30
C. 40 D. 50

40. A man sends some money to his friend at 5% per annum of interest rate. After 2 years, the difference between the Simple and the Compound interest on money is ₹ 50. What will be the value of the amount at the end of 3 years if compounded annually?
A. ₹ 24512.5 B. ₹ 22252.7
C. ₹ 23152.5 D. ₹ 21325.6

41. Sumit was doing a multiplication, but by mistake instead of 25 as one of the multiplier, he took 34. Now, because of this mistake, the answer was 405 more than the correct answer. Find out the answer that Sumit arrived at?
A. 1620 B. 1530
C. 1450 D. 1350

42. If the ratio of sides of a rectangle is 2 : 7 and perimeter is 360m, then find out its length and breadth.
A. 144m, 36m B. 140m, 40m
C. 135m, 45m D. 160m, 20m

43. How many different words can be formed with the word CUSTOM with a condition that the word should begin with M?
A. 720 B. 120
C. 540 D. 180

44. If h, c, v are the height, the curved surface area and the volume of a cone.
Then $3\pi vh^3$ is ______.
A. $c^2h^2 - 9v^3$ B. $9v^2 - c^2h^3$
C. $c^2h^2 - 9v^2$ D. $c^2h^2 - 16v^2$

45. A is able to do a piece of work in 10 days and B can do the same work in 15 days. If they can work together for four days, what is the fraction of work left?
A. 3/2 B. 1/3
C. 1/2 D. 2/3

46. If 20% of $x = y$, then y% of 20 is same as:
A. 15% of x B. 20% of x
C. 4% of x D. 10% of x

47. $\left\{\dfrac{2^{\frac{1}{2}} \times 3^{\frac{1}{3}} \times 4^{\frac{1}{4}}}{10^{-\frac{1}{5}} \times 5^{\frac{3}{5}}} \div \dfrac{3^{\frac{4}{3}} \times 5^{-\frac{7}{5}}}{4^{\frac{3}{5}} \times 6}\right\} \times 2 =$
A. 20 B. 40
C. 10 D. 30

48. If a labor output function for a laundry service is described by the following equaiton:
$O = (a^0 - 4L^0)^6$, where L denotes Labor and O denotes output. Then, output ______.
A. Decreases as labor increases
B. Remains constant irrespective of the amount of labor
C. None of the option is correct
D. Increases as labor increases

49. Excluding the halts, the speed of the bus is 50 km/hr and including the halts it is 40 km/hr. For how many minutes does the bus stop per hour?
A. 15 B. 11
C. 12 D. 10

50. ABC Paints Ltd. is planning to create different combination of dyes. The research team has decided they will be using five different green dyes, three different red dyes and four different blue dyes. How many combinations of dyes can be created by ABC Paints Ltd. by including at least one blue and one green dye?
A. None of the options is correct
B. 31
C. 60
D. 3720

51. The expressions $(a - b)^3 + (b - c)^3 + (c - a)^3$ can be factorized as:
A. $(a - b)(b - c)(c - a)$
B. $3(a - b)(b - c)(c - a)$
C. None of the option is correct
D. $3(a + b)(b - c)(c + a)$

52. Pune taxi services has a fixed charge plus a variable charge based on the distance covered. For travelling 10 km the total fare paid by Somu is ₹ 150 and for a journey of 15 km the total fare paid by Ramu is ₹ 220. Then Tomy will pay _____ as the total fare for travelling a distance of 25 km.

A. ₹ 350 B. ₹ 360
C. ₹ 380 D. ₹ 400

53. In the figure given below, a Cylinder is inserted into a cone and the vertical height of the cone is 30 cm. The vertical angle of the cone is 60°. The diameter of the cylinder is $8\sqrt{3}$ cm. What is the volume of the cone?
Note: The base of the cylinder and the base of the cone are on the same plane.

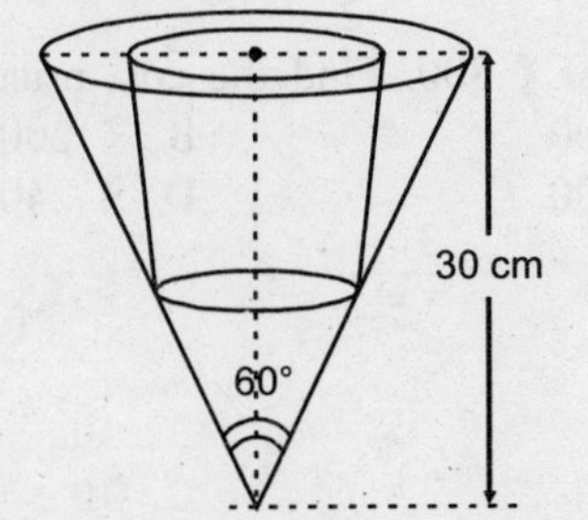

A. 2020 π cm³ B. 3000 π cm³
C. 2890 π cm³ D. 2500 π cm³

54. The number of ways that 5 Marathi, 3 English and 3 Tamil Books be arranged if the books of each language are to be kept together is _______.
A. 2480 B. 5920
C. 9240 D. 29520

55. An umbrella is made by stitching 10 triangular pieces of cloth of two different colours each piece measuring 30 cm, 60 cm, 60 cm. How much cloth of each colour is required for umbrella?
A. $1015\sqrt{45}$ B. $1235\sqrt{15}$
C. $1220\sqrt{45}$ D. $1125\sqrt{15}$

56. If $\log_{10} 11 = a$ then $\log_{10}(1/110)$ is equal to _____.
A. $1/10a$ B. $-a$
C. $-(a + 1)$ D. $(1 + a)^{-1}$

57. Sonal and Meenal appear in a interview for same post having two vacancies. If 1/7 is Sonal's probability of selection and 1/5 is Meenal's probability of selection then what is the probability that only one of them is selected?
A. 2/7 B. 1/5
C. 1/7 D. 3/5

58. Alok and Atul are competing in 800m race. Alok gives Atul a lead of 100m. Initially, Alok runs at x times of Atul's speed (x is an integer > 1) but after crossing 600m, he slows down to a speed which is $\frac{1}{x}$ time of Atul's speed. Atul continues to run at his original speed. If both of them meet for the second time at a distance of 50m from the finish line, then who wins the race and by what distance?
A. Atul, 33.33 m B. Alok, 37.5 m
C. Atul, 37.5 m D. Alok, 25 m

59. Nisha went to buy three types of stationery products, each of them were priced at ₹ 5, ₹ 2 and ₹ 1 respectively. She purchased all three types of products in more than one quantity and gave ₹ 20 to shopkeeper. Since the shopkeeper had no change with him/her; He/she gave Nisha three more products of price ₹ 1 each. Find out the number of products with Nisha at the end of the transaction.
A. 10 B. 12
C. 11 D. 8

60. During the placement, season of a Class, 21 students got shortlisted for company A, 26 got shortlisted for company B and 29 got shortlisted for company C. 14 students got shortlisted for both A and B, 12 students got shortlisted for A and C and 15 for both B and C. All the companies shortlisted 8 students from the class. Then what is the ratio of the number of students who got shortlisted for only 8 and number of students who got shortlisted for only C?
A. 2 : 3 B. 3 : 2
C. 1 : 1 D. 1 : 2

61. The capital of a company is made up 50,000 preferred shares with dividends of 20% and 20,000 common shares, the par value of each type of share being ₹ 10. The Company had a total profit of ₹ 1,80,000 out of which ₹ 30,000 was kept in reserve fund and the remaining distributed to shareholders. Find the dividend per cent to the common share holders.
A. 20% B. 24%
C. 25% D. 30%

62. Sham is trying to solve the expression:
Log tan 1° + log tan2° + log tan 3° + ___ + log tan 89°.

A. $\frac{1}{\sqrt{2}}$ B. 0

C. –1 D. 1

63. If the price of sugar increases by 20% and Salman intends to spend only an additional 5% on sugar, then find out the percentage decrease in his sugar consumption.

A. 10.25 B. 12.5

C. 11.75 D. 5.25

64. If $R_c = m \ln\left(1+\frac{R_m}{m}\right)$ then R_m is equal to

A. $R_m = m\left(e^{R_c/m}-1\right)$ B. $R_m = \ln\left(1+\frac{R_c}{m}\right)$

C. Cannot be determined D. None of the above

65. Mr. Sandeep was placing an order with his carpenter to make wooden boxes to pack his cotton toys which were to be delivered to his friend. Sandeep wanted 2 types of boxes. The larger one with a dimension of 25 cm * 20cm * 5 cm and the smaller box with a dimension of 15 cm * 12 cm * 5 cm. Sandeep also told the carpenter that 5% of total surface area is required as extra to cover the overlaps. If the cost of the wood is ₹ 4/- cm^2, the cost of wood required for supplying 250 boxes of each type would be ₹ ______.

A. 2184000

B. 3184000

C. 4184000

D. None of these

66. David is trying to solve the expression.

$$\frac{(4)^2 \times 2^{\pi+1} - 4\times 2^{\pi}}{(4)^{-1}\times 2^{\pi+2} - 2\times 2^{\pi+2}}$$

And you help him to do the same and finally arrive at the answer with correct to one decimal which would be:

A. +4

B. –4

C. –2

D. None of these

67. Mahesh has a toy which has the shape of a trapezium. The two sides which are parallel have a length of 60 cm and 77 cm and the other sides are 25 cm and 26 cm. He requests you to calculate the area and you help him to arrive at the correct answer which is _____ sq. cm.

A. 1622 sq.cm. B. 1644 sq.cm.

C. 1620 sq.cm. D. None of these

68. Ramesh is trying to simplify the expression. $(p + q)^3 - (p - q)^3 - 6q(p^2 - q^2)$ and if $q = 1$. You helped him and the solution arrived was _____.

A. $2(p^3 + 2)$ B. $2(p^3 - 3)$

C. $2(p^3 + 3)$ D. None of these

69. In Roman Numerals, a number has been written as MMXVIII. In Arabic numbers, it will be ______. (Note:

A. 1118 B. 2018

C. 4018 D. None of these

70. A shopkeeper raised the market price of an article by 60% and allows three successive discounts of $12\frac{1}{2}$%, $14\frac{2}{7}$% and 20%. If shopkeeper would allowed only two discounts of $12\frac{1}{2}$% and 20% on new mp, he made a profit of ₹ 800. Find the cost price of article?

A. ₹ 1800 B. ₹ 2000

C. ₹ 1600 D. ₹ 2400

SECTION-III

CURRENT AFFAIRS

71. How many persons were awarded the Nobel Prize in Economic Sciences in 2018?

A. William D. Nordhaus

B. Arthur Ashkin

C. Tasuku Honja

D. None of these

72. From the beginning of 2018 which of these sectors has been allowed for automatic hundred per cent foreign direct investment (FDI) in India?

A. Multi Brand Retailling

B. Banking

C. Single Brand Retailing

D. Defence Manufacturing

73. In 2017 Facebook acquired the video editing technology company named ______.

A. Survival Vision B. Fayteq

C. BenQ D. Instagram

74. In August, 2018, Intel was overtaken by ______ as the world's largest chipmaker, has made a major move into the autonomous driving space, with its acquisition of Israeli visual sensor company Mobileye.

A. Samsung B. AMD

C. Qualcomm D. Google

75. In October 2018, _____ received the champions of the Earth, a top United Nations honour that recognizes contribution in the field of Environment and environment protection.
A. Shri James Hansen
B. Shri Bill Mckibben
C. Shri Narendra Modi
D. Smt. Vandana Shiva

76. In September 2018, ______ was appointed as the Vice Chairman of the University Grants Commission.
A. Prof. Rais Ahmed
B. Prof. Bhushan Patwardhan
C. Prof. B. Ramachandra Rao
D. Prof. Satish Chandra

77. The Nobel Peace Prize 2018 was awarded jointly to _____ and _____ for "their efforts to end the use of sexual violence as a weapon of war and armed conflict."
A. Denis Margo, Nadia Moratis
B. Denis Levis, Nadia Johnes
C. Denis Mukwege, Nadia Murad
D. Denis Gurang, Nadia Posche

78. Assembly polls were held in ______ between November 2018 to December 2018.
A. Madhya Pradesh, Rajasthan, Chhattisgarh, Mizoram and Telangana only
B. Madhya Pradesh, Rajasthan, Chhattisgarh and Mizoram only
C. Madhya Pradesh, Rajasthan and Chhattisgarh only
D. Madhya Pradesh, Rajasthan, Chhattisgarh and Telangana only

79. The Prime Minister of India on September 23, 2018 rolled out the ______, touted as the world's largest healthcare scheme. The insurance scheme provides an annual health cover of Rs. 5 lakhs per family for secondary and tertiary care hospitalization.
A. Ayushman Bharat-Pradhan Mantri Jan Aarogya Yojana
B. Ayushman Desh-Pradhan Mantri Jan Aarogya Yojana
C. Ayushman Parivar-Pradhan Mantri Jan Aarogya Yojana
D. Ayushman Jeevan-Pradhan Mantri Jan Aarogya Yojana

80. Almost six months after chief economic advisor Arvind Subramanian left the finance ministry, the government has recently appointed ______ to the post.
A. Krishnamurthy Subramanian
B. Shaktikanta Das
C. Krishnan Subramanian
D. Krishnakumar Subramanian

81. In June 2018, ______ was appointed MD and CEO of IDBI Bank in place of ______ who took charge as Deputy Governor of Reserve Bank of India.
A. D Shriram, Rahul Kumar Jain
B. B Sriram, Mahesh Kumar Jain
C. R Raghuram, Vinnet Kumar Jain
D. K. Sridharan, Shyam Kumar Jain

82. In October 2018, __________, ex-director of Amazon joined as a` Chief Technology and Product Officer (CTPO) for chqbook.com, the Gurgaon based financial technology startup.
A. Rohit Bansal
B. Prakash Malhotra
C. Sachin Arora
D. Suresh Wadhwani

83. Name Indians who won the Ramon Magsaysay Award 2018?
A. Bharat Vatwani
B. Sonam Wangchuk
C. Option A and B
D. None of these

84. In September, 2018, _____ and _____ were all set to acquire Aditya Birla Group's Food and grocery retail "More", even as Investment Bank _____ exited the consortium.
A. Samara capital, Alibabamazon, Barclays
B. Softbank, Amazon, Deutsche Bank
C. Soft Bank, Amazon, JP Morgan
D. Samara Capital, Amazon, Goldman Sachs

85. Recently the ______ of the RBI Act came into spotlight amid the war between the Central Government and the Reserve Bank of India (RBI). The provision in the RBI Act empowers the government to issue directions to the RBI.
A. Section 9 B. Section 8
C. Section 7 D. Section 10

86. In December 2017, ______ bought Foodpanda India from its German parent Delivery _______ in a stock deal pegged at about $30 Million, entering a resurgent food-delivery market that has matured and become more competitive.
A. Ola, Hero B. Swiggy, Siemens
C. Uber, Volkswagen D. Zomato, BOSCH

87. The political party floated by the actor turned politician Kamal Hasan is
A. Makkal Needhi Maiam
B. Makkal Kazahgam
C. Makkal Munnetra Kazhagam
D. Makkal Munnetra Maiam

88. Recently, Tata Motor's first SUV built on Land Rover's DB architecture was developed in collaboration with Jaguar Land Rover. It is called _______.
A. Tata Hexa B. Tata Harrier
C. Tata Crux D. Tata Sierra

89. The Statue of Unity, built as a tribute to the first _____ was inaugurated on Oct. 31, 2018.

A. Deputy Chief Minister of Gujarat
B. President of India
C. Vice President of India
D. Deputy Prime Minister of India

90. The world's first holographic screen phone _____ has a holographic screen that 3-D visuals without needing special glasses.
A. Nitrogen One B. Carbon One
C. Hydrogen One D. Oxygen One

91. In October 2018, Dish TV: India's first and largest direct-to-home (DTH) company, signed up _____ as its brand ambassador as it is planning to woo the younger audience with its largest brand campaign.
A. Tiger Shroff B. Ayushman Khurana
C. Ranveer Singh D. Shahrukh Khan

92. The 2018 Asian Games were held in how many cities of Indonesia?
A. 3 B. 4
C. 2 D. 1

93. The Ministry of Agriculture and Farmer Welfare launched the Krishi Kalyan Abhiyan from the __th month of 2018 which could lead and advise farmers on the means to improve their farming techniques.
A. 5 B. 9
C. 6 D. 8

94. What was the rank of India in 2018 United Nation World Happiness Index Report?
A. 133 B. 140
C. 116 D. 130

95. _____ was the number of seats won by the party who got majority in Chhattisgarh in recent 2018 Vidhan Sabha elections.
A. 68 B. 60
C. 52 D. None of these

SECTION-IV

ANALYTICAL AND LOGICAL REASONING

96. Find the missing number in the series.
16, __, 68, 88
A. 08 B. 36
C. 46 D. 06

97. Aman, Raman, Saman, and Chaman are standing on the four corners of a square carpet as shown in the fig. here.

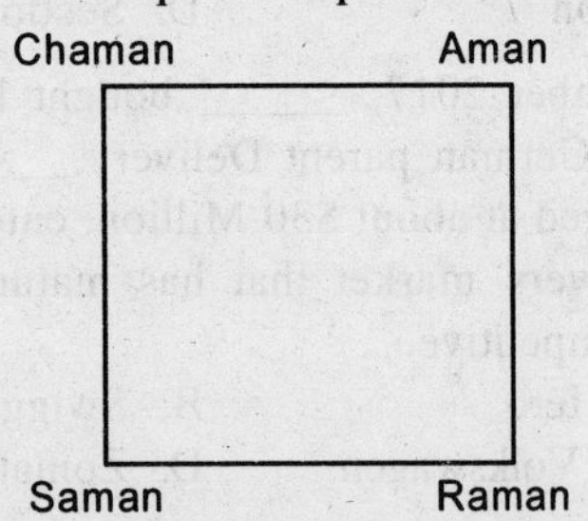

Aman starts walking towards Saman diagonally. After walking half the diagonal he turns to the left. What direction is Aman facing now?
A. North west
B. East
C. South
D. South east

98. If Golu is son of Molu, Molu and Polu are sisters, Pillu is Polu's mother, Gannu is son of Pillu. Which of the following statements is definitely true?
A. Golu's nephew is Gannu
B. None of the options is correct
C. Golu and Gannu are cousins
D. Golu's maternal uncle is Gannu

99. In a pile of 10 books there are 3 of History, 3 of Hindi, 2 of Maths, and 2 of English. Taking from above there is an English book between a History and Maths book, a History book between a Maths and an English book, a Hindi book between an English and a Maths book, a Maths book between two Hindi books, and two Hindi books between a Maths and a History book. Book of which subject is at the sixth position from the top?
A. English B. Hindi
C. History D. Maths

100. Lectures by industry experts have to be scheduled in Marketing, Finance, Operations, Organisational Behaviour and Information Systems. This has to be scheduled on a weekday (Monday to Friday). Only one lecture can be scheduled per day. Marketing cannot be scheduled on Tuesday as it is inconvenient for the expert. Expert for Finance is available only on Tuesday. As Information Systems and Finance are always considered important, Information Systems has to be scheduled immediately on the day following the Marketing Lecture. The Dean has decided that Organisational Behaviour lecture has to be scheduled immediately before the day of Marketing lecture. Finally the schedule is prepared which satisfies all condiditons.
Which lecture is scheduled on Monday?
A. Information System B. Operations
C. Marketing D. Finance

101. Ramu has several friends and all of them are well settled in India. 1/5th of Ramu's friends went to Mumbai and 1/3rd of his friends went to Delhi, three times the difference of these two went to Chennai and only one went to Pune. How many of his friends went to Mumbai?

A. 15 B. 10
C. cannot be determined D. 3

102. Messenger was returning to his base station which was in front of him to the north. When his base station was 100 m away from him he turned to the left and moved 50 m to deliver the last message to the Peshwa's troops. He then moved in the same direction for 40m, turned to his right and moved 100m. How many meters away he was now from his base station?

A. 0 B. 150
C. 90 D. 100

103. If Pig : Farrow and Lion : Cub, then Bear : _____

A. Foal B. Cub
C. Cygnet D. Goat

104. Four classes, Rural Marketing, Digit Marketing, International Marketing and Service Marketing were taught in four consecutive periods of one hour each starting from 8.00 am on a week day.

Read the statements and find out the probable opinion which can be sufficient to answer the question. At what time was the Digital Marketing class scheduled?

Statement I : International marketing class ended at 10.00 a.m. which was preceded by Service Marketing.

Statement II : Rural Marketing was scheduled as the last class.

Statement III : International Marketing was immediately followed by Digital Marketing.

A. II only
B. II and III together
C. I only
D. I and II together or I and III together

105. Find the missing number (?)

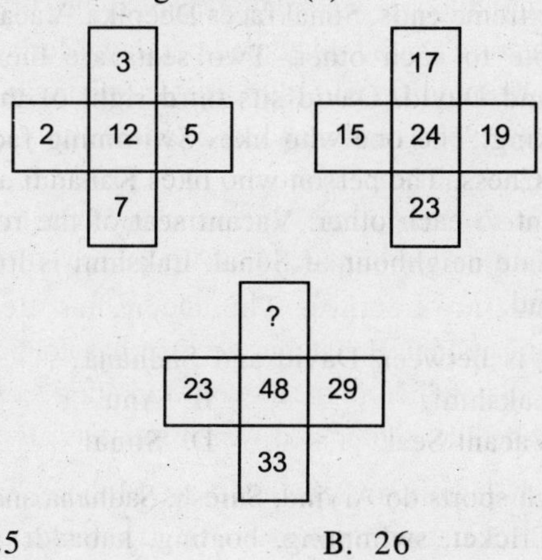

A. 25 B. 26
C. 27 D. None of these

106. Babu is younger than Gopu. Somu is older than Pappu. Read the statements and find out the probable option which can be sufficient to answer the question. Who among them is the oldest?

Statement I : Somu is older than Babu.

Statement II : Gopu is older than Somu.

Statement III : Pappu is the youngest of all.

A. II only B. I and II together
C. I, II and III together D. III only

107. A Men's singles Tennis tournament is being held out at Mumbai. 30 players participated in the tournament. There is a rule that is implemented. One rule states that if a player loses a match then he is eliminated from the tournament. How many matches have to be played to decide the winner of the tournament?

A. 15 B. 29
C. 10 D. 30

108. A fast train leaves Chennai to Bengaluru and at the same point in time a slow train leaves from Bengaluru to Chennai. The fast train moves at a speed of 60 km/hr and the slow train moves at half of the speed of the fast train. Both these trains meet at a point after some time. Which of these trains are further away from Chennai when they meet? (Do not consider the length of the train)

A. The slow train is further from Chennai
B. The fast train is further from Chennai
C. None of the option is correct
D. Both the trains are equidistant from Chennai.

109. If A and B are represented by the above diagrams as two circles then shaded area is represented as?

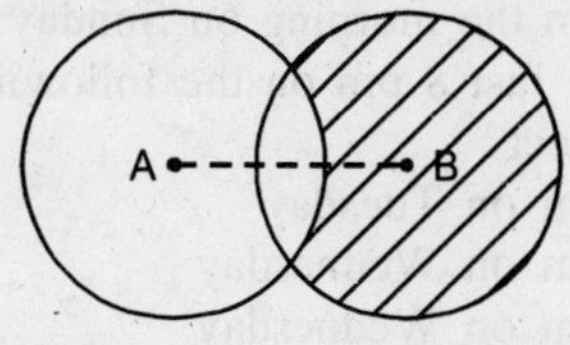

A. $(A \cup B) - A$ B. $(A \cap B) \cup A$
C. $(A \cup B) - B$ D. None of these

110. Read the statements and find out the probable option which can be sufficient to answer the question. What is the total monthly salary of Kakde?

Statement I : Kakade's basic salary is ₹ 100 more than Madan's salary, who also serves in Kakade's company.

Statement II : Other allowance taken by Madan besides his basic salary is 1/10th of Kakade's salary.

Statement III : Madan's basic salary is ₹ 15000/- per month.

A. II and III together B. Data insufficient
C. I, II and III together D. I and III together

111. There are 3 closed cartons in a room. One of the cartons contains cash. There is a printed message that is displayed outside each carton. Only one message is true and the other two messages are false. The first carton has the message 'Cash is not in the carton'. The second carton has the message 'No cash in the carton'. The third carton has the message, 'Cash is in the second carton'. Which Carton has the cash?

A. Third B. Second
C. First D. Cannot be determined

112. Given the series : Z, Y, X, V, S, ... The next alphabet will be?

A. L B. N
C. O D. K

113. Which is the seventh number in the sequence?

5 9 16 29 54

A. 103 B. 300
C. 330 D. 200

114. Nath places three identical cones of radius '*r*' on a table in such a way that each cone base touches the other two, also the attitude of the cone is perpendicular to the table. The radius of the circle drawn through the cone vertices would be?

A. Smaller than *r* B. Cannot be determined
C. Larger than *r* D. Equal to *r*

115. If in a code language, MASTER is written as 79711512131109
Then LAUGH will be written as ______ ?

A. 7371274349
B. 7371274943
C. 7127734349
D. 4943127773

116. A watch which gains uniformly, is 5 minutes slow at 8 o'clock in the morning on Sunday and is 5 minutes 48 seconds fast 8 pm on the following Sunday When was it correct?

A. 7:20 pm on Tuesday
B. 9:20 pm on Wednesday
C. 7:20 pm on Wednesday
D. 9:20 pm on Tuesday

117. If January = 9 and July = 14 then June = ?

A. 5 B. 6
C. 15 D. 25

118. In the figure given below there are maximum ______ number of triangles that can be identified.

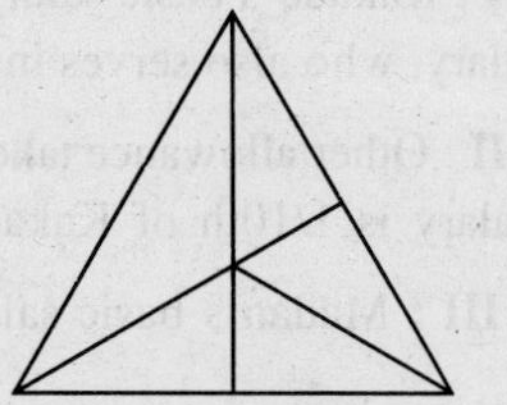

A. 11 B. 13
C. 15 D. 17

119. Rohan (Coded as 1), Pawan (coded as 2), Mohan (coded as 3), Sohan (coded as 4) are waiting for a group photo to be taken. The photographer arrives to take the photo. They decide to stand in a straight row and inform the photographer to take photo by facing them. Mohan is to the left of Pawan and Sohan is to the right of Pawan. Rohan is between Sohan and Pawan.

The code of the person who would be second from the left for the photographer is ______ ?

A. 1 B. 2
C. 3 D. 4

Directions (Qs. Nos. 120 to 126): *Please read the following information and answer the questions:*

There are 12 seats in total which are arranged as 6 in a row facing each other. Ten people have occupied the seats in such a way that 5 are seated in each row and there is equal distance between adjacent seats. In row 1 : Sadhana, David, Lakshmi, Sonal and Anu are seated and all of them are facing south. In row 2-Lily, Suresh, Deepika, Mahesh and Arvind are sitting and all of them are facing north. One seat is vacant in each row. In the given seating arrangement described above, each person seated in a row faces another member of the other row or a vacant seat. Each member likes only one activity or sport namely Kabaddi, Cricket, Baseball, Chess, Wrestling, Boating, Walking, Running, Swimming and Skating.

Mahesh sits third to the right of Deepika and likes Chess. Only two people sit between Suresh and the vacant seat. Suresh sits at one of the extreme end. Deepika does not like Kabaddi and Running. Suresh does not like Wrestling and Baseball. Anu is not an immediate neighbour of Lakshmi. David likes Skating. The one who likes Baseball faces the other one who likes Running. Vacant seat of Row 1 does not face Mahesh and he does not sit at any extreme ends. The one who likes Baseball sits opposite to the one, who sits third right of the one, who sits opposite to Mahesh. Lakshmi is not an immediate neighbor of Sonal. Arvind, who neither likes Wrestling nor Boating does not face vacant seat and sits opposite to the person who likes Kabaddi. Deepika does not sit at extreme ends. Sonal faces Deepika. Vacant seats are not opposite to each other. Two seats are there between Lakshmi and David. David sits third right of the one who likes Walking. The one who likes Swimming faces the one who likes Chess. The person who likes Kabaddi and running are adjacent to each other. Vacant seat of the row 1 is not an immediate neighbour of Sonal. Lakshmi is located at an extreme end.

120. Who is between David and Sadhana?

A. Lakshmi B. Anu
C. Vacant Seat D. Sonal

121. Which sports do Arvind, Suresh, Sadhana and Anu like?

A. Cricket, swimming, boating, kabaddi
B. Baseball, boating, swimming, kabaddi

C. Cricket, boating, swimming, kabaddi
D. Cricket, boating, kabaddi, swimming

122. What sport does Deepika like?
A. Boating B. Wrestling
C. Cricket D. Baseball

123. What sport does Lily like?
A. Boating B. Baseball
C. Wrestling D. Cricket

124. Who face vacant seats?
A. David, Lily B. David, Lakshmi
C. Anu, David D. Lily, Sadhana

125. Find the statement which is false.
A. Deepika likes baseball
B. Sonal likes running
C. Mahesh likes chess
D. Sadhana likes boating

126. Who amongst the following sits to the immediate right of the person who faces the vacant seat?
A. The person who likes Running
B. The person who likes Cricket
C. The person who likes Boating
D. The person who likes Walking

Directions (Qs. Nos. 127 to 130): *Answer the questions based on the following information.*

Four persons (1) Mohit, (2) Monohar, (3) Prasant and (4) Dinesh each had some initial money with them. They all were playing bridge in a way that the loser doubles they money of each of the other three persons from his share. They played four rounds and each person lost one round in the order 1, 2, 3 and 4 as mentioned above. At the end of the fourth round each person had ₹ 32000/-.

127. What was the amount with Mohit to start with?
A. ₹ 60,000 B. ₹ 34,000
C. ₹ 66,000 D. ₹ 28,000

128. What was the amount with Monohar at the end of the first round?
A. ₹ 68,000 B. ₹ 72,000
C. ₹ 64,000 D. ₹ 80,000

129. Who has the lowest amount at any round of play throughout the tournament?
A. Mohit B. Monohar
C. Prasant D. Dinesh

130. What was the amount with Prasant at the end of the second round?
A. 36000 B. 72000
C. 16000 D. 68000

ANSWERS

1	2	3	4	5	6	7	8	9	10
B	A	A	D	B	C	D	C	D	C
11	**12**	**13**	**14**	**15**	**16**	**17**	**18**	**19**	**20**
B	A	B	B	C	A	D	B	B	B
21	**22**	**23**	**24**	**25**	**26**	**27**	**28**	**29**	**30**
C	B	A	C	C	B	C	B	A	D
31	**32**	**33**	**34**	**35**	**36**	**37**	**38**	**39**	**40**
A	D	D	A	B	C	D	A	C	C
41	**42**	**43**	**44**	**45**	**46**	**47**	**48**	**49**	**50**
B	B	B	C	B	C	A	A	C	D
51	**52**	**53**	**54**	**55**	**56**	**57**	**58**	**59**	**60**
B	B	B	C	D	C	A	A	A	D
61	**62**	**63**	**64**	**65**	**66**	**67**	**68**	**69**	**70**
C	B	B	A	A	B	B	C	B	B
71	**72**	**73**	**74**	**75**	**76**	**77**	**78**	**79**	**80**
A	C	B	A	C	B	C	A	A	A
81	**82**	**83**	**84**	**85**	**86**	**87**	**88**	**89**	**90**
B	C	C	D	C	A	A	B	D	C
91	**92**	**93**	**94**	**95**	**96**	**97**	**98**	**99**	**100**
C	C	C	A	A	D	D	D	B	B
101	**102**	**103**	**104**	**105**	**106**	**107**	**108**	**109**	**110**
D	C	B	D	A	C	B	D	A	B
111	**112**	**113**	**114**	**115**	**116**	**117**	**118**	**119**	**120**
C	C	A	C	A	C	A	A	A	C
121	**122**	**123**	**124**	**125**	**126**	**127**	**128**	**129**	**130**
C	D	C	A	D	A	C	A	A	B

EXPLANATORY ANSWERS

QUANTITATIVE, DATA INTERPRETATION

36. Required shaded region Area

$$= \frac{\pi(28)^2}{2\times4} + \frac{\pi(14)^2}{2\times4} - \frac{\pi(7)^2}{2\times4} - \frac{\pi(7)^2}{2\times4}$$

$$= \frac{784\pi}{8} + \frac{196\pi}{8} - \frac{49\pi}{8} - \frac{49\pi}{8}$$

$$= \frac{980\pi}{8} - \frac{98\pi}{8}$$

$$= \frac{980\pi - 98\pi}{8} = \frac{882\pi}{8}$$

$$= 110.25\pi$$

38.

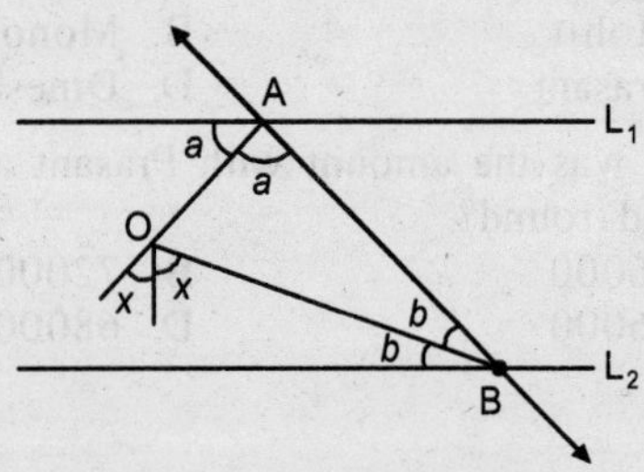

$$2a + 2b = 180 \quad (\because L_1 | L_2)$$

$$a + b = \frac{180}{2} = 90°$$

$a + b = 2x$ (Exterior Angle)

$$x = \frac{a+b}{2}$$

$$x = \frac{90}{2} = 45°$$

39. $$\frac{\text{Price of TV}}{\text{Price of Fridge}} = \frac{150\%}{100\%} = \frac{3}{2}$$

Required percentage $= \frac{2}{3+2} \times 100 = 40\%$

40. $\because$ CI and SI 2 years difference $= P\left(\frac{r}{100}\right)^2$

$$50 = P\left(\frac{5}{100}\right)^2$$

$$50 = P\left(\frac{1}{20}\right)^2$$

$$50 = P \times \frac{1}{400}$$

$$P = 50 \times 400 = 20{,}000$$

Amount after 3 years $= P\left(1 + \frac{r}{100}\right)^3$

$$= 20{,}000\left(1 + \frac{5}{100}\right)^3$$

$$= 20{,}000\left(\frac{21}{20}\right)^3$$

$$= 20{,}000 \times \frac{21\times21\times21}{20\times20\times20}$$

$$= ₹\ 23152.5$$

41. Let the second multiplier $= x$

then, $34 \times x - 25 \times x = 405$

$9x = 405$

$x = 45$

$\therefore$ Sumit's answer $= 34 \times 45$

$= 1530$

42. Let the side of Rectangle is $2x$ and $7x$ respectively.

$2(2x + 7x) = 360$

$18x = 360$

$x = 20$

$\therefore \quad b = 2x = 2 \times 20 = 40\text{m}$

$l = 7x = 7 \times 20 = 140\text{m}$

43.

No. of word can be formed $= 5! \times 1!$

$= 120$

44. $C = \pi rl = \pi r\sqrt{r^2 + h^2}$

$$C^2 = \pi^2r^2(r^2 + h^2) = \pi^2r^4 + \pi^2r^2h^2$$

$$C^2h^2 = \pi^2r^4h^2 + \pi^2r^2h^4$$

$$V = \frac{1}{3}\pi r^2h \quad \Rightarrow V^2 = \frac{1}{9}\pi^2r^2h^2 \quad \Rightarrow 9V^2 = \pi^2r^4h^2$$

$$\therefore \quad 3\pi vh^3 = 3\pi \times \frac{1}{3}\pi r^2h\,(h^3)$$

$$= \pi^2r^2h^4$$

$$C^2h^2 - 9V^2 = \pi^2r^4h^2 + \pi^2r^2h^4 - \pi^2r^4h^2$$

$$= \pi^2r^2h^4$$

$$= 3\pi Vh^3$$

45. Work done in 4 days $= 4\left(\frac{1}{10}+\frac{1}{15}\right)$

$$= 4\left(\frac{3+2}{30}\right) = \frac{20}{30} = \frac{2}{3}$$

Remaining work $= 1-\frac{2}{3} = \frac{1}{3}$

46. 20% of $x = y$

$$y = \frac{x}{5}$$

y% of 20 $= \frac{x}{5}$% of 20 $= x$% of 4 = 4% of x.

47.
$$\left\{\frac{2\tfrac{1}{2} \times 3^{\frac{1}{3}} \times 4^{\frac{1}{4}}}{10^{-\frac{1}{5}} \times 5^{\frac{3}{5}}} \div \frac{3^{\frac{4}{3}} \times 5^{-\frac{7}{5}}}{4^{-\frac{3}{5}} \times 6}\right\} \times 2$$

$$= \left\{\frac{2^{\frac{1}{2}} \times 3^{\frac{1}{3}} \times 4^{\frac{1}{4}} \times 4^{-\frac{3}{5}} \times 6}{10^{-\frac{1}{5}} \times 5^{\frac{3}{5}} \times 3^{\frac{4}{3}} \times 5^{-\frac{7}{5}}}\right\} \times 2$$

$$= \left\{\frac{2^{\frac{1}{2}} \times 3^{\frac{1}{3}} \times 2^{\frac{1}{2}} \times 2^{-\frac{6}{5}} \times 2 \times 3}{5^{-\frac{1}{5}} \times 2^{-\frac{1}{5}} \times 5^{\frac{3}{5}} \times 3^{\frac{4}{3}} \times 5^{-\frac{7}{5}}}\right\} \times 2$$

$$= \left\{\frac{2^{\frac{1}{2}+\frac{1}{2}-\frac{6}{5}+1} \times^{\frac{1}{3}+1}}{5^{-\frac{1}{5}+\frac{3}{5}-\frac{7}{5}} \times 2^{-\frac{1}{5}} \times 3^{\frac{4}{3}}}\right\} \times 2$$

$$= \left\{\frac{2^{\frac{5+5-12+10}{10}} \times 3^{\frac{4}{3}}}{5^{-1} \times 2^{-\frac{1}{5}} \times 3^{\frac{4}{3}}}\right\} \times 2$$

$$= \left\{2^{\frac{4}{5}} \times 2^{\frac{1}{5}} \times 5^{1}\right\} \times 2$$

$= (2 \times 5) \times 2$

$= 20$

49. Required stoppage time per hour =

$$\frac{\text{Speed Excluding Halts} - \text{Speed Including Halts}}{\text{Speed Excluding Halts}} \times 60$$

$$= \frac{50-40}{50} \times 60 = \frac{60}{5} = 12$$

50. The number of ways of choosing of at least one blue and one green dyes

$= (2^5 - 1)(2^4 - 1) \times 2^3$

$= (32 - 1)(16 - 1) \times 8$

$= 31 \times 15 \times 8$

$= 3720$

51. $(a - b)^3 + (b - c)^3 + (c - 1)^3$

$\because (a - b) + (b - c) + (c - a) = 0$

$\therefore (a - b)^3 + (b - c)^3 + (c - a)^3$

$= 3(a - b)(b - c)(c - a)$

[$\because$ We know that if $a + b + c = 0$ then $a^3 + b^3 + c^3 = 3abc$]

52. Fare from 0 km to 10 km $= \frac{150}{10} =$ ₹ 15

Fare from 10 km to onward $= \frac{70}{5} =$ ₹ 14

Tom's total fare for travelling a distance

25 km $= 15 \times 10 + 15 \times 14$

$= 150 + 210$

$= 360$

53.

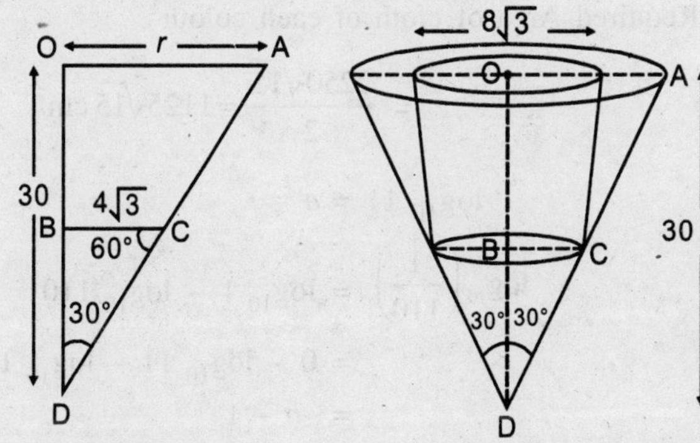

In Δ AOD,

$$\tan 30° = \frac{OA}{OD}$$

$$\frac{1}{\sqrt{3}} = \frac{OA}{30}$$

$$OA(r) = \frac{30}{\sqrt{3}} = 10\sqrt{3} \text{ cm}$$

$\therefore$ Volume of cone $= \frac{1}{3}\pi r^2 h$

$$= \frac{1}{3}\pi\left(10\sqrt{3}\right)^2 \times 30$$

$$= 3000\,\pi \text{ cm}^3$$

54. Required ways $= \frac{(5+3+3)!}{5! \times 3! \times 3!}$

$$= \frac{11!}{5! \times 3! \times 3!}$$

$$= \frac{11 \times 10 \times 9 \times 8 \times 7 \times 6}{6 \times 6}$$

$= 11 \times 5 \times 3 \times 8 \times 7$

$= 9240$

55. 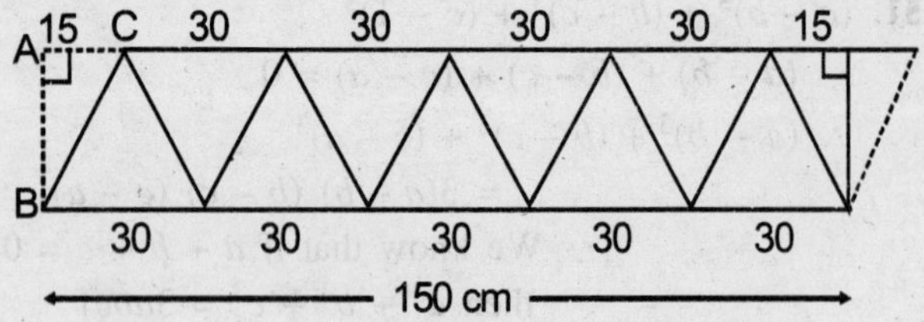

In ΔABC,

$$AB^2 = 60^2 - 15^2$$
$$= 3600 - 225 = 3375$$
$$AB = 15\sqrt{15}$$
$$\text{Total} = l \times b$$
$$= 150 \times 15\sqrt{15}$$
$$= 2250\sqrt{15}$$

Required Area of cloth of each colour

$$= \frac{2250\sqrt{15}}{2} = 1125\sqrt{15}\text{ cm}^2$$

56. $\log_{10} 11 = a$

$$\log_{10}\left(\frac{1}{110}\right) = \log_{10} 1 - \log_{10} 110$$
$$= 0 - \log_{10} 11 - \log_{10} 10$$
$$= -a - 1$$
$$= -(a + 1)$$

57. Required Probability $= \left(1-\frac{1}{7}\right) \times \left(\frac{1}{5}\right) + \frac{1}{7} \times \left(1-\frac{1}{5}\right)$

$$= \frac{6}{7} \times \frac{1}{5} + \frac{1}{7} \times \frac{4}{5}$$
$$= \frac{10}{35} = \frac{2}{7}$$

58. It is clear that Atul wins the race.
So (a) and (c) are left since they meet at a distance of 50 m from the finish line, to distance covered by Atul = 650 m and Alok = 750 m

$$\Rightarrow 100 + \frac{600}{x} + 150x = 750$$
$$\Rightarrow (3x-4)(x-3) = 0$$
$$\Rightarrow x = 3 \text{ or } \frac{4}{3}$$

Since, x is an integer, it has to be 3.

So, when Atul covers the remaining 50m, Alok covers only $\frac{50}{3}$, i.e. 16.67 m.

Therefore, Atul wins by 33.33 m.

59.

Price =	₹ 5	₹ 2	₹ 1	
Proudct =	2	2	6	
Total	10	4	6	= 20

$\therefore$ Total product = 2 + 2 + 6 = 10

60. 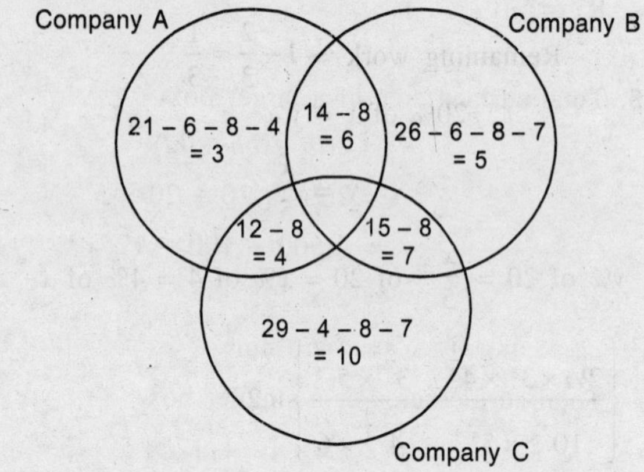

No. of students who got shortlisted for only B = 5
No. of students who got shortlisted for only C = 10
Required Ratio = 5 : 10 = 1 : 2

61. Dividend for preferred share

$$= 50{,}000 \times 10 \times \frac{20}{100}$$
$$= ₹\ 1{,}00{,}000$$

Rest dividend = 1,80,000 – 1,00,000 – 30,000
= ₹ 50,000

$$\%\text{ dividend for common share} = \frac{50{,}000 \times 100}{20{,}000 \times 10} = 25\%$$

62. Log tan 1° + log tan 2° + log tan 3° + ... + log tan 89°
= log (tan 1° × tan 2° × tan 3° × .... × tan 89°)
= log 1
= 0

63. Price × Consumption = Expenditure

Assume →	10	×	10	=	100
	20% ↑				↑ 5% ↑
	12	×	$\frac{105}{12} = 8.75$		= 105

$$\text{Decrease in consumption} = \frac{10 - 8.75}{10} \times 100 = 12.5\%$$

64. We have,

$$R_c = m\ln\left(1 + \frac{R_m}{n}\right)$$
$$\frac{R_c}{m} = \ln\left(1 + \frac{R_m}{m}\right)$$

$$e^{\frac{R_c}{m}} = 1 + \frac{R_m}{m}$$

$$\frac{R_m}{m} = e^{\frac{Rc}{m}} - 1$$

$$R_m = m\left(e^{\frac{R_c}{m}} - 1\right)$$

65. Total surface area of a larger box

$= 2(lb + bh + hl)$

$= 2(25 \times 20 + 20 \times 5 + 5 \times 25)$

$= 2(500 + 100 + 125)$

$= 2 \times 725$

$= 1450 \text{ cm}^2$

Total surface area of a smaller box

$= 2(15 \times 12 + 12 \times 5 + 5 \times 15)$

$= 2(180 + 60 + 75)$

$= 2 \times 315 = 630 \text{ cm}^2$

$\therefore$ Total surface Area of 250 boxes of each

$= 1450 \times 250 + 630 \times 250$

$= 362500 + 157500 = 520000 \text{ cm}^2$

5% extra to cover envelop $= 520000 \times 105\%$

$= 546000 \text{ cm}^2$

Then total cost of wood $= 4 \times 54{,}6000$

$=$ ₹ 21,84,000

66. $$\frac{(4)^2 \times 2^{\pi+1} - 4 \times 2^{\pi}}{4^{-1} \times 2^{\pi+2} - 2 \times 2^{\pi+2}}$$

$$= \frac{2^{\pi}(2^5 - 2^2)}{2^{\pi}(2^{-2} \times 2^2 - 2 \times 2^2)}$$

$$= \frac{2^5 - 2^2}{1 - 2^3}$$

$$= \frac{32 - 4}{1 - 8} = -\frac{28}{7} = -4$$

67. Area of ΔECB

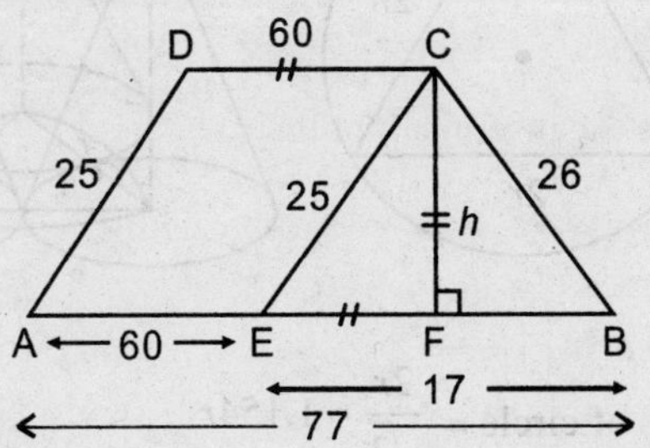

$$\frac{1}{2} \times \text{base} \times \text{height} = \sqrt{s(s-a)(s-b)(s-c)}$$

$$s = \frac{25 + 26 + 17}{2} = \frac{68}{2} = 34$$

$$\frac{1}{2} \times 17 \times h = \sqrt{34(34-17)(34-25)(34-26)}$$

$$\frac{1}{2} \times 17 \times h = \sqrt{34 \times 17 \times 9 \times 8}$$

$$\frac{1}{2} \times 17 \times h = 17 \times 4 \times 3$$

$$\frac{1}{2} \times h = 12$$

$h = 24$

Area of Trapezium $= \frac{1}{2} \times$ Sum of parallel side $\times$ Height

$= \frac{1}{2}(60 + 77) \times 24$

$= 1644$ sq.cm

68. $(p + q)^3 - (p - q)^3 - 6q(p^2 - q^2)$

$2(p^3 + 3p^2q) - 6q(p^2 - q^2)$

$[\because (a + b)^3 - (a - b)^3 = 2(b^3 + 3a^2b)]$

put $q = 1$,

$= 2(p^3 + 3p^2) - 6(p^2 - 1)$

$= 2p^3 + 6p^2 - 6p^2 + 6$

$= 2p^3 + 6$

$= 2(p^3 + 3)$

69. Roman Numerical = MMXVIII

In Arabic Numbers M = 1000

$= 1000 + 1000 + 10 + 8$

$= 2018$

70. Let original marked price be ₹ $100x$

Then, New market price of article $= 160x$

Selling price of article $= 160x \times \frac{7}{8} \times \frac{6}{7} \times \frac{4}{5} =$ ₹ $96x$

CP of article $= 96x \times \frac{5}{6} =$ ₹ $80x$

2nd Selling price $= 160x \times \frac{7}{8} \times \frac{4}{5} =$ ₹ $112x$

Given, $112x - 80x = 800$

$32x = 800$

$x = 25$

Cost price $= 80 \times 25 =$ ₹ 2000

ANALYTICAL AND LOGICAL REASONING

97.

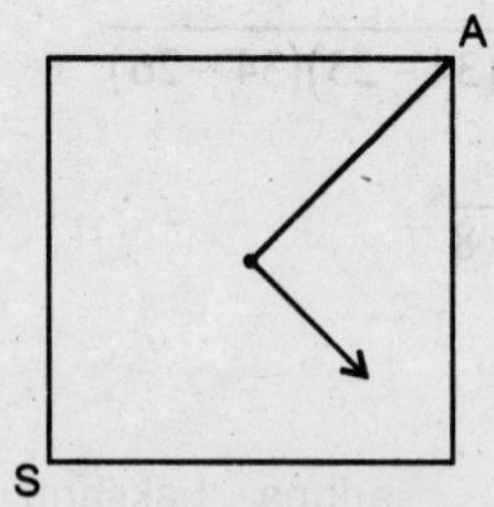

He would be facing South-east.

98.

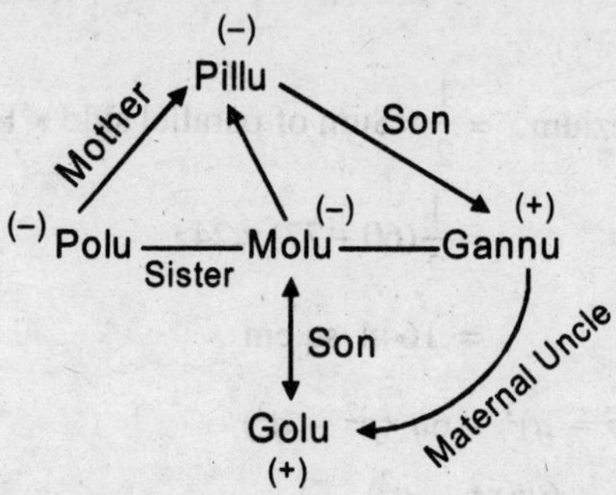

100. *Lectures*	*Day*
Operations	Monday
Finance	Tuesday
Organisational Behaviour	Wednesday
Marketing	Thursday
Information systems	Friday

101. Let Ramu's total friends = x

Then No. of Ramu's friends went to Mumbai = $\frac{x}{5}$

No. of Ramu's friends went to Delhi = $\frac{x}{3}$

No. of Ramu's friends went to Chennai = $3\left(\frac{x}{3}-\frac{x}{5}\right) = \frac{2x}{5}$

No. of Ramu's friends went to Pune

$$= x-\left(\frac{x}{5}+\frac{x}{3}+\frac{2x}{5}\right)$$

$$= \frac{15x-3x-5x-6x}{15} = \frac{x}{15}$$

Given that $\frac{x}{15}=1$

$x = 15$

No. of Ramu's friend went to mumbai = $\frac{x}{5} = \frac{15}{5} = 3$

102.

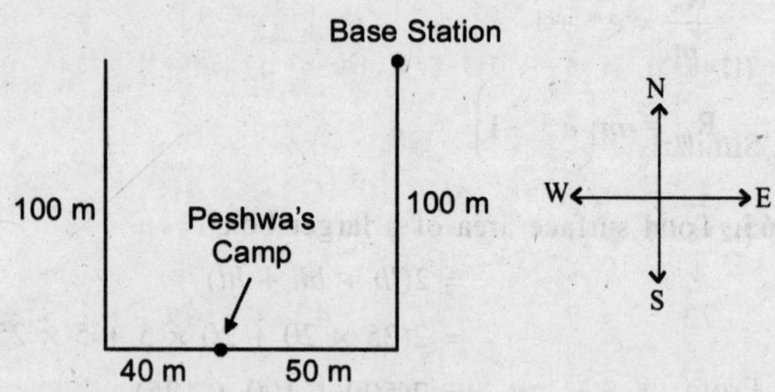

105. As $(5-2)\times(7-3) = 3\times 4 = 12$

$(19-15)\times(23-17) = 4\times 6 = 24$

Similarly

$(29-23)\times(33-?) = 48$

$6\times(33-?) = 48$

$? = 33-8 = 25$

106. Babu < Gopu

Pappu < Somu

Statement I : Babu < Somu

Statement II : Somu < Gopu

From I and II

Babu < Somu < Gopu

Statement III : Pappu is youngest

From I, II and III

Pappu < Babu < Somu < Gopu

↓ youngest ↓ oldest

113.

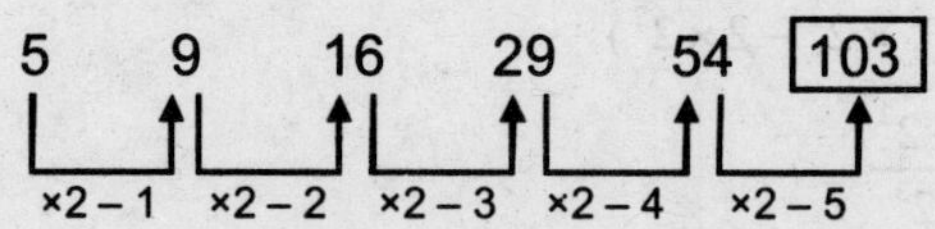

114.

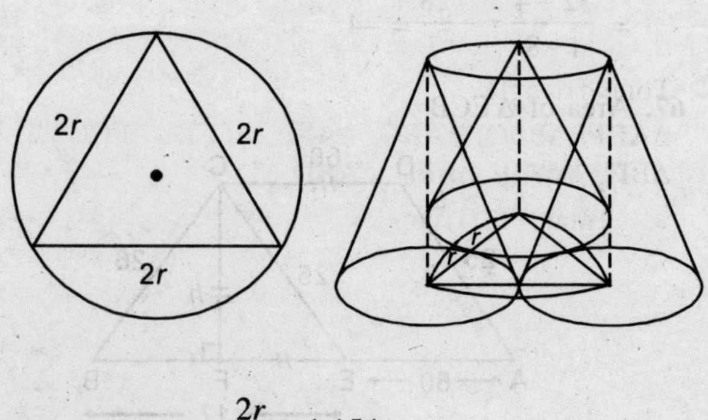

Radius of circle = $\frac{2r}{\sqrt{3}} = 1.154r$

$\therefore$ Radius is Larger than r.

115. As,

M	A	S	T	E	R
79	7	115	121	31	109
↓	↓	↓	↓	↓	↓
(13×6+1)	(1×6+1)	(19×6+1)	(20×6+1)	(5×6+1)	(18×6+1)

Similarly,

L	A	U	G	H
(12×6+1)	(1×6+1)	(21×6+1)	(7×6+1)	(8×6+1)
↓	↓	↓	↓	↓
73	7	127	43	49

116. From 8 AM on Sunday to 8 PM on the following Sunday the watch will be 348 seconds fast.

∴ There will be difference of 648 seconds in the duration of 180 hours.

∴ Difference of 800 seconds (5 min), so that the watch shows the correct time will be there after $\frac{180}{648} \times 300$ hours *i.e.* $\frac{250}{3}$ hours *i.e.*, $33\frac{1}{3}$ hours *i.e.* at 7:20 pm on Wednesday.

117. As January = 10 + 1 + 14 + 21 + 1 + 18 + 25
= 90
= 9 + 0 = 9

July = 10 + 21 + 12 + 25
= 68
= 6 + 8 = 14

Similarly

June = 10 + 21 + 14 + 5
= 50
= 5 + 0
= 5

118.

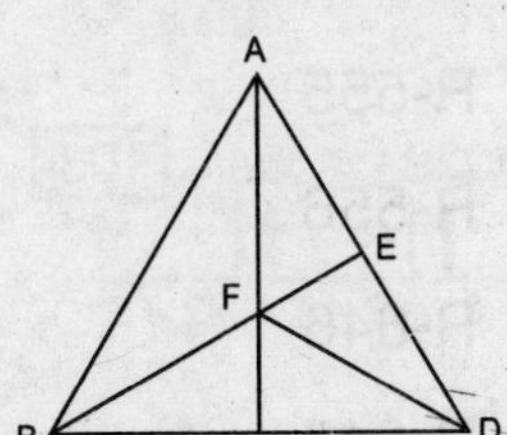

Total Triangle

ΔABF, ΔBFC, ΔABC, ΔAFE, ΔFED, ΔFCD, ΔACD, ΔBFD, ΔABE, ΔBED, ΔABD

∴ Total = 11 Triangles

119.

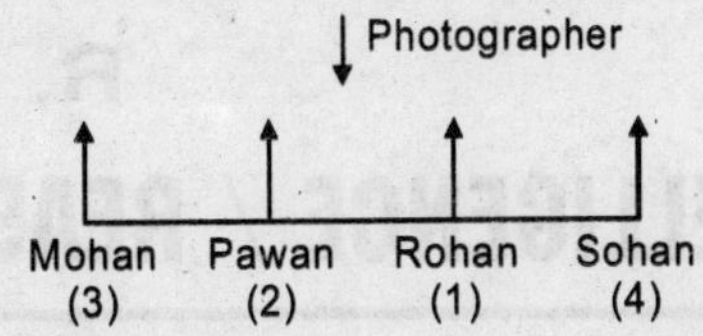

Rohan (1) is second from the left for the photographer.

For Answers (Qs. 120 to 126):

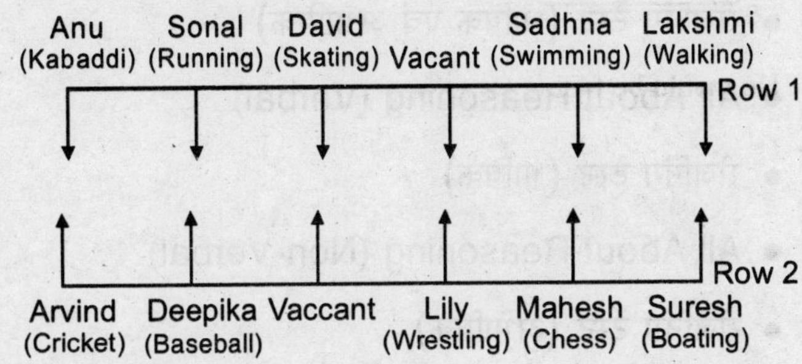

For Answers (Qs. 127 to 130):

These kind of questions are best solved in the reverse order by progressively doubling the money of the individuals depending on who lost the game. The following table is self explanatory:

After	Mohit	Monohar	Prasant	Dinesh	Loser in Round
4	32000	32000	32000	32000	Dinesh
3	16000	16000	16000	80000	Prasant
2	8000	8000	72000	40000	Monohor
1	4000	68000	36000	20000	Mohit
Starting Values	66000	34000	18000	10000	

The solutions can then be read off the table:

127. Mohit had ₹ 66000 to start with.
Option (C) is correct.

128. Monohar had ₹ 68000 at the end of the first round.
Option (A) is correct.

129. Mohit had ₹ 4000 at the end of the first round and that is the lowest value throughout the tournament.
Option (A) is correct.

130. Prasant had ₹ 72000 at the end of the second round.
Option (B) is correct.

Previous Paper (Solved)

Symbiosis National Aptitude Test (SNAP), 2017*

SECTION- I

CURRENT AFFAIRS

1. The Ex-German Chancellor who was credited with Germany unification died in 2017.
A. Gerhard Schroeder B. Angela Merkel
C. Helmut Schmidt D. Helmut Kohl

2. Which of the following was detected in the year 2016 by scientists?
A. Magnetic wave B. Levitational wave
C. Electric wave D. Gravitational wave

3. e-Nivaran is a new facility launched by Government of India. What is its purpose?
A. To help the farmers to avail easy loans
B. To help the RTI petitions
C. To submit and track Income Tax grievances
D. To get help for health related issues

4. The technology that shares the limelight of Bitcoin by providing technological support to the said Bitcoin is called:
A. Blockcurrency B. Blockcash
C. Blockchain D. Blockcipher

5. Who won the Hridayanath Mangeshkar award in 2016?
A. Vishwanathan Anand
B. Aamir Khan
C. Javed Akhtar
D. Anu Malik

6. The last person to 'walk the moon' passed away this year. He was:
A. Buzz Aldrin B. Neil Armstrong
C. Yuri Gagrin D. Eugene Cernan

7. In January, 2017 .... banned FM Broadcasting.
A. France B. Sweden
C. Denmark D. Norway

8. Columbia's Juan Manuel Santos won the 2016 Nobel Prize for peace. What was his contribution that was recognised?
A. For his efforts to bring an end to the exploitation of agricultural labourers
B. For his efforts to end the country's civil war
C. For his efforts to bring peace between Latin American countries
D. For his efforts to eradicate the use of land mines in war

9. Which is India's longest tunnel?
A. Chenani-Nashri Tunnel
B. Jawahar Tunnel
C. Zoji-La Tunnel
D. Banihal-Nashri Tunnel

10. How many Men's Single Wimbledon titles does Roger Federer have to his name including the 2017 title?
A. 9 B. 7
C. 8 D. 6

11. Sweden based Furniture maker IKEA acquired this online (and mobile) marketplace for finding freelance labour to carry out local on demand day-to-day tasks.
A. HandyMan B. UrbanClap
C. ManFriday D. Taskrabbit

12. Which of the following personalities is a member of the committee of administrations of Indian Cricket during October, 2017?
A. Anjum Chopra B. Sonika Kaliraman
C. Diana Eduji D. Mithaly Raj

13. The Flipkart backed startup is into a social network platform exclusively for new parents.
A. Newparents B. Post9Months
C. Tinystep D. Babystep

14. Who are the last two (present and the immediate past) Attorney Generals?
A. Goolam Vahanvati and Milon Banerji
B. K.K. Venugopal and Goolam Vahanvati
C. Mukul Rohatgi and Milon Banerji
D. K.K. Venugopal and Mukul Rohatgi

15. Walmart acquired Men's fashion retailer .... for $310 million in 2017.
A. GAP B. BONOBOS
C. ZARA D. BEBE

16. Which of the following is the cyclone that hit India in December 2016?
A. Eona B. Nada
C. Gloria D. Ockhi

* Online exam held on 17/12/2017

17. The RP-Sanjiv Goenka Group has roped in .... as the brand ambassador for its package snacking brand .... on a bid to scale up its FMCG business.

A. M.S. Dhoni, E-Vita
B. Virender Singh, Yummee
C. P.V. Sindhu, Hungree
D. Virat Kohli, Too Yumm

18. Identify the leading business personality in the picture:

A. Alice Vaidyan—CMD General Insurance Company
B. Aarthi Subramanian—Exec. Director (Global Head-DEGC) TCS
C. Kaku Nakhate—President and Country Head Bank of America Merill Lynch
D. Chitra Ramakrishna—MD & CEO National Stock Exchange

19. The Eurasia Tunnel that was opened for traffic in December 2016 connecting Europe and Asia passes underneath the ....

A. Bosphorus Strait
B. Dover Strait
C. Bering Strait
D. Hormuz Strait

20. What was the name of the mascot in the recently concluded Asian Athletic Championship 2017 held at Bhubaneshwar?

A. Shera B. Coby
C. Olly D. Vinicious

21. One of Olympics longest official sponsors recently ended its 41 year long sponsorship. This organization was ....

A. Pepsi B. General Motors
C. Adidas D. McDonalds

22. On 15th February 2017, ISRO made history by launching 104 satellites through a single rocket. What was the rocket?

A. PSLV 3104
B. GSLV-IRS-P3
C. PSLV-C37
D. GSLV-C38

23. In the world of Information Technology, SMAC has become very popular. What does the A stand for in the abbreviation?

A. Advanced B. Analytics
C. Analysis D. Artificial Intelligence

24. Who owns the recently launched New Channel called Republic TV?

A. Anil Ambani
B. Subramaniam Swami
C. Rajeev Chandrashekhar
D. Arnab Goswami

25. In a move that will reverberate from Hollywood and Silicon Valley to TVs and smartphones around the world, the .... said Thursday (14th Dec 2017) that it had reached a deal to buy most of the .... , the empire controlled by ...., in an all-stock transaction valued at approximately ....

A. Walt Disney Company, 20th Century Fox, Rupert Murdoch, $70 billion
B. Paramount Pictures, 20th Century Fox, Rupert Murdoch, $50 billion
C. Walt Disney Company, 21st Century Fox, Rupert Murdoch, $50 billion
D. Paramount Pictures, 21st Century Fox, Rupert Murdoch, $70 billion

26. The language creator who constructed artificial language 'Dothraki' for the television series Game of Thrones is ....

A. David Henchman
B. David Peterson
C. Robert Peterson
D. Robert Henchman

27. The human organ was discovered in 2017 in our digestive system.

A. Mesentery B. Lacteal
C. Enterocyte D. Ileocecal Valve

28. This comic book company was acquired by Netflix in 2017.

A. Millerworld B. DC Comics
C. Marvel D. Doton

29. The country was recently boycotted by a majority of its neighbours.

A. Oman B. Qatar
C. Bahrain D. Kuwait

30. In Jan 2017, Japan launched its first ever military communications satellite. It was called:

A. Fuji-1 B. Kirameki-2
C. Suzuki-1 D. Takahiro-2

SECTION-II

GENERAL ENGLISH, READING COMPREHENSION, VERBAL REASONING, VERBAL ABILITY

31. Give synonym for:

Artless

A. Pretentious B. Innocent
C. Devious D. Hypocritical

32. Choose a word from the given alternative which is opposite in meaning to the word:

Hiatus

A. Sardonic B. Pessimist
C. Sentimental D. Continuation

33. Choose the right meaning of the Idiom:

A pyrrhic victory

A. A victory that is not worth winning
B. Joyous victory
C. Undeserving victory
D. None of the mentioned options

34. Choose the correct option based on the sentence:

Akhil fared poorly in the entrance test. He found it ....

A. Fathomable B. Esoteric
C. Manifest D. Coherent

35. Choose one word from the options given, that can fit into all sentences.

1. You should not .... to his demands.
2. Ravi needed a loan. However, he did not .... to the loan agreement.
3. The state will .... to the union.
4. The Prince will .... to the throne.

A. Accede B. Precede
C. Recede D. Secede

36. Choose the correct word(s) for the underlined word(s):

You cannot trust a person who plays false and loose with others.

A. fastloose B. false or loose
C. fast or loose D. fast and loose

37. Choose the correct meaning of the phrase.

The team of doctors sent to the village to investigate the epidemic were exasperated. The conditions there were *'a perfect storm'*.

A. Torrential rains
B. A worst possible scenario
C. A wonderful situation
D. A risky proposition

38. Choose the correct meaning of the phrase.

When playing the Clarinet, Jerry is 'all thumbs'.

A. Perfunctory B. Dancing
C. An expert D. Clumsy

39. Identify the Compound Relative Adverb in the sentence.

I do not know the venue where the match will be played.

A. will be B. where
C. know D. played

40. Choose one word from the options given, that can fit into all sentences.

1. I cannot .... this principled man.
2. These religious songs will .... you.
3. Better suppress than .... your desires.
4. Only delicious food can .... my hunger.

A. Gratify B. Edify
C. Occupy D. Glorify

41. Select the pair that best expresses the relationship similar to that expressed in this pair.

convivial : jocund

A. lyrical : prosaic B. umbrage : happiness
C. timid : brave D. folly : silliness

42. Rearrange the sentences to make it a readable paragraph:

1. There is a difference between foreign equity stakes and that of domestic promoters.
2. Apart from it, many other differences exist.
3. The hike in the former was important as a signal.
4. These have to be treated at different levels.
5. A similar signal is not required in respect of domestic industry.

A. 15234 B. 14352
C. 23145 D. 43215

43. Find a correct match of grammatical function with the usage of the word REBEL.

Function	**Usage**
1. Noun	(*a*) Her rebellious moods are due to the side effects of prescription drugs
2. Adjective	(*b*) Most teenagers rebel against their elders
3. Verb	(*c*) Armed rebels moved towards the city
4. Adverb	(*d*) His body shivering with anger, Kunal's eyes fixed rebelliously

	1	2	3	4
A.	(*a*)	(*c*)	(*b*)	(*d*)
B.	(*d*)	(*b*)	(*c*)	(*a*)
C.	(*c*)	(*a*)	(*b*)	(*d*)
D.	(*c*)	(*d*)	(*a*)	(*b*)

44. Choose a word from the given alternatives which is opposite in meaning to the word—SABLE.

A. Ebony B. Chromatic
C. Swarthy D. Raven

45. From the given alternatives, choose the one which best expresses the given sentence in PASSIVE voice.

Who ate the last piece?

A. The last piece was eaten by who?
B. Who was eating the last piece?
C. The last piece who ate?
D. The last piece was eaten by whom?

46. Give antonym for

Prevaricate

A. Gubble B. Unambiguous
C. Equivocate D. Tergiversate

47. Give antonym for:

Dilatory

A. Tardy B. Remiss
C. Indolent D. Diligent

48. Rearrange the sentences to make it a readable paragraph:

1. Rather, it is the consequence of passive public policy.
2. It follows that the output of economy must grow.
3. High unemployment is not the inevitable result of the pace of technological change.
4. This will prevent the unemployment rate from rising.
5. We can anticipate a moderate increase in labour force, accompanied by a slow decline in the working hours.

A. 35421 B. 42351
C. 24351 D. 31524

49. From the given alternatives, choose the one which best expresses the given sentence in ACTIVE voice.

Sugarcane is raised by some people in the Hinterland.

A. Some people in the Hinterland raised sugarcane.
B. Sugarcane is raised in the Hinterland by some people.
C. In the Hinterland is raised sugarcane by some people.
D. Some people raise sugarcane in the Hinterland.

50. Find a correct match of grammatical function with the usage of the word SHADE.

Function	Usage
1. Noun	(*a*) He shaded his eyes and smiles.
2. Adjective	(*b*) She saw a shady garden.
3. Verb	(*c*) He slunk shadily in the shadows.
4. Adverb	(*d*) The shade of the tree is welcome.

	1	2	3	4
A.	(*d*)	(*b*)	(*a*)	(*c*)
B.	(*d*)	(*b*)	(*c*)	(*a*)
C.	(*a*)	(*c*)	(*d*)	(*b*)
D.	(*a*)	(*b*)	(*c*)	(*d*)

51. Choose the correct word based on the sentence.

Raj, normally an amicable person, gets .... towards the unjust.

A. Affable B. Benign
C. Belligerent D. Genial

52. In the following question choose the word which best expresses the opposite meaning of the word in capital:

She was very INDUSTRIOUS by temperament.

A. Assiduous B. Off-the-cuff
C. Indolent D. Apathetic

53. Give antonym for:

Fleece

A. Satiate B. Bilk
C. Mulct D. Rook

54. Give synonym for:

Perfunctory

A. Precise B. Conscientious
C. Deliberate D. Negligent

55. Identify the Compound Relative Pronoun in the sentence:

Whosoever diggeth a pit shall fall therein.

A. therein B. fall
C. diggeth D. whosoever

56. Choose the correct word based on the sentence:

Rupert drained all his trust funds. He was a ....

A. Miser B. Churl
C. Scrooge D. Wastrel

57. Select the pair that best expresses the relationship similar to that expressed in this pair:

vacillate : be indecisive

A. vacuous : empty B. vindictive : vague
C. vigilant : sleepy D. vociferous : silent

58. Choose the correct word(s) for the underlined word(s):

If I was you, I would do it at once.

A. am B. should be
C. would be D. were

Directions (Qs. 59-64): *Read the following passage and answer the questions that follow.*

The first thing to know when talking about privacy in India is that a majority of the population does not always understand what it means. It is at times confused with shame. It's also confused with the emotion we feel when we do something that does not meet our standards or our sense of what is right. Modern Indian languages do not seem to have an exact word which captures the meaning of privacy; they are usually some variation of the words for isolation, intimacy or secret, once again hinting at a conceptual confusion. This explains the reactions of many who wonder what's the big deal about privacy because they have nothing to hide from the government anyway. Privacy, however, is not only about hiding something or keeping it secret. It is, at its core, the right to be left alone. It does not mean that one is withdrawing from society. It is an expectation that society will not interfere in the choices made by the person so long as they do not cause harm to others. It means that one's right to eat whatever one chooses, the right to drink what one chooses, the right to love and marry whom one chooses, to wear what one chooses, among others, are rights which the state cannot interfere with.

In a society, where adults do not necessarily exercise most of these choices of their own free will (either because of family, caste or societal pressures), it is natural that the very concept of privacy seems incomprehensible. If you have grown up in a society where everything is dictated by someone else, and the cost of disobedience is high, to have the freedom to choose what you will in such important matters sounds like fantasy. But it is also a common misconception that the non-well-off in India do not know or care about privacy. Millions of men and women push back daily against the oppressive hold of their families and communities, and fight for the freedom to make their own choices. They may not have the right word for it, but they are creating space for themselves to exercise the right to privacy.

It is in this context that one must understand the hearings in the Supreme Court on the right to privacy. Although the nine-judge bench has been constituted to decide whether there is a fundamental right to privacy protected under the Constitution in the specific context of the Aadhaar case, privacy has many more dimensions than just data protection or surveillance by the state. A fundamental right to privacy, enshrined and protected in the Constitution, would mean that all persons have the right to be left alone by the state, unless such intrusion is necessitated by a just, reasonable, and fair law.

The nine-judge bench was necessitated in the first place because while multiple judgements have held that there is common law right to privacy (claimed against other individuals and entities); there was doubt as to whether such a right could be claimed against government. Obviously, the Constitution does not use the word "privacy" or we wouldn't be having these hearings. Where, then, does the right to privacy find a place in the Constitution? To answer that it is necessary to go deep into what is meant by a fundamental right. At their core, such rights can be said to be the lines drawn by the Constitution delineating boundaries for the government's actions. Such boundaries necessarily imply, the petitioners' counsels have argued, that the Constitution guarantees that individuals have a right to be left alone by the state on matters of individual choice. They have argued that the earlier decisions in MP Sharma vs Satish Chandra (1954) and Kharak Singh v State of UP (1962) were rendered relying in a narrow and pedantic interpretation of fundamental rights—an approach that has been discarded by the Supreme Court since the 1970s.

The Union government has argued that it does not think that the right to privacy is a fundamental right protected under the Constitution. Attorney General K.K. Venugopal has argued that while the right to privacy may be protected as a common law right or some element of it part of a fundamental right, by itself, it could not per se be guaranteed as a fundamental right. The argument of the Union government and state government supporting it have been premised on an "originalistic" interpretation of the Constitution—that the framers never intended privacy to be a fundamental right available to citizens. Given the Supreme Court's recent approach where it has not been hesitant to depart from the narrow interpretation of the Constitution when the situation demands it (such as appointment of judges), perhaps this approach may not find much Judicial favour.

Far more worrisome is the argument that privacy is only the preserve of the well-off and the elite, and protecting it through law and legal institutions may stand in the way of "development" and poverty alleviation. This not only misunderstands what the right to privacy means but underplays its role in allowing individuals to make free choices. It is an argument of a paternalistic and patriarchal state that knows what's good for you and will not let you make your own choices. It also clashes with the ethos of a limited government enshrined in the Constitution.

A nine-judge bench of the Supreme Court holding that the Constitution guarantees a right to privacy will, however, only settle one issue—that there is a right to privacy guaranteed against state intervention. To what extent this right can be claimed and in what circumstances the state may be allowed to intrude will have to be decided on a case by case basis. At most, the court's judgement may outline the principles on the basis of which judicial review will be carried out, but it cannot be expected to prescribe an answer for every foreseeable situation. Whatever the final judgement, the implications will go far beyond just the Aadhar scheme and law. The law laid down by the Supreme Court on privacy could affect the course of development of the law governing reproductive rights, gay rights, beef bans, prohibition, among a host of other issues that the Indian state and society are grappling with.

59. According to the author, what is meant by a fundamental right?

A. Fundamental rights can be said to be the lines drawn by the Constitution delineating boundaries for the government's actions

B. Privacy is not a fundamental right

C. All the mentioned options are correct
D. Fundamental rights is the preserve of the well-off and the elite

60. The tone of this passage is:
A. Social B. Philosophical
C. Moral D. Reflective

61. When the author says.... 'to have the freedom to choose what you will, in such matters, sounds like fantasy', he means ....
A. To be able to do what you desire in essential matters is like a dream come true.
B. You can do what you like and not pay for it
C. To care two hoots for somebody
D. None of the mentioned options are correct

62. The rights that the state cannot interfere with are:
A. The right to drink whatever one chooses
B. The right to wear what one chooses
C. The right to love and marry whom one chooses
D. All the mentioned options are correct

63. The author says people usually continue privacy with:
A. Emotion B. Shame
C. Hatred D. Despondency

64. The most suitable title for this passage would be:
A. The right to be left alone
B. Only the elite can enjoy privacy
C. The Constitution
D. Aadhaar card

Directions (Qs. 65-70): *Read the following passage and answer the questions that follow.*

Poor Elizabeth Beckford. She was 70 years old and her crime was stealing 12lb of Gloucester cheese. For that she could have hanged. Hundreds did in those violent, vengeful days, dancing "the Tyburn frisk" in the words of those who crammed around the gallows to watch this favourite spectator sport of the 18th century. But the state, in its mercy, saved her life—and gave her a punishment that some would see as worse than death. She was an unwilling passenger on a fleet of 11 ships that set out from England in 1787, the first of the convoys of the criminal underclass—as the ruling elite of Georgian England saw them—sent in chains to colonise new and dangerous shores on the other side of the world.

Those 736 sad souls on that pioneering voyage would establish a new world. Though she didn't know it—and the thought would have given her no consolation as she lay crammed with others in cell-like spaces below decks—Elizabeth was a founder member of a new country, Australia.

On Thursday, more than 200 years later, those who made those dreadful voyages—163,000 in all over the years to come—are feted. Twenty-first century Australians celebrate their convict past, taking their lead from premier John Howard, a descendant of transported folk on both sides of his family. The shipping and court registers of the banished have long lain in the National Archive in London. Now, in the knowledge that two million of us in Britain probably have blood links with Australia's criminal forebears, they have been put online for the hundreds of thousands of amateur genealogists in this country, eager to find out more about their roots.

The history they hide may not be pleasant. Elizabeth, incredibly, was not the oldest on that first ark of despair. Dorothy Handland, a dealer in rags and old clothes, was 82. How she was expected to contribute to empire-building in a virgin land whose hardships could only be guessed at is a mystery as great as the place she was being sent to. But nonetheless she was among the waggon-loads of prisoners dragged down to the docks in Portsmouth from the sunless ship hulks at Woolwich where they had been held because the prisons were all full. They were dressed in rags, their faces pale from imprisonment, louse-ridden and thin as rakes from the slops they had been forced to live on.

Alongside the grannies were 120 other women, mostly young, like 22-year-old Elizabeth Powley. Penniless at home in Norfolk she had raided someone's kitchen for a few shillings' worth of bacon, flour and raisins and "24 ounces weight of butter valued 12d". The death sentence on this starving girl was commuted and, as Robert Hughes, historian of the transportations, notes wryly in his book, The Fatal Shore, "she was sent to Australia, never to eat butter again".

At least the youngest of the "passengers", John Hudson, would never be pushed up another chimney. The nine-year-old sweep was condemned to seven years' exile for theft. All on board were small-time criminals whose punishment, by the standards of later generations, in no degree fitted the crime. James Grace, 11, had taken some ribbon and a pair of silk stockings. John Wisehammer, 15, snatched some snuff from a shop counter in Gloucester. For that, they would never see home again. The most extraordinary crime was that of William Francis, who stole a book about 'the flourishing state of the island of Tobago' from a gentleman in London. If he had had time to read it before he was caught, perhaps he had an inkling of what now lay ahead of him in a British colony far rawer than the West Indies.

There were no political prisoners, however, no rabble-rousing, hay stackburning activists or trades unionists sentenced for their subversive activities, as some of today's anti-Pom Australians like to think. The women, however, were treated poorly. They arrived at the gangplank of their vessel, the Lady Penrhyn, "in a situation that stamps them with infamy", according to the officer in command of the expedition, Captain Arthur Phillip. He was appalled at their treatment by the magistrates who had sentenced them and the jailers who had held them. Whether he could guarantee them better lives at the end of their nine-month voyage was yet to be seen.

What they were about to embark on was the longest journey ever attempted by such a large group of people. Where they were going might as well have been the moon. Crewmen, let alone convicts, believed they would never see home or their loved ones again. "Oh my God," wrote one officer of Marines in his journal, "all my hopes are over of seeing my beloved wife and son."

As for the country they were going to, almost nothing was known except for the promise of Captain James Cook, its discoverer, that this 'New South Wales' as he chose to call it, was now British. But, to some observers of the hang 'em tendency, the thought that the felons might be better off than if they had languished in jail provoked bitter reproach.

65. Why was Dorothy Handland sent to the "New World"?
A. She dealt in rags and old clothes
B. She was a criminal
C. She was old
D. There is no clear reason

66. Who discovered Australia?
A. John Haward B. Arthur Phillip
C. James Cook D. Robert Hughes

67. What was the crime of William Francis?
A. He stole some snuff
B. He stole ribbon and silk stockings
C. He stole a book
D. He stole 12 lb of butter

68. Lady Penrhyn refers to?
A. A surgeon B. A judge
C. A criminal D. A ship

69. On the basis of the passage, why was Australia developed in 1787?
A. To send ill people to a warmer island
B. To serve as a penal colony
C. To employ people
D. To serve as a military-colony

70. As per the passage, what does the world "Pom" denote?
A. Aboriginals B. Australian
C. Englishmen D. Pomegranate

SECTION-III

QUANTITATIVE, DATA INTERPRETATION AND DATA SUFFICIENCY

71. Find out next number in the series:
2, 7, 17, 34, 60, ...
A. 97 B. 108
C. 95 D. 121

72. Kiran, Kumar and Kedar were watching cricket and eating groundnuts. They pop a groundnut in their mouth every 40 seconds, 72 seconds and 108 seconds respectively and munch it. If all of them pop a groundnut simultaneously at 8:20:00 hours, then at what time will they again pop groundnuts again in their mouth simultaneously?
A. 8:22:20 hours B. 8:27:12 hours
C. 8:38:00 hours D. 8:25:15 hours

73. Find the odd one in the series:
10, 16, 24, 28, 30
A. 16 B. 28
C. 30 D. 24

74. A pharmacist stocks a sophisticated life saving injection vial which expires at the end of the day. The cost of procuring that drug is ₹ 3000 and it is solf at a market price of ₹ 10000 per vial. The pharmacist observes that the probability of not selling any vial on a day is 10 per cent, that of selling one vial is 20 per cent, two vials is 30 per cent, three vials is 20 per cent, four vials is 10 per cent and five vials is 10 per cent. How many vials should the pharmacist stock, so that the expected profit is maximized?
A. 0 B. 5
C. 2 D. 4

75. Mr. Murthy invested ₹ 25000 in a mutual fund three years ago. The value of the account increased by 15% in the first year, it decreased by 15% in the second year and again increased by 10% during the third year. What is the value of the investment at the end of the three year period?
A. 26880 B. 28890
C. 27500 D. 26500

76. Two congruent triangles Δ ABC and Δ DEF each with area 198 cm^2 are placed over each other as shown in the figure. AB is parallel to DE and the centres of both the triangles coincide with each other. The area of region PQRSTU is ....

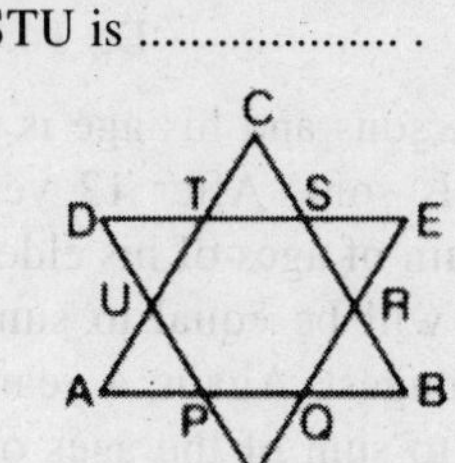

A. 199 cm^2 B. Cannot be determined
C. 198 cm^2 D. 132 cm^2

77. An Olympic size swimming pool requires 6 hours to fill up the brim when water is supplied simultaneously by three pipes A, B and C. The pipes start filling at 12 O'clock in the night but if pipe C is shut off at 3 am, the swimming pool fill up to the brim at 7 am. The swimming pool is drained in the evening. That night, if only pipe C is used to fill up the swimming pool and the filling starts at 12 O'clock in the night, what will be the time in the clock when the swimming pool will be filled up to the brim?

A. 12 O'clock night B. 9 am morning
C. 6 pm evening D. 12 O'clock noon

78. In an international school, 80% of the students know French, 85% know German and 75% know both French and German. If 40 students don't know both, then find total number of students.

A. 400 B. 350
C. 460 D. 450

79. The price of an item was increased during festival season by 10%. Later the new price was decreased by 10%. If the latest price is 1000, what was the approximate original price?

A. 1234.50 B. 1020.50
C. None of these D. 1232.75

80. John was writing an entrance test. Every question has multiple choice questions with four probable answers and one correct answer. The probability that he knows the answer is 80%. If he correctly answered a question, then what is the probability that he was guessing?

A. $\frac{1}{5}$ B. $\frac{1}{17}$
C. None of these D. $\frac{1}{27}$

81. In an India-Pakistan 20 over cricket match, Pakistan played first and India had to chase a high score. Rain created havoc and the average number of overs was reduced to 18, India needed 144 runs to win. In the initial 5 overs, the average scoring rate was 6 but in next 5 overs it increased to 8. It then declined to 5 in next 5 overs. Now, what is the average needed by India in the remaining overs to win the match?

A. 16.33 B. 13.33
C. 12.33 D. 14.33

82. Ravi has three sons and his age is equal to the sum of the ages of his sons. After 12 years his age will be equal to the sum of ages of his eldest two sons. 4 years later, his age will be equal to sum of the age of the oldest and youngest. Again, 4 years after that, his age will be equal to sum of the ages of the two youngest. Find the present age of Ravi and his sons.

A. 44, 18, 16, 12 B. 48, 18, 16, 10
C. 48, 20, 16, 12 D. 48, 20, 12, 8

83. If $\log_x y = 100$ and $\log_3 x = 100$, then the value of y is:

A. 3^{10000} B. 3^{100}
C. 3^{1000} D. 3^{10}

84. The population of a state is consistently decreasing due to migration owing to lack of opportunities in the state. It has been observed that the population has decreased at a rate of 10 per thousand each year in the last few years. If today, the population of the state is 100 million, what will be the population of the state after 3 years?

A. None of these
B. 83 million
C. 73 million
D. 63 million

85. In the figure given below, the semi circle with centre O has a diameter AB = 2*r*. PQRS is a Square of maximum possible area. P and Q are lie on the diameter AB. R and S are on the arc of the semi-circle. There are two more squares of maximum possible area EFGP and CDQH. Then RC + FS = ...

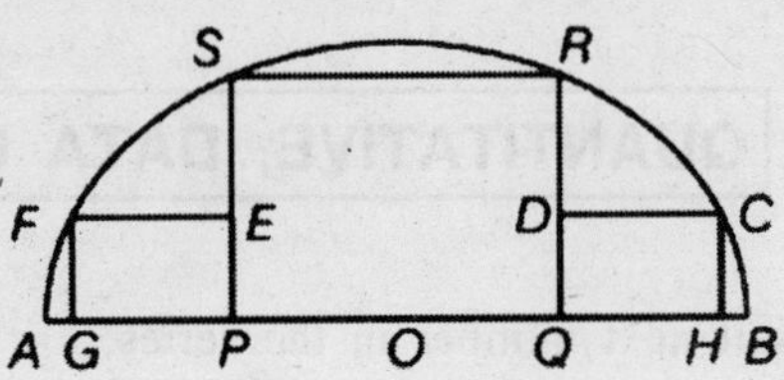

A. $\frac{\sqrt{2r}}{5\sqrt{3}}$ B. $\frac{\sqrt{2r}}{5}$
C. None of these D. $2r\sqrt{\frac{2}{5}}$

86. In the given figure CD = 10 cm. CT is the tangent at C, making an angle of $\frac{\pi}{4}$ with CD. O is the centre of the circle. The perimeter of the shaded region is Δ AOC is:

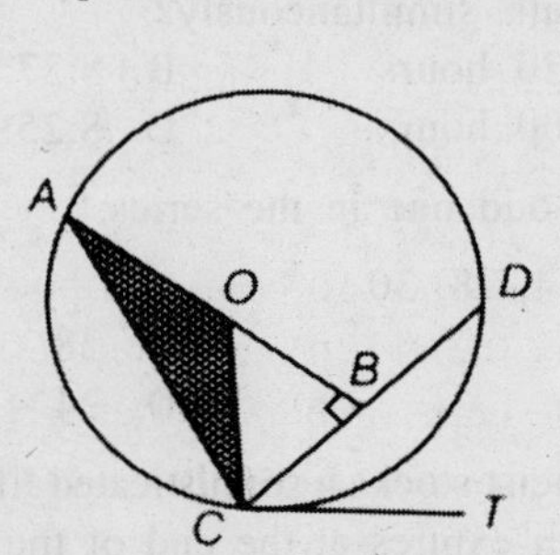

A. 27 cm B. 2 cm
C. 30 cm D. 13 cm

87. In how many ways can we arrange the word CAPSTONE so that all the vowels come together?

A. 1440 B. 4320
C. 2160 D. 3480

88. The area of minor segment PQ is:

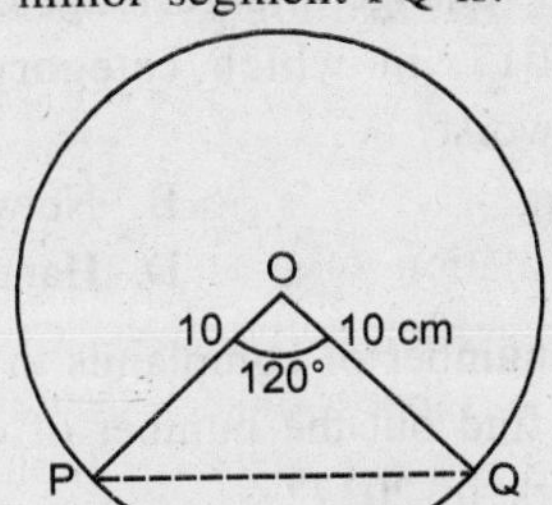

A. $\frac{100\pi}{3} - 25\sqrt{3}$ B. $\left(\frac{100\pi}{3}\right) - \left(\frac{45}{3}\right)$

C. $\left(\frac{100\pi}{3}\right) - \left(\frac{15}{3}\right)$ D. $\left(\frac{100\pi}{3}\right) - \left(\frac{75}{3}\right)$

89. A company employs 35 people whose average age is 45 years. The 10 most aged people have an average age that is 20% more than the average of all 35 people in the company. Three new employees of aged 32, 35 and 30 joined the company. Now what is the difference in the average age of the same 10 most aged people and the average age to all the people in the company?

A. 10 years B. 9 years
C. 12 years D. 11 years

90. On her birthday, Vrinda gets a gift box which contains 6 chocolates and 8 cup cakes. She eats 7 of them randomly. What is the probability that she ate 3 chocolates and 4 cup cakes?

A. 0.16231 approximately
B. 0.05655 approximately
C. 0.40792 approximately
D. 0.1165 approximately

91. To cross a man running at 9 km/hr in the direction of train, a train takes 40 seconds and to cross another man running at 6 km/h in same direction it takes 30 seconds. Find the speed of the train.

A. 20 km/hr
B. Cannot be determined
C. 14 km/hr
D. 18 km/hr

92. A library has 10 copies of book A, 3 copies each of books B and C, 2 copies each of books D, E, F and a single copy of books G and H. The total number of ways in which these books can be arranged on a shell is:

A. $\frac{(24!)}{10!(3!)^2(2!)^3}$ B. $\frac{(24!)}{(10!3!2!)}$

C. $\frac{(24!)}{10!(3!)^2(2!)^2}$ D. $\frac{(12!)}{10!(3!)^2(2!)^2}$

93. What is the value of $(p-1)^{q+1}$, if $p^q = 169$ and p and q are whole numbers?

A. 1728 B. 2187
C. 169 D. 144

94. The population of a city increased by 5%, but decreases by 0.25% each year due to emigration. Find the net increase in per cent in 3 years?

A. 11.24% B. 13.23%
C. 14.93% D. None of these

95. How many factors of $2^8 \times 3^6 \times 5^4 \times 10^5$ are multiples of 120?

A. None of these B. 660
C. 540 D. 594

96. The remainder obtained when 2^{97} is divided by 97 is:

A. 2 B. 0
C. 97 D. None of these

97. Out of all the MBA aspirants interviewed, 50% of those who attended the first interview received the call for second interview. 90% of the people who got a call for second interview felt good about their first interview. 80% of people who did not receive a call for second interview also felt good about their first interview. If a person left good after his/her first interview, what is the probability that he/she will receive a call for the second interview?

A. Approximately 73%
B. Approximately 89%
C. Approximately 53%
D. Approximately 63%

98. The sum of three numbers is 138. The ratio of first to the second number is 3 : 4 and that of second to the third is 8 : 9. The first number is

A. 48 B. 36
C. 54 D. 42

99. If A is the set of odd integers less than 50, and B is the set of the squares of integers in A, how many elements does the intersection of A and B contain?

A. 4
B. 5
C. 3
D. Cannot be determined

100. A Bank lent a certain sum of money at a certain rate of simple interest for 2 years to David. He lent this sum at the same rate of interest but compounded annually for the same period to Pankaj, who has a bad credit history. At the end of two years, David received ₹ 2400 as compound interest but paid ₹ 2000 only as simple interest. Find the rate of interest.

A. 20% B. 10%
C. 30% D. None of these

101. A fish observes a submarine from the bottom of ocean, when the submarine starts moving up vertically. The angle of elevation of the submarine changes from 25° to 60° in 2 minutes. The point of observation of the fish is situated 300 m (displacement) away from the takeoff point. If the submarine is going up in straight line at a constant speed, what is the speed of the submarine? (Here, tan 25° = 0.47)

A. 4.16 m/s
B. 2.22 m/s
C. 5.22 m/s
D. 3.16 m/s

102. Gopal is planning to sell two radios. The first radio is marked ₹ 200 over and above the cost price and is sold at a discount of 20% on it. The other radio is marked 50% over and above the cost price and sold at discount of ₹ 160. If his profit is 15% on both radios combined and 5% on the first alone, what is the percentage of profit on the second radio?

A. 25%
B. 20%
C. Cannot be determined
D. 7%

103. A factory has 900 employees, 25% of whom are men. The factory has to hire 200 additional employees to increase production. What percentage of the additional employees hired must be men in order to raise the percentage of men employees in the factory to 35%?

(Assume transgender for his situation)

A. 70% B. 90%
C. 75% D. 80%

Directions (Qs. 104-105): *Use information provided in the below given figure to solve the questions.*

Percentage of consumer complaints in different categories

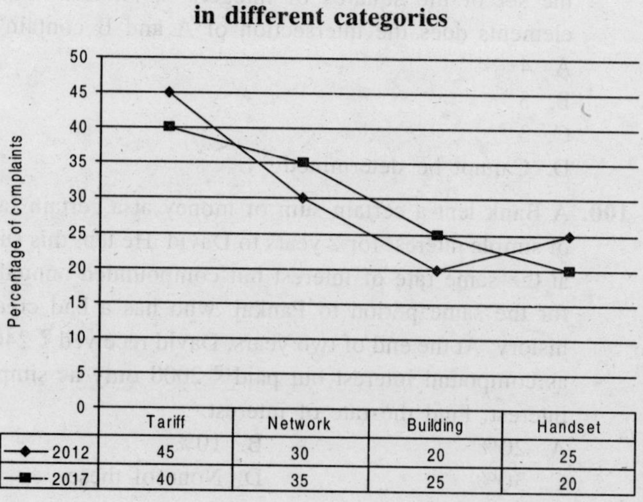

	Tariff	Network	Building	Handset
2012	45	30	20	25
2017	40	35	25	20

104. If we see the change in percentage of complaints from 2012 to 2017, in which category complaints has performed worst?

A. Building B. Network
C. Tariff D. Handset

105. If the total number of complaints in the year 2017 are 2500, then find out the number of complaints related with network in 2017?

A. 900 B. 695
C. 850 D. None of these

Directions (Qs. 106-110): *Data in the pie chart describes the population engaged in various occupations in a German city for a year Based on the information answer questions.*

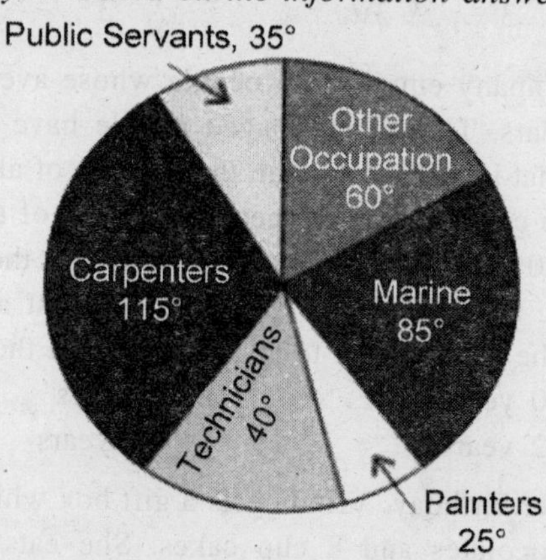

106. What percent of wage earners in the city re-engaged in Marine Engineering?

A. 13.6% B. 19.6%
C. 23.6% D. 8.6%

107. If the number of persons engaged in painting is 2160, the number of persons engaged as technicians would be:

A. 1456 B. 5456
C. 2458 D. 3456

108. If the total earnings from carpenters is twice that from technicians, the ratio of average wages in these occupations is:

A. 23 : 8 B. 5 : 4
C. 16 : 23 D. 23 : 16

109. If the average earning of a wage earner from carpenters category was twice that from technicians, the ratio of the total earnings from these occupations is

A. 4 : 3 B. 5 : 4
C. 4 : 5 D. 23 : 4

110. If the total number of wage earners increases by 10% and the number of wage earners engaged in carpentry increases by 15% in the next year, the nearest central angle of carpenters in the pie chart for that year will be:

A. 120° B. 125°
C. 130° D. None of these

SECTION-IV

ANALYTICAL & LOGICAL REASONING

111. What is the missing number?

2691 3870 ? 9108

A. 1629 B. 9018
C. 5823 D. 7830

112. Four women P, Q, R, S and three men X, Y, Z play bridge, a card game for four players.

- The men and women consist of three married couples and a widow.
- The members of each married couple are never partners in a game.
- No more than one married couple ever plays in the same bridge game.

One night they played four bridge games in which partners were as follows :

Game 1 : P and X versus Q and Y

Game 2 : P and Z versus S and Y

Game 3 : Q and R versus Y and Z

Game 4 : R and X versus S and Z

Who is the widow?

A. P B. S
C. R D. Q

113. David pans a small tea estate in Assam, all of which is shared by him with Rahul and his friends. Rahul has a share of 3/5th of it, and his friend Santosh has twice as much as John, who is also a close friend of Rahul. What fraction of the land belongs to John?

A. Cannot be determined B. $\frac{2}{5}$

C. $\frac{2}{10}$ D. $\frac{2}{15}$

114. What numbers should replace the question marks?

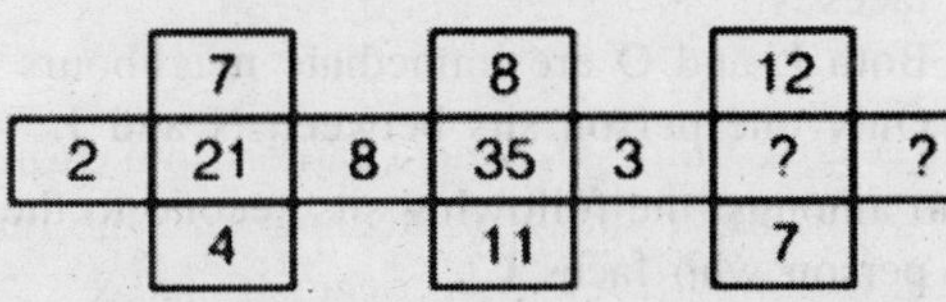

A. 31 and 4 B. 49 and 21
C. 29 and 2 D. None of these

115. Study the rules employed in the following example and answer the question.

Input: from 22 61 tea India east 55 74

Step I: east from 22 61 tea India 55 74

Step II: east 74 from 22 61 tea India 55

Step III: east 74 tea from 22 61 India 55

Step IV: east 74 tea 61 from 22 India 55

Step V: east 74 tea 61 India from 22 55

Step VI: east 74 tea 61 India 55 from 22

Identify the input if the result of Step II of an input is

Money 56 to 11 66 people notes 24

A. Money 56 to 15 66 people notes 24
B. To 15 Money 56 66 people notes 24
C. To 15 66 Money people notes 24 56
D. None of these

116. Mr. Golu, the farmer is desperate to make some money by selling his livestock. He has many animals and has a unique set of customer requirements. The customers do not want to buy animals with disabilities or strange features. Mr. Golu has a Stock of animals which include Monkeys with two tails, Sheep with six legs, cows with two udders, horses with two right hooves and chicken with two toes on each foot. Which animals will he be able to sell?

A. He will not be able to sell any animal
B. Cow
C. Monkey
D. Horse

117. Which of the following is the greatest?

I. log (2 + 4)
II. (log 2 + log 4)
III. log(6 – 3)
IV. log 6 – log 3

A. Only III B. Only IV
C. Only II D. Both I and II

118. In a village located in remote part of the world 5% of the population have a birth defect and hence have only one leg, and half of the others go barefoot as they never wear shoes. What is the least number of shoes needed in the village if the total population is 10000?

A. 5000 B. 4750
C. 9500 D. 10000

119. At 12:00 noon, my watch was correct. However, it then started to lose 18 minutes per hour. Four hours ago, it stopped completely and is now showing the time as 15:30. What is the correct time?

A. 17:00 B. 21:00
C. 20:33 D. 19:00

120. 3182596 is to 65283 and 6742835 is to 53476, therefore 7496258 is to ………?

A. 98765 B. 76574
C. 65987 D. 85647

121. If 645 = 10 and 316 = 12, then 765 = ?
A. 18 B. 5
C. 7 D. 210

122. A villager places a map of his farm in a grampanchayat meeting for resolving his dispute. He places the map in a way that South-East becomes North and North-East becomes West. The Sarpanch is confused. His assistant comes to help and turns the anti-clockwise by 90°. If the Sarpanch has to see the South side in the map, which side will he look?
A. North B. South-West
C. South D. West

123. A, B and C played some card games.
- Each game had exactly one winner.
- No player won 2 game in succession.
- When a player was the dealer for a game he/she did not win the game.
- The dealer position proceeded from A to B to C, then this order was repeated until they stopped.
- The only player to win more than 2 games did not win the first game.

Who was the only player to win more than 2 games?
A. Cannot be determined
B. B
C. C
D. A

124. There are several aspirants who are made to sit in single row for an entrance examination. Rajesh is fifteenth from the left and Mohan is eight from the right end. If there are four aspirants between them and Mohan is to the right of Rajesh, then the total number of aspirants sitting in the row are ....
A. None of these B. 28
C. 27 D. 26

125. Study the following information and answer the question.

In a recent performance approval done by Company X, more than 70% of the employees were found to be underperforming.

'I think that the restructuring done by the company in the previous year is responsible for the under-performance of the employees'-HR manager of company X.

Which of the following Statements weakens the Statement of HR Manager of Company X?
A. The incentives linked to performance were abolished upon restructuring creating dissatisfaction among employees.
B. After restructuring the decision making power was taken away from employees thus delaying the work by long hours.
C. The number of projects in Company X increased by 60% this year thereby increasing the burden on the existing employees this year.
D. After restructuring the employees were required to report to multiple bosses leading to ambiguity at the work place.

Directions (Qs 126 to 130): *Study the following information to answer the given questions.*

Twelve people are sitting in two parallel rows containing six people each in such a way that there is an equal distance between adjacent persons. In row 1—J, K, L, M, N and O are seated (but not necessarily in the same order) and all of facing South. In row 2—U, V, W, X, Y and are seated (but not necessarily in the same order) and all of them are facing North. Therefore, in the given seating arrangement, each member seated in a row faces another member of the other row.

M sits fourth to the left of J. The one facing J sits third to the left of Y. Only one person sits between Y and U. U does not sit at any of the extreme ends of the line. The one facing Z sits second to the right of K. Z does not sit at any of the extreme ends of the line. Only two people sit between K and O. The one facing K sits second to the left of X. V is not an immediate neighbour of Z. L is not immediate neighbour of M.

126. Which of the following groups of people represents the people sitting at extreme ends of both the rows?
A. M, O, X, W B. M, K, V, W
C. N, K, V, Y D. J, N, U, V

127. Who amongst the following faces V?
A. M B. L
C. J D. K

128. Which of the following is true with respect to the given information?
A. K faces one of the immediate neighbours of X.
B. V sits exactly between W and U.
C. None of the given options is true
D. J is an immediate neighbour of K

129. Which of the following is true regarding N?
A. K sits second to right of N.
B. V is an immediate neighbour of the person who faces N.
C. Both L and O are immediate neighbours of N.
D. Only one person sits between N and J.

130. Who amongst the following sits second to the right of the person who faces L?
A. V B. Z
C. W D. Y

131. If '4' is subtracted from each odd digit and '1' is added to each even digit in the number 9476582, which of the following numbers will appear twice in the new number thus formed?
A. Only 2 B. Both 1 and 7
C. Both 3 and 5 D. None of these

132. Read the given information and answer the question.

The government of State D was recently criticised for accepting a proposal as per which schools in the State will not conduct exams up the standard nine. Rather, schools will promote all the students up to standard nine and from standard nine onwards the students would either be passed or failed depending upon their marks in the exams.

Which of the following may not be reason for the criticism of the government of State D for accepting the proposal?

A. Several parents have raised concern that students would not be willing to pay attention in class and retain the information because they do not have to write exams.

B. Students learn to write exam papers in a stipulated period of time with practice of solving exam papers over the years and many students may not be able to grasp this directly in the standard nine.

C. Many teachers are of the opinion that students have become very competitive and are mainly focused on acquiring marks rather than gaining knowledge.

D. Studies suggest that students become less prone to exam anxiety and exam fear when they get used to passing exams over the years as compared to when they do not appear for an exam at all.

Directions (Qs 133 to 138): *Study the following information to answer the given questions.*

Seven friends, namely P, Q, R, S, T, U and V visit seven different countries namely Japan, Germany, China, India, Nepal, Australia and Malaysia, not necessarily in the same order, starting from Monday to Sunday (of the same week).

R visits on Thursday. Only two people visit between R and the one who visits Germany. Only four people visit between the one who visits Germany and V. The one who visits Malaysia visits immediately before V. Only two people visit between the one who visits Malaysia and P. S visits on one of the days after the one who visits Malaysia.

U visit immediately after the one who visit Japan. U does not visit Malaysia. Only three people visit between the one who visits Japan and the one who visits Nepal. The one who visits Australia visits immediately before the one who visits China. Q does not visit on Monday.

133. Which of the following countries does Q visit?

A. China B. Malaysia
C. Japan D. Nepal

134. On which of the following days does U visit a country?

A. Friday B. Saturday
C. Sunday D. Wednesday

135. Which of the following is true about T?

A. T visits immediately before P
B. T visits of Friday
C. T visits China
D. All the given options are true

136. Who amongst the following visits India?

A. S B. T
C. P D. G

137. As per the given arrangement, P is related to the one who visits Japan in a certain way and V is related to the one who visits Nepal in the same way. To which of the following is R related to in the same way?

A. The one who visits Australia
B. The one who visits China
C. The one visits India
D. The one who visits Malaysia

138. Four of the following five are alike in a certain way and thus form a group as per the given arrangement. Which of the following does not belong to that group?

A. U-Friday B. Q-Thursday
C. S-Saturday D. V-Sunday

Directions (Qs 139 to 142): *Study the following information to answer the given questions.*

When a word and number arrangement machine is given an input line of words and numbers, it arranges them following a particular rule. The following is an illustration of input and rearrangement. (All the numbers are two digit numbers)

Input: 23 kinetic amount 64 nature 71 58 opium verdict 96 elderly 15

Step I: opium 23 kinetic amount 64 nature 71 verdict 96 elderly 15 58

Step II: elderly opium 23 kinetic amount nature 71 verdict 96 15 58 64

Step III: amount elderly opium 23 kinetic nature 71 verdict 15 58 64 96

Step IV: 15 amount elderly opium 23 kinetic nature 71 58 64 96 verdict

Step V: 23 15 amount elderly opium kinetic 71 58 64 96 verdict nature

Step VI: 71 23 15 amount elderly opium 58 64 96 verdict nature kinetic

Step VI is the last step of the above arrangement and as he intended arrangement is obtained.

As per the rules followed in the given steps, find out the appropriate steps for the given input.

Input: adverb 59 36 salient 81 idioms bakery 14 launch 47 umpire 62

139. Which elements come exactly between '59' and 'bakery' in Step II of the given input?

A. Only launch B. Only 62
C. Only idioms D. Both 81 and salient

140. Which of the following combination represents the first two and the last two element of the step VI of the given input?

A. 81, 59, launch, bakery
B. 81, 62, salient, launch
C. 50, 47, bakery, salient
D. 62, 14, idioms, umpire

141. If in the step III, 'idioms' interchanges its position with '81' and 'salient' also interchanges its position with '14', then which element will be third to the left of '47'?

A. 14 B. salient
C. adverb D. idioms

142. Which step are the elements 'bakery launch 47 14' found in the same order?

A. Sixth
B. Third
C. Fourth
D. The given order of elements is not found in any step

143. The question consist of information and two Statements numbered I and II given below it. You have to decide which of the given Statements weakens or strengthens the information and decide the appropriate answer.

Information: Due to increased cases of kidnapping in its vicinity, school M has made it compulsory for parents or guardians of the students to give a duly signed authority letter to the person picking up the students from school.

(I) Most working parents rely on their domestic help, for picking up their children from school, who can easily coerced into forging or misusing the authority letter for monetary plans.

(II) There is no photograph on the authority letter making it difficult to identify the person who comes to pick up the child.

A. Both Statement I and Statement II strengthen the information.
B. Statement I strengthens the information while Statement II weaken the information.
C. Both Statements I and II weaken the information.
D. Both Statements I and II are neutral Statements.

144. How many such pairs of letters are there in the word 'PAINTED' each of which has as many letters between them in the word in both forward and backward directions, as there are between them in the English alphabetical series?

A. More than three B. Two
C. Three D. None

Directions (Qs 145 to 149): *Study the following information to answer the given questions.*

In a certain code language, 'urban people prefer cars' is written as 've fm ab eg' 'profit for urban areas' is written as 'ab ep zi so' 'people demand for hike' is written as 'zi qr cd ve' 'hike in profit margin' is written as 'al jn ep cd' (All codes are two letter codes only)

145. What will be the possible code for 'urban food demand' in the given code language?

A. qr ab zi B. qr cr ab
C. nj qr cd D. qr ab jn

146. What is the code for areas in the given code language?

A. ab B. zi
C. ep D. so

147. What does the code 'jn' stand for in the given code language?

A. either 'hike' or 'people'
B. either 'in' or 'margin'
C. profit
D. hike

148. In the given code language, if 'small' is coded as 'wy', then who will 'prefer small cars' be coded as?

A. wy eg ab B. fm ve wy
C. eg wy fm D. a beg fm

149. What will be the code the 'hike' in the given code language?

A. xl
B. zi
C. qr
D. other than those given as options

150. Read the given information and answer the question.

Long term usage of antibiotics causes the disease 'Cretosis' as it decreases the secretion of hormone X. While body can endure the level of hormone X dropping to half the required number of micrograms in levels dropping to 23-micrograms needs immediate medical attention.

Which of the following can be concluded from the given Statement?

A. A patient can be said to have Cretosis only if his/her hormone 'X' levels are 23 micrograms or low.
B. Usage of antibiotics on a short term cannot cause 'Cretosis' ever to a minor extent.
C. All micrograms is exactly half of the amount of hormone X required daily by the body.
D. In a patient with hormone X level of 21 microgram, if administrated another 23 micrograms would bring the level to absolute normal.

ANSWERS

1	2	3	4	5	6	7	8	9	10
D	D	C	C	A	D	D	B	A	C
11	**12**	**13**	**14**	**15**	**16**	**17**	**18**	**19**	**20**
D	C	C	D	B	B	D	D	A	C
21	**22**	**23**	**24**	**25**	**26**	**27**	**28**	**29**	**30**
D	C	B	D	C	B	A	A	B	B
31	**32**	**33**	**34**	**35**	**36**	**37**	**38**	**39**	**40**
B	D	A	B	A	D	B	D	B	A
41	**42**	**43**	**44**	**45**	**46**	**47**	**48**	**49**	**50**
D	B	C	B	D	B	D	D	D	A
51	**52**	**53**	**54**	**55**	**56**	**57**	**58**	**59**	**60**
C	C	A	D	D	D	A	D	A	D
61	**62**	**63**	**64**	**65**	**66**	**67**	**68**	**69**	**70**
A	D	A	A	D	C	C	D	B	C
71	**72**	**73**	**74**	**75**	**76**	**77**	**78**	**79**	**80**
A	C	A	A	A	D	A	A	B	A
81	**82**	**83**	**84**	**85**	**86**	**87**	**88**	**89**	**90**
A	C	A	A	D	A	B	A	A	C
91	**92**	**93**	**94**	**95**	**96**	**97**	**98**	**99**	**100**
B	A	A	C	D	A	C	B	A	A
101	**102**	**103**	**104**	**105**	**106**	**107**	**108**	**109**	**110**
D	A	D	A	D	C	D	D	B	A
111	**112**	**113**	**114**	**115**	**116**	**117**	**118**	**119**	**120**
C	C	A	B	D	A	C	B	B	D
121	**122**	**123**	**124**	**125**	**126**	**127**	**128**	**129**	**130**
B	B	C	C	C	A	D	C	B	D
131	**132**	**133**	**134**	**135**	**136**	**137**	**138**	**139**	**140**
C	C	B	D	A	A	B	A	D	A
141	**142**	**143**	**144**	**145**	**146**	**147**	**148**	**149**	**150**
D	B	C	C	B	D	B	B	D	A

EXPLANATORY ANSWERS

71.

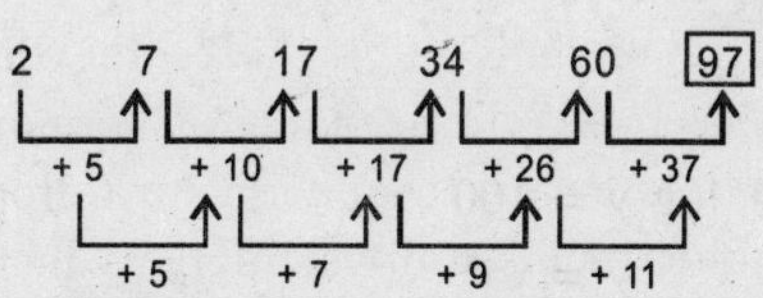

Therefore 97 will be the next number of the given series.

72. We have, 40 second, 72 second, 108 second

They again pop groundnuts in their mouth simultaneously after LCM (40 sec, 72 sec, 108 sec)

$\therefore$ LCM (40 sec, 72 sec, 108 sec)

= 1080 second = 18 minute

Time = 8:20:00 hours + 18 minute = 8:38:00 hours.

73.

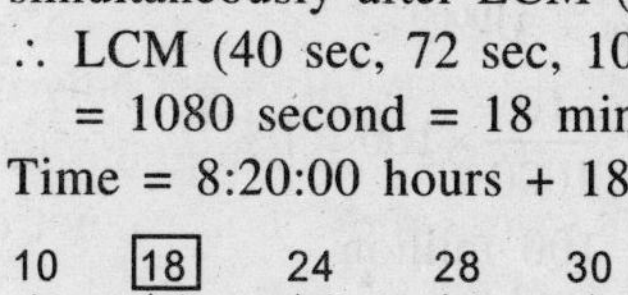

So '16' will be '18'.

Hence, 16 is odd number.

75. Value of investment after three years

$$= 25000\left(1+\frac{15}{100}\right)\left(1-\frac{15}{100}\right)\left(1+\frac{10}{100}\right)$$

$$= 25000 \times \frac{115}{100} \times \frac{85}{100} \times \frac{110}{100}$$

$$= 26881.25 \sim 26880$$

76.

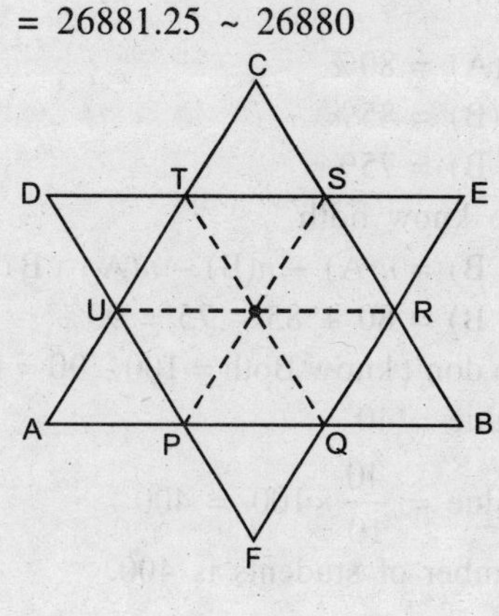

We have, Ar (Δ ABC) = Ar (Δ DEF) = 198 cm^2

There are 12 identical trangles, and 9 identical triangles of each triangle Δ ABC and Δ DEF

Out of 9 triangles we have taken 6 triangles

So, the area of region PQRSTU = $\frac{6}{9} \times$ Ar (Δ ABC)

$= \frac{6}{9} \times 198 \text{ cm}^2 = 132 \text{ cm}^2.$

77. We have, (A + B + C) pipes fill the pool in 6 hours

⇒ 3AM – 12 O'clock = 3 hours

In 3 hours (A + B + C) pipes fill the pool

$= \frac{3}{6} = \frac{1}{2}$ part

Rest part = $1 - \frac{1}{2} = \frac{1}{2}$ [This part fill by A + B)

⇒ 7AM – 3 AM = 4 hour

So, $\frac{1}{2}$ part fill by (A + B) = 4 hours

1 part fill by (A + B) = 8 hours

Now,

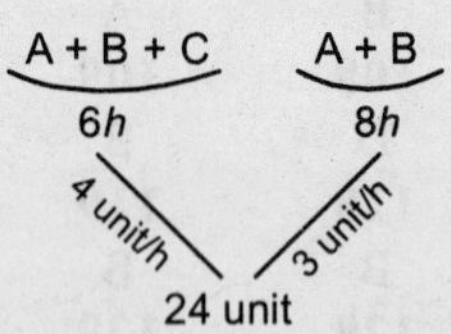

C efficiency = 4 – 3 = 1 unit/h

Time taken by 'C' = $\frac{24}{1}$ = 24 h

So the swimming pool will be filled in 24 h by 'C' only. Time in the clock will be 12 O'clock night.

78.

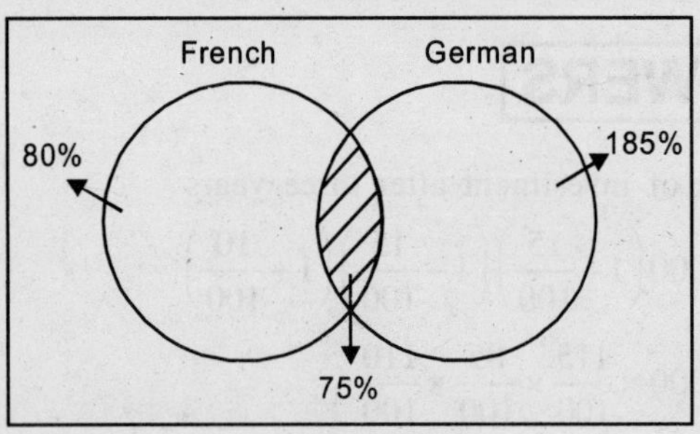

We have,

French $n(A)$ = 80%

German $n(B)$ = 85%

Both $n(A \cap B)$ = 75%

Total students who know both

$$n(A \cup B) = n(A) + n(B) - n(A \cap B)$$

$$n(A \cup B) = 80 + 85 - 75 = 90\%$$

These students who don't know Both = 100 – 90 = 10%

10% value = 40

$$100\% \text{ value} = \frac{40}{10} \times 100 = 400$$

Therefore total number of students is 400.

79. Let the original price was ₹ x

Then,

$$1000 = x\left(1+\frac{10}{100}\right)\left(1-\frac{10}{100}\right) = x \times \frac{110}{100} \times \frac{90}{100}$$

$$\therefore \quad x = \frac{1000 \times 100 \times 100}{110 \times 90} = 1010.10 \sim 1020.50.$$

81. Total run needed to win = 144 (in 18 overs)

Total run score by India

= (5 × 6 + 5 × 8 + 5 × 5) = 30 + 40 + 25 = 95

India score 95 run in 15 overs

Remaining run = 144 – 95 = 49

Remaining over = 18 – 15 = 3

Required average = $\frac{49}{3}$ = 16.33.

82. Let the present age of Ravi and his three sons are P, Q, R and S respectively.

Then, according to question,

P = Q + R + S ...(*i*)

After 12 years,

P + 12 = (Q + 12) + (R + 12)

P = (Q + R) + 12 ...(*ii*)

Put in eq (*i*) we get S = 12 years

Again 4 years later,

P + 16 = (Q + 16) + (S + 16)

P = Q + 28 [∵ S = 12 years] ...(*iii*)

Put in eq (*ii*), we get R = 16 years

Again 4 years after

P + 20 = (R + 20) + (S + 20)

P = (R + S) + 20

Put in eq (*i*) we get Q = 20 years

Put, Q = 20, R = 16, S = 12 in eq (*i*) we get

P = 20 + 16 + 12 = 48 years

So their ages are 48 years, 20 years, 16 years, 12 years respectively.

83. We have,

$$\log_x y = 100$$

$$\therefore \quad y = x^{100}$$

$$\text{and} \quad \log_3 x = 100$$

$$\therefore \quad x = 3^{100}$$

$$\text{So,} \quad y = (3^{100})^{100}$$

$$\therefore \quad y = 3^{10000}.$$

84. Rate of decreasing = $\frac{10}{1000} \times 100 = 1\%$

Present population = 100 million

Population after 3 year

$$= 100\left(1-\frac{1}{100}\right)\left(1-\frac{1}{100}\right)\left(1-\frac{1}{100}\right)$$

$$= 100 \times \frac{99}{100} \times \frac{99}{100} \times \frac{99}{100} \text{ million} = 97.029 \text{ million.}$$

85.

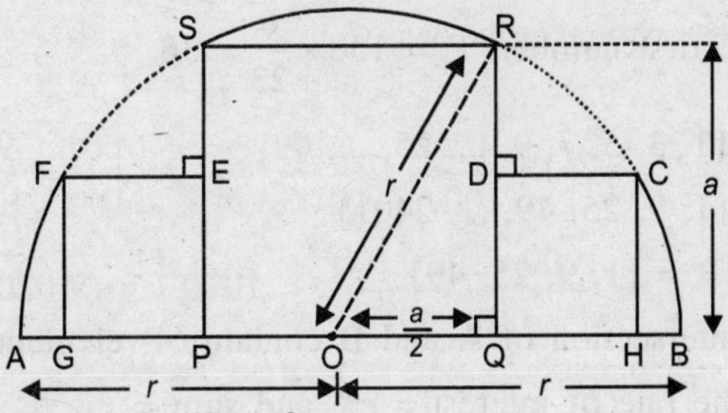

We have, AB = 2r
Radius of semicircle = 'r'
Let side of Bigger Square is 'a'
Then,
In ΔOQR,

$$OR^2 = OQ^2 + QR^2$$

$$r^2 = \frac{a^2}{4} + a^2$$

$$\frac{5a^2}{4} = r^2$$

$$\Rightarrow \quad a = \frac{2r}{\sqrt{5}}$$

Side of smaller square = $\frac{a}{2} = \frac{r}{\sqrt{5}}$

$\therefore$ DR = DC = $\frac{r}{\sqrt{5}}$

In ΔRDC,

$$RC^2 = RD^2 + DC^2$$

$$RC^2 = \left(\frac{r}{\sqrt{5}}\right)^2 + \left(\frac{r}{\sqrt{5}}\right)^2$$

$$RC^2 = \frac{2r^2}{5}$$

$$RC = \sqrt{\frac{2}{5}}\, r$$

$$\because \quad FS = RC$$

$$\therefore \quad FS + RC = 2r\sqrt{\frac{2}{5}}$$

86.

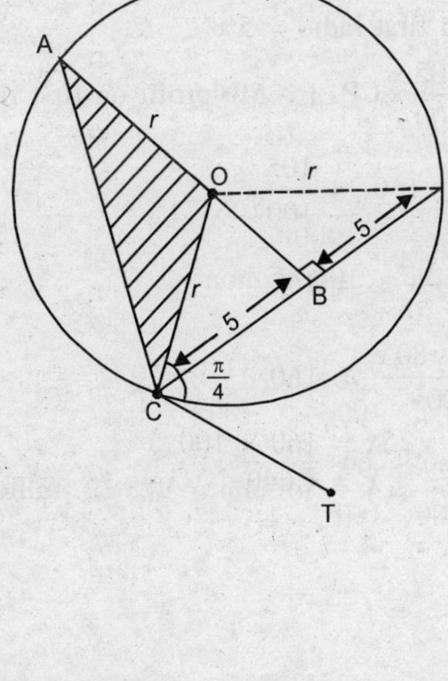

We have CD = 10 cm,

$$\angle DCT = \frac{\pi}{4}$$

$\therefore$ CB = BD = 5 cm and
OA = OC = OD ≡ r (radius)

$$\because \quad \angle OCT = \frac{\pi}{2}$$

$$\therefore \quad \angle OCB = \frac{\pi}{2} - \frac{\pi}{4} = \frac{\pi}{4}$$

ΔOCB,

$$\cos\frac{\pi}{4} = \frac{5}{r}$$

$$\frac{1}{\sqrt{2}} = \frac{5}{r}$$

$$r = 5\sqrt{2}$$

In ΔOBD,

$$OB^2 = OD^2 - BD^2$$

$$OB^2 = \left(5\sqrt{2}\right)^2 - (5)^2$$

$$OB^2 = 25$$

$$\Rightarrow \quad OB = 5$$

$$AB = AO(r) + OB = \left(5\sqrt{2} + 5\right)$$

In ΔABC,

$$AC^2 = AB^2 + BC^2$$

$$AC^2 = \left(5\sqrt{2} + 5\right)^2 + (5)^2$$

$$AC^2 = 50 + 25 + 50\sqrt{2} + 25$$

$$AC^2 = 100 + 50\sqrt{2} \sim 170$$

$$AC \sim 13$$

So, we get OA = $5\sqrt{2}$ ~ 7, OC = $5\sqrt{2}$ ~ 7, AC = 13
$\therefore$ Perimeter of ΔAOC = OA + OC + AC
= 7 + 7 + 13 = 27 cm

87. There are three vowels (A, O, E) in the given word A, O and E should be together, so we will assume that these three letters to be tied up with each other

Now we have 6 distinct things to be arranged

C P S T N [A O E]

6 letter

So, the number of
Arrangements = 6!

However, the place of A, O and E can be interchanged between themselves.

So, the total number of arrangement
= 6! × 3! = 4320

89. 35 people average age = 45 years

Average age of 10 most aged people = 45 × 120%
= 54 years

New average age of (35 + 3) = 38 people of the company

$= \dfrac{35 \times 45 + (32 + 35 + 30)}{38} = \dfrac{1575 + 97}{38} = \dfrac{1672}{38}$

= 44 years

Required difference = 54 – 44 = 10 years.

92. A → 10 copy | E → 2 copy
B → 3 copy | F → 2 copy
C → 3 copy | G → 1 copy
D → 2 copy | H → 1 copy

Total 24 copies

$$\frac{24!}{10!(3!)^2 \times (2!)^3 \times (1!)^2} = \frac{24}{10!(3!)^2 \times (2!)^3}.$$

93. We have $P^2 = 169$, where p and q whole number

There are two set of p, q value

(*a*) $(13)^2 = 169 \Rightarrow (p, q) = (13, 2)$

(*b*) $(169)^1 = 169 \Rightarrow (p, q) = (169, 1)$

So the value of $(p - 1)^{2+1} = (13 - 1)^{2+1} = (12)^3 = 1728$.

94. Net increase or decrease in a year

$= 5\% + (-0.25\%) + \dfrac{5\% \times (-0.25)}{100} = 5 - 0.25 - 0.0125$

= 5 – 0.2625 = 4.74% increase

Net increase in I and IInd years

$= 4.74 + 4.74 + \dfrac{4.74 \times 4.74}{100} = 9.48 + 0.22 = 9.70$

Now, net increase with IIIrd year

$= 9.7 + 4.74 + \dfrac{9.7 \times 4.74}{100} = 14.44 + 0.46$

= 14.9% ~ 14.93%.

95. $2^8 \times 3^6 \times 5^4 \times 10^5$

We can write this into prime factors

$2^8 \times 3^6 \times 5^4 \times (2 \times 5)^5$

$2^{13} \times 3^6 \times 5^9$

Prime factor of 120 = $2^3 \times 3^1 \times 5^1$

Multiples of 120 = $2^{10} \times 3^5 \times 5^8$

Total factor = (10 + 1)(5 + 1)(8 + 1) = 11 × 6 × 9 = 594.

96. We know, Fermat's Remainder Theorem $\dfrac{A^{p-1}}{p}$, remainder is 1.

We have, $\dfrac{2^{97}}{97} = \dfrac{2^{96}.2}{97} = \dfrac{2^{97-1}.2}{97} = 1 \times 2 = 2.$

98. We have I + II + III = 138

I : II = 3 : 4 = 6 : 8

II : III = 8 : 9

I : II : III = 3 : 4 : 8 : 9

24 : 32 : 36

I : II : III = 6 : 8 : 9

Then, first number 'I' = $138 \times \dfrac{6}{23} = 36$.

99. A = {1, 3, 5, 7, 9, ..., 25, ..., 49}

B = {1, 9, 25, 49, ..., 2401}

A ∩ B = {1, 9, 25, 49}

So, Intersection of A and B contain '4' elements.

100. Let the rate of interest = $r\%$ and sum = x

SI given to Bank = ₹ 2000

$$2000 = \frac{x \times 2 \times r}{100}$$

$$x \times r = 200000 \quad ...(i)$$

We know, 2 year CI and SI difference = $P\left(\dfrac{r}{100}\right)^2$

$$2400 - 2000 = \frac{xr^2}{10000}$$

Put $x \times r$ = 2,00,000

$$400 = \frac{r \times 200000}{10000}$$

$$r = 20\%.$$

102. Let the cost price of both the radio are ₹ x and ₹ y respectively

Then,

First Radio

CP = ₹ x

MP = $x + 200$

SP = $(x + 200) \times \dfrac{4}{5} = \dfrac{4}{5}x + 160$

Second Radio

CP = ₹ y

MP = $\dfrac{3}{2}y$

SP = $\left(\dfrac{3}{2}y - 160\right)$

Given, profit on first radio = 5%

SP = $\dfrac{105}{100} \times$ CP [∵ 5% profit on first radio]

$$\frac{4}{5}x + 160 = \frac{105}{100}x$$

$$\frac{105x}{100} - \frac{4}{5}x = 160$$

$$\frac{105x - 80x}{100} = 160$$

$$25x = 160 \times 100$$

$$x = ₹\ 640$$

Total CP = $x + y$

Total SP = $\frac{4}{5}x + 160 + \frac{3}{2}y - 160 = \frac{4}{5}x + \frac{3}{2}y$

$$SP = \frac{115}{100} \times CP \quad [\because \text{ 15\% profit on both radio}]$$

$$\frac{4}{5}x + \frac{3}{2}y = \frac{23}{20}(x+y)$$

$$\frac{4}{5} \times 640 + \frac{3}{2}y = \frac{23}{20}(640 + y) \quad [\because x = ₹\,640]$$

$$512 + \frac{3}{2}y = 736 + \frac{23}{20}y$$

$$\frac{3}{2}y - \frac{23}{20}y = 224$$

$$\frac{30y - 23y}{20} = 224$$

$$\frac{7y}{20} = 224$$

$$y = ₹\,640$$

CP of second radio = ₹ 640

SP of second radio = $\frac{3}{2}y - 160 = 960 - 160 = ₹\,800$

$$\%P = \frac{SP - CP}{CP} \times 100 = \frac{800 - 640}{640} \times 100 = \frac{160}{640} \times 100$$

$= 25\%$.

103. Total Employees = 900

Male = 25% = 225

Total Employees after hiring = 900 + 200 = 1100

Male employee in the factory = 1100 × 35% = 385

Additional male = 385 − 225 = 160

% male in the additional employee = $\frac{160}{200} \times 100 = 80\%$.

104. (*a*) Tariff = $\frac{40-45}{45} \times 100 = -11\frac{1}{9}\% \downarrow$

(*b*) Network = $\frac{35-30}{30} \times 100 = \frac{50}{3}\% = 16\frac{2}{3}\% \uparrow$

(*c*) Building = $\frac{25-20}{20} \times 100 = 25\% \uparrow$

(*d*) Handset = $\frac{20-25}{25} \times 100 = -20\% \downarrow$

We can see in Building category complaints are increased more than the other, so, Building performed the worst.

105. Complaints related with network in 2017

$= 2500 \times 35\% = 2500 \times \frac{35}{100} = 875.$

106. % Marine Engineering = $\frac{85}{360} \times 100 = 23.6\%$.

107. Painters *i.e.*, 25° value = 2160

Technicians *i.e.*, 40° value = $\frac{2160}{25} \times 40 = 3456.$

108. Let the total population of the city = x

Population engaged in corpentry = $\frac{115}{360} \times x = \frac{23}{72}x$

Population engaged in technicians = $\frac{40}{360} \times x = \frac{x}{9}$

Let total earnings from technicians = ₹ y

Then total earnings from carpentry = ₹ $2y$

$$\text{Required average} = \frac{\dfrac{2y}{\left(\frac{23}{72}x\right)}}{\dfrac{y}{\left(\frac{x}{9}\right)}} = \frac{2y \times 72 \times x}{23x \times y \times 9}$$

$$= \frac{16}{23} = 16 : 23$$

109. Population engaged on painters = $\frac{25}{360} \times x = \frac{5}{72}x$

Population engaged in technician = $\frac{x}{9}$

Let the total earnings of painter and technicians are ₹A and ₹B respectively

Then,

$$\frac{\dfrac{A}{\left(\frac{5}{72}x\right)}}{\dfrac{B}{\left(\frac{x}{9}\right)}} = \frac{2}{1}$$

$$\frac{A \times 72 \times x}{B \times 9 \times 5x} = \frac{2}{1}$$

$$\frac{A}{B} = \frac{10}{8} = 5 : 4$$

110. Now total wage earners = $110\%x = \frac{11}{10}x$ [10% ↑]

population engaged in carpentry (initially) = $\frac{23}{72}x$

population engaged in carpentry (next year)

$$= 115\%\left(\frac{23}{72}x\right) = \frac{529}{1440}x$$

Central Angle of carpenters in the next year

$$= \frac{\frac{529}{1440}x}{\left(\frac{11}{10}\right)x} \times 360° = \frac{529 \times 10 \times 360}{1440 \times 11} = 120.22 \sim 120°.$$

113. We have,

$$\frac{\text{Rahul}}{\text{Total}} = \frac{3}{5}, \quad \frac{\text{Santosh}}{\text{John}} = \frac{2}{1}$$

∵ We cannot Bridging the element of ratio, so we cannot calculate the Fraction of the land belongs to John.

117. I. $\log(2 + 4) = \log 6 = \log 3 + \log 2$

II. $\log 2 + \log 4 = \log 8 = 3 \log 2$

III. $\log(6 - 3) = \log 3$

IV. $\log 6 - \log 3 = \log 2$

So, $3 \log 2 > \log 3 + \log 2 > \log 3 > \log 2$

∴ Only II $\log 2 + \log 4$ is the greatest.

118. One leg population = 10000 × 5% = 500

Remaining population = 10,000 – 500 = 9500

$$\text{Barefoot population} = \frac{9500}{2} = 4750$$

Population who have two legs = 4750

4750, least shoes needed in the village.

119. The time when the watch was stop

= 15 : 30 – 4 hour = 11 : 30

Watch was correct at 12 : 00 noon

Watch loose 18 min/hour

Right time	Watch shows time
1 : 00 pm	12 : 42 (–18 min)
2 : 00 pm	12 : 24 (–18 min)
3 : 00 pm	12 : 06 (–18 min)
4 : 00 pm	11 : 48 (–18 min)
5 : 00 pm	11 : 30 (–18 min)

So, correct time at 11 : 30 = 17 : 00

Correct time of the watch = 17 : 00 + 4 hour = 21 : 00.

121. As, 645 = (6 – 4) × 5 = 10

316 = (3 – 1) × 6 = 12

Similarly,

765 = (7 – 6) × 5 = 5

So, 765 coded as 5.

122.

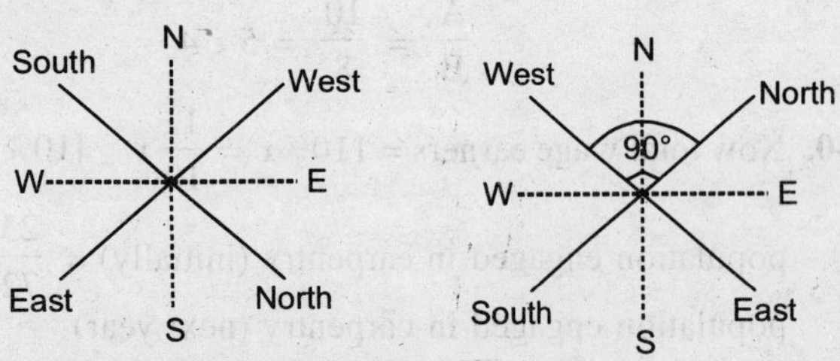

Map after 90° anticlockwise

South side ⇒ South West direction

124.

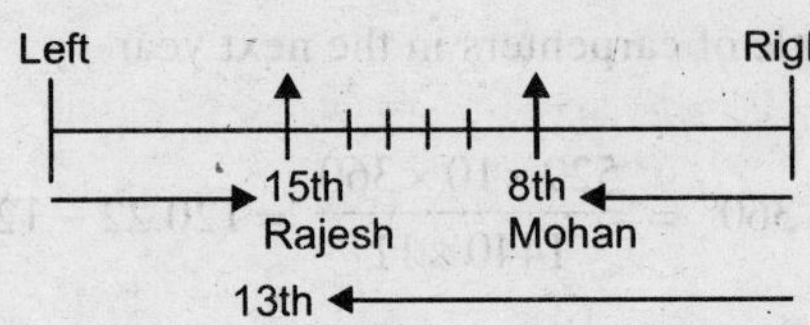

∵ 4 aspirants are sit between them

So, Rajesh is 13th from Right

Total = Left + Right – 1

= 15 + 13 – 1 = 27.

131.

9	4	7	6	5	8	2
–4↓	+1↓	–4↓	+1↓	–4↓	+1↓	+1↓
5	5	3	7	1	9	3

∴ Both 3 and 5 appears twice.

For Q.No. 133 to 138

Country	Friend's Name	Day
Germany	T	Monday
Japan	P	Tuesday
Australia	U	Wednesday
China	R	Thursday
Malaysia	Q	Friday
Nepal	V	Saturday
India	S	Sunday

For Q.No. 139 to 142

Input: adverb 59, 36 salient 81 idioms bakery 14 launch 47 umpire 62

Step I: umpire adverb 59 36 salient 81 idioms bakery launch 47 62 14

Step II: Idioms umpire adverb 59 salient 81 bakery launch 47 62 14 36

Step III: Idioms umpire adverb 59 salient 81 bakery launch 47 14 36 62

Step IV: 47 Idioms umpire adverb 59 81 bakery launch 14 36 62 salient

Step V: 59 47 Idioms umpire adverb 81 bakery 14 36 62 salient launch

Step VI: 81 59 47 Idioms umpire adverb 14 36 62 salient launch bakery

144.

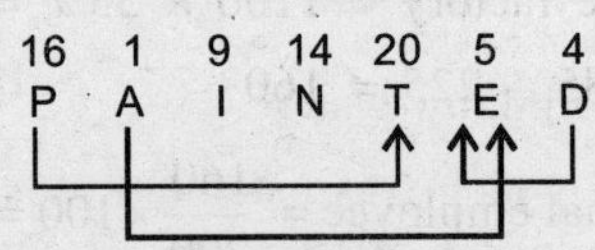

For Q.No. 145 to 149

Urban people prefer cars — ve fm ab eg

Profit for urban Areas — ab ep zi so

People demand for hike — zi qr cd ve

hike in profit margin — al jn ep cd

From eq (1) and eq (2) urban code is "ab"

From eq (1) and eq (3) people code is "ve"

From eq (3) and eq (4) hike code is "id"

From eq (2) and eq (3) for code is "zi"

From eq (3) demand code is "qr"

From eq (2) and eq (4) profit code is "ep"

From eq (2) Areas code is "so"

From eq (1) prefer/cars code is "ve/fm"

From eq (4) in/margin code is "jn/al"

Previous Paper (Solved)

Symbiosis National Aptitude Test (SNAP), 2016*

SECTION- I

GENERAL AWARENESS

1. Football club which has been promoted to the Premier League Season 2016-17, is ....

A. Burnley B. Real Madrid
C. FC Barcelona D. Liverpool F.G.

2. Who won the 2016 Nobel Prize for Literature?

A. Duncan Haldane B. Bob Dylan
C. Oliver Hart D. Ben Feringa

3. Who will take over as the new UN Secretary General?

A. Antonio Guterres B. Ban Ki-Moon
C. Antonio Hardy D. Simon Philip

4. Who beat Michael Phelops in Rio to claim Singapore's first ever Gold Medal in the Olympics?

A. Michael Shultz B. Joseph Schooling
C. Henery Walker D. James Hardy

5. Which country did Germany defeat to clinch the FIFA World Cup trophy in 2014?

A. Spain B. Brazil
C. Argentina D. Italy

6. The Kigali Agreement is related to ....

A. Phasing down the use of Hydroflurocarbons
B. Biosafety
C. Global Terrorism
D. Availability of Drinking water

7. The Indian company which was listed among the Top 50 global R&D spenders in 2015, is ....

A. Reliance B. Tata Motors
C. Hindustan Motors D. ONGC

8. Which tournament is known as Women's "Davis Cup"?

A. Nehru Cup B. Ezra Cup
C. Rovers Cup D. Fed Cup

9. Who retired in December 2015 as Chairperson of HSBC India?

A. Arundhati Bhattachary
B. Smriti Irani
C. Naina Lal Kidwai
D. Sujata Mehta

10. Which project has been launched in to make National Highways free of Railway Crossings?

A. Setu Bharatam B. Setu Nagaram
C. Setu Samudram D. None of these

11. Who is the first African-American whose portrait will appear on the new series US$20 currency notes?

A. Harriet Tubman B. Simon Tardy
C. Mitchell Weiser D. None of these

12. The Oscar 2016 Best Movie was ....

A. Moonlight B. Spotlight
C. Twenty Years Back D. No Mans Land

13. Who is the youngest cricketer to score 10000 test runs?

A. Ravindra Jadeja B. Mohammad Shami
C. Virat Kohli D. Alistair Cook

14. Indian scientist who was honored by the International Astronautical Federation's Hall of Fame award in 2016, is ....

A. A.P.J. Abdul Kalam B. Nilesh Kukreja
C. U.R. Rao D. Kiran Reddy

15. Who is the first Muslim mayor of London?

A. Sadiq Khan B. Gulam Ahmad
C. Gulam Rashul D. Tariq Khan

16. The organization which signed an agreement to acquire LinkedIn, is ....

A. Google B. Microsoft
C. Amazon D. Flipkart

17. Flipkart acquired which fashion portal in July 2016?

A. Jabong B. Zivame
C. Koovs D. Limeroad

18. Which operation was launched to evacuate Indians stranded in South Sudan in July 2016?

A. Rakshak B. Vikalp
C. Tarun D. Sankat Mochan

19. The Lata Mangeshkar Award for Lifetime Achievement in 2016 is given to ....

A. Uttam Singh B. Anu Malik
C. Rajesh Roshan D. Udit Narayan

20. The world's largest solar power plant at a single location is in ....

* Based on Memory.

A. Karnataka B. Tamil Nadu
C. Madhya Pradesh D. Haryana

21. The name of Google's instant messenger is ....
A. Milly B. Jolly
C. Allo D. Timmy

22. Which organization took control of Time Warner Cable?
A. AB Communications
B. Warner Communications
C. Simon Communications
D. Charter Communications

23. The current Vice President of USA is ....
A. Joe Biden B. Michael Wrest
C. Admond Walker D. None of these

24. Which technology company did Dell Inc, acquire in 2016 and create the largest technology merger in history?
A. IBM B. EMC Corp
C. Microsoft D. Foxconn

25. Which one is correctly matched.
A. Janet Yellen – Chair of US Fed Reserve
B. Urjit Patel – Chairman, SEBI
C. Mark Carney – President, European Central Bank
D. Mario Draghi – Governor, Bank of England

26. The Paralympian who was sentenced to six and half years of imprisonment in 2016 for the murder of his girlfriends was ....
A. Oscar Pistorius B. Chris Clarke
C. Richard Chiassaro D. Mickey Bushell

27. The mini satellite developed by engineering students of Pune, and launched by ISRO, is ....
A. Param B. Swayam
C. Kalamsat D. Hamsafar

28. Which animal carried out an attack at Disneyland, killing a 2 years old boy?
A. Bear B. Tiger
C. Alligator D. Elephant

29. British MP was shot to death in 2016, was ....
A. Gerry Adams B. Robert Adley
C. Nick Ainger D. Jo Cox

30. The most realistic humanoid robot did China unveil recently?
A. Jia Jia B. Asimo
C. Petman D. Nao

SECTION-II

GENERAL ENGLISH

Directions (Qs. 31-34): *Fill in the blanks with suitable words:*

31. Last year our PM .... the festival with troops.
A. Expand B. Spent
C. Went D. Restrained

32. Marine algae .... a key role in ocean ecology.
A. Sports B. Draws
C. Plays D. Tries

33. The drone rose even ....
A. Lower B. Higher
C. Wider D. Deeper

34. The students were awaiting .... results.
A. a B. an
C. the D. None of these

Directions (Qs. 35 and 36): *Select the Synonyms of the underlined words:*

35. Rich merchant was loved for his <u>magnanimous</u> attitude.
A. Miserly B. Callous
C. Seductive D. Generous

36. Father had little cause to <u>regret</u> dot com boom to doom.
A. Refuse B. Regress
C. Rue D. Relish

Directions (Qs. 37-39): *Select the correct options:*

37. Doubtful about making a judgement:
A. Vacillating B. Agnostic
C. Incisive D. Decoy

38. Deja vu
A. See off B. Never seen
C. See back D. Already seen

39. Sunflower, moonlight etc. are .... words.
A. Single B. Blended
C. Compound D. Complex

Directions (Qs. 40-42): *Select the Antonyms of the given words:*

40. NADIR
A. Zero B. Zenith
C. Villain D. Nasty

41. ABSTRUSE
A. Occult B. Abstract
C. Overt D. Obscure

42. EBULLIENT
A. Silent B. Jovial
C. Candid D. Blunt

Directions (Qs. 43 and 44): *Fill in the blanks with right combination of words:*

43. Please .... me so with your valuable ....

A. Tell, Story B. Ask, Questions
C. Advise, Advice D. Advise, Mind

44. .... the police approaching to ....

A. Eluding, Allude B. Alerting, Flee
C. Alluding, Elude D. Fleeing, Alert

Directions (Qs. 45 and 46): *Select the right combination of words for the meanings of the given pair of words:*

45. DISCRETE/DISCREET

A. Different/Separate B. Distinct/Subtle
C. Disgust/Disdain D. Disgrace/Dishonest

46. INGENIOUS/INGENUOUS

A. Clever/Innocent B. Candid/Bright
C. Crafty/Clumsy D. Naive/Shrewd

Directions (Qs. 47-49): *Select the nearest opposites of the given words:*

47. GREGARIOUS

A. Greedy B. Generous
C. Reserved D. Smart

48. BRUSQUE

A. Agile B. Agog
C. Alert D. Affable

49. LUCID

A. Distinct B. Explicit
C. Obscure D. Limpid

Directions (Qs. 50-53): *Select the right option to answer the questions:*

50. Achilles' heel meaning?

A. Highpoint B. Vulnerable point
C. Low point D. Recovery point

51. Wealthy : Indigent : : Graceful : ?

A. Gauche B. Doughty
C. Lofty D. Churlish

52. Almond : Nut : : Cheddar : ?

A. Fruit B. Vegetable
C. Drink D. Cheese

53. Chaff : Wheat : : Dregs : ?

A. Rice B. Pulse
C. Wine D. Cloth

Directions (Qs. 54 and 55): *Each of these questions has a group of sentences marked 1, 2, 3, 4 and 5. Arrange these to form a logical sequence.*

54. 1. But as emotions give way to cold calculations and analyses, two points are emerging powerfully.
2. One, there's no reason for India to be paranoid as the economics of outsourcing is too compelling for American companies and the US government can do little to stem it.
3. Two, Indian companies should increase their employee strength in the US with more locals, a strategy that some companies have already said they would adopt.
4. This will help to calm American nerves, and also eventually make Indian companies a more global force with a better understanding of the US market.
5. India reacted strongly to the US increase in work visa costs and the Ohio governor's statements against offshoring.

A. 4-5-2-1-3 B. 1-2-5-3-4
C. 5-1-2-3-4 D. 1-2-3-4-5

55. 1. It occurs widely in the tropics, including continental USA and increasingly in southern China and almost 100 other countries.
2. Unlike malaria, dengue is just as prevalent in the urban districts of its range as in rural areas.
3. It is also known as 'breakbone fever', since it can be extremely painful.
4. Dengue fever and dengue hemorrhagic fever (DHF) are acute febrile diseases transmitted by mosquitoes, which occur in the tropics, can be life-threatening, and are caused by four closely related virus serotypes of the genus *Flavivirus,* family Flaviviridae.
5. The mosquitoes that spread dengue usually bite at dusk and dawn but may bite at any time during the day, especially indoors, in shady areas, or when the weather is cloudy.

A. 4-3-1-2-5 B. 4-2-5-3-1
C. 5-4-3-2-1 D. 3-4-5-1-2

Directions (Qs. 56-67): *Read the passage and answer the questions that follow each passage:*

Passage-I

The Indian Economy is expected to grow by over eight per cent per annum until 2020. It can become the second largest in the world, ahead of the United States, by 2050, and the third largest after China and the US by 2032. However, Industry experts believe this target would be difficult to achieve without the active participation of the Small and Medium Enterprises (SMEs). At present, the SME sector contributes around 15 to 16 per cent to India's GDP and is expected to touch a figure of 22 per cent in the next three years. However, a lot is expected and much needs to be done if the figure needs to be achieved. Nearly 60 per cent of the SMEs in India fall in the unorganized sector. Once this untapped potential becomes the source for growth of these units, the size of Indian GDP can surpass that of developed nations. The SME sector at times fails on achieving its full potential due to lack of direction. A small entrepreneur faces several difficulties including access to finance to expand their operation. In addition, the loans that they get are not on preferential rates but are available around 15-18 per cent rate of interest which adds to their cost of operation. In several Asian countries, loans are available on interest as low as three per cent annually. India should take such steps. Small Industries Development Bank of India (SIDBI) offers loans to SMEs without collateral to the tune of around ₹ 1 crore. However, many are of the opinion that this limit needs to be increased to ₹ 2.5 crore. The government is also taking several steps to boost manufacturing sector growth as it would lead to job creation and boost the country's economy. At present the SMEs share in the manufacturing sector is just to the tune of around seven per cent. This is a very important sector and creates lot of employment opportunities both directly and indirectly. The Central government on its part has also set up National Manufacturing Competitiveness Council to suggest ways to enhance competitiveness in the manufacturing sector. The government has already announced a National Manufacturing Policy that aims at raising the share of manufacturing to 25 per cent of GDP by 2022.

56. What is the target of the Indian Economy?
A. Active participation of SME
B. Organization of SME
C. Growth by 8% per annum until 2020
D. Decrease the interest rate on loan

57. The growth of manufacturing sector would
A. boost the country's economy
B. surpass the GDP of developed nations
C. fail on achieving its full potential
D. set up national manufacturing competitiveness Council

58. According to the passage, SME sector
A. doesn't lead to job creation
B. plays a vital role in the growth of Indian economy
C. is a setback to manufacturing sector
D. provides loans on preferential rates

59. What step can be taken by Indian government to enhance the potential of Indian economy?
A. Provide loans on interest as low as 3% per year
B. Shift focus to other sectors
C. Decrease competitiveness in manufacturing sector
D. Ban small entrepreneurship

Passage-II

Sugar lowering drugs prescribed to patients with diabetes may pose an increased risk of heart failure in these patients, a comprehensive study examining 95,000 patients has claimed. Dr. Jacob Udell, the study's principal investigator states; patients randomized to new or more intensive blood sugar lowering drugs or strategies to manage diabetes showed an overall 14% increased risk for heart failure. This increased risk was directly associated with the type of diabetes therapy that was chosen, with some drugs more likely to cause heart failure than others, compared with placebo or standard care. While some drugs showed an increased risk, other strategies tested, such as intensive weight loss to control blood sugar, showed a trend towards a lower risk for heart failure.

Overall, the study found that for every one kilogram of weight gain attributed to a sugar lowering diabetes medication or strategy, there was an associated 7% increased risk of heart failure directly linked to that medication or strategy. The results of this study could prove to be the catalyst for how diabetes patients at risk for heart disease are managed moving forward. The study reviewed and analyzed data from 14 trails involving 95,502 patients. Four per cent of patients suffered a heart failure event while being followed in the individual trails, while 9.8 of patients in the study endured a major adverse cardiovascular event. The relative increase in the risk of heart failure outweighed a 5% reduction in heart attacks. The numbers show that for about every 200 patients treated, one additional hospital admission for heart failure was seen after an average of 4 years of follow-up. Heart failure events are common occurrences for patients with type 2 diabetes and have a major impact on one's life expectancy and quality of life as well as representing a major driver of healthcare hosts.

60. What primary issue is being discussed in the passage?
A. Types of diabetes therapy & drugs
B. Intensive weight loss in patients
C. Treatment of 92,502 patients
D. Increasing risk of cardiovascular event due to diabetic drugs

61. Who is the principal investigator of the study?
A. The Lancet Diabetes & Endocrinology
B. Dr. Michael Farkouh
C. Dr. Jacob Udell
D. Dr. Barry Rubin

62. Sugar controlling drugs can increase the risk of heart failure by ....
A. 9.8% B. 14%
C. 5% D. 7%

63. According to the study, the increased cardiovascular risk is directly associated with
A. Blood sugar level
B. Diabetes drugs and therapy
C. Weight gain attribution
D. None of these

Passage-III

Antibiotic resistance data from four big hospitals is being studied and analysed for the first time in India to help forecast drug resistance patterns that may surface in future. The Indian Council for Medical Research (ICMR) has been on the resistance trail since last year. It had set up six nodal centres at four hospitals—two at PGI, Chandigarh, and Christian Medical College, Vellore, and one each at All India Institute of Medical Sciences (AIIMS) in Delhi and JIPMER, Puducherry. The ICMR's antimicrobial surveillance and research network has been tracking resistance pattern of six pathogens causing diarrhoea, typhoid, fungal infections and infections that patients acquire in hospitals. "We initiated these projects in early 2014 and we are now analysing the first year's data. We did not have a nationally representative data on pathogens that cause 80% of infections. It is important to get a handle on what is the drug resistance prevalence in these six pathogenic groups," said Dr. Kamini Walia from the division of Epidemiology and communicable diseases, ICMR. The data will be used to forecast the drug resistance pattern to guide clinicians, she added. There is alarm over rising resistance to antibiotics in India, largely attributed to their indiscriminate use to treat routine infections like cough, cold and fever and their over-the-counter availability. Using antibiotics inappropriately, such as stopping the dosage mid-course has added to the resistance. There are broad estimates and observations so far—resistance to carbapenem (the strongest class of antibiotic) is 50-60% and in the case of acenotobacter (a hospital bug) seen among ICU patients, the resistance is 70-80%, however, there is no accurate resistance data, with most hospitals not reporting such figures, fearing an impact on patient's footfall. "The project would help design a standard treatment guideline and physicians will be able to decide what antibiotic should be given," Dr. Walia said. The surveillance and research network will expand to 15 more hospitals across the country, which will feed their data into these centres that will in turn map the resistance pattern. "In India, there are no statistics. Anyone who comes to the clinic or OPD is not tested for microbiological examination. A test is conducted only in the case of in-patients," V Balaji, Head of Microbiology, CMC, Vellore said. Though there is an antibiotic policy in place since 2012, officials said that surveys have revealed the compliance was in only a few hospitals.

64. The study conducted in various hospitals will
- A. divide epidemiology and communicable diseases
- B. be used to forecast the drug-resistance patterns
- C. help to set up six nodal centres
- D. be used for microbiological examination

65. The study on antibiotic resistance was initiated in ....
- A. 2011
- B. 2014
- C. 2010
- D. 2012

66. Which of the following statements is not true?
- I. The project would help design a standard treatment guideline.
- II. The research will cover 15 more hospitals in India.
- III. There is no antibiotic policy in India
- IV. Rising resistance to antibiotics is a matter of concern in India.

- A. I & II are not true
- B. III & IV are not true
- C. Only I is not true
- D. Only III is not true

67. Which disease is not covered in the research?
- A. Diarrhoea
- B. Heart disease
- C. Typhoid
- D. Fungal infections

Directions (Qs. 68-70): *Identify the best way of writing the sentence in the context of correct usage of standard of written english.*

68.
- A. John said that he will visit me now.
- B. John said that he will visit me then.
- C. John said that he would visit me now.
- D. John said that he would visit me then.

69.
- A. Excess knowledge is a dangerous thing.
- B. Excessive knowledge is a dangerous thing.
- C. A little knowledge is a dangerous thing.
- D. A less knowledge is a dangerous thing.

70.
- A. His house consists three bed rooms and a toilet.
- B. His house comprises of three bed rooms and a toilet.
- C. His house comprises three bed rooms and a toilet.
- D. His house consists with three bed rooms and a toilet.

SECTION-III

QUANTITATIVE, DATA INTERPRETATION AND DATA SUFFICIENCY

71. When the equation $\{(x+y)^3-(x-y)^3-6y(x^2-y^2)\}\times y$ is simplified, it gives the expression:
- A. $6y^2$
- B. $6y^3$
- C. $8y^2$
- D. None of the above

72. The sum of two numbers is 216 and their HCF is 27. These numbers are:
- A. 60 and 90
- B. 81 and 135
- C. 64 and 128
- D. 30 and 84

73. In the figure given below AB || CD, a° would be:

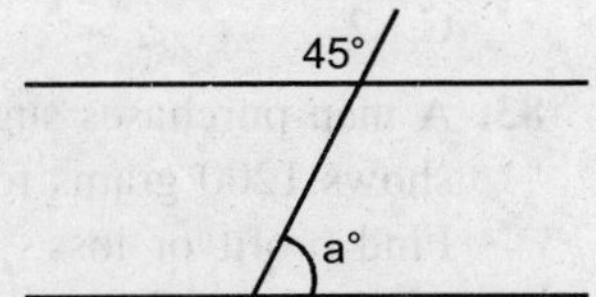

- A. 140°
- B. 130°
- C. 145°
- D. 135°

74. The H.C.F. and L.C.M. of two numbers are 50 and 250. On dividing one of these numbers by 2, 50 is obtained as quotient. The numbers are:
- A. 100, 125
- B. 80, 100
- C. 125, 200
- D. 200, 250

75. A number of dogs get together and decided to kill between the 999919 cats. Each dog killed equal number of cats. How many dogs were there?

A. 919 B. 996
C. 989 D. None of these

76. The circumference of the front wheels of a vehicle is 3.5 metres and that of the back wheels is 3 metres. If the vehicle is moving at a speed of 54 km/hr, the shortest time in which both the wheels will make a whole number of turns is:

A. 6.4 seconds B. 4 seconds
C. 2.1 seconds D. 1.4 seconds

77. A started to chase B, when B was 27 steps ahead of A. After how many steps did A catch up with B?

A. 25 B. 28
C. 30 D. 24

78. The capacity of a cuboidal tank is 50,000 litres. Find the breadth of the tank if the length and depth are respectively 2.5 m and 10 m

A. 2 m B. 3 m
C. 4 m D. 5 m

79. Three bells ring respectively at an interval of 15 seconds, 20 seconds and 24 seconds. If they ring continuously for 12 minutes, then how many times during this period, will they ring together?

A. 2 times B. 6 times
C. 5 times D. 3 times

80. In a software company average age of all female employees is 21 years and male average is 32 years and average of group is 28. What can be total number of people?

A. 240 B. 225
C. 231 D. 245

81. There are 12 balls in which some are red. If 3 red balls were added the probability of selecting red balls doubled. Find the number of red balls

A. 3 B. 4
C. 5 D. None of these

82. log tan 1° + log tan 2° + ... + log tan 89° = ?

A. 0 B. 1
C. 2 D. 3

83. A man purchases sugar on a weighing machine which shows 1200 grams for 1 kg and sells at 20% mark up. Find profit or loss

A. 20% profit B. 30% loss
C. No profit no loss D. 40% profit

84. Ram is two times as efficient as Shyam. Shyam can complete a Job assignment in 15 days. For how many days Ram and Shyam work together?

A. 2 days B. 3 days
C. 4 days D. 5 days

85. If unit digit of 129 × 206 × 47$ × 1484 is 2. What is the value of $?

A. 5 B. 7
C. 3 D. 4

86. Everybody in a room shakes hands with everybody else. The total number of hand shakes is 66. The total number of person in the room is:

A. 11 B. 12
C. 8 D. 14

87. Rohan speaks truth 60% of the cases and Ramesh speaks truth in 80% of the cases. The probability that they will say the same thing which could be a truth or could be a lie while describing a single event would be:

A. 0.42 B. 0.48
C. 0.56 D. 0.65

88. A clock which is running 12 minutes late on a Sunday noon is 16 minutes ahead of the time at 12 noon on the next Sunday. When is the clock 8 minutes ahead of the time?

A. Monday noon
B. Tuesday noon
C. Friday noon
D. Saturday noon

89. In a triangle ABC, the measure of an angle A is 50° and the external bisectors of angle B and angle C meet at O. The measure of angle BOC is:

A. 140° B. 40°
C. 105° D. 65°

90. The salary of Raju and Ram is 20% and 30% less than the salary of Saroj respectively. By what per cent is the salary of Raju more than the salary of Ram?

A. 33.33% B. 50%
C. 15.18% D. 14.28%

91. A man sold $\frac{3}{5}$ of his articles at a gain of 20% and the remaining at cost price. Find the gain earned in the transaction.

A. 8 B. 10
C. 12 D. 14

92. The sum of all two digit numbers that give a remainder 2 when they are divided by 7 is:

A. 552 B. 654
C. 658 D. 684

93. A person has a bag which contains 9 bulbs out of which 2 are fused and can not be used to lighten the room. Two bulbs are selected at random. What is the probability that all the two bulbs chosen can be used to lighten the room?

A. $\frac{5}{12}$ B. $\frac{7}{12}$

C. $\frac{9}{12}$ D. $\frac{10}{12}$

94. Consider the equation

$y = (5!)\ 5! + (10!)\ 10! + (50!)\ 50! + (100!)\ 100!$

The number of zeros at the end of the whole expression would be:

A. 120 B. 125

C. 130 D. 110

95. A bouncing ball rebounds to a height equal to one half of height of the previous bounce each time. Find the total distance travelled when it hits the ground for the 8th time

A. 45.25 B. 48.45

C. 47.75 D. 44.85

96. There are 9 humans in a ship, each human has 9 cages and each cage has 9 huge lions and each lion has 9 cubs. How many legs are there in the ship?

(Human have two legs, lions have four legs, cubs have four legs)

A. 747 B. 3258

C. 29178 D. 26561

97. If $2^x + 2^{x+1} = 48$, then the value of x^x is equal to:

A. 04 B. 64

C. 256 D. 16

98. A dishonest seller sells his grocery items using a false weight and thus gains 5% for a kilogram, he uses the weight of approximately:

A. 940.251

B. 943.121

C. 948.238

D. 952.381

99. If a man walks from A to B and then takes a bike from B to A, he takes 5 hours. How long would he take to walk from A to B and then back from B to A if he saves time 3 hrs and rests at B for 30 minutes?

A. 6.5 hrs B. 5.5 hrs

C. 8.5 hrs D. 7.5 hrs

100. A merchant marks up his goods by 80% and gives discount of 25%. What is the net profit of the merchant?

A. 40% B. 50%

C. 35% D. 45%

101. The value of $\log_{\left(-\frac{1}{4}\right)} 256$ is equal to:

A. 64 B. 16

C. –4 D. None of these

102. If 60% of $(x - y)$ = 40% of $(x + y)$, then what per cent of x is y?

A. 20% B. 25%

C. 22% D. None of these

103. $\left(1-\frac{1}{2}\right)\left(1-\frac{1}{3}\right)\left(1-\frac{1}{4}\right)\left(1-\frac{1}{5}\right)\ldots\ldots\ldots\left(1-\frac{1}{40}\right)=\frac{x}{40}$

What is the value of x?

A. 13 B. 39

C. 21 D. 1

104. If $\left(\frac{a}{b}\right)^{2x-3} = \left(\frac{b}{a}\right)^{x-9}$ then what is the value of x?

A. –6 B. –2

C. 12 D. 4

105. There are 30 railway stations between A and B, which includes A and B. If 30 tickets were issued for each station, then number of tickets issued at station A is:

A. 480 B. 770

C. 870 D. 680

106. There are 6 varieties of fruits. A box can contain minimum 10 and maximum 15 fruits. The maximum number of fruits can be put in 15 boxes are:

A. 375 B. 400

C. 360 D. 450

107. The HCF of two expression P and Q is 1. The LCM is:

A. $p + q$ B. $\frac{P}{Q}$

C. PQ D. P – Q

108. The number of triangles whose vertices are the vertices of an octagon but none of the sides of such triangles are taken from the sides of the octagon is:

A. 12 B. 16

C. 18 D. 14

109. One side of an equilateral triangle is 24 cm. The mid points of its sides are joined to form another triangle whose mid points are in turn joined to form still another triangle. This process continues indefinitely. Find the sum of the perimeters of all the triangles.

A. 144 cm

B. 72 cm

C. 536 cm

D. 676 cm

110. A and B entered in to a partnership with capital investment in the ratio 4 : 5. After 3 months A withdrew $\frac{1}{4}$ of his capital and B withdrew $\frac{1}{5}$ of his capital. The gain at the end of 10 months was ₹ 760. The share of profit A is:

A. ₹ 220 B. ₹ 330

C. ₹ 340 D. ₹ 290

SECTION-IV

ANALYTICAL & LOGICAL REASONING

111. Brijesh, Jayesh, Amar and Praveer are playing a game of cards. Amar is to the right of Jayesh who is to the right of Brajesh. Who is to the right of Amar?

A. Brijesh B. Praveer
C. Brijesh or Praveer D. Jayesh

112. Six friends A, B, C, D, E and F are standing in a circle. B is between F and C; A is between E and D; F is to the left of D. Who is between A and F?

A. C B. B
C. D D. E

Directions (Q. 113) : *P, Q, R and S are standing on four corners of a square piece of plot as shown in the given figure. They start moving, and the movements are explained in each of the questions. Read the question and select the right alternative.*

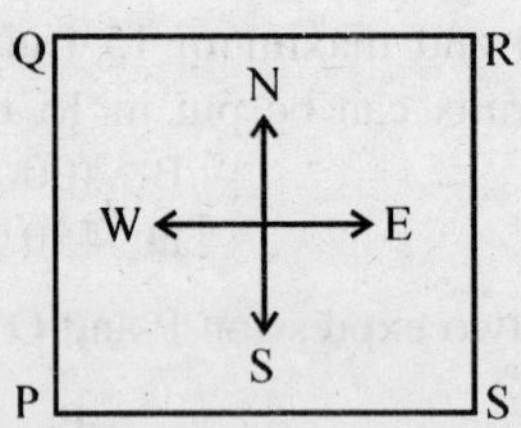

113. P and S walk one and a half length of the side clockwise and anticlockwise respectively. Which one of the following statements is true?

A. P is at midpoint between Q and R and S at the corner originally occupied by P
B. P and S are both at the midpoint between R and S
C. P and S are both at the midpoint between Q and R
D. S is at midpoint between Q and R and P is at the midpoint between original side of R and S

114. Rahim started from point X and walked straight 5 km West, then turned left and walked straight 2 km and again turned left and walked straight 7 km. In which direction is he from point X?

A. North-East B. South-West
C. South-East D. North-West

115. Five persons A, B, C, D and E are sitting in a row facing you such that D is on the left of C and B is on the right of E. A is on the right of C and B is on the left of D. If E occupies a corner position, then who is sitting in the centre?

A. A B. B
C. C D. D

116. Pointing to Snehal, Mahesh said, "Her mother's only daughter is my daughter". How is Mahesh related to Snehal?

A. Brother B. Uncle
C. Son D. Father

117. Is D brother of F? To find out the answer which of the following informations given in the statements (*a*) and (*b*) is/are sufficient?

(*a*) B has two sons of which F is one.
(*b*) D's mother is married to B.

A. Both (*a*) and (*b*) together are needed
B. Both (*a*) and (*b*) together are not sufficient
C. Only (*a*) is sufficient
D. Only (*b*) is sufficient

118. If M + N means 'M' is the brother of 'N', M – N means 'M' is the sister of 'N', M × N means 'M' is the mother of 'N' and M ÷ N means 'M' is the father of 'N'. Which of the following would mean 'E' is the paternal Aunt of 'F'?

A. E – G ÷ F B. E + G × F
C. E × F – G D. F × G + E

119. A train started from station 'A' and proceeded towards station 'B' at a speed of 48 km/hr. Forty-five minutes later another train started from station 'B' and proceeded towards station 'A' at 50 km/hr. If the distance between the two stations is 232 km, at what distance from station 'A' will the trains meet?

A. 132 km B. 144 km
C. 108 km D. 160 km

120. A toy cube is painted orange on all sides. It is cut into 64 smaller cubes of equal size. How many smaller cubes are not painted at all?

A. 4 B. 8
C. 16 D. 20

121. What will be the number of dots on the top of the following dice, if it is placed as in the second figure?

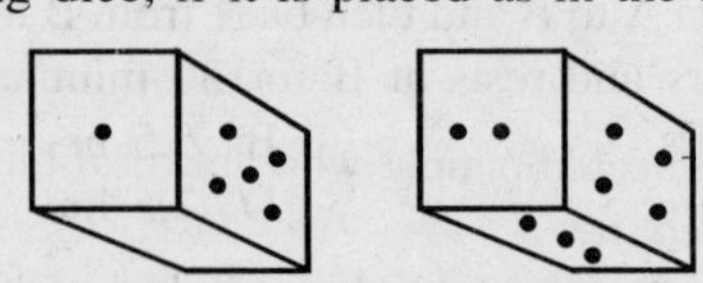

A. 5 B. 1
C. 6 D. Can't be said

Directions (Qs 122 and 123): *Each of the questions below consist of a question and two statements numbered I and II given below it. You have to decide whether data provided in the statements are sufficient to answer the question. Read both the statements and give answer—*

A. if the data in statement I alone are sufficient to answer the question, while the data in statement II alone are not sufficient to answer the question.

B. if the data in statement II alone are sufficient to answer the question, while the data in statement I alone are not sufficient to answer the question.
C. if the data even in both statements I and II together are not sufficient to answer the question.
D. if the data in both statements I and II together are necessary to answer the question.

122. Ramesh, Nanda and Rohit together wanted to meet on Wednesday afternoon between 12.00 Noon to 5.00 p.m. At what time can they meet?
I. Ramesh has his tuitions between 12.00 Noon to 5.00 p.m. every day except on Wednesday and Saturday.
II. Nanda goes to swimming between 2.00 p.m. to 4.00 p.m. on all days except Sunday and Rohit attends craft class between 3.00 p.m. to 5.00 p.m. on all days except Thursday.

123. Who among P, Q, R and S is the shortest?
I. L is not as tall as P or R but is taller than S and Q, Q being shorter than R.
II. S is shorter than R who is shorter than P who is taller than Q.

Directions (Qs. 124-126): *The questions are based on the following diagram in which the circle stands for educated, the square stands for lazy, the triangle stands for rural and the rectangle stands for corrupt people. Different regions in the diagram are numbered. Study the diagram and choose the answer of each question from the given alternatives.*

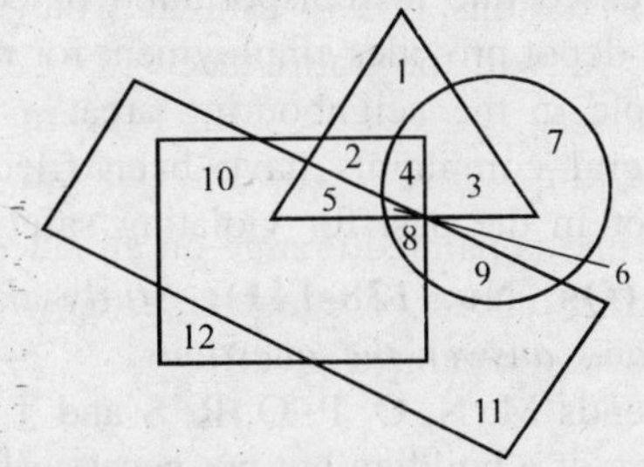

124. Non-educated rural people who are lazy but not corrupt are indicated by
A. 1 B. 2
C. 5 D. 4

125. Lazy educated people who are rural but not corrupt are indicated by
A. 3 B. 4
C. 2 D. 6

126. Non-rural educated people who are corrupt but not lazy are indicated by
A. 9 B. 8
C. 10 D. 6

127. Mohsin is carrying 8 fruits .... orange, mango, pineapple strawberry, guava, blueberry, apples and watermelon to the farmhouse. However, he has only two boxes of equal weight and each will carry same number of fruits. Therefore, he has to ensure:
(*a*) Orange and pineapple will be together, but not with watermelon.
(*b*) Mango and strawberry cannot be with pineapple.
(*c*) Blueberry will be in the same box with orange and guava but not with the apple. One more fruit will be with blueberry.

If Apple and watermelon are in the same box which are the other two fruits in the same box?
A. Strawberry and pineapple
B. Mango and guava
C. Mango and strawberry
D. indeterminent

128. Seven army recruits Laxman, Mahendar, Narendar, Omprakash, Preetam, Qutub and Raman are standing in a column for physical inspection in order of height, shorter first. Omprakash is standing between Laxman and Qutub. Narendar is taller than Laxman. Raman is shorter than Narendar. Mahendar and Preetam are not the two tallest. Preetam is two places ahead of Qutub. Which of the following is true?
A. Narendar is tallest
B. Qutub is taller than Laxman
C. Raman is ahead of Omprakash
D. Mahendar is shortest

129. In a small organization, Saurav and Kartik plays Tennis and Basketball very well. Manpreet and Lokesh are very good in Athletics, Tennis and Swimming. Kartik and Nikhil are good in Athletics and Badminton. Ganesh and Lokesh are good in Badminton and Hockey. Prakash and Rohit are good in Hockey and Basketball. Ganesh, Prakash, and Saurav are good in swimming. Manpreet and Nikhil also play Hockey well. If an inter-organizational event is arranged and only one person can be selected from one organization, who will be selected from this organization if Tennis, Swimming and Badminton has to be played in the competition?
A. Saurav B. Kartik
C. Manpreet D. Lokesh

130. One day five friends Amar, Bittu, Chintan, Darshan and Ehsaan having bikes Pulsor, Splendor, Apache, Unicorn and Hayabusa, but not in the same order, decided to race. Chintan came second last and was not driving Pulsor or Hayabusa. Ehsaan came just after Amar and was driving Unicorn. The person who came last was driving Apache. Darshan did not come last and was driving Hayabusa. Amar had neither came first nor last and was not driving Apache or Splendor. Who was driving Splendor?
A. Darshan B. Amar
C. Chintan D. Bittu

131. In a pile of 10 books there are 3 of History, 3 of Hindi, 2 of Maths, and 2 of English. Taking from above there is an English book between a History and Maths book, a History book between a Maths and an English book, a Hindi book between an English and a Maths book, a Maths book between two Hindi books, and two Hindi books between a Maths and a History book. Book of which subject is at the sixth position from the top?

A. English B. Hindi
C. History D. Maths

132. A team of six members is selected from a bunch of boys B1, B2, B3, B4, B5 and B6 and four girls G1, G2, G3 and G4.

Some of the criteria are as follows:
B2 and B4 have to go together
B3 cannot go with G2
B6 cannot go with B1 and B4
G1 and G3 have to be together
G1 cannot go with G4
B1 and B5 have to be together
If four of the members including B5 have to be boys and one of the girls has to be G3 then the team members are:

A. B1, B4, B5, B6, G1, G3
B. B2, B4, B5, B6, G1, G3
C. B1, B2, B4, B5, G3, G4
D. B1, B2, B4, B5, G1, G3

133. In a certain code 'nee tim see' means 'who is it'. 'ble nee see' means 'where is it'. 'ble see tee' means 'where is what'. What is the code for 'who is where'?

A. tim ble see B. tim nee see
C. tim ble nee D. nee ble see

134. Its dinner time and snow white is serving soup to the seven dwarfs namely grumpy, happy, bashful, dopey, sneezy, sleepy and dreamy. The dwarfs are standing in a line where sneezy is between bashful and dopey. Grumpy is just before happy and dreamy is just before sleepy. Bashful is not the last and he is behind sleepy. Dopey is also not last. Who is standing just before grumpy?

A. sleepy B. dreamy
C. happy D. dopey

135. Mac is brother of Tina and brother-in-law of Tim. Sharon has two children and her son-in-law is Tim. Sheeba's grandmother is Sharon. Teena is Sheeba's aunt. Which of the following statements is true?

A. Sheeba is daughter of Tim
B. Tina is Tim's sister-in-law
C. Sheeba is niece of Tim
D. Sheeba is niece of Mac

136. Mr. Biswas is planning for 2 weeks holiday with his family to UK, Germany, France, Switzerland, Belgium, Italy, Norway and Austria. However his family has imposed certain conditions for travel:

(*i*) Will travel to Austria and Switzerland in different weeks
(*ii*) Germany must be visited after Belgium in the same week
(*iii*) Switzerland must be visited after France in the same week
(*iv*) Will travel to Germany and Norway in different weeks
(*v*) Will travel to UK and France in different weeks
(*vi*) Will travel to Austria and Belgium in the same weeks
(*vii*) Italy must be visited in the 2nd week
(*viii*) Equal numbers of places must be visited in week 1 and 2

Which places are visited in a week 2, if all places are visited only once?

A. Italy, UK, Switzerland, Norway
B. Italy, France, Germany, Norway
C. Italy, UK, Austria, Switzerland
D. Italy, France, Switzerland, Norway

Directions (Qs. No. 137): *Read the following information and answer the given question.*

Despite strong opposition from the social activists of college X, Pollution Control Board renewed the license of a coal depot to operate coal handling barely one kilometre away from the village.

137. Which of the following statements supports the decision of Pollution Control Board to some extent?

A. Several crops were destroyed due to disposal of waste from the coal depot in the previous year.
B. Maintenance cost of roads in Village X has increased due to transportation of coal.
C. The depot provides employment for nearly 70% of people in the neighbouring area.
D. Several complaints have been filed against the depot in the past for violating safety standards.

Directions (Qs. No. 138-141): *Study the following information and answer the questions.*

Eight friends M, N, O, P, Q, R, S and T live on eight different floors of a building but not necessarily in the same order. The lowermost floor of the building is numbered 1, the one above that is numbered 2 and till the topmost floor is numbered 8.

Q lives on an odd numbered floor but not on the floor numbered 7. T lives immediately below Q.

Only three people live between T and M. O lives on an even numbered floor immediately above S.

Only two people live between N and R. N lives on one of the floors above R.

P does not live on floor numbered five.

138. Who amongst the following lives on the floor numbered 5?

A. Q B. N
C. R D. M

139. Which of the following is true about N as per the given arrangement?

A. S lives on one of the floors above N.
B. Only two people live between N and Q.
C. N lives on the topmost floor.
D. N lives immediately below P.

140. If M and S interchange their positions and so do P and T, who amongst the following will live between T and Q as per the new arrangement?
A. No one
B. Both S and P
C. Only N
D. Both R and M

141. Four of the following five are alike as per the given arrangement and thus form a group. Which of the following does not belong to that group?
A. NR
B. QO
C. TP
D. TR

Directions (Qs. No. 142 & 143): *Study the following information and answer the questions.*

A chef is trying to recipe for a tasty ice cream using four ingredients. He can choose from three liquids for taste which are labelled A, B and C which are stable in nature and the choice for flavour can be from four liquids which are labelled W, X, Y and Z. For the new ice cream recipe to be tasty there must be two liquids from the taste giving liquids. Also certain liquids cannot be mixed because of their reactions which make it unhealthy for human consumption as the same is given below:

B cannot be mixed with W
C cannot be mixed with Y
Y cannot be mixed with Z

142. If the chef calculated that Y is the most important flavour and must be used in the recipe, which other ingredients must be part of the recipe?
A. A, B and W
B. A, B and X
C. A, B and C
D. B, C and X

143. Which of the following combination of liquids is impossible?
I. Using Y and W together
II. Using B and C together
III. Using W, X and Z together
A. I only
B. II only
C. III and I only
D. II and I only

144. Which letter from the options given will replace the "?" in the table given below?

CAT	389376	DOG
RAT	1758276	MAT
CAB	15876	FAN
CAN	571536	?

A. MAN
B. GIFT
C. PAN
D. SOFT

145. A producer of drama theatre is creating his weekend schedule. The producer has six plays to choose from: "Made in India", "Laugh for a While", "The Life is your choice", "MBA–Maha Budhiman Aadmi", "Placements—my goal", "MBA Go Getters", The producers sets a schedule based upon the following criterion.
I. "Made in India" must be shown before "The Life is your choice" and "Laugh for a while".
II. "Laugh for a while" must be shown before "MBA–Maha Budhiman Aadmi".
III. "Placements—My Goal" must be shown after "The Life is your choice" and "MBA Go Getters".

Which of the following weekend schedules are consistent with producer's criterion?
A. Made in India, Laugh for a while, Life is your choice, MBA—Maha Budhiman Aadmi, Placements—my goal and MBA Go Getters.
B. MBA Go Getters, Made in India, Laugh for a while, The Life is your choice, Placements—my goal and MBA—Maha Budhiman Aadmi.
C. Made in India, MBA Go Getters, Placements—my goal, Laugh for a while, MBA—Maha Budhiman Aadmi and Life is your choice
D. Made in India, MBA Go Getters, MBA—Maha Budhiman Aadmi, The Life is your choice, Placements—My Goal and Laugh for a while.

146. Assertion (A) : The steam engine was invented by James Watt.
Reasoning (R): There was a problem of taking out water from flooded mines.
A. Both A and R are true and R is the correct explanation of A.
B. Both A and R are true but R is not the correct explanation of A.
C. A is true but R is false.
D. A is false but R is true.

147. Assertion (A) : Inside the earth, metals are present in molten state.
Reasoning (R): Earth absorbs the sun's rays.
A. Both A and R are true and R is the correct explanation of A.
B. Both A and R are true but R is not the correct explanation of A.
C. A is true but R is false.
D. A is false but R is true.

148. In the question that follows, fill up the blanks in the given series out of the given choices.
c_bba_cab_ac_ab_ac
A. bcbac
B. cabcb
C. accbc
D. acbcb

149. Vertex : Pyramid : .... : ....
A. Strand : Hair
B. Rung : Ladder
C. Summit : Mountain
D. Frame : Picture

150. Piety : Pilgrim : Curiosity : ....
A. Explorer
B. Pinchpenny
C. Gamester
D. Monk

ANSWERS

1	2	3	4	5	6	7	8	9	10
A	B	A	B	C	A	B	D	C	A
11	**12**	**13**	**14**	**15**	**16**	**17**	**18**	**19**	**20**
A	B	D	C	A	B	A	D	A	B
21	**22**	**23**	**24**	**25**	**26**	**27**	**28**	**29**	**30**
C	D	A	B	A	A	B	C	D	A
31	**32**	**33**	**34**	**35**	**36**	**37**	**38**	**39**	**40**
B	C	B	D	D	C	A	D	C	B
41	**42**	**43**	**44**	**45**	**46**	**47**	**48**	**49**	**50**
C	B	C	C	B	A	C	D	C	B
51	**52**	**53**	**54**	**55**	**56**	**57**	**58**	**59**	**60**
A	D	C	C	B	C	A	B	A	D
61	**62**	**63**	**64**	**65**	**66**	**67**	**68**	**69**	**70**
C	B	B	B	B	D	B	D	C	C
71	**72**	**73**	**74**	**75**	**76**	**77**	**78**	**79**	**80**
D	B	D	A	D	D	C	A	B	C
81	**82**	**83**	**84**	**85**	**86**	**87**	**88**	**89**	**90**
D	A	C	D	B	B	C	C	D	D
91	**92**	**93**	**94**	**95**	**96**	**97**	**98**	**99**	**100**
C	B	B	A	C	C	C	D	C	C
101	**102**	**103**	**104**	**105**	**106**	**107**	**108**	**109**	**110**
C	A	D	D	C	D	C	B	A	B
111	**112**	**113**	**114**	**115**	**116**	**117**	**118**	**119**	**120**
B	C	C	C	D	D	A	A	C	B
121	**122**	**123**	**124**	**125**	**126**	**127**	**128**	**129**	**130**
C	C	D	B	B	A	C	A	D	C
131	**132**	**133**	**134**	**135**	**136**	**137**	**138**	**139**	**140**
B	D	A	D	C	D	C	A	D	C
141	**142**	**143**	**144**	**145**	**146**	**147**	**148**	**149**	**150**
D	B	C	C	B	C	B	D	C	A

EXPLANATORY ANSWERS

71. $\left[(x+y)^3-(x-y)^3-6y(x^2-y^2)\right]\times y$

$= [x^3+y^3+3x^2y+3xy^2-x^3+y^3+3x^2y-3xy^2$
$-6x^2y+6y^3]\times y$

$= (6y^3)y = 6y^4$.

72. Let the numbers are x and y

$$x + y = 216 = 27 \times 8$$

$\because$ HCF of these two numbers is 27, hence, common factor of these two numbers is 27.

These numbers are in the form of 27×1, 27×2, 27×3, ... etc.

According to the condition, sum of these two numbers is 216.

The numbers will be $27 \times 3 = 81$ and $27 \times 5 = 135$

Hence, numbers are 81 and 135.

73. 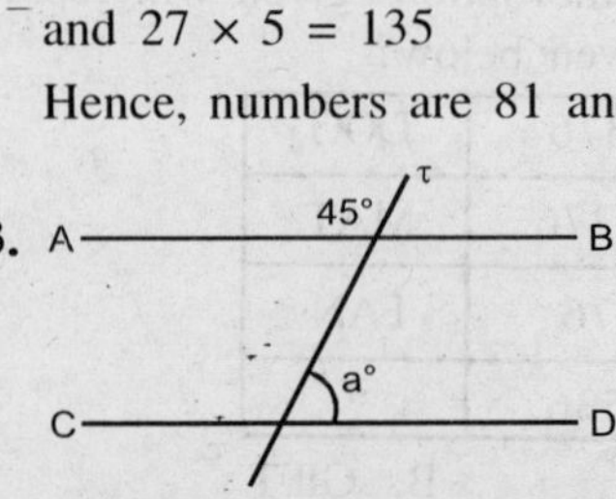

$\because$ AB $\parallel$ CD and τ is transversal line

$\therefore$ $a = 180 - 45° = 135°$.

74. According to the condition of the problem, 50 is obtained on dividing one of the numbers by 2

One number = 50 × 2 = 100

Now, Ist number × IInd number = HCF × LCM

$$100 \times \text{2nd no.} = 250 \times 50$$

$$\text{2nd number} = \frac{250 \times 50}{100} = 125$$

∴ Numbers are 100 and 125.

75. Number of dogs = $\sqrt{999919}$ = 999.

76. Speed of vehicle in m/s = $54 \times \frac{5}{18} = 15$ m/s

Time taken both the wheels make a whole number of turns = LCM of 3.5 m and 3 m = 21 m

Time required to cover 21 m = $\frac{21}{15}$ = 1.4 seconds.

77. After 30 steps A catched up with B.

78. Let the breadth of the tank = x m.

V = Capacity of the tank = $\frac{50000}{1000}$ m^3 = 50 m^3

$$\text{Breadth} = \frac{\text{Volume}}{l \times h} = \frac{50}{2.5 \times 10}$$

$$\Rightarrow \quad x = \frac{50}{25} = 2 \text{ m.}$$

Hence, breadth of the tank = 2 m.

80. Let no. of female = x and no. of male = y

$$28(x + y) = 21x + 32y$$

$$\Rightarrow \quad 28x + 28y = 21x + 32y$$

$$\Rightarrow \quad 7x = 4y \Rightarrow \frac{x}{y} = \frac{4}{7}$$

$$\therefore \quad \text{No. of female} = \frac{4}{11} \times 231 = 4 \times 21 = 84$$

$$\text{No. of male} = \frac{7}{11} \times 231 = 7 \times 21 = 147$$

Hence, total no. of people = 84 + 147 = 231

81. Let no. of red balls = x

According to the question,

$$2\left(\frac{x}{12}\right) = \frac{x+3}{15}$$

$$\Rightarrow \quad \frac{x}{6} = \frac{x+3}{15}$$

$$\Rightarrow \quad 15x = 6x + 18$$

$$\Rightarrow \quad 9x = 18 \Rightarrow x = 2$$

Hence, no. of red balls = 2.

82. ∵ $\log(mn) = \log m + \log n$

∴ log tan 1° + log tan 2° + ... + log tan 89°

= log (tan 1°· tan 2°·tan 3° .... · tan 89°)

= log (tan 1°· tan 89°·tan 2°·tan 88° .... tan 45°)

= log [tan 1°· cot 1°·tan 2°·cot 2° .... tan 45°)

= log [1] = 0

84. Ratio of efficiency of Ram and Shyam = 2 : 1

Hence, ratio of their time taken = 1 : 2

∵ Shyam can finish the job in 15 days

(Ram + Shyam)'s 1 day work = $\frac{2}{15} + \frac{1}{15} = \frac{3}{15} = \frac{1}{5}$

Hence, they together will finish the job in 5 days.

85. 129 × 206 × 47$ × 1484

The product of unit digit = 2

$$\Rightarrow \quad 9 \times 6 \times \$ \times 4 = 2$$

$$\Rightarrow \quad 9 \times 6 \times 7 \times 4 = 2$$

Hence, value of $ = 7.

87.

$$P(E_1) = 60\%, \; P(\bar{E}_1) = 40\%$$

$$P(E_2) = 80\%, \; P(\bar{E}_2) = 20\%$$

$$P(E) = 60\% \times 20\% + 40\% \times 80\%$$

$$= \frac{60}{100} \times \frac{20}{100} + \frac{40}{100} \times \frac{80}{100}$$

$$= \frac{1200}{10000} + \frac{3200}{10000}$$

$$= 12\% + 32\% = 44\%$$

$$P(E') = 56\% = \frac{56}{100} = 0.56$$

89.

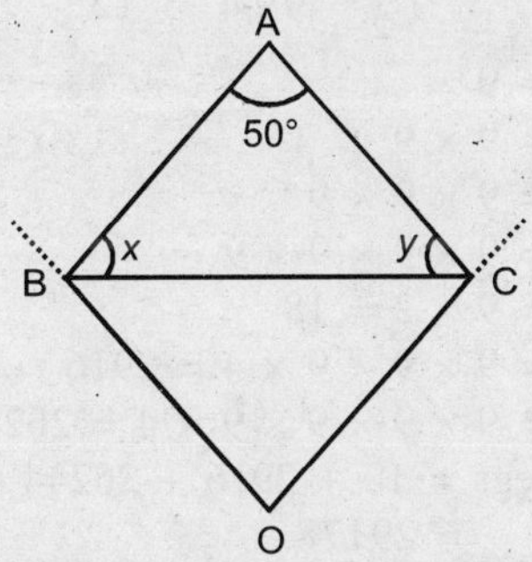

$$\angle BOC = 180° - \angle OBC - \angle OCB$$

$$= 180° - \left(\frac{50 + x}{2}\right) - \left(\frac{50 + y}{2}\right)$$

$$= 180 - 50 - \left(\frac{x + y}{2}\right)$$

$$= 130 - \left(\frac{180 - 50}{2}\right) = 65°.$$

90. Let the salary of Saroj = ₹ 100

Salary of Raju and Ram is 20% and 30% less than Saroj's salary.

Hence, Raju's salary = ₹ 80

and Ram's salary = ₹ 70

Thus, Raju's salary is greater than Ram's salary by

$$\frac{80 - 70}{70} \times 100 = \frac{10}{70} \times 100$$

$$= \frac{100}{7} = 14.28\%.$$

91. Let CP = ₹ 100

$$SP = \frac{3}{5} \times 120 = ₹\ 72$$

$$SP = \frac{2}{5} \times 100 = ₹\ 40$$

Total CP = ₹ 100
Total SP = 72 + 40 = ₹ 112
Profit = 112 – 100 = ₹ 12.

92. 16, 23, 30, ... 93

nth term $= a + (n - 1)d$

$$93 = 16 + (n - 1) \times 7$$
$$93 = 16 + 7n - 7$$
$$93 = 7n + 9$$
$$n = 12$$

$$\text{Sum of } n \text{ terms} = \frac{n}{2}\{2a + (n-1)d\}$$
$$= \frac{12}{2}\{2 \times 16 + (12-1)7\}$$
$$= 6\{32 + 77\} = 6(109) = 654.$$

93. Number of bulb that can be used to lighten the room = 9 – 2= 7

$$\text{Required probability} = \frac{{}^7C_2}{{}^9C_2} = \frac{7 \times 3}{9 \times 4} = \frac{7}{12}.$$

96. Let no. of men = 9
No. of cages = 9 × 9
No. of lions = 9 × 9 × 9
No. of cubs = 9 × 9 × 9 × 9
No. of legs, Men = 9 × 2 = 18
Lions = 9 × 9 × 9 × 4 = 2916
Cubs = 9 × 9 × 9 × 9 × 4 = 26244
∴ Total number of legs = 18 + 2916 + 26244 = 29178.

97. $2^x + 2^{x+1} = 48$

Let $2^x = y$, then $y + 2y = 48$

$\Rightarrow 3y = 48 \Rightarrow y = 16$

$\therefore 2^x = 16 \Rightarrow 2^x = 2^4$

$\Rightarrow x = 4$

Value of $x^x = (4)^4 = 256.$

98. $$\text{Required weight} = \frac{100}{105} \times 1000 = \frac{20000}{21} = 952.381$$

99. Walking + bike = 5 hrs ...(*i*)
bike + bike = 2 hrs
⇒ 2 bike = 2 hrs ...(*ii*)
Equation (*i*) multiply by 2
2 walking + 2 bike = 10 hrs
2 bike = 2 hrs

2 Walking = 8 hrs.

rest = 30 minutes = $\frac{1}{2}$ hrs

Required time = $8\frac{1}{2}$ hrs.

100. Let CP = ₹ 100
MP = 100 + 80 = ₹ 180

$$\text{Discount} = \frac{25}{100} \times 180 = ₹\ 45$$

SP = ₹(180 – 45) = ₹ 135
Profit = 135 – 100 = 35

$$\text{Profit\%} = \frac{35}{100} \times 100 = 35\%$$

101. Let $\log_{\left(-\frac{1}{4}\right)} 256 = x$

$$\Rightarrow \left(-\frac{1}{4}\right)^x = 256$$
$$\Rightarrow (-4)^{-x} = 16 \times 16$$
$$\Rightarrow (-4)^{-x} = (4)^4$$
$$\Rightarrow (-4)^{-x} = (-4)^4$$
$$\Rightarrow -x = 4$$
$$\Rightarrow x = -4.$$

102. 60% $(x - y)$ = 40% of $(x + y)$

$$\Rightarrow \frac{60}{100} \times (x-y) = \frac{40}{100} \times (x+y)$$
$$\Rightarrow 6(x - y) = 4(x + y)$$
$$\Rightarrow 6x - 6y = 4x + 4y$$
$$\Rightarrow 2x = 10y$$
$$\Rightarrow x = 5y$$
$$\Rightarrow y = \frac{1}{5}x$$
$$\Rightarrow y = \frac{1}{5} \times 100\% \text{ of } x$$
$$\Rightarrow y = 20\% \text{ of } x.$$

103. $$\left(1-\frac{1}{2}\right)\left(1-\frac{1}{3}\right)\left(1-\frac{1}{4}\right)\left(1-\frac{1}{5}\right)\ldots\ldots\ldots\left(1-\frac{1}{40}\right) = \frac{x}{40}$$

$$\Rightarrow \frac{1}{2} \times \frac{2}{3} \times \frac{3}{4} \times \frac{4}{5} \times \ldots\ldots \times \frac{39}{40} = \frac{x}{40}$$

$$\Rightarrow \frac{1}{40} = \frac{x}{40} \Rightarrow x = 1.$$

104. $$\because \left(\frac{a}{b}\right)^{2x-3} = \left(\frac{b}{a}\right)^{x-9}$$

$$\Rightarrow \left(\frac{a}{b}\right)^{2x-3} = \left(\frac{a}{b}\right)^{-x+9}$$
$$\Rightarrow 2x - 3 = -x + 9$$
$$\Rightarrow 3x = 9 + 3$$
$$\Rightarrow 3x = 12$$
$$\Rightarrow x = 4$$

105. Required number of tickets = 29 × 30 = 870.

106. LCM of 10 and 15 = 30
Maximum no. of fruits can put in 15 boxes = 30 × 15 = 450.

107. $\because$ HCF of P and Q = 1

$\therefore$ LCM of P and Q = P × Q = PQ.

109. Perimeter of the largest equilateral triangle is 72 cm. Now, triangle formed by joining the mid points will have perimeter half of that of original triangle.

Hence, the required sum

= 72 + 36 + 18 + ... up to infinite terms

This is nothing but an infinite GP, with common ratio $= \frac{1}{2}$

Hence, the required sum $= \dfrac{72}{1-\frac{1}{2}} = \dfrac{72}{\frac{1}{2}} = 72 \times 2 = 144.$

110. Ratio of their capitals

$$= \left[4\times 3+\left(4-\frac{1}{4}\times 4\right)\times 7\right] : \left[5\times 3+\left(5-\frac{1}{5}\times 5\right)\times 7\right]$$

= 33 : 43

Hence, A's profit $= \dfrac{33}{76}\times 760 = ₹\ 330.$

111. The order from left to right is:
Brijesh, Jayesh, Amar, Praveer.

112. The pattern of standing is:

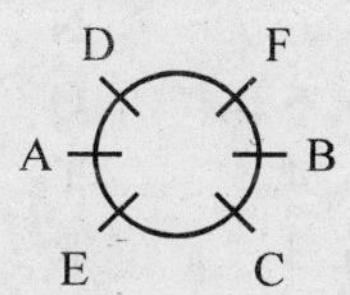

113. The movements are:

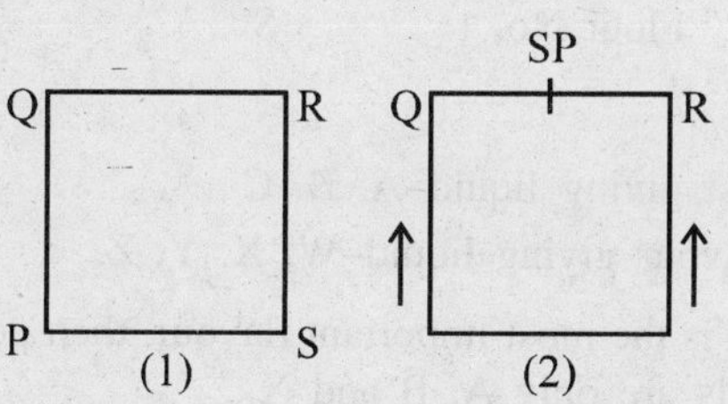

114.

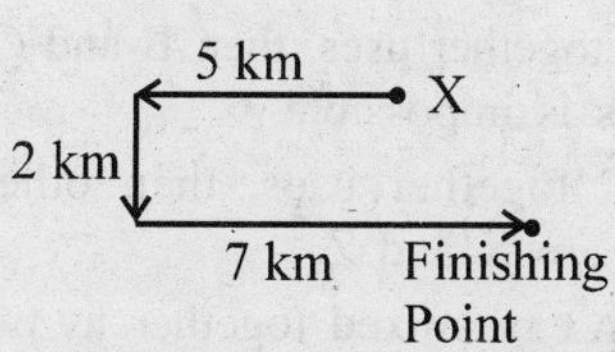

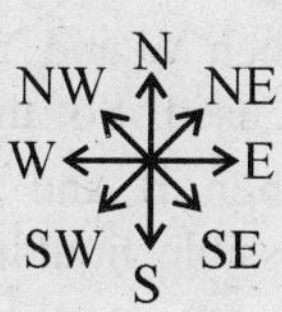

115. Sitting order while facing us is :
A, C, D, B, E.

116.

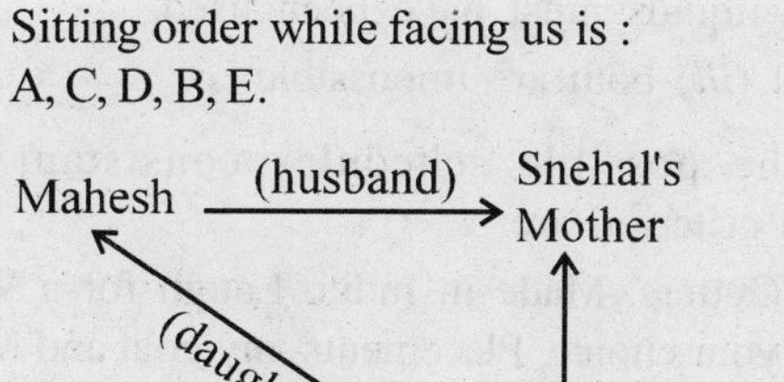

'Her (Snehal's) mother's only daughter' is Snehal herself. So, Snehal is the daughter of Mahesh who in turn is her father.

117.

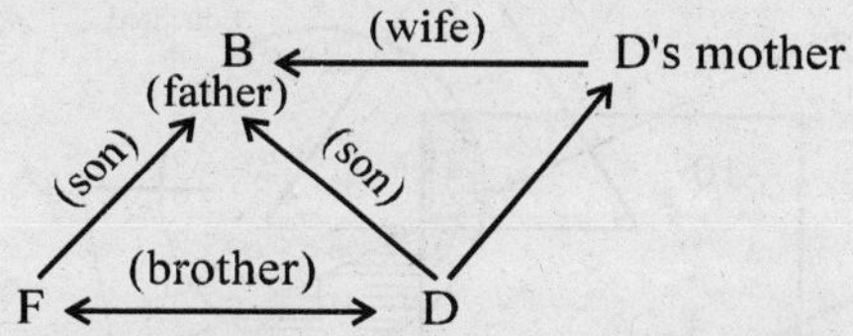

Both the statements are needed to know that 'D' is the brother of 'F'.

118.

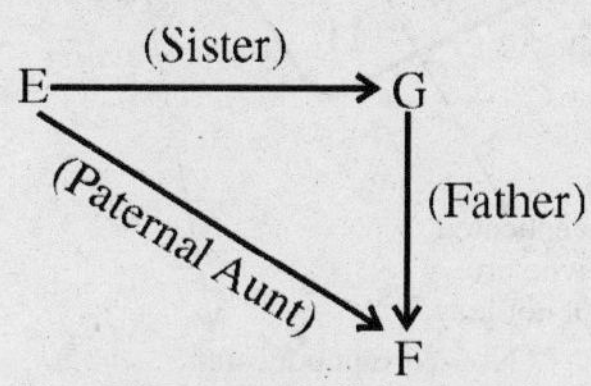

E – G means 'E' is the sister of 'G' and G ÷ F means 'G' is the father of F.

119.

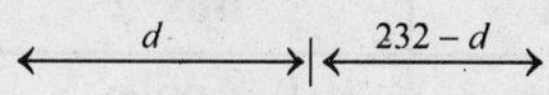

i.e., $\dfrac{d}{48} = \dfrac{232-d}{50} + \dfrac{3}{4}$

$\dfrac{d}{48} = \dfrac{464-2d+75}{100}$

$100d = (539 - 2d)\,48$

$100d = 25872 - 96d$

$196d = 25872$

$d = 25872 \div 196 = 132$ km.

120. The smaller 64 pieces will be cut in the manner that:

1. 8 pieces will be painted on 3 sides,
2. 24 pieces on 2 sides.
3. 24 pieces on 1 side, and
4. 8 will not have paint at all.

Diagrammatically, the explanation taking one side of the cube will be :

a	b	b	a
b	c	c	b
b	c	c	b
a	b	b	a

1. 'a' are the corner pieces [4 × 2 = 8]
2. 'b' are the centre pieces of the cornered sides [8 × 3 = 24]
3. 'c' are the centre pieces [4 × 6 = 24]
4. Remaining interior pieces
$= 64 - (8 + 24 + 24) = 64 - 56 = 8.$

121. The numbers adjacent to 2 are 1, 5, 3 and 4. So, the number of dots opposite to the side with 2 dots will be 6.

123. According to both the statements the ascending order will be P < R < L < Q/S. Either Q or S can be the shortest. So, even both the statements together are not sufficient to answer the question.

124-126.

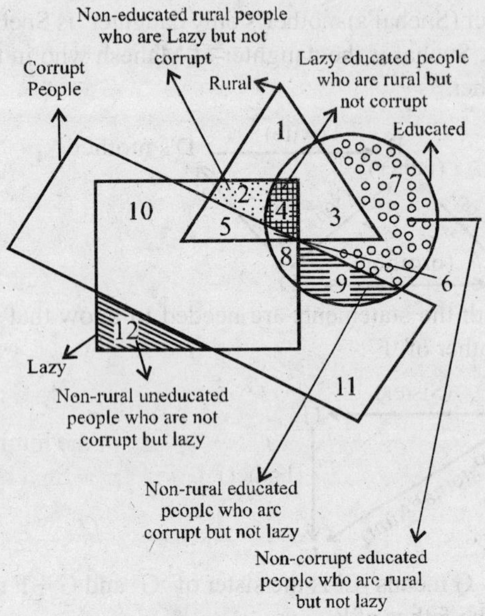

127.

	Box 1	Box 2
Fruits	(*i*) Orange	(*i*) Apple
	(*ii*) Pineapple	(*ii*) Watermelon
	(*iii*) Blueberry	(*iii*) Mango
	(*iv*) Guava	(*iv*) Strawberry

129. From given information we tabulated players with their games:

Tennis	Saurav, Kartik, Manpreet, Lokesh
Swimming	Manpreet, Lokesh, Ganesh, Prakash, Saurav
Badminton	Kartik, Nikhil, Ganesh, Lokesh

Hence, player one who plays above three games is Lokesh.

130. Rankwise arrangement of Riders with their bikes

Darshan	Amar	Ehsaan	Chintan	Bittu
Hayabusa	Pulsor	Unicorn	Splendor	Apache

132. Six members team with four boys and two girls is possible with following combination:

As B5 is included, so B1 is also included.

When B1 is included, B6 is excluded from team.

So, Team of boys are B1, B2, B4, B5.

and Team of girls are G1 and G3 as G3 is included together with G1.

Hence, Team includes : B1, B2, B4, B5, G1, G3.

133.

nee tim see – who is it

ble nee see – where is it

ble see tee – where is what

Hence, Code for 'who' is 'tim' and for 'where' is 'ble'

Hence, 'who is where' is coded as 'tim ble see'.

134. Standing Position of Seven dwarfs are as follows:

Happy Grumpy Dopey Sneezy Bashful Sleepy Dreamy

Hence, Dopey is just before Grumpy.

135.

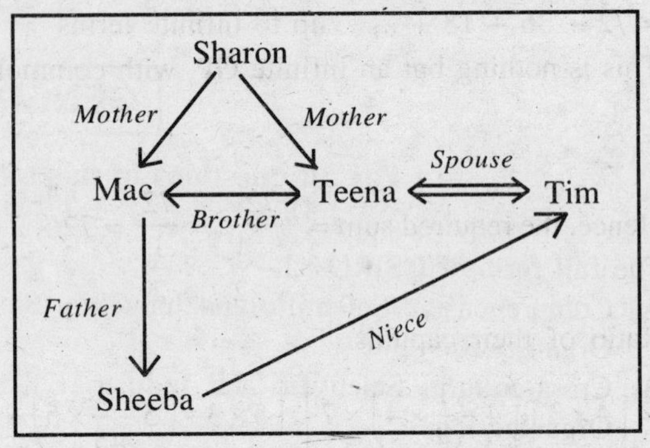

136.

Week 1	Week 2
Austria, Belgium	France, Norway
Germany, U.K.	Switzerland, Italy

138-141.

Floor No. 8	M
Floor No. 7	P
Floor No. 6	N
Floor No. 5	Q
Floor No. 4	T
Floor No. 3	R
Floor No. 2	O
Floor No. 1	S

142-143.

Test giving liquid–A, B, C

Flavour giving liquid–W, X, Y, Z.

142. As Y is the most important flavour, then possible three liquids are only A, B and X.

143. (*i*) When W and Y together uses, then B and C can't be used. So, this is impossible.

(*ii*) When B and C together uses, then other two possible liquids are X and Z.

(*iii*) W, X and Z can't be mixed together as two test giving liquids must have been used.

So, (*i*) and (*iii*) both are impossible.

145. One of the possible schedules consistant with producer's criterion are:

MBA Go Getters, Made in India, Laugh for a While, The life is your choice, Placements–my goal and MBA–Maha Budhiman Aadmi.

148. The Series is cabbac, cabbac, cabbac.

Previous Paper (Solved)

Symbiosis National Aptitude Test (SNAP), 2015

SECTION- I

GENERAL AWARENESS

For all questions in this section, each correct answer carries 1 mark

1. The full form of CSIRO is
A. Comprehensive Scientific and Industrial Research Organization
B. Cross-cultural Scientific and Industrial Research Organization
C. Commonwealth Scientific and Industrial Research Organization
D. Council of Scientific and Industrial Research Organization

2. The full form of UNRISD is
A. United Nations Research Institute for Soviet Development
B. United Nations Research Institute for Scientific Development
C. United Nations Research Institute for Socio-economic Development
D. United Nations Research Institute for Social Development

3. *Bhaona* is a presentation of the *Ankia Naat* of Assam. In *Bhaona* the cultural glimpses of .... is/are reflected.
A. Assam and Orissa B. Bengal
C. Mathura and Brindavan D. All of the above

4. Under which Act does the Archaeological Survey of India (ASI) protect monuments, sites and remains of national importance?
A. AMASR Act, 1958 B. AMASR Act, 1968
C. AMASR Act, 1978 D. AMASR Act, 1948

5. In 2015 the President of India approved conferment of Padma Awards to all of the following but for ....
A. Kharag Singh Valdiya
B. Mohammad Yusuf Khan
C. Lakshmi Gopala Naidu
D. Saichiro Misumi

6. In Grammy Awards 2015 the best folk album was won by ....
A. Pharrell Williams - "Happy"
B. Old Crow Medicine Show – Remedy
C. Rosanne Cash – The River & The Thread
D. None of the above

7. Scientists discovered a new species in the human family tree which is a small creature with a tiny brain. The new species has been named as ....
A. Homo Naledi B. Dwarf Sapien
C. Dwarf Homo Sapien D. None of the above

8. The ministry of External Affairs has recently decided to change the nomenclature of Indian Based Domestic Assistance (IBDA) to ....
A. Government Serving Based Domestic Assistance (GSBDA)
B. Swadesh Swatcha Domestic Assistance (SSDA)
C. Service Staff (SS)
D. Bharath Based Service Assistance (BBSA)

9. The Foreign Exchange Reserve of India consists of ....
A. The foreign currency assets held by RBI and the gold holding of RBI
B. The gold holding of RBI and special drawing rights
C. The gold holding of RBI, the foreign currency assets held by RBI and special drawing rights
D. Only the foreign currency assets held by RBI

10. .... is a charge for converting bullion into coins where free coinage is permitted. This charge is equal to the cost of bullion to coins transformation.
A. Bull-Coin B. Brassage
C. Bit Coin D. Coin Levy

11. Microsoft introduced several new products for education customers, this includes a note-taking app called ....
A. Office Note classroom
B. Micro Note class Notebook
C. Soft Note for classroom
D. None Note class Notebook

12. The National Merit Scholarship Scheme, which provided financial assistance to meritorious students from Class XI to Post-Graduation level in Government Schools/Colleges/Universities has been discontinued from the year ....
A. 2009 B. 2008
C. 2007 D. None of the above

13. The Symbol used for reusable microwaveable plastic ware is ....

A. 

B. 

C. 

D. 

14. Which is the National Emblem of Iran, used from 1979 to 1980?

A. 

B. 

C. 

D. 

15. Super Blood Moon, which is a rare astronomical phenomenon hasn't happened since .... and won't happen again until ....

A. 1982, 2033 B. 1980, 2035
C. 1978, 2032 D. 1975, 2031

16. The galaxy ...., was initially discovered with NASA's Spitzer Space Telescope in infrared light and is believed to be at least 9 billion years old.

A. ANDROMEDA B. COSMOS REDSHIFT 7
C. SAGE0536AGN D. SUNFLOWER

17. In the year 2014 Facebook bought Whatsapp for .... US Dollars.

A. 17 billion B. 18 billion
C. 19 billion D. 20 billion

18. Shri Narendra Modi was sworn in as the Prime Minister of India on .... at the Rashtrapati Bhavan in New Delhi.

A. 25th May 2014 B. 19th May 2014
C. 26th May 2014 D. 27th May 2014

19. Where is the doldrums belt located?

A. Near the Equator
B. Near the Poles
C. Near the Tropic of Cancer
D. Near the Tropic of Capricorn

20. The air quality in Singapore deteriorated to a hazardous level in September 2015, forcing the city-state to shut its schools for the first time in 12 years. This was due to ....

A. Haze created mainly due to companies in Malaysia
B. Haze created mainly due to companies in Indonesia
C. Haze created mainly due to companies in Philippines
D. Haze created mainly due to companies in Vietnam

21. In 1347 during the reign of Muhammed Tuglak an Afghan officer named .... setup an independent kingdom called Bahmani Kingdom.

A. Mahmud Gawan Bahmani
B. Mohammad Yusuf Bahmani
C. Hasan Gangu Bahmani
D. Khwaja Tuglak Bahmani

22. The task of consolidating Mughal Kingdom was left to Akbar who was only .... years old at the time of his accession of the throne.

A. Eleven Years B. Twelve Years
C. Thirteen Years D. Fourteen Years

23. Smoking in public places was prohibited nationwide from ....

A. 15th August, 2008 B. 2nd October, 2008
C. 15th August, 2009 D. 2nd October, 2009

24. A constitutional right can be ...., recognized and established by a Sovereign State or Union of States.

A. a prerogative or a duty or a restraint of power
B. a prerogative or a duty, a power or a restraint of power
C. a prerogative or a duty, a power but not a restraint of power
D. a power but not a prerogative or a duty

25. In Nepal the festival of lights i.e. Diwali is celebrated by some Buddhists as ....

A. Tihar
B. Swanti
C. Both options A and B
D. Neither of the options above

26. The term "Vrajapati" used in Indian Mythology denoted ....

A. The Head of the Village
B. The Head of the Family
C. The Head of a Society
D. The Head of Merchants

27. One of India's most distinguished constitutional lawyers who had received brickbats for arguing in favour of Dow Chemicals in the Bhopal gas disaster case is ....

A. Ram Jethmalai B. Fali Nariman
C. Mukul Rohatgi D. Pramila Nesargi

28. The Nobel Peace Prize winner who gave up freedom and a life with her family in Britain, to protest against the military rule at another country, who is also the Chair of the National League for democracy is ....

A. Angela Merkel B. Christine Lagarde
C. Brito Polman D. Aung San Suu Kyi

29. In September 2015 it was revealed that .... is the country which was exporting drone components worth hundreds of millions to countries that include Saudi Arabia and South Korea, to regain lost ground in a global arms race.
A. France
B. United Kingdom
C. Japan
D. United State of America

30. In the Indian general assembly elections 2014 the BJP-led NDA won .... seats out of .... Lok Sabha seats that were announced.
A. 335, 546 B. 334, 545
C. 336, 543 D. 332, 544

31. The BS EN 16001 solutions from SBI is applicable for ....
A. Energy Management Systems
B. Environment Management Systems
C. Energy Process Systems
D. Environment Standard Systems

32. The WEEE (Waste from Electrical and Electronic Equipment) is a directive that controls ....
A. how electric and electronic equipment is handled and recycled
B. how electric and electronic equipment is manufactured and handled
C. how electric and electronic equipment is recycled
D. how electric and electronic equipment is manufactured, handled and recycled.

33. The Sensex and Nifty are both indices. The base years for the BSE Sensex and Nifty are .... and .... respectively.
A. 1980-81 and 1990
B. 1990-91 and 2000
C. 1978-79 and 1995
D. 2000-01 and 2004

34. The artist who painted Irises, Sunflowers, Red Poppies, Pink Roses was ....
A. Vincent van Gogh B. Sandro Botticelli
C. Leonardo da Vinci D. Michelangelo

35. Which Indian satellite was launched that has a fantastic timing and records 1000th of a second?
A. GSAT-16 B. IRNSS-1D
C. GSAT-6 D. Astrosat

36. W3C stands for ....
A. Triple Web Consortium
B. Triple Web Consolidation Council
C. World Wide Web Consortium
D. World Wide Web Company

37. The country/countries that has/have resorted to Quantitative easing in the last decade is/are
A. United States of America
B. United Kingdom
C. Japan
D. All of the above

38. Who is the Indian badminton player who after spending an year after his shoulder injury earned a final appearance at the Korean Open in 2015?
A. Chetan Anand B. Ajay Jayaram
C. Parupalli Kashyap D. Sameer Verma

39. *Tamaasha*, the traditional folk theatre form of Maharashtra has evolved from the folk forms of ....
A. Gondhal, Jagran and Kirtan
B. Only Kirtan
C. Only Gondhal, Jagran
D. Only Gondhal

40. The last series of wall painting in India are from .... near Hindupur belonging to 16th century A.D.
A. Lepakshi temple
B. Shiva temple
C. Sri Venkateswara Swami temple
D. Ganesh temple

SECTION-II

ANALYTICAL & LOGICAL REASONING

For all questions in this section, each correct answer carries 2 marks

41. A bungalow has one of its room located on the first floor and there are three identical 100 watt electrical bulbs fixed in the room. Each bulb is connected to a specific switch located at the basement. There are only three switches in the basement. All the bulbs are switched off at present and you are also at the basement area. The first floor cannot be seen from the basement area. If you are allowed to use your common prudence, what is the minimum number of times that you will have to go from basement to first floor to identify which switch goes to which bulb?
A. 03 times B. 20 times
C. 01 time D. 06 times

42. Shyam is running a start-up. His initial investments are high and he is trying hard to manage and increase his cash flow. The sundry expenses that his firm incurs is negligible. He found from his accountant that the amount of pre-paid expenses in the balance sheet, which were booked from the previous year to the current year was increased. Shyam also ensured that his cash funded by the venture capitalists did not reduce when compared to the previous year. The interest that he gets from his fixed deposits increases, which is in tune with his sundry expenses. The final effect on cash for this year would be ....

A. Cash flow marginally increases
B. Cash flow exponentially increases
C. Cash flow remains the same
D. None of the above

43. A producer of a drama theatre is creating his weekend schedule. The producer has six plays to choose from: "Made in India", "Laugh for a while", "The Life is your choice", "MBA–Maha Budhiman Aadmi", "Placements – My goal", "MBA Go Getters". The producer sets a schedule based upon the following criteria.

I. "Made in India" must be shown before "The Life is your choice" and "Laugh for a while"

II. "Laugh for a while" must be shown before "MBA – Maha Budhiman Aadmi"

III. "Placements – my goal" must be shown after "The Life is your choice" and "MBA Go Getters".

Which of the following weekend schedules are consistent with the producer's criteria?

A. Made in India, Laugh for a while, The Life is your choice, MBA – Maha Budhiman Aadmi, Placements – my goal and MBA Go Getters.
B. MBA Go Getters, Made in India, Laugh for a while, The Life is your choice, Placements my goal and MBA – Maha Budhiman Aadmi.
C. Made in India, MBA Go Getters, Placements – my goal, Laugh for a while, MBA – Maha Budhiman Aadmi and The Life is your choice.
D. Made in India, MBA Go Getters, MBA – Maha Budhiman Aadmi, The Life is your choice, Placements – my goal and Laugh for a while.

44. Mr. Peter gave his eldest son David a bag with 1000 gold coins. David took 230 gold coins from the bag and gave the rest to his younger brothers John, Joe and Jonathan, and advised them to distribute the balance left in the bag amongst themselves in proportion to their age which together amounted to 17.5 years. After a lot of deliberation and discussion John, Joe and Jonathan came to a conclusion to distribute the gold coins. Their methodology was as follows: As often John took 4 gold coins, Joe took 3. As often John took 6 gold coins Jonathan took 7. What was the age of John, Joe and Jonathan?

A. 6 years, 4.5 years and 7 years respectively
B. 5 years, 5.5 years and 7 years respectively
C. 6 years, 5 years and 6.5 years respectively
D. 5 years, 6.5 years and 6 years respectively

45. Mohan has an antique clock which strikes and makes loud gong sound every hour. It strikes the exact number of times indicating the time of the day or night. His clock takes seven seconds to strike Seven O'clock, how many seconds will his clock take to strike Eleven O'clock?

A. 11.22222227 seconds B. 11 seconds
C. 11.66666667 seconds D. None of these

46. Find the missing number : 2, 6, 20, 42, 110, —

A. 126 B. 156
C. 176 D. 196

47. Fifteen years back, Ms. Kalpana had three sons Ramesh, Suresh and Rajesh. The sum of the age of Ramesh, Suresh and Rajesh was equal to half of the age of their mother. It was during the next five years when Mahesh was born. Then the age of Ms. Kalpana was equal to the sum of the ages of all her children. Time went on and years passed and Dinesh was born and age of Ramesh equalled the sum of the ages of Rajesh and Mahesh. Now, it so happened that the sum of the ages of Ramesh, Suresh, Rajesh, Mahesh and Dinesh was double the age of their mother and was also equal to the sum of the ages of Suresh and Ramesh. Also Ramesh's age was equal to sum of the ages of Mahesh and Dinesh. What is the age of Ms. Kalpana?

A. 39 years B. 42 years
C. 41 years D. None of these

48. Find the missing number : –1, 0, 0, –, 8

A. 1 B. 2
C. 3 D. 4

49. Ceiling Fan : Table Fan : abcdefg : ....

A. abcdefg B. abcdefg
C. gfedcba D. None of these

Directions (Qs. 50 & 51): *There is a statement followed by two arguments. Choose the correct option which gives the decision on the arguments, which is derived from the statement.*

50. Given Statement: The new amendment of Corporate Social Responsibility (CSR) in India refers to bringing an overall positive impact on the communities, cultures, societies and environments. The fundamentals of CSR rest on the fact that not only public policy but even corporate should be responsible enough to address social issues.

Argument I: Government should not enforce companies to take up CSR.

Argument II: Companies moral responsibility is to take up CSR for a long run benefit.

A. Argument II is correct but Argument I is wrong
B. Argument I is correct but Argument II is wrong
C. Both the arguments are false
D. Both the arguments are correct

51. LPG subsidy is a stand taken by the government. Linking of Aadhar card to a bank account has been made mandatory for receiving the subsidy.

Argument I: All people should give up LPG subsidies.

Argument II: To fill the gap between poor and rich.

A. Argument II is the output of Argument I
B. Argument I is a byproduct of Argument II
C. Argument I and Argument II are complementing to each other
D. No logical link between Argument I and Argument II.

52. If CAB is coded as 723 –5 58 in a coded language then how will DAD be coded?

A. 4023 –5 4023 B. 4090 –5 4090
C. 1024 –5 1024 D. 1246 –5 1189

53. If MONEY is coded as 144 200 171 0 600 then DOLLAR is coded as ....

A. –9 200 119 119 –24 299
B. –8 200 106 106 –20 200
C. 122 200 102 102 10 154
D. 120 200 101 101 08 156

54. There are 100 MBA aspirants in a classroom and 99% of them are engineers. How many engineers must leave the classroom in order to reduce the percentage of engineers in the classroom to 98%?

A. 1 B. 2
C. 50 D. 90

55. Vikas was showing a photograph to his friend and pointed to a boy and told the following statement "His name is Atul and his maternal grandfather's brother is my maternal grandfather's sister's son." How is Atul related to Vikas?

A. They are brothers
B. Vikas is the uncle of Atul
C. They are distant cousins
D. None of the above

56. There was a party organized and the following members attended the party : Sheela, Amruta, Rohit, Rahul, Ajey, Ranveer and Gauri.

Sheela is mother-in-law of Amruta, who is sister-in-law of Rohit. Rahul is the father of Ajey. Ajey is the brother of Rohit. Ranveer is the only brother of Rahul and the father-in-law of Gauri. Gauri was married to Rohit. How is Sheela related to Ranveer?

A. Cousin B. Mother
C. Sister D. None of these

Directions (Qs. 57-60): *Using the information given below and answer the question.*

A chef is trying a recipe for a tasty ice cream using four ingredients. He can choose from three liquids for taste which are labelled A, B and C which are stable in nature and the choice for flavor can be from four liquids which are labelled W, X, Y and Z. For the new ice cream recipe to be tasty, there must be two liquids from the taste giving liquids. Also certain liquids cannot be mixed because of their reactions which makes it unhealthy for human consumption and the same is given below:

B cannot be mixed with W
C cannot be mixed with Y
Y cannot be mixed with Z

57. If the chef calculated that Y is the most important flavour and must be used in the recipe, which other ingredients must be part of the recipe?

A. A, B and W B. A, B and X
C. A, B and Z D. B, C and X

58. If the chef rejected B because of its possible side effects but decided to use Z, which is a possible combination of the four ingredients in the recipe?

A. A, C, W and Z B. A, X, Y and Z
C. A, W, X and Z D. A, C, Y and Z

59. Which of the following combination of liquids is impossible?

I. Using Y and W together
II. Using B and C together
III. Using W, X and Z together

A. I only B. II only
C. III and I only D. II and I only

60. Which of the following must always be true?

I. If C is used W must be added
II. If Y is used B must be added
III. If C is not used W cannot be added

A. I and II only B. II and III only
C. I, II and III only D. II only

61. Which letter from the options given will replace the " ? " in the table given below:

CAT	389376	DOG
RAT	1758276	MAT
CAB	15876	FAN
CAN	571536	?

A. MAN B. GIFT
C. PAN D. SOFT

62. Which word from the given options will replace the " ? " in the table given below:

WOE	1089	MISERY
VANQUISH	1308	SUBDUE
TACITURNITY	1547	SILENCE
SPLEEN	1806	?

A. MALEVOLENCE
B. JOIN
C. LINK
D. DRIP

63. Which number from the options given will replace the " ? " in the table given below:

MBA	728	NCB
BCA	24	CDB
BCS	24	CDT
FCA	?	GDB

A. 48 B. 168
C. 64 D. 132

64. What is the least number of straight lines needed to draw the following diagram?

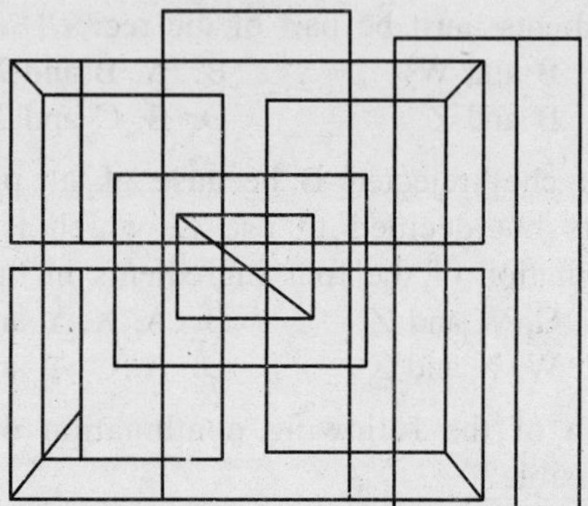

A. 39 B. 40
C. 42 D. None of these

65. If  THEN

 : ?

A. 

B. 

C. 

D. 

Directions (Qs. 66-70): *Based on the information given and below answer the questions.*

Ten friends Matt, Sam, Pat, Tom, Sid, Alex, Kat, Jim, Jane and John are having dinner on a rectangular table. Eight facing each other along the length of the table while two facing each other along the smaller side of the table. Pat is sitting diagonally opposite to Kat, Alex is facing Jim, John is to the right of Jane, Tom is sitting between Sam and Jim, Pat is to the extreme left of Jane who is sitting on the extreme right along the length of the table. Sid is facing Tom and Matt is on the right of Sam.

66. Who do you think is sitting in front of each other along the small sides of the table?
A. Matt and John
B. Matt and Sam
C. Matt and Kat
D. John and Sam

67. Sid is sitting between which of the following two?
A. Sam and Kat B. Pat and Alex
C. Alex and Jane D. Kat and Jane

68. Who are sitting diagonally opposite to each other?
A. Sid and Tom B. Sam and Jane
C. Alex and Jim D. Jane and Kat

69. Who is sitting to the immediate right of Jim?
A. Sam B. Jane
C. Alex D. Tom

70. Who is sitting two places left to Jane?
A. Pat B. Alex
C. Sid D. Sam

SECTION-III

GENERAL ENGLISH

For all questions in this section, each correct answer carries 1 mark

71. "To catch a tartar" means ....
A. To trap wanted criminal with great difficulty
B. To catch a dangerous person
C. To catch a person who is more than one's match
D. To meet with disaster

72. A leopard can't change its ....
A. Dots B. Stripes
C. Colour D. None of these

73. "So sober sometimes serious Sam smiles on silly things" is a/an ....
A. Hyperbole B. Assonance
C. Anaphora D. Alliteration

74. "The strength given by my mother is bigger than the cosmic energy in this cosmos" is ....
A. Rhyme B. Metaphor
C. Personification D. Hyperbole

75. The buzzing of bees is an example of ....

A. Simile
B. Metonymy
C. Onomatopoeia
D. Paradox

76. The word CACTI is of Latin origin. It can also be replaced by ....

A. Cactus
B. Cats
C. Cactuses
D. Cactusus

77. The word TROUSSEAUX is of French origin. It can also be replaced by ....

A. Troussers
B. Trousseaus
C. Troussears
D. None of these

78. The singular of the word SCARVES is spelt as ....

A. Scarve
B. Scarfe
C. Scarv
D. None of these

79. The "Drawing Pins" in British English is referred to as .... in American English.

A. Thumb pins
B. Board pins
C. Broad pins
D. Thumbtacks

80. "Aubergine" in Britain is referred to as .... in United States of America.

A. Migraine
B. Eggplant
C. Margarine
D. Egg

81. Base ball in American English is commonly referred to as .... in British English.

A. Run ball
B. Strike ball
C. Rounders ball
D. Rounders

Directions (Qs. 82-84): *Complete the collocation words.*

82. .... weapon

A. Nuclear
B. Atomic
C. Molecular
D. Electronic

83. Seminal ....

A. News
B. River
C. Nuance
D. Research

84. .... Percentage

A. Huge
B. Big
C. More
D. Large

85. The idiom "Against the clock" means

A. Break the Rules
B. Rushed and short on time
C. Go back to the Past
D. Look at the Past

86. The idiom "Buy a lemon" means ....

A. A superstitious way to say 'good luck'
B. A lie which is propaganda for people to believe
C. An unbelievable story which is told for people to believe
D. To purchase a vehicle that constantly gives problems or stops running after you drive it away

87. Kedar uses the following sentence to introduce himself. Choose the correct option.

A. Myself Kedar, I belong to Mumbai
B. Myself Kedar and I am from Mumbai
C. Myself Kedar from Mumbai
D. None of the above

88. Ram uses the following sentence to tell the time of the day to Shyam. Which one is the correct sentence?

A. It is 2 pm in the afternoon
B. It is 2 pm
C. It is 2 pm in the noon
D. It is 2 o'clock in the afternoon

Directions (Qs. No. 89-91): *In the questions from 89 to 91 parts have been underlined. If any rule of correct English is violated then it could be only in the UNDERLINED part, marked as (a), (b), (c) or (d). Choose the option, which violates usage of correct English.*

89. My Parents/*(a)* are Indians but/*(b)* I am/*(c)* born in Sydney/*(d)*

A. Only *(a)*
B. *(a)* and *(c)*
C. Only *(c)*
D. Only *(d)*

90. Standing/*(a)* on the top of the tower/*(b)* the whole city could be seen/*(c)*

A Only *(a)*
B. Only *(b)*
C. *(a)* and *(b)*
D. None of these

91. Ganesh is taller than/*(a)* Ramesh but/*(b)* Anoop is/*(c)* more taller/*(d)*.

A. Only *(a)*
B. Only *(b)*
C. Only *(c)*
D. Only *(d)*

Directions (Qs. No. 92-96): *Read the following passage and answer the questions.*

"There are several factors that contribute to wisdom. Of these I should put first a sense of proportion; the capacity to take account of all the important factors in a problem and to attach to each its due weight. This has become more difficult than it used to be owing to the extent and complexity of the specialized knowledge required of various kinds of technicians. Suppose, for example, that you are engaged in research in scientific medicine. The work is difficult and is likely to absorb the whole of your intellectual energy. You have no time to consider the effect which your discoveries or invention may have outside the field of medicine. You succeed (let us say), as modern medicine has succeeded, in enormously lowering the infant death-rate, not only in Europe and America, but also in Asia and Africa. This has the entirely unintended result of making the food supply inadequate and lowering the standard of life in the most populous parts of the world. To take an even more spectacular example, which is in everybody's mind at the present time-you study the composition of the atom from a disinterested desire for knowledge and incidentally place in the hands of powerful lunatics the means of destroying the human race. In such ways the pursuit of knowledge may become harmful unless it is combined with wisdom; and wisdom in the sense of comprehensive vision is not necessarily present in specialists in the pursuit of knowledge.

Comprehensiveness alone, however, is not enough to constitute wisdom. There must be, also, certain awareness of ends of human life. This may be illustrated by the study of history. Many eminent historians have done more harm than good because they viewed facts through the distorting medium of their own passions. Hegel had a philosophy of history which did not suffer from any lack of comprehensiveness, since it started from earliest time and continued into an indefinite future. But the chief lesson of history which he sought to inculcate was that from the year A.D. 400 down to his own time, Germany had been the most important nation and the standard bearer of progress in the world. Perhaps one could stretch the comprehensiveness that constitutes wisdom to include not only intellect but also feeling. It is by no means uncommon to find men/women whose knowledge is wide but those feelings are narrow. Such men/women lack what I am calling wisdom.

I think the essence of wisdom is emancipation, as far as possible, from the tyranny of the here and the now. We cannot help the egoism of our senses. Sight, sound and touch are bound up with our own bodies and cannot be made impersonal. Our emotions start similarly from ourselves. An infant feels hunger or discomfort; gradually with the years his horizon widens, and, in proportion as his thoughts and feelings become less personal and less concerned with his own physical states, he achieves growing wisdom. This is of course a matter of degree. No one can view the world with complete impartiality; however, it is possible to make a continual approach towards impartiality, on the one hand, by knowing things somewhat remote in time or space, and, on the other hand, by giving to such things their due weight in our feelings. It is this approach towards impartiality that constitutes growth in wisdom.

Perhaps in this sense the wisdom can be taught. I think that this teaching should have a larger intellectual element than has been customary in what has been thought of as moral instruction. I think that the disastrous result of hatred and narrow mindedness to those who fed them can be pointed out incidentally in the course of giving knowledge. Knowledge and morals ought not to be too much separated. It is true that the kind of specialized knowledge which is required for various kinds of skills has very little to do with wisdom. But it should be supplemented in education by wider surveys calculated to put it in its place in the totality of human activities. Even the best technicians should also be good citizens, *i.e.*, citizens of the world and not of any one nation. With every increase of knowledge and skill, wisdom becomes more necessary for every such increase augments our capacity of realizing our purposes, and therefore augments our capacity for evil, if our purposes are unwise. The world needs wisdom as it has never needed it before; and if knowledge continues to increase, the world will need wisdom in the future even more than it does now.

92. According to the author what results in growth of wisdom?
A. Widening knowledge and narrowing feelings
B. Acquiring specialized knowledge which is required for various kinds of skills
C. Viewing the world with complete impartiality
D. None of the above

93. According to the author the essence of wisdom is
A. Deliverance from the oppression of here and now
B. Subduing from the oppression of here and now
C. Captivity from the oppression of here and now
D. All of the above

94. What according to the author is the relationship between knowledge and wisdom?
A. As human wisdom increases there is increase in knowledge created
B. As knowledge keeps on increasing there is lesser need of wisdom
C. As knowledge keeps on increasing there is a higher need for wisdom
D. As growth in wisdom stops, knowledge creation stagnates

95. The example used by the author to explain the ways in which the pursuit of knowledge can be harmful, unless combined with wisdom, is
A. the space mission
B. medicine that lowers infant mortality across the world
C. the progress of Germany
D. None of the above

96. What factors according to the author, contribute to wisdom?
A. a sense of proportion, giving knowledge, study of history, emancipation
B. a sense of proportion, dignity, knowledge, skill
C. comprehensiveness, a sense of proportion, awareness of the end of human life, emancipation from the tyranny of the present
D. None of the above

Directions (Qs. No. 97 & 98): *Read each of the components of the given sentences and mark the components with grammatical error.*

97. (*i*) He is capable at
(*ii*) twisting any fact
(*iii*) without any suspicion
(*iv*) at any time.
A. Only (*i*) B. Only (*ii*)
C. Only (*iii*) D. Only (*iv*)

98. (*i*) My cousin brother, who lives
(*ii*) in Goa, is eager to visit us
(*iii*) in Mumbai and aspires to have
(*iv*) a glimpse of the city
A. Only (*i*) B. Only (*ii*) and (*iii*)
C. Only (*iv*) D. None of the above

Directions (Qs. No. 99 & 100): *Choose the correct word which best fits for the sentence to be complete and grammatically correct.*

99. It was no wonder that after the roads were closed with continuous snow fall, hotels started .... off the tourists.

A. Raking B. Taking
C. Beating D. Looting

100. When the penalty corner was saved, the players .... in toward the goal keeper to congratulate him.

A. closed B. went
C. crashed D. pooled

101. The synonym for the word "Inclement" is ....

A. Stormy B. Intimate
C. Advocacy D. Immediate

102. The antonym for the word "Taciturn" is ....

A. Garrulous B. Energetic
C. Ephemeral D. Enigmatic

103. Complete with the appropriate collocation word: .... activism.

A. Judicial B. Legal
C. Prosecutional D. Lawful

104. If Propensity : Tendency then ....

A. Prologue : Epilogue B. Master : Slave
C. Audacity : Impudence D. Conduct : Immorality

105. If Tepid : Hot then ....

A. Jealousy : Envy B. Hatred : Antipathy
C. Unity : Harmony D. Joy : Ecstasy

106. From the following words pick the odd word out.

A. Lampoon B. Satire
C. Ridicule D. Parable

107. From the following words pick the odd word out.

A. Euphemishm B. Maxim
C. Aphorism D. Dictum

108. From the following words pick the odd word out.

A. Force B. Intimidation
C. Shakedown D. Bleak

Directions (Qs. No. 109 & 110): *The questions 109 and 110 have sentences which are incomplete. Pick up one phrase/clause from the options given, that will complete the sentence logically.*

109. To ensure success in a difficult task ....

A. one needs to be persistent
B. persistence is needed
C. you need a person of persistence
D. persistence is what one needs

110. The more we looked at the piece of modern art ....

A. we liked it less
B. better we liked it
C. the less we liked it
D. we liked it more and more

SECTION-IV

QUANTITATIVE, DATA INTERPRETATION AND DATA SUFFICIENCY

For all questions in this section, each correct answer carries 1 mark

111. A batsman was having 32 runs per innings as his average after 15th innings. His average increased by 2 runs after 16th inning. Then what was his score in the 16th inning?

A. 64 B. 60
C. 46 D. 62

112. The least number which is a perfect square and is divisible by each of the numbers 14, 16, 18 is ....

A. 6048 B. 7056
C. 1008 D. 2046

113. Four people clap after every 20 minutes, 30 minutes, 40 minutes and 50 minutes respectively. All of them clapped together at 10.00 am. Then they will again clap together at ....

A. 3 pm B. 5 pm
C. 6 pm D. 8 pm

114. Three candidates "A", "B". "C" participated in an election. "A" gets 40% of the votes more than "B". "C" gets 20% votes more than "B". "A" also overtakes "C" by 4000 votes. If 90% voters voted and no invalid or illegal votes were cast, then what will be the number of voters in the voting list?

A. 72000 B. 80000
C. 70000 D. 78500

115. In a competitive exam there were 5 sections. 10% of the total number of students cleared the cut off in all the sections and 5% cleared none of the sections. From the remaining candidates 30% cleared only section 1, 20% cleared only section 2, 10% cleared only section 3 and remaining 1020 candidates cleared only section 4. How many students appeared in the competitive exam?

A. 2550 B. 2800
C. 3000 D. 3200

116. A man sold $3/5^{th}$ of his articles at a gain of 20% and the remaining at cost price. Find the gain earned in the transaction.

A. 8 B. 10
C. 12 D. 14

117. A trader sells 20 articles at ₹ 54 per article after giving 10% discount and gains 50% profit. If the discount is not given, the profit gained is ....

A. 56.76% B. 66.66%
C. 62.66% D. 63.66%

118. A bottle contains 50 liters of milk. From this bottle 5 liters of milk was taken out and replaced by water. This process was repeated further for three times. How much milk is now contained in the bottle?

A. 32.8 liters B. 34.4 liters
C. 36.8 liters D. 46.5 liters

119. A ball is dropped from a height of 200 meters. After striking the floor it re-bounces to 4/5th of the height from where it fell. The total distance it travels before coming to rest is ....

A. 1200 meters B. 1600 meters
C. 1800 meters D. 1820 meters

120. The sum of all two digit numbers that give a remainder 2 when they are divided by 7 is ....

A. 552 B. 654
C. 658 D. 684

121. A man covers half of his journey by train at 90 km/hr, one-third of the remainder by bus at 30 km/hr and the rest by cycle at 10 km/hr. The average speed during the entire journey is ....

A. 22.5 km/hr B. 28.5 km/hr
C. 30.0 km/hr D. 32.5 km/hr

122. John's grandfather was five times older to him 5 years ago. He would be two times of his age after 25 years from now. What is the ratio of John's age to that of his grandfather?

A. 7:11 B. 5:11
C. 3:11 D. 4:11

123. A number when successively divided by 5 and 6 gives remainders 3 and 2 respectively. What will be the remainders if the number is successively divided by 3 and 4?

A. 2, 3 B. 2, 1
C. 1, 2 D. 3, 4

124. How many zeros would be there in 1024!

A. 240 B. 248
C. 256 D. 252

125. If $x = 3 + 2\sqrt{2}$ what will be the value of $x^2 + \frac{1}{x^2}$?

A. 35 B. 32
C. 36 D. 34

126. The unit digit in the final solution when, 13*27*63*51*98*46 is ....

A. 4 B. 8
C. 2 D. None of the above

127. A dishonest seller sells his grocery items using a false weight and thus gains 5% for a kilogram, he uses the weight of approximately ....

A. 940.251 B. 943.123
C. 948.238 D. 952.381

128. A, B and C can do a work in 6, 8 and 12 days respectively. If they do the work together and earn ₹ 2700, what is the share of C in that amount?

A. 600 B. 900
C. 1000 D. 700

129. How many words each of two vowels and three consonants can be formed from the letters of the word "UNIVERSAL"?

A. 7000 B. 7200
C. 7400 D. 7800

130. Three pipes A, B and C can fill a tank in 12 hours. All the pipes started working together and after 3 hours, C is closed. If A and B can fill the remaining part in 10 hours, then the number of hours taken by C alone to fill the tank is ....

A. 100 hours B. 110 hours
C. 120 hours D. 130 hours

131. If the numbers between 1 to 65 which will be divisible by 4 are taken and then if the number present in the units places and tens places is swapped, post which they are written in ascending order, then which of the following number will be at 10th place from the last?

A. 40 B. 24
C. 44 D. 25

132. Ajit, Ravi and Hari were trying to hit a target. If Ajit hits the target 5 times in 8 attempts, Ravi hits it 3 times in 5 attempts and Hari hits it 2 times in 4 attempts. What is the probability that the target is hit by at least 2 persons?

A. 31/80 B. 49/80
C. 09/80 D. 03/16

133. If $\frac{1}{2}\log x + \frac{1}{2}\log y + \log 2 = \log(x + y)$ then ....

A. $x = -y$ B. $x = y + 1$
C. $x = y$ D. $y = x + 1$

134. $\log_5 25 + \log_2 (\log_3 81)$ is ....

A. 1 B. 2
C. 3 D. 4

135. Peter was standing on the top of a rock cliff facing the sea. He saw a boat coming towards the shore. As he kept seeing time just flew. Ten minutes less than half of an hour, the angle of depression changed from 30 to 60. How much more time in minutes will the boat take to reach the shore?

A. 5 B. 10
C. 15 D. 20

136. In a school where there was a compulsion to learn at least one foreign language from the choice given to them, namely German, French and Spanish. Twenty eight students took French, thirty took German and thirty two took Spanish. Six students learnt French and German, eight students learnt German and Spanish, ten students learnt French and Spanish. Fifty four students learnt only one foreign language while twenty students learnt only German. Find the number of students in the school.

A. 60 B. 62
C. 70 D. None of these

137. Sonali can solve 70% of the problems in a competitive exam and Nirali can solve only 60% in the same exam. What is the probability that at least one of them will solve a problem, provided selection of questions is done randomly from the same exam?

A. 0.82 B. 0.88
C. 0.62 D. 0.72

138. ₹ XYZ was deposited at simple interest at a specific rate for 3 years. Had it been deposited at 2% higher rate, it would have fetched ₹ 360 more. Find ₹ XYZ.

A. ₹ 5500 B. ₹ 5000
C. ₹ 6000 D. ₹ 4500

139. A man invests certain amount at 6% per annum simple interest and another amount at 7% per annum simple interest. His income from the interest after 2 years was ₹ 348. The ratio of first amount to second is 4:5. Find the total amount invested.

A. ₹ 2600 B. ₹ 2900
C. ₹ 2700 D. None of these

140. A person has a bag which contains 9 bulbs out of which 2 are fused and cannot be used to lighten the room. Two bulbs are selected at random. What is the probability that all the two bulbs chosen can be used to lighten the room?

A. 5/12 B. 7/12
C. 9/12 D. 10/12

141. The value of $(p - a) * (p - b) * (p - c)$ .... $* (p - z)$ is ....

A. A complex polynomial which starts with p^{24}
B. Zero
C. A complex polynomial which starts with p^{26}
D. A complex polynomial which has several variables including p^{26} and p^{24}

142. There are nine humans in a ship, each human has nine cages and each cage has nine huge lions and each lion has nine cubs. How many legs are there in the ship? (Human have two legs, lions have four legs, cubs have four legs.)

A. 747 B. 3258
C. 29178 D. 26561

143. As shown in the figure, there is a square of 24 cm. A circle is inscribed inside the square. Inside the circle are four circles of equal radius which are inscribed. The total area of the shaded region in the figure given below is ....

A. 576 – 196 π
B. 584 – 196 π
C. 864 – 196 π
D. None of the above

144. In the figure given below the value of x and y would be ....

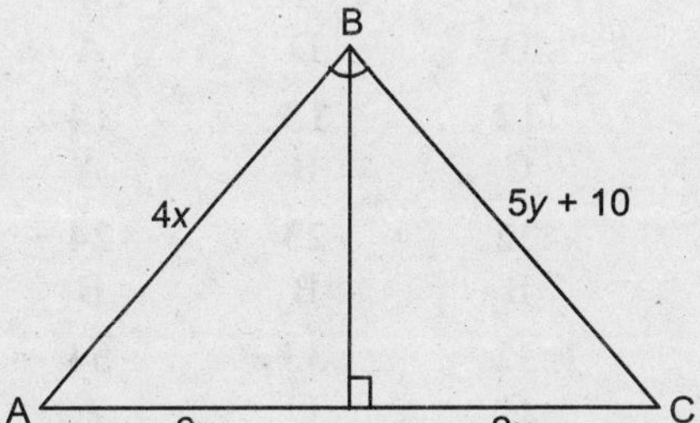

A. 10 and 15 B. 15 and 10
C. 06 and 12 D. 12 and 06

145. Mr. Suresh went into a restaurant and had a happy lunch for ₹ 162. He paid for the same using a 500 rupee note. He was so happy with the meal and hence purchased a mini snack box for ₹ 37 and paid for the same using a 100 rupee note. Mr. Suresh ensured that he collected the balance from the cashier in both cases. Although he was satisfied with the service, he did not pay any tips to the server. The next day when the cashier went and deposited the money in the bank, it was found by the banker that only the two currency notes which were given by Mr. Suresh were counterfeit notes. As per policy, the bank immediately tore both the notes using their shredding machine. What is the total loss to the restaurant in this transaction?

A. 1200
B. 600
C. 7991
D. None of the above

146. If $2^x + 2^{x+1} = 48$ then the value of x^x is ....

A. 04 B. 64
C. 256 D. 16

Directions (Qs. 147-150): *Following table shows the percentage population of six states below poverty line and proportion of male female.*

State	% Population below Poverty line	Proportion of male & female: Below Poverty line M : F	Proportion of male & female: Above Poverty line M : F
A	16	2:3	3:4
B	10	4:3	5:2
C	14	3:4	2:3
D	20	5:2	4:3
E	25	4:1	2:1
F	20	2:3	4:1

147. If the total population of state A is 5000 then what is the number of females above poverty line in State A?

A. 2000 B. 2400
C. 2600 D. Data inadequate

148. If the population of C & D together is 20,000. What is the total numbers of females below poverty line in the above states?

A. 5000 B. 6000
C. 7200 D. None of the above

149. If the population of males below poverty line in state C is 6000 & State E is 10,000, then what is the ratio of the total population of state C & E is:

A. 2 : 1 B. 3 : 5
C. 1 : 5 D. None of the above

150. If in State F population of females below poverty line is 16,000, then what is the population of males below poverty line in that state?

A. 8000 B. 6000
C. 12000 D. None of the above

ANSWERS

1	**2**	**3**	**4**	**5**	**6**	**7**	**8**	**9**	**10**
C	D	D	A	C	B	A	A	C	A
11	**12**	**13**	**14**	**15**	**16**	**17**	**18**	**19**	**20**
D	C	B	A	A	C	C	C	A	B
21	**22**	**23**	**24**	**25**	**26**	**27**	**28**	**29**	**30**
C	B	B	B	C	B	B	D	B	C
31	**32**	**33**	**34**	**35**	**36**	**37**	**38**	**39**	**40**
A	C	C	A	C	C	D	B	A	A
41	**42**	**43**	**44**	**45**	**46**	**47**	**48**	**49**	**50**
C	A	B	A	C	B	D	C	B	A
51	**52**	**53**	**54**	**55**	**56**	**57**	**58**	**59**	**60**
A	B	A	B	A	D	B	A	C	B
61	**62**	**63**	**64**	**65**	**66**	**67**	**68**	**69**	**70**
C	D	B	A	D	A	B	B	D	C
71	**72**	**73**	**74**	**75**	**76**	**77**	**78**	**79**	**80**
B	D	D	D	C	C	B	D	D	B
81	**82**	**83**	**84**	**85**	**86**	**87**	**88**	**89**	**90**
D	A	D	D	B	C	D	D	B	D
91	**92**	**93**	**94**	**95**	**96**	**97**	**98**	**99**	**100**
D	C	A	C	D	C	A	B	A	D
101	**102**	**103**	**104**	**105**	**106**	**107**	**108**	**109**	**110**
A	A	A	C	D	D	A	D	A	C
111	**112**	**113**	**114**	**115**	**116**	**117**	**118**	**119**	**120**
A	B	C	B	C	C	B	C	C	B
121	**122**	**123**	**124**	**125**	**126**	**127**	**128**	**129**	**130**
A	C	C	D	D	A	D	B	B	
131	**132**	**133**	**134**	**135**	**136**	**137**	**138**	**139**	**140**
A	B	C	D	B	C	B	C	D	B
141	**142**	**143**	**144**	**145**	**146**	**147**	**148**	**149**	**150**
	C	A	B	B	C	B	D	D	D

EXPLANATORY ANSWERS

43. Producer's Criteria : MBA Go Getters, Made in India, Laugh for a while, The Life is your choice, Placements–My Goal and MBA – Maha Budhiman Aadmi.

44. Ratio of Coins : $\frac{\text{John}}{\text{Joe}} = \frac{4}{3}$

Ratio of Coins : $\frac{\text{John}}{\text{Jonathan}} = \frac{6}{7}$

Ratio of Coins ⇒ John : Joe : Jonathan : : 6 : 4.5 : 7

Sum of years = 17.5

i.e., $6x + 4.5x + 7x = 17.5$

$17.5x = 17.5$

$x = 1$

Hence, Age of John = $6x$ = 6 years

Age of Joe = $4.5x$ = 4.5 years

Age of Jonathan = $7x$ = 7 years.

46.

2	6	20	42	110	156
↓	↓	↓	↓	↓	↓
1×2	2×3	4×5	6×7	10×11	12×13

47. Let present age of Ramesh, Suresh, Rajesh and Kalpana are w, x, y and z years respectively.

From question, 15 years back,

$$w + x + y - (15 + 15 + 15) = \frac{z-15}{2}$$

$$2(w + x + y) - 90 = z - 15$$

$$2\,(w + x + y) - z = 75 \qquad ...(i)$$

Again, 10 years back,

$$(w + x + y) - (10 + 10 + 10) = z - 10$$

$$(w + x + y) - z = 20 \qquad ...(ii)$$

From (*i*) and (*ii*), $z = 75 - 40$

= 35 years

So, Ms. Kalpana's age = 35 years

52.

C	A	B
|	|	|
$(3)^6 - 6 = 723$	$(1)^6 - 6 = -5$	$(2)^6 - 6 = 58$

Similarly,

D	A	D
|	|	|
$(4)^6 - 6 = 4090$	$(1)^6 - 6 = -5$	$(4)^6 - 6 = 4090$

53.

	M	O	N	E	Y
Letter Position:	13	15	14	5	25
	$(13)^2 - 25$ = 144	$(15)^2 - 25$ = 200	$(14)^2 - 25$ = 171	$(5)^2 - 25$ = 0	$(25)^2 - 25$ = 600

Similarly,

	D	O	L	L	A	R
Letter Position:	4	15	12	12	1	18
	$(4)^2 - 25$ = –9	$(15)^2 - 25$ = 200	$(12)^2 - 25$ = 119	$(12)^2 - 25$ = 119	$(1)^2 - 25$ = –24	$(18)^2 - 25$ = 299

54. Engineers – 99

Percentage of Engineers in the classroom

$$= 99 \times \frac{98}{100} = 96.02 \approx 97$$

Number of Engineer in order to reduce

= 99 – 97 = 2.

56.

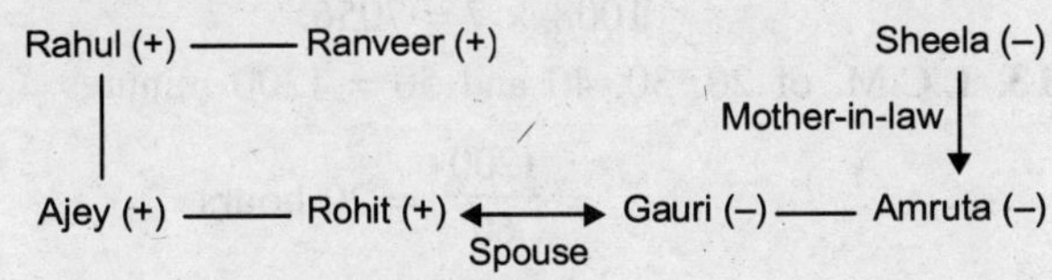

Hence, Correct option is D.

57. Liquid for taste : A, B and C

Liquid for flavour : W, X, Y and Z

If Y is selected, then C & Z are not be selected. So possible choice is A, B, X only.

58. If B is not selected then W may be selected. But if Z is selected, Y is not selected so, possible combination A, C, W and Z.

59. I. If Y is selected, C and Z is not selected. A and B should must be included. So W can't be selected.

II. Possible combination is B, C, X and Z.

III. W, X and Z can't be together.

Hence, correct answer is (C).

60. I. If C is used, W can't be used.

II. If Y is used, C and Z can't be used. So, B can be used.

III. If C is not used, A and B must be used so, W can't be used.

Hence, correct answer is (C).

63.

M	B	A	= $(13 + 14)^2 - 1$ = 728	N	C	B
13	2	1		14	3	2
B	C	A	= $(2 + 3)^2 - 1$ = 24	C	D	B
2	3	1		3	4	2
B	C	S	= $(2 + 3)^2 - 1$ = 24	C	D	T
2	3	19		3	4	20
F	C	A	= $(6 + 7)^2 - 1$ = 168	G	D	B
6	3	1		7	4	2

64. Least number of Straight lines needed to draw the diagram is 39.

66-70.

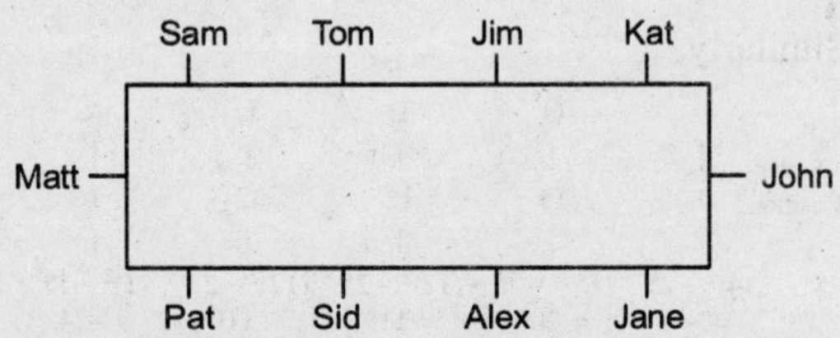

111. Let the Score in the 16th inning is n runs

Then, Sum of runs in 15 inning = 32 × 15 = 480 runs

Sum of runs in 16 inning = 34 × 16 = 544

Run made in 16th inning = 544 – 480 = 64 runs.

112. L.C.M. of 14, 16 and 18 = 1008

Least perfect square number divisible by

$$1008 \times 7 = 7056.$$

113. L.C.M. of 20, 30, 40 and 50 = 1200 minutes

$$= \frac{1200}{60} = 20 \text{ hours}$$

Time when all clap together

$$= 10.00 + 20 \text{ hours} = 6 \text{ p.m.}$$

114. Let number of voters in the voter list is x and number of votes B's got is n

Then, number of votes A got = 1.4 n

Number of votes C got = 1.2 n

From question, $1.4n - 1.2n = 4000$

$$0.2n = 4000$$

$$n = 20000$$

Again, total poll $= 0.9x$

$$1.4n + 1.2n + n = 0.9x$$

$$3.6n = 0.9x$$

$$x = \frac{3.6n}{0.9} = 4n$$

$$x = 4 \times 20000 = 80000.$$

115. Let number of students appeared in the competitive exam is x

Number of students cleared all sections

$$= x \times \frac{10}{100} = 0.1x$$

Number of students cleared none of the section

$$= x \times \frac{5}{100} = 0.05x$$

Remaining students $= x - 0.1x - 0.05x = 0.85x$

Number of students, cleared Section–1

$$0.85x \times \frac{30}{100} = 0.85x \times 0.3$$

Number of students, cleared Section–2

$$0.85x \times \frac{20}{100} = 0.85x \times 0.2$$

Number of students, cleared Section–3

$$0.85x \times \frac{10}{100} = 0.85x \times 0.1$$

Number of students, cleared Section–4

$$= 0.85x(1 - 0.3 - 0.2 - 0.1) = 1020$$

$$= 0.85x \times 0.4 = 1020$$

$$x = \frac{1020 \times 1}{0.85 \times 0.4} = 3000.$$

116. Let the cost price of an article = x

$$\text{Selling Price} = x \times \frac{120}{100} = 1.2x$$

Total selling price $= 3 \times 1.2x + 2x = 5.6x$

$$\text{Profit earned} = \left(\frac{5.6x - 5x}{5x}\right) \times 100 = 12.$$

117. Selling price of 20 articles = 20 × 54 = 1080

$$\text{Selling price after discount} = 1080 \times \frac{90}{100} = 972$$

$$\text{Cost price} = 972 \times \frac{100}{150} = 648$$

When discount is not offered, then

$$\text{Profit \%} = \left(\frac{1080 - 648}{648}\right) \times 100 = 66.66\%.$$

118. Ratio of milk and water after first replacement

$$= 45 : 5 \Rightarrow 9 : 1$$

On second replacement

$$\text{Ratio} = \left(45 - \frac{5 \times 9}{10}\right) : \left(5 - \frac{5}{10}\right) + 5$$

$$= 40.5 : 9.5 \Rightarrow 81 : 19$$

On third replacement

$$\text{Ratio} = \left(40.5 - 5 \times \frac{81}{100}\right) : \left(9.5 - 5 \times \frac{19}{100}\right) + 5$$

$$= 36.45 : 13.55$$

Milk contained in the bottle = 36.45 ≈ 36 liters

119. Total distance covered

$$= 200 + 2\left[200 \times \frac{4}{5} + 200 \times \left(\frac{4}{5}\right)^2 + \ldots.\infty\right]$$

$$= 200 + 2 \times 200 \times \frac{4}{5}\left[1 + \frac{4}{5} + \ldots.\infty\right]$$

$$= 200 + \frac{1600}{5}\left[\frac{1}{1 - \frac{4}{5}}\right] = 200 + \frac{1600 \times 5}{5} = 1800.$$

120. First two digit number that gives remainder 2 when divided by 7 is 16

and Last two digits number = 93

Common difference between two consecutive number

$$= 7$$

Number of such numbers

$$93 = 16 + (n - 1) \times 7$$

$$\therefore \quad n = 12$$

Sum of such numbers $= \frac{n}{2}(16+93)$

$$= \frac{12}{2}[16+93] = 6 \times 109 = 654.$$

121. Let total distance is d km

Time taken to cover half of his journey by train

$$t_1 = \frac{\frac{d}{2}}{90} = \frac{d}{180} \text{ hr.}$$

Time taken to cover one-third of the remainder by bus

$$t_2 = \frac{\frac{d}{2} \times \frac{1}{3}}{30} = \frac{d}{180} \text{ hr.}$$

Time taken to cover remaining distance by cycle

$$t_3 = \frac{\frac{d}{3}}{10} = \frac{d}{30} \text{ hr.}$$

$$\text{Average speed} = \frac{\text{total distance}}{\text{total time taken}}$$

$$= \frac{d}{\frac{d}{180} + \frac{d}{180} + \frac{d}{30}} = \frac{180}{8} = 22.5 \text{ km/hr.}$$

122. Let present age of John and his grandfather is x and y respectively.

From question,

5 years ago, $(y - 5) = 5(x - 5)$

$\Rightarrow \quad 5x - y = 20 \quad ...(i)$

and, 25 years from now

$y + 25 = 2(x + 25)$

$2x - y = -25 \quad ...(ii)$

From (i) and (ii), we have,

$x = 15$ years

and $y = 55$ years

Ratio $x : y = 15 : 55 = 3 : 11.$

125.

$$x = 3 + 2\sqrt{2}$$

$$\frac{1}{x} = 3 - 2\sqrt{2}$$

$$\left(x + \frac{1}{x}\right)^2 = x^2 + \frac{1}{x^2} + 2$$

$$\left(3+2\sqrt{2}+3-2\sqrt{2}\right)^2 = x^2 + \frac{1}{x^2} + 2$$

$$(6)^2 = x^2 + \frac{1}{x^2} + 2$$

$$\therefore \quad x^2 + \frac{1}{x^2} = 34.$$

126. $13 \times 27 \times 63 \times 51 \times 98 \times 46 = 5083955604$

$\therefore$ Unit digit in the final solution is 4.

127. The weight he uses $= 1000 \times \frac{100}{105} = 952.381.$

129. Vowel in the given word 'UNIVERSAL' is A, E, I, U

Consonant in the given word 'UNIVERSAL' is N, V, R, S and L.

Required number of words thus formed

$$= \left({}^4C_2 \times {}^5C_3\right) \times 5!$$

$$= (6 \times 10) \times 120 = 7200.$$

131. Numbers between 1 to 65 divisible by 4 are

04, 08, 12, 16, 20, 24, 28, 32, 36, 40, 44, 48, 52, 56, 60, 64

Numbers after swapping its digits

40, 80, 21, 61, 02, 42, 82, 23, 63, 04, 44, 84, 25, 65, 06, 46

Numbers in ascending order

02, 04, 06, 21, 23, 25, 40, 42, 44, 46, 61, 63, 65, 80, 82, 84.

132. Probability for hitting target by

Ajit, P(A) $= \frac{5}{8}$; Ravi, P(R) $= \frac{3}{5}$; Hari, P(H) $= \frac{2}{4} = \frac{1}{2}$

Probability for hitting the target by at least two

$$= \sum P(A).P(R).P'(H) + \sum P(A).P(R).P(H)$$

$$= \frac{5}{8} \times \frac{3}{5} \times \left(1 - \frac{1}{2}\right) + \frac{5}{8}.\left(1 - \frac{3}{5}\right).\frac{1}{2} + \left(1 - \frac{5}{8}\right).\frac{3}{5}.\frac{1}{2} + \frac{5}{8}.\frac{3}{5}.\frac{1}{2}$$

$$= \frac{3}{16} + \frac{1}{8} + \frac{9}{80} + \frac{3}{16} = \frac{15+10+9+15}{80} = \frac{49}{80}.$$

133.

$$\frac{1}{2}\log x + \frac{1}{2}\log y + \log 2 = \log (x + y)$$

$$\log\sqrt{x} + \log\sqrt{y} + \log 2 = \log (x + y)$$

$$\log 2\left(\sqrt{xy}\right) = \log(x + y)$$

$$e^{(x+y)} = e^{2\left(\sqrt{xy}\right)}$$

$$x + y = 2\sqrt{xy}$$

$$\left(\sqrt{x}\right)^2 + \left(\sqrt{y}\right)^2 - 2\sqrt{x}.\sqrt{y} = 0$$

$$\left(\sqrt{x} - \sqrt{y}\right)^2 = 0$$

$$\therefore \quad x = y.$$

134. $\log_5 25 + \log_2(\log_3 81)$

$= \log_5 5^2 + \log_2 (\log_3 3^4)$

$= 2\log_5 5 + \log_2 4$

$= 2 + \log_2 2^2 = 2 + 2 = 4.$

135. 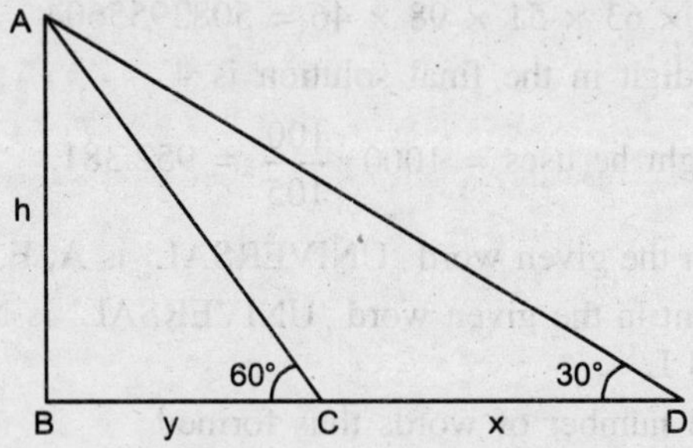

From, ΔABD, $\tan(30°) = \frac{h}{x+y}$

$$\frac{1}{\sqrt{3}} = \frac{h}{x+y}$$

$$\sqrt{3}h = x + y \quad ...(i)$$

Again, from ΔABC,

$$\tan(60°) = \frac{h}{y} \Rightarrow \sqrt{3} = \frac{h}{y}$$

$$h = \sqrt{3}y \quad ...(ii)$$

From (*i*) and (*ii*), $3y = x + y \Rightarrow x = 2y$

From question, time taken to cover distance x is 20 minutes.

Time taken to cover distance $y = \frac{x}{2} = \frac{20}{2} = 10$ minutes.

136. Let, number of students took German, $n(G)$
number of students took French, $n(F)$
number of students took Spanish, $n(S)$
then, $n(F) = 28$, $n(G) = 30$, $n(S) = 32$
$n(F \cap G) = 6$, $n(G \cap S) = 8$, $n(F \cap S) = 10$
only $n(G) = 20$
only one language = 54

$$n(F \cap G \cap S) = \text{only } n(G) - n(S) + n(F \cap G) + n(G \cap S)$$
$$= 20 - 30 + 6 + 8 = 4$$

$$n(F \cup G \cup S) = n(S) + n(F) + n(G) - n(S \cap F) - n(F \cap G) - n(G \cap S) + n(S \cap F \cap G)$$
$$= 32 + 28 + 30 - 10 - 6 - 8 + 4$$
$$= 70.$$

137. Probability for solving a problem by

$$\text{Sonali, } P(S) = \frac{70}{100} = 0.7$$

$$\text{Nirali, } P(N) = \frac{60}{100} = 0.6$$

Probability for not solving a problem by

$$\text{Sonali, } P'(S) = 1 - \frac{70}{100} = 0.3$$

$$\text{Nirali, } P'(N) = 1 - \frac{60}{100} = 0.4$$

Probability for non of them solve a problem

$$P'(S).P'(N) = 0.3 \times 0.4 = 0.12$$

Probability for at least one of them solve a problem

$$= 1 - 0.12 = 0.88$$

138. Let amount XYZ is P, rate is $r\%$ then,

$$\text{S.I.} = \frac{P \times r \times 3}{100}$$

when, rate is increases by 2%,

then, $\text{S.I.} = \frac{P \times (r+2) \times 3}{100}$

From question,

$$\frac{3P(r+2)}{100} - \frac{3Pr}{100} = 360$$

$$\frac{6P}{100} = 360 \Rightarrow P = 6000$$

$\therefore$ ₹ XYZ = 6000.

139. Total S.I. = 348

$$\frac{4x \times 6 \times 2}{100} + \frac{5x \times 7 \times 2}{100} = 348$$

$$\frac{48x + 70x}{100} = 348$$

$$\frac{118x}{100} = 348 \Rightarrow x = \frac{348 \times 100}{118}$$

Total amount invested = $9x$

$$= \frac{9 \times 348 \times 100}{118} = ₹\ 2654.2$$

140. Number of bulb that can be used to lighten the room

$$= 9 - 2 = 7$$

$$\text{Required probability} = \frac{{}^7C_2}{{}^9C_2} = \frac{7 \times 3}{9 \times 4} = \frac{7}{12}.$$

142. No. of Men = 9
No. of Cages = 9 × 9
No. of Lions = 9 × 9 × 9
No. of Cubs = 9 × 9 × 9 × 9

No. of Legs, Men = 9 × 2 = 180
Lions = 9 × 9 × 9 × 4 = 2916
Cubs = 9 × 9 × 9 × 9 × 4 = 26244

$\therefore$ Number of legs = 29178.

145. Loss of the restaurant = 500 + 100 = ₹ 600.

146. $2^x + 2^x + 1 = 48$

Let $2^x = y$ then, $y + 2y = 48$

$$3y = 48 \Rightarrow y = 16$$
$$2^x = 16$$
$$2^x = 2^4 \Rightarrow x = 4$$
$$x^x = 4^4 = 256.$$

Previous Paper (Solved)

Symbiosis National Aptitude Test (SNAP), 2014

SECTION- I

GENERAL ENGLISH

For all questions in this section, each correct answer carries 1 mark

1. Which of these is not a compound word?
A. Hard-wired B. Run-of-the-mill
C. Bottleneck D. Minestrone

2. Find the antonym of 'Native' is
A. Tarnished B. Characteristic
C. Intrinsic D. Exotic

3. A shortfall in supplies would knock the bottom out of the engineering industry. What does "knock the bottom out of" stand for?
A. Make it baseless
B. Shake its foundation
C. Cause it to collapse
D. None of the above

4. Choose the word with correct spellings
A. Demurrage
B. Demurage
C. Demmurrage
D. Demurragge

Directions (Qs. Nos. 5 to 9): *Fill in the blanks with proper answer:*

5. .... the stable door after the horse has bolted.
A. Lock B. Shut
C. Close D. Open

6. There were .... viewers than expected.
A. Fewer B. Less
C. Many D. Lesser

7. .... when, after waiting for hours, we were told the tickets were sold out.
A. We were at the end of our tether
B. We were on edge
C. Tempers frayed
D. Nerves frayed

8. Your proficiency in the French language depends on how hard you work. .... if you live in the culture of your target language, you will learn even more quickly.
A. Likewise B. Hence
C. Furthermore D. In fact

9. The government is planning to sell PSUs .... BHEL, SAIL and NTPC .... first wants a regulatory agency .... the IRDA to assess their worth.
A. like/and/like B. like/but/such as
C. such as/so/like D. such as/but/like

10. Choose the sentence which is grammatically correct
A. I was struck in trafficjam
B. I struck in traffic jam
C. I was struck up in traffic jam
D. I was struck upon traffic jam

11. Choose the option having correct word order in the following statements:
A. Immigrant communities in the U.S. have brought renewed pressure after the catastrophic terrorist attacks on the World Trade Center in New York in 2001.
B. The aftermath of the catastrophic terrorist attacks on the World Trade Center in New York in 2001 has brought renewed pressure on immigrant communities in the U.S.
C. The World Trade Center, after the catastrophic terrorist attacks in New York in 2001, has brought renewed pressures on immigrant communities in the U.S.
D. In 2001, the aftermath of the catastrophic terrorist attacks on the World Trade Center in New York has brought renewed pressures on immigrant communities in the U.S.

12. The football team travelled .... Train.
A. in B. by
C. from D. on

13. Choose the correct synonym for: **Attenuate**
A. Enervate B. Attention
C. Awaken D. Trigger

14. Choose the antonym for: **Cajole**
A. Dissuade B. Beguile
C. Flatter D. Blandish

15. Choose the correct synonym for: **Consanguinity**
A. Affinity B. Corpulent
C. Estrangement D. Anarchy

Directions (Qs. Nos. 16 to 22): *Fill in the blanks with proper answer:*

16. Autotomy is the ability of an animal to .... a part, of its body at will to save itself from predators and attackers.

A. Cast off B. Cast away
C. Cut off D. Cut down

17. They all stared .... as the door of the dilapidated house swung open.

A. with bated breath
B. in beeline
C. in the same boat
D. a bone of contention

18. But at a time when business values are falling, stakeholders tend not to take kindly to .... at a knock-down price.

A. Selling the family silver
B. Selling one's soul
C. Selling yourself short
D. Selling someone a pup

19. Although this political party has reputation for ...., its behaviour towards the public has always been ...., suggesting it may not be as insolent as has been projected by its competitors.

A. Vendibility, Impertinent
B. Secrecy, Uniform
C. Temerity, Courteous
D. Nepotism, Varied

20. The company believes in getting value for money and it is wrong of the employees to view its .... as ....

A. Forehandedness, Stinginess
B. Reserve, Levity
C. Insolence, Shortage
D. Indolence, Destitution

21. In the midst of radiantly beautiful women, this homely women looked strikingly .... but it transpired later that she was an arbitrator and not ....

A. unattractive, an adjudicator
B. incongruous, a contestant
C. dejected, a raconteuse
D. personable, an audience

22. This author is not outspoken on political issues and her writings are more famous for their aesthetic qualities than for their .... ones.

A. Polemical
B. Commercial
C. Contemporary
D. Dramatic

23. Choose the word with correct spellings

A. Perniceous B. Pernicious
C. Perniciuos D. Pernecious

Directions (Qs. Nos. 24 to 26): *Read the given passage and answer the questions.*

Read not to contradict and confute, nor to believe and take for granted, nor to find talk and discourse, but to weigh and consider. Some books are to be tasted, others to be swallowed, and some few to be chewed and digested; that is, some books are to be read only in parts; others to be read, but not curiously; and some few to be read wholly, and with diligence and attention. Some books may also be read by deputy, and extracts made of them by others; but that would be only in the less important arguments and the meaner sort of books; else distilled books are like common distilled waters, flashy things. Reading makes a full man, conference a ready man, and writing an exact man. And therefore, if a man writes little, he need have a good memory; if he confers little, he need have a present wit; and if he reads little, he need have much cunning to seem to know that he doth not.

Histories make men wise, poets witty, the mathematics subtitle, natural philosophy deep, moral grave, logic and rhetoric able to contend.

(Source : Of-studies - Francis Bacon)

24. What should be the real object of reading?

A. To distil the contents
B. To understand author's views and their application in day-to-day life
C. To have a good memory
D. To weigh and consider

25. The word 'confute' in the passage means

A. To argue B. To refuse
C. To disapprove D. To doubt

26. What can be done with regard to the 'meaner' sort of books?

A. To be read but not to contradict and confute
B. To be read curiously
C. To be read, but not curiously
D. They can be "read by deputy and extracts made of them by others."

Directions (Qs. Nos. 27 to 32): *Read the given passage and answer the questions.*

Problems Facing Potential Exporters

Many firms fail because when they begin exporting, they have not researched the target markets or have not developed an international marketing plan to be successful. A firm must clearly define goals, objectives and potential problems. Secondly it must develop a definitive plan to accomplish its objective regardless of the problems involved. Unless the firm is fortunate enough to possess a staff with considerable expertise, it may not be able to take this crucial first step without qualified outside guidance.

Often top management is not committed enough to overcome the initial difficulties and financial requirements of exporting. It can often take more time and effort to establish a firm in a foreign market than in the domestic one. Although the early delays and costs involved in exporting may seem difficult to justify when compared to established domestic trade the exporter should take a more objective view of this process and carefully monitor international marketing efforts through these early difficulties. If a good foundation is laid for export business the benefits derived should eventually outweigh the investment.

Another problem area is in the selection of the foreign distributor. The complications involved in overseas communications and transportation require international distributors to act with greater independence than their domestic counterparts also since a new exporter's trademarks and reputation is usually unknown in the foreign market. Foreign customers may buy on the strength of the distributing agent's reputation. A firm should therefore conduct a thorough evaluation of the distributor's facilities, the personnel handling its account and the management methods employed.

Another common difficulty for the new exporter is the neglect of the export market once the domestic one booms: too many companies only concentrate on exporting when there is a recession. Others may refuse to modify products to meet the regulations or cultural preferences of other countries. Local safety regulations cannot be ignored by exporters. If necessary modifications are not made at the factory the distributor must make them, this is usually at a greater cost and probably not as satisfactorily as it should be. It should also be noted that the resulting smaller profit margin makes the account less attractive.

If exporters expect distributing agents to actively promote their accounts they must be trained and their performance must be continually monitored. This requires a company's marketing executive to be located permanently in the distributor's geographical areas until there is sufficient business to support the representative. The distributor should also be treated on an equal basis with domestic counterparts for e.g. special discount offers, sales incentive programmes and special credit terms should be available.

Considering a joint-venture or licensing agreement is another option for new exporters. However, many companies still dismiss international marketing as unviable. There are a number of reasons for this. There may be import restrictions in the target market. The company may lack sufficient financial resources or its product line may be too limited. Yet, many products that can compete on a national basis can be successful in the majority of world markets. In general, all that is needed for success is flexibility in using the proper combinations of marketing techniques.

27. In the first paragraph, the writer suggests that firms thinking about exporting should
A. get professional advice
B. study international marketing
C. identify the most profitable markets
D. have different objectives to other exporters

28. The writer believes that if sufficient preparation is undertaken
A. initial difficulties can be easily avoided.
B. the costs can be recovered quite quickly.
C. management will become more committed.
D. the exporter will be successful in the long term.

29. An exporter should choose a distributor who
A. has experienced personnel.
B. has good communication skills.
C. is well established in the target market.
D. is not financially dependent on the import business.

30. New exporters often make the mistake of ignoring the export market when
A. distribution costs are too high.
B. their product is selling well at home.
C. there is a global economic recession.
D. distributors cannot make safety modifications.

31. For a distributor to be successful, the exporter must
A. focus on one particular region.
B. finance local advertising campaigns.
C. give the same support as to domestic agents.
D. make sure there are sufficient marketing staff locally.

32. The writer states that some companies are reluctant to export because
A. there is little demand for their products.
B. the importation of certain goods is controlled.
C. they do not have good marketing techniques.
D. they are not able to compete with local businesses.

33. The kids were all coming to the pool. The pool attracted kids like .....
A. rain to the desert sands
B. elephants to water
C. a balloons to helium
D. flies to honey

34. If thieves ..... our house, I ..... frightened to death.
A. will enter, would be
B. enter, would be
C. would enter, would be
D. enter, would have been

35. The enemy turned ..... and ran away as soon as it saw the brave warrior.
A. back B. tail
C. face D. feet

36. Vodafone ..... and initiated a fresh style in television advertising.
A. broke new ground B. cut the ground
C. cut corners D. prepared the ground

37. He took to ..... his impatience indiscriminately, on whoever happened to be at hand.
A. giving free rein to
B. airing
C. venting
D. All of the above

38. He provides a .... sketch of the political values, strategy and practices of these diverse civilizations.
A. thumbnail
B. thumbling
C. thumbprint
D. None of the above

39. "This desk is more square than that one" is an erroneous statement, the correct statement would be
A. This desk is squarer than that one.
B. This desk is more nearly square than that one.
C. This desk is most square than that one.
D. This desk is much more square than that one.

40. "Half a loaf is better than no bread" means
A. God helps those who help themselves
B. When a bad thing happens, there is always a positive side to it
C. You should be happy with what you have even if it is less than what you want
D. Don't be upset by trivialities

SECTION-II

QUANTITATIVE, DATA INTERPRETATION AND SUFFICIENCY

For all questions in this section, each correct answer carries 1 mark

41. The line joining the points (5, 2) and (6, –15) subtends a right angle at the point:
A. (–1, 2) B. (0, 7)
C. (0, 0) D. Insufficient Data

42. Find the least positive integer which when divided by 2 gives remainder 1, when divided by 3 gives remainder 2, when divided by 4 gives remainder 3 and when divided by 5 gives remainder 4, what would the integer be?
A. 33 B. 59
C. 61 D. 49

43. How many numbers exist between 100 and 1000 which are divisible by 7?
A. 142 B. 138
C. 60 D. 128

44. Two athletes A and B are running on a 900 m long track from opposite ends, their speeds are in the ratio of 4 : 5, when will they cross each other:
A. 500 m from A's end
B. 400 m from B's end
C. 400 m from A's end
D. 100 m from A's end

45. Find the domain of the expression Y = $\sin x + \log x$
A. $(-\infty, \infty)$ B. $(-\infty, 0)$
C. $(0, \infty)$ D. $(0, 1)$

46. The expression $x^4 + 5x^2 + 36$ can be factorized as
A. $(x^2 + x + 1)(x^2 - x + 36)$
B. $(x^2 - x + 6)(x^2 + 4x + 6)$
C. $(x^2 - x + 6)(x^2 + x + 6)$
D. None of the above

47. There are 4 members in a committee. If one of them is replaced by a younger member such that the average age of the committee is as it was 6 years ago. The new member is how many years younger than the member replaced?
A. 42 years B. 24 years
C. 6 years D. None of the above

48. The probability that a builder will get a construction contract is $\frac{3}{4}$ and the probability that he will not get the electrical contract is $\frac{7}{9}$. If the probability of getting at least one contract is $\frac{5}{6}$, what is the probability that he will get both?
A. $\frac{5}{36}$ B. $\frac{5}{6}$
C. $\frac{5}{9}$ D. None of the above

49. In how many ways can 21 books on Literature and 19 books on Nature be placed in a row on a shelf so that two books on Nature may not be together?
A. 4620 B. 3672
C. 1540 D. None of the above

50. A sphere of radius 4 cm is dropped into a cylindrical vessel partly filled with water. The radius of the vessel is 8 cm. If the sphere is submerged completely, by how much will the surface of the water be raised?
A. $\frac{3}{4}$ units B. $\frac{4}{3}$ units

C. $\frac{1}{4}$ units D. None of the above

51. A solid metal cylinder of radius 5 cm and height 12 cm is melted to make solid balls of radius 1 mm. How many such balls are produced (in lakhs)?

A. 2.5 B. 2.25
C. 1.5 D. None of the above

52. The value of $\log_{(-1/4)} 256$ is equal to

A. 64 B. 16
C. –4 D. None of the above

53. A salesman's commission is 6% on all sales upto ₹ 15,000 and 4% on all sales exceeding this amount. He remits ₹ 30000 to his parent company after deducting his commission. Find the total sales.

A. ₹ 31562.5
B. ₹ 31500.5
C. ₹ 32570.5
D. None of the above

54. If 60% of $(x - y)$ = 40% of $(x + y)$, then what per cent of x is y?

A. 20% B. 25%
C. 22% D. None of the above

55. The investments of A and B in a business is in ratio 4 : 5 and the period of their investments is in the ratio 9 : 8. Their profit will be in the ratio?

A. 9 : 10 B. 9 : 8
C. 8 : 11 D. None of the above

56. A new watch was purchased and it was showing 6 o'clock. The purchaser was soon informed that the hour hand and minute hand had been interchanged and attached to the wrong axes. This resulted in the hour hand to move with a speed which is twelve times higher than the minute hand. Hence the watch was taken to the mechanic for repair, at that moment the watch showed the right time! If the clock started at 6 o'clock in the correct position, then what was the first moment that the watch showed the correct time?

A. $6\frac{5}{11}$ minutes past 9 B. $5\frac{5}{11}$ minutes past 7

C. $7\frac{5}{11}$ minutes past 7 D. None of the above

57. If $x + y + z = \pi$, then $\cot\frac{x}{2} + \cot\frac{y}{2} + \cot\frac{z}{2}$ will be equal to

A. $\tan\frac{x}{2} * \tan\frac{y}{2} * \tan\frac{z}{2}$ B. $\cos\frac{x}{2} * \cos\frac{y}{2} * \cos\frac{z}{2}$

C. $\cot\frac{x}{2} * \cot\frac{y}{2} * \cot\frac{z}{2}$ D. None of the above

58. If $y^2 + 7ay + 40 = 0$ and $y^2 + 2ay - 60 = 0$ have a common root, then the value of a is

A. ± 4 B. ± 2
C. ± 5 D. ± 6

Directions (Qs. Nos. 59 to 63): *These questions are based on the following pie chart which shows the viewership of different knowledge based channels (in degrees) in the month of December, 2013. There are no overlaps in viewership of channels.*

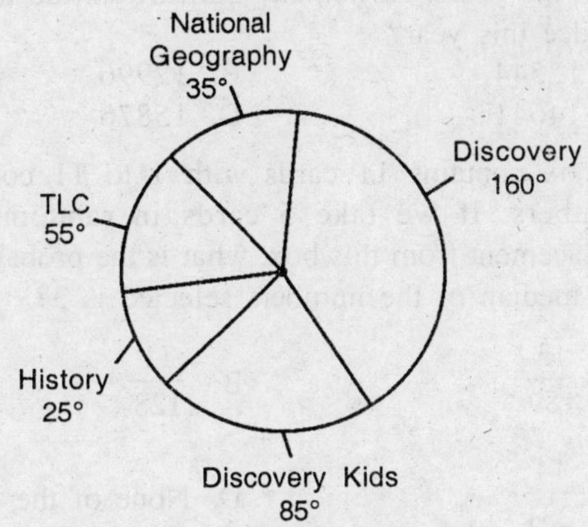

59. During the given period for how many channels is the viewership more than 25% of the total viewership?

A. 1 B. 2
C. 3 D. None of the above

60. If 70,000 people watched National Geographic on an average per day in December, 2013 then how many more people watched Discovery than History on an average per day for the same period?

A. 3,60,000 B. 36,000
C. 2,70,000 D. None of the above

61. If the viewership of TLC on an average was 55,000 per day, then what was the viewership for all the channels on an average per day?

A. 3,60,000 B. 5,40,000
C. 6,30,000 D. None of the above

62. If the viewership of Discovery Kids channel for the first half of December is half of that of second half of December, then what is the ratio of the viewership of Discovery Kids for the second half to that of TLC for the whole month?

A. 8 : 9 B. 10 : 11
C. 9 : 10 D. None of the above

63. By mistake the viewership of Discovery Kids has been underquoted by 25%. If this mistake is corrected then what is the correct share of viewership of Discovery?

A. 45%
B. 41.55%
C. 39.15%
D. None of the above

64. How many three digit positive integers exist that when divided by 6 leave a reminder of 4?
A. 149 B. 150
C. 151 D. 152

65. A person grows sunflowers in his Square Shaped farm. One sunflower takes 1 square feet of area in his farm. This year he has increased his output by 241 Sunflowers as compared to last year. The shape of the farm used for growing the Sunflowers has remained the same, both the years. How many Sunflowers did he in total prodce this year?
A. 12544 B. 12996
C. 14641 D. 15876

66. A box contains 11 cards with 1 to 11 consecutive numbers. If we take 5 cards in random without replacement from this box, what is the probability that the median of the numbers selected is 3?
A. $\frac{13}{189}$ B. $\frac{11}{128}$
C. $\frac{14}{231}$ D. None of the above

67. In the given figure if A, B, C, D are Centers of the Circles and the radius of each Circle is $7\sqrt{3}$ units then what is the area of the shaded region?

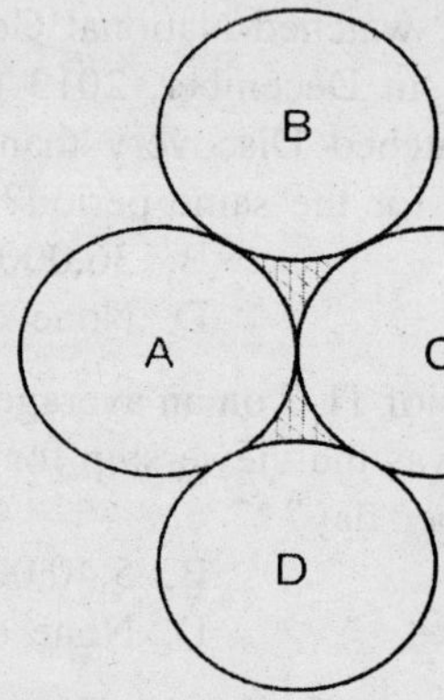

A. $74\left[\sqrt{3}-\frac{\pi}{2}\right]$ B. $49\left[\sqrt{3}-\frac{\pi}{2}\right]$
C. $300\left[\sqrt{3}-\frac{\pi}{2}\right]$ D. $294\left[\sqrt{3}-\frac{\pi}{2}\right]$

68. Instead of dividing ₹ 180 among A, B, C in the ratio $\frac{1}{3}:\frac{1}{4}:\frac{1}{6}$ respectively, it was divided in the ratio 3 : 4 : 6 by mistake. Who gained in this transaction?
A. Only A
B. Only B
C. Only C
D. Both B and C

69. The sum of ages of mother and daughter is 42 years. After three years the age of mother will be three times that of the daughter. Then the mother's age after three years will be?
A. 28 B. 33
C. 36 D. 38

70. Find the missing number in the series 2, 5, 11, 20, 32, 47, ..., 86
A. 60 B. 65
C. 58 D. 62

71. One number in the given series is wrong. Find out the wrong one.

91, 74, 59, 47, 35, 26, 19,
A. 47 B. 35
C. 26 D. 19

72. A player can take a maximum of 4 chances to hit a bottle with a flying disc. The probability of hitting the bottle at the first, second, third and fourth shots are 0.1, 0.2, 0.35 and 0.45 respectively. What is the probability that the player hits the bottle with the flying disc?
A. 0.6573
B. 0.2574
C. 0.7426
D. None of the above

73. If log 2 = 0.30103 and log 3 = 0.4771, then number of digits in $(10368)^6$ will be approximately equal to
A. 15 digits B. 25 digits
C. 30 digits D. 40 digits

74. What is the measure of the angle between the hands of the clock at 8:24 p.m.?
A. 100 degree B. 102 degree
C. 106 degree D. 108 degree

75. A train after travelling for a certain distance develops a problem and decreases its speed to half its original speed and reaches its destination 45 minutes late. Had the problem occurred 30 km further on, it would have reached its destination 15 minutes earlier. What is the speed of the train?
A. 72 km/hr B. 90 km/hr
C. 120 km/hr D. 150 km/hr

76. A manufacturer sells a good to a wholesale dealer at a profit of 18%. The wholesaler sells the same to a retailer at a profit of 20%. The retailer in turn, sells it to the customer for ₹ 30.09 (Rupees thirty and paise nine only) thereby earning profit of 25%. What is the cost of the manufacturer?
A. ₹ 15.30
B. ₹ 17.00
C. ₹ 20.00
D. ₹ 22.00

77. If ₹ 1000 is invested at an interest rate of 5% and the interest earned is added to the principal every 10 years, then after how many years will the amount become ₹ 2000?

A. $\frac{20}{3}$ B. $\frac{50}{3}$

C. $\frac{25}{4}$ D. None of the above

78. Two pipes of P and Q can fill a tank A in 10 hrs and 12 hrs respectively. A third pipe R can empty a tank B in 40 hrs. Tank B is two times in volume compared to tank A. If P and Q are kept open to fill tank A, if R is connected to empty tank A at the same time such that all the pipes operate simultaneously, how long will take for the tank A to be fully filled?

A. 6 hours 40 minutes
B. 8 hours 30 minutes
C. 7 hours 30 minutes
D. None of the above

79. ABCD is a trapezium and AB is parallel to CD. AB has a length, which is two times the length of CD. If the diagonals of the trapezium intersect at O, then the ratio of AOB triangle and COD triangle is

A. 1 : 4 B. 4 : 1
C. 1 : 2 D. 2 : 1

80. The MRP of a chocolate is ₹ 65. A customer purchased the chocolate for ₹ 56.16 (Rupees fifty six and paise sixteen only). He got two successive discounts of which one is 10%. Find the other discount of the discount scheme that was given by the seller.

A. 3%
B. 4%
C. 5%
D. None of the above

SECTION-III

GENERAL AWARENESS

For all questions in this section, each correct answer carries 1 mark

81. What is the full-form of RSVP?
A. rigorous session of voice protocol
B. répondez, s'ilvous plait
C. répondez, sans vous plaint
D. response sought vous plaint

82. The Callanish Stones, which are considered to be petrified giants, as per a folklore, who did not convert to Christianity are located in which country:
A. Norway B. Wales
C. Scotland D. Ireland

83. Which of the following sites or monuments is not in the UNESCO list of World Heritage Sites?
A. Humayun's Tomb at Delhi
B. Kaziranga National Park in Assam
C. Tirupati Temple
D. Agra Fort

84. Mr. Ramon Magsaysay, in whose name the prestigious Magsaysay award is given, belonged to:
A. Malaysia B. Thailand
C. Philippines D. Indonesia

85. Indian musician noted for his exemplary skills at playing mandolin who died in 2014:
A. Uppalapu Srinivas B. Amjad Ali Khan
C. Zarin Daruwalla D. Ustad Usman Khan

86. In the Fifth meeting of SAARC Energy Ministers, which from the following has been the key area of discussion:
A. Telecom B. Environment
C. Power Grid D. Education

87. Malaysian Airlines Flight MH17 was a scheduled international passenger flight, presumed to have been shot down, crashed on 17 July 2014 killing all 283 passengers and 15 crew on board was flying,
A. from Amsterdam to Kuala Lumpur
B. from Kuala Lumpur International Airport to Beijing Capital International Airport
C. from Kuala Lumpur to Amsterdam
D. from Beijing Capital International Airport to Kuala Lumpur International Airport

88. Ukraine has been in news recently on account of conflict with Russia. The name of the Ukrainian President is:
A. Petro Poroshenko
B. Viktor Yushebenko
C. Yulia Timoshenko
D. None of the above

89. Oldest brand of the Tata Group is
A. The Taj Palace and Tower
B. Ginger
C. Taj Cape Town
D. Quilon

90. 'Gold fixing' refers to setting the price of gold. Where is the value of gold set?
A. New York B. London
C. Tokyo D. Singapore

91. Which of the following nations has declared issuance of digital currency?

A. Ecuador
B. Estonia
C. United States of America
D. United Kingdom

92. What is the significance of EPS in the stock market parlance?

A. Equilibrium of Private Securities
B. Earning Per Share
C. Elongated Period of Settlement
D. Equity Per Share

93. Which committee was appointed by RBI to provide recommendations on financial services for small businesses and low income households in the year 2013?

A. Nachiket Mor Committee
B. Bimal Jalan Committee
C. Tarapore Committee
D. Damodaran Committee

94. Dadabhai Naoroji, in his book "Poverty and Un-British rule in India" in 1867 explained about

A. Brain Drain Theory
B. Economic Drain Theory
C. Wants and Means of India
D. Benefits of British Rule

95. He is a famous playwright, socialist and one of the Co-Founders of London School of Economics, name the person:

A. Beatrice Webb
B. George Bernard Shaw
C. Arnold Bennett
D. Richard Brome

96. Auto didacticism is a form of education where

A. One automatically learns from experience of peers
B. A set of logical deductions are taught by the teacher
C. One teaches oneself about a subject
D. One learns automatically from elders

97. The author of the book "An inquiry into the nature and causes of The Wealth of Nations" is,

A. Adam Smith B. Karl Marx
C. Amartya Sen D. Alfred Marshall

98. The Universities of Delhi, Madras and Calcutta were set up in the same year. Which was the year?

A. 1818 B. 1857
C. 1947 D. 1860

99. 'Shaurya-Dridata-Karmanishtha' is the motto of

A. Border Security Force
B. The Indian Coast Guard
C. Indo Tibetan Border Police
D. National Security Guard

100. The following is the logo of

A. Cannes International Film Festival
B. Berlin International Film Festival
C. London Film Festival
D. Amnesty International Film Festival

101. The Giant Panda, world's most threatened species is a symbol of:

A. IFAW B. WWF
C. CEE D. UNEP

102. Acid rain is caused mainly due to emissions of

A. Calcium oxide and sulphur dioxide
B. Nitrogen oxide and sulphur dioxide
C. Sodium oxide and nitrogen dioxide
D. Calcium oxide and carbon dioxide

103. Which of the given rivers does not flow into River Ganga?

A. Brahmaputra B. Gomti
C. Gandak D. Tapti

104. The International agreement for control of Green House Gas (GHG) emissions is named

A. Adelaide Agreement B. Kyoto Protocol
C. Manila Convention D. Kuala Lumpur Treaty

105. The International Standard which checks the compliance of the environmental practices of an organization against the best practices followed in the industry is

A. ISO 9001 B. TS 16949
C. ISO 14001 D. ISO 9002

106. Comiket the world's largest comic festival is held in which country?

A. Japan B. France
C. Korea D. Germany

107. In .... Infosys was listed in NASDAQ and in .... Infosys became a part of NASDAQ 100

A. 1998, 2009 B. 1997, 2008
C. 1999, 2006 D. None of the above

108. The Indian team finished with 8th position in the recent Asian games 2014 held at South Korea with a total .... tally of medals.

A. 57 B. 37
C. 56 D. 58

109. A full moon occurs once every

A. 14 days B. 30 days
C. 21 days D. 15 days

110. The River Danube empties in
A. The Dead Sea B. The Mediterranean Sea
C. The Black Sea D. The Red Sea

111. Cartography is the study of:
A. Cartilages and Bones B. Cartoons
C. Maps D. Carrots

112. Regional rural banks are jointly owned by
A. Central government, a commercial bank and Central bank of a country
B. Central government, a commercial bank and a state government
C. Central government, Central bank and a state government
D. A commercial bank, Central bank of a country and a state government

113. The famous Meenakshi Temple in Madurai was built by .... and beautified by the ....
A. Vakatakas, Nayaks B. Chalukyas, Chollas
C. Satavahanas, Chollas D. Pandyas, Nayaks

114. Which of the given statements is not correct as per Indian constitution?
A. There can be a single person as the Governor for two or more states.
B. Governor is the Constitutional or nominal head of the State Executive and is bound by the advice of the State Council of Ministers.
C. All executive actions of the State shall be expressed to be taken in the name of the Chief Minister.
D. The President of India consults the Governor on appointment of the High Court judges.

115. Kwan Yin has a special place in Buddhist tradition and is known:
A. To represent healing nature
B. To bring in good luck charm
C. As a Goddess of compassion
D. As Related to higher psychological-scientific teachings from Buddha

116. "Bachpan Bachao Andolan" was initiated by
A. Jawaharlal Nehru B. Kailash Satyarthi
C. Malala Yusefzai D. Benzeer Bhutto

117. Dr. Ram Manohar Lohia, the great freedom fighter and thinker, established a political party named
A. Samajwadi Party
B. Samyukta Socialist Party
C. Janata Party
D. Praja Socialist Party

118. The Nobel Prize in Physics 2014 was awarded jointly to three scientists "for the invention of efficient blue light-emitting diodes which has enabled bright and energy-saving white light sources". Who among the given four names was not in the team?
A. Isamu Akasaki B. Hiroshi Amano
C. Shuji Nakamura D. Takada Ekimano

119. Chandrasekhar Venkat (CV) Raman was awarded the Nobel Prize for investigation into 'Raman Effect', a phenomenon that deals with
A. Change in the speed of light as it passes through a transparent medium
B. Change in the atomic structure of the medium impacted by light
C. Change in the wavelength of the light as it passes through a transparent medium
D. Change of light in liquid crystals alone

120. "Working towards a world free of poverty" is a motto of one of these International Organizations:
A. UNESCO B. UNICEF
C. WORLD BANK D. UNCTAD

SECTION-IV

ANALYTICAL & LOGICAL REASONING

For all questions in this section, each correct answer carries 2 marks

121. Vertex : Pyramid : .... : ....
A. Strand : Hair B. Rung : Ladder
C. Summit : Mountain D. Frame : Picture

122. Piety : Pilgrim : Curiosity : ....
A. Explorer B. Pinchpenny
C. Gamester D. Monk

Directions (Qs. Nos. 123 and 124): *Read the paragraphs carefully and determine the main point the author is trying to make. What conclusion can be drawn from the argument? Each paragraph is followed by four statements. One statement supports the author's argument better than the others do.*

123. Nowadays people are so busy in their work that they do not take care of their health. A lot of diseases like diabeties, arthritis, heart attack, blood pressure are increasing due to unhealthy lifestyle. People do not find time to exercise. Stress is continuously increasing. The fast food is quick, easy and tastes better but it is not healthy. Habits like smoking and drinking alcohol have many side effects. Polluted city environment has also negative impact on the health of citizens. To live happy and healthy life; it is important to take care of oneself.
A. Nowadays everybody smokes and drinks.
B. Diseases are increasing only because of pollution.

C. People are happy and stress free because their life is too busy.
D. Current city environment, life style and habits are affecting health of citizens.

124. Diwali is a festival of lights; which signifies that good will always be better than evil. In the present day many people have developed a lot of negativity in their minds and hearts, which has led to high corruption and higher crime rates. Value based education is the need of the day to eliminate darkness of cruelty, selfishness and greed and bring the light of wisdom, peace, kindness, truthfulness, right attitude, hard work and co-operation. Mahatma Gandhi also said that we should eradicate sins from hearts not the sinners. If a society has people with good heart, it will definitely flourish.
A. Nowadays nobody is good.
B. Nowadays everybody is corrupt and criminal.
C. Value based education can save the society from and corruption.
D. It is not possible to fight against corruption as everybody is corrupt.

125. If X + Y means X is the mother of Y; X – Y means X is the brother of Y; X % Y means X is the father of Y and X × Y means X is the sister of Y, which of the following shows that P is the maternal uncle of Q?
A. Q – N + M × P B. P + S × N – Q
C. P – M + N × Q D. Q – S % P

126. In the following question, three statements numbered– I, II and III and four conclusions are given. Select the correct answer from the options given.
Statements:
I. Some Newspaper are ACs.
II. Some ACs are Trees.
III. All Books are ACs.
Conclusions:
1. Some Books are Newspapers.
2. All Books are Trees.
3. All Trees are Books.
4. Some Books are not Newspapers.
Options:
A. Only 1 follows
B. Only 1 or 3 follows
C. Only either 1 or 4 follows
D. Only 2 and 3 follow

Directions (Qs. Nos. 127 and 128): *A statement is given followed by several assumptions. An assumption is something assumed or taken for granted. Read the statement carefully and decide which of the assumptions are implicit in the statement.*

127. Statement: "I think that she has gone mad. Last night I saw her barking like a dog."– A tells B about Madhu.
Assumptions:
I. Madhu is a friend of A.
II. B Knows Mahdu.
III. Some people do not bark like dogs.
A. I and II are implicit.
B. II and III are implicit.
C. I and III are implicit.
D. All are implicit.

128. Statement: "The Prime Minister is expected to announce an expansion in his ministry shortly."–A newspaper report.
Assumptions:
I. The newspaper has quoted authentic sources.
II. The newspaper has reliable sources for the news.
III. The Prime Minister has the power to expand his cabinet.
A. I and II are implicit.
B. II and III are implicit.
C. I and III are implicit.
D. All are implicit.

129. How many times in a day, do the hands of a clock coincide with each other?
A. 12 B. 24
C. 22 D. 44

Directions (Qs. Nos. 130 to 134): *These questions are based on the data given below.*

There are two circular tables in a room. Six Americans–A1, A2, A3, A4, A5, A6 are sitting at one table and six Germans– G1, G2, G3, G4, G5, G6 are sitting at the other table. A1 and A4 are sitting opposite to each other. A2 and A5 are also sitting opposite to each other. A3 sits to the right of A4. A2 is the only person who can translate English to German language; A3 is the only person who can translate German language to English, and none of them does the vice-versa, unless so stated. A5 and A6 sit adjacent to each other. Similarly, G1 sits opposite to G4; G6 sits opposite to G3 and G5 sits to the right of G6. G2 and G4 sit adjacent to G3. On the table, any person can talk to another person, only as stated below. The only conversations that took place are as given below:

A1 spoke to A2; A2 to G6; G6 to A3; G6 to G5; G5 to G4; G4 to G3; G3 to G2; G2 to G1; A3 to A4; A4 to A5; A6 to A1; G1 to G6; and A5 to A6.

The conversations are one-sided *i.e.,* A1 spoke to A2 implies that A1 is the speaker and A2 is the listener and not vice-versa.

130. If the person to the right of A1 wants to send a message to the person sitting to the left of G2, then the message has to pass through how many persons? (Excluding the first and the last)
A. 7 B. 5
C. 6 D. None of the above

131. If A3 wants to send a message to G1, then how many persons must the message pass through? (excluding the first and the last)

A. 6 B. 4
C. 10 D. None of the above

132. If the person sitting to the right of A2 wants to send a message to the person sitting two places to the left of A4, then what is the minimum number of people involved between them? (excluding the two people)

A. 6 B. 10
C. 4 D. None of the above

133. If every person is allowed to speak to the person who sits opposite to him directly, then what is the least number of persons involved in passing a message from A3 to G1? (excluding the two people)

A. 3 B. 6
C. 7 D. None of the above

134. If the order of conveying messages is reversed at both the tables *i.e.,* "A1 speaks to A2", now becomes "A2 speaks to A1", and so on) and also A2 and A3 exchange their interpreting skills, then which of the following must be true?

A. A1 can send a message to G5 involving only 2 persons.
B. The person sitting to the right of A6 can send a message to the person sitting to the left of G2, by involving only five persons.
C. The person sitting to the right of A3, on the same table, can translate English to German language.
D. The maximum number of persons involved in the longest message in this new arrangement is more than that in the previous arrangement.

Directions (Qs. Nos. 135 and 136): *These questions are based on the data given below.*

There are 3 females A, B and E and 4 males C, D, F and G standing in a straight line.

No two females are together. B is to right of C, F and D are not together as A is placed between them. G is not near B or E but E and F are together. D is not to the right of B.

135. Who is in the extreme end?

A. G and B B. C and F
C. B and D D. None of them

136. Who is exactly in the middle?

A. A B. F
C. E D. None of the above

137. Look carefully at the sequence of letters and symbols to find the pattern. Select correct pattern.

N_2WF LT_3G JQH_4 .... FK_6J

A. H_5NI B. H_5SI
C. HR_2I_3 D. H_5NJ

Directions (Qs. Nos. 138 to 141): *Answer the questions based on the details given in the paragraph.*

There were 200 students who gave a competitive exam. The exam had three sections English, Mathematics and General Knowledge (GK). Ten students could not pass in any of the subjects. Number of students passed in English was 125, Mathematics was 95 and General Knowledge was 95. 55 students passed in English and Mathematics. 45 students passed in GK and Mathematics and 60 students passed in GK and English. 20 students passed in English and Mathematics but failed in GK.

138. How many students passed in GK and Mathematics, but failed in English?

A. 30 B. 60
C. 45 D. 10

139. How many students passed in English and GK but failed in Mathematics?

A. 60 B. 25
C. 20 D. 35

140. How many students passed only in Mathematics?

A. 30 B. 20
C. 25 D. 35

141. How many students passed in all three Subjects?

A. 25 B. 35
C. 15 D. 20

Directions (Qs. Nos. 142): *Answer the question based on the details given in the paragraph.*

In a family of six, there are three men A, B and C and three women R, S and T. The six persons are Architect, Lawyer, CA, Professor, Doctor and Engineer by profession but not in the same order. There are two married couples and two unmarried persons. C is not R's husband. The Doctor is married to a Lawyer. R's grandfather is a Professor. B is neither A's son, nor he is an Architect or a Professor. The Lawyer is T's daughter-in-law. C is T's son and the Engineer's father. A is married to the CA.

142. Which among the following is the correct pair of married couples?

A. AS and CR B. AC and SR
C. AT and CS D. AT and RB

143. Five friends A, B, C, D and E have different jobs. They work with either Bank or Cinema or Pharma or Sales or Jewellery. Pick the right job in which they are employed if:

B is neither concerned with Bank nor with Sales. E has nothing to do with Cinema or Jewellery. C is not in Bank or Cinema. D has nothing to do in Sales or Pharma. Cinema has not given a chance to B or D nor E is a bank employee. C is not in Sales nor B is in Jewellery.

A. A–Cinema; B–Sales; C–Jewellery: D–Pharma; E–Bank
B. A–Cinema; B–Sales; C–Pharma: D–Bank; E–Jewellery
C. A–Cinema; B–Sales; C–Jewellery: D–Bank; E–Pharma
D. A–Cinema; B–Pharma; C–Jewellery: D–Bank; E–Sales

Directions (Qs. Nos. 144): *In the following question, a statement or two are followed by two conclusions:*

144. Statement: Of all the television sets manufactured in India, the 'Solar' brand has the largest sale.

Conclusions:

I. The volume of sales of all the brands of televeision sets manufactured in India is known.

II. The production of no other television set in India is as large as that of 'Solar'.

A. Conclusion I follows
B. Conclusion II follows
C. Both I and II follow
D. Either I or II follows

Directions (Qs. Nos. 145 to 148): *Read the following information carefully and answer the questions.*

Radhika, Neha, Rupa, Mahima, Sayali and Shobha joined six different renowned MBA institutes–A, B, C, D, E and F and opted for six different specializations in Management–Finance, IT, Operations, HR, Marketing and General Management. Each of them has a different educational qualification out of BCA, B.Com, CA, BBM, B.Sc., and B.Tech.

The following information is available about them.

(*i*) Mahima joined institute B and did not take up Marketing or IT and is not a B.Tech.

(*ii*) The person who joined institute E opted for HR and is not a CA or a B.Sc.

(*iii*) The person, who is a BCA graduate, joined the institute D and took IT. Whereas, the person, who is B.Com graduate, took up General Management.

(*iv*) Shobha is a CA and has taken Finance.

(*v*) The person, who is a B.Tech graduate has taken Operations Management and did not join institute A or C.

(*vi*) Radhika joined institute F, Rupa is a BCA graduate and Neha took up HR.

145. What is the educational qualification of Mahima and what is the course she opted for?
A. B.Sc. and Finance
B. B.Com. and General Management
C. BBM and HR
D. None of the above

146. Which Institute did the B.Tech Graduate join?
A. F B. D
C. A D. None of these

147. The educational qualification of Sayali is
A. B.Sc B. BCA
C. BBM D. CA

148. If Shobha joined the institute C, then the student who joined the institute A opted for
A. Finance B. Marketing
C. General Marketing D. HR

Directions (Qs. Nos. 149 to 150): *Read the following information carefully and answer the questions.*

In a village, there used to live three types of tribes SIM, LIM and JIM. SIMs always tells the truth, JIM always lie and LIMs tell the truth and lie alternating. (They can tell truth and lie in any sequence). Three persons (of different tribes) from the village gave following statements.

NICE says "RICE is of SIM tribe; I am of LIM tribe".

MICE says "NICE is of JIM tribe; I am of SIM tribe".

RICE says "MICE is of JIM tribe; I am of LIM tribe".

149. Who is telling the truth always?
A. NICE B. MICE
C. RICE D. None of the above

150. NICE belongs to which of the following tribe?
A. SIM B. JIM
C. LIM D. None of the above

ANSWERS

1	2	3	4	5	6	7	8	9	10
D	D	C	A	C	A	A	C	D	A
11	**12**	**13**	**14**	**15**	**16**	**17**	**18**	**19**	**20**
A	B	A	A	A	A	A	B	C	B
21	**22**	**23**	**24**	**25**	**26**	**27**	**28**	**29**	**30**
B	A	B	D	C	D	A	D	C	B
31	**32**	**33**	**34**	**35**	**36**	**37**	**38**	**39**	**40**
C	B	D	B	A	A	D	A	B	C

41	42	43	44	45	46	47	48	49	50
C	B	D	C	A	D	B	A	C	B
51	**52**	**53**	**54**	**55**	**56**	**57**	**58**	**59**	**60**
B	C	A	A	A	C	C	B	A	C
61	**62**	**63**	**64**	**65**	**66**	**67**	**68**	**69**	**70**
A	D	C	B	C	C	D	C	C	B
71	**72**	**73**	**74**	**75**	**76**	**77**	**78**	**79**	**80**
A	C	B	D	C	B	B	C	B	B
81	**82**	**83**	**84**	**85**	**86**	**87**	**88**	**89**	**90**
B	C	C	C	A	C	A	A	A	B
91	**92**	**93**	**94**	**95**	**96**	**97**	**98**	**99**	**100**
B	B	A	B	B	C	A	B	C	A
101	**102**	**103**	**104**	**105**	**106**	**107**	**108**	**109**	**110**
B	B	D	B	A	A	C	A	B	C
111	**112**	**113**	**114**	**115**	**116**	**117**	**118**	**119**	**120**
C	B	B	C	C	B	D	D	C	C
121	**122**	**123**	**124**	**125**	**126**	**127**	**128**	**129**	**130**
C	A	D	C	C	C	A	B	C	B
131	**132**	**133**	**134**	**135**	**136**	**137**	**138**	**139**	**140**
C	D	B	B	A	B	A	D	B	A
141	**142**	**143**	**144**	**145**	**146**	**147**	**148**	**149**	**150**
B	C	D	A	B	A	A	B	B	B

SOME SELECTED EXPLANATORY ANSWERS

41. Two points A = (5, 2), B = (6, –15)

Let, third point is origin C (0, 0)

$$\text{Slope of line AC} = \frac{2-0}{5-0} = \frac{2}{5}$$

$$\text{Slope of line BC} = \frac{-15-0}{6-0} = \frac{-15}{6} = \frac{-5}{2}$$

$$\text{Product of the slopes} = \frac{2}{5} \times \left(\frac{-5}{2}\right) = -1$$

$\therefore$ Line AC $\perp$ BC

Hence, points (5, 2) and (6, –15) subtends a right angle at the origin.

43. First No. b/w 100 and 1000 which is divisible by 7 is $a = 105$

Last No. b/w 100 and 1000 which is divisible by 7 is $l = 994$ and common difference is $d = 7$

$\Rightarrow \quad l = a + (n-1) \times d$

$\Rightarrow \quad 994 = 105 + (n-1) \times 7$

$\Rightarrow \quad 994 - 105 = 7(n-1)$

$\Rightarrow \quad 889 = 7(n-1)$

$\Rightarrow \quad (n-1) = \dfrac{889}{7} = 127$

$\Rightarrow \quad (n-1) = 127$

$\Rightarrow \quad n = 127 + 1$

$\Rightarrow \quad n = 128$

Hence, 128 numbers exist b/w 100 and 1000 which are divisible by 7.

44. Given as,
$$\frac{s_A}{s_B} = \frac{4}{5}$$

or
$$\frac{s_B}{s_A} = \frac{5}{4}$$

Let, C be the meeting point of A and B at this point,

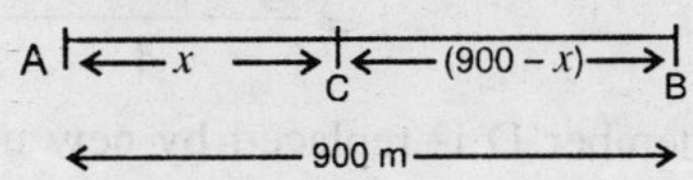

$$t_A = t_B$$

$$\frac{t_A}{t_B} = 1 \quad \ldots(i)$$

$D_{AC} = x$ metre and $D_{BC} = (900 - x)$ metre

$$\text{Time} = \frac{\text{Distance}}{\text{Speed}}$$

$$\frac{t_A}{t_B} = \frac{D_{AC} / s_A}{D_{BC} / s_B}$$

$$1 = \frac{D_{AC} / s_B}{D_{BC} / s_A}$$

[∵ from equation (*i*)]

$$1 = \frac{x}{(900 - x)} \times \frac{5}{4}$$

[∵ Given as and from let]

$$1 = \frac{5x}{3600 - 4x}$$

$$3600 - 4x = 5x$$

$$9x = 3600$$

$$x = \frac{36000}{9} = 400$$

$$\Rightarrow \quad x = 400 \text{ m}$$

400 m from A's end.

45. $y = \sin x + \log x$

$y = (-\infty, +\infty) \cup (0, +\infty)$

$y = (-\infty, +\infty)$.

46. Factorization of the given expression:

$= x^4 + 5x^2 + 36$

$= x^4 + 12x^2 + 36 - 7x^2$

$$= \left(x^2 + 6\right)^2 - \left(\sqrt{7x}\right)^2$$

$$= \left(x^2 - \sqrt{7x} + 6\right) \times \left(x^2 + \sqrt{7}x + 6\right)$$

Hence, None of the above is correct.

47. Let, present age of four members on a committee are A, B, C and D and present age of a younger member is E.

Average age of 4 members, A, B, C and D is

$$\frac{A + B + C + D}{4}$$

Average age of above members before 6 years

$$= \frac{A + B + C + D}{4} - 6$$

when member D is replaced by new member E,

$$\text{their average Age} = \frac{A + B + C + E}{4}$$

from question,

$$\frac{A + B + C + E}{4} = \frac{A + B + C + D}{4} - 6$$

or, $\quad E = D - 24$

or, $\quad D - E = 24$

Hence, Difference of age = 24 years.

48. Probability for getting construction contract

$$P(C) = \frac{3}{4}$$

Probability for not getting electrical contract

$$P(E)' = \frac{7}{9}$$

Probability for getting electrical contract

$$P(E) = 1 - P(E')$$

$$= 1 - \frac{7}{9} = \frac{2}{9}$$

Probability for getting atleast one contract

$$P(C \cup E) = \frac{5}{6}$$

∴ Probability for getting both contract

$$P(C \cap E) = P(C) + P(E) - P(C \cup E)$$

$$= \frac{3}{4} + \frac{2}{9} - \frac{5}{6}$$

$$= \frac{27 + 8 - 30}{36} = \frac{5}{36}.$$

49. 19 books on nature can be placed between 21 + 1 = 22 vacant places between 21 books on literature in $^{22}C_{19}$ ways.

$$^{22}C_{19} = \frac{22 \times 21 \times 20}{3 \times 2}$$

$$= 1540 \text{ ways.}$$

50. Given as,

Radius of sphere, r_s = 4 cm

Radius of cylinder, r_c = 8 cm

$$\text{volume of sphere} = \frac{4}{3}\pi r_s^3$$

$$= \frac{4}{3}\pi(4)^3 \text{ cm}^3$$

The height occupied by $\frac{4}{3}\pi(4)^3$ cm^3 of water in a cylinderical vessel of 8 cm radius

$$= \frac{\frac{4}{3}\pi(4)^3}{(8)^2 \pi}$$

$$= \frac{4 \times 64}{3 \times 64} = \frac{4}{3} \text{cm}.$$

51. Given as,

Radius of cylinder, $r_c = 5$ cm

Height of cylinder, $h = 12$ cm

and Radius of sphere, $r_s = 1$ mm

$$r_s = \frac{1}{10}\text{cm}$$

$$\text{No. of balls} = \frac{\text{Volume of cylinder}}{\text{Volume of each sphere}}$$

$$= \frac{\pi r_c^2 h}{\frac{4}{3}\pi r_s^3}$$

$$= \frac{5\times5\times12\times3}{4\times\frac{1}{10}\times\frac{1}{10}\times\frac{1}{10}}$$

$$= 225 \times 1000 = 225000$$

$$= 2.25 \text{ lakhs.}$$

52. Let, $\log_{(-1/4)}256 = x$

$$\Rightarrow \left(-\frac{1}{4}\right)^x = 256 \qquad [\because \log_a N = x \Rightarrow a^x = N]$$

$$\Rightarrow (-4)^{-x} = 16 \times 16 \qquad [\because (16)^2 = 256]$$

$$\Rightarrow (-4)^{-x} = 4 \times 4 \times 4 \times 4 \qquad [\because (4)^2 = 16]$$

$$\Rightarrow (-4)^{-x} = (4)^4$$

$$\Rightarrow (-4)^{-x} = (-4)^4$$

comparing power both sides, we get

$$\Rightarrow -x = 4$$

$$\Rightarrow x = -4.$$

53. Let total sales is ₹ x (> 15000)

then, amount remitted to the company

$$= 15000\times\left(1-\frac{6}{100}\right)+(x-15000)\times\left(1-\frac{4}{100}\right) = 30000$$

$$= 15000\times\frac{94}{100}+(x-15000)\times\frac{96}{100} = 30000$$

$$14100 + 0.96x - 14400 = 30000$$

$$0.96x = 30000 + 300$$

$$= 30300$$

$$x = \frac{30300}{0.96}$$

$$= ₹\ 31562.5.$$

54. Given as,

60% of $(x - y)$ = 40% of $(x + y)$

$$\Rightarrow \frac{60}{100}\times(x-y) = \frac{40}{100}\times(x+y)$$

$$\Rightarrow 36(x - y) = 4^2(x + y)$$

$$\Rightarrow 3(x - y) = 2(x + y)$$

$$\Rightarrow 3x - 3y = 2x + 2y$$

$$\Rightarrow 3x - 2x = 2y + 3y$$

$$\Rightarrow x = 5y$$

$$\Rightarrow y = \frac{1}{5}x$$

$$\Rightarrow y = \frac{1}{5}\times 100\% \text{ of } x$$

$$\Rightarrow y = 20\% \text{ of } x$$

57.

$$x + y + z = \pi$$

$$\frac{x+y}{2} = \frac{\pi}{2}-\frac{z}{2}$$

$$\tan\left(\frac{x+y}{2}\right) = \tan\left(\frac{\pi}{2}-\frac{z}{2}\right)$$

$$\frac{\tan\frac{x}{2}+\tan\frac{y}{2}}{1-\tan\frac{x}{2}.\tan\frac{y}{2}} = \cot\frac{z}{2}$$

$$\frac{\cot\frac{y}{2}+\cot\frac{x}{2}}{\cot\frac{x}{2}.\cot\frac{y}{2}-1} = \cot\frac{z}{2}$$

$$\cot\frac{y}{2}+\cot\frac{x}{2} = \cot\frac{x}{2}.\cot\frac{y}{2}.\cot\frac{z}{2}-\cot\frac{z}{2}$$

$$\therefore \cot\frac{x}{2}+\cot\frac{y}{2}+\cot\frac{z}{2} = \cot\frac{x}{2}.\cot\frac{y}{2}.\cot\frac{z}{2}.$$

58. $(y^2 + 7ay + 40 = 0) \times 2$

$(y^2 + 2ay - 60 = 0) \times 7$

$$2y^2 + 14ay + 80 = 0$$

$$7y^2 + 14ay - 420 = 0$$

$$-\quad -\quad +$$

$$-5y^2 + 500 = 0$$

$$\Rightarrow -5y^2 = -500$$

$$\Rightarrow y^2 = \frac{500}{5} = 100$$

$$\Rightarrow y^2 = 100$$

$$\Rightarrow y = \pm 10$$

Putting $y = \pm 10$ in $y^2 + 7ay + 40$ so, we get

$$\Rightarrow (\pm 10)^2 + 7 \times a\ (\pm 10) + 40 = 0$$

$$\Rightarrow 100 \pm 70a + 40 = 0$$

$$\Rightarrow 140 \pm 70a = 0$$

$$\Rightarrow \pm 70a = 140$$

$$70a = 140$$
$$a = 2$$
$$-70a = 140$$
$$a = -2$$
$$a = \pm 2.$$

59. Total viewership (in degree)
$= 160 + 85 + 25 + 55 + 35 = 360°$

Per cent of viewership in:

$$\text{Discovery kids} = \frac{85}{360} \times 100 = 23.6\%$$

$$\text{History} = \frac{30}{360} \times 100 = 8.33\%$$

$$\text{TLC} = \frac{55}{360} \times 100 = 15.3\%$$

$$\text{National Geographic} = \frac{35}{360} \times 100 = 9.72\%$$

$$\text{Discovery} = \frac{165}{360} \times 100 = 45.83\%$$

Hence, Number of channels having viewership more than 25% is one *i.e.,* Discovery.

60. Let, total number of viewer on an average per day in December 2013 is x.

Then, $x \times \frac{35}{360°} = 70000$

$$x = 720000$$

Now, Difference of No. of viewer between Discovery and History on an average per day

$$= 720000 \times \left[\frac{160°}{360°} - \frac{25°}{360°}\right]$$

$$= 720000 \times \frac{135}{360°}$$

$$= 2,70,000.$$

61. Total Number of viewer on an average per day

$$= 55000 \times \frac{360°}{55}$$

$$= 3,60,000.$$

62. Let, the viewership of discovery kids channel for the first half of December is x

Then, it is on second half $= 2x$

Total number of discovery kids viewership in december $= x + 2x = 3x$

Now, total number of viewership of all channels

$$= \frac{3x \times 360}{85}$$

Number of TLC viewership

$$= \frac{3x \times 360 \times 55}{85 \times 360}$$

$$= \frac{3x}{85} \times \frac{360 \times 55}{360}$$

$$= \frac{33x}{17}$$

$$\therefore \quad \text{Required ratio} = \frac{2x}{\frac{33x}{17}}$$

$$= \frac{34}{33}$$

64. First three digit no. which is divided by 6 leave a reminder of 4 is

$$a = 100$$

last three digit no. which is divided by 6 leave a reminder of 4 is

$$l = 994$$

and common difference,

$$d = 6$$
$$l = a + (n - 1)d$$
$$\Rightarrow \quad 994 = 100 + (n - 1)6$$
$$\Rightarrow \quad 994 - 100 = (n - 1)6$$
$$\Rightarrow \quad 894 = (n - 1)6$$
$$\Rightarrow \quad (n - 1) = \frac{894}{6} = 149$$
$$\Rightarrow \quad n = 149 + 1$$
$$\Rightarrow \quad n = 150.$$

66. Box :

1	2	3	4	5	6	7	8	9	10	11

As the median of the selected number is 3, so, two of the selected numbers must be less than 3 and other two selected numbers must be greater than 3 *i.e.,* out of five selected numbers, three numbers would be 1, 2 and 3 and other two numbers would be any two numbers out of remaining eight numbers.

Two of way of selection of 2 numbers from 8 numbers $= {}^8C_2$

Total ways of selecting 5 numbers out of 11 $= {}^{11}C_5$

$$\text{Required probability} = \frac{{}^8C_2}{{}^{11}C_5} = \frac{14}{231}.$$

67.

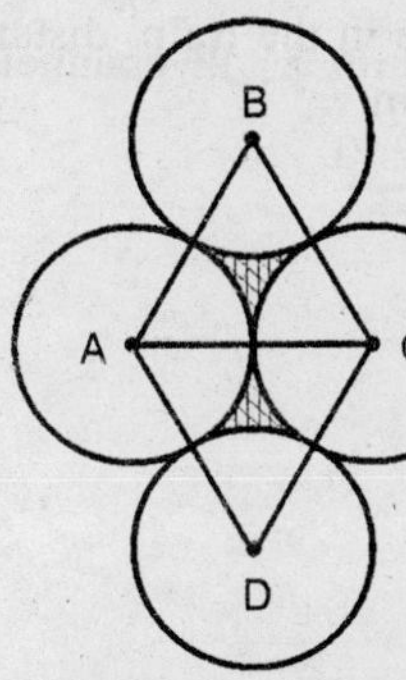

On connecting points A, B, C and D, we get two equilateral triangles ABC and ACD.

Required area = 2[Area of equilateral triangle ABC – $3\times\frac{60}{360}\times$ Area of circle]

Side length AB $= 2r = 14\sqrt{3}$

$$\therefore \text{ Required Area } = 2\left[\frac{\sqrt{3}}{4}\times\left(14\sqrt{3}\right)^2-\frac{\pi}{2}\left(7\sqrt{3}\right)^2\right]$$

$$= 2\left[147\sqrt{3}-147\frac{\pi}{2}\right]$$

$$= 294\left[\sqrt{3}-\frac{\pi}{2}\right].$$

68. Total amount = ₹ 180

Let, A got = ₹ $\frac{x}{3}$

B got = ₹ $\frac{x}{4}$

C got = ₹ $\frac{x}{6}$

By mistake, A got = ₹ $3x$

B got = ₹ $4x$

C got = ₹ $6x$

According to statement,

$$\frac{x}{3}+\frac{x}{4}+\frac{x}{6} = 180$$

$$\frac{4x+3x+2x}{12} = 180$$

$$9x = 180 \times 12$$

$$x = \frac{180\times 12}{9}$$

$$x = 20 \times 12$$

$$x = ₹\ 240$$

Since, A got $= \frac{240}{3}$ = ₹ 80

B got $= \frac{240}{4}$ = ₹ 60

C got $= \frac{240}{6}$ = ₹ 40

By mistake,

$$3x + 4x + 6x = 180$$

$$13x = 180$$

$$x = \frac{180}{13}$$

Since, A got $= 3\times\frac{180}{13}$

= ₹ 41.538

B got $= 4\times\frac{180}{13}$ = ₹ 55.384

C got $= 6\times\frac{180}{13}$ = ₹ 83.077

Hence, only C gained in this transaction.

69. Let, present age of daughter = x years

Present age of mother = $(42 - x)$ years

After three years,

Daughter's age = $(x + 3)$ years

Mother's age = $(42 - x + 3)$ years

= $(45 - x)$ years

According to statement,

$\Rightarrow$ $(45 - x) = 3(x + 3)$

$\Rightarrow$ $45 - x = 3x + 9$

$\Rightarrow$ $3x + x = 45 - 9$

$\Rightarrow$ $4x = 36$

$\Rightarrow$ $x = \frac{36}{4} = 9$

$\Rightarrow$ $x = 9$ years

Mother's age after 3 years will be (45 – 9) years

= 36 years.

70. 2 5 11 20 32 47 [65] 86

3×1 =3, 3×2 =6, 3×3 =9, 3×4 =12, 3×5 =15, 3×6 =18, 3×7 =21

Required no. = 47 + 18 = 65.

71. 91 74 59 [46] 35 26 19

–17 –15 –13 –11 –9 –7

Hence, wrong number in the series is 47.

72. Probability of not hitting a bottle in

first chance = 1 – 0.1 = 0.9

second chance = 1 – 0.2 = 0.8

third chance = 1 – 0.35 = 0.65

fourth chance = 1 – 0.45 = 0.55

Now, Probaility of not hitting the bottle in all the four chance = 0.9 × 0.8 × 0.65 × 0.55 = 0.2574

∴ Probability of hitting the bottle in any chance

= 1 – 0.2574

= 0.7426.

73. From Logarithm, we know that log of single digit number have 0 digit before decimal

Log of two digits number have 1 digit before decimal and 80 on.

Now, $\text{Log } (10368)^6 = 6 \text{ Log } (10368)$

$= 6 \text{ Log } (2^7 \times 3^4)$

$= 6 [7 \text{ Log } 2] + 6 [4 \text{ Log } 3]$

$= 42 \text{ Log } 2 + 24 \text{ Log } 3$

$= 42 [0.30103] + 24[0.4771]$

$= 12.6433 + 11.4504$

$= 24.09366$

Hence, Number of digit in $(10368)^6$ are 25.

74. Rate of rotation of hour hand

$$= \frac{360°}{12} = 30°/\text{hour}$$

Rate of rotation of minute hand

$$= \frac{360°}{60} = 6°/\text{minute}$$

Angle rotated by hour hand in 8.24 minutes @ 30° per hour

$$= 8 \times 30° + 30° \times \frac{24}{60}$$

$= 240° + 12°$

$= 252°$

Angle rotated by minute hand in 24 minutes @ 6° per minute.

$= 6 \times 24 = 144°$

Angle between two hands

$= 252° – 144° = 108°$.

75.

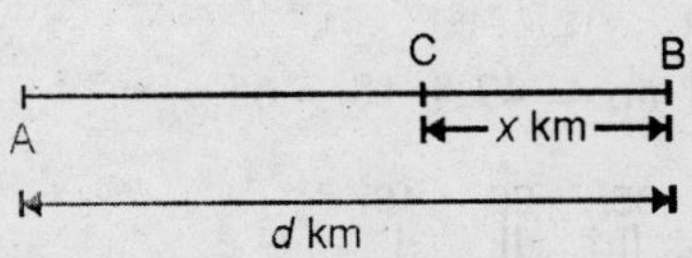

Let, the distance, train need to cover AB = d km and speed of the train is S km/hr.

At point C, some problem occure in the train, distance AC = $(d – x)$ km and BC = x km

Now, from question,

$$\left(\frac{d-x}{S}\right) + \frac{2x}{S} = \frac{d}{S} + \frac{45}{60}$$

$$\therefore \quad \frac{x}{S} = 0.75 \quad ...(i)$$

and
$$\left(\frac{d-x+30}{S}\right) + \frac{2(x-30)}{S} = \frac{d}{S} + \frac{(45-15)}{60}$$

$$\frac{x}{S} - \frac{30}{S} = 0.5 \quad ...(ii)$$

From (i) and (ii), we get

$$0.75 - \frac{30}{S} = 0.5$$

$$\frac{30}{S} = 0.25$$

⇒ S = 120 km/hr.

76. Cost of manufacturer

$$= 30.09 \times \frac{100}{118} \times \frac{100}{120} \times \frac{100}{125}$$

= ₹ 17.00.

78. Tank B is double in volume than tank A.

So, Pipe R can empty tank A in $\frac{40}{2} = 20$ hours.

Now, when all the three pipes P, Q and R open simultaneously,

Portion of tank A filled in 1 hour

$$= \frac{1}{10} + \frac{1}{12} - \frac{1}{20}$$

$$= \frac{6+5-3}{60} = \frac{8}{60} = \frac{2}{15} \text{ tanks}$$

so, time required to fill the tank A

$$= \frac{15}{2}$$

= 7 hours 30 minutes.

80. Let, other discount is x %

then,
$$65 \times \left(\frac{100-10}{100}\right) \times \left(\frac{100-x}{100}\right) = 56.16$$

$$65 \times 0.9 \times (1 - 0.01x) = 56.16$$

$$1 - 0.01x = \frac{56.16}{65 \times 0.9}$$

$$1 - 0.01x = 0.96$$

$$0.01x = 0.04$$

$$x = 4\%.$$

121. Vertex : Pyramid :: Summit : Mountain

The vertex is the highest point of a pyramid, the summit is the highest point of a mountain.

122. Piety : Pilgrim :: Curiosity : Explorer

A pilgrim travels because of his piety; an explorer travels because of his curiosity.

123. Bad habits like smoking & drinking, laying stressful life with to much work load and polluted city environment are serious cause to affect the health of citizens.

124. The conclusion can be drawn from the above paragraph is value based education can save the society from crime & corruption.

125. Decoding of option:

A. Q – N + M × P means

Q is the brother of N, N is the mother of M and M is the sister of P. Here Q is the maternal uncle of P.

B. P + S × N – Q means

P is the mother of S, S is the sister of N, and N is the brother of Q. Here P is the mothe of S, N and Q.

C. P – M + N × Q means

P is the brother of M, M is the mother of N and N is sister of Q. Here, P is the maternal uncle of Q and N.

D. Q – S % P means

Q is the brother of S and S is father of P. Here Q is uncle of P.

Hence, correct option is (C).

130-134. Sitting arrangement of 6 Americans and 6 Germans around the circular table

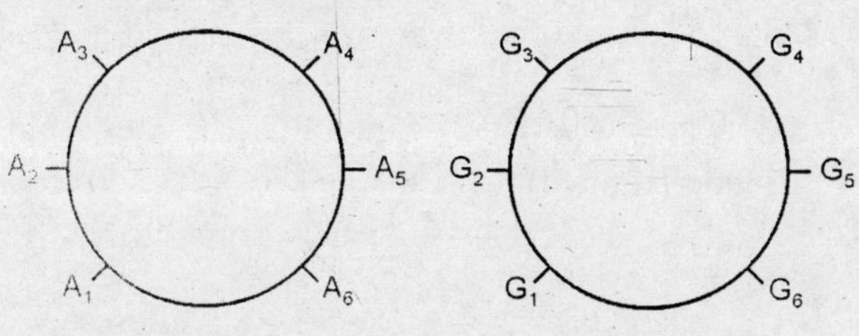

130. Right of A_1 is A_6 and Left of G_2 is G_3.

Now, Message from A_6 to G_3 forwarded via

$A_6 \rightarrow A_1 \rightarrow A_2 \rightarrow G_6 \rightarrow G_5 \rightarrow G_4 \rightarrow G_3$

Hence, number of Persons through which message passes = 5.

131. Possible path to send the message from A_3 to G_1 is

$A_3 \rightarrow A_4 \rightarrow A_5 \rightarrow A_6 \rightarrow A_1 \rightarrow A_2 \rightarrow G_6 \rightarrow G_5 \rightarrow G_4 \rightarrow G_3 \rightarrow G_2 \rightarrow G_1$

Hence, message passes through 10 persons.

132. Possible path to send the message from A_1 to A_6 is

$A_1 \rightarrow A_2 \rightarrow G_6 \rightarrow A_3 \rightarrow A_4 \rightarrow A_5 \rightarrow A_6$

Hence, message has to passes through 5 persons.

133. Possible path to send the message from A_3 to G_1 is

$A_3 \rightarrow A_4 \rightarrow A_1 \rightarrow A_2 \rightarrow G_6 \rightarrow G_5 \rightarrow G_4 \rightarrow G_1$

Hence, 6 is the minimum number of persons through which message has to passes.

134. Message from A_5 to G_3 can follow the way:

$A_5 \rightarrow A_4 \rightarrow A_3 \rightarrow G_6 \rightarrow G_1 \rightarrow G_2 \rightarrow G_3$

Hence, it has to passes through 5 persons.

135-136. Based on the given information, Position of seven persons in a straight line is as follows:

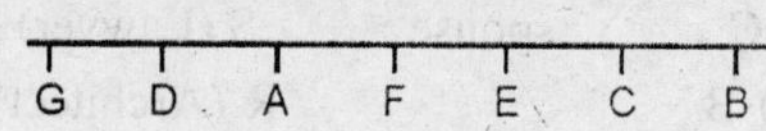

135. G and B are in the extreme end.

136. F is exactly in the middle.

137.

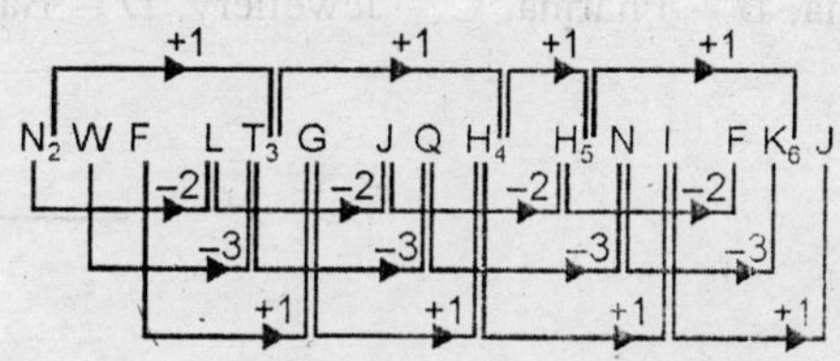

138-141. Total number of students $n(s) = 200$

Total number of students Passed in Math

$n(M) = 95$

Total number of students Passed in G.K.

$n(G) = 95$

Total number of students Passed in English

$n(E) = 125$

Total number of students Passed in English & Math

$n(M \cap E) = 55$

Total number of students Passed in G.K. & Math

$n(G \cap M) = 45$

Total number of students Passed in G.K. & English

$n(G \cap E) = 60$

Total number of students Passed in English & Math but failed in G.K.

$n(E \cap M \cap G') = 20$

Number of Students passes in either of subjects

$= 200 - 10 = 190$

$n\,(E \cup M \cup G)$

Number of students passed in all three subjects

$n(E \cap M \cap G)$

$= n(E \cup M \cup G) + n(M \cap E) + n(G \cap M) + n(G \cap E) - n(E) - n(M) - n(G)$

$= 190 + 55 + 45 + 60 - 95 - 95 - 125 = 35$

Now, $n(E \cap M \cap G')$

$= n(E \cap M) - n(E \cap M \cap G) = 55 - 35 = 20$ (given).

138. $n(M \cap G \cap E') = n(M \cap G) - n(M \cap G \cap E)$

$= 45 - 35 = 10.$

139. $n(E \cap G \cap M') = n(E \cap G) - n(M \cap G \cap E)$

$= 60 - 35 = 25.$

140. Number of students passed only in Math

$= n(M) - n(M \cap E) - n(M \cap G) + n(M \cap E \cap G)$

$= 95 - 55 - 45 + 35 = 30.$

142.

Male		**Female**
(Professor) A	spouse	T (CA)
(Doctor) C	spouse	S (Lawyer)
(Engineer) B		R (Architect)

143. Based on given data, employment of five persons is like this:

A – Cinema, B – Pharma, C – Jewellery, D – Bank, E – Sales.

145-148.

Names	Institutes	Specializations	Qualifications
Radhika	F	B.Tech	Operation Management
Neha	E	BBM	H.R.
Rupa	D	BCA	I.T.
Mahima	B	B.Com	General Management
Sayali	A or C	B.Sc.	Marketing
Shobha	C or A	C.A.	Finance

145. Mahima has specialization in B.Com and joined the institute B for General management qualifications.

146. B.Tech graduate Radhika joined the institute F in operation management.

147. Educational qualification of Sayali is B.Sc.

148. If Shoba joined the institute C, then Sayali joined the institute A for qualification in Marketing.

149-150. SIM – Always true

JIM – Always lie

LIM – True & lie (Alternatively).

Previous Paper (Solved)

Symbiosis National Aptitude Test (SNAP), 2013

SECTION- I

GENERAL AWARENESS

For all questions in this section, each correct answer carries 1 mark

1. The economically diverse region of MENA nations refers to the countries in
A. Middle East and North African Countries
B. Middle East and North American Countries
C. Mediterranean Europe and North Atlantic Countries
D. None of the above

2. Open market operations refers to the sale and purchase by the RBI of
A. Foreign currency B. Gold
C. Government bonds D. All the above

3. The agency that estimates national income in India is
A. RBI
B. Planning commission
C. Central Statistics Organisation
D. None of the above

4. Panchatantra was written by
A. Bhramabhatt B. Vishnu Sharma
C. Jaya Deva D. None of the above

5. IRDA is the Indian regulator for the
A. Banking sector B. Mutual funds
C. Rural Development D. None of the above

6. The Speaker of the Lok Sabha is elected by:
A. Members of Lok Sabha
B. President of India
C. Leader of the majority party in the Lok Sabha
D. None of the above

7. The Finance Minister's Budget speech presented to the Indian parliament every year usually has two parts. What does the Part A relate to:
A. Taxation Proposals
B. General Economic Survey
C. Both A and B
D. None of the above

8. Which two institutions are popularly referred to as the 'Bretton Woods Twins?
A. International Monetary Fund and World Bank
B. World Bank and Asian Development Bank
C. Asian Development Bank and International Monetary Fund
D. United Nations and World Trade Organisation

9. Based on the statements given below, Special Drawing Rights is best described to include which of the following statements:
1. It is a basket of currencies reviewed every 5 years
2. It was created to reduce the dependencies on gold and US dollar
3. It is a supplementary foreign exchange reserve assets maintained by IMF

A. 1 and 2 B. 2 and 3
C. 1, 2 and 3 D. only 3

10. Which NATO country's parliament was the first to vote in favour of the move to make military service compulsory for women
A. Italy B. Germany
C. Norway D. None of the above

11. The market for government securities is also referred to as
A. Bullion market B. Gilt-edge market
C. Secondary market D. Money market

12. From April 2010 the savings bank interest is calculated
A. at 4% per annum
B. on daily balance method based on the closing balance maintained every day
C. lowest balance available between 10th and the final day of the month
D. None of the above

13. Financial Action Task Force (FATF) is an Inter-governmental Body established in 1989. It is concerned with
A. Money laundering
B. Stock markets
C. Foreign Direct Investment
D. Foreign Institutional Investors

14. Which of the following sectors in India was not nationalised by the government at any point of time
A. Banking B. Insurance
C. Coal D. None of these

15. Which of the following is not a Palindrome
A. A car, a man, a maraca
B. A man, a plan, a canal, Panama
C. As you sow, so you reap
D. None of the above

16. The movie 'Life of Pi' is based on an award winning novel written by
A. Peter Carey B. Alan Hollinghurst
C. Yann Martel D. None of the above

17. LIDL, Walgreen, ASDA and Kroger are
A. Automobile companies
B. Retailers
C. Electronic product manufacturers
D. None of the above

18. Which bank was the first to introduce the Automated Teller Machine (ATM) in India
A. Hongkong and Shanghai Banking Corporation (HSBC)
B. Citibank
C. Standard Chartered Bank
D. HDFC Bank

19. The Three Gorges Dam, is the world's largest hydropower project and is located in
A. China B. Brazil
C. Russia D. None of the above

20. India signed the Kyoto Protocol in the year:
A. 2000 B. 1998
C. 2002 D. 1995

21. Given below are the creators of some well-known cartoon characters. Identify the creator–cartoon pair that does not match
A. Elzie Crisler Segar – Popeye
B. Dan Yaccarino – Oswald the Octopus
C. William Hanna – Richie Rich
D. Walt Disney – Donald Duck

22. The goals of monetary policy do not include
A. full employment
B. price stability
C. maximising tax revenue
D. Any of the above

23. The Corporate Head Quarters of Tata Group is
A. Bombay House B. Tata Towers
C. JRD Centre D. None of the above

24. Dronacharya Award for Coaching for the year 2013 was received by Narinder Singh Saini for which discipline
A. Archery B. Boxing
C. Hockey D. None of the above

25. This company signed an agreement with the Government of India for the Passport Seva Project in 2008
A. Infosys Technologies
B. Mahindra Satyam
C. Tata Consultancy Services
D. Wipro

26. Recommendations of which committee led to the reforms in the banking sector in India in the nineties
A. Malhotra Committee
B. Narasimham Committee
C. Mayaram Committee
D. Damodaran Committee

27. Which country is the largest producer of newsprint?
A. Indonesia B. Finland
C. Canada D. None of the above

28. India's Direct Cash Transfer Scheme draws similarities from models of which country
A. Brazil B. Germany
C. Australia D. Sri Lanka

29. Who put forth the theory of 'Zero Defects':
A. W. Edward Deming B. Philip Crosby
C. Joseph Juran D. Martin Hall

30. Logic of induction is very close to
A. the logic of observation
B. the logic of sampling
C. the logic of the controlled variables
D. None of the above

31. How many pairs of chromosomes exist in a human cell?
A. 12 B. 23
C. 34 D. 45

32. International Date Line roughly follows the longitude of
A. 180 degrees B. 90 degrees
C. 360 degrees D. None of the above

33. Daylight saving time (DST) implies adjusting the clock by one hour:
A. forward near the start of spring and backward in autumn
B. forward near the start of autumn and backward in spring
C. forward near the start of spring and forward in autumn
D. backward near the start of autumn and backward in spring

34. Who among the following sports persons are twins
A. Serena Williams and Venus Williams
B. Mark Waugh and Steve Waugh
C. Irfan Pathan and Yusuf Pathan
D. None of the above

35. Individuals with this blood type are universal red cell donors
A. A^- B. B^+
C. AB D. O^-

36. 'Relationship beyond banking' is the tag line of
A. State Bank of India
B. Central Bank of India
C. Bank of India
D. Bank of Maharashtra

37. Find the odd man out:
A. Sarod B. Sitar
C. Shehenai D. Santoor

38. Which instrument measures Blood Pressure?
A. Sphygmomanometer B. Hygrometer
C. Saccharimeter D. Spherometer

39. Find the odd person out:
A. Shikha Sharma B. Chanda Kochhar
C. Shubhalakshmi Panse D. Kiran Mazumdar Shaw

40. Extradition is a process whereby
A. one country transfers a suspected or convicted criminal to another country
B. olden traditions are rejuvenated as a part of cultural events
C. additional diction efforts are taken for language improvisation
D. extraction of the tar is carried out from the mines

SECTION-II

ANALYTICAL & LOGICAL REASONING

For all questions in this section, each correct answer carries 2 marks

41. What will be the measurement of the angle made by the hands of a clock when the time is 8.35?
A. 45 degrees B. 32.5 degrees
C. 47.5 degrees D. 30 degrees

42. Find the odd man out.
A. DIRAMLOG B. ERENIGEN
C. UIHCSSBI D. AFOLDFID

43. Raghu is at point A. He walks 3 km to the north and then turns to his left. He walks 4 km in this direction. He turns left again and walks 6 km. If he wishes to reach point A again, in which direction should he be walking and what distance will he have to cover?
A. South east – 5 km B. South east – 4 km
C. North east – 5 km D. North east – 4 km

Directions (Q. No. 44 to 45): *Read the following information carefully and answer the questions.*

Six students A, B, C, D, E and F participated in a dancing competition wherein they won prizes 12000, 10000, 8000, 6000, 4000, 2000 according to the position secured. The following information is known to us:

(*a*) A won less money than B
(*b*) The differences between the winning of C and F was ₹ 2000
(*c*) The differences between the winning of D and F was at least ₹ 4000
(*d*) E won the ₹ 8000 prize.

44. Which of the following could be the ranking from first place to sixth places of students?
A. A, D, E, B, F, C B. B, A, E, C, F, D
C. F, B, E, A, C, D D. B, A, E, D, C, F

45. If A won ₹ 4000, how much in total did C and F win?
A. ₹ 6000 B. ₹ 10000
C. ₹ 22000 D. ₹ 18000

46. Where should the number 17 be placed to fit into the sequence?
18, 11, 15, 14, 19, 16, 10, 12, 20
A. Between 14 and 19 B. Between 18 and 11
C. Between 19 and 16 D. Between 12 and 20

47. There are three facts and you are given three more statements. You have to determine which of these, if any is also a fact.

Fact 1. All kids like to play.
Fact 2. Some kids like ice cream.
Fact 3. Some kids look like their mothers.
Statement 1. All kids who like ice cream look like their mothers.
Statement 2. Kids who like ice cream also like to play.
Statement 3. Kids who like to play do not look like their mothers.
A. 1 only B. 2 only
C. 2 and 3 only D. None can be a fact

48. Insert the missing letter

A	F	K
H	M	R
	T	Y

A. O B. P
C. Q D. R

Direction (Q. 49): *Six statements are followed by options consisting of three statements put together in a specific order. Choose the option which indicates a valid argument, that is, the third argument is a conclusion drawn from the earlier two statements.*

A. Amar is unhappy
B. Amar is honest
C. Some magicians are honest
D. No honest man is a magician
E. No magician is happy
F. Amar is not a magician
A. FED B. BFA
C. AEF D. BDF

Directions (Q. Nos. 50 and 51): *The table below presents data on Gross National Product (GNP) and Net National Product (NNP) for India for six years. Answer questions based on these data.*

Gross National Product and Net National Product

	GNP (in Rs. Crores)	NNP (in Rs. Crores)
1988-89	348210	309289
1989-90	402931	357285
1990-91	470252	418057
1991-92	541593	479001
1992-93	615831	543566
1993-94	695342	615273

50. The year during which both GNP and NNP have shown almost equal rate of growth:
A. 1989-90 B. 1990-91
C. 1991-92 D. 1992-93

51. Which of the following statements could be stated to be true:
A. GNP has been growing during the period under review
B. NNP has been growing during the period under review
C. The growth pattern has been different for both indicators during period under review
D. None of these

52. If PEACE is coded as RGCEG then how is MICKEY coded in that code?
A. MOUSE B. OKEMA
C. LHBJX D. JDLJ

53. A set of symbols is given indicating the terms for which they will be used in the questions following them.

Θ = equal to, φ = not equal to, Δ = greater than, x = less than, + = not greater than, \$ = not less than

A Δ B Δ C does not imply which of the following options?
A. A \$ B Δ C B. C + B x A
C. A x B Θ C D. C + B φ A

54. Given below are 4 pictures of a cube

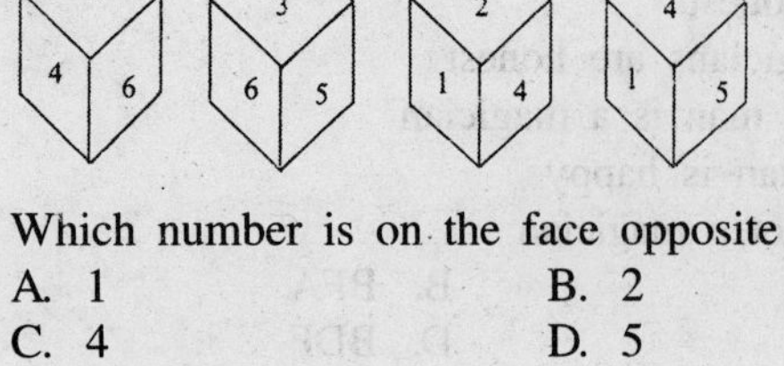

Which number is on the face opposite to 3?
A. 1 B. 2
C. 4 D. 5

Directions (Q. Nos. 55 and 56):

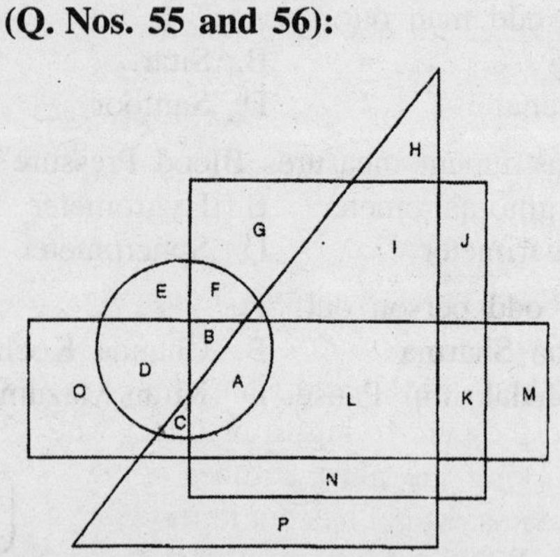

Rectangle indicates Faculty members, Square indicates Research scholars, Circle indicates Indians and Triangle indicates Americans.

55. In the given figure show the area which indicates faculty members who are research scholars but are neither Americans nor Indians
A. L B. N
C. K D. G

56. In the given figure show the area which indicates Indian Americans who are faculty members but not research scholars
A. A B. B
C. C D. F

57. While facing East you turn to your left and walk 10 yards. Then turn to your left and walk 10 yards, and now turn 45° to your right and go straight to cover 50 yards. Now in what direction are you with respect to the starting point?
A. North-East B. North
C. South-East D. North-West

58. If P – Q means Q is son of P, P × Q means P is brother of Q, P ÷ Q means Q is sister of P and P + Q means P is mother of Q, which of the following is definitely true about N × K – M ÷ L?
A. K is father of L and M
B. L is daughter of K and niece of uncle N
C. K is father of L and M – his son and daughter respectively
D. M is uncle of K's brother N

Directions (Q. Nos. 59 and 60): *Five girls Rama, Sudha, Tara, Uma and Veena share an apartment and have distributed the task as one girl making breakfast per day, Monday through Friday, one of the five dishes upama, dosa, idli, uttapa and paratha.*

Veena does not make uttapa and does not cook on Tuesday.
Sudha makes parathas but not on Monday or Friday.
Upama is made on Thursday.
Tara makes her dish which is not uttapa on Wednesday.
Idli is made on Friday but not by Uma.
Rama cooks on Monday.

59. What does Tara cook on Wednesday?

A. Upama B. Paratha
C. Idli D. Dosa

60. What day does Uma prepare her dish?

A. Monday B. Tuesday
C. Thursday D. Friday

Direction (Q. 61): *Answer the following question on the basis of the following data:*

Kunal, Rahul, Mohit and Charan are constables on duty in Delhi, and they are trying to recollect who amongst them has been working the most number of days without leave. Here are the facts: Kunal has put in more days of consecutive duty than has Rahul, who has got in five days of consecutive duty. Mohit has gone fifteen days straight without leave, more than Kunal and Rahul combined. Charan has put in eight consecutive days of duty, which is less than what Kunal has done.

61. How many days has Kunal gone without leave?

A. 6 B. 9
C. 12 D. 4

Directions (Q. Nos. 62 to 64): *Five hundred school children were asked cricketers they liked. 80 said they like Sachin Tendulkar, 60 said they liked Sourav Ganguly, 90 said they liked Mahendra Singh Dhoni. 10 said they liked all three. 40 said they liked Sachin and Sourav, 50 liked Sachin and Dhoni, while 30 said they liked Sourav and Dhoni.*

62. How many, out of the 500, didn't choose Sachin, Sourav or Dhoni?

A. 280 B. 380
C. 182 D. 480

63. How many like Sourav but not Tendulkar?

A. 21 B. 22
C. 23 D. None of the above

64. How many like fewer than two cricketers?

A. 410 B. 400
C. 390 D. 395

65. If 7 is coded as CBRT343, then 9 is

A. CBRT27 B. SQRT81
C. CBRT729 D. CBRT6561

66. If HELLO is coded as 15/12/12/5/8, then WHERE is

A. 16/13/13/6/9 B. 5/18/5/8/23
C. 5/5/18/23/8 D. 5/5/18/8/23

67. Complete the following series 3 10 59

A. 402 B. 403
C. 404 D. 405

68. If WOOD is coded as 23/225/4 then MEET is

A. 13/5/20 B. 13/10/20
C. 13/25/20 D. None of the above

Directions (Q. Nos. 69 and 70): *The following table gives data on Sectoral Investment during the Eight plan. Read the data and answer the questions.*

Sr. No.	Sector	Public Sector (Figures in Rs. Crores)	Private Sector (Figures in Rs. Crores)	6th Plan (Figures in %)	7th Plan (Figures in %)
1.	Agriculture	52000	96800	15.2	11.2
2.	Mining and Quarrying	28500	11100	6.1	6.7
3.	Manufacturing	47100	141300	23.6	26.0
4.	Electricity, Gas and Water	92000	10120	12.1	13.7
5.	Construction	3300	17240	2.7	1.9
6.	Transport	49200	38710	9.4	9.9
7.	Communication	25000	1000	1.5	2.0
8.	Services	63900	120730	29.4	28.6

69. Which of the following sectors has shown a steady increase in investments during last three plan periods?

A. Agriculture
B. Communication
C. Services
D. Construction

70. During the eight plan period, in which sector the percentage share of Public Sector investment has been the highest?

A. Mining and Quarrying
B. Electricity, Gas and Water
C. Communication
D. None of these

SECTION-III

QUANTITATIVE & DATA INTERPRETATION AND SUFFICIENCY

For all questions in this section, each correct answer carries 1 mark

71. Dinesh can do a piece of work in days, Harish can do it in 12 days, Dinesh, Harish and Subodh can together do it in 4 days. In how many days can Subodh alone do this work?
A. 12 B. 15
C. 10 D. 8

72. A village has 2000 people out of which 1100 are male and 900 female. The overall literacy rate is 95%, while only 3% male are not literate. What percentage of the female population is literate?
A. 92.55% B. 93.65%
C. 94% D. None of the above

73. A sum of money is sufficient to pay Ravi's wage for 18 days or Ajit's wage for 12 days. The same money is sufficient to pay the wages of both for how many days?
A. 9 days B. 8.5 days
C. 7.2 days D. 5.5 days

74. What is the number to be added to 1/4 of 30% of 120 so that it is 50% more than 25% of 40?
A. 6 B. 7
C. 5 D. 4

75. 10 years ago the average age of all 25 teachers of a school was 45 years. 4 years ago one teacher retired from her post at the age of 58. After one year a new teacher whose age was 54 years was recruited from outside. The present average age of all the teachers is:
A. $55\frac{17}{25}$ years B. $55\frac{4}{5}$ years
C. $54\frac{18}{25}$ years D. $49\frac{1}{2}$ years

76. X can complete a task in 15 days and Y can complete the same task in 20 days. X started the task alone and after 5 days Y joins X. After working together for three days, X had to leave the work. How long does it take for Y to complete the remaining work?
A. 4.5 days B. 5.3 days
C. 2.5 days D. 6.3 days

77. Find out the value of $[(2.093 + 0.089)^2 - (2.093 - 0.089)^2] / [(2.093)(0.089)]$
A. 4 B. 3.79
C. 4.12 D. 2.09

78. In a triangle ABC the measure of angle A is 50° and the external bisectors of angle B and angle C meet at O. The measure of angle BOC is
A. 140° B. 40°
C. 105° D. 65°

79. The average salary of all the 60 employees in an office is ₹ 12000/- per month. If the number of executives is twice the number of non-executive employees, then the average salary of all the non-executive employee is:
A. ₹ 9000/- B. ₹ 8000/-
C. ₹ 6000/- D. Cannot be determined

80. Two pots A and B contain syrup and water in the ratio 3:7 and 5:2 respectively. Find the ratio in which these mixtures be mixed to get a new mixture containing syrup and water in the ratio of 2:3?
A. 22:7 B. 4:3
C. 16:9 D. None of the above

81. A truck carrying goods has travelled for 3 days. The 1st day it travelled for 12 hours at an average speed of 40 km/hour, the 2nd day it travelled for 10 hours at an average speed of 50 km/hour, the third day it completed 360 kilometres travelling for 8 hours. The average speed of the truck in km/hour for all the three days is:
A. 45.22 B. 49.58
C. 43.73 D. 44.66

82. The H.C.F. of two number is 8 and their L.C.M. is 96. If one of the numbers is 32, then the other is
A. 24 B. 42
C. 16 D. None of the above

83. The ratio between present age of A and B is 3:4. The present age of B is 10 years more than the age of A 5 years ago. Find out the present age of B.
A. 12 B. 17
C. 20 D. None of the above

84. The circumference of the front wheels of a vehicle is 3.5 metres, and that of the back wheels is 3.0 metres. If the vehicle is moving at a speed of 54 km/hr, the shortest time in which both the wheels will make a whole number of turns is
A. 6.4 seconds B. 4 seconds
C. 2.1 seconds D. 1.4 seconds

85. If p is 81% of q, what percentage of p is q?
A. 81% B. 101.9%
C. 112.3% D. 123.45%

86. The length of a rectangular plot is 15 metres more than its breadth. If the cost of fencing the plot is ₹ 2700 at the rate of ₹ 30 per metre, what is the length of the plot in metres?
A. 25 B. 35
C. 30 D. None of the above

87. In a class, 70% of the students opted for History, 45% of the students opted for Geography, and 20% opted for neither History nor Geography. What percentage of the class opted for both the subjects?
A. 35% B. 25%
C. 30% D. 20%

88. A boy lying on the ground 1000 ft away from a multistoried building, can see the top of the building exactly in line with the top of a 10 ft tall flag post. If the flag post is 20 ft away from the boy, what is the height of the building?
A. 500 ft B. 600 ft
C. 800 ft D. 1000 ft

89. 21 boys turn out for an NCC parade. Of these 19 are wearing NCC caps and 11 are wearing NCC shoes. There are no boys without one or the other. How many boys are wearing the full uniform?
A. 2 B. 6
C. 9 D. 10

90. A number is increased by 20% and then is reduced by 20%. The number
A. does not change B. decreases by 4%
C. increases by 4% D. increases by 8%

91. In a class of 42, there are 22 boys and 20 girls. The class teacher has to appoint a girl monitor and a boy monitor. In how many ways can the selection be made?
A. 22 B. 420
C. 42 D. 440

92. Arvind and Mohan are 20 km apart. If they travel in opposite directions, they meet each other after half an hour. If they travel in the same direction, they meet after 2 hours. If Arvind travels faster than Mohan, his speed is
A. 15 km/hr B. 20 km/hr
C. 25 km/hr D. 30 km/hr

93. Let A = {1, 2, 3, 4, 5, 6}
How many subsets of A can be formed with just two elements, one even and one odd?
A. 6 B. 8
C. 9 D. 10

94. A grocer sells a soap marked ₹ 30/- at 15% discount and gives a shampoo sachet costing ₹ 1.5 free with each soap. He then makes 20% profit. His cost price per soap is
A. ₹ 15 B. ₹ 16
C. ₹ 18 D. ₹ 20

95. Which one of the following is correct?
A. $\log (m + n) = \log m + \log n$
B. $\log (m - n) = \log m - \log n$
C. $\log (m \times n) = \log m + \log n$
D. $\log m^n = (\log m)^n$

96. The sum of two numbers is 15; their product is 30. The sum of their reciprocals is
A. 2 B. 1/3
C. 1/2 D. 4/15

97. Three persons Amar, Akbar and Anthony agree to pay their hotel bills in the ratio of 3:4:5. Amar pays the first day's bill which amounts to ₹ 26.85. Akbar pays the second day's bill which amounts to ₹ 42.75 and Anthony pays the third day's bill which amounts to ₹ 53. When they settle their accounts, which of the following happens?
A. Amar gives ₹ 3 to Anthony
B. Akbar gives ₹ 2 to Amar
C. Amar gives Akbar ₹ 1.95 and ₹ 2 to Anthony.
D. None of the above

98. Inspired by the 'Golden quadrilateral project', the UP Government recently accomplished a diamond triangular project. Under this project the State Government laid down 6 lane roads connecting three cities Ayodhya, Banaras and Chitrakoot, which are equally separated from each other, *i.e.*, in terms of geometry they form an equilateral triangle. Angad and Bajrang start simultaneously from Ayodhya and Banaras respectively, towards Chitrakoot. When Angad covers 100 kms, Bajrang covers such a distance that the distance between Angad and Bajrang makes 90° angle with the road joining Banaras to Chitrakoot. When Bajrang reaches Chitrakoot, Angad is still 150 km away from Chitrakoot.
What is the distance between Ayodhya and Chitrakoot?
A. 250 km B. 450 km
C. 300 km D. None of the above

99. (1 – 1/2) (1 – 1/3) (1 – 1/4) (1 – 1/5) ... (1 – 1/40) = X/40
What is the value of X?
A. 13 B. 39
C. 21 D. 1

100. The smallest number which when divided by 12, 16, 18 gives 7 as remainder in each case, is
A. 103 B. 135
C. 151 D. 187

101. If $(a/b)^{2x-3} = (b/a)^{x-9}$, then x is equal to
A. –6 B. –2
C. 12 D. 4

102. The population of vultures in a locality decreases by a certain rate (compounded annually). If the current population of vultures 29160 and the ratio of the decrease in population for the second year and the third year be 10:9, what was the population of vultures 3 years ago?
A. 30000 B. 35000
C. 40000 D. 50000

103. Anna sold his car to Boney at a profit of 20% and Boney sold it to Chakori at a profit of 10%. Chakori sold it to a mechanic at a loss of 9.09%. The mechanic spent 10% of his purchase price on repairs and then sold it at a profit of 8.33% to Anna once gain.

A. Anna's loss is 23% B. Anna's loss is 29%
C. Anna's gain is 5% D. Anna's gain is 4%

104. A circle and a square have the same perimeter. Then

A. The area of the circle is greater than that of the square
B. The area of the square is $1/\pi$ times that of the circle
C. The area of the square is greater than that of a circle
D. The areas are equal

105. Two taps A and B can fill a cistern in 15 hours and 10 hours respectively. Tap C can empty full cistern in 30 hours. All the three taps were opened for 2 hours, when it was remembered that the emptying tap had been left open. It was then closed.

How many hours more would it take for the cistern to be filled?

A. 4.4 hours B. 1.2 hours
C. 4 hours D. 3.5 hours

106. At Atharva Prakashan every book goes through 3 phases —typing, composing and binding. There are 16 typists, 10 composers and 15 binders. A typist can type 8 books in each hour, a composer can compose 12 books in each hour and a binder can bind 12 books in each hour. All the people at Atharva Prakashan work for 10 hours a day and each person is trained to do only one type of job. How many books can be prepared in each day?

A. 1500 B. 1200
C. 1440 D. 1380

107. A contractor employed a certain number of workers to finish constructing a road in a certain scheduled time. Sometime later, when a part of work had been completed, he realized that the work would delayed by three-fourth of the scheduled time. So, he at once doubled the number of workers and thus he managed to finish the road on the scheduled time. How much work had been completed before increasing the number of workers?

A. $\frac{77}{9}\%$ B. $\frac{100}{7}\%$
C. 20% D. None of the above

Directions (Q. Nos. 108 to 110): *A train enters into a tunnel AB at A and B. A jackal is sitting at O in another by pass tunnel AOB, which is connected to AB at A and B, where OA is perpendicular to OB. A cat is sitting at P inside the tunnel AB making the shortest possible distance between O and P, such that AO : PB = 30 : 32. A train before entering into the tunnel AB blows the whistle (or siren) somewhere before A. On hearing the whistle, if the jackal and the cat run towards A, they would meet with an accident with the train exactly at A, the entrance to the tunnel. Further if the jackal moves towards B instead of A (on hearing the whistle) it would again meet with an accident exactly at B, the exit of the tunnel, with the same train coming from the same direction.*

108. What is the ratio of speeds of the Jackal and the Cat?

A. 4:3 B. 5:3
C. 1:1 D. None of the above

109. The ratio of speeds of the Jackal to the train is:

A. 5:1 B. 3:5
C. 1:5 D. None of the above

110. If the tunnels were to be connected from O to P and the Jackal moves from O to A through P, it will meet with train at M_1. Then AM_1 is:

A. 20 kms B. 16 kms
C. 10 kms D. None of the above

SECTION-IV

GENERAL ENGLISH

For all questions in this section, each correct answer carries 1 mark.

Directions (Q. Nos. 111 to 116): *Choose the correct synonym of the word in capital letters from the four options given below.*

111. ENTRENCHED

A. Filled up B. Fortified
C. Followed by D. Kept down

112. RECTITUDE

A. renew B. resort
C. integrity D. liberate

113. INVIDIOUS

A. unacceptable B. enticing
C. breathtaking D. inverted

114. AFICIONADO

A. affection B. good friend
C. tour guide D. ardent fan

Direction (Q. 115): *Choose the correct alternative for the following sentence:*

115. Despite immense development in almost all the fields, humans still cannot control nature and .... to do so.

A. probably not be able
B. probably will never able
C. probably will never be able
D. perhaps be never having ability

Direction (Q. 116): *The sentence given below is jumbled and each **bold** part is numbered. You are required to arrange it in the most logical and grammatically acceptable sequence.*

116. 1. **a path**
2. **put up a notice.**
3. **to another**
4. **from one place**
5. **the public**
6. **Halton Borough Council**
7. **about its**
8. **plans to move**
9. **to tell**

A. 5, 2, 9, 6, 7, 8, 1, 4, 3
B. 6, 8, 1, 4, 3, 2, 9, 7, 5
C. 5, 8, 1, 4, 3, 2, 9, 7, 6
D. 6, 2, 9, 5, 7, 8, 1, 4, 3

Directions (Q. Nos. 117 to 118): *Choose the correct spelling:*

117. Disgraphia is a deficiency in the ability to write, spell or put down thoughts on paper. The underlined word is spelled incorrectly. Which of the following is the correct spelling?

A. Disgrephia B. Disgrephia
C. Dysgraphya D. Dysgraphia

118. PANGERYRIC means "pay tribute to". The underlined word is spelt incorrectly. Mark the correct spelling

A. panegyric B. penagyric
C. penagiryc D. none of the above

Directions (Q. Nos. 119 to 120): *Identify two most appropriate options from those given below that can complete the sentence without changing the meaning.*

119. I .... scorpions, they are diabolic.

(*i*) Hate (*ii*) Scare
(*iii*) Loath (*iv*) Wary
(*v*) Cagey

A. (*i*), (*v*) B. (*v*), (*iii*)
C. (*i*), (*iii*) D. (*ii*), (*iv*)

120. To fully understand (A) .... work such as George Orwell's Animal Farm, one must be able to differentiate (B) .... events of the plot from the abundant extended metaphors.

Blank A	Blank B
(*i*) an allegorical	(*iv*) literal
(*ii*) a tragic	(*v*) superficial
(*iii*) a parabolic	(*vi*) Figurative

A. (*ii*), (*iv*) B. (*i*), (*iv*)
C. (*iii*), (*iv*) D. (*i*), (*v*)

Direction (Q. 121): *Find out the two words that are nearly the same or opposite in meaning and indicate the correct combination provided in the four options.*

121. I. attenuate II. repent
III. Make thin IV. Force

A. I-III B. III-II
C. II-IV D. I-II

Direction (Q. 122): *Choose the word nearest in meaning to the word underlined.*

122. A **rapier** is a

A. baton B. rascal
C. quick sharp hit D. long thin light sword

Directions (Q. Nos. 123 and 124): *Choose the correct alternatives for the following idioms.*

123. Raise Cain

A. re-incarnate B. cause trouble
C. bring up D. look after

124. Go to seed.

A. harvesting
B. planting
C. become unfit and run down
D. financial ruin

Directions (Q. Nos. 125 and 128): *Read the following passage and answer the questions that follow.*

The next ingredient is a very remarkable one: Good Temper. "Love is not easily provoked". Nothing could be more striking than to find this here. We are inclined to look upon bad temper as a very harmless weakness. We speak of it as a mere infirmity of nature, a family failing, a matter of temperament, not a thing to take into very serious account in estimating a man's character. And yet here, right in the heart of this analysis of love, it finds a place; and the Bible again and again returns to condemn it as one of the most destructive elements in human nature. The peculiarity of ill temper is that it is the vice of the virtuous. It is often the one blot on an otherwise noble character. You know men who are all but perfect, and women who would be entirely perfect, but for an easily ruffled quick-temper or "touchy" disposition. This compatibility of ill temper with high moral character is one of the strongest and saddest problems of ethics. The truth is there are two great classes of sins—sins of the Body, and sins of disposition. The prodigal son may be taken as a type of the first, the Elder Brother of the second. Now society has no doubt whatever as to which of these is the worse. Its brand falls, without a challenge, upon the Prodigal. But are we right? We have no balance to weigh one another's sins, and coarser and finer are but human words; but faults in the higher nature may be less venial than those in the lower, and to the eye of Him who is love, a sin against love may seem a hundred times more base. No form of vice, not worldliness, not greed of gold, not drunkenness itself does more to un-christianise society than evil temper. For embittering life, for breaking up communities, for destroying the most sacred relationships, for devasting homes, for withering up men and women, for taking the bloom off childhood; in short for sheer gratuitous misery-producing power, this influence stands alone. Jealousy, anger, pride, uncharity, cruelty, self-righteousness, touchiness, doggedness, sullenness-in varying properties these are the ingredients of all ill temper. Judge if such sins of disposition are not worse to live in, and for others to live with than sins of the body. There is really no place in Heaven for a disposition like this. A man with such a mood could only make Heaven miserable for all the people in it.

Excerpts from passages by Henry Drummond.

125. What is the popular notion about bad temper?

A. it is a serious malady
B. it is a harmless weakness
C. it is a peculiarity
D. it shows strength of character

126. What does the word "venial" mean?

A. forgivable sin or mistake
B. deer meat
C. wicked
D. insect

127. The Bible condemns which of the following as one of the most destructive elements in human nature?

A. anger B. laziness
C. procrastination D. splurging

128. Find the exact word in the passage which means easily or quickly offended.

A. uncharity B. self righteousness
C. touchiness D. pride

Direction (Q. 129): *Fill in the blank with the correct option.*

129. "If you are **morose**," you feel ....

A. excited B. sleepy
C. peaceful D. gloomy

Directions (Q. Nos. 130 and 131): *Name the part of speech of the **bold** word in the following sentences.*

130. We scored as many goals as **them**.

A. conjunction B. pronoun
C. adverb D. preposition

131. There is more evidence **yet** to be offered.

A. verb B. adverb
C. preposition D. none of the above

Direction (Q. 132): *Identify a synonym to replace the **bold** word(s).*

132. The baby is in the **carrying basket**.

A. bassinet B. tub
C. baby bag D. perambulator

Direction (Q. 133): *Which of the following does not denote the **bold** word(s)?*

133. The festival of **Christmas**.

A. Noel B. Yule
C. Crimbo D. Halloween

134. Choose the correct option to fill in the blank with the correct grammatical usage.

He has his finger .... the pulse of the nation.

A. in B. below
C. on D. over

135. Identify the error in the following sentence by choosing from the given alternatives

"Where do you live" asked the stranger?

A. an exclamation should be used instead of a question mark
B. the question mark should be used inside the quotation marks
C. the question mark should be used outside the quotation marks
D. none of the above

Directions (Q. Nos. 136 and 137): *In each of the following questions, the part of the sentence which is in capitals and has been **bold** may have an error. Correct the error.*

136. Unless he **DOES NOT GIVE UP** smoking, we cannot be sure of any improvement in his health

A. has not given up B. gives up
C. did not give up D. no improvement

137. A recently carried out meta-analysis of two decades of published research **DOES NOT SUGGEST THAT THERE SHOULD BE AN ASSOCIATION BETWEEN COFFEE DRINKING AND CORONARY AILMENTS.**

A. suggests that there is no association between coffee drinking and coronary ailments.
B. any association between and coronary ailments is not suggested.
C. shows that coffee drinking should not be suggested to have an association with coronary ailments.
D. suggests the association between coffee drinking and coronary ailments on a macro level.

138. In the following sentences, the order of the sentences has been jumbled up. Make the best choice to find out the correct sequence so as to form a meaningful paragraph

1. The following represents a condensed statement of what we think we now know about the relationship between education and employment
2. Although the linkages between education and employment are complex, and in the past, often analysed with simplistic notions of casualty, recent research results have yielded new insights about the nature of these linkages.

3. In the interests of brevity, the argument is put forward as a series of major propositions and derivatives strategies relating to the education employment nexus.
4. Intensive research efforts are currently being supported through the developing world both by national governments and international donor agencies in the hope of improving, understanding of the nature and causes of rising unemployment.
5. The massive problems of widespread and chronic unemployment and underemployment in less developed nations will remain among the most serious challenges to the development policy during the next several decades.

A. 1 3 2 4 5　　B. 2 1 3 4 5
C. 3 1 2 4 5　　D. 4 2 5 1 3

139. Given below are four sentences, each of which may or may not have an error. Following them are four options. Select the option indicating the error free statement(s).

1. Fruits from Madhavs' garden were stolen.
2. Fruit's are sold here
3. All the students of the students' council agreed to raise funds.
4. The dress's texture was coarse

A. 1 and 2 are correct　　B. 1 and 3 are correct
C. Only 3 is correct　　D. 2 and 4 are correct

140. Choose the odd word out

A. esoteric　　B. exigent
C. occult　　D. mystical

141. Choose the word or phrase that best completes the sentence:

The rebels sought to overcome the .... strength of the police forces by engaging in .... tactics

A. lack of .... divisionary
B. augmented .... peaceful
C. preponderant .... guerilla
D. power .... foolish

Directions (Q. Nos. 142 and 144): *Read the following passage and answer the questions that follow.*

My new mistress proved to be all she appeared when I first met her at the door—a woman of the kindest heart and feelings. She had never had a slave under her control previously and prior to her marriage she had been dependent upon her own industry for a living. She was by trade a weaver, and by constant application to her business, she had been in a good degree preserved from the lighting and dehumanizing effects of slavery. I was utterly astonished at her goodness. I scarcely knew how to behave towards her. My early instruction was all out of place. The crouching servility, usually so acceptable a quality in a slave, did not answer when manifested toward her. Her favour was not gained by it; she seemed to be disturbed by it. She did not deem it to be impudent or unmannerly for a slave to look in the face. The meanest slave was put fully at ease in her presence, and none left without feeling better for having seen her. But alas! This kind heart had but a short time to remain such. The fatal poison of irresponsible power was already in her hands, and soon commenced its internal work.

Very soon I went to live with Mr. and Mrs. Auld; she very kindly commenced to teach me the A, B, C. After I had learnt this, she assisted me in learning to spell words of three and four letters. Just at this point of my progress, Mr. Auld found out what was going on, and at once forbade Mrs. Auld to instruct me further, telling her that it was unlawful, as well as unsafe, to teach a slave to read. Further he said, "If you give a slave an inch, he will take a mile. A slave should know nothing but to obey his master—to do as he is told to do. Learning would spoil the best slave in this world. "Now," said he, "If you teach that boy (speaking of myself) how to read, there would be no keeping him." It would forever unfit him to be a slave. He would at once become unmanageable, and of no value to his master. As to him, it could do him no good, but a great deal of harm. It would make him discontented and unhappy. "These words sank deep into my heart, stirred up sentiments within that lay slumbering, and called into existence an entirely new train of thought." I now understood what had been to me a most perplexing difficulty—the white man's power to enslave the black man. From that moment, I understood that pathway from slavery to freedom. Though conscious of the difficulty of learning without a teacher, I set out with high hope, and a fixed purpose, at whatever cost of trouble, to learn how to read. The very decided manner with which my master spoke, and strove to impress his wife with the evil consequences of giving me instruction, served to convince me that he was deeply sensible of the truths he was uttering. It gave me the best assurance that I might rely with the utmost confidence on the results which, he said, would flow from teaching me to read. What he most dreaded, that I most desired. What he most loved, that I most hated. That which to him was great evil, to be carefully shunned, was to me a great good, to be diligently sought; and the argument which he so warmly urged, against my learning to read, only served to inspire me with a desire and determination to learn. In learning to read, I owe almost as much to the bitter opposition of my master, as to the kindly aid of my mistress. I acknowledge the benefit of both.

Excerpt from Fredrick Douglass.

142. The author's main purpose in this passage is to

A. describe a disagreement between a woman and her husband
B. analyze the reason for prohibiting the education of slaves
C. describe a slave's discovery of literacy as a means of freedom
D. dramatize a slave's change in attitude towards his mistress

143. For which of the following reasons does Mr. Auld forbid his wife to educate the slave?
1. Providing slaves with an education violates the law
2. He believes slaves lack the capacity for education
3. He fears education would leave the slave less submissive

A. Only 1 B. Only 3
C. 1 and 2 D. 1 and 3

144. The tone of author in acknowledging his debt to his master can be best described as
A. sentimental and nostalgic
B. cutting and ironic
C. petulant and self-righteous
D. resigned but wistful

145. Replace the bold phrase with the one which is closest to it in meaning.
I DROPPED A CLANGER when I mentioned her ex-husband.
A. Inadvertently caused embarrassment
B. Let go of a clanger
C. Toppled a lampshade
D. None of the above

Directions (Q. Nos. 146 to 148): *Fill in the blanks in the sentences with the appropriate choices.*

146. I should not have .... to talk in such a .... strain especially when I had not studied the man to whom I was talking.
A. daring .... commanding
B. tried .... bold
C. ventured .... peremptory
D. emboldened .... Reckless

147. The .... of Agatha Christie's argument is that human nature remains the same, wherever you are.
A. acne B. pinnacle
C. nadir D. crux

148. The .... play caused me to squirm in my seat, but she began to .... her eyes in a way that irritated me.
A. delirious .... spot
B. grin .... wet
C. disastrous .... damp
D. maudlin .... daub

Directions (Q. Nos. 149 and 150): *Select the choice that best replaces the bold part of the sentence given in the question:*

149. There they stood on the top shelf, all 20 volumes, **their red binding and gold lettering undoubtedly making them the most attractive between all the books** on the bookshelf
A. their red binding and gold lettering undoubtedly making them the most attractive between all the books on
B. their red binding and gold lettering undoubtedly making them the most attractive among all the books on
C. their red binding and gold lettering undoubtedly made them the most attractive among the books in
D. their red binding and gold lettering undoubtedly making them seem the most attractive among all the books on

150. Everything in the world **decays, it comes to an end and also dies inevitably**.
A. decays, it comes to an end and also dies inevitably
B. decays, comes to an end and dies inevitably
C. decays and as it comes to an end it also dies inevitably
D. decay, they come to an end and then they die inevitably

ANSWERS

1	2	3	4	5	6	7	8	9	10
A	C	A	B	B	A	B	A	C	C
11	**12**	**13**	**14**	**15**	**16**	**17**	**18**	**19**	**20**
B	B	A	B	C	C	B	A	A	C
21	**22**	**23**	**24**	**25**	**26**	**27**	**28**	**29**	**30**
C	A	A	C	C	B	C	A	A	B
31	**32**	**33**	**34**	**35**	**36**	**37**	**38**	**39**	**40**
B	A	A	B	D	C	C	A	D	A
41	**42**	**43**	**44**	**45**	**46**	**47**	**48**	**49**	**50**
C	*	C	D	C	C	A	A	D	A
51	**52**	**53**	**54**	**55**	**56**	**57**	**58**	**59**	**60**
A	B	C	C	D	C	A	B	D	C

61	62	63	64	65	66	67	68	69	70
B	B	D	B	C	B	C	C	B	B
71	72	73	74	75	76	77	78	79	80
C	A	C	A	C	D	A	D	D	A
81	82	83	84	85	86	87	88	89	90
D	A	C	D	D	C	A	A	C	B
91	92	93	94	95	96	97	98	99	100
D	C	C	D	C	C	C	C	D	C
101	102	103	104	105	106	107	108	109	110
D	C	A	A	A	B	A	B	C	D
111	112	113	114	115	116	117	118	119	120
B	C	A	D	C	D	D	A	C	B
121	122	123	124	125	126	127	128	129	130
A	D	B	C	B	A	A	C	D	B
131	132	133	134	135	136	137	138	139	140
B	A	D	D	B	B	A	B	C	B
141	142	143	144	145	146	147	148	149	150
C	C	D	B	A	C	B	D	B	B

SOME SELECTED EXPLANATORY ANSWERS

41. $\frac{\text{Angle rotated}}{\text{minute}}$ by minute hand $= \frac{360^\circ}{60} = 6^\circ$

Angle rotated by minute hand in 35 minutes $= 6 \times 35 = 210^\circ$.

$\frac{\text{Angle rotated}}{\text{hour}}$ by hour hand $= \frac{360^\circ}{12} = 30^\circ$

Angle rotated by hour hand in 8.35 hours.

$= \left(8 + \frac{35}{60}\right) \times 30^\circ = 257.5^\circ$.

Angle between two hands

$= 257.5 - 210^\circ = 47.5^\circ$

43. From ΔABC,

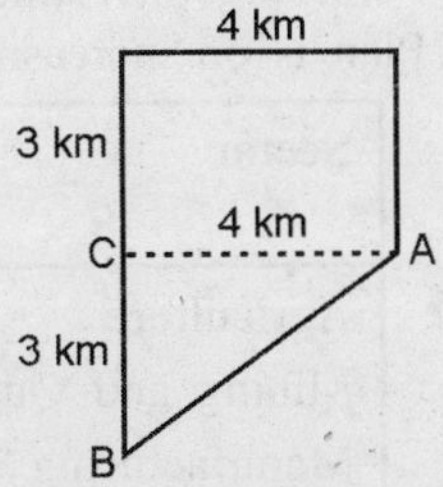

$AB = \sqrt{(AC)^2 + (BC)^2}$

$= \sqrt{(3)^2 + (4)^2}$

$= \sqrt{25}$

= 5 km North-East.

44. Based on given data, ranking from first place to sixth places of students could be in many ways. One of them is B, A, E, D, C, F.

45. If A won ₹4000, then their are two way of their ranking that is

C, F, E, B, A, D or F, C, E, B, A, D

In both cases total amount win by C and F is 12000 + 10000 = 22000.

48.

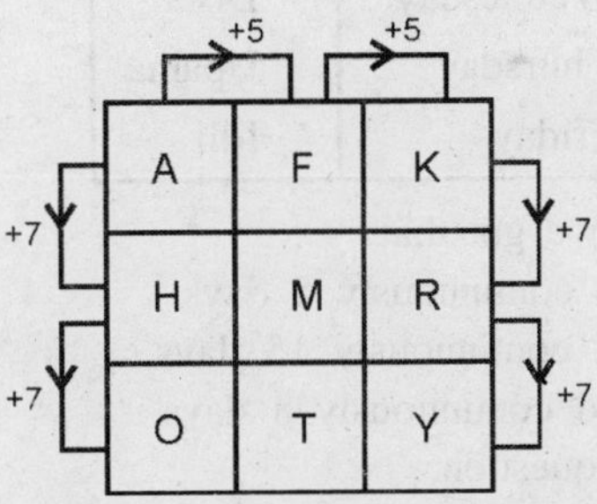

Row 1 : $A \xrightarrow{+5} F \xrightarrow{+5} K$

Row 2 : $H \xrightarrow{+5} M \xrightarrow{+5} R$

Row 3 : $O \xrightarrow{+5} T \xrightarrow{+5} Y$

49. From statement B : Amar is honest and from Statemen D : No honest man is magician.

We can draw conclusion from above two statements that Amar is not a magician.

52.

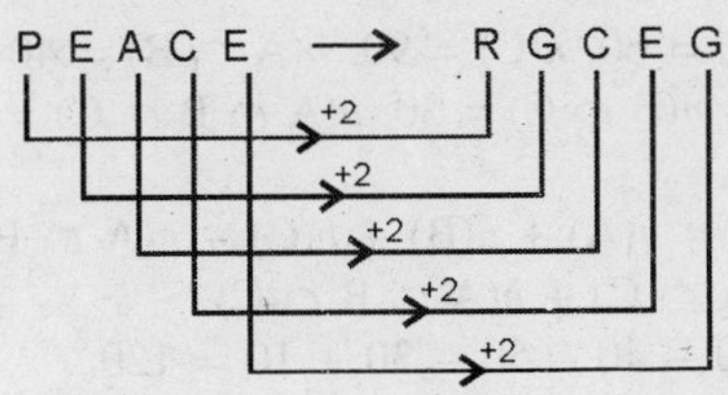

Similarly,

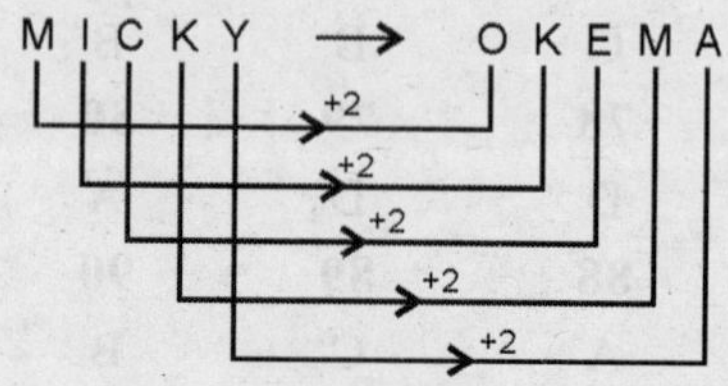

53. A > B > C

But A × B Θ C   means A < B = C.

57. Person is in North-East direction

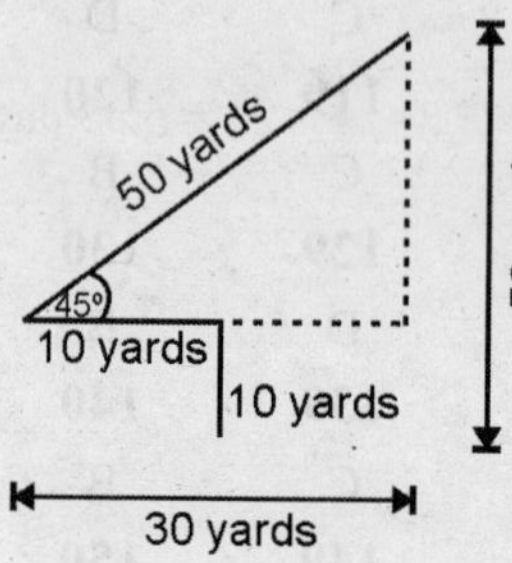

58. There is no data specify gender of K. So, we can not be as certain about K (male or female) gender. Hence correct option is B.

59-60.

Name	Day	Dish
Rama	Monday	Uttapa
Sudha	Tuesday	Parathas
Tara	Wednesday	Dosa
Uma	Thursday	Upama
Veena	Friday	Idli

61. From given data, we got that
Rahul has worked continuously 5 days.
Mohit has worked continuously 15 days.
Charan has worked continuously 8 days.
According to the question,

$X + 5 < 15$

and $X > 18$

$\therefore$ $X = 9$

here X is number of days Kunal has worked continuously.

62-64.

Notation: For
Sachin Tendulkar — A
Sourav Ganguly — B
Mahendra Singh Dhoni — C

$n(A) = 80$, $n(B) = 60$, $n(C) = 90$, $n(A \cap B) = 40$

$n(A \cap C) = 50$, $n(B \cap C) = 30$, $n(A \cap B \cap C) = 10$.

From formula

$n(A \cup B \cup C) = n(A) + n(B) + n(C) - n(A \cap B) - n(B \cap C) - n(A \cap C) + n(A \cap B \cap C)$

$= 80 + 60 + 90 - 40 - 50 - 30 + 10 = 120.$

62. Number of students didn't choose any
$= 500 - 120 = 380.$

63. Student who likes Sourav but not Sachin
$= n(B) - n(A \cap B)$
$= 60 - 40 = 20.$

64. Students like fewer than two cricketers
= Students like none + Students like only one
$= 380 + (100 - 50 - 30) = 400.$

65. 7 — CBRT $(7)^3$

9 — CBRT $(9)^3$ = CBRT 729.

66.

	H	E	L	L	O
Alphabet	↓	↓	↓	↓	↓
Position:	8	5	12	12	15

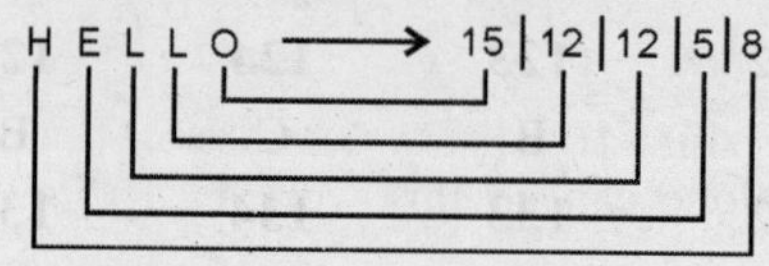

Similarly,

	W	H	E	R	E
Alphabet	↓	↓	↓	↓	↓
Position:	23	8	5	18	5

W H E R E ⟶ 5 | 18 | 5 | 8 | 23

68. W O O D   Similarly M E E T

Alphabet Position: 23 | 15 × 15 = 225 | 4   13 | 5 × 5 = 25 | 20

69. 8th Plan investment on communication in public sector (figure in %)

$$= \frac{25000}{361000} \times 100 = 6.9\%$$

Investment in private sector

$$= \frac{1000}{437000} \times 100 = 0.228\%$$

8th Plan investment on communication (figure in %)

$= 6.9\% + 0.228\%$

$= 7.128\%$

Percentage investment on communication for last three Plan is on increasing roll.

70.

Sector	8th Plan (figure in %)
Agriculture	14.4
Mining and Quarrying	7.89
Manufacturing	13.04
Electricity, Gas & Water	25.48
Construction	0.914
Transport	13.62
Communication	6.92
Services	17.7

Hence, highest percentage share on Electricity, Gas and Water.

71. Dinesh's one day work $= \frac{1}{15}$

Harish's one day work $= \frac{1}{12}$

Work done by all three together in one day $= \frac{1}{4}$

$\therefore$ Work done by Subodh in one day

$$= \frac{1}{4} - \frac{1}{12} - \frac{1}{15}$$

$$= \frac{15-5-4}{60} = \frac{6}{60} = \frac{1}{10}$$

Hence, Subodh alone can do this work in 10 days.

72. Number of literate people $= 2000 \times \frac{95}{100} = 1900$

Number of literate male $= 1100 \times \frac{97}{100} = 1067$

Number of literate female = 1900 – 1067 = 833

Literacy rate of female $= \frac{833}{900} \times 100 = 92.55\%$.

73. Let a sum of money 'X' is sufficient to pay Ravi's wage for 18 days or Ajit's wage for 12 days.

Money to pay Ravi and Ajit's wage per day

$$= \frac{X}{18} + \frac{X}{12} = \frac{5X}{36}$$

Let presume money X can pay them for n days.

Then, $\frac{n \cdot 5X}{36} \le X$

or, $n \le \frac{36X}{5X}$

$n \le 7.2$

$\therefore$ Number of days = 7.2 days.

74. Let the number to be added = x.

Then,

$$x + \frac{1}{4} \times 120 \times \frac{30}{100} - 40 \times \frac{25}{100} = \frac{50}{100} \times \frac{25}{100} \times 40$$

$x + 9 - 10 = 5 \quad \Rightarrow \quad x = 6.$

76. x's one day work $= \frac{1}{15}$

y's one day work $= \frac{1}{20}$

In 5 day x completed $5 \times \frac{1}{15} = \frac{1}{3}$ work

Workd done by x and y together in one day

$$= \frac{1}{15} + \frac{1}{20} = \frac{7}{60}$$

Work done by together in 3 days

$$= 3 \times \frac{7}{60} = \frac{7}{20}$$

Remaining work

$$= 1 - \frac{1}{3} - \frac{7}{20} = \frac{19}{60}$$

Remaining work completed by y in $\frac{19}{60} \times 20 = 6.33$ days.

77. $$\frac{\left[(a+b)^2 - (a-b)^2\right]}{ab} = 4$$

Here a = 2.093, b = 0.089.

78. $\angle BOC = 180^\circ - \angle OBC - \angle OCB$

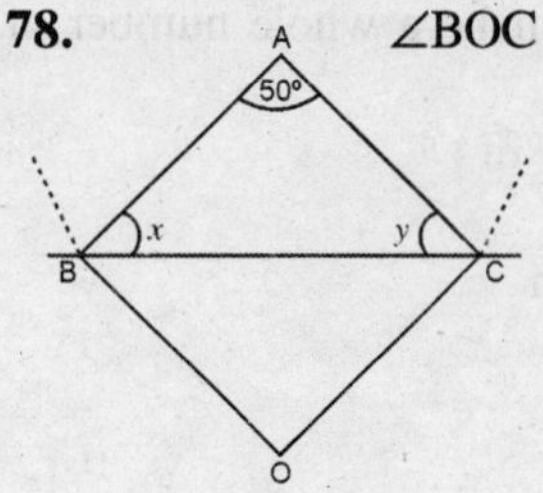

$$= 180^\circ - \left(\frac{50+x}{2}\right) - \left(\frac{50+y}{2}\right)$$

$$= 180^\circ - 50 - \left(\frac{x+y}{2}\right)$$

$$= 130^\circ - \left(\frac{180^\circ - 50^\circ}{2}\right)$$

$$= 65^\circ.$$

79. As the number of executives is twice the number of non-executive

So, number of executives = 40

and number of non-executives = 20

Total Salary of 60 employees

= 12000 × 60

= 7,20,000

Due to lack of data we can't determine required value.

80. Fraction of syrup in pot A $= \frac{3}{10}$

fraction of syrup in pot B $= \frac{5}{7}$

fraction of syrup in new mixture $= \frac{2}{5}$

Now from alligation

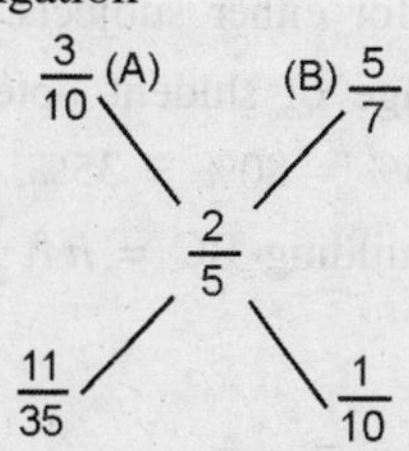

Hence, required ratio is $\frac{11}{35} : \frac{1}{10} = 22 : 7$.

81. Distance travelled on first day = 40 × 12 = 480 km

Distance travelled on second day = 50 × 10 = 500 km

Total distance travelled in 3 days

$= 480 + 500 + 360 = 1340$ km

Total time of travel $= 12 + 10 + 8 = 30$ hours

Average speed of the truck $= \frac{1340}{30} = 44.66$ km/hr.

82. Product of two numbers = L.C.M. × H.C.F.

Required number $= \frac{8 \times 96}{32} = 24$.

83. Ratio of age of A and B $= 3x : 4x$.

According to the question,

$4x = 10 + 3x - 5$

$4x = 3x + 5 \Rightarrow x = 5$

Hence, Present age of B = 20 years.

84. Speed of vehicle in m/sec. $= 54 \times \frac{5}{18} = 15$ m/sec

Time when both the wheels make a whole number of turns

= L.C.M. of 3.5 m and 3 m

= 21 m

Time required to cover 21 m

$= \frac{21}{15} = 1.4$sec.

85. $p = \frac{81}{100}q$ or, $q = \frac{180}{81}p$

Required percentage $= \frac{\frac{100}{81} \times p \times 100}{p}$

$= \frac{10000}{81} = 123.45\%$

86. Perimeter of the plot $= \frac{2700}{30} = 90$ metre

Let Length of the plot $= x$ m

Width of the plot $= (x - 15)$ m

$90 = 2(x + x - 15)$

$2x - 15 = 45$

$x = 30$ m

87. As 20% opted for none subjects so, 80% opted for either subjects.

Hence, Percentage of student opted both subjects

$= 70\% + 45\% - 80\% = 35\%$.

88. Let height of building BC $= h$ ft

From ΔADE,

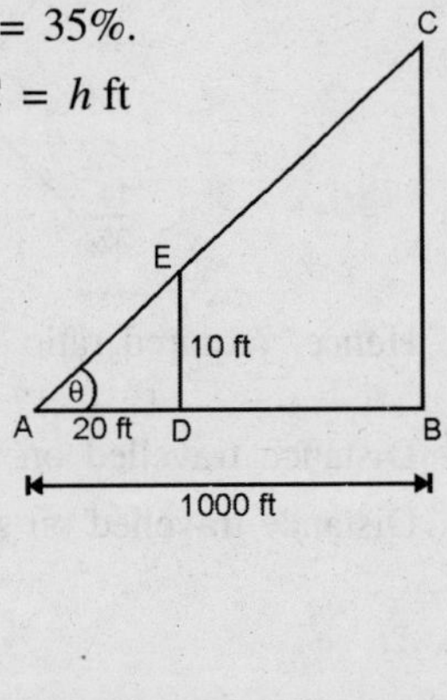

$\tan\theta = \frac{10}{20} = \frac{1}{2}$

Again from ΔABC,

$\tan\theta = \frac{h}{1000}$

or, $\frac{1}{2} = \frac{h}{1000}$

Height of building = 500 ft.

89. Total number of boys = 21

boys with NCC caps = 19

boys with NCC shoes = 11

Boys with ful uniform $= 19 + 11 - 21 = 9$

90. Let number is 100,

Increase by 20% $= 100 + 100 \times \frac{20}{100} = 120$

Decrease by 20% $= 120 - 120 \times \frac{20}{100} = 96$

Decrease % of the number $= \frac{(100 - 96)}{100} \times 100 = 4\%$.

91. Number of ways in which a boy elected for monitor

$= {}^{22}C_1 = 22$

Number of ways in which a girl elected for monitor

$= {}^{20}C_1 = 20$

Total number of ways for the selection $= 22 \times 20 = 440$

92. Let speed of Arvind and Mohan are x km/hr and y km/hr respectively.

According to the question,

$\frac{x}{2} + \frac{y}{2} = 20$

or, $x + y = 40$ ...(*i*)

Again, $2x - 2y = 20$ ...(*ii*)

From (*i*) and (*ii*), we get $x = 25$ km/hr

Speed of Arvind = 25 km/hr.

93. Required number of subject with one even and one odd $= 3 \times 3 = 9$.

94. Let cost price of each soap $= x$.

Total cost price including shampoo $= (x + 1.5)$

According to the question,

$(x + 1.5) \times \frac{120}{100} = 30 \times \frac{85}{100}$

or, $(x + 1.5) \times \frac{6}{5} = 25.5$

or, $(x + 1.5) = \frac{25.5 \times 5}{6}$

$= 21.25$

$x = 19.75$

96. Let two numbers are x and y.

According to the question,

$x + y = 15$ and $x \cdot y = 30$

Sum of their reciprocals

$$= \frac{1}{x} + \frac{1}{y} = \frac{x+y}{xy}$$

$$= \frac{15}{30} = \frac{1}{2}.$$

98.

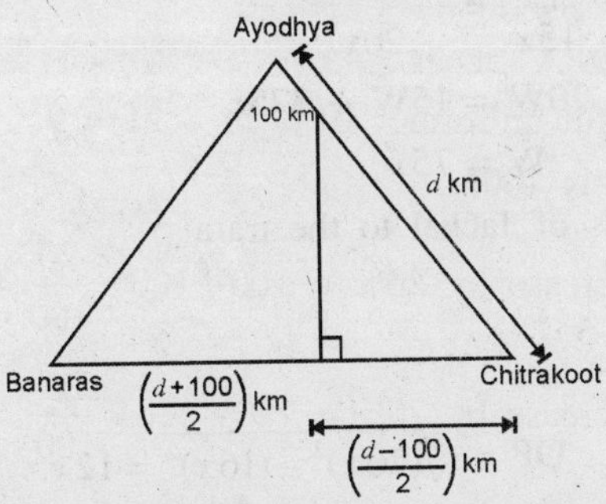

Let the distance between Ayodhya and Chitrakoot is d km. and Speed of Angad and Bajrang is u km/hr and v km/hr respectively.

According to the question,

$$\frac{d+100}{2v} = \frac{100}{u} \Rightarrow \frac{v}{u} = \frac{d+100}{200} \quad ...(i)$$

Again,

$$\frac{d}{v} = \frac{d-150}{u} \Rightarrow \frac{v}{u} = \frac{d}{d-150} \quad ...(ii)$$

From (*i*) and (*ii*), we get

$$\frac{d+100}{200} = \frac{d}{d-150}$$

or, $d^2 - 50d - 15000 = 200d$

or, $d^2 - 250d - 15000 = 0$

$\therefore$ $d = 300$ km.

99. $$\left(1-\frac{1}{2}\right)\left(1-\frac{1}{3}\right)\left(1-\frac{1}{4}\right)\left(1-\frac{1}{5}\right)...\left(1-\frac{1}{40}\right) = \frac{x}{40}$$

$$\frac{1}{2}\cdot\frac{2}{3}\cdot\frac{3}{4}\cdots\frac{39}{40} = \frac{1}{40}$$

$\therefore$ $x = 1$

100. L.C.M. of 12, 16, 18 = 144

Desired number = 144 + 7 = 151.

101. $$\left(\frac{a}{b}\right)^{2x-3} = \left(\frac{b}{a}\right)^{x-9}$$

or, $$\left(\frac{a}{b}\right)^{2x-3} = \left(\frac{a}{b}\right)^{9-x}$$

$\therefore$ $2x - 3 = 9 - x$

$3x = 9 + 3$

$x = 4.$

102. Let rate of decrease in vulture population is r %.

According to the question,

$$\frac{29160\left(1-\frac{r}{100}\right)^2}{29160\left(1-\frac{r}{100}\right)^3} = \frac{10}{9}$$

or, $$10\left(1-\frac{r}{100}\right) = 9$$

or, $0.1\,r = 1 \quad \Rightarrow r = 10\%.$

Population of vultures 3 years ago

$$\frac{29160}{\left(1-\frac{10}{100}\right)^3} = \frac{29160}{0.729}$$

$= 40000.$

104. Let perimeter of circle and square = P.

$$\text{Side length of square} = \frac{P}{A}$$

$$\text{Radius of circle} = \frac{P}{2\pi}$$

$$\text{Area of square} = \frac{P^2}{16} = 0.0625P^2$$

$$\text{Area of circle} = \pi r^2 = \frac{P^2}{4\pi}$$

$$= 0.079P^2$$

Hence, circle area is greater than square area.

105. When all the three taps are open, the portion of tank filled in 2 hours

$$= 2\times\left(\frac{1}{15}+\frac{1}{10}-\frac{1}{30}\right) = \frac{4}{15} \text{ tank}$$

Remaining empty portion of the tank

$$= 1-\frac{4}{15} = \frac{11}{15} \text{ tank}$$

Now, in 1 hour, tap A and B together can fill

$$\frac{1}{15}+\frac{1}{10} = \frac{1}{6} \text{ tank}$$

Time required to fill $\frac{11}{15}$ tank

$$= \frac{\frac{11}{15}}{\frac{1}{6}} = \frac{22}{5} = 4.4 \text{ hours.}$$

106.

	Typing	Composing	Binding
Books/hour	8	12	12
No. of empolyees	16	10	15
Working hour/day	10	10	10
Books/day	8 × 16 × 10 = 1280	12 × 10 × 10 = 1200	12 × 15 × 10 = 1800

Hence, books prepared in each day = 1200.

108. Let AP = z and OB = y and OP = D

According to the question,

AO : PB = $15x$: $16x$

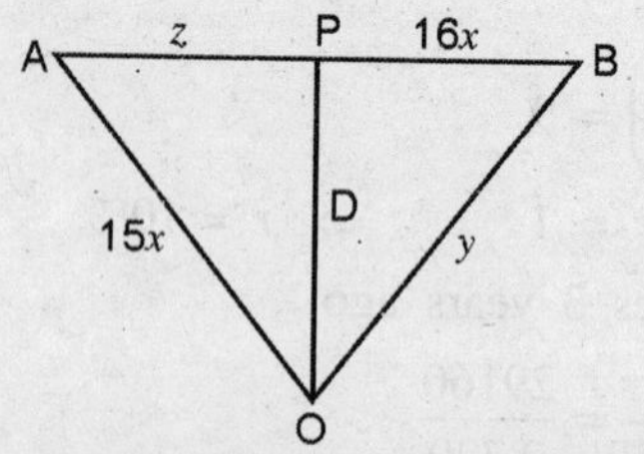

From ΔAOP, $(15x)^2 = z^2 + D^2$ ...(*i*)

From ΔOPB, $(y)^2 = (16x)^2 + D^2$ ...(*ii*)

From ΔAOB, $(16x + z)^2 = (15x)^2 + y^2$ ...(*iii*)

From (*ii*) and (*iii*), we get

$$(16x + z)^2 = (15x)^2 + (16x)^2 + D^2 \quad ...(iv)$$

From (*i*) and (*iv*), we get

$(16x + z)^2 = 2(15x)^2 - z^2$

or, $z^2 + 16 \times z - (15x)^2 = 0$

$$z = \frac{-16 \pm \sqrt{1156x}}{2} = \frac{-16x \pm 34x}{2} = 9x$$

Ratio of speed of Jackal and the Cat

$= 15x : 9x = 5 : 3.$

109. Let train blows the whistle at distance W before A.

According to the question,

$$\frac{W}{15x} = \frac{W + 25x}{20x}$$

or, $20W = 15W + 375x$

or, $W = 75x$

Ratio of speeds of Jackal to the train

$$= \frac{15x}{75x} = 1:5.$$

110. From ΔOPB, $OP = \sqrt{(20x)^2 - (16x)^2} = 12x$

When train reached at point A, Jackal covered distance $15x$.

At that moment distance between Jackal and train

$= 9x - 15x + 12x = 6x$

Hence, distance AM_1 = 10 kms.

Here we consider only prime factor.

Model Paper (Solved)

Symbiosis National Aptitude Test (SNAP), 2012

SECTION- I

GENERAL AWARENESS

1. Julia Gillard is currently the Prime Minister of
A. UK B. Holland
C. France D. None of these

2. World Post Day is celebrated on
A. 9 October B. 30 January
C. 2 November D. 27 November

3. Who among the following became the President of Venezuela for the third consecutive term?
A. Henrique Capriles
B. Hugo Chavez
C. Tanvir Mahmud
D. None of these

4. Britain and Scotland have signed an agreement to hold a referendum in ________ on the latter's independence.
A. 2015 B. 2014
C. 2016 D. 2017

5. Who among the following is the teenaged Pakistani girl activist who become famous world wide after a near fatal attack on her?
A. Kainat B. Shazia
C. Malala D. None of these

6. Monisha Kaltenborn is associated with which sport?
A. Golf B. Cycling
C. Formula One D. Lawn Tennis

7. Who was captain of the West Indies cricket team that won the -20 Cricket World Cup 2012 held in Sri Lanka?
A. Kieron Pollard
B. Chris Gayle
C. Dwayne Bravo
D. Darren Sammy

8. Wayne Ronney of England is a well known player of
A. Polo B. Football
C. Lawn Tennis D. Chess

9. Which of the following cities will be hosting the next Olympic Games to be held in 2016?
A. Moscow
B. Melbourne
C. Rïo de Janeiro
D. Toronto

10. Which car company has recently launched SUV segment vehicle 'Quanto' in India?
A. BMW
B. Skoda
C. Mahindra and Mahindra
D. Ford

11. The Diesel Loco Modernization Works of Indian Railways is situated in
A. Ludhiana B. Chennai
C. Patiala D. Kolkata

12. The government recently cancelled the permission given to________ for setting up of a special economic zone (SEZ) in Mundra in Gujarat due to charges of violation of the SEZ Act.
A. Reliance Group
B. Aditya Birla Group
C. Adani Group
D. Godrej Group

13. Which Indian IT company has recently acquired consulting firm Lodestone?
A. TCS
B. Wipro
C. Infosys
D. Mahindra-Satyam

14. Which of the following is an Indian origin banking firm?
A. HSBC B. Citibank
C. Yes Bank D. RBS

15. POSCO is a major steel of which country?
A. Korea B. China
C. Indonesia D. None of these

16. Steve Ballmer is the CEO of which corporate?
A. Vodafone B. Microsoft
C. Google D. None of these

17. Which of the following brand of cars is owned by the Japanese automobile major Nissan
A. Rapid B. Brio
C. Sunny D. EON

18. The tag line 'The power to do more' is linked with
A. Dell B. Lenovo
C. HP D. Sony

19. Timothy geithner is currently the treasure secretary of which country?
A. France B. US
C. England D. None of these

20. Information technology corporate giant Apple recently won a $1.05 billion patent case against
A. Samsung B. LG
C. Hyundai D. Philips

21. World Bank has recently appointed ______ as its chief economist & senior vice president
A. Montek Singh Ahluwalia
B. Kaushik Basu
C. Amartya Sen
D. C. Rangarajan

22. Which of the following corporate houses is the India Partner of global coffee chain 'starbucks'
A. Tatas
B. Aditya Birla
C. Godrej
D. None of these

23. Which state has the largest urban population in India?
A. Maharashtra
B. Kerala
C. Chattisgarh
D. West Bengal

24. Mr. Hamid Ansari was re-elected as India's ______ vice president
A. 13th B. 14th
C. 12th D. 11th

25. 'Dream Yuga' is a motorbike launched by which company?
A. Hero B. Honda
C. TVS D. Yamaha

26. Which Insurance Company uses the tag line 'jiyo befikar'?
A. LIC
B. ICICI Prudential
C. TATA-AIG
D. Bajaj Allianz

27. Which Indian city has an IIT and IIM both situated there?
A. Indore B. Raipur
C. Kozikhode D. Banglore

28. Who became the first woman speaker of lok sabha in India?
A. Sarojini Naidu
B. Meira Kumar
C. Aruna Asaf ali
D. Indira Gandhi

29. Kargil Vijay Divas is celebrated on ______ every year in honour of kargil war heroes
A. 2 November B. 26 July
C. 27 November D. 15 July

30. CBDT stands for
A. Central Boards of Direct Tarrifs
B. Central Boards of Direct Taxation
C. Central Boards of Direct Taxes
D. Central Boards of Daily Taxes

31. Which is the most mobile chess piece?
A. Rook B. Horse
C. Queen D. Bishop

32. Indian Institute of Sugarcane Research is located at
A. Nasik B. Lucknow
C. Meerut D. Pune

33. Telecom giant 'NOKIA' comes from
A. Germany B. Finland
C. Poland D. Australia

34. Which Indian state projects itself as 'God's Own Country'?
A. Uttarakhand
B. Uttar Pradesh
C. Kerala
D. Madhya Pradesh

35. Which country includes its income from cocaine and narcotics in its GDP?
A. Columbia
B. Venezuela
C. Afghanistan
D. Peru

36. Which of the following sectors is not the part of infrastructure?
A. Railways
B. Power
C. Food processing
D. Roads

37. Which among the following is the oldest development financial institution of India?
A. UTI B. IDBI
C. ICICI D. IFCI

38. 'Prima' is a newly launched
A. car B. tractor
C. Truck D. motor-cycle

39. Dudhwa National Park is located on
A. Uttar Pradesh
B. Uttarakhand
C. Madhya Pradesh
D. West Bengal

40. Jaffna is situated in the ______ part of Sri lanka
A. Western B. southern
C. Northern D. Eastern

SECTION-II

GENERAL ENGLISH

Directions (Qs. 41 to 46): *Read the following passage and answer within its context.*

The world dismisses curiosity by calling it idle, or mere idle curiosity - even though curious persons are seldom idle. Parents do their best to extinguish curiosity in their children because it makes life difficult to be faced every day with a string of answerable questions about what makes fire hot or why grass grows. Children whose curiosity survives parental discipline are invited to join our university. Within the university, they go on asking their questions and trying to find the answers. In the eyes of a scholar, that is mainly what a university is for. Some of the questions that scholars ask seem to thė world to be scarcely worth asking let alone answering. They ask questions too minute and specialized for you and me to understand without years of explanation. If the world inquires of one of them why he wants to know the answer to a particular question he may say, especially if he is a scientist, that the answer will in some obscure way make possible a new machine or weapon or gadget. He talks that way because he knows that the world understands and respects utility. But to you who are now part of the university, he will say that he wants to know the answer simply because he does not know it. The way a mountain climber wants to climb a mountain simply because it is there. Similarly a historian when asked by outsiders why he studies history may come out with argument that he has learnt to repeat on such occasions. Something about knowledge of the past making it possible to understand the present and mould the future. But if you really want to know why a historian studies the past, the answer is much simpler: something happened, and he would like to know what. All this does not mean that the answers which scholars find to their questions have no consequences. They may have enormous consequences but these seldom form the reason for asking the question or pursuing the answers. It is true that scholars can be put to work answering questions for the sake of the consequences as thousands are working now, for example, in search of a cure for cancer. But this is not the primary function of the scholar, for the consequences are usually subordinate to the satisfaction of curiosity.

41. Common people consider some of the questions asked by scholars as unimportant
A. since they are not worth asking of answering.
B. because the question is related to new machines and gadgets.
C. because the common man does't understand questions without years of explanations.
D. scholars ask very minute, specialized questions beyond the comprehension of the common man.

42. In the statement 'that is mainly what a university is for' 'that' refers to
A. parents refusal to answer questions.
B. children's curiosity that survives parents structures.
C. questions not worth answering.
D. the aim and scope of the university to provide an opportunity to curious minds to find out the answers to their questions.

43. According to the passage the general public respects
A. new inventions.
B. any useful invention.
C. any invention that makes life easier for them.
D. a scientist who invents gadgets and machines for them.

44. The writer compares the scientist to
A. a historian and mountain climber.
B. a historian.
C. a mountain climber.
D. a scholar.

45. The primary function of a scholar is different from the search for a cure for cancer because
A. the answers to the scholar's question have no consequence unlike the results of the research involving a cure for cancer.
B. the answer sought by the scholar is selfish unlike the consequences of cancer research which are for the common weal.
C. the primary function of a scholar is satisfaction of his mental curiosity, while research involving a cure for cancer demands a constant, systematic and planned pursuit by several scholars.
D. several scholars work for a cancer cure while a single scholar works with a selfish motive.

46. Idle curiosity means
A. curiosity is lazy. B. idle people are curious.
C. curiosity is apt. D. casual curiosity.

47. Find the correct match of definition/meaning with usage for the word: WOULD

Definition/meaning	Usage
1. willingness	5. He would go for a walk even when it was raining.
2. obstinacy persistence	6. He would do as you say.
3. determination	7. He beat the ox, but it wouldn't move.
4. improbable or real condition	8. If you came across a snake what would you do?

A. 1-8, 2-5, 3-6, 4-6 B. 1-6, 2-7, 3-5, 4-8
C. 1-5, 2-7, 3-8, 4-6 D. 1-7, 2-5, 3-6, 4-8

48. Choose the most appropriate passive construction of the sentence:
'He is doing his job well.'
A. His job is done well by him.
B. His job is being done well.
C. He has been doing his job well.
D. His job is being well done.

49. There are three underlined words below, followed by their usages. Determine the sentences, in which the use of words is correct or appropriate:
Pray. Prey. Prying
A. If you pray with faith, they say, it will be answered.
B. He has fallen a prey to cheats
C. Prying into the affairs of others is bad.
A. A and B B. A and C
C. A, B and C D. B and C

Directions (Qs. 50 to 54): *The following is an excerpt from a recent article by David Ewing Duncan. Read the passage and answer the questions within its context.*

Eye surgeon Virendar Sangwan has perfected a procedure so cutting-edge that most who have tried it have failed. In an operating theatre in the central Indian city of Hyderabad, he surgically implants corneas grown in a petri dish from stem cells by his colleague Geeta Vemuganti in patients with damaged eyes. Together they perform about 80 corneal regeneration procedures a year, making the L.V. Prasad Eye Institute, where they work, one of the most prolific facilities in the world using stem cells to regenerate tissues of any kind.

The Sangwan-Vemuganti team uses stem cells found in the tissues of living adults, not ones derived from embryos. Teams all over the world are working with adult stem cells, trying to coax them to regrow cells in hearts, brains, livers and other organs, but progress is slow. Besides corneas, scientists have had some success regrowing skin cells and bone tissues, but those procedures remain experimental. "A number of programs around the world have tried to perfect this treatment, but they have had bad outcomes," says University of Cincinnati eye surgeon and stem cell specialist Edward Holland. "It is impressive what they are doing at Prasad."

In addition to the Hyderabad project, only Holland's program and a half-dozen others in the world conduct operations using corneas grown from stem cells.

The treatment uses stem cells harvested from the limbus, located where the cornea touches the white of the eye. For those with damaged corneas, these cells – called "limbic" and "conjunctiva" – are harvested from a patients good eye, if he has one, or from a close relative. They are placed in a petri dish and chemically tweaked to grow into the lower layer of a cornea, called epithelium. It is then transplanted into the eye of the patient where in most cases it takes hold and grows. In 56% of the cases at the Prasad Institute, patient could still see clearly after 40 months later.

Indians are well known for reverse engineering, meaning they can deduce how drugs are made in order to produce generic versions. But in this case, Sangwan and Vemuganti, a pathologist, developed the technique on their own from reading papers and running experiments in the lab. Sangwan says he had a number of patients with burned eyes who could not be helped with standard corneal transplants from cadavers, so he persuaded Vemuganti to try growing corneas in her lab. "You know how to grow cells, and I know how to do the transplant surgery." Vemuganti recalls him saying. "Why don't we work together?" She smiles and shakes her head. "I had no clue if this was going to work."

Vemuganti's major innovation was developing a platform on which to grow corneas. First she designed a circular glass tube about the size of a stack of coins. Then she overlaid the glass with tissue from a human placenta which is "a good surface to grow corneas on." She says. After that she placed stem cells in four places around a circle, added a growth medium, and watched the corneas begin to grow.

Commercial interests among stem cell companies for the procedure has been scant because of the perceived small volume of patients, says venture capitalist Antoun Nabhan of Bay Capital, who sits on the board of Cellerant, a leading stem cell company in San Carlos, Calif. But corneal stem cell treatment may have wider applications, say ophthalmologist Ivan Schwab of University of California at Davis. "These stem cells are similar to others in the body that make mucous membrane," he say. "These techniques of growing stem cells might one day be used to treat mucousmembrane tissue in the sinuses, bladder, and other organs."

50. According to the article Sangwan-Vemuganti team's cutting-edge procedure of implanting cornea grown from stem cells is considered a major advancement by the experts because
A. They derive stems cells from embryos.
B. Their labs are customised to grow stem cells.
C. They regrow cells in hearts, brains, livers with stems cells from tissues of living adults.
D. They derive stem cells from tissues of living adults and grow cells in labs.

51. Sangwan-Vemuganti procedure is carried out on
A. Patients requiring any corneal transplant
B. Patients with damaged corneas
C. Patients with damaged eyes of any kind
D. None of the above

52. The world recognises this Indian innovation because Indian scientists are normally known
A. To be good at analysing and finding out a method of how an existing drug is made.
B. As they are good researchers of drugs.
C. As they are good at carrying out experiments to create generic drugs.

D. As they are able to carry-out drug trials on large samples.

53. The pathologist, Vemuganti, started growing cornea in a petri dish

A. By following procedures published in research papers

B. By inventing a totally new procedure

C. By experimenting with procedures published in journals

D. By following the instruction of the transplant surgeon.

54. In the context of the passage choose the correct set of meanings for the words: PLATFORM and GENERIC

A. Lab table; related to genes

B. Method; related to genes

C. Lab experiments; without a brand name

D. Methodology; without a brand name

55. Choose the option which is closest in meaning to the word SUBTLE

A. Innocent B. Elusive

C. Dangerous D. Insidious

56. The following sentence has a missing punctuation mark, choose the right answer.

My mother who is from the village is very superstitious.

A. Brackets B. Comma

C. Semicolon D. Apostrophe

57. For the pair of sentences below choose the right option.

1. Those are them.
2. Those are they.

A. The first sentence has an error.

B. The second sentence is erroneous.

C. Both sentences are incorrect.

D. Both sentences are correct.

Directions (Qs. 58-62): *Read the following passage and answer within its context.*

TRIPs agreement provides a comprehensive set of global trade rules for the protection of copyright patents, trademarks, industrial designs, trade secrets, semiconductor lay out designs, and geographical indications, that apply to all the member-countries irrespective of their levels of development, natural and human endowments and history. Every member-country has been asked by the WTO to amend its national patent law to conform to that universal globalized format for legislation relating to pharmaceutical, agrochemical, food, alloys, etc.

Under Article 65, the developed countries have been asked to change their laws within another five years, and the less developed countries within an additional five years. The least developed countries have been asked to make those changes by 2005 AD.

This attempt at global standardisation and uniformity by way of TRIP's agreement is in conflict with the main thrust of the Rio Earth Summit of 1992 that set out the conditions for sustainable development. These two reveal two contrasting types of international approaches and norms.

While the 1992 Earth Summit and the 1993 convention on biodiversity (CBD) focused on 'diversity' as being fundamental to sustain life and development, TRIPs and WTO are pushing for 'conformity' to international standardized norms on patents, services, labour, investment and what not irrespective of their history, ecology, level of economic development, etc. But despite their diametrically opposed viewpoints, 170 countries signed CBD upholding the need for diversity, and 50 countries signed the TRIPs agreement in 1994 claiming the urgency of uniformity, with a very large element of common names (130) in both.

The convention on bio-diversity (CBD) in its Article 16.5 specifically asserts that intellectual property right must not be in conflict with conservation and sustainable use of bio-diversity, a provision that has been totally ignored by those who composed the TRIPs agreement. While in case of agriculture the higher yield of patented products induces the farmers to switch from a more varied production pattern, the resulting narrowing of genetic base makes the economy and society more vulnerable to plant disease and epidemics. It is true that the move towards cultivation of a smaller number of higher yielding varieties and the uniform spread of the same variety over a large space predates the present debate on patent, particularly since the introduction of the green revolution technology in the mid-sixties, but there can be no doubt that the latter has brought about a qualitative change in the scenario and has created possibility of a vast quantitative change too in that direction. So far no attempt has been made to reconcile the two conflicting approaches of CBD and TRIPs. If diversity is so important for sustaining life, how can WTO demand conformity to standardised global formats?

58. The author points out that intellectual property rights and their administration mechanism

A. is throttling the interest of global bio-diversity

B. is working to help sustain global bio-diversity

C. is being sustained by global bio-diversity

D. is what the global bio-diversity needs

59. Which of the following has been said by the author in the passage?

A. A high number of countries have signed both CBD and TRIPs, two conflicting treaties.

B. A narrow genetic base, if stuck to for long, is fraught with danger.

C. Although a non-discriminatory approach has been followed in the applicability of TRIPs, there has been a confessional attitude in prescribing a timeframe for Transition, as per needs of the respective countries.

D. The author is supportive of international conventions and treaties such as TRIPs, CBD etc.

60. Out of the countries that signed CBD, the percentage of those that signed the TRIPs also, is

A. 76.5 B. 74.5
C. 78.5 D. 80.2

61. According to the author, a higher-yield seed variety is not always welcome as it also ultimately leads to

A. diseases among the consumers.
B. diseases among the plants.
C. monopoly of developed countries.
D. monopoly of developing countries.

62. As per the TRIPs agreement not much differentiation is made between a developed country such as the USA and an undeveloped country such as Sudan. This is

A. definitely true B. probably true
C. probably false D. definitely false

63. A single word equivalent for the statement 'Speak falsely with deliberate intent' is

A. repudiate B. prevaricate
C. disparage D. equivocate

64. Identify the option with the correct spelling.

A. paraphernalia B. programme
C. pediatrics D. beserk

65. Which two sentences in the following convey the same idea?

Choose from the combinations listed below:

1. Wasn't there any checking at the airport?
2. I want to know if there was any checking at the airport.
3. I wonder if there should have been any checking at the airport?
4. There should have been checking at the airport.

A. 2, 3 B. 1, 4
C. 3, 4 D. 2, 4

66. Choose the grammatically correct sentence.

A. You should at once report it to the concerned authority.
B. You should report it at once, to the authority concerned.
C. You should report it at once, to the concerned authority.
D. You should at once report it to the authority concerned.

67. Find the correct match of grammatical function with usage for the word: AFTER

Grammatical function	**Usage**
1. adjective	5. You may go after having your lunch
2. adverb	6. It appears to be the after effect of the disease
3. conjunction	7. Many graduates are hankering after jobs
4. preposition	8. He came soon after

A. 1-8, 2-5, 3-6, 4-7 B. 1-6, 2-5, 3-8, 4-7
C. 1-5, 2-8, 3-7, 4-6 D. 1-6, 2-8, 3-5, 4-7

68. NEOPHYTE is the opposite of

A. student B. clown
C. veteran D. professional

69. Find the maximum number of times that any one of the given words fits the set of sentences.

disabled **flimsy** **crippled** **lame**

1. Don't make .... excuses.
2. Liberalization may have .... smaller manufacturers.
3. Being a defaulter at the stock exchange makes him a .... duck.
4. A .... person may limp.

A. in all the four sentences
B. in three sentences
C. in two sentences
D. in only one

70. Which of the following does not make a sensible word/phrase when added to the given word?

FIRE

A. fly B. engine
C. stick D. escape

71. Arrange the sentence 1,2,3,4 to form a logical sequence between sentences I and II. Choose the alternative where the four combinations make a meaningful sentence.

I. We all value having the freedom
1. which many of us fail to honour
2. to make the choices we want in our careers
3. but with great freedom comes great responsibility
4. so most companies fall prey to the policies which become rigid

II. and that's probably one reason we find most companies not following what they preach

A. 1, 3, 4, 2 B. 2, 3, 1, 4
C. 1, 4, 2, 3 D. 3, 2, 1, 4

Directions (Qs. 72-74): *Read the edited excerpt of an article by NELSON VINOD MOSES and answer the questions in this context.*

A successful non-resident Indian imployed in the United states returns to a backward Indian village and transforms the lives of the villagers. Sounds familiar. At 31, Ashwin Naik is pacing through the path Shah Rukh Khan traced in his off beat Bollywood movie, Swades. Naik had just quit his cushy job in a genomics firm in the US to join MIT Sloan School of Business. With a month in hand, he headed home and travelled through the remote areas of Bagalkot district in Karnataka. The woeful social conditions he saw moved him. Naik chucked the MBA course and in six months set up Vaatsalya Healthcare, a rural healthcare delivery system.

In February 2005, Vaatsalya's first hospital opened in Hubli. Two more centres were opened in Gadag and Karwar to offer specialist services of surgeons and facilities such as physiotherapy for children suffering from cerebral palsy. "We introduced paediatric surgery for infants below six months," says Naik. "Else, patients would have to be taken to distant cities of Hubli or Bangalore." Naik plans 100 more units in five states in the next three years. Mere charity by an affluent, middle-class professional? Far from it. Vaatsalya is one among rapidly spreading 'for profit' social enterprises that serves the poor and brings in profit. Mumbai based Ziqitza, an imbalance services company, is another. It never refuses a patient for money, and charges ₹ 50 to 200.

Done fleetingly in India and elsewhere till now, entrepreneurial minds with a social conscience are methodically creating such models at a greater pace. "There has been a boom in the past two years," says Varun Sahni, country director of Acumen Fund, a US-based social fund that invests in companies that target low income communities. "Currently, there are about 1,000 in India."

The timing seems perfect. There is a wide market acceptance and funding has been coming in easily. These enterprises work across a swathe of areas including healthcare, education, rural energy, agriculture, arts and crafts, banking and more. 'For profit' entrepreneurs are obsessed with social and environmental impact in addition to the financial returns. Since they are answerable to the investors, they try expanding the business rapidly. SKS Microfinance, for instance, started in 1998 and has now over 9,00,000 customers, 440 branches and an outstanding loan disbursement of over ₹ 452 crore as of August 2007.

72. Identify the appropriate business model of the kind of enterprise described by the author.

A. Servicing societies at no profit
B. Profiting from poor people
C. Setting up enterprises for masses of low-income groups on experimental basis.
D. Setting up enterprises for social causes for profit and expand rapidly

73. Which of the following companies does not illustrate the idea explained by the author?

A. SKS Microfinance
B. Acumen Fund
C. Ziqitza
D. Vaatsalya Healthcare

74. According to the author, which of the following options describes 'for profit' entrepreneurs most appropriately?

A. NRI's paying back to their motherland.
B. Those affluent, middle-class professionals treating it as charity.
C. Those who work towards getting financial returns on social business by expanding quickly.
D. Those who have sympathetic investors for their business ideas for poor.

75. A contextual usage is provided for the word below. Pick the word that is most inappropriate.
MALINGER: The young man made it a point to malinger inspite of the assigned work load.

A. Wander B. Laze
C. Evade D. Argue

76. The following is a scrambled sentence with the segments marked 1, 2, 3 and 4. Choose the alternative with the order of segments that best reconstruct the sentence.

1. For all the padre's rhetoric about the English as God's Chosen People, the padre had a whole tribe of Anglo-Indian first cousins.
2. Padre Rotton was an even more striking case.
3. By various Indian wives, all of whom were at that moment engaged in fighting on the rebel side in Avadh, where they took an active part in besieging the British Residency in Lucknow.
4. These included James Rotton who could not speak English and the twenty two Muslim sons of his convert cousin, Felix Rotton

A. 1, 2, 3, 4 B. 2, 1, 4, 3
C. 1, 4, 2, 3 D. 2, 4, 1, 3

77. Choose the sentence in which the given word is used correctly (grammatically and semantically) ALMOST

A. As I crossed the road a scooterist almost hit me.
B. Crossing the road a scooterist hit me almost.
C. A scooterist across the road almost hit me.
D. A scooterist almost hit me crossing the road.

78. In the following sentence choose the erroneous segment.

We took a taxi (A) so we would be on time (B) for the meeting (C)

A. Error in segment A
B. Error in segment B
C. Error in segment C
D. No error

79. Find the ODD one out from the group of words which are related in some way or the other

A. din B. cacophony
C. racket D. cadence

80. Fill in the blanks with the correct alternative.
Caw is to crows as ....is to cows.

A. bleat B. snort
C. low D. bellow

SECTION-III

QUANTITATIVE & DATA INTERPRETATION AND SUFFICIENCY

81. In 4 years, the SI on a certain sum of money is 7/25 of the principal. What is the annual rate of interest?

A. 4% B. 4.5%
C. 7% D. 9%

82. Thirty days are in September, April, June and November. Some months are of thirty one days. A month is chosen at random.
Then its probability of having exactly three days less than maximum of 31 is

A. 15/16 B. 1
C. 3/48 D. None of these

83. Two persons are climbing up on two moving escalators which have 120 steps. The ratio of 1st person's speed to that of 1st escalator is 2:3 (steps). The ratio of 2nd person's speed to that of 2nd escalator is 3:5 (steps). Find the total number of steps they both have taken together.

A. 85 B. 93
C. 80 D. 75

84. The table below shows the temperatures in degrees Fahrenheit in a city over one week.

Noontime Temperature!						
Mon	Tue	Wed	Thu	Fri	Sat	Sun
66	78	75	69	78	77	70

If m represents the median temperature, f represents the temperature that occurs most often, and a represents the average (arithmetic mean) of the seven temperatures, which of the following is the correct order of m, f, and a?

A. $a < m > f$ B. $a < m < f$
C. $m < a < f$ D. $m < f < a$

85. Inside a triangular park, there is a flower bed forming a similar triangle. Around the flower bed runs a uniform path of such a width that the sides of the park are exactly double the corresponding sides of the flower bed. The ratio of areas of the path to the flower bed is:

A. 1 : 1 B. 1 : 2
C. 1 : 3 D. 3 : 1

86. Consider the following piecewise definition of the function f.

$$f(x) = \begin{cases} 3 - x, \text{ if } x \leq 0 \\ x^2 + 2, \text{ if } x \geq 0 \end{cases}$$

Evaluate $f(-3)$.

A. 6 B. 0
C. 11 D. –7

87. A special lottery is to be held to select a student who will live in the only deluxe room in a hostel. There are 100 Year-III, 150 Year-II, and 200 Year-I students who applied. Each Year-III's name is placed in the lottery 3 times; each Year-II's name, 2 times; and each Year-I's name, 1 time. What is the probability that a Year-III's name will be chosen?

A. 1/8 B. 2/9
C. 2/7 D. 3/8

88. The average of nine numbers is M and the average of three of these is P. If the average of remaining numbers is N, then

A. M = N + P B. 2M = N + P
C. 3M = 2N + P D. 3M = 2P + N

89. The intersection of two cubes cannot be

A. cube B. triangle
C. a rectangle D. None of these

90. In a factory, each day the expected number of accidents is related to the number of overtime hour by a linear equation. Suppose that on one day there were 1000 overtime hours logged and 8 accidents reported and on another day there were 400 overtime hours logged and 5 accidents. What is the expected number of accidents when no overtime hours are logged?

A. 2 B. 3
C. 4 D. 5

91. A rainy day occurs once in every 10 days. Half of the rainy days produce rainbows. What per cent of all the days do not produce rainbow?

A. 95% B. 10%
C. 50% D. 5%

92. Bags I, II and III together have ten balls. If each bag contains at least one ball, how many balls does each bag have? Decide whether the data given in the statements are sufficient to answer the question.

Statement (1): Bag I contains five balls more than bag III.

Statement (2): Bag II contains half as many balls as bag I.

A. Statement (1) ALONE is sufficient, but statement (2) alone is not sufficient to answer the question asked.

B. Statement (2) Alone is sufficient, but statement (1) alone is not sufficient to answer the question asked.

C. BOTH statements (1) and (2) TOGETHER are sufficient to answer the question asked, but NEITHER statement ALONE is sufficient to answer the question asked

D. EACH statement ALONE is sufficient to answer the question asked.

93. Area of a square natural lake is 50 sq. kms. A diver wishing to cross the lake diagonally, will have to swim a distance of

A. 10 miles B. 12 miles
C. 15 miles D. None of these

94. If $n = 1 + x$, where x is the product of 4 consecutive positive integers, then which of the following is/are true?

1. n is odd;
2. n is prime
3. n is a perfect square

A. 1 and 3 only B. 1 and 2 only
C. 1 only D. None of these

95. How many arrangements can be formed out of the letters of the word EXAMINATION so that vowels always occupy odd places?

A. 72000 B. 86400
C. 51600 D. 64000

96. In a school drill, a number of children are asked to stand in a circle. They are evenly spaced and the 6th child is diametrically opposite the 16th child. How many children are made to stand in the circle?

A. 16 B. 20
C. 22 D. None of these

97. A man purchased 40 fruits; apples and oranges for ₹ 17. Had he purchased as many oranges as apples and as many apples as oranges, he would have paid ₹ 15. Find the cost of one pair of an apple and an orange.

A. 70 paise B. 60 paise
C. 80 paise D. 1 rupee

98. The number 311311311311311311311 is

A. divisible by 3 but not by 11
B. divisible by 11 but not by 3
C. divisible by both 3 and 11
D. neither divisible by 3 nor by 11

99. A man earns 6% SI on his deposits in Bank A while he earns 8% simple interest on his deposits in Bank B. If the total interest he earns is ₹ 1800 in three years on an investment of ₹ 9000, what is the amount invested at 6%?

A. 3000 B. 6000
C. 4000 D. 4500

100. Two identical trains A and B running in opposite directions at same speed take 2 minutes to cross each other completely. The number of bogies of A are increased from 12 to 16. How much more time would they now require to cross each other?

A. 40s B. 50s
C. 60s D. 20s

101. A ladder is lying against a wall which is 5 metres high. If the ladder slips 2 metres away from the wall, the top of the ladder touches the foot of the wall. The length of the ladder is

A. 5 m B. 5.25 m
C. 7.25 m D. 4 m

Directions (Qs. 102–104): *The following line graph shows the ratio of the wheat production to rice production of a state in India over a period 7 years from 1999–2006. Answer the questions based on this data.*

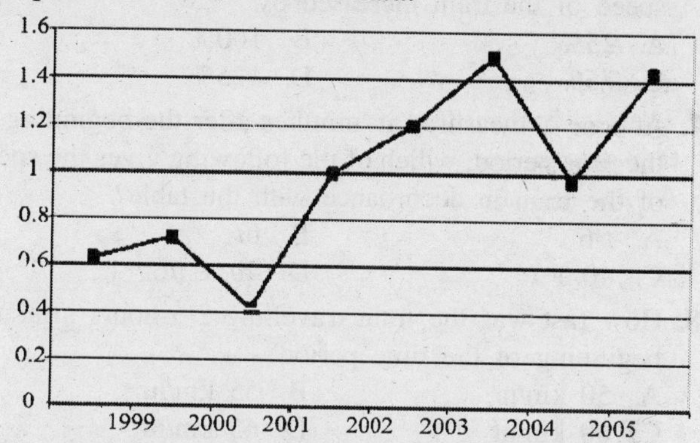

102. If the rice production in 2003 was 4 lac tons what was its wheat production during 2003?

A. 1.2 lac tons B. 3.6 lac tons
C. 4.8 lac tons D. 5.0 lac tons

103. In how many years wheat production was more than the rice production?

A. 2 B. 3
C. 4 D. 5

104. If the total annual grain requirement (wheat and rice together) of the state is 5 lac tons and if the rice production during 2001 was 3 lac tons, how much grain had to be imported from the other state of India to fulfill the requirement?

A. 2.00 lac tons B. 1.55 lac tons
C. 1.05 lac tons D. 0.65 lac tons

105. The profits of Biscuits India Ltd soared by 32% in the year 2006–07 as compared to year 2005–06. By what % did Biscuits India's Sales increase in 2006–07 compared to the previous year?

(Assume: Profit = Sales – Expenses)

Decide whether the information given in the two statements is sufficient to solve the problem.

Statement (1): Expenses in 2006–07 were ₹ 1,400 crores, as compared to ₹ 1,220 crores in 2005–06.

Statement (2): Sales in 2006–07 were ₹ 4,300 crores

A. Any one of the two statements (1) or (2) taken alone, is sufficient to answer the question.
B. Each of the statement (1) or (2) taken alone, is sufficient to answer the question.
C. Both statements, taken together, are sufficient to answer the question, but neither of them alone is sufficient.
D. Both the statements together are insufficient to answer the question.

Directions (Qs. 106–108): *Use the table about the speed of a train over a 3-hour period. The time count does not begin from starting of train.*

Time (Minutes)	0	30	45	60	90	120	150	180
Speed at given time (in km/hour)	40	45	47.5	50	55	60	65	70

106. During the three-hour period shown in the table, the speed of the train increased by

A. 25% B. 100%
C. 75% D. 125%

107. At time t, measured in minutes after the beginning of the time period, which of the following gives the speed of the train in accordance with the table?

A. $t/6$ B. $6t$
C. $40 + t$ D. $40 + t/6$

108. How fast was the train travelling 2½ hours after the beginning of the time period?

A. 50 km/hr B. 55 km/hr
C. 60 km/hr D. 65 km/hr

109. In March 2007, Computers Ltd. made a bundled offer of its Laptops together with Deskjet printers to boost Sales, though both the Laptop and the printer were also available individually. What is the price of the Printer, if purchased separately? Decide whether the information given in the two statements is sufficient to solve the problem.

Statement (1): The bundled offer price was ₹ 42,600

Statement (2): The Laptop, without the bundle offer was priced at ₹ 39,400.

A. Any one of the two statements (1) or (2) is sufficient to answer the question.
B. Each of the statement (1) or (2) taken alone, is sufficient to answer the question.
C. Both statements, taken together, are sufficient to answer the question, but neither of them alone is sufficient.
D. Both the statements together are insufficient to answer the question.

Directions (Qs. 110–113): *The following chart presents the data collected for an online job portal InfiniteJobs.com. For each question, there are two statements, each of which could be either True or False. Choose one of the following options as an answer:*

A. Only [X] is True
B. Only [Y] is True
C. Both [X] and [Y] are True
D. Neither [X] nor [Y] is True

Year	Category	Number of Registrations	Number of Candidates who posted their CV's	Number of Candidates shortlisted by Employers	Number of offered jobs
2004	Technical	61,205	59,981	684	181
	Managerial	19,236	15,389	138	48
2005	Technical	63,298	60,133	637	115
	Managerial	45,292	40,763	399	84

110. Statement X: The percentage increase in number of Registrations from 2004 to 2005 is higher for Managerial candidates than for Technical candidates.

Statement Y: The overall Registrations have grown by more than 25%.

111. Statement X: The percentage of drop-outs (from Registration stage to posting CV's) had decreased from 2004 to 2005 for Managerial category.

Statement Y: The percentage of drop-outs was higher for Technical than for Managerial category in 2005.

112. Statement X: The success rate of candidates getting shortlisted based on CV's is higher for Managerial category than for Technical category in 2005.

Statement Y: The success rate of candidates getting shortlisted based on CV's is better for Managerial category in 2005 than in 2004.

113. Statement X: In 2004, the number of candidates offered jobs as a proportion of the number of CV's posted was higher for Technical than for Managerial candidates

Statement Y: In 2004, among those short-listed by Employers, the Technical category had a higher success rate in securing jobs than the Managerial category.

Directions (Qs. 114–116): *The following table gives cost data of select stock prices on 3rd Dec 2003 in two markets, BSE of India and NQE of Kya Kya island. Closing stock refers to the price at the close of trading hours and opening stock to the price at the beginning of the day. The currency of Kya Kya is # and the exchange rate is # = ₹11.*

3rd Dec 2003				
	Closing		Opening	
	BSE (₹)	NQE (in #)	BSE (₹)	NQE (in #)
SIFY	247	22	232	21
INFY	115	10.5	105	9.5
WIPRO	65	6.5	60	5.5
TCS	444	40	450	40.5

114. Which share and which market showed the highest % increase on 3rd Dec 2003?

A. SIFY on BSE B. INFY on NQE
C. Wipro on NQE D. TCS on NQE

115. Arbitrage means buying in one market and selling simultaneously in another market. If Mr. Ghosh Babu were to buy a share at the opening price on one exchange and be able to sell it at the opening price on the other exchange, on which share does he make maximum % profit?

A. SIFY B. INFY
C. Wipro D. TCS

116. In an M&A deal SIFY is purchased by its parent company SATYAM which purchases 15% of SIFY's equity shares. Total of SIFY's equity shares is 1 million. How much does Satyam pay in ₹ for the stake if 50% of its purchases were on BSE's operating price and balance on BSE's closing price?

A. 36 million B. 3.5 million
C. 363 million D. 217 million

Directions (Qs. 117–120): *The following pie-charts give the percentage distribution of employees in different Departments in XYZ Company Ltd. during the years 2005 and 2006. Answer the questions using this data.*

A : Administration **B : Operations**
C : Sales & Marketing **D : Finance & Accounts**
E : Corporate HQ

2005
(No. of employees - 18,000)

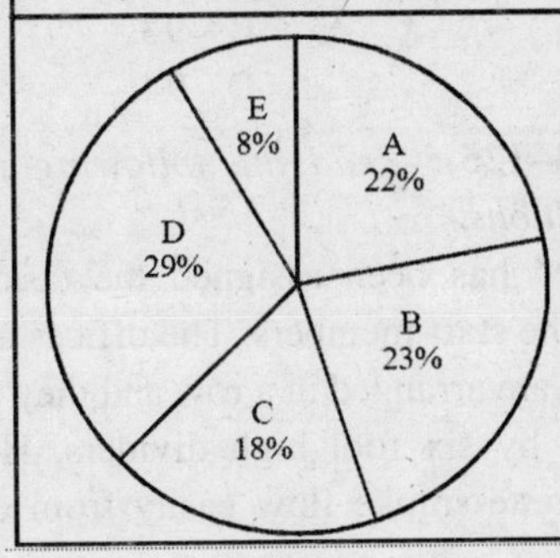

2006
(No. of employees - 20,000)

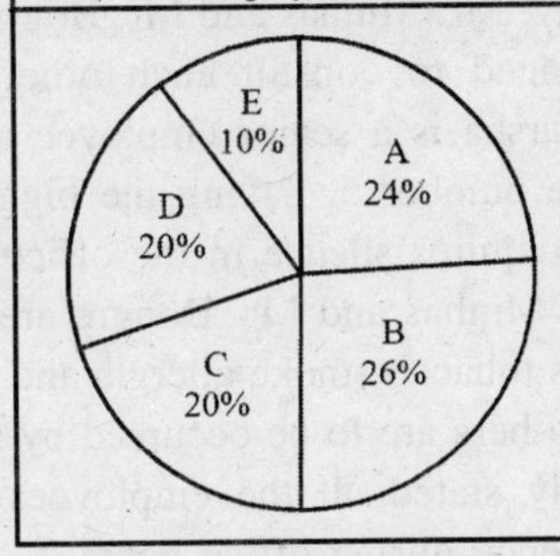

117. If the average monthly salary of employees in Administration was ₹ 12,000 in 2005, what was the approximate total salary expense of Administration in 2005?

A. ₹ 57 lakhs B. ₹ 4.7 crores
C. ₹ 66 crores D. ₹ 66 lakhs

118. What is the percentage increase in No. of employees in Sales & Marketing?

A. 2% B. 2.34%
C. 23.45% D. None of these

119. In which Dept is the variation in strength the maximum in 2006?

A. A B. B
C. D D. E

120. If 300 employees left Operations at the end of 2005, how many joined in 2006?

A. 480 B. 960
C. 1360 D. None of these

SECTION-IV

ANALYTICAL & LOGICAL REASONING

121. You have three chests in front of you. The first chest is labeled "GOLD", the second is labeled "SILVER" and the third is labeled "GOLD OR SILVER". You have been told that all the labels are on the wrong chests and that one chest contains gold coins, one contains silver coins and one contains bronze coins. How many chests do you need to open to deduce which label goes on which chest?

A. 0
B. 1
C. 2
D. Cannot deduce

122. How many minutes before 12 noon is it when it is 27 minutes past 10 a.m.?

A. 30 B. 93
C. 49 D. 94

Directions (123–125): *Read the following passage and answer the questions.*

An employee has been assigned the task of allotting offices to six of the staff members. The offices are numbered 1 - 6. The offices are arranged in a row and they are separated from each other by six foot high dividers. Hence voices, sounds and cigarette smoke flow easily from one office to another.

Miss Ruby needs to use the telephone quite often throughout the day. Mr. Minhas and Mr. Brar need adjacent offices as they need to consult each other often while working. Miss Harsha is a senior employee and has to be allotted the office number 5, having the biggest window.

Mr. Dongre requires silence in the offices next to his. Mr. Tanjore, Mr. Minhas and Mr. Dongre are all smokers. Miss Harsha finds tobacco smoke allergic and consequently the offices next to hers are to be occupied by non-smokers. Unless specifically stated all the employees maintain an atmosphere of silence during office hours.

123. The ideal candidate to occupy the office furthest from Mr. Brar would be

A. Miss Harsha B. Mr. Minhas
C. Mr. Tanjore D. Mr. Dongre

124. The three employees who are smokers should be seated in the offices.

A. 1, 2 and 4 B. 2, 3 and 6
C. 1, 2 and 6 D. 1, 2 and 3

125. In the event of what occurrence, within a period of one month since the assignment of the offices, would a request for a change in office be put forth by one or more employees?

A. Mr. Dongre quitting smoking
B. The installation of a noisy teletype machine by Miss Harsha in her office.
C. Miss Ruby needs silence in the office(s) next to her own.
D. Mr. Tanjore taking over the duties formerly taken care of by Miss Ruby.

126. Shankar and Jwala are both members of a Youth club, though they are not speaking to each other and refuse to work with each other. Chhaya, the club president, is appointing members to the fundraising committee, but she has resolved that she will not appoint anyone without his or her explicit consent. Shankar tells Chhaya, "I will not consent to appointment on that committee unless I know whether Jwala is to be a member of it." And Jwala says, "I will not consent to be a member of that committee unless I know whether Shankar will be appointed to it." If all three of these people stick by these resolutions, then:

A. Neither of them can be appointed to the committee.
B. They must either both be appointed or both be left out.
C. The committee may finally have one of them, both of them, or neither of them as members.
D. Either one of them can be appointed, but not both.

127. A bank customer had ₹ 100 in his account. He then made 6 withdrawals, totaling ₹ 100. He kept a record of these withdrawals, and the balance remaining in the account, as follows:

Withdrawals	Balance left
₹ 50	₹ 50
₹ 25	₹ 25
₹ 10	₹ 15
₹ 8	₹ 7
₹ 5	₹ 2
₹ 2	₹ 0
₹ 100	₹ 99

So, why are the totals not exactly right?

A. There is a mistake in the total of withdrawals.
B. There is a mistake in the total of the balance.
C. The two totals need not be equal.
D. The bank has cheated the customer.

128. Which image from the bottom row should replace the question mark?

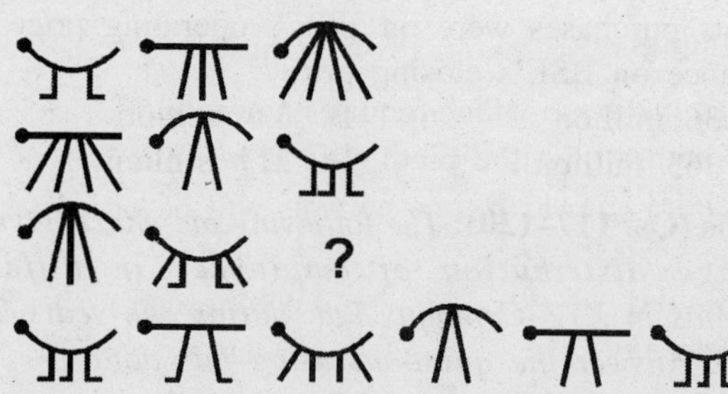

A. 3 B. 6
C. 2 D. 5

129.

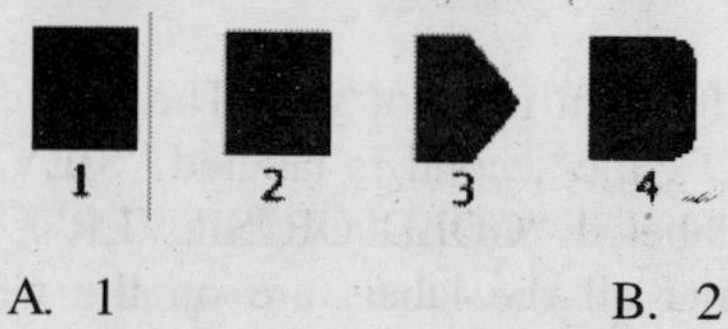

A. 1 B. 2
C. 3 D. 4

130. Which pattern from the bottom line (A, B, C, D or E) is missing from the top line?

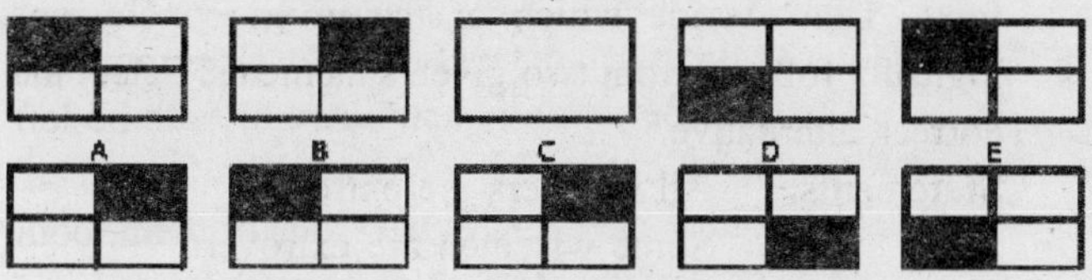

A. A B. B
C. D D. E

Directions (Qs. 131-133): *These questions are based on the following information.*

During their school Silver Jubilee Reunion, four alumni were discussing their starting annual salaries back in 1981. The salaries in question were Rupees 40, 50, 60 and 70 thousand per year. Of course the present MD of a private company earned the most. Arvind earned more than Biswajeet, and the doctor earned more than Dhruv the engineer. Chinmay could not remember what he started on. Biswajeet the lawyer did not start on Rs.50,000, nor did Dhruv.

131. What is Chinmay's current profession?
A. MD B. Lawyer
C. Doctor D. Engineer

132. What was the Lawyer's starting salary?
A. 40,000 B. 50,000
C. 60,000 D. 70,000

133. Who received the highest starting salary?
A. Arvind B. Biswajeet
C. Chinmay D. Dhruv

134. A man has a job which requires him to work 8 straight days and rest on the ninth day. If he started work on a Monday, the 12th time he rests will be on what day of the week?
A. Sunday B. Wednesday
C. Tuesday D. Friday

135. From the images below (1-7) find 3 odd ones out.

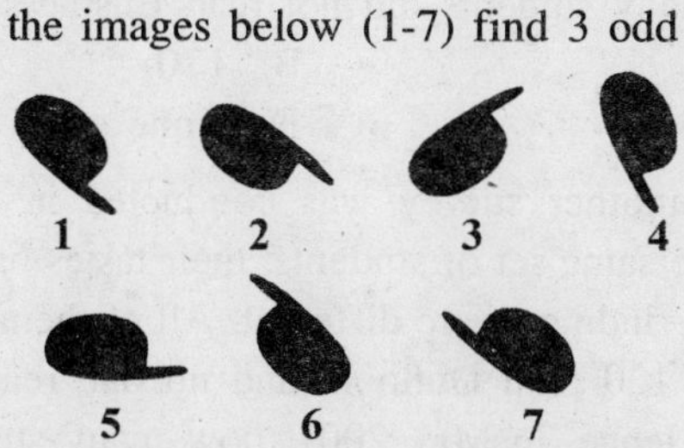

A. 1, 2, 3 B. 2, 3, 7
C. 3, 7, 4 D. 4, 1, 6

136. Babloo and Bunty were excitedly describing the result of the First Annual Running Race at Damapur High School. Snehal, Tanmay and Waman had been the three contestants. "Tanmay won the race; Waman was in second place," reported Babloo. Bunty disagreed. "It was Snehal who won. Tanmay came second." In fact, neither Babloo nor Bunty had given a correct version of the result as each had made one true and one false statement.

What was the actual placing of the three contestants?
A. Snehal, Waman, Tanmay.
B. Snehal, Tanmay, Waman.
C. Waman, Snehal, Tanmay.
D. Tanmay, Waman, Snehal.

137. What is the number of routes from P to Q?

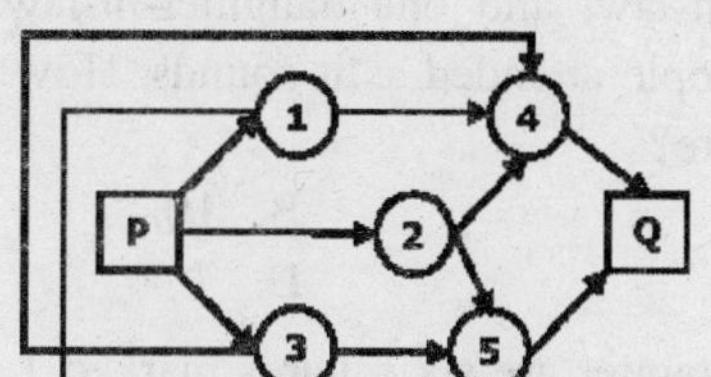

A. 5 B. 6
C. 9 D. 12

138. Find the circle, which does not include a word using all letters.

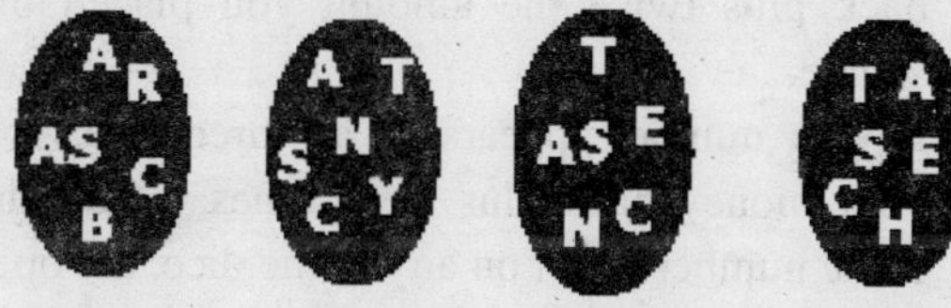

A B C D

A. A B. B
C. C D. D

139. It is possible to arrange eight of nine numbers 2, 3, 4, 5, 7, 10, 11, 12, 13 in the vacant squares of the 3 by 4 array shown below so that the arithmetic average of the numbers in each row and column is the same integer.

1			15
	9		
		14	

Which of the nine numbers must be left out when completing the array?
A. 4 B. 10
C. 15 D. 7

140. What number will replace the question mark?

A. 40 B. 18
C. 19 D. 24

141. Where should the letter 'W' be placed?

B C D E J K L M

A F G H I N

A. Above B. Below
C. Both D. None

142. At a family reunion were the following people: one grandfather, one grandmother, two fathers, two mothers, four children, three grandchildren, one brother, two sisters, two sons, two daughters, one father-in-law, one mother-in-law, and one daughter-in-law. But not as many people attended as it sounds. How many person were there?

A. 5 B. 10
C. 15 D. 7

143. On the counter are six squares marked 1, 2, 3, 4, 5, 6. Players are invited to place as much money as they wish on any one square. Three dice are then thrown.

- If your number appears on one die only, you get your money back plus the same amount.
- If two dice show your number, you get your money back plus twice the amount you placed on the square.
- If your number appears on all three dice, you get your money back plus three times the amount.
- If the number is not on any of the dice, the operator gets your money.

For example, suppose that you bet one Rupee on square No. 6. If one die shows a 6, you get your Rupee back plus another Rupee. If two dice show 6, you get back your Rupee plus two Rupees. If three dice show 6, you get your Rupee back plus three Rupees.

From a player's point of view, the chance of his number showing on one die is 1/6, but since there are three dice, the chances must be 3/6 or 1/2, therefore the game is a fair one. Of course this is the way the operator of the game wants everyone to reason, for it is quite fallacious.

What is the probable story?

A. Operator gets a profit of 6% on each Rupee bet.
B. Operator suffers a loss of 7.8% on each Rupee bet.
C. Operator gets a profit of 7.8% on each Rupee bet.
D. The player suffers a loss of 6% on each Rupee bet.

144. Jaideep was given some money by his mother on his birthday. Jaideep spent all of it in five stores. In each store he spent one rupee more than half of what he had when he came in.

How much did he get from his mother?

A. ₹ 25 B. ₹ 30
C. ₹ 46 D. ₹ 52

145. Consider the following two statements to be true if they seem to be at variance from commonly known facts. Then decide which of the given conclusions logically follows from two given statements. Select the correct alternative.

Statements: All Lawyers are extrovert.
Some wisemen are extrovert.

Conclusions: (*i*) All lawyers are wisemen.
(*ii*) All wisemen are lawyers.
(*iii*) Some extrovert are wisemen.
(*iv*) All extrovert are lawyers

A. Only (*i*) follows.
B. Only (*ii*) and (*iii*) follows.
C. Only (*i*) and (*iii*) follows.
D. Only (*iii*) follows.

Directions (Qs. 146-148): *Use the data given below to answer the questions.*

The following are the results of a survey conducted on a small cross-section of students from Symbiosis Group of institutes, to determine the readership of three magazines. This survey was conducted in Dec. 2006.

- Number of students who read only Business India was 40
- 60 students read only Outlook
- 110 students read only India Today
- 30 students read all three magazines
- 20 read Business India and India Today, but not Outlook
- 50 read Business India and Outlook, but not India Today
- 40 read Outlook and India Today, but not Business India

146. What was the total no. of students surveyed?

A. 210 B. 350
C. 390 D. None of these

147. How many students did not read Business India?

A. 40 B. 170
C. 240 D. None of these

148. When another survey was conducted in May 2007 with the same set of students, their tastes had changed and the findings were different. All of them read India Today. 120 read Outlook, and no one read Business India. Hence, in May 2007, how many students read only India Today?

A. 60 B. 110
C. 230 D. None of these

149. An ingredient in coffee, known as RTC, has been found to inactivate common cold viruses in experiments. In previous experiments, researchers found that inactivated common cold viruses can convert healthy cells into cancer cells. It can be concluded that the use of coffee can cause cancer.

Which one of the following, if true, most seriously weakens the argument?

A. Several teams of scientists performed the various experiments, and all of the teams had similar results.
B. The carcinogenic effect of RTC could be neutralized by the other ingredients found in coffee.
C. When RTC kills common cold viruses it weakens the immune system, and it might thus diminish the body's ability to fight other viruses, including viruses linked to cancers.
D. If chemists modify the structure of RTC, RTC can be safely incorporated into medications to prevent the common cold.

150. A census taker approaches a house and asks the woman who answers the door, "How many children do you have, and what are their ages?"

Woman: "I have three children, the product of their ages is 36, the sum of their ages is equal to the address of the house next door."

The census taker walks next door, comes back and says, "I need more information."

The woman replies, "I have to go, my oldest child is sleeping upstairs."

Census taker: "Thank you, I now have everything I need."

What are the ages of each of the three children?

A. 9, 2, 2
B. 6, 6, 1
C. 12, 3, 1
D. 6, 3, 2

ANSWERS

1	2	3	4	5	6	7	8	9	10
D	A	B	B	C	C	D	B	C	C
11	**12**	**13**	**14**	**15**	**16**	**17**	**18**	**19**	**20**
C	C	C	C	A	B	C	A	B	A
21	**22**	**23**	**24**	**25**	**26**	**27**	**28**	**29**	**30**
B	A	A	B	B	D	A	B	B	C
31	**32**	**33**	**34**	**35**	**36**	**37**	**38**	**39**	**40**
C	B	B	C	D	C	D	C	A	C
41	**42**	**43**	**44**	**45**	**46**	**47**	**48**	**49**	**50**
D	D	B	A	C	D	B	A	C	D
51	**52**	**53**	**54**	**55**	**56**	**57**	**58**	**59**	**60**
C	A	C	D	D	B	A	A	D	A
61	**62**	**63**	**64**	**65**	**66**	**67**	**68**	**69**	**70**
B	D	D	C	A	D	D	C	C	C
71	**72**	**73**	**74**	**75**	**76**	**77**	**78**	**79**	**80**
B	D	B	C	A	B	A	B	D	D
81	**82**	**83**	**84**	**85**	**86**	**87**	**88**	**89**	**90**
C	D	B	B	D	A	D	C	A	B
91	**92**	**93**	**94**	**95**	**96**	**97**	**98**	**99**	**100**
A	C	D	C	C	B	C	D	B	D
101	**102**	**103**	**104**	**105**	**106**	**107**	**108**	**109**	**110**
C	C	A	D	C	C	D	D	D	C
111	**112**	**113**	**114**	**115**	**116**	**117**	**118**	**119**	**120**
A	B	D	C	D	A	B	C	C	C
121	**122**	**123**	**124**	**125**	**126**	**127**	**128**	**129**	**130**
A	B	D	D	D	D	C	D	C	C
131	**132**	**133**	**134**	**135**	**136**	**137**	**138**	**139**	**140**
C	C	A	B	B	A	C	A	D	B
141	**142**	**143**	**144**	**145**	**146**	**147**	**148**	**149**	**150**
B	D	C	B	D	B	D	C	C	D

SOME SELECTED EXPLANATORY ANSWERS

81. Simple Interest = $\frac{P\times4\times r}{100}$

(where P is principal and r is rate of interest in percentage.)

$\Rightarrow \frac{7P}{25} = \frac{P\times4\times r}{100} \Rightarrow \frac{7}{25} = \frac{4r}{100}$

$\Rightarrow r = 7\%$.

82. February is the only month having 28 days which is three days less than the maximum of 31.
Prob. of choosing February is 1 month from 12 months

$= \frac{1}{12}$.

83. 2 steps of 1st person = 3 steps of escalators
Hence steps for 1st person

$= \left(\frac{2}{3}\times\frac{120}{1+\frac{2}{3}}\right) = 120\times\frac{3}{5}\times\frac{2}{3} = 72\times\frac{2}{3} = 48$

Similarly, steps for 2nd person

$= \frac{3}{5}\times\frac{120}{1+\frac{3}{5}} = \frac{3}{5}\times120\times\frac{5}{8} = 45$

Total steps taken together = 48 + 45 = 93.

84. Arranging data in ascending order, we get 66, 69, 70, 75, 77, 78, 78
So, median, $m = 75$

Arithmatic mean = $\frac{66+69+70+75+77+78+78}{7}$
= 73.29

Mode = 78.
Median, lies between mean and mode i.e., $a < m < f$.

85. Let side of the park be x then Park area = $\frac{\sqrt{3}}{4}x^2$

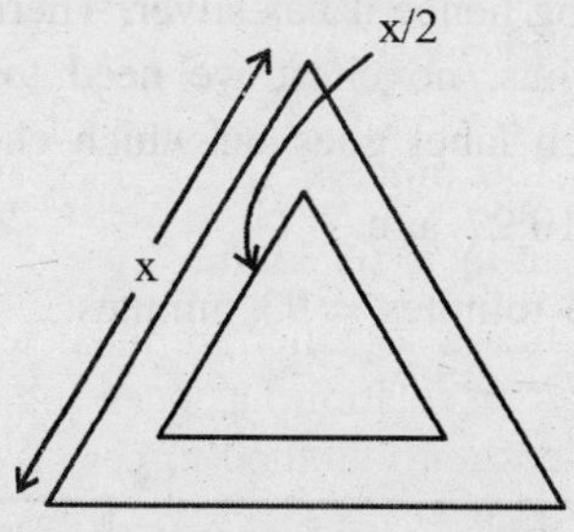

Flower bed area = $\frac{\sqrt{3}}{4}\left(\frac{x}{2}\right)^2 = \frac{\sqrt{3}}{4}.\frac{x^2}{4}$

Path area = $\frac{\sqrt{3}}{4}x^2\left(1-\frac{1}{4}\right)$, Ratio = $\frac{\frac{\sqrt{3}}{4}\left(\frac{3}{4}\right)}{\frac{\sqrt{3}}{4}\left(\frac{1}{4}\right)}$ = 3 : 1.

86. $f(-3) = 3 - (-3) = 6$ [Since $-3 < 0, f(x) = 3 - x$ bonds.]

87. Total names in the lottery

$= 3 \times 100 + 2 \times 150 + 200 = 800$

No. of year-III names = 3×100

Probability $= \frac{300}{800} = \frac{3}{8}$.

88. Let $A_1, A_2, \ldots\ldots\ldots A_9$ be the numbers and their average = M.(given)

So, $A_1 + A_2 + \ldots\ldots\ldots + A_9 = 9M$

$A_1 + A_2 + A_3 = 3P$

(since average of A_1, A_2 and A_3 = P)

$A_4 + A_5 + \ldots\ldots\ldots A_9 = 6N$

(since average of A_4 to A_9 = N)

$\Rightarrow$ 9M = 3P + 6N

3M = 2N + P.

89. Cubes are 3-dimensional. So, their intersection will always be two-dimensional. Hence, it cannot be a cube.

90. Suppose no. of accidents $y = a + bx$
Where x is no. of overtime hours
From the given data $y = 8$ for $x = 1000$ and $y = 5$ for $x = 400$

$\Rightarrow 8 = a + 1000b$ and $5 = a + 400b$,

$\Rightarrow b = \frac{1}{200}$, $a = 3$

So, the relation, $y = 3+\frac{x}{200}$

For $x = 0$, $y = 3$.

91. Two rainy days occur in 20 days. So, rainbow will occur once in 20 days.
Rest 19 days will have no rainbow.

% of not producing rainbows = $\frac{19}{20}\times100$ = 95%.

92. Suppose, the bag contains x, y and z balls.

$x + y + z = 10$

From statement 1, $x = z + 5 \Rightarrow z = x - 5$

From statement 2, $y = \frac{1}{2}x$

Solving, we get $x+\frac{x}{2}+x-5 = 10$

$x = 6$, $y = 3$, $z = 1$.

So, both statements together are sufficient but neither statement alone is sufficient.

93. A = 50 sq km.

Length of each side = $\sqrt{50} = 5\sqrt{2}$

Diagonal Length = $\sqrt{2}\times5\sqrt{2}$ = 10 km

Which does not match with option A to C.

94. 4 consecutive positive integers will have two even numbers and two odd numbers.

So, x will always be even.

$\Rightarrow n = 1 + x$ will always be odd.

Let 4 consecutive integers be $(p - 1)(p)(p + 1)(p + 2)$

$n = 1 + p(p + 2)(p^2 - 1) = 1 + p^4 + 2p^3 - p^2 - 2p$

Which is not necessarily a perfect square.

$n = 1 + x$ may not be a prime number in all the cases.

But $1 + x$ is definitely an odd number.

95. No. of vowels = E, A, A, I, I, O

No. of non-vowels = X, M, N, N, T

Six vowels are to be placed at six odd places in 11-letter word while 5 non-vowels will occupy the rest places.

There are two pairs of same vowels. No. of arrangements

$= \frac{\lfloor 6}{\lfloor 2 \times \lfloor 2} \times \frac{\lfloor 5}{\lfloor 2} = 180 \times 60 = 10800.$

96. Since 6 and 16th are standing diametrically opposite number of children on one side between 6th & 16th child = 9

So, total children = 9 + 9 + 2 = 20.

97. Let the man buys x apples at p price and y oranges at q price, then $x + y = 40$

$$px + qy = 17 \quad \text{... (1)}$$

Also $$qx + py = 15 \quad \text{... (2)}$$

Adding (1) & (2)

$$(p + q)(x + y) = 17 + 15$$

$$p + q = \frac{32}{40} = ₹\,0.80.$$

98. 311 is repeated seven times in the number, 311 is not divisible by 3 but 311 repeated twice is not divisible by 3, but divisible by 11.

Similarly 311 repeated thrice is divisible by 3, but not by 11.

As 311 is repeated seven times, which is neither multiple of 2 nor 3.

So, number is not divisible by 3 or 11.

99. Suppose, the man invests x in 6% and y in 8%

$$x + y = 9000 \quad \text{....(1)}$$

$$3\times\left(\frac{x\times 6}{100} + \frac{y\times 8}{100}\right) = 1800$$

$$3x + 4y = 3000 \quad \text{....(2)}$$

From (1) & (2),

$$x = 6000.$$

100. Total bogies initially is 12 + 12 = 24, since trains move in opposite direction.

Additional bogies = 16 – 12 = 4, so total bogies = 28

24 bogies take 2 minutes, so, 28 bogies will take

$\frac{2\times 60}{24}\times 28 = 140$ seconds,

so, 20 seconds more are required.

101. Let the length be x

then $$x^2 = (x - 2)^2 + (5)^2$$

$$x^2 = x^2 - 4x + 4 + 25$$

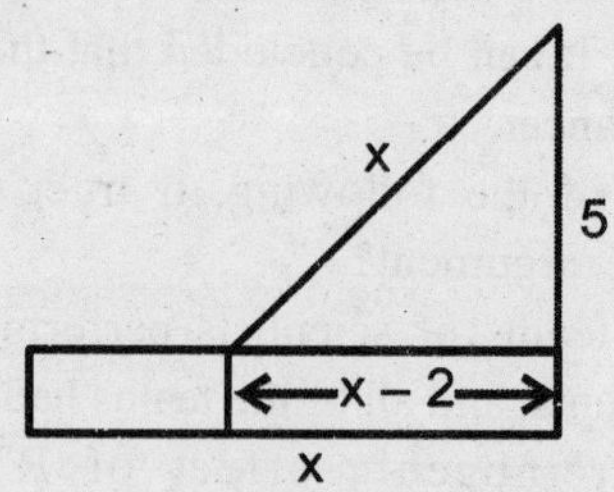

$$4x = 29 \Rightarrow x = \frac{29}{4} = 7.25 \text{ m.}$$

102. Wheat production = 4 × 1.2 = 4.8 lac tons.

103. Only for year 2003 & 2004, it is higher.

104. Wheat production during 2001 = 0.45(3) = 1.35

Total grain production = 3 + 1.35 = 4.35

Import needed = 5 – 4.35 = 0.65 lac tons.

105. E_1 = 1220. Expenses, 2005-06

E_2 = 1400, Expenses for 2006-07

S_2 = 4300, Sales for 2006-07

P_2 = 4300 – 1400 = 2900, profit for 2006-07

Since there is 32% increase

$P_2 = 1.32\ P_1 = 2900$

$P_2 = \frac{2900}{1.32} = 2196.37$

$S_1 = E_1 + P_1$ (sales in 2005-06)

= 1220 + 2196.37 = 3416.91

So using both statements, % increase in sales can be determined.

106. Per cent increase in speed $= \frac{70 - 40}{40}\times 100 = 75\%.$

107. Le the equation be of the form $v = u + at$

For $t = 30$ min

$v = 45$ and initial speed = 40

So, $45 = 40 + a \times 30$

$\Rightarrow a = \frac{t}{6}$

So, $40 + \frac{t}{6}$.

108. $2\frac{1}{2}$ hours = 150 min.

Speed of the train is 65 km/hr.

109. Both the statements together are insufficient to answer the question as no relation has been indicated between individual prices and bundled prices. We cannot draw the inference that

Printer price = bundled offer – individual laptop price.

110. For statement X,

$$\frac{45292 - 19236}{45292} > \frac{63298 - 61205}{63298}$$

$\frac{26066}{45292} > \frac{2093}{63298}$ which is correct.

For statement Y,

$$\left(\frac{26066}{45292}+\frac{2093}{63298}\right)\times 100 (= 61\%)$$

$\Rightarrow$ 61725 which is correct.

111. For statement X,

$$\frac{19236-15389}{15389} > \frac{45292-40763}{45292}$$

or, $\frac{3847}{15389} > \frac{4529}{45292}$ which is correct

For statement Y,

$$\frac{63298-60133}{63298} > \frac{45292-40763}{45292}$$

or, $\frac{3165}{63298} > \frac{4529}{45292}$ which is incorrect.

112. As per statement X,

$\frac{399}{40763} > \frac{637}{60133}$ which is not correct

As per statement Y, $\frac{399}{40763} > \frac{138}{15333}$

Which is true, only Y is true.

113. As per statement X,

$\frac{181}{59981} > \frac{48}{15389}$ Which is not true.

As per statement Y,

$\frac{181}{684} > \frac{48}{138}$ which is not true.

114. By observation, % increase of Wipro in NQE market is the highest as $\frac{6.5-5.5}{5.5}$ is the highest among all.

115. For SIFY, % difference = $\frac{21\times 11-232}{21\times 11} = \frac{-1}{21\times 11}$

For INFY = $\frac{9.5\times 11-105}{9.5\times 11} = \frac{-0.5}{9.5\times 11}$

For Wipro = $\frac{5.5\times 11-60}{5.5\times 11} = \frac{-0.5}{5.5\times 11}$

For TCS = $\frac{40.5\times 11-450}{40.5\times 11} = \frac{-4.5}{40.5\times 11}$

Among all figures $\frac{-4.5}{40.5\times 11}$ is the highest for TCS.

116. Total price of 15% share

$$= \frac{7.5\times 10^6}{100}\times 232+\frac{7.5\times 10^6}{100}\times 247$$

$= 7.5 \times 10^4(232 + 247)$

$= 7.5 \times 479 \times 10^4 \approx 36$ million.

117. For 22% employees

Salary = $12,000 \times 22 \times 180$

$= 216 \times 2.2$ lakhs ≈ 4.75 crores.

118. Per cent increase = $\frac{C_2-C_1}{C_1}\times 100$, C_2 = 20% of 20000 and C_1 = 18% of 18000.

$$= \frac{20\times 200-18\times 180}{18\times 180}\times 100$$

$$= \frac{760\times 100}{18\times 180} = \frac{1900}{81} = 23.45\%.$$

119. For year 2005,

$A_1 = 22 \times 180$, $B_1 = 23 \times 180$

$C_1 = 18 \times 180$, $D_1 = 29 \times 180$

$E_1 = 8 \times 180$

For year 2006,

$A_2 = 24 \times 200$, $B_2 = 26 \times 200$

$C_2 = 20 \times 200$, $D_2 = 20 \times 200$

$E_2 = 10 \times 200$

Variation of strength is maximum for D as strength dips from 29 to 20.

120. $B_2 - (B_1 - 300) = 26 \times 200 - (23 \times 180 - 300)$

$= 5200 - (4140 - 300) = 1360.$

121.

GOLD	SILVER	GOLD or SILVER
1st chest	2nd chest	3rd chest

If all the labels are on wrong chests then 3rd chest has neither gold nor silver. Hence it has bronze.

Now, first chest has either gold or gold or silver. But label is wrong hence it has silver. Therefore 2nd chest has gold. Thus, no chest we need to be opened to deduce which label goes on which chest.

122. 12 noon – 10:27 a.m.

= 1 hour 33 minutes = 93 minutes.

For (Qs. 123–125) : The arrangement of offices is as follows:

1	2	3	4	5	6
Mr Dongre	Mr Tanjore	Mr Minhas	Mr Brar	Miss Harsha	Miss Ruby
Requires silence	Needs adjacent office			Office no. 5 smoke allergic	Uses telephone (Noise)
Smokers					

123. Miss Harsha has been alloted the office number 5. So office number 4 and 6 have to be alloted to non-smokers. Mr. Brar and Miss Ruby are non-smokers. Mr. Brar has to be alloted office number 4 and Mr. Minhas number 3 because they need to consult each other often while working. Mr. Dongre requires silence. So he has to be alloted office number 1. Hence the ideal candidate to occupy the office furthest from Mr. Brar is Mr. Dongre.

126. Shankar and Jwala do not speak to each other and refuse to work with each other. So either one of them can be appointed, but not both.

128. The number of feet in each row's and column's designs are same. Top portions take any one of design among,

,  and  in each column and row without repetitions.

129. From figure first to second the number of sides is increasing by one. Triangle is followed by a quadrilateral. So rectangle will be followed by a pentagon.

130. The dark square moves one place in clockwise direction in each step.

For (Qs. 131 - 133)

Alumni	Profession	Starting salary
Arvind	MD	₹ 70,000
Biswajeet	Lawyer	₹ 60,000
Dhruv	Engineer	₹ 40,000
Chinmay	Doctor	₹ 50,000

134.

1st	2nd	3rd	4th	5th	6th	7th	8th	9th	10th	11th	12th
↓	↓	↓	↓	↓	↓	↓	↓	↓	↓	↓	↓
Tue	Thu	Sat	Mon	Wed	Fri	Sun	Tue	Thu	Sat	Mon	Wed

Every next rest will occur on two days later from previous rest.

135. All others are in anticlockwise direction.

136.

Place	Babloo's Statements	Bunty's Statements
1st	Tanmay (False*)	Snehal (True*)
2nd	Waman (True*)	Tanmay (False*)

* Each has made one true and one false statement. Hence, actual placing of the three contestants was Snehal, Waman and Tanmay.

137. Probable routes are
P-1-4-Q, P-3-5-Q, P-2-4-Q, P-2-5-Q, P-3-5-2-4-Q
P-1-4-2-5-Q, P-1-4-3-5-Q, P-3-5-1-4-Q, P-2-5-1-4Q

138. From option B. we get SCANTY (too little in amount for what is needed)
From option C. we get ASCENT (the out of climbing or moving up)
From option D. we get CHASTE (not having sex with any one).

139.

1	13	3	15
11	9	7	5
12	2	14	4

The arithmetic average of numbers in each row (32 ÷ 4 = 8) and column (24 ÷ 3 = 8) is 8.

140. The product of two numbers is double than the product of rest two numbers.

Example : $12 \times 8 = 2(8 \times 6)$
$6 \times 10 = 2(6 \times 5)$
$\boxed{18} \times 4 = 2(12 \times 3)$

141.

B C D E J K L M | R S T U
A F G H I N | O P Q V [W]

142. Let A be the grandfather and B be the grandmother. C is the son of A and B. D is the wife of C. E is the son and F and G are daughters of C and D. Here A and C are two fathers. B and D are two mothers. C, E, F and G are four children. E, F and G are three grand-children. E is one brother. F and G are two sisters, C and E are two sons. F and G are two daughters. A, B and D are father-in-law, mother-in-law and daughter-in-law respectively. So there were 7 persons.

145.

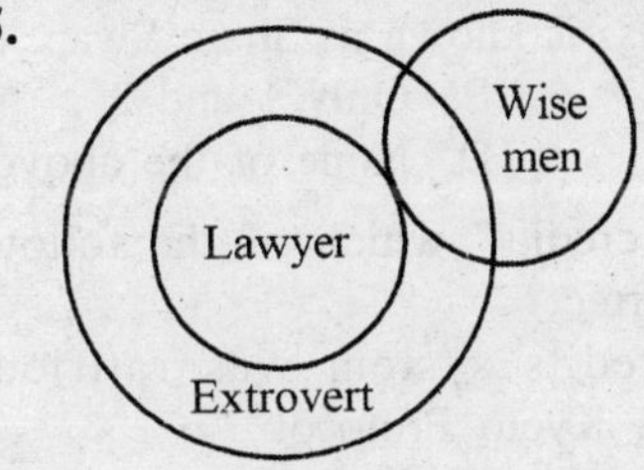

Case I

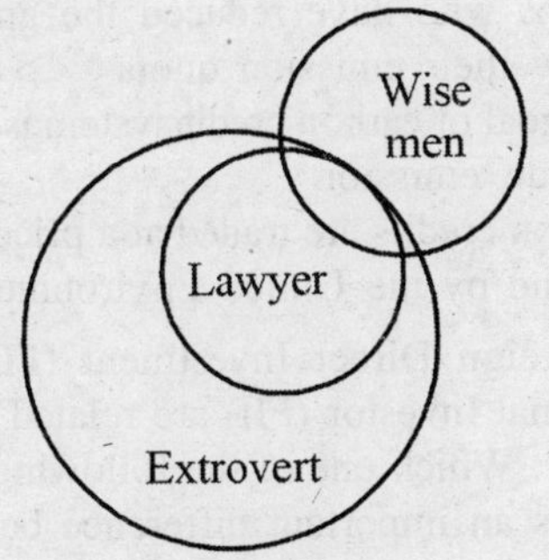

Case II

(i) ✗ *(ii)* ✗ *(iii)* ✔ *(iv)* ✗

Hence only *(iii)* follows.

146. Total no. of students surveyed
= 40 + 60 + 110 + 30 + 20 + 50 + 40 = 350.

147. No. of students who did not read Business India
= 60 + 110 + 40 = 210.

148. No. of students who read only India Today
= 350–120 = 230.

Previous Paper (Solved)

Symbiosis National Aptitude Test (SNAP), 2011

SECTION- I

GENERAL AWARENESS

1. "Swayamsidha" is a scheme launched by the Government of India to help

A. School Children only

B. Health workers only

C. Senior citizen without any regular income

D. Women only

2. What is 'Share swap'?

1. A business takeover in which acquiring company uses its own stock to pay for the acquired company.
2. When a company uses its own share to get some short term loan for working capital requirement
3. When companies are required to float a new issue to earn capital for their expansion programmes, each shareholder gets some additional preferential share. The process is known as Share Swap.

A. Only 1 B. Only 1 and 2

C. Only 3 D. None of the above

3. Regarding "carbon credits" which of the following statements is not correct?

A. The carbon credits system was ratified in conjunction with Kyoto Protocol

B. Carbon credits are awarded to those countries or groups who have reduced the green house gases below their emission quota

C. The goal of carbon credit system is to reduce carbon dioxide emission

D. Carbon credits are traded at a price fixed from time to time by the United Environment Programme

4. Both Foreign Direct Investment (FDI) and Foreign Institutional Investor (FII) are related to investment in a country. Which one of the following statements best represents an important difference between the two?

A. FII helps bring better management skills and technology while FDI only bring capital

B. FII helps in increasing capital availability in general, while FDI only targets specific sectors

C. FDI flows only into secondary markets while FII targets primary markets

D. FII is considered to be more stable than FDI

5. In context of global oil prices, "Brent crude oil" is frequently referred to in the news. What does this term imply?

1. It is a major classification of crude oil
2. It is sourced from North sea
3. It does not contain sulphur

Which of the statement given above is/are correct?

A. 2 only B. 1 and 2 only

C. 1 and 3 only D. 1, 2 and 3

6. With reference to "Look East Policy" of India consider the following statements:

1. India wants to establish itself as an important regional player in the East Asian affairs.
2. India wants to plug the vacuum created by the termination of Cold war.
3. India wants to restore the historical and cultural ties with its neighbors in Southeast and East Asia.

Which of the statements given above is/are correct?

A. 1 only B. 1 and 3 only

C. 3 only D. 1, 2 and 3

7. Recently "oilzapper" was in news. What is it?

A. It is eco friendly technology for the remediation of oily sludge and oil spills

B. It is the latest technology developed under sea oil exploration

C. It is genetically engineered high biofuel–yielding maize variety

D. It is the latest technology to control accidentally caused flames from oil wells.

8. Why is the offering of "teaser loans" by commercial banks a cause of economic concern?

1. The teaser loans are considered to be an aspect of sub – prime lending and banks may be exposed to the risk of defaulters in future.
2. The teaser loans are given to inexperienced entrepreneurs to set up manufacturing or export units.

Which of the statements given above is/are correct?

A. 1 only B. 2 only

C. Both 1 and 2 D. Neither 1 nor 2

9. Why is the government of India disinvesting its equity in the Central Public Sector Enterprises (CPSE)?

1. The Government intends to use the revenue earned from disinvestment mainly to pay the External debt
2. The Government no longer intends to retain the management control of CPSEs

Which of the statements given above is/are correct?
A. 1 only B. 2 only
C. Both 1 and 2 D. Neither 1 nor 2

10. The book "Unto this Last" which influenced Gandhiji, was authored by
A. Boris Yeltsin B. Pushkin
C. Ruskin Bond D. John Ruskin

11. Besides USA, India has signed, with which of the following countries the agreement being named as "Cooperation Agreement on the development of Peaceful Uses of Nuclear Energy"?
A. Italy B. Germany
C. France D. Australia

12. Which of the following country is not a member of the Nuclear Suppliers Group (NSG)?
A. Russia B. USA
C. Italy D. Iran

13. Many times we read in the newspaper that several companies are adopting the FCCBs route to raise capital. What is the full form of FCCB?
A. Foreign Currency Convertible Bonds
B. Foreign Convertible Credit Bonds
C. Financial Consortium and Credit Bureau
D. None of these

14. As per the reports published in various newspapers, RBI has asked banks to make plan to provide banking services in villages having population of 2000. This directive issued by the RBI will fall in which of the following categories?
A. Plan for financial inclusion
B. Efforts to meet Priority sector lending
C. Extension of Internet and Branchless banking
D. None of these

15. Which of the following is/ are treated as artificial currency?
A. ADR B. GDR
C. Both ADR and GDR D. SDR

16. Which of the following terms indicates a mechanisms used by commercial banks to provide credit to the Government?
A. Cash Credit Ratio
B. Debit Service Obligation
C. Liquidity Adjustment Facility
D. Statutory Liquidity Ratio

17. Who among the following was not part of the drafting committee of the Lokpal Bill?
A. Salman Khurshid B. Anna Hazare
C. Arvind Kejriwal D. Kiran Bedi

18. Criminal procedure is a subject of which of the following lists?
A. State List B. Concurrent List
C. Union List D. Both A and C

19. One of the world's biggest Uranium resources was recently found in
A. Andhra Pradesh B. Maharashtra
C. Jharkhand D. Uttarakhand

20. Which of the following was the first venture of Kishore Biyani?
A. Pantaloons
B. Big Bazaar
C. Central
D. Future Capital Holdings

21. Who among the following is often referred to as father of India's "Green Revolution"?
A. Dr. Verghese Kurien
B. Dr. Amrita Patel
C. M.S. Swaminathan
D. Gurudev Khush

22. Who among the following directed the movie Peepli Live?
A. Ashutosh Gowarikar B. Kiran Rao
C. Anusha Rizvi D. Seema Chisti

23. The point at which soild, liquid and gaseous forms of a substance co-exist is called
A. sublimation B. distillation point
C. triple point D. melting point

24. The limit beyond which the stars suffer internal collapse is called the
A. Raman Effect B. Chandrashekhar limit
C. Aurora Borealis D. Quasan Zone

25. UNDP reports publishes every year the Human Development Index. Which of the following is not a criteria used to measure the Human Development Index?
A. Health B. Education
C. Living Standards D. Human Rights

26. The software company I–flex Solutions was originally a division of which famous financial services company?
A. Citicorp B. ICICI
C. HSBC D. ABN Amro Bank

27. Which of the following is not a member of SAARC?
A. Bhutan B. Bangladesh
C. Burma D. Maldives

28. Which article of the Indian constitution recognizes Hindi in Devanagari Script as thé official language of India?
A. Art 345 B. Art 343
C. Art 348 D. Art 334

29. India has finally woken up to the needs of the country's elderly. With the number of people in the 60–plus age group in India expected to increase to 100 million in 2013 and to 198 million in 2030, the health ministry is all set to roll out the

A. National Programme for Health care of the Elderly
B. National Programme for Senior Citizens
C. National Programme for Old Aged
D. Rashtriya Vriddha Swasthya Yojana

30. Which of the following statements is/are correct?
1. Non-resident Indians (NRIs) can now cast votes in their home constituencies in India.
2. The NRI can cast his vote by postal balloting.

A. 1 only B. 2 only
C. Both 1 and 2 D. Neither 1 nor 2

31. In a landmark move, which of these State governments has set up a Savarna Aayog, a commission to identify the deprived and underprivileged families among upper castes?
A. Uttar Pradesh B. Bihar
C. Madhya Pradesh D. Rajasthan

32. Which of these countries has announced one billion US dollar aid for the reconstruction of Nalanda University?
A. Japan B. Singapore
C. Malaysia D. China

33. What is the name of the in-house magazine to be published by the Indian Railways, to be distributed to the passengers on high end trains?
A. Rail Vishwa B. Rail Bandhu
C. Rail Mail D. Rail Patra

34. The government, in February 2011, set up a task force to create a way to provide direct subsidies to the ultimate beneficiaries on petroleum working gas and fertilizers. The task force will be headed by
A. Nandan Nilekani B. Apoorva Sinha
C. Sangam Chitra D. All of these

35. Which of these countries has been added in the 'BRIC' (Brazil–Russia–India–China) grouping of emerging economies?
A. South Korea B. Venezuela
C. South Africa D. Malaysia

36. Which of the following countries was the host of First Asian Yoga Championship held recently?
A. Thailand B. Cambodia
C. Vietnam D. Laos

37. Who among the following has been defeated by Saina Nehwal, who has won the Swiss Open Grand Prix Gold title recently, becoming the First Indian to do so?
A. Ji Hyun Sung B. Wang Lin
C. Kamilla Rytter Juhl D. Shinta Mulia Sari

38. On March 6, 2011, the Supreme Court held that has ample powers to enact legislation with respect to extra – territorial aspects for the security of India?
A. President B. Prime Minister
C. Supreme Court D. Parliament

39. Which country has appointed Brigadier General Ravinder Singh as its next Army chief, the first Sikh in nearly 30 years to be given the force's batton
A. Thailand B. Singapore
C. Malaysia D. Indonesia

40. Anant Pai, better known as 'Uncle Pai', who recently died was the creator of which among the following comic series?
A. Chacha Chaudhary B. Chandamama
C. Indrajaal D. Amar Chitra Katha

SECTION-II

ANALYTICAL & LOGICAL REASONING

For all questions in this section, correct answers carry 2 marks each.

Directions (Q.Nos. 41 and 42): *Each question consists of a set of numbered statements. Assume that each one of these statements is individually true. Each of the four choices consists of a subset of these statements. Choose the subset as your answer where the statements therein are logically consistent among themselves:*

41. A. Only if the water level in the coastal areas rises, then the people change their lifestyle.
B. People change their lifestyle only if they are rewarded.
C. If people are rewarded, then they will not change their lifestyle.
D. If the temperature rises, then the water level in the coastal areas rises.
E. Whenever the water level in the coastal area rises, then the temperature rises.
F. Unless the people change their lifestyle, temperature rises.
G. People are rewarded.
H. Water level in the coastal areas does not rise.

A. C, D, F, G and H B. G, F, D, B and H
C. E, F, G, H and B D. None of the above

42. A. If Kumar sings, then the audiences sleep.
B. If Kumar sings, then the audiences dance.
C. Unless audience do not dance, the concert will be successful.
D. Only if the audience dance, the concert will be successful.

E. If Vina dances, then Kumar sings.
F. Kumar sings only if Vina dances.
G. Vina dances.
H. The concert is successful.

A. C, F, G, B and H B. A, C, F, G and H
C. E, C, G, B and H D. Both B and C

Directions (Q. No. 43 to 45): *These questions are based on the data given below. There are only four members of a family viz., A, B, C and D and there is only one couple among them. When asked about their relationships, following were their replies:*

a. A: B is my son. D is my mother.
b. B: C is my wife. D is my father.
c. C: D is my mother-in-law. A is my daughter.
d. D: A is my grand-daughter. B is my daughter-in-law.

43. Who always speaks the truth?
A. A B. B
C. C D. D

44. How is B related to C?
A. Father B. Mother
C. Wife D. Husband

45. Which of the following statements must be true?
A. A's grandmother alternates between the truth and lie.
B. C's wife always speaks the truth.
C. A's grandfather always speaks the truth.
D. B's daughter always tells lies.

Directions (Q. No. 46 to 48): *These questions are based on the following information. A cube of 7cm × 7cm × 7cm is kept in the corner of a room and painted in three different colours, each face in one colour. The cube is cut into 343 smaller but identical cubes.*

46. How many smaller cubes do not have any face painted?
A. 125 B. 180
C. 144 D. 216

47. How many smaller cubes have exactly one colour on them?
A. 108 B. 72
C. 36 D. 24

48. How many smaller cubes have at the most two faces painted?
A. 343 B. 342
C. 256 D. 282

Directions (Q. Nos. 49 and 50): *Amit was driving in New Town, where all roads either north-south or east-west forming a grid. Roads were at a distance of 1 km from each other in parallel.*

49. Amit started at the intersection of streets no. 7 and 8. He drove 3 km north, 3 km west and 4 km south. Which further route could bring him back to his starting point?
I. 3 km east, then 2 km south
II. 1 km north, then 3 km east
III. 1 km north, then 2 km west

A. I only B. II only
C. I and II only D. II and III only

50. After driving as stated in question no. 49 above, Amit did not return to his starting point, but instead drove 4 km east and 1 km north. How far is he from his starting point?
A. 5 km B. 4 km
C. 1 km D. 7 km

Directions (Q. Nos. 51 to 55): Refer to the following data and answer the questions that follow:

A numerical machine accepts two values X and Y. Then it updates these values as X = XY and Y = Y + 1 in every step. The machine stops at $X \geq N$.

51. For X = 3, Y = 2 and N = 100, how many steps are performed before the machine stops?
A. 2 B. 3
C. 4 D. 5

52. In the above question (51), what is the final value of X?
A. 6 B. 20
C. 72 D. 360

53. In the above question (51), what is the final value of Y?
A. 4 B. 5
C. 6 D. 20

54. If the value of N is changed to 500, what would be the final value of X?
A. 360 B. 500
C. 560 D. 2160

55. If X = 2 and Y = 3, what should be the minimum value of N such that final value of Y is 7?
A. 300 B. 360
C. 720 D. 860

Directions (Q. Nos. 56 to 60): Refer to the following statements and answer the questions:

Seven students Priya, Ankit, Raman, Sunil, Tony, Deepak and Vicky take a series of tests. No two students get similar marks. Vicky always scores more than Priya. Priya always scores more than Ankit. Each time either Raman scores the highest and Tony gets the least, or alternatively Sunil scores the highest and Deepak or Ankit scores the least.

56. If Sunil is ranked sixth and Ankit is ranked fifth, which of the following can be true?
A. Vicky is ranked first or fourth
B. Raman is ranked second or third
C. Tony is ranked fourth or fifth
D. Deepak is ranked third or fourth

57. If Raman gets the highest, Vicky should be ranked not lower than:

A. Second B. Third
C. Fourth D. Fifth

58. If Raman is ranked second and Ankit is ranked first, which of the following must be true?

A. Sunil is ranked third
B. Tony is ranked third
C. Priya is ranked sixth
D. None of these

59. If Sunil is ranked second, which of the following can be true?

A. Deepak gets more than Vicky
B. Vicky gets more than Sunil
C. Priya gets more than Raman
D. Priya gets more than Vicky

60. If Vicky is ranked fifth, which of the following must be true?

A. Sunil scores the highest
B. Raman is ranked second
C. Tony is ranked third
D. Ankit is ranked second

61. In 2002, according to a news poll, 36% of the voters had leaning towards party "Y". In 2004, this figure rose to 46%. But in another survey the percentage was down to 40%. Therefore, the party "Z" is likely to win the next election. Which of the following, if true, would seriously weaken the above conclusion?

A. People tend to switch their votes at the last minute.
B. It has been showed that 85% of the voters belonging to the party "Y" vote in an election as compared to 80% of the voters belonging to party "Z".
C. 35% of people favour party "Z".
D. No one can predict how people will vote.

62. Inflation rose by 5% over the second quarter, by 4% during the first quarter and higher than 3% recorded during the same time last year. However, the higher price index did not seem to alarm National Stock Index as stock prices remain steady.

Which of the following, if true, could explain the reaction of National Stock Index?

A. RBI announced that it will take necessary corrective measures
B. Stock prices were steady because of a fear that inflation would continue.
C. Economists warned that inflation would continue.
D. Much of the quarterly increase in the price level was due to a summer drought effect on food price.

Direction (Q. No. 63): Pick up the appropriate analogy.

63. Birth : Dirge

A. sunset : sunrise
B. security check : arrival
C. marriage : alimony
D. welcome address : vote of thanks

64. Beautiful beaches attract people, no doubt about it. Just look at this city's beautiful beaches, which are among the most overcrowded beaches in the state.

Which of the following exhibits a pattern of reasoning most similar to the one exhibited in the argument above?

A. Moose and bear usually appear at the same drinking hole at the same time of day. Therefore, moose and bear must grow thirsty at about the same time.
B. Children who are scolded severely tend to misbehave more often than other children. Hence if a child is not scolded severely that child is less likely to misbehave.
C. This software programme helps increase the work efficiency of its users. As a result, these users have more free time for other activities.
D. During weather my dog suffers from fleas more than during cooler weather. Therefore, fleas must thrive in a warm environment.

65. No national productivity measures are available for underground industries that may exist but remain unreported. On the other hand, at least some industries that are run entirely by self-employed industrialists are included in national productivity measures.

From the information given above, it can be validly concluded that

A. there are at least some industries run entirely by self-employed industrialists that are underground industries
B. no industries that are run entirely by self-employed industrialists operate underground.
C. there are at least some industries other than those run entirely by self-employed industrialists that are underground industries.
D. there are at least some industries run entirely by self-employed industrialists that are not underground industries

66. Nilu has never received a violation from the Federal Aviation Administration during her 16-year flying career. Nilu must be a great pilot.

Which of the following can be said about the reasoning above?

A. The definitions of the terms create ambiguity
B. The argument uses circular reasoning
C. The argument is built upon hidden assumptions
D. The argument works by analogy

67. Many people argue that the death penalty deters murder. However, the notorious killer Tom Hanks deliberately moved to a state that imposes the death penalty just before embarking on a series of ferocious murders. Thus, it seems clear that the existence of the death penalty does not serve as a deterrent to murder.

The argument above may best be characterized as:

A. an appeal to emotion.
B. a flawed analogy.
C. a general conclusion based on a specific example.
D. circular reasoning.

68. What number should replace the question mark?

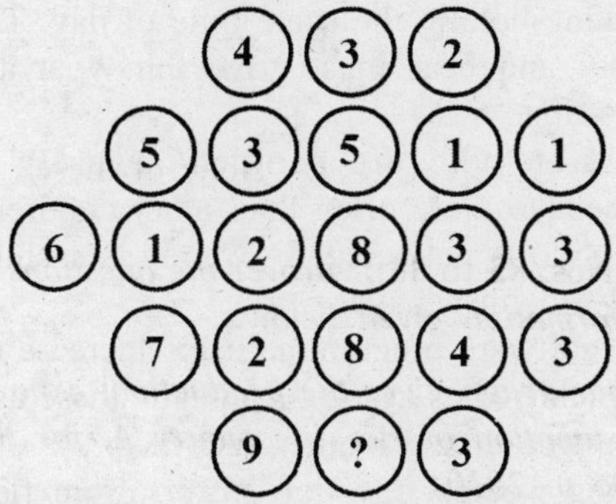

A. 1 B. 4
C. 12 D. 6

69. The fewer restrictions there are on the advertising of legal services, the more lawyers there are who advertise their services, and the lawyers who advertise a specific service usually charge less for that service than lawyers who do not advertise. Therefore if the state removes any of its current restrictions, such as the one against advertisements that do not specify fee arrangements, overall consumer legal costs will be lower than if the state retains its current restrictions.

If the statements above are true, which of the following must be true?

A. Some lawyers who now advertise will charge more for specific services if they do not have to specify fee arrangements in the advertisements.
B. More consumers will use legal services if there are fewer restrictions on the advertising of legal services.
C. If the restrictions against advertisements that do not specify fee arrangements is removed, more lawyers will advertise their services.
D. If more lawyers advertise lower prices for specific services, some lawyers who do not advertise will also charge less than they currently charge for those services.

70. Which of the following, if true, would most seriously weaken the argument concerning overall consumer legal costs?

A. The state is unlikely to remove all of the restrictions that apply solely to the advertising of legal services.
B. Lawyers who do not advertise generally provide legal services of the same quality as those provided by lawyers who do advertise.
C. Most lawyers who now specify fee arrangements in their advertisements would continue to do so even in the specification were not required.
D. Most lawyers who advertise specific services do not lower their fees for those services when they begin to advertise.

SECTION-III

QUANTITATIVE & DATA INTERPRETATION AND SUFFICIENCY

For all questions in this section, correct answers carry 1 mark each.

71. A train travelling at 36 kmph crosses a platform in 20 seconds and a man standing on the platform in 10 seconds. What is the length of the platform in meters?

A. 240 meters B. 100 meters
C. 200 meters D. 300 meters

72. By walking at 4/5th of his usual speed, a man reaches office 10 minutes later than usual. What is his usual time?

A. 20 min B. 40 min
C. 30 min D. 50 min

73. A man and a woman 81 miles apart from each other, start travelling towards each other at the same time. If the man covers 5 miles per hour to the women's 4 mile per hour, how far will the woman have travelled when they meet?

A. 27 B. 36
C. 45 D. None of these

74. Two people were walking in opposite directions. Both of them walked 6 miles forward then took right and walked 8 miles. How far is each from starting positions?

A. 14 miles and 14 miles
B. 10 miles and 10 miles
C. 6 miles and 6 miles
D. 12 miles and 12 miles

75. Four men and three women can do a job in 6 days. When 5 men and 6 women work on the same job, the work gets completed in 4 days. How long will 2 women and 3 men take to do the job?

A. 18 B. 10
C. 8.3 D. 12

76. Ram completes 60% of a task in 15 days and then takes the help of Rahim and Rachel. Rahim is 50% as efficient as Ram is and Rachel is 50% as efficient as Rahim is. In how many more days will they complete the work?

A. $\frac{121}{3}$ B. $\frac{51}{7}$
C. $\frac{40}{7}$ D. $\frac{65}{7}$

77. A and B can do a piece of work in 21 and 24 days respectively. They start the work together and after some days A leaves the work and B completes the remaining work in 9 days. After how many days did A leave?

A. 5 B. 7
C. 8 D. 6

78. A trader makes a profit equal to the selling price of 75 articles when he sold 100 of the articles. What % profit did he make in the transaction?

A. 33.33% B. 75%
C. 300% D. 150%

79. In a 100 M race, if A gives B a start of 20 meters, then A wins the race by 5 seconds. Alternatively, if A gives B a start of 40 meters the race ends in a dead heat. How long does A take to run 200 M?

A. 10 seconds B. 20 seconds
C. 30 seconds D. 40 seconds

80. A 4 cm cube is cut into 1 cm cubes. What is the percentage increase in the surface area after such cutting?

A. 4% B. 300%
C. 75% D. 400%

81. A number G236G0 can be divided by 36 if G is:

A. 8
B. 6
C. 1
D. More than one values are possible.

82. Amit can do a work in 12 days and Sagar in 15 days. If they work on it together for 4 days, then the fraction of the work that is left is:

A. $\frac{3}{20}$ B. $\frac{3}{5}$
C. $\frac{2}{5}$ D. $\frac{2}{20}$

83. A rectangular park 60 m long and 40 m wide has two concrete crossroads running in the middle of the park and rest of the park has been used as a lawn. If the area of the lawn is 2109 sq. m, then what is the width of the road?

A. 2.91 m B. 3 m
C. 5.82 m D. None of these

84. A bag contains 5 white and 3 black balls; another bag contains 4 white and 5 black balls. From any one of these bags a single draw of two balls is made. Find the probability that one of them would be white and other black ball.

A. $\frac{275}{504}$ B. $\frac{5}{18}$
C. $\frac{5}{9}$ D. None of these

Directions (Q. Nos. 85 to 88): *Answer the questions on the basis of the information given below.*

The following bar graph gives the production, exports and per capita consumption of rice in country A, for the five years from 2006 to 2010.

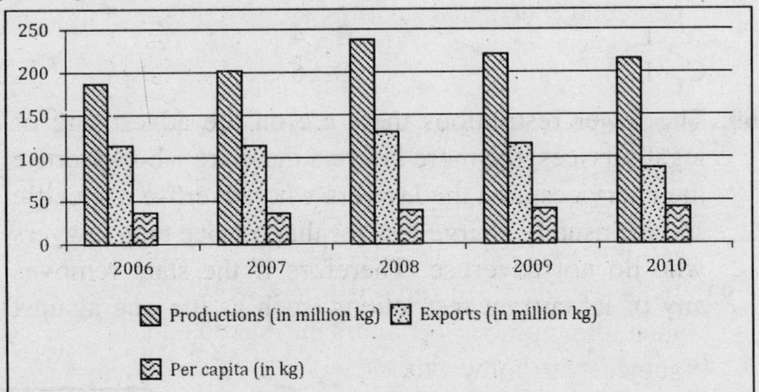

Consumption = Production – Exports
Per Capita Consumption = (Consumption) ÷ (Population)

85. In which year was the percentage increase in the consumption of rice over the previous year, the highest?

A. 2007 B. 2008
C. 2009 D. 2010

86. What is the population of country A in the year 2008 (in million)?

A. 2.64 million B. 2.72 million
C. 2.79 million D. 2.85 million

87. The ratio of exports to consumption in the given period was the highest in the year

A. 2006 B. 2007
C. 2008 D. 2009

88. In which of the given years was the population of country A, the highest?

A. 2007 B. 2008
C. 2009 D. 2010

Directions (Q. Nos. 89 to 93): *The following pie chart shows the hourly distribution (in degrees) of all the major activities of a student.*

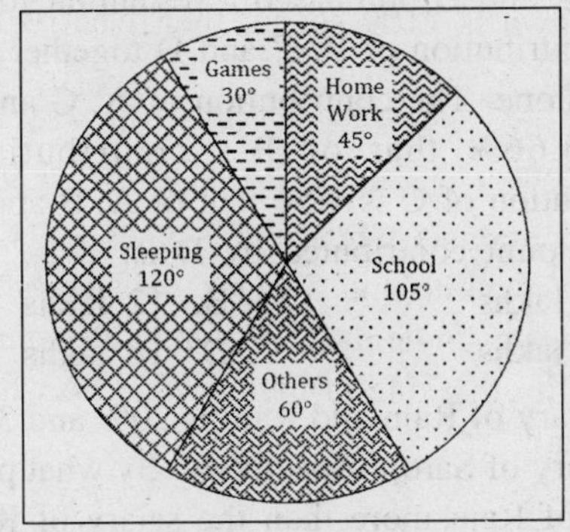

89. The percentage of time which he spends in school is:
A. 38% B. 30%
C. 40% D. 25%

90. How much time (in per cent) does he spend in games in comparison to sleeping?
A. 30% B. 40%
C. 25% D. None of these

91. If he spends the time in games equal to the home work and remains constant in other activities, then the percentage decrease in time of sleeping:
A. 15% B. 12.5%
C. 20% D. None of these

92. What is the difference in time (in hours) spent in school and in home work?
A. 2 B. 3
C. 4 D. 8

93. If he spends 1/3rd time of home work in Mathematics then the number of hours he spends in rest of the subjects in home work:
A. 1 B. 2
C. 3 D. 4

94. Three parallel lines are cut by two transversals as shown in the given figure. If AB = 2 cm, BC = 4 cm and DE = 1.5 cm, then the length of EF is:

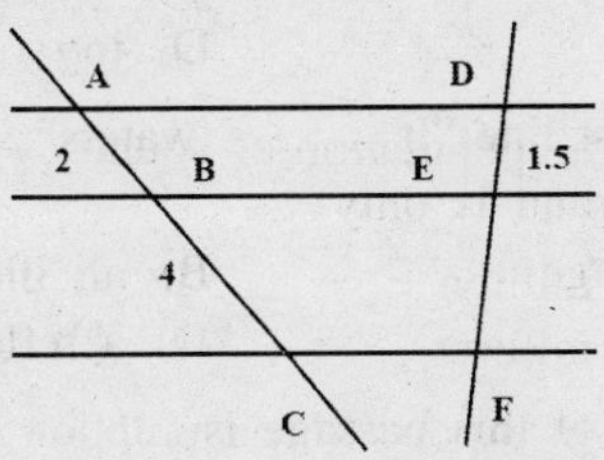

A. 2 cm B. 3 cm
C. 3.5 cm D. 4 cm

95. $\log_{10}10 + \log_{10}10^2 + \ldots\ldots + \log_{10}10^n$
A. $n^2 + 1$ B. $n^2 - 1$
C. $\frac{(n^2+n)}{2} \cdot \frac{n(n+1)}{3}$ D. $\frac{n^2+n}{2}$

96. The sum of a number and its reciprocal is thrice the difference of the number and its reciprocal. The number is:
A. $\pm\sqrt{2}$ B. $\pm\frac{1}{\sqrt{2}}$
C. $\pm\frac{1}{\sqrt{3}}$ D. $\pm\sqrt{3}$

97. The total number of Natural numbers that lie between 10 and 300 and are divisible by 9 is
A. 32 B. 30
C. 33 D. 34

98. If ${}^nC_x = 56$ and ${}^nP_x = 336$, find n and x.
A. 7, 3 B. 8, 4
C. 8, 3 D. 9, 6

99. One side of an equilateral triangle is 24 cm. The midpoints of its sides are joined to form another triangle whose midpoints are in turn joined to form still another triangle. This process continues indefinitely. Find the sum of the perimeters of all the triangles.
A. 144 cm B. 72 cm
C. 536 cm D. 676 cm

100. The probability that a leap year selected at random contains either 53 Sundays or 53 Mondays, is:
A. $\frac{17}{53}$ B. $\frac{1}{53}$
C. $\frac{3}{7}$ D. None of these

101. Find the intercepts made by the line $3x + 4y - 12 = 0$ on the axes:
A. 2 and 3 B. 4 and 3
C. 3 and 5 D. None of these

102. The average of 4 distinct prime numbers a, b, c, d is 35, where $a < b < c < d$. a and d are equidistant from 36 and b and c are equidistant from 34 and a, b are equidistant from 30 and c and d are equidistant from 40. The difference between a and d is:
A. 30 B. 14
C. 21 D. Cannot be determined

103. Ramsukh bhai sells rasgulla (a favourite Indian sweets) at ₹ 15 per kg. A rasgulla is made up of flour and sugar in the ratio 5 : 3. The ratio of price of sugar and flour is 7 : 3 (per kg). Thus he earns $66\frac{2}{3}$ profit. What is the cost price of sugar?
A. ₹ 10/kg B. ₹ 9/kg
C. ₹ 18/kg D. ₹ 14/kg

104. A reduction of 20% in the price of sugar enables a person to purchase 6 kg more for ₹ 240. What is the original price per kg of sugar?
A. ₹ 10/kg B. ₹ 8/kg
C. ₹ 6/kg D. ₹ 5/kg

105. A solid sphere is melted and recast into a right circular cone with a base radius equal to the radius of the sphere. What is the ratio of the height and radius of the cone so formed?
A. 4 : 3 B. 2 : 3
C. 3 : 4 D. None of these

106. The speed of scooter, car and train are in the ratio of 1 : 4 : 16. If all of them cover equal distance then the ratio of time taken/velocity for each of the vehicle is:
A. 256 : 16 : 1 B. 1 : 4 : 16
C. 16 : 4 : 1 D. 16 : 1 : 4

107. B is twice efficient as A and A can do a piece of work in 15 days. A started the work and after a few days B joined him. They completed the work in 11 days, from the starting. For how many days did they work together?
A. 1 day B. 2 days
C. 6 days D. 5 days

108. A, B, C and D purchased a restaurant for ₹ 56 lakhs. The contribution of B, C and D together is 460% that of A, alone. The contribution of A, C and D together is 366.66% that of B's contribution and the contribution of C is 40% that of A, B and D together. The amount contributed by D is:
A. 10 lakhs B. 12 lakhs
C. 16 lakhs D. 18 lakhs

109. The salary of Raju and Ram is 20% and 30% less than the salary of Saroj respectively. By what per cent is the salary of Raju more than the salary of Ram?
A. 33.33% B. 50%
C. 15.18% D. 14.28%

110. The radius of a wire is decreased to one-third and its volume remains the same. The new length is how many times the original length?
A. 2 times B. 4 times
C. 5 times D. 9 times

SECTION-IV

GENERAL ENGLISH

For all questions in this section, correct answers carry 1 mark each.

Directions (Q. Nos. 111 to 116): *Read the following passage and answer the questions that follow.*

D.H. Lawrence—1885-1930: The Supreme Triumph

For man, the vast marvel is to be alive. For man as for flower and beast and bird, the supreme triumph is to be most vividly, most perfectly alive. Whatever the unborn and the dead may know, they cannot know the beauty, the marvel of being alive in the flesh. The dead may look after the afterwards. But the magnificent here and now of life in the flesh is ours, and ours alone, and ours only for a time. We ought to dance with rapture, that we should be alive and in the flesh, and part of the living, incarnate cosmos. I am part of the sun as my eye is the part of me. That I am part of the earth my feet below know the perfectly, and my blood is part of the sea. My soul knows that I am a part of the human race, my soul is an inorganic part of the great human soul, as my spirit is a part of my nation. In my own very self, I am part of my family. There is nothing of me that is alone and absolute except my mind, and we shall find that the mind has no existence by itself, it is only the glitter of the sun on the surface of the waters. – Apocalypse, 1931.

111. By *triumph* the author means—
A. sin B. loss
C. sorrow D. victory

112. When the dead look after the afterwards the living should look at life
A. forever
B. for some months
C. for only a short while
D. in the past

113. By *rapture* the author means an emotion involving great
A. trepidation B. thrill
C. fear D. joy

114. By the last line "It .... waters", the author means that the mind is only
A. a mirage B. an illusion
C. magic D. a reflection

115. The tone of this passage is
A. social B. moral
C. reflective D. philosophical

116. The most suitable title for this passage would be—
A. The Surface of the Waters
B. My Mind
C. The Human Race
D. Alive and Kicking

Directions (Q. Nos. 117 to 119): *Choose the correct option.*

117. The repetition of messages or the use of superfluous expressions is called ....

A. redundancy B. hyperbole
C. alliteration D. allegory

118. Ink : pen : paper

A. watch : dial : strap
B. book : paper : words
C. farmer : plough : field
D. colour : brush : canvas

119. realia

A. theoretical constructs
B. fabricated examples
C. objects from real life
D. based on reality

Directions (Q. Nos. 120 to 125): *Choose the grammatically correct option from the following.*

120. A. 'Are these gloves belonging to you?' she asked.
B. 'Does this gloves belong to you?' she asked.
C. 'Do these gloves belongs to you?' she asked.
D. 'Do these gloves belong to you?' she asked.

121. A. I live in a house in a street in the countryside. The street is called "Bear Street" and the house is old – more than 100 years old!
B. I live in the house in the street countryside. The street is called "Bear Street" and the house is old – more than 100 years old!
C. I live in a house in the street in the countryside. The street is called "Bear Street" and the house is old – more than 100 years old!
D. I live in a house in a street in the countryside. The street is called "Bear Street" and a house is old – more than 100 years old!

122. A. The teachers will be able to visit our schools and compare our teaching methods to their own.
B. The teachers will be able to pay a visit to our schools and compare teaching methods for their own.
C. The teachers will be able to visit our schools and compare our teaching methods with their own.
D. The teachers will be able to visit our school and compare their teaching method with their own.

123. A. Could you give me the amount that you filled out in the cheque which was sent?
B. Could you give me the amount what you filled out in the cheque you sent?
C. Could you give me the amount for which you filled out in the cheque you sent?
D. Could you give me the amount wherein you filled out in the cheque you sent?

124. A. I have completed the work yesterday.
B. I did completed the work yesterday.
C. I have had completed the work yesterday.
D. I completed the work yesterday.

125. A. The train couldn't stop in time and crashed with the truck.
B. The train couldn't stop in time and crashed into the truck.
C. The train couldn't stop in time and crashed against the truck.
D. The train couldn't stop in time and crashed before the truck.

Directions (Q. Nos. 126 to 130): *Choose the correct synonymous word or description for each italicized word.*

The Jan Lokpal Bill, also ***referred*** to ....(126).... as the citizens' ombudsman bill, is a proposed ***independent*** ....(127).... anti-corruption law in India. Anti-corruption social activists proposed it as a more effective improvement on the original Lokpal Bill, which is currently being proposed by the government of India. The Jan Lokpal Bill aims to effectively ***deter*** ....(128).... corruption, redress ***grievances*** ....(129).... of citizens, and protect whistle – blowers. If made into law, the bill would create an independent ***ombudsman*** ....(130).... body called the Lokpal. It would be empowered to register and investigate complaints of corruption against politicians and bureaucrats without prior government approval. *(Source: Wikipedia)*

126. A. described as B. included in
C. supported for D. reformed as

127. A. self-centered B. impartial
C. self-seeking D. self-possessed

128. A. swell B. prevent
C. propel D. lucubrate

129. A. complaints B. dishonesty
C. committees D. opinions

130. A. a government appointee who investigates complaints by private persons against bureaucrats and/or politicians.
B. a government appointee who investigates complaints by government against common citizens.
C. a government appointee who investigates complaints by citizens against citizens.
D. a government appointee who investigates complaints by government against government officials.

Directions (Q. Nos. 131 to 134): *In the following sentences, one word or a phrase is used wrongly. Choose the word that must be changed or modified or deleted to make the sentence correct. There are sentences without any errors too.*

131. The committee **came to** a decision **to discuss** in detail **about** assorted problems that people have been facing for **a long time**.
A. came to B. to discuss
C. about D. a long time

132. I know you must not **see eye to eye** with the philosophy of Ramkrishna but you must admit that he **had had** tremendous **influence over** a great many followers.
A. see eye to eye B. had had
C. influence over D. no error

133. The Principal **inquired with** the students if they would like their teacher **to repeat** the lesson **again**.
A. inquired with B. to repeat
C. again D. no error

134. One of the **security guards** rushed **forward**; unlocked the gate and asked whether I had anything **objectionable**.
A. security guards B. forward
C. objectionable D. no error

Directions (Q. Nos. 135 and 136): Rearrange the following sentence fragment (P, Q, R and S) to make meaningful sentences.

135. I always told them .... and consideration for me.
P. How for me she
Q. Always exuded warmth
R. Was like a family's
S. Elder and how she has
A. PRSQ B. PQSR
C. PRQS D. SQPR

136. A year or .... foreign languages
P. I picked up a liking for learning
Q. the largest urban area and primary
R. so after leaving Bangkok
S. city of Thailand and a place of unique beauty
A. RPQS B. RQSP
C. RSPQ D. SPQR

Directions (Q. Nos. 137 to 139): Choose the correct option.

137. This book .... five sections.
A. comprises of B. comprises
C. consists D. comprises to

138. My grandfather left most of his money to an NGO; the rest went directly to my daughter and ....
A. I B. me
C. myself D. myself too

139. Parts of a country behind the coast or a river's banks
A. Isthumus B. Archipelago
C. Hinterland D. Swamps

Directions (Q. Nos. 140 and 141): In the questions below, the sentences have been given in Active/Passive voice. From the given alternatives, choose the one which best expresses the given sentence in Passive/Active voice.

140. Who is creating this mess?
A. Who has been created this mess?
B. By whom has this mess been created?
C. By whom this mess is being created?
D. By whom is this mess being created?

141. You should open the wine about three hours before you use it.
A. Wine should be opened about three hours before use.
B. Wine should be opened by you three hours before use.
C. Wine should be opened about three hours before you use it.
D. Wine should be opened about three hours before it is used.

Direction (Q. No. 142): *In the questions below, the passage consist of six sentences. The first and the sixth sentences are given. (S1, S6) The middle four sentences in each have been removed and jumbled up. These are labelled as P, Q, R and S. Find out the proper order for the four sentences.*

S1: Metals are today being replaced by polymers in many applications.
P: Above all, they are cheaper and easier to process making them a viable alternative to metals.
Q: Polymers are essentially a long chains of hydrocarbons molecules.
R: Today polymers as strong as metals have been developed.
S: These have replaced the traditional chromium – plated metallic bumpers in cars.
S6: Many Indian Institutes of Science and Technology run special programmes on polymer science.

The proper sequence should be:

142. A. QRSP B. RSQP
C. RQSP D. QRPS

Directions (Q. Nos. 143 and 144): *Choose the word which best expresses the meaning of the given word.*

143. FRUGALITY
A. Foolishness B. Extremity
C. Enthusiasm D. Economy

144. HARBINGER
A. Massenger B. Steward
C. Forerunner D. Pilot

Directions (Q. Nos. 145 and 146): *Choose the word which is the exact OPPOSITE of the given words.*

145. EXODUS
A. Influx B. Home-coming
C. Return D. Restoration

146. EQUANIMITY
A. Resentment B. Dubiousness
C. Duplicity D. Excitement

Direction (Q. No. 147): *Find the Odd one out from the group of words.*

147. A. Bludgeon B. Dragon
C. Black Jack D. Order

Direction (Q. No. 148): *Each pair of capitalized words given is followed by four pair of words. Select the pair that DOES NOT expresses a relationship similar to that expressed by the capitalized pair.*

148. KERNAL : SHELL
A. Caterpillar : Pupa
B. Larva : Cacoon
C. Lassitude : Syncope
D. Passenger : Car

149. Both of them ………… since their childhood.
A. are working here
B. work here
C. have been working
D. are liking to work

150. Although initial investigations pointed towards him …………
A. the preceding events corroborated his involvement in the crime.
B. the additional information confirmed his guilt
C. the subsequent events established that he was guilty
D. the subsequent events proved that he was innocent

ANSWERS

1	**2**	**3**	**4**	**5**	**6**	**7**	**8**	**9**	**10**
D	A	D	B	B	B	A	A	A	D
11	**12**	**13**	**14**	**15**	**16**	**17**	**18**	**19**	**20**
C	D	A	A	D	D	D	B	A	A
21	**22**	**23**	**24**	**25**	**26**	**27**	**28**	**29**	**30**
C	C	C	C	D	A	C	B	A	A
31	**32**	**33**	**34**	**35**	**36**	**37**	**38**	**39**	**40**
B	D	B	A	C	A	A	D	B	D
41	**42**	**43**	**44**	**45**	**46**	**47**	**48**	**49**	**50**
C	C	D	C	D	D	A	B	B	C
51	**52**	**53**	**54**	**55**	**56**	**57**	**58**	**59**	**60**
C	D	C	D	A	D	C	D	A	A
61	**62**	**63**	**64**	**65**	**66**	**67**	**68**	**69**	**70**
C	D	C	D	D	C	C	D	C	D
71	**72**	**73**	**74**	**75**	**76**	**77**	**78**	**79**	**80**
B	B	B	B	C	C	B	C	C	B
81	**82**	**83**	**84**	**85**	**86**	**87**	**88**	**89**	**90**
A	C	B	A	B	C	A	D	B	C
91	**92**	**93**	**94**	**95**	**96**	**97**	**98**	**99**	**100**
B	C	B	B	D	A	A	C	A	C
101	**102**	**103**	**104**	**105**	**106**	**107**	**108**	**109**	**110**
B	B	A	D	D	A	B	D	D	D
111	**112**	**113**	**114**	**115**	**116**	**117**	**118**	**119**	**120**
D	C	D	D	D	D	A	B	C	D
121	**122**	**123**	**124**	**125**	**126**	**127**	**128**	**129**	**130**
A	C	A	D	B	A	B	B	A	A
131	**132**	**133**	**134**	**135**	**136**	**137**	**138**	**139**	**140**
C	B	C	D	A	B	B	B	C	D
141	**142**	**143**	**144**	**145**	**146**	**147**	**148**	**149**	**150**
D	A	D	C	A	D	D	D	C	D

SOME SELECTED EXPLANATORY ANSWERS

41. Let us solve the problem by eliminating the options.

G: People are rewarded.

C: If people are rewarded then they will not change their lifestyle.

F: Unless people change their lifestyle, temperature rises.

This can also be put in the following manner.

If people do not change their lifestyle then temperature rises.

This is an example of binary logic – If not A then B which can also be expressed as If not B then A.

Thus if the temperature doesn't rise then people change their lifestyle.

Here, according to G and C, people do not change their lifestyle and so temperature rises.

D: If temperature rises, then water level in the coastal area rises.

H: Water level in the coastal area doesn't rise.

Here, D is in accordance with G, C and F but H is not.

So option A is incorrect.

B: People change their lifestyle only if they are rewarded.

This means that if people have changed their lifestyle then they have been rewarded.

Note that if people are rewarded, they may or may not have changed their lifestyle.

G: People are rewarded.

From B and G, we don't know whether people have changed their lifestyle or not.

F: Unless the people change their lifestyle, temperature rises.

This can also be put as – If the people don't change their lifestyle then temperature rises.

But as we don't know whether people have changed their lifestyle or not, we cannot conclude whether temperature has risen or not.

D: If temperature rises, then water level in the coastal area rises.

H: Water level in the coastal area doesn't rise.

As we don't know whether temperature has risen or not, H cannot be concluded using D.

Hence, option B is also incorrect.

E: Whenever the water level in the coastal areas rises, then the temperature rises.

H: Water level in the coastal area doesn't rise.

Thus from these two statements the temperature may or may not rise.

While discussing the previous option, we have proved that from statements B, F and G we also come to the same conclusion that the temperature may or may not rise.

Hence, statements E, F, G, H and B are consistent with each other.

Hence, option C is correct.

42. In option B, statement A says – If Kumar sings then the audiences sleep.

Out of the other 4 statements C, F, G and H, none of the statements mention anything about sleep and hence option B is incorrect.

If option B is incorrect, option D is also incorrect as it mentions both B and C.

C: Unless audience does not dance, the concert will be successful.

This can also be put as – If the audience dance, the concert will be successful.

F: Kumar sings, only if Vina dances.

This means that – If Kumar has sung, then Vina has danced.

But if Vina has danced, Kumar may or may not have sung.

B: If Kumar sings, then the audience dances.

We don't know whether Kumar has sung or not and so we don't know whether the audience have danced or not.

Hence we cannot conclude that the concert is successful as given in statement H.

So option A is also logically incorrect.

E: If Vina dances, then Kumar sings.

G: Vina dances.

Thus from these two statements we know that Kumar sings.

B: If Kumar sings, then audiences dance.

Thus audiences have danced.

C: Unless audience does not dance, the concert will be successful.

This can also be put as – If the audience dance, the concert will be successful.

Thus the concert is successful as mentioned in statement H.

Hence, option C is correct.

43. The person with both statements true will be the one whose statements help establish the relationships between all 4 members and especially between B and C. This is because one of the questions explicitly asks for the relationship between B and C and there is no answer option such as "None of the above" or "Data Insufficient".

Consider A: B is A's son and D is A's mother. However, there is no information about C. Therefore, the relationship between B and C cannot be established. Therefore, both of A's statements cannot be true.

Consider B: B is C's husband and D is B's father.

However, there is no information on A. A could be B and C's child or B's sibling. Therefore, both of B's statements cannot be true.

Consider C: D is C's mother-in-law and A is C's daughter. Since the family has only four members, B has to be C's spouse. Thus, the overall relationship between the four members is known. It is also known that B and C are a couple. However, it cannot be said whether B is the husband or wife of C. Therefore, both of C's statements also cannot be considered true.

Consider D: A is D's grand-daughter and B is D's daughter-in-law. Again, since there are only 4 members in the family, C should be B's husband, D's son and A's father.

Thus, D is the grand-parent, B and C are the couple where C is the husband and A is the grand-daughter. Thus, D's statements give the overall relationships as well as the specific relationship between B and C.

Therefore, D is the person who always speaks the truth.

Hence, option D is correct.

44. Consider the solution to the previous solution. B is C's wife.

Hence, option C is correct.

45. Consider each option.

D is either A's grandfather or grandmother. Since D always speaks the truth, option C may or may not be true (depending on the gender of D) while option A is definitely false.

B is C's wife. Based on the fact that both of D's statements are true, both of B's statements turn out to be false. Therefore, option B is also false.

A is B's daughter. Based on the fact that both of D's statements are true, both of A's statements turn out to be false.

Thus, B's daughter always tells lies.

Hence, option D is correct.

46. Observe that the question states that the cube is kept in the corner of the room. Assume that opposite faces are painted red, blue and green. Since the cube is in the corner, one red, blue and green face each will be adjacent to the two walls and floor and will not be painted. Thus, there are only three faces that are painted. Since the cube is cut into 343 identical cubes, it means that each face is cut into 7 equal layers along each dimension. Thus, there are 7 similar layers along the length, breadth as well as height. Since only one face is painted along each dimension, if we remove one layer each along the length, breadth as well as height, the remaining 6 layers in each dimension will be unpainted.

Therefore, number of cubes that have no face painted $= 6 \times 6 \times 6 = 216$

Hence, option D is correct.

47. Consider the 3 faces that are painted. Let us first count the number of cubes that have at least one colour painted on them.

Consider the top face. Assume it is painted blue. It will have $7 \times 7 = 49$ cubes with at least one colour on them.

Now, consider the front face. Assume it is painted red. It will also have 49 cubes with at least one colour. However, the top 7 cubes on this face are also common to the top face and have been counted once.

So, number of cubes to be considered from front face $= 7 \times 6 = 42$

Finally, consider the side face which will be painted green and which will also have 49 cubes with at least one colour.

However, it also has 7 cubes common to the top face, 7 cubes common with the front face and 1 cube common to both the top and front face.

So, number of cubes to be considered from side face $= 49 - (7 + 7 - 1) = 36$

So, number of cubes with at least one colour on them $= 49 + 42 + 36 = 127$

Now, there are 7 cubes common to the top and front face. Similarly, there are 7 cubes common to the front and side face as well as 7 cubes common to the top and side face. Here, the cube at the intersection of all three faces will have red, blue as well as green colour, while the remaining 6 faces in each case will have a red-blue or red-green or blue-green combination.

Thus, there is only 1 cube that is painted in 3 colours while there are 6×3 i.e. 18 cubes that are painted in exactly 2 colours.

So, number of cubes that have exactly one colour on them $= 127 - (1 + 18) = 108$

Hence, option A is correct.

48. Number of cubes that have at the most 2 faces painted = Total number of cubes – Number of cubes that have 3 faces painted.

From the solution to the previous question, number of cubes that have at the most 2 faces painted $= 343 - 1 = 342$

Hence, option B is correct.

49. After driving 3 km north, 3 km west and 4 km south, Amit is 1 km to the south and 3 km to the west of his starting point. So he will have to move 1 km north and 3 km east to reach the point from where he started.

Hence, option B is correct.

50. As stated in the solution of Q. 49, Amit will reach the starting point if he moves 1 km north and 3 km east. If he moves 1 km north and 4 km east, he will be 1 km away from his starting point.

Hence, option C is correct.

51. $X = XY$ and $Y = Y + 1$ is the process that takes place in every step and the machine stops at $X \geq N$

Input: X = 3, Y = 2 and N = 100.
Step 1: X = 6, Y = 3
Step 2: X = 18, Y = 4
Step 3: X = 72, Y = 5
Step 4: X = 360, Y = 6
Thus it takes 4 steps before the machine stops.
Hence, option C is correct.

52. In Q. 51, the value of X becomes 360.
Hence, option D is correct.

53. In Q. 51, the value of Y becomes 6.
Hence, option C is correct.

54. If N = 500, then by using the solution of question 51, we can further calculate Step 5 as follows:
Step 5: X = 2160, Y = 7.
Thus value of X will be 2160.
Hence, option D is correct.

55. Input: X = 2, Y = 3
Step 1: X= 6, Y = 4
Step 2: X = 24, Y = 5
Step 3: X = 120, Y = 6.
Thus for any value of N greater than 120, the value of Y will become 7.
The correct answer should be 121.
Since, the answer closest to 121 here is 300, it is the correct answer.
Hence, option A is correct.

56. Vicky always score more than Priya and Priya always scores more than Ankit.
Thus Vicky's rank is always numerically lower than Priya's and Priya's rank is always numerically lower than Ankit's.
Either of the following two situations take place.
If Raman scores the highest, then Tony scores the least or if Sunil scores the highest, then Deepak or Ankit scores the least.
Thus if Raman is ranked 1st, then Tony is ranked 7th and if Sunil is ranked 1st then either Deepak or Ankit is ranked 7th.
The question states that Sunil is ranked 6th and Ankit is ranked 5th.
Thus Raman is ranked 1st and Tony is ranked 7th.
So the following situations are possible for ranks 2, 3 and 4.
2 – Vicky, 3 – Priya, 4 – Deepak or
2 – Vicky, 3 – Deepak, 4 – Priya or
2 – Deepak, 3 – Vicky, 4 – Priya.
Vicky can never be 1st, Raman will never be 2nd or 3rd,
Tony is ranked 7th and hence cannot be 4th or 5th.
Deepak can be 3rd or 4th according to the above obtained conditions.
Hence, option D is correct.

57. If Raman gets the highest, he will be ranked 1st. Accordingly, Tony will be ranked 7th.
Now Priya and Ankit are always ranked numerically higher than Vicky.
Hence, Vicky can never be ranked lower than 4th as if he is ranked 5th then there will not be enough positions to accommodate both Ankit and Priya.
Hence, option C is correct.

58. The question states that Ankit is ranked 1st.
This is not possible as he always scores less than Vicky and Priya.
Hence, option D is correct.

59. If Sunil is ranked 2nd, then Raman is ranked 1st and Tony is ranked 7th.
Vicky's lowest numerical rank will be 3rd and so he will always get less than Sunil.
Raman is ranked 1st and so Priya will always get less than Raman.
Vicky scores more than Priya is given in the question data.
Hence by eliminating the options, we can conclude that it is possible for Deepak to get more than Vicky.
Hence, option A is correct.

60. If Vicky is ranked 5th, then Priya is ranked 6th and Ankit is ranked 7th.
Thus Sunil is ranked 1st.
Hence, option A is correct.

61. The conclusion is that Party Z is likely to win the next election. A weakening statement would have to show party Y having a favourable advantage or party Z having a clear disadvantage.
Option A is loose and does not provide anything concrete.
Option B shows the percentage of people voting from each party but fails to bring out the vantage point of a single party.
Option D does not weaken the conclusion.
Option C clearly states a percentage of people in favour of party Z, which is less than 50%.
Hence, the correct answer is option C.

62. The information given to us is about inflation rising in quarters of a year.
Option A talks about the RBI taking "necessary measures" which is ambiguous.
Option B and C do not provide a strong explanation for the question.
Option D, is right on target. It talks about the quarterly increase in inflation and is in line with the information given.
Hence, the correct answer is option D.

63. Birth (positive connotation)—The act or process of bearing young; Dirge (negative connotation)—A funeral hymn or lament.

Similarly: Marriage - is a social union or legal contract; alimony-allowance made to a wife out of her husband's estate or income for her support, upon her divorce. Options A, B and D do not have the same relation between the two words.
Hence, the correct answer is option C.

64. The reasoning in the main statement is that, 'because beaches are beautiful, they are crowded', *i.e.,* it is a 'Because A therefore B' type of a statement.
Option A only states that Moose and bear grow thirsty at the same time, it does not give a reason for this.
Option B shows an 'If not A then B' type of relationship.
Option C does not show a relationship between two variables.
Option D, shows a similar relation to the one described in the main statement.
Hence, the correct answer is option D.

65. Option A is incorrect as there is no available measure for underground industries.
Option B assumes that all self-employed industries are not underground, which is erroneous.
Option C states that industries 'except for self-employed ones' may be running underground. This is a wrong conclusion.
Option D is correct. The latter statement states that some self-employed industries are included in national productivity measures and only non-underground industries are included in national productivity measures.
Hence, the correct answer is option D.

66. This statement has an underlying assumption that 'receiving a violation from the FAA is a hindrance to an excellent performance record'.
Hence, the correct answer is option C.

67. The example of Tom Hanks is specific. It is just one case upon which the broad conclusion of 'death penalty not serving as a deterrent to murder' has been made.
Hence, the correct answer is option C.

68. The number at the center of each row is the sum of the rest of the numbers divided by 2.
In the first row the central number is 3 which is $\frac{(4+2)}{2}$
In the second row the central number is 5 which is $\frac{(5+3+1+1)}{2}$
Similarly, the central number in the last row is $\frac{(9+3)}{2}$ = 6
Hence, option D is correct.

69. The latter part of the information states that removal of restrictions such as the ones including fee arrangements for advertisements will cause reduction in overall consumer legal costs and the former part of the information states that fewer restrictions on advertising will lead to an increase in lawyers advertising for their services.
Only option C captures the essence of the information.
Hence, the correct answer is option C.

70. The information clearly mentions that those lawyers who advertise for their services charge lesser than those who do not advertise. Option D contradicts this and weakens the argument concerning consumer legal costs.
Hence, the correct answer is option D.

71. Speed of train = 36 kmph = $\frac{36\times 5}{18}$ = 10 m/s
Let the length of the platform be P and length of the train be T.
Now, the train crosses the platform in 20 seconds and the man in 10 seconds.
Hence, the train travelled P + T in 20 seconds and T in 10 seconds.
Hence, it travelled P distance in 20 – 10 = 10 seconds.
Hence, P = 10 × 10 = 100 meters.

72. Let t be the usual time taken by the man.
Now, if he travels at $\frac{4}{5}$ th of his usual speed then time taken by him will be $\frac{5}{4} \times t$
Hence, $\frac{5}{4} \times t = t + 10$
Hence, $t = 40$

73. Relative speed of man and woman is 9 kmph.
Hence, time required to cover 81 km = $\frac{81}{9}$ = 9 hours.
In 9 hours, the woman will travel 9 × 4 = 36 km.

74. Assume that one of them walked 6 km towards east and then 8 km south.
Hence, he is 6 km east and 8 km south of the original position.
Hence, he is $\sqrt{(6^2+8^2)}$ = 10 km from his original position.
Similarly, we can prove that the other person is also 10 km from his original position.

75. Let a man take m days and a woman takes w days to complete the work.
Hence, we have,

$$\frac{4}{m}+\frac{3}{w} = \frac{1}{6} \qquad ...(i)$$

$$\frac{5}{m}+\frac{6}{w} = \frac{1}{4} \qquad ...(ii)$$

Solving (*i*) and (*ii*), we get,
$m = 36$ and $w = 54$
Hence, 2 women and 3 men will complete;

$$\frac{2}{54}+\frac{3}{36} = \frac{13}{108}$$

Hence, they will complete the work in $\frac{108}{13} = 8.3$ days

Note: In actual paper, all the options were incorrect. Hence, we have changed one of the options to include the correct answer.

76. Ram completes 60% of the work in 15 days.
Hence, he will take 25 days to complete the work.
Now, Rahim is 50% as efficient as Ram, and Rachel is 50% as efficient as Rahim, hence Ram, Rahim and Rachel together will complete;

$\frac{1}{25}+\frac{1}{50}+\frac{1}{100}=\frac{7}{100}$ of the work in one day.

Hence, they will complete the work in $\frac{100}{7}$ days.
Hence, they can complete the remaining 40% work in $0.4\times\frac{100}{7}=\frac{40}{7}$ days.

Note: In actual paper, all the options were incorrect. Hence, we have changed one of the options to include the correct answer.

77. In 9 days, B will do $\frac{9}{24} = \frac{3}{8}$ of the work.

Hence, A and B together did $1 - \frac{3}{8} = \frac{5}{8}$ of the work together.
Now, A and B can complete;

$\frac{1}{21}+\frac{1}{24}=\frac{5}{56}$ of the work in one day.

Hence, they will complete, $\frac{5}{8}$ of the work in;

$\frac{5}{8}\times\frac{56}{5}=7$ days.

78. Let selling price of each article be 1.
Hence, selling price of 100 article = 100
Hence, profit = 75
Hence, cost price = 100 – 75 = 25

Hence, percentage profit = $\frac{75}{25} \times 100 = 300\%$

79. Let V_a and V_b be the speed of A and B respectively.

$$\therefore \quad \frac{V_a}{100}+5 = \frac{80}{V_b} \quad ...(i)$$

$$\frac{V_a}{100} = \frac{60}{V_b} \quad ...(ii)$$

Solving (*i*) and (*ii*), we get,

$$V_b = 4,\ V_a = \frac{100}{15}$$

Hence, A will take;

$$\frac{200}{\frac{100}{15}} = 30 \text{ seconds}$$

80. Surface area of the original cube = 4 × 4 × 6
= 6 × 16 cm^2
Now, there will 64 new cubes of side 1 cm.
Hence, surface area of these 64 cubes = 64 × 6 cm^2
Hence, percentage increase in surface area is;

$$\frac{64\times6-16\times6}{16\times6}\times100 = 300\%$$

81. G236G0 is divisible by 36
Hence, G236G0 is divisible by 9 and G is an even number.
Hence, G + 2 + 3 + 6 + G + 0 = 2G + 11 is divisible by 9.
Only G = 8 satisfies the given condition.

82. In one day, Amit and Sagar can do;

$\frac{1}{12}+\frac{1}{15}=\frac{3}{20}$ of the work in one day.

Hence, in 4 days they will complete $\frac{3}{5}$th of the work.
Hence, amount of work remaining after 4 days

$$= 1-\frac{3}{5}=\frac{2}{5}$$

83. Let the width of the road be *d*.
Hence, Area of lawn = Area of park – Area of road
= 2400 – (40*d* + 60*d* – d^2)
Hence, 2109 = 2400 – 100*d* + d^2
Hence, d^2 – 100*d* + 297 = 0
Hence, *d* = 97 or *d* = 3.

84. Probability of choosing either bag is $\frac{1}{2}$.

Now, from the first bag, we can choose one white and one black ball in 5 × 3 = 15 ways.
While, two balls can be selected in $^8C_2$ = 28 ways
Hence, probability that the two ball chosen are of different colour = $\frac{15}{28}$

Similarly, we can prove that the probability that two balls chosen from the second bag are of different colour

$$= \frac{20}{36}$$

$$= \frac{1}{2}\times\frac{15}{28}+\frac{1}{2}\times\frac{20}{36}=\frac{275}{504}$$

85. Consumption = Production – Exports

$$\text{Per Capita Consumption} = \frac{\text{Consumption}}{\text{Population}}$$

$$\therefore \text{Population} = \frac{\text{Consumption}}{\text{Per Capita Consumption}}$$

Thus, the consumption and population for each year can be calculated as shown in the table below:

Year	Production	Exports	Consumption	Per Capita consumption	Population	% Increase in Consumption	Exports/ Consumption
2006	186.5	114	72.5	36.25	2.00	—	1.57
2007	202	114	88	35.2	2.50	21.38	1.30
2008	238	130	108	38.7	2.79	22.73	1.20
2009	221	116	105	40.5	2.59	(2.78)	1.10
2010	215	88	127	42	3.02	20.95	0.69

Thus, the percentage increase in the consumption of rice was the highest in 2008.

86. Consider the table obtained in the solution to the first question.
The population of country A in 2008 was 2.59 million.

87. Consider the table obtained in the solution to the first question.
As can be seen, the ratio of consumption to exports in the given period was highest in 2006 (1.57).

88. Consider the table obtained in the solution to the first question.
The population of country A was the highest in 2010 (3.02 million).

89. The percentage of time spent in school is

$$\frac{105}{360}\times 100 = 29.17\% \approx 30\%$$

90. Since the angle subtended by "games" is 30° while that subtended by "sleeping" is 120°, the required percentage is $\frac{30}{120}\times 100$ i.e. 25%

91. Since the student spends an equal time in games and home work, he can either increase the time spent on games or decrease the time spent on home work.
It is also given that the time spent on other activities remains constant while the decrease in time spent on sleeping is to be found.
This can only happen if the time spent on games has increased.
So, the angle subtended for games should increase from 30° to 45°. If this happens, the angle subtended for sleeping will decrease from 120° to 105°.

$$\therefore \text{Percentage decrease} = \frac{(120-105)}{120}\times 100 = 12.5\%$$

92. Difference in time spent in school and home work is

$$\frac{(105-45)}{360}\times 24 = 4 \text{ hours}$$

93. Time spent on home work $= \frac{45}{360}\times 24 = 3$ hours

Out of this, one-third *i.e.* one hour is spent on Mathematics.
So, the remaining two hours are spent on the rest of the subjects.

94. In the given diagram,

$$\frac{AB}{BC} = \frac{DE}{EF}$$

Substituting value of AB, BC and DE, we get,
EF = 2 × 1.5 = 3

95. $\text{Log } 10 + \log 10^2 + \ldots + \log 10^n = 1 + 2 + \ldots + n$

$$= \frac{n(n+1)}{2}$$

96. Let a be the required number.
Hence, we have;

$$\left(a+\frac{1}{a}\right) = 3\left(a-\frac{1}{a}\right)$$

Hence, $a^2 = 2$

Hence, $a = \pm\sqrt{2}$

97. Smallest and largest numbers, in the given region, which are divisible by 9 are 18 and 297 respectively.

Hence, the required number $= \frac{(297-18)}{9}+1$

= 31 + 1 = 32

98. $^{n}C_x = \frac{n!}{x!(n-x)!} = 56$

$^{n}P_x = \frac{n!}{(n-x)!} = 336$

Hence $x! = 6$
Hence, $x = 3$

Hence, from the given options, n can be either 7 or 8.
But $^8C_3 = 56$.

99. Perimeter of the largest equilateral triangle is 72 cm
Now, triangle formed by joining the midpoints will have perimeter half of that of original triangle.
Hence, the required sum is;
$72 + 36 + 18 + ...$ upto infinite terms
This is nothing but an infinite GP, with
common ratio $= \frac{1}{2}$

Hence, the required sum $= \left(\frac{72}{1-\frac{1}{2}}\right) = 144$

100. In a leap year, there are 366 days.
Now, $366 \pmod 7) \equiv 2 \pmod 7)$
This implies that if year started on Monday then it will end on Tuesday, and so on.
And also the first day of the year will definitely appear for 53 times in the calendar.
Hence, if year starts on Saturday, Sunday or Monday then it will definitely contain either 53 Sundays or 53 Mondays.
Now, a leap year chosen can start from any of the seven day.
But it will contain 53 Monday or Sunday only if it starts on Saturday, Sunday or Monday.

Hence the required probability $= \frac{3}{7}$

101. $3x + 4y = 12$

Hence, $\frac{x}{4} + \frac{y}{3} = 12$

This is nothing but the intercept form of the equation of a line.
Hence, intercept are 4 and 3.

102. a and d are equidistant from 36.
This implies that average of a and d is 36.
Hence, $a + d = 72$
Similarly, we have,
$b + c = 68$, $a + b = 60$ and $c + d = 80$
Now, $a < b$ and $a + b = 60$
Hence, $a < 30$ and $b > 30$
Similarly,
$b < 34$ and $c > 34$
And $c < 40$ and $d > 40$
Now, only primes which satisfy this conditions are,
$a = 29$, $b = 31$, $c = 37$ and $d = 43$
Hence, $d - a = 43 - 29 = 14$

103. After selling at ₹ 15/kg, Ramsukh earns a profit of 66.66%.
Hence, cost price of Rasgulla is ₹ 9/kg.
Now, ratio of flour and sugar is 5 : 3.

Hence, 1 kg of rasgulla is made up of $\frac{5}{8}$ kg of flour and $\frac{3}{8}$ kg of sugar.
Let price of 1 kg of flour $= 3k$
Hence, price of 1 kg of sugar $= 7k$
Hence price of 1 kg of Rasgulla is;

$\frac{3}{8} \times 7k + \frac{5}{8} \times 3k = 9$

Hence, $k = 2$
Hence, cost price of sugar $= 7k = 7 \times 2 =$ ₹ 14/kg

104. Let cost price of sugar be x.
Hence, reduced cost price $= 0.8x$

Hence, $\frac{240}{x} + 6 = \frac{240}{0.8x}$

Hence, $\frac{300}{x} = \frac{240}{x} + 6$

Hence, $\frac{60}{x} = 6$

Hence, $x = 10$

105. Volume of a solid sphere is given as $\frac{4\pi r^3}{3}$ where r is the radius of the sphere.

Volume of a right circular cone is $\frac{\pi r^2 h}{3}$ where r is the radius of the base of the cone and h is the height of the cone. The solid sphere is recast into a right circular cone with base radius equal to the radius of the sphere.

Thus $\frac{4\pi r^3}{3} = \frac{\pi r^2 h}{3}$

Hence, $h = 4r$
Thus the ratio of the height to the radius is 4 : 1.

106. Assume that the scooter, the car and the train travel 16 km.
Let their speeds be 1 kmph, 4 kmph and 16 kmph respectively.
Thus time taken by the scooter, the car and the train is 16 hours, 4 hours and 1 hour respectively.
Ratio of time taken to the velocity of each vehicle will be

$\left(\frac{16}{1}\right) : \left(\frac{4}{4}\right) : \left(\frac{1}{16}\right) = 16 : 1 : \left(\frac{1}{16}\right) = 256 : 16 : 1$

107. A can do a piece of work in 15 days.
B is twice as efficient as A.
So B can do the same piece of work in 7.5 days.
Assume that there is 15 units or work to be done.
A does 1 unit per day and B does 2 units per day.
For the first x days, A worked alone and hence work done per day is $1 \times x = x$ units.

For the next $(11 - x)$ days, A and B worked together.

In 1 day, they do $(1 + 2) = 3$ units.

So in $11 - x$ days, they do $33 - 3x$ units.

$33 - 3x + x = 15$

Thus $2x = 18$ and $x = 9$.

Thus A worked alone for 9 days the A and B worked together for 2 days.

108. Let the contributions of A, B, C and D in purchasing a restaurant be a, b, c and d respectively.

Thus $a + b + c + d = 5600000$ ...(1)

$b + c + d = 4.6a$ ...(2)

$a + c + d = 3.666b$ ...(3)

$0.4\,(a + b + d) = c$

$a + b + d = 2.5c$ ...(4)

Using (1) and (2) we get

$$a = \frac{5600000}{5.6} = 1000000 \quad ...(5)$$

Using (1) and (3) we get

$$b = \frac{5600000}{4.66} = 1201716.73 \quad ...(6)$$

Using (1) and (4) we get

$$c = \frac{5600000}{3.5} = 1600000 \quad ...(7)$$

Using (1), (5), (6) and (7), we get $d \approx 1800000$

109. Let the salary of Saroj be ₹100.

Salary of Raju and Ram is 20% and 30% less than Saroj's salary.

Hence, Raju's salary is Rs. 80 and Ram's salary is Rs. 70.

Thus Raju's salary is greater than Ram's salary by

$$\frac{80-70}{70}\times 100 = 14.28\%$$

110. Let the radius of the wire be r and its length be l.

A wire is cylindrical in nature where the length is considered to be its height.

Hence volume of this wire is $\pi r^2 l$

The modified radius is one-third of the original radius.

Thus $r' = \frac{r}{3}$

Let the modified length be l'

As the volume remains constant,

$$\pi r^2 l = \pi\left(\frac{r}{3}\right)^2 l'$$

Hence, $l' = 9l$

Thus the modified length is 9 times the original length.

111. Victory is the synonym of triumph.

112. Consider the fourth sentence in the paragraph: 'But the magnificent here and now ... ours only for a time'. This line validates what is mentioned in option C.

113. Rapture means 'extreme joy; being transported by a lofty emotion; an expression of ecstatic feeling'.

114. The sun's reflection is seen on the surface of the water.

115. The passage is pertaining to 'the rational investigation of truth and principles of being, knowledge or conduct'. This is characteristic of a philosophical approach/tone.

116. The passage talks about the gift of life and how one must make the most of it. Option D is a befitting title.

117. Redundancy can be defined as: 1. The state of being repetitive or no longer needed. 2. Superfluous repetition or overlapping, especially of words.

118. The relationship conveyed is that of 'material—object that utilizes the material—where the final result is seen'

This correlation is seen in option D.

119. Realia can be defines as: 1. Objects like coins, tools, etc. used by a teacher to illustrate everyday living. 2. Things which are real.

120. In sentence A, 'Are' is used erroneously. It is the present indicative plural of 'be'.

In sentence B, 'Does' and 'this' are incorrect as gloves are plural. 'Does this **pair** of gloves' is correct

In sentence C, 'belongs' is incorrect. It should be used with singular nouns/pronouns eg. It belongs to you.

Sentence D is grammatically correct.

121. In the first sentence, "I live…..countryside", the nouns 'house' and 'street' require an indefinite article (the nouns are not particular or specific ones). Hence, options B and C are ruled out.

Between sentences A and D, A is error-free. The noun 'house' in the latter part of the sentence requires a definite article as it is talking about a specific 'old house'.

122. In sentence A, the preposition 'to' is incorrect. One uses 'to' only in case of similarities. For example: She is similar to me. Her teaching method is similar to mine.

In sentence B, 'for' is erroneous. It is incorrect usage.

Between options C and D, option D mentions 'compare **their** teaching method with **their** own'. This is erroneous.

Option C is grammatically correct.

123. Option A uses the word 'that' which is appropriate to introduce the restrictive relative clause "you filled out……which was sent?"

124. In sentence A, 'have' denotes the simple present tense which is incorrect as the word 'yesterday' is in the past.

When the word 'did' is used in a sentence, it must be followed with the stem of the verb, i.e. 'Did complete'. Hence, this statement is erroneous.

In sentence C, 'have had' indicates the perfect present tense which is incorrect.

125. 'Crashed into' is the correct phrase.
Hence, the correct answer is option B.

126. Here, 'referred to' indicates something 'also known as' or 'described as' something else.
Hence, the correct answer is option A.

127. In this context the word to replace independent would be impartial which means 'unbiased; independent of bias; neutral'.
Self-centred means concerned solely or chiefly with one's own interests
Self-seeking means the seeking of one's own interest or selfish ends
Self-possessed means having full possession of ones faculties
Hence, the correct answer is option B.

128. To 'deter' means to 'discourage or prevent'.
Hence, the correct answer is option B.

129. 'Grievance' can mean the following:
1. suffering; distress
2. a cause of distress (as an unsatisfactory working condition) felt to afford reason for complaint or resistance
3. the formal expression of a grievance : complaint

Hence, the correct answer is option A.

130. 'Ombudsman' can mean the following:
1. a government appointed to receive and investigate complaints made by individuals against abuses or capricious acts of public officials
2. one that investigates, reports on, and helps settle complaints

Hence, the correct answer is option A.

131. The sentence does not require the word 'about'. It can be written as '...to discuss in detail the assorted problems that people have been facing for a long time'.
We discuss a/the problem and not discuss about a problem.
Hence, the correct answer is option C.

132. The sentence requires a single 'had'.
Hence, the correct answer is option B.

133. The word 'again' is redundant as it is placed after 'to repeat' which means the same.
Hence, the correct answer is option C.

134. This sentence has no error.
Hence, the correct answer is option D.

135. The RS link—'was like a family's elder....'—is essential which is only seen in option A.
Hence, the correct answer is option A.

136. R has to be the first sentence as 'A year or so' is the correct phrase.
Q must follow R as it describes Bangkok. S takes this further. The correct order is RQSP.
Hence, the correct answer is option B.

137. Comprises means 'consists of' and is the appropriate word for the blank.

138. The blank needs to be filled by a pronoun in the subjective case. 'I' is ruled out. It is appropriate to use "myself" when you have used "I" earlier in the same sentence.

139. Isthmus— a narrow strip of land connecting two larger land areas
Archipelago—an expanse of water with many scattered islands
Hinterland—a region lying inland from a coast
Swamp—a wetland often partially or intermittently covered with water

140. The tense is present-continuous and must be retained in the passive voice.

141. Option A misses the pronoun 'it'. Option B misses the detail 'about' and distorts what is mentioned in the actual statement. Option C mentions 'you', in the passive voice 'by you/him/her' is understood and need not be mentioned.

142. Q expands the idea of polymers which is introduced in the statement.
R talks about the variants of polymers. S talks about 'those polymers' having replaced the chromium-plated ones, which is introduced in R.
P gives a fitting conclusion to the paragraph.

143. Frugal means characterized by or reflecting economy in the use of resources.

144. A harbinger is the one that pioneers in or initiates a major change.

145. Exodus—a mass departure
Influx—a coming in

146. Equanimity—evenness of mind especially under stress
Excitement—something that rouses

147. Dragon—a violent, combative, or very strict person- is the odd man out.
The other words are related to 'issuing commands' or 'giving orders'.

148. Shell is "built around" a particular component like a kernal.
In the same way a Pupa is the outer layer of a growing Caterpillar and the Cocoon is the outer layer of the larva.
Syncope is an "umbrella term" and technical word for faint and lassitude means a state or feeling of weariness, diminished energy, or listlessness.

149. The tense required here is the present perfect continuous.

150. The latter part of the sentence must continue with the word that juxtaposes 'initial', as the word 'although' in the beginning suggests a turn of events in the latter part of the sentence.
Also, the initial part of the sentence states that 'the investigations pointed towards him', the only option that juxtaposes this is option D.

Previous Paper (Solved)

Symbiosis National Aptitude Test (SNAP), 2010

SECTION- I

GENERAL KNOWLEDGE

For all questions in this section, correct answers carry 1 mark each.

1. Name the jailed Chinese pro-democracy activist who won this year's Nobel Peace Prize for his 'long and non-violent struggle for fundamental human rights in China'.

A. Went Xiabo B. Wen Jiabao
C. Liu Jiabao D. Liu Xiaobo

2. Which government is behind the 'The Nalanda Proposal'; proposing Nalanda as an ideal site for establishing a 21st century learning institution.

A. Singapore B. India
C. U.K. D. China

3. Jimmy Wales and Larry Sanger are known for founding

A. Facebook B. Orkut
C. Wikipedia D. Google

4. The year 2010 is represented in Roman Numerals as

A. LLX B. MMX
C. IIXX D. CCXX

5. In which country is the seat of the United Nations International Court of Justice?

A. France B. Norway
C. Britain D. Netherlands

6. In 1965 Gordon Moore, Co-founder of Intel, made a prediction about the future of computer processing. What does his prediction, known as Moore's Law say?

A. As technology continues to advance, computer chips will become obsolete.
B. Computer-processing power will double every 18 months to two years.
C. There will eventually be no need for transistors in high-tech electronics.
D. As the number of transistors increases, computer-processing power will be reduced by half in every two years.

7. 'At 60 miles an hour the loudest noise in this new Rolls-Royce comes from the electric clock.' Who wrote this famous advertising headline?

A. David Ogilvy B. Walter Thompson
C. Leo Burnett D. Salman Rushdie

8. His life's motto was 'simple living and high thinking'. He was one of the greatest intellectual and activists of the 19th century and one of the pillars of the Bengal Renaissance. He was a polymath, Sanskrit pundit, educator, social reformer, writer and philanthropist. Due to his relentless efforts, on 26th July 1856, widow re-marriage was legalized by the then Government of India. Who is this towering personality?

A. Ishwar Chandra Vidyasagar
B. Raja Ram Mohan Roy
C. Rabindranath Tagore
D. Bankim Chandra Chatterjee

9. Eustace Fernandes, who passed away in early 2010 was the much admired creator of

A. Tom and Jerry B. Amul Girl
C. Air India Maharaja D. Snow White

10. Find the mismatch

A. Somdev Devvaraman : Badminton
B. Gagan Narang : Shooting
C. Arjun Atwal : Golf
D. Anita Sood : Swimming

11. Who designed the new rupee symbol?

A. Dilip Chhabria B. D. Udaya Kumar
C. Tarun Tahiliani D. S. Arun Kumar

12. What exactly is cloud computing?

A. A way to organize desktop computers
B. Lightweight software that takes up little space on a hard drive
C. Computing resources that can be accessed on demand, like electricity from a utility
D. The World Wide Web

13. What is NDM-1?

A. National Defence Missile 1, developed by Defence Research and Development Organisation (DRDO) as a part of the comprehensive missile shield for India.
B. A bacterial gene called New Delhi Mettallo-lactamase-1, dubbed the 'superbug' because of it being resistant to most antibiotics.
C. New Directions in Management 1—the first among a series of international conferences on Management, to be inaugurated by Bill Gates in Mumbai in January 2012.
D. A vision document on Disaster Management, released by Planning Commission.

14. 'Niyamgiri Hills' was in the news because of
A. Vedanta's failed mining proposal in the area inhabited by Dongria Knodh tribals.
B. The helicopter crash and death of Y.S. Rajasekhara Reddy.
C. Headquarters of the Naxalite Red corridor.
D. Conde Nast Traveller magazine selected it as the best trekking holiday spot in the world.

15. Match the following husband-and-wife team with the awards they have received for exemplary work:

1. Bill and Melinda Gates	(*i*) Right to Livelihood Awards
2. Sankaralingam and Krishnammal	(*ii*) UN Population Award
3. Prakash and Mandakini Amte	(*iii*) Kyoto Prize
4. Peter and Rosemary Grant	(*iv*) Magsaysay Award

A. 1–(*ii*), 2–i, 3–(*iii*), 4–(*iv*)
B. 1–(*ii*), 2–(*i*), 3–(*iv*), 4–(*iii*)
C. 1–(*iv*), 2–(*iii*), 3–(*ii*), 4–(*i*)
D. 1–(*ii*), 2–(*iv*), 3–(*iii*), 4–(*i*)

16. What is Renminbi?
A. It is the official currency of the People's Republic of China (PRC), whose principal unit is the Yuan.
B. Low cost car being developed by Volkswagen in China, at prices lower than Nano.
C. New aircraft Company floated by Brazil to challenge Boeing and Airbus.
D. New currency mooted for the entire ASEAN region like Euro for Europe.

17. Embraer is one of the world leaders in the manufacturing of corporate/business jets. Embraer belongs to which country?
A. Germany B. Japan
C. Brazil D. France

18. The Unique Identification (UID) Project, headed by Nandan Nilekani has been renamed as
A. Adhaar B. Sambhav
C. Sambhandh D. Alekh

19. India's first Special Economic Zone dedicated to the Aerospace Industry has been launched at
A. Hyderabad B. Hallargi
C. Shimla D. Ahmedabad

20. The prime purpose of WTO is to promote:
A. Financial Support B. Global Peace
C. Unilateral Trade D. Multilateral Trade

21. On March 5, 2010, which of the following personalities from India is among 19 members chosen by UN chief Ban Ki-moon for a high-level advisory group on Climate Change Financing tasked with mobilizing funds pledged during the Copenhagen meet to tackle global warming?
A. Shyam Saran
B. Montek Singh Ahluwalia
C. Chandrashekhar Dasgupta
D. Pradipto Ghosh

22. Prime Minister Manmohan Singh has termed the 11th Five Year Plan as:
A. India's health plan
B. India's poverty eradication plan
C. India's rural prosperity plan
D. India's education plan

23. Which of the following milestones was achieved by New Zealand Cricket Team Captain Daniel Vettori recently?
A. 2,000 runs and 200 wickets in Test Cricket.
B. 3,000 runs and 300 wickets in One-day International (ODIs)
C. 3,000 runs and 300 wickets in Test Cricket.
D. 2,000 runs and 200 wickets in One-day International (ODIs)

24. What is the full form of the term 'NPA' as used in banking environment?
A. Not Profitable Assets
B. New Potential Accounts
C. Non Performing Assets
D. Net Performing Assets

25. Which of the following contributes to the highest share of revenue earned by the Government of India?
A. Income Tax B. Excise Duty
C. Value Added Tax D. Corporate Tax

26. Which country was world's largest exporter with merchandise export worth \$1.47 trillion in 2008, according to the World Trade Organisation?
A. USA B. China
C. Germany D. Russia

27. Where in India, recently has the Clinton Foundation, founded by former US President Bill Clinton, firmed up its plans to set up world's largest solar park (3,000 to 5,000 MW capacity)?
A. Bihar B. Orissa
C. Gujarat D. Rajasthan

28. Which three public sector lenders have entered recently into a joint venture agreement for setting up a banking subsidiary, India BIA Bank (Malaysia) Bhd, in Malaysia?
A. Bank of India, Indian Overseas Bank and Andhra Bank
B. Bank of Baroda, Indian Overseas Bank and Bank of India
C. Bank of Baroda, Indian Overseas Bank and Andhra Bank
D. Bank of Baroda, Indian Bank and Andhra Bank

29. What is the compaign of Union and State Governments against the Naxalite movement called?
A. Operation Red Alert
B. Operation Green Hunt
C. Operation Cobra Den
D. Operation Clean Corridor

30. The instrument used to measure the speed of the wind is
A. Altimeter B. Anemometer
C. Chronometer D. Dosimeter

31. Which of the following countries is the first in the world to propose a carbon tax for its people to address global warming?
A. Finland B. Japan
C. Germany D. Australia

32. Amino acids are found in
A. Carbohydrates B. Fats
C. Proteins D. Vitamins

33. Which among the following is the world's largest milk producing country?
A. India B. China
C. The US D. Germany

34. A group of words that share the same spelling and the same pronunciation but have different meaning, e.g. left (opposite of right) and left (past tense of leave)
A. Synonyms B. Homonyms
C. Heteronyms D. Acronyms

35. The National Flag of India was designed by
A. Mahatma Gandhi
B. Jawarharlal Nehru
C. Rabindra Nath Tagore
D. Pingali Venkayya

36. In 1679, Denis Papin, a French physicist, who assisted Robert Boyle, used the latter's scientific discoveries and invented what is today one of the most commonly found kitchen equipment. His invention earned him a membership of the Royal Society of England. What was the invention?
A. Knife B. Fork
C. Pressure Cooker D. Stove

37. Mr. Ratan Tata refused to job with one of the following companies to join Tata Steel in 1962.
A. IBM B. HUL
C. Seimens D. SKF International

38. Which of the following have owned Land Rover branch before Tata Motors?
A. BMW B. British Leyland
C. British Aerospace D. All of the above

39. Who is known as 'The Man Who Broke the Bank of England' after he made a reported $1 billion during the 1992 Black Wednesday UK currency crisis?
A. George Soros B. George W Bush
C. Paul Volcker D. Ben Bernake

40. Who is the current dean of Harvard Business School?
A. Nitin Nohria B. Deepak Jain
C. Kim Clark D. None of the above

SECTION-II

GENERAL ENGLISH

For all questions in this section, correct answers carry 1 mark each.

Directions (Q. Nos. 41-44): *Read the passage and then determine the best choice for an answer from the questions given below. Base your choice on what this passage states directly or implies, not on any information you may have got from elsewhere.*

"The emancipation of women", James Joyce told one of his friends, "has caused the greatest revolution in our time."

Other modernists agree: Virginia Woolf, claiming that in about 1910 "human character changed" and illustrating the new balance between the sexes, urged, "Read the 'Agamemon' and see whether your sympathies are not almost entirely with Clytemnestra". D.H. Lawrence wrote "perhaps the deepest fight for 200 years and more has been the fight for women's independence".

But if modernist writers considered women's revolt against men's domination as one of their "greatest" and "deepest" themes, only recently, perhaps in the past 15 years has literary criticism begun to catch up with it. Not that the images of sexual antagonism that abound in modern literature have gone unremarked—far from it. We are able to see in literary works the perspective we bring to them and now that women are enough to make a difference in reforming canons and interpreting literature, the landscapes of literary history and the features of individual books have begun to change.

41. According to the passage, modernists are changing literary criticism by:-
A. Noting instances of hostility between men and women
B. Seeing literature from fresh points of view

C. Studying the works of early twentieth-century writers
D. Reviewing books written by feminists

42. The author quotes James Joyce, Virginia Woolf and D.H. Lawrence primarily in order to show that:-
A. These were feminist writers
B. Although well-meaning, they were ineffectual
C. Before the twentieth century, there was little interest in women's literature
D. None of the above

43. The author's attitude towards women's reformation of literary canons can best be described as one of.-
A. Ambivalence B. Antagonism
C. Indifference D. Endorsement

44. Which of the following titles best describes the contents of the passage?
A. Modernist Writers and the Search for Equality
B. The meaning of Literature, from 1910 onwards
C. Transforming Literature
D. None of the options

45. Choose the correct sentence
A. A anthropologist by profession, he is also a trained classical singer.
B. The anthropologist by profession, he is also a trained classical singer.
C. As anthropologist by profession, he is also a trained classical singer.
D. An anthropologist by profession, he is also a trained classical singer.

Directions (Q. Nos. 46-49): *Fill in the blanks and choose the correct definition for the punctuation.*

46. __________ is used to indicate possession.
A. Hyphen B. Apostrophe
C. Semi-colon D. Period

47. __________is used to mark the end of declarative and imperative sentences.
A. Semi-colon B. Comma
C. Dash D. Period

48. When a subordinate clause is followed by the main clause, _______ is required.
A. Dash B. Semi-colon
C. Comma D. Colon

49. When no connecting word is used to connect two independent clauses, one should use______.
A. Comma B. Semi-colon
C. Period D. Colon

Directions (Q. Nos. 50-51): *Which is the correct proverb?*

50. A. Sleeping dogs tell lies.
B. Dogs sleeping lie till late.
C. Lie sleeping dogs till the dawn comes.
D. Let the sleeping dogs lie.

51. A. A fool is always parted from his money.
B. A fool and his money are parted easily.
C. Money and the fool must part ways.
D. You can always part a fool from his money.

Directions (Q. Nos. 52-57): *Choose the correct meaning of the following idioms.*

52. If someone said, "You are the bomb!" she or he probably would be telling you:
A. You have a bad temper.
B. You are a war weapon.
C. You are exceptional and/or wonderful.
D. You are dangerous.

53. When someone is described as being "flighty", the person described is probably:
A. Light.
B. Indecisive and irresponsible.
C. Someone who loves flying.
D. Someone who flies kites.

54. What does "to take down the enemy" mean?
A. To take the enemy's pictures off the wall.
B. To kill the enemy.
C. To make friends with the enemy.
D. To ignore the enemy.

55. What does, "Dime a dozen" mean?
A. For one dime you get a dozen
B. All dozens cost a dime
C. Anything that is common and easy to get
D. It is difficult to get people

56. "Throw the baby out with the bath water" means,
A. Clean out everything
B. Throw out the good things with the unwanted
C. Being thorough
D. Create the impression of an accident

57. "Bark up the wrong tree" means,
A. Skin of another animal
B. Behave like a dog
C. Purposely make an error
D. Make the wrong choice

58. I ______ my bike yesterday, so my legs are sore.
A. road B. rode
C. rhode D. ride

59. Insulation was fitted to ______ further heat loss from the building.
A. guard B. protect
C. save D. prevent

60. A _______ rate of inflation makes exports difficult.
A. great B. high
C. large D. tall

61. My boat has two _____.
A. sales B. sails
C. sailes D. sells

62. Can you give me _____ details, please?
A. faster B. further
C. farther D. furthur

63. A baby deer is called a _____.
A. Foal B. Fawn
C. Calf D. Joe

Directions (Q. Nos. 64-65): *Complete the sentence by choosing the correct words from the given alternatives.*

64. The greatest _________ of my generation is that a human being can alter his life by _________ his attitude.
A. gift ... gifting
B. discovery ... altering
C. misgiving ... elevating
D. thing ... flaunting

65. When it comes to staying ____ , a mind-lift beats a _________ any day.
A. young ... face-lift B. at home ... egg
C. light ... elevator D. away ... sleep

66. None are so _______ as those who are full of themselves
A. empty B. important
C. vital D. indispensible

67. Your most unhappy customers are your greatest source of ___________
A. earning B. irritation
C. worry D. learning

68. Choose the correct **antonym** for the word below from the options provided 'Eulogize'
A. Extol B. Criticize
C. Emulate D. Amulet

69. Correct synonym for **Pedantic** is,
A. Referring to small children
B. Teaching Methodology
C. Finicky
D. Angry

70. **Pyrophobia** means
A. Fear of Pythons B. Fear of funeral pyres
C. Fear of fever D. Fear of fire

Directions (Q. Nos. 71-72): *Choose the kangaroo word that carries a smaller version of the word with a very similar meaning.*

71. A. Masculine B. Woman
C. Man D. Child

72. A. Sleep B. Respite
C. Walk D. Talk

Directions (Q. No. 73): *Choose the word that completes the first and begins the second word.*

73. Paper __________ and __________ lifter
A. cone B. weight
C. light D. fly

Directions (Q. Nos. 74-75): *Choose the word that cannot be coupled with the given word to form a new word.*

74. Out
A. Shine B. Number
C. Bug D. Run

75. News
A. Letter B. Week
C. Stand D. Paper

76. I did not see you _____ the office party.
A. in B. for
C. at D. on

Directions (Q. Nos. 77-78): *Fill in the blanks with the correct simile.*

77. As cool as
A. a Cucumber B. the winter night
C. an ice cream D. a rock star

78. As fresh as
A. a daisy B. a rose
C. milk D. dew

79. Choose the option that does not belong with the rest:
A. Consort B. Spouse
C. Partner D. Clear

80. Nerd means
A. Genius B. Uninteresting person
C. Worm D. Arthropod

SECTION-III

QUANTITATIVE & DATA INTERPRETATION AND SUFFICIENCY

For all questions in this section, correct answers carry 1 mark each.

81. In a Retail outlet the average revenue was ₹ 10,000 per day over a 30 day period. During this period the average daily revenue on weekends (total 8 days) was ₹ 20,000 per day. What was the average daily revenue on weekdays?
A. 6364 B. 5250
C. 6570 D. 8060

82. Two different prime numbers X and Y, both are greater than 2, then which of the following must be true?
A. X – Y = 23 B. X + Y ≠ 87
C. Both A and B D. None of the above

83. It takes 6 hours for pump A, used alone, to fill a tank of water. Pump B used alone takes 8 hours to fill the same tank. A, B and another pump C all together fill the tank in 2 hours. How long would pump C take, used alone, to fill the tank?

A. 4.8 B. 6
C. 5.6 D. 3

84. A swimming pool can be filled by pipe A in 3 hours and by pipe B in 6 hours, each pump working on its own. At 9 a.m. pump A is started. At what time will the swimming pool be filled if pump B is started at 10 a.m.?

A. 11.20 a.m. B. 11.05 a.m.
C. 11.10 a.m. D. 10.50 a.m.

85. The sum of prime numbers that are greater than 60, but less than 70 is:

A. 128 B. 191
C. 197 D. 260

86. Find out the appropriate next number in the series from the options below 0, 2, 6, 12, 20, 30, 42, ?

A. 56 B. 62
C. 49 D. 5

87. A bakery opened with its daily supply of 40 dozen rolls. Half of the rolls were sold by noon, and 60% of the remaining rolls were sold between noon and closing time. How many dozen rolls were left unsold?

A. 6 B. 8
C. 10 D. 12

88. Stuart, Jack and Leo are colleagues working in a plant. Stuart and Jack can do a work in 10 days, Jack and Leo can do the same work in 15 days while Stuart and Leo can do it in 12 days. All of them started the work together. After two days, Leo was shifted to some other work. How many days will Stuart and Jack take to finish the rest of the work?

A. 9 B. 12
C. 8 D. 7.5

89. The missing numbers in the below series would be 1 : 1, 8 : 4, 9 : 27, 64 : 16, 25 : 125, ? : ?, 49 : 343,

A. 36 : 316 B. 216 : 36
C. 316 : 16 D. 32 : 316

90. The difference between the value of a number increased by 25% and the value of the original number decreased by 30% is 22. What is the original number?

A. 70 B. 65
C. 40 D. 90

91. Running at the same constant rate, 6 identical machines can produce a total of 180 bottles per hour. How many bottles could 15 such machines produce in 30 minutes?

A. 225 B. 300
C. 250 D. 350

92. A number whose fifth part increased by 4 is equal to its fourth part diminished by 10, is :

A. 240 B. 260
C. 270 D. 280

93. Which one of the following numbers will completely divide (461 + 462 + 463 + 464)?

A. 3 B. 10
C. 11 D. 13

Directions (Q. Nos. 94–100): *Each of the following questions is followed by two statements.*

Mark (a) if statement I alone is sufficient to answer the question.
(b) if statement II alone is sufficient to answer the question.
(c) if both statement I and II together are necessary to answer the question.
(d) if both statements I and II together are not sufficient to answer the question.

94. Is 'b' positive?

(I) a + b is positive
(II) a – b is positive

95. In a general body election, 3 candidates' p, q and r were contesting for a membership of the board. How many votes did each receive?

(I) p received 17 votes more than q and 103 votes more than r.
(II) Total votes cast were 1703

96. If C1 and C2 are the circumferences of the outer and inner circles respectively, What is C1 : C2?

(I) The two circles are concentric
(II) The area of the ring is 2/3 the area of greater circle

97. What is the middle number of 7 consecutive whole numbers?

(I) Product of number is 702800
(II) Sum of the numbers is 105

98. Total marks obtained by P, Q, R and S in Mathematics is 360. How many marks did P secure in Mathematics?

(I) P secured one-third marks of the total of Q, R and S
(II) Average marks obtained by Q and R are 20 more than that secured by S.

99. How many ice cubes can be accommodated in a container?

(I) The length and breadth of the container is 20 cm and 15 cm respectively
(II) The edge of the ice cube is 2 cm

100. Ram got ₹ 1500 as dividend from a company. What is the rate of interest given by the company?

(I) The dividend paid last year was 10%
(II) Ram has 350 shares of ₹ 10 denomination

Directions (Q. Nos. 101-104): *Refer to the following bar diagram showing the sales (in ₹Crores) of top market brands among pain killers in India and answer the questions*

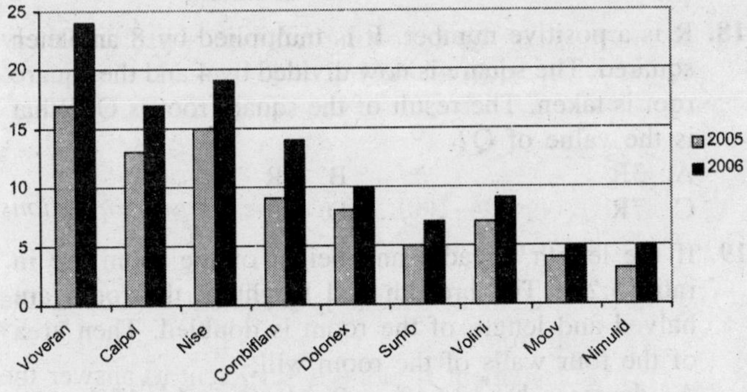

101. What is the difference between the sales of Voveran in 2006 and those of Calpol in 2005 (in Rs lacs)?

A. 1000 B. 50
C. 100 D. 500

102. Percentage of increase in sales from 2005 to 2006 is the highest for which brand of a pain killer?

A. Voveran B. Volini
C. Dolonex D. Sumo

103. Percentage increase in sales from 2005 to 2006 is the lowest for which brand of a pain killer?

A. Voveran B. Volini
C. Moov D. Nise

104. What is the approximate percentage of increase in the sales of Voveran from 2005 to 2006?

A. 35% B. 40%
C. 45% D. 50%

105. Which of the designs best completes the following sequence?

A. 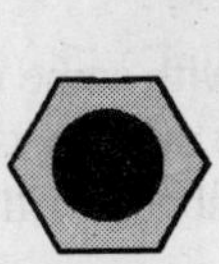 B. 

C. 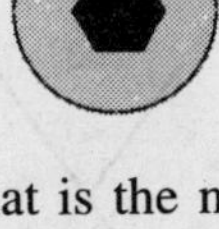 D. 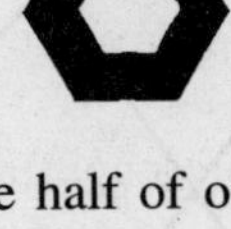

106. What is the number that is one half of one quarter of one tenth of 400?

A. 2 B. 5
C. 8 D. 10

107. Consider a square ABCD with midpoints E, F, G and H of sides AB, BC, CD and DA. Let L denote the line passing through F and H. Consider points P and Q on the line L inside the square such that the angles APD and BQC are both equal 120°. What is the ratio ABCQPD to the remaining area of ABCD?

A. $4+\frac{\sqrt{2}}{3}$ B. $2\sqrt{3}-1$

C. $2+\sqrt{3}$ D. $\frac{10-3\sqrt{3}}{9}$

108. The Price of Darjeeling Tea (in rupees Kilogram) is $100 + 0.1\ n$, on the nth day of a non-leap year (n = 1, 2, 3 ..., 100) and then remains constant. On the other hand the price of Ooty tea (in rupees per kilogram) is $85 + 0.15n$, on the nth day (n = 1, 2, ..., 365). On which date of that year will the prices of these two varieties of the tea be equal?

A. 27th October B. 16th June
C. 15th June D. 28th October

Directions: *Each of questions 109–112 consist of question followed by two statements numbered I and II*

Answer (A) if data in Statement I alone is sufficient to answer the question but the data in Statement II alone is not sufficient to answer the question. Answer (B) if data in Statement II alone is sufficient to answer the question but the data in Statement I alone is not sufficient to answer the question. Answer (C) if data in Statement I and II together are necessary to answer the question. Answer (D) if data in Statement I and II together are not sufficient to answer the question.

109. ΔABC and ΔPQR are congruent

(I) Area of ΔABC and ΔPQR are same
(II) ΔABC and ΔPQR are right angle Triangles

110. Salary of A and B is in ratio 3:4 and expenditure is in ratio 4:5. What is the ratio of their saving?

(I) B's saving is 25% of his salary
(II) B's salary is ₹ 2500

111. What is the average height of the class?

(I) Average height of the class decreases by 1 cm if we exclude the tallest person of the class whose height is 56 cm
(II) Average height of the class increases by 1 cm if we exclude the shortest person of the class whose height is 42 cm

112. Ram is taller than Shyam and Jay is shorter than Vikram. Who is the shortest among them?

(I) Ram is the tallest.
(II) Shyam is taller than Vikram.

113. In September 2009, the sales of a product were 2/3rd of that in July 2009. In November 2009, the sales of the product were higher by 5% as compared to September 2009. How much is the percentage of increase in sales in November 2009 with respect to the base figure in July 2009?

A. + 40% B. – 20%
C. – 30% D. + 25%

114. For what range of values of 'x', will be the inequality

$$15x - \frac{2}{x} > 1 \, ?$$

A. $x > 0.4$

B. $x < \frac{1}{3}$

C. $-\frac{1}{3} < x < 0.4,\ x > \frac{15}{2}$

D. $-\frac{1}{3} < x < 0,\ x > \frac{2}{5}$

115. How many litres of a 30% alcohol solution should be added to 40 litres of a 60% alcohol solution to prepare a 50% solution?

A. 30 B. 20
C. 24 D. 32

116. 66 cubic centimetres of silver is drawn into a wire of 1 mm diametre. The length of the wire in metres will be :

A. 84 B. 90
C. 168 D. 336

117. A train 108 m long moving at a speed of 50 km/hr crosses a train 112 m long coming from opposite direction in 6 seconds. The speed of the second train is,

A. 48 km/hr B. 54 km/hr
C. 66 km/hr D. 82 km/hr

118. R is a positive number. It is multiplied by 8 and then squared. The square is now divided by 4 and the square root is taken. The result of the square root is Q. What is the value of Q?

A. 3R B. 4R
C. 7R D. 9R

119. If the length, breadth and height of the room are in ratio 3:2:1. The breadth and height of the room are halved and length of the room is doubled. Then area of the four walls of the room will,

A. decrease by 13.64% B. decrease by 15%
C. decrease by 18.75% D. decrease by 30%

120. A survey was conducted of 100 people whether they have read recent issues of 'Golmal', a monthly magazine. Summarized Information is presented below:

Only September: 18
September but not August: 23
September and July: 8
September: 28
July: 48
July and August: 10
None of the three months: 24

What is the number of surveyed people who have read exactly for two consecutive months?

A. 7 B. 9
C. 12 D. 14

SECTION-IV

ANALYTICAL & LOGICAL REASONING

For all questions in this section, correct answers carry 2 marks each.

121. Four children A, B, C & D are having some chocolates each.

A gives B as many as he already has, he gives C twice of what C already has and he gives D thrice of what D already has.

Now, D gives 1/8th of his own chocolates to B.

Then A gives 10% chocolates he now owns to C and 20% to B.

Finally, all of them have 35 chocolates each.

What is the original number of chocolates each had in the beginning?

A. A–110, B–10, C–10, D–10
B. A–90, B–20, C–20, D–10
C. A–70, B–25, C–25, D–20
D. A–125, B–5, C–5, D–5

122. There are two similar figures below with some numbers. The left one is complete whereas one number is missing in the right one. Find a suitable number to fill in place of the question mark.

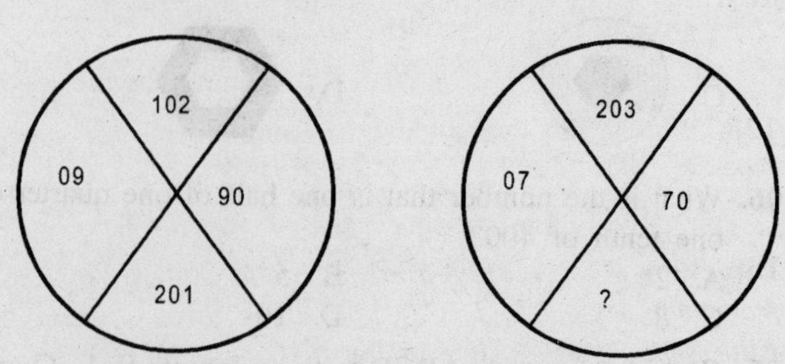

A. 280 B. 303
C. 362 D. 382

123. Complete the following series by replacing the?: (TBLD, VEPI, XHTN, ?)

A. ZJVP B. ZVJP
C. ZKXS D. ZKXP

124. In a cricket team, three batsmen Ricky, Sachin and Brian are the top three run-scorers in any order. Each of them gives two replies to any question, one of which is true and the other is false, again, in any order. When asked about who the top scorer was, following were the replies they gave:

Sachin: I got the top score. Ricky was second.
Brian: I got the top score. Sachin was second.
Ricky: I got the top score. Sachin was third.

Which of the following is the correct order of batsmen who got the top score, second best and third best score respectively?

A. Brian, Ricky, Sachin
B. Brian, Sachin, Ricky
C. Ricky, Sachin, Brian
D. Sachin, Brian, Ricky

125. 60 employees in an office were asked about their preference for tea and coffee. It was observed that for every 3 people who prefer tea, there are 2 who prefer coffee. For every 6 people who prefer tea, there are 2 who drink both of tea and coffee. The number of people who drink both is the same as those who drink neither.

How many people drink both tea and coffee?

A. 10 B. 12
C. 14 D. 16

126. A clock strikes once at 1 o'clock, twice at 2 o'clock and so on. If it takes 6 seconds to strike at 3 o'clock, how much time will it take to strike at 9 o'clock?

A. 24 seconds B. 18 seconds
C. 20 seconds D. None of these

Direction (Q. Nos. 127–128): *E-1, E-2 and E-3 are three engineering students writing their assignments at night. Each of them starts at a different time and completes at a different time. The digit in their name and the order of their starting and completing the assignment is certainly not the same. The last student to start is the first to complete the assignment.*

127. Who is the first student to start writing the assignment?

A. E-1 B. E-2
C. E-3 D. Cannot be decided

128. Who is the last student to complete the assignment?

A. E-1 B. E-2
C. E-3 D. Cannot be decided

Directions (Q. Nos. 129–130): *A, B and C are three students from Don School and P, Q and R are three students from Elite School. Q is brighter than R but duller than the Don School student who is brighter than A. The same Don School student is duller than P but is brighter than C.*

129. Who is brightest amongst all?

A. B B. P
C. R D. Cannot be decided

130. Who is the dullest amongst the three students from Elite School?

A. P B. Q
C. R D. Cannot be decided

131. When Rafael entered the class, there were already 10 students in the class. 5 students entered the class between Roger and Rafael. Total 10 students entered after Roger. Exactly how many students are in the class finally?

A. 15 B. 25
C. 27 D. Cannot be decided

Directions (Q. Nos. 132-134): *Arijit, Biplab, Chintan, Debashish, Elangovan, Frederick, Gautam and Himadri are sitting around a circular table. Some information about the order in which they are sitting is available as follows:*

(1) Debashish is sitting opposite to Himadri and to the immediate right of Gautam.
(2) Elangovan is sitting to the immediate right of Biplab.
(3) Arijit is sitting opposite Chintan who is not immediately next to Frederick on either side.

132. Who is sitting to the immediate right of Himadri?

A. Arijit B. Debashish
C. Elangovan D. Frederick

133. Who is sitting opposite Biplab?

A. Arijit B. Debashish
C. Frederick D. Himadri

134. Who is to the immediate right of Chintan?

A. Arijit B. Biplab
C. Elangovan D. Himadri

135. Select the alternative that logically follows the two given statements:

Some rocks are not tables.
Some rocks are balloons.

A. Some tables are not balloons
B. Some tables are balloons
C. Some balloons are not tables
D. None of the above

Directions (Q. Nos. 136-137): *A, B, C, D and E sit on a long bench. C does not sit next to A or E. A and E have three persons sitting between them.*

136. Who is sitting in the middle of the bench?

A. B B. C
C. D D. None of these

137. Who are sitting at the extreme ends of the bench?

A. A & E B. B & D
C. C & E D. None of these

Directions (Q. Nos. 138–139): *Observe the chart below and answer the following questions.*

1 Yearly Commission Earned by Five Salesmen.

Figures in Rupees

Year / Salesman	1990	1991	1992	1993	1994	1995
A	27350	28500	25200	29800	24600	27000
B	26850	27900	27400	28000	28500	29000
C	26200	27900	28200	29100	29400	30000
D	27850	30040	29800	30060	29800	32000
E	28640	29000	28750	30000	29750	29700

138. In the year 1994, the commission earned by salesman D was approximately what per cent more of the commission earned by A?

A. 18 B. 82.5
C. 21 D. 17

139. In the year 1993, the commission of B was approximately what per cent of the total commission earned by five salesmen that year?

A. 30 B. 20
C. 40 D. 80

140. Find the Missing Number in the following set

2	4	6	8	10
2	14	34	??	98

A. 30 B. 62
C. 42 D. 78

141. There are 6 volumes of books on a rack kept in order (such as, Vol. 1, Vol. 2 and so on).

After some readers used them, their order got disturbed. The changes showed as follows:
Vol. 5 was directly to the right of Vol. 2
Vol. 4 has Vol. 6 to its left and both were not at Vol. 3's place
Vol. 1 has Vol. 3 on right and Vol. 5 on left
An even numbered volume is at Vol. 5's place

Find the order in which the books are kept now, from the 4 given alternatives.

A. 6, 3, 5, 1, 4, 2 B. 4, 6, 3, 5, 1, 2
C. 3, 4, 1, 6, 5, 3 D. 2, 5, 1, 3, 6, 4

142. All German philosophers, excepts for Marx, are idealist.

From which of the following can the statement above be most properly inferred?

A. Except for Marx, if someone is an idealist philosopher, then he or she is German.
B. Marx is the only non-German philosopher who is an idealist.
C. If a German is an idealist, then he or she is a philosopher, as long as he or she is not Marx.
D. Marx is not an idealist German philosopher.

143. Ramaswami was studying for his examinations and the lights went off. It was around 1:00 a.m.: He lighted two uniform candles of equal length but one thicker than the other. The thick candle is supposed to last six hours and the thin one two hours less. When he finally went to sleep, the thick candle was twice as long as the thin one.

For how long did Ramaswami study in candle light?

A. 2 hours B. 3 hours
C. 2 hours 45 minutes D. 4 hours

144. The numerator and denominator of a fraction is in the ratio 2:3. If 6 are subtracted from the numerator the value of the fraction becomes 2/3 of the original fraction. The numerator of the original fraction is,

A. 16 B. 21
C. 18 D. 30

145. A person wanted to withdraw X rupees and Y paise from the bank. But cashier made a mistake and gave him Y rupees and X paise. Neither the person nor the cashier noticed that.

After spending 20 paise, the person counts the money. To his surprise, he has double the amount he wanted to withdraw.

Find X and Y. (1 Rupee = 100 Paise)

A. X = 3, Y = 6 B. X = 26, Y = 53
C. X = 15, Y = 30 D. X = 9, Y = 36

146. A drawer contains 10 black and 10 brown socks which are all mixed up. What is the fewest number of socks you can take from the drawer without looking and be sure to get a pair of the same colour?

A. 7 pairs B. 7 pieces only
C. 10 pieces only D. 3 pieces only

147. A placement company has to assign 1000 SW personnel who are skilled in Java and Dot Net to a prospective outsourcing company. He finds that 750 are have Dot Net skills and 450 have Java skills. Some have skills in both Java and Dot Net. Find the numbers who have skills in both Java and Dot Net.

A. 250 B. 200
C. 350 D. 100

148. All good athletes who want to win are disciplined and have a well balanced diet. Therefore athletes who do not have well balanced diets are bad athletes.

Based on the sentence above which of the statement below strongly supports the view:

A. No bad athlete wants to win
B. No athlete who does not eat a well balanced diet is good athlete
C. Every athlete who eats a well balanced diet is good athlete
D. All athletes who want to win are good athletes.

149. The numbers in these series are arranged in a triangle which has a logic as shown below. Find the missing numbers shown as (?) from the choices given below:

```
              2
           2     2
        2     4     2
     2     8     ?     2
  2    16    64    16    2
2   32   1024   ?    ?    2
```

A. {16, 32, 64} B. {8, 1024, 32}
C. {24, 1024, 64} D. {24, 32, 64}

150. If for a particular value of the variable x, the following holds good, $17 = \frac{17x}{1-x}$, then compute the value of $(2x)^*x$.

A. 17 B. 1
C. 2 D. $\frac{1}{2}$

ANSWERS

1	**2**	**3**	**4**	**5**	**6**	**7**	**8**	**9**	**10**
D	A	C	B	D	B	A	A	B	A
11	**12**	**13**	**14**	**15**	**16**	**17**	**18**	**19**	**20**
B	C	B	A	B	A	C	A	B	D
21	**22**	**23**	**24**	**25**	**26**	**27**	**28**	**29**	**30**
B	D	C	C	D	C	C	C	B	B
31	**32**	**33**	**34**	**35**	**36**	**37**	**38**	**39**	**40**
A	C	A	B	D	C	A	D	A	A
41	**42**	**43**	**44**	**45**	**46**	**47**	**48**	**49**	**50**
B	D	D	D	D	B	D	C	B	D
51	**52**	**53**	**54**	**55**	**56**	**57**	**58**	**59**	**60**
B	C	B	B	C	B	B	B	D	B
61	**62**	**63**	**64**	**65**	**66**	**67**	**68**	**69**	**70**
B	B	B	B	A	A	D	B	B	D
71	**72**	**73**	**74**	**75**	**76**	**77**	**78**	**79**	**80**
A	B	B	C	B	C	A	A	D	B
81	**82**	**83**	**84**	**85**	**86**	**87**	**88**	**89**	**90**
A	B	A	A	A	A	B	D	B	C
91	**92**	**93**	**94**	**95**	**96**	**97**	**98**	**99**	**100**
A	D	B	D	C*	B	B	A	D	B
101	**102**	**103**	**104**	**105**	**106**	**107**	**108**	**109**	**110**
A	C	D	B	A	B	B*	B*	D*	A
111	**112**	**113**	**114**	**115**	**116**	**117**	**118**	**119**	**120**
C	B	C	D	B	A	D	B	D	B
121	**122**	**123**	**124**	**125**	**126**	**127**	**128**	**129**	**130**
A	A	A	A	B	A	C	A	B	C
131	**132**	**133**	**134**	**135**	**136**	**137**	**138**	**139**	**140**
D	A	C	B	D	B	A	C	B	B
141	**142**	**143**	**144**	**145**	**146**	**147**	**148**	**149**	**150**
D	C*	B	C	B	D	B	B	B	D

SOME SELECTED EXPLANATORY ANSWERS

41. Refer to the last paragraph—"Not that the images .... we are able to see in literary works ..... change." Author says that modernist writers have left an impact on literature by works that have a newer perspective on the theme.

42. By quoting James Joyce and Virginia Woolf Author wants to emphasize on the fact that the fight for women's independence is not a new theme and that it has been going on for several decades now.

43. Author appreciates the change that is visible in the literature with reference to the discussed theme.

44. Author has discussed how modernist writers have impacted the issue of Women's Emancipation. None of the given options are correct.

45. "Anthropologist" starts with a Vowel sound and hence should use "An" before it. Hence the only correct answer choice is option D.

46. Apostrophe is used as the marking of possessive case (as in the cat's whiskers).

47. Period is used to mark the end of declarative and imperative sentences. (Hand in the poster essays no later than noon on Friday.)

48. Comma (I am, as you have probably noticed, very nervous about this.)

49. Semi-colon is used between closely related independent clauses not conjoined with a coordinating conjunction (I went to the basketball court; I was told it was closed for cleaning.)

50. "Let the sleeping dogs lie" means 'Do not instigate trouble'.

51. "A fool and his money are parted easily" means that foolish people lose wealth easily.

52. "You are a bomb" means exceptionally wonderful.

53. "Flighty" means Indecisive and irresponsible.

54. "Take down the enemy" means to kill/defeat the enemy.

55. "Dime a dozen" means very common.

56. "Throw the baby out with the bath water" means to discard the useful along with the useless.

57. "Bark up the wrong tree" means to make a wrong choice.

68. Eulogize means to praise highly. Opposite will be to Criticize.

69. Pedantic means ostentatious in one's learning.

70. Pyrophobia is the fear of fire.

71. Masculine has "man".

72. Respite has "rest".

73. Weight can be clubbed with both paper (Paper-weight) and lifter (weight-lifter).

74. "out" can be clubbed with all the words except "bug". 'Out-shine', 'out number' and 'out run'.

75. News can be clubbed with all the words except Week. 'News letter', 'news stand', 'news paper'.

81. Total revenue for 30 days = 10,000 × 30 = ₹ 3,00,000
Total revenue for weekends = 20,000 × 8 = ₹ 1,60,000
Total revenue for weekdays = 3,00,000 – 1,60,000 = ₹ 1,40,000

$\therefore$ Average daily revenue for weekdays = $\frac{140000}{22}$

= 6363.6 = ₹ 6364.

82. As both the prime numbers are greater than 2. So, both must be odd. Sum of any two odd numbers must be even. Hence, $x + y \neq 87$.

83. Let total units to be filled be 24.
Units filled by A in one hour = 4
Units filled by B in one hour = 3
Units filled by A, B and C in one hour = 12
So, units filled by C in one hour = 5

Hence, C can fill the complete tank in $\frac{24}{5}$ = 4.8 hours.

84. Let total units to be filled be 6
Units filled by A in one hour = 2
Units filled by B in one hour = 1
Units filled in the swimming pool upto 10:00 a.m. = 2
Units filled in the swimming pool upto 11:00 a.m. = 2 + 3 = 5
Now as 2 units are being filled in one hour.

So, 1 unit will be filled in $\frac{1}{3}$ hr *i.e.* 20 minutes.

Hence, swimming pool will be completely filled at 11 : 20 a.m.

85. Prime numbers between 60 and 70 are 61 and 67. So, sum = 61 + 67 = 128

86. It is an A.P. in second form.
The difference between the terms is in A.P.
So, the missing term is 42 + 14 = 56

87. Rolls at the time of opening = 40 dozens
Rolls sold by noon = 20 dozens
So, rolls remaining = 20 dozens
Rolls sold between noon and closing time = 12 dozens
So, rolls left unsold = 8 dozens.

88. Let total units of work to be done be 120.

Now, units done by Stuart and Jack in one day = 12
Units done by Jack and Leo in one day = 8
Units done by Leo and Stuart in one day = 10
So, units done by all three in 2 days = 30
Remaining units to be done = 90

So, time taken by Stuart and Jack = $\frac{90}{12}$ = 7.5 days.

90. Let original number be y

Now, increased number = $\frac{5}{4}y$

And decreased number = $\frac{7}{10}y$

According to questions:

$$\frac{5}{4}y - \frac{3}{10}y = 22$$

$$\Rightarrow \frac{25-14}{20}y = 22$$

$$\Rightarrow \frac{11}{20}y = 22 \Rightarrow y = 40.$$

91. Bottles produced by 6 machines in one hour = 180
Bottles produced by 6 machines in 30 minutes = 90
Bottles produced by 1 machine in 30 minutes = 15
Bottles produced by 15 machines in 30 minutes = 15 × 15 = 225

92. Let the number be x.
Now, according to condition

$$\frac{x}{5}+4 = \frac{x}{4}-10$$

$$\Rightarrow \frac{x+20}{5} = \frac{x-40}{4}$$

$$\Rightarrow 4x + 80 = 5x - 200$$

$$\Rightarrow x = 280.$$

93. $461 + 462 + 463 + 464 \equiv 1 + 2 + 3 + 4 \equiv 10 \equiv \frac{10}{10}$

∴ The sum will end in 0.
∴ The sum will be divisible by 10.

94. From (I) statement 'b' can be either negative or positive. Also from (II) statement 'b' can be either negative or positive. (I) and (II) cannot be combined.
So, the answer can not be found out.

95. From statement (I)
$p = q + 17$ and $p = r + 103$
From statement (II)
$p + q + r = 1703$
Combining statement (I) and statement (II)
$p + p - 17 + p - 103 = 1703$
$\Rightarrow 3p = 1583 \Rightarrow p = 527.66$
Which is not possible since number of votes cannot be fraction.
So, value of p, q and r can not be found out.

96. From statement (I) nothing can be said.
From statement (II)

$$\pi\left(r_1^2 - R_2^2\right) = \frac{2}{3}\pi R_1^2$$

So, ratio of $R_1 : R_2$ can be find out.
Hence, ratio of $C_1 : C_2$ can be found out.

97. From statement (I) we can not find out the answer.
From statement (II)
$a - 3 + a - 2 + a - 1 + a + a + 1 + a + 2 + a + 3 = 105$
$\Rightarrow 7a = 105$
$\Rightarrow a = 15$
Hence, middle number can be determined.

98. From statement (I)

$$P = \frac{1}{3}(Q+R+S)$$

and we know that $P + Q + R + S = 360$.
Hence, marks of P can be determined.
From statement (II) we can not say anything.

99. Even after combining both statement the volume of container cannot be determined. Hence, number of ice cubes that can be accommodated cannot be determined.

100. Statement (I) tells only about the previous years rate.
Statement (II)
Dividend from 350 shares = ₹ 1500

Dividend from 1 share = $\frac{150}{35} = \frac{30}{7} = 4.28$

So, rate of interest can be find out

as $4.28 = \frac{10 \times R \times T}{100}$

$\Rightarrow$ R = 42.8%.

101. Sales of Voveran in 2006 = ₹ 23,00,00,000
Sales of Calpol in 2005 = ₹ 13,00,00,000
So, difference = ₹ 10,00,00,000 = ₹ 1,000 lakhs.

102. Percentage increase in sales of Voveran

$= \frac{6.5}{16.5} \times 100 = 39.39\%$

Percentage increase in sales of Volini

$= \frac{2}{7} \times 100 = 28.57\%$

Percentage increase in sales of Dolonex

$= \frac{3.5}{7} \times 100 = 50\%$

Percentage increase in sales of Sumo

$= \frac{2.25}{5.25} \times 100 = 42.85.$

103. Percentage increase in sales of Voveran

$= \frac{6.5}{16.5} \times 100 = 39.39\%$

Percentage increase in sales of Volini

$= \frac{2}{7} \times 100 = 28.57\%$

Percentage increase in sales of Moov = $\frac{1}{4} \times 100 = 25\%$

Percentage increase in sales of Nise = $\frac{3}{15} \times 100 = 20\%$

104. Percentage increase in the sales of Voveran

$= \frac{6.5}{16.5} \times 100 = 39.39\% = 40\%$ (appox.)

106. $\frac{1}{10} \times 400 = 40$

$\frac{1}{4} \times 40 = 10$

$\frac{1}{2} \times 10 = 5$

Hence, required answer is 5.

107. 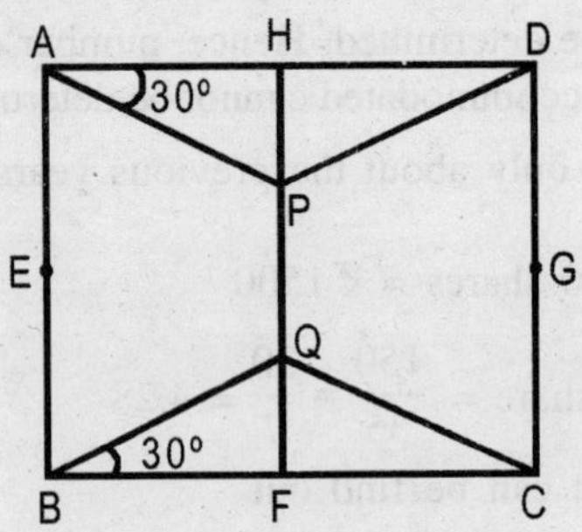

Let the edge of the square measures $2a$ units.
In right triangle BFQ:

QF = BF tan 30° = a tan 30° = $\frac{a}{\sqrt{3}}$

Area of triangle BQC = 2(Area of traingle BFQ)

$= 2\left(\frac{1}{2}\right)(a)\left(\frac{a}{\sqrt{3}}\right) = \frac{a^2}{\sqrt{3}}$ unit square

Area (ABCQPD) = Area (ABFH) + 2 Area (QFC)

$= \left(\frac{1}{2}\right)$ Area (ABCD) + Area (BQC)

$= 2a^2 + \frac{a^2}{\sqrt{3}} = a^2\left(2 + \frac{1}{\sqrt{3}}\right)$

Remaining area of ABCD

$= 4a^2 - \left(2a^2 + \frac{a^2}{\sqrt{3}}\right) = a^2\left(2 - \frac{1}{\sqrt{3}}\right)$

Answer = $a^2\left(2 + \frac{1}{\sqrt{3}}\right) : a^2\left(2 - \frac{1}{\sqrt{3}}\right)$

$= \frac{(2\sqrt{3}+1)}{(2\sqrt{3}-1)}$

$= \frac{(13+4\sqrt{3})}{11}$ which is none of the given options.

Hence the question is wrong.

Assumption:

Had it been polygon ABQCPD in the question instead of ABCQPD:

answer = $4a^2$ – 2(area of ΔBQC) : 2(area of ΔBQC)

$$= \frac{\left(4a^2 - \frac{2a^2}{\sqrt{3}}\right)}{\left(\frac{2a^2}{\sqrt{3}}\right)} = 2\sqrt{3} - 1$$

which is option (B).

108. Price of Darjeeling tea on every day after 100th day = 100 + 10 = 110

So, according to question 110 = 85 + 0.15x

$\Rightarrow x = \frac{25}{15} \times 100 = \frac{500}{3} = 166.67$

i.e. on 167th day price will be equal.

Hence, on 16th June prices will be equal.

109. From statement (I) we can not say that triangles must be congruent. Also from statement II we can not say that every right angle triangles are congruent to each other.

Even on combining both the statements we can not arrive at the conclusion whether the triangles are congruent or not.

110. Let salary of A be $3x$ and B be $4x$ and expenditure of A be $4y$ and that of B be $5y$.

From statement (I)

B's saving = x

$4x - 5y = x$

$\Rightarrow \quad 3x = 5y$

A's saving $3x - 4y = 3x - \frac{3}{5}x = \frac{12}{5}x$

Hence, ratio of saving can be determined.

From statement (II) we can not say anything.

111. Let average height of class be A and number of persons in the class be N.

From statement (I)

$\frac{AN - 56}{N - 1} = A - 1 \Rightarrow AN - 56$

$= AN - A - N + 1 \quad \Rightarrow A + N = 55$

From statement (II)

$\frac{AN - 42}{N - 1} = A + 1 \quad \Rightarrow AN - 42$

$= AN - A + N - 1 \quad \Rightarrow A - N = 43$

Combining (I) and (II)

A + N = 55

$A - N = 43$

$2A = 98$

$\Rightarrow A = 49$

Hence, answer can be determined.

112. R > S

J < V

From statement (I)

Ram is tallest but we can not know who is taller among Shyam and Vikram.

From statement (II)

S > V

Hence, Jay is the shortest among them.

113. Let the sales in July be 150 units.

So, the sales in September = 100 units

and the sales in November = 105 units

Hence, required percentage change $= \frac{-45}{150} \times 100\%$

$= -30\%$

114. $15x - \frac{2}{x} > 1$

$$\Rightarrow 15x - \frac{2}{x} - 1 > 0$$

$$\Rightarrow \frac{15x^2 - x - 2}{x} > 0$$

$$\Rightarrow \frac{15x^2 - 6x + 5x - 2}{x} > 0$$

$$\Rightarrow \frac{3x(5x-2) + 1(5x-2)}{x} > 0$$

$$\Rightarrow \frac{(3x+1)(5x-2)}{x} > 0$$

$$\Rightarrow \frac{\left(x+\frac{1}{3}\right)\left(x-\frac{2}{5}\right)}{(x-0)} > 0$$

$$\therefore\ x \in \left(-\frac{1}{3}, 0\right) \cup \left(\frac{2}{5}, \infty\right)$$

115.

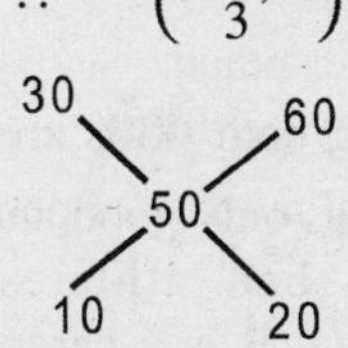

i.e. 1 : 2

So, with 40 litres of a 60% alcohol solution 20 litres of a 30% alcohol solution should be added to prepare a 50% solution.

116. Let length of the wire be x mm.

So, according to condition

$$\pi\left(\frac{1}{2}\right)^2 \times x = 66000$$

$$\Rightarrow \frac{22}{7} \times \frac{1}{4} \times x = 66000$$

$\Rightarrow x = 84{,}000$ mm

So, length of wire in metres = 84m

117. Let speed of second train be x km/hr.

Length of both the trains = 220m

Now, $\frac{220}{6} = (50 + x) \times \frac{5}{18}$

$\Rightarrow 50 + x = 132 \quad \Rightarrow x = 82$ km/hr.

118. $(8R)^2 = 64R^2$

After dividing by 4 = $16R^2$

Its square root = 4R

So, Q = 4R

119. Let length, breadth and height of room be $6x$, $4x$ and $2x$.

So, area of four walls = $24x^2 + 16x^2 = 40x^2$.

Dimensions after changes are $12x$, $2x$ and x.

So, area of four walls = $24x^2 + 4x^2 = 28x^2$.

Hence, percentage reduction in area = $\frac{12}{40} \times 100 = 30\%$.

120.

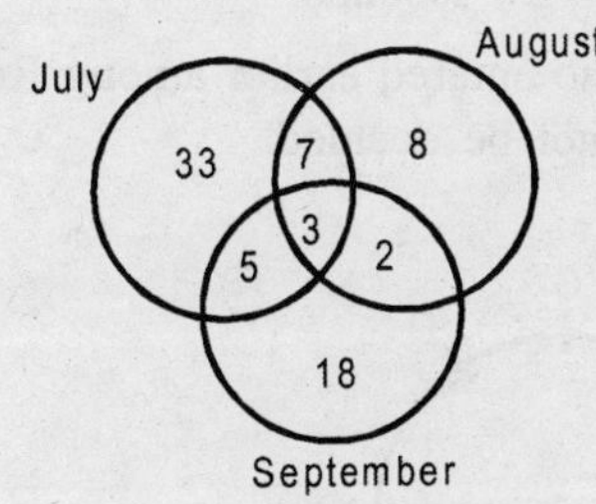

Hence, 7 + 2 = 9 people read magzine for exactly two consecutive months.

121. Let number of chocolates with A, B, C and D be a, b, c and d.

After 1st transfer chocolates with A = a – b – 2c – 3d with B = 2b, C = 3c, D = 4d.

After 2nd transfer B = $2b + \frac{d}{2}$, D = $\frac{7}{2}d$

After 3rd transfer number of chocolates with

$$A = \frac{9}{10}\ (a - b - 2c - 3d)$$

$$B = 2b + \frac{d}{2} + \frac{2}{10}(a - b - 2c - 3d) = 35,\ D = \frac{7}{2}d = 35,$$

$$C = 3c + \frac{1}{10}(a - b - 2c - 3d) = 35$$

Solving these equations we get the values of A – 110, B – 10, C – 10 and D – 10.

122. 201 is the sum of all three numbers.

So, question mark should be 07 + 203 + 70 = 280.

123. The first letter is being incremented by two letters in each element of series, the second is being incremented by three letters and so on.

So, the answer must be ZKXS.

124. As one statement of each has to be wrong and one has to right.
So first statement of any one of the three must be right while that of other two must be wrong.
If we assume first statement of Sachin and Ricky to be right, we get some contradicting results.
So, Brian must be top scorer and Sachin cannot be second.
Hence, correct order is Brian, Ricky, Sachin.

125. Let number of employees who prefer tea be T and who prefer coffee be C, while those who drink both and those who drink neither be B and N respectively.
Now, T = 3B and 2T = 3C ⇒ C = 2B.
Also, B = N
So, T + C – B + N = 60
⇒ 3B + 2B – B + B = 60 ⇒ 5B = 60 ⇒ B = 12

126. As the clock strikes three times in 6 seconds i.e. at 0 seconds, 3 seconds and 6 seconds, means taking 3 seconds in between two strikes.
Now, clock has to strike 9 times at 9 o'clock.
So, it will take 8 × 3 = 24 seconds.

131. As we do not know who entered earlier among Roger and Rafael. So, it cannot be decided.

For questions 132 to 134:

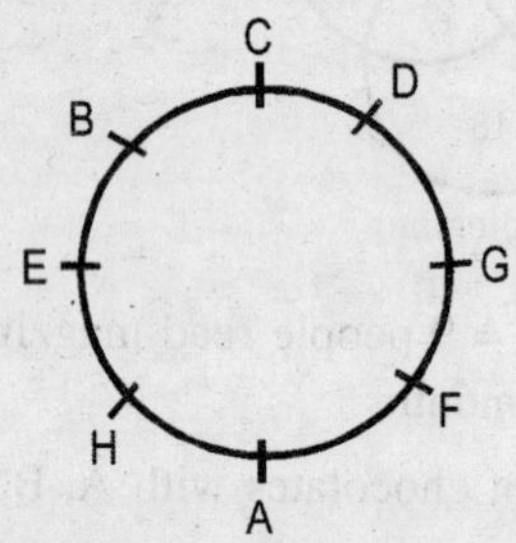

135. There is no evidence given to support or deny a relationship between balloons and tables. Hence none of the first three given options can be conclusively deduced.

138. Commission earned by D in 1994 = ₹ 29,800
Commission earned by A in 1994 = ₹ 24,600

So, required percentage = $\frac{5200}{24600} \times 100 = 21.14\%$
= 21% approximately.

139. Commission earned by B in 1993 = ₹ 28,000
Total commission earned by all salesmen in 1993 = ₹ 1,46,960

So, required percentage = $\frac{28000}{146960} \times 100 = 19.05\%$
= 20% approximately.

140. The numbers in the second row are squares of numbers in first row decreased by 2 i.e. $n^2 - 2$
So, missing number must be $(8)^2 - 2 = 64 - 2 = 62$.

141. The order will 2, 5, 1, 3, 6, 4

142. Options B and D can be eliminated as they are specific in nature. Option C is the best answer amongst the given although this can act only as one premise while concluding the given statement. We need more information to conclusively deduce the given statement.

143. Let the length of each candle be L.
And n is the number of hours.

Now, $2\left(L - \frac{nL}{4}\right) = L - \frac{nL}{6}$

⇒ $2L - \frac{nL}{2} = L - \frac{nL}{6}$

⇒ $L = \frac{nL}{3}$

⇒ $n = 3$.

So, after three hours length of both the candles will be same.

144. Let the numerator and denominator of the fraction be $2x$ and $3x$.

So, $\frac{2x-6}{3x} = \frac{2}{3} \times \frac{2x}{3x}$

⇒ $18x - 54 = 12x$

⇒ $6x = 54$

⇒ $x = 9$

So, the numerator = 2 × 9 = 18

145. According to question $100y + x - 20 = 2(100x + y)$
$100y + x - 20 = 200x + 2y$
$98y - 199x = 20$
Solving we get $x = 26$ and $y = 53$

146. If we take two pieces then they can either be both of same colour, or of different colours. But third one must be of the one of the two colours. So, by taking out 3 pieces we must have at least one pair.

147.

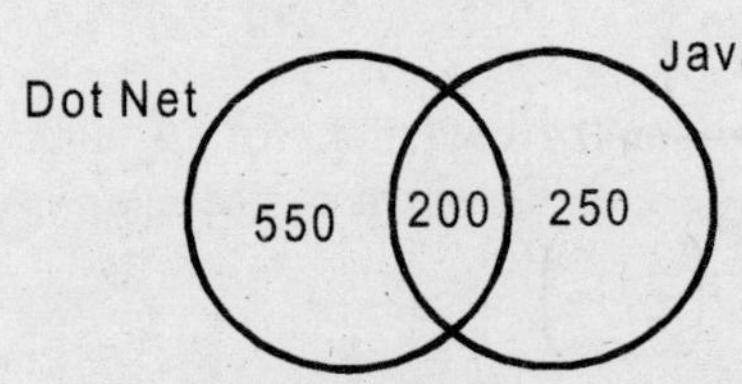

Number of SW personnels having both skills are 200.

148. If we assume that the set athlete has only good and bad athletes and that good and bad athletes are complimentary sets, then Statement B is a paraphrase of the conclusion statement and hence the answer.

149. The missing numbers are 8, 1024 and 32.

150. $17 = \frac{17x}{1-x}$

⇒ $17 - 17x = 17x$ ⇒ $34x = 17$

⇒ $x = \frac{1}{2}$

So, $(2x) \times x = 2 \times \frac{1}{2} \times \frac{1}{2} = \frac{1}{2}$.

Previous Paper (Solved)

Symbiosis National Aptitude Test (SNAP), 2009

SECTION-I

QUANTITATIVE & DATA INTERPRETATION & DATA SUFFICIENCY

For all questions in this section, correct answers carry 1 mark each.

1. A dice is rolled three times and sum of three numbers appearing on the uppermost face is 15. The chance that the first roll was a four is

A. $\frac{2}{5}$ B. $\frac{1}{5}$

C. $\frac{1}{6}$ D. None of these

2. A boat covers a distance of 30 kms downstream in 2 hours while it takes 6 hours to cover the same distance upstream. What is the speed of the boat in kms per hour?

A. 5 B. 7.5

C. 13 D. 18

3. A five digit number is formed by using the digits 1, 2, 3, 4 and 5 without repetitions. What is the probability that the number is divisible by 4?

A. $\frac{1}{5}$ B. $\frac{5}{6}$

C. $\frac{4}{5}$ D. None of these

4. If the algebraic sum of deviations of 20 observations measured from 23 is 70, mean of these observations would be

A. 24 B. 25

C. 26 D. None of these

5. An alloy of gold and silver weighs 50 gms. It contains 80% gold. How much gold should be added to the alloy so that percentage of gold is increased to 90?

A. 50 gms B. 60 gms

C. 30 gms D. 40 gms

6. Weekly incomes of two persons are in the ratio of 7 : 3 and their weekly expenses are in the ratio of 5 : 2. If each of them saves ₹ 300 per week, then the weekly income of the first person is

A. ₹ 7500 B. ₹ 4500

C. ₹ 6300 D. ₹ 5400

7. Wheat is now being sold at ₹ 27 per kg. During last month its cost was ₹ 24 per kg. Find by how much per cent a family reduces its consumption so as to keep the expenditure fixed.

A. 10.2% B. 12.1%

C. 12.3% D. 11.1%

8. There are 10 stations on a railway line. The number of different journey tickets that are required by the authorities is

A. 92 B. 90

C. 91 D. None of these

9. The radius of circle is so increased that its circumference increased by 5%. The area of the circle then increases by

A. 12.5% B. 10.25%

C. 10.5% D. 11.25%

10. In how many ways can the letters of the word ABACUS be rearranged such that the vowels always appear together?

A. 6!/2! B. 3!*3!

C. (3!*3!)/2! D. (4!*3!)/2!

11. In 4 years, ₹ 6000 amounts to ₹ 8000. In what time at the same rate will ₹ 525 amount to ₹ 700?

A. 2 years B. 3 years

C. 4 years D. 5 years

Direction (Q. Nos. 12–14): *At the start of a game of cards, J and B together had four times as much money as T, while T and B together had three times as much as J. At the end of the evening, J and B together had three times as much money as T, while T and B together had twice as much as J. B lost ₹200.*

12. What fraction of the total money did T have at the beginning of the game?

A. $\frac{1}{3}$ B. $\frac{1}{8}$

C. $\frac{2}{9}$ D. $\frac{1}{5}$

13. What fraction of the total money did J win/lose?

A. Won $\frac{1}{12}$ B. Lost $\frac{1}{6}$

C. Lost $\frac{1}{3}$ D. Won $\frac{1}{5}$

14. What amount did B start with?

A. ₹ 575 B. ₹ 375
C. ₹ 825 D. ₹ 275

15. If a and b are negative, and c is positive, which of the following statement/s is/are true?

I. $a - b < a - c$

II. if $a < b$, then $\frac{a}{c} < \frac{b}{c}$

III. $\frac{a}{b} < \frac{a}{c}$

A. I only B. II only
C. III only D. II and III only

16. The diagonal of a square is $4\sqrt{2}$ cm. The diagonal of another square whose area is double that of the first square is

A. 8 cm B. $8\sqrt{2}$ cm
C. $4\sqrt{2}$ cm D. 16 cm

17. The maximum length of a pencil that can be kept in a rectangular box of dimensions 8 cm × 6 cm × 2 cm, is

A. $2\sqrt{13}$ cm B. $2\sqrt{14}$ cm
C. $2\sqrt{26}$ cm D. $10\sqrt{2}$ cm

Directions (Q. Nos. 18–19): *Answer the questions based on the information given below. The Venn diagram given below shows the estimated readership of 3 daily newspapers (X, Y & Z) in a city. The total readership and advertising cost for each of these papers is as below:*

Newspapers	Readership (lakhs)	Advertising cost (₹ per sq. cm)
X	8.7	6000
Y	9.1	6500
Z	5.6	5000

The total population of the city is estimated to be 14 million. The common readership (in lakhs) is indicated in the given Venn diagram

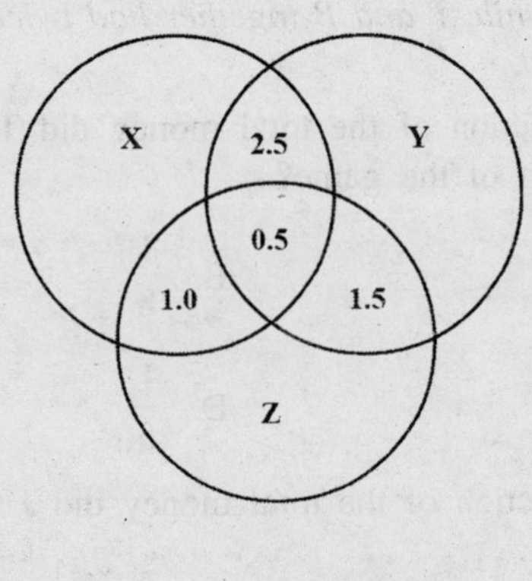

18. The number of people (in lakhs) who read at least one newspaper is

A. 4.7 B. 11.9
C. 17.4 D. 23.4

19. The number of people (in lakhs) who read only one newspaper is

A. 4.7 B. 11.9
C. 17.4 D. 23.4

20. Sonali invests 15% of her monthly salary in insurance policies. She spends 55% of her monthly salary in shopping and on household expenses. She saves the remaining amount of ₹ 12,750. What is Sonali's monthly income?

A. ₹ 42,500 B. ₹ 38,800
C. ₹ 40,000 D. ₹ 35,500

21. How many kgs of tea worth ₹ 25 per kg must be blended with 30 kgs of tea worth ₹ 30 per kg so that by selling the blended variety at ₹ 30 per kg there should be a gain of 10%?

A. 36 kgs B. 40 kgs
C. 32 kgs D. 42 kgs

22. In an examination, out of 80 students 85% of the girls and 70% of the boys passed. How many boys appeared in the examination if total pass percentage was 75%?

A. 370 B. 340
C. 320 D. 360

23. 300 gms of salt solution has 40% salt in it. How much salt should be added to make it 50% in the solution?

A. 40 gms B. 60 gms
C. 70 gms D. 80 gms

24. What number should replace the question mark in the image below?

6	12	?
48	96	192

A. 18 B. 20
C. 22 D. 24

25. What image from bottom row should replace the question mark?

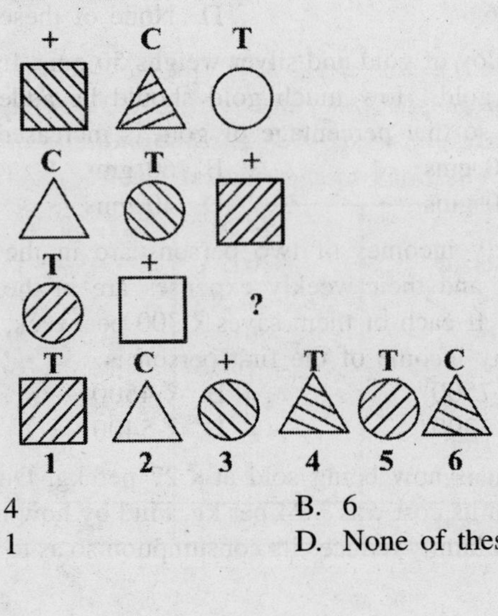

A. 4 B. 6
C. 1 D. None of these

26. A five-digit number divisible by 3 is to be formed using numerical 0, 1, 2, 3, 4 and 5 without repetition. The total number of ways this can be done is:

A. 122 B. 210
C. 216 D. 217

27. A contract is to be completed in 50 days and 105 men were set to work, each working 8 hours a day. After 25 days, 2/5th of the work is finished. How many additional men be employed so that the work may be completed on time, each man now working 9 hours a day?

A. 34 B. 36
C. 35 D. 37

28. A can built up a structure in 8 days and B can break it in 3 days. A has worked for 4 days and then B joined to work with A for another 2 days only. In how many days will A alone build up the remaining part of the structure?

A. 10 days B. 9 days
C. 12 days D. None of these

29. The first two terms of a geometric progression add up to 12. The sum of the third and the fourth terms is 48. If the terms of the geometric progression are alternately positive and negative, then the first term is

A. –2 B. –4
C. –12 D. 8

30. The mean of the numbers *a*, *b*, 8, 5, 10 is 6 and the variance is 6.80. Then which one of the following gives possible values of *a* and *b*?

A. $a = 0, b = 7$ B. $a = 5, b = 2$
C. $a = 3, b = 4$ D. $a = 2, b = 4$

31. An agent sells goods of value ₹ 15,000. The commission which he receives at the ratio of 12 ½ % is

A. ₹ 1875 B. ₹ 2125
C. ₹ 2000 D. ₹ 2700

32. $\sqrt{110.25} \times \sqrt{0.01} \div \sqrt{0.00025} - \sqrt{420.25}$ equal to

A. 0.75 B. 0.50
C. 0.64 D. 0.73

Directions (Q. Nos. 33 – 35): *The table below shows the number of people who responded to a survey about their favourite style of music. Use this information to answer the following questions to the nearest whole percentage.*

Age	15–20	21–30	31+
Classical	6	4	17
Pop	7	5	5
Rock	6	12	14
Jazz	1	4	11
Blues	2	3	15
Hip-Hop	9	3	4
Ambient	2	2	2
Total	33	33	68

33. What percentage of respondents under 31 indicated that Blues is their favourite style of music?

A. 7.1 B. 7.6
C. 8.3 D. 14.1

34. What percentage of respondents aged 21–30 indicated a favourite style other than Rock music?

A. 64% B. 60%
C. 75% D. 36%

35. What percentage of the total sample indicated that Jazz is their favourite style of music?

A. 6% B. 8%
C. 22% D. 12%

Directions (Q. Nos. 36 – 40): *The pie charts below show the percentage of students in each faculty at North West University and the number of non-US students in the Arts faculty. These percentages have been rounded to the nearest whole number. There are a total of 1049 students in Arts faculty. Use this information to answer the following questions.*

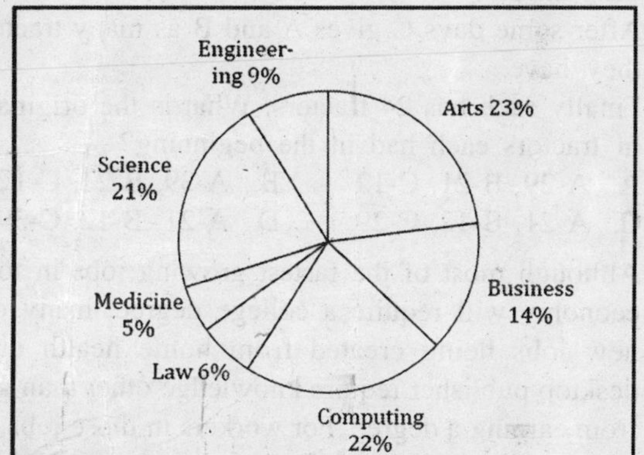

Students by Faculty

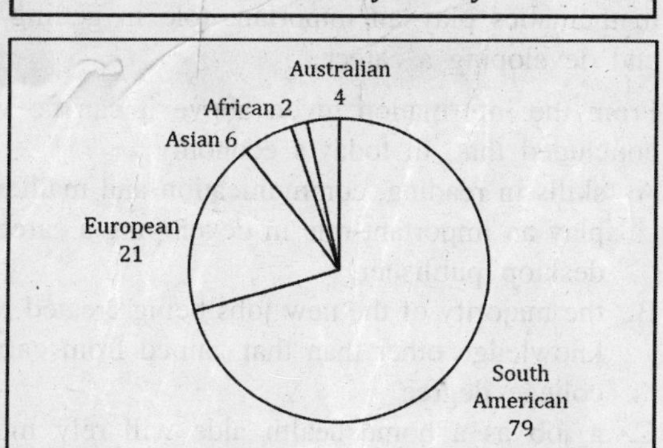

Arts Students (Non-US)

36. What percentage of students in the Arts faculty are non-US students?

A. 14% B. 9%
C. 30% D. 11%

37. How many students are there in the Engineering faculty?

A. 420 B. 410
C. 390 D. 440

38. How many students are there at the University?
A. 4650 B. 4560
C. 4640 D. 4450

39. If six per cent of Science students are Asian. How many Asian students are studying Science?
A. 48 B. 66
C. 120 D. 57

40. There are 34 European medical students. What percentage of the faculty does this represent?
A. 14% B. 18%
C. 16% D. 15%

SECTION-II

ANALYTICAL & LOGICAL REASONING

For all questions in this section, correct answers carry 2 marks each.

41. There are 3 societies A, B, C having some tractors each.
A gives B and C as many tractors as they already have.
After some days B gives A and C as many tractors as they have.
After some days C gives A and B as many tractors as they have.
Finally each has 24 tractors. What is the original No. of tractors each had in the beginning?
A. A-29, B-21, C-12 B. A-39, B-21, C-12
C. A-21, B-12, C-29 D. A-21, B-12, C-39

42. Although most of the fastest growing jobs in today's economy will require a college degree, many of the new jobs being created from home health aide to desktop publisher require knowledge other than gained from earning a degree. For workers in those jobs, good basic skills in reading, communication, and mathematics play an important role in getting a job and developing a career.

From the information given above it can be validly concluded that, in today's economy
A. skills in reading, communication and mathematics play an important role in developing a career as a desktop publisher.
B. the majority of the new jobs being created require knowledge other than that gained from earning a college degree
C. a job as a home health aide will rely more on communication skills than on basic skills in reading and mathematics
D. if a job is one of the fastest growing jobs, it will require a college degree.

43. According to the National Agricultural Aviation Society (NAAS), without the use of crop protection products to control insects, weeds, and diseases, crop yields per acre will drop by more than 50 per cent. The first aerial application of insecticide occurred in 1921, and it was a huge success. By contrast, in today's economy all aircraft that are classified as aerial applicators do more than just apply insecticide; today, they also spread sees and apply fertilizer.

From the information given above it CANNOT be validly concluded that
A. according to the NAAS, if crop yields per acre never drop by more than 50 per cent, then crop protection products have been used to control insects, weeds, and diseases.
B. in today's economy any aircraft that cannot be used to apply fertilizer cannot be classified as an aerial applicator
C. in today's economy, if an aerial applicator is used, then it will be able to spread seed and to apply fertilizer
D. according to the NAAS, if crop yields per acre drop by more than 50 per cent, then crop protection products have not been used to control insects, weeds, and diseases.

44. Lou observes that if flight 409 is cancelled, then the manager could not possibly arrive in time for the meeting. But the flight was not cancelled. Therefore, Lou concludes, the manager will certainly be on time. Evelyn replies that even if Lou's premises true, his argument is fallacious. And therefore, she adds, the manager will not arrive on time after all.

Which of the following is the strongest thing that we can properly say about this discussion?
A. Evelyn is mistaken in thinking Lou's argument to be fallacious, and so her own conclusion is unwarranted.
B. Evelyn is right about Lou's argument, but nevertheless her own conclusion is unwarranted.
C. Since Evelyn is right about Lou's argument, her own conclusion is well supported.
D. Since Evelyn is mistaken about Lou's argument, her own conclusion must be false.

45. Cars are safer than planes. Fifty per cent of plane accidents result in death, while only one per cent of car accidents result in death.

Which of the following, if true, would most seriously weaken the argument above?

A. Planes are inspected more often than cars.
B. The number of car accidents is several hundred thousand times higher than the number of plane accidents.
C. Pilots never fly under the influence of alcohol, while car drivers often do.
D. Plane accidents are usually the fault of air traffic controllers, not pilots.

Directions (Q. Nos. 46–48): *Refer to the chart below showing annual production and answer the questions that follow.*

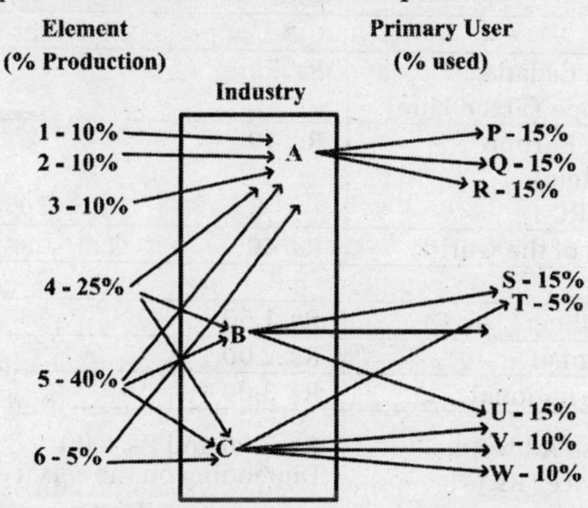

100% = 100,000 tons **Total – 100%**

46. Which industry/industries contribute/contributes to company S?

A. Industry A and B only
B. B and C only
C. A and C only
D. B only

47. Industry B processes what percentage of the total production of listed elements?

A. 25% B. 65%
C. 40% D. Cannot be determined

48. Of the listed elements processed by the Industry A, how many tons are produced annually?

A. 30,000 B. 50,000
C. 1,00,000 D. 55,000

Directions (Q. Nos. 49–52): *In each of the following questions there are two blanks marked I & II. The words to fill in these blanks are given against I as (A, B, C, D) and II as (P, Q, R, S) RESPECTIVELY. The right words to fill in these blanks are given as four alternatives. The words on either side of the sign (::) have a similar relationship. That alternative which signifies this relationship is your answer.*

49. I : Increase : : Descend : II

I. (A) Grow (B) Ascend
(C) Rise (D) Price

II. (P) Reduce (Q) Down
(R) Decrease (S) Mountain

A. AR B. RB
C. CP D. DQ

50. Modern : I : : II : Old

I. (A) Ancient (B) Death
(C) Famous (D) Civilization

II. (P) Industrialisation (Q) Young
(R) Fashion (S) Western

A. AQ B. AS
C. BP D. CR

51. Part : I : : Class : II

I. (A) Section (B) Whole
(C) School (D) Students

II. (P) Student (Q) School
(R) Teachers (S) Room

A. AR B. BQ
C. CP D. DS

52. Summit : Apex : : I : II

I. (A) Beautiful (B) Picture
(C) Attractive (D) Enhancing

II. (P) Comfortable (Q) Pretty
(R) Healthy (S) Brave

A. AQ B. BP
C. DS D. CR

53. If in a certain code BEAUTIFUL is coded as 573041208, BUTTER as 504479, how is FUTURE coded in that code?

A. 201497 B. 204097
C. 704092 D. 204079

54. How many Mondays are there in particular month of a particular year if the month ends on Wednesday?

A. 4 B. 5
C. 3 D. Cannot be specified

Directions (Q. Nos. 55–56): *There are two rows of numbers in each question. The upper row is complete and in the lower row one number is missing. Find a suitable number to fill the blank space so that the symmetry with the upper row is maintained.*

55.

17	102	12
15	?	10

A. 211 B. 75
C. 125 D. 117

56.

12	336	14
15	?	16

A. 220 B. 480
C. 125 D. 450

Directions (Q. Nos. 57–58): *Seven People A, B, C, D, E, F, G are planning to enjoy boating. There are only two boats, and the following conditions are to be kept in mind.*

(I) A will go in the same boat in which E is to go.

(II) F cannot go in the same boat in which C is, unless D is also accompanying.

(III) Neither B nor C can be given the boat in which G is.

(IV) The maximum number of persons in one boat can be four only.

57. If F and B are in one boat, which of the following statements is true?

A. G is in the other boat
B. D is in the other boat
C. C is in the other boat
D. E is with F and B in one boat

58. If E gets the boat with F, which of the following is the complete and accurate list of the people who must be sitting in other boat?

A. F and E
B. G and A
C. D and A
D. C, D and B

Directions (Q. Nos. 59–60): A series of figures has been shown on the left. Find the figure in the place of (?) from the figures on the right.

59. 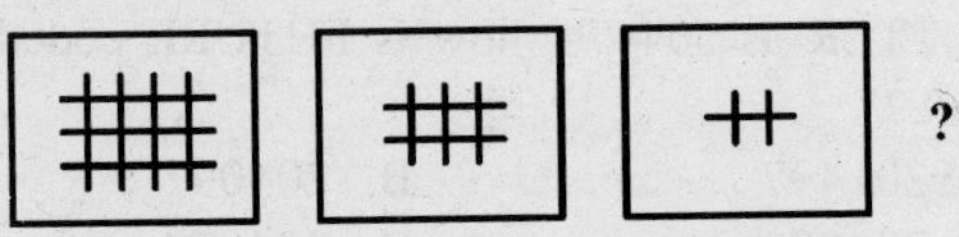

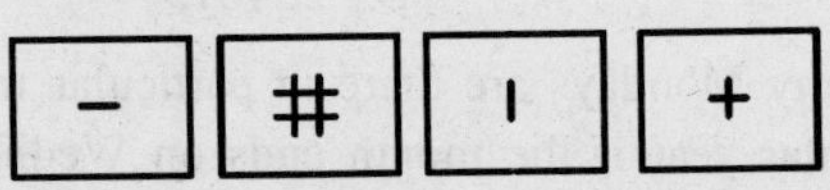

A. B. C. D.

60. 

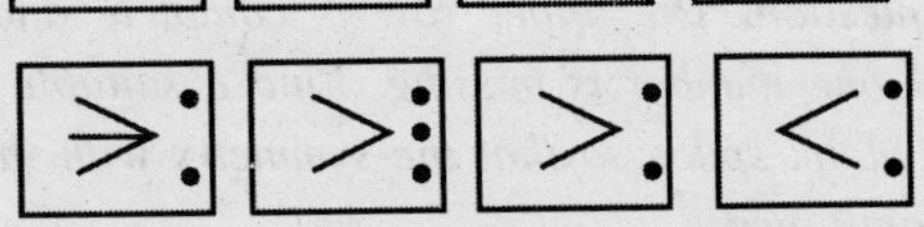

A. B. C. D.

61. Complete the following series.
(GMSY, IOVA, KQWC, ?)

A. MSYE
B. NSYE
C. MYTE
D. MSYF

Directions (Q. Nos. 62 – 65): Following is a new plan called 'Airtel One', introduced by Airtel (GSM Service Provider) for its post-paid customers.

Pulse Rate	60 Secs		
Price of Plan	Rs. 99		
Free Airtime on Plan	Nil		
Incoming Calls (Rs.)	Free		
Outgoing Calls (Rs./minute)	Airtel	GSM/ CDMA (10 Digit)	Landline/ WLL
LOCAL RATES	Rs. 1	Rs. 1	Rs. 2
STD RATES			
50-200 KM	Rs. 1.50	Rs. 1.50	Rs. 2
200-500 KM	Rs. 1.50	Rs. 2.50	Rs. 2.50
500+ KM	Rs. 1.50	Rs. 3.00	Rs. 3.50
ISD			
USA, Canada, Europe (Fixed Line)	Rs. 7		
Gulf, Europe (Mobile) SAARC	Rs. 10		
Rest of the world	Rs. 40		
SMS			
Local	Rs. 1.50		
National	Rs. 2.00		
International	Rs. 5.00		
Value Added Service (Rs.)	Rs. 3.00 and Rs. 6.00 Depending on the service		
Monthly bill = Call charges + SMS charges + Price of Plan			

62. If Mohan made STD calls (within 100 kms) for 30 min and spent 100 min on local calls with 30% on landline, 40% in GSM and 30% on Airtel, and Rohan spent 18 min on STD within 150 Kms and spent 120 mins on local calls with 30%, 40% and 30% on GSM, landline and Airtel respectively, then who spent more?

A. Mohan
B. Rohan
C. Both spent the same amcunt.
D. Cannot be determined.

63. A new SMS scheme was introduced @60p/local SMS, with an additional monthly charge of ₹ 35. Who will not benefit from the scheme? A Person sending

A. 38 local SMS a month
B. 40 local SMS a month
C. 60 local SMS a month
D. 59 local SMS a month

64. A bill of ₹ 199/Month is definitely not possible when you make only

A. 30 calls to Airtel on STD (500+Kms) and 55 local calls in a month
B. 16 calls to Airtel on STD (15 Kms and 76 local calls in a month)

C. 10 STD calls (250 Kms) each to Airtel, GSM and landline respectively and 30 locals in a month

D. 8, 4 and 7 calls to Airtel, GSM and landline respectively on STD (500+Kms) and 55 local calls in a month

65. A person makes 12 minutes of ISD calls in a month. If 80% of his ISD bill came from calling the rest of the world, then the ISD calls for the minimum duration were made to which of the following group of countries?

A. USA, Canada, Europe (Fixed Line)
B. Gulf, Europe (Mobile), SAARC
C. Rest of the world
D. Cannot be determined.

Directions (Q. Nos. 66 – 67): *These questions are based on the graph which shows the demand and production statistics of 5 T.V. companies.*

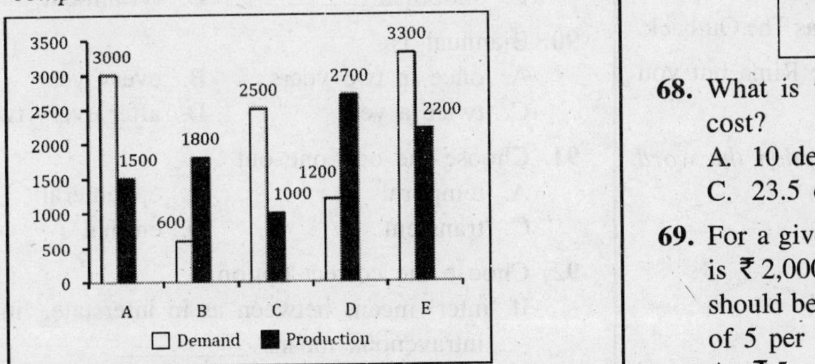

66. What is the difference between the average demand and the average production of the five companies taken together?

A. 1400 B. 400
C. 280 D. 138

67. The production of the company D is how many times of the production of the company A?

A. 1.8 B. 1.5
C. 2.5 D. 1.11

Directions (Q. Nos. 68 – 69): *The following questions are based on the pie-chart given below which gives the expenditure incurred in printing a magazine:*

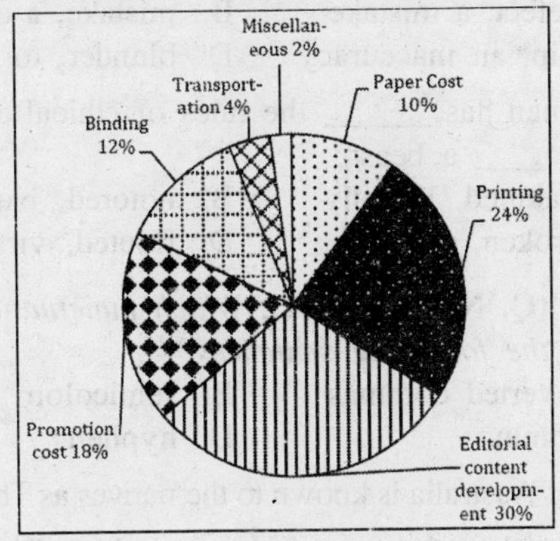

68. What is the angle for the sector representing paper cost?

A. 10 degrees B. 36 degrees
C. 23.5 degrees D. 45 degrees

69. For a given issue of the magazine, the 2% of the cost is ₹ 2,000 and the print-run is 12,500 copies. What should be the sale price if the publisher desires a profit of 5 per cent?

A. ₹ 5 B. ₹ 7.5
C. ₹ 8 D. ₹ 8.40

70. If 'water' is called 'food', 'food' is called 'tree', 'tree' is called 'sky', 'sky' is called 'wall', on which of the following does a 'fruit' grow?

A. Water B. Food
C. Tree D. Sky

SECTION-III

GENERAL ENGLISH

For all questions in this section, correct answers carry 1 mark each.

Directions (Q. Nos. 71– 72): *Choose the correct option to fill the blank for correct grammatical use.*

71. She is a close friend of ________

A. my B. I
C. me D. mine

72. He took care of her ________ she was restored to health.

A. when B. before
C. after D. till

73. Choose the correct option: The man ________ his appearance completely since then.

A. is changing B. changed
C. has changed D. is changed

74. From the options below, choose the word with the incorrect spelling:

A. Diarrhea B. Diaper
C. Dichotomy D. Dias

75. Choose the word spelt correctly :

A. superintendant B. seive
C. alloted D. dissipate

Directions (Q. Nos. 76 – 79): *Fill in the blanks with the correct pair of words:*

76. He lives ____ Bangaluru ____115, Richmond Road.

A. at, in B. in, at
C. at, on D. in, on

77. 'Aurally challenged' is a ______________ for the ______________.

A. metaphor, blind B. euphemism, deaf
C. simile, disabled D. synonym, dumb

78. If error is a ______________ , fault is ______________.

A. defect, a mistake B. mistake, a defect
C. slip, an inaccuracy D. blunder, to blame

79. The man has ______ the rules of ethical conduct; he is ______ a beast.

A. flaunted, literally B. ignored, basically
C. broken, as D. flouted, virtually

Directions (Q. Nos. 80 – 81): *Which punctuation mark is missing in the following sentences?*

A. inverted commas B. semicolon
C. comma D. hyphen

80. Part of Australia is known to the natives as The Outback.

81. I know that you want to learn to drive Rima but you are too young.

Directions (Q. Nos. 82 – 83): *Select as option the word closest in meaning to the given word:*

82. Veracious

A. False B. Varied
C. Image D. Truthful

83. Perturb

A. Stipulate B. Turn around
C. Disturb greatly D. Compatible

Directions (Q. Nos. 84 – 85): *Choose the option that is the closest in meaning to the phrases in bold:*

84. He was a king who ruled his subjects **with a high hand**.

A. sympathetically B. oppressively
C. democratically D. generously

85. **I racked my brains** to solve this difficult problem.

A. I read a large number of books.
B. I consulted several people.
C. I used my commonsense.
D. I subjected my mind to hard thinking.

Directions (Q. Nos. 86 – 87): *Select the most appropriate meaning of the underlined idiomatic phrase:*

86. I was supposed to give a speech to my English class, but <u>I got cold feet</u> and I didn't go.

A. My English class-room makes my feet feel too cold.
B. My feet got cold so I didn't go.
C. I got too nervous and I didn't go.
D. It was winter time and I had no socks for my feet.

87. Take care of what u say! <u>You will have to eat your words!</u>

A. You have no food to eat.
B. You will have to take back what you have said.
C. You are not good with your language.
D. None of the above.

88. Which two sentences convey the same idea?

1. Shouldn't they have checked your tickets?
2. I wonder if they should have checked your tickets.
3. I want to know if they checked your tickets.
4. They should have checked your tickets.

A. 2, 3 B. 1, 2
C. 1, 4 D. 2, 4

Direction (Q. No 89): *Choose a single word which is opposed to the meaning contained in the sentence.*

89. He was magnanimous and his benevolence made him give to charity.

A. mean B. cruel
C. snobbish D. tyrannical

90. Biannual is

A. once in two years B. every year
C. twice a year D. after every two years

91. Choose the odd one out :

A. temporal B. ephemeral
C. transient D. eternal

92. Choose the correct option :
If 'inter' means between as in interstate, 'intra' as in intravenous means ________.

A. into B. onto
C. within D. without

93. Match the part of speech (in bold) to their usage **(down)**

1. Adjective 5. The fire engine came rushing **down** the hill.
2. Preposition 6. He has seen the ups and **downs** of life.
3. Noun 7. **Down** with the tyrant!
4. Verb 8. The porter was hit by the **down** train.

A. 1-8, 2-5, 3-6, 4-7 B. 1-6, 2-5, 3-8, 4-7
C. 1-5, 2-6, 3-7, 4-8 D. 1-5, 2-7, 3-6, 4-8

Direction (Q. No. 94): *Match the items in the column I with those in column II. Choose the correct answer combination given below:*

94.

I	II
1. As deaf as	5. gall
2. As bitter as	6. an eel
3. As unpredictable as	7. a post
4. As slippery as	8. the weather

A	B	C	D
1 – 7	1 – 7	1 – 8	1 – 5
2 – 6	2 – 5	2 – 6	2 – 7
3 – 8	3 – 8	3 – 5	3 – 6
4 – 5	4 – 6	4 – 7	4 – 8

Direction (Q. No. 95): *Match the part of speech (in bold) to their usage.*

95. ABOVE

1. Adjective	5. Rain comes from **above**.
2. Adverb	6. His conduct is **above** suspicions.
3. Noun	7. Look **above** the mantel piece.
4. Preposition	8. The **above** information is for the public.

A	B	C	D
1 – 8	1 – 7	1 – 8	1 – 6
2 – 7	2 – 6	2 – 5	2 – 8
3 – 5	3 – 5	3 – 6	3 – 5
4 – 6	4 – 8	4 – 7	4 – 7

Directions (Q. Nos. 96 - 97): *Choose the segment with the error. If there is no error, choose D.*

96. (A) Idli and Sambar/(B)make the breakfast/(C) in the South./(D) No Error.

97. (A) Your account should/(B) have been credited/(C) with three months' interest./(D) No Error.

98. Choose the sentence where the underlined word is used appropriately.

A. Who's car are you planning to borrow?
B. Whose planning to have a birthday party?
C. He is the person who you met at the junction.
D. Does he know whom to talk about the hotel reservations?

Directions (Q. Nos. 99–100): Choose

A. if the first sentence is correct
B. if the second sentence is correct
C. if both sentences are correct
D. if both sentences are wrong.

99. 1. Can you imagine his forgetting his own birthday?
2. Can you imagine him forgetting his own birthday?

100. 1. Recently I read about a unique wedding that took place in the newspaper.
2. Recently I read in the newspaper about a unique wedding that took place.

101. Complete the given sentence by choosing the correct phrase.

You cannot succeed unless

A. you do not work hard.
B. you shall not work hard.
C. you will work hard.
D. you work hard.

102. Re-arrange the scrambled segments in logical order to make a complete sentence.

1. to place where more opportunities are available
2. and so there is a great demand for English
3. for professional and economic growth
4. because it takes one outside one's own community
5. English is language of opportunities

A. 1, 2, 5, 3, 4 B. 5, 4, 1, 3, 2
C. 3, 1, 4, 2, 5 D. 2, 1, 3, 5, 4

103. Choose the best construction:

A. If you would have taken care you wouldn't have got typhoid.
B. If you took care you wouldn't have got typhoid.
C. If you take care you wouldn't have got typhoid.
D. If you had taken care you wouldn't have got typhoid.

Directions (Q. Nos. 104–105): Choose the correct option.

104. At times, we are all _______ to be mistaken.

A. apt B. likely
C. considered D. able

105. He hardly cares, _______?

A. does he B. doesn't he
C. will he D. won't he

Directions (Q. Nos. 106–107): *Identify the figures of speech in the following sentences:*

106. As proud as a peacock.

A. Metaphor B. Simile
C. Apostrophe D. Epigram

107. Death lays his icy hand on kings.

A. Personification B. Exclamation
C. Simile D. Anticlimax

Directions (Q. Nos. 108–110): *Read the passage carefully and answer within the context.*

"A way to deal with frozen feelings"

Every child experiences all that happens around him with total awareness. In the first seven years the child's brain is like a sponge, taking in all sensory inputs and building his idea of his surroundings. As long as the environment is safe, the child learns with incredible speed. However, when the environment is scary or stressful, the child unlearns past learning just as rapidly.

In the early years of every child's life, whenever there is shock, violence, fear or pain, these intense emotions are imprinted deeply into memory. Whenever the same activity or situation is repeated, the nervous system and body subconsciously re-experience the memory of that trauma.

Any emotional situation that takes us out of the present and into the past means that whenever the same kind of emotion crops up later in our life we return to the past for our reference point. If that point was at age three, we find ourselves behaving like a three-year-old. We feel childish and we behave childishly. Our feelings are the cause of this 'glitch' in our learning process. We know we should be able to make a positive change, but that doesn't change anything.

The process of change need not be traumatic. We couldn't have done any better because we didn't know how to. But we should realise that was then and this is now! We can choose to choose again. It's up to us. It's our movie!

108. The "Frozen Feelings" being talked about are about
A. negative childhood experiences
B. childhood learning patterns
C. inability to learn as an adult
D. None of the above.

109. A 'glitch' is
A. a ditch
B. uneasy emotions
C. sudden malfunction or breakdown
D. learning patterns

110. Identify the correct sentence, based on the paragraph
A. The process of change needs to be traumatic.
B. We feel childish and we behave childishly.
C. Both sentences are incorrect.
D. Both the sentences are correct.

SECTION- IV

GENERAL KNOWLEDGE

For all questions in this section, correct answers carry 1 mark each.

111. Dendrochronology is
A. The method of scientific dating based on the analysis of tree-ring growth patterns.
B. A core sample from the accumulation of snow and ice over many years that have re-crystallized and have trapped air bubbles from previous time periods.
C. The study of glaciers or more generally ice and natural phenomena that involve ice.
D. The scientific study of landforms and the processes that shape them.

112. Which company launched the first mobile phone operation in India?
A. Bharti Airtel B. Essar
C. Max Touch D. Modi Telstra

113. Which is the longest running English language TV cartoon in history, first launched by Warner Brothers in 1969 exclusively for television, and is now popular fare on the Cartoon Network?
A. Scooby Doo B. Tom and Jerry
C. Popeye D. Johnny Bravo

114. The expansion for BIFR, in the context of the Indian Industry is:
A. Board for Industrial and Financial Reconstruction
B. Bureau for Industrial and Financial Reconstruction
C. Board for Investment and Financial Reconstruction
D. Bureau for Investment and Financial Reconstruction

115. What is Ekistics?
A. The water sports related study
B. The science of body mass
C. The art of card tricks
D. The science of human settlements

116. What is Red Herring in an IPO?
A. Prospectus
B. Submission of Form
C. Funds Generated during IPO
D. Minimum Offer per Share

117. Eight O'clock Coffee, a US based coffee marketer, is a group company of
A. Tata B. ITC
C. Britannia D. Unilever

118. Bio-diesel is extracted from the following plant:
A. Hibiscus B. Aloe Vera
C. Jatropha D. Chamomile

119. Which woman tennis player won the maximum number of Grand Slam singles titles in the history of tennis?
A. Steffi Graf B. Martina Navratilova
C. Billi Jean King D. Margaret Court

120. What is the nickname of New York city?
A. Big Sweet Lime B. Big Apple
C. Big Grape D. None of the above

121. Which was the first Indian company to be listed in NASDAQ?
A. TCS B. Wipro
C. Infosys D. Cognizant

122. "TAREGNA" – a village in Bihar was in the news because of
A. Rahul Gandhi's election campaign
B. Viewing of the solar eclipse
C. A visit by Hillary Clinton
D. The Naxal massacre of tribals

123. The driver for Formula One's Force India Team is
A. Mark Webber B. Adrian Sutil
C. Felipe Massa D. Jensen Button

124. (MDRT) – Million Dollar Round Table consists of
A. World's richest industrialists
B. World's most powerful golfers
C. Global Association of Life Insurance professionals
D. World's richest sports persons

125. Who declares the "Credit Policy" of India?
A. The Finance Ministry
B. Reserve Bank of India
C. The Commerce Ministry
D. The Indian Bank's Association

126. Name of the body which regulates the Stock Exchanges in India is
A. CBI B. CID
C. SHCIL D. SEBI

127. Rafflesia arnoldii is its botanical name and it holds a world record in the plant world. What's the record?
A. Tallest cacti
B. Largest bloom
C. Longest surviving tree
D. Smallest plant

128. Temujin was a famous historical character of the world, but we know him by a different name. Which one?
A. Babur B. Genghis Khan
C. Kublai Khan D. Emperor Nero

129. What is Short Selling in stock market lingo?
A. Selling the shares which you do not own.
B. Selling a share after owning it for a short period of time.
C. Selling all the shares in your booty which are not productive.
D. Selling of shares which are quoted for a short period of time.

130. Mahatma Gandhi was nominated for the Nobel Peace prize for the first time in the year 1937 and last time in the year 1948. How many times in between these two years was he nominated?
A. 1 B. 3
C. 2 D. 0

131. Which organization instituted the Nobel Prize in Economics?
A. Svenska Handelsbanken
B. International Monetary Fund
C. Economics Department of Oxford University
D. Sveriges Riksbank

132. Film stars have been endorsing Lux beauty soap since ages. Who was the first actress of Bollywood to do so?
A. Madhubala B. Leela Naidu
C. Suraiya D. Leela Chitnis

133. India's interstate cricket tournament is named after Kumar Shri Ranjitsinghji Vibhaji Jadeja or in short Ranji, the great cricket legend. He played his cricket in England and also played for English national team. Which county he used to play for in the league cricket of England?
A. Sussex B. Essex
C. Somerset D. Surrey

134. As a practice, all major states' capitals have offices of Reserve Bank of India with only one exception. Which city of India has an office of RBI without being a state capital?
A. Pune B. Allahabad
C. Baroda D. Nagpur

135. This distinguished Indian was born on 28th December 1937 in Surat. He has completed the Advanced Management Program at Harvard Business School, USA. He turned down a job offer from IBM to join his family business and today he is one of the top business leaders of India. Name the tycoon.
A. Ratan Tata B. Anil Ambani
C. Azim Premji D. Anand Mahindra

136. It's known to be used in Egypt in 3000 BC, but its name had been derived from an old French expression. It is made from the fermentation of ethanol in a process that yields its key ingredient, ethanoic acid. What is it?
A. Tabasco Sauce B. Phenol
C. Vinegar D. Soya Sauce

137. The name of this Japanese company literally means "three diamonds". Which one?
A. Sanyo B. Mutsuhito
C. Mitsubishi D. Suzuki

138. Who were the first twins to play test cricket?
A. Merv and Kim Hughes
B. Ian and Greg Chappell
C. Surinder and Mohinder Amarnath
D. Steve and Mark Waugh

139. Which country has the largest rail network in the world?
A. India B. China
C. U.S.A D. U.K.

140. Who won the 2009 French open tennis ladies title?
A. Svetlana Kuznetsova
B. Kristina Mladenovic
C. Maria Sharapova
D. Venus Williams

141. Who invented the microchip?
A. Robert Noyce and Jack Kilby
B. Larry Page and Sergey Brin
C. Sabeer Bhatia
D. Gordon Moore & Robert Noyce

142. Which is the largest landlocked country in the world?
A. Switzerland B. China
C. Kazakhstan D. Mongolia

143. Which country makes Panama hats?
A. Ecuador B. Panama
C. Cuba D. Bolivia

144. Who was the Prime Minister of India when the 42nd Amendment Bill bringing in the changes in the Constitution was passed?

A. Rajiv Gandhi B. V.P. Singh
C. Indira Gandhi D. Morarji Desai

145. Which one of the following planets lost its planet status recently?

A. Moon B. Neptune
C. Pluto D. Saturn

146. One barrel of oil is approximately equivalent to how many litres?

A. 200 B. 150
C. 120 D. 160

147. Five year plans in India are finally approved by the :

A. Union Cabinet
B. President on the advice of prime minister
C. National Development Council
D. Planning Commission

148. Many a time we read in the newspapers about Southern Cone of South America which includes Argentina, Chile, Paraguay, Uruguay and Peru. Which is the most spoken language of this geographical area?

A. Portuguese B. Spanish
C. French D. English

149. Which one of the following is not a member of the organization of petroleum exporting countries (OPEC)?

A. Algeria B. Brazil
C. Ecuador D. Nigeria

150. Who invented the 'computer laptop'?

A. Arthur Wynna – USA
B. Q. Daimler – Germany
C. Sinclair – Britain
D. None of the above

ANSWERS

1	2	3	4	5	6	7	8	9	10
B	?	A	D	A	C	D	B	B	D
11	**12**	**13**	**14**	**15**	**16**	**17**	**18**	**19**	**20**
C	D	A	C	D	A	C	C	B	A
21	**22**	**23**	**24**	**25**	**26**	**27**	**28**	**29**	**30**
A	?	B	D	B	C	C	D	C	C
31	**32**	**33**	**34**	**35**	**36**	**37**	**38**	**39**	**40**
A	B	B	A	D	D	B	B	D	D
41	**42**	**43**	**44**	**45**	**46**	**47**	**48**	**49**	**50**
B	A	D	B	B	B	D	C	B	A
51	**52**	**53**	**54**	**55**	**56**	**57**	**58**	**59**	**60**
B	A	B	D	B	B	A	D	C	C
61	**62**	**63**	**64**	**65**	**66**	**67**	**68**	**69**	**70**
A	B	A	D	D	C	A	B	D	D
71	**72**	**73**	**74**	**75**	**76**	**77**	**78**	**79**	**80**
D	D	C	D	D	B	B	B	D	A
81	**82**	**83**	**84**	**85**	**86**	**87**	**88**	**89**	**90**
C	D	C	B	D	C	B	C	A	C
91	**92**	**93**	**94**	**95**	**96**	**97**	**98**	**99**	**100**
D	C	A	B	A	B	D	D	B	B
101	**102**	**103**	**104**	**105**	**106**	**107**	**108**	**109**	**110**
D	B	D	B	A	B	A	A	C	B
111	**112**	**113**	**114**	**115**	**116**	**117**	**118**	**119**	**120**
A	D	?	A	D	A	A	C	D	B
121	**122**	**123**	**124**	**125**	**126**	**127**	**128**	**129**	**130**
C	B	B	C	B	D	B	B	A	B
131	**132**	**133**	**134**	**135**	**136**	**137**	**138**	**139**	**140**
D	D	A	D	A	C	C	D	C	A
141	**142**	**143**	**144**	**145**	**146**	**147**	**148**	**149**	**150**
A	C	A	C	C	D	D	B	B	D

SOME SELECTED EXPLANATORY ANSWERS

1. The sum of numbers can be 15 in the following three ways.

Case (i):

15 = 3 + 6 + 6

The first, second and third throws can be (3, 6, 6), (6, 3, 6) and (6, 6, 3) respectively.

∴ Total number of ways in which 3, 6 and 6 can be obtained is 3.

Case (ii):

15 = 4 + 5 + 6

The first, second and third throws can be either of 4, 5 and 6.

∴ Total number of ways in which 4, 5 and 6 can be obtained is 6.

Case (iii):

15 = 5 + 5 + 5

The first, second and third throws can be 5, 5 and 5.

∴ Total number of ways in which 5, 5 and 5 can be obtained is 1.

∴ The total number of ways in which the sum of throws can be 15 is 3 + 6 + 1 = 10

The total number of ways in which the first roll will be a 4 is 2

∴ Required chance = $\frac{2}{10} = \frac{1}{5}$

2. Let b and s be the speed of the boat and stream respectively.

As per the given conditions,

$$\frac{30}{b+s} = 2 \quad ...(i)$$

and

$$\frac{30}{b-s} = 6 \quad ...(ii)$$

Dividing (i) and (ii),

$$\frac{b-s}{b+s} = \frac{1}{3}$$

∴ $b = 25$

Substituting in (i)

$$\frac{30}{2s-s} = 2$$

∴ $s = 5$

∴ $b = 10$ kmph

The correct answer is not there in the options.

3. A number divisible by 4 formed using the digits 1, 2, 3, 4 and 5 has to have the last two digits 12 or 24 or 32 or 52.

In each of these cases, the five digit number can be formed using the remaining 3 digits in 3 × 2 × 1 = 6 ways.

∴ A number divisible by 4 can be formed in 6 × 4 = 24 ways.

Total numbers that can be formed using the digits 1, 2, 3, 4 and 5 without repetitions = 5! = 120

∴ Required probability = $\frac{24}{120} = \frac{1}{5}$.

4. Let a, b, c, d, e, and f be six numbers, such that

$$a = e + m$$
$$b = e - k$$
$$c = e + l$$
$$d = e - j$$
$$f = e + p$$

Then, the sum of deviations of a, b, c, d, e, and f from e is $(m - k + l - j + p)$

The means of a, b, c, d, e, and f is given by

$$\frac{a+b+c+d+e+f}{6}$$

$$= \frac{e+m+e-k+e-l+e-j+e+p}{6}$$

$$= \frac{6e+(m-k-l-j+p)}{6}$$

$$= \frac{6e + \text{sum of deviations}}{6}$$

Using the similar logic, the required mean

$$= \frac{20 \times 23 + 70}{20} = 26.5.$$

5. Let x gms of gold be added to the alloy to increase the percentage of gold in it to 90%.

∴ $\frac{80\% \text{ of } 50+x}{50+x} = \frac{90}{100}$

∴ $\frac{40+x}{50+x} = \frac{90}{100}$

∴ $x = 50$ gms.

6. Let the incomes be $7x$ and $3x$ and let the expenses be $5y$ and $2y$.

∴ $7x - 5y = 300 \quad ...(i)$

and $3x - 2y = 300 \quad ...(ii)$

Solving (i) and (ii) simultaneously, we get

∴ A's income = $7x = 7 \times 900$ = ₹ 6300.

7. Assume the family consumes 1 kg usually.

To keep expenditure at ₹ 24, its new consumption should be $\frac{24}{27} = \frac{8}{9}$ kg

∴ Percentage decrease in consumption = $\frac{\left(1-\frac{8}{9}\right)}{1} \times 100$

= 11.1%.

8. From a certain station, there will be a ticket for each of the other 9 stations and there are 10 stations on the railway line.

∴ The number of different journey tickets = $9 \times 10 = 90$.

9. Circumference of a circle = $2\pi r$, where r is the radius.
Circumference $\propto r$
As the circumference increases by 5%, the radius also increases by 5%
∴ New radius = $1.05r$
∴ As area $\propto$ (radius)^2
∴ New area = $(1.05)^2 \times$ old area
= $1.1025 \times$ old area
∴ Percentage increase in area = 10.25%.

10. The three vowels in ABACUS are A, A and U. These three can be arranged among themselves in $3!/2! = 3$ ways.
As the three vowels are to appear together, we consider them as one entity. Thus we have four letters; (AAU), B, C and S to be arranged.
This can be done in 4! ways.

∴ Required number of ways = $\dfrac{4! \times 3!}{2}$.

11. We can see that $\dfrac{8000}{6000} = \dfrac{700}{525} = \dfrac{4}{3}$

∴ The time that 525 requires to grow to 700 will be the same as the time that 6000 requires to grow to 8000.
∴ Required number of years = 4.

12. At the start of the game J and B together had 4 times the money that T had.
Let the total money be m.
Then $j + b + t = 4t + t = 5t = m$
∴ The total money is divisible by 5.
Similarly from the other three statements in the data, we find that the total money is divisible by 4 and 3 as well.
∴ The total money is divisible by $4 \times 3 \times 5 = 60$
Let the total money be $60y$
At the start of the game,

$j + b = 4t$
∴ $j + b + t = 5t = 60y$
∴ $t = 12y$
∴ $j + b = 48y$ ...(*i*)

Also,

$t + b = 3j$
∴ $4j = 60y$
∴ $j = 15y$

Substituting in (*i*),

$15y + b = 48y$
∴ $b = 33y$

∴ The amounts with J, B and T at the start were $15y$, $33y$ and $12y$ respectively.

Now, at the end of the game,

$j + b = 3t$
∴ $4t = 60y$
∴ $t = 15y$
∴ $j + b = 45y$ ...(*ii*)

Also, $t + b = 2j$
∴ $3j = 60y$
∴ $j = 20y$

Substituting in (*ii*), we get

$b = 25y$

∴ The amounts with J, B and T at the end were $20y$, $25y$ and $15y$ respectively.
∴ The fraction of total money that T had at the beginning was

$$\frac{12y}{60y} = \frac{1}{5}.$$

13. J won $20y - 15y = 5y$

∴ Required fraction = $\dfrac{5y}{60} = \dfrac{1}{12}$.

14. As B lost ₹ 200,

∴ $33y - 25y = 200$
∴ $8y = 200$
∴ $y = 25$
∴ $33y = 825$

∴ B started with ₹ 825.

15. Here a and b are negative and c is positive.
Let $a = -4$, $b = -3$ and $c = 2$
(I) $a - b = -4 + 3 = -1$
and $a - c = -4 - 2 = -6$
So, $a - b > a - c$
So, (I) is not true.
(II) Taking the same values of a, b and c,

we have $\dfrac{a}{c} = -\dfrac{4}{2} = -2$

$\dfrac{b}{c} = -\dfrac{3}{2} = -1.5$

Since $-2 < -1.5$, $\dfrac{a}{c} < \dfrac{b}{c}$

So, (II) is true.
(III) For any value of b and c,

$\dfrac{1}{b} < \dfrac{1}{c}$ since b is negative and c is positive.

So, III is true.
So, statements II and III are true.

16. Let the length of the side of the original square be x.

Then, the length of the diagonal = $\sqrt{x^2 + x^2} = 4\sqrt{2}$

∴ $2x^2 = 32$
∴ $x^2 = 16$
∴ $x = 4$

Now, the area of the other square is twice that of the first square.

So, the area of the second square = $2x^2 = 32$

If the length of the second square is y, then $y^2 = 32$

$\therefore$ Length of the diagonal of this square = $\sqrt{2y^2} = \sqrt{64}$

= 8 cm.

17. Let the length, breadth and height of the rectangular box be l, b and h respectively. Then the maximum length of the pencil that can be accommodated in the box will be equal to the length of its greatest diagonal, which is the body diagonal.

This length is given as $\sqrt{l^2 + b^2 + h^2}$

Dimensions of the box are given as 8 × 6 × 2.

So the maximum length of the pencil that can be kept in the box = $\sqrt{8^2 + 6^2 + 2^2} = \sqrt{104} = 2\sqrt{26}$ cm.

18. Total readership of X is 8.7 lakhs.

$\therefore X + 2.5 + 0.5 + 1 = 8.7$

$\therefore \quad X = 4.7$

Also, total readership of Y is 9.1 lakhs

$\therefore 2.5 + 0.5 + 1.5 + Y = 9.1$

$\therefore \quad Y = 4.6$

Also, total readership of Z is 5.6 lakhs

$\therefore Z + 1 + 0.5 + 1.5 = 5.6$

$\therefore \quad Z = 2.6$

Therefore, we have

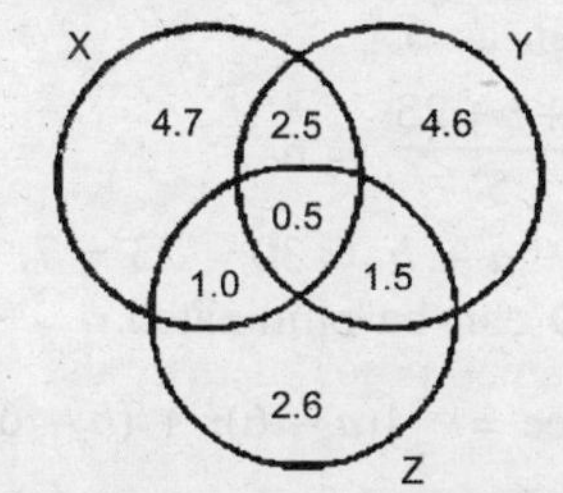

$\therefore$ Number of people who read at least one paper = 4.7 + 4.6 + 2.6 + 1 + 1.5 + 0.5 + 2.5 = 17.4.

19. From the solution to the previous question,

Number of people who read only one newspaper = 4.6 + 4.7 + 2.6 = 11.9.

20. Percentage of Sonali's salary invested in insurance policies = 15%

Percentage of salary spent in shopping and household expenses = 55%

$\therefore$ Percentage of salary saved = 100 – (15 + 55) = 30%

Let Sonali's monthly salary be x

$\therefore \quad 30\%$ of $x = 12750$

$$x = \frac{12750 \times 100}{30}$$

₹ 42,500.

21. There are two varieties of tea, one worth ₹ 25 per kg and the other worth ₹ 30 per kg.

Now, by selling the blended variety at ₹ 30 Kg, profit should be 10%

So, the cost price of the blended tea should be

$$\frac{30}{1.1} = ₹\ \frac{300}{11} \text{ kg}$$

This problem can be depicted through alligation as below

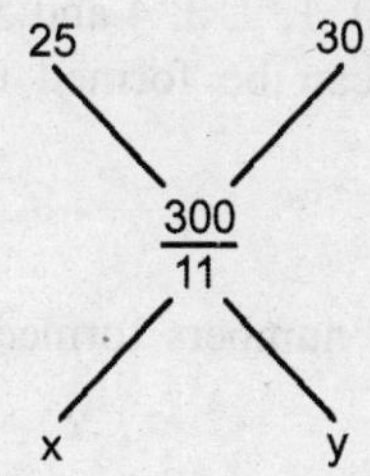

We get $\dfrac{x}{y} = \dfrac{30}{25}$

Amount of tea of ₹ 30 kg to be used is 30 kg. If the amount of tea worth ₹ 25 kg is, a

$$\therefore \quad \frac{30}{25} = \frac{a}{30}$$

$$\therefore \quad a = \frac{30 \times 30}{25} = 36 \text{ kgs.}$$

22. Of the total 80 students, let there be b boys and g girls.

$\therefore \quad b + g = 80 \quad \ldots(i)$

Total pass percentage was 75%.

So number of students who passed = 0.75 × 80 = 60

Now, it is given that 85% of girls and 70% of boys passed.

$\therefore \quad 0.7b + 0.85g = 60 \quad \ldots(ii)$

Solving (i) and (ii), we get

$g = 26.67$ and $b = 53.33$

None of the options has the correct answer.

23. 300 gms solution has 40% salt in it.

So, the amount of salt in it = $\dfrac{40 \times 300}{100}$ = 120 gms

For the salt percentage to be 50%,

$$\frac{120 + x}{300 + x} = \frac{1}{2}$$

$\therefore \quad 240 + 2x = 300 + x$

$\therefore \quad x = 60$ gms.

24. Here, each number in the second row is obtained by multiplying the corresponding number of the first row by 8.

$\therefore$ The number that should replace the question mark

$= \dfrac{192}{8} = 24.$

25. The rectangle, triangle and the circle each move one step to the left in every subsequent row.

Thus, the rightmost figure in the third row is the triangle. We can also see that all the three colours, white, grey and black are present in each row. So the third row should have a black triangle. In the first and second rows, the rectangle is labeled "+", the triangle "C" and the circle "T". Thus the black triangle in row three should also be labeled "C".

26. Using the digits 0, 1, 2, 3, 4 and 5, five digit numbers divisible by 3, can be formed using the following combinations.

Case (i):

1, 2, 3, 4, 5

Total number of numbers formed using these digits = 5! = 120

Case (ii):

0, 1, 2, 4, 5

Total number of numbers formed using these digits = 4 × 4 × 3 × 2 = 96

Thus, total numbers = 120 + 96 = 216.

27. Since $\frac{2}{5}$th of the work is completed in 25 days, remaining $\frac{3}{5}$th of the work is to be completed in 25 days.

Let x men work for 25 days to complete $\frac{3}{5}$th of the work.

$$\frac{M_1D_1H_1}{W_1} = \frac{M_2D_2H_2}{W_2}$$

$$\therefore \quad \frac{25\times105\times8\times5}{2} = \frac{x\times25\times9\times5}{3}$$

$$\therefore \quad x = \frac{105\times8}{2\times3} = 140$$

∴ Additional men employed = 140 – 105 = 35.

28. A can build the structure in 8 days.

∴ Fraction of structure built in a day by A = $\frac{1}{8}$

Similarly, fraction of structure broken by B in a day $= \frac{1}{3}$

Amount of work done by A in 4 days $= \frac{4}{8} = \frac{1}{2}$

Now, both A and B work together for 2 days.

So, the fraction of structure built in 2 days

$$= 2\left(\frac{1}{8}-\frac{1}{3}\right) = \frac{2\times(-5)}{24} = \frac{-5}{12}$$

Fraction of structure still to be built

$$= \frac{1}{2}+\frac{5}{12} = \frac{11}{12}$$

If A takes x days, to build up the remaining structure, then

$$\frac{x}{8} = \frac{11}{12}$$

$$\therefore \quad x = \frac{22}{3}.$$

29. Let the first term and the ratio of the Geometric Progression be a and r respectively.

$\therefore \quad a + ar = 12$

$\therefore \quad a(1 + r) = 12$ ...(*i*)

Also, $ar^2 + ar^3 = 48$

$\therefore \quad ar^2(1 + r) = 48$ ...(*ii*)

Dividing (*ii*) by (*i*),

$$\frac{ar^2(1+r)}{a(1+r)} = \frac{48}{12} = 4$$

$\therefore \quad r^2 = 4$

$\therefore \quad r = \pm\sqrt{4} = \pm 2$

Since the terms of the Geometric Progression are alternately positive and negative,

$r = -2$

∴ From (*i*), $a(1 - 2) = 12$

$\therefore \quad a = -12.$

30. Mean of the given numbers $= \frac{a+b+23}{5}$

Mean is given as 6.

$$\therefore \quad \frac{a+b+23}{5} = 6$$

$\therefore \quad a + b = 30 - 23 = 7.$

So, option D can be eliminated.

Now, variance $= \frac{1}{5}[(a - 6)^2 + (b - 6)^2 + (5 - 6)^2 + (10 - 6)^2 + (8 - 6)^2]$

$\therefore \frac{1}{5}[(a - 6)^2 + (b - 6)^2 + 21] = 6.8$

$\therefore (a - 6)^2 + (b - 6)^2 + 21 = 34$

$\therefore (a - 6)^2 + (b - 6)^2 = 13$

Only option C fits into the above equation.

31. Value of goods = ₹ 15,000

Commission he receives $= 12\frac{1}{2}\%$

∴ Commission received $= \frac{12.5\times15000}{100} = 1875.$

32. $\sqrt{110.25} = 10.5$

$\sqrt{0.01} = 0.1$

$\sqrt{0.0025} = 0.05$

$\sqrt{420.25} = 20.5$

$\therefore \sqrt{110.25} \times \sqrt{0.01} \div \sqrt{0.0025} \times \sqrt{420.25}$

$= 10.5 \times \frac{0.1}{0.05} - 20.5$

$= \frac{1.05}{0.05} - 20.5$

$= 21 - 20.5 = 0.5.$

33. The total number of respondents under the age of 31 = 33 + 33 = 66

The number of respondents under the age of 31 who indicated that Blues is their favourite style of music = 2 + 3 = 5

$\therefore$ The required percentage = $\frac{5}{66} \times 100 \approx \frac{15}{2} \approx 7.5\%$.

34. Therewere 33 respondents in all in the 21–30 age group.

Therewere 12 respondents in this group who preferred Rock music.

$\therefore$ Therewere 33 – 12 = 21 students who favoured a style other than Rock music.

$\therefore$ The required percentage = $\frac{21}{33} \times 100 \approx 64\%$.

35. The total number of respondents = 33 + 33 + 68 = 134

The number of respondents who indicated that Jazz is their favourite style of music = 1 + 4 + 11 = 16

$\therefore$ The required percentage = $\frac{16}{134} \times 100 \approx 12\%$.

36. The total number of students in the Arts faculty = 1049

The number of non-US students in the Arts faculty = 79 + 21 + 6 + 2 + 4 = 112

$\therefore$ The required percentage = $\frac{112}{1049} \times 100 \approx 10.67\%$

The closest option is 11%.

37. Let the total number of students in the University be x.

There are 1049 students in the Arts faculty, which represents 23% of the total students.

$\therefore \quad 1049 = 0.23x$

The number of students in the Engineering faculty is 9% of the total.

$\therefore$ The total number of Engineering students

$= \frac{9}{100} \times \frac{1049}{0.23} \approx 410$ students.

38. 1049 students in Arts represent 23% of the total students.

Let the total number of students in the University be x.

$\therefore \quad 1049 = 0.23x$

$\therefore \quad x = \frac{1049 \times 100}{23} = 4560$

$\therefore$ The total number of students is 4560.

39. The Science students are 21% of the total number of students.

6% out of these are Asians.

The total number of students in the University = 4560

$\therefore$ The number of Asian students studying Science =

$\frac{6}{100} \times \frac{21}{100} \times 4560 \approx 57.$

40. The number of medical students is 5% of the total students.

$\therefore$ The number of medical students = $\frac{5}{100} \times 4560 = 228$

Out of these, 34 are from Europe.

This represents $\frac{34}{228} \times 100 = 14.9\% \approx 15\%$ of the faculty.

41. A gives B as many tractors that they already have.

$\therefore$ Number of tractors with A > Number of tractors with B and C.

42. Option B can be eliminated as "many" is not the same as "majority".

The passage states that "...good basic skills in reading, communication, and mathematics play an important role..." and not reliance on communication skills over basic skills in reading and mathematics. Eliminate option C.

The passage states "most" of the fastest growing jobs. Option D implies "all" the fastest growing jobs. Eliminate option D.

The passage mentions that good basic skills in reading, communication and mathematics are important prerequisites for jobs that do not require a college degree-jobs such as that of a desktop publisher, as mentioned in option A.

Hence, the correct answer is option A.

43. Options A, B and C can be concluded from the data given in the passage.

With the help of binary logic, we arrive at the following to be valid conclusions:

If X, then Y and If not Y, then not X

The following conclusions may or may not be valid:

If Y, then X and If not X, then not Y

The passage states that "If there are no crop protection products (X), then crop yield per acre will drop by more than 50% (Y)."

Option D states If crop yield per acre drops by more than 50% (Y), then crop protection products have not been used (X).

Hence, the conclusion in option D may or may not be valid.

Hence, the correct answer is option D.

44. This is a case of If X, then Y where X is a sufficient but not a necessary condition for Y to occur.

If flight 409 is cancelled (X), then the manager cannot arrive on time (Y).
The other valid conclusion would be Not Y, therefore Not X—The manager can arrive on time, therefore flight 409 is not cancelled.
There are other two conclusions that may or may not be valid—they may or may not be sufficient conditions:
If Y, then X—The manager cannot arrive on time, therefore flight 409 is cancelled.
If not X, then not Y—Flight 409 is not cancelled, therefore the manager can arrive on time.
Lou's conclusion is based on Not X, therefore not Y, which may or may not be true—hence, Lou's argument is fallacious.
Evelyn states that the manager will not arrive at time—there is no possibility that the manager can arrive on time. But, based on data, if flight 409 is not cancelled (not X), then the manager can arrive on time (not Y) may or may not happen. Hence, Evelyn's conclusion that the manager will not arrive on time is unwarranted.
Hence, the correct answer is option B.

45. The passage assumes that a higher percentage of accident related deaths in planes compared to cars implies cars are safer.
We need an argument that attacks the premise that percentage is the only way to determine safety; or we need an argument that brings in another element that shows cars are not as safe as percentages make them to be.
Options A, B and D touch upon other aspects which merely add data to the main statement.
If the number of car accidents is several hundred thousand times higher than the number of plane accidents, it nullifies the premise of percentages being the deciding factor as well as presents data that supports that cars are not as safe as they are made out to be vis-a-vis planes.
Hence, the correct answer is option B.

46. From the figure, it is clear that only Industry B and C contribute to company S.
Hence, option B.

47. Observe that Industry B processes element 4 and 5. But elements 4 and 5 are also processed at Industry A and C.
Hence, the proportion of element 4 that gets processed in Industry B is unknown.
Similarly, the proportion of element 5 that gets processed in Industry B is unknown.
∴ The percentage of total production of listed elements processed by Industry B cannot be determined.
Hence, option D.

48. Industry A processes all the listed elements, i.e. Element A, B, C, D and E.
The total annual production of all these elements is 1,00,000 tons.
Hence, option C.

49. RB shows the correct analogy with exact antonyms in both the pairs- Ascend : Increase :: Descend : Decrease.
Hence, the correct answer is option B.

50. AS introduces 'Western' which does not logically fit in the analogy.
AQ gives the perfect antonymous analogous relationship—Modern : Ancient :: Young : Old.
Hence, the correct answer is option A.

51. BQ gives the correct relationship—
Part : Whole :: Class : School.
Hence, the correct answer is option B.

52. Summit and Apex are synonyms. The only option which comes close to being a synonym is AQ-Beautiful : Pretty.
Hence, the correct answer is option A.

53. The code for B is 5, code for E is 7 and so on. Based on this coding system, the code for FUTURE is 204097.
Hence, option B.

54. If the month that ends on Wednesday have 31 days, there will be 5 Mondays otherwise will have 4 Mondays. Therefore, cannot be specified.
Hence, option D.

55. From the first row, $\frac{(17 \times 12)}{2} = 102$
∴ From the second row, $\frac{(15 \times 10)}{2} = 75$
Hence, option B.

56. From the first row, $12 \times 14 \times 2 = 336$
∴ From the second row, $15 \times 16 \times 2 = 480$
Hence, option B.

57. From condition (III), B and G are in different boats.
∴ If F and B are in one boat, G is in the other boat.
Hence, option A.

58. From condition (IV), the maximum number of persons in one boat can be four only.
∴ There must be at least three people in the complete list of people who must be sitting in either boat.
Hence, option D.

59. In the first figure there are 4 vertical and 3 horizontal lines.
In the following figures both the vertical and horizontal lines reduce by 1.
∴ In the fourth figure there will be only one vertical line.
Hence, option C.

60. In the first figure there are 3 shaded balls on 3 lines. In the second figure the shaded balls remain in the same places and the lines are turned in the opposite direction.

In second figure there are 2 shaded balls on two sticks.
∴ In the fourth figure the lines are turned in the opposite direction.
Hence, option C.

61. In the alphabetic order I is second to the right of G and K is second to the right of I, similarly M is second to the right of K.
Therefore, the first alphabet of the next element in the series will be M.
Therefore, we can eliminate option B.
Now, O is second to the right of M, Q is second to the right of O and similarly S is second to the right of Q.
∴ We can eliminate option C.
In option A and D only the last letter is different.
Consider the last letters. A is second to the right Y, C is second to the right of A and similarly E is second to the right of C.
Hence, option A.
[**Note:** There is a misprint in the second word. The third letter should be U instead of V.]

62. Mohan spent 30% on landline, 40% on GSM and 30% on Airtel and spent 30 minutes and 100 minutes on STD and landline calls respectively.
This breakup can also be shown in a tabular manner as follows:

Mohan	Airtel	GSM	Landline
STD (minutes)	9	12	9
Local (minutes)	30	40	30

Mohan falls in the category of 50–200 km.
Hence, Mohan's STD cost is (9 × 1.5) + (12 × 1.5) + (9 × 2) = 13.5 + 18 + 18 = ₹ 49.5
His cost of local calls is (30 × 1) + (40 × 1) + (30 × 2) = ₹ 130
Price of plan = ₹ 99
Hence, Mohan's total cost = 99 + 49.5 + 130 = ₹ 278.5
Rohan spent 40% on landline, 30% on GSM and 30% on Airtel and spent 18 minutes and 120 minutes on STD and landline calls respectively.
This breakup can also be shown in a tabular manner as follows:

Rohan	Airtel	GSM	Landline
STD (minutes)	5.4	5.4	7.2
Local (minutes)	36	36	48

Rohan falls in the category of 50 – 200 km.
Hence, Rohan's STD cost is (5.4 × 1.5) + (5.4 × 1.5) + (7.2 × 2) = 8.1 + 8.1 + 14.4 = ₹ 30.6
His cost of local calls is (36 × 1) + (36 × 1) + (48 × 2) = ₹ 168
Price of plan = ₹ 99
Hence, Rohan's total cost = 99 + 30.6 + 168 = ₹ 297.6
Hence, Rohan spent more.
Hence, option B.

63. Solve this problem by comparing the charges incurred while sending 38, 40, 60 and 59 local SMSs respectively under the old scheme and the new scheme.

Old Scheme			New Scheme			
Number of SMS	Rate	Total Cost	Monthly Charge	Number of SMS	Rate	Total Cost
38	1.5	57	35	38	0.6	57.8
40	1.5	60	35	40	0.6	59
60	1.5	90	35	60	0.6	71
59	1.5	88.5	35	59	0.6	70.4

Thus, a person sending 38 local SMS per month actually pays 80 paise more.
Hence, option A.

64. Note that this question actually asks for a plan which is not possible at all.
The other three plans may be possible in some cases.
The fixed price of the plan is ₹ 99.
Hence, the call charges should be exactly ₹ 100 in three out of four cases.

Consider option A:
30 calls to Airtel on STD (500 + kms) cost the user ₹ 1.5 per call.
Hence, the total STD cost = 30 × 1.5 = ₹ 45.
If the person makes all his local calls to Airtel numbers or to GSM phones, he will be charged ₹ 1 per call.
Hence, his total local call cost = 55 × 1 = ₹ 55
Hence, the total calling cost = 45 + 55 = ₹ 100.
Hence, a bill of 99 + 100 = ₹ 199 is possible.

Consider option B:
16 calls to Airtel on STD (15 kms) cost the user ₹ 1.5 per call.
Hence, the total STD cost = 16 × 1.5 = ₹ 24.
If the person makes all his local calls to Airtel numbers or to GSM phones, he will be charged ₹ 1 per call.
Hence, his total local call cost = 76 × 1 = ₹ 76
Hence, the total calling cost = 24 + 76 = ₹ 100.
Hence, a bill of 99 + 100 = ₹ 199 is possible.

Consider option C:
10 STD calls (250 kms) each to Airtel, GSM and landline respectively.
Hence the total STD cost = 10 × (1.5 + 2.5 + 2.5) = ₹ 65
He makes 30 local calls in all.
Hence, if the person makes 25 calls to Airtel and 5 call to a landline, he gets charged as ₹ 1 and ₹ 2 respectively i.e. (25 × 1) + (5 × 2) = ₹ 35
Hence, the total calling cost = 65 + 35 = ₹ 100
Hence, a bill of 99 + 100 = ₹ 199 is possible.

Consider option D:
The person makes 8, 4 and 7 calls to Airtel, GSM and landline respectively on STD (500+ kms).
Hence, his total STD cost = (8 × 1.5) + (4 × 3) + (7 × 3.5) = ₹ 49.5

He makes 55 local calls a month.
Even if he makes all the calls to Airtel, he gets charged at ₹ 1
Hence, his minimum local call charges are ₹ 55
Hence, his minimum total call charges are 49.5 + 55 = ₹ 104.5
Hence, a total of ₹ 199/month cannot be obtained for this breakup.
Hence, option D.

65. 80% of the ISD bill came from calling the rest of the world.
Hence, 20% of the bill came from the other regions.
The total time spent on ISD calling was 12 minutes.
Since the ISD rates for each region vary, the 12 minutes cannot be proportionately divided based on the bill.
Consequently, it is not possible to find out the calling duration for the other two regions.
Hence, the answer cannot be determined.
Hence, option D.

66. Total demand = 3000 + 600 + 2500 + 1200 + 3300 = 10600 units.

$$\text{Average demand} = \frac{\text{Total demand}}{\text{Number of T.V. companies}}$$

$$= \frac{10600}{5} = 2120 \text{ units.}$$

Total production = 1500 + 1800 + 1000 + 2700 + 2200 = 9200 units.

$$\text{Average production} = \frac{\text{Total production}}{\text{Number of T.V. companies}}$$

$$= \frac{9200}{5} = 1840 \text{ units.}$$

Difference between the average demand and the average production = 2120 − 1840 = 280
Hence, option C.

67. Production of company D = 2700.
Production of company A = 1500.

$$\frac{\text{Production of company D}}{\text{Production of company A}} = \frac{2700}{1500} = 1.8$$

i.e. Production of company D is 1.8 times the production of company A.
Hence, option A.

68. Paper cost is 10% of the total cost.
Angle for sector representing paper cost =

$$\left(\frac{10}{100}\right) \times 360 \text{ degrees} = 36 \text{ degrees.}$$

Hence, option B.

69. 2% of the cost is ₹ 2000 for 12500 copies.
∴ 1% of the cost is ₹ 1000.
Using this information,
The total expenditure = ₹ 1,00,000
∴ Cost for 1 copy = (100000/12500) = ₹ 8.
S.P for a profit of 5 percent = 8 × 1.05 = ₹ 8.40.
Hence, option D.

70. 'Tree' is called 'sky'. Hence, logically 'fruit' grows on 'sky'.
Hence, the correct answer is option D.

Previous Paper (Solved)

Symbiosis National Aptitude Test (SNAP), 2008

SECTION- I

GENERAL ENGLISH

Direction (Q. No. 1): *Match the following idiomatic references to parts of the human anatomy.*

1. 1. palm — 5. could not tolerate the insult
2. foot — 6. to look at with envy and desire
3. eye — 7. to put the blame on someone else
4. stomach — 8. forced to pay the bill

A. 1-8, 2-6, 3-5, 4-7 B. 1-5, 2-7, 3-6, 4-8
C. 1-7, 2-8, 3-6, 4-5 D. 1-6, 2-8, 3-5, 4-7

Directions (Q. Nos. 2–5): *Answer the following questions based on the information given below:*

2. Find the maximum number of times any one of the given words fits the sets of sentences

RAISE ARISE
AROSE RISE

(*a*) Opportunities will ________, and you must grab them.
(*b*) A hot wind ________ from the desert.
(*c*) I ________ at dawn on most days.
(*d*) A mood of optimism ________ among the people.

A. in all four sentences B. in 3 sentences
C. in 2 sentences D. in 1 sentence

3. Which two sentences in the following convey the same idea? Choose from the combinations listed below:

(*a*) He is in a fool's paradise
(*b*) He can't see the wood for the trees
(*c*) He can't distinguish between reality and fancy.
(*d*) He is unable to separate unimportant details from the really important ones

A. 2, 3 B. 2, 4
A. 1, 4 D. 1, 3

4. Find the correct match of grammatical function with usage for the word THEN.

Function	
1. Noun	5. He was the then King of Nepal.
2. Adjective	6. I have not heard about him since then.
3. Adverb	7. He is not feeling well, then how can he possibly go for the picnic.
4. Conjunction	8. He was not a graduate then.

A. 1-8, 2-5, 3-7, 4-7 B. 1-6, 2-5, 3-8, 4-7
C. 1-7, 2-5, 3-6, 4-8 D. 1-8, 2-7, 3-6, 4-5

5. We can never make our beliefs, regarding the world, certain. Even scientific theory of a most rigorous and well-confirmed nature is likely to change over a decade or even tomorrow. If we refuse to even try to understand, then it is like resigning from the human race. Undoubtedly life of an unexamined kind is worth living in other respects—as it is no mean thing to be a vegetable or an animal. It is also true that a man wishes to see this speculation domain beyond his next dinner. From the above passage it is clear that the author believes that

A. men would do well not to speculate
B. progress in the scientific field is impossible
C. one should live life with the dictum 'what will be will be'
D. men are different from animals as far as their reasoning abilities are concerned.

Directions (Q. No. 6–11): *Answer the questions based on the passage given below:*

Rajendra K. Pachauri, head of the Intergovernmental Panel on Climate Change, is getting nightmares because of the Nano, Tata's soon-to-be-launched Rs. One lakh car. Sunita Narain of the Centre for Science and Environment (CSE) says that it isn't the Nano by itself but cars overall that give her nightmares. The villains in my nightmares are neither the Nano nor cars overall, but stupid government policies that subsidize and encourage pollution, adulteration and congestion.

Sanctimonious greens call the Nano disastrous because of its affordability—millions more will now clog roads and consume more fossil fuel. This is elitism parading as virtue. Elite greens own cars, but cannot stand the poorer masses becoming mobile, since the consequent congestion will eat into the time of the elite!

More logical would be a protest against big cars that use more space and fuel, or highly polluting old cars. Instead, green hypocrites aim at a new car with the lowest cost, best mileage and least emissions. The Nano will not burden us with too many cars. India has very few cars per person by world standards. London and New York have ultra-high car densities, yet have clearer air than Delhi. Our problem is too many bad policies, not too many cars.

We subsidize vehicles on a gargantuan scale invisible to lay folk. Roads and flyovers cost crores to build and maintain, yet road use is free (save on a few toll roads). Traffic police and lights are costly, yet are provided free. These invisible subsidies starve cities of funds to expand roads and public transport.

Land in cities now costs lakhs per square metre. Yet parking is free in the suburbs, and often costs just ₹ 10 day per day in city centres. A single parking space of 23 square meters occupies land worth ₹ 40 lakhs. A car occupies more space than an office desk, yet the desk space pays full commercial rent while parking space costs just about ₹ 10 per day.

Daily parking charges range from $30 (₹ 630) in Washington to $60 (₹ 1260) in New York. CSE launched a sensible campaign to raise parking fees in Delhi to ₹ 120 per day, but was foiled. So, parking space now exceeds green space, a scathing comment on priorities.

The world price of crude oil has risen 13 fold since 1998 to over $139 per barrel, but Indian petrol prices have barely doubled. Left Front politicians, who once wanted to soak the rich, now want to subsidize them. Under-recoveries of oil companies' total may be ₹ 2,00,000 crore, even after a recent price hike. This is far more than the cost of Sarva Shiksha Abhiyan (education for all) and the Employment Guarantee Scheme put together.

We sanctimoniously lecture rich countries to reduce their green house emissions, yet subsidize our own. Diesel is subsidized to be cheaper than petrol. So, Indian car makers produce the highest proportion of diesel cars in the world. Diesel fumes contain suspended particles that are highly toxic. This subsidy kills.

So does kerosene provided at throwaway prices, ostensibly to benefit poor villagers. One third of all kerosene is used to adulterate petrol and diesel. This causes horrendous pollution even in the greenest of cars.

What's the way forward? We must abolish subsidies and raise taxes on vehicles and fuels to reflect their full social cost. The biggest but least visible subsidy is for parking, and we should start there.

Many car owners in the West take public transport to work since parking space downtown is costly and scarce. We should levy parking fees on an hourly, not daily, basis. ₹ 10 per hour could be a starting point in the metros.

In parts of Tokyo, you cannot own a car unless you own a private parking space. This is too extreme for India, but indicates the future path. If we charge owners the full social cost of parking, people will buy smaller and perhaps fewer vehicles, and fewer still will take them to work. That will slash congestion and pollution.

Cities should levy stiff annual taxes on vehicles, not a one-time tax, and use the revenue to constantly expand public transport and roads. This will create economic synergy: Private transport will finance public transport. London and New York have high density public transport as well as high car density.

Apart from underground rail, cities need elevated roads to ease congestion and pollution. Lata Mangeshkar helped kill a proposal for an elevated road near her Mumbai flat: perhaps she felt her throat and singing would be affected. She did not care that the throats of poor people living on the pavements were far worse affected by fumes, and might get relief if some fumes were diverted to a higher level. What elitism!

Next, some medicine that will be really bitter, politically. The excise duty on all automotive vehicles should be raised to reflect their social costs. Fuel subsidies should be abolished. Price differentials between petrol, diesel and kerosene should be removed, ending incentives for adulteration. Diesel cars should bear a heavy additional cess to finance improved healthcare for those affected by their emission of harmful particulate matter.

That is a long, politically difficult agenda. Only part of it will ever be achieved. Yet that is the way to go, rather than agitate the Nano.

6. By 'Sanctimonious greens' the writer refers to
A. aristocratic environmentalists
B. the rich
C. environmentalists with a 'holier than thou' attitude
D. those who decry deforestation

7. The elite are
A. jealous of Nano owners
B. afraid of traffic jams and depletion of fossil fuel
C. afraid of reaching their destinations late
D. full of disdain that the poor can afford cars

8. The paradox of the situation is that
A. bigger cars mean more fuel, more space and more pollution
B. though India has fewer cars the Nano will bring more pollution
C. London and New York have more cars and less pollution
D. though India is smaller than the US its cars cause more pollution

9. In saying 23 square metres of parking space costs 40 lakhs, the writer is ______
A. caustic B. exaggerating
C. sarcastic D. ironical

10. The writer blames India for
A. subsidizing kerosene whereby greenhouse emissions are indirectly subsidized
B. subsidizing diesel

C. for increasing the cost of parking by the hour
D. for not making it mandatory for car owners to own parking space

11. The most suitable title for this passage is
A. Polluting Politics
B. No No Nano
C. Submerge Subsidies
D. More Cars, Less Pollution

Directions (Q. Nos. 12–21): *Answer the following questions based on the information given below:*

12. The plural of Virus is—
A. Viruses B. Virae
C. Virii D. Virus

13. If the following segments of a sentence are to be rearranged in logical order as A, B, C, D where would '3' be placed?
1. to see that students do not altogether forget to write especially during exam time
2. the education groups are now asking for hand writing classes
3. thanks to mobile testing and computer literacy
4. writing in long hand is becoming a vanishing art
A. A B. B
C. C D. D

14. If leaf is to leaves and knife is to knives, then belief is to—
A. beliefs B. believes
C. belief D. believing

15. Choose the sentence where the underlined word is used correctly.
A. This latest novel is a <u>pedestrian</u> story about spies.
B. The exam paper is not <u>pedestrian</u> but difficult.
C. This is the <u>pedestrian</u> highway.
D. None of these.

16. When the fire alarm rang ______ left the building immediately.
A. all B. everyone
C. all the people D. every person

17. In the following sentence choose the erroneous segment/s.
(A) He is one of those people /(B) who thinks /(C) he owns the world
A. Error in segment A
B. Error in segment B
C. Error in segment A & C
D. Error in segment B & C

18. Choose the correct meaning for the word: CYNIC
A. the person who is selfish
B. the person who is concerned about others
C. the person who isn't misanthropic
D. the person who believes that people always act from selfish motives

19. Choose the word with correct spelling.
A. catagories B. diarhoea
C. ommission D. inagurate

20. Pick out the right sentences.
1. I will go with you.
2. There was nobody I could go with.
3. I have a glass with painting on it.
4. The curtains do not match with the furniture.
A. 1 & 2 B. 2 & 3
C. 1 & 4 D. All

21. About the following pair of phrases, choose the correct option.
i. A two days' visit
ii. A two day's visit
A. The first phrase is erroneous
B. The second phrase is erroneous
C. Both phrases are erroneous
D. Both phrases are correct

Directions (Q. Nos. 22 – 24): *Answer the questions based on the passage given below:*

Nearly two thousand years have passed since a census decreed by Caesar Augustus became part of the greatest story ever told. Many things have changed in the intervening years. The hotel industry worries more about overbuilding than overcrowding, and if they had to meet an unexpected influx, few inns would have managed to accommodate the weary guests. Now it is the census taker that does the travelling in the fond hope that a highly mobile population will stay put long enough to get a good sampling. Methods of gathering, recording and evaluating information have presumably been improved a great deal. And where then it was the modest purpose of Rome to obtain a simple head count as an adequate basis for levying taxes, now batteries of complicated statistical series furnished by governmental agencies and private organizations are eagerly scanned and interpreted by sages and seers to get a clue for future events.

The Bible does not tell us how the Roman census takers made out, and as regards our more immediate concern, the reliability of present-day economic forecasting, there are considerable differences of opinion. They were aired at the celebration of the 125th anniversary of the American Statistical Association. There was the thought that business forecasting might well be on its way from an art to a science, and some speakers talked about new-fangled computers and high-faulting mathematical systems in terms of excitement and endearment, which we, at least in our younger years when these things mattered, would have associated more readily with the description of a fair maiden.

But others pointed to a deplorable record of highly esteemed forecasts and forecasters with a batting average below that of the Mets and the President-elect of the Association cautioned that "high-powered statistical methods are usually in order where the facts are crude and inadequate, statisticians assume".

We left his birthday party somewhere between hope and despair and with the conviction, not really newly acquired, that proper statistical methods applied to ascertainable facts have their merits in economic forecasting as long as neither forecaster nor public is deluded into mistaking the delineation of probabilities and trends for a prediction of certainties of mathematical exactitude.

22. According to the passage, taxation in Roman times was based on

A. mobility
B. wealth
C. population
D. census takers

23. The author refers to the Mets primarily in order to

A. show that sports do not depend on statistics
B. contrast verifiable and unverifiable methods of record keeping
C. indicate the changes in attitudes from Roman days to the present
D. illustrate the failure of statistical predictions.

24. The author's tone can best be described as

A. jocular
B. scornful
C. pessimistic
D. humanistic

Directions (Q. Nos. 25-27): *Answer the following questions based on the information given below:*

25. Disinterested is closest in meaning to

A. bored
B. unbiased
C. not interested
D. indifferent

26. Choose the option which is nearly the opposite in meaning to BERATE

A. grant
B. praise
C. refer
D. purchase

27. Arrange the following in the right order to make a complete sentence

(*i*) with interconnected vibrating balls and springs
(*ii*) in a naïve sense, a field in physics may be envisioned as if space were filled
(*iii*) as the displacement of a ball from its rest position
(*iv*) and the strength of the field can be visualized

A. (*ii*), (*i*), (*iv*), (*iii*)
B. (*i*), (*ii*), (*iii*), (*iv*)
C. (*iv*), (*iii*), (*ii*), (*i*)
D. (*iii*), (*iv*), (*i*), (*ii*)

Direction (Q. No. 28): *Select the odd man out from the given alternatives.*

28. Find the odd one out

A. latent
B. natural
C. inborn
D. inherent

Direction (Q. No. 29): *In each of the following sentences, parts of the sentence are left blank. Beneath each sentence, different ways of completing the sentence are indicated. Choose the best alternative among them.*

29. He told the teacher that ______.

A. he was liked by the whole class
B. you are liked by the whole class
C. he is liked by the whole class
D. you were liked by the whole class

Directions (Q. Nos. 30 and 31): *Answer the following questions based on the information given below:*

30. Match the several meanings of the word COMPLEX with their appropriate usages.

Meaning	Usage
1. complicated	5. A new sports complex is coming up for the Common Wealth Games.
2. abnormal state of mind	6. Culture is a complex whole of many things.
3. group of structures	7. She has a complex about being overweight.
4. mixture	8. His motives in carrying out the crime were complex.

A. 1-6, 2-8, 3-7, 4-5
B. 1-8, 2-7, 3-5, 4-6
C. 1-5, 2-7, 3-6, 4-8
D. 1-8, 2-5, 3-6, 4-7

31. Which does not make a sensible word/phrase when added to the word:
HONEY

A. pot
B. suckle
C. comb
D. taste

Directions (Q. Nos. 32-34): *Answer the question based on the passage given below:*

In September of 1929, traders experienced a lack of confidence in the stock market's ability to continue its phenomenal rise. Prices fell. For many inexperienced investors, the drop produced a panic. They had all their money tied up in the market, and they were pressed to sell before the prices fell even lower. Sell orders were coming in so fast that the ticker tape at the New York Stock Exchange could not accommodate all the transactions.

To try to reestablish confidence in the market, a powerful group of New York bankers agreed to pool their funds and purchase stock above current market values. Although the buy orders were minimal, they were counting on their reputations to restore confidence on the part of the smaller investors, thereby affecting the number of sell orders. On Thursday, October 24, Richard Whitney, the Vice President of the New York Stock Exchange and a broker for the J.P. Morgan Company, made the effort on their behalf. Initially it appeared to have been successful, then, on the following Tuesday, the crash began again and accelerated. By 1932, stocks were worth only twenty percent of their value at the

1929 high. The results of the crash had extended into every aspect of the economy, causing a long and painful depression, referred to in American history as the Great Depression.

32. The New York bankers counted on—

A. Current market values
B. The number of sell orders
C. Confidence
D. Their reputation

33. The cause of downfall of share market was—

A. Inexperienced investors
B. Phenomenal decrease
C. Lack of confidence in stock market's ability
D. Panic amongst investors

34. Choose the word in the passage that is an antonym of 'MINIMAL'

A. Negligible B. Minimum
C. Maximal D. Significant

35. Identify the correct sentence.

A. The office is opposite to the bank.
B. The office is opposite the bank.
C. The office is opposite from the bank.
D. The office is opposite of the bank.

SECTION-II

ANALYTICAL & LOGICAL REASONING

36. A band passes around all the wheels so that they can all be turned by the driving wheel. When the driving wheel turns in the direction shown, which way will the wheel B turn?

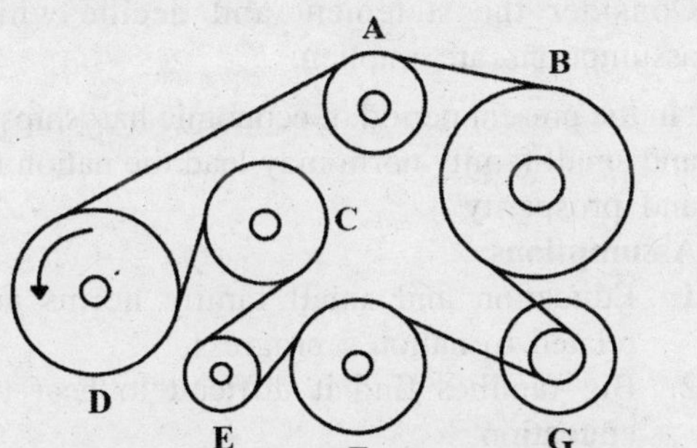

A. Clockwise B. Cannot move
C. Anti-clockwise D. Either way

37. In a certain language, (A) 'Sun shines brightly' is written as 'ba lo sul'; (B) Houses are brightly lit' is written as 'kado udo ari ba'; and (C) 'Light comes from sun' as 'dapi kup lo nro'. What words will be written for 'sun' and 'brightly'?

A. lo, ba B. ba, lo
C. sul, lo D. ba, sul

38. Given are the following three equations:

1. 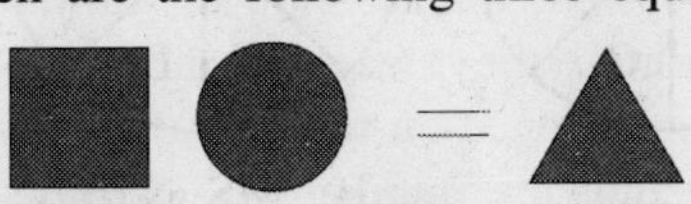

2. 

3. 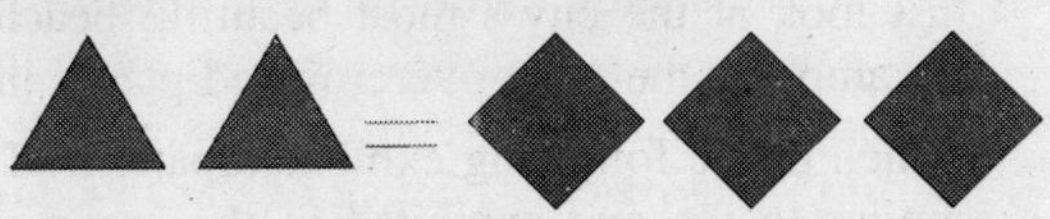

How many circles are equivalent to a square, as per the ratios in the three given equations?

A. 

B. 

C. 

D. 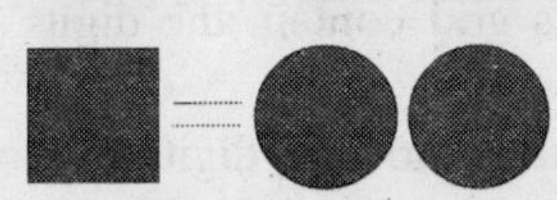

39. Each child in a family has at least 4 brothers and 3 sisters. What is the smallest number of children the family might have?

A. 7 B. 8
C. 9 D. 10

40. In the following question two statements are followed by two conclusions numbered I and II. Assume the two statements are true even if they are at variance with commonly known facts. Then pick the correct answer from the choices given below:

1. Only conclusion I follows
2. Only conclusion II follows
3. Both conclusions I & II follow
4. Neither conclusion I nor conclusion II follows

Statements: Some doctors are fools. Joshi is a doctor.

Conclusions: I. Joshi is a fool.
II. Some fools are doctors.

A. 1 B. 2
C. 3 D. 4

41. Debu walks towards the East then towards North and turning 45° right walks for a while and lastly turns towards left. In which direction is he walking now?

A. North B. East
C. South-East D. North-West

Directions (Q. Nos. 42-44): *Answer the following questions based on the information given below:*

I. There is a rectangular wooden block of length 4 cm, height 3 cm and breadth 3 cm.
II. The two opposite surfaces of 4 cm × 3 cm are painted yellow on the outside.
III. The other two opposite surfaces of 4 cm × 3 cm are painted red on the outside.
IV. The remaining two surfaces of 3 cm × 3 cm are painted green on the outside.
V. Now, the block is cut in such a way that cubes of 1 cm × 1 cm × 1 cm are created.

42. How many cubes will have only one colour?

A. 10 B. 12
C. 14 D. 18

43. How many cubes will have no colour?

A. 1 B. 2
C. 4 D. 8

44. How many cubes will have any two colours?

A. 34 B. 24
C. 16 D. 12

45. Read the following about the grid given below and answer.

- The cells in this grid contain the digits 1 to 9 in random order.
- Column A contains no odd digits.
- Cell C3 minus Cell C2 equals 4.
- The sum of three digits in Row 1 is 17.
- Number 7 is in Column B; its left hand neighbour is not 4.
- The digits of Column C add up to 14.
- 2 is not in the same horizontal row as 8; and 9 is not immediately below 3.

Which cell holds the number 9?

	A	B	C
1			
2			
3			

A. B1 B. B3
C. C2 D. C1

46. Replace the question mark with the right option.

4, 32, 288, ?, 31680

A. 25600 B. 2880
C. 7420 D. 10000

47. In the Sunday bazzar, Jamuna sells her lemons at ₹ 0.50 for two. Her neighbour Seema has a little smaller lemons; she sells hers at ₹ 0.50 for three. After a while, when both ladies have the same number of lemons left, Seema is called away. She asks her neighbour to take care of her goods. To make things simple, Jamuna puts all lemons in one big pile, and starts selling five lemons per one rupee. When Seema returns, at the end of the day, all lemons have been sold. But when they start dividing the money, there appears to be a shortage of ₹ 3.50. Supposing they divide the money equally, how much does Jamuna lose with this deal?

A. ₹ 10.50 B. ₹ 11.50
C. ₹ 42.00 D. ₹ 52.50

48. There are two cups, one containing orange juice and one containing an equal amount of lemonade. One teaspoon of the orange juice is taken and mixed with the lemonade. Then a teaspoon of this mixture is mixed back into the orange juice. Is there more lemonade in the orange juice or more orange juice in the lemonade?

A. More orange juice in the lemonade
B. More lemonade in the orange juice
C. Equal amount of each juice between the two cups
D. None of the above

49. Consider the statement and decide which of the assumptions are implicit:

"In the present period of economic hardships, education and small family norm may lead the nation to progress and prosperity"

Assumptions:

1. Education and small family norms are directly related to nation's progress.
2. Big families find it difficult to bear the cost of education.

A. Only 1 is implicit.
B. Only 2 is implicit.
C. Both 1 and 2 are implicit.
D. Neither 1 nor 2 is implicit.

50. Fill in the blanks to find out two words that are synonyms

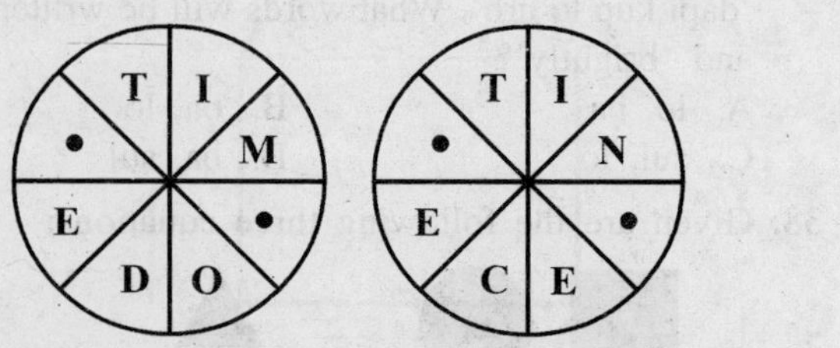

A. KS, ST B. MS, NT
C. ST, DN D. MS, DN

51. Beautiful beaches attract people, no doubt about that. Just look at the city's most beautiful beaches, which are amongst the most overcrowded places in the state. Which of the following exhibits a pattern of reasoning similar to the one exhibited in the argument above?

A. Moose and bear usually appear at the same drinking hole at the same time of the day. Therefore, moose and bear must be feeling thirsty at about the same time.
B. Children who are scolded severely tend to misbehave more often than other children. Hence if a child is not scolded severely, that child is less likely to misbehave.
C. During warm weather my dog suffers more fleas than during cool weather. Therefore, fleas must thrive in a warm environment.
D. Tally accounting software helps increase the work efficiency of its users. As a result, these users have more time for other activities.

52. Abdul, Mala and Chetan went bird watching. Each of them saw one bird that none of the others did. Each pair saw one bird that the third did not. And one bird was seen by all three. Of the birds Abdul saw, two were yellow. Of the birds Mala saw, three were yellow. Of the birds Chetan saw, four were yellow. How many yellow birds were seen in all? How many non-yellow birds were seen in all?
A. 7 yellow birds and 3 non yellow birds
B. 5 yellow birds and 2 non yellow birds
C. 4 yellow birds and 2 non yellow birds
D. 3 yellow birds and 2 non yellow birds

53. In each of the following two sets I & II, find the word or pair of words different from the other three words or pair of words:

I:	J. Lake	K. Brook
	L. Stream	M. River
II:	J. Weighty-Heavy	K. Broad-Wide
	L. Big-Large	M. Tiny-Small

A. I-J, II-J B. I-K, II-M
C. I-K, II-J D. I-J, II-K

54. A, B, C and D are standing on the four corners of a square field as shown in the figure. From the positions shown in the figure, A walks to North position and B walks to the East position while C decides to walk two sides in anticlockwise direction. B walks to North and then changes his mind to take the previous position. Identify the choice with correct positions.

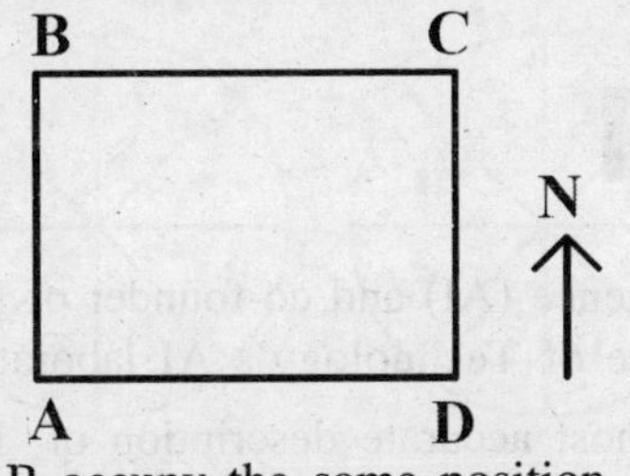

A. A & B occupy the same position.
B. C & D occupy the same position.
C. D & B are in their original positions.
D. B & C are diagonally opposite positions.

55. A gambler bet on a horse race, but the bookie wouldn't tell him the results of the race. The bookie gave clues as to how the five horses finished – which may have included some ties – and wouldn't pay the gambler off unless the gambler could determine how the five horses finished based on the following clues:
1. Penn Fe finished before Night Marvel and after Wish Bones.
2. If Hallelujah is not tied with Sundae, then Wish Bones is tied with Penn Fe.
3. Penn Fe finished as many places after Sundae as Sundae finished after Wish Bones if and only if Wish Bones finished before Night Marvel.

The gambler thought for a moment, then answered correctly. How did the five horses finish the race?
A. Sundae came in first. Wish Bones and Hallelujah tied for second place. Penn Fe came in fourth. Night Marvel came in fifth.
B. Wish Bones came in first. Sundae and Penn Fe tied for second place. Hallelujah came in fourth. Night Marvel came in fifth.
C. Wish Bones came in first. Sundae and Hallelujah tied for second place. Penn Fe came in fourth. Night Marvel came in fifth.
D. Penn Fe came in first. Night Marvel and Hallelujah tied for second place. Wish Bones came in fourth. Sundae came in fifth.

56. In a school drill, a number of children are asked to stand in a circle. They are evenly spaced and the 6th child is diametrically opposite the 16th child. How many children are made to stand in the circle?
A. 16 B. 20
C. 22 D. None of the above

57. In this question insert the missing number at the sign of interrogation.

8	4	9	5
5	7	3	4
3	4	5	8
39	44	60	?

A. 62 B. 72
C. 60 D. 70

58. Steel cylinders are made so that each one has a large and small hole through the middle. In the drawing six cylinders have been stacked on top of each other. To stop the cylinders from rolling on the smooth floor they are wedged by heavy blocks at each side of bottom row. If the heavy blocks are removed what would be the position of the cylinder when they stopped rolling?

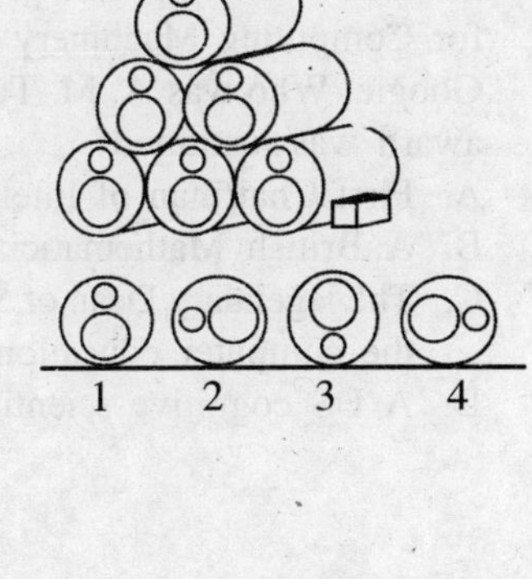

A. 1
B. 2
C. 3
D. 4

Directions (Q. Nos. 59-61): *Answer the following questions based on the information given below:*

I. There is a group of 5 persons A, B, C, D and E
II. In the group there is one badminton player, one chess player and one tennis player
III. A and D are unmarried ladies and they do not play any games
IV. No lady is a chess player or a badminton player
V. There is a married couple in the group of which E is the husband
VI. B is the brother of C and is neither a chess player nor a tennis player

59. Which of the groups has only ladies?
A. ABC B. BCD
C. CDE D. None of the above

60. Who is the tennis player?
A. B B. C
C. D D. E

61. Who is the wife of E?
A. A B. B
C. D D. None of the above

62. Consider the following statements and answer the question.
M, N, O and P are all different individuals.
M is the daughter of N.
N is the son of O.
O is the father of P.
Which among the following statements is contradictory to above premises?
A. P is the father of M.
B. O has three children.
C. M has one brother.
D. M is the granddaughter of O.

63. The drawing shows a cross section where the land meets the sea. The section covered is 5 kilometers. On a hot day, in which direction, indicated by four arrows, is the wind most likely to blow?

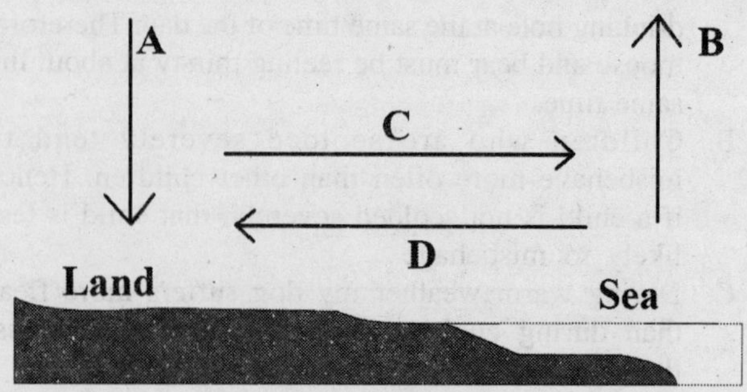

A. A B. B
C. C D. D

Direction (Q. Nos. 64-65): *Answer the following questions based on the information given below:*

In the diagram below, the circle stands for 'educated', square stands for 'hard working', triangle for 'urban people' and rectangle for 'honest'. The different regions of the diagram are numbered from 1 to 12.

Study the diagram carefully and answer the questions:

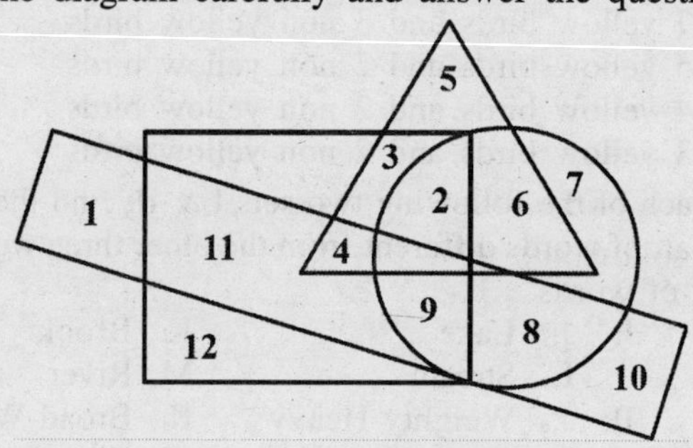

64. Uneducated urban hard-working and honest people are indicated by:
A. 3 B. 11
C. 9 D. 4

65. Non-urban educated people who are neither hard-working nor honest are indicated by:
A. 5 B. 7
C. 10 D. 11

SECTION-III

GENERAL AWARENESS

66. A. M. Turing award is considered as the Nobel Prize in the field of computers, given annually by Association for Computing Machinery co-sponsored by Intel and Google. Who was A. M. Turing in whose memory the award was instituted?
A. First Chairman of Intel
B. A British Mathematician
C. The legendary Dean of Stanford who revolutionized the computer education in the US.
D. A US cognitive scientist in the field of Artificial Intelligence (AI) and co-founder of Massachusetts Institute of Technology's AI laboratory.

67. Find the most accurate description of 'Bt Cotton'.
A. A variety of cotton that grows fast
B. A genetically modified crop of cotton whose DNA has been altered by introducing an extra gene.
C. A variety of cotton that is immune to diseases.
D. A genetically modified crop of cotton whose DNA has been designed in the lab.

68. Las Vegas, US-based tour operator AMX Company has filed a trademark patent for which of the following Tagline?
A. "Gods Own Country"
B. "Heaven on Earth"
C. "Discover Incredible India"
D. "Come Alive with India"

69. Which out of the following holds the highest number of shares of ICICI Bank?
A. Banks Financial Institutes and Insurance
B. FIIs
C. General public
D. Private Corporate Bodies

70. Indian Standard time is based on the longitude of 82.5 degrees passing through which of the following places?
A. Mirzapur, U.P. B. Jabalpur, M.P.
C. Nagpur, Maharashtra D. Jaipur, Rajasthan

71. The Black Box of an aircraft is an important part as all the conversations and data are recorded therein. Although it is named black box, but its colour is not black. What is the colour of the black box in a commercial airplane?
A. Brown B. Yellow
C. Red D. Orange

72. Find the person who plays the odd sport out of the following:
A. Navaratilova B. Michael Phelps
C. David Beckham D. Prakash Padukone

73. Sariska and Ranthambore are the reserves for which of the following animals?
A. Lion B. Deer
C. Tiger D. Bear

74. March, 2008 witnessed a turning point in the history of which of the following Himalayan Kingdoms when democracy was ushered in replacing monarchy?
A. Bhutan B. Nepal
C. Sikkim D. None of these

75. Neil Armstrong brought back a rock from the moon. On earth
A. The mass of the rock will change
B. The weight of the rock will change
C. Mass and weight will change
D. None of the above

76. Milk, Cheese and Eggs are the source of
A. Vitamin C & A B. Vitamin A & D
C. Vitamin C & D D. Vitamin B & C

77. In August 2008, India's longest runway for the passenger aircraft was commissioned in:
A. Mumbai B. New Delhi
C. Hyderabad D. Bangalore

78. The first ever public hearing in India, almost like a referendum, on the fate of SEZ was held during the month of September 2008 in/at ______.
A. Verna IT/ITeS SEZ in Goa.
B. Haryana for Reliance Haryana SEZ.
C. Pen, Maharashtra.
D. POSCO India, Steel SEZ Phase II, Orissa.

79. Al Ahram is
A. A University in Damascus.
B. A Newspaper in Cairo.
C. A Mosque in Riyadh.
D. A Small mountain range in Jordan.

80. India's largest and first multi-national pharmaceutical giant Ranbaxy is being bought over by ______.
A. Matrix Pharma. B. GVK Bio-sciences.
C. Merck. D. None of these.

81. Find the odd product out of the following:
A. Pamper B. Dove
C. Tide D. Pantene

82. In the internet sphere, 'Opera' is the name of a:
A. Web conference site
B. Web advertising firm.
C. Software for webinars
D. Web browser.

83. Which of the following pairs is not correct?
A. Shahid Parvez: Sitar
B. Rashid Khan: Vocal Music
C. Ayaan Ali Khan: Sarod
D. Romu Muzumdar: Guitar

84. At what frequency SENSEX calculation is carried out?
A. Every half an hour. B. Every hour.
C. Real time. D. 6 times a day.

85. The second largest manufacturer of CD's, DVD's and other optical media in the world is:
A. Sony B. Intel
C. Philips D. Moser Baer

86. Which of the following is a legal right and not a fundamental right
A. Right to Liberty of thought and expression
B. Right against exploitation
C. Right to property
D. Right to practice religion

87. The term Net Shot is associated with
A. Badminton B. Tennis
C. Table Tennis D. Volley Ball

88. What is Dry Ice?
A. Solidified Carbon Dioxide
B. Transparent plastic decoy which looks like ice
C. Sulfur dioxide at –57 degree Celsius
D. Chemically treated ice to make it dry.

89. The rail-based mass rapid transit system in Mumbai has been awarded to a consortium of companies led by:
A. Reliance Infrastructure B. Reliance Industries
C. GMR Infrastructure D. Lanco Infrastructure

90. If bilirubin is high in a human body, which organ is most affected?

A. Pancreas B. Liver
C. Kidney D. Large Intestine

91. Who of the following is one of the most celebrated Photo Journalists in India?

A. Gautam Rajadhyaksha
B. Raghu Rai
C. Anjii Reddy
D. Sudhir Das

92. Inflation implies

A. Rise in budget deficit
B. Rise in general price index
C. Rise in price of consumer goods
D. Rise in money supply

93. The India-US Nuclear Deal is called 123 Agreement. What does 123 denote?

A. The deal governed by the clause number 123 of International Atomic Energy Agency, in the context of peaceful use of atomic energy.
B. This number denotes a cluster of clauses of Hyde Act of US Government
C. This is the simple number related to chain reaction for creation of atomic energy.
D. Section 123 of the US Atomic Energy Act.

94. In the United States of America, the President is elected

A. By the electorate
B. By the electoral college
C. By the two national political parties
D. By the U.S. congress

95. The term "Uruguay Round" is associated with an important world organization. Which one?

A. UNDP B. GATT
C. WTO D. World Bank

96. Siebel is a software firm that is now taken over by:

A. IBM B. Accenture
C. Gartner D. Oracle

97. First Indian motion picture insured by a General Insurance company is

A. Hum Apke Hain Kaun
B. Taal
C. Dil Se
D. Ashoka

98. India's first coalition government in New Delhi was formed under the leadership of:

A. V. P. Singh B. P. V. Narasimharao
C. Morarji Desai D. Atal Behari Vajpayee

99. WiMax stands for:

A. Wireless internet for microwave access
B. Wireless inter-operability for microwave access
C. Worldwide internet for microwave access
D. Worldwide inter-operability for microwave access

100. El Nino is

A. A Sea storm
B. A warm ocean current
C. A tropical disturbance
D. Another name of typhoon

SECTION-IV

QUANTITATIVE & DATA INTERPRETATION & DATA SUFFICIENCY

101. A factory is to commission two production lines. Production line 1 is to use existing technology. Production line 2 is to use the latest innovation in technology and, while promising to achieve considerable advances in productivity, it will take longer to start and is likely to experience teething problems. Graph indicates the productive record of each product line. Refer to the graph to answer the following:

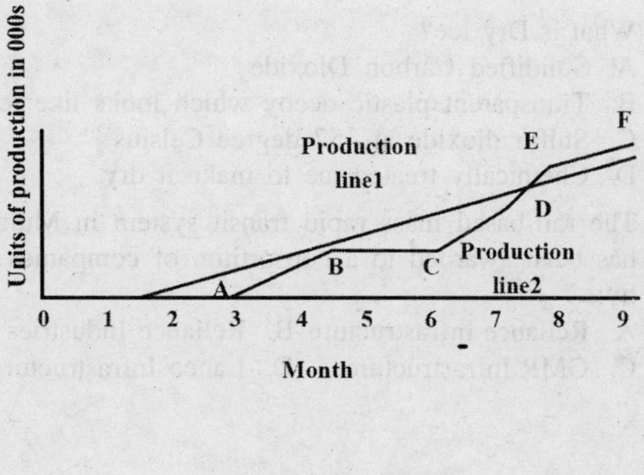

1. Can the duration of reported breakdown be established?
2. Can the loss of production be quantified?

A. Both questions 1 and 2 can be answered
B. Only 1 question can be answered
C. Only 2 question can be answered
D. Neither question can be answered

102. From a book, a number of consecutive pages are missing. The sum of the page numbers of these pages is 9808. Which pages are missing?

A. The page 9808 is missing.
B. The pages 291 up to 322 are missing.
C. The pages 291 up to and including 322 are missing
D. Either A or C

103. In the following series find the one number that is wrong

2, 3, 13, 37, 86, 167, 288

A. 3 B. 13
C. 37 D. 86

104. Two sea trawlers left a sea port simultaneously in two mutually perpendicular directions. Half an hour later, the shortest distance between them was 17 km and another 15 minutes later, one sea trawler was 10.5 km farther from the original than the other. Find the speed of each sea trawler.

A. 16 km/hr, 30 km/hr
B. 18 km/hr, 24 km/hr
C. 20 km/hr, 22 km/hr
D. 18 km/hr, 36 km/hr

105. The image below indicates the number of residents at 5 hotels on 1 Feb and 1 July 1998.

Which hotel had the greatest increase in the total number of adult residents on 1 July 1998 compared with 1 Feb of that year?

1-Jul-98

Hotel	Female Adults	Male Adults	Female Children
V	41	73	16
W	18	37	7
X	47	92	12
Y	37	59	19
Z	81	112	26

1-Feb-98

Hotel	Female Adults	Male Adults	Female Children	Male Children
V	32	64	8	11
W	11	28	3	2
X	47	84	14	7
Y	22	42	5	6
Z	63	104	18	23

A. V　　B. W
C. Y　　D. Z

106. This question consists of a question and two statements numbered I and II. Decide whether the data given in the statements are sufficient to answer the question.

What is the 57th number in a series of numbers?

I. Each number in the series is three more than the preceding number.
II. The tenth number in the series is 29.

A. The data in Statement I alone is sufficient to answer the question while the data in Statement II alone is not sufficient to answer the question.
B. The data in Statement II alone is sufficient to answer the question, while the data in Statement I alone is not sufficient to answer the question.
C. If the data either in Statement I or Statement II alone are sufficient to answer the question.
D. If the data in both Statements I and II together is necessary to answer the question.

107. The cost of levelling and turfing a square field at ₹ 160 per hectare is ₹ 2624.40. The cost of surrounding it with a railing costing 25 paise per metre is:

A. ₹ 350　　B. ₹ 375
C. ₹ 400　　D. ₹ 405

108. When the Sun ray's inclination increases from 30° to 60°, the length of the shadow of a tower decreases by 60 m. Find the height of the tower.

A. 50.9 m　　B. 51.96 m
C. 48.8 m　　D. None of these

Directions (Q. Nos. 109–112): *Answer the following questions based on the information given below:*

A survey of movie goers from five cities A, B, C, D and E is summarized below. The first column gives the percentage of viewers in each city who watch less than one movie a week. The second column gives the total number of viewers who view one or more movies per week.

City	I	II
A	60	2400
B	20	3000
C	85	2400
D	55	2700
E	75	8000

109. How many viewers in the city C watch less than one movie a week?

A. 2040　　B. 13600
C. 16000　　D. 3600

110. Which city has the highest number of viewers who watch less than one movie a week?

A. City E　　B. City D
C. City B　　D. City C

111. A city with second lowest number of movie watchers is

A. City E　　B. City D
C. City B　　D. City C

112. The total number of all movie goers in the five cities who watch less than one movie per week is

A. 50000　　B. 36000
C. 18500　　D. 45250

113. The 288th term of the sequence *a, b, b, c, c, c, d, d, d, d* ..... is

A. *u*　　B. *v*
C. *w*　　D. *x*

114. The inequality of $p^2 + 5 < 5p + 14$ can be satisfied if:

A. $p \leq 6, p > -1$　　B. $p = 6, p = -2$
B. $p \leq 6, p \leq 1$　　D. $p \geq 6, p = 1$

115. Answer the question based on the following table:

Relative Sweetness of Different			
Lactose	0.16	Maltose	0.32
Glucose	0.74	Sucrose	1.00
Saccharin	675.00		

What is the ratio of glucose to lactose in a mixture as sweet as maltose?

A. 8 : 21 B. 1 : 3
C. 3 : 2 D. 16 : 9

116. A colourless cube is painted blue and then cut parallel to sides to form two rectangular solids of equal volume. What percentage of surface area of each of new solids is not painted blue?

A. 25 B. 16
C. 20 D. 18

117. There are 10 stations on a railway line. The number of different journey tickets that are required by the authorities is

A. 10! B. 90
C. 81 D. 10

118. A and B throw one dice for a stake of ₹ 11, which is to be won by the player who first throws a six. The game ends when stake is won by A or B. If A has the first throw, what are their respective expectations?

A. 5 and 6 B. 6 and 5
C. 11 and 0 D. 10 and 1

119. Which investment gives a better return, assuming the face value of shares to be ₹ 10?

1. 5% stock at 75, subject to 30% income tax
2. 4% stock at 90, tax free

A. B
B. A
C. Both A and B
D. None of these

120. Four stacks containing equal number of chips are to be made from 11 orange, 9 white, 13 black and 7 yellow chips. If all of these chips are used and each stack contains at least one chip of each colour, what is the maximum number of white chips in any one stack?

A. 3 B. 4
C. 5 D. 6

121. A 14.4 kg gas cylinder runs for 104 hours when the smaller burner on the gas stove is fully opened while it runs for 80 hours when the larger burner on the gas stove is fully opened.

Which of these values are closest to the percentage difference in the usage of gas per hour, of the smaller burner over the larger burner?

A. 26.23% B. 30%
C. 32.23% D. 23.07%

122. Study the questions and the statements given below. Decide whether any information provided in the statement(s) is redundant and/or can be dispensed with, to answer it.

If 7 is added to numerator and denominator each of fraction *a*/*b*, will the new fraction be less than the original one?

Statement I: $a = 73$, $b = 103$
Statement II: The average of *a* and *b* is less than *b*
Statement III: $a - 5$ is greater than $b - 5$

A. II and either I or III
B. Only II or III
C. Any two of them
D. Any one of them

123. Coefficient of variation is useful to study

A. Risk B. Disparity
C. Consistency D. All of the above

124. A cyclist drove one kilometer, with the wind in his back, in three minutes and drove the same way back, against the wind in four minutes. If we assume that the cyclist always puts constant force on the pedals, how much time would it take him to drive one kilometer without wind?

A. $2\frac{1}{3}$ B. $3\frac{3}{7}$
C. $2\frac{3}{7}$ D. $3\frac{7}{12}$

125. A, B and C started a business by investing 1/2, 1/3rd and 1/6th of the capital respectively. After 1/3rd of the total time, A withdrew his capital completely and after 1/4th of the total time B withdrew his capital. C kept his capital for the full period. The ratio in which total profit is to be divided amongst the partners is

A. 1 : 2 : 1 B. 4 : 1 : 4
C. 2 : 1 : 2 D. 1 : 2 : 2

126. A number lock consists of 3 rings each marked with 10 different numbers. In how many cases the lock cannot be opened?

A. 3^{10} B. 10^3
C. 30 D. 999

127. A person buys 18 local tickets for ₹ 110. Each first class ticket costs ₹ 10 and each second class ticket costs ₹ 3. What will another lot of 18 tickets in which the number of first class and second class tickets are interchanged cost?

A. 112 B. 118
C. 121 D. 124

128. A clock loses 12 minutes every 24 hours. It is set right at 7:25 p.m. on Monday. What will be the time when clock shows 1:45 p.m. the following day?

A. 1:20:35 p.m. B. 1:35:50 p.m.
C. 1:25:35 p.m. D. None of these

129. In a row at a bus stop, A is 7th from the left and B is 9th from the right. They both interchange their positions. A becomes 11th from the left. How many people are there in the row?

A. 18 B. 19
C. 20 D. 21

130. A merchant wants to make profit by selling food grains. Which of the following will maximize his profit?

I. Sell product at 30% profit

II. Increase the price by 15% over the cost price and reduce weight by 15%

III. Use 700 gm of weight instead of 1 kg.

IV. Mix 30% impurities in grains and sell it at cost price

A. III B. II and I

C. II D. All give the same profit

Directions (Q. Nos. 131-134): *Answer the following questions based on the information given below:*

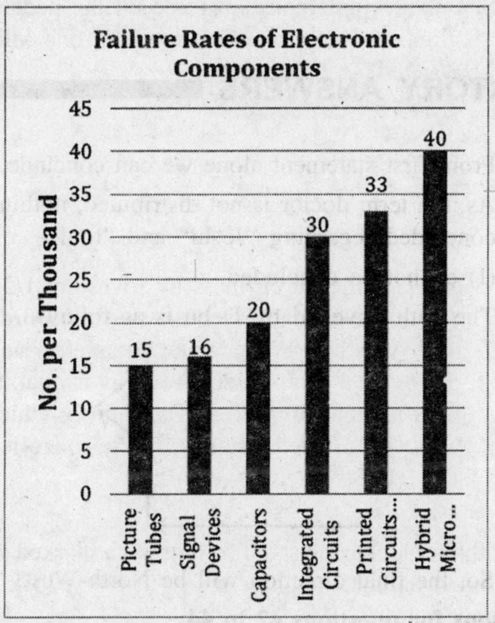

131. From the failure data of electronics components presented here, which statement is true?

A. Integrated circuits and Printed circuit boards are more reliable.

B. Capacitors are more reliable than pictures tubes.

C. Higher number of component failures is likely to be due to picture tubes.

D. The least reliable component is Hybrid Micro circuits.

132. Which of the following components has a failure rate 25% more than that of signal devices?

A. Pictures tubes B. Capacitors

C. Integrated circuits D. Printed Circuits boards

133. Lowest priority for investing in any changes or additions to the component manufacturing units, in the company's investment plans, may be given to the following:

A. Printed circuits boards and hybrid micro circuits

B. Capacitors and integrated circuits

C. Pictures tubes and signal devices

D. Signal devices and capacitors

134. For the equipments using Integrated Circuit Board: 400, Capacitors: 240 and Printed Circuit Boards: 120 to run with minimum downtime, how many spares should be kept in the store respectively?

A. 12, 4, 3 B. 12, 5, 4

C. 5, 12, 4 D. 12, 4, 4

135. The water from a roof, 9 sq. metres in area, flows down to a cylindrical container of 900 cm^2 base. To what height will the water rise in cylinder if there is a rainfall of 0.1 mm?

A. 0.1 cm B. 0.1 metre

C. 0.11 cm D. 1 cm

ANSWERS

1	2	3	4	5	6	7	8	9	10
C	C	B	B	C	C	C	C	C	A
11	**12**	**13**	**14**	**15**	**16**	**17**	**18**	**19**	**20**
C	A	B	A	D	B	D	D	B	A
21	**22**	**23**	**24**	**25**	**26**	**27**	**28**	**29**	**30**
C	C	D	B	B	B	A	A	A	B
31	**32**	**33**	**34**	**35**	**36**	**37**	**38**	**39**	**40**
D	D	D	D	B	B	A	A	C	B
41	**42**	**43**	**44**	**45**	**46**	**47**	**48**	**49**	**50**
D	A	B	C	B	B	A	C	A	D
51	**52**	**53**	**54**	**55**	**56**	**57**	**58**	**59**	**60**
C	B	A	D	C	B	B	C	D	B
61	**62**	**63**	**64**	**65**	**66**	**67**	**68**	**69**	**70**
D	A	D	D	B	B	B	C	B	A
71	**72**	**73**	**74**	**75**	**76**	**77**	**78**	**79**	**80**
D	B	C	A	B	B	B	C	B	D

81	82	83	84	85	86	87	88	89	90
B	D	D	C	D	C	A	A	A	B
91	**92**	**93**	**94**	**95**	**96**	**97**	**98**	**99**	**100**
B	B	D	B	B	D	B	C	D	B
101	**102**	**103**	**104**	**105**	**106**	**107**	**108**	**109**	**110**
B	C	B	A	C	D	D	B	B	A
111	**112**	**113**	**114**	**115**	**116**	**117**	**118**	**119**	**120**
B	D	D	A	A	A	B	B	B	D
121	**122**	**123**	**124**	**125**	**126**	**127**	**128**	**129**	**130**
D	B	D	B	C	D	D	D	B	A
131	**132**	**133**	**134**	**135**					
D	B	C	B	D					

SOME SELECTED EXPLANATORY ANSWERS

36. Look at the wheel B from the top. As D is rotating in anticlockwise direction, it is evident that the wheel should move in anticlockwise direction. Now look at the wheel B from the bottom, here it should rotate in clockwise direction. This is contradictory, hence wheel B cannot move.

37. From (A) and (B) 'brightly' is common in both the phrases and also "ba" is the only word common in the code.

So, "brightly" is written as "ba".

Similarly from (A) and (C), "sun" is coded as "lo".

38. Let us represent square as '*s*', triangle as '*t*' circle as '*c*' and rhombus as '*r*'.

(1) $s + c = t$

(2) $s = c + r$

(3) $2t = 3r$.

From (1) and (2) substituting *s* in (1) we get,

$$2c + r = t$$

From (3)

$$2c + r = \frac{3}{2}r$$

$$\Rightarrow \quad 2c = \frac{1}{2}r$$

$$\Rightarrow \quad 4c = r$$

Now from (2)

$s = c + r = c + 4c = 5c$

$\therefore$ There should be 5 circles.

39. Assume, a person is male, he should have 4 brothers and three sisters. This implies that there are at least 5 male children.

This implies that in that family every female child has 5 brothers and two sisters.

Therefore there must be at least 3 female children, hence there must be one more female.

Hence, there must be at least 9 children.

40. From first statement alone we can conclude (II).

As, the term doctor is not distributed, nothing can be concluded regarding "Joshi" and "fool".

(I) cannot be concluded.

41. The path travelled by Debu is as follows

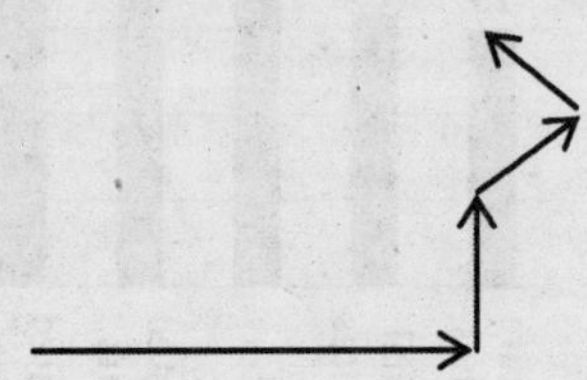

So, the final direction will be North-West.

Solutions for questions 42 to 44:

42. Given that there are four 4 cm × 3 cm faces.

The middle two cubes on each of these faces will have only one colour on them.

$\therefore$ 4 × 2 = 8 cubes

Also given, there are two 3 cm × 3 cm faces. The middle cube on either of these faces will have one colour on it.

$\therefore$ 2 × 1 = 2 cubes

$\therefore$ A total of 10 cubes are there with only one colour.

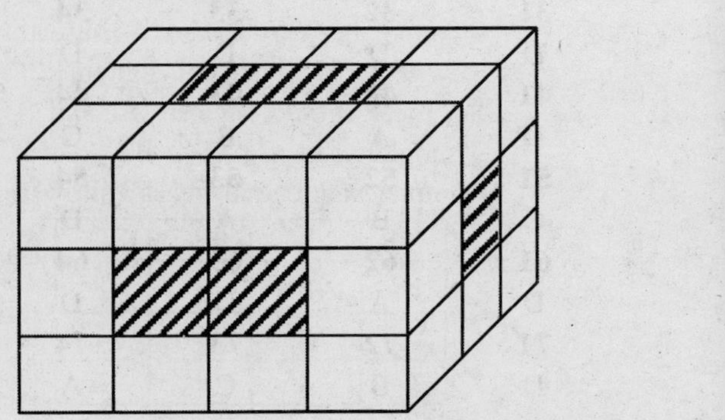

43. The cuboid consists of four layers along first dimension, three layers along second dimension and three layers along third dimension.

As we need cubes without any colour, we are not interested in the outer most layers along each of the three dimensions.

∴ The number of cubes without any colour
$= 2 \times 1 \times 1 = 2$.

44. There are a total of 36 cubes and from the above two solutions, we have the number of cubes with one colour as 10, the number of cubes without any colour as 2. As each colour is painted in opposite pairs of faces, each of the 8 corner cubes will have three colours.

∴ The number of cubes with any two colours
$= 36 - 10 - 2 - 8 = 16$.

Alternative solution:

From the given information, we can say that the dimensions of the cube are 4 cm, 3 cm and 3 cm.

∴ We have four edges of 4 cubes, four edges of 3 cubes and four edges of 3 cubes.

On an edge with four cubes, only the middle two cubes have two colours on them.

On an edge with three cubes, only the middle cube has two colours on it.

∴ Total number of cubes with two colours on them = $4 \times 2 + 8 \times 1 = 16$.

45. As cell C3 – cell C2 = 4

The possibilities are—(1, 5), (2, 6), (3, 7), (4, 8), (5, 9) as column A cannot have any odd number it should have even numbers, i.e., three numbers among – 2, 4, 6 and 8.

So, in C, there cannot be two even numbers and also as, number 7 is in column B.

Cell C3 and C2 can have (1, 5) or (5, 9).

Again given, the sum of the digits of column C is 14.

So, (5, 9) is not possible in C2 and C3.

∴ C3 is 5.

C2 is 1.

So, C1 must be 8. [as sum is 14]

So, 4, 6 and 2 will be in column A.

Sum of the three digits in row 1 is 17.

9 cannot be in row 1 as we already have 8.

Row 1 should also not contain 2.

So, there should be 6 in A1 and 3 in B1.

9 should be in B3. 7 in B2, 2 is A2 and 3 in A3.

Complete table will be as follows:

	A	B	C
1	6	3	8
2	2	7	1
3	4	9	5

46. The series is as follows:

4, 32, 288, <u>2880</u>, 31680
(×8, ×9, ×10, ×11)

2880 is the missing number.

47. Let us assume each one of them have 6 lemons.

Selling 6 lemons Jamuna would have got ₹ 1.5 and Seema would gave got Re.1.

After combining cost of (6 + 6), i.e., 12 lemons

$= \frac{12}{5} = 2.4$

∴ There is ₹ 0.1 loss.

But at the end they found a shortage of ₹ 3.5.

∴ Total number of lemons $= 3.5 \times \frac{12}{0.1} = 420$

∴ For her 210 lemons Jamuna receives at original rate

$= \frac{0.5}{2} \times 210 = 52.50$

If they divide equally the amount after selling @5 lemons per ₹ 1 Jamuna receives $\frac{210}{5} = 42$

∴ Jamuna receives (52.50 – 42) = 10.50 less.

48. Let the lemonade and the orange juice be contained in cups A and B respectively.

Let the quantity of each juice be 1 litre.

After both the transformations the quantity of the mixture in each cup is the same, i.e., 1 litre.

Now, suppose x ml orange juice is present in cup A.

∴ The quantity of lemonade in cup A is $(1000 - x)$ ml. It implies that the remaining x ml of lemonade must be in cup B.

So, the quantity of orange juice in cup A is equal to the quantity of lemonade in cup B.

The above illustration can be explained through following diagrams.

<u>Initial</u>

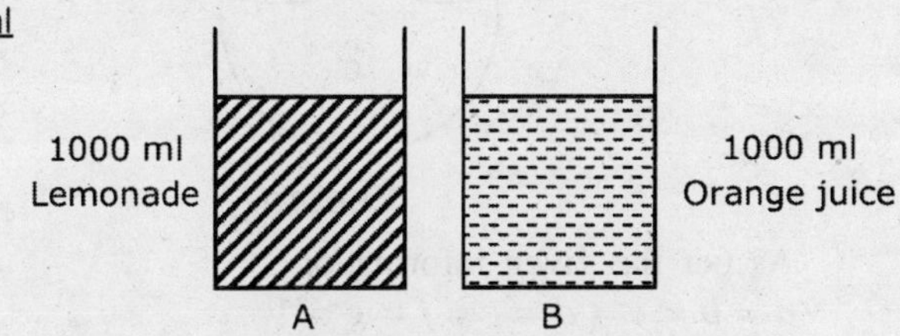

<u>Final</u>

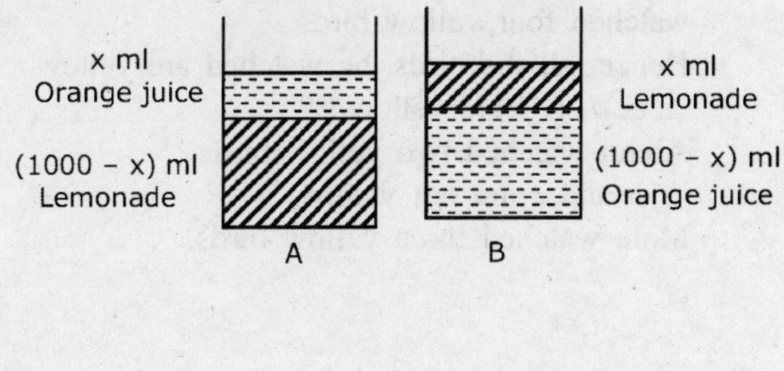

49. As the author says, "progress and prosperity" can be obtained by "education and small family norms" he mean to say that these events are related events.

So, I is implicit.

There is no reference of big families so, II is irrelevant.

50. MS and DN are the missing letter. The synonyms words that can be obtained from these letters are, IMMODEST and INDECENT.

51. The question statement present a situation in which numbers of species (people) are drawn to a locale because of its attribute (beauty) therefore the assumption in the statement is that the crowds on the beach are there to enjoy the beauty.

Choice (B) and (D) are easily eliminated.

Choice (B) woks at an entirely different logical relationship – one between behavioural aspects – where increase/decrease in scolding leads to increase/ decrease in misbehaviour.

Choice (D) presents a totally different argument–where increased efficiency leads to increase in free time.

Choice (A) is not similar – it speaks of 2 species and the species share an attribute.

The attribute of the locale is not the focus of the statement.

Choice (C) is appropriate. Numbers of fleas – (the tiny insects that live in the fur of animals) are drawn to an environment when it is warm (the animal's fur exposed to warm weather). The assumption then is that the fleas are there (and proliferating) because of the warmth. This choice is the closest, in reasoning to the question statement.

52. Let us represent the birds watched by the three in the venn diagram.

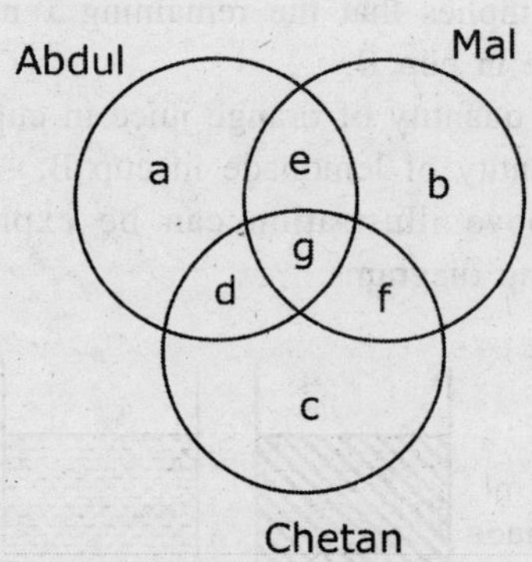

As per the given information

$a = b = c = d = e = f = g = 1$.

So, Chetan watched 4 birds and given that Chetan watched four yellow birds.

Hence, all the birds, he watched are yellow.

$\therefore$ *c, d, g,* f are yellow.

Abdul watched two yellow birds.

$\therefore$ *a* and *c* are not yellow.

Mala watched three yellow birds.

$\therefore$ *b* is yellow.

$\therefore$ There are 5 (*b, e, d, g, f*) yellow and 2 (*a* and *c*) non-yellow birds.

53. Lake is a closed water body whereas others are open water bodies.

Except Weighty – Heavy, others are related to size.

54. The initial positions of A, B, C and D is as follows:

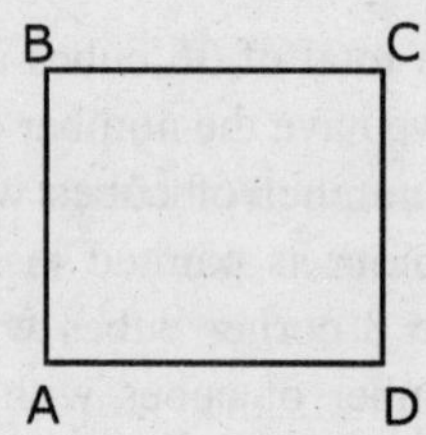

The final positions of A, B, C and D is as follows:

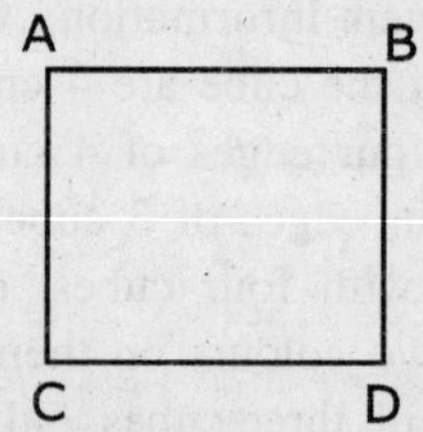

(A) is false

(B) cannot be determined

(C) cannot be determined

(D) is definitely true.

55. It is given that Penn Fe finished before Night Marvel and after Wish Bones.

i.e., Wish Bones

Penn Fe

Night Marvel

from (3), we have Penn Fe finished as many places after Sundae as Sundae finished after Wish Bones if and only if Wish Bones finished before Night Marvel.

From (1), we have Wish Bones finished before Night Marvel.

Hence, from (1) and (3), we have

Wish bones

Sundae

Penn Fe

Night Marvel

From (2), Wish Bones is not tied with Penn Fe, hence Hallelujah tied with Sundae.

The final order is

1 – Wish Bones

2, 3 } Sundae, Hallelujah

4 – Penn Fe

5 – Night Marvel

56. It is given that in a school drill all the students are made to stand in a circle. In the circle 6th child is diametrically opposite to the 16th child.

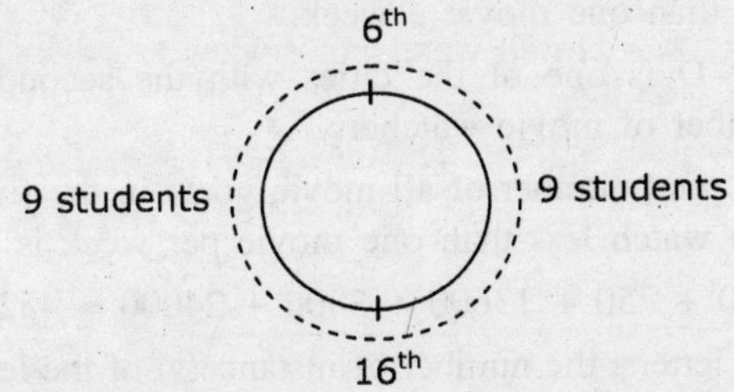

Hence there are 9 students in between these two students on either side.
In total there are 9 + 9 + 2 = 20 students.

57. In each column, the sum of first two numbers is multiplied with the third number to get the last number in each column

i.e., (8 + 5) × 3 = 39

(4 + 7) × 4 = 44

(9 + 3) × 5 = 60

(5 + 4) × 8 = 72.

58. The cylinders when they stopped rolling, in each cylinder the larger hide (less weight) goes up while the smaller hoe (more weight) comes down. Hence, (c) is the position.

Solutions for questions 59 to 61:

It is given that – A, B, C, D and E are five persons. Among them, there is one tennis player, one chess player and one badminton player.

From (III), A and D are spinsters and do not play any games.

Hence, B, C, E are the players.

Among B, C and E there is one married couple.

From (V) and (VI),

B and E are males.

Hence, C is a female who is married to E.

From (IV) C plays tennis

From (VI) B plays badminton

E plays chess.

59. A, C and D are the females in the group.

60. C is the tennis player.

61. C is the wife of E.

62. The given information can be represented as follows.

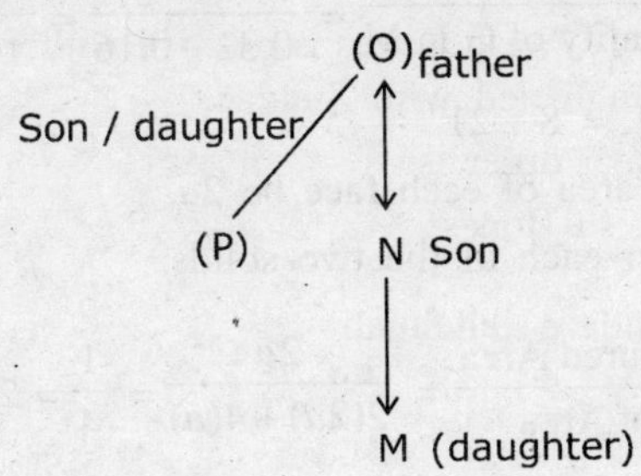

(A) P is the sibling of M's father. Hence, P could be either Aunt or uncle to M but not father.

63. On a hot day, the heat from the land goes up and there will be vacuum space formed on the earths atmosphere. Hence the wind from the sea blows towards this vacuum space. The wind blows from sea towards land i.e., D.

Solutions for questions 64 and 65:

64. Urban, hardworking honest people but not educated is the region which is common to all the figures except circle and is represented by 4.

65. Non-urban educated people who are neither hard working nor honest is the region belongs to only circle, and is represented by 7.

101. As the graph representing production line 1 is showing a steady increase, the given graphs represent cumulative productions. The graph representing production line 2 is showing no change from the 4th month to the 6th month.

∴ It was the period of breakdown. As the graph representing production line 2 is not following any pattern, we cannot quantify the loss of production as the production in those months cannot be determined.

102. A single leaf (folio) has two pages. It is not possible to have one page in isolation.

If a consecutive number of pages are missing, the number of missing pages has to be even. It can be 291 up to and including 322 (*i.e.,* 32 pages). We can verify that the sum of these numbers $\left[32\left(\frac{291+322}{2}\right)\right]$ is equal to 9808.

103. The 7 numbers, t_i (i = 1 to 7) and the differences of successive numbers $t_i - t_{i-1}$ (for i = 2 to 7) are tabulated below:

i	1	2	3	4	5	6	7
t_i	2	3	13	37	86	167	288
$t_i - t_{i-1}$		1	10	24	49	81	121

We see that if t_3 = 12 (instead of 13), the successive differences would become the squares of successive odd numbers. So, 13 is the odd one.

104. The positions of the 2 trawlers after 30 minutes and 45 minutes are shown below:

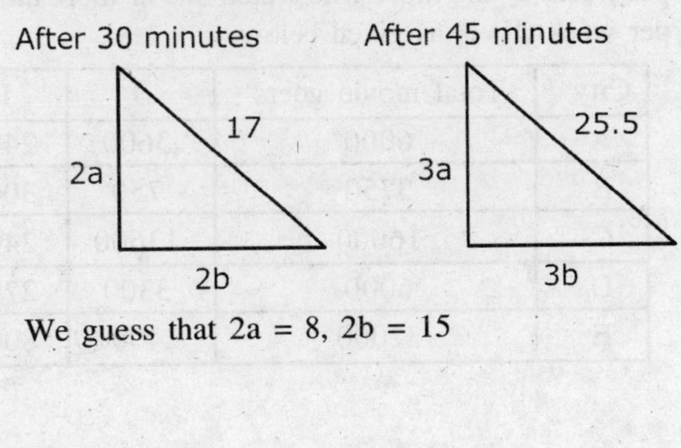

We guess that 2a = 8, 2b = 15

$\therefore$ 3a = 12 and 3b = 22.5 as 3b – 3a = 10.5, our guess is right (*i.e.*, 2a = 8 and 2b = 15)

$\therefore$ The speeds are 16 km/hr and 30 km/hr.

105. The hotel which had the greatest increase in the number of adult residents on 1st July 1998 compared to 1st February of that year is Y.

106. From statement I, $t_i + 1 - t_i = 3$ for all integer values of *i*. But we don't know the value of any term. This is not sufficient.

From statement II, $t_{10} = 29$. But we can't determine t_{57} as we don't know any thing about the series.

Combining the two statements, we get $t_{57} = t_{10} + 47(3) = 29 + 141 = 170$.

107. Cost of levelling and turfing = ₹ 2624.40

Unit cost = ₹ 160/hectare

$\therefore$ Area (in hectares) = $\frac{2624.40}{160} = \frac{26244}{1600} = 16.4025$

or 164025 m^2

$\therefore$ Side of square = 405 m

Cost of fencing = perimeter (cost per unit length)

= ₹ 4(405) $\left(\frac{1}{4}\right)$ = ₹ 405.

108.

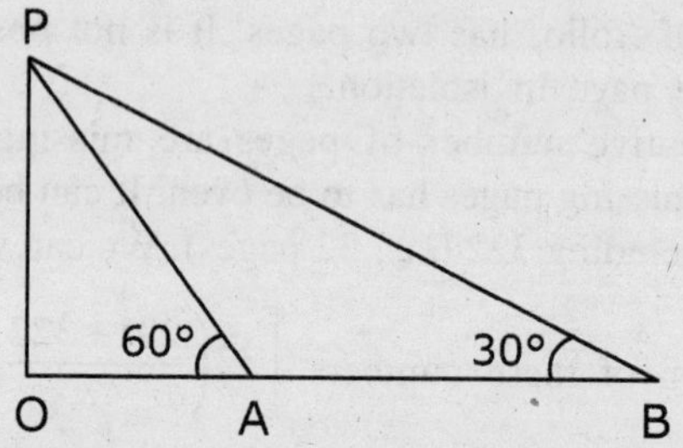

When the angle of inclination of the sun's rays (θ) = 60°, the shadow of the tower OP is OA.

When θ = 30°, the shadow is OB.

Let OA = *x*, $\therefore$ OP = $\sqrt{3}x$ and OB = $\sqrt{3}$ (OP) = 3*x*, *i.e.*, AB = 2*x*

Given AB = 60 m

$\therefore$ *x* = 30 m and OP = $\sqrt{3}x$ = 30(1.732)m = 51.96 m.

Solutions for questions 109 to 112:

The total number of movie goers in the five cities and the number of people who watch less than one movie per week (I) and those who watch one or more movies per week (II) is as given below:

City	Total movie goers	I	II
A	6000	3600	2400
B	3750	750	3000
C	16000	13600	2400
D	6000	3300	2700
E	32000	24000	8000

109. 13,600 viewers in city C watch less than one movie a week.

110. City E has the highest number of viewers who watch less than one movie a week.

111. City D is one of the cities with the second lowest number of movie watchers.

112. The total number of all movie goers in the five cities who watch less than one movie per week is

3600 + 750 + 13600 + 3300 + 24000 = 45250.

113. The letters, the number of instance(s) of the letter, the total number of instances of all the letters are tabulated as:

Letter	No. of instances n	Total S_n
a	1	1
b	2	3
c	3	6
d	4	10
e	5	15
f	6	21
g	7	28
h	8	36

The last instance of the nth letter is $\frac{n(n+1)}{2}$th letter of the series.

We see that S_{23} = (23) (12) = 276 and

S_{24} = (12) (25) = 300

$\therefore$ All the terms from the 277th to the 300th are the 24th letter of the alphabet, *i.e.*, *x*.

114. $p^2 + 5 < 5p + 14$

$\Rightarrow p^2 - 5p - 9 < 0$

$\Rightarrow \frac{5-\sqrt{61}}{2} < p < \frac{5+\sqrt{61}}{2}$

p lies between –1.4 and 6.4.

None of the choices satisfy.

115. The relative sweetness of the base sugars [(lactose (L), glucose (G)] and that of the mixture (M) are tabulated below:

Relative sweetness	0.16	0.32	0.74
Sugar/mixture	L	M	G

$\therefore \frac{\text{Quantity of L in M}}{\text{Quantity of G in M}} = \frac{0.74-0.32}{0.32-0.16} = \frac{42}{16} = \frac{21}{8}$

$\therefore$ G : L = 8 : 21.

116. Let the area of each face be 2*a*.

Consider each of the two solids.

$\frac{\text{Uncoloured Area}}{\text{Total Area}} = \frac{2a}{2(2a)+4(a)} = \frac{1}{4} = 25\%.$

GENERAL ENGLISH

1 Comprehension Passages

In the English language paper, questions on comprehension test are very important for the examinees appearing in the competitive examinations.

Comprehension means the act of comprehending or the capacity of the mind to understand. In the examination papers, questions on comprehension test are included to judge the ability of the examinees to understand the given passage.

Therefore, they should understand how to solve these questions. Practice of solving these questions will greatly help them in the examination.

PASSAGE-1

Cyber crime is the branded stigma defacing the culture and magnanimity of computer technology. It is up keeping the flag with indomitable triumph against developing computer technology worldwide.

Modern age is striding with marching steps of technology revolution beating the past decade of ancestral belief with ultimate care. Computer invention has unfolded the mystery of quick access with the objective of minimum manpower and cutting the time consumption parameters.

As each coin has dual face of its portrait, likewise computer technology is sick of creeping virus. Synthetic manmade dilemma of site hackers activation is causing setback to the expanding anchor of revolutionary device with great loss of time, economy and data profile as suffered by consumers. A recent report from Internet security firm Websense estimates that 85.6 per cent of all the unwanted e-mails contained links to spam sites. The company's data suggests that the number of malicious sites grew 233 per cent in the last six months and saw 671 per cent growth in the number of malicious sites during the last year. In June alone, the total number of e-mails detected as containing viruses increased by 600 per cent compared to May.

Chat rooms, blogs and message boards where the users post comments have been identified as the top targets of hackers and spammers due to the high traffic these attract. According to Websense, 95 per cent of user-generated comments to blogs, chat rooms and message boards during the first half of 2009 were malicious.

It is advisable not to click on spurious links and stay away from keying in passwords at unknown sites as they are most likely to be spammed. Hackers can steal your passwords and log in to your account and access critical information like account numbers and contact details among other things.

1. The above passage is:
A. an advisory for the computer users.
B. an advisory for the Internet users.
C. a warning against possible threat to the Internet users.
D. related to chat rooms, blogs and message boards.

2. People who use chat rooms and blogs,
A. are safe and have no threats from spam.
B. are more prone to malicious e-mails.
C. create virus and hack the accounts of others.
D. are unsafe.

3. According to the writer, it is not safe to:
A. log on to spurious links.
B. access one's own account frequently.
C. have essential information stored in a computer.
D. include data-stealing code.

4. Hackers and spammers, according to the writer, are:
A. a new threat to the Internet users and the economy.
B. only pranksters and not serious threat to the system.
C. trained, professional technocrats who are an asset.
D. not expert professionals.

5. Which word in the passage is synonym of *weblog*?
A. blog B. password
C. site D. e-mail

6. Which agency has assessed the data record with spam sites?
A. Microsoft
B. Internet security from Google
C. Websense
D. Spammers

7. Which month dominates the e-mails detection data record?
A. May B. June
C. Last six months D. Last year

PASSAGE-2

South Asian tropical weather reels valuable geographical aesthetic beauty surrounded by wildlife, mountains, steep elevations with dense forest. Human sapiens has been old and innocent victim of natural calamities branded as flood, hurricane, storm, earthquake and tsunami since ages.

Biological data depicts the loss of valuable men and material suffered by mankind owing to the chain reaction by such fatal natural incidents. Global warming has turned an eyesore to the rapid glory of rising civilsation with great mystery yet to be unfolded. UNO data reveals the rate of mortality owned due to natural disaster with great sway of cruel time factor.

Of the 245 natural disasters in the first 11 months of 2009, 224 or 91.4 per cent were due to weather, a UN study has found.

Of the 58 million people affected, 55 million were affected by weather-related disasters, which accounted for $15 billion (₹ 7,05,000 crore) spent in coping with these disasters.

A study has found that half of these disasters, mainly storms and floods, have taken place in Asia, Debarati Guha Sapir of Brussels-based CRED said.

And this data does not include the loss due to drought because that takes longer to account for. Guha Sapir said, "We have found that this is due to global warming, weak political systems, urbanisation, weak infrastructure, population increase and deforestation."

"We have to cap the emissions of greenhouse gases because they produce greater risk of disasters," WMO head Michel Jerraud said. Greenhouse gases, mainly carbon dioxide, are causing the climate change that is already affecting farm output, making droughts, floods and storms more frequent and more severe, and raising the sea level.

1. Mark the correct statement(s).
1. Weather is responsible for most of the natural disasters.
2. There is no need to worry about the climatic changes.
3. All the nations are trying to have a common pact to tackle natural disasters.

A. 1 is correct
B. Only 2 is correct
C. Only 3 is correct
D. 1 and 2 are correct

2. Which one of the following statements is not true?
A. More than 58 million people have been affected by natural disasters.
B. $15 billion have been spent in coping with these disasters.
C. Most of the natural disasters have taken place in Asia.
D. The report does not include the loss caused by droughts.

3. The main cause of natural disasters is :
A. global warming
B. weak political system
C. deforestation
D. all of these

4. The first priority before the nations and the scientists across the world is :
A. to cap the emission of greenhouse gases.
B. to reach an agreement on the nuclear weapons.
C. to help one another when natural disasters occur.
D. to find out the causes of natural disasters.

5. Which word in the passage means the same as *serious*?
A. Reduce B. Severe
C. Underlined D. Underscore

6. Which region is mainly affected by these disasters?
A. United Nation B. Africa
C. Asia D. Brussels

7. Which statement is true?
A. Sea level rise is an alarming incident.
B. Rapid urbanisation is a curse to the global warming.
C. Deforestation with climatic change is responsible for green house gaseous effect.
D. All of these.

PASSAGE-3

Time has brought forth different classical and instrumental range of device that serves healing element to the aweful and mentally distressed state of affair. Temperament compatibility matching is a reasonable trend of motion adhered with the concept of body movement termed as dance therapy.

Dance has mobilised different acknowledgement of its noted culture proclaimed by experts and sensational aspirants. Vedas describe the dance origin from aryan culture to the creeping span of modern India. Archaeological Survey of India has claimed the civilsation existence issue owing to excavation of past events traced in the stone idols displayed in dance "mudra".

Dance is an art form that generally refers to movement of the body, usually rhythmic and to music, used as a form of expression, social interaction or presented in a spiritual or performance setting. Gymnastics, figure skating and synchronised swimming are sports that incorporate dance, while martial arts Kata are often compared to dances. Motion in ordinarily inanimate objects may also be described as dances (the leaves dance in the wind). Every dance, no matter what style, has something in common. It not only involves flexibility and body movement, but also physics. If the proper physics is not taken into consideration, injuries may occur. Choreography is the art of creating dances. The person who creates (*i.e.* choreographs) a dance, is known as the choreographer. Dance has certainly been an important part of ceremony, rituals, celebrations and entertainment since the birth of the earliest human civilisations. Archaeology delivers trace of dance from prehistoric times such as the 9,000 years old Rock Shelters of Bhimbetka painting in India and Egyptian tomb painting depicting dancing figures from circa 3300 BC. One of the earliest structured uses of dance may have been in the performance and in the telling of myths. Before the production of written languages, dance was one of the methods of passing these stories down from generation to generation.

1. Dance does not necessarily involve:
   A. movement of the body.
   B. face expression.
   C. rhythmic movements.
   D. costumes.
2. According to the passage, which one of the following sports does not incorporate dance?
   A. Figure skating B. Gymnastics
   C. Chess D. Swimming
3. Which one, according to the passage, is not common in all the dances?
   A. Flexibility B. Body movement
   C. Music D. Physics
4. Physics in every dance is an important element to:
   A. master the art of dance.
   B. to perform it scientifically.
   C. to save oneself from injury.
   D. to get an award.
5. In prehistoric times dance was:
   A. a substitute of written language.
   B. a source of entertainment.
   C. a necessary ritual.
   D. not allowed in royal palaces.
6. Choreographer is a term addressed to:
   A. a sports trainer B. a gym coach
   C. a painting teacher D. a dance teacher
7. What does the term 'the telling of myths' refer to?
   A. Telling the beads B. Vedic convention
   C. Spiritual recitation D. Protest march

PASSAGE-4

Personal sentiments exposure portray in sober class of distinct feature with fair and pleasant expression. As the cruelty can be overpowered by the amicable terms likewise, impossible review let with expression can be publically announced by either electronic device or diary writing habit. On line conversation such as chattering, message convey are the rampant tool of the time to manage in communication congestion.

Blogs, online versions of personal diaries, have been a craze for quite some time now. Some of us, before the Internet age, used to maintain a small diary to jot down random thoughts, reflections, sayings, poems and other such stuff. Their online versions, however, are much more than that. The biggest difference between a blog and a personal diary is while the latter was a very private affair accessible not even to family members, a blog is a public account. In fact the whole world connected to the internet gets to read its contents if they come across it.

What a blogger should keep in mind while blogging is that by displaying our thoughts, emotions, opinions and personal life in full public glory, we are revealing a part of our private life. We are giving outsiders, strangers, friends, colleagues and relatives a full view of what's happening inside us and without lives, something we would hesitate to do in real life. Once blogging begins, it's difficult to control, and we end up sharing our personal thoughts and emotions that we would otherwise have kept to ourselves. The blog entices us, and we fall to be bait.

Blogs don't preserve our anonymity. They let us publish our pictures, professional and personal details, physical address and locations, etc. The biggest precaution to be observed therefore is to think twice before publishing anything, and ask oneself if one would like to make such matters public. The blog readers try to judge us, understand us, get information about us and take pleasure while peeping into our private lives without realising it.

1. Blogs are ______.
   A. diaries B. personal diaries
   C. online diaries D. official diaries
2. Earlier the personal diaries were not available even to family members but blog can be read by:
   A. family and friends
   B. friends only
   C. anyone who connects to the Internet
   D. relatives
3. Which one of the following statements is not true?
   A. Blogging does not affect our privacy.
   B. Blogging displays our private life in public.
   C. Blogging cannot be controlled easily.
   D. Blogging allows outsiders to peep into our personal life.
4. The writer in the above passage:
   A. favours blogging.
   B. criticises the practices.
   C. does not want the reader to become a victim of blogging.
   D. favours blogging but with some precautions.
5. Which word in the passage stands for *an account from on point of view*?
   A. version B. block
   C. affair D. account
6. The word Anonymity in the passage refers to:
   A. hostility B. friendship
   C. unacknowledged D. differences

7. What precaution is stressed by the narrator while releasing personal information?
   A. Suggest Comprehensive issue only
   B. Reader friendly blog appreciation
   C. Hiding personal facts
   D. Diary maintenance

PASSAGE-5

Pleasure and sorrow being twin face of a coin suffers the sign of commend and condemn since long before the historic ages. Warriors outcry with victory decorated with deep throat laugh has marked the sentimental coverage of battle filed victory. Baffled war victims, war widows, dependents are the oppressed classes. They seek relief against those causing sabotage to their established settlement.

The history of literature really began long before man learned to write. Dancing was the earliest of the arts. Man danced for joy round his primitive camp fire after the defeat and slaughter of his enemy. He yelled and shouted as he danced and gradually the yells and shouts became coherent and caught the measure of the dance and thus the first war song was sung. As the idea of God developed, prayers were framed. The song and prayers became traditional and were repeated from one generation to another, each generation adding something of its own. As man slowly grew more civilsed, he was compelled to invent some method of writing by three urgent necessities. There were certain things that were dangerous to forget and which, therefore, had to be recorded.

It was often necessary to communicate with persons who were some distance away and it was necessary to protect one's property by making tools and so on, in some distinctive manner. So man taught himself to write and having learned to write purely for utilitarian reasons, he used this new method for preserving his songs and his prayers. Of course, among these ancient people, there were only a very few individuals who learned to write, and only a few could read what was written.

1. Liguistic comment and condemn refers to:
   A. physical grudge against sorrows
   B. a temporary verbal communication
   C. slip of tongue with praise and denounce
   D. a communal fury against the joys
2. Which one of the following statements is true?
   A. Study reveals the historic onslaught with cruelty only.
   B. Suffers pledge to reform the existing tyranny.
   C. A permanent code of sympathetic temper is sought by the suffers.
   D. Language dominates with great passion during mental aggression.
3. Before man invented writing, ______.
   A. literature was passed on by word of mouth
   B. prayres were considered literature
   C. literature was just singing and dancing
   D. all of the above
4. According to the passage, dance originated from
   A. literature
   B. yelling and shouting
   C. prayers
   D. war
5. Writing was invented because
   A. there were certain things which had to be remembered.
   B. there were certain things which had to be communicated to the people who were away.
   C. there were certain rituals which had to be preserved.
   D. all of the above
6. Which one of the following statements is not true?
   A. Since only a few individuals had learnt to write and a few could read what was written.
   B. Prayers were developed after the idea of god developed.
   C. People learned writing for utilitarian reasons.
   D. Dancing was the earliest form of art.
7. Which one of the following words in the passage stands for *ancient*?
   A. primitive B. traditional
   C. coherent D. utilitarian

PASSAGE-6

Quality dominates the world ahead of its spherical, classical, racial or any undue discrimination ever exercised by any individual. Physical characteristics of a produce differs its structural attribution in quality from substandard produce with richness of prolong usage features declared by the claimant. Talent Hunt initiated by quality seekers have turned the table of established brand on account of innovative additions presented by their hardships.

The simple message that quality matters because it is cheaper to do things right the first time is yet to sink into the thinking of Indian managers and workers who seem quite content to live in their cloistered world where mediocrity reigns supreme. It is only the ignorant who believe that quality consciousness is a cultural trait. Barely half a century back no one associated Japan with high quality, but today 'Made in Japan' is a symbol of assured quality. The daily drill at school is far more important than the seasonal seminar and an obsession with quality at home is the best way of convincing the world that we value good quality. Indian exports will never be recognised for quality unless Indians are known to insist on good quality at home.

Global appreciation surrounds the exceptional brand by keeping up the physical standard neatly and efficiently. It is designed to meet the international standards' numerous logos and certificates to upkeep the label of quality consciousness. It has prime motto of swaying clean chit observation to the sanctioned produce without any objection. Modern age has been proved a turning point to the old established traditional belief. They are using tiring house hold produce to industrial device responsible to deface the status of rapid development. We breath into the cool and fresh air of healthy surrounding devised by mankind monitored by quality management with rigorous span or struggle tackled with due care.

1. The theme of the passage is:
A. quality consciousness
B. export promotion
C. quality management
D. promoting Indian culture

2. According to the passage, the Indian exports will never be recognised for quality:
A. unless there is technical advancement.
B. unless there is competition.
C. unless the Indians are known to insist on good quality.
D. unless government supports the exporters.

3. Which one of the following statements is true?
A. Indians are known for their insistence on quality.
B. Indians are not quality conscious.
C. Japan is not known for its quality product.
D. 'Made in Japan' is not a symbol of assured quality.

4. The best way to convince the world about quality is:
A. to conduct seminars on quality control.
B. to advertise the products.
C. to value for quality products at home.
D. all of the above

5. According to the passage, the main fault with the Indian managers is that:
A. they are not hard-working.
B. they are not quality conscious.
C. they are not well trained in quality control.
D. they lack technical knowledge.

6. How quality consciousness is termed as a cultural trait with the ignorant?
A. The mature opinion of quality.
B. Quality dominates ahead of traditional belief.
C. Cultural trait justifies any quality standard.
D. Quality consciousness leads to management growth.

7. Why 'Made in Japan' is a symbol of assured quality?
A. Traditional belief
B. Quality consciousness
C. Age old negligence suffering
D. High export offer

PASSAGE-7

Creamy outlook with soft speech presentation is a prima facie attraction that brings joy on the withering faces with a mixture of sensational touch. Smile with grace is the ultimate boon to the mankind keeping away any melancholy at bay. Old age is often miscalculated as liability with little attention paid to them. Their talents and learned potential is buried alive with undue care by oblivious approach due to disfigured or deformed physical structure taken over by aged talents of the society.

It happens to us all, however hard we may try to delay the process. We grow old. Cosmetic surgery may remove the wrinkles, skin which has sagged may be tightened by means of a facelift and hair dresser may dye grey hair a more youthful colour, but we cannot remain young forever.

However, what is important is the quality of life. Some people are lucky to be taken care of at home whereas others may have to move to residential homes. The worst part of ageing is that often the mind becomes less alert. As people grow older, they experience short term memory. Later some may suffer from dementia, often in the form of Alzheimer's disease.

By no means all people are in this category. Many senior citizens are in the possession of all their faculties and see retirement as a time of freedom. Not only that if they have a generous retirement pension, they are likely to be quite well off with money to be spent on a holiday and other luxuries. Because of this, both businesses and government have a new respect for what is known as grey power.

It is unfortunate that many people regard old people as geriatrics who have one foot in the grave. Someone should remind them that they too would be old one day!

They are honoured with senior citizen label granted by the authority with due regard paid from distinct possession of living cadres. One should always bear in mind the old saying "Tit for Tat". Which denotes its symbolic regulation of periodical sufferings. The aforesaid misdoings of emotional disparity, done by striding steps is a curse to the mankind.

1. Cosmetic surgery :
A. helps in the retention of youth forever.
B. cannot stop the natural process of ageing.
C. is the solution to ageing.
D. is the fancy of the rich.

2. While ageing man's greatest worry is :
A. his growing wrinkles.
B. his short term memory.
C. his quality of life.
D. moving to a residential home.

3. The government considers grey power as :
A. a liability
B. an asset
C. neither an asset nor a liability
D. a spent force

4. Retirement is seen as a time of freedom because people:
A. no longer have to go to office.

B. can enjoy their earnings.
C. can spend time with their families.
D. finally move into residential homes.

5. *geriatrics* in the passage refers to :
A. handicapped people
B. doctors for the aged
C. a branch of medicine
D. aged people

6. What is the theme of the passage?
A. Old is gold
B. The grey power
C. Youth domination in the society
D. Aged suffering

7. What does the phrase 'suffer from dementia' refer to?
A. Harmful
B. Short term memory loss
C. Lunacy
D. Paralysis

PASSAGE-8

The social Darwinist conception of evolution is not that of the biologist but of the propagandist looking for justification for his political theories. The simplistic formulation of ferocious antisocial struggle finds no place in the theories of modern biologists. Even on the animal level the scientific picture is not that of "struggle" and ruthless elimination. If in the tropics a pigmented skin is an advantage, no one is swept to destruction in a fight for blackness. Modifications of that sort, which prove advantageous, gradually establish themselves without trouble. Even living on other animals for food does not imply savagery—fishermen are not unpleasant and aggressive persons; tribes living, as they once did, on shellfish are not haters of their kind; even pig breeders and chicken farmers can be kind to their neighbours and no worse than the rest of us. Lorenz, constantly quoted to defend innate aggression, points out that the predatory carnivores are not angry when bringing down an antelope. It is simply a matter of going to fetch the dinner. A lion can be angry, but not when going out to kill. One might go on and take the whole case to pieces bit by bit, but enough has been said to indicate that the picture of "nature red in tooth and claw" is the poet's view—the phrase was Tennyson's—not the scientist's: it is tendentious, controversial exaggeration, not objective science. Even the notion of the "survival -of the fittest" begs the question. If we say that the fittest survive, we only mean that they do survive. It implies no other quality than survival ability. It holds just as well for the oyster or the flourishing flea as for the beautifully adapted camel or flying fox. It does not follow that the survivor is the fittest even in being the finest specimen of its kind, let alone of the kind that we would prefer to see flourish.

1. The instinct to kill in carnivores comes from their
A. need to survive
B. need to prove their might
C. inherently brutal nature
D. liking for flesh .

2. The writer's attitude towards poets is that of
A. awe B. scorn
C. anger D. indifference

3. The term 'survival of the fittest' suggests the survival of the
A. mightiest B. finest
C. best D. prepared

4. By citing various examples the writer tries to establish the fact that savagery does not entail
A. killing of any kind
B. killing of lesser animals
C. killing dispassionately to serve an end
D. ruthless killing

5. The concept of ferocious antisocial struggle finds credence with
A. biologist B. scientists
C. Lorenz D. Darwin

PASSAGE-9

The difficulty in the education of young infants is largely the *delicate balance* required in the parents. Constant watchfulness and much labour are needed to avoid injury to health : these qualities will hardly exist in the necessary degree except where there is strong parental affection. But where this exists, it is very likely not to be wise. To the devoted parents, the child is immensely important. Unless care is taken, the child feels this and judges himself as important as his parents feel but later in life his social environment will not regard him so fondly and the habit of assuming that he is the centre of other people's universe will lead to disappointment. It is, therefore, necessary, not only in the first year, but afterwards also, that parents should be breezy, cheerful and rather matter-of-fact where the child's possible ailments are concerned. In old days, infants were too much petted, sung to, rocked and dandled. This was wrong since it turned them into helpless pampered parasities. The right rule is : encourage spontaneous activities but discourage demands on others. Do not let the child see how much you do for him or how much trouble you take. Let him, wherever possible, taste the joy of success, achieved by his own efforts, not extracted by tyrannising over the grown-ups.

1. 'Delicate balance' required in parents in sentence one refers to
A. cautiousness B. possessiveness
C. indifference D. matter-of-factness

2. In the past the excessively fondled children became
A. troublesome B. mischievous
C. parasites D. abnormal

3. For the possible ailments of the child, the writer advised the parents to be
A. indifferent B. demanding
C. unsympathetic D. matter-of-fact

4. The author wishes the parents to let the child
A. manage his own things
B. be absolutely free
C. work under their guidance
D. seek anybody's help if needed

5. A child pampered by his parents is likely to become
A. disappointed B. demoralised
C. dependent D. independent

PASSAGE-10

The world dismisses curiosity by calling it idle or mere idle curiosity-even though curious persons are seldom idle. Parents do their best to extinguish curiosity in their children because it makes life difficult to be faced everyday with a string of unanswerable questions about what makes fire hot or why grass grows. Children whose curiosity survives parental discipline are invited to join our university. With the university, they go on asking their questions and trying to find the answers. In the eyes of a scholar, that is what a university is for. Some of the questions which the scholars ask seem to the world to be scarcely worth asking, let alone answering. They asked questions too minute and specialised for you and me to understand without years of explanation. If the world inquires of one of them, why he wants to know the answer to a particular question, he may say, especially if he is a scientist, that the answer will in some obscure way make possible a new machine or weapon or gadget. He talks that way because he knows that the world understands and respects utility.

But to you who are now part of the university, he will say that he wants to know the answer simply because he does not know it, the way the mountain climber wants to climb a mountain, simply because it is there. Similarly, a historian asked by an outsider, why he studies history, may come out with the argument that he has learnt to repeat on such occasions, something about knowledge of the past making it possible to understand the present and mould the future. But if you really want to know why a historian studies the past, the answer is much simpler, something happened and he would like to know what. All this does not mean that the answers which scholars find to their questions have no consequences. They may have enormous consequences but these seldom form the reason for asking the question or pursuing the answers. It is true that scholars can be put to work answering questions for the sake of the consequences as thousands are working now, for example, in search of a cure for cancer. But this is not the primary function of the scholars. The consequences are usually subordinate to the satisfaction of curiosity.

1. The common people consider some of the questions that the scholars ask unimportant
A. as they are too lazy and idle
B. as they are too modest
C. as it's beyond their comprehension
D. as it is considered a waste of time

2. According to the passage, parents do their best to discourage curiosity in their children
A. because they have no time
B. because they have no patience to answer them
C. because they feel that their children ask stupid questions continuously
D. because they are unable to answer all their questions

3. According to the passage, the children make life difficult for their parents
A. by their ceaseless curiosity
B. by unceasing bombardment of questions
C. by asking irrelevant questions
D. by posing profound questions

4. Children whose curiosity survives parental discipline means
A. children retaining their curiosity in spite of being discouraged by their parents
B. children pursuing their mental curiosity
C. children's curiosity subdued due to parents' intervention
D. children being disciplined by their parents

5. A historian really studies the past
A. to comprehend the present and to reconstruct the future to
B. to explain the present and plan the future
C. to understand the present and make fortune
D. to understand the present and mould the future

PASSAGE-11

Shames and delusions are esteemed for soundest truths, while reality is *fabulous*. If men would steadily observe realities only, and not allow themselves to be deluded, life, to compare it with such things as we know, would be like a fairy tale and the Arabian Night's entertainments. If we respect only what is inevitable and has a right to be, music and poetry would resound along the streets. When we are unhurried and wise, we perceive that only great and worthy things have any permanent and absolute existence—that petty fears and petty pleasures are but the shadow of the reality. This is always exhilarating and sublime. By closing the eyes and slumbering and consenting to be deceived by shows, men establish and confirm their daily life of routine and habit everywhere, which still is built on purely illusory foundations. Children, who play life, discern its true law and relations more clearly than men, who fail to live it worthily, but who think that they are wiser by experience; that is by failure.

I have read in a Hindu book that there was a king's son who, being expelled in infancy from his native city, was brought up by a forester and growing up to maturity in that state, imagined himself to belong to the barbarous race with which he lived. One of his father's ministers, having discovered him, revealed to him what he was and the misconception of his character was removed and he knew himself to be a prince, "So soul," continues the Hindu philosopher, "from the circumstances in which it is placed, mistakes its own character, until the truth is revealed to it by some holy teacher and then it knows itself to be Brahma."

We think that is which appears to be. If a man should give us an account of the realities he beheld, we should not recognize the place in his description. Look at a meeting house, or a court house, or a jail, or a shop, or a dwelling house and say what that thing really is before a true gaze and they would all go to pieces in your account of them. Men esteem truth remote, in the outskirts of the system, being the farthest star, before Adam and after the last man. In entirety, there is indeed something true and sublime. But all these times and places and occasions are now and here. God himself culminates in the present moment and will never be more divine in the lapse of all ages. And we are enabled to apprehend at all what is sublime and noble only by the perpetual instilling and drenching of the reality that surrounds us. The universe constantly and obediently answers to our conceptions : Whether we travel fast or slow; the track is laid for us. Let us spend our lives in conceiving, then. The poet or the artist never yet had so fair and noble a design but some of his posterity at least could accomplish it.

1. The author believes that children are often more acute than adults in their appreciation of life's relations because
A. children know more than adults
B. children can use their experience better
C. children's eyes are unclouded by failure
D. experience is the best teacher

2. The writer's attitude towards the arts is one of
A. indifference B. suspicion
C. admiration D. repulsion

3. The passage implies that human beings
A. cannot distinguish the true from the untrue
B. are immoral if they are lazy
C. should be bold and fearless
D. believe in fairytales

4. The author is primarily concerned with urging the reader to
A. mediate on the meaninglessness of the present
B. appraise the present for its true value
C. look to the future for enlightessment
D. spend more time in leisure activities

5. The word 'fabulous' in the passage means
A. wonderful B. delicious
C. birdlike D. illusion

PASSAGE-12

The stress-relieving power of humour is well-established. Laughter relieves tension, breaks negative 'holding patterns,' and helps out our problems in perspective. Humour is listed as a major stress-coping mechanism among healthy men. It has been credited with reducing the coronary risk of a high-stress lifestyle. Not only does it help diffuse anxiety and anger, as a heart attack victim points out, it acts as a blocking agent against the ravages of panic. The panic produced by a heart attack can be just as dangerous as the heart attack. It constricts blood vessels and destablises the heart. But through humour, reassurance and positive thinking you can control the panic and thereby enhance your prospects for recovery. Laughter creates an environment for healing. And that goes for all illnesses—not only coronary artery disease. Negative emotions, such as depression, weaken the immune system, making us more susceptible to a host of illnesses ranging from cold to cancer. It stands to reason that positive emotions such as humour and laughter may have the opposite effect, keeping our immune system strong and defending us against sickness. Sick people, especially those who are depressed, take themselves much too seriously. Even though controlled scientific studies for measuring the chemistry of laughter are rare, yet what is there does support the theory that good humour breeds good health.

1. Humour helps fight heart attacks by
A. alerting us against pain
B. beginning a fitness programme
C. introducing us to high-stress life
D. removing our worries

2. Panic caused by a heart attack
A. makes the blood vessels narrower
B. builds our confidence
C. makes us active and self-sufficient
D. makes us seek medical help

3. It has been proved that humour is
A. a negative trait
B. a solution to all problems
C. helpful in reducing tensions
D. present in all healthy men

4. Humour helps us become
A. richer B. healthier
C. ambitious D. scientific

5. Negative feelings lead to
   A. stronger defences B. increased opposition
   C. many illnesses D. bad reputation

PASSAGE-13

In the technological systems of tomorrow—fast, fluid and self-regulating—machines will deal with the flow of physical materials; men with the flow of information and insight. Machines will increasingly perform tasks. Machines and men both, instead of being concentrated in gigantic factories and factory cities, will be scattered across the globe, linked together by amazingly sensitive, *near-instantaneous communications.* Human work will move out of the factory and mass office into the community and the home. Machines will be synchronised, as some already are to the billionth of a second; men will be de-synchronised. The factory whistle will vanish. Even the clock, "the key machine of the modern industrial age," as Lewis Mumford called it a generation ago, will lose some of its power over humans, as distinct from technological affairs. Simultaneously, the organisation needed to control technology will shift from bureaucracy to Ad-hocracy, from permanence to transience and from a concern with the present to a focus on the future.

In such a world, the most valued attributes of the industrial age become handicaps. The technology of tomorrow requires not millions of lightly lettered men, ready to work in unison at endlessly repetitive jobs, it requires not men who take orders in unblinking fashion, aware that *the price of bread is mechanical submission to authority,* but men who can make critical judgments, who can weave their way through novel environments, who are quick to spot new relationships in the rapidly changing reality. It requires men who, in C.R Snow's compelling terms, "have the future in their bones."

1. The future man, according to this passage, must be
   A. most adaptative and intelligent
   B. most capable of dealing with the changing reality
   C. more concerned with the present than the future
   D. trained and obedient
2. The technological system of tomorrow will be marked by
   A. dehumanisation B. perfection
   C. automation D. unpredictability
3. The type of society which the author has mentioned makes a plea for
   A. a mind assimilative of modern scientific ideas
   B. a critical mind having insight into future
   C. a mind well-versed in cultural heritage
   D. a mind with firm principles of life
4. 'Near-instantaneous communications' may be regarded as a symbol of
   A. anachronisation B. mischronisation
   C. desynchronisation D. synchronisation
5. If a person believes that 'the price of bread is mechanical submission to authority,' he is
   A. a believer in devotion to duty
   B. a believer in taking things for granted
   C. a believer in doing what he is told, right or wrong
   D. a believer in the honesty of machines

PASSAGE-14

Foot-racing is a popular activity in the United States. It is seen not only as a competitive sport but also as a way to exercise, to enjoy the *catneraderie* of like-minded people and to donate money for a good cause. Though serious runners may spend months training to compete, other runners and walkers might not train at all. Those not competing might run in an effort to beat their own time or simply to enjoy the fun and exercise. People of all ages, from those less than one year of age (who may be pushed in *strollers*) to those in their eighties, enter into this sport. The races are held in city streets, on college campuses, through parks and in suburban areas and they are commonly 5 to 10 kilometres in length.

The largest foot-race in the world is the 12 kilometre Bay to Breakers race that is held in San Francisco every spring. This race begins on the east side of the city near San Francisco Bay and ends on the west side at the Pacific Ocean. There may be 80,000 or more people running in this race through the streets and hills of San Francisco. In the front are the serious runners who compete to win and who might finish in as little as 34 minutes. Behind them are the thousands who take several hours to finish. In the back of the race are those who dress in costumes and come just for fun. One year, there was a group of men who dressed like El vis Presley and another group consisted of fire-fighters who were tied together in a long line carrying a fire-hose. There was even a bridal party, in which the bride and the groom threw flowers to bystanders and they were actually married at some point along the route.

1. The word "camaraderie" could best be replaced by which of the following words?
   A. Games B. Jokes
   C. Companionship D. Views
2. The main purpose of the passage is
   A. make fun of runners in costumes
   B. give reasons for the popularity of foot-races
   C. describe a popular activity
   D. encourage people to exercise
3. As used in the passage the word "strollers" refers to:
   A. carriages B. wheelchairs
   C. wagons D. cribs
4. Which of the following is not the reason why people enter foot-races?
   A. To exercise B. To enjoy
   C. As a compulsion D. For charity

5. Which of the following is NOT implied by the author?
   A. Foot-races appeal to a variety of people
   B. Running is a good way to strengthen the heart
   C. Workers can compete for prizes
   D. Age and profession is no bar to enter footraces

PASSAGE-15

The conservative is not an extreme individualist. He may be willing to concede numerous arguments of the unqualified individualists, for his own respect for the dignity of the individual is not surpassed by that of any man. Yet he cannot agree to the full implications of individualism, which is based, so he thinks, on an incorrect appraisal of man, society, history, and government. In his own way, the individualist is as much a perfectionist as the socialist, and with perfectionism the conservative can have no *truck*.

In particular, the conservative refuses to go all the way with economic individualism. His distrust of unfettered man, his recognition to groups, his sense of the complexity of the social process, his iecognition of the real services that government can perform-all these sentiments make it impossible for him to subscribe to the dogmas and shibboleths of economic individualism : laissez-faire, the negative state, enlightened self-interest, the law of supply and demand, the profit motive. The conservative may occasionally have kind words for each of these nations, but he is careful to qualify his support by stating other, more important social truths. For example, he does not for a moment deny the prominence of the profit motive, but he insists that it be recognised for the selfish thing it is and be kept within socially imposed reasonable limits.

1. Which of the following words can replace the word 'truck?'
   A. Dealing B. Bargain
   C. Debate D. Transport
2. The conservative is
   A. A perfectionist B. An economist
   C. A socialist D. None of these
3. Which of the following statements is true?
   A. The socialist and the individualist tend to be broadly similar in their views
   B. The conservative believes that profit motive originates in selfishness
   C. The conservative is also an extreme individualist
   D. None of these
4. The conservative is against economic individualism for all the following reasons except
   A. he does not trust free men
   B. he believes in the authority of the government
   C. he believes in groups
   D. he feels that social processes are important
5. The author mentions all the following catchwords of economic individualism except
   A. free trade
   B. the profit motive
   C. balance of trade
   D. the negative state

PASSAGE-16

Every survey ever held has shown that the image of an attractive woman is the most effective advertising gimmick. She may sit astride the mudguard of a new car, or step into it ablaze with jewels, she may lie at the man's feet stroking his new socks, she may hold the petrol pump in a challenging pass, or dance through woodland glades in slow motion in all the glory of a new shampoo. Whatever she does, her image sells. The *gynolatry* of our civilization is written large upon its face, upon hoardings, cinema screens, television, newspapers, magazines, tins, packets, cartons, bottles, all consecrated to the reigning deity, the female fetish. Her dominion must not be thought to entail the rule of women, for she is not a woman. Her glossy lips and matt complexion, her unfocused eyes and flawless fingers, her extraordinary hair all floating and shining, curling and gleaming, reveal the inhuman triumph of cosmetics, lighting, focussing and printing. She sleeps unruffled, her lips red and juicy and closed, her eyes as crisp and black as if newly painted and her false lashes immaculately curled. Even when she washes her face with a new and creamier toilet soap her expression is as tranquil and vacant, and the paint as flawless as ever. If ever she should appear tousled and troubled, her features are miraculously smoothed to their proper veneer by a new washing powder on a buillon cube. For, she is a doll : weeping, pouting or sinking, running or reclaiming, she is a doll.

1. The author's primary purpose in this passage is
   A. to ridicule women
   B. to show the dominance of women in advertising
   C. to portray the obsession of women with trivial thing
   D. to depict the emancipation of women
2. What point is the writer trying to make when he says "she may lie at a man's feet stroking his new socks"?
   A. Women like being subservient
   B. Women are observed with clothes
   C. This is a typical posture of women in advertising
   D. Women enjoy this kind of intimacy
3. The 'gynolatry' of one civilization would suggest all the following except that
   A. women enjoy immense power in modern society
   B. the image of women boost sales as few other things can
   C. women worship is all pervasive in advertising
   D. glamorous and attractive women are the forte of modern advertising

4. By saying that women depicted in an advertisement is 'not a woman' the author implies that
   A. in real life women are less attractive
   B. the depiction of women in advertisement is grossly artificial and unreal
   C. in-real life women are more dominant
   D. in advertisement, a woman is a mere commercial symbol
5. In the last sentence of the paragraph, the word 'doll' is meant to express
   A. tenderness B. delicacy
   C. contempt D. beauty

PASSAGE-17

Everything that men do or think concerns either the satisfaction of the needs they feel or the need to escape from pain. This must be kept in mind when we seek to understand spiritual or intellectual movements and the way in which they develop, for feeling and longing are the motive forces of all human striving and productivity-however nobly the latter may display themselves to us.

What, then, are the feelings and the needs which have brought mankind to religious thought and to faith in the widest sense? A moment's consideration shows that the most varied emotions stand at the cradle of religious thought and experience.

In primitive peoples it is, the fear that awakens religious ideas—fear of hunger, of wild animals, of illness and death. Since the understanding of causal connections is usually limited on this level of existence, the *human soul forges a being,* more or less like itself, on whose will and activities depends the experiences which it fears. One hopes to win the favour of this being by deeds and sacrifices which according to the tradition of the race are supposed to appease the being or to make him well disposed to man. I call this the religion of fear.

This religion is considerably stabilized, though not caused, by the formation of the priestly caste which claims to mediate between the people and the being they fear and so attains a position of power. Often a leader or a despot will combine the functions of the priesthood with his own temporal rule for the sake of greater security; or an alliance may exist between the interests of the political power and the priestly caste.

1. Who motivates man's action or thinking?
   A. To satisfy his needs or to escape pain
   B. His desire for progress or to rule
   C. His spiritual urge
   D. To carry out the dictates of his religious faith
2. What feeling prompted primitive man to create religion?
   A. Love B. Anger
   C. Fear D. Spiritual revelation
3. How did priests come to acquire political power?
   A. By joining hand with the despotic rulers
   B. By protecting the believers against despotic rulers
   C. By generating fear of the unknown in the minds of rulers
   D. By giving religious blessing to political movements
4. How did religion become firmly established?
   A. Through the constant fear of death
   B. Through the perpetuation of faith in God
   C. Through the establishment of religious practices
   D. Through the growth of a priestly class
5. "Human soul forges a being" means
   A. that ghosts and witches are a creation of human mind
   B. that the concept of God is a creation of human mind
   C. Both (A) and (B)
   D. Neither (A) nor (B)

PASSAGE-18

Several years ago my parents, my wife, my son and I aie at a restaurant. After a wonderful dinner, the waiter set the bill in the middle of the table. That's what it happened, my father did not reach for the bill.

Conversation continued. Finally, it dawned on me. I was supposed to pay the bill. After hundreds of restaurant meals with my parents after a lifetime of thinking of my father as the one with the money, it had all changed. I reached for the bill and my view of myself suddenly altered. 1 was an adult.

Some people mark off their lives in years, I measure mine in small events—in rites of passage. I did not become a young man at a particular age, like 13, but rather when a boy strolled into the shop where I worked and called me "mister". The realization hit me like a punch ; I was suddenly a mister.

I never thought that I would fall asleep in front of the television set as my father did. Now it's what I do best. I never thought I would prefer to stay at home, but now I find myself foregoing parties. I used to think that people who watched birds were weird, but this summer I found myself watching them, and may be I'll get a book on the subject. I yearn for a religious conviction that I never thought I'd want to feel close to my ancestors long gone, and echo my father in arguments with my son. I still lose.

One day I became a father, and not too long after that I picked up the bill for my own father. I thought then it was a rite of passage for me. But one day, when I was little older, I realized it was one for him too.

1. Which of the following does the author consider a rite of passage for him?

A. Becoming thirteen years old
B. Going to a restaurant with his parents, wife and son
C. Being called 'mister' by the boy in the shop
D. Working in a shop as a young boy

2. The author's father expected his son to pay the restaurant bill because
A. he did not have enough money
B. he acknowledged that his son was now an adult
C. it was the first time the family was eating out together
D. the waiter had set the bill in the middle of the table

3. "Rites of passage" for the author generally refers to
A. events that mark the passage of time
B. imitating his father as he grows older
C. slowing down in his pace of living
D. the act of growing old with time

4. In the fourth paragraph the author makes the point that as people grow older they
A. become lazier and less adventurous in their behaviour
B. have to take on many more financial responsibilities
C. turn to their children more for help and advice
D. behave in the same way as their elders even if they never thought they would

5. In the fourth paragraph, 'foregoing parties' means
A. going early to parties
B. hosting many parties
C. giving up parties
D. going to many parties

PASSAGE-19

Mountaineering is now looked upon as the king of sports. But men have lived amongst the mountains since prehistoric times and in some parts of the world, as in the Andes and Himalayas, difficult mountain journeys have inevitably been part of their everyday life. However, some of the peaks were easily accessible from most of the cities of Europe. It is quite interesting that while modern mountaineers prefer difficult routes for the greater enjoyment of sport, the early climbers looked for the easiest ones, for the summit was the prize they all set their eyes on. Popular interest in mountaineering increased considerably after the ascent of the Alpine peak of Matterhorn in 1865 and Edward Whymper's dramatic account of the climb and fatal accident which occurred during the descent.

In the risky sport of mountaineering, the element of competition between individuals or teams is totally absent. Rather one can say that the competition is between the team and the peaks themselves. The individuals making up a party must climb together as a team, for they depend upon one another for their safety. Mountaineering can be dangerous unless reasonable precautions are taken. However, the majority of fatal accidents happen to parties which are inexperienced or not properly equipped. Since many accidents are caused by bad weather, the safe climber is the man who knows when it is time to turn back, however, tempting it may be to press on and try to reach the summit.

1. People living in the Andes and the Himalayas made mountain journeys because
A. it was a kind of sport
B. they had to undertake them in their day-to-day life
C. they lived in prehistoric times
D. of the challenge offered by the difficult journey

2. Mountaineering is different from other sports because
A. it is risky and dangerous
B. it can be fatal
C. it is most thrilling and exciting
D. there is no competition between individuals

3. "..... the summit was the prize they all set their eyes on." In the context of the passage, this means
A. reaching the top was their exclusive concern
B. they kept their eyes steadily on reaching the summit
C. they cared for nothing but the prize of reaching the summit
D. they chose a route from which they could see the summit clearly

4. Mountaineers climb as a team because
A. the height is too much for one individual
B. the competition is between the team and the peak
C. they have to rely on each other for safety
D. there is no competition among them

5. "To press on" in the last sentence of the passage means
A. to struggle in a forceful manner
B. to force upon others
C. to work fearlessly
D. to continue in a determined manner

PASSAGE-20

Human analytical abilities remain vastly superior to anything demonstrated elsewhere in the animal kingdom. Virtually in all studies of animal intelligence and language skills, performance plummets as more elements are added to a task and as an animal has to remember these elements for long periods. By contrast, humans can call on vast working memory.

Many evolutionary scholars suspect that as ancient human groups became larger, the need to keep track of ever more complex social interactions was what really pushed the human brain toward superiority. Both dolphins and chimps have very complex interactions, but the intricacy of their social world pales beside the lattice of entanglements that characterised human society as early. Homosapiens bonded together to gather food and defend themselves. In Somalia today, warring clans identify friend or foe by demanding that those accosted recite their ancestry going back many generations. It is easy to see how similar challenges in antiquity might have driven the development of brainpower.

It does not lessen the grandeur of the human intellect to argue that it evolved partly in response to social pressures or that these pressures also produced similar abilities in "lesser" creatures. Instead, the fact that nature may have broadly sown the seeds of consciousness, suggests a world enlivened by many different minds. There may even be practical applications. Studies of animal cognition and language have yielded new approaches to communicating with handicapped and autistic children. Some scientists are pondering ways to turn intelligent animals like sea lions and dolphins into research assistants in marine studies or into lifeguards who can save the drowning upon command.

If the notion that animals might actually think poses a problem, it is an ethical one. The great philosopher, such as Descartes, used their belief that animals cannot think as a justification for arguing that they do not have moral rights. It is one thing to treat animals as mere resources if they are presumed to be little more than living robots, but it is entirely different if they are recognised as fellow sentient beings. Working out the moral implication makes a perfect puzzle for a large-brained, highly social species like our own.

1. Does the author believe that the grandeur of the human intellect is lessened by the fact that the same evolutionary pressure produced similar abilities in animals as in humans?
   A. Perhaps yes B. Perhaps no
   C. Yes D. No

2. What could be some of the practical applications of animal intelligence?
   A. To use dolphins for solving difficult mathematical problems
   B. To use some animals for guarding and supervising other animals
   C. To develop new approaches to communicate with handicapped and mentally ill children
   D. All of these

3. In many studies of animal intelligence and language skills, the performance significantly drops
   A. as we move from humans to animals
   B. membered for long periods
   C. as we move from animals with large brains to those with small brains
   D. as the time allowed to develop these skills is reduced

4. What makes a perfect puzzle for a large-brained highly social species like humans, according to the author?
   A. Determining the moral implications of the fact that animals might actually think
   B. discovering the real reason why nature developed the ability to think in animals
   C. Deciding what should be the ideal relationship between humans and animals
   D. To answer the question if animals are different from humans

5. What really pushed the human brain towards superiority?
   A. The need to keep track of ever more complex social interactions
   B. The need to develop efficient methods of gathering food
   C. The need to find better methods of defending themselves
   D. The need to evolve faster than other species

6. What is the ethical problem posed by the notion that animals might actually think?
   A. It is difficult to accept that humans are also animals
   B. animals should perhaps not be used in laboratory experiments
   C. Animals cannot perhaps be treated as mere resources
   D. It has to correctly assessed as to how much can an animal actually think

7. Why did early Homosapiens bond together?
   A. They formed social groups as they were social animals
   B. To gather food and to defend themselves
   C. To maintain genetic purity
   D. They were more comfortable with similar people

ANSWERS

PASSAGE-1

1	2	3	4	5	6	7
A	D	A	A	A	C	B

PASSAGE-2

1	2	3	4	5	6	7
A	A	D	A	B	C	D

PASSAGE-3

1	2	3	4	5	6	7
D	C	D	C	A	D	B

PASSAGE-4

1	2	3	4	5	6	7
C	C	A	D	A	D	C

PASSAGE-5

1	2	3	4	5	6	7
B	D	D	B	D	C	A

PASSAGE-6

1	2	3	4	5	6	7
A	C	B	C	B	C	B

PASSAGE-7

1	2	3	4	5	6	7
B	C	C	B	D	B	C

PASSAGE-8

1	2	3	4	5
A	D	A	C	D

PASSAGE-9

1	2	3	4	5
A	C	D	A	C

PASSAGE-10

1	2	3	4	5
C	D	A	A	D

PASSAGE-11

1	2	3	4	5
C	C	A	B	D

PASSAGE-12

1	2	3	4	5
D	A	C	B	C

PASSAGE-13

1	2	3	4	5
B	C	B	D	C

PASSAGE-14

1	2	3	4	5
C	C	B	C	C

PASSAGE-15

1	2	3	4	5
A	C	C	C	C

PASSAGE-16

1	2	3	4	5
B	C	B	B	D

PASSAGE-17

1	2	3	4	5
A	C	A	D	D

PASSAGE-18

1	2	3	4	5
C	A	A	D	C

PASSAGE-19

1	2	3	4	5
B	C	A	B	D

PASSAGE-20

1	2	3	4	5	6	7
A	C	B	A	A	C	B

2

Spelling Errors

There are thousands of words in English language. It is difficult to remember the spellings and meanings of all at once. Try to learn as many as you can. Use a dictionary regularly.

MULTIPLE CHOICE QUESTIONS

Directions: *In each of the following questions, four words are given of which only one is correctly spelt. Find the correctly spelt word.*

1. A. Haphazard B. Hapahazard C. Haphzard D. Haphazrd

2. A. Peevishness B. Pievishness C. Peivishness D. Peeveshness

3. A. Gelactic B. Galectic C. Galactic D. Galactec

4. A. Imbibision B. Imbision C. Embibition D. Imbibition

5. A. Zoparasite B. Zooparaset C. Zooparasite D. Zoprasite

6. A. Neuralgia B. Nuraliga C. Neurolgia D. Neuraltia

7. A. Survellance B. Surveilance C. Surveillance D. Survaillance

8. A. Schedule B. Schdule C. Schedale D. Schedeule

9. A. Occurad B. Occurred C. Ocurred D. Occured

10. A. Licentious B. Licontious C. Licenttious D. Licientious

11. A. Chancelery B. Chancellery C. Chancellary D. Chancelary

12. A. Itenerary B. Itinarery C. Itinarary D. Itinerary

13. A. Grieff B. Grief C. Grieef D. Grrief

14. A. Indipenseble B. Indispansible C. Indispensable D. Indipensable

15. A. Guarantee B. Garuntee C. Guaruntee D. Gaurantee

16. A. Meddicine B. Medicine C. Medicene D. Medicinne

17. A. Benefeted B. Benefitted C. Benifited D. Benefited

18. A. Kleptomonia B. Kleptemonia C. Kleptomania D. Klaptomania

19. A. Sepalchrle B. Sepalchral C. Sepulchrle D. Sepulchral

20. A. Acommodation B. Acomodation C. Accomodation D. Accommodation

21. A. Teracherous B. Treacherous C. Treacheraus D. Treachereans

22. A. Querrelsome B. Quarrelsame C. Quarrelsome D. Queralsome

23. A. Farmament B. Farmement C. Fermament D. Fremament

24. A. Sympathetic B. Smypathetic C. Sympothetic D. Sympethetic

25. A. Pecification B. Pacification C. Pecifacation D. Pecefication

26. A. Meritricious B. Merefrecious C. Meretricious D. Merritricious

27. A. Hypocritical B. Hypocretical C. Hypocriticel D. Hypocirticel

28. A. Vaingloriaus B. Vaniglorious C. Vaniglerious D. Vaingloreus

29. A. Imprecticability B. Impracticebility C. Impracticibility D. Impracticability

30. A. Prograssive B. Progressive C. Progresive D. Prograsive

31. A. Incradulous B. Incredulous C. Incridulous D. Incredalous

32. A. Uncivilized B. Uncevilized
C. Uncivillized D. Uncevelized

33. A. Osttentatious B. Ostentetious
C. Ostentatious D. Ostenttatious

34. A. Extravagant B. Extreragent
C. Extreregant D. Extravegent

35. A. Missunderstood B. Miesunderstood
C. Misunderstood D. Misunderstod

36. A. Belligerent B. Beligirent
C. Belligarant D. Belligerrent

37. A. Pasiveness B. Passiveness
C. Passevenis D. Passivines

38. A. Astonished B. Astronished
C. Astoneshed D. Asstonished

39. A. Obnosious B. Obnoxeous
C. Obnoxious D. Obnoseous

40. A. Uncivilised B. Uncivelized
C. Uncevelized D. Uncivilized

41. A. Sincerely B. Sencerely
C. Sincerelly D. Sincerrely

42. A. Efflorascence B. Eflorescene
C. Effllorescence D. Efflorescence

43. A. Vulnarable B. Valnerable
C. Velnerable D. Vulnerable

44. A. Rigourous B. Rigerous
C. Rigorous D. Regerous

45. A. Omenous B. Ominous
C. Ommineous D. Omineous

46. A. Puerille B. Puerrile
C. Puerile D. Purrile

47. A. Satellite B. Sattellite
C. Satelite D. Sattelite

48. A. Pesanger B. Passenger
C. Pessenger D. Pasanger

49. A. Humurous B. Humorous
C. Humoreus D. Humorrous

50. A. Equinimity B. Equanimmity
C. Equannimity D. Equanimity

51. A. Exeggerate B. Exaggerate
C. Exadgerate D. Exagerate

52. A. Fariegn B. Forein
C. Foriegn D. Foreign

53. A. Rennaissance B. Rennaisance
C. Renaisance D. Renaissance

54. A. Excesive B. Excessive
C. Exccessive D. Exccesive

55. A. Forcaust B. Forcast
C. Forecast D. Forecaste

56. A. Valuptuous B. Volluptous
C. Voluptuous D. Volupttuous

57. A. Entrepraneur B. Entreprenuer
C. Entrapreneur D. Entrepreneur

58. A. Paralleted B. Paralelled
C. Parralleled D. Parallelled

59. A. Ocasion B. Occassion
C. Occasion D. Ocassion

60. A. Boquet B. Bouquet
C. Bouquete D. Bouquette

61. A. Loquacieous B. Lequacious
C. Loquacious D. Lequocious

62. A. Cataclysm B. Cataclism
C. Catacilysm D. Cataclesm

63. A. Chettering B. Chaterring
C. Chattering D. Chatering

64. A. Fulfilment B. Fulffilment
C. Fulfilmient D. Fullfilment

65. A. Conjectore B. Conjecture
C. Cenjecture D. Cenjectire

66. A. Endication B. Endicaltion
C. Indication D. Indicetion

67. A. Splendour B. Spllendour
C. Splandour D. Splendeur

68. A. Excllusion B. Exclussion
C. Exclusion D. Exclosion

69. A. Discourage B. Disscourage
C. Discourege D. Discaurage

70. A. Curageous B. Courageous
C. Courrageous D. Couregeous

71. A. Abandon B. Abanddon
C. Abendon D. Abbandon

72. A. Embarassment B. Emberrassement
C. Embarrassment D. Embbaresment

73. A. Veneration B. Venration
C. Venneration D. Venerration

74. A. Tacciturnity B. Taciturnity
C. Taciturrnity D. Tacitturnity

75. A. Despponding B. Despending
C. Desponding D. Dessponding

76. A. Bueyant B. Buoyant
C. Bueeyant D. Buoyent

77. A. Eccintric B. Eccentrie
C. Eccentric D. Eccintrie

78. A. Corrigible B. Corigible
C. Corregible D. Corrigiblle

79. A. Amandable B. Amendable
C. Amenddable D. Amendablle

80. A. Occasional B. Occassional
C. Occesional D. Occessional

81. A. Exttirpate B. Extirpete
C. Extirpate D. Exterpate

82. A. Vorecious B. Varacious
C. Voarcious D. Voracious

83. A. Insolvent B. Insolvant
C. Insollvent D. Insolvvent

84. A. Querrel B. Querral
C. Quarrel D. Quarel

85. A. Contrebution B. Contribution
C. Contributtion D. Conterbution

86. A. Desgrace B. Disgrece
C. Disgrice D. Disgrace

87. A. Harassment B. Herassment
C. Harasment D. Harassmient

88. A. Tranquilitty B. Tranquility
C. Trenquility D. Tranquillity

89. A. Imaginative B. Imeginative
C. Imagenative D. Imaginetive

90. A. Suficient B. Suficiant
C. Sufficient D. Sufficiant

91. A. Adequate B. Edequate
C. Adaquete D. Edaquete

92. A. Exparienced B. Experianced
C. Experienced D. Experrienced

93. A. Flatering B. Fletering
C. Flattering D. Fletaring

94. A. Cuttiveted B. Culltrivated
C. Cultivated D. Caltivated

95. A. Materialistic B. Materielistic
C. Meterialistic D. Matterialistic

96. A. Praiceworthy B. Peiseworthy
C. Praiseworthy D. Praisaworthy

97. A. Tranquil B. Trenquil
C. Tranquel D. Trinquil

98. A. Neggardly B. Nigardly
C. Niggerdly D. Niggardly

99. A. Magnanimity B. Megnanimity
C. Magnenimity D. Magnanimety

100. A. Darogate B. Derogate
C. Derogeta D. Deragate

101. A. Philanthropist B. Philenthropist
C. Phelanthropist D. Philanthropest

102. A. Ephameral B. Ephimeral
C. Ephemeral D. Ephemerel

103. A. Fanaticism B. Fenaticism
C. Faneticism D. Fanatecism

104. A. Profesional B. Professionel
C. Professional D. Profissional

105. A. Ameteur B. Amateur
C. Amataur D. Amateor

106. A. Philenderer B. Philandarer
C. Philanderer D. Philonderer

107. A. Honorary B. Henorary
C. Honerary D. Honorery

108. A. Darmatologist B. Dermatologist
C. Dermetologist D. Dermatologest

109. A. Compansation B. Compensetion
C. Compensotion D. Compensation

110. A. Quadruplets B. Quedruplets
C. Quadroplets D. Quadruplats

111. A. Econemist B. Economest
C. Economist D. Econimist

112. A. Catholic B. Cetholic
C. Cathelic D. Cathilic

113. A. Unfevourable B. Unfevaurable
C. Unfavourable D. Unfivourable

114. A. Bombastic B. Bombestic
C. Bombostic D. Bombastie

115. A. Posthumaus B. Posthomous
C. Posthumous D. Pasthumous

116. A. Epitaph B. Epetaph
C. Epiteph D. Epitoph

117. A. Falicitate B. Felicitate
C. Falecitate D. Falecetate

118. A. Schezophrenia B. Schizaphrenia
C. Schizophrenia D. Schizophrania

119. A. Claustrophobia B. Cloustrophobia
C. Claustraphobia D. Claustrophabia

120. A. Connoiseur B. Connaisseur
C. Connaissour D. Connoisseur

121. A. Transperency B. Transparency
C. Transpirency D. Tranporency

122. A. Superanuation B. Superennuation
C. Superannuation D. Superannuetion

123. A. Antiquated B. Antiqueted
C. Antequated D. Antiquatted

124. A. Adalescent B. Addescant
C. Adolescent D. Adolascent

125. A. Newrologist B. Neurologist
C. Neurologest D. Newrologest

126. A. Obstatricion B. Obstitricion
C. Obstetricion D. Obstetracion

127. A. Illiterate B. Iliterate
C. Illitarate D. Illiterete

128. A. Thioretical B. Theoretical
C. Theoratical D. Theoritical

129. A. Mersenary B. Marcenary
C. Marcenery D. Mercenary

130. A. Belligerent B. Beligerent
C. Belligarent D. Belligerant

131. A. Extrevagant B. Extravegant
C. Extravagant D. Extravagent

132. A. Aristocracy B. Arestocracy
C. Arestocracy D. Aristocrecy

133. A. Lagislature B. Legislature
C. Lageslature D. Legesleture

134. A. Dormetory B. Dormitery
C. Dormitory D. Dermitory

135. A. Garulous B. Garrulaus
C. Gorrulous D. Garrulous

136. A. Regecide B. Regicide
C. Rigecide D. Ragicide

137. A. Epicurean B. Epecurean
C. Epicurian D. Epicorean

138. A. Commetti B. Commitee
C. Committee D. Comitte

139. A. Inemitable B. Inimitable
C. Inimetable D. Inimitabli

140. A. Pletocray B. Plitocracy
C. Plotocracy D. Plutocracy

141. A. Immortal B. Emortal
C. Immortel D. Immortol

142. A. Perrhic B. Pyrrhic
C. Pyrrhec D. Porrhec

143. A. Chauvinism B. Chouvinism
C. Chevinism D. Cauvinism

144. A. Rumenate B. Ruminote
C. Rimunote D. Ruminate

145. A. Teciturn B. Tociturn
C. Tecitorn D. Taciturn

146. A. Verteran B. Viteran
C. Veterain D. Veteran

147. A. Antitode B. Antedote
C. Antidote D. Autitote

148. A. Audeble B. Audible
C. Audcble D. Audebli

149. A. Bellicose B. Billicose
C. Bellecose D. Belicose

150. A. Pilferage B. Peliferage
C. Pilfarege D. Poliferage

151. A. Arestocracy B. Aristocracy
C. Aristocracy D. Aristocy

152. A. Sychologist B. Psycologist
C. Psychologist D. Psychogist

153. A. Neucteus B. Neucleus
C. Nucleus D. Neocleus

154. A. Negotcate B. Negociate
C. Negotiate D. Nagotiate

155. A. Skelled B. Skilled
C. Skiled D. Skillid

156. A. Logecal B. Logical
C. Lagical D. Logicle

157. A. Effeciency B. Efficiency
C. Eficiency D. Efficency

158. A. Cumpartment B. Compartment
C. Compartant D. Comportent

159. A. Continuous B. Cuntinuous
C. Continus D. Contineuous

160. A. Valuable B. Valueable
C. Valuabel D. Valoble

161. A. Courteous B. Curteous
C. Corteous D. Courtus

162. A. Prophecy B. Phropecy
C. Propecy D. Propheci

163. A. Lonesome B. Lonsome
C. Lonesum D. Lonsum

164. A. Formedable B. Formidable
C. Formidabul D. Formdable

165. A. Inibriated B. Enibrited
C. Inebriated D. Inubrited

166. A. Guellible B. Guiellible
C. Guillibel D. Guillible

167. A. Subsequent B. Subsiquent
C. Subsequint D. Subquent

168. A. Alein B. Alien
C. Abein D. Abien

169. A. Misogynesm B. Mesogynism
C. Misogynism D. Misosogesim

170. A. Sanguini B. Sangune
C. Sanguine D. Sanguine

171. A. Exemplary B. Eximplary
C. Exemplari D. Examplary

172. A. Sycophants B. Sicophants C. Sycopants D. Sycophints

173. A. Petchy B. Patchy C. Patchi D. Patche

174. A. Impassi B. Impasse C. Impase D. Inpassi

175. A. Aceelerate B. Accilerate C. Accelerate D. Accelerati

ANSWERS

1	**2**	**3**	**4**	**5**	**6**	**7**	**8**	**9**	**10**
A	A	D	D	C	A	C	A	B	A
11	**12**	**13**	**14**	**15**	**16**	**17**	**18**	**19**	**20**
C	D	B	C	A	B	B	C	D	D
21	**22**	**23**	**24**	**25**	**26**	**27**	**28**	**29**	**30**
B	C	C	A	B	C	A	B	D	B
31	**32**	**33**	**34**	**35**	**36**	**37**	**38**	**39**	**40**
B	A	C	A	C	A	B	A	C	D
41	**42**	**43**	**44**	**45**	**46**	**47**	**48**	**49**	**50**
A	D	D	C	B	C	A	B	B	D
51	**52**	**53**	**54**	**55**	**56**	**57**	**58**	**59**	**60**
B	D	D	B	C	C	D	A	C	B
61	**62**	**63**	**64**	**65**	**66**	**67**	**68**	**69**	**70**
C	A	C	A	B	C	A	C	A	B
71	**72**	**73**	**74**	**75**	**76**	**77**	**78**	**79**	**80**
A	C	A	B	C	B	C	A	B	A
81	**82**	**83**	**84**	**85**	**86**	**87**	**88**	**89**	**90**
C	D	A	C	B	D	A	B	A	C
91	**92**	**93**	**94**	**95**	**96**	**97**	**98**	**99**	**100**
A	C	A	C	A	C	A	D	A	B
101	**102**	**103**	**104**	**105**	**106**	**107**	**108**	**109**	**110**
A	C	A	C	B	C	A	B	D	A
111	**112**	**113**	**114**	**115**	**116**	**117**	**118**	**119**	**120**
C	A	C	A	C	A	B	C	A	D
121	**122**	**123**	**124**	**125**	**126**	**127**	**128**	**129**	**130**
B	C	A	C	B	C	A	B	D	A
131	**132**	**133**	**134**	**135**	**136**	**137**	**138**	**139**	**140**
C	A	B	C	D	B	A	C	B	D
141	**142**	**143**	**144**	**145**	**146**	**147**	**148**	**149**	**150**
A	B	A	D	D	D	C	B	D	A
151	**152**	**153**	**154**	**155**	**156**	**157**	**158**	**159**	**160**
C	C	C	C	B	B	B	B	A	A
161	**162**	**163**	**164**	**165**	**166**	**167**	**168**	**169**	**170**
A	A	A	B	C	D	A	B	C	D
171	**172**	**173**	**174**	**175**					
A	A	B	B	C					

3

Spot The Errors

The most common errors in English are of spellings, grammar and usage of words. By regular practice, the errors can be easily spotted and minimised. The best method of improving your usage of English with this guide is to study the sample sentences, then do the practice exercises at the end of the section. Practise carefully.

I. COMMON ERRORS WITH NOUNS AND NOUN-PHRASES

	Incorrect	Correct
1.	I have bought new *furnitures.*	I have bought new *furniture.*
2.	The wages of sin *are* death.	The *wages* of sin is death.
3.	She told these *news* to her mother.	She told her mother this *news.*
4.	He took *troubles* to do his work.	He took *trouble* (or pains) over his work.
5.	The *cattles* were grazing.	The *cattle* were grazing.
6.	He showered *many abuses* on me.	He showered *much abuse* on me.
7.	I spent the holidays with my *family members.*	I spent the holidays with my *family.*
8.	There is no *place* in this compartment.	There is no *room* in this compartment.
9.	Write this new *poetry* in your *copy.*	Write this new *poem* in your *note-book.*
10.	He took *insult* at this.	He took *offence* at this.
11.	Put your *sign* here.	Put your *signatures* here.
12.	She is my *cousin sister.*	She is my *cousin.*
13.	*Sunil's* my *neighbour's* house was burgled.	*Sunil,* my *neighbour's* house was burgled.
14.	I lost a *ten-rupees* note.	I lost *a ten-rupee* note.
15.	Road closed for *repair.*	Road closed for *repairs.*
16.	His house is out of *repairs.*	His house is out of *repair.*
17.	What is the *reason* of an earthquake?	What is the *cause* of an earthquake?
18.	This building is made of *stones.*	This building is made of *stone.*
19.	I disapprove of *these kinds* of games.	I disapprove of *this kind* of games.
20.	Veena's and Sheela's father is ill.	Veena and Sheela's father is ill.
21.	His *son-in-laws* are doctors.	His *sons-in-law* are doctors.
22.	*Alms* is given to the *poor.*	*Alms* are given to the poor.
23.	He always keeps his words.	He always keeps his *word.*
24.	I carried the *luggages.*	I carried the *luggage.*
25.	*Two-third* of the work is left.	*Two-thirds* of the work is left.

II. COMMON ERRORS WITH PRONOUNS

	Incorrect	Correct
1.	Both did not go.	Neither went.
2.	We all did not go.	None of us went.
3.	Each of these boys play.	Each of these boys plays.
4.	Whoever does best he will get a prize.	Whoever does best will get a prize.

	Incorrect	Correct
5.	One should not waste his time.	A man should not waste his time.
6.	I and she are sisters.	She and I are sisters.
7.	He is wiser than me.	He is wiser than I.
8.	Between you and I, Anil is not to be trusted.	Between you and me, Anil is not to be trusted.
9.	Nobody was there but I.	Nobody was there but me.
10.	Who is there ? It is me.	Who is there? It is I.
11.	Only he and me can use this card.	Only he and I can use this card.
12.	Let you and I go now.	Let you and me go now.
13.	Everyone got one's pay.	Everyone got his pay.
14.	Everyone is frightened when they see a tiger.	Everyone is frightened when he sees a tiger.
15.	These two friends are fond of one another.	These two friends are fond of each other.
16.	I did not like him coming at that hour.	I did not like his coming at that hour.
17.	Who do you think I met?	Whom do you think I met?
18.	You should avail this opportunity.	You should avail yourself this opportunity.
19.	When you have read these books, please return the same to me.	When you have read the books, please return them to me.
20.	They that are humble need fear no fall.	Those that are humble need fear no fall.

III. COMMON ERRORS WITH ADJECTIVES

	Incorrect	Correct
1.	These all oranges are good.	All these oranges are good.
2.	He held the book in the both hands.	He held the book in both hands.
3.	Both men have not come.	Neither man has come.
4.	That man should do some or other work.	That man should do some work or other.
5.	He is elder than I.	He is older than I.
6.	Shakespeare is greater than any other poets.	Shakespeare is greater than any other poet.
7.	He is a coward man.	He is a cowardly man.
8.	Many villagers cannot write his own name.	Many villagers cannot write their own name.
9.	Each of us loves our home.	Each of us loves his home.
10.	Much efforts bring their reward.	Much effort brings its reward.
11.	He found hundred rupees.	He found a hundred rupees.
12.	He had leave of four days.	He had four days leave.
13.	This is a worth seeing sight.	This is a sight worth seeing.
14.	He will spend his future life here.	He will spend the rest of his life here.
15.	There is a best teacher in that class.	There is a very good teacher in that class.
16.	Of the two plans this is the best.	Of the two plans this is the better.
17.	He is becoming strong every day.	He is becoming stronger every day.
18.	He is worst than I.	He is worse than I.
19.	Jaipur is hot than Delhi.	Jaipur is hotter than Delhi.
20	In our library the number of books is less.	In our library the number of books is small.
21.	From the three he is more clever.	He is the cleverest of the three.
22.	India is the first peace-loving country in the world.	India is the foremost peace-loving country in the world.
23.	Verbal instruction will not do.	Oral instruction will not do.
24.	Her command over French is most excellent.	Her command over French is excellent.
25.	He has not some money with him.	He has not any money with him.

Incorrect	Correct
26. I have visited Bombay many a times.	I have visited Bombay many a time.
27. Death is more preferable to dishonour.	Death is preferable to dishonour.
28. I gave him a few books I had.	I gave him the few books I had.
29. If he wants farther help send him to me.	If he wants further help, send him to me.
30. She is so cunning as a fox.	She is as cunning as a fox.

IV. COMMON ERRORS WITH VERBS

Incorrect	Correct
1. He asked had we taken our luggage.	He asked if we had taken our luggage.
2. She asked what are you doing.	She asked what we were doing.
3. Rama asked to Anil why he is angry.	Rama asked Anil why he was angry.
4. He does not care for his money.	He does not take care of his money.
5. He does not care for his work.	He takes no care over his work.
6. No one cared for him after his mother died.	No one took care of him after his mother died.
7. He got angry before I said a word.	He got angry before I had said a word.
8. I met a man who was my tutor 20 years ago.	I met a man who had been my tutor twenty years ago.
9. I had been for walking yesterday.	I went for a walk yesterday.
10 If I shall do this I shall be wrong.	If I do this I shall be wrong.
11. I have left trekking.	I have given up trekking.
12. I came to know as to how he did this.	I learnt how he did this.
13. I came to know why he was sad.	I found out why he was sad.
14. He knows to swim.	He knows how to swim.
15. The criminal's head was cut.	The criminal's head was cut off.
16. I said to him to go.	I told him to go.
17. I told the teacher to excuse me.	I asked the teacher to excuse me.
18. He is troubling me.	He is giving me trouble.
19. I have got a hurt on my leg.	I have hurt my leg.
20. She gave a speech.	She made a speech.
21. He has given his examination.	He has sat for his examination.
22. He took out his shoes.	He took off his shoes.
23. I have ordered for a new car.	I have ordered a new car.
24. He would not hear me.	He would not listen to me.
25. I struck a blow on his face.	I struck him in the face.
26. He denied to come.	He refused to come.
27. He lived there for a day.	He stayed there for a day.
28. The book is not found.	The book is lost.
29. Shut the light.	Turn off the light.
30. I must revenge my brother.	I must avenge my brother.

V. COMMON ERRORS IN SUBJECT-VERB AGREEMENT

Incorrect	Correct
1. The owners of this factory is very rich.	The owners of this factory *are* very rich.
2. The pleasures of nature that one can experience at Shimla is beyond description.	The pleasures of nature that one can experience at Shimla *are* beyond description.

	Incorrect	Correct
3.	There is no street lights in our colony.	There *are* no street lights in our colony.
4.	He and I am entrusted with the job.	He and I *are* entrusted with the job.
5.	Rice and curry are his favourite dish.	Rice and curry *is* his favourite dish.
6.	The honour and glory of our country are at stake.	The honour and glory of our country *is* at stake.
7.	Time and tide waits for none.	Time and tide *wait* for none.
8.	All the passengers with the driver was killed.	All the passengers, with the driver, *were* killed.
9.	The teacher, with her students, were going out.	The teacher, with her students, *was* going out.
10.	I am as well as they are tired.	I as well as they *are* tired.
11.	Not only the soldiers but their captain also were captured.	Not only the soldiers but their captain also was captured.
12.	Neither you nor I were selected.	Neither you nor I *was* selected.
13.	Either of these two applicants are fit for the job but neither want to accept it.	Either of these two applicants *is* fit for the job but neither wants to accept it.
14.	One of these students are sure to stand first.	One of these students *is* sure to stand first.
15.	Everyone of these workers want a raise.	Everyone of these workers wants a raise.
16.	None of these letters has been answered so far.	None of these letters *have* been answered so far.
17.	None of the girls were present at the party.	None of the girls *was* present at the party.
18.	Many a battle were fought on Indian soil.	Many a battle *was* fought on Indian soil.
19.	A lot of work remain to be done.	A lot of work *remains* to be done.
20.	The majority of these girls likes music.	The majority of these girls *like* music.
21.	The number of admissions are encouraging.	The number of admissions *is* encouraging.
22.	A large number of boys was present.	A large number of boys *were* present.
23.	A variety of books was on display.	A variety of books *were* on display.
24.	Variety are the spice of life.	Variety *is* the spice of life.
25.	If my estimates are correct, I will need another hundred rupees.	If my estimate is correct, I will need another hundred rupees.
26.	Mathematics are my favourite subject.	Mathematics *is* my favourite subject.
27.	'Gulliver's Travels' are written by Swift.	'Gulliver's Travels' *is* written by Swift.
28.	Ten miles are a long distance to cover on foot.	Ten miles *is* a long distance to cover on foot.
29.	A new pair of shoes are to be purchased.	A new pair of shoes is to be purchased.
30.	The Committee have issued its report.	The Committee *has* issued its report.
31.	I, who am your friend, has always been on your side.	I, who am your friend, *have* always been on your side.
32.	I am the person who have always stood by you.	I am the person who *has* always stood by you.
33.	This is one of the best novels that has been published this year.	This is one of the best novels that *have* been published this year.
34.	Less than half the amount have been wasted.	Less than half the amount *has* been wasted.
35.	A lot of people has turned up for the show.	A lot of people *have* turned up for the show.
36.	Much of their honour are undeserved.	Much of their honour *is* undeserved.
37.	More than a decade have passed since this house was built.	More than a decade *has* passed since this house was built.
38.	Either she or he are mistaken.	Either she or he *is* mistaken.
39.	Plenty of information are available on the subject.	Plenty of information *is* available on the subject.
40.	Plenty of pamphlets is available on the subject.	Plenty of pamphlets *are* available on the subject.

VI. COMMON ERRORS IN USE OF WILL, SHALL, WOULD, SHOULD, MAY, MIGHT, MUST

	Incorrect	Correct
1.	When I shall see him I shall tell him this.	When I *see* him, I shall tell him this.
2.	If I should do wrong, he would punish me.	If I *did* wrong, he would punish me.
3.	Until he will have confessed his fault, he will be kept in prison.	Until he *has* confessed his fault, he will be kept in prison.
4.	She will obey me.	She *shall* obey me.
5.	You would work hard.	You *should* work hard.
6.	You shall find him in the garden.	You *will* find him in the garden.
7.	He must have died of exposure, but we cannot be certain.	He *might* have died of exposure, but we cannot be certain.
8.	You might not show disrespect to your elders.	You *must* not show disrespect to your elders.
9.	You may take exercise in order to maintain good health.	You *must* take exercise in order to maintain good health.
10.	He must be a crook for all we know.	He *may* be a crook for all we know.

VII. COMMON ERRORS IN THE USE OF ADVERBS

(Very, Much, Too, Enough, Quite, Hardly, Scarcely, Before, Ago, Since, Yet, Still, etc.)

	Incorrect	Correct
1.	He is very much angry.	He is *very* angry.
2.	She was very good enough to help me.	She was *good enough* to help me.
3.	She runs much fast.	She runs *very* fast.
4.	She runs very faster than Seema.	She runs *much* faster than Seema.
5.	It is bitter cold today.	It is *bitterly* cold today.
6.	He is a much learned man.	He is a very learned man.
7	She is thinking very hardly.	She is thinking very hard.
8.	To tell in brief the film was boring.	*In short* the film was boring.
9.	He told the story in details.	He told the story *in detail.*
10.	I did it anyhow.	I *managed to do* it somehow.
11.	Aeroplanes reach Europe soon.	Aeroplanes reach Europe quickly.
12.	Before long there were dinosaurs on the earth.	*Long ago,* there were dinosaurs on the earth.
13.	This book is too interesting.	This book is *very* interesting.
14.	He lives miserly.	He lives in *a miserly* way.
15.	Just I had gone when she came.	I had just gone when she came.
16.	He sings good.	He sings *well.*
17.	He sings good than I.	He sings *better* than I.
18.	Really speaking it is cold.	*As a matter of fact* it is cold.
19.	He is enough tall to reach the ceiling.	He is *tall enough* to reach the ceiling.
20.	He went directly to his college.	He went *direct* to his college.
21.	He is presently at Delhi.	He is at Delhi *at present.*
22.	Last night she returned lately.	Last night she *returned late.*
23.	He was even blamed by his friends.	He was *blamed even* by his friends.
24.	I only employed him for a week.	I employed him for a week only.

Incorrect	Correct
25. I met him four months before.	I met him four *months ago.*
26. Anil seldom ever goes to school.	Anil *seldom goes* to school.
27. I will wait here until you do not go.	I will wait here until *you go.*
28. I never remember having met her before.	I *do not remember* having met her before.
29. She has not been here too long to have many friends.	She has not *been here long* enough to have many friends.
30. Hardly I have had any rest since one week.	Hardly have *I had* any rest for a week.
31. Scarcely the water crossed the danger level, the warning signals were sounded.	Scarcely *had the* water *crossed the* danger level, when the warning signals were sounded.
32. She is neat but fairly slow.	She is *neat* but rather slow.
33. It is a rather good film.	It is a *fairly good* film.
34. We yet have time to catch the bus.	We *still have* time to catch the bus.
35. She has not still spent all her money.	She has *not yet* spent all her money.

VIII. COMMON ERRORS IN THE USE OF CONJUNCTIONS

Incorrect	Correct
1. As he is fat so he runs slowly.	As he is fat *he* runs slowly.
2. If he is fat then he will run slowly.	If he is fat, he will run slowly.
3. Though he is fat still he runs fast.	Though he is fat, *he runs* fast.
4. *As* I pulled the trigger at the sametime he shook my arm.	As I pulled the trigger, he shook my arm.
5. No sooner I had spoken than he left.	No sooner *had* I spoken than he left.
6. Not only he will go, but also he will stay there.	Not only *will he* go, but he *will also* stay there.
7. Neither he comes nor he writes.	Neither *does he* come nor *does he* write.
8. Scarcely he entered the room than the telephone rang.	Scarcely *had* he entered the room *when the* telephone rang.
9. Hardly she had left the house than it began to rain.	Hardly *had she* left the house *when* it began to rain.
10. He is the fastest runner and he comes last.	He is the fastest runner *but* he comes last.
11. She is as innocent as if she looks.	She is as innocent as she looks.
12. Until he does not try he must be punished.	He must be punished unless he tries.
13. I want to know as to why you are late.	I want to know why you are late.
14. I am fond of Chinese food as for example sweet and sour prawns.	I am fond of Chinese food, for example, sweet and sour prawns.
15. He was angry therefore I ran away.	He was angry so I ran away.
16. I was trying to work, at that time he was disturbing me.	While I was trying to work, he was disturbing me.
17. Supposing if he is late, what will happen?	Supposing he is late (or if he is late) what will happen?
18. He asked me that why I was late.	He asked me why I was late.
19. Let us catch a taxi lest we should not get late.	Let us catch a taxi lest we should get late.
20. She dresses herself like the teacher does.	She dresses herself as the teacher does.
21 Wait while I come.	Wait *until* (or *till)* I come.
22. Until, there is corruption in India, there can be little progress.	*As long* as there is corruption in India there can be little progress.
23. I have never told a lie nor cheated anybody.	I have never told a lie *nor have I* cheated anybody.
24. Both Mohan as well as Arun are responsible for this action.	Both Mohan *and* Arun are responsible for this action.
25. Hindus and Muslims both are to blame for the riots.	*Both Hindus* and Muslims are to blame for the riots.
26. I have bought paintings, books, records, and etc.	I have bought paintings, books *and records etc.*

	Incorrect	Correct
27.	He as well as you is a fool.	He as well as you *are* a fool.
28.	He is so poor and he cannot save anything.	He is so *poor that* he cannot save anything.
29.	Such a book that you want is not available.	Such a book *as* you want is not available.
30.	Such was her condition as everyone was moved to pity.	Such was her condition that everyone was moved to pity.

IX. COMMON ERRORS IN THE USE OF PREPOSITIONS

	Incorrect	Correct
1.	I will not listen him.	I will not listen *to* him.
2.	Copy this word by word.	Copy this word *for* word.
3.	He enquired from her where she lived.	He enquired *of* her where she lived.
4.	Sign here with ink.	Sign here *in* ink.
5.	Has she come in train or by foot?	Has she come *by* train or *on* foot?
6.	She said this at his face.	She said this *to* his face.
7.	Open the book on page one.	Open the book *at* page one.
8.	I was invited for lunch.	I was *invited to* lunch.
9.	I am ill since three months.	I have been *ill for* three months.
10.	This paper is inferior than that.	This paper is inferior *to* that.
11.	This resembles to that.	This *resembles* that.
12.	My brother is superior than you in strength.	My brother is superior *to* you in strength.
13.	He wrote me.	He wrote *to* me.
14.	I shall explain them this.	I shall explain this *to* them.
15.	Send this letter on my address.	Send this letter *to* my address.
16.	He suggested me this.	He suggested this *to* me.
17.	He goes on his work.	He goes *to his* work.
18.	He reached to Nagpur.	He *reached* Nagpur.
19.	He told to me to go.	He told *me* to go.
20	The term begins from July 1st.	The term begins *on* July 1st.
21.	There are many advantages from this.	The advantages *of* this are many.
22.	We waste much time in trifles.	We waste much time *on* (or *over)* trifles.
23.	He sat on a tree.	He sat *in* a tree.
24.	This is a comfortable house to live.	This is a comfortable house to live in.
25.	This is the road to go.	This is the road to go *by*.
26.	He married with an Indian lady.	He *married an Indian* lady.
27.	He accompanied with his friend.	He *accompanied his* friend.
28.	He went for doing some business.	He went away *on* business.
29.	He went for riding.	He went *for a ride*.
30.	I pitied on him.	I *pitied* him.
31.	When this was searched it was found.	When this was searched *for* it was found.
32.	I shall inform them this.	I shall inform them *of* this.
33.	Due to illness I cannot go to school.	*Owing* to illness I cannot go to school.
34.	He went to the back side of the house.	He went behind (or to the back of) the house.
35.	I must go; there is no help.	I must go; there is no help *for it*.
36.	I met with your friend there.	I *met your* friend there.
37.	The First World War was fought during 1914-1918.	The First World War was fought *between* 1914-1918.

Incorrect	Correct
38. England grew prosperous between Queen Victoria's reign.	England grew prosperous *during* Queen Victoria's reign.
39. He asked a holiday.	He asked *for* a holiday.
40. I am obliged of you for this good turn.	I am obliged *to* you for this good turn.
41. There is no harm to try.	There is no harm *in trying*.

X. MISCELLANEOUS ERRORS
(Including Ambiguities and Indianisms)

Incorrect	Correct
1. Many *homes* are lying vacant.	Many *houses* are lying vacant.
2. It is cool in the shadow of the tree.	It is cool in the *shade* of the tree.
3. She *keeps* good health.	She *enjoys* good health.
4. My leg is paining.	*I am feeling pain* in my leg.
5. See this word in the dictionary.	*Look up* this word in the dictionary.
6. The train will arrive *just now*.	The train will arrive *shortly*.
7. They are *pulling* on well.	They are *getting* on well.
8. The river has *over flown* its bank.	The river has *over flown* its banks.
9. He was appointed *on* the post.	He was appointed *to* the post.
10. Last but not *the least,* we have to discuss the problem of overpopulation.	Last but not *least,* we have to discuss the problem of overpopulation.
11. *Cities* after *cities* fell.	*City* after *city* fell.
12. What is the use Munir going there?	What is the use of Munir going there?
13. He *did many mischief.*	He *made much mischief.*
14. It is exact five *in* my watch.	It is exact five *by* my watch.
15. I will dine with them on *next Sunday.*	I will dine with them *Sunday next.*
16. Misfortunes when faced bravely and *manly* become less troublesome.	Misfortunes when faced bravely and *manfully* become less troublesome.
17. I am *laid down* with fever.	I am *laid up* with fever.
18. He is habituated to smoking.	He is *addicted* to smoking.
19. *According to my opinion* he is right.	*In my opinion* (or according *to me)* he is right.
20. Could you please *open* this knot?	Could you please *untie* this knot?
21. When five *years old* his father died.	*When he was five years old* his father died.
22. I made him *to* do this work.	I *made him do* this work.
23. What is the *cost* of this camera?	What is the *price of* this camera?
24. He wants *as many as* five kilograms of sugar.	He wants *as much as* five kilograms of sugar.
25. I have come to a final *conclusion.*	I have come to a *conclusion* (or *to a final decision).*
26. Do you wish me to teach *you or the principal?*	Do you wish me or the principal to teach you?
27. The tree was *loaded with* fruit.	The tree was *laden with* fruit.
28. What sort of a man is he?	What sort of man is he?
29. My views are different *than* you.	My views are different *from yours.*
30. I take this opportunity *to thank* you.	I take this opportunity *of thanking* you.

MULTIPLE CHOICE QUESTIONS

Directions: *In this section, each sentence has three parts, indicated by (A), (B) and (C). Read each sentence to find out whether there is an error. If you find an error in any of the parts (A, B, C), indicate your response by marking the letter related to that part. If a sentence has no error, indicate this by marking '(D)' which stands for "No error". Errors may belong to grammar, usage or idiom. Ignore errors of punctuation, if any.*

1. (A) The reason why/(B) he was rejected/(C) was because he was too young./(D) No error.

2. (A) The scientist must follow/(B) his hunches and his data/(C) wherever it may lead./(D) No error.

3. (A) Firstly you should/(B) think over the meaning of the words/(C) and then use them./(D) No error.

4. (A) Scarcely had/(B) I arrived than/(C) the train left./(D) No error.

5. (A) Unless you stop to make noise at once/(B) I will have no option but to/(C) bring the matter to the attention of the police./(D) No error.

6. (A) He couldn't but help/(B) shedding tears at the plight of the villagers/(C) rendered homeless by a devastating cyclone./(D) No error.

7. (A) Since it was his first election campaign, the candidate was confused;/(B) none could clearly understand/(C) either the principles he stood for or the benefits he promised./(D) No error.

8. (A) It is an established fact that the transcendental American poets and philosophers,/(B) who lived in the latter half of the nineteenth century,/(C) were more influenced by Indian philosophy, in particular by Upanishadic Philosophy./(D) No error.

9. (A) The crew were on board/(B) and they soon busied themselves/(C) in preparing to meet the storm./(D) No error.

10. (A) One of the members/(B) expressed doubt if/(C) the Minister was an athiest./(D) No error.

11. (A) Beware of/(B) a fair-weather friend/(C) who is neither a friend in need nor a friend indeed./(D) No error.

12. (A) Copernicus proved/(B) that Earth/(C) moves round the Sun./(D) No error.

13. (A) Seldom we have been treated/(B) in such a rude manner/(C) by the police personnel./(D) No error.

14. (A) Some men are born great,/(B) some achieve greatness/(C) and some had greatness thrust on them./(D) No error.

15. (A) The property/(B) was divided/(C) among the two brothers./(D) No error.

16. (A) I am quite certain/(B) that the lady is not only greedy/(C) but miserly./(D) No error.

17. (A) The aircraft overloaded/(B) there was something wrong of the battery/(C) and the engine was making a queer noise./(D) No error.

18. (A) A thorough inquiry of the misappropriation of funds/(B) is now imperative/(C) to bring the guilty to book./(D) No error.

19. (A) The brilliant success in the examination/(B) as well as his record in sports/(C) deserves high praise./(D) No error.

20. (A) While travelling by a train/(B) on a cold winter night/(C) an argument rose between two passengers in our compartment./(D) No error.

21. (A) I cannot find/(B) where has he gone/(C) though I have tried may best./(D) No error.

22. (A) If I was/(B) the Prime Minister of India/(C) I would work wonders./(D) No error.

23. Amit's severe bout of flu/(B) debilitated him so much/(C) that he was too tired to do for work for a week./(D) No error.

24. (A) This is the crux of the entire problem;/(B) everything centres on/(C) it being resolved./(D) No error.

25. (A) One of the major aims of the Air Force/(B) was the complete demolition of all means of transportation/(C) by the bombing of rail lines and terminals./(D) No error.

26. (A) His strong voice cut over/(B) the hum of conversation/(C) like a knife through butter./(D) No error.

27. (A) Even though they weren't expecting us/(B) they managed to knock up/(C) a marvellous meal./(D) No error.

28. (A) The celebrated singer was/(B) surrounded by the usual crowd/(C) of lackeys and hanger-ons./(D) No error.

29. (A) If it weren't/(B) for you,/(C) I wouldn't be alive today./(D) No error.

30. (A) He looked like a lion/(B) baulked from/(C) its prey./(D) No error.

31. (A) Widespread flooding/(B) is effecting/(C) large areas of the villages./(D) No error.

32. (A) She regards/(B) negotiating prices with customers/(C) as her special preserve./(D) No error.

33. (A) Often in political campaigns, a point is reached at which/(B) the candidates take out their gloves,/(C) and start slugging with bare fists./(D) No error.

34. (A) If we really set to/(B) we can get the whole house/(C) cleaned in an afternoon./(D) No error.

35. (A) Pieces of rock plummeted/(B) down the mountainside/(C) in the ground below./(D) No error.

36. (A) Since the two parties each won/(B) the same number of seats,/(C) the minority party holds the balance of power./(D) No error.

37. (A) It's arrogant for you/(B) to assume you'll/(C)win every time./(D) No error.

38. (A) We've paid for our travel and accommodation,/(B) so we need only to take/(C) some pocket-money with us./(D) No error.

39. (A) There's no evidence to show/(B) that information technology secrets are more/(C) vulnerable in India than Britain or the US./(D) No error.

40. (A) It is shameful that hunting/(B) is still considered sport/(C) by some unscrupulous people in the civilized world./(D) No error.

41. (A) The Prime Minister's good looks won him/(B) the election but he has still to prove/(C) that he's not a just pretty face./(D) No error.

42. (A) The two books are the same/(B) except for the fact that this/(C) has an answer in the back./(D) No error.

43. (A) He estimated his income tax bill/(B) by extrapolation over figures/(C) submitted in previous years./(D) No error.

44. (A) The modern office block/(B) sticks out like a sore thumb/(C) among the old buildings in the area./(D) No error.

45. (A) I will try to put over/(B) some feelers to gauge/(C) people's reactions to our proposal./(D) No error.

46. (A) A major contribution of Mathura sculptors/(B)of that period were the creation and popularization/(C) of the Buddha's image in human form./(D) No error.

47. (A) Amit has been deceiving Mona/(B) for many years but she/(C) has not still tumbled to it. (C)/No error. (D)

48. (A) Mahavira was an advocate of non-violence and vegetarianism,/(B) who revived and reorganized the Jain doctrine/(C) and established rules for their monastic order. (D) No error.

49. (A) Microwaves are the principle carriers/(B) of television, telephone and data transmissions/(C) between stations on earth and between the earth and satellites./(D) No error.

50. (A) An unit is an abstract idea,/(B) defined either by reference to/(C) a randomly chosen material standard or to a natural phenomenon./(D) No error.

51. (A) With the crisis deepening,/(B) the critics sense an opportunity/(C) about putting in place a more radical strategy./(D) No error.

52. (A) The salesman gave us/(B) a big spiel about why/(C) we should buy his product./(D) No error.

53. (A) I will need several weeks/(B) to invent the lie of the land before/(C) I can make any decision about the future of the business./(D) No error.

54. (A) You should be cautious/(B) and make a few discrete enquiries about/(C) the firm before you sign anything./(D) No error.

55. (A) Your husband doesn't/(B) believe that you are older/(C) than I./(D) No error.

56. (A) There is a beautiful moon out tonight/(B) and Neeta and I are going for a stroll/(C) —would you like to come along with she and I?/(D) No error.

57. (A) The data on/(B) the divorce case is/(C) on the judge's desk./(D) No error.

58. (A) She stood off/(B) from the crowd/(C) because of her height and flaming red hair./(D) No error.

59. (A) It's stupid to go/(B) to the expense of taking/(C) music lessons if you never practise./(D) No error.

60. (A) You will find it difficult/(B) to explain of your use/(C) of such offensive language./(D) No error.

61. (A) Because of the/(B) extenuating circumstances/(C) the court acquitted him out of the crime./(D) No error.

62. (A) The carpet was badly stained/(B) to such an extent that/(C) you couldn't tell its original colour./(D) No error.

63. (A) It is greatly to Amit's credit/(B) that he gave back the money he found/(C) his honesty does for him credit./(D) No error.

64. (A) A terrific hue and cry/(B) was raised/(C) at the new tax proposals./(D) No error.

65. (A) The former General was/(B) exiled of his country because of/(C) his part in the plot against the government./(D) No error.

66. (A) The company has/(B) set off itself some stiff production/(C) goals for this year./(D) No error.

67. (A) The music was so loud/(B) that we had to bellow over each/(C) other to be heard./(D) No error.

68. (A) When this beautiful girl arrived, (A) all the men in the room/(C) gravitated over her./(D) No error.

69. (A) The children are/(B) really in their element/(C) playing on the beach./(D) No error.

70. (A) The refugees are/(B) badly off for blankets,/(C) and even worse for food./(D) No error.

71. (A) From their vintage-point on the cliff,/(B) the children could watch/(C) the ships coming and going./(D) No error.

72. (A) A cogent remark/(B) compels acceptance because/(C) of their sense and logic./(D) No error.

73. (A) Credit cards have/(B) brought about a revolution/(C) in people's spending habits./(D) No error.

74. (A) In financial matters/(B) it is important to/(C) get disinterested advice./(D) No error.

75. (A) Some women admit that/(B) their principle goal in life/(C) is to marry a wealthy man./(D) No error.

76. (A) Take two spoonsful/(B) of this medicine/(C) every three hours./(D) No error.

77. (A) The film was so disjointed/(B) that I could not tell you/(C) what the story was about./(D) No error.

78. (A) He had been/(B) saved of death as if/(C) by divine intervention./(D) No error.

79. (A) I informed the principal/(B) that I was running temperature/(C) and, therefore, could not attend the meeting./(D) No error.

80. (A) The lady was broken with grief/(B) when she heard the sad news of the train disaster/(C) in which her brother was killed./(D) No error.

81. (A) The farmer is irrigating/(B) his fields/(C) since morning./(D) No error.

82. (A) I could not/(B) answer to/(C) the question./(D) No error.

83. (A) Two years passed/(B) since/(C) my cousin died./(D) No error.

84. (A) He hesitated to accept the post/(B) as he did not think/(C) that the salary would be enough for a man with a family of three./(D) No error.

85. (A) Have you gone through/(B) either of these three chapters/(C) that have been included in this volume?/(D) No error.

86. (A) I am learning English/(B) for ten years/(C) without much effect./(D) No error.

87. (A) Ramesh has agreed/(B) to marry with the girl/(C) of his parent's choice./(D) No error.

88. (A) The pity is that/(B) no sooner he had left the place/(C) than the fire broke out./(D) No error.

89. (A) When he was arriving/(B) the party was/(C) in full swing./(D) No error.

90. (A) The Dean wrote he constituted a committee of experts/(B) comprising of five members/(C) before the next meeting took place./(D) No error.

91. (A) Inflation and shortages/(B) have made it very difficult for him/(C) to make his both ends meet./(D) No error.

92. (A) The most studious boy/(B) in the class/(C) was made as the captain./(D) No error.

93. (A) I am participating/(B) in the two-miles race/(C) tomorrow morning./(D) No error.

94. (A) The sum and substance/(B) of his speech/(C) were essentially anti-establishment./(D) No error.

95. (A) It has been such a wonderful evening/(B) I look forward to meet you again/(C) after the vacations./(D) No error.

96. (A) When the boy committed a mistake/(B) the teacher made him to do/(C) the sum again./(D) No error.

97. (A) Unless the government does not revise its policy of liberalization/(B) the growth of the indigenous technology/(C) will be adversely affected./(D) No error.

98. (A) Supposing if you get/(B) a seat in the plane/(C) you will not take more than two hours to reach Mumbai./(D) No error.

99. (A) Whenever a person lost anything/(B) the poor folk around/(C) are suspected./(D) No error.

100. (A) Still impressive is that/(B) we achieve this selective attention/(C) through our latent ability to lip-read./(D) No error.

101. (A) The brakes and steering failed/(B) and the bus ran down the hill/(C) without anyone being able to control it./(D) No error.

102. (A) The polling was marred/(B) at many a place/(C) by attempts at rigging./(D) No error.

103. (A) He wanted to work all right/(B) but we saw that he was completely worn/(C) and so we persuaded him to stop./(D) No error.

104. (A) When a whale is washed ashore by the tide,/(B) the people flock together to see it,/(C) wondering how so huge an animal can swim about in the water./(D) No error.

105. (A) Few scientists changed/(B) people's ideas as much as/(C) Darwin with his Theory of Evolution./(D) No error.

106. (A) Were he/(B) to see you,/(C) he would have been surprised./(D) No error.

107. (A) The number of marks carried by each question/(B) are indicated/(C) at the end of the question./(D) No error.

108. (A) An animal/(B) can be just as unhappy in a vast area/(C) or in a small one./(D) No error.

109. (A) It is time/(B) we did something/(C) to stop road accidents./(D) No error.

110. (A) A free press is not a privilege/(B) but the organic necessity/(C) in a free society./(D) No error.

111. (A) The Indian radio,/(B) which was previously controlled by the British rulers,/(C) is free now from the narrow vested interests./(D) No error.

112. (A) Because of the emergency help/(B) that the patient received,/(C) he would have died./(D) No error.

113. (A) At present juncture,/(B) however, the super-computer/(C) would be a costly toy./(D) No error.

114. (A) Students should not take part/(B) in party politics and political demonstrations/(C) as they interfere in serious study./(D) No error.

115. (A) Wherever they go/(B) Indians easily adapt to/(C) local circumstances./(D) No error.

116. (A) According to the Bible/(B) it is meek and humble/(C) who shall inherit the earth./(D) No error.

117. (A) I was there/(B) many a time/(C) in the past./(D) No error.

118. (A) As much as I admire him for his sterling qualities./(B) I cannot excuse him for/(C) being unfair to his friends./(D) No error.

119. (A) Were you/(B) given a choice/(C) or you had to do it?/(D) No error.

120. (A) When he was asked what is wrong with him,/(B) he said that he was not well,/(C) and asked for leave of absence for one day./(D) No error.

121. (A) At the end of the year/(B) every student who had done adequate work/(C) was automatically promoted./(D) No error.

122. (A) Many times the news has been published/(B) in the papers that the end of the world will be certain/(C) if a nuclear war breaks out./(D) No error.

123. (A) Happily, zoos were/(B) unwilling to cooperate/(C) in a scheme that was potentially harmful to animal welfare./(D) No error.

124. (A) We discussed about the problem so thoroughly/(B) on the eve of the examination/(C) that I found it very easy to work it out./(D) No error.

125. (A) She reluctantly said that/(B) if nobody else was doing it/(C) she will do it./(D) No error.

126. (A) He will end up his work/(B) in the city/(C) by the end of the year./(D) No error.

127. (A) Though child marriage/(B) has been banned,/(C) the custom still prevailed among some groups in India./(D) No error.

128. (A) Supposing if/(B) there is no bus,/(C) how will you get there?(D) No error.

129. (A) At the moment the house/(B) was burgled the family/(C) attended a night party in the neighbourhood./(D) No error.

130. (A) On a holiday/(B) Sudha prefers reading/(C) than going out visiting friends./(D) No error.

131. (A) Neither he/(B) nor his father is interested/(C) in joining the party./(D) No error.

132. (A) A group of friends/(B) want to visit/(C) the new plant as early as possible./(D) No error.

133. (A) May I/(B) know who you want/(C) to see please./(D) No error.

134. (A) Myself and Gopal/(B) will take care of/(C) the function on Sunday./(D) No error.

135. (A) I could not put up in a hotel/(B) because the boarding and lodging charges/(C) were exorbitant./(D) No error.

136. (A) He is not coming tomorrow/(B) as he is having a pain in the chest/(C) and has to see a doctor./(D) No error.

137. (A) They have been/(B) very close friends/(C) until they quarrelled./(D) No error.

138. (A) Since India has gained Independence/(B) 49 years ago,/(C) much progress has been made in almost every field./(D) No error.

139. (A) The party chief made it a point to state that/(B) the Prime Minister and the Union Home Minister should also come/(C) and they see what his party men had seen./(D) No error.

140. (A) Due to me being a newcomer/(B) I was unable to get a house/(C) suitable for my wife and me./(D) No error.

141. (A) We are meeting today afternoon/(B) to discuss the matter/(C) and reach a compromise./(D) No error.

142. (A) Either Ram or/(B) you is responsible/(C) for this action./(D) No error.

143. (A) The student flatly denied/(B) that he had copied/(C) in the examination hall./(D) No error.

144. (A) By the time you arrive tomorrow/(B) I have finished/(C) my work./(D) No error.

145. (A) The speaker stressed repeatedly on/(B) the importance of improving/(C) the condition of the slums./(D) No error.

146. (A) The captain with the members of his team/(B) are returning/(C) after a fortnight./(D) No error.

147. (A) After returning from/(B) an all-India tour/(C) I had to describe about it./(D) No error.

148. (A) The teacher asked his students/(B) if they had gone through/(C) either of the three chapters included in the prescribed text./(D) No error.

149. (A) Although they are living in the country/(B) since they were married/(C) they are now moving to the town./(D) No error.

150. (A) Do you know/(B) how old were you/(C) when you came here?/(D) No error.

ANSWERS

1	2	3	4	5	6	7	8	9	10
C	C	A	B	A	A	D	C	C	B
11	12	13	14	15	16	17	18	19	20
D	B	A	C	C	C	B	A	D	C
21	22	23	24	25	26	27	28	29	30
B	A	C	C	B	A	A	C	C	C
31	32	33	34	35	36	37	38	39	40
C	A	A	A	C	A	A	B	D	B
41	42	43	44	45	46	47	48	49	50
C	C	C	D	A	B	C	C	A	A
51	52	53	54	55	56	57	58	59	60
C	D	D	D	C	C	D	A	B	B
61	62	63	64	65	66	67	68	69	70
C	A	C	A	B	B	B	C	B	D
71	72	73	74	75	76	77	78	79	80
A	C	D	D	D	A	B	B	B	A
81	82	83	84	85	86	87	88	89	90
A	B	A	D	B	A	B	B	A	B
91	92	93	94	95	96	97	98	99	100
C	C	B	C	B	B	A	A	A	C
101	102	103	104	105	106	107	108	109	110
C	D	B	C	A	A	B	C	D	B
111	112	113	114	115	116	117	118	119	120
C	A	A	C	B	B	A	A	A	A
121	122	123	124	125	126	127	128	129	130
D	D	C	A	C	A	C	A	A	C
131	132	133	134	135	136	137	138	139	140
D	B	B	A	A	C	A	A	C	A
141	142	143	144	145	146	147	148	149	150
A	B	D	A	A	B	C	C	B	D

4 Filling the Blanks

Filling the blanks or sentence completion is such an exercise that starts with the primary education and continues in the highest level of examinations. One must practise it regularly to score well.

MULTIPLE CHOICE QUESTIONS

Directions: *Fill in the blanks with the appropriate words given in choice.*

1. While the contest in Kashmir has international ....., the outcome in Gujarat will be of enormous significance at the national level.
A. ravening B. ramifications
C. reconnaissance D. renunciation

2. No amount of post facto sophistry can conceal the ..... and inflammable crudity of this exhortation.
A. cruel B. shameful
C. horrible D. horrendous

3. There were ..... voices coming out from the amalgam.
A. discordant B. sneering
C. impalpable D. disquisite

4. Online learning can also offer a ..... of exciting and technology driven courses.
A. spectrum B. spume
C. spate D. stolidity

5. Work experience does carry some weight, the extent of which is a matter of ..... .
A. erudition B. decision
C. determination D. conjecture

6. The curricula of education has to respond and reshape with ..... modification according to the needs and aspirations of the dynamic society.
A. ambient B. alliterate
C. allotropic D. ample

7. It may look ..... to pronounce but this is the reality that the situation is that of the largest concentration of illiterate population is in existence.
A. surprising B. contradictory
C. paradoxical D. contrary

8. If we take the global, developmental ....., education is not only a master key but also acts as a catalyst in the positive direction.
A. context B. pretext
C. progress D. continuum

9. It is mainly the utmost utilization of knowledge which emanates from the deep-delved ..... in every human being.
A. acquisition B. core
C. mind D. self

10. The book is a rare ..... of facts and figures illustrating every aspect of education.
A. chronicle B. compendium
C. combination D. presentation

11. The ...... reason behind the terrorist attack on the temple is obvious.
A. devious B. diabolical
C. factual D. original

12. He had the ..... to reject plans.
A. disquisition B. gibe
C. gumption D. grudge

13. Unable to ..... the election process in Kashmir, the terrorists have spread out all over India.
A. scrounge B. scruff
C. scuttle D. scuffle

14. Their hopes of a communal ..... is not going to come true.
A. conflagration B. plenipotentiary
C. podium D. coterie

15. Fundamentalism refers to an ideology which takes certain articles of faith to be ....., beyond question.
A. inviolable B. invulnerable
C. irrefragable D. inerrant

16. The separation of Church and State has its origins in the vicious and ..... 16th century sectarian wars within European Christianity.
A. debacle B. debilitating
C. debase D. dastard

17. It's time to look beyond the simplistic ..... between secularism and fundamentalism.
A. dichotomy B. disarray
C. diabolical D. fetid

18. When unemployment is rampant, the future is ..... and fraught with worry, depression and frustration.
A. fluster B. foible
C. occlude D. ominous

19. We are worried by the ..... tendencies prevailing in our social and political life.
A. flotsam B. fissiparous
C. perfidy D. aphoristic

20. The intellectual experience had culminated new methodology in ..... .
A. assuage B. value
C. pedagogy D. pedantic

21. Human rights awareness is ignited to propagate in conflagration which could ..... down to the masses.
A. perfidy B. kindle
C. percolate D. exude

22. It is no use living a wretched life, where human dignity is an ..... .
A. abyss B. abhor
C. abnegation D. chastise

23. Upheavals and ..... de-humanisation are the paralytic traumas which degrade and nullify glamourisation of the human values.
A. profilgate B. entropic
C. obdurate D. intransigence

24. The driver tried to ..... the fears of the passenger by announcing that the accident has been prevented.
A. overcome B. remove
C. eradicate D. allay

25. ..... should be removed among the people and equal chances should be given to those who have been denied equality so long.
A. Distinction B. Diaphanous
C. Differences D. Disparities

26. It is uncertainity and ..... of purpose which create defects in our present system of education.
A. confusion B. ambivalence
C. vagueness D. doubt

27. The golden jubilee of the Independence of India is being celeberated in an ..... and jubilant manner.
A. joyful B. pleasant
C. ecstatic D. eclat

28. Education has become capable of meeting all the changes of the ..... society by broadening the new paths of learning.
A. dynamic B. growing
C. developing D. modern

29. The world and inevitably India, are in the ..... of major economic and technological changes.
A. gulf B. tendentious
C. insularity D. throes

30. ..... officer finally surrendered before the court yesterday, after being on the run for two months.
A. Crustacean B. Complacent
C. Fugitive D. Beguile

31. Gujarat riots have attracted national and international ..... .
A. opprobrium B. oposum
C. repugnance D. reconnoitre

32. The political and administrative leadership may have been determined to prevent any mischief makers from running ..... .
A. gauche B. amok
C. catharic D. berserk

33. After her long illness she is suffering from general ..... .
A. fawning B. debility
C. lethargic D. inchoate

34. They are so poor that they have to .....
A. slack B. slander
C. skimp D. bicker

35. ..... about in the corridors.
A. Fecund B. Droll
C. Carp D. Skulk

36. The Queen's request was ..... to a command.
A. tantamount B. tenet
C. nexus D. nascent

37. We cannot ..... base metals into gold.
A. transmute B. trek
C. sodden D. impede

38. ..... an official tour to Switzerland.
A. Warble B. Wane
C. Wangle D. Warren

39. His inattention in the classroom is a clear ..... of failure.
A. pristine B. propriety
C. prognosis D. prognostic

40. I find singing ..... challenging.
A. intrinsically B. incredibly
C. extrinsically D. preposterously

41. Educating people to think your way is ....
A. prodigious B. presumptuous
C. sagacious D. unpretentious

42. The birth of United Nations Organisation in 1945 was a ..... device to achieve specific objectives in the realm of world order.
A. subjugate B. precarious
C. pragmatic D. unabated

43. Inefficiency and delaying habits are the ..... of Indian offices.
A. averse B. assuagement
C. banal D. bane

44. It is a most ..... situation for a young person to find a place under the sun and lead a comfortable though hard-working life.

A. exhilarating B. barbarous
C. bucolic D. cantankerous

45. The hot weather created a feeling of ..... and encouraged drowsiness.

A. uninterested B. lethargic
C. lassitude D. exhaustion

46. One prefers to be alone than to be in his company because of his ..... temper.

A. stoic B. smattering
C. sleight D. irascible

47. The boss scolded him in a ..... voice.

A. preposterous B. loud
C. suave D. strident

48. The ..... provided by the government to those affected by the flood were not sufficient.

A. ensconce B. victuals
C. baroque D. accrue

49. Zakir Hussain is a tabla ....

A. abettor B. adroit
C. virago D. virtuoso

50. He took.....at his remarks and resigned immediately.

A. umbrage B. surly
C. unseemly D. valour

ANSWERS

1	2	3	4	5	6	7	8	9	10
B	D	A	A	D	C	C	D	A	B
11	**12**	**13**	**14**	**15**	**16**	**17**	**18**	**19**	**20**
B	C	C	A	A	B	A	D	B	C
21	**22**	**23**	**24**	**25**	**26**	**27**	**28**	**29**	**30**
C	A	B	D	D	B	C	A	D	C
31	**32**	**33**	**34**	**35**	**36**	**37**	**38**	**39**	**40**
A	B	B	C	D	A	A	C	D	B
41	**42**	**43**	**44**	**45**	**46**	**47**	**48**	**49**	**50**
B	C	D	A	C	D	D	B	B	A

5

Synonyms & Antonyms

There are thousands of words in English language. No one can remember their meanings easily but with regular practice one can memorise most of them. A number of words with their Synonyms and Antonyms are compiled here. Try to learn as many as you can and answer the questions thereafter.

WORDS	SYNONYMS	ANTONYMS
Abandon	Cease, Forsake	Continue
Abhor	Hate, Loathe, Detest	Like, Love
Abiding	Enduring, Durable	Fleeting
Able	Proficient, Competent	Incompetent, Unfit
Ability	Skill, Power	Disability, Inability
Abortive	Fruitless, Futile	Fruitful, Successful
Abolish	Destroy, Undo	Restore, Revive
Abridge	Shorten, Curtail	Lengthen, Expand
Absolve	Forgive, Pardon, Excuse	Condemn
Accelerate	Hasten	Retard
Accord	Agreement, Harmony	Discord, Disagreement
Accumulate	Collect, Store, Amass	Distribute, Scatter
Adamant	Hard, Inflexible	Flexible
Adversity	Misfortune, Distress	Prosperity
Adept	Expert, Skilful	Inexpert, Unskillful
Aggravate	Heighten, Intensify	Quell, Suppress
Agile	Nimble	Clumsy, Undeft
Alert	Vigilant	Heedless
Allay	Calm, Soothe, Assuage	Arouse
Ameliorate	Improve, Advance, Amend	Worsen, Deteriorate
Ambiguous	Vague, Unclear	Clear
Amiable	Lovable, Agreeable	Disagreeable
Annihilate	Destroy	Create
Arduous	Hard, Strenuous	Easy
Attack	Assault	Defend
Audacity	Boldness	Cowardice
Auspicious	Favourable, Propitious, Lucky	Ominous, Inauspicious, Unlucky
Austere	Harsh, Severe, Rigorous	Easy-going
Authentic	True, Genuine	Spurious, False
Avarice	Greed	Generosity
Averse	Unwilling, Loth, Disinclined	Willing, Inclined
Aversion	Hostility, Hatred	Affinity, Liking

WORDS	SYNONYMS	ANTONYMS
Base	Low, Mean, Ignoble	Noble, Exalted
Boisterous	Noisy, Stormy	Calm, Quiet
Brave	Courageous, Daring, Bold, Plucky	Cowardly, Dastardly, Timid
Brief	Short, Concise, Laconic	Lengthy, Diffuse
Bright	Vivid, Radiant	Dull, Dark
Brutal	Savage, Cruel	Humane, Kindly
Callous	Hard, Cruel, Indifferent	Soft, Tender, Concerned
Cautious	Careful, Wary	Rash, Reckless, Foolhardy
Censure (*n*)	Blame, Condemnation	Praise
Censure (*vb*)	Blame, Condemn	Praise, Commend
Circumscribed	Restricted, Confined, Limited	Unconfined, Unrestricted
Civil	Polite, Courteous, Gracious, Urbane	Rude, Uncivil, Impolite, Ungracious
Coerce	Compel, Force	Volunteer
Compassionate	Pitiful, Sympathetic, Merciful	Unsympathetic, Merciless, Cruel
Compress	Condense, Abbreviate	Expand, Lengthen
Conspicuous	Noticeable, Manifest	Inconspicuous
Constant	Steady, Steadfast, Uniform	Inconstant, Variable
Cordial	Friendly, Warm, Hearty	Cold, Unfriendly
Covert	Hidden, Secret	Overt, Open
Cruel	Savage, Ruthless, Vicious	Kind, Gentle, Benevolent
Cursory	Rapid, Superficial	Thorough, Exhaustive, Intensive
Credible	Believable, Probable, Plausible	Incredible, Unbelievable, Fantastic
Crafty	Cunning, Sly	Artless, Simple, Ingenuous
Costly	Expensive, Dear	Cheap, Inexpensive
Confidence	Trust, Reliance	Distrust, Doubt
Death	Decease, Demise	Existence, Life
Dearth	Scarcity, Lack, Want, Paucity, Shortage	Plenty, Abundance
Decay	Dissolution, Decline, Decomposition, Disintegration	Regeneration
Deference	Respect, Reverence	Disrespect, Irreverence
Deficient	Lacking, Inadequate	Complete, Sufficient
Desolate	Lonely, Deserted	Crowded, Occupied
Destitute	Wanting, Needy	Rich, Affluent
Diligence	Industry, Perseverance	Idleness
Disgrace	Dishonour, Discredit	Honour, Credit
Dwindle	Decrease, Shrink	Grow, Increase
Earthly	Terrestrial, Mundane	Celestial, Heavenly, Unearthly
Eligible	Qualified, Suitable	Ineligible, Unsuitable
Emancipate	Liberate, Free	Enslave
Excited	Impassioned, Stimulated	Composed, Cool, Impassive
Extraordinary	Uncommon, Remarkable, Marvellous	Commonplace, Ordinary
Extravagant	Lavish, Prodigal, Wastrel, Spendthrift	Thrifty, Economical, Frugal
Fabricate	Construct, Make	Destroy
Fabulous	Fictitious, Mythical	Actual, Real
False	Untrue, Mendacious	True, Genuine

WORDS	SYNONYMS	ANTONYMS
Famous	Well-known, Renowned	Obscure, Unknown
Fantastic	Fanciful, Imaginative, Visionary	Practical, Down to earth
Fearful	Nervous, Anxious, Afraid, Scared	Fearless, Dauntless
Felicity	Happiness	Sorrow
Gaiety	Joyousness, Hilarity	Mourning, Dullness
Garrulous	Talkative, Loquacious	Taciturn, Silent, Reserved
Generous	Liberal, Magnanimous	Stingy, Miserly
Gigantic	Huge, Colossal	Minute, Small
Graphic	Vivid, Pictorial, Meaningful	Vague
Guest	Visitor	Host
Guile	Fraud, Trickery	Artlessness, Ingenuousness
Gratitude	Gratefulness	Ingratitude, Ungratefulness
Gratuitous	Voluntary, Spontaneous, Unwarranted	Involuntary, Forced
Hamper	Hinder, Obstruct	Facilitate, Ease
Haughty	Arrogant, Proud	Humble, Modest
Hazardous	Dangerous, Perilous	Safe, Secure, Protected
Headstrong	Obstinate, Stubborn	Weak-willed, Flexible
Hope	Belief, Conviction, Expectation	Despair, Hopelessness
Improvident	Prodigal, Carelessness	Provident, Economical
Incessant	Unceasing, Continuous	Discontinuous
Indolent	Slothful, Lethargic	Active, Energetic
Joy	Delight, Pleasure	Sadness, Gloom
Jolly	Jovial, Merry	Gloomy, Sad
Judicious	Discreet, Prudent	Indiscreet, Injudicious
Knowledge	Enlightenment, Learning	Ignorance, Stupidity
Laborious	Industrious, Assiduous	Slothful, Lazy
Laxity	Slackness, Looseness	Firmness
Lenient	Mild, Forbearing	Strict, Stern
Lethal	Deadly, Fatal, Mortal	Life-giving, Vital, Vivifying
Liberal	Generous, Tolerant	Intolerant, Illiberal
Liberty	Freedom, Independence	Slavery, Bondage
Lively	Animated, Active	Dull, Listless
Loyal	Faithful, Devoted	Treacherous, Disloyal, Unfaithful
Lucky	Fortunate	Unlucky, Unfortunate
Lucrative	Profitable	Unprofitable
Magnanimous	Generous, Largehearted	Ungenerous, Stingy
Malady	Illness, Ailment	Health
Manifest	Noticeable, Obvious	Obscure, Puzzling
Meagre	Small	Plentiful, Large
Mean	Low, Abject	Noble, Exalted
Mendacious	False, Untruthful	Truthful
Misery	Sorrow, Distress	Happiness, Joy
Morbid	Sick, Diseased	Healthy
Mournful	Sorrowful, Sad	Joyful, Happy
Negligent	Careless, Heedless	Careful

WORDS	SYNONYMS	ANTONYMS
Notorious	Infamous, Disreputable	Reputable
Obedient	Submissive, Compliant, Docile	Disobedient, Recalcitrant, Wayward
Obsolete	Antiquated, Out-of-Date	Current, Modern
Opportune	Timely, Seasonable	Inopportune
Opulence	Wealth, Riches	Penury, Poverty
Onerous	Heavy, Burdensome	Light, Easy
Palatable	Tasty, Delicious	Unpalatable
Pathetic	Touching	Joyous, Cheery
Persuade	Urge, Induce	Dissuade
Praise (*vb*)	Applaud, Eulogise	Condemn
Praise (*n*)	Applause, Eulogy	Condemnation
Precarious	Risky, Uncertain	Safe, Certain
Pretence	Pretext, Excuse	Candour, Frankness
Propagate	Breed, Circulate	Terminate, Restrict
Quaint	Odd, Singular	Usual, Ordinary
Quell	Suppress, Subdue	Agitate, Arouse
Rare	Uncommon, Scarce	Common, Ordinary
Refined	Polished, Elegant	Crude, Coarse
Remote	Distant	Near, Close
Renown	Fame, Reputation	Infamy, Notoriety
Rigid	Stiff, Unyielding	Flexible, Yielding
Remorseful	Regretful, Repentant	Unrepentant
Rebellion	Revolt, Mutiny, Insurgency	Loyalty
Scared	Holy, Consecrated	Profane, Unholy
Sane	Sensible, Sound	Insane
Scold	Chide, Rebuke	Praise
Serious	Grave, Earnest	Frivolous
Shy	Bashful	Bold, Impudent
Simple	Plain, Artless	Complex, Cunning, Shrewd
Solitary	Single, Lonely, Secluded	Numerous, Multitude
Shallow	Superficial	Deep
Solace	Comfort, Relief	Discomfort, Grief
Spurious	Sham, False	Genuine, Authentic
Stagnant	Still, Motionless	Moving
Surplus	Excess	Deficit, Shortage
Tame	Gentle, Mild, Domesticated	Savage, Wild
Teacher	Instructor, Educator	Student, Pupil
Tedious	Wearisome, Monotonous	Agreeable, Lively
Temporal	Worldly, Secular	Spiritual
Temperate	Moderate	Immoderate, Intemperate
Tortuous	Winding, Circuitous	Straight, Direct
Tough	Hard, Strong	Tender, Soft, Flexible
Transient	Temporary, Fleeting	Lasting, Durable, Permanent
Trustworthy	Reliable	Unreliable, Untrustworthy
Tranquil	Calm	Agitated

WORDS	SYNONYMS	ANTONYMS
Ugly	Unsightly, Repulsive	Beautiful, Attractive
Useful	Advantageous, Serviceable	Useless
Vehemence	Passion, Force	Apathy, Indifference
Vindictive	Revengeful	Forgiving
Wholesome	Healthy	Unwholesome, Morbid, Unhealthy, Diseased
Wicked	Evil, Impious	Pious, Good
Wise	Sagacious, Erudite	Foolish, Stupid
Wrath	Anger, Fury, Rage	Love, Peace, Calm
Wreck	Ruin, Destroy	Create, Construct
Yield	Surrender, Submit	Resist, Revolt
Yielding	Submissive, Supple	Inflexible, Intractable
Yoke	Oppression, Bondage	Freedom
Zeal	Passion, Fervour	Apathy, Indifference
Zest	Relish, Enthusiasm	Distaste, Disrelish

MULTIPLE CHOICE QUESTIONS

Directions (Qs. 1 to 50): *In the following questions choose the word which best expresses the MEANING of the given word.*

1. ADVERSITY
A. Failure B. Helplessness
C. Misfortune D. Crisis

2. TURN UP
A. Land up B. Show up
C. Crop up D. Come up

3. DEIFY
A. Flatter B. Challenge
C. Worship D. Face

4. ERROR
A. Misadventure B. Misgiving
C. Ambiguity D. Blunder

5. SHALLOW
A. Artificial B. Superficial
C. Foolish D. Worthless

6. MASSACRE
A. Murder B. Stab
C. Assassinate D. Slaughter

7. COMBAT
A. Conflict B. Quarrel
C. Feud D. Fight

8. VORACIOUS
A. Wild B. Hungry
C. Angry D. Quick

9. IMPROMPTU
A. Offhand B. Unimportant
C. Unreal D. Effective

10. RABBLE
A. Mob B. Noise
C. Roar D. Rubbish

11. TEPID
A. Hot B. Warm
C. Cold D. Boiling

12. MAYHEM
A. Jubilation B. Havoc
C. Excitement D. Defeat

13. TIMID
A. Fast B. Slow
C. Medium D. Shy

14. CANTANKEROUS
A. Quarrelsome B. Rash
C. Disrepectful D. Noisy

15. PRECARIOUS
A. Cautious B. Critical
C. Perilous D. Brittle

16. TACITURNITY
A. Dumbness B. Changeableness
C, Hesitation D. Reserve

17. INEBRIATE
A. Dreamy B. Stupefied
C. Unsteady D. Drunken

18. HARBINGER
A. Massenger B. Steward
C. Forerunner D. Pilot

19. INTIMIDATE
A. To hint B. Frighten
C. Bluff D. Harass

20. IRONIC
A. Inflexible B. Bitter
C. Good-natured D. Disguisedly sarcastic

21. STRINGENT
A. Tense B. Stringy
C. Strict D. Causing to shrink

22. ECSTATIC
A. Animated B. Bewildered
C. Enraptured D. Illful

23. COMMENSURATE
A. Measurable B. Proportionate
C. Beginning D. Appropriate

24. DESTITUTION
A. Humility B. Moderation
C. Poverty D. Beggary

25. ASCEND
A. Leap B. Grow
C. Deviate D. Mount

26. UNCOUTH
A. Ungraceful B. Rough
C. Slovenly D. Dirty

27. LYNCH
A. Hang B. Madden
C. Killed D. Shoot

28. LAUD
A. Lord B. Eulogy
C. Praise D. Extolled

29. CORRESPONDENCE
A. Agreements B. Contracts
C. Documents D. Letters

30. VENUE
A. Place B. Agenda
C. Time D. Duration

31. STERILE
A. Barren B. Arid
C. Childless D. Dry

32. SYNOPSIS
A. Index B. Mixture
C. Summary D. Puzzle

33. GERMANE
A. Responsible B. Logical
C. Possible D. Relevant

34. PONDER
A. Think B. Evaluate
C. Anticipate D. Increase

35. CANNY
A. Obstinate B. Handsome
C. Clever D. Stout

36. ABUNDANT
A. Ripe B. Cheap
C. Plenty D. Absent

37. CONSEQUENCES
A. Results B. Conclusions
C. Difficulties D. Applications

38. SHIVER
A. Shake B. Rock
C. Tremble D. Move

39. DILIGENT
A. Progressive B. Brilliant
C. Inventive D. Hard-working

40. DISTANT
A. Far B. Removed
C. Reserved D. Separate

41. FORAY
A. Excursion B. Contest
C. Ranger D. Intuition

42. FRUGALITY
A. Foolishness B. Extremity
C. Enthusiasm D. Economy

43. GARNISH
A. Paint B. Garner
C. Adorn D. Abuse

44. VIGOUR
A. Strength B. Boldness
C. Warmth D. Enthusiasm

45. CANDID
A. Apparent B. Explicit
C. Frank D. Bright

46. BRIEF
A. Limited B. Small
C. Little D. Short

47. GARRULITY
A. Credulity B. Senility
C. Loquaciousness D. Speciousness

48. FURORE
A. Excitement B. Worry
C. Flux D. Anteroom

49. NEUTRAL
A. Unbiased B. Non-aligned
C. Undecided D. Indifferent

50. LAMENT
A. Complain B. Comment
C. Condone D. Console

Directions (Qs. 51 to 100): *In the following questions choose the word which is the exact OPPOSITE of the given words.*

51. REPEL
A. Attend B. Concentrate
C. Continue D. Attract

52. ARTIFICIAL
A. Real B. Natural
C. Truthful D. Solid

53. CAPACIOUS
A. Limited B. Caring
C. Foolish D. Changeable

54. PROVOCATION
A. Vocation B. Pacification
C. Peace D. Destruction

55. METICULOUS
A. Mutual B. Shaggy
C. Meretricious D. Slovenly

56. ABLE
A. Disable B. Inable
C. Unable D. Misable

57. COMFORT
A. Uncomfort B. Miscomfort
C. Discomfort D. None of these

58. GAIN
A. Loose B. Fall
C. Lost D. Lose

59. SYNTHETIC
A. Affable B. Natural
C. Plastic D. Cosmetic

60. ACQUITTED
A. Freed B. Burdened
C. Convicted D. Entrusted

61. STRINGENT
A. General B. Vehement
C. Lenient D. Magnanimous

62. FLIMSY
A. Frail B. Filthy
C. Firm D. Flippant

63. BUSY
A. Occupied B. Engrossed
C. Relaxed D. Engaged

64. ADAPTABLE
A. Adoptable B. Flexible
C. Yielding D. Rigid

65. LOVE
A. Villainy B. Hatred
C. Compulsion D. Force

66. BALANCE
A. Disbalance B. Misbalance
C. Debalance D. Imbalance

67. RELINQUISH
A. Abdicate B. Renounce
C. Possess D. Deny

68. MOUNTAIN
A. Plain B. Plateau
C. Precipice D. Valley

69. FICKLE
A. Courageous B. Sincere
C. Steadfast D. Humble

70. PERENNIAL
A. Frequent B. Regular
C. Lasting D. Rare

71. RARELY
A. Hardly B. Definitely
C. Frequently D. Periodically

72. STARTLED
A. Amused B. Relaxed
C. Endless D. Astonished

73. ADHERENT
A. Detractor B. Enemy
C. Alien D. Rival

74. QUIESCENT
A. Indifferent B. Troublesome
C. Weak D. Unconcerned

75. CONDENSE
A. Expand B. Distribute
C. Interpret D. Lengthen

76. BENIGN
A. Malevolent B. Soft
C. Friendly D. Unwise

77. OBSCURE
A. Implicit B. Obnoxious
C. Explicit D. Pedantic

78. HYPOCRITICAL
A. Gentle B. Sincere
C. Amiable D. Dependable

79. EVASIVE
A. Free B. Honest
C. Liberal D. Frank

80. INDUSTRIOUS
A. Indifferent B. Indolent
C. Casual D. Passive

81. EXTRICATE
A. Manifest B. Palpable
C. Release D. Entangle

82. LUCID
A. Glory B. Noisy
C. Obscure D. Distinct

83. INSIPID
A. Tasty B. Stupid
C. Discreet D. Feast

84. OBEYING
A. Ordering B. Following
C. Refusing D. Contradicting

85. VICTORIOUS
A. Defeated B. Annexed
C. Destroyed D. Vanquished

86. COMMISSIONED
A. Started B. Closed
C. Finished D. Terminated

87. VANITY
A. Pride B. Humility
C. Conceit D. Ostentious

88. ZENITH
A. Acme B. Top
C. Nadir D. Pinnacle

89. TANGIBLE
A. Ethereal B. Concrete
C. Actual D. Solid

90. REPRESS
A. Inhibit B. Liberate
C. Curb D. Quell

91. EPILOGUE
A. Dialogue B. Prelude
C. Post script D. Epigram

92. FRAUDULENT
A. Candid B. Direct
C. Forthright D. Genuine

93. LOQUACIOUS
A. Reticent B. Talkative
C. Garrulous D. Verbose

94. NIGGARDLY
A. Frugal B. Thrifty
C. Stingy D. Generous

95. PERTINENT
A. Irrational B. Irregular
C. Insistent D. Irrelevant

96. FAINT-HEARTED
A. Warm-hearted B. Full-blooded
C. Hot-blooded D. Stout-hearted

97. VIOLENT
A. Humble B. Harmless
C. Gentle D. Tame

98. STATIONARY
A. Active B. Mobile
C. Rapid D. Busy

99. HONORARY
A. Dishonourable B. Reputed
C. Paid D. Official

100. COMMON
A. Rare B. Small
C. Petty D. Poor

ANSWERS

1	2	3	4	5	6	7	8	9	10
C	B	C	D	B	D	D	B	A	A
11	**12**	**13**	**14**	**15**	**16**	**17**	**18**	**19**	**20**
B	B	D	A	B	D	D	C	B	D
21	**22**	**23**	**24**	**25**	**26**	**27**	**28**	**29**	**30**
C	C	B	C	D	A	C	C	D	A
31	**32**	**33**	**34**	**35**	**36**	**37**	**38**	**39**	**40**
A	C	D	A	C	C	A	C	D	A
41	**42**	**43**	**44**	**45**	**46**	**47**	**48**	**49**	**50**
A	D	C	A	C	D	C	A	A	A
51	**52**	**53**	**54**	**55**	**56**	**57**	**58**	**59**	**60**
D	B	A	B	D	C	C	D	B	C
61	**62**	**63**	**64**	**65**	**66**	**67**	**68**	**69**	**70**
C	C	C	D	B	D	C	D	C	D
71	**72**	**73**	**74**	**75**	**76**	**77**	**78**	**79**	**80**
D	B	B	A	C	A	C	B	B	B
81	**82**	**83**	**84**	**85**	**86**	**87**	**88**	**89**	**90**
D	C	A	A	A	D	B	C	A	B
91	**92**	**93**	**94**	**95**	**96**	**97**	**98**	**99**	**100**
B	D	B	D	D	D	C	B	C	A

6

One Word Substitutions

There are many words in English language which can be perfectly used for a number of words. These words help in expressing ideas in a short and correct manner for the right occasion. Such words not only enhance the vocabulary but also enable you to economise in the use of words to a great extent.

PERTAINING TO GOVERNMENT

To give up a throne or other office of dignity	*abdicate*
Absence of government	*anarchy*
Government by Sovereign of uncontrolled authority	*autocracy, despotism*
Government by departments of states	*bureaucracy*
Government of the people, for the people and by the people	*democracy*
Government by the nobility	*aristocracy*
The right of self-government	*autonomy*
Government by a few	*oligarchy*
Government by the wealthy	*plutocracy*
Government by divine guidance	*theocracy*
To decide a political question by the direct vote of the whole electorate	*referendum*
Sweeping governmental change	*revolution*
The science of government	*politics*

PERTAINING TO RELIGION

One who believes that man can have no knowledge of God but only of natural phenomena	*agnostic*
One who renounces his religious vows or forsakes his religious principles	*apostate*
One who believes in existence of God	*theist*
One who does not believe in existence of God	*atheist*
One intolerantly devoted to a particular creed	*bigot*
To utter profane language against God or anything holy	*blaspheme*
A breaker of church images	*iconoclast*
Worship of images or idols	*idiolatory*
One who believes in one God	*monotheist*
One who believes in many Gods	*polytheist*

PERTAINING TO MARRIAGE AND CHILDREN

One who marries a second wife or husband while the legal spouse is alive	*bigamist*
One vowed to a single or unmarried life	*celibate*
One engaged to be married	*fiancé, fiancee*
A child whose parents are dead	*orphan*
A hater of marriage	*misogamist*
One who has more than two wives at a time	*polygamist*

PERTAINING TO DEATH

Dead and decaying flesh (esp. of animals)	*carrion*
A monument set up for persons who are buried elsewhere	*cenotaph*
To preserve a dead body from putrefaction	*embalm*
Words inscribed on a tomb	*epitaph*
An examination of dead body	*postmortem, autopsy*
An account in the newspaper of the funeral of one deceased	*obituary*
The property left to someone by a will	*legacy*
Occurring after death	*posthumous*
The act of killing a human being	*homicide*
Murder of a new born child	*infanticide*
Murder of a brother	*fratricide*
Murder of a sister	*sororicide*
Murder of a mother	*matricide*
Murder of a father	*patricide*
Murder of a parent	*parricide*
Murder of a king	*regicide*

PERTAINING TO THE LITERARY SPHERE

A work whose writer is unknown	*anonymous*
A record of one's life written by oneself	*autobiography*
The history of the life of a person	*biography*
The heading or short description of a newspaper article, chapter of a book etc.	*caption*
A humorous play, having a happy ending	*comedy*
A list of books in a library	*catalogue, bibliography*
A book in which the events of each day are recorded	*diary*
A books containing the words of a language with their definitions, in alphabetical order	*dictionary*
A book of names and addresses	*directory*
A short speech by a player at the end of a play	*epilogue*
A brief summary of a book	*epitome*
A book containing information on all branches of knowledge	*encyclopaedia*
To remove the offensive portions of a book	*expurgate*
A speech delivered without earlier preparation	*extempore*
A noisy or vehement speech intended to excite passions	*harangue*
A written account, usually in book form of the interesting and memorable experiences of one's life	*memoirs*
A note to help the memory	*memorandum*
A declaration of plans and promises put forward by a candidate for election, political party or a sovereign	*manifesto*
A short speech by a player at the beginning of a play	*prologue*
Literary theft, or passing off an author's original work as one's own	*plagiarism*
Speaking aloud to oneself	*soliloquy*
A play with a sad or tragic end	*tragedy*

PERTAINING TO SCIENCES AND ARTS

The study of all heavenly bodies and the earth in relation to them	*astronomy*
The science of land management	*agronomics*

The study of mankind	*anthropology*
The study of physical life or living matter	*biology*
The study of plants	*botany*
The art of beautiful hand-writing	*calligraphy*
The science which deals with the varieties of human race	*ethnology*
The study of the origin and history of words	*etymology*
The study of coins	*numismatics*
The study of human face	*physiognomy*
The art of making fireworks	*pyrotechnics*
The study of birds	*ornithology*
The study of languages	*philology*
At home equally on land or in water	*amphibious*
The inside of a nut	*kernel*
The central or innermost part of fruit	*core*
The animals of a certain region	*fauna*
The plants and vegetation of a certain region	*flora*
Absence of rain for, a long time	*drought*
To supply land with water by artificial means	*irrigate*
One who studies plant and animal life	*naturalist*
A cud-chewing animal, *e.g.* the cow	*ruminant*
A gnawing animal, *e.g.* the rat	*rodent*
A four-footed animal	*quadruped*
Animals which carry their young in a pouch, *e.g.* kangaroo	*marsupials*
Soil composed largely of decayed vegetable matter	*humus*
A preparation for killing insects	*insecticide*
A plant or animal growing on another	*parasite*
Living for many years	*perennial*

PERTAINING TO MEDICINE

A substance which destroys or weakens germs	*antiseptic*
Any medicine which produces insensibility	*anaesthetic*
A medicine to counteract poison	*antidote*
Want or poorness of blood	*anaemia*
A medicine which alleviates pain	*anodyne*
To cut off from a person's body a part which is infected	*amputate*
One who is recovering from illness	*convalescent*
To be able to tell the nature of disease by its symptoms	*diagnose*
A disease affecting many persons at the same place and time	*epidemic*
A disease confined to a particular district or place	*endemic*
To disinfect by smoke	*fumigate*
Free or exempt from infection	*immune*
A person who is sick	*invalid*
A cure for all diseases	*panacea*
A disease widely epidemic	*pandemic*
Confinement to one place to avoid spread of infection	*quarantine*

PERTAINING TO WAR

An unprovoked attack by an enemy	*aggression*
Shells, bombs, military stores	*ammunition*
A place where naval or military weapons are made or stored	*ordnance*
An agreement between belligerents to stop fighting	*armistice*
A general pardon of offenders	*amnesty*
To reduce to nothing	*annihilate*
Nations carrying on warfare	*belligerents*
To surround a place with the intention of capturing	*besiege*
To camp in the open air without tents or covering	*bivouac*
To seize for military use	*commandeer*
A person who is forced by law to become a soldier	*conscript*
An order prohibiting ships to leave the ports	*embargo*
A number of firearms being discharged continuously	*fusillade*
To make an examination or preliminary survey of enemy territory or military objectives	*reconnoitre*

PERTAINING TO PROFESSIONS

One who attends to the diseases of the eye	*oculist*
One who tests eyesight and sells spectacles	*optician*
One who attends to the teeth	*dentist*
One skilled in the care of hands and feet	*chiropodist*
A physician who assists women at child-birth	*obstetrician*
One who drives a motor-car	*chauffeur*
One who makes or sells candles	*chandler*
One who preserves the skins of animals and mounts them so as to resemble the living animals	*taxidermist*
One who compiles a dictionary	*lexicographer*
One who writes books	*author*
One skilled in the treatment of diseases of animals	*veterinarian*
A tradesman who manages funerals	*undertaker*
One who draws up contracts and lends money on interest	*scrivener*
One who lends money and keeps goods as security	*pawnbroker*
A teacher who travels from place to place to give instruction	*peripatetic*
One who travels from place to place selling miscellaneous articles	*hawker; pedlar*
One who collects postage stamps	*philatelist*
One who lends money at exorbitant interest	*usurer*
One who takes care of a building	*janitor*
One who sells sweets and pastries	*confectioner*
One who works in a coal-mine	*collier*
One who flies an aeroplane	*pilot, aviator*
One who studies rocks and soils	*geologist*
One who shoes horses	*farrier*
A professional rider in horses races	*jockey*
One who deals in silks, cotton, woollen, and linen goods	*mercer*
One who deals in wines	*vintner*
One who deals in fish	*fishmonger*

One who deals in iron and hardware	*ironmonger*
One who sells fruits, vegetables, etc. from a barrow	*costermonger*
One who sets type (in a printing office)	*compositor*

PERTAINING TO CHARACTERISTICS AND ACTIONS

One who devotes his life to the welfare and interests of other people	*altruist*
One who can use both hands with equal case	*ambidexterous*
One who fishes with a rod	*angler*
One who kills secretly or by surprise	*assassin*
A person who collects things belonging to ancient times	*antiquary*
One who is always finding faults	*censorious*
One living at the same time as another	*contemporary*
One who sneers at the aims and beliefs of his fellowmen	*cynic*
One who delights in speaking about oneself	*egotist*
One who exalts his own opinion	*egoist*
One who dies for a noble cause	*martyr*
One who retires from society to live a solitary life	*recluse, hermit*
One who maliciously sets fire to buildings	*incendiary*
One who is banished from his home or his country	*exile*
One who takes refuge in a foreign country	*refugee, alien*
One who runs away from justice or the law	*fugitive*
One who walks in his sleep	*somnambulist*
One who looks on the bright side of things	*optimist*
One who looks on the dark side of things	*pessimist*
A hater of mankind	*misanthropist*
One who knows everything	*omniscient*
One who is all powerful	*omnipotent*
One who is present everywhere	*omnipresent*
One who devotes his service or wealth for the love of mankind	*philanthropist*
One new to anything	*novice, tyro, neophyte*
One who engages in any pursuit for the love of it, and not for gain	*amateur*
One who journeys to a holy place	*pilgrim*
A leader of the people who can sway his followers by his oratory	*demagogue*
One who has special skill in judging art, music, tastes, etc.	*connoisseur*
One whose reasoning is clever yet false	*sophist*
One who makes a display of his learning	*pedant*
One who is indifferent to pain or pleasure	*stoic*
One who loves his country and serves it devotedly	*patriot*
One devoted to the pleasures of eating and drinking	*epicure*
One who poses to be what he is not	*hypocrite, impostor*
One who foretells events	*prophet*
One who pretends to know a great deal about everything	*mountebank, charlatan, quack*
One versed in many languages	*linguist*
One who cannot pay one's debts	*insolvent*
One who takes over after another in office or employment	*successor*
One who has been before another in office or employment	*predecessor*
One who is opposed to intellectual progress	*obscurant*

One who abstains from alcoholic drinks	*teetotaller*
One who hides away on a ship to obtain a free passage	*stowaway*
One who spends very little	*miser*
One who spends too much	*spendthrift*

DENOTING NUMBERS

A collection of poems	*anthology*
A number of merchant ships protected by warships (in war-time)	*convoy*
A number of stars grouped together	*constellation*
A number of hired applauders, *i.e.* persons paid to clap	*claque*
A number of people at church	*congregation*
A number of people gathered together for some common purpose	*gathering, assembly, society*
A group of people who get together to work for some cause of common interest	*coterie*
A number of workmen, prisoners, thieves	*gang*
A number of sheep	*flock*
A number of geese	*gaggle*
A number of leopards	*leap*
A number of lions, monkeys	*pride, troop*
A number of herrings, mackerel	*shoal*
A number of fish taken in a net	*catch, haul*
A number of whales, porpoises	*school, gam*
A number of oxen or horses (two or more) harnessed together	*team*
A number of ships	*fleet*

DENOTING PLACES

A place where bees are kept	*apiary*
A place where birds are kept	*aviary*
A place where fishes are kept	*aquarium*
A dwelling-place of an animal underground	*burrow*
A squirrel's home	*drey*
A nest of a bird of prey	*eyrie, aerie*
A place where spirituous liquors are produced	*distillery*
A place where clothes are washed and ironed	*laundry*
A place where Government records are kept	*archives*
A place where treasures of art, curiosities, etc. are preserved or exhibited	*museum*
A place where fruit trees are grown	*orchard*

MISCELLANEOUS

Loud enough to be heard	*audible*
Not distinct enough to be heard	*inaudible*
Fit for food	*edible*
Unfit for human consumption	*inedible*
Fit to be chosen or selected	*eligible*
Not having the qualities for being chosen	*ineligible*
(Writing) that is easy to read	*legible*
Writing that is difficult to decipher	*illegible*

Able to read	*literate*
Unable to read	*illiterate*
Born of married parents	*legitimate*
Born of unmarried parents	*illegitimate*
To send back a person to his own country	*repatriate*
To banish from one's country	*expatriate*
To move from one country to another	*migrate*
One who leaves his country to settle in another	*emigrant*
One who comes into a foreign country to settle there	*immigrant*
Incapable of being redeemed from evil, *i.e.* beyond correction	*incorrigible*
That which cannot be rubbed out or blotted out	*ineffaceable, indelible*
That which cannot be conquered	*invincible*
Incapable of making errors	*infallible*
That which cannot be avoided or prevented	*inevitable*
Incapable of being burnt	*incombustible*
That which easily catches fire	*inflammable, flammable*
That which cannot be seen	*invisible*
Living for ever	*immortal*
Increase the gravity of an offence	*aggravate*
Ordinary or commonplace remark	*platitude*
That which cannot be satisfied	*insatiable*
That which cannot be repaired	*irreparable*
That which cannot be imitated	*inimitable*
Persons (or efforts) that cannot be wearied	*indefatigable*
One who eats too much	*glutton*
To destroy completely	*annihilate*
A statement open to more than one interpretation	*ambiguous*
A round about way of speaking	*circumlocution*
Cautious observation of events, etc.	*circumspection*
That which cannot be hurt	*invulnerable*

MULTIPLE CHOICE QUESTIONS

Directions: *Below are given some statements each of which is capable of being expressed by one word only. Each statement is followed by four such words of which only one is correct. Out of the given alternatives A, B, C and D choose the correct word which very closely fits each definition.*

1. Pour out secrets:
   A. disclose B. unbosom
   C. uncouth D. defiled
2. Too deep to reach the bottom of:
   A. tedious B. unapproachable
   C. unfathomable D. discursive
3. One who shows to people their seats in a theatre:
   A. liasion B. guide
   C. tutlage D. usher
4. Person whose business is lending money for profit:
   A. utilitarian B. usurer
   C. usurp D. jew
5. One who tries to plan a perfect society:
   A. upholsterer B. stoic
   C. idealist D. utopian
6. Influenced by or done for bribery:
   A. venal B. unsavoury
   C. vertigo D. vile
7. Reversal of one's attitude towards something:
   A. voodoo B. vicissitude
   C. volte-face D. willy-nilly
8. Person who becomes successful very quickly:
   A. vainglorious B. whiz-kid
   C. wind-fall D. jackpot

9. Person anxious for or trying to provoke war:
A. warhead B. wary
C. warmonger D. extremist

10. Young and ambitious professional-person, working in a city:
A. yokel B. yankee
C. zippy D. yuppie

11. A set of papers giving information about a person or event:
A. editorial B. agenda
C. dossier D. clippings

12. List and explanations of technical, difficult, etc. terms:
A. preface B. glossary
C. bibliography D. appendix

13. One who pays too much care to his clothes and personal appearance:
A. conscious B. dandy
C. cynosure D. leer

14. A child born after death of father:
A. effeminate B. immature
C. fatherless D. posthumous

15. A person who is hard to please:
A. insensible B. hypocrite
C. egoist D. fastidious

16. Complete failure in something attempted:
A. fiasco B. vain
C. falsies D. flop

17. A word which can be interpreted in any way:
A. confusing B. misnomer
C. ambiguous D. precise

18. A group of persons united by common interests:
A. society B. class
C. clique D. coterie

19. A person who is inexperienced and easily deceived:
A. fool B. immature
C. blackhorn D. greenhorn

20. Instruction by question and answer:
A. interrogatory B. catechism
C. bulletin D. alibi

21. Record of events by someone with first hand knowledge:
A. novice B. report
C. memoir D. itinerary

22. A mode of talk familiar to a particular group or profession:
A. jargon B. coloquial
C. slang D. abuses

23. Person who is crazy:
A. lackadaisical B. tartar
C. scrupulous D. loony

24. A person who readily believes others:
A. hasty B. immature
C. credulous D. credible

25. One who eats human flesh:
A. beast B. cannibal
C. maneater D. carnivorous

26. Intentional destruction of racial groups:
A. genocide B. homicide
C. fratricide D. communal riot

27. One with weakness of mind caused by old age:
A. dastard B. dotard
C. fop D. philanderer

28. Study of insects:
A. zoology B. philology
C. entomology D. etymology

29. A critical judge of any art particularly fine arts:
A. connoisseur B. veteran
C. philistine D. dilettante

30. One who introduces items of an entertainment:
A. consort B. compere
C. director D. introducer

31. A combination of business firms:
A. enterprise B. franchise
C. cartel D. autonomous

32. Freeing of suppressed emotions:
A. catastrophe B. catharsis
C. epilogue D. chargin

33. One who believes in the power of fate:
A. optimist B. pessimist
C. fearful D. fatalist

34. The murder of brother:
A. genocide B. homicide
C. salvage D. fratricide

35. Additional to proper payment:
A. premium B. overtime
C. ex-gratia D. fringe

36. A person or thing used to tempt somebody into a position of danger:
A. decoy B. provocateur
C. jinx D. scapegoat

37. Custom of having more than one husband at the same time:
A. polygamy B. polyandry
C. bigamy D. sexagenary

38. A person of refined taste in food and drink:
A. emeritus B. epicure
C. glutton D. gourmand

39. A tombstone inscription:
A. epithet B. encomium
C. epistle D. epitaph

40. An office which has no work but high salary:
A. fortuitous B. gratuitous
C. sinecure D. ceremonial

41. One who is indifferent to pains and pleasure of life:
A. stoic B. metaphysical
C. pessimist D. sceptic

42. List of names of persons showing duties to be performed by each in turn:
A. notice B. bulletin
C. roster D. agenda

43. Long angry speech of criticism or accusation:
A. track record B. tirade
C. taut D. outburst

44. One who is in favour of never drinking alcoholic drinks:
A. idealist B. uxorious
C. teetotaller D. unalcoholic

45. Trying to win favour by flattery and charm:
A. smarmy B. shrewish
C. toady D. loquacious

46. A decision on which one cannot go back:
A. incorrigible B. irrevocable
C. rigid D. invulnerable

47. The science that studies the varieties of human race:
A. phrenology B. physiology
C. sociology D. ethnology

48. A person who hates women:
A. philogynist B. philatelist
C. misogamist D. misogynist

49. A group of people in a particular grade in an organisation:
A. colleague B. staff
C. echelon D. commission

50. Honourably discharged from service:
A. emeritus B. latitudinarian
C. infallible D. meritorious

51. Incapable of being wounded:
A. inevitable B. invulnerable
C. indomitable D. indefatigable

52. A short description of a picturesque scene or event of rural life:
A. narration B. panorama
C. idyll D. passage

53. The act of violating the sanctity of church:
A. sacrilege B. dissenter
C. immorality D. desecration

54. A dishonest dealing:
A. hanky-panky B. hocus-pocus
C. hoity-toity D. hilly-billy

55. A short and stout person:
A. lilliput B. hussy
C. humpty-dumpty D. paragon

56. Word that reads the same backwards as forwards (like did, dad, noon):
A. paradox B. palindrome
C. kowtow D. ragtag

57. Walking in sleep:
A. delusion B. somniloquism
C. somnabulism D. lethargic

58. Glamour and excitement:
A. razzle-dazzle B. razzmatazz
C. kudos D. winsome

59. Violent and bad tempered woman:
A. virago B. moll
C. concubine D. profane

60. One who pays excessive attention to preserving his health:
A. pagan B. bellicose
C. sycophant D. fop

61. Talking to one's own self:
A. pessimism B. introvert
C. soliloquy D. brooding

62. One who pretends to have more knowledge than he really has:
A. hypocrite B. charlaton
C. paragon D. lackadaisical

63. Brief biography of a person:
A. profile B. bibliography
C. synopsis D. resume

64. Very small allowance or wage:
A. lumpsum B. consolidated
C. pittance D. paltry

65. Purely spiritual love:
A. devotion B. platonic
C. passionate D. divine

66. Place where people often meet:
A. common room B. discotheque
C. roster D. rendezvous

67. A completely perfect person:
A. upright B. straightforward
C. paragon D. intellectual

68. Commencement of words with the same alphabet:
A. symmetry B. pun
C. phonetic D. alliteration

69. A person who loves everybody:
A. ascetic B. philander
C. philistine D. cosmopolitan

70. A government where the powers are concentrated in the hands of one person:
A. dictatorship B. autocracy
C. plutocracy D. anarchy

71. That which is outdated or out of use or fashion:
A. discarded B. obsolete
C. classic D. antic

72. Calendar giving information about sun, moon, stars and sometimes also the future:
A. almanac B. appendix
C. horoscope D. chronology

73. Collection in one volume of works by various authors:
A. panorama B. anthology
C. autobiography D. bibliography

74. One who takes part in a study, game, etc. for the love of it and not for professional gain:
A. apologist B. august
C. amateur D. avant-garde

75. A thing that causes annoyance:
A. bugbear B. arduous
C. anathema D. shrill

76. An allowance for support made to a wife by her husband after their legal separation:
A. alimony B. compensation
C. settlement D. ex-gratia

77. Place for hiding treasure, food, weapons, etc.:
A. store B. cache
C. stock D. armoury

78. A person who is stubborn in resisting change:
A. irresistible B. die-hard
C. dead beat D. terse

79. Able to use the left hand or the right hand equally well:
A. ambidextrous B. epicurean
C. apostle D. arbiter

80. A man who rarely speaks the truth:
A. liar B. unreliable
C. hypocrite D. bellicose

81. Partial or total loss of memory:
A. alexia B. anorexia
C. anexia D. amnesia

82. A word made by rearranging the letters of another word like eat-tea:
A. acronym B. anagram
C. byword D. dossier

83. A mass of snow and ice sliding down from a mountain:
A. avalanche B. glacier
C. iceberg D. azure

84. To speak irreverently of god:
A. impious B. irreligious
C. immoral D. blasphemy

85. Sudden increase in business:
A. bonanza B. boom
C. bounce D. booze

86. Object of laughter or ridicule:
A. butt B. but
C. divert D. hurtle

87. A pioneer of a reform movement:
A. sinecure B. sycophant
C. gallant D. apostle

88. Art of handwriting:
A. cartography B. choreography
C. calligraphy D. graph

89. One who takes a low view of human nature:
A. coquette B. cynic
C. crony D. coy

90. Document that is handwritten by the author:
A. autograph B. holograph
C. editorial D. preface

91. A person who forsakes religion:
A. irreligious B. non-religious
C. falalist D. apostle

92. Speaker or agitator who can stir the emotions of the masses:
A. demagogue B. dexterous
C. diligent D. reformist

93. Uncontrollable craving for alcohol:
A. dyspepsia B. delirium
C. edict D. dipsomania

94. A lady who remains unmarried:
A. concubine B. alien
C. spinster D. damsel

95. State payment to an unemployed person:
A. pension B. dole
C. increment D. grant in aid

96. Person slow at learning:
A. adroit B. dubious
C. dunce D. agile

97. One who has a good taste for food and enjoys it:
A. gourmet B. sensitive
C. deleterious D. debonair

98. One who is fond of gossip:
A. flippant B. flibbertigibbet
C. fuddy-duddy D. gregarious

99. A long, loud and angry speech:
A. incongruous B. hyperbole
C. hullabaloo D. harangrue

100. A person chosen by the opposing parties to settle their differences:
A. mediator B. prude
C. paramour D. arbitrator

101. Study of population statistics:
A. topography B. demography
C. entomology D. dermatology

102. One who explains a subject:
A. preacher B. theologist
C. exponent D. exquisite

103. Person living outside his country:
A. expatriate B. exodus
C. statesman D. emigrant

104. Looking or sounding strange:
A. outlandish B. unique
C. peculiar D. paradox

105. Thing that excites a person's rage:
A. risque B. red rag
C. excruciating D. feat

106. Short story or poem conveying a moral:
A. sonnet B. prelude
C. fable D. farce

107. A person filled with unreasonably great enthusiasm for something:
A. fanciful B. emotional
C. fanatic D. factitious

108. One who hinders the passage of a law in a parliament by making a very long speech:
A. verbose B. garrulous
C. freebooter D. filibuster

109. Temperature at which a substance will ignite:
A. break point B. flashpoint
C. flamboyant D. saturation point

110. A speech delivered without any previous preparation:
A. abrupt B. spontaneous
C. maiden D. extempore

111. A person with full discretionary powers to act on behalf of a country:
A. ambassador B. plenipotentiary
C. diplomat D. emirate

112. State of continuous movement:
A. frenzied B. vibration
C. inertia D. flux

113. To utter senseless or rapid unintelligible sounds:
A. stammer B. yell
C. gibber D. glib

114. To grind the teeth in rage:
A. gnash B. gnaw
C. gobble D. goggle

115. Freedom from care:
A. unburden B. unconcerned
C. insouciance D. nonchalance

116. To turn out of society or fellowship:
A. rusticate B. expel
C. ostracize D. extradite

117. Too great to be described in words:
A. implicit B. ineffable
C. hyperbole D. metaphor

118. At the point of death:
A. end B. climax
C. moribund D. extreme

119. View of a wide area:
A. background B. landscape
C. pagan D. panorama

120. Private conversation between two persons:
A. tete-a-tete B. tip-top
C. razzle-dazzle D. rat-tat

121. Wealthy and powerful businessman:
A. tycoon B. wizard
C. versatile D. voracious

122. Easy to mould:
A. flexible B. tractable
C. malleable D. detachable

123. A new word coined by an author:
A. inculcation B. neologism
C. syllogism D. neolithic

124. The giving of special favour by a person in high position to his relatives:
A. recommendation B. partiality
C. honorarium D. nepotism

125. Things that exist only in imagination:
A. nonpareil B. nonentity
C. nonsensical D. nonsequitur

126. Conclusion which does not follow from the premises:
A. fallacious B. irrelevant
C. circuitous D. nonsequitur

127. Longing for something one has known in the past:
A. craving B. nostalgia
C. yearn D. kleptomania

128. One who is breaker of idols:
A. jailbird B. prude
C. iconoclast D. tenacious

129. Highflown speech or style of writing:
A. grandiloquence B. quintessence
C. harangue D. epigram

130. Parliamentary measure to curtail time in debating:
A. quorum B. guillotine
C. ordinance D. habeas-corpus

131. Philosophic belief that pleasure is the highest good:
A. plagiarism B. egalitarianism
C. aestheticism D. hedonism

132. In haste and confusion:
A. helter-skelter B. hullabaloo
C. harum-scarum D. hectic

133. Doctrine contrary to the teachings of one of the recognised religions:
A. heresy B. diction
C. soliloquy D. fanatic

134. A formal written charge against a person for some crime or offence:
A. chargesheet B. indictment
C. warranty D. conviction

135. System in which each person belongs to a specific rank:
A. ordinal B. allocation
C. hierarchy D. organization

136. Land lying behind the coastline or behind a sea port:
A. island B. hamlet
C. hunterland D. hinterland

137. Having a rough, husky voice:
A. shrill B. yell
C. hoarse D. horrid

138. Difficult or impossible to read:
A. vague B. obsolete
C. obscene D. illegible

139. Noisy and energetic activity of many people together:
A. din and bustle B. humdrum
C. hugger-mugger D. hurly-burly

140. Great destruction of life:
A. holocaust B. hurricane
C. devastation D. debris

141. Captive whose safety is used as a bargaining point:
A. warranter B. mortgage
C. hostage D. humdinger

142. State in which one believes mistakenly that one is ill:
A. hysteria B. hypochondria
C. mania D. kleptomania

143. Moving from place to place:
A. peripatetic B. vagabond
C. indecisive D. patrolling

144. External boundary or surface:
A. adjacent B. periphery
C. complementary D. circumscribe

145. Person who knowingly make a false statement after taking an oath to tell the truth:
A. lackadaisical B. camouflage
C. intrigue D. perjurer

146. Act of reading carefully:
A. perusal B. petulance
C. review D. pun

147. Study of fishes:
A. aquaculture B. ichthyology
C. numismatics D. holograph

148. Court instruction which can be enforced by law:
A. rulings B. judgement
C. injunction D. mandatory

149. Guess which is based on instinctive knowledge:
A. telepathy B. innate
C. inherent D. intuition

150. One who is excessive in his expressions of patriotism:
A. jubilant B. jingoist
C. dainty D. compatriot

151. A money liberally given:
A. honorarium B. bonus
C. largesse D. laurels

152. Statement damaging to a person's reputation:
A. litigation B. encroachment
C. libel D. infringement

153. Commonplace remarks:
A. platitudes B. lampoon
C. meiosis D. magniloquent

154. To wind up the affairs of a company:
A. levitate B. liquidate
C. manoeuvre D. orientate

155. Passion for appearing important:
A. magalomania B. monomania
C. narcissism D. nymphomania

156. Person authorised to draw up and attest deeds:
A. attorney B. notary
C. incharge D. proliferate

157. Object of pure thought:
A. niche B. notion
C. onus D. noumenon

158. Something that can be taken for granted:
A. trivial B. postulate
C. prodigal D. paltry

159. Scene of din and uproar:
A. hurly-burly B. hugger-mugger
C. pandemonium D. pacificism

160. One who fully depends on others:
A. protocol B. parasite
C. scrounge D. sceptic

161. Making a show of piety:
A. savage B. sanctimonious
C. salacious D. licentious

162. A fort where an enemy cannot enter:
A. impregnable B. impenetrable
C. inescapable D. indiscernible

163. Withdraw from an organization to form a new independent body or state:
A. autonomous B. secede
C. secular D. sovereign

164. That which cannot be understood or known:
A. indescriptable B. inaccessible
C. insensible D. inscrutable

165. In a careless and inefficient way:
A. shoddy B. slovenly
C. uninterestingly D. seriatim

166. Paying great attention to details:
A. concerned B. sceptic
C. scrupulous D. self-righteous

167. A person who cannot be easily approached:
A. introvert B. indifferent
C. inaccessible D. impractical

168. Planning of evil acts:
A. skullduggery B. sensational
C. deception D. introspection

169. Content with one's way of life:
A. improvident B. smug
C. self-centered D. sophist

170. Doing harm secretly:
A. heinous B. incapacitate
C. incineration D. insidious

171. Person setting fire to property unlawfully and with an evil purpose:
A. incendiary B. incense
C. indemnity D. indent

172. Person who flatters another in the hope of gain or advantage:
A. selfish B. misdemeanor
C. scurrilous D. toady

173. Person who changes from one party or side to another:
A. opportunist B. turncoat
C. turncourt D. selfish

174. One who apes the upper classes:
A. proud B. sophisticated
C. snob D. malicious

175. Temporary stay or residence:
A. adjourn B. sojourn
C. deputation D. unstable

176. Legal order requiring attendance at a court:
A. summons B. writ
C. obligatory D. rulings

177. Dulling of the senses caused by alcohol or fatigue:
A. stupor B. numb
C. symphilis D. obliteration

178. Art of politics:
A. trickery B. statecraft
C. dexterity D. polity

179. Less than the speed of sound:
A. sub-sonic B. subsist
C. substratum D. subsume

180. Evasive play in argument or discussion:
A. subtlety B. subvert
C. subterfuge D. succinct

181. Be of help in distress:
A. cooperative B. generous
C. suppliant D. succour

182. Make agree in time:
A. suffice B. synchronise
C. complacence D. conciliation

183. Philosophical or other discussion among several speakers:
A. symposium B. discourse
C. discernment D. suffrage

184. Strongly and effectively expressed:
A. verbose B. trenchant
C. terse D. reasonable

185. Violent outburst of temper:
A. hypertension B. tantrum
C. tenor D. shrill

186. One who reveals secrets:
A. propellant B. conjurer
C. telltale D. manipulator

187. Extremely violent wind storm:
A. tempest B. tornado
C. torrent D. turmoil

188. Person who does something new or original:
A. novice B. neolithic
C. top-notch D. trail-blazer

189. Person who deliberately avoids his duties:
A. truant B. escapist
C. laggard D. wag

190. Remaining cool in a crisis:
A. unflappable B. unflagging
C. innocuous D. unaffected

191. One who does not believe in the existence of God:
A. atheist B. altruist
C. agnostic D. flippant

192. The large money paid for an early retirement:
A. gold mine B. golden handshake
C. gratuity D. lion's share

193. A hater of learning and knowledge:
A. misologist B. misogynist
C. philologist D. philanthropist

194. One who lives and acts for the welfare of others:
A. misanthrope B. credulous
C. altruist D. gregarious

195. Mental weariness for want of occupation:
A. desperate B. depression
C. fatigue D. ennui

196. That which makes it difficult to recognise the presence of real of somebody or something:
A. cover B. mask
C. make-up D. camouflage

197. To bear a hand in one's new business:
A. partnership B. interfere
C. assist D. join

198. One who renounces the world and practices self-discipline in order to attain salvation:
A. devotee B. ascetic
C. theist D. parsimonious

199. A place where gambling is practiced:
A. hotel B. motel
C. casino D. stadium

200. A sad song is a:
A. dirge B. elegy
C. ditty D. knell

ANSWERS

1	2	3	4	5	6	7	8	9	10
B	C	D	B	D	A	C	B	C	D
11	12	13	14	15	16	17	18	19	20
C	B	B	D	D	A	C	C	D	B
21	22	23	24	25	26	27	28	29	30
C	A	D	C	B	A	B	C	A	B
31	32	33	34	35	36	37	38	39	40
C	B	D	D	D	A	B	B	D	C
41	42	43	44	45	46	47	48	49	50
A	C	B	C	A	B	D	D	C	A
51	52	53	54	55	56	57	58	59	60
B	C	D	A	C	B	C	B	A	A
61	62	63	64	65	66	67	68	69	70
C	B	A	C	B	D	C	D	D	B
71	72	73	74	75	76	77	78	79	80
B	A	B	C	A	A	B	B	A	C
81	82	83	84	85	86	87	88	89	90
D	B	A	D	B	A	D	C	B	B
91	92	93	94	95	96	97	98	99	100
D	A	D	C	B	C	A	B	D	D
101	102	103	104	105	106	107	108	109	110
B	C	A	A	B	C	C	D	B	D
111	112	113	114	115	116	117	118	119	120
B	D	C	A	C	C	B	C	D	A
121	122	123	124	125	126	127	128	129	130
A	C	B	D	B	D	B	C	A	B
131	132	133	134	135	136	137	138	139	140
D	A	A	B	C	D	C	D	D	A
141	142	143	144	145	146	147	148	149	150
C	B	A	B	D	A	B	C	D	B
151	152	153	154	155	156	157	158	159	160
A	C	A	B	A	B	D	B	C	B
161	162	163	164	165	166	167	168	169	170
B	A	B	D	B	C	C	A	B	D
171	172	173	174	175	176	177	178	179	180
A	D	B	C	B	A	A	B	A	C
181	182	183	184	185	186	187	188	189	190
D	B	A	B	B	C	B	D	A	A
191	192	193	194	195	196	197	198	199	200
A	B	A	C	D	D	C	B	C	A

7

Kangaroo Words

A Kangaroo word is one which carries within itself a smaller word which is a synonym of the word itself. The letters of the synonym are found often in the same order as in the given word, but some letters are omitted.

The kangaroo carries its young in a pouch unlike other animals. Likewise, a kangaroo word carries its young in itself. For example, Encourage and Evacuate are kangaroo words because Urge and Vacate that appear in order are synonymous with Encourage and Evacuate respectively. The kangaroo word is also known as marsupial.

A list of kangaroo words is as follows:

Kangaroo	Child	Kangaroo	Child
accustomed	used	destruction	ruin
aeroplane	plane	devilish	evil
amend	mend	disappointed	sad
amicable	amiable	diversified	divers
appropriate	apt	domesticate	tame
appropriate	proper	earthquake	quake
architecture	art	especial	special
assured	sure	espy	spy
blossoms	blooms	esquire	squire
boundaries	bounds	exhilaration	elation
calumines	lies	exists	is
capable	able	fabrication	fiction
catacomb	tomb	facade	face
clue	cue	facetiousness	fun
container	can	falsities	lies
curtail	cut	fatigue	fag
deceased	dead	feasts	eats
deliberate	debate	festival	feast

Kangaroo	Child	Kangaroo	Child
hamburger	burger	prosecute	sue
helicopter	copter	rampage	rage
hostelry	hotel	rapscullion	rascal
human	man	recline	lie
hurried	hied	regulates	rules
hurries	hies	respite	rest
illuminated	lit	reverence	revere
indolent	idle	revolution	revolt
instantaneously	instantly	rotund	round
instructor	tutor	salvage	save
joviality	joy	satisfied	sated
lineage	line	separate	part
market	mart	signal	sign
masculine	male	skilfulness	skill
matches	mates	slithered	slid
meadow	mead	splotches	spots
nourished	nursed	spoiled	soiled
observe	see	supervisor	superior
omnibus	bus	surpeme	super
pantaloons	pants	telephone	phone
parambulate	ramble	utilise	use
perambulate	ambulate	variegated	varied
perimeter	rim	violincello	cello
periwig	wig	wealth	weal
pinioned	pinned	whimsies	whims
prattle	prate	whinge	whine
precipitation	rain		

MULTIPLE CHOICE QUESTIONS

Directions: *Select the kangaroo-child word that carries a smaller version of the given word with a similar meaning:*

1. AMICABLE
   A. Amicable B. Able
   C. Cable D. Amiable

2. APPROPRIATE
   A. Appropriate B. Private
   C. Priority D. Apt

3. SALVAGE
   A. Salvage B. Sale
   C. Wage D. Save

4. HOSTELRY
A. Hostelry B. Total
C. Motel D. Hotel

5. ENCOURAGE
A. Encourage B. Rage
C. Urge D. Verge

6. RECLINE
A. Recline B. Lie
C. Clinch D. Record

7. BEFORE
A. Before B. Bore
C. Ere D. For

8. INSTRUCTOR
A. Instructor B. Tractor
C. Cutter D. Tutor

9. SEPARATE
A. Separate B. Part
C. Spurt D. Rate

10. CATACOMB
A. Catacomb B. Cat
C. Comb D. Tomb

11. INDOLENT
A. Indolent B. Indent
C. Idle D. Dent

12. PRECIPITATION
A. Precipitation B. Rain
C. Recipe D. Citation

13. DELIBERATE
A. Deliberate B. Date
C. Debate D. Liberate

14. OBSERVE
A. Observe B. Serve
C. Look D. See

15. TRANSGRESSION
A. Transgression B. Sin
C. Trans D. Agression

16. UNSIGHTLY
A. Unsightly B. Sigh
C. Ugly D. Sight

17. DESTRUCTION
A. Destruction B. Damage
C. Destroy D. Ruin

18. MASCULINE
A. Masculine B. Muscle
C. Male D. Line

19. UMPTEENTH
A. Umpteenth B. Ten
C. Teen D. Nth

20. BECAUSE
A. Because B. As
C. Be D. Cause

21. ALONE
A. Alone B. All
C. One D. Lane

22. CONTAINER
A. Container B. Can
C. Cone D. Coin

23. FRANGIBLE
A. Frangible B. Feeble
C. Able D. Frail

24. FALSITIES
A. Falsities B. Fail
C. Lies D. Ties

25. PERAMBULATE
A. Perambulate B. Ram
C. Late D. Ramble

26. EXPURGATED
A. Expurgated B. Exude
C. Pure D. Exurb

ANSWERS

1	2	3	4	5	6	7	8	9	10
D	D	D	D	C	B	C	D	B	D
11	12	13	14	15	16	17	18	19	20
C	B	C	D	B	C	D	C	D	B
21	22	23	24	25	26				
C	B	D	C	D	C				

8

Odd Words Out

MULTIPLE CHOICE QUESTIONS

Directions : *In each of the following questions, four words/ pair of words are given. Three of them are alike in some manner. Spot the odd one out.*

1. A. Green B. Red C. Colour D. Orange
2. A. Stable B. Hole C. Canoe D. Sty
3. A. Nose B. Eyes C. Skin D. Teeth
4. A. Venus B. Moon C. Pluto D. Mars
5. A. Happy B. Gloomy C. Lively D. Cheerful
6. A. Cone B. Circle C. Triangle D. Rectangle
7. A. Lead B. Mercury C. Copper D. Iron
8. A. Kite B. Bird C. Radar D. Jet
9. A. Knee B. Shoulder C. Ankle D. Palm
10. A. Deluge B. Calamity C. Catastrophe D. War
11. A. Cub B. Chicken C. Pig D. Pup
12. A. Rabbit B. Crocodile C. Earthworm D. Snail
13. A. Tree B. Leaf C. Bush D. Herb
14. A. Doctor B. Teacher C. Engineer D. Diver
15. A. Trot B. Equestrian C. Derby D. Grunt
16. A. Ornate B. Pleasant C. Decorate D. Beautify
17. A. Polo B. Chess C. Ludo D. Squash
18. A. Tutor B. Principal C. Pupil D. Professor
19. A. Pond B. River C. Stream D. Brook
20. A. Quotation B. Duty C. Tax D. Octroi
21. A. Root B. Tree C. Branch D. Flower
22. A. Mumbai B. Chandigarh C. Lucknow D. Hyderabad
23. A. Immortal B. Eminence C. Perpetual D. Everlasting
24. A. Spinach B. Potato C. Carrot D. Ginger
25. A. Van B. Aeroplane C. Helicopter D. Transport
26. A. Fathom B. Marine C. Deep D. Lacuna
27. A. Attorney B. Lawyer C. Judge D. Liquidator
28. A. Sparrow B. Kingfisher C. Kiwi D. Parrot
29. A. Arrow B. Daggar C. Knife D. Sword
30. A. Mathematics B. Algebra C. Trigonometry D. Geometry
31. A. Irish B. Iranian C. Eastern D. Chinese
32. A. December B. June C. January D. March
33. A. Boxer B. Wrestler C. Jockey D. Player
34. A. Mature B. Outdo C. Ripen D. Bloom
35. A. Kanpur B. Haridwar C. Varanasi D. Lucknow
36. A. Adore B. Like C. Love D. Covet
37. A. Greedy B. Rapacious C. Endear D. Avaricious

38. A. Club B. Heart
C. Spade D. Ace

39. A. Permit B. Allow
C. Agree D. Confess

40. A. Stool B. Wood
C. Table D. Chair

41. A. Chair - Furniture B. Shirt - Garment
C. Necklace - Jewellery D. Bogie - Engine

42. A. Crayon - Paper B. Pencil - Lead
C. Pen - Ink D. Brush - Paint

43. A. War - Peace B. Real - Natural
C. Premiere - First D. Wrath - Anger

44. A. Finger - Thimble B. Head - Cap
C. Waist - Tiara D. Foot - Shoe

45. A. Day - Night B. Clever - Foolish
C. Clear - Blurred D. Arrive - Come

46. A. Quintal - Gallon B. Bouquet - Flowers
C. Book - Pages D. Parliament - Members

47. A. Birds - Chirp B. Horses - Hum
C. Lions - Roar D. Snakes - Hiss

48. A. Niece - Nephew B. Husband - Wife
C. Brother - Sister D. Father - Mother

49. A. Petrol - Car B. Oil - Lamp
C. Diesel - Wood D. Wax - Candle

50. A. Ganga - Narmada B. Thar - Gobi
C. Stomach - Hands D. Everest - Mountain

51. A. Medicine - Doctor B. Flower - Artist
C. Shoes - Cobbler D. Skin - Dermatologist

52. A. Authority - Sanction B. Repel - Attract
C. Finicky - Choosy D. Breath - Entity

53. A. Polo - Rink B. Golf - Lawn
C. Tennis - Court D. Chess - Board

54. A. Ecstacy - Fantasy B. Price - Thrice
C. Crime - Prune D. Both - Oath

55. A. Army - General B. College - Principal
C. Ship - Captain D. Navy - Lieutenant

56. A. Train - Tracks B. Birds - Fly
C. Aeroplane - Sky D. Submarine - Sea

57. A. Jelly - Gentle B. Stone - Hard
C. Fur - Soft D. Glass - Smooth

58. A. Branch - Tree B. Minute - Hour
C. Sentence - Paragraph D. Student - Teacher

59. A. Sum - Total B. Now - Present
C. Big - Notion D. Yes - Agree

60. A. Four - Foursome B. Three - Triplet
C. Two - Double D. One - Single

61. A. Brahmaputra - River B. Sahara - Desert
C. Aravalli - Ocean D. Europe - Continent

62. A. Much - Most B. Bad - Worse
C. Simple - Simpler D. Little - Less

63. A. Ripe - Crude B. Close - Near
C. Warm - Tender D. Believe - Trust

64. A. Giraffe - Tall B. Snail - Heavy
C. Jet - Fast D. Ant - Industrious

65. A. Nine - Twine B. Fire - Liar
C. Honey - Money D. Rust - Crest

ANSWERS

1	2	3	4	5	6	7	8	9	10
C	C	D	B	B	A	B	C	D	D
11	**12**	**13**	**14**	**15**	**16**	**17**	**18**	**19**	**20**
C	A	B	D	D	B	A	C	A	A
21	**22**	**23**	**24**	**25**	**26**	**27**	**28**	**29**	**30**
A	B	B	A	D	D	D	C	A	A
31	**32**	**33**	**34**	**35**	**36**	**37**	**38**	**39**	**40**
C	B	D	B	D	D	C	D	D	B
41	**42**	**43**	**44**	**45**	**46**	**47**	**48**	**49**	**50**
D	A	A	C	D	A	B	A	C	D
51	**52**	**53**	**54**	**55**	**56**	**57**	**58**	**59**	**60**
B	B	A	C	D	B	A	D	C	B
61	**62**	**63**	**64**	**65**					
C	A	A	B	D					

9

Word Analogy

Like different people words also have different relationships in comparison to each other. This is called word analogy. Try and master it in the following questions.

MULTIPLE CHOICE QUESTIONS

Directions (Qs. 1 to 100): *Each question consists of two words which have a certain relationship to each other followed by four pairs of related words. Select the pair which has the same relationship.*

1. Lunatic : Asylum

A. The poor : palace
B. Soldiers : barracks
C. Flowers : fields
D. Crops : garden

2. Soldier : Rifle

A. Blacksmith : anvil
B. Thief : ladder
C. Watchman : uniform
D. Barber : hair

3. Frog : Tadpoles

A. Man : sons
B. Woman : daughters
C. Cat : kids
D. Elephant : calves

4. Lizards : Reptiles

A. Frogs : amphibians
B. Tree : fruit
C. Earth : moon
D. Man : woman

5. Earthworm : Segments

A. Tree : wood
B. Crocodiles : scales
C. Apple : juice
D. Earth : moisture

6. Mosquito : Malaria

A. Rabid dog : rabies
B. House fly : AIDS
C. Cockroach : influenza
D. Virus : diabetes

7. Fish : Water

A. Man : air
B. Cat : mouse
C. Lion : flesh
D. Tree : roots

8. Impulsive : Cautious

A. Liberty : bondage
B. Tyrant : bully
C. Tremble : quake
D. Vulgar : indecent

9. Weighty : Voluminous

A. Jolly : dismal
B. Loyalty : perfidy
C. Wealth : opulence
D. Odd : normal

10. Innocent : Guilty

A. Kill : produce
B. Suspect : doubt
C. Depress : discourage
D. Contract : condense

11. Useless : Cheap

A. Late : unpunctual
B. Real : illusive
C. Just : unfair
D. Humble : proud

12. Loyal : Faithful

A. Treacherous : devoted
B. Just : fair
C. Yield : suppress
D. Uncertain : sure

13. Cancel : Confirm

A. Certain : ambiguous
B. Vulgar : crude
C. Reliable : trustworthy
D. Obscure : anonymous

14. Gloomy : Bright
A. Pleasant : delightful
B. Funny : comic
C. Condemn : blame
D. Hard : flexible

15. Modern : Ancient
A. Happy : joyful
B. Service : slavery
C. Mild : harsh
D. assumed : unreal

16. Dexterous : Clumsy
A. Cheap : worthless
B. Usual : common
C. Zenith : acme
D. True : false

17. Luscious : Sweet
A. Above : below
B. Absolve : penalise
C. Lucrative : profitable
D. Adopt : abandon

18. Fraud : Duplicity
A. Inferior : cheap
B. Delicious : abhorrent
C. Desire : detest
D. Diligent : lackadiasical

19. Weired : Ordinary
A. Vacant : void
B. Vivid : clear
C. Vivid : dim
D. Wicked : unvirtuous

20. Build : Demolish
A. Abide : persist
B. Compel : coax
C. Odious : hateful
D. Claim : demand

21. Passionate : Impassive
A. Deter : discourage
B. Folly : stupidity
C. Polite : rude
D. Fanciful : whimsical

22. Gauche : Attractive
A. True : exact
B. Tough : hard
C. Tasteful : palatable
D. Insipid : delicious

23. Existence : Life
A. Silence : noise
B. Risk : shield
C. Enormous : immense
D. Rural : urban

24. Belief : Suspicion
A. Unravel : unweave
B. Profuse : insufficient
C. Urge : spur
D. Use : occupy

25. Violent : Calm
A. Tranquil : peaceful
B. Reticent : quiet
C. Even : smooth
D. Disdain : reverence

26. Lawyer : Court
A. businessman : market
B. chemist : laboratory
C. labourer : factory
D. athlete : olympics

27. Corporeal : Spirtual
A. mesa : plateau
B. moron : savant
C. foreigner : immigrant
D. pedagogue : teacher

28. Ten : Decimal
A. seven : septet
B. four : quartet
C. two : binary
D. five : quince

29. Grain : Salt
A. shard : pottery
B. shred : wood
C. blades : grass
D. chip : glass

30. Fish : Shoal
A. audience : theatre
B. elephant : flock
C. whale : herd
D. shark : school

31. Wan : Colour
A. copulent : weight
B. insipid : flavour
C. pallid : complexion
D. enigmatic : puzzle

32. Thrust : Spear
A. mangle : iron
B. scabbard : sword
C. bow : arrow
D. fence : epee

33. Indigent : Wealthy
A. angry : rich
B. native : affluent
C. gauche : graceful
D. scholarly : erudite

34. Diva : Opera
A. producer : theatre
B. director : drama
C. conductor : bus
D. thespian : play

35. Army : Logistics
A. business : strategy
B. soldiers : students
C. war : logic
D. team : individual

36. Distance : Mile
A. liquid : litre
B. bushel : corn
C. weight : scale
D. fame : television

37. After : Before
A. first : second
B. present : past
C. contemporary : historic
D. successor : predecessor

38. Mundane : Spiritual
A. common : ghostly
B. worldly : unworldly
C. routine : novel
D. secular : clerical

39. Filter : Water
A. curtail : activity
B. expurge : book
C. edit : text
D. censor : play

40. Pork : Pig
A. rooster : chicken
B. mutton : sheep
C. steer : beef
D. lobster : crustacean

41. Pain : Sedative
A. comfort : stimulant
B. grief : consolation
C. trance : narcotic
D. ache : extraction

42. Hope : Aspires
A. love : elevates
B. film : flam
C. fib : lie
D. fake : ordinary

43. Gravity : Pull
A. iron : metal
B. north pole : directions
C. magnetism : attraction
D. dust : desert

44. Light : Blind
A. speech : dumb
B. language : deaf
C. tongue : sound
D. voice : vibration

45. Symphony : Composer
A. Leonardo : music
B. Fresco : painter
C. colours : pallet
D. art : appreciation

46. Astringent : Pucker
A. car : engine
B. spark : ignition
C. quell : allay
D. sophisticated : rural

47. Saga : Heroic
A. epigram : pithy
B. tragedy : comedy
C. editorial : newspaper
D. picture : book

48. Obdurate : Flexibility
A. stagger : amaze
B. adamant : submissiveness
C. semite : jew
D. redicule : scorn

49. Percept : Instructive
A. ribald : decent
B. metaphor : figurative
C. institution : study
D. library : calmness

50. Hermit : Gregarious
A. sanguine : optimistic
B. repose : rest
C. ruthless : merciless
D. ascetic : hedonostic

51. Braggart : Modesty
A. profound : intelligence
B. opinionated : over confident
C. buffoon : dignity
D. arrogant : acuity

52. Posthumous : Death
A. selection : nomination
B. product : factory
C. clay : mould
D. cloudy : sunny

53. Prophet : Foreteller
A. precursor : forerunner
B. prtentious : ambitious
C. prodigy : wonderful
D. teacher : intelleclual

54. Indigent : Wealth
A. yelp : cry
B. yearn : disire
C. wrangle : argue
D. emociated : nourishment

55. Senorita : Woman
A. sophisticated : self assured
B. monologue : discourse
C. semite : jew
D. general : military

56. Yeoman : Farming
A. clerk : office
B. waiter : restaurant
C. hawker : business
D. apprentice : engineer

57. Genuine : Authentic
A. light : reflection
B. image : shadow
C. mirage : illusion
D. vaccum : transparency

58. Colour : Bleach
A. buzz : hum
B. echo : sound
C. paper : burn
D. blush : faint

59. Labour : Wages
A. capital : interest
B. doctor : medicine
C. ink : writing
D. agree : oppose

60. Brusque : Unceremoniousness
A. apparatus : experiment
B. auditor : lecture
C. malaria : relapse
D. obstinate : intractability

61. Mature : Regressed
A. intruder : privacy
B. varied : monotonus
C. sufficiency : adequacy
D. train : compartment

62. Celerity : Snail
A. curosity : cat
B. cunning : crow
C. humility : peacock
D. obstinacy : mule

63. Dilatory : Expeditious
A. direct : circuitous
B. skirmish : mar
C. language : grammar
D. ice : coolness

64. Molt : Feathers
A. flutter : wings
B. slough : skin
C. sharpen : talons
D. bare : fangs

65. Jeopardy : Peril
A. plead : insist
B. promote : reward
C. rely : support
D. jealousy : envy

66. Surfeit : Appetite
A. refine : taste
B. sniff : nose
C. cloy : palate
D. cling : touch

67. Aeroplane : Cockpit
A. train : engine
B. train : compartment
C. train : coach
D. train : wagon

68. Rider : Bill
A. meter : electricity
B. endorsement : policy
C. purchase : receipt
D. violation : law

69. Wax : Wane
A. earth : orbit
B. depth : shallow
C. fall : bottom
D. zenith : nadir

70. Painting : Artist
A. wood : furniture
B. analysis : critic
C. symphone : composer
D. symphone : poet

71. Windfall : Jubilation
A. grief : sorrow
B. pratfall : embarrassment
C. bravery : courageous
D. calm : melancholy

72. Foresight : Anticipation
A. insomnia : sleeplessness
B. intuition : happening
C. inddent : incredible
D. impromptu : preparation

73. Novice : Learner
A. voters : politicians
B. gigolo : dance
C. harbinger : forerunner
D. roster : duty

74. Life : Autobiography
A. witness : truth
B. witness : documents
C. culprit : punishment
D. jailor : prison

75. Grieve : Sorrow
A. wallow : misery
B. smart : pain
C. weaken : strength
D. afflict : torment

76. Cloud : Scud
A. fog : disperse
B. blood : clot
C. mist : fall
D. water : race

77. Fire : Ashes
A. heart : breath
B. school : education
C. exphesion : debris
D. almamater : certificate

78. Ocean : Water
A. mountain : peak
B. cave : hollow
C. glacier : ice
D. river : current

79. Unctuous : Sincerity
A. urban : society
B. pleasant : jovial
C. wrest : amusing
D. sardonic : sympathy

80. Audible : Hear
A. fascinating : see
B. palpable : touch
C. distinctive : smell
D. unique : personality

81. Article : Newspaper
A. scene : play
B. food : diet
C. account : diary
D. maize : field

82. Beginning : Culmination
A. tragedy : catastrophe
B. language : grammar
C. stanza : poetry
D. puologue : denouncement

83. Ballerina : Ballet
A. stage : performance
B. protagonist : play
C. grief : tragedy
D. ballad : poem

84. Bellicose : Quarrelsome
A. boost : boom
B. bonny : unhealthy
C. babel : babble
D. bane : ruin

85. Annihilation : Fire
A. ordinance : parliament
B. emergency : failure
C. cataclysm : flood
D. cataclysm : earthquake

86. Shovel : Lift
A. break : pieces
B. amend : join
C. thrust : dig
D. push : fall

87. Jade : Green
A. garnet : red
B. rainbow : sky
C. mercury : shine
D. white : peace

88. Gamut : Completeness
A. square : perimeter
B. gas : volume
C. conclusion : end
D. radius : circle

89. Truth : Sham
A. wariness : caution
B. honesty : deceit
C. cant : jargon
D. abut : border

90. Glass : Crockery
A. hat : millinery
B. books : school
C. operation theatre : hospital
D. poetry : books

91. Install : Officer
A. campaign : candidate
B. promote : worker
C. succeed : prince
D. inangurate : president

92. Quark : Particle
A. proton : molecule
B. zine : element
C. light : photosynthesis
D. uranium : fission

93. Doctor : Disease
A. judge : culprit
B. politician : parliament
C. teacher : ignorance
D. policeman : criminal

94. Line : Segment
- A. big : small
- B. part : whole
- C. square : rectangle
- D. circle : arc

95. Breeze : Cyclone
- A. drizzle : downpour
- B. snowfall : cold
- C. earthquake : debris
- D. heat : sunstroke

96. Jail : Jailor
- A. museum : curator
- B. school : board
- C. life : wealth
- D. couple : marriage

97. Dimwit : Intelligent
- A. dingy : decent
- B. felon : guilty
- C. hoary : old
- D. harry : ravage

98. Grim : Stern
- A. laud : blame
- B. zest : gusto
- C. morbid : healthy
- D. overt : concealed

99. Trap : Spring
- A. prison : escape
- B. sentence : period
- C. alarm : trigger
- D. convict : corner

100. Orgy : Drunken
- A. wine : bar
- B. obese : fast food
- C. oasis : desert
- D. quicksilver : mercury

ANSWERS

1	2	3	4	5	6	7	8	9	10
B	A	D	A	B	A	A	A	C	A
11	**12**	**13**	**14**	**15**	**16**	**17**	**18**	**19**	**20**
A	B	A	D	C	D	C	A	C	B
21	**22**	**23**	**24**	**25**	**26**	**27**	**28**	**29**	**30**
C	D	C	B	D	A	B	C	D	D
31	**32**	**33**	**34**	**35**	**36**	**37**	**38**	**39**	**40**
C	D	C	D	A	A	D	B	D	B
41	**42**	**43**	**44**	**45**	**46**	**47**	**48**	**49**	**50**
B	C	C	A	B	B	A	B	B	D
51	**52**	**53**	**54**	**55**	**56**	**57**	**58**	**59**	**60**
C	A	A	D	B	C	C	D	A	D
61	**62**	**63**	**64**	**65**	**66**	**67**	**68**	**69**	**70**
B	C	A	B	D	C	A	B	D	C
71	**72**	**73**	**74**	**75**	**76**	**77**	**78**	**79**	**80**
B	A	C	A	B	D	C	C	D	B
81	**82**	**83**	**84**	**85**	**86**	**87**	**88**	**89**	**90**
A	D	B	D	C	C	A	C	B	A
91	**92**	**93**	**94**	**95**	**96**	**97**	**98**	**99**	**100**
D	B	C	D	A	A	A	B	C	D

10

Idioms, Phrases & Proverbs

An idiom is a group of words established by usage as having a meaning different from the individual words.
A phrase is a small group of words standing together as an idiomatic expression.
A proverb is a short pithy saying in general use, stating a general truth or piece of advice.
The use of idioms and phrases makes your expression more effective and powerful.
Learn and practise as many as you can.

A wet blanket: A person who discourages others or prevents them from enjoying what they are doing

A stag party: A party where only men are allowed to be present

A hen party: A party where only women are allowed to be present

A working lunch/dinner: A meal at which the people who attend discuss business

A working knowledge: Enough practical knowledge to do something

A willing horse: A helpful person who often gets all the work to do

A wild goose chase: A search for something which has no chance of being successful

A trend-setter: A person who starts a new fashion which becomes popular

A thick-skinned person: A person who is not easily offended

Teething troubles: Troubles and difficulties which occur during the early stages of an activity or enterprise, but which will lessen with time

A sweeping statement: A statement which is not careful or correct in its details; a generalisation

A spending spree: An outing during which one spends a lot of money

A tall order: A request which is unreasonably difficult to grant or carry out

A soft spot: A feeling of special fondness for something or someone

A sore point: A matter which irritates or hurts when it is mentioned

The small hours: The early morning hours just after midnight

A sleeping partner: A business partner who provides a share of the capital for a venture but does not take an active part in its operation

Small talk: Trivial, light conversation on unimportant topics

Second nature: Some acquired habit or skill which seems perfectly natural to someone; a very firmly fixed habit

A shady character: A person of very doubtful honesty or character

Second thoughts: A change of mind, attitude or decision after thinking about the matter

A security risk: A person who is a risk to a country's security because of his political activities; one whose loyalty is doubted

A jail bird: A man who has spent a lot of time in prison

A ladies' man: A man who likes to be with women or who likes to please them

A last fling: The last opportunity for pleasure or amusement before having to stop

An inside job: A theft committed by someone inside a building, i.e., by an employee, not a stranger

A knowing look: A look which suggests that, the person is well-informed of a matter

A hushed-up affair: An affair or matter that is kept secret by forcing silence about it

Hush money: A bribe paid to keep a matter secret from the public

A hot line: A direct telephone line between heads of governments, to be used at times of great difficulty

Hot news: Very recent, important or sensational news

High time: At the most important point of time

Fast living: A person who spends too much money in enjoying life

A fishy story: An untrue tale; an unlikely story

Half-hearted: Having no enthusiasm for a particular task

An iron will: A very strong will

Hot water: Trouble

A laughing stock: A person who is unkindly laughed at by everyone

A light-fingered person: A person who is in the habit of stealing small things

Pin money: A wife's allowance for her personal needs

Plain sailing: A plan or action that is simple and free from trouble

A pep talk: A friendly talk to give encouragement to win a game, etc. or to complete something well

A package tour: A completely planned holiday at a fixed price arranged by a company

A passing fancy: A temporary liking for something or someone

An open secret: Something supposed to be a secret but which in fact is known to everyone

A package deal: An offer or agreement which includes a number of things all of which must be taken

Plain dealing: Truthfulness and honesty, especially in business

A pet aversion: Something or someone much disliked

A sideline: Work one does apart from one's regular job

Second sight: An ability to foresee future events

A rolling stone: Person who travels around a lot without staying in any one place or job for long

A raw deal: An unjust or cruel treatment

In a rash moment: Doing something hastily without thinking of the results

A practical joke: A trick played on a person to make him look silly and to give amusement to others

A sugar daddy: An older man who has a relationship, especially sexual, with a younger woman, providing her with money and presents

Take-home pay/wages: The amount a person receives after taxes have been deducted

Backstairs influence: Secret and usually unfair influence

Bad blood: Bad feelings or unfriendliness between two persons

A bad egg/a bad penny: A bad character

A false alarm: A warning of something bad, which does not happen

Elbow room: Space, in which to move and act freely

A double agent: A person who works secretly for two opposing sides without either of the sides knowing

A dead-end job: A job without prospects; a job that leads to nothing further

A dark horse: A person who competes successfully against another although little is known about him

A confirmed bachelor: A man who has no desire to marry

A burning question: A matter having to be dealt with at once; an urgent important matter

A big shot: An important person, usually someone who is wealthy and influential

A blind date: A social meeting of two persons, usually a boy and a girl who have not met before

Broken English: Imperfect, ungrammatical English

A chicken-hearted person: A timid, cowardly person

A close-fisted person : Someone who is stingy; a miser

A country cousin: A person who is simple and inexperienced and not used to city life

A close shave: A narrow escape from danger, an accident, etc.

A light sleeper: A person who is easily awakened from sleep

A cock-and-bull story: An invented story

A cold war: A serious political struggle between countries which does not actually result in fighting

A dog-eared book: A book where the corners of the pages have been turned down with use

An easy victim/mark: Somebody who can easily be cheated or treated badly

A bosom friend: A very close, trusted friend

Crocodile tears: Pretended sorrow; insincere tears

A fast colour: A colour which does not run out of the material when washed or dried

A fair copy: A neat, legible copy

An armchair critic: A person who critically judges others' work and gives advice, but who has not himself experienced doing the work

Apple-pie order: In perfect order

An absent-minded person: A forgetful, inattentive person

A flying visit: A very short visit

Forty winks: A short nap or sleep

Hard labour: Hard, manual work usually done as a punishment

Hard cash: Actual money, not a cheque, etc.

A hard drinker: A person who drinks large quantities of alcohol

A golden opportunity: A very good opportunity

A going concern: A successful, thriving business

A foregone conclusion: A result that is certain, sure

The naked eye: The eye unaided by any instrument

An odd-job man: A man who does various types of work, usually manual, for pay

Old hat: No longer fashionable

A grass widow: A wife who is alone because her husband is away temporarily

A good samaritan: A person who helps a stranger in difficulties

French leave: Absence from work that is taken without permission

The generation gap: The difficulties arising when younger and elder people do not understand each other's way of life

The happy medium: The middle course of doing something when opposite ways are suggested

A henpecked husband: A husband who is dominated and nagged by his wife

A hard and fast rule: A rule that cannot be changed; fixed, unchangeable rule

One-track mind: A person's mind that is limited and thinks of only one thing at a time

A short cut: An easier method of doing something or getting somewhere; a quicker, more direct way

A small-time business: A business limited in activity and profits

A splitting headache: A very severe and painful headache

A square deal: A fair bargain or fair treatment

A square meal: A meal with adequate good; a satisfying meal

Strong language: Angry language consisting of many swear words

Tall talk: Boastful talk

A tight spot/corner: A difficult situation

Tough luck: Bad luck; hard luck

Ready money: Money which is immediately available in coins or notes

Second best: Next to the best; second in value or importance

The writing on the wall: An indication that something bad or difficult is about to happen

A storm in a teacup: A big fuss made over a small, trivial event

The talk of the town: A person or event that is causing great excitement and is the subject of much discussion

A snake in the grass: A cunning, harmful person who pretends to be a friend; a false friend

A shot in the dark: A guess which is not supported by arguments; a wild guess

The tricks of the trade: The best and most successful ways of doing good business

The bottom of the ladder: In the least important position

A shot in the arm: Something which acts to bring back a happy state or condition

The man in the street: The ordinary, average man

The gift of the gab: A person fluent in speech; a persuasive speaker

A man about town: A man who is used to living in the city and associating with well-known fashionable people

The life and soul: The person who is the centre of attaction because he is amusing and lively

The lion's share: The greater part; an unfairly large share

The long arm of the law: The police

A bear hug: A rough, tight hug

A brain wave: A clever idea; sudden inspiration

A bone of contention: Something that causes argument

Birds of a feather: People of the same kind (often bad) who like each other's company

A feather in one's cap: A deserved honour that one is proud of

Castles in the air: Plans which cannot be carried out or which will probably not succeed

A fish out of water: A person who feels uncomfortable because he is in a strange place, or among people who are very different from himself

The ups and downs: The good and bad times, the happy and sad times

A stone's thorw: A short distance

A slip of the tongue: Something said unintentionally

A slap in the face: A rebuff

The rank and file: Ordinary people without special positions; the masses

A man of means: A rich man

A man of iron: A cruel, hard person

A jack of all trades: A person who can do many different kinds of work but who may not be good at any of them

By hook or by crook: By any means possible

By fits and starts: At irregular intervals

A pack of lies: A whole lot of untruths

A clean bill of health: A certificate or announcement that someone is healthy

A happy-go-lucky person: Someone who does not think or plan carefully

A stay-at-home person: A person who is in the habit of staying at home, and not liking to travel

A smash-and-grab robbery: A robbery done quickly, usually by breaking windows or showcases taking away valuables, and running away

A hit-and-run driver: A driver who drives away after an accident without stopping to find out about damage or injuries

A run-of the-mill job: An ordinary unexciting job

A round the-clock service: A twenty-four hour service

An out-of the way place: A place that is distant and far away from people and places

A much-talked-of affair: Something which is the subject of much discussion

A nine to six job: A job in which the working hours are from nine in the morning to six in evening

An open and shut case: A case which is easy to settle or solve

Bumper-to-bumper traffic: Cars that are very close together one after another

A hand-me-down dress: A dress used by someone after belonging to another

A well-to-do man/woman: A rich person

An angel: A person who is gentle and kind to others
A devil: A particularly wicked or mischievous person
A giant: A person who is very large
A wizard: A clever and skillful person
An Adonis: A very handsome man
An Apollo: A man with a perfect physique or body shape
A Hercules: A very strong man
A Goliath: An extremely large man
A Judas: A mean traitor
A Samson: A very strong man
A Solomon: A very wise man
A dog: A worthless person; someone who is despised
A cat: A malicious, person. usually referring to females
A bear: A clumsy, rough person
A fox: A crafty, cunning person
A goose: A silly person
A lamb: An innocent harmless person
A monkey: A child who is lively and full of mischievous tricks
A mule: A very stubborn person
A wolf: A greedy and cunning person
A snake/serpent: A treacherous, dangerous person
A parrot: A person who learns things off by heart without understanding the real meaning
A pest: A troublesome person; one who annoys and pesters others
A pig: A greedy and dirty person
A shark: A greedy and cunning person; one who swindles others
A parasite: A person who lives off others and gives nothing in return
An ape: A person who foolishly imitates others
An ass/donkey: A stupid person
An elephant: A huge, clumsy, ungainly person
Again and again: Repeatedly; continually; very often
Back and forth/backward: Moving first in one direction and then in the opposite direction and forwards
By and by: Soon; before long
By and large: On the whole; in general; all things considered
Far and away: Very much
Far and near/far and wide: Everywhere
First and foremost: In the first place; firstly
Here and there: In various places; firstly
Here, there and everywhere: In all possible places; everywhere
In and out: Sometimes inside and sometimes outside
Up and about: On one's feet; up again after an illness
To-and fro: Backwards and forwards; from side to side

Through and through: Completely; entirely; in every way
Out and out: Complete, thorough; total
Out and away: By far
Over and over again: Again and again; repeatedly
Round and round: Repeatedly moving round
Once and for all: Finally; for the last time
On and on: Without stopping
Off and on: From time to time; occasionally
Every now and then/again: From time to time; at times
Now and again: Sometimes; occasionally; not very often
Nowadays: In these times
To backbite a person: To speak ill of someone
To back chat: To be cheeky; to talk back to someone
To be a bag of bones: To be extremely thin
To be a blockhead: To be an idiot
To be a great hand at: To be an expert
To be a sight for sore eyes: Something or someone who appears before another and is very welcome
To be all ears: To be eager to listen
To be armed to the teeth: Fully armed or protected
To be at one's wits' end: Completely puzzled
To be close-fisted: To be mean; miserly
To be double-faced: Insincere
To be down in the mouth: To feel depressed and unhappy
To be head and shoulders above someone else: To be very much better or superior
To be out of one's mind: To be mad; crazy
To be some one's blue-eyed boy: To be someone's favourite
To be the apple of someone's eye: To be someone's favourite
To bite someone's head off: To speak abruptly or sharply to another
To burn one's fingers: To get into trouble
To catch someone's eye: To suddenly exchange glances with another
To change hands: To pass from one owner to another
To dig someone in the ribs: To make unkind remarks or insinuations about another
To eat one's heart out for another: To long for someone who is unattentive
To foot the bill: To pay the bill, especially for some social outing
To get on someone's nerves: To irritate someone very greatly
To get something off one's chest: To confess to something
To get the upper hand: To gain mastery over an opponent
To give someone the cold shoulder: To ignore someone; to slight another

To give someone the sharp: To thoroughly scold another person with edge of one's tongue

To go on all hands: To crawl on hands and knees

To have a chip on one's shoulder: To bear a grudge against the world/society

To have a finger in every pie: To have many business interests

To have a head like a sieve: To have a very poor memory

To have a skeleton in the cupboard: To have a guilty secret, especially a family scandal

To have at one's fingertips: To know thoroughly

To have cold feet: To be afraid

To have no backbone: To be a weak, spineless person

To have one's head screwed on: To be very sensible

To have someone under one's thumb: To dominate another

To have the upper hand: To be in command/charge; to be someone in power

To have two left hands: Someone who is clumsy

To hold one's tongue: To keep silent even though there is an urge to talk

To hold up one's head: To be able to look people in the face

To jump out of one's skin: To get a sudden fright

To keep a cool head: To keep calm, especially in an emergency

To keep a stiff upper lip: To remain calm and courageous

To keep one's fingers crossed: To hope for luck

To keep one's hair on: To keep cool and patient

To keep one's head above water: To avoid getting into debt or trouble

To keep someone at arm's length: Not to be too friendly with another

To knock something on the head: To suddenly stop an idea/project

To know something by heart: To be able to say something from memory

To lay hands on someone: To handle someone roughly

To lend a hand: To help

To let the grass grow under one's feet: To be inactive and uninvolved

To live from hand to mouth: To live in hardship with the barest essentials

To lose face: To lose one's reputation/honour

To lose heart: To be discouraged

To lose one's head: To be very angry; to lose control of oneself

To make a clean breast of something: To confess to something; to reveal all

To make neither head nor tail of something: To understand nothing of a particular issue

To pat someone on the back: To give praise and encouragement to another

To pay through the nose for something: To pay an excessiveiy high price for something

To poison someone's mind: To turn someone against another person

To pull one's weight: To do one's share of work

To pull someone's leg: To play a joke on someone

To pull the wool over someone's eyes: To deceive another

To put one's back into it: To work hard

To put one's foot down: To be firm about an issue

To put one's shoulder to the wheel: To work hard and earnestly

To put someone's nose out of joint: To make someone jealous by taking his place as the centre of attraction

To rack one's brains: To try to recall or remember something

To receive with open arms: To welcome warmly

To save face: To save one's honour/reputation/name

To show a clean pair of heels: To escape by running

To snap someone's head off: To speak abruptly

To stand on one's own feet: To do things without help; to be independent

To sweep someone off his/her feet: To flatter and woo someone, usually a girl

To take matters into one's own hands: To do something independently without help

To take something to heart: To take something seriously

To take to one's heels: To run away

To throw dust in someone's eyes: To try to deceive someone

To tread in someone's foot steps: To follow the example of another, usually of talent/occupation

To tread on someone's toes: To offend another person

To try one's hand at: To try doing something for the first time

To turn a blind eye: To refuse to see

To turn a deaf ear: To refuse to hear

To turn up one's nose at something: To treat with contempt

To twist someone's arm: To force someone into doing something for you

To wash one's hands of something: To have nothing to do with a particular matter

The green-eyed monster: Jealousy

To scream blue murder: To complain loudly

Out of the blue: Suddenly; without warning

Once in a blue moon: Very rarely

To feel blue/have the blues: To be sad or depressed

A bolt from the blue: A complete surprise
Blue blood: Royal or noble blood
To blackmail someone: To threaten to reveal someone's misdeeds unless he pays money
In someone's black books: Out of favour with someone
A black sheep: A bad character who disgraces his family
A blacklist: A list of people groups, etc. who have done something wrong
Black and blue: Badly bruised after a beating/a fall
In black and white: In writing or print
A greengrocer: A person who sells fruit and vegetables
A greenhorn: An inexperienced person
The green light: The signal to proceed
Green with envy: Tilled with jealousy
Grey matter: Common sense; brains
In the pink of health: In excellent health
Purple with rage: So angry that the face becomes purple
To be caught red-handed: To be caught in the act of committing a crime
A red carpet welcome: A special ceremonial welcome to guest
The red flag: A flag used as a danger signal
A red-letter day: A notable, fortunate day that will be remembered
The red light: A signal to stop; a danger signal
Red tape: Official rules which cause delay in settling one's problems
To jump out of the frying-pan into the fire: To get out of one bad situation and fall into another worse situation
To see red: To lose control of oneself in anger
A white-collar job: An office job
A white elephant: A worthless possession which the owner wants to get rid of
The white flag: A sign that one accepts defeat or surrenders
To show the white feather: To show signs of cowardice
Yellow at heart: Cowardly
To be/feel off colour: To be unwell or rather ill
To come off with flying colours: To achieve great success
To have a high colour: To appear very red in the face
To show one's true colours: To show what one is really like; to show one's true character
A blue-collar job: A factory job
To act the goat: To behave foolishly
To air one's opinion: To express one's feelings in public
To be a chip off the old block: A son who is very much like his father in appearance, character, habits, etc
To be as fit as a fiddle: To be in excellent health
To be at a loose end: To have nothing to do
To be at a loss for words: Not knowing what to say; to be speechless
To be at death's door: To be so ill as to be near death
To be at sixes and sevens: To be in confusion and discord
To be at the end of one's teeter: To be at a loss as to what to do
To be born under a lucky star: To have considerable luck in whatever one does
To be born with a silver spoon in one's mouth: To be born into a rich family
To be dead beat: To be worn out by fatigue
To be in hot water: To be in serious trouble
To be in someone's bad/good books: To be out of/in favour with someone
To be in the doldrums: To be in low spirits
To be in the same boat: To have the same troubles/ circumstances
To be mad about someone: To like someone very much
To be made of money: To have plenty of money; to be well-off
To be on tenterhooks: To be in a state of suspense and anxiety
To be on the ball: To be alert, up-to-date, prepared
To be on the horns of a dilemma: Not knowing what to do from several courses of action
To be on the rocks: To be in danger of failing, collapse, especially a business, a marriage, etc
To be on top of the world: To be very happy; to be in high spirits
To be out of the woods: To be out of danger, especially a serious illness
To be quick on the uptake: To be alert and quick to do or say something
To be rolling in money: To be rich; to have plenty of money
To be scared to death: To be extremely frightened of something
To be slow on the uptake: Slow to do or say something
To be too big for one's boots: To think very highly of oneself
To bear the brunt of: To bear the main force or strain of some action
To beat about the bush: To be indecisive; not coming to the point
To blow hot and cold: To do one thing at one time and the opposite soon after
To blow one's own trumpet: To boast about one's abilities, achievements etc.

To break the ice: To take the first step; to be the first to begin, especially a conversation

To bring down the house: To cause loud and long applause through one's skill

To bring to light: To disclose, reveal

To build castles in the air: To think of things impossible to realize

To burn the candle at both ends: To use one's energy in two directions at the same time

To burn the midnight oil: To work till late at night

To call a spade a spade: To speak out bluntly and frankly

To clip one's wings: To deprive one of power and/or freedom

To cook an account: To tamper with/falsify an account

To come down on someone like a ton of bricks: To harshly criticise another

To come of age: To reach adulthood

To come to grief: To meet with danger

To come to light: To become known

To count-sheep: To try various methods to get to sleep

To cry over spilt milk: To be unhappy about something that can no longer be remedied

To curry favour: To seek/win favour by flattery or gifts

To cut off in its prime: To destroy something when it is at its best

To cut one's coat according to one's cloth: To live within one's income

To cut someone down to size: To put someone in his place especially a person who thinks too highly of himself

To do a city/the sights: To visit a city and see its well-known places

To do a thing by fits and starts: To work intermittently on something

To do a thing by hook or by crook: To do something by any means, fair or foul, direct or indirect

To do something on the spur of: To suddenly decide to do something the moment

To drop a brick: To say or do something very tactless and unexpected

To drop a hint: To make veiled and subtle suggestions

To eat like a horse: To eat a lot

To eat one's words: To take back what one has said; to apologise

To egg on: To spur on to further action

To face the music: To meet the worst; to face punishment for doing something wrong

To fall out with someone: To quarrel with someone

To fall through: To fail, especially a plan, project, etc.

To fancy someone/something: To take a great liking to

To fight shy of: To avoid

To fire someone: To dismiss someone from employment

To flog a dead horse: To try fruitlessly to revive interest in something

To follow the crowd: To follow unthinkingly; to believe or act as most people do

To follow suit : To do the same thing as another person

To gain ground: To make progress in an undertaking

To get away with something: To escape detection despite doing something wrong

To get blood out of stone: To get more than is fair out of someone

To get down to work: To start on the task in hand

To get even with someone: To take revenge of someone

To get hitched: To get married

To get into a flap: To get worried and nervous

To get into hot water: To be in/to get into trouble

To get off scot-free: To escape punishment; to be let off payment

To get out of bed on the wrong side: To start the day in a bad temper

To get the sack: To be dismissed from employment

To get wind of: To hear a rumour of

To get the wind up: To get worried and nervous; to get a fright

To give someone a telling-off: To scold someone

To give someone the slip: To manage to avoid someone who is looking for you

To give the game away: To let out a secret

To give up the ghost: To die; to cease trying

To go a long way: To excel; to go far especially in a career

To go off at a tangent: To change suddenly to a different course of thought or action

To go like a bomb: To go fast; to function well, especially a vehicle

To go to rack and ruin: To fall into a ruined or disorganised state

To grin and bear it: To make the best of a bad job; to put up with something

To haul someone over the coals: To reprimand someone

To have a bee in one's bonnet: To be strong-minded over certain issues

To have a soft spot for: To be very fond of someone or something

To have an axe to grind: To have a selfish motive for doing something

To have one's knife in someone: To be always finding fault with another person

To hit below the belt: To attack an opponent unfairly

To hit the nail on the head: To do or say exactly the right thing

To jog someone's memory: To remind someone of something

To keep a person in the dark: To keep something hidden from someone

To keep open house: To encourage visitors to come at any time

To keep one's pecker up: To keep one's courage up

To keep one's shirt on: To keep cool and calm in a dispute

To keep the ball rolling: To keep a conversation going

To keep up appearances: To pretend to be well-off and to live in a certain style

To kick the bucket: To die

To kid someone: To tease

To kill two birds with one stone: To accomplish two things with one action

To knock off work: To stop working

To know the ropes: To be thoroughly familiar with a particular situation

To knock someone down with a feather: To surprise someone very much

To kowtow to someone: To act in a very servile manner hoping for favour

To lay down the law: To speak in an authoritative way

To lead someone up the garden path: To deceive someone by hiding the real intention

To leave someone in the lurch: To desert someone in time of difficulties

To let bygones be bygones: To let things that are past and gone remain untouched and forgotten

To let the cat out of the bag: To reveal a secret; to expose a trick

To look like a wet week: To look miserable and depressed

To make a go of something: To try one's best to succeed in something

To make a hash/a mess of something: To fail at a task

To make believe: To pretend

To make both ends meet: To be able to manage on one's income

To make hay while sun shines: To take advantage of an opportunity

To make much ado about nothing: To make a great fuss about something unimportant

To make oneself scarce: To disappear

To nip something in the bud: To stop something before it can develop into something serious

To pass the buck: To pass on the responsibility

To pick holes in something: To find fault with

To pick to pieces: To analyse critically

To pick and choose: To make a careful selection

To play a double game: To do one thing openly and a different thing in secret

To play fast and loose: To disregard one's promises

To play second fiddle: To take a subordinate position

To pooh-pooh an idea: To express contempt for an idea

To pull strings: To use friends in influential positions, or to use one's own influence, to secure favours

To put the cart before the horse: To do the wrong thing first

To put two and two together: To arrive at a conclusion

To put a spoke in one's wheel: To hinder; to obstruct progress

To rain cats and dogs: To rain very heavily

To read between the lines: To detect the hidden meaning

To rule the roost: To domineer over others

To run in the blood: Inherited qualities

To see something through coloured spectacles: To see only the best points about a thing

To see the light: To begin to understand

To sit on the fence: To take neither side in a dispute; to remain neutral

To smell a rat: To become suspicious about something

To steer clear of: To avoid

To take French leave: To be absent from work without permission

To take someone for a ride: To deceive someone

To take the bull by the horns: To meet a difficulty with courage

To take the law into one's own hands: To punish a person, usually by force, without his being tried in a law court

To take with a pinch of salt: To accept what someone has said with doubt and misgiving

To talk nineteen to the dozen: To speak a lot

To talk the hind legs of a donkey: To talk a great deal at great length

To tear a strip off someone: To scold harshly

To throw cold water upon: To discourage effort

To turn over a new leaf: To give up one's bad ways, and lead a better life

To turn the tables on someone: To defeat someone who has previously defeated you

To upset the apple cart: To disturb the routine of something

MULTIPLE CHOICE QUESTIONS

Directions: *Some idioms/phrases are given below with their probable meanings. Select the options with their correct meanings.*

1. Pay off old scores
A. To repay the debt B. To have revenge
C. To invite D. Secretly pay

2. Turn turtle
A. To cheat B. To be lopsided
C. To frustrate D. To dance to the tune

3. Wash one's hands of
A. To refuse B. To assist
C. To abuse D. To refuse to be

4. Under duress
A. Under compulsion
B. Willing
C. To elicit information
D. To demand

5. To turn the tables
A. To ruin someone
B. To turn the situation to one's own side
C. To reverse the situation
D. To move from one point to another

6. On the cards
A. Possibly B. Probably
C. Openly D. Likely

7. To leave someone in the lurch
A. To come to compromise with someone
B. Constant source of annoyance to someone
C. To put someone at ease
D. To desert someone in his difficulties

8. To play second fiddle
A. To be happy, cheerful and healthy
B. To reduce importance of one's senior
C. To support the role and view of another person
D. To do back seat driving

9. Day and night, he **'yearns for'** his beloved, means
A. He weeps for his beloved
B. He remembers his beloved
C. He admires his beloved
D. He keenly desires to meet his beloved

10. 'Call off' means
A. To finish B. To withdraw
C. To postpone D. To cry

11. 'Carry out' means
A. To take from one place to another
B. To continue
C. To obey
D. To make efforts

12. In the same boat
A. A worn out choice
B. Indifferent
C. In identical circumstances
D. Carry off

13. In one's good book
A. A costly book B. A priceless treasure
C. In one's favour D. An enchanting beauty

14. Keep a straight face
A. To do make up
B. To change clothes
C. Assume responsibility
D. To remain serious

15. To be above board
A. To have a good height
B. To be honest in any business deal
C. To have no debts
D. To try to be beautiful

16. On the face of it
A. To agree B. From an action
C. More than enough D. Apparently

17. Let the bygones be bygones
A. In one's favour B. To pretend
C. To forget the past D. Other choice

18. To split hairs
A. Major distinctions
B. Hair with two ends
C. To make minute distinction
D. Without distinction

19. 'Bread and butter' means
A. Both bread and butter
B. Something essential
C. Livelihood
D. Relevant things

20. 'To bell the cat' means
A. To catch a cat and tie a bell round its neck
B. To make an effort
C. To be quick
D. To face a risk

21. 'Hard and fast' means
A. Strict B. Solid
C. Fast moving D. Some hard surface

22. 'Part and parcel' means
A. The part of a parcel
B. An essential part
C. A missing parcel
D. Some part of a machine sent by parcel

23. 'Null and void' means
A. Something invalid
B. Something that can be avoided

C. Something that can be nullified
D. Something evil

24. To make clean breast of
A. To gain prominence
B. To praise oneself
C. To confess without reserve
D. To destroy before it blooms

25. 'Trump card' means
A. A powerful means of achieving an object
B. Resourcefulness
C. The best gamble to attain success
D. None of these

26. 'Tall talk' means
A. A discussion continued for a long time
B. A high sounding talk
C. A meaningful talk
D. A useless talk

27. 'Small talk' means
A. Gossip
B. A discussion carried on for a long time
C. A brief discussion
D. None of these

28. Throw out of gear
A. To replace B. Hinder, disturb
C. To decide D. Take up tune

29. To and fro
A. Back and forth B. Puzzled
C. Amazed D. Reprove

30. To bell the cat
A. To do an easy job
B. To be indifferent to
C. To undertake a difficult job
D. To clarify

31. To be under cloud
A. Puzzle B. Enjoy the favour
C. Talk thoughtlessly D. To be under suspicion

32. A labour of love
A. A tragic end
B. A funny thing
C. Not fruitful
D. Work done without payment

33. Follow suit
A. Follow an example B. Wear a new dress
C. Irrelevant D. A gay person

34. 'Foul play' means
A. Bad intentions
B. A play not well acted
C. A play not liked by the audience
D. none of these

35. To pick holes
A. To find some reason to quarrel
B. To destroy something
C. To criticise someone
D. To cut some part of an item

36. To smell a rat
A. To see signs of plague epidemic
B. To get bad smell of a dead rat
C. To suspect foul dealings
D. To be in a bad mood

37. To put a spoke in one's wheel
A. To encourage B. Act without restraint
C. Risk something D. To obstruct one's progress

38. To pull one's leg
A. To give up B. Take care of
C. To befool D. To know

39. To play with fire
A. Grasp the truth
B. To handle something dangerous
C. To ridicule
D. To flee away

40. To reckon with
A. Take up time B. Make an inventory
C. To deal with D. Submit to punishment

41. To run short
A. Talk until one is tired at
B. Apply to oneself
C. To get rid of
D. To have or be too little

42. 'A man of letters' means
A. A postman B. A learned man
C. A hypocrite D. An ignorant man

43. 'A maiden speech' is
A. A speech made in the parliament
B. A speech made before unmarried girls
C. A speech made by a political leader
D. A speech made for the first time

44. 'Order of the day' means
A. An order passed on a particular day
B. A current law
C. Something common or general
D. None of these

45. To end in smoke
A. To make completely understand
B. To ruin oneself
C. To excite great applause
D. None of these

46. To give vent to
A. To allow to flow forth
B. To prove a failure
C. To amass wealth
D. To evade

47. To eat a humble pie
A. To apologise or confess
B. To order

C. To flatter
D. To get rid of

48. A black sheep
A. An unlucky person B. A negro
C. An ugly person D. None of these

49. To catch a tartar
A. To trap wanted criminal with great difficulty
B. To catch a dangerous person
C. To meet with disaster
D. To deal with a person who is more than one's reach

50. Sit on fence
A. To remain neutral
B. To show contempt
C. To enjoy the surroundings
D. To become fond of

51. To beg the question
A. To refer to B. To take for granted
C. To raise objections D. To be discussed

52. To drive home
A. To find one's roots
B. To return to place of rest
C. Back to original position
D. To emphasise

53. Take a cue
A. A thoughtful action B. Cautious act
C. To gather a hint D. To sharpen weapon

54. Turn a blind eye
A. To refuse to see B. Become blind
C. To act D. To hesitate

55. Take it upon oneself
A. To abuse B. To set on fire
C. To talk foolishly D. Assume responsibility

56. To take the cake
A. To eat
B. To defeat
C. To carry off the prize
D. To steal

57. 'Crocodile tears' means
A. False tears
B. Wrong impression
C. Tears shed by a crocodile
D. None of these

58. 'At sea' means
A. On a ship
B. On a sea journey
C. Weak and perplexed
D. Something vague

59. 'At loggerheads' means
A. To be enemies
B. To be friends
C. Criticising each other
D. None of these

60. 'To burn the midnight oil' means
A. To work lazily
B. To work very hard till late at night
C. To work by fits and starts
D. None of these

61. To eat one's heart out
A. To cry B. To laugh
C. To suffer silently D. Pretend to suffer

62. Null and void
A. Not concerned with
B. To avenge
C. To avoid taking sides
D. Not valid

63. To cry wolf
A. To listen eagerly
B. To give false alarm
C. To turn pale
D. To keep off starvation

64. A man of straw
A. A man of no substance
B. A very active person
C. A worthy fellow
D. An unreasonable person

65. 'A white lie' means
A. A lie told with a good intention
B. A lie told by a white man
C. A harmful lie
D. None of the above

66. 'To while away' means
A. To go away
B. To spend the time uselessly
C. To use the time
D. To calculate the time

67. 'A square meal' means
A. A good diet B. A full meal
C. A timely meal D. A proper diet

68. A wet blanket
A. Inflammable item B. Orthodox
C. One who kills joy D. To speak bluntly

69. Yeoman's service
A. A hostile service B. Help in need
C. To fulfil D. To guess right

70. Wild goose chase
A. A futile and foolish search
B. An emergency
C. Conscious of something
D. Interest in something

71. Wear and tear
A. Contempt
B. Damage caused by constant use
C. Festive occasion
D. Be leisurely

72. To get the wind of
A. To face danger B. To run away
C. To fight stubbornly D. To know something

73. Gift of the gab
A. Dull person B. An artist
C. A witty person D. A liar

74. Hobson's choice
A. A choice for the rich
B. Very good choice
C. A choice that is really no choice
D. An excuse

75. Carry the day
A. Win B. Lose
C. Evening D. Carry in time

76. To do oneself justice
A. To dispense justice on our own
B. To treat others with due respect
C. To defend one's point of view
D. None of these

77. 'Lend your ear to what I say' means
A. You should cut your ear and give it to me
B. You should listen to me
C. You should not pay any attention to what I say
D. You should know my intentions

78. 'A lion's share' means
A. The part of a lion B. The share of a lion
C. Major part D. Prey

79. 'A scape goat' means
A. A goat that has escaped from the herd
B. A person who is made to suffer for the misdeeds of another
C. A prisoner who has run away from custody
D. A person whose company is not desirable

80. To give a piece of one's mind
A. Scold B. Praise
C. Abandon D. Discharge

81. Capital punishment
A. Object to B. Death penalty
C. Release D. Balance

82. A burning question
A. A present
B. An old important issue
C. A matter hotly discussed in public
D. Life imprisonment

83. A feather in one's cap
A. Something to be ashamed of
B. Something to be proud of
C. Creditable success
D. Keep silent

84. Chicken-hearted
A. Cowardly B. Fearlessly
C. Festive occasion D. Sincerely

85. Come off with flying colours
A. Compensate
B. Take care of
C. Achieve creditable success
D. Keep silent

86. Cut a sorry figure
A. To beat severely B. Make a poor show
C. To venture D. To discourage

87. Eat one's words
A. Withdraw one's words
B. To continue on
C. To persist
D. Remain valid

88. Eat a humble pie
A. Feel hungry B. Eat greedily
C. To face humiliation D. To face jubilation

89. Get into hot water
A. Take bath B. Get into trouble
C. Without aim D. A difficult problem

90. 'To miss the bus' means
A. Being not able to catch the bus
B. To miss the chance
C. To lose a thing
D. To forget something

91. 'To miss the boat' means
A. Being not able to catch the boat
B. To lose a thing
C. To miss a chance
D. None of these

92. He is a **'man of word'** means
A. He is a boastful fellow
B. He talks very less
C. A reliable person
D. None of these

93. Tooth and nail
A. Secretly B. Swiftly
C. Completely D. With utmost effort

94. 'A bone of contention' means
A. A cause of quarrel B. A broken bone
C. A disputed matter D. None of these

95. 'To carry the day' means
A. To work hard
B. To win
C. To work throughout the day
D. None of these

96. Scot-free
A. Unpunished B. Respected
C. Give up D. Continue on

97. Red-letter day
A. A day of great joy or importance
B. A miserable day
C. Ripe time
D. In perfect order

98. Make hay while the sun shines
A. To dance happily
B. To hurry
C. Take advantage of favourable condition
D. To destroy

99. 'Well-to-do' means
A. A work excellently done
B. To do something in a good way
C. Well done
D. Rich

100. He is a **'bookworm'** means
A. He is in the habit of reading much
B. He is in the habit of reading less
C. He tears the pages of the book
D. He crams

101. 'Wait upon' means
A. To serve
B. To wait for someone
C. To contact a man at his place
D. To call a man

102. To pull along with
A. To insult B. To abuse
C. To cooperate D. To fight

103. Raw deal
A. Unripe fruits B. Hostility
C. Harsh treatment D. Opportunistic

104. See eye to eye
A. Enmity B. Friendship
C. Contempt D. To agree in certain matters

105. Show the white feather
A. To decorate B. To act like a coward
C. To hate D. To admire

106. To die in harness
A. Talk thoroughly B. To die while in work
C. To retire D. Tit for tat

107. Dark horse
A. Unexpected winner
B. To be indifferent
C. Natives of Africa
D. Decide for good reasons

108. To egg on
A. To make omelette B. To instigate
C. To act as traitor D. To revolt

109. Fall through
A. To begin B. To rise up
C. Collapse D. To chatter

110. To grease the palm of
A. To polish B. To repair
C. To bribe D. To reward

111. 'To make bare' means
A. To sleep without clothes
B. To be frank
C. To remove clothes
D. To expose

112. 'To put off' means
A. To remove B. To extinguish
C. To postpone D. None of the above

113. 'Turn-down' means
A. To reject B. To dismiss
C. To disappear D. To waste

114. 'An underdog' is
A. A notorious fellow B. A starving person
C. The poor and needy D. The neglected class

115. Burn one's fingers
A. Become reconciled
B. To hurry
C. To get into trouble
D. Make it difficult for

116. Burn the candles at both ends
A. A wealthy person
B. A lazy person
C. Treat as important
D. Use up too much energy

117. Face the music
A. Love music
B. To avoid
C. Face the consequence of one's action
D. To delay

118. Keep abreast of
A. To be informed of something
B. A crisis
C. To welcome
D. To turn hostile

119. Kick the bucket
A. To drop the bucket
B. To destroy a thing
C. To breathe one's last
D. To reveal a secret

120. Make a clean sweep of
A. To face a problem
B. To amass wealth
C. To rob
D. To admit one's guilt fully

121. 'A hard nut to crack' means
A. A problem difficult to solve
B. A nut which cannot be broken with hands
C. A nut which is not easily available
D. None of these

122. 'To hold good' means
A. To be useful B. To be valid
C. To be good D. To be clear

123. 'Hallmark' means
A. A big mark B. A spot
C. Genuine excellence D. Grand

124. Smell a rat
- A. To act unfairly
- B. To talk boastfully
- C. To have reason to suspect
- D. To discourage

125. Pick holes in another's coat
- A. To insult
- B. To applaud
- C. To waste time
- D. To find fault with another

126. In vogue
- A. In action
- B. In fashion
- C. In play
- D. In full cooperation

127. In memoriam
- A. In memory of
- B. To forget
- C. Without happiness
- D. With happiness

128. Via media
- A. Right course
- B. Left course
- C. Extreme course
- D. Middle course

129. Vis-a-vis
- A. In good faith
- B. Face to face
- C. Contrary to the fact
- D. Similar

130. In toto
- A. Unknown
- B. On the whole
- C. Isolated
- D. Part by part

ANSWERS

1	2	3	4	5	6	7	8	9	10
B	B	D	A	C	D	D	C	D	B
11	**12**	**13**	**14**	**15**	**16**	**17**	**18**	**19**	**20**
C	C	C	D	B	D	C	C	D	D
21	**22**	**23**	**24**	**25**	**26**	**27**	**28**	**29**	**30**
A	B	A	C	C	B	A	B	A	C
31	**32**	**33**	**34**	**35**	**36**	**37**	**38**	**39**	**40**
D	D	A	A	C	C	D	C	B	C
41	**42**	**43**	**44**	**45**	**46**	**47**	**48**	**49**	**50**
D	B	D	C	D	A	A	D	B	A
51	**52**	**53**	**54**	**55**	**56**	**57**	**58**	**59**	**60**
B	D	C	A	D	C	A	C	A	B
61	**62**	**63**	**64**	**65**	**66**	**67**	**68**	**69**	**70**
C	D	B	A	A	B	B	C	B	A
71	**72**	**73**	**74**	**75**	**76**	**77**	**78**	**79**	**80**
B	D	C	C	A	D	B	C	B	A
81	**82**	**83**	**84**	**85**	**86**	**87**	**88**	**89**	**90**
B	C	B	A	C	B	A	C	B	B
91	**92**	**93**	**94**	**95**	**96**	**97**	**98**	**99**	**100**
C	C	D	A	B	A	A	C	D	A
101	**102**	**103**	**104**	**105**	**106**	**107**	**108**	**109**	**110**
A	C	C	D	B	B	A	B	C	C
111	**112**	**113**	**114**	**115**	**116**	**117**	**118**	**119**	**120**
D	C	A	C	C	D	C	A	C	D
121	**122**	**123**	**124**	**125**	**126**	**127**	**128**	**129**	**130**
A	B	C	C	D	B	A	D	B	B

11

Ordering of Words

The words form a sentence and convey their meaning only when they are arranged in a proper order. One must study and practise it regularly.

MULTIPLE CHOICE QUESTIONS

Directions: *In the following questions, some parts of the sentence have been jumbled up. You are required to rearrange these parts which are labelled P, Q, R and S to produce the correct sentence. Choose the option with proper sequence.*

1. We are doing
P : to the people
Q : to give relief
R : all we can
S : but more funds are needed
The correct sequence should be
A. P Q R S B. R Q P S
C. Q P R S D. S P Q R

2. The man
P : when he was
Q : in the office last evening
R : could not finish
S : all his work
The correct sequence should be
A. P Q R S B. Q R S P
C. R Q P S D. R S P Q

3. The people decided
P : they were going
Q : how much
R : to spend
S : on the construction of the school building
The correct sequence should be
A. Q P R S B. P Q R S
C. P R Q S D. S Q P R

4. The man said that
P : those workers
Q : would be given a raise
R : who did not go on
S : strike last month
The correct sequence should be
A. P Q R S B. P R S Q
C. Q P R S D. R S P Q

5. I think
P : the members
Q : are basically in agreement
R : of the group
S : on the following points.
The correct sequence should be
A. R Q P S B. S Q R P
C. P R Q S D. P Q S R

6. While it was true that
P : I had
Q : to invest in industry
R : some lands and houses
S : I did not have ready cash
The correct sequence should be
A. P Q R S B. P R S Q
C. S Q P R D. Q P R S

7. P : But your help
Q : to finish this work
R : it would not have been possible
S : in time
The correct sequence should be
A. P R Q S B. S P Q R
C. R P Q S D. P Q R S

8. The boy
P : in the competition
Q : who was wearing spectacles
R : won many prizes
S : held in our college
The correct sequence should be
A. P Q R S B. R P S Q
C. Q R P S D. Q P S R

9. About 200 years ago,
P : in the south of India
Q : an old king
R : ruled over a kingdom
S : called Rajavarman.
The correct sequence should be
A. Q S R P B. P Q R S
C. Q P S R D. Q S P R

10. P : his land
Q : a wooden plough
R : the Indian peasant still uses
S : to cultivate.
The correct sequence should be
A. R Q P S B. Q P S R
C. S R Q P D. R Q S P

11. He was a man,
P : even if he had to starve
Q : who would not beg
R : borrow or steal
S : from anyone.
The correct sequence should be
A. P Q R S B. P R Q S
C. Q R S P D. Q P R S

12. P : in the progress of
Q : universities play a crucial role
R : our civilization
S : in the present age.
The correct sequence should be
A. S Q P R B. Q R S P
C. Q R P S D. S Q R P

13. P : far out into the sea
Q : for the next two weeks there were further explosions
R : which hurled
S : ashes and debris.
The correct sequence should be
A. Q R P S B. R S P Q
C. Q R S P D. S R P Q

14. William Shakespeare,
P : in his lifetime
Q : the great English dramatist
R : wrote thirty-five plays
S : and several poems.
The correct sequence should be
A. P Q R S B. R S P Q
C. Q S R P D. Q R S P

15. Whenever I am,
P : with an old friend of mine
Q : in New Delhi
R : to have dinner
S : I always try.
The correct sequence should be
A. S Q P R B. Q S R P
C. R P S Q D. P R Q S

16. P : I don't know
Q : must have thought
R : what people sitting next to me
S : but I came away.
The correct sequence should be
A. R S Q P B. R Q S P
C. P Q R S D. P R Q S

17. P : in estimating the size of the earth
Q : but they were hampered by the lack of instruments of precision
R : ancient astronomers
S : used methods which were theoretically valid
The correct sequence should be
A. R P Q S B. P R Q S
C. R S Q P D. R P S Q

18. P : It is a pity that
Q : by offering a handsome dowry
R : a number of parents think that
S : they will be able to ensure the happiness of their daughters
The correct sequence should be
A. S Q R P B. P R S Q
C. P S R Q D. P R Q S

19. The common man
P : in nurturing
Q : a more active role
R : communal harmony
S : should play
The correct sequence should be
A. P R S Q B. S Q P R
C. S Q R P D. P R Q S

20. The doctor
P : able to find out
Q : what has caused
R : the food poisoning
S : has not been
The correct sequence should be
A. S P R Q B. P R Q S
C. P R S Q D. S P Q R

21. P : was suspended
Q : the officer being corrupt
R : before his dismissal
S : from service
The correct sequence should be
A. Q P S R B. Q P R S
C. R S Q P D. R S PQ

22. With an unsteady hand
P : on my desk
Q : from his pocket
R : he took an envelope
S : and threw it

The correct sequence should be
A. Q R P S B. Q R S P
C. R Q P S D. R Q S P

23. P : she gave her old coat
Q : to a beggar
R : the one with the brown fur on it
S : shivering with cold

The correct sequence should be
A. S Q R P B. S P R Q
C. P R Q S D. P S Q R

24. It is a privilege
P : to pay tax
Q : of every citizen
R : as well as the duty
S : as well as the duty who is well-placed

The correct sequence should be
A. R P S Q B. S P R Q
C. R Q S P D. S Q R P

25. It is not good
P : of the wicked persons
Q : to overthrow
R : to accept the help
S : the righteous persons

The correct sequence should be
A. R S Q P B. Q S R P
C. R P Q S D. Q P R S

26. Life is judged
P : and not by
Q : of work done
R : the longevity of years
S : by the quality

The correct sequence should be
A. Q S P R B. S Q R P
C. Q S R P D. S Q P R

27. P : When he learns that
Q : you have passed the examination
R : in the first division
S : your father will be delighted

The correct sequence should be
A. Q P S R B. S P Q R
C. Q R S P D. S R Q P

28. P : The journalist
Q : saw
R : countless number of the dead
S : driving across the field of battle

The correct sequence should be
A. P Q S R B. P Q R S
C. P S Q R D. S R Q P

29. P : Jane planned
Q : some stamps
R : to buy
S : this afternoon

The correct sequence should be
A. P R Q S B. P S Q R
C. Q R P S D. Q S P R

30. Her mother
P : when she was
Q : hardly four years old
R : began to teach Neha
S : English

The correct sequence should be
A. R S Q P B. S R P Q
C. R S P Q D. S R Q P

31. P : Bill had
Q : a friend
R : an appointment
S : to meet

The correct sequence should be
A. P S R Q B. P R S Q
C. Q S R P D. Q R S P

32. For fear
P : that may or may not affect them perhaps at first
Q : of upsetting young people
R : only healthy people over 80 should be sequenced
S : about their genetic propensities

The correct sequence should be
A. S Q P R B. Q S R P
C. S Q R P D. Q S P R

33. While traditional
P : under made-up Americans aliases pretending familiarity with a culture and climate
Q : India sleeps a dynamic young cohort of highly skilled articulate professionals
R : they've never actually experienced earning salaries that were undreamt of by their elders
S : work through the night in the call centres functioning on US time

The correct sequence should be
A. P R Q S B. Q S P R
C. P S Q R D. Q R P S

34. IITs are
P : of great self-confidence and competitive advantage for India today
Q : in science and technology which has become a source
R : as they epitomize his creation of an infrastructure for excellence
S : perhaps Jawaharlal Nehru's most consequential legacy

The correct sequence should be
A. Q P S R B. S R Q P
C. Q R S P D. S P Q R

35. As India
P : from nearly 250 years of the British rule in India
Q : first major struggle for independence from the British rule
R : celebrates the Diamond Jubilee of its independence
S : it also observes simultaneously the 150th Anniversary of the Great Indian Mutiny

The correct sequence should be

A. R S P Q B. Q P S R
C. R P S Q D. Q S P R

36. There have been
P : a day after high intensity violence left at least 50 persons
Q : sporadic clashes between
R : dead in the northern city of Tripoli
S : the Lebanese army and militants

The correct sequence should be

A. Q S R P B. S Q R P
C. Q S P R D. S Q P R

37. Although
P : of non-owner managers came to be widely appreciated
Q : political freedom from the British masters
R : came to us in 1947 it was not until
S : well into the following decade that the role

The correct sequence should be

A. S P Q R B. Q R S P
C. S R Q P D. Q P S R

38. Conditions
P : for marketing in the U.S. and Canada
Q : Mexico as a manufacturing base
R : that Indian companies aspiring to tap
S : would have to fulfil include the complex rules of origin

The correct sequence should be

A. R Q P S B. S P Q R
C. R P Q S D. S Q P R

39. Aside
P : of the same three-storey building in the military academy
Q : from eating in the same dining hall
R : half to the north of the entrance half to the south
S : the 206 troops live side by side on the ground floor

The correct sequence should be

A. R P S Q B. Q S P R
C. R S P Q D. Q P S R

40. Russia's test firing
P : to US steps that have sparked an arms race
Q : of an intercontinental ballistic missile on
R : and undermined world security
S : Tuesday was in response

The correct sequence should be

A. S Q P R B. Q S R P
B. S Q R P D. Q S P R

41. Marks, cities, civilization —
P : on the verge of globalization; poised to
Q : the slow ascent to where he is today, poised
R : it is in this order that primitive man made
S : achieve universal prosperity and abundance

The correct sequence should be

A. R Q P S B. P S R Q
B. R S P Q D. P Q R S

42. I bow my head
P : for their sense of the beautiful in
Q : nature and for their foresight in investing beautiful
R : manifestations of nature with a religious significance
S : in reverence to our ancestors

The correct sequence should be

A. Q R S P B. S P Q R
B. Q P S R D. S R Q P

43. With all the crime and sleaze
P : I am not sure how many parents will be able to
Q : how many will have the courage to satisfy the child's uncomfortable queries
R : that dominates the front page of the newspapers today
S : read out the headlines to their children and if they do so

The correct sequence should be

A. R P S Q B. S Q R P
C. R Q S P D. P R S Q

44. The way
P : processes that govern their actions
Q : nutrients become integral parts
R : depends on the physiological and biochemical
S : of the body and contribute to its functions

The correct sequence should be

A. Q R S P B. P S R Q
C. Q S R P D. S P R Q

45. Thus,
P : international surveys would hence forth record
Q : if dirt-poor people in the developing world
R : their wealth of happiness alongside their material poverty
S : display a general sense of well-being

The correct sequence should be

A. S Q R P B. Q S P R
C. S Q P R D. Q S R P

46. It's
P : someone who's grieving but
Q : natural to feel uncomfortable
R : that prevent you from being there
S : or awkward when you have to help

The correct sequence should be

A. Q P S R B. R S P Q
C. Q S P R D. R P S Q

47. Developing countries
P : along the equator, which
Q : could become leaders in energy production
R : are expected to face the brunt of global warming
S : with a solar energy breakthrough

The correct sequence should be

A. Q S P R B. P R Q S
C. Q R P S D. P S Q R

48. A diversified
P : use as a heating or power generation fuel by converting gas into
Q : adding a new dimension to the traditional use of gas
R : of natural gas is emerging
S : amongst other products, high quality diesel transportation fuel virtually free of sulphur

The correct sequence should be

A. R P Q S B. S Q P R
C. R Q P S D. S P Q R

49. As things stand
P : but a majority still does not have access to English
Q : linguistic edge they are equipped with
R : after globally because of the
S : Indian professionals are much sought

The correct sequence should be

A. R S P Q B. S R Q P
C. R S Q P D. S R P Q

50. While advocates
P : of its provisions with the
Q : there is some misguided concern about a possible clash of some
R : of social reform have generally hailed the new legislation
S : religious and customary practices in vogue in the country

The correct sequence should be

A. R Q P S B. Q R S P
C. R Q S P D. Q R P S

51. He has
P : while has in a reverie
Q : found the book
R : at the bus-stop
S : he lost

The proper sequence should be:

A. Q R S P B. P R Q S
C. Q S R P D. P Q S R

52. Then the women
P : lamenting their evil desire
Q : that had brought
R : wept loudly
S : this sorrow upon them

The proper sequence should be:

A. R P Q S B. R Q P S
C. P Q S R D. P R Q S

53. It is easy to excuse
P : but it is hard
Q : in a boy of fourteen
R : the mischief of early childhood
S : to tolerate even unavoidable faults

The proper sequence should be:

A. R P Q S B. Q R S P
C. Q R P S D. R P S Q

54. I don't remember
P : I saw a man dying in front of a hospital
Q : but when I left Lucknow in 1984
R : hit apparently by a fast moving car
S : the exact date

The proper sequence should be:

A. S Q R P B. S Q P R
C. Q R P S D. S P R Q

55. Since the beginning of history
P : have managed to catch
Q : the Eskimos and Red Indians
R : by a very difficult method
S : a few specimens of this acquatic mammal

The proper sequence should be:

A. Q P R S B. S Q P R
C. S Q R P D. Q P S R

56. I saw that
P : but seeing my host in this mood
Q : I deemed it proper to take leave
R : as I had frequently done before
S : it had been my intention to pass the night there

The proper sequence should be:

A. Q P S R B. Q R P S
C. S P Q R D. S R P Q

57. It was to be
P : before their school examination
Q : which was due to start
R : the last expedition
S : in a month's time

The proper sequence should be:

A. S R Q P B. R Q S P
C. R P Q S D. S P R Q

58. They felt safer
P : to watch the mountain
Q : of more than five miles
R : as they settled down
S : from a distance

The proper sequence should be:

A. R P S Q B. R S Q P
C. P Q S R D. P R S Q

59. If you need help
P : promptly and politely
Q : ask for attendants
R : to help our customers
S : who have instructions
The proper sequence should be:
A. S Q P R B. Q P S R
C. Q S R P D. S Q R P

60. He was so kind and generous that
P : he not only
Q : made others do so
R : but also
S : helped them himself
The proper sequence should be:
A. P S R Q B. S P Q R
C. P R S Q D. Q P R S

61. People
P : at his dispensary
Q : went to him
R : of all professions
S : for medicine and treatment
The proper sequence should be:
A. Q P R S B. R P Q S
C. R Q S P D. Q R P S

62. When it began to rain suddenly on the first of January
P : to celebrate the new year
Q : we ran for shelter
R : to the neighbouring house
S : where many people had gathered
The proper sequence should be:
A. Q R P S B. P S Q R
C. P R S Q D. Q R S P

63. The master
P : who was very loyal to him
Q : punished the servant
R : without giving any valid reason
S : when he left the work unfinished
The proper sequence should be:
A. R Q P S B. R Q S P
C. Q P S R D. Q R P S

64. The appearance
P : this dinosaurs were at their peak
Q : of the first mammals on the earth
R : at the time when
S : went almost unnoticed
The proper sequence should be:
A. S R P Q B. Q S R P
C. Q R P S D. R P Q S

65. It is easier
P : to venture into space
Q : for men
R : beneath their feet
S : than to explore
The proper sequence should be:
A. Q R P S B. Q P S R
C. P S R Q D. P Q S R

66. It is very easy
P : a great deal more than one realises
Q : may mean
R : that a phrase that one does not quite understand
S : to persuade oneself
The proper sequence should be:
A. R S Q P B. S P Q R
C. S R Q P D. R Q P S

67. The national unity of a free people
P : to make it impracticable
Q : for there to be an arbitrary administration
R : depends upon a sufficiently even balance of political power
S : against a revolutionary opposition that is irreconciably opposed to it.
The proper sequence should be:
A. Q R P S B. Q R S P
C. R P Q S D. R S P Q

68. He told us that
P : and enjoyed it immensely
Q : in a prose translation
R : he had read Milton
S : which he had borrowed from his teacher
The proper sequence should be:
A. R S Q P B. Q R P S
C. R Q S P D. R Q P S

69. This time
P : exactly what he had been told
Q : the young man did
R : beyond his dreams
S : and the plan succeeded
The proper sequence should be:
A. Q P R S B. Q P S R
C. P Q S R D. Q S R P

70. As a disease
P : and breaks up marriages
Q : accidents and suicides
R : alcoholism leads to
S : affecting all ages
The proper sequence should be:
A. S R P Q B. R P S Q
C. S R Q P D. R Q P S

71. This majestic mahogany table
P : belongs to an old prince
Q : which has one leg missing
R : who is no impoverished
S : but not without some pride

The proper sequence should be:
A. P Q S R B. Q R S P
C. P R S Q D. Q P R S

72. We have to
P : as we see it
Q : speak the truth
R : there is falsehood and darkness
S : even if all around us

The proper sequence should be:
A. R Q S P B. Q R P S
C. R S Q P D. Q P S R

73. He sat
P : through the Town Hall Park
Q : which flanked a path running
R : under the boughs
S : of a spreading tamarind tree

The proper sequence should be:
A. P Q S R B. R S Q P
C. R S P Q D. P R S Q

74. We went
P : along the railway line
Q : and had a right to
R : where other people were not allowed to go
S : but daddy belonged to the railway

The proper sequence should be:
A. R P Q S B. P R S Q
C. R S Q P D. P R Q S

75. In the darkness
P : the long, narrow beard
Q : was clearly visible with
R : the tall stooping figure of the doctor
S : and the aquiline nose

The proper sequence should be:
A. R Q P S B. P S Q R
C. R S Q P D. Q P R S

76. It is foolish
P : of those who posses them
Q : to believe that
R : will result in victory
S : the use of nuclear weapons.

The correct sequence should be:
A. R S P Q B. Q S R P
C. P R Q S D. S Q P R

77. A distressing fact is that
P : social accountability
Q : are dominated only by greed
R : many people today
S : and there is hardly any

The correct sequence should be:
A. S R P Q B. Q S R P
C. P R Q S D. R Q S P

78. I once had
P : every morning
Q : a client who swore
R : for the past four years
S : she had a headache

The correct sequence should be:
A. P R S Q B. Q S P R
C. R P Q S D. S Q R P

79. People know
P : not only of the smokers themselves,
Q : that smoking tobacco
R : but also of their companions
S : is injurious to the health

The correct sequence should be:
A. P S Q R B. R P S Q
C. Q P R S D. Q S P R

80. He had
P : finished his lunch
Q : hardly
R : at the door
S : when someone knocked

The correct sequence should be:
A. Q P R S B. P Q R S
C. Q P S R D. R P Q S

81. Mr. Sexena was a profound scholar who
P : was held in high esteem by all those
Q : who read his books and visited him regularly
R : till his untimely death
S : though not popular with the general public

The correct sequence should be:
A. P Q R S B. R P Q S
C. S R Q P D. S P Q R

82. The Government wants that
P : by the veterinary surgeons
Q : by the butchers
R : all the goats slaughtered
S : must be medically examined

The correct sequence should be:
A. R P S Q B. Q S R P
C. R Q S P D. P R S Q

83. The general line about television
P : is that it is very exciting,
Q : but also potentially very dangerous
R : immensely powerful
S : that I took myself

The correct sequence should be:
A. P Q R S B. S P R Q
C. P R Q S D. R P Q S

84. The second test of good government is that
P : to every man and woman
Q : and act only with their consent

R : it should give a lot of freedom
S : and should treat their personalities with respect and sympathy

The correct sequence should be:

A. Q S P R B. S R Q P
C. R P S Q D. P Q R S

85. The teacher warned that
P : he would not let
Q : go home
R : those students
S : who do not finish the class work

The correct sequence should be:

A. P Q R S B. P R Q S
C. P R S Q D. R S P Q

86. Towards the end of the eighteenth century, quite a number of economists
P : in the near future
Q : at the possibility of
R : were seriously perturbed
S : the world facing starvation

The correct sequence should be:

A. P R Q S B. R Q S P
C. Q S P R D. R P Q S

87. The best way of understanding our own civilization
P : is to examine
Q : an ordinary man
R : in the life of
S : an ordinary day

The correct sequence should be:

A. P Q R S B. R Q P S
C. P S R Q D. R S P Q

88. What greater thing is there
P : for two human souls to feel
Q : to rest on each other in all sorrow,
R : that they are joined for life,
S : to strengthen each other in all labour

The correct sequence should be:

A. S Q R P B. R P Q S
C. Q R S P D. P R S Q

89. Fame
P : by showing off
Q : to the best advantage
R : one's ability and virtue
S : is earned

The correct sequence should be:

A. P Q R S B. S P R Q
C. P R S Q D. P Q S R

90. When he was a child
P : passed his happiest hours
Q : the boy who was to become Britain's Baron Haden
R : staring out of his apartment window
S : living in New York

The correct sequence should be:

A. Q S P R B. P R Q S
C. S Q P R D. R S Q P

91. P : The teacher had to be specially careful
Q : because he enjoyed the confidence
R : about how he faced up to this problem
S : of all the boys

The correct sequence should be:

A. P R Q S B. Q P S R
C. S P R Q D. P S R Q

92. Movies made in
P : all around the globle
Q : Hollywood in America
R : by people
S : are seen at the same time

The correct sequence should be:

A. Q S R P B. Q R P S
C. P S R Q D. Q P S R

93. P : The foundations of the prosperity of a state
Q : primary health and education but also
R : involves the creation of job oppotunities
S : does not merely rest on

The correct sequence should be:

A. P S Q R B. P Q R S
C. P R Q S D. P S R Q

94. I am pure
P : and will be happy
Q : sooner or later
R : a day will come
S : when all will be equal

The correct sequence should be:

A. Q P R S B. Q S R P
C. R Q S P D. R S Q P

95. P : To do his/her work properly
Q : it should be the pride and honour
R : without anybody forcing him/her
S : of every citizen in India

The correct sequence should be:

A. Q S R P B. P R Q S
C. Q S P R D. P Q R S

96. The person who can state
P : correct than the person who cannot
Q : is more likely to be
R : his antagonist's point of view
S : to the satisfaction of the antagonist

The correct sequence should be:

A. R S Q P B. R Q P S
C. P Q R S D. S Q R P

97. The time has come
P : for future generations to come
Q : that the ideal of peace is a distant ideal
R : or one which can be postponed
S : when man must no longer think

The correct sequence should be:
A. P Q R S B. S Q R P
C. Q R S P D. R S P Q

98. I had been staying with
P : at his cottage among the Yorkshire fells
Q : a friend of mine
R : a delightfully lazy fellow
S : some ten miles away from the railway station
The correct sequence should be:
A. P Q R S B. Q R P S
C. Q R S P D. R Q P S

99. All the evil in this world is brought about by person
P : when they ought to be up
Q : but do not know
R : nor what they ought to be doing
S : who are always up and doing
The correct sequence should be:
A. P Q S R B. Q P R S
C. S Q P R D. P Q R S

100. If all the countries
P : of mankind and agree to obey
Q : work together for the common good
R : with each other and there will be no more war
S : the laws, then they will never fight
The correct sequence should be:
A. P Q R S B. Q S P R
C. Q P S R D. R Q P S

ANSWERS

1	2	3	4	5	6	7	8	9	10
B	D	A	B	C	C	A	C	A	D
11	**12**	**13**	**14**	**15**	**16**	**17**	**18**	**19**	**20**
C	A	C	D	B	D	C	B	B	D
21	**22**	**23**	**24**	**25**	**26**	**27**	**28**	**29**	**30**
B	D	C	C	B	D	B	C	A	B
31	**32**	**33**	**34**	**35**	**36**	**37**	**38**	**39**	**40**
B	D	B	B	C	C	C	A	B	D
41	**42**	**43**	**44**	**45**	**46**	**47**	**48**	**49**	**50**
A	B	A	C	B	C	B	A	B	A
51	**52**	**53**	**54**	**55**	**56**	**57**	**58**	**59**	**60**
C	A	D	B	D	D	C	A	C	A
61	**62**	**63**	**64**	**65**	**66**	**67**	**68**	**69**	**70**
C	D	C	C	B	B	D	C	B	C
71	**72**	**73**	**74**	**75**	**76**	**77**	**78**	**79**	**80**
D	D	B	B	A	B	D	B	D	C
81	**82**	**83**	**84**	**85**	**86**	**87**	**88**	**89**	**90**
D	C	B	C	B	B	C	D	B	C
91	**92**	**93**	**94**	**95**	**96**	**97**	**98**	**99**	**100**
A	A	A	C	C	A	B	B	C	C

12

Ordering of Sentences

A paragraph is formed from sentences, it will convey its true meaning and purpose only when the sentences are arranged in a proper manner. Try and practise it in this exercise.

MULTIPLE CHOICE QUESTIONS

Directions: *In questions below, each passage consists of six sentences. The first and the sixth sentences are given in the beginning. The middle four sentences in each have been removed and jumbled up. These are labelled P, Q, R and S. Find out the proper order for the four sentences.*

1. S_1: Since the sixties there has been an increasing interest in human neurophysiology, which deals with the neural bases of mental activity and behaviour.

S_6: So far the journal has published a mixture of articles including reports and investigations.

P: It has format which is very similar to that of *Brain and Language*, a sister journal.

Q: Since then, a number of journals devoted entirely to this area of research have appeared.

R: Before the 1960's when this field was the concern of a small number of investigators, research articles were scattered in various neurological journals.

S: *Brain and Cognition* is one such journal.

The proper sequence should be:

A. R Q S P B. Q R S P
C. Q S P R D. R S P Q

2. S_1: She said on the phone that she would report for duty next day.

S_6: Eventually we reported to the police.

P: We waited for a few days, then we decided to go to her place.

Q: But she did not.

R: We found it locked.

S: Even after that we waited for her for quite a few days.

The proper sequence should be:

A. P R S Q B. Q P S R
C. Q P R S D. S Q P R

3. S_1: A force of attraction exists between everybody in the universe.

S_6: The greater the mass, the greater is the earth's force of attraction on it—we call this force of attraction gravity.

P: Normally it is very small but when one of the bodies is a planet, like the earth, the force is considerable.

Q: It has been investigated by many scientists including Galileo and Newton.

R: Everything on or near the surface of the earth is attracted by the mass of the earth.

S: This gravitational force depends on the mass of the bodies involved.

The proper sequence should be:

A. P R Q S B. P R S Q
C. Q S R P D. Q S P R

4. S_1: Metals are today being replaced by polymers in many applications.

S_6: Many Indian Institutes of Science and Technology run special programmes on polymer science.

P: Above all, they are cheaper and easier to process, making them a viable alternative to metals.

Q: Polymers are essentially long chains of hydrocarbon molecules.

R: Today polymers as strong as metals have been developed.

S: These have replaced the traditional chromium-plated metallic bumpers in cars.

The proper sequence should be:

A. Q R S P B. R S Q P
C. R Q S P D. Q R P S

5. S_1: Biological evolution has not fitted man to any specific environment.

S_6: That brilliant sequence of cultural peaks can most appropriately be termed the ascent of Man.

P: It is by no means a biological evolution, but it is a cultural one.

Q: His imagination, his reason, his emotional subtlety and toughness, make it possible for him not to accept the environment but to change it.
R: And that series of inventions by which man from age to age has reshaped his environment is a different kind of evolution.
S: Among the multitude of animals which scamper, fly, burrow and swim around us, he is the only one who is not locked into his environment.

The proper sequence should be:
A. Q P R S   B. S R Q P
C. Q R S P   D. S Q R P

6. S_1: Growing up means not only getting larger, but also using our senses and our brains to become more aware of the things around us.
S_6: In other words, we must develop and use our ability to reason, because the destruction or the preservation of the places in which we live depends on us.
P: Not only does he have a memory but he is able to think and reason.
Q: In this, man differs from all other animals.
R: Before we spray our roadside plants or turn sewage into our rivers, we should pause to think what the results of our actions are likely to be.
S: That is to say, he is able to plan what he is going to do in the light of his experience before he does it.

The proper sequence should be:
A. Q R S P   B. S P Q R
C. S P R Q   D. Q P S R

7. S_1: It is regrettable that there is widespread corruption in the country at all levels.
S_6: This is indeed a tragedy of great magnitude.
P: So there is hardly anything that the government can do about it now.
Q: And there are graft and other malpractices too.
R: The impression that corruption is a universal phenomenon persists and the people do not cooperate in checking this evil.
S: Recently several offenders were brought to book, but they were not given deterrent punishment.

The proper sequence should be:
A. Q S R P   B. S Q R P
C. R S Q P   D. P Q S R

8. S_1: Smoke oozed up between the planks.
S_6: Most people bore the shock bravely.
P: Passengers were told to be ready to quit the ship.
Q: The rising gale fanned the smouldering fire.
R: Everyone now knew there was a fire on board.
S: Flames broke out here and there.

The proper sequence should be:
A. S R Q P   B. Q P S R
C. R S P Q   D. Q S R P

9. S_1: It was a dark moonless night.
S_6: They all seemed to him to be poor and ordinary—mere childish words.
P: He turned over the pages, reading passages here and there.
Q: He heard them on the floor.
R: The poet took down his books of poems from his shelves.
S: Some of them contained his earliest writings which he had almost forgotten.

The proper sequence should be:
A. R P Q S   B. R Q S P
C. R S P Q   D. R P S Q

10. S_1: A noise started above their heads.
S_6: Nearly two hundred lives were lost on the fateful day.
P: But people did not take it seriously.
Q: That was to show everyone that there was something wrong.
R: It was a dangerous thing to do.
S: For, within minutes the ship began to sink.

The proper sequence should be:
A. P Q S R   B. P R Q S
C. Q P R S   D. Q P S R

11. S_1: The cooperative system of doing business is a good way of encouraging ordinary workers to work hard.
S_6: The main object is to maintain the interest of every member of the society and to ensure that the members participate actively in the projects of the society.
P: If the society is to be well run, it is necessary to prevent insincere officials being elected to the committee which is solely responsible for the running of the business.
Q: They get this from experienced and professional workers who are not only familiar with the cooperative system, but also with efficient methods of doing business.
R: To a large extent, many cooperative societies need advice and guidance.
S: The capital necessary to start a business venture is obtained by the workers' contributions.

The proper sequence should be:
A. S Q P R   B. P Q S R
C. S R Q P   D. P S R Q

12. S_1: American private lives may seem shallow.
S_6: This would not happen in China, he said.
P: Students would walk away with books they had not paid for.
Q: A Chinese journalist commented on a curious institution: the library.
R: Their public morality, however, impressed visitors.
S: But in general they returned them.

The proper sequence should be:

A. P S Q R B. Q P S R

C. R Q P S D. R P S Q

13. S_1: The *Bhagavadgita* recognises the nature of man and the needs of man.

S_6: A man who does not harmonise them, is not truly human.

P: All these three aspects constitute the nature of man.

Q: It shows how the human being is a rational one, an ethical one and a spiritual one.

R: More than all, it must be a spiritual experience.

S: Nothing can give him fulfilment unless it satisfies his reason, his ethical conscience.

The proper sequence should be:

A. P S R Q B. R S P Q

C. Q P S R D. P S Q R

14. S_1: I usually sleep quite well in the train, but this time I slept only a little.

S_6: It was shut all night, as usual.

P: Most people wanted it shut and I wanted it open.

Q: As usual, I got angry about the window.

R: The quarrel left me completely upset.

S: There were too many people and too much luggage all around.

The proper sequence should be:

A. R S Q P B. S Q P R

C. S Q R P D. R S P Q

15. S_1: For decades, American society has been called a melting pot.

S_6: In recent years, such differences—accentuated by the arrival of immigrants from Asia and other parts of the world in the United States—have become something to celebrate and to nurture.

P: Differences remained—in appearance, mannerisms, customs, speech, religion and more.

Q: The term has long been a cliche, and a half-truth.

R: But homogenisation was never achieved.

S: Yes, immigrants from diverse cultures and traditions did cast off vestiges of their native lands and become almost imperceptibly woven into the American fabric.

The proper sequence should be:

A. Q R S P B. S Q R P

C. S Q P R D. Q S R P

16. S_1: While talking to a group, one should feel self-confident and courageous.

S_6: Any man can develop his capacity if he has the desire to do so.

P: Nor is it a gift bestowed by Providence on only a few.

Q: One should also learn how to think calmly and clearly.

R: It is like the ability to play golf.

S: It is not as difficult as most men imagine.

The proper sequence should be:

A. S Q P R B. Q S P R

C. Q R S P D. R S Q P

17. S_1: In 1934, William Golding published a small volume of poems.

S_6: *But Lord of the Flies* which came out in 1954 was welcomed as "a most absorbing and instructive tale".

P: During the World War II (1939-45) he joined the Royal Navy and was present at the sinking of the *Bismarck.*

Q: He returned to teaching in 1945 and gave it up in 1962, and is now a full-time writer.

R: In 1939, he married and started teaching at Bishop Wordsworth's School in Salisbury

S: At first his novels were not accepted.

The proper sequence should be:

A. R P Q S B. R P S Q

C. S R P Q D. S Q P R

18. S_1: Our ancestors thought that anything which moved itself was alive.

S_6: Therefore some scientists think that life is just a very complicated mechanism.

P: The philosopher Descartes thought that both men and animals were machines.

Q: But a machine such as a motorcar or a steamship moves itself, and as soon as machines which moved themselves had been made, people asked, "Is man a machine?"

R: And before the days of machinery that was a good definition.

S: He also thought that the human machine was partly controlled by the soul action on a certain part of the brain, while animals had no souls

The proper sequence should be:

A. P R S Q B. R P Q S

C. P S Q R D. R Q P S

19. S_1: But how does a new word get into the dictionary?

S_6: He sorts them according to their grammatical function, and carefully writes a definition.

P: When a new dictionary is being edited, a lexicographer collects all the alphabetically arranged citation slips for a particular word.

Q: The dictionary makers notice it and make a note of it on a citation slip.

R: The moment a new word is coined, it usually enters the spoken language.

S: The word then passes from the realm of hearing to the realm of writing.

The proper sequence should be:

A. P Q R S B. P R S Q

C. R Q P S D. R S Q P

20. S_1: There is a touching story of Professor Hardy visiting Ramanujan as he lay desperately ill in hospital at Putney.

S_6: It is the lowest number that can be expressed in two different ways as the sum of two cubes."

P: 'No, Hardy, that is not a dull number in the very least.

Q: Hardy, who was a very shy man, could not find the words for his distress.

R: It was 1729.

S: The best he could do, as he got to the bedside was: "I say, Ramanujan, I thought the number of the taxi I came down in was a very dull number".

The proper sequence should be:

A. P R S Q B. Q S R P
C. Q S P R D. S Q R P

21. S_1: The heart is the pump of life.

S_6: All this was made possible by the invention of the heart-lung machine.

P: They have even succeeded in heart transplants.

Q: Nowadays surgeons are able to stop a patient's heart and carry out complicated operations.

R: A few years ago it was impossible to operate on a patient whose heart was not working properly.

S: If the heart stops we die in about five minutes.

The proper sequence should be:

A. S R Q P B. S P R Q
C. S Q P R D. S R P Q

22. S_1: On vacation in Tangier, Morocco, my friend and I sat down at a street cafe.

S_6: Finally a man walked over to me and whispered, "Hey buddy ..... this guy's your waiter and he wants your order."

P: At one point, he bent over with a big smile, showing me a single gold tooth and a dingy fez.

Q: Soon I felt the presence of someone standing alongside me.

R: But this one wouldn't budge.

S: We had been cautioned about beggars and were told to ignore them.

The proper sequence should be:

A. S Q R P B. S Q P R
C. Q S R P D. Q S P R

23. S_1: There is only one monkey we can thoroughly recommend as an indoor pet.

S_6: Finally, let me say that no other monkey has a better temper or more winning ways.

P: They quickly die from colds and coughs after the first winter fogs.

Q: It is the beautiful and intelligent Capuchin monkey.

R: The lively little Capuchins, however, may be left for years in an English house without the least danger to their health.

S: The Marmosets, it is true, are more beautiful than the Capuchins and just as pleasing, but they are too delicate for the English climate.

The proper sequence should be:

A. P Q R S B. Q R P S
C. Q S P R D. R P S Q

24. S_1: Hungry, with a population of about ten million, lies between Czechoslovakia to the north and Yugoslavia to the south.

S_6: The new industries derive mainly from agricultural production.

P: Here a great deal of grain is grown.

Q: In recent years, however, progress has been made also in the field of industrialisation.

R: Most of this country consists of an extremely fertile plain, through which the river Danube flows.

S: In addition to grain, the plain produces potatoes, sugar, wine and livestock.

The proper sequence should be:

A. Q R S P B. R P S Q
C. P R S Q D. R Q S P

25. S_1: Throughout history man has used energy from the sun.

S_6: This energy comes from inside atoms.

P: Today, when we burn wood or use electric current we are drawing on energy.

Q: However, we now have a new supply of energy.

R: All our ordinary life depends on the sun.

S: This has come from the sun.

The proper sequence should be:

A. S Q P R B. R Q P S
C. Q S R P D. P S R Q

26. S_1: In India marriages are usually arranged by parents.

S_6: She felt she was a modern girl and not a subject for bargaining

P: Sometimes girls and boys do not like the idea of arranged marriages.

Q: Most young people accept this state of affairs.

R: Shanta was like that.

S: They assume their parents can make good choices.

The proper sequence should be:

A. S P R Q B. P S R Q
C. Q S P R D. R Q P S

27. S_1: I had halted on the road.

S_6: I decided to watch him for a while and then go home.

P: As soon as I saw the elephant I knew I should not shoot him.

Q: It is a serious matter to shoot a working elephant.

R: I knew that his 'must' was already passing off.

S: The elephant was standing eighty yards from the road.

The proper sequence should be:
A. S P Q R B. P Q S R
C. R Q P S D. S R P Q

28. S_1: A man can be physically confined within stone walls.
S_6: No tyranny can intimidate a lover of liberty.
P: But his mind and spirit will still be free.
Q: Thus his freedom of action may be restricted.
R: His hopes and aspirations still remain with him.
S: Hence, he will be free spiritually if not physically.

The proper sequence should be:
A. P Q R S B. S R Q P
C. Q P R S D. Q P S R

29. S_1: The dictionary is the best friend for your task.
S_6: Soon you will realize that this is an exciting task
P: That may not be possible always.
Q: It is wise to look it up immediately.
R: Then it must be firmly written on the memory and traced at the first opportunity.
S: Never allow a strange word to pass unchallenged.

The proper sequence should be:
A. P Q R S B. S P Q R
C. Q R P S D. S Q P R

30. S_1: Far away in a little street there is a poor house.
S_6: His mother has nothing to give but water, so he is crying
P: Her face is thin and worn and her hands are coarse, pricked by a needle, for she is a seam-stress.
Q: One of the windows is open and through it I can see a poor woman.
R: He has a fever and a asking for oranges.
S: In a bed in a corner of the room her little boy is lying ill.

The proper sequence should be:
A. S R Q P B. P Q S R
C. Q P S R D. R S P Q

31. S_1: Calcutta unlike other cities, kept its trams.
S_6: The foundation stone was laid in 1972.
P: As a result, there is horrendous congestion.
Q: It was going to be the first in South Asia.
R: They run down the centre of the road.
S: To ease in the city decided to build an underground railway line.

The proper sequence should be:
A. P R S Q B. P S Q R
C. S Q R P D. R P S Q

32. S_1: We now know that oceans are very deep.
S_6: This reaches from India to the Antarctic.
P: For example, the Indian Ocean has a range called the Indian Ridge.
Q: Much of it is fairly flat.
R: However, there are great mountain ranges as well.
S: On average the bottom is two and a half to three and a half miles down.

The proper sequence should be:
A. S Q P R B. P Q S R
C. R S Q P D. Q P R S

33. S_1: As he passed beneath her he heard the swish of her wings.
S_6: The next moment he felt his wings spread outwards.
P: He was not falling head long now.
Q: Then monstrous terror seized him
R: But it only lasted a minute.
S: He could hear nothing.

The proper sequence should be:
A. P S Q R B. Q S P R
C. Q S R P D. P R Q S

34. S_1: When a satellite is launched, the rocket begins by going slowly upwards through the air.
S_6: Consequently, the rocket still does not become too hot.
P: However, the higher it goes, the less air it meets.
Q: As the rocket goes higher, it travels faster.
R: For the atmosphere becomes thinner.
S: As a result there is less friction.

The proper sequence should be:
A. Q P R S B. Q S P R
C. P Q R S D. P Q S R

35. S_1: Sunbirds are among the smallest of Indian birds.
S_6: Our common sunbirds are the purple sunbird, the glossy black species and purplerumped sunbird, the yellow and maroon species
P: Though they are functionally similar to the hummingbirds of the New World, they are totally unrelated.
Q: They do eat small insects too.
R: They are also some of the most brilliantly-coloured birds.
S: Sunbirds feed on nectar mostly and help in pollination.

The proper sequence should be:
A. S Q P R B. R P S Q
C. Q P R S D. P S R Q

36. S_1: Venice is a strange and beautiful city in the north of Italy.
S_6: This is because Venice has no streets.
P: There are about four hundred old stone bridges joining the island of Venice.
Q: In this city there are no motor cars, no horses and no buses.
R: These small islands are near one another.
S: It is not an island but a hundred and seventeen islands.

The proper sequence should be:
A. P Q R S B. P R Q S
C. S R P Q D. P Q S R

37. S_1: A ceiling on urban property.

S_6: Since their value would exceed the ceiling fixed by the Government.

P: No mill-owner could own factories or mills or plants.

Q: And mass circulation papers.

R: Would mean that.

S: No press magnate could own printing presses.

The proper sequence should be:

A. Q S R P B. R P S Q
C. S R P Q D. Q P S R

38. S_1: The weather-vane often tops a church spire, tower or high building.

S_6: The weather-vane can, however, give us some indication of the weather.

P: They are only wind-vanes.

Q: Neither alone can tell us what the weather will be.

R: They are designed to point to the direction from which the wind is coming.

S: Just as the barometer only tells us the pressure of the air, the weather-vane tells us the direction of the wind.

The proper sequence should be:

A. P Q R S B. P S R Q
C. P R S Q D. S P Q R

39. S_1: Most of the universities in the country are now facing financial crisis.

S_6: The Government should realise this before it is too late.

P: Cost benefit yardstick thus should not be applied in the case of the univesities.

Q: The current state of affairs cannot be allowed to continue for long.

R: Universities cannot be equated with commercial enterprises.

S: Proper development of universities and colleges must be ensured

The proper sequence should be:

A. Q R P S B. Q S P R
C. Q R S P D. Q P R S

40. S_1: I keep on flapping my big ears all day.

S_6: Am I not a smart, intelligent elephant?

P: They also fear that I will flap them all away.

Q: But children wonder why I flap them so.

R: I flap them so to make sure they are safely there on either side of my head.

S: But I know what I am doing.

The proper sequence should be:

A. S R Q P B. Q P S R
C. Q P R S D. P S R Q

ANSWERS

1	2	3	4	5	6	7	8	9	10
A	C	D	A	C	D	A	A	D	C
11	**12**	**13**	**14**	**15**	**16**	**17**	**18**	**19**	**20**
A	B	B	B	B	B	A	C	A	B
21	**22**	**23**	**24**	**25**	**26**	**27**	**28**	**29**	**30**
A	C	B	B	D	C	B	A	D	C
31	**32**	**33**	**34**	**35**	**36**	**37**	**38**	**39**	**40**
D	A	C	A	A	C	B	D	A	B

13

Punctuation

Punctuation Marks help us to make grammatically correct sentences.

The following pairs of sentences will bring home to you the importance of punctuation marks:

(*i*) Jack, my neighbour, is going abroad.

(*ii*) Jack, my neighbour is going abroad.

The sentence no. (*i*) means 'Jack' is my neighbour and he is going abroad. The sentence no. (*ii*) means that I'm telling Jack that my neighbour is going abroad.

The following are some most commonly used punctuation marks:

I. Full Stop	II. Note of Interrogation
III. Note of Exclamation	IV. Comma
V. Inverted commas	VI. Apostrophe
VII. Capital Letters	VIII. Colon
IX. Semicolon	X. Dash
XI. Oblique	XII. Parentheses
XIII. Hyphen	

I. Full Stop (.)

(A) At the end of a statement or an imperative sentences the full stop is applied.

(*i*) Mohan sings a song. (Positive Statement)

(*ii*) She does not tell a lie. (Negative Statement)

(*iii*) Open the door. (Imperative Sentence)

(B) To mark abbreviations:

(*i*) M.L.A. (*ii*) M.B.B.S.

(*iii*) B.A. (*iv*) Mr R.S. Gill

Note: Now a days after Mr, Mrs, Miss, a full stop is not applied.

II. Note of Interrogation (?)

It is used to make direct questions as well as tags.

(*i*) What are you doing?

(*ii*) Where is your friend?

(*iii*) You like eggs, don't you?

(*iv*) She doesn't obey her elder brother, does she?

Note: The note of interrogation is not used after indirect questions.

(*i*) She was sure whether her brother would come there to see her or not.

(*ii*) I asked him when he would start a confectionary shop.

III. Note of Exclamation (!)

(A) It is used after an optative sentence

(*i*) May you live long!

(*ii*) May God bless you!

(B) After an exclamatory sentence

(*i*) What a beautiful butterfly it is!

(*ii*) How wonderful the rainbow is!

(C) After an Interjection

(*i*) Hurrah! We have won the match.

(*ii*) Alas ! I am done.

IV. Comma (,)

It is used:

(A) To separate identical words.

(*i*) Mohit was tall, thin and handsome.

(*ii*) I bought apples, bananas and guava yesterday.

(*iii*) This shopkeeper sells cakes, pastries and biscuits.

(B) In sentences of direct speech.

(*i*) She said, "I wish to be a doctor."

(*ii*) "Don't move," the man shouted.

(C) To separate pairs of words when they are joined by 'and'.

(*i*) I like rice and curry, idli and dosa.

(*ii*) In the fair were present all kinds of persons rich and poor, weak and strong, foolish and wise.

(D) To join two coordinate clauses when the latter is the consequence of the former.

(*i*) He was rejected by his friends, and he decided never to meet any of them in future.

(*ii*) He was beaten by his opponent, and he decided to leave boxing.

(E) To separate a noun or phrase in apposition.

(*i*) The Mahabharata, the longest epic, was written by Ved Vyas.

(*ii*) Hari, the tallest boy in our class, is a good player of basketball.

(F) To separate an adverbial clause from the rest of the sentences.

(*i*) If you tell me the reality, you won't be punished.

(*ii*) As soon as the thief saw the policeman, he ran away.

(*iii*) When he laughs, the dog in his house will bark.

(G) To separate a nominative of address

(*i*) Where are you, Mr Smith?

(*ii*) How are you, Seema?

(H) To separate a noun clause from the rest of the sentence.
  (*i*) When she will get up, nobody knows.
  (*ii*) Why he is crying so loudly, is known to his parents only.
  (*iii*) What he says, no one cares.

V. Inverted Commas (" ")

Inverted commas are used in sentences of direct speech.

(*i*) She says, "I can drive a car."
(*ii*) He said to me, "I'll convey your message to your uncle."

VI. Apostrophe(')

It is used:

(A) To make Genitive Case when the noun is singular
  (*i*) Mohan's shirt
  (*ii*) James' shoes or James's shoes.

(B) To makes Genitive Case when the noun is plural
  (*i*) Boys' college (*ii*) Girls' hostel
  (*iii*) Men's club (*iv*) Women's hostel

(C) To make plural form of letters and figures
  (*i*) H's (*ii*) A's
  (*iii*) 4's (*iv*) 8's

(D) To denote the omission of letter or letters from words.
  (*i*) can't (*iii*) won't
  (*ii*) don't

(E) To express a possessive pertaining to a period of time:
  (*i*) a week's holiday.
  (*ii*) a year's programme.

VII. Capital Letters

Capital letters are used as under:

(A) The first word of a sentence should start with a capital letter.
  (*i*) She does not know how to play the piano.
  (*ii*) We are going to perform a dance in London.

(B) A proper name should be started with a capital letter.
  (*i*) Mohan is my fast friend.
  (*ii*) I like all students of my class but I like Ajay, Rajesh and Vinod in particular.
  (*iii*) Delhi is the capital of India.

(C) The important words in the title of a book should start with a capital letter each:
  (*i*) The Cat and the Mouse.
  (*ii*) Love and Life.

(D) The first letter of the days of a week and the months of year should be written in capital.
  (*i*) Sunday, Monday, Tuesday.
  (*ii*) January, February, March.

(E) The names of all rivers, oceans, mountains, bays, channels, peaks, etc., should start with a capital letter.
  (*i*) The Ganga (*ii*) The Himalayas
  (*iii*) The Bay of Bengal (*iv*) Mount Everest
  (*v*) The English Channel (*vi*) The Indian Ocean

(F) The names of all holy books, magazines, newspapers, etc. should start with a capital letter.
  (*i*) The Gita (*ii*) The Bible
  (*iii*) The Tribune (*iv*) The Sun

VIII. Colon (:)

Colon represents a long pause. It is not used so frequently but in some cases it is applied.

(A) To start the names of certain items both proper and common nouns:
  (*i*) Some of Shakespeare's plays are: Macbeth, Tempest, Julius Caesar, etc.
  (*ii*) The following items are required by our school urgently:
    (*a*) Five Hockey Sticks (*b*) Ten Cricket Bats
    (*c*) Four Football (*d*) Eight Basketballs

(B) To introduce a quotation.
  (*i*) Keats says: A thing of beauty is a joy forever.
  (*ii*) Shakespeare says: All that glitters is not gold.

(C) Before some examples:
  (*i*) I know several languages:
  Hindi, Punjabi, French and English.
  (*ii*) Mohan is: honest, wise, handsome, rich and kind.

IX. Semicolon (;)

It represents a pause which is longer than a comma but is of less duration than a colon. A semicolon is used:

1. To mark off a series of clauses:
  (*i*) We love our young ones; they too in return love us.
  (*ii*) This forest is a witness to good and bad days of mankind; this is here since ancient times.
2. To break up long sentences

My brother reads in 10th standard; he always goes to school late; he is almost everyday punished by the teacher.

X. Dash (—)

It is used:

(A) To mark a sudden change in our thoughts
  (i) I think you like berries—but your brother told me that you spend no money on fruit.
  (ii) You must read this book—but how can you read it when you're writing a letter to your friends?

(B) To mark a sudden interruption in a sentence.
  I know him—I met him five years ago—where? I can't recall.

(C) To divulge a secret in a dramatic way.
  (*i*) The most famous modern novelist is—Sydney Sheldon.
  (*ii*) The best singer is—Mohammad Rafi.

XI. Oblique (/)

(A) It is used to represent abbreviations
  (*i*) D/o (Daughter of) (*ii*) S/o (Son of)
  (*iii*) W/o (Wife of)

(B) To represent masculine gender or feminine gender
 (*i*) He/She can join the army.
 (*ii*) I'll ask questions to him/her.

(C) To represent possibilities.
 (*i*) The girl/girls should work even harder.
 (*ii*) The boy/boys cannot be given any special rights.

XII. Parentheses (or Brackets) ( )

These are used:

(A) To explain something additional which is usually not of much significance.
 (*i*) Manoj (a five year old boy) saved the life of his friend.
 (*ii*) Meena (a girl of 5th standard) was called by the President for her remarkable paintings.

(B) To separate an after thought
 (*i*) All his sons (including the eldest one, who is an addict to alcohol) are going abroad.
 (*ii*) All the ingredients (excluding milk) should be put in a pan.

XIII. Hyphen (-)

It is shorter than a dash in length.

It is most often used to make compound words.
 (*i*) sun-rise
 (*ii*) sun-set

MULTIPLE CHOICE QUESTIONS

Directions: *In the questions given below, which one alternative out of four has been applied with correct punctuation marks?*

1. Hurry up the train leaves at six o'clock
 A. Hurry up; the train leaves at six o'clock.
 B. Hurry up, the train leaves at six O'clock.
 C. Hurry up! the train leaves at six o'clock.
 D. Hurry up, the train leaves at six o'clock.

2. It is a sin to waste food said the father
 A. It is a sin to waste food said the father.
 B. 'It is a sin to waste food' said the father.
 C. "It is a sin to waste food" said the father.
 D. "It is a sin to waste food," said the father.

3. This is his fathers car
 A. This is his fathers' car
 B. This is his father's car
 C. This is his father's car.
 D. This is his fathers' car.

4. The man asked the taxi driver will you take me to the airport
 A. The man asked the taxi driver, will you take me to the airport?
 B. The man asked the taxi driver, "Will you take me to the airport?"
 C. The man asked the taxi driver, "Will you take me to the airport."
 D. The man asked the taxi driver, "will you take me to the airport!"

5. Bring me a glass of water said the teacher I am feeling very thirsty
 A. "Bring me a glass of water," said the teacher, "I am feeling very thirsty."
 B. "Bring me a glass of water" said the teacher, "I am feeling very thirsty."
 C. "Bring me a glass of water" said the teacher "I am feeling very thirsty."
 D. "Bring me a glass of water, said the teacher, "I am feeling very thirsty."

6. Ram said I am in a hurry I cannot spare time
 A. Ram said, "I am in a hurry" "I cannot spare time."
 B. Ram said "I am in a hurry I cannot spare time."
 C. Ram said, "I am in a hurry, I cannot spare time."
 D. Ram said, "I am in a hurry I cannot spare time."

7. How have you found so much wealth asked Anil's wife
 A. "How have you found so much wealth!" asked Anil's wife.
 B. "How have you found so much wealth?" asked Anil's wife.
 C. "How have you found so much wealth." asked Anil's wife.
 D. "How have you found so much wealth?", asked Anil's wife.

8. What is the matter why are you crying he asked the child
 A. "What is the matter, why are you crying?" he asked the child.
 B. "What is the matter? why are you crying?" he asked the child
 C. "What is the matter" "why are you crying?" he asked the child.
 D. "What is the matter? Why are you crying?" he asked the child.

9. I am going to marry and I want a house to live in said the young man
 A. "I am going to marry and I want a house to live in", said the young man.
 B. "I am going to marry" and "I want a house to live in" said the young man.
 C. "I am going to marry and I want a house to live in" said the young man.
 D. "I am going to marry and I want a house to live in." said the young man.

10. What harm is there if we admit the error the lawyer said
 A. "What harm is there, if we admit the error?" the lawyer said.

B. "What harm is there if, we admit the error?" the lawyer said.

C. "What harm is there if we admit the error?" the lawyer said.

D. "What harm is there? if we admit the error". the lawyer said.

11. The more we hate one another the worse it will be for everybody

A. The more we hate one another the worse it will be for everybody.

B. The more we hate one another, the worse it will be for everybody.

C. The more we hate one another; the worse it will be for everybody.

D. The more we hate one another – the worse it will be for everybody.

12. Anita's father Mr Anil Kumar is a teacher

A. Anita's father Mr Anil Kumar is a teacher.

B. Anita's father, Mr Anil Kumar is a teacher.

C. Anita's father Mr Anil Kumar, is a teacher.

D. Anita's father, Mr Anil Kumar, is a teacher.

13. Considering all aspects of the matter they decided to give up the plan.

A. Considering all aspects of the matter, they decided to give up the plan.

B. Considering, all aspects of the matter, they decided to give up the plan.

C. Considering all aspects, of the matter, they decided to give up the plan.

D. Considering all aspects of the matter they decided to give up the plan.

14. The Narmada the Ganga and the Yamuna are three of the largest Indian rivers

A. The Narmada, the Ganga and the Yamuna are three of the largest Indian rivers

B. The Narmada, the Ganga, and the Yamuna are three of the largest Indian rivers

C. The Narmada, the Ganga and the Yamuna are three of the largest Indian rivers.

D. The Narmada, the Ganga and, the Yamuna are three of the largest Indian rivers.

15. What a remarkable victory

A. What, a remarkable victory.

B. What a remarkable victory.

C. What a remarkable victory?

D. What a remarkable victory!

16. The election many people said would spring a big surprise and they were proved right

A. The election, many people said, "would spring a big surprise and they were proved right."

B. "The election," many people said, "would spring a big surprise and they were proved right."

C. "The election" many people said, "Would spring a big surprise and they were proved right."

D. "The election" many people said "would spring a big surprise and they were proved right."

17. Are you going he asked we are not I replied

A. "Are you going?" he asked, "We are not." I replied.

B. "Are you going?" he asked, "we are not" I replied.

C. "Are you going?" he asked. "We are not," I replied.

D. "Are you going?" he asked, "we are not" I replied.

18. Here I am master I am ready to go wherever you want me to go he said

A. "Here I am, master, I am ready to go," wherever you want me to go he said.

B. "Here I am master, I am ready to go wherever you want me to go" he said.

C. "Here I am master, I am ready to go wherever you want me to go", he said.

D. "Here I am, master, I am ready to go wherever you want me to go", he said.

19. What is your reason for refusing to marry this girl asked his mother

A. "What is your reason for refusing to marry this girl?" asked his mother.

B. "What is your reason, for refusing to marry this girl?" asked his mother.

C. "What is your reason for refusing to marry this girl?" asked his mother

D. "What is your reason for refusing to marry this girl," asked his mother.

20. He said my brother had an accident on Sunday the 28th April 2013 and is still confined to bed

A. He said "my brother had an accident on Sunday the 28th April 2013 and is still confined to bed."

B. He said "My brother had an accident on Sunday the 28th April 2013 and is still confined to bed."

C. He said, "My brother had an accident on Sunday, the 28th April 2013, and is still confined to bed."

D. He said," My brother had an accident on Sunday, the 28th April 2013 and is still confined to bed."

21. How many r's are there in referred I asked

A. How many r's are there in referred? I asked.

B. "How many r's are there, in referred" I asked.

C. "How many r's are there in referred?" I asked.

D. "How many rs' are there in referred?" I asked.

22. Friends colleagues relatives none stood by him

A. Friends, colleagues, relatives none stood by him.

B. Friends, colleagues, relatives, none stood by him.

C. Friends, colleagues, relatives, none, stood by him

D. Friends, colleagues, relatives none, stood by him.

23. He said I am sorry I could not reach in time

A. He said, "I am sorry I could not reach in time."

B. He said "I am sorry. I could not reach in time."

C. He said, "I am sorry. I could not reach in time."
D. He said "I am sorry I could not reach in time."

24. Never listen to his stories don't believe what he says and always distrust him
A. Never listen to his stories, don't believe what he says? and always distrust him.
B. Never listen to his stories, don't believe, what he says? and always distrust him.
C. Never listen to his stories, do'nt believe what he says and always distrust him.
D. Never listen to his stories, don't believe what he says and always distrust him.

25. Man proposes God disposes
A. Man proposes God disposes.
B. Man proposes, God disposes.
C. Man proposes, God disposes!
D. Man proposes God disposes!

26. He asked me what I wanted
A. He asked me, what I wanted?
B. He asked me "what I wanted."
C. He asked me what I wanted.
D. He asked me "what I wanted!"

27. He said to me lets wait for him
A. He said to me, "Let's wait for him."
B. He said to me 'Lets wait for him.'
C. He said to me, "let's wait for him."
D. he said to me, "Let's wait for him."

28. In india the ganga and the yamuna are regarded as sacred rivers
A. In India, the Ganga and the Yamuna are regarded as sacred rivers.
B. In India, the ganga and the Yamuna are regarded as sacred rivers.
C. In India, the ganga and the yamuna are regarded as sacred rivers.
D. In India, the Ganga and the yamuna are regarded as sacred rivers.

29. To err is human to forgive divine
A. To err is human to forgive divine.
B. To err is human, to forgive divine.
C. To err, is human to forgive, divine.
D. To err is human; to forgive, divine.

30. He was born on August 16, 1944 in Howrah West Bengal
A. He was born on August 16, 1944 in Howrah West Bengal.
B. He was born on August 16, 1944, in Howrah West Bengal.
C. He was born on, August 16 1944 in Howrah West Bengal.
D. He was born on August, 16 1944, in Howrah West Bengal.

31. He is to tell you the truth is a fool
A. He is, to tell you the truth, is a fool.
B. He is to tell you, the truth is a fool.
C. He is to tell you the truth, is a fool.
D. He is to tell you the truth is a fool.

32. This in brief is what he said
A. This in brief is what he said.
B. This in brief, is what he said.
C. This, in brief, is what he said.
D. This, in brief is what, he said.

33. If you need my help write to me
A. If you need my help write to me.
B. If, you need my help, write to me.
C. If you need, my help, write to me.
D. If you need my help, write to me.

34. They started running for they were getting late
A. They started running, for they were getting late.
B. They started running for they were getting late.
C. They started running for, they were getting late.
D. They started, running for, they were getting late.

35. The sun having set it became pleasant
A. The sun having set it became pleasant.
B. The sun, having set, it became pleasant.
C. The sun having set, it became pleasant.
D. The sun, having set it, became pleasant.

36. Mr Sushil please listen to me
A. Mr Sushil please listen to me.
B. Mr Sushil, please listen to me.
C. Mr Sushil please, listen to me.
D. Mr Sushil, please, listen to me.

37. Wherever you go you face the same difficulties
A. Wherever you go you face the same difficulties.
B. Wherever, you go, you face the same difficulties.
C. Wherever you go, you face, the same difficulties.
D. Wherever you go, you face the same difficulties.

38. Tell me what I can do for you
A. Tell me what I can do for you.
B. Tell me what I can do for you!
C. Tell me what I can do for you?
D. Tell me, what I can do for you?

39. May you live long
A. May you live long.
B. May you live long?
C. May you live long!
D. May, you live long!

40. May God the Greatest bless you
A. May God the Greatest bless you.
B. May God, the Greatest bless you!
C. May God, the Greatest, bless you.
D. May God, the Greatest, bless you!

41. The sun having risen the fog disappeared
A. The sun having risen the fog disappeared.
B. The sun having risen, the fog disappeared.
C. The sun having risen; the fog disappeared.
D. The sun, having risen, the fog disappeared!

42. Having done his homework he went to bed
A. Having done his homework he went to bed.
B. Having done, his homework he went to bed.
C. Having done, his homework, he went to bed.
D. Having done his homework, he went to bed.

43. He is indeed a foolish boy
A. He is indeed, a foolish boy.
B. He is indeed a foolish boy.
C. He is, indeed, a foolish boy.
D. He, is indeed, a foolish boy.

44. Naresh my younger brother has gone to Patna
A. Naresh, my younger brother, has gone to Patna.
B. Naresh, my younger brother has gone to Patna.
C. Naresh my younger brother, has gone to Patna.
D. Naresh! my younger brother has gone to Patna.

45. Well I shall look into the matter
A. Well I shall look into the matter.
B. Well, I shall look into the matter.
C. Well I shall, look into the matter.
D. Well I shall, look into, the matter.

46. All right you can go
A. All right you can go.
B. All right you, can go.
C. All right, you can go.
D. All right! you can go.

47. Carry on your work I am watching you
A. Carry on your work I am watching you.
B. Carry on, your work, I am watching you.
C. Carry on, your work, I am watching you.
D. Carry on your work, I am watching you.

48. He was a kind bold man and we all praised him.
A. He was a kind bold man and we all praised him
B. He was a kind, bold man and we all praised him.
C. He was a kind bold, man and we all praised him.
D. He was, a kind bold man, and we all praised him.

49. He said Honesty is the best policy
A. He said Honesty is the best policy.
B. He said, Honesty is the best policy.
C. He said "Honesty is the best policy."
D. He said, "Honesty is the best policy."

50. What a beautiful flower it is
A. What a beautiful flower it is.
B. What a beautiful flower it is?
C. What a beautiful flower it is!
D. What! a beautiful flower it is.

ANSWERS

1	2	3	4	5	6	7	8	9	10
D	D	C	B	A	C	B	D	A	C
11	**12**	**13**	**14**	**15**	**16**	**17**	**18**	**19**	**20**
B	D	A	C	D	B	C	D	A	D
21	**22**	**23**	**24**	**25**	**26**	**27**	**28**	**29**	**30**
C	B	A	D	B	C	A	A	D	B
31	**32**	**33**	**34**	**35**	**36**	**37**	**38**	**39**	**40**
A	C	D	A	C	B	D	A	C	D
41	**42**	**43**	**44**	**45**	**46**	**47**	**48**	**49**	**50**
B	D	C	A	B	C	D	B	D	C

14

Figures of Speech

It is a common saying that a picture is worth a thousand words. Same goes with the figures of speech as they elaborate the meaning of words.

FIGURATIVE LANGUAGE

One meaning of "figure" is "drawing" or "image" or "picture". Figurative language creates figures (pictures) in the mind of the reader or listener. These pictures help convey the meaning faster and more vividly than words alone.

We use figures of speech in "figurative language" to add colour and interest, and to awaken the imagination. Figurative language is everywhere, from classical works like Shakespeare or the Bible, to every day speech, pop music and television commercials. It makes the reader or listener use their imagination and understand much more than the plain words.

Figurative language is the opposite of literal language. Literal language means exactly what it says. Figurative language means something different to (and usually more than) what it says on the surface. *e.g.*,

- He ran fast. (literal)
- He ran like the wind. (figurative)

In the above example "like the wind" is a figure of speech (in this case, simile). It is important to recognize the difference between literal and figurative language. There are many figures of speech that are commonly used and which you can learn by heart. At other times, writers and speakers may invent their own figures of speech. If you do not recognize them as figures of speech and think that they are literal, you will find it difficult to understand the language.

In this chapter we will discuss four common types of figures of speech:

- Simile
- Metaphor
- Hyperbole
- Oxymoron

Simile

A simile is a figure of speech that says that one thing is like another different thing. We can use similes to make descriptions more emphatic or vivid.

We often use the words as....as and like with similes. Common patterns for similes, with example sentences, are:

- Something AS adjective AS something
  His skin was as cold as ice.
  It felt as hard as rock.
  She looked as gentle as a lamp.
- Something LIKE something
  My love is like a red, red rose.
  These cookies taste like garbage.
  He had a temper (that was) like a volcano.
  Something LIKE something
  He eats like a pig.
  He smokes like a chimney.
  They fought like cats and dogs.

Here are some more examples of well known similes:

AS adjective AS something	Meaning
as blind as a bat	completely blind
as cold as ice	very cold
as flat as a pancake	completely flat
as gentle as a lamb	very gentle
as light as a feather	very light
as old as the hills	very old
as sharp as a knife	very sharp
as strong as a bull	very strong
as white as snow	pure white
as wise as an owl	very wise

LIKE something	Possible meaning (depending on context)
like a rose	beautiful
like a volcano	explosive
like garbage	disgusting

like an animal	inhuman
like spaghetti	entangled
like dewdrops	sweet and pure
like golddust	precious
like a tip	very untidy (tip = garbage dump)
like a dream	wonderful, incredible
like stars	bright and beautiful

LIKE something	**Meaning**
to drink like a fish	to drink a lot
to eat like a bird	to eat very little
to eat like a horse	to eat a lot
to eat like a pig	to eat impolitely
to fight like cats and dogs	to fight fiercely
to sing like an angel	to sing beautifully
to sleep like a log	to sleep well and soundly
to smoke like a chimney	to smoke heavily, all the time
to soar like an eagle	to fly high and free
to work like a dog	to work very hard

Note that with the AS...AS pattern, the first AS is sometimes suppressed, for example :

His skin was cold as ice.

The above patterns of simile are the most common, but there are others made with adverbs or words such as than and as if, for example :

- He ran as fast as the wind.
- He is larger than life.
- They ran as if for their lives.

Similes can include other figures of speech. For example. "He ran like greased lightning" is a simile that includes hyperbole (greased lightning).

Similes often make use of irony or sarcasm. In such cases they may even mean the opposite of the adjective used. Look at these examples :

- His explanation was as clear as mud. (not clear at all since mud is opaque)
- The film was about as interesting watching a copy of Windows download. (long and boring)
- Watching the show was like watching paint dry. (very boring)

Similes are often found (and they sometimes originate) in poetry and other literature. Here are a few examples :

- A woman without a man is like a fish without a bicycle–Irina Dunn
- Dawn breaks open like a wound that bleeds afresh–Wilfred Owen
- Death has many times invited me : it was like the salt invisible in the waves–Pablo Neruda
- Guiltless forever, like a tree–Robert Browning
- Happy as pigs in mud–David Eddings
- How like the winter hath my absence been–William Shakespeare
- As idle as a painted ship upon a painted ocean–Samuel Taylor Coleridge
- Jubilant as a flag unfurled–Dorothy Parker
- So are you to my thoughts as food to life–William Shakespeare
- Yellow butterfiles flickered along the shade like flecks of sun–William Faulkner

Metaphor

A metaphor is a figure of speech that says that one thing is another different thing. This allows us to use fewer words and forces the reader or listener to find the similarities.

The word metaphor comes from the Greek word *metapherin* (meaning "transfer").

The simplest form of metaphor is : "The [first thing] is a [second thing]."

Look at this example :

Her home was a prison.

In the above sentence, we understand immediately that her home had some of the characteristics of a prison. Mainly, we imagine, she could not leave her home. She was trapped inside. Why it was a prison we do not know, but that would be clear from the context–perhaps her husband forced her to stay at home, perhaps she was afraid of the outside. We don't know, but the rest of the story would tell us. What is important here is that in five simple words we understand a lot about her environment, how she felt and how she behaved. In this sentence, "prison" is a metaphor.

Look at another example :

George is a sheep.

What is one characteristic of sheep? They follow each other. So we can imagine that George is a follower, not a leader. In this sentence "sheep" is a metaphor.

Metaphors are very common in everyday language. But poets also like to use metaphors.

Look at these examples of metaphors with sample sentences and meanings :

Metaphor example	Metaphorical sense	Original sense
I'm not an angel, but I wouldn't behave like that.	exemplary person	a spiritual being believed to be a messenger of God
America is a melting pot.	place where different peoples, styles and cultures are mixed together	a container in which metals or other materials are melted and mixed
John is a real pig when he eats.	greedy person	a four-legged animal kept for meat (pork)
My father is a rock.	very strong or reliable person	a hard, mineral material made of stone
How could she marry a snake like that!	traitor	a long, limbless reptile (e.g., cobra, python, viper)
The policeman let him off with a yellow card. cautioning them	warning	(in soccer) a yellow card that the referee shows to players when

All the above metaphors (the simplest form) are nouns. But there are other ways of making metaphors, for example with verbs or adjustives. Here are some examples :

Metaphor example	Original sense of the word (example)
The committee **shot** ideas **down** one by one.	Anti-aircraft guns that shoot down planes.
The private detective **dug** up enough evidence to convince the police to act.	Dugs like to bury bones and dig them up later.
He **broke into** her conversation	Burglars break into buildings.
The new movie was very popular. People **flocked** to see it.	Birds flock together before they migrate.
His head was **spinning** with ideas.	Some computer hard drives spin at over 10,000 revolutions per minute.
Reading that book **kindled** my interest in politics.	You need to start with twigs and small branches when you kindle a camp fire.
Tim lost his job after a **heated** argument with his boss.	We have a heated swimming pool.
The new car's **sexy** design increased sales for the company.	Some women think that lipstick makes them look sexy.
He was dressed rather vulgarly in a **loud** checked suit.	I can't hear you because the radio is too loud.
It wasn't long before their relationship turned **sour**.	Sour food has an acid taste like lemon or vinegar.

Difference Between Metaphor and Simile

Both similes and metaphors link one thing to another. A simile usually uses "as" or "like". A metaphor is a condensed simile, a shortcut to meaning, which omits "as" or "like". A metaphor creates a relationship directly and leaves more to the imagination. With simile, A is like B. With metaphor, A is B.

Simile	Metaphor
Your eyes are like the sun.	You are my sunshine.
He eats like a pig.	He is a pig.
He lives like a pig.	He is a pig.

Dead Metaphors

In the phrase "to grasp the concept" the physical action "to grasp" is used as a metaphor for "to understand" (which is non-physical). But this phrase has been used so often that most English speakers do not have an image of the physical action in their mind. This metaphor is a "dead metaphor".

Mixed Metaphors

The awkward use of two or more different metaphors at the same time is normally best avoided. It creates conflicting images in the reader or listener's mind, reduces each metaphor's impact, and generally causes confusion. Look at this example :

America is a **melting pot** where new ideas are **kindled**.

Hyperbole

Hyperbole is a figure of speech that uses an exaggerated or extravagant to create a strong emotional response. As a figure of speech it is not intended to be taken literally. Hyperbole is frequently used for humour. Examples of hyperbole are :

- They ran like greased lightning.
- He's got tons of money.
- Her brain is the size of a pea.
- He is older than the hills.
- I will die if she asks me to dance.
- She is as big as an elephant!
- I'm so hungry I could eat a horse.
- I have told you a million times not to lie!

The media and the advertising industry often use hyperbole (which may then be described as hyper or media hype).

Oxymoron

An oxymoron is a figure of speech that deliberately uses two contradictory ideas. This contradiction creates a paradoxical image in the reader or listener's mind that generates a new concept or meaning for the whole. Some typical oxymorons are :

- a living death
- sometimes you have to be cruel to be kind
- a deafening silence
- bitter-sweet
- The Sounds of Silence (song title)
- make haste slowly
- he was conspicuous by his absence

Pseudo Oxymorons

In the standard meaning of oxymoron the contradiction is deliberate. However, in popular usage oxymoron is sometimes used to mean "contradiction in terms", where the contradiction is unintentional. Such expressions, unlike real oxymorons, are commonly used without any sense of paradox in everyday language, for example :

- anecdotal evidence
- friendly fire
- pretty ugly

A common attempt at humour is to describe a certain phrase as an oxymoron, implying that the two parts of the phrase are mutually exclusive and that consequently the phrase as a whole must be nonsensical :

- airline food
- American culture
- eco-tourism
- Microsoft security
- military intelligence

The following are the other major Figures of Speech :

1. **Personification**—"In Personification inanimate objects and abstract notions are spoken of as having life or intelligence."
2. **Onomatopoeia**—"Onomatopoeia is that artifice of language by which the sound of words is made to suggest or echo the sense."
3. **Apostrophe**—"An Apostrophe is a direct address to the dead, to the absent, or to a personified object or idea."
4. **Alliteration**—"Alliteration consists in the repetition of the same sound or syllable at the beginning of two or more words."
5. **Irony**—"Irony is a mode of speech in which the real meaning is exactly the opposite of that which is literally conveyed."
6. **Pun**—"This consists in a play on the various meanings of a word, and is seldom used except as a joke."
7. **Metonymy**—"In Metonymy an object is designated by the name of something which is generally associated with it."
8. **Synecdoche**—"In Synecdoche a part is used to designate the whole, or the whole to designate a part."
9. **Antithesis**—"In Antithesis a striking opposition or contrast of words or sentiments is made in the same sentence. It is employed to secure emphasis."
10. **Transferred Epithet**—"In this figure an epithet is tranferred from its proper word to another that is closely associated with it in the sentence."

MULTIPLE CHOICE QUESTIONS

Directions (Qs. 1 to 40) : *Fill in the blanks with the correct similes.*

1. As black as ....
A. coal B. colour
C. coat D. cloud

2. As blind as a ....
A. cat B. rat
C. dog D. bat

3. As bold as ....
A. grass B. brass
C. ass D. gas

4. As brave as a ....
A. horse B. elephant
C. lion D. bull

5. As busy as a ....
A. man B. animal
C. bird D. bee

6. As cheap as ....
A. dirt B. shirt
C. trouser D. shoes

7. As clear as ....
A. shirt B. picture
C. crystal D. sky

8. As cold as ....
A. ice B. water
C. milk D. breeze

9. As cunning as a ....
A. ox B. fox
C. crow D. man

10. As dead as ....
A. button B. air
C. water D. mutton

11. As deaf as a ....
A. man B. bird
C. stone D. animal

12. As dry as ....
A. cloth B. bone
C. air D. sand

13. As easy as ....
A. ABC B. 123
C. PQR D. XYZ

14. As free as a ....
A. kite B. aircraft
C. bird D. tiger

15. As gentle as a ....
A. child B. woman
C. man D. lamb

16. As good as ....
A. silver B. gold
C. jewellery D. ornaments

17. As good as ....
A. new B. old
C. ancient D. modern

18. As green as ....
A. grass B. colour
C. tree D. life

19. As hard as a ....
A. cock B. glass
C. rock D. ice

20. As light a ....
A. kite B. feather
C. bulb D. hair

21. As mild as ....
A. silk B. water
C. feather D. milk

22. As neat as a new ....
A. sin B. kin
C. pin D. tin

23. As old as the ....
A. air B. water
C. river D. hills

24. As pissed as a ....
A. rat B. dog
C. donkey D. cat

25. As playful as a ....
A. toy B. animal
C. kitten D. bird

26. As pretty as a ....
A. breeze B. ice
C. river D. picture

27. As proud as a ....
A. crow B. peacock
C. vulture D. eagle

28. As quick as a ....
A. flash B. light
C. air D. sound

29. As silent as a ....
A. engine B. machine
C. women D. grave

30. As sly as a ....
A. cat B. fox
C. dog D. crow

31. As smooth as ....
A. ice B. floor
C. silk D. road

32. As sober as a ....
A. judge B. driver
C. cop D. convict

33. As solid as a ....
A. wood B. ice
C. glass D. rock

34. As straight as an ....
A. arm B. ass
C. arrow D. abacus

35. As strong as ....
A. a fox B. wood
C. a camel D. an ox

36. As stubborn as a ....
A. dog B. mule
C. horse D. elephant

37. As sweet as ....
A. pie B. coconut
C. cucumber D. gourd

38. As tough as old ....
A. suit B. man
C. boots D. books

39. As ugly as ....
A. man B. woman
C. moon D. sin

40. As worm as ....
A. milk B. toast
C. water D. air

Directions (Qs. 41 to 60) : *Select the correct option for each of the following questions.*

41. Recognize the figure of speech in the following statement.
He is the Newton of our class.
A. Simile
B. Hyperbole
C. Metaphor
D. Climax

42. Which of the following statements shows oxymoron:
A. He is leading a busy life.
B. He is leading a highly busy-idle life.
C. He is leading an idle life.
D. How happy I am!

43. Recognize the figure of speech in the following:
The Child is father of the Man.
A. alliteration B. hyperbole
C. paradox D. metaphor

44. Which of the following shows Metonymy:
A. Have you read the whole of George Eliot?
B. Which building in Delhi do you like most?
C. Whose acting do you like most among the Bollywood actors?
D. Why are day and night equal on 21 March?

45. Recognize the figure of speech in the following statement.
I wandered lonely as a cloud.
A. Simile B. Metaphor
C. Alliteration D. Hyperbole

46. Which of the following phrases shows onomatopoeia:
A. living like a king.
B. looking beyond the horizon.
C. burning the candle at both ends.
D. murmuring of innumerable bees.

47. Recognize the figure of speech in the following:
Five miles meandering in mazy motion.
A. oxymoron B. climax
C. anticlimax D. alliteration

48. Which of the following shows 'climax':
A. I walk, I jump, I dance.
B. I run, I walk, I sit.
C. He lies, runs and sits.
D. She thinks, writes and observes.

49. Recognize the figure of speech in the following:
Lightly, O lightly, they bear her along.
A. Hyperbole B. Onomatopoeia
C. Climax D. Repetition

50. Which of the following shows 'Hyperbole':
A. You will have to speak for a hundred thousand years to describe her beauty.
B. Love begets love.
C. Better alone than in bad company.
D. Nothing can be decided by war.

51. What Figure of Speech is used when a statement is made emphatic by over statement?
A. Apostrophe B. Hyperbole
C. Personification D. Antithesis

52. What Figure of Speech is used when two contradictory qualities are predicted at once of the same thing?
A. Antithesis B. Onomatopoeia
C. Oxymoron D. Hyperbole

53. In which Figure of Speech the sense is conveyed by the sound of words?
A. Metonymy B. Synecdoche
C. Alliteration D. Onomatopoeia

54. In which Figure of Speech is there a play on the various meanings of a word?
A. Pun B. Metaphor
C. Transferred Epithet D. Irony

55. Which Figure of Speech is produced when the same sound is repeated more than twice at the beginning of a word?
A. Onomatopoeia B. Alliteration
C. Metonymy D. Synecdoche

56. Which of the following is an example of Onomatopoeia?
A. Sweet-bitter tears flowed from her eyes
B. The beetle wheels his droning flight
C. A load of learning lumbering in his head
D. Fair is that fair does

57. Which of the following is an example of Oxymoron?
A. There is kind cruelty in the surgeon's knife
B. Ruin seize thee, ruthless king!
C. And Brutus is an honourable man!
D. Kalidas is the Shakespeare of India

58. Which of the following is an example of Antithesis?
A. Who is to blame but you!
B. Fair is foul, and foul is fair!
C. As you sow, so will you reap
D. Forget me not!

59. Which of the following is an example of Apostrophe?
A. Barking dogs seldom bite
B. My misfortune is your fortune
C. O Solitude! Where are thy charms?
D. I am the monarch of all I survey

60. Which of the following is an example of Alliteration?
A. Sweet are the uses of adversity!
B. Thus idly busy rolls their world away
C. He is the heir apparent
D. With beaded bubbles winking at the brim

Directions (Qs. 61 to 90) : *In each of the following sentences or verses a Figure of Speech has been used. You have to choose the correct Figure of Speech out of the four choices given under each :*

61. Rivers of blood flowed on the battlefied :
A. Simile B. Metaphor
C. Hyperbole D. Alliteration

62. Camel is the ship of the desert :
A. Metaphor B. Simile
C. Pun D. Personification

63. 'O my love's liek a red, red rose, that's newly sprung in June':
A. Alliteration B. Metaphor
C. Simile D. Hyperbole

64. "Opportunity knocks at the door but once."
A. Hyperbole B. Metaphor
C. Antithesis D. Personification

65. "Death lays his icy hand on Kings" :
A. Personification B. Antithesis
C. Metaphor D. Oxymoron

66. "The murmurous haunt of flies on summer eves."
A. Hyperbole B. Alliteration
C. Onomatopoeia D. Personification

67. "Milton! thou should'st be living at this hour. "
A. Personification B. Apostrophe
C. Alliteration D. Irony

68. How high his honour holds his haughty head!
A. Oxymoron B. Personification
C. Alliteration D. Irony

69. "Yet Brutus says he was ambitious. And Brutus is an honourable man."
A. Irony B. Oxymoron
C. Apostrophe D. Pun

70. Is life worth living? That depends upon the liver.
A. Pun B. Alliteration
C. Irony D. Metaphor

71. The pen is mightier than the sword.
A. Simile B. Metaphor
C. Metonymy D. Irony

72. A reeling road, a rolling road, that rambles round the shire.
A. Alliteration B. Metonymy
C. Personification D. Apostrophe

73. "The ploughman homeward plods his weary way."
A. Oxymoron B. Synecdoche
C. Transferred Epithet D. Pun

74. To err is human, to forgive divine :
A. Oxymoron B. Antithesis
C. Transferred Epithet D. Pun

75. He passed a sleepless night :
A. Transferred Epithet B. Antithesis
C. Synecdoche D. Oxymoron

76. The best brains assembled there :
A. Oxymoron B. Synecdoche
C. Irony D. Pun

77. I have many mouths to feed :
A. Synecdoche B. Irony
C. Pun D. Oxymoron

78. Man proposes, God disposes :
A. Synecdoche B. Antithesis
C. Pun D. Irony

79. O Death! Where is thy sting?
A. Metaphor B. Oxymoron
C. Personification D. Apostrophe

80. "Life is a tale told by an idiot,
Full of sound and fury ...."
A. Simile B. Personification
C. Metaphor D. Apostrophe

81. "Thy soul was like a star, and dwelt apart."
A. Metaphor B. Personification
C. Simile D. Onomatopoeia

82. He can devour mountains of food, and drink rivers of whisky.
A. Onomatopoeia B. Hyperbole
C. Metaphor D. Personification

83. "There Honour comes a pilgrim grey" :
A. Metaphor B. Apostrophe
C. Personification D. Onomatopoeia

84. An ambassador is a man who lies abroad for the good of his country :
A. Metonymy B. Personification
C. Pun D. Irony

85. "Husbands had she five on the holy altar" :
A. Hyperbole B. Metaphor
C. Irony D. Pun

86. His honour rooted in dishonour stood :
A. Antithesis B. Oxymoron
C. Irony D. Personification

87. "Sceptre and crown must tumble down."
A. Metonymy B. Antithesis
C. Irony D. Oxymoron

88. A lie has no legs :
A. Apostrophe B. Pun
C. Personification D. Irony

89. Youth is full of pleasure; Age is full of care :
A. Antithesis B. Pun
C. Metaphor D. Personification

90. The righteous shall flourish as the palm trees :
A. Simile B. Transferred Epithet
C. Antithesis D. Synecdoche

ANSWERS

1	2	3	4	5	6	7	8	9	10
A	D	B	C	D	A	C	A	B	D
11	12	13	14	15	16	17	18	19	20
C	B	A	C	D	B	A	A	C	B
21	22	23	24	25	26	27	28	29	30
D	C	D	A	C	D	B	A	D	B
31	32	33	34	35	36	37	38	39	40
C	A	D	C	D	B	A	C	D	B
41	42	43	44	45	46	47	48	49	50
C	B	C	A	A	D	D	A	D	A
51	52	53	54	55	56	57	58	59	60
B	C	D	A	B	B	A	B	C	D
61	62	63	64	65	66	67	68	69	70
C	A	C	D	A	C	B	C	A	A
71	72	73	74	75	76	77	78	79	80
C	A	C	B	A	B	A	B	D	C
81	82	83	84	85	86	87	88	89	90
C	B	C	C	C	B	A	C	A	A

15

Literary Terms

Abridged edition: Condensed or shortened version of a work.

Acrostic : Poem in which the initial letters of the word of each line make a word or words when read downwards.

Act: Major division in a dramatic work.

Adage: Maxim or proverb.

Addendum: Addition or appendix to a book.

Allegory: Story or tale in which characters and events have a second, symbolic, meaning.

Alliteration: Figure of speech where consonants at the beginning of words, or stressed syllables, are repeated.

Almanac: Book or table comprising a calendar of days, weeks and months indicating special occasions or events.

Anagram: Letters or word or phrase which, when re-arranged, form a new word.

Anecdote: Brief account or story about an individual.

Annotation: Textual comment on a book.

Anthology: Collection of poetry or prose from diverse sources.

Antithesis: In Antithesis a striking opposition or contrast of words or sentiments is made in the same sentence. It is employed to secure emphasis.

Aphorism: Short statement of a truth or dogma couched in memorable terms.

Apostrophe: An Apostrophe is a direct address to the dead, to the absent, or to a personified object or idea.

Argot: Slang or coarse vernacular language.

Assonance: Repetition of similar vowel sounds in speech or writing to achieve a particular effect.

Autobiography: An account of a person's life by himself or herself.

Avant-garde: Advanced (in style or thought); ahead of its time.

Ballad: Story in verse, a narrative song, usually treating a single event.

Belles lettres: Essays on literary studies and the aesthetics of literature.

Bibliography: List of works on a particular subject or author.

Biography: An account of a person's life.

Blank verse: Poetry with unrhymed line endings (usually written in iambic pentameters).

Blurb: Brief description of a book printed on the dust cover or jacket.

Bound verse: Verse based on metrical pattern.

Bowdlerize: To 'cleanse' a work by omitting or cutting out indecent passages, phrases or words.

Cadence: Rhythm and phrasing of language.

Caesura: Pause in line of verse.

Canon: Collection of works established as genuine.

Canto: Division of an epic or narrative poem.

Caption: Short description accompanying an illustration.

Catalogue: List of books or works.

Chronicle: A compilation of events in sequence, either actual or fictional.

Cipher: Writing that employs substitution or transposition of letters.

Cliche: Over-used expression that has become trite.

Colloquialism: Word, phrase or expression in everyday use.

Comedy: Humorous dramatic piece; *high comedy* has welldrawn characterization and witty dialogue; *low comedy*, or *slapstick* has absurd situations and boisterous action; *farce* is exaggerated comedy; *tragi-comedy* is a blend of the tragic and comic; *satire* uses sarcasm and wit to ridicule people's follies and vices (originally in verse).

Concordance: Alphabetical index of words in a work, or works by a single author.

Copyright: Protective law to prevent pirating or plagiarism of an author's work.

Couplet: Two successive rhyming lines in poetry.

Cycle: Group of works united by an overall theme.

Denouement: Final unwinding of complex plot.

Dialect: Manner of speaking pertaining to a particular class or geographical region.

Dialogue: The speech of characters in a story or play; part of work in which characters speak.

Diary: Sequential private record of personal and other events kept by an individual.

Dictionary: Book containing the words of a language, and their definitions, alphabetically; two-language dictionaries contain the corresponding words of both languages and their meanings, also alphabetically; any book that gives words and phrases about a particular subject, arranged alphabetically.

Digest: Publication where works are abridged.

Documentary: Form of fiction or drama based on documentary evidence provided by newspapers, recent historical reports or other contemporary or recent factual evidence.

Doggerel: Rough, ill-constructed verse.

Double entendre: Ambiguity in a word or expression, one of whose meanings is usually bawdy or frivolous.

Drama: Work to be performed on a stage by actors.

Dramatic irony: Situation or remark that has a significance unperceived by the character involved.

Dramatis Personae: Characters in a play.

Duologue: Conversation between two characters in a play, story or poem.

Edition: Total number of copies of a book printed from unchanged set of type.

Elegy: Serious meditative poem; often lamenting death.

Encyclopedia: Comprehensive work encompassing and describing many aspects of knowledge; specialist work covering comprehensively a subject or discipline.

Epic: Long narrative poem incorporating myth, legend, folklore and history, about the deeds of heroes, warriors, important people; a grandiose treatment of an individual's or nation's story.

Epigram: Short witty statement in verse or prose.

Epistle: Direct address to another person, a 'letter' in the form of verse.

Epitaph: Valediction to dead person or persons; inscription on a tomb or grave.

Eponymous: Person whose name becomes the title of the work.

Essay: Short composition, usually in prose, discussing a topic or variety of topics.

Euphemism: Substitution of a bland expression for a harsh or blunt one.

Exposition: Summary of plot and events given at the beginning of play.

Fable: A short narrative in prose or verse, often with animals, that points a moral.

Fiction: General term for an imaginative work, usually in prose.

First person: Style of novel, etc., in which narrator is a character.

Folk tale: Story handed down by word of mouth from generation to generation.

Foot: In poetry, rhythmic unit of two or three syllables.

Foreword: Short introduction to a work.

Free verse (*vers libre*): Poetry free from mechanical restrictions such as metre and rhyme, cadenced according to meaningful stress.

Gazetteer: Geographical dictionary or index.

Ghost writer: Person who undertakes literary work for another who takes the credit.

Glossary: Alphabetical list of unfamiliar or uncommon words, usually appended to work in which they appear.

Haiku: Japanee verse form composed of 17 syllables (5, 7, 5).

Heroic Couplet: Rhymed couplet in *horoic lines* (iambic pentameters).

Heroic verse : That used in *epic poetry.*

Hyperbole: Figure of speech that contains exaggeration for emphasis.

Idiom: Form of expression of phrase peculiar to its language and possessing a meaning other than its literal one.

Idyll: Short pastoral poem conveying mood of happy innocence.

Imagism: School of poetry (early 1900s) concerned with precise language, direct treatment, and freedom of form.

Introduction: Essay stating author's intention to reader.

Irony: Irony is a mode of speech in which the real meaning is exactly the opposite of that which is literally conveyed.

Jingle: Verse or verses with strong rhyme and rhythm.

Journal: Paper, periodical or magazine.

Legend: Story about a particular person containing myth and historical fact; explanation of symbols on a map or chart.

Lexicon: Dictionary, especially for Classical and Middle Eastern languages.

Limerick: Humorous verse form of five lines rhyming *a-a-b-b-a* (*a* in trimeter, *b* in diameter).

Linguistics: The scientific study of language.

Lyric: Usually short, song-like poem in which the poet expresses personal feelings.

Manuscript: Book or work written by hand.

Maxim: Short statement or even sentence containing a general truth about human conduct and human nature.

Melodrama: Exaggeratedly and emotionally written story, usually tragic.

Metaphor: Figure of speech in which one thing is described in terms of another by implication.

Meter: Patterns of stressed and unstressed syllables in verse.

Metonymy: In Metonymy an object is designated by the name of something which is generally associated with it.

Monologue: Single person addressing an audience alone, in drama, verse or prose. *Interior monologue*: unspoken but thought speech in verse or prose.

Myth: Non-factual story that embodies explanation of how something comes to exist, usually involving supernatural or superhuman creatures.

Narrative: Story or tale in prose or verse.

Neoclassicism: Movement of late 17th and 18th century reviving classical values in English literature, emphasizing discipline, reason and clarity.

Neologism: Newly-coined word.

New Wave (*la Nouvelle Vague*): Movement in literature and cinema originating in France in late 1950s that attempted to eschew fixed values, revealing a character by the way he experienced objects and events, which were often meticulously described.

Nom de plume: Term used to indicate a fictitious name used by a writer to represent his work.

Novel: Extended piece of prose fiction.

Novelette: Long short story of some 15,000 words.

Novella: Prose fiction longer than a short story, shorter than a novel (about 30,000 words), concentrating upon a single event.

Ode: Lyric poem describing an event or feelings about a person, object or event. *Pindaric Ode* (Greece) written for choral recitation in units of three stanzas; *Horatian Ode* (Rome) consisted of succession of stanzas following pattern of first one.

Onomatopoeia: Onomatopoeia is that artifice of language by which the sound of words is made to suggest or echo the sense.

Parable: A story which conveys a moral lesson is known as a parable, as for instance, the biblical parables.

Paradox: Apparently self-contradicting statement that, on closer examination, is found to contain a truth reconciling the two opposites.

Parody: Comic imitation of serious words, style, sense, subject of a writer to make them appear ridiculous.

Pastoral: Literature depicting idealized rural life.

Periodical: Magazine or journal published regularly.

Personification: In Personification inanimate objects and abstract notions are spoken of as having life or intelligence.

Philology: Study of literature, language and linguistics.

Picaresque: Chronicle of adventures of a rogue; originated in 16-century Spain.

Plagiarism: Wrongful appropriation and publication of another's work as one's own.

Plot: Plan or organization of events and characters in a play, work of fiction or poem as to induce curiosity and suspense in the audience or reader.

Poem: Literary work which may be in rhyme, blank verse or a combination of the two.

Poetry: Any kind of metrical composition in a literary work.

Pornography: Work in which there is a deliberate emphasis on the sexual behaviour of characters, in order to arouse sexual excitement.

Potboiler: Work written essentially to gain the author a livelihood.

Precis: Concise statement; short summary of a work.

Preface: Introduction to a work.

Prologue: Opening section of the work itself.

Propaganda: Work devoted to the dissemination of an idea or belief, usually biased.

Prose: Writings not in verse.

Prosody: Study of the handling of language in poetry.

Pseudonym: Name other than the true one used by an author to represent his work.

Pun: Figure of speech that involves a play upon the words used.

Realism: Literature that attempts to depict life objectively and faithfully.

Refrain: Phrase, line or lines repeated at intervals or at the end of a stanza in poetry.

Review: Notice or critical article on musical, artistic or literary work.

Revue: Theatrical entertainment made up of sketches, dances, songs, recitals and improvised pieces, usually humorous, satirical or topical.

Rhetoric: Art of using language, in speaking or writing, to persuade. The persuasive style (rather than the content) of a work.

Rhyme: Echoing sound or audible similarity in two or more words, especially at endings. Device used to construct much poetry.

Riddle: Puzzle, question or enigma; conundrum.

Roman a clef: Novel based upon actual people under disguised names.

Romance: Tale of chivalry, originally written in verse (Medieval times); term applied now to any wonderful or mysterious tale far removed from reality.

Romanticism: Movement originating in late 18th century as revolt against *neoclassicism*, emphasizing imaginative style.

Saga: Lengthy prose work, sometimes in several parts, describing the history and events surrounding medieval kings, warriors and, recently, fictitious families.

Satire: Any work in which wickedness, folly, vice and corruption in individuals and society is exposed by ridicule and scorn.

Science fiction: Literature based on scientific fact or fantasy, on earth or on other worlds in space, often set in periods in the distant future.

Semantics: Branch of linguistics dealing with meaning, and the change of meaning, in words.

Semiology: General study or science of signs with which humans communicate with each other (including words and their use in every context).

Sermon: Verbal instruction of some length, usually religious.

Short story: Work of fiction usually revolving around single event.

Sic: Included in brackets after a printed word or quoted passage to indicate that it is quoted accurately, however actually incorrect.

Simile: Figure of speech where one thing is likened to another explicitly.

Slang: Colourful and vernacular language of the street, marketplace, barrackroom`, workplace, sportsplace and playground.

Soliloquy: Kind of monologue in which a stage character expresses his thoughts and feelings.

Song: Poem or verse set to music.

Sonnet: 14-line poem of set rhyme-scheme; Elizabethan (Shakespearean) *ababcdcdefefgg*, Italian *abbaabba cdcdcd* (or *cdecde*).

Stanza: Group of lines of verse in definite pattern, usually repeated in a poem.

Style: Characteristic manner of expression in prose or verse; the way in which a thing is written by an author.

Symbolism: French poetic movement of late 19th century that developed as revolt against realism, concentrating on evoking emotions by use of indirect suggestion (symbol and metaphor); flourished in Russia at turn of century as literary movement, and later in British novel.

Synecdoche: In synecdoche a part is used to designate the whole, or the whole to designate a part.

Synonym: A word similar in meaning to another.

Synopsis: Outline of the main points of a work; a summary.

Tale: Spoken or written narrative.

Tautology: Overuse of synonyms or repetition of ideas in sentence.

Theme: Central idea of a work, rather than its subject.

Thesis: Long essay or treatise on a subject, usually expository; work presented to examiners for academic qualification.

Third person: Style of novel, etc., in which narrator is outside action.

Tract: Short printed treatise on religious or political subject.

Tragedy: Serious dramatic composition in prose or verse in which the principal character or characters are victims, often fatally so, of circumstance, or of a fatal flaw of character, of a crime of violence, especially one involving the death of an innocent person.

Transferred Epithet: In this figure an epithet is transferred from its proper word to another that is closely associated with it in the sentence.

Translation: Rendering of a work into another language.

Treatise: Formal work examining a subject and the principles underlying it.

Vade mecum: Manual or handbook carried for frequent reference.

Vernacular: Domestic or native language.

Versification: Study of how traditional verse is constructed.

Yarn: Story or tale, sometimes improbable and far-fetched.

Quantitative, Data Interpretation & Data Sufficiency

1 Simplification

BODMAS-RULE

This rule is very important for the arithmetical simplification. When vinculum, bracket, of, division, multiplication, addition, subtraction all or two or more than two operations are present in any question, then we can find out the result (answer) with the help of BODMAS - Rule. Details of BODMAS-Rule is given below :

Order	Abbreviated Letter Used in rule	Meaning	Notation
1.	V	Vinculum or Bar	——
2.	B	Brackets	[ ], { }, ( )
3.	O	Of	of
4.	D	Division	÷
5.	M	Multiplication	×
6.	A	Addition	+
7.	S	Subtraction	–

Note :

(*i*) Order of the letter which is used in BODMAS Rule is always fixed.

(*ii*) Absence of any operation or more than one operations does not changes the order of BODMAS.

(*iii*) 'of' means multiplication.

BRACKETS

When all brackets are present in a question, in that condition **ViCiCuSq-Rule** is applied. This ViCiCuSq-Rule stands for brackets and represents the order of calculation of brackets. Details are given below :

Order	Abbreviated Letter Used in rule	Meaning	Notation
1.	Vi	Vinculum	——
2.	Ci	Circular Bracket	( )
3.	Cu	Curly Bracket	{ }
4.	Sq	Square Bracket	[ ]

Note : This order of brackets (ViCiCuSq) is also fixed and not variable.

SOLVED EXAMPLES

Example 1: $\frac{3}{5}$ of $\frac{5}{7}$ of 357 = ?

A. 51 B. 153

C. 255 D. 159

Solution: $357 \times \frac{5}{7} \times \frac{3}{5} = 153$

Example 2: 125% of 860 + 75% of 480 = ?

A. 1415 B. 1385

C. 1435 D. 1365

Solution: $\frac{860 \times 125}{100} + \frac{480 \times 75}{100}$

$= 1075 + 360$

$= 1435$

Example 3: 1121.21 + 121.021 + 21.0021 = ?

A. 1263.2331

B. 1263.2121

C. 1163.2121

D. 1163.2331

Solution: 1121.21 + 121.021 + 21.0021

= 1263.2331

Example 4: $\left(\sqrt{5}-\sqrt{6}\right)^2 + \left(\sqrt{3}+\sqrt{10}\right)^2 = (?)^3 - 40$

A. 8 B. 4

C. 3 D. 6

Solution: Let ? = x

$\therefore \left(\sqrt{5}-\sqrt{6}\right)^2 + \left(\sqrt{3}+\sqrt{10}\right)^2 = (x)^3 - 40$

$\Rightarrow 5+6-2\sqrt{30}+3+10+2\sqrt{30} = x^3 - 40$

$\Rightarrow 24 = x^3 - 40$

$\Rightarrow x^3 = 40 + 24$

$= 64 = 4^3$

$\Rightarrow x = 4$

MULTIPLE CHOICE QUESTIONS

1. 68% of $\sqrt{2916} \times 25 = ? + 189$
A. 728 B. 718
C. 729 D. 739

2. $\sqrt{5^2 \times 41 \times 5 - 17^2 - 75} = ?$
A. 69 B. 61
C. 71 D. 79

3. 8.88 × 88.8 × 88 = ?
A. 68301.142 B. 79391.642
C. 65365.824 D. None of these

4. $\dfrac{9 \div 2 \times 27 \div 9}{18 \div 7.5 \times 5 \div 4} = ?$
A. 4.5 B. 5.7
C. 2.5 D. 6.8

5. $\sqrt{\sqrt{2500} + \sqrt{961}} = (?)^2$
A. 81 B. 3
C. 6561 D. 9

6. $64^{12} \div 4^{15} = 64^{?}$
A. 9 B. 3
C. 12 D. 7

7. $\sqrt{97344} = ?$
A. 302 B. 322
C. 292 D. None of these

8. 14% of 255 + ?% of 405 = 124.8
A. 22 B. 24
C. 18 D. 15

9. 19.25 × 16.4 × ? = 3472.7
A. 15 B. 17
C. 12 D. None of these

10. 83.04 – 66.46 – 11.71 = ?
A. 5.63 B. 2.92
C. 3.87 D. None of these

11. $\sqrt{529} + 46 \times 7.4 + (8)^3 - 251 = ?$
A. 265.1 B. 246.1
C. 256.4 D. None of these

12. $(16 \times 4)^3 \div (4)^5 \times (2 \times 8)^2 = (4)^{?}$
A. 5 B. 6
C. 3 D. 8

13. 534.58 – 386.89 + 221.45 – 195.42 = ?
A. 162.65 B. 109.65
C. 173.72 D. None of these

14. $\sqrt{5625} \times \dfrac{3}{5} + (85)^2 = ?$
A. 7270 B. 7280
C. 7260 D. 7250

15. $(7)^3 \times (3)^3 \div (6)^2 = ?$
A. 12348 B. 85.75
C. 1543.5 D. 257.25

16. $\dfrac{17^{6.87}}{17^{4.23}} = 17^{(?)}$
A. 1.62 B. 2.64
C. 11.1 D. 29.06

17. $\sqrt{2401} \times \sqrt{3249} = ?$
A. 2793 B. 2842
C. 2679 D. 2891

18. (5967 – 2437 – 1910) ÷ ? = 27
A. 60 B. 50
C. 65 D. 45

19. $\sqrt[3]{12167} \times \sqrt{?} = 621$
A. 841 B. 27
C. 625 D. None of these

20. 4.5% of 800 ÷ 0.5% of 640 = ?
A. 11.75 B. 12
C. 112.05 D. 11.25

21. 383 ÷ 25 × 2.5 + 12 = ?
A. 50.30 B. 222.14
C. 68.30 D. 124.24

22. 5083 + 25% of ? + 289 = 6385.5
A. 4044 B. 4054
C. 4154 D. 4104

23. $0.2 \times 10^{-2} = ?$
A. 2 B. 0.02
C. 0.002 D. 0.0002

24. 15 – 10 + 5 × 2 ÷ 5 = ?
A. 70 B. 4
C. 5 D. None of these

25. 37% of 150 – 0.05% of 1000 = ?
A. 50 B. 55
C. 55.5 D. 55.55

26. $(15)^{98} \times (15)^{-95} = ?$
A. 3375 B. 225
C. 45 D. 75

27. $\sqrt[3]{?} = (36 \times 24) \div 9$
A. 884736 B. 804036
C. 854734 D. 814736

28. (43% of 2750) – (38% of 2990) = ?
A. 49.3 B. 44.7
C. 43.6 D. 46.3

29. $40 \times 2 \div 10 + 5 - 4 = ?$

A. 5 B. 8
C. 11 D. 9

30. $28 \times 104 \div (18 + 6) + 3 = ?$

A. $124\frac{1}{3}$ B. $104\frac{1}{3}$
C. $125\frac{1}{3}$ D. 128

31. $0.99 \times 14 \div 11 \div 0.7 = ?$

A. 2.9 B. 1.6
C. 1.8 D. 2.8

32. $8\frac{1}{3}+5\frac{1}{4}\times13\frac{1}{5} \div 6\frac{3}{5} = ?$

A. $19\frac{1}{3}$ B. $18\frac{1}{2}$
C. $21\frac{1}{3}$ D. None of these

33. $\frac{14}{3}$ of $\frac{5}{8}$ of $72 = ?$

A. 209 B. 217
C. 210 D. 199

34. $10 \times 10 \times 10 \div (20 \div 10 \times 10 - 10) + 6 = ?$

A. 108 B. 111
C. 106 D. 114

35. $\dfrac{\frac{1}{3}\times 20 \div 4}{\frac{1}{4}\times 25 \div 5} = ?$

A. $\frac{7}{3}$ B. $\frac{4}{3}$
C. $\frac{5}{3}$ D. $\frac{10}{3}$

36. $\dfrac{2.70\times2.70+4.30\times4.30+8.60\times2.70}{2.70+4.30} = ?$

A. 6.8 B. 7.0
C. 7.6 D. 8.5

37. $1150 \div 50 \div 23 + 15 = ?$

A. 16 B. 20
C. 22 D. 18

38. $\dfrac{(0.08)^3+(0.011)^3}{(0.08)^2-0.08\times0.011+(0.011)^2} = ?$

A. 0.087 B. 0.091
C. 0.077 D. 0.067

39. $22 \div \left[(28-13)\div\{(32-8)\div\left(5+\frac{1}{3}\right)\}\right] = ?$

A. 7.9 B. 6.8
C. 6.6 D. 5.7

40. $\dfrac{(20^2-10^2)+5\times3+10}{\frac{1}{3}\text{ of }27+10+2+1} = ?$

A. $10\frac{2}{3}$ B. $14\frac{1}{3}$
C. 15 D. $12\frac{2}{3}$

41. $11^2 - 6^2 \div 6 \times \frac{5}{2} = 2$ of $10 = ?$

A. 126 B. 108
C. 110 D. 125

42. $\frac{25}{3}-\frac{4}{7}$ of $\frac{7}{5}+\frac{11}{3}+\frac{2}{3}-4 = ?$

A. $8\frac{1}{15}$ B. $9\frac{1}{30}$
C. $7\frac{1}{30}$ D. $9\frac{1}{5}$

43. $35 \times 0.07 - 21 \times 0.03 = ?$

A. 2.75 B. 1.72
C. 1.82 D. 2.13

44. $60 \times [35 - \{25 - (18 - \overline{9-3}) \div 11\}] = ?$

A. $566\frac{5}{7}$ B. $665\frac{5}{11}$
C. $665\frac{8}{11}$ D. $765\frac{5}{11}$

45. $\dfrac{140-44\times9\div3}{\frac{1}{2}\text{ of }18\div9+2} = ?$

A. $2\frac{2}{3}$ B. $3\frac{1}{3}$
C. $4\frac{2}{3}$ D. $6\frac{4}{5}$

46. $4\frac{1}{3} \times 5\frac{1}{3} \div 6\frac{1}{3} \div \frac{1}{3} + 5\frac{1}{3} - 2\frac{1}{3} = ?$

A. $14\frac{14}{19}$ B. $13\frac{19}{21}$
C. $13\frac{18}{19}$ D. $17\frac{8}{11}$

47. $1+\dfrac{1}{1+\dfrac{1}{1+\frac{1}{3}}} = ?$

A. $1\frac{4}{7}$ B. $2\frac{4}{7}$
C. $3\frac{4}{7}$ D. $4\frac{4}{7}$

48. $3 - [9 + \{14 - (6 - \overline{3-21})\}] = ?$

A. 0 B. 4
C. 18 D. 6

49. (8 ÷ 88) × 8888088 = ?
A. 8008008 B. 808088
C. 808080 D. 808008

50. 1260 ÷ 15 ÷ 7 = ?
A. 588 B. 122
C. 58 D. 12

51. 2 – [2 – {2 – 2(2 + 2)}] = ?
A. 6 B. – 6
C. 4 D. – 4

52. 100 × 10 – 100 + 2000 ÷ 100 = ?
A. 979 B. 920
C. 780 D. 29

53. 3640 ÷ 14 × 16 + 340 = ?
A. 3500 B. 4500
C. 3525 D. 4480

54. 25 – 5 [2 + 3 {2 – 2 (5 – 3) + 5} –10] ÷ 4 = ?
A. 5 B. 25
C. 23.25 D. 23.75

55. $(-5)\ (4)\ (2)\left(-\frac{1}{2}\right)\left(\frac{3}{4}\right) = ?$
A. 15 B. –15
C. 30 D. –30

ANSWERS

1	2	3	4	5	6	7	8	9	10
C	A	D	A	B	D	D	A	D	D
11	**12**	**13**	**14**	**15**	**16**	**17**	**18**	**19**	**20**
D	D	D	A	D	B	A	A	D	D
21	**22**	**23**	**24**	**25**	**26**	**27**	**28**	**29**	**30**
A	B	C	D	B	A	A	D	D	A
31	**32**	**33**	**34**	**35**	**36**	**37**	**38**	**39**	**40**
C	D	C	C	B	B	A	B	C	D
41	**42**	**43**	**44**	**45**	**46**	**47**	**48**	**49**	**50**
A	B	C	B	A	C	A	B	D	D
51	**52**	**53**	**54**	**55**					
B	B	B	D	A					

SOME SELECTED EXPLANATORY ANSWERS

1. $\sqrt{2916} \times 68\% \times 25 = ? + 189$

$\Rightarrow 54 \times \frac{68}{100} \times 25 = ? + 189$

$\Rightarrow 918 = ? + 189$

$\Rightarrow ? = 918 - 189 = 729$

2. $? = \sqrt{25 \times 41 \times 5 - 17^2 - 75}$

$= \sqrt{5125 - 289 - 75} = \sqrt{4761} = 69$

3. 8.88 × 88.8 × 88 = 69391.872

4. $\frac{9 \div 2 \times 27 \div 9}{18 \div 7.5 \times 5 \div 4} = \frac{\frac{9}{2} \times \frac{27}{9}}{\frac{18}{7.5} \times \frac{5}{4}} = \frac{\frac{27}{2}}{3} = 4.5$

5. $\sqrt{\sqrt{2500 + 961}}$

$= \sqrt{\sqrt{50 + 31}} = \sqrt{9} = \sqrt{(3)^2} = 3.$

6. Let $(64)^x = 64^{12} \div 4^{15}$

$\Rightarrow \quad 4^{3x} = 4^{36-15} = 4^{21}$

$\therefore \quad 3x = 21 \quad \therefore x = 7$

7. $\sqrt{97344} = 312$

8. ∵ 14% of 255 + ?% of 405 = 124.8

$\Rightarrow \quad 35.7 + 405 \times ? = 124.8$

$\Rightarrow \quad \frac{405 \times ?}{100} = 124.8 - 35.7 = 89.1$

$\therefore \quad ? = \frac{98.1 \times 100}{405} = \frac{891 \times 10}{405}$

$= \frac{891 \times 2}{81} = \frac{1782}{81} = 22.$

9. ∵ 19.25 × 16.4 × ? = 3472.7

$\therefore \quad ? = \frac{3472.7}{19.25 \times 16.4} = 11$

10. 83.04 – 66.46 – 11.71 = 4.87

11. $\sqrt{529} \div 46 \times 7.4 + (8)^3 - 251$

$= \frac{23}{46} \times 7.4 + 512 - 251$

$= 3.7 + 512 - 251 = 264.7$

12. $(4)^? = (16 \times 4)^3 \div (4)^5 \times (2 \times 8)^2$

$(4)^? = \frac{(4)^9}{(4)^5} \times (4)^4 = (4)^{9+4-5}$

$\therefore \quad ? = 8.$

13. $534.58 - 386.89 + 221.45 - 195.42 = 172.72$

14. $\sqrt{5625} \times \frac{3}{5} + (85)^2 = 75 \times \frac{3}{5} + 7225 = 7270$

15. $343 \times 27 \div 36 = 257.25$

16. $17^{(?)} = \frac{17^{6.87}}{17^{4.23}} = 17^{6.87-4.23} = 17^{2.64}$

$\therefore \quad ? = 2.64.$

17. $\sqrt{2401} \times \sqrt{3249} = 49 \times 57 = 2793.$

18. $(5967 - 2437 - 1910) \div ? = 27$

$\therefore \quad \frac{1620}{?} = 27$

$\therefore \quad ? = \frac{1620}{27} = 60$

19. $\because \sqrt[3]{12167} \times \sqrt{?} = 621$

$\Rightarrow \quad 23 \times \sqrt{?} = 621$

$\Rightarrow \quad \sqrt{?} = \frac{621}{23} = 27$

$\therefore \quad ? = (27)^2 = 729$

20. $\frac{4.5}{100}$ of $800 \div \frac{0.5}{100}$ of 640

$= 36 \div 3.2 = 11.25$

21. $383 \div 25 \times 2.5 + 12$

$= \frac{383}{25} \times 2.5 + 12 = 50.30.$

22. $\because 5083 + \frac{25}{100}$ of $? + 289 = 6385.5$

$\Rightarrow 0.25 \times ? = 6385.5 - 5083 - 289$

$\therefore \quad ? = \frac{1013.5}{0.25} = 4054.$

23. $0.2 \times 10^{-2} = 0.2 \times \frac{1}{100} = 0.002$

24. $15 - 10 + 5 \times 2 \div 5$

$= 15 - 10 + \frac{5 \times 2}{5} = 15 - 10 + 2 = 7$

25. 37% of 150 – 0.05% of 1000

$= \frac{37}{100}$ of $150 - \frac{0.05}{100}$ of 1000

$= \frac{111}{2} - 0.5 = 55.5 - 0.5 = 55.$

26. $(15)^{98} \times (15)^{-95} = (15)^3 = 3375$

27. $(36 \times 24) \div 9 = \sqrt[3]{?}$

$\Rightarrow 4 \times 24 = \sqrt[3]{?}$

$\therefore \; ? = 96^3 = 884736$

28. (43% of 2750) – (38% of 2990)

$= 0.43 \times 2750 - 0.38 \times 2990 = 46.3$

29. $\because \; 40 \times 2 \div 10 + 5 - 4 = 40 \times \frac{2}{10} + 5 - 4$

$= 8 + 5 - 4 = 9$

30. $\because \; 28 \times 104 \div (18 + 6) + 3$

$= 28 \times 104 \div 24 + 3$

$= 28 \times \frac{104}{24} + 3 = 28 \times \frac{13}{3} + 3$

$= \frac{364}{3} + 3 = \frac{373}{3} = 124\frac{1}{3}$

31. $\because \; 0.99 \times 14 \div 11 \div 0.7$

$= 0.99 \times \frac{14}{11} \div 0.7 = \frac{0.99 \times 14}{11 \times 0.7} = 1.8$

32. $\because \; 8\frac{1}{3} + 5\frac{1}{4} \times 13\frac{1}{5} + 6\frac{3}{5}$

$= \frac{25}{3} + \frac{21}{4} \times \frac{66}{5} + \frac{33}{5} = \frac{25}{3} + \frac{21}{4} \times \frac{66}{5} \times \frac{5}{33}$

$= \frac{25}{3} + \frac{21}{2} = \frac{50+63}{6} = \frac{113}{6} = 18\frac{5}{6}$

33. $\because \; \frac{14}{3}$ of $\frac{5}{8}$ of $72 = \frac{14 \times 5 \times 72}{3 \times 8} = 210$

34. $10 \times 10 \times 10 \div (20 \div 10 \times 10 - 10) + 6$

$= 10 \times 10 \times 10 \div \left(\frac{20}{10} \times 10 - 10\right) + 6$

$= 10 \times 10 \times 10 \div (20 - 10) + 6$

$= 10 \times 10 \times 10 \div 10 + 6 = 10 \times 10 \times \frac{10}{10} + 6$

$= 10 \times 10 \times 1 + 6 = 100 + 6 = 106$

35. $\because \; \frac{\frac{1}{3} \times 20 \div 4}{\frac{1}{4} \times 25 \div 5} = \frac{\frac{1}{3} \times 5}{\frac{1}{4} \times 5} = \frac{\frac{5}{3}}{\frac{5}{4}} = \frac{4}{3}$

36. $\frac{2.70 \times 2.70 + 4.30 \times 4.30 + 8.60 \times 2.70}{2.70 + 4.30}$

$= \frac{(2.70)^2 + (4.30)^2 + 2 \times 4.30 \times 2.70}{2.70 + 4.30} = \frac{(2.70 + 4.30)^2}{2.70 + 4.30}$

$[\because a^2 + b^2 + 2ab = (a + b)^2]$

$= 2.70 + 4.30 = 7.00$

37. $1150 \div 50 \div 23 + 15 = \frac{1150}{50} \div 23 + 15$

$= 23 \div 23 + 15 = \frac{23}{23} + 15 = 1 + 15 = 16.$

38. $\frac{(0.08)^3 + (0.011)^3}{(0.08)^2 - 0.08 \times 0.011 + (0.011)^2}$

$\frac{(0.08 + 0.011)\left[(0.08)^2 - 0.08 \times 0.011 + (0.011)^2\right]}{(0.08)^2 - 0.08 \times 0.011 + (0.011)^2}$ $[\because a^3 + b^3 = (a + b)(a^2 - ab + b^2)]$

$= 0.08 + 0.011 = 0.091.$

39. $22 \div \left[(28-13)+\left\{(32-8)+\left(5+\frac{1}{3}\right)\right\}\right]$

$= 22 \div \left[15+\left\{24+\frac{16}{3}\right\}\right]$

$= 22 \div \left[15+\frac{24}{\frac{16}{3}}\right] = 22 \div \left[15+\frac{9}{2}\right]$

$= 22 \div \frac{15}{\frac{9}{2}} = 22 \div \frac{10}{3} = \frac{22\times 3}{10} = 6.6$

40. $\frac{(20^2-10^2)\div 5\times 3+10}{\frac{1}{3}\text{ of } 27+10\div 2+1}$

$= \frac{(400-100)\div 5\times 3+10}{9+10\div 2+1} = \frac{300\div 5\times 3+10}{9+5+1}$

$= \frac{60\times 3+10}{15} = \frac{190}{15} = \frac{38}{3} = 12\frac{2}{3}$

41. $11^2 - 6^2 \div 6 \times \frac{5}{2} + 2 \text{ of } 10$

$= 11^2 - 6^2 \div 6 \times \frac{5}{2} + 20$

$= 11^2 - 6 \times \frac{5}{2} + 20 = 11^2 - 15 + 20$

$= 121 + 5 = 126.$

42. $\frac{25}{3} - \frac{4}{7}\text{ of }\frac{7}{5} + \frac{11}{3} \div \frac{2}{3} - 4$

$= \frac{25}{3} - \frac{4}{5} + \frac{11}{3} \div \frac{2}{3} - 4 = \frac{25}{3} - \frac{4}{5} + \frac{11}{2} - 4$

$= \frac{83}{6} - \frac{24}{5} = \frac{415-144}{30} = \frac{271}{30} = 9\frac{1}{30}$

43. $35 \times 0.07 - 21 \times 0.03 = 2.45 - 0.63 = 1.82$

44. $60 \times [35 - \{25 - (18 - \overline{9-3}) \div 11\}]$

$= 60 \times [35 - \{25 - (18 - 6) \div 11\}$

$= 60 \times [35 - \{25 - 12 \div 11\}]$

$= 60 \times \left[35 - \left\{25 - \frac{12}{11}\right\}\right] = 60 \times \left[35 - \frac{263}{11}\right]$

$= 60 \times \frac{122}{11} = 665\frac{5}{11}$

45. $\frac{140-44\times 9\div 3}{\frac{1}{2}\text{ of } 18\div 9+2} = \frac{140-44\times 3}{9\div 9+2}$

$= \frac{140-132}{1+2} = \frac{8}{3} = 2\frac{2}{3}$

46. $4\frac{1}{3}\times 5\frac{1}{3} \div 6\frac{1}{3} \div \frac{1}{3} + 5\frac{1}{3} - 2\frac{1}{3}$

$= \frac{13}{3} \times \frac{16}{3} \div \frac{19}{3} \div \frac{1}{3} + \frac{16}{3} - \frac{7}{3}$

$= \frac{13}{3} \times \frac{16}{19} + \frac{1}{3} + \frac{16}{3} - \frac{7}{3}$

$= \frac{13}{3} \times \frac{48}{19} + \frac{16}{3} - \frac{7}{3} = \frac{208}{19} + \frac{16}{3} - \frac{7}{3}$

$= 10\frac{18}{19} + 3 = 13\frac{18}{19}$

47. This type of questions is solved starting from the bottom.

$1+\frac{1}{1+\frac{1}{1+\frac{1}{3}}} = 1+\frac{1}{1+\frac{1}{\frac{4}{3}}}$

$= 1+\frac{1}{1+\frac{3}{4}} = 1+\frac{1}{\frac{7}{4}} = 1+\frac{4}{7} = \frac{11}{7} = 1\frac{4}{7}$

48. $3 - [9 + \{14 - (6 - \overline{3-21})\}]$

$= 3 - [9 + \{14 - (6 + 18\}]$

$= 3 - [9 + \{14 - 24\}] = 3 - [9 - 10]$

$= 3 + 1 = 4.$

49. $(8 \div 88) \times 888808 = \frac{8}{88} \times 8888088 = 808008$

50. $1260 \div 15 \div 7 = 1260 \times \frac{1}{15} \times \frac{1}{7} = 12$

51. $2 - [2 - \{2 - 2(2 + 2)\}]$

$= 2 - [2 - \{2 - 8\}] = 2 - [2 + 6]$

$= 2 - 8 = -6$

52. $100 \times 10 - 100 + 2000 \div 100$

$= 100 \times 10 - 100 + 20$

$= 1000 - 100 + 20 = 920$

53. $3640 \div 14 \times 6 + 340$

$= 3640 \times \frac{1}{14} \times 16 + 340 = 4160 + 340 = 4500$

54. $25 - 5\,[2 + 3\,\{2 - 2\,(5 - 3) + 5\} - 10] \div 4$

$= 25 - 5\,[2 + 3\,\{2 - 4 + 5\} - 10] \div 4$

$= 25 - 5\,[2 + 9 - 10] \div 4$

$= 25 - 5 \times 1 \times \frac{1}{4} = 25 - \frac{5}{4} = \frac{95}{4} = 23.75$

55. $(-5)\,(4)\,(2)\left(-\frac{1}{2}\right)\left(\frac{3}{4}\right)$

$= -5 \times 4 \times 2 \times -\frac{1}{2} \times \frac{3}{4} = 15$

2

L.C.M. & H.C.F.

LCM

Multiple : A number is said to be multiple of other when it is exactly divisible by the other.

Common multiple : A common multiple of two or more numbers is a number which is exactly divisible by each of them. For example, for 2 and 3, common multiples are 6, 12, 18, 24 and so on.

Explanation:

Consider the two numbers 2 and 3

Multiple of 2 are 2, 4, 6, 8, 10, 12, ...

Multiple of 3 are 3, 6, 9, 12, 15, ...

$\therefore$ Common multiples are 6, 12, 18, ...

Least Common Multiple (L.C.M.) : L.C.M. of two or more given numbers is the least number which is exactly divisible by each of them. For example,

6 is a common multiple of 2 and 3

12 is also common multiple of 2 and 3

18 is also common multiple of 2 and 3

But 6 is the least common multiple (L.C.M.) of 2 and 3.

Methods to find out L.C.M.

L.C.M. of two or more given numbers is determined by following two methods:

1. By prime factorization method
2. By division method

1. By prime factorization method : Resolve the given numbers into their prime factors and then find the product of the highest power of all the factors that occur in the given numbers. This product will be the L.C.M.

Example 1 : Find the L.C.M. of 40, 50, 60 and 80.

Sol :

$$40 = 2 \times 2 \times 2 \times 5 = 2^3 \times 5$$
$$50 = 2 \times 5 \times 5 = 2 \times 5^2$$
$$60 = 2 \times 2 \times 3 \times 5 = 2^2 \times 3 \times 5$$
$$80 = 2 \times 2 \times 2 \times 2 \times 5 = 2^4 \times 5$$

Here, the prime factors that occur in the given numbers are 2, 3 and 5 and their highest powers are respectively 2^4, 5^2 and 3.

Hence, the required L.C.M. $= 2^4 \times 3 \times 5^2 = 1200$.

2. By division method : This is the quicker method to find the prime factors and hence L.C.M.

For determining L.C.M. of the numbers 40, 50, 60 and 80, following process of division is adopted:

2	40, 50, 60, 80
2	20, 25, 30, 40
2	10, 25, 15, 20
5	5, 25, 15, 10
	1, 5, 3, 2

Now required L.C.M. $= 2 \times 2 \times 2 \times 5 \times 5 \times 3 \times 2 = 1200$

L.C.M. of decimal : First of all, we find out the L.C.M. of numbers without decimal and then, we see the number in which decimal is given in the minimum digit from right to left. We put the decimal in our result which is equal to that number of digits.

Example : Find the L.C.M. of 0.6, 9.6 and 0.36.

Sol : The given numbers are equivalent to 0.60, 9.60 and 0.36.

Now, we find out the L.C.M. of 60, 960 and 36, which is equal to 2880.

Hence, the required L.C.M. = 2880.

L.C.M. of fraction : The L.C.M. of two or more fractions is the least fraction or integer which is exactly divisible by each of them.

If $\frac{a}{b}, \frac{c}{d}, \frac{e}{f}$ be the proper fractions, then their L.C.M. is given by

$$\frac{\text{L.C.M. of numerators } a, c, e}{\text{H.C.F. of denominators } b, d, f}$$

An example : Find the L.C.M. of $\frac{1}{2}, \frac{3}{5}, \frac{4}{7}$ and $\frac{5}{12}$

Sol : $\text{L.C.M.} = \frac{\text{L.C.M. of } 1, 3, 4, 5}{\text{H.C.F. of } 2, 5, 7, 12}$

$$= \frac{60}{1} = 60$$

HCF

Factor : One number is said to be a factor of other when it divides the other exactly. Hence, 5 and 7 are factors of 35.

Common factor : A common factor of two or more numbers is a number that divides each of them exactly. Hence, 5 is a common factor of 15, 25, 35 and 55.

Highest Common Factor (H.C.F.) : H.C.F. of two or more numbers is the largest number by which each given number is divisible without leaving any remainder.

An example : It is required to find the H.C.F. of 6 and 8.

Factors of 6 are 1, 2, 3, 6 and

Factors of 8 are 1, 2, 4, 8.

The common factors are 1, 2, but highest of these is 2. Hence, 2 is the H.C.F.

Note : *The terms Highest Common Divisor (H.C.D.) and Greatest Common Measure (G.C.M.) are often used in the sense of Highest Common Factor (H.C.F.)*

We can find H.C.F. by two methods:

1. By prime factorization method :

An example : Find the H.C.F. of 144, 336 and 2016.

Sol : $144 = 2 \times 2 \times 2 \times 2 \times 3 \times 3 = 2^4 \times 3^2$

$336 = 2 \times 2 \times 2 \times 2 \times 3 \times 7 = 2^4 \times 3 \times 7$

$2016 = 2 \times 2 \times 2 \times 2 \times 2 \times 7 \times 3 \times 3 = 2^5 \times 7 \times 3^2$

$\therefore$ H.C.F. of given numbers $= 2^4 \times 3 = 48$.

2. By division method : Divide the greater number by the smaller number, divide the divisor by the remainder, divide the remainder by the next remainder, and so on until no remainder is left. The last divisor is the required H.C.F.

An example : Find the H.C.F. of 48, 168 and 324.

Sol :

```
48 ) 168 ( 3
     144
     ---
      24 ) 48 ( 2
           48
           --
           ××
```

Thus, the H.C.F. of 48 and 168 is 24.

Now, we find out the H.C.F. of 24 and 324

```
24 ) 324 (13
     24
     --
     84
     72
     --
     12 ) 24 (2
          24
          --
          ××
```

$\therefore$ Required H.C.F. = 12.

H.C.F. of Decimals : First of all find the H.C.F. of the given numbers ignoring decimals and then put decimal at maximum digits from right to left.

An example : Find the H.C.F. of 0.0012, 1.6 and 2.8.

Sol : First we find the H.C.F. of 12, 16 and 28, which comes to 4.

So, H.C.F. of 0.0012, 1.6 and 2.8 will be 0.0004.

H.C.F. of Fractions : If $\frac{a}{b}, \frac{c}{d}, \frac{e}{f}, \ldots$ be the proper fraction, their H.C.F. is equal to $\frac{\text{H.C.F. of numerators}}{\text{L.C.M. of denominators}}$

An example : Find the H.C.F. of $\frac{54}{9}, 3\frac{9}{17}$ and $\frac{36}{51}$

Sol : Here, $\frac{54}{9} = \frac{6}{1}$; $3\frac{9}{17} = \frac{60}{17}$ and $\frac{36}{51} = \frac{12}{17}$

Hence, the fractions are $\frac{6}{1}, \frac{60}{17}$ and $\frac{12}{17}$

$\therefore$ H.C.F. $= \frac{\text{H.C.F. of 6, 60, 12}}{\text{L.C.M. of 1, 17, 17}} = \frac{6}{17}$.

Relationship Between Two Numbers and Their L.C.M. and H.C.F.

Product of the H.C.F. and the L.C.M. of two numbers is equal to the product of the given numbers.

i.e., 1st number × 2nd number = H.C.F. × L.C.M.

MULTIPLE CHOICE QUESTIONS

1. The largest three-digit number, when divided by 6, 9 and 12 leaves 1 as remainder in each case, will be

A. 887 B. 987
C. 973 D. 730

2. Two numbers are in the ratio of 8 : 15. If their H.C.F. is 4, the numbers are

A. 32 and 60 B. 16 and 30
C. 80 and 150 D. 64 and 120

3. The largest number that will divide 226 and 272 leaving 1 and 2 as remainders respectively, is

A. 36 B. 45
C. 55 D. 59

4. Which of the following is the greatest common divisor of 1170 and 102?

A. 8 B. 4
C. 6 D. 3

5. A number, when 3 is added to it, becomes divisible by 36, 45 and 50. The smallest such number is

A. 987 B. 798
C. 986 D. 897

6. The least perfect square number which is completely divisible by 10, 20, 30 and 40 is

A. 4800 B. 3600
C. 4400 D. 2500

7. The product of two numbers is 2160 and their H.C.F. is 12. How many such pairs of numbers can be possibly formed?

A. 3 B. 1
C. 2 D. None

8. The greatest number that will divide 366, 513 and 324 leaving the same remainder in each case is
A. 21 B. 18
C. 27 D. 42

9. The sum of two numbers is 216 and their H.C.F. is 27. These numbers are
A. 60 and 90 B. 81 and 135
C. 64 and 128 D. 30 and 84

10. The L.C.M. of two numbers is 45 times their H.C.F. If the sum of the L.C.M. and the H.C.F. of these two numbers is 1150 and one of the numbers is 125, then the other number is
A. 256 B. 225
C. 250 D. 255

11. The H.C.F. and the L.C.M. of two numbers are 50 and 250 respectively. On dividing one of these numbers by 2, 50 is obtained as quotient. The numbers are
A. 100, 125 B. 80, 100
C. 125, 100 D. 200, 250

12. The largest three-digit number, which when successively divided by 6, 9 and 12, leaves 3 as remainder in each case, is
A. 575 B. 795
C. 975 D. 525

13. The greatest number that will divide 33, 64 and 80 leaving 3, 4 and 5 as remainders respectively, is
A. 10 B. 20
C. 15 D. 22

14. The H.C.F. of three numbers is 12. If the three numbers are in the ratio of 1 : 2 : 3, then the numbers are
A. 14, 28, 42 B. 12, 24, 36
C. 15, 30, 45 D. 24, 48, 72

15. The greatest four-digit number completely divisible by 2, 3, 4 and 5 is
A. 9960 B. 9690
C. 8990 D. 9980

16. Three bells ring respectively at an interval of 15 seconds, 20 seconds and 24 seconds. If they ring continuously for 12 minutes then how many times, during this period, will they ring together?
A. 2 times B. 6 times
C. 5 times D. 3 times

17. The smallest number, on being successively divided by 5, 6, 8, 9 and 12 leaves 1 as remainder in each case and is completely divisible by 13, will be
A. 4603 B. 6305
C. 4503 D. 3601

18. If in the process of finding H.C.F. of two numbers by continued division method, 49 is the last divisor and quotients obtained (from the beginning) are 17, 3 and 2 respectively, then the numbers are
A. 432 and 4929 B. 343 and 5829
C. 388 and 5880 D. 472 and 5930

19. A, B and C start running together in a particular direction from a particular point on a 12 kms long circular path. If the speeds of A, B and C are 3 kms/h, 7 kms/h and 13 kms/h respectively, then after how many hours will they meet together again?
A. 8 B. 6
C. 12 D. 10

20. Which of the following has most numbers of divisors?
A. 182 B. 176
C. 101 D. 99

ANSWERS

1	2	3	4	5	6	7	8	9	10
C	A	B	C	D	B	C	A	B	B
11	**12**	**13**	**14**	**15**	**16**	**17**	**18**	**19**	**20**
A	C	C	B	A	B	D	B	C	B

SOME SELECTED EXPLANATORY ANSWERS

1. L.C.M. of 6, 9 and 12

$$\begin{array}{r|l} 3 & 6,\ 9,\ 12 \\ \hline 2 & 2,\ 3,\ 4 \\ \hline & 1,\ 3,\ 2 \end{array}$$

$\therefore$ LCM = 3 × 2 × 3 × 2 = 36.

$\because$ Largest 3-digit number = 999

$$36\ )\ 999\ (\ 27$$
72
279
252
× 27

$\therefore$ Largest 3-digit number which is exactly divisible by 6, 9 and 12

= 999 − 27 = 972

$\therefore$ The required number = 972 + 1 = 973

2. Let the numbers be $8x$ and $15x$

$8x = 2 \times 2 \times 2 \times x$

$15x = 3 \times 5 \times x$

$\therefore$ LCM of $8x$ and $15x = 2 \times 2 \times 2 \times x \times 3 \times 5$
$= 120x$

Now, 1st number × 2nd number = HCF × LCM

$\Rightarrow$ $8x \times 15x = 4 \times 120x$

$\Rightarrow$ $120x^2 = 4 \times 120x$

⇒ $x = 4$

∴ Numbers are 8 × 4 = 32 and 15 × 4 = 60

3. 226 – 1 = 225 and 272 – 2 = 270

Now, HCF of 225 and 270

225) 270 (1
225
45) 225 (5
225
×

∴ The required number is 45.

4. 1170 = 2 × 5 × 3 × 3 × 13

102 = 2 × 3 × 17

∴ Greatest common divisor = 2 × 3 = 6

5. It is clear from the given conditions of the problem that the smallest number will be 3 less than the LCM of 36, 45 and 50.

∴ LCM of 36, 45 and 50

3	36, 45, 50
2	12, 15, 50
3	6, 15, 25
5	2, 5, 25
	2, 1, 5

∴ LCM = 3 × 2 × 3 × 5 × 2 × 5 = 900

∴ Required number = 900 – 3 = 897.

6. LCM of 10, 20, 30 and 40

2	10, 20, 30, 40
5	5, 10, 15, 20
2	1, 2, 3, 4
	1, 1, 3, 2

∴ LCM = 2 × 2 × 5 × 3 × 2

But the number is a perfect square number

∴ Required number = 2 × 2 × 5 × 5 × 3 × 3 × 2 × 2 = 3600.

7. Product of two numbers = their HCF × their LCM

2160 = 12 × LCM

$$\therefore \quad \text{LCM} = \frac{2160}{12} = 180$$

Therefore such pairs of numbers in which product of numbers is 2160 and HCF is 12 will be only 2.

i.e., 12 × 180 and 36 × 60.

8. Difference between 366 and 513

= 513 – 366 = 147

and difference between 513 and 324

= 513 – 324 = 189

∴ HCF of 147 and 189

147) 189 (1
147
× 42) 147 (3
126
× 21) 42 (2
42
×

∴ The required largest number is 21.

9. Let the numbers be x and y.

∴ $x + y = 216 = 27 \times 8$

Since HCF of these two numbers is 27, hence, common factor of these two numbers is 27.

Therefore these numbers will be of the form of 27 × 1, 27 × 2, 27 × 3, ... etc.

According to the condition of the problem sum of these numbers is 216.

∴ The numbers will be 27 × 3 = 81

and 27 × 5 = 135

10. LCM of the two numbers = 45 × HCF

and LCM + HCF = 1150

⇒ 45 × HCF + HCF = 1150

⇒ HCF(45 + 1) = 1150

$$\Rightarrow \quad \text{HCF} = \frac{1150}{46} = 25$$

∴ LCM = 45 × 25 = 1125

∵ 1st number × 2nd number = LCM × HCF

∴ 125 × 2nd number = 1125 × 25

$$\therefore \quad \text{2nd number} = \frac{1125 \times 25}{125} = 225.$$

11. According to the condition of the problem, 50 is obtained on dividing one of the numbers by 2

∴ One of the numbers = 50 × 2 = 100

Now, 1st number × 2nd number = LCM × HCF

∴ 100 × 2nd number = 250 × 50

$$\therefore \quad \text{2nd number} = \frac{250 \times 50}{100} = 125$$

Hence, numbers are 100 and 125.

12. LCM of 6, 9 and 12 = 36 and largest 3-digit number = 999

∴ HCF of 36 and 999

36) 999 (27
72
279
252
27

∴ Largest three-digit number completely divisible by 6, 9 and 12

= 999 – 27 = 972

Since 3 is left as remainder in each case,

∴ Required largest number of 3-digit

= 972 + 3 = 975.

13. 33 – 3 = 30, 64 – 4 = 60 and 80 – 5 = 75

Now, HCF of 30, 60 and 75:

30 = 2 × 3 × 5

60 = 2 × 2 × 3 × 5

75 = 3 × 5 × 5

∵ Common factor = 3 × 5 = 15

∴ Required largest number = 15.

14. Let the numbers be x, $2x$ and $3x$ respectively. Here, HCF of the three numbers is 12, therefore 12 is the largest common factor of these three numbers. Thus, it is clear that $x = 12$

$\therefore$ These numbers are 12, 24 and 36 respectively.

15. LCM of 2, 3, 4 and 5 = 60

Largest 4-digit number = 9999

$$\text{Now, } 60\ )\ 9999\ (\ 166 \\ \underline{60} \\ 399 \\ \underline{360} \\ \times\ 399 \\ \underline{360} \\ 39$$

$\therefore$ Largest 4-digit number completely divisible by 2, 3, 4 and 5

$$= 9999 - 39 = 9960$$

$\therefore$ Required number = 9960

16. LCM of 15, 20 and 24

5	15,	20,	24
4	3,	4,	24
3	3,	1,	6
	1,	1,	2

LCM = 5 × 4 × 3 × 2 = 120

$\because$ 12 minutes = 12 × 60 = 720 seconds

$\therefore$ Number of times the bells will ring together during 12 minutes

$$= \frac{720}{120} = 6 \text{ times.}$$

17. LCM of 5, 6, 8, 9 and 12 = 360. Hence, the smallest number which when divided by 5, 6, 8, 9 and 12 leaves 1 as remainder in each case will be 360 + 1 = 361. But 361 is not completely divisible by 13. Hence, on considering 360 × 2 + 1, 360 × 3 + 1,... , 360 × n + 1, we conclude that 360 × 10 + 1 = 3601 is the required number.

18. This question is based on finding HCF by continued division method. Now according to the condition of the problem,

Last divisor = 49

and Last quotient = 2

$\therefore$ dividend = 49 × 2 = 98

Now, divisor = 98 and quotient = 3

and remainder = 49

$\therefore$ dividend = 98 × 3 + 49 = 343

Now, divisor = 343

quotient = 17

and remainder = 98

$\therefore$ Dividend = 343 × 17 + 98 = 5829

Therefore, these numbers are 343 and 5829.

19. A, B and C will take 4 hours, $\frac{12}{7}$ hours and $\frac{12}{13}$ hours respectively in completing one round on the circular path.

$$\therefore \text{LCM of } \frac{4}{1}, \frac{12}{7} \text{ and } \frac{12}{13}$$

$$= \frac{\text{LCM of 4,12 and 12}}{\text{HCF of 1,7 and 13}}$$

$\because$ LCM of 4, 12 and 12= 12 and HCF of 1, 7 and 13 = 1

$$\therefore \quad \text{Required LCM} = \frac{12}{1} = 12$$

Hence, they will meet together again after 12 hrs.

20.

Numbers		***Their divisors***
182	→	1, 2, 7, 13, 14, 26, 91 and 182
176	→	1, 2, 4, 8, 16, 22, 44, 88 and 176
101	→	1 and 101
99	→	1, 3, 9, 11, 33 and 99

Therefore, 176 has the most number of divisors.

3 Ratio & Proportion

Some Important Facts:

1. If a and b are two quantities, then
   (a) Duplicate ratio of $a : b = a^2 : b^2$
   (b) Sub-duplicate ratio of $a : b = \sqrt{a} : \sqrt{b}$
   (c) Triplicate ratio of $a : b = a^3 : b^3$
   (d) Sub-triplicate ratio $a : b = \sqrt[3]{a} : \sqrt[3]{b}$
   (e) Inverse or reciprocal ratio of $a : b = \frac{1}{a} : \frac{1}{b}$
2. If $a : b = c : d$, then
   (a) Third proportional (c) to 'a' and 'b' $= \frac{b^2}{a}$
   (b) Fourth proportional (d) $= \frac{bc}{a}$
   (c) Mean proportional between a and $b = \sqrt{ab}$
   (d) If $\frac{a}{b} = \frac{c}{d}$ then $\frac{a+b}{a-b} = \frac{c+d}{c-d}$
   (Componendo and dividendo)
3. (a) Comparision of ratios $(a : b) > (c : d) \Leftrightarrow \frac{a}{b} > \frac{c}{d}$
   (b) Compounded ratio of $(a : b)$, $(c : d)$, $(e : f) = \frac{ace}{bdf}$
4. If $A : B = x : y$ and $B : C = p : q$, then
   (a) $A : C = \frac{x \times p}{y \times q}$
   (b) $A : B : C = px : py : qy$
5. In what *ratio* the two kinds of tea must be mixed together one at ₹ x per kg and another at ₹ y per kg, so that the mixture may cost ₹ z per kg?
   $$\text{Ratio} = \frac{z-y}{x-z}$$
6. A grey hound persues a hare as takes J_1 leaps for every J_2 leaps of the hare. If K_1 leaps of the hound are equal to K_2 leaps of the hare, then the Ratio of speeds of the hound and hare is
   $$\frac{J_2 \times K_1}{J_1 \times K_2}$$
7. The incomes of two persons are in the ratio of $a : b$ and their expenditure are in the ratio of $x : y$. If the saving of each person is ₹ s, then income of each is ₹ $\frac{as(y-x)}{ay-bx}$ and ₹ $\frac{bs(y-x)}{ay-bx}$ respectively.
8. In a mixture of z litre, the ratio of milk and water is $x : y$. If another p litres of water is added to the mixture, the ratio of milk and water in the resulting mixture $= \frac{xz}{yz + p(x+y)}$
9. There are four members a, b, c and d, then formula for
   (a) What should be added to each of these numbers so that the remaining numbers may be proportional $= \frac{ad - bc}{(b+c) - (a+d)}$.
   (b) What should be subtracted from each of these numbers so that the remaining numbers may be proportional $= \frac{ad - bc}{(a+d) - (b+c)}$.
10. In a mixture the ratio of milk and water is $a : b$. If in this mixture another K litre of water is added, then the ratio of milk and water in the resulting mixture becomes $a : m$. Then the quantity of milk in the original mixture $= \frac{ak}{m-b}$ and quantity of water $= \frac{bk}{m-b}$.

SOLVED EXAMPLES

Example 1: The ratio between the three angles of a quadrilateral is 6 : 7 : 8 respectively. The value of the fourth angle of the quadrilateral is 108°. What is the difference between the largest and the smallest angles of the quadrilateral?

Solution: $6x + 7x + 8x = 360° - 108°$

$\Rightarrow \quad 21x = 252°$

$\Rightarrow \quad x = 12°$

$\therefore$ Required difference $= 108° - 6 \times 12°$
$= 108° - 72° = 36°$.

Example 2: Three-fifth of a number is equal to 70% of another number. What is the ratio between the first number and the second number respectively?

Solution: Let the numbers are x and y respectively.

$$\therefore \quad x \times \frac{3}{5} = y \times \frac{70}{100}$$

$$\Rightarrow \quad x \times \frac{3}{5} = y \times \frac{7}{10}$$

$$\Rightarrow \quad \frac{x}{y} = \frac{7}{10} \times \frac{5}{3} = 7 : 6.$$

Example 3: The respective ratio between the speeds of a car, a train and a bus is 5 : 9 : 4. The average speed of the car, the bus and the train is 72 km/hr together. What is the average speed of the car and the train together?

Solution: Let the speed of car, train and bus be $5x$, $9x$ and $4x$ km/hr respectively.

$$\because \quad 5x + 9x + 4x = 72 \times 3$$

$$\therefore \quad x = 12$$

$\therefore$ Average speed of car and train

$$= \frac{5x+9x}{2} = \frac{14 \times 12}{2} = 84 \text{ km/hr}$$

Example 4: The ratio between the present ages of Meena and Priya is 3 : 4 respectively. Ten years ago the ratio between their ages was 4 : 7 respectively. What will be Meena's age after 5 years?

Solution: Let the present age of Meena be $3x$ years

$\therefore$ The present age of Priya = $4x$ years

$$\frac{3x-10}{4x-10} = \frac{4}{7}$$

$$21x - 70 = 16x - 40$$

$$x = \frac{70-40}{5} = 6$$

$\therefore$ Meena's age after 5 years = $3 \times 6 + 5$

= 23 years.

MULTIPLE CHOICE QUESTIONS

1. An amount of rupees 4500 is to be distributed among Shrinath, Ramnath and Raghunath in the ratio of 3 : 7 : 5 respectively. If ₹ 500 is added to each of the share, the new ratio will become ......?

A. 7 : 10 : 13 B. 9 : 16 : 13
C. 2 : 4 : 3 D. 4 : 5 : 2

2. The total number of students studying in a school is 10396. The ratio of the total number of girls to the total number of boys studying in the school is 52 : 61 respectively. What is the total number of girls studying in the school?

A. 4874 B. 4784
C. 5216 D. 5612

3. In a box the numbers of one rupee, half rupee and quarter rupee are in the ratio 3 : 4 : 5 and the total value of the coins is ₹ 100. Find the number of each kind of coins respectively—

A. 48, 64, 80 B. 48, 60, 70
C. 55, 78, 90 D. 44, 68, 80

4. The ratio of the ages of two persons is 4 : 7 and the age of one of them is more than the other by 30 years. What is the sum of their ages in years?

A. 100 B. 110
C. 120 D. 130

5. The ratio of the amounts of A and B is 4 : 3 and the sum of their amount is ₹ 700. What amount in rupees should A give B to make the ratio 1 : 1?

A. 100 B. 200
C. 150 D. 50

6. The ratio of the volumes of water and glycerine in 240 c.c. glycerine mixed with water is 1 : 3. What additional volume of water should be mixed with it so that the ratio of the volumes of water and glycerine becomes 2 : 3?

A. 70 c.c. B. 80 c.c.
C. 60 c.c. D. 50 c.c.

7. If 15% of A is the same as 20% of B, then A : B is :

A. 9 : 11 B. 4 : 3
C. 3 : 4 D. 6 : 7

8. The fourth proportional to 1, 3 and the third proportional to 1 and 3 is :

A. 9 B. 81
C. 3 D. 27

9. The ratio of third proportional to 12 and 30 and the mean proportional to 9 and 25 is :

A. 5 : 1 B. 2 : 1
C. 9 : 14 D. 7 : 15

10. If $a + b : b + c : c + a = 6 : 7 : 8$ and $a + b + c = 14$ then the value of c is :

A. 14 B. 7
C. 8 D. 6

11. ₹ 120 are divided among P, Q, R such that P's share is ₹ 20 more than Q's and ₹ 20 less than R's share. What is Q's share?

A. ₹ 20 B. ₹ 15
C. ₹ 25 D. ₹ 10

12. If four numbers are in the ratio 5 : 7 : 11 : 13 be such that the sum of their cubes is 3996, find the average of four numbers.

A. 13 B. 11
C. 12 D. 9

13. If 2A = 3B = 4C then A : B : C is :

A. 2 : 3 : 4 B. 4 : 3 : 2
C. 6 : 4 : 3 D. 3 : 4 : 6

14. If $\frac{1}{5}:\frac{1}{x}=\frac{1}{x}:\frac{1}{1.25}$, then the value of x is :

A. 1.25 B. 1.5
C. 2.5 D. 2.25

15. A fraction bears the same ratio to $\frac{1}{27}$ as $\frac{3}{7}$ does to $\frac{5}{9}$. The fraction is :

A. $\frac{7}{45}$ B. $\frac{1}{35}$
C. $\frac{45}{7}$ D. $\frac{5}{21}$

16. The fourth proportional to 0.2, 0.12 and 0.3 is:

A. 0.13 B. 0.15
C. 0.18 D. 0.8

17. A bag contains 25 paise, 10 paise and 5 paise coins in the ratio 1 : 2 : 3. If their total value is ₹ 30, the number of 5 paise coins is :

A. 50 B. 100
C. 150 D. 200

18. In a class, the number of boys is more than the number of girls by 12% of the total strength. The ratio of boys to girls is :

A. 11 : 14 B. 14 : 11
C. 25 : 28 D. 28 : 25

19. A right cylinder and a right circular cone have the same radius and the same volume. The ratio of the height of the cylinder to that of the cone is :

A. 3 : 5 B. 2 : 5
C. 3 : 1 D. 1 : 3

20. The speeds of three cars are in the ratio 3 : 4 : 5. The ratio between times taken by them to travel the same distance is :

A. 3 : 4 : 5 B. 5 : 4 : 3
C. 12 : 15 : 20 D. 20 : 15 : 12

21. A and B are two alloys of gold and copper prepared by mixing metals in proportions 7 : 2 and 7 : 11 respectively. If equal quantities of the alloys are melted to form a third alloy C, the proportion of gold and copper in C will be :

A. 5 : 9 B. 5 : 7
C. 7 : 5 D. 9 : 5

22. If $a : b = c : d$, then $\frac{ma+nc}{mb+nd}$ is equal to :

A. $m : n$ B. $na : mb$
C. $a : b$ D. $md : nc$

23. Vijay got thrice as many marks in Maths as in English. The proportion of his marks in Maths and History is 4 : 3. If his total marks in Maths, English and History are 250, what are his marks in English?

A. 120 B. 90
C. 40 D. 80

24. A mixture contains milk and water in the ratio 5 : 1. On adding 5 litres of water, the ratio of milk and water becomes 5 : 2. The quantity of milk in the mixture is:

A. 16 litres B. 25 litres
C. 32.5 litres D. 22.75 litres

25. Two equal glasses are respectively $\frac{1}{3}$ and $\frac{1}{4}$ full of milk. They are then filled up with water and the contents mixed in a tumbler. The ratio of milk and water in the tumbler is:

A. 7 : 5 B. 7 : 17
C. 3 : 7 D. 11 : 23

26. The ratio between two numbers is 3 : 4 and their L.C.M. is 180. The first number is :

A. 15 B. 20
C. 45 D. 60

27. A man has some hens and cows. If the number of heads be 48 and number of feet equals 140, the number of hens will be :

A. 22 B. 23
C. 24 D. 26

28. The ratio between the ages of Kamal and Savita is 6 : 5 and the sum of their ages is 44 years. The ratio of their ages after 8 years will be

A. 5 : 6 B. 7 : 8
C. 8 : 7 D. 14 : 13

29. One year ago the ratio between Laxmi's and Gopi's salary was 3 : 4. The ratios of their individual salaries between last year's and this year's salaries are 4 : 5 and 2 : 3 respectively. At present the total of their salary is ₹ 4160. The salary of Laxmi now, is :

A. ₹ 1040 B. ₹ 1600
C. ₹ 2560 D. ₹ 3120

30. If 15% of x = 20% of y, then $x : y$ is :

A. 3 : 4 B. 4 : 3
C. 17 : 16 D. 16 : 17

31. If $\frac{x}{5}=\frac{y}{8}$, then $(x + 5) : (y + 8)$ is equal to :

A. 3 : 5 B. 13 : 8
C. 8 : 5 D. 5 : 8

32. The salaries of A, B, C are in the ratio 2 : 3 : 5. If the increments of 15%, 10% and 20% are allowed respectively in their salaries, then what will be the new ratio of their salaries?

A. 3 : 3 : 10 B. 10 : 11 : 20
C. 23 : 33 : 60 D. Cannot be determined

33. A sum of ₹ 1300 is divided amongst P, Q, R and S such that $\frac{\text{P's share}}{\text{Q's share}} = \frac{\text{Q's share}}{\text{R's share}} = \frac{\text{R's share}}{\text{S's share}} = \frac{2}{3}$.

Then, P's share is :

A. ₹ 140 B. ₹ 160

C. ₹ 240 D. ₹ 320

34. The ratio of the number of boys and girls in a college is 7 : 8. If the percentage increase in the number of boys and girls be 20% and 10% respectively, what will be the new ratio?

A. 8 : 9 B. 17 : 18

C. 21 : 22 D. Cannot be determined

35. The third proportional to 0.36 and 0.48 is :

A. 0.64 B. 0.1728

C. 0.42 D. 0.94

36. An alloy is to contain copper and zinc in the ratio 9 : 4. The zinc required to be melted with 24 kg of copper is :

A. $10\frac{2}{3}$ kg B. $10\frac{1}{3}$ kg

C. $9\frac{2}{3}$ kg D. 9 kg

37. 15 litres of mixture contains 20% alcohol and the rest water. If 3 litres of water be mixed with it, the percentage of alcohol in the new mixture would be :

A. 15% B. $16\frac{2}{3}$%

C. 17% D. $18\frac{1}{2}$%

38. The sides of a triangle are in the ratio $\frac{1}{2}:\frac{1}{3}:\frac{1}{4}$ and its perimeter is 104 cm. The length of the longest side is :

A. 52 cm B. 48 cm

C. 32 cm D. 26 cm

39. x varies inversely as square of y. Given that $y = 2$ for $x = 1$. The value of x for $y = 6$ will be equal to :

A. 3 B. 9

C. $\frac{1}{3}$ D. $\frac{1}{9}$

40. Which of the following ratios is greatest?

A. 7 : 15 B. 15 : 23

C. 17 : 25 D. 21 : 29

ANSWERS

1	2	3	4	5	6	7	8	9	10
E	B	A	B	D	C	B	D	A	D
11	**12**	**13**	**14**	**15**	**16**	**17**	**18**	**19**	**20**
A	D	C	C	B	C	C	B	D	D
21	**22**	**23**	**24**	**25**	**26**	**27**	**28**	**29**	**30**
C	C	C	B	B	C	D	C	B	B
31	**32**	**33**	**34**	**35**	**36**	**37**	**38**	**39**	**40**
D	C	B	C	A	A	B	B	D	D

SOME SELECTED EXPLANATORY ANSWERS

2. From question,

The total number of students in the school = 10396 and ratio of number of girls : boys = 52 : 61

∴ The total number of girls in that school

$$= \frac{10396 \times 52}{(52+61)} = \frac{10396 \times 52}{113}$$

$= 92 \times 52 = 4784$.

3. $3x + 4x \times \frac{1}{2} + 5x \times \frac{1}{4} = 100$

$12x + 8x + 5x = 400$

$x = 16$

Number of 1 rupee = 48

Number of 1/2 rupee = 64

Number of quarter rupee = 5 × 16 = 80

4. ∵ $7x - 4x = 30$,

$x = 10$

∴ Sum of their ages = (4 + 7) × 10

= 11 × 10

= 110 years

5. ∵ $4x + 3x = 700$

⇒ $x = 100$

400 300

Ⓐ Ⓑ

∵ $\frac{400 - x}{300 + x} = \frac{1}{1}$

⇒ $400 - x = 300 + x$

∴ $x = 50$

6. Quantity of water in 240 cc glycerine and water

$= \frac{1}{4} \times 240$ cc = 60 cc

Quantity of glycerine = (240 – 60) cc

= 180 cc

$\because \quad \frac{60+x}{180} = \frac{2}{3}$

$\Rightarrow \quad x = 60$ cc.

8. First we find the third proportional to 1 and 3. Let the third proportional to 1 and 3 is

$\Rightarrow \quad 1 : 3 :: 3 : x$

$\Rightarrow \quad x = 3 \times 3$

$\Rightarrow \quad x = 9$

The fourth proportional to 1, 3 and the third proportional to 1 and 3

= The fourth proportional to 1, 3 and 9. Let the fourth proportional to 1, 3 and 9 is y, then $1 : 3 :: 9 : y$

$\Rightarrow \quad 1 \times y = 3 \times 9$

$\Rightarrow \quad y = 27$

∴ Required answer is 27.

9. Let the third proportional to 12 and 30 be y, then $12 : 30 :: 30 : y$

$\Rightarrow \quad 12y = 30 \times 30$

$y = \frac{30 \times 30}{12} = 75$

Mean proportional to 9 and 25

$= \sqrt{9 \times 25} = 3 \times 5 = 15$

∴ The required ratio is 75 : 15 = 5 : 1

10. Given

$$\left.\begin{aligned} a+b &= 6x \\ b+c = 7x \text{ and } c+a &= 8x \end{aligned}\right\} \quad ...(1)$$

and $\quad a + b + c = 14 \quad ...(2)$

adding equations in (1) we get

$2(a + b + c) = 21x$

using (2) we get

$2 \times 14 = 21x$

$\Rightarrow \quad x = \frac{2 \times 14}{21}$

$= \frac{2 \times 2}{3} = \frac{4}{3}$

$a + b = 6x = 6 \times \frac{4}{3} = 8.$

Using $a + b + c = 14$ and $a + b = 8$

We get $c = 6$.

11. Given P's share = Q's share + ₹ 20

P's share = R's share – ₹ 20

let Q's share be y.

then P's share = y + 20

y + 20 = R's share – ₹ 20

$\Rightarrow$ R's share = y + 40

P's share + Q's share + R's share = 120

$y + 20 + y + y + 40 = 120$

$3y + 60 = 120$

$3y = 120 - 60 = 60$

$y = 20$

∴ Q's share = y = ₹ 20.

12. Let the four numbers be $5x$, $7x$, $11x$ and $13x$.

Given $(5x)^3 + (7x)^3 + (11x)^3 + (13x)^3 = 3996$.

$125x^3 + 343x^3 + 1331x^3 + 2197x^3 = 3996$

$\Rightarrow \quad 3996x^3 = 3996$

$\Rightarrow \quad x^3 = 1 \quad \Rightarrow \quad x = 1$

∴ The four numbers are 5, 7, 11 and 13. Average of these four numbers

$= \frac{5+7+11+13}{4} = \frac{36}{4} = 9.$

13. Let 2A = 3B = 4C = x.

Then, A = $\frac{x}{2}$, B = $\frac{x}{3}$ and C = $\frac{x}{4}$.

∴ A : B : C = $\frac{x}{2} : \frac{x}{3} : \frac{x}{4}$ = 6 : 4 : 3.

Hence, A : B : C = 6 : 4 : 3.

14. $\frac{1}{x} \times \frac{1}{x} = \frac{1}{5} \times \frac{1}{1.25}$

or, $\quad \frac{1}{x^2} = \frac{4}{25}$

or, $\quad x^2 = \frac{25}{4}$

$\therefore \quad x = \frac{5}{2} = 2.5.$

15. $x : \frac{1}{27} = \frac{3}{7} : \frac{5}{9} \Rightarrow \frac{5}{9}x = \frac{1}{27} \times \frac{3}{7}.$

$\therefore \quad \frac{5}{9}x = \frac{1}{63}$ or $x = \left(\frac{1}{63} \times \frac{9}{5}\right) = \frac{1}{35}.$

16. Let 0.2 : 0.12 :: 0.3 : x.

Then, $0.2x = 0.12 \times 0.3$ or

$x = \frac{0.12 \times 0.3}{0.2} = 0.18.$

17. Ratio of their values = $\frac{1}{4} : \frac{2}{10} : \frac{3}{20}$

= 5 : 4 : 3.

$\therefore$ Value of 5-paise coins = ₹ $\left(30 \times \frac{3}{12}\right)$

= ₹ 7.50.

$\therefore$ Number of 5-paise coins = $\frac{750}{5} = 150$.

19. Let the heights of the cylinder and cone be h and H respectively. Then, $\pi r^2 h = \frac{1}{3}\pi r^2$ H or $\frac{h}{\text{H}} = \frac{1}{3}$.

So, their heights are in the ratio 1 : 3.

20. Ratio of time taken = $\frac{1}{3} : \frac{1}{4} : \frac{1}{5}$ = 20 : 15 : 12.

21. Gold in C = $\left(\frac{7}{9} + \frac{7}{18}\right) = \frac{21}{18} = \frac{7}{6}$.

Copper in C = $\left(\frac{2}{9} + \frac{11}{18}\right) = \frac{15}{18} = \frac{5}{6}$.

$\therefore$ Gold : Copper = $\frac{7}{6} : \frac{5}{6}$ = 7 : 5.

22. Let $\frac{a}{b} = \frac{c}{d} = k$. Then, $a = bk$ and $c = dk$.

$\therefore \frac{ma + nc}{mb + nd} = \frac{mbk + ndk}{mb + nd} = k\left(\frac{mb + nd}{mb + nd}\right)$

$= k = \frac{a}{b}$.

23. M = 3E and $\frac{\text{M}}{\text{H}} = \frac{4}{3}$.

$\therefore$ H = $\frac{3}{4}$M = $\frac{3}{4} \times 3\text{E} = \frac{9}{4}$E.

Now, M + E + H = 250

$\Rightarrow$ 3E + E + $\frac{9}{4}$E = 250

$\therefore$ 25E = 1000 or E = 40.

24. Let quantity of milk and water be $5x$ and x litres.

Then, $\frac{5x}{x + 5} = \frac{5}{2}$ or, $10x = 5x + 25$

or, $x = 5$.

$\therefore$ Quantity of milk = $5x$ = 25 litres.

26. Let the numbers be $3x$ and $4x$.

Then, their L.C.M. = $12x$.

$\therefore$ $12x = 180$ or $x = 15$.

Hence, the first number = 45.

27. Let the number of hens = x and number of cows = y.

Then, $x + y = 48$ and $2x + 4y = 140$.

Solving these equations, we get, $2y = 44$

or $y = 22$.

So, $x = (48 - 22) = 26$.

$\therefore$ Number of hens = 26.

28. Let their ages be $6x$ and $5x$ years.

$\therefore$ $6x + 5x = 44$ or, $x = 4$.

So, their present ages are 24 years and 20 years.

Ratio of their ages after 8 years = 32 : 28

= 8 : 7.

29. Let the salaries of Laxmi and Gopi one year before be x_1, y_1 respectively and now x_2, y_2 respectively. Then,

$\frac{x_1}{y_1} = \frac{3}{4}, \frac{x_1}{x_2} = \frac{4}{5}, \frac{y_1}{y_2} = \frac{2}{3}$ and $x_2 + y_2 = 4160$.

Solving these equations, we get $x_2 = 1600$.

30. 15% of x = 20% of y

$\Rightarrow \frac{15x}{100} = \frac{20y}{100}$

$\Rightarrow \frac{x}{y} = \left(\frac{20}{100} \times \frac{100}{15}\right) = \frac{4}{3}$

$\Rightarrow x : y = 4 : 3$.

31. Let $\frac{x}{5} = \frac{y}{8} = k$. Then, $x = 5k$ and $y = 8k$.

$\therefore \frac{x + 5}{y + 8} = \frac{5k + 5}{8k + 8} = \frac{5(k + 1)}{8(k + 1)} = \frac{5}{8}$

$\Rightarrow (x + 5) : (y + 8) = 5 : 8$.

32. Let A = $2k$, B = $3k$ and C = $5k$

A's new salary = $\frac{115}{100}$ of $2k$

$= \left(\frac{115}{100} \times 2k\right) = \frac{23}{10}k$

B's new salary = $\frac{110}{100}$ of $3k$

$= \left(\frac{110}{100} \times 3k\right) = \frac{33}{10}k$

C's new salary = $\frac{120}{100}$ of $5k$

$= \left(\frac{120}{100} \times 5k\right) = 6k$.

$\therefore$ New ratio = $\frac{23k}{10} : \frac{33k}{10} : 6k$ = 23 : 33 : 60.

34. Originally, let the number of boys and girls in the college be $7x$ and $8x$ respectively. Their increased number is (120% of $7x$) and (110% of $8x$)

i.e., $\left(\frac{120}{100} \times 7x\right)$ and $\left(\frac{110}{100} \times 8x\right)$

i.e. $\frac{42x}{5}$ and $\frac{44x}{5}$.

$\therefore$ Required ratio $= \frac{42x}{5} : \frac{44x}{5} = 21 : 22.$

35. Let the third proportional to 0.36 and 0.48 be *x*.
Then, 0.36 : 0.48 :: 0.48 : *x*

$\Rightarrow x = \left(\frac{0.48 \times 0.48}{0.36}\right) = 0.64.$

36. Let the required quantity of copper be *x* kg.
Then, 9 : 4 :: 24 : *x* $\Rightarrow$ 9*x* = 4 × 24

$\Rightarrow x = \frac{4 \times 24}{9} = 10\frac{2}{3}.$

Hence, the required quantity of copper is $10\frac{2}{3}$ kg.

37. Alcohol in 15 litres of mixture = 20% of 15 litres = $\left(\frac{20}{100} \times 15\right)$ litres = 3 litres.

Water in it = (15 – 3) litres = 12 litres.

New quantity of mixture = (15 + 3) litres = 18 litres.

Quantity of alcohol in it = 3 litres.

Percentage of alcohol in new mixture

$= \left(\frac{3}{18} \times 100\right)\% = 16\frac{2}{3}\%.$

38. Ratio of sides $= \frac{1}{2} : \frac{1}{3} : \frac{1}{4} = 6 : 4 : 3.$

Largest side $= \left(104 \times \frac{6}{13}\right)$ cm = 48 cm.

39. Given $x = \frac{k}{y^2}$, where *k* is a constant.

Now, *y* = 2 and *x* = 1 gives *k* = 4.

$\therefore x = \frac{4}{y^2} \Rightarrow x = \frac{4}{6^2}$,

when *y* = 6 $\Rightarrow x = \frac{4}{36} = \frac{1}{9}.$

40. $\frac{7}{15} = 0.466, \frac{15}{23} = 0.652,$

$\frac{17}{25} = 0.68$ and $\frac{21}{29} = 0.724.$

Clearly, 0.724 is greatest and hence, 21 : 29 is greatest.

4

Age and Average

IMPORTANT FACTS

1. The present age of the father is x times the age of his son. T years hence, the father's age become y times the age of his son. Then the present ages of the father and his son are:

   Present age of the son = $\frac{(y-1)T}{x-y}$ years and age of the

   father = $\frac{(y-1)T \times x}{x-y}$ years.

2. T years ago, the father's age was x times that of his son and at present the father's age is y times the age of his son. Then their present ages are:

   Son's age = $\frac{T(x-1)}{x-y}$ years;

   Father's age = $\frac{T(x-1)}{x-y} \times y$ years.

3. Average Speed = $\frac{\text{Total travelled distance}}{\text{Total taken time}}$

4. If equal distances are travelled at the rate X and Y,

   then Average Speed = $\frac{2XY}{X+y}$

5. When a person leaves a group and another person joins the group in the place of person left, then
   I. In the case of increasing of average age, Age of the newcomer = Age of person left + no. of persons in the group X increase in average age.
   II. In the case of decreasing of average age, Age of the newcomer = Age of person left – no. of persons in the group X decrease in average age.

6. When a person joins a group without replacing any previous person from that group, then
   I. In the case of increasing of average age, Age of the newcomer = Previous of the age + no. of all persons (including newcomer) × increaser in average age.
   II. In the case of decreasing of average age, Age of the newcomer = Previous average age – no. of all persons (including new comer) × decrease in average age.

7. When a person leaves the group but nobody joins this group, then
   I. In the case of increasing of average age, Age of man left = Previous Average of the age – no. of present persons × decrease in the average age.
   II. In the case of decreasing of average age, Age of man left = Previous Average age + no. of present persons × increase in average age.

8. Average related to Natural Numbers

   I. Average of consecutive natural numbers till n $= \frac{(n+1)}{2}$

   II. Average of squares of numbers till n $= \frac{(n+1)(2n+1)}{6}$

   III. Average of cubes of consecutive natural numbers till $n = \frac{n(n+1)^2}{4}$

9. Average related to Even Numbers

   I. Average of n = consecutive even numbers $= n + 1$

   II. Average of consecutive even number till n $= (n/2 + 1)$

   III. Average of squares of n consecutive even numbers $= \frac{2(n+1)(2n+1)}{3}$

   IV. Average of squares of consecutive even numbers till $n = \frac{(n+1)(2n+1)}{3}$

10. Average related to Odd Numbers

    I. Average of n consecutive odd numbers = n

    II. Average of consecutive odd numbers till n $= \frac{n(n+1)}{3}$

    III. Average of squares of consecutive odd numbers till $n = \frac{n(n+1)}{3}$

SOLVED EXAMPLES

Example 1: Ankur's present age is 12 years more than Raman's age 5 years ago. Presently Raman is 24 years older than Raju's present age who is 12 years younger than Amit. If Amit's present age is 42 years, what is Ankur's age 3 years ago?

Solution: Amit's present age = 42 years
Raju's present age = 42 – 12 = 30 years
Raman's present age = 30 + 24 = 54 years
Ankur's present age = 54 – 5 + 12
= 61 years
∴ Ankur's age 3 years ago = 58 years

Example 2: Present ages of Leela and Shivani are in the ratio of 4 : 5 respectively. Three years hence this ratio becomes 5 : 6. What is the difference between their present ages?

Solution: Let Leela's and Shivani's present ages be $4x$ and $5x$ years respectively.
After 3 years,

$$\frac{4x+3}{5x+3} = \frac{5}{6}$$

$$\Rightarrow \quad 25x + 15 = 24x + 18$$

$$\Rightarrow \quad 25x - 24x = 18 - 15 = 3$$

$\therefore x = 3$ = Required difference

Example 3: Radha's present age is three years less than twice her age 12 years ago. Also the respective ratio between Raj's present age and Radha's present age is 4 : 9. What will be Raj's age after 5 years?

Solution: Let the present age of Radha be x years. As per question,

$$\because \quad x = (x - 12) \times 2 - 3$$

$$\Rightarrow \quad x = 2x - 27$$

$$\therefore \quad x = 27$$

∵ Present age of Raj

$$= \frac{4}{9} \times 27 = 12 \text{ years}$$

∴ Raj's age after 5 years = 12 + 5 = 17 years

Example 4: If $12a + 12b = 45$, what is the average of a and b?

Solution: $\because \quad 12a + 12b = 45$

$$\Rightarrow \quad 12(a + b) = 45$$

$$\therefore \quad \frac{a+b}{2} = \frac{45}{12 \times 2} = 1.875.$$

Example 5: The sum of five numbers is 555. The average of first two numbers is 75 and the third number is 115. What is the average of last two numbers?

Solution: Average of last two numbers

$$= \frac{555 - 2 \times 75 - 115}{2}$$

$$= \frac{555 - 150 - 115}{2} = \frac{290}{2} = 145.$$

Example 6: The average marks of nine students in a group is 63. Three of them scored 78, 69 and 48 marks. What are the average marks of remaining six students?

Solution: The average marks of remaining 6 students

$$= \frac{9 \times 63 - (78 + 69 + 48)}{6}$$

$$= \frac{567 - 195}{6} = 62.$$

MULTIPLE CHOICE QUESTIONS

1. The sum of 8 consecutive odd numbers is 656. Also average of four consecutive even number is 87. What is the sum of the smallest odd number and second largest even number?
A. 165 B. 175
C. 163 D. Cannot be determined

2. The ages of Meena, Tina and Seema are in the ratio of 6 : 4 : 7 respectively. If the sum of their ages is 34 years, what is Seema's age?
A. 12 years B. 10 years
C. 18 years D. None of these

3. The average of 4 positive integers is 52.5. The highest integer is 76 and the lowest integer is 34. The difference between the remaining two integers is 4. Which of the following integers is higher of the remaining two integers?
A. 52 B. 54
C. 48 D. Cannot be determined

4. The present age of Deepa is 3 times the present age of his son. Five years hence the ratio of their ages will be 34 : 13 respectively. What is the present age of Deepa?
A. 63 years B. 68 years
C. 58 years D. 60 years

5. The average of 5 consecutive even numbers A, B, C, D and E is 66. What is the product of B and E?
A. 4352 B. 4340
C. 4620 D. None of these

6. The present age of father is 34 years more than that of his son. Twelve years ago father's age was 18 times the age of his son. What is the present age of the son in years?
A. 16 B. 14
C. 12 D. 18

7. A father is older than his first son by 30 years, the first son is older than the second son by 10 years.

If the sum of their ages by 95 years, find the age of father—

A. 65 years B. 55 years
C. 45 years D. 50 years

8. In 10 years, Alok will be twice as old as Babul was 10 years ago. If Alok is now 9 years older than Babul, the present age of Babul is :

A. 19 years B. 29 years
C. 39 years D. 49 years

9. The difference between the ages of two persons is 10 years. 15 years ago, the elder one was twice as old as the younger one. The present age of the elder person is :

A. 25 years B. 35 years
C. 45 years D. 55 years

10. Ratio of Anshu's age to Jolly's age is equal to 4 : 3. Anshu will be 26 years old after 6 years. How old is Jolly now?

A. 12 years B. 15 years
C. $19\frac{1}{2}$ years D. 21 years

11. The ratio between the ages of Asha and Bhushan at present is 2 : 3. Five years hence the ratio of their ages will be 3 : 4. What is the present age of Asha?

A. 10 years B. 15 years
C. 25 years D. data inadequate

12. Three years ago the average age of A and B was 18 years. With C joining them, the average becomes 22 years. How old is C now?

A. 24 years B. 27 years
C. 28 years D. 30 years

13. Ten years ago A was half of B in age. If the ratio of their present ages is 3 : 4, what will be the total of their present ages?

A. 8 years B. 20 years
C. 35 years D. 45 years

14. Out of three numbers, the first is twice the second and is half of the third. If the average of the three numbers is 56, the three numbers in order are :

A. 48, 96, 24 B. 48, 24, 96
C. 96, 24, 48 D. 96, 48, 24

15. The average of 50 numbers is 38. If two numbers, namely 45 and 55 are discarded, the average of the remaining numbers is :

A. 36.5 B. 37.0
C. 37.5 D. 37.52

16. The average temperature of the first three days is 27° C and that of the next three is 29° C. If the average of the whole week is 28.5° C, the temperature of the last day of the week is:

A. 10.5° C B. 21° C
C. 31.5° C D. 42° C

17. The average of an adult class is 40 years. 12 new students with an average age of 32 years join the class, thereby decreasing the average by 4 years. The original strength of the class was :

A. 10 B. 11
C. 12 D. 15

18. The average weight of 8 persons is increased by 2.5 kg when one of them whose weight is 56 kg is replaced by a new man. The weight of the new man is :

A. 66 kg B. 75 kg
C. 76 kg D. 86 kg

19. In a TV factory, an average of 60 TVs are produced per day for the first 25 days of the month. A few workers fell ill for the next five days, reducing the daily average for the month to 58 sets per day. The average production per day for the last 5 days is :

A. 45 B. 48
C. 52 D. 58

20. The average expenditure of a man for the first five months is ₹ 120 and for the next seven months it is ₹ 130. If he saves ₹ 290 in that years, his monthly average income is :

A. ₹ 140 B. ₹ 150
C. ₹ 160 D. ₹ 170

21. The average age of 24 students in a class is 10. If the teacher's age is included, the average increases by one. The age of the teacher is :

A. 25 years B. 30 years
C. 35 years D. 40 years

22. A batsman has a certain average runs for 11 innings. In the 12th inning he made a score of 90 runs and thereby decreased his average by 5. His average after 12th inning is :

A. 127 B. 145
C. 150 D. 217

23. The average weight of three men A, B and C is 84 kg. Another man D joins the group and the average now becomes 80 kg. If another man E, whose weight is 3 kg more than that of D, replaces A, then the average weight of B, C, D and E becomes 79 kg. The weight of A is :

A. 70 kg B. 72 kg
C. 75 kg D. 80 kg

24. A man whose bowling average is 12.4 takes 5 wickets for 26 runs and thereby decreases his average by 0.4. The number of wickets, taken by him, before his last match, is :

A. 85 B. 78
C. 72 D. 64

25. The present ages of three persons are in proportions 4 : 7 : 9. Eight years ago, the sum of their ages was 56. Find their present ages (in years).

A. 8, 20, 28 B. 16, 28, 36
C. 20, 35, 45 D. Cannot be determined

26. A is two years older than B who is twice as old as C. If the total of the ages of A, B and C be 27, then how old is B?

A. 7 B. 8
C. 9 D. 10

27. A person's present age is two-fifth of the age of his mother. After 8 years, he will be one-half of the age of his mother. How old is the mother at present?

A. 32 years B. 36 years
C. 40 years D. 48 years

28. Quasab is as much younger than Rahim as he is older than Tanveer. If the sum of the ages of Rahim and Tanveer is 50 years, what is definitely the difference between Rahim's and Quasab's age?

A. 1 year B. 2 years
C. 25 years D. Can not be determined

29. A father said to his son, "I was as old as you are at present at the time of your birth." If the father's age is 38 years now, the son's age five years back was :

A. 14 years B. 19 years
C. 33 years D. 38 years

30. Asha's father was 38 years of age when she was born while her mother was 36 years old when her brother four years younger to her was born. What is the difference between the ages of her parents?

A. 2 years B. 4 years
C. 6 years D. 8 years

31. A student was asked to find the arithmetic mean of the numbers 3, 11, 7, 9, 15, 13, 8, 19, 17, 21, 14 and x. He found the mean to be 12. What should be the number in place of x?

A. 3 B. 7
C. 17 D. 31

32. If a, b, c, d, e are five consecutive odd numbers, their average is :

A. $5\ (a + 4)$

B. $\dfrac{abcde}{5}$

C. $5\ (a + b + c + d + e)$

D. None of these

33. Of the four numbers, the first is twice the second, the second is one-third of the third and the third is 5 times the fourth. The average of the numbers is 24.75. The largest of these numbers is :

A. 9 B. 25
C. 30 D. 45

34. Of the three numbers, the average of the first and the second is greater than the average of the second and the third by 15. What is the difference between the first and the third of the three numbers?

A. 15 B. 45
C. 60 D. None of these

35. The average temperature of the town in the first four days of a month was 58 degrees. The average for the second, third, fourth and fifth days was 60 degrees. If the temperatures of the first and fifth days were in the ratio 7 : 8, then what is the temperature on the fifth day?

A. 64 degrees B. 62 degrees
C. 56 degrees D. 46 degrees

36. The captain of a cricket team of 11 members is 26 years old and the wicket keeper is 3 years older. If the ages of these two are excluded, the average age of the remaining players is one years less than the average age of the whole team. What is the average age of the team?

A. 23 years B. 24 years
C. 25 years D. 32 years

37. A cricketer has a certain average for 10 innings. In the eleventh inning, he scored 108 runs, thereby increasing his average by 6 runs. His new average is :

A. 48 runs B. 52 runs
C. 55 runs D. 60 runs

38. The average weight of 3 men A, B and C is 84 kg. Another man D joins the group and the average now becomes 80 kg. If another man E, whose weight is 3 kg more than that of D, replaces A, then the average weight B, C, D and E becomes 79 kg. The weight of A is :

A. 70 kg B. 72 kg
C. 75 kg D. 80 kg

39. The average age of husband, wife and their child 3 years ago was 27 years and that of wife and the child 5 years ago was 20 years. The present age of the husband is :

A. 35 years B. 40 years
C. 50 years D. 36 years

40. In an examination, a pupil's average marks were 63 per paper. If he had obtained 20 more marks for his Geography paper and 2 more marks for his History paper, his average per paper would have been 65. How many papers were there in the examination?

A. 8 B. 9
C. 10 D. 11

ANSWERS

1	2	3	4	5	6	7	8	9	10
C	D	A	A	D	B	B	C	B	B
11	**12**	**13**	**14**	**15**	**16**	**17**	**18**	**19**	**20**
A	A	C	B	C	C	C	C	B	B
21	**22**	**23**	**24**	**25**	**26**	**27**	**28**	**29**	**30**
C	B	C	A	B	D	C	D	A	C
31	**32**	**33**	**34**	**35**	**36**	**37**	**38**	**39**	**40**
B	D	D	D	A	A	A	C	B	D

SOME SELECTED EXPLANATORY ANSWERS

1. Let the smallest odd and smallest even numbers are x and y respectively. Then,

$\because\ x + (x + 2) + (x + 4) + (x + 6) + (x + 8) + (x + 10) + (x + 12) + (x + 14) = 656$

$\Rightarrow\quad 8x + 56 = 656$

$\therefore\quad x = \dfrac{656-56}{8} = 75$

$\because\ y + (y + 2) + (y + 4) + (y + 6) = 4 \times 87$

$\Rightarrow\quad 4y + 12 = 348$

$y = \dfrac{348-12}{4} = 84$

$\therefore$ Reqd. sum of the smallest odd number and second largest even number

$= x + (y + 4) = 75 + (84 + 4) = 163$

2. Seema's age $= \dfrac{7 \times 34}{(6+4+7)} = 14$ years.

3. Let the 4 positive integers, kept in descending order be a, b, c and d respectively. Then from question,

$\because\quad a + b + c + d = 4 \times 52.5 = 210$

$\Rightarrow 76 + (b + c) + 34 = 210$

$\Rightarrow\quad (b + c) = 100 \quad ...(i)$

Again from question,

$b - c = 4 \quad ...(ii)$

Solving (*i*) and (*ii*), we get

$b = 52$

and $\quad c = 48$

Hence, higher integer $= b = 52$

4. Let the present age of Deepa's son $= x$ years.

Then from question,

Deepa's present age $= 3x$ years

$\because\quad \dfrac{3x+5}{x+5} = \dfrac{34}{13}$

$\Rightarrow\quad 39x + 65 = 34x + 170$

$\Rightarrow\quad 5x = 105$

$\therefore\quad x = 21$

Deepa's present age $= 3x = 3 \times 21 = 63$ years

5. From question,

$A + B + C + D + E = 66 \times 5 = 330$

Let the 1st even number A be n, then

$\because\ n + (n + 2) + (n + 4) + (n + 6) + (n + 8) = 330$

$\Rightarrow\quad 5n + 20 = 330$

$\Rightarrow\quad 5n = 310$

$\therefore\quad n = 62$

Thus the required product

$= B \times E = (n + 2) \times (n + 8)$

$= 64 \times 70 = 4480$

7. Age of 1st son $= x$

Father's age $= x + 30$

Age of 2nd son $= x - 10$

$\because\ x + (x + 30) + (x - 10) = 95$

$\Rightarrow\quad 3x + 20 = 95$

$\Rightarrow\quad 3x = 75$

$\therefore\quad x = 25$

$\therefore$ Father's age $= (25 + 30)$ years $= 55$ years

8. Let the present ages of Babul and Alok be x years and $(x + 9)$ years respectively.

Then, $(x + 9 + 10) = 2(x - 10)$ or $x = 39$.

9. Let the present age of the elder person be x years. Then, the present age of the other person

$= (x - 10)$ years

$(x - 15) = 2(x - 10 - 15)$ or $x = 35$

$\therefore$ The present age of the elder person is 35 years.

10. Let Anshu's age $= 4x$ and Jolly's age $= 3x$ years.

$\therefore\ 4x + 6 = 26 \quad \Rightarrow \quad x = 5$

$\therefore$ Jolly's age $= 3x = 15$ years.

11. Let the ages of Asha and Bhushan be $2x$ and $3x$ years.

$\dfrac{2x+5}{3x+5} = \dfrac{3}{4}$

$\Rightarrow\quad 4(2x + 5) = 3(3x + 5)$

$\Rightarrow \quad x = 5.$

$\therefore$ Asha's present age = $2x$ = 10 years.

12. Sum of ages of A and B, 3 years ago = (18 × 2) = 36 years.
Sum of ages of A, B and C = (22 × 3) = 66 years.
Sum of ages of A and B = (36 + 6) years = 42 years.
$\therefore$ C's age = (66 – 42) years = 24 years.

14. Let second number = x. Then, first one = $2x$ and third number = $4x$.

$$\therefore \frac{x+2x+4x}{3} = 56 \Rightarrow 7x = 168 \text{ or } x = 24.$$

So, the numbers are 48, 24, 96.

15. Total of 50 numbers = (50 × 38) = 1900.
Total of 48 numbers = [1900 – (45 + 55)] = 1800.

$$\therefore \text{Required average} = \frac{1800}{48} = \frac{225}{6} = 37.5.$$

16. $3 \times 27 + 3 \times 29 + x = 7 \times 28.5$
$\therefore \quad x = 31.5$

17. Let original strength = x. Then,
$40x + 12 \times 32 = (x + 12) \times 36$
or $40x + 384 = 36x + 432$ or $4x = 48$
or $\quad x = 12.$

18. Total increase = (8 × 2.5) kg = 20 kg.
Weight of new man = (56 + 20) kg = 76 kg.

19. Production during these 5 days = (30 × 58 – 25 × 60) = 240.

$$\therefore \text{Average for 5 days} = \frac{240}{5} = 48.$$

20. Total income = ₹ 120 × 5 + 130 × 7 + 290) = ₹ 1800.

$$\text{Average monthly income} = ₹ \left(\frac{1800}{12}\right) = ₹\ 150.$$

21. Age of the teacher = (25 × 11 – 24 × 10) years = 35 years.

22. $11x + 90 = (x - 5) \times 12$ or $x = 150.$
$\therefore$ Average after 12th inning = (150 – 5) = 145.

23. A + B + C = 3 × 84 = 252, A + B + C + D = (4 × 80) = 320.
$\therefore$ D = (320 -- 252) = 68 and so E = (68 + 3) = 71.
Now, B + C + D + E = (4 × 79) = 316
So, A = (320 – 245) = 75 kg.

24. Let the number of wickets taken before the last match = x.

$$\text{Then, } \frac{12.4x+26}{x+5} = 12 \Rightarrow x = 85.$$

25. Let their present ages be $4x$, $7x$ and $9x$ years respectively.
Then, $(4x - 8) + (7x - 8) + (9x - 8) = 56$
$\Rightarrow 20x = 80 \Rightarrow x = 4.$
$\therefore$ Their present ages are 16 years, 28 years and 36 years respectively.

26. Let C's age be x years.
Then, B's age = $2x$ years.
A's age = $(2x + 2)$ years.
$\therefore (2x + 2) + 2x + x = 27$
$\Rightarrow 5x = 25 \Rightarrow x = 5.$
Hence, B's age = $2x$ = 10 years.

27. Let the mother's present age be x years. Then, the person's present age = $\left(\frac{2}{5}x\right)$ years.

$$\therefore \left(\frac{2}{5}x+8\right) = \frac{1}{2}(x+8)$$

$\Rightarrow 2(2x + 40) = 5(x + 8) \Rightarrow \quad x = 40.$

28. Rahim – Quasab = Rahim – Tanveer
$\Rightarrow$ Quasab = Tanveer
Also, Rahim + Tanveer = 50
$\Rightarrow$ Rahim + Quasab = 50.
So, (Rahim – Quasab) cannot be determined.

29. Let the son's present age be x years.
Then, $(38 - x) = x$
$\Rightarrow 2x = 38$
$\Rightarrow x = 19.$
$\therefore$ Son's age 5 years back = (19 – 5) years = 14 years.

30. Mother's age when Asha's brother was born = 36 years.
Father's age when Asha's brother was born = (38 + 4) years = 42 years.
$\therefore$ Required difference = (42 – 36) years = 6 years.

31. Clearly, we have

$$\left(\frac{3+11+7+9+15+13+8+19+17+21+14+x}{12}\right) = 12$$

or, $137 + x = 144$
or, $x = 144 - 137 = 7.$

32. Clearly, $b = a + 2$, $c = a + 4$, $d = a + 6$ and $e = a + 8$.
$\therefore$ Average

$$= \frac{a+(a+2)+(a+4)+(a+6)+(a+8)}{5}$$

$$= \left(\frac{5a+20}{5}\right) = (a + 4).$$

34. Let the numbers be x, y and z.

Then, $\left(\frac{x+y}{2}\right) - \left(\frac{y+z}{2}\right) = 15$

or, $(x + y) - (y + z) = 30$

or, $x - z = 30$.

35. Sum of temperatures on 1st, 2nd, 3rd and 4th days

$= (58 \times 4) = 232$ degrees ...(*i*)

Sum of temperatures on 2nd, 3rd, 4th and 5th days

$= (60 \times 4) = 240$ degrees ...(*ii*)

Subtracting (*i*) from (*ii*), we get :

Temperature on 5th day – Temperature on 1st day

= 8 degrees.

Let the temperatures on 1st and 5th days be $7x$ and $8x$ degrees respectively.

Then, $8x - 7x = 8$ or $x = 8$.

$\therefore$ Temperature on the 5th day $= 8x = 64$ degrees.

36. Let the average age of the whole team be x years.

$\therefore\ 11x - (26 + 29) = 9\ (x - 1)$

$\Rightarrow\ 11x - 9x = 46 \Rightarrow\ 2x = 46$

$\Rightarrow\ x = 23.$

So, average age of the team is 23 years.

37. Let average for 10 innings be x. Then,

$$\frac{10x+108}{11} = x + 6$$

$\Rightarrow\ 11x + 66 = 10x + 108$

$\Rightarrow\ x = 42.$

$\therefore$ New average $= (x + 6) = 48$ runs.

39. Sum of the present ages of husband, wife and child = $(27 \times 3 + 3 \times 3)$ years = 90 years.

Sum of the present ages of wife and child

$= (20 \times 2 + 5 \times 2)$ years = 50 years.

$\therefore$ Husband's present age $= (90 - 50)$ years

= 40 years.

40. Let the number of papers be x.

Then, $63x + 20 + 2 = 65x$

or, $2x = 22$

or, $x = 11.$

5

Percentage

IMPORTANT FACTS

To solve quickly the problems related to percentage, remember following rules:

Rule 1:

(*a*) Of the given two numbers if the first is $x\%$ more than the second, then the second will be $\left(\frac{100 \times x}{100 + x}\right)\%$ less than the first.

(*b*) Of the given two numbers if the first is $x\%$ less than the second, then the second will be $\left(\frac{100 \times x}{100 - x}\right)\%$ more than the first.

(*c*) If two numbers are respectively $x\%$ and $y\%$ more than a third number, then the first number will be $\left(\frac{100 + x}{100 + y} \times 100\right)\%$ of the second.

(*d*) If two numbers are respectively $x\%$ and $y\%$ less than a third number, then the first number will be $\left(\frac{100 - x}{100 - y} \times 100\right)\%$ of the second.

Rule 2 :

(*a*) If a number or quantity is increased by $x\%$ then in order to restore its original value it must be decreased by $\left[\frac{100 \times x}{100 + x}\right]\%$.

(*b*) If a number or quantity is decreased by $x\%$ then in order to restore its original value it must be increased by $\left[\frac{100 \times x}{100 - x}\right]\%$.

Rule 3 :

(*a*) If a number is successively increased by $x\%$ and $y\%$ then a single equivalent increase in that number will be $\left(x + y + \frac{xy}{100}\right)\%$

(*b*) If two successive discounts of $x\%$ and $y\%$ are allowed on a particular amount, then a single discount that is equivalent to the two successive discounts will be $\left(x + y - \frac{xy}{100}\right)\%$

(*c*) If a number is successively increased by $x\%$, $y\%$ and $z\%$, then a single equivalent increase in that number will be

$$\left[(x + y + z) + \left(\frac{xy + yz + zx}{100}\right) + \frac{(xyz)}{10000}\right]\%$$

(*d*) If three successive discounts of $x\%$, $y\%$ and $z\%$ are allowed on an amount, then a single discount that is equivalent to the three successive discounts will be

$$\left[x + y + z - \frac{(xy + yz + zx)}{100} + \frac{xyz}{10000}\right]\%$$

Rule 4 :

(*a*) If a number is increased by $x\%$ and thereafter reduced by $x\%$, then the number will be reduced by $\left(\frac{x^2}{100}\right)$ per cent.

(*b*) If a number is reduced by $x\%$ and thereafter increased by $x\%$ then the number will be reduced by $\left(\frac{x^2}{100}\right)$ per cent.

(*c*) If due to an increase of $x\%$ in the selling price of certain commodity the sell/consumption of the commodity decreases by $y\%$, then gross receipts on account of sale of that commodity will be increased or decreased by $\left(x - y - \frac{xy}{100}\right)\%$, where $x > y$ and will be decreased by $\left(y - x + \frac{xy}{100}\right)\%$, where $y > x$.

SOLVED EXAMPLES

Example 1: If the numerator of a fraction is increased by 200% and the denominator is increased by 400%. The resultant fraction is $1\frac{1}{20}$. What was the original fraction?

Solution: Let original fraction = $\frac{x}{y}$; then

$$\frac{x+2x}{y+4y} = \frac{21}{20}$$

$$\frac{3x}{5y} = \frac{21}{20}$$

$$\frac{x}{y} = \frac{21\times5}{20\times3} = \frac{7}{4} = 1\frac{3}{4}.$$

Example 2: In a class of 65 students and 4 teachers, each student got sweets that are 20% of the total number of students and each teacher got sweets that are 40% of the total number of students. How many sweets were there?

Solution: The share of each student = $\frac{20}{100}\times 65$

= 13 sweets

The share of each teacher = $\frac{40}{100}\times 65$

= 26 sweets

Total number of sweets = 65 × 13 + 4 × 26

= 845 + 104 = 949.

Example 3: Bovina spent ₹ 44,668/- on her air tickets, ₹ 56,732/- on buying gifts for the family members and the remaining 22% of the total amount had as cash with her. What was the total amount?

Solution: Let total amount = ₹ x ; then

$$\frac{78}{100}\times x = 44668 + 56732$$

$$\Rightarrow x = \frac{101400\times100}{78} = ₹\ 1{,}30{,}000.$$

Example 4: Rubina decided to donate 16% of her monthly salary to an NGO. On the day of donation she changed her mind and donated ₹ 6,567/- which was 75% of what she had decided earlier. How much is Rubina's monthly salary?

Solution: Let her monthly salary be ₹ x ; then

$$\frac{16}{100}\times\frac{75}{100}\times x = 6567$$

$$\Rightarrow \frac{3}{25}x = 6567$$

$$\Rightarrow x = \frac{6567\times25}{3} = ₹\ 54725$$

Thus, her monthly salary = ₹ 54725.

MULTIPLE CHOICE QUESTIONS

1. Mr. Rajesh spent 20% of his monthly income on grocery and household expenses. Out of the remaining he spent 25% on children's education 20% on transport and 15% on entertainment. He is left with an amount of ₹ 14,400 after incurring above expenses. What was his monthly income?

A. ₹ 45,000 B. ₹ 24,000
C. ₹ 36,000 D. ₹ 32,000

2. Number of students in a school in June every year increases by 10% from the previous year. If the school had 2,400 students in 2009, how many students did it have in June 2011?

A. 2,840 B. 2,880
C. 2,904 D. 2,920

3. On a test consisting of 150 questions, Reema answered 40% of the first 75 questions correctly. What per cent of the other 75 questions does she need to answer correctly for her grade on the entire exam to be 60%?

A. 80 B. 70
C. 40 D. 50

4. If the numerator of a fraction is increased by 200% and the denominator is increased by 300%. The resultant fraction is $\frac{9}{17}$. What was the original fraction?

A. $\frac{11}{17}$ B. $\frac{9}{11}$
C. $\frac{13}{17}$ D. None of these

5. One-sixth of a number is 53. What will 57% of that number be?

A. 181.26 B. 136.74
C. 197.16 D. 149.46

6. Mrs. Sudha Jain invests 14% of her monthly salary, *i.e.* ₹ 7,014 in Insurance Policies. Later she invests 21% of her monthly salary on Family Mediclaim Policies; also she invests another 6.5% of her salary on Mutual Funds. What is the total annual amount invested by Mrs. Sudha Jain?

A. ₹ 25,050 B. ₹ 50,100
C. ₹ 2,42,550 D. ₹ 2,49,498

7. A sum of ₹ 817 is divided among A, B and C such that 'A' receives 25% more than 'B' and 'B' receives 25% less than 'C'. What is the 'A's share in the amount?

A. ₹ 228 B. ₹ 247
C. ₹ 285 D. ₹ 304

8. Ms. Sujan invests 7% *i.e.* ₹ 2170 of her monthly salary in mutual funds. Later she invests 18% of her monthly salary in recurring deposits also, she invests 6% of her salary on NSC's. What is the total annual amount invested by Ms. Sujan?
A. ₹ 1,25,320 B. ₹ 1,13,520
C. ₹ 1,35,120 D. ₹ 1,15,320

9. What is the 30% of 40% of $\frac{2}{5}$th of 2500?
A. 500 B. 400
C. 360 D. 120

10. A sum of ₹ 731 is divided among A, B and C, such that 'A' receives 25% more than 'B' and 'B' receives 25% less than 'C'. What is C's share in the amount?
A. ₹ 172 B. ₹ 200
C. ₹ 262 D. None of these

11. Thirty five per cent of 740 is 34 more than a number. What is two-fifth of the number?
A. 45 B. 90
C. 180 D. 120

12. Ramu scored 456 marks in an exam and Shama got 54 per cent marks in the same exam which is 24 marks less than Ramu. If the minimum passing marks in the exam is 34 per cent, then how much more marks did Ramu score than the minimum passing marks?
A. 184 B. 196
C. 190 D. 180

13. Shreyas bought an article and sold for 125% of its cost price. What was the cost price of the article, if Shreyas sold it for ₹ 30,750?
A. ₹ 24,600 B. ₹ 25,640
C. ₹ 24,250 D. ₹ 23,200

14. When 20% of a number is added to another number the number increases by 50%. What is the respective ratio between the first and the second number?
A. 3 : 2 B. 2 : 3
C. 5 : 2 D. Cannot be determined

15. If a number is added to seven-ninth of thirty five per cent of 900, the value so obtained is 325. What is the number?
A. 60 B. 120
C. 90 D. None of these

16. The difference between 60% of a number and 20% of the same number is 316. What is 35% of that number?
A. 270.5 B. 285.5
C. 276.5 D. 275

17. In an examination it is required to get 36% of the aggregate marks to pass. A student gets 174 marks and is declared failed by 96 marks. What is the maximum aggregate marks a student can get?
A. 750 B. 650
C. 600 D. Cannot be determined

18. There is a rebate of 15%, if electric bills are paid in time. A man got a rebate of ₹ 54 by paying the bill in time. What was his electric bill in rupees?
A. 300 B. 350
C. 360 D. 380

19. A number is increased by 20% then again by 20%. By what per cent the original number is changed?
A. 40% B. 50%
C. 44% D. 36%

20. In an examination 62% of the examinees got plucked in Mathematics, 52% passed in English and 22% failed in both the subjects. If 108 is the number of examinees who passed in both the subjects then find out the total number of candidates—
A. 1200 B. 900
C. 800 D. 1000

21. Ramesh gets 10% more than Mohan. How much per cent will Mohan get less than Ramesh?
A. 10% B. 9%
C. $9\frac{1}{11}$% D. 11%

22. By what quantity the use of coal is to be reduced to keep one's family budget same when the price of coal is increased by 25%?
A. 25% B. 20%
C. 30% D. 15%

23. The percentage equivalent to the fraction 3/4 is :
A. 75% B. 25%
C. 60% D. 80%

24. If each side of a square be increased by 10%, what is the percentage increase of the area?
A. 21% B. 10%
C. 20% D. 100%

25. Suppose 35% of the males are married. Find the percentage of unmarried females in the total population, if $\frac{5}{7}$ part of the population in a village are males.
A. 21% B. 25%
C. $3\frac{4}{7}$% D. 28%

26. In an examination, 80% of the students failed in English, 85% in Maths and 75% in both English and Maths. If 40 students passed in both the subjects, the total number of students is
A. 200 B. 400
C. 600 D. 800

27. If the ratio of Rohan's expenditure on food and clothes be 3 : 2 and the prices register an increase of 20% on food and 25% on clothes of expenditure. By how much should his salary be increased in order that he may maintain the same level of consumption as before, his present salary being ₹ 1750?

A. ₹ 2135 B. ₹ 1200
C. ₹ 387 D. ₹ 385

28. A hockey team lost 20% of the total number of matches it played during a year If it won 70% of the matches played by the team and 40 matches were drawn. What was the total number of matches played by the team during the year.

A. 400 B. 200
C. 500 D. 300

29. A rise of 25% in the price of groundnuts compels a person to buy 1.5 kg of groundnuts less for ₹ 240, original price of groundnuts per kg. is :

A. ₹ 40 B. ₹ 100
C. ₹ 32 D. ₹ 140

30. In an election between two candidates A and B, A got 55% of the total valid votes. 20% of the votes were declared invalid. If the total votes were 7500, the number of valid votes polled in favour of the candidate B is :

A. 2700 B. 3300
C. 7500 D. 1500

31. The population of a city is 35000. On an increas of 6% in the number of men and an increase of 4% in the number of women, the population would become 36700. The number of men and women separately in thousand is:

A. 15, 16 B. 16, 17
C. 17, 18 D. 18, 17

32. After 30 kg of water were evaporated from a solution of salt and water which had 15% of salt, the remaining solution had 20% salt. The weight of the original solution is :

A. 30 kg B. 60 kg
C. 120 kg D. 600 kg

33. X's salary is half that of Y. If X got a 50% rise in his salary and Y got a 25% rise in his salary, then find the percentage increase in combined salaries of both.

A. 25% B. $16\frac{2}{3}\%$
C. 30% D. $33\frac{1}{3}\%$

34. In a particular school in a rural area, 80% of students with age below 14 years in the area are enrolled; of them, 75% come regularly to the school. If 270 students are regular, how many students are there in the area who are below 14 years?

A. 270 B. 370
C. 450 D. 550

35. A's weight is 25% that of B's and 40% that of C's. What percentage of C's weight is B's weight?

A. 80% B. 60%
C. 160% D. 90%

ANSWERS

1	2	3	4	5	6	7	8	9	10
A	C	A	D	A	D	C	D	D	D
11	12	13	14	15	16	17	18	19	20
B	A	A	C	D	C	A	C	C	B
21	22	23	24	25	26	27	28	29	30
C	B	A	A	C	B	D	A	C	A
31	32	33	34	35					
D	C	D	C	C					

SOME SELECTED EXPLANATORY ANSWERS

1. Let Rajesh's monthly income be ₹ 100.

∴ Expenditure on grocery and household expenses = ₹ 20.

Expenditure on education etc.

$$= \frac{80 \times 60}{100} = ₹\ 48$$

Saving = 100 – 20 – 48 = ₹ 32

Now,

when saving ₹ 32 then income = ₹ 100

when saving ₹ 14400 then income

$$= \frac{100}{32} \times 14400 = ₹\ 45000$$

2. Required number of students

$$= 2400\left(1 + \frac{10}{100}\right)^2 \qquad \left[P = P_0\left(1 + \frac{R}{100}\right)^T\right]$$

$$= 2400 \times \frac{11}{10} \times \frac{11}{10} = 2904.$$

3. Required % $= \dfrac{150\times\frac{60}{100}-75\times\frac{40}{100}}{75}\times 100\%$

$= \dfrac{90-30}{75}\times 100\% = 80\%$

4. Let the fraction be $\dfrac{x}{y}$

$\therefore \quad \dfrac{x+2x}{y+3y} = \dfrac{9}{17}$

$\Rightarrow \quad \dfrac{3x}{4y} = \dfrac{9}{17}$

$\therefore \quad \dfrac{x}{y} = \dfrac{9}{17}\times\dfrac{4}{3} = \dfrac{12}{17}$

5. Let the number be x

$\because \quad \dfrac{x}{6} = 53$

$\Rightarrow \quad x = 53 \times 6$

$\therefore \quad x\times\dfrac{57}{100} = \dfrac{53\times 6\times 57}{100} = 181.26$

6. Let the monthly salary be ₹ x

$\because \quad x\times\dfrac{14}{100} = 7014$

$\Rightarrow \quad x = \dfrac{7014\times 100}{14} =$ ₹ 50100

∴ Total amount of annual investment

$= \dfrac{(14+21+6.5)}{100}\times 50100\times 12 =$ ₹ 249498

8. Monthly salary of Ms. Sujan $= \dfrac{2170\times 100}{7}$

= ₹ 31,000.

∴ Total money invested by her in a year

$= 31000\times\dfrac{(7+18+6)}{100}\times 12$

= ₹ 9610 × 12

= ₹ 115320

9. $2500\times\dfrac{30}{100}\times\dfrac{40}{100}\times\dfrac{2}{5} = 120.$

10. Let the share of C be ₹ x

∴ Share of B $= \dfrac{75x}{100} =$ ₹ $0.75x$

and share of A $= 0.75x\times\dfrac{125}{100} =$ ₹ $0.9375x$

$\because 0.9375x + 0.75x + x = 731$

$\therefore \quad x = \dfrac{731}{2.6875} =$ ₹ 272

11. Number $= 740\times\dfrac{35}{100} - 34 = 225$

$\therefore \dfrac{2}{5}$ of the number $= 225\times\dfrac{2}{5} = 90.$

12. Let the maximum marks of the eximination be x. Then, from question—

$\because \quad 54\%$ of $x = 456 - 24 = 432$

$x = \dfrac{432\times 100}{54} = 800$

⇒ Minimum passing marks $= \dfrac{34}{100}\times 800$

$= 272$

∴ Required difference = 456 – 272 = 184.

13. C.P. of the article $= \dfrac{30750\times 100}{125} =$ ₹ 24600.

14. Let the two numbers be x and y respectively. Then,

$\because \quad \dfrac{20x}{100} + y = y + \dfrac{50y}{100}$

$\Rightarrow \quad \dfrac{x}{5} + y = y + \dfrac{y}{2}$

$\Rightarrow \quad \dfrac{x}{5} = \dfrac{y}{2}$

$\Rightarrow \quad \dfrac{x}{y} = \dfrac{5}{2}$

$\therefore \quad x : y = 5 : 2$

15. Let the number be x.

$\because \quad 900\times\dfrac{35}{100}\times\dfrac{7}{9} + x = 325$

$\therefore \quad x = 325 - 245 = 80$

16. Let the number be 'n', then from question,

$\because \quad (60\% - 20\%)$ of $n = 316$

$\Rightarrow \quad 40\%$ of $n = 316$

$\Rightarrow \quad 35\%$ of $n = \dfrac{316}{40}\times 35$

$= 276.5$

17. Let the maximum marks of the exam. paper $= n$. Then from question,

$n\times\dfrac{36}{100} = 174 + 96$

$\Rightarrow \quad n = \dfrac{270\times 100}{36}$

$\therefore \quad n = 750$

18. $15\% = ₹\ 54$

$$100\% = ₹\ \frac{54}{15}\times 100$$

$$= ₹\ 360$$

19. Let the first number be 100.

After 20% increase, the number becomes = 120

After again 20% increase, the number becomes

$$= 120\times\frac{120}{100} = 144$$

∴ Original number is changed by (144 – 100)% = 44%

20. Failed in Mathematics = 62%

Failed in English = (100 – 52)% = 48%

Failed in both = 22%

Failed only in Maths = (62 – 22)% = 40%

Failed only in English = (48 – 22)% = 26%

∴ Total failed = (40 + 26 + 22)%
= 88%

∴ Passed in both = (100 – 88)% = 12%

12% = 108

$$100\% = \frac{108}{12}\times 100 = 900$$

21. Ramesh gets = 110,

Mohan = 100

$$\text{Required percentage} = \frac{10}{110}\times 100 = 9\frac{1}{11}\%.$$

22. $$\text{Required percentage} = \frac{25}{(100+25)}\times 100$$

$$= \frac{25}{125}\times 100 = 20\%.$$

23. $$\text{Required percentage} = \frac{3}{4}\times 100\% = 75\%.$$

24. Percentage increase in area

$$= \frac{(110)^2-(100)^2}{(100)^2}\times 100\%$$

$$= \frac{210\times 10}{100\times 100}\times 100\% = 21\%.$$

26. Let the total number of students be x. Number of student failed in one or both is given by

$n(A \cup B) = n(A) + n(B) - n(A \cap B)$

= 80% of x + 85% of x – 75% of x

$$= \frac{90x}{100} = \frac{9x}{10}$$

$$\text{passed in both} = \left(x-\frac{9x}{10}\right) = \frac{x}{10}$$

From question, $\frac{x}{10} = 40$

⇒ $x = 400$

27. Let expenditure on food and clothes be ₹ $3x$ and ₹ $2x$.

From question, $3x + 2x = 1750$

$x = 350$

Food = 3 × 350 = 1050

Clothes = 2 × 350 = 700

New expenditure = 120% of 1050 + 125% of 700

$$= \left(\frac{120}{100}\times 1050\right)+\left(\frac{125}{100}\times 700\right)$$

$$= ₹\ 2135$$

∴ Required increase = 2135 – 1750 = ₹ 385.

28. Let the total number of matches played be x.

From question, 20% of x + 70% of x + 40 = x

$$\Rightarrow \frac{20x}{100}+\frac{70x}{100}+40 = x$$

⇒ $x = 400.$

30. Total votes = 7500

$$\text{Invalid votes} = 20\% \text{ of } 7500 = \frac{20}{100}\times 7500$$

= 1500

∴ Valid votes = 7500 – 1500 = 6000

Valid votes polled in favour of the candidate A

$$= 55\% \text{ of } 6000 = \frac{55}{100}\times 6000 = 3300$$

∴ Valid votes polled in favour of the candidate B = 6000 – 3300 = 2700.

31. Increase in the population = 36760 – 35000
= 1760

Let the number of men in the population be x

∴ Increase in the men population = 6% of x

$$= \frac{6x}{100}$$

Population of women = 35000 – x

Increase in the women population

$$= 4\% \text{ of } (35000 - x) = \frac{4}{100}(35000-x)$$

Now From question,

$$\frac{6x}{100}+\frac{4}{100}(35000-x) = 1760$$

⇒ $6x + 4(35000 - x) = 176000$

⇒ $2x = 176000 - (35000 \times 4)$

⇒ $2x = 176000 - 140000$

⇒ $2x = 36000;\ x = \frac{36000}{2} = 18000$

∴ Numbere of men in the population of city = 18000 and Number of women in the population of city = 35000 − 18000 = 17000.

32. Let the weight of the original solution be x kg.

The weight of salt at 15% = $\frac{15}{100} \times x$

$= \frac{15x}{100}$ kg.

Weight of the remaining solution when 30 kg of water from the solution is evaporated

= $(x - 30)$ kg.

∴ Percentage of salt in the remaining solution

$$= \frac{15x/100}{x-30} \times 100\% = \left(\frac{15x}{x-30}\right)\%$$

It is given that the remaining solution had 20% salt (Given)

$$\therefore \quad \frac{15x}{x-30} = 20$$

$$\Rightarrow \quad 15x = 20(x - 30)$$

$$\Rightarrow \quad 20x - 15x = 600$$

$$\therefore \quad 5x = 600;$$

$$x = 600 \div 5 = 120$$

Hence, the weight of the original solution is 120 kg.

33. Suppose the initial salary of Y = ₹ 100

Then, initial salary of X = $\frac{1}{2}$ × ₹ 100 = ₹ 50

Increase in Y's salary = 25% of ₹ 100 = ₹ 25

Increase in X's salary = 50% of ₹ 50

$= ₹\left(50 \times \frac{50}{100}\right) = ₹\ 25$

∴ Total increase = ₹ (25 + 25) = ₹ 50

∴ Percentage increase in combined salaries of both

$$= \frac{50}{(100+50)} \times 100$$

$$= \frac{50}{150} \times 100 = 33\frac{1}{3}\%.$$

34. Let total number of students with age below 14 years in the area be 100

Number of students enrolled = 80% of 100 = 80

Number of students who come regularly = 75% of 80

$$= 80 \times \frac{75}{100} = 60$$

If number of regualr students is 60, total number = 100

If number of regular students is 270, total number

$$= \frac{100}{60} \times 270 = 450.$$

6 Profit & Loss

GENERAL FORMULAE

1. $\text{Gain \%} = \left(\frac{\text{Gain}}{\text{C.P}} \times 100\right)$
2. $\text{Loss \%} = \left(\frac{\text{Loss}}{\text{C.P}} \times 100\right)$
3. $\text{S.P} = \left(\frac{100 + \text{Gain \%}}{100}\right) \times \text{C.P}$
4. $\text{S.P} = \left(\frac{100 - \text{Loss \%}}{100}\right) \times \text{C.P}$
5. $\text{C.P} = \left(\frac{100}{100 + \text{Gain \%}}\right) \times \text{S.P}$
6. $\text{C.P} = \left(\frac{100}{100 - \text{Loss \%}}\right) \times \text{S.P}$

Formulae For Short Cut Solution

Type 1 :

Conditional Tricks

If cost price of X goods = Selling price of Y goods then,

I. $\text{Gain \%} = \frac{X - Y}{Y} \times 100$ (In case of $X > Y$)

II. $\text{Loss \%} = \frac{Y - X}{Y} \times 100$ (in case of $X < Y$)

Type 2 :

I. If X_1 and X_2 both are the rate of gain or both are the rate of loss, then, $\text{C.P.} = \left(\frac{100}{X_1 - X_2}\right) \times$ amount of difference between S.Ps.

II. If in X_1 and X_2 one is the rate of gain and another is the rate of loss, then $\text{C.P.} = \left(\frac{100}{X_1 + X_2}\right) \times$ amount of difference between S.Ps

Type 3 :

Miscellaneous Tricks

I. When a man buys two things on equal price and in those things one is sold on the profit of X % and another is sold on the loss of X %, then there is no loss or gain percent.

II. When a man sells two things at the same price each and in this process his loss on first thing is X % and gain on second thing is X % then in such type of questions, there is always a loss.

$$\text{Loss \%} = X\text{ \% of } X = \frac{X^2}{100} = \left(\frac{X}{10}\right)^2$$

III. Dishonest dealer and less weight

$$\text{Gain} = \frac{\text{Error}}{\text{True value} - \text{Error}} \times 100$$

Where, error = 100 gm – used weight of goods

IV. If A sells a thing to B at a gain of R_1 %, B sells it to C at a gain of R_2 % and C sells it to D at a gain of R_3 % Then, cost price of D = Cost Price of $A\ (1 + R_1/100)\ (1 + R_2/100)\ (1 + R_3/100)$

V. If A sells a thing, to B at a loss of R %, B sells it to C at a loss of R % and C sells it to D at a loss of R % then, Cost price of D = Cost price of $A\ (1 - R_1/100)\ (1 - R_2/100)\ (1 - R_3/100)$

SOLVED EXAMPLES

Example 1: Suman purchased an item of ₹ 72,000 and sold it at a loss of 27 percent. With that amount she purchased another item and sold it at a gain of 30 percent. What is her overall profit/loss?

Solution: First S.P. $= \frac{72000 \times 73}{100} =$ ₹ 52560

Second S.P. $= \frac{52560 \times 130}{100} =$ ₹ 68328

Loss = 72000 − 68328 = ₹ 3672.

Example 2: Mr. Babulal purchased 100 pieces of an article at the rate of ₹ 480 per piece. He then listed the price so as to gain a profit of 25%. While selling the articles he offered a discount of 5%. What is the percentage of profit earned in the deal?

Solution: Profit per cent = $\left(x+y+\frac{xy}{100}\right)\%$

Here, $x = 25\%,\ y = -5\%$

$= \left(25-5-\frac{25\times5}{100}\right)\% = 18.75\%.$

Example 3: The profit earned after selling an article for ₹ 625 is the same as loss incurred after selling the article for ₹ 435. What is the cost price of the article?

Solution: C.P. of the article = $\frac{625+435}{2}$ = ₹ 530.

Example 4: A man sold an item for ₹ 6,750 at a loss of 25%. What will be the selling price of same item if he sells it at a profit of 15%?

Solution: C.P. of the item = $\frac{6750\times100}{75}$ = ₹ 9000

∴ Required S.P. of the item = $\frac{9000\times115}{100}$ = ₹ 10350.

Example 5: A man sold a wristwatch for ₹ 2,400 at a loss of twenty-five per cent. At what rate should he have sold the wristwatch to earn a profit of twenty five per cent?

Solution: C.P. of the wristwatch = $\frac{100\times2400}{100-25}$ = ₹ 3200

∴ Required S.P. of the wristwatch

= $\frac{125\times3200}{100}$ = ₹ 4000.

MULTIPLE CHOICE QUESTIONS

1. What profit / loss per cent did Raveena earn if she purchased an item of ₹ 5,600 and sold it at three-fourth of its cost price?
A. Loss of 20 per cent B. Gain of 25 per cent
C. Neither gain nor loss D. Loss of 15 per cent

2. By what per cent must the cost price be raised in fixing the sale price in order that there may be a profit of 20% after allowing a commission of 10%?
A. 30% B. $33\frac{1}{3}\%$
C. 25% D. 40%

3. A sold a watch to B at a loss of 10%. B sold it to C at a gain of 10%. If A had sold at the same price which C paid for it, what would have been his gain or loss per cent?
A. 1% loss B. 10% gain
C. 12% gain D. 11% loss

4. A shopkeeper purchased two cycles at ₹ 1600 and sold the first at 10% profit and second one at 20% profit. If he could sell the first cycle at a profit of 20% and the second one at a profit of 10% then he could get ₹ 5 more. What will be the cost price of each in rupees?
A. 925, 675 B. 1000, 600
C. 825, 775 D. 900, 700

5. A person makes a profit of 25%, by selling an article for ₹ 35.50. By how much should he increase his selling price so as to make a profit of 40%?
A. ₹ 4.60 B. ₹ 4.26
C. ₹ 4 D. ₹ 3.90

6. A man bought a pen and sold it at a loss of 20%. If he had bought it for 30% less and sold it for ₹ 60 more, he would have made a profit of 50%. Find the C.P. of the article.
A. ₹ 200 B. ₹ 140
C. ₹ 240 D. ₹ 210

7. The cost price of 15 article is equal to S.P. of 10 articles. The gain percent is:
A. 50% B. 25%
C. $16\frac{2}{3}\%$ D. 100%

8. Raghav bought a radio with 20% discount on the labelled price. Had he bought it with 25% discount, he would have saved ₹ 500. At what price did he buy the radio?
A. ₹ 1000 B. ₹ 500
C. ₹ 5000 D. ₹ 10000

9. A man buys mangoes at ₹ 5 a dozen and an equal number at ₹ 4 a dozen. He sells them at ₹ 5.50 a dozen and makes a profit of ₹ 50. How many mangoes does he buy?
A. 30 dozens B. 40 dozens
C. 50 dozens D. 60 dozens

10. How many for a rupee must a shopkeeper sell to gain 25%, if he bought toys at 10 for a rupee?
A. 10 B. 9
C. 8 D. 6

11. Suppose the ratio of two investment be 3 : 2. What is the gain or loss on the two investments taken together if a man loses 20% on an investment but gains 25% on another investment.
A. 2.5% B. 0.2%
C. 2% D. 3%

12. A shopkeeper gives 12% additional discount after giving an initial discount of 20% on the labelled price of a T.V. If the sale price of the T.V. is ₹ 704. What is the labelled price of the T.V?

A. ₹ 844.80 B. ₹ 929.28
C. ₹ 1000 D. ₹ 900

13. By selling 12 oranges for a rupee, a man loses 10%. How many oranges for a rupee be sold to gain 20%?

A. 5 B. 4
C. 9 D. 12

14. A dishonest shopkeeper professes to sell pulses at his cost price but uses a false weight of 950 gm for each kilogram. His gain per cent is :

A. $5\frac{5}{19}\%$ B. 19%
C. $4\frac{5}{19}\%$ D. 95%

15. Sushil bought two boxes for ₹ 1300. He sold one box at a profit of 20% and the other box at a loss of 12%. If the selling price of both boxes is the same. Then the cost price of each box is :

A. ₹ 850, ₹ 650 B. ₹ 750, ₹ 750
C. ₹ 550, ₹ 750 D. ₹ 750, ₹ 650

16. A producer sells his articles to a wholesaler at 10% profit. The wholesaler sells it to a retailer at 20%. The retailer sells it to a customer at ₹ 41.25 making 25% profit. Then the cost of production is :

A. ₹ 52 B. ₹ 132
C. ₹ 110 D. ₹ 25

17. A bought two oxen for ₹ 1080. He sold one at a loss of 5% and the other at a gain of 7%. He neither gained nor lost in this transaction. C.P. of each is:

A. ₹ 630, ₹ 450 B. ₹ 680, ₹ 400
C. ₹ 600, ₹ 480 D. ₹ 540, ₹ 540

18. A businessman makes a profit of 20% in the first year. In the subsequent year, he had a loss of 25% on the capital which he had at the beginning of the second year. His overall gain or loss per cent in the capital is:

A. 120% B. 90%
C. 100% D. 10%

19. A co-operative society manufactures note books. It makes a profit of ₹ 9.00 when it sells 100 copies to a local shop. The local shop, in turn, makes a profit of ₹ 2 on selling 50 copies at 69 paisa per copy. How much did it cost the society to manufacture 100 note books?

A. ₹ 65 B. ₹ 55
C. ₹ 56 D. ₹ 32.50

20. A man marks his goods at a price that would give him 40% profit. He sells three-fourths of his goods at the marked price. For the remaining, he has to lower the marked price by 40%. His per cent profit on the entire transaction is:

A. 21% B. 84%
C. 52% D. 26%

21. By reducing the selling price of an article by ₹ 80 a gain of 5% turns into a loss of 5%. Then the original selling price of the article is:

A. ₹ 1050 B. ₹ 525
C. ₹ 510 D. ₹ 515

22. A man bought a horse and a carriage for ₹ 3000. He sold the horse at a gain of 20% and carriage at a loss of 10%, thereby gaining 2% on the whole. Then the cost of the horse is:

A. ₹ 3060 B. ₹ 1200
C. ₹ 2100 D. ₹ 3000

23. If 20% more would be gained by selling a price of cloth for ₹ 45.50 than by selling it for ₹ 39. What per cent would be gained by selling if for ₹ 42.25?

A. 32% B. 42%
C. 30% D. 35%

24. There would be 10% loss if rice is sold at ₹ 5.40 per kg. At what price per kg should it be sold to earn a profit of 20%?

A. ₹ 6 B. ₹ 6.48
C. ₹ 7.02 D. ₹ 7.20

25. A fruitseller buys lemons at 2 for a rupee and sells them at five for three rupees. His gain per cent is:

A. 10% B. 15%
C. 20% D. 25%

26. By selling 100 bananas, a fruitseller gains the selling price of 20 bananas. His gain per cent is :

A. 10% B. 15%
C. 20% D. 25%

27. A shopkeeper mixes two varieties of tea, one costing ₹ 35 per kg and another at ₹ 45 per kg in the ratio 3 : 2. If he sells the mixed variety at ₹ 41.60 per kg, his gain or loss per cent is :

A. $6\frac{2}{3}\%$ gain B. $6\frac{2}{3}\%$ loss
C. 4% gain D. 4% loss

28. A dishonest dealer professes to sell his goods at cost price. But he uses a false weight and thus gains $6\frac{18}{47}\%$. For a kg, he uses a weight of :

A. 953 gms B. 940 gms
C. 960 gms D. 947 gms

29. By selling toffees at 20 for a rupee, a man loses 4%. To gain in 20%, for one rupee he must sell

A. 16 toffees B. 20 toffees
C. 25 toffees D. 24 toffees

30. A man sold two houses for ₹ 7.81 lakhs each. On one he gained 5% and on the other he lost 5%. What per cent is the effect of the sale on the whole?
A. 0.25% loss B. 0.25% gain
C. 25% loss D. 25% gain

31. By selling 45 oranges for ₹ 40, a man loses 20%. How many should he sell for ₹ 24 so as to gain 20% in the transaction?
A. 16 B. 18
C. 20 D. 22

32. A man gains 10% by selling an article for a certain price. If he sells it at double the price, the profit made is :
A. 20% B. 60%
C. 100% D. 120%

33. A man sells two horses for ₹ 4000 each, neither losing nor gaining in the deal. If he sold one horse at a gain of 25%, the other horse is sold at a loss of:
A. $16\frac{2}{3}\%$ B. $18\frac{2}{9}\%$
C. 25% D. 20%

34. Due to an increase of 30% in the price of eggs, 3 eggs less are available for ₹ 7.80. The present rate of eggs per dozen is :
A. ₹ 8.64 B. ₹ 8.88
C. ₹ 9.36 D. ₹ 9.25

35. The per cent profit when an article is sold for ₹ 78 is twice as when it is sold for ₹ 69. The C.P. of the article is :
A. ₹ 49 B. ₹ 51
C. ₹ 57 D. ₹ 60

36. 6% more is gained by selling a radio for ₹ 475 than by selling for ₹ 451. The C.P. of the radio is :
A. ₹ 400 B. ₹ 434
C. ₹ 446.50 D. ₹ 500

37. A tradesman by means of a false balance defrauds to the extent of 8% in buying goods and also defrauds to 8% in selling. His gain per cent is :
A. 15.48% B. 16%
C. 16.64% D. 20%

38. While selling a watch, a shopkeeper gives a discount of 5%. If he gives a discount of 7%, he earns ₹ 15 less as profit. The marked price of the watch is :
A. ₹ 697.50
B. ₹ 712.50
C. ₹ 787.50
D. ₹ 750.00

39. The marked price is 10% higher than the cost price. A discount of 10% is given on the marked price. In this kind of sale, the seller
A. bears no loss, no gain
B. gain 1%
C. loss 1%
D. gain 10%

40. The difference between a discount of 40% on ₹ 500 and two successive discounts of 36% and 4% on the same amount is :
A. 0 B. ₹ 2
C. ₹ 1.93 D. ₹ 7.20

ANSWERS

1	2	3	4	5	6	7	8	9	10
A	B	A	C	B	C	A	D	C	C
11	**12**	**13**	**14**	**15**	**16**	**17**	**18**	**19**	**20**
C	C	C	A	C	D	A	D	C	D
21	**22**	**23**	**24**	**25**	**26**	**27**	**28**	**29**	**30**
B	B	C	D	C	D	A	B	A	B
31	**32**	**33**	**34**	**35**	**36**	**37**	**38**	**39**	**40**
B	D	A	C	D	A	C	D	C	D

SOME SELECTED EXPLANATORY ANSWERS

1. S.P. of the item $= 5600 \times \frac{3}{4} = ₹\ 4200$

$\therefore$ Loss % $= \frac{5600 - 4200}{5600} \times 100 = 25$

2. Let x% must the cost price be raised

$\therefore$ S.P. = (100 + x)

$\because \quad (100 + x)\frac{90}{100} = 120$

$\Rightarrow \quad 3x = 100$

$\therefore \quad x = 33\frac{1}{3}\%.$

3. Let A's C.P. = ₹ 100

A's S.P. $= 100 \times \frac{90}{100}$;

B's S.P. $= 100 \times \frac{90}{100} \times \frac{110}{100}$

C paid = B's S.P. = $100 \times \frac{90}{100} \times \frac{110}{100}$

= 99

If A had sold at ₹ 99

Then, his loss = (100 − 99)% = 1%.

4. Let the C.P. of 1st cycle = ₹ x

and C.P. of 2nd cycle = ₹ $(1600 - x)$

Total S.P. in 1st case

$$= x \times \frac{110}{100} + (1600 - x) \times \frac{120}{100}$$

$$= \frac{12 \times 1600 - x}{10}$$

Total S.P. in 2nd case

$$= x \times \frac{120}{100} + (1600 - x) \times \frac{110}{100}$$

$$= \frac{11 \times 1600 + x}{10}$$

$$\because \frac{11 \times 1600 + x}{10} - \frac{12 \times 1600 - x}{10} = 5$$

$$\therefore x = \frac{1650}{2} = 825$$

C.P. of 1st cycle = 825

C.P. of 2nd cycle = 1600 − 825 = 775.

5. Let new S.P. be ₹ x

Then (100 + Gain%) : (1st S.P.)

= (100 + Gain%) : (2nd S.P.)

125 : 35.50 = 140 : x

$$\Rightarrow x = \frac{35.50 \times 140}{125} = ₹\ 39.76$$

∴ Increased in S.P. = ₹ (39.76 − 35.50)

= ₹ 4.26.

6. Let original C.P. be ₹ x

$$\text{S.P.} = 80\% \text{ of } x = \frac{80x}{100} = \frac{4x}{5}$$

$$\text{New C.P.} = 70\% \text{ of } x = \frac{70x}{100} = \frac{7x}{10}$$

$$\text{New S.P.} = 150\% \text{ of } \frac{7x}{10} = \frac{21x}{20}$$

$$\therefore \frac{21x}{20} - \frac{4x}{5} = 60$$

$$\Rightarrow \frac{x}{4} = 60$$

$$x = 240$$

7. Let C.P. of each article be ₹ 1

Then C.P. of 10 article = ₹ 10

S.P. of 10 article = ₹ 15

$$\text{Gain \%} = \left(\frac{15-10}{10} \times 100\right)\% = \left(\frac{5}{10} \times 100\right)\%$$

= 50%.

8. Let the labelled price be x

$$\therefore \text{S.P.} = 80\% \text{ of } x = \frac{4x}{5}$$

$$\text{New S.P.} = 75\% \text{ of } x = \frac{3x}{4}$$

$$\therefore \frac{4x}{5} - \frac{3x}{4} = 500$$

$$\Rightarrow x = ₹\ 10000.$$

9. C.P of 2 dozens of mangoes = ₹ (5 + 4)

= ₹ 9

S.P. of 2 dozens of mangoes = ₹ 11

2 dozens of mangoes are bought at a profit of ₹ 2.

If profit is ₹ 50, mangoes bought = $\frac{2}{2} \times 50$

= 50 dozens.

10. C.P of 10 toys = ₹ 1

Gain = 25%

$$\text{S.P of 10 toys} = \frac{125}{100} \times 1 = ₹\ \frac{5}{4}$$

Thus S.P of 10 toys is ₹ $\frac{5}{4}$

∴ For ₹ 1, Number of toys = $10 \times \frac{4}{5} = 8$

So he must sell 8 toys for a rupee.

11. Let the investments are $3x$ and $2x$

Then total investment = $5x$

Receipt = (80% of $3x$ + 125% of $2x$)

= $(2.4x + 2.5x) = 4.9x$

$$\therefore \text{loss} = \left(\frac{0.1x}{5x} \times 100\right)\% = 2\%$$

12. Let the labelled price of the T.V be ₹ x

$$\text{Initial S.P} = 80\% \text{ of } x = \frac{80x}{100}$$

$$\text{New S.P} = 88\% \text{ of } \frac{80x}{100}$$

$$= \frac{88}{100} \times \frac{80x}{100} = \frac{704x}{1000}$$

From question,

$$\frac{704x}{1000} = 704$$

$704x = 704 \times 1000$

$\Rightarrow \quad x = 1000$

$\therefore$ The labelled price of the T.V = ₹ 1000.

14. Let C.P. of 1 gm pulses be ₹ 1

$\therefore$ C.P. of 950 gm pulses = ₹ 950

S.P. of 950 gm = C.P. of 1 kg

= C.P. of 1000 gm = ₹ 1000

$\therefore$ Gain = ₹ (1000 – 950) = ₹ 50

$\therefore$ Gain % = $\frac{50}{950} \times 100 = 5\frac{5}{19}\%$.

17. From question,

$\therefore$ Loss on one = Gain on other

$\therefore$ 5% of the C.P. of one = 7% of the C.P. of the other

$\Rightarrow \frac{\text{C.P. of 1st}}{\text{C.P. of 2nd}} = \frac{7\%}{5\%} = \frac{7}{5}$

or, C.P. of 1st : C.P. of 2nd = 7 : 5

Divide ₹ 1080 in the ratio of 7 : 5

$\therefore$ C.P. of the 1st = $\frac{7}{12} \times 1080$ = ₹ 630.

and C.P. of the 2nd = ₹ $\frac{5}{12} \times 1080$ = ₹ 450.

18. Let his capital at the beginning of first year be ₹ 100

Profit during first year = 20% of ₹ 100 = ₹ 20

$\therefore$ Capital at the beginning of the second year

= ₹ (100 + 20) = ₹ 120

Loss during second year = 25% of ₹ 120

= ₹ $\left(120 \times \frac{25}{100}\right)$ = ₹ 30

$\therefore$ Capital at the end of second year

= ₹ (120 – 30) = ₹ 90

$\therefore$ Loss = ₹ (100 – 90) = ₹ 10

Overall Loss % on his capital = $\left(\frac{10}{100} \times 100\right)\%$

= 10%.

19. The local shop sells 50 copies @ 69 paisa per copy

= ₹ (50 × 0.69) = ₹ 34.50

His profit = ₹ 2

$\therefore$ The local shop purchases 50 copies for

= S.P. – Profit = ₹ (34.50 – 2) = ₹ 32.50

$\therefore$ The local shop purchases 100 copies for

= ₹ $\left(\frac{32.50}{50} \times 100\right)$ = ₹ 65

$\therefore$ Co-operative society sells 100 copies for ₹ 65

Profit = ₹ 9

$\therefore$ Co-operative society manufactures 100 copies for

= ₹ (65 – 9) = ₹ 56.

23. Let C.P. of cloth be ₹ x. Then $x \times \frac{20}{100}$

= ₹ (45.50 – 39) = ₹ 6.50

$\Rightarrow \quad x = \frac{6.50 \times 100}{20}$ = ₹ 32.50

Gain = ₹ (42.25 – 32.50) = ₹ 9.75

$\therefore$ Gain% = $\frac{9.75}{32.50} \times 100 = 30\%$.

24. 90 : 5.40 :: 120 : x

$\therefore x = \frac{5.40 \times 120}{90} = 7.20$.

So, the rice must be sold at ₹ 7.20 per kg.

25. Suppose he buys 10 lemons.

C.P. = ₹ $\left(\frac{10}{2}\right)$ = ₹ 5, S.P. = ₹ $\left(\frac{3}{5} \times 10\right)$ = ₹ 6.

Gain% = $\left(\frac{1}{5} \times 100\right)\% = 20\%$.

26. Gain = (S.P. of 100 bananas) – (C.P. of 100 bananas)

(S.P. of 20) = (S.P. of 100) – (C.P. of 100)

₹ S.P. of 80 = C.P. of 100.

Let C.P. of each = ₹ 1.

C.P. of 80 = ₹ 80

S.P. of 80 = ₹ 100.

$\therefore$ Gain % = $\left(\frac{20}{80} \times 100\right)\% = 25\%$

27. Suppose he purchases 3 kg and 2 kg of tea of first and second kind respectively.

C.P. = ₹ (3 × 35 + 2 × 45) = ₹ 195.

S.P. = ₹ (5 × 41.60) = ₹ 208.

Gain % = $\left(\frac{13}{195} \times 100\right)\% = 6\frac{2}{3}\%$.

28. Let the error be x gms. Then,

$\frac{x}{1000 - x} \times 100 = \frac{300}{47}$ or $\frac{x}{1000 - x} = \frac{3}{47}$

$\therefore 47x = 3000 - 3x$ or $x = 60$.

So, he uses a weight = (1000 – 60) gm = 940 gm for a kg.

29. Suppose he sells 20 toffees.

Then, S.P. = ₹ 1 and Loss = 4%.

$\therefore$ C.P. = $\left(\frac{100}{96} \times 1\right)$ = ₹ $\frac{25}{24}$

Gain = 20%

$\therefore$ S.P. = $₹\left(\frac{120}{100}\times\frac{25}{24}\right) = ₹\frac{5}{4}$.

For $₹\frac{5}{4}$, toffees sold = $\left(20\times\frac{4}{5}\right) = 16$.

So, he must sell 16 toffees for a rupee.

30. Loss % = $\left(\frac{5}{10}\right)^2 = \frac{1}{4}\% = 0.25\%$.

31. S.P. of 1 orange = $₹\left(\frac{40}{45}\right) = ₹\frac{8}{9}$.

80% of C.P. = $\frac{8}{9}$ or C.P. = $\left(\frac{8}{9}\times\frac{100}{80}\right) = ₹\frac{10}{9}$.

S.P. = $\left(120\% \text{ of } ₹\frac{10}{9}\right) = ₹\frac{4}{3}$.

For $₹\frac{4}{3}$, he sells 1 orange

For ₹ 24, he would sell $\left(\frac{3}{4}\times 24\right) = 18$ oranges.

32. Let C.P. = ₹ x.

First S.P. = 110% of $x = ₹\left(\frac{11}{10}x\right)$

Second S.P. = $\frac{22}{10}x$

New gain = $\left(\frac{22}{10}x - x\right) = \frac{12x}{10}$.

New gain % = $\left(\frac{12x}{10\times x}\times 100\right)\% = 120\%$.

33. C.P. of two horses = ₹ 8000.

S.P. of one horse = ₹ 4000, gain = 25%.

$\therefore$ C.P. of this horse = $₹\left(\frac{100}{125}\times 4000\right)$

= ₹ 3200.

C.P. of another horse = ₹ (8000 − 3200)

= ₹ 4800.

S.P. of this horse = ₹ 4000.

$\therefore$ Loss% = $\left(\frac{800}{4800}\times 100\right)\% = 16\frac{2}{3}\%$.

35. Let the C.P. be ₹ x.

Then, $\frac{2(69-x)}{100} = \frac{78-x}{100}$

or, $138 - 2x = 78 - x$

or, $x = 60$

$\therefore$ C.P. = ₹ 60.

36. Difference between two selling prices = ₹ 24

$\therefore$ 6% of C.P. = ₹ 24

Hence, C.P. = $₹\left(\frac{24\times 100}{6}\right) = ₹400$.

37. Rule. In such questions we adopt the rule :

Gain % = $\frac{(100 + \text{common gain\%})^2}{100} - 100$

$= \left\{\frac{(108)^2}{100} - 100\right\}\% = 16.64\%$.

38. Let the marked price be ₹ x.

Then, (7% of x) − 15 = 5% of x

or, $\frac{7x}{100} - \frac{5x}{100} = 15$ or $x = 750$.

39. Let C.P. = ₹ 100.

Marked Price = ₹ 110.

S.P. = 90% of ₹ 110 = ₹ 99.

$\therefore$ Loss = 1%.

40. Price after 40% discount = 60% of ₹ 500

= ₹ 300.

Price after 36% discount = 64% of ₹ 500

= ₹ 320.

Price after next 4% discount = 96% of ₹ 320

= ₹ 307.20.

Difference in two prices = ₹ 307.20 − ₹ 300.00

= ₹ 7.20.

7 Mixture & Alligation

Alligation deals with calculation of values or properties of a mixture. Alligation is the rule that enables us—

(1) to find the proportion in which the two or more ingredients at the given prices must be mixed to yield a mixture at the given price. This is termed as "Alligation Alternate".

(2) to calculate the average or mean value of a mixture when the prices of two or more ingredients which are to be mixed together and proportion in which they are to be mixed are given. This is termed as "Alligation Medial".

1. Rule of Alligation:

$$\frac{\text{A of Cheap mixture}}{\text{A of Dearer mixture}} = \frac{\text{CP of Dearer} - \text{Mean Price}}{\text{Mean Price} - \text{CP of Cheaper}}$$

Here cost price of unit quantity of the mixture is called the *Mean Price.*

The above rule may be represented schematically as under:

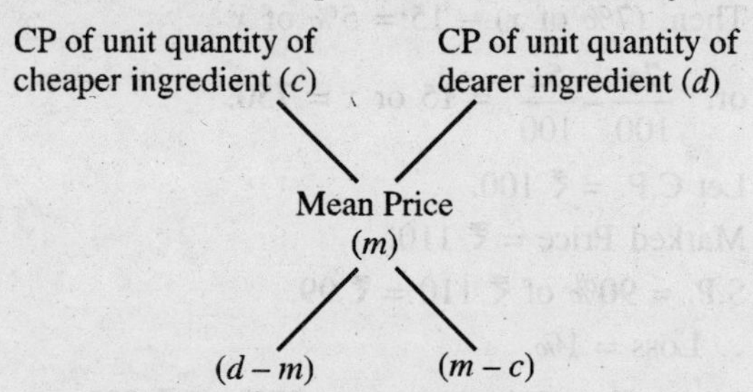

(Cheaper quantity) : (Dearer quantity)

$= (d - m) : (m - c)$

This relationship is very helpful in solving problems on mixture involving percentage values, rates, prices, speeds etc.

2. m gm of sugar solution has x % sugar in it. To increase the sugar content in the solution to y %,

quantity of sugar need to be added $= \dfrac{m(y-x)}{100-y}$

3. A vessel contains x litres of liquid A. y litres are withdrawn and replaced by liquid B. Next y litres of the mixture is withdrawn and again replaced by liquid B.

This operation is repeated n times.

$$\frac{\text{Quantity of liquid A left after } n\text{th operation}}{\text{Whole quantity of liquid A initially present}}$$

$$= \left(\frac{x-y}{x}\right)^n \text{ or } \left(1-\frac{y}{x}\right)^n$$

SOLVED EXAMPLES

Example 1: Find the quantities of two types of wheat, one @ ₹ 6 per kg and the other @ ₹ 4 per kg. to get 20 kgs. of wheat mixture worth ₹ 4.5 per kg.

Solution:

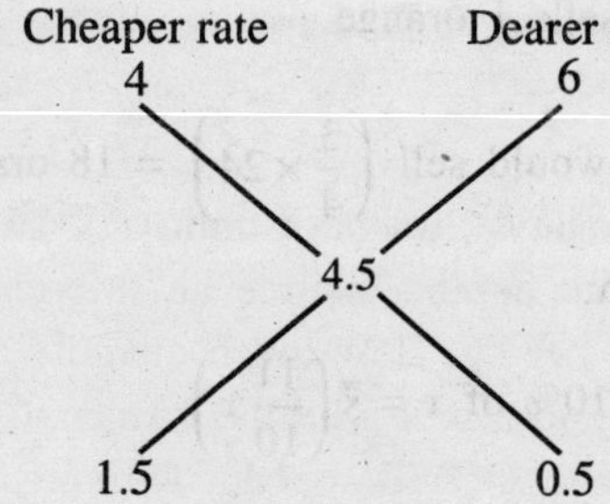

$$\frac{\text{Quantity of cheaper wheat}}{\text{Quantity of dearer wheat}} = \frac{1.5}{0.5} = \frac{3}{1}$$

Quantity of cheaper wheat

$= \dfrac{3}{3+1} \times 20 = 15$ kgs

Quantity of dearer wheat

$= \dfrac{1}{3+1} \times 20 = 5$ kgs.

Example 2: In what proportion water be mixed with pure milk in order to make a profit of 20% by selling it at cost price?

Solution: Let cost price of pure milk be Re. 1 per litre.

The SP of mixture = ₹ 1 per litre

Profit = 20%

So, CP of 1 litre of mixture

$= ₹ \left(1 \times \dfrac{100}{120}\right) = ₹ \dfrac{5}{6}$

We assume that CP of 1 litre of water is zero. Using the rule of alligation on 1 litre,

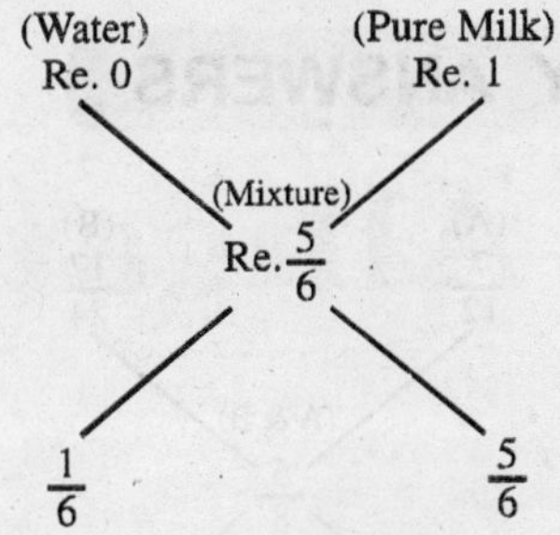

$$\frac{\text{Quantity of water}}{\text{Quantity of pure milk}} = \frac{\frac{1}{6}}{\frac{5}{6}} = \frac{1}{5}$$

or, Ratio of water to pure milk in the mixture = 1 : 5.

MULTIPLE CHOICE QUESTIONS

1. In what proportion must tea at ₹ 62 per kg be mixed with tea at ₹ 72 per kg in order to obtain the mixture worth ₹ 65 per kg?

A. 4 : 6 B. 7 : 3
C. 2 : 3 D. 4 : 7

2. Find the quantity of rice @ ₹ 10 per kg. which should be mixed with 25 kgs of rice @ ₹ 8 per kg, so that on selling the mixture @ ₹ 15 per kg there is 80% profit.

A. 6 kgs B. 7 kgs
C. 3 kgs D. 5 kgs

3. A shopkeeper buys 26 kgs of milk @ ₹ 16 per kg. He also buys from another source an inferior quality of milk @ ₹ 10 per kg. How much quantity of the latter should he buy to mix it with the former so that he can sell the mixture @ ₹ 14 per kg without making any loss?

A. 13 kgs B. 12 kgs
C. 14 kgs D. 16 kgs

4. Two vessels A and B contain milk and water in the ratio 7 : 5 and 17 : 7 respectively. In what ratio mixtures from two vessels should be mixed to get a new mixture containing milk and water in the ratio 5 : 3?

A. 1 : 2 B. 2 : 1
C. 2 : 3 D. 3 : 2

5. Two vessels A and B contain mixture of milk and water in the ratio 4 : 1 and 9 : 11 respectively. They are mixed in the ratio of 3 : 2. Find the ratio of milk : water in the resulting mixture.

A. 34 : 16 B. 33 : 17
C. 16 : 34 D. 17 : 33

6. A person has two solutions of sugar with 30% and 50% concentration respectively. In what proportion should he mix two solutions to get 45% concentration in the resulting mixture?

A. 1 : 3 B. 3 : 1
C. 2 : 3 D. 3 : 2

7. 6 litres of milk and water mixture has 75% milk in it. How much milk should be added to the mixture to make it 90% pure?

A. 8 litres B. 9 litres
C. 10 litres D. 12 litres

8. In what ratio must water be added to spirit to gain 25% by selling it at cost price?

A. 1 : 4 B. 4 : 1
C. 3 : 4 D. 4 : 3

9. A shopkeeper has 50 kgs of rice. He sells a part of it at 20% profit and the rest at 40% profit. If he gains 25% on the whole, find the quantity of each part.

A. 12.5 kgs and 37.5 kgs
B. 37.5 kgs and 12.5 kgs
C. 23.5 kgs and 21.5 kgs
D. 21.5 kgs and 23.5 kgs

10. A shopkeeper has 100 kgs of tea. He sells a part of it at 20% profit and the rest at 5% loss. If his overall profit is 10%, find the quantity for each part.

A. 20 kgs B. 25 kgs
C. 30 kgs D. 40 kgs

ANSWERS

1	2	3	4	5	6	7	8	9	10
B	D	A	B	B	A	B	A	A	D

SOME SELECTED EXPLANATORY ANSWERS

1. CP of 1 kg. Cheaper tea Rs. 62 — CP of 1 kg. Dearer tea Rs. 72

Mean Price per kg Rs. 65

$d - m = 7$ — $m - c = 3$

Using Alligation rule,

$$\frac{\text{Quantity of cheaper tea}}{\text{Quantity of dearer tea}} = \frac{d-m}{m-c} = \frac{7}{3}$$

Therefore, they must be mixed in the ratio of 7 : 3.

2. Cost price of the mixture $= 15 \times \frac{100}{180}$

$= ₹\ \frac{25}{3}$ per kg

8 — 10

$\frac{25}{3}$

$\frac{5}{3}$ — $\frac{1}{3}$

$$\frac{\text{Quantity of rice @ ₹ 8 per kg}}{\text{Quantity of rice @ ₹ 10 per kg}} = \frac{5/3}{1/3} = \frac{5}{1}$$

Quantity of rice @ ₹ 10 per kg $= 25 \times \frac{1}{5}$

= 5 kgs.

3. 16 — 10

14

4 — 2

or 2 : 1

$$\frac{\text{Quantity of milk @ ₹ 10 per kg}}{\text{Quantity of milk @ ₹ 16 per kg}} = \frac{1}{2}$$

So, quantity of milk @ ₹ 10 per kg. $= \frac{26}{2}$

= 13 kgs.

4. First of all we write the fraction of milk present in three mixtures.

In A : $\frac{7}{12}$ In B : $\frac{17}{24}$

In combination of A and B : $\frac{5}{8}$

We now apply alligation rule on these fractions.

(A) $\frac{7}{12}$ — (B) $\frac{17}{24}$

(A & B) $\frac{5}{8}$

$\frac{2}{24}$ — $\frac{1}{24}$

or, 2 : 1

So, Ratio of A : B = 2 : 1.

5. Fraction is *Milk* *Water*

A : $\frac{4}{5}$ $\frac{1}{5}$

B : $\frac{9}{20}$ $\frac{11}{20}$

$(3A + 2B) = A \text{ and } B : \left(\frac{12}{5} + \frac{9}{10}\right) \left(\frac{3}{5} + \frac{11}{10}\right)$

$\frac{33}{10}$ $\frac{17}{10}$

So, Ratio of milk : water in the resulting mixture = 33 : 17.

6. 30% — 50%

45%

5 — 15

He should mix 30% and 50% in the ratio 5 : 15 or 1 : 3.

$$\frac{\text{30\% Solution}}{\text{50\% Solution}} = \frac{1}{3}$$

or 1 : 3

7. The given solution has 75% milk.

Milk to be added has 100% milk.

75% — 100%

90%

10 — 15

Milk should be added to the given mixture in the ratio 15 : 10 or 3 : 2.

$\therefore$ Quantity of milk to be added $= \frac{3}{2} \times 6$

$= 9$ litres.

8. Let cost price of spirit be Re. 1 per litre.

Then SP of mixture = Re. 1 per litre

Gain = 25%

So, CP of mixture $= 1 \times \frac{100}{125} = \text{Re. } \frac{4}{5}$

We assume that CP of water is zero.

Using alligation rule on cost price,

Water should be mixed to spirit in the ratio $\frac{1}{5} : \frac{4}{5}$ or $1 : 4$.

9.

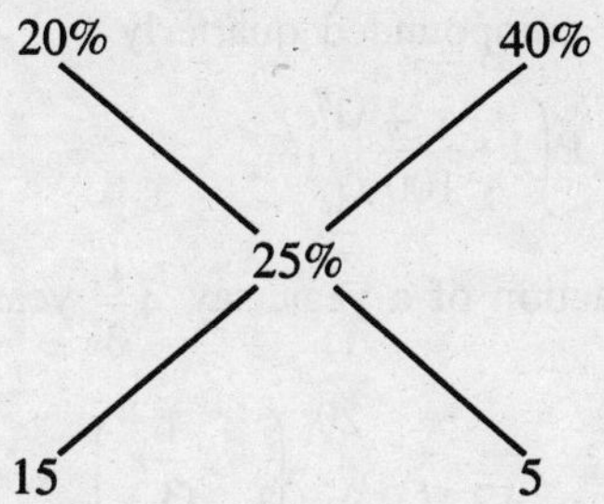

or 3 : 1

Quantity sold at 20% profit $= \frac{3}{3+1} \times 50$

$= 37.5$ kgs.

Quantity sold at 40% profit $= (50 - 37.5)$

$= 12.5$ kgs.

10.

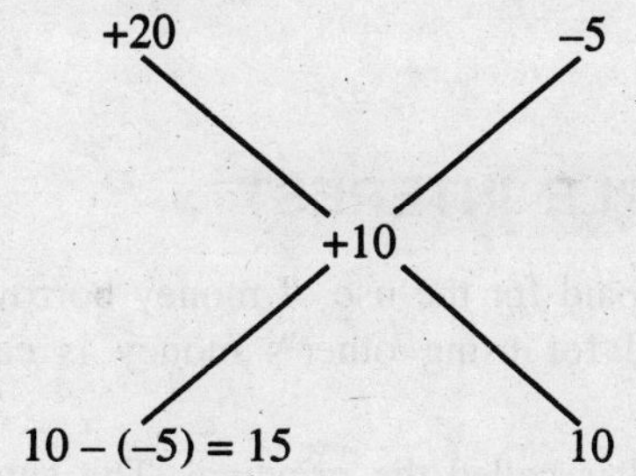

or 3 : 2

Quantity sold at 20% profit $= \frac{3}{3+2} \times 100$

$= 60$ kgs.

Quantity sold at 5% loss $= (100 - 60)$

$= 40$ kgs.

8 Simple & Compound Interests

SIMPLE INTEREST

Interest is the money paid for the use of money borrowed, *i.e.*, extra money paid for using other's money is called *interest.*

The sum borrowed is called the principal. The sum of interest and principal is called the *Amount.*

If the interest on a certain sum borrowed for a certain period is reckoned uniformly, then it is called simple interest, denoted by S.I.

Thus, if A = Amount, P = Principal,
I = Interest, T = Time (in year),
R = Rate per cent per annum, then

(*a*) $I = \frac{P \times R \times T}{100}$ (*b*) $P = \frac{100 \times I}{R \times T}$

(*c*) $T = \frac{100 \times I}{P \times R}$ (*d*) $R = \frac{100 \times I}{P \times T}$

(*e*) $P = \frac{100\,A}{100 + RT}$ (*f*) $A = P + I.$

COMPOUND INTEREST

In business transaction if interest as it becomes due is not paid to the lender but is added on to the principal, the money is said to be lent at *compound interest* and the total sum owed after a given time is called the amount at compound interest for that time.

After a certain period, the difference between the amount and the original principal is called the compound Interest (C.I.).

Some Important Formulae :

Let Principal = P
Time = *n* years
Rate = *r*% p.a.

then, the amount,

(*a*) when interest is compounded annually:

$$\text{then, Amount} = P\left(1 + \frac{r}{100}\right)^n$$

(*b*) when interest is compounded half yearly:

$$\text{then, Amount} = P\left(1 + \frac{r/2}{100}\right)^{2n}$$

(*c*) when interest is compounded quarterly :

$$\text{then, Amount} = P\left(1 + \frac{r/4}{100}\right)^{4n}$$

(*d*) when time is fraction of a year, say $4\frac{1}{3}$ years,

$$\text{then, Amount} = P\left(1 + \frac{r}{100}\right)^4 \times \left(1 + \frac{\frac{1}{3}r}{100}\right)$$

(*e*) when rates are $r_1\%$, $r_2\%$ and $r_3\%$ for Ist, IInd and IIIrd year respectively;

$$\text{then, Amount} = P\left(1 + \frac{r_1}{100}\right)\left(1 + \frac{r_2}{100}\right)\left(1 + \frac{r_3}{100}\right)$$

SOLVED EXAMPLES

Example 1: A man gets a simple interest of ₹ 1,000 on a certain principal at the rate of 5% p.a. in 4 years. What compound interest will the man get on twice the principal in two years at the same rate?

Solution: $\because$ Principal $= \frac{1000 \times 100}{4 \times 5} = ₹\ 5000$

$$\therefore \text{Required C.I.} = 1000\left[\left(1 + \frac{5}{100}\right)^2 - 1\right]$$

$$= 1000\left[\frac{441 - 400}{400}\right]$$

$$= 1000 \times \frac{41}{400}$$

$= ₹\ 102.5.$

Example 2: The simple interest accrued on an amount of ₹ 25,000 at the end of four years is ₹ 8,000. What would be the compound interest accrued on the same amount at the same rate in the same period?

Solution: Rate $= \frac{8000 \times 100}{25000 \times 4}\% = 8\%$

$\therefore$ C.I. $= 25000\left[\left(1+\frac{8}{100}\right)^4 - 1\right]$

$= 25000\left(\frac{27^4 - 25^4}{25^4}\right)$

$= 25000\left[\frac{531441 - 390625}{390625}\right]$

$=$ ₹ 9012.224

Example 3: The difference between the amount of compound interest and simple interest accrued on an amount of ₹ 26,000 at the end of 3 years is ₹ 2994.134. What is the rate of interest p.c.p.a.?

Solution: $\because$ $2994.134 = \frac{26000r^2}{(100)^2}\left(\frac{r}{100}+3\right)$

$= \frac{26r^2}{10}\left(\frac{r+300}{100}\right)$

$\Rightarrow$ $2994134 = 26r^3 + 7800r^2$

$\Rightarrow$ $115159 = r^3 + 300r^2$

$r = 19$

Example 4: What would be the compound interest accrued on an amount of ₹ 45,400/- at the end of two years at the rate of 15% p.a.?

Solution: C.I. $= 45400\left[\left(1+\frac{15}{100}\right)^2 - 1\right]$

$= 45400\left[\frac{23}{20} \times \frac{23}{20} - 1\right]$

$= 45400 \times \frac{129}{400}$

$= \frac{29283}{2} =$ ₹ 14641.5

MULTIPLE CHOICE QUESTIONS

1. Anshul borrowed some money at the rate of 4% p.a. for the first three years, at the rate of 8% p.a. for the next two years and at the rate of 9% p.a. for the period beyond 5 years. If he pays a total simple interest of ₹ 19,550 at the end of 7 years, how much money did he borrow?

A. ₹ 39,500 B. ₹ 42,500
C. ₹ 41,900 D. ₹ 43,000

2. The simple interest accrued on an amount at the end of five years @ 12.5% p.a. ₹ 1,575. What is the amount?

A. ₹ 2,050 B. ₹ 2,550
C. ₹ 2,250 D. ₹ 2,520

3. Siyaram invested equal sum of money in two schemes. Under scheme X, the compound interest rate was 10% p.a. and under scheme Y, the compound interest rate was 12% p.a. The interest after two years on the sum invested in Scheme X was ₹ 63. How much is the interest earned under Scheme Y after two years?

A. ₹ 79.0272 B. ₹ 70.56
C. ₹ 76.32 D. Cannot be determined

4. What would be the compound interest accrued on an amount of ₹ 15,000 at the rate of 15% at the end of three years?

A. ₹ 7813.125 B. ₹ 7762.50
C. ₹ 7762.125 D. ₹ 11235.09375

5. If the difference between the simple interest and compound interest earned on an amount @ 15% p.a. at the end of 3 years is ₹ 595.35, what is the amount?

A. ₹ 8,400 B. ₹ 9,200
C. ₹ 6,800 D. Cannot be determined

6. What will be the simple interest earned on an amount of ₹ 988 @ 18% p.a., at the end of 5 years?

A. ₹ 711.36 B. ₹ 898.2
C. ₹ 799.25 D. None of these

7. The simple interest accrued on a sum of certain principal is ₹ 1,200 in four years at the rate of 8% p.a. What would be the simple interest accrued on thrice of that principal at the rate of 6% p.a. in 3 years?

A. ₹ 2,025 B. ₹ 3,025
C. ₹ 2,250 D. ₹ 2,150

8. What amount of compound interest can be obtained on an amount of ₹ 18,000 at the rate of 10% p.a. at the end of three years?

A. ₹ 5,628 B. ₹ 5,400
C. ₹ 5,940 D. None of these

9. Shrivastava invests ₹ 14500 for 6 years at the certain rate of simple interest. After 6 years his invested sum becomes ₹ 21460. Find out the rate of interest per annum—

A. 4% p.a. B. 10% p.a.
C. 6% p.a. D. 8% p.a.

10. A sum of ₹ 12,500 amounts to ₹ 15,500 in 4 years at the rate of simple interest. What is the rate of interest?

A. 3% B. 4%
C. 5% D. 6%

11. A sum fetched a total simple interest of ₹ 4016. 25 at the rate of 9% p.a. in 5 years. What is the sum?

A. ₹ 4462.50 B. ₹ 8032.50
C. ₹ 8900 D. ₹ 8925

12. ₹ 800 becomes ₹ 956 in 3 years at a certain rate of simple interest. If the rate of interest is increased by 4%, what amount will ₹ 800 become in 3 years?
A. ₹ 1020.80 B. ₹ 1025
C. ₹ 1052 D. Data inadequate

13. A certain amount earns simple interest of ₹ 1750 after 7 years. Had the interest been 2% more, how much more interest would it have earned?
A. ₹ 35 B. ₹ 245
C. ₹ 350 D. Cannot be determined

14. Neelam borrowed some money at the rate of 6% p.a. for the first three years, 9% p.a. for the next five years and 13% p.a. for the period beyond eight years. If the total interest paid by him at the end of eleven years is ₹ 8160, how much money did he borrow?
A. ₹ 8000 B. ₹ 10,000
C. ₹ 12000 D. Data inadequate

15. The simple interest on a sum of money will be ₹ 600 after 10 years. If the principal is trebled after 5 years, what will be the total interest at the end of the tenth year?
A. ₹ 600 B. ₹ 900
C. ₹ 1200 D. ₹ 1500

16. Simple interest on a certain sum at a certain annual rate of interest is $\frac{1}{9}$ of the sum. If the numbers representing rate percent and time in years be equal, then the rate of interest is :
A. $3\frac{1}{3}$% B. 5%
C. $6\frac{2}{3}$% D. 10%

17. A borrowed some money from B at 12% p.a. S.I. for 3 years. He then added some more money to the borrowed sum and lent it to C for the same period at 14% p.a. rate of interest. If A gains ₹ 93.90 in the whole transaction, how much money did he add from his side?
A. ₹ 35 B. ₹ 55
C. ₹ 80 D. ₹ 105

18. A sum of ₹ 725 is lent in the beginning of a year at a certain rate of interest. After 8 months, a sum of ₹ 362.50 more is lent but at the rate twice the former. At the end of the year, ₹ 33.50 is earned as interest from both the loans. What was the original rate of interest?
A. 3.6% B. 4.5%
C. 5% D. None of these

19. The difference between the simple interest received from two different sources on ₹ 1500 for 3 years is ₹ 13.50. The difference between their rates of interest is:
A. 0.1% B. 0.2%
C. 0.3% D. 0.4%

20. Priya invested an amount of ₹ 12,000 at the rate of 10% p.a. simple interest and another amount at the rate of 20% p.a. simple interest. The total interest earned at the end of one year on the total amount invested became 14% p.a. Find the total amount invested.
A. ₹ 20,000 B. ₹ 22,000
C. ₹ 24,000 D. ₹ 25,000

21. If the rate increases by 2%, the simple interest received on a sum of money increases by ₹ 108. If the time period is increased by 2 years, the simple interest on the same sum increases by ₹ 180. The sum is:
A. ₹ 1800 B. ₹ 3600
C. ₹ 5400 D. Data inadequate

22. A sum of ₹ 2600 is lent out in two parts in such a way that the interest on one part at 10% for 5 years is equal to that on another at 9% for 6 years. The sum lent out at 10% is:
A. ₹ 1150 B. ₹ 1250
C. ₹ 1350 D. ₹ 1450

23. A sum of 1550 was lent partly at 5% and partly at 8% p.a. simple interest. The total interest received after 3 years was ₹ 300. The ratio of the money lent at 5% to that lent at 8% is :
A. 5 : 8 B. 8 : 5
C. 16 : 15 D. 31 : 6

24. Deepak invested certain amount in three different schemes A, B and C with the rate of interest 10% p.a., 12% p.a. and 15% p.a. respectively. If the total interest accrued in one years was ₹ 3200 and the amount invested in Scheme C was 150% of the amount invested in Scheme A and 240% of the amount invested in Scheme B, what was the amount invested in Scheme B?
A. ₹ 5000 B. ₹ 6500
C. ₹ 8000 D. Cannot be determined

25. Shyam invested ₹ 15,000 @ 10% per annum for one year. If the interest is compounded half-yearly, then the amount received by Shyam at the end of the year will be :
A. ₹ 16,500 B. ₹ 16,525.50
C. ₹ 16,537.50 D. ₹ 18,150

26. The difference between simple interest and compound interest on ₹ 1200 for one year at 10% per annum reckoned half-yearly is :
A. ₹ 2.50 B. ₹ 3
C. ₹ 3.75 D. ₹ 4

27. There is 60% increase in an amount in 6 years at simple interest. What will be the compound interest of ₹ 12,000 after 3 years at the same rate?
A. ₹ 2160 B. ₹ 3120
C. ₹ 3972 D. ₹ 6240

28. The difference between compound interest and simple interest on an amount of ₹ 15,000 for 2 years is ₹ 96. What is the rate of interest per annum?
A. 8 B. 10
C. 12 D. Cannot be determined

29. The difference between the simple interest on a certain sum at the rate of 10% per annum for 2 years and compound interest which is compounded every 6 months is ₹ 124.05. What is the principal sum?
A. ₹ 6000 B. ₹ 8000
C. ₹ 10,000 D. ₹ 12,000

30. Mr Dubey invested money in two schemes A and B offering compound interest @ 8% p.a. and 9% p.a. respectively. If the total amount of interest accrued through two schemes together in two years was ₹ 4818.30 and the total amount invested was ₹ 27,000, what was the amount invested in Scheme A?
A. ₹ 12,000 B. ₹ 13,500
C. ₹ 15,000 D. Cannot be determined

31. A sum of money becomes ₹ 13,380 after 3 years and ₹ 20,070 after 6 years on compound interest. The sum is:
A. ₹ 8800 B. ₹ 8890
C. ₹ 8920 D. ₹ 9040

32. A sum of ₹ 12,000 deposited at compound interest becomes double after 5 years. After 20 years, it will become:
A. ₹ 96,000 B. ₹ 1,20,000
C. ₹ 1,24,000 D. ₹ 1,92,000

33. The least number of complete years in which a sum of money put out at 20% compound interest will be more than doubled is:
A. 3 B. 4
C. 5 D. 6

34. What annual payment will discharge a debt of ₹ 1025 due in 2 years at the rate of 5% compound interest?
A. ₹ 550 B. ₹ 551.25
C. ₹ 560 D. ₹ 560.75

35. A sum of money is borrowed and paid back in two annual instalments of ₹ 882 each allowing 5% compound interest. The sum borrowed was:
A. ₹ 1620 B. ₹ 1640
C. ₹ 1680 D. ₹ 1700

ANSWERS

1	2	3	4	5	6	7	8	9	10
B	D	C	A	A	D	A	D	D	D
11	**12**	**13**	**14**	**15**	**16**	**17**	**18**	**19**	**20**
D	C	D	A	C	A	D	D	C	A
21	**22**	**23**	**24**	**25**	**26**	**27**	**28**	**29**	**30**
D	C	C	A	C	B	C	A	B	A
31	**32**	**33**	**34**	**35**					
C	D	B	B	B					

SOME SELECTED EXPLANATORY ANSWERS

2. Required amount $= ₹\dfrac{1575\times100}{12.5\times5} = ₹\ 2520.$

3. Suppose Sriram invested ₹ P in each scheme.

$$\because \quad 63 = P\left[\left(1+\frac{10}{100}\right)^2-1\right]$$

$$\Rightarrow \quad 63 = P\left(\frac{121}{100}-1\right) = \frac{21P}{100}$$

$$\therefore \quad P = \frac{63\times100}{21} = ₹\ 300$$

Compound interest of 2 years under scheme Y

$$= 300\left[\left(1+\frac{12}{100}\right)^2-1\right]$$

$$= \frac{300\times159}{625} = ₹\ 76.32.$$

5. $\because$ Required Difference $= \dfrac{P\times r^2}{(100)^2}\left(\dfrac{r}{100}+3\right)$

$$\Rightarrow \quad 595.35 = \frac{P\times15\times15}{100\times100}\left(\frac{15}{100}+3\right)$$

$$= \frac{225P}{10000}\times\frac{315}{100}$$

$$\therefore \quad P = \frac{595.35\times1000000}{225\times315} = ₹\ 8400$$

6. S.I. $= \dfrac{988\times18\times5}{100} = ₹\ 889.20$

7. Principal $= \dfrac{1200\times100}{4\times8} = ₹\ 3750$

$$\therefore \quad S.I. = \frac{3750\times3\times3\times6}{100} = ₹\ 2025$$

9. From question,

$$(21460 - 14500) = \frac{14500 \times r \times 6}{100}$$

$$\Rightarrow \frac{6960}{145 \times 6} = r$$

$$\therefore r = 8$$

Hence the rate of interest per annum = 8%.

10. S.I. = ₹ (15500 − 12500) = ₹ 3000.

$$\text{Rate} = \left(\frac{100 \times 3000}{12500 \times 4}\right)\% = 6\%.$$

11. $$\text{Principal} = ₹\left(\frac{100 \times 4016.25}{9 \times 5}\right)$$

$$= ₹\left(\frac{401625}{45}\right) = ₹\ 8925.$$

12. S.I. = ₹ (956 − 800) = ₹ 156.

$$\text{Rate} = \left(\frac{100 \times 156}{800 \times 3}\right)\% = 6\frac{1}{2}\%.$$

$$\text{New rate} = \left(6\frac{1}{2} + 4\right)\% = 10\frac{1}{2}\%.$$

$$\text{New S.I.} = ₹\left(800 \times \frac{21}{2} \times \frac{3}{100}\right) = ₹\ 252.$$

∴ New amount = ₹ (800 + 252) = ₹ 1052.

13. We need to know the S.I., principal and time to find the rate. Since the principal is not given, so data is inadequate.

14. Let the sum be ₹ x. Then,

$$\left(\frac{x \times 6 \times 3}{100}\right) + \left(\frac{x \times 9 \times 5}{100}\right) + \left(\frac{x \times 13 \times 3}{100}\right) = 8160$$

$$\Rightarrow 18x + 45x + 39x = (8160 \times 100)$$

$$\Rightarrow 102x = 816000 \Rightarrow x = 8000.$$

15. Let the sum be ₹ x. Now, S.I. = ₹ 600, T = 10 years.

$$\text{Rate} = \left(\frac{100 \times 600}{x \times 10}\right)\% = \left(\frac{6000}{x}\right)\%.$$

$$\text{S.I. for first 5 years} = ₹\left(\frac{x \times 5 \times 6000}{x \times 100}\right) = ₹\ 300.$$

$$\text{S.I. for last 5 years} = ₹\left(3x \times 5 \times \frac{6000}{x \times 100}\right) = ₹\ 900.$$

∴ Total interest = ₹ 1200.

16. Let sum = x. Then, S.I. = $\frac{x}{9}$.

Let rate = R% and time = R years.

$$\therefore \left(\frac{x \times R \times R}{100}\right) = \frac{x}{9} \Rightarrow R^2 = \frac{100}{9}$$

$$\Rightarrow R = \frac{10}{3} = 3\frac{1}{3}.$$

Hence, rate = $3\frac{1}{3}\%$.

17. Let the money added be ₹ x Then,

$$\frac{(830 + x) \times 14 \times 3}{100} - \frac{830 \times 12 \times 3}{100} = 93.90$$

$$\Rightarrow 830 \times 42 + 42x - 830 \times 36 = 9390$$

$$\Rightarrow 42x + 830 \times (42 - 36) = 9390$$

$$\Rightarrow 42x = 9390 - 4980 \Rightarrow x = \frac{4410}{42} = 105.$$

∴ Money added = ₹ 105.

18. Let the original rate be R%. Then, new rate = (2R)%.

$$\therefore \left(\frac{725 \times R \times 1}{100}\right) + \left(\frac{36250 \times 2R \times 1}{100 \times 3}\right) = 33.50$$

$$\Rightarrow (2175 + 725)\ R = 33.50 \times 100 \times 3 = 10050$$

$$\Rightarrow R = \frac{10050}{2900} = 3.46.$$

∴ Original rate = 3.46%.

19. $$\left(\frac{1500 \times R_1 \times 3}{100}\right) - \left(\frac{1500 \times R_2 \times 3}{100}\right) = 13.50$$

$$\Rightarrow 4500\ (R_1 - R_2) = 1350$$

$$\Rightarrow R_1 - R_2 = \frac{1350}{4500} = 0.3\%.$$

20. Let the second amount be ₹ x. Then,

$$\left(\frac{12000 \times 10 \times 1}{100}\right) + \left(\frac{x \times 20 \times 1}{100}\right) = \left[\frac{(12000 + x) \times 14 \times 1}{100}\right]$$

$$\Rightarrow 120000 + 20x = 168000 + 14x$$

$$\Rightarrow 6x = 48000 \Rightarrow x = 8000.$$

∴ Total investment = ₹ (12000 + 8000) = ₹ 20000.

21. Let the sum be ₹ x, rate be R% p.a. and time be T years.

$$\text{Then, } \left[\frac{x \times (R + 2) \times T}{100}\right] - \left(\frac{x \times R \times T}{100}\right) = 108$$

$$\Rightarrow 2xT = 10800 \quad ...(i)$$

$$\text{And, } \left[\frac{x \times R \times (T + 2)}{100}\right] - \left(\frac{x \times R \times T}{100}\right) = 180$$

$$\Rightarrow 2xR = 18000 \quad ...(ii)$$

Clearly, from (*i*) and (*ii*), we cannot find the value of x. So, the data is inadequate.

22. Let the sum lent at 10% be ₹ x and that lent at 9% be ₹ $(2600 - x)$. Then,

$$\left(\frac{x \times 10 \times 5}{100}\right) = \frac{(2600 - x) \times 9 \times 6}{100}$$

$\Rightarrow 50x = (2600 \times 54) - 54x$

$\Rightarrow \quad x = \left(\frac{2600 \times 54}{104}\right) = 1350.$

$\therefore$ Sum lent at 10% = ₹ 1350.

23. Let the sum lent at 5% be ₹ x and that lent at 8% be ₹ $(1550 - x)$. Then,

$$\left(\frac{x \times 5 \times 3}{100}\right) + \left[\frac{(1550 - x) \times 8 \times 3}{100}\right] = 300$$

$\Rightarrow 15x - 24x + (1550 \times 24) = 30000$

$\Rightarrow 9x = 7200 \quad \Rightarrow \quad x = 800.$

$\therefore$ Required ratio = 800 : 750 = 16 : 15.

24. Let x, y and z be the amounts invested in schemes A, B and C respectively. Then,

$$\left(\frac{x \times 10 \times 1}{100}\right) + \left(\frac{y \times 12 \times 1}{100}\right) + \left(\frac{z \times 15 \times 1}{100}\right) = 3200$$

$\Rightarrow \quad 10x + 12y + 15z = 320000 \quad ...(i)$

Now, $\quad z = 240\%$ of $y = \frac{12}{5}y \quad ...(ii)$

And, $\quad z = 150\%$ of $x = \frac{3}{2}x$

$\Rightarrow x = \frac{2}{3}z = \left(\frac{2}{3} \times \frac{12}{5}\right)y = \frac{8}{5}y \quad ...(iii)$

From (i), (ii) and (iii), we have :

$16y + 12y + 36y = 320000 \Rightarrow 64y = 320000$

$\Rightarrow y = 5000.$

$\therefore$ Sum invested in Scheme B = ₹ 5000.

25. P = ₹ 15000; R = 10% p.a. = 5% per half-year; T = 1 year = 2 half-years.

$\therefore$ Amount = ₹$\left[15000 \times \left(1 + \frac{5}{100}\right)^2\right]$

= ₹$\left(15000 \times \frac{21}{20} \times \frac{21}{20}\right)$

= ₹ 16537.50.

26. S.I. = ₹$\left(\frac{1200 \times 10 \times 1}{100}\right)$ = ₹ 120.

C.I. = ₹$\left[1200 \times \left(1 + \frac{5}{100}\right)^2 - 1200\right]$ = ₹ 123.

$\therefore$ Difference = ₹ (123 – 120) = ₹ 3.

27. Let P = ₹ 100. Then, S.I. = ₹ 60 and T = 6 years.

$\therefore \quad R = \frac{100 \times 60}{100 \times 6} = 10\%$ p.a.

Now, P = ₹ 12000, T = 3 years and R = 10% p.a.

$\therefore \quad$ C.I. = ₹$\left[12000 \times \left\{\left(1 + \frac{10}{100}\right)^3 - 1\right\}\right]$

= ₹$\left(12000 \times \frac{331}{1000}\right)$ = ₹ 3972.

28. $\left[15000 \times \left(1 + \frac{R}{100}\right)^2 - 15000\right] - \left(\frac{15000 \times R \times 2}{100}\right) = 96$

$\Rightarrow \quad 15000\left[\left(1 + \frac{R}{100}\right)^2 - 1 - \frac{2R}{100}\right] = 96$

$\Rightarrow \quad 15000\left[\frac{(100 - R)^2 - 10000 - 200R}{10000}\right] = 96$

$\Rightarrow \quad R^2 = \frac{96 \times 2}{3} = 64 \Rightarrow R = 8.$

$\therefore$ Rate = 8%.

30. Let the investment in scheme A be ₹ x.

Then, investment in scheme B = ₹ $(27000 - x)$

$\therefore \left[x \times \left\{\left(1 + \frac{8}{100}\right)^2 - 1\right\} + (27000 - x)\left\{\left(1 + \frac{9}{100}\right)^2 - 1\right\}\right]$

= 4818.30.

$\Rightarrow \quad \left(x \times \frac{104}{625}\right) + \frac{1881(27000 - x)}{10000} = \frac{481830}{100}$

$\Rightarrow \quad 1664x + 1881\ (27000 - x) = 48183000$

$\Rightarrow \quad (1881x - 1664x) = (50787000 - 48183000)$

$\Rightarrow \quad 217x = 2604000$

$\Rightarrow \quad x = \frac{2604000}{217} = 12000.$

32. $\quad 12000 \times \left(1 + \frac{R}{100}\right)^5 = 24000$

$\Rightarrow \quad \left(1 + \frac{R}{100}\right)^5 = 2$

$\therefore \quad \left[\left(1 + \frac{R}{100}\right)^5\right]^4 = 2^4 = 16$

$\Rightarrow \quad \left(1 + \frac{R}{100}\right)^{20} = 16$

$\Rightarrow \quad P\left(1 + \frac{R}{100}\right)^{20} = 16P$

$\Rightarrow 12000\left(1 + \frac{R}{100}\right)^{20} = 16 \times 12000 = 192000.$

33. $P\left(1+\frac{R}{100}\right)^n > 2P$ or $\left(\frac{6}{5}\right)^n > 2$

Now, $\left(\frac{6}{5}\times\frac{6}{5}\times\frac{6}{5}\times\frac{6}{5}\right) > 2$. So, $n = 4$ years.

34. Let each instalment be ₹ x. Then,

$$\frac{x}{\left(1+\frac{5}{100}\right)}+\frac{x}{\left(1+\frac{5}{100}\right)^2} = 1025$$

$$\Rightarrow \quad \frac{20x}{21}+\frac{400x}{441} = 1025$$

$$\Rightarrow \quad 820x = 1025 \times 441$$

$$\Rightarrow \quad x = \left(\frac{1025\times 441}{820}\right) = 551.25.$$

So, value of each instalment = ₹ 551.25

35. Principal = (P.W. of ₹ 882 due 1 year hence) + (P.W. of ₹ 882 due 2 years hence)

$$= \left[\frac{882}{\left(1+\frac{5}{100}\right)}+\frac{882}{\left(1+\frac{5}{100}\right)^2}\right]$$

$$= \left(\frac{882\times 20}{21}+\frac{882\times 400}{441}\right) = ₹\ 1640.$$

* * * * *

9

Time & Work

IMPORTANT FACTS

1. If A can do a piece of work in n days, then work done by A in 1 day $= \frac{1}{n}$.
2. If work done by A in 1 day $= \frac{1}{n}$; then A can finish the whole work in n days.
3. If A is twice as good a workman as B then; Ratio of work done by A and B = 2 : 1
   Ratio of times taken by A and B to finish a work = 1 : 2.
4. If a pipe can fill a tank in x hours, then the part filled in 1 hour $= \frac{1}{x}$.
5. If a pipe can empty a tank in y hour, then the part of the tank emptied in 1 hour $= \frac{1}{y}$.
6. If two pipes can fill a tank in x and y hours respectively and both the pipes are opened simultaneously then, the part filled in 1 hour $= \frac{1}{x} + \frac{1}{y}$.
7. (*a*) A tap fills a cistern in x hours and the other can empty the cistern in y hours. If both the taps are opened simultaneously, then the net part of the tank filled in hour $= \frac{1}{x} - \frac{1}{y}$; when $y > x$.

   (*b*) A tap can fill a tank in x hours and other can empty in y hours. If both the taps are opened simultaneously, then the net part of tank emptied in 1 hour $= \frac{1}{y} - \frac{1}{x}$; when $x > y$.

SOLVED EXAMPLES

Example 1: A and B together can complete a piece of work in 4 days. If A alone can complete the same piece of work in 5 days, in how many days will B alone complete it?

Solution: (A + B)'s 1 day's work $= \frac{1}{4}$

A's 1 day's work $= \frac{1}{5}$

$\therefore$ B's 1 day work $= \frac{1}{4} - \frac{1}{5} = \frac{5-4}{20} = \frac{1}{20}$

$\therefore$ B alone will complete the work in 20 days.

Example 2: Four examiners can examine a certain number of answer papers in 10 days by working for 5 hours a day. For how many hours in a day would 2 examiners have to work in order to examine twice the number of answer papers in 20 days?

Solution:

No. of exam	Hrs. per day	No. of papers	Days
4 ↓	5 ↓	1 ↓	10 ↓
2	x	2	20

$$\left.\begin{array}{l} 2 : 4 \\ 1 : 2 \\ 20 : 10 \end{array}\right\} :: 5 : x$$

$$\therefore \quad x = \frac{4 \times 2 \times 10 \times 5}{2 \times 20} = 10 \text{ hrs./day}$$

Example 3: 6 women and 6 men together can complete a piece of work in 6 days. In how many days can 15 men alone complete the piece of work if 9 women alone can complete the work in 10 days?

Solution: $\because$ Work of 1 woman for 1 day

$$= \frac{1}{9 \times 10} = \frac{1}{90}$$

$\Rightarrow$ Work of 6 women for 6 days

$$= \frac{1}{90} \times 6 \times 6 = \frac{2}{5}$$

$\Rightarrow$ Work of 6 men for 6 days

$$= 1 - \frac{2}{5} = \frac{3}{5}$$

$\Rightarrow$ Work of 1 man for 1 day

$$= \frac{3}{5 \times 6 \times 6} = \frac{1}{60}$$

∴ Work of 15 men for 1 day

$$= \frac{1}{60} \times 15 = \frac{1}{4}.$$

Hence, 15 men will complete the work in 4 days.

Example 4: Six boys or four men can complete a piece of work in 24 days. In how many days will 3 boys and 10 men together complete the same piece of work?

Solution: 4 men = 6 boys

∴ 10 men $= \frac{6}{4} \times 10$

= 15 boys

∴ 10 men + 3 boys = 15 + 3

= 18 boys

∴ Required number of days $= \frac{6 \times 24}{18} = 8.$

MULTIPLE CHOICE QUESTIONS

1. 24 men can complete a piece of work in 16 days. The same work can be completed by 8 women in 72 days, whereas 24 children take 32 days to complete it. If 10 men, 15 women and 24 children work together, in how many days can the work be completed?

A. 18 B. 8
C. 22 D. 12

2. 18 Men alone can complete a piece of work in 24 days. 54 Women alone can complete the same piece of work in 12 days and 16 children alone can complete the same piece of work in 54 days. In how many days can 3 men, 9 women and 6 children together complete the piece of work?

A. 18 B. 36
C. 27 D. Cannot be determined

3. If 80 persons can finish a work within 16 days by working 6 hours a day then how many hours per day should 64 persons work to complete that very job within 15 days?

A. 7 hrs B. 8 hrs
C. 6 hrs D. 5 hrs

4. A and B can do a piece of work in 12 days. B and C in 15 days and C and A in 20 days. In how many days A, B and C together can do the work?

A. 8 days B. 12 days
C. 10 days D. 14 days

5. 5 labourers can make 5 mats in 5 hours. How many mats will 10 labourers make in 10 hours?

A. 10 B. 20
C. 30 D. 35

6. A can do $\frac{1}{2}$ of a work in 9 days while B can do $\frac{1}{3}$ of the same work in 6 days. How long would it take for A and B together to complete the work?

A. 18 B. 6
C. 12 D. 9

7. A can do a piece of work in 40 days. He works on it for 5 days and then B completes it in 21 days. How long will A and B together take to complete the work?

A. 20 days B. 15 days
C. 30 days D. 10 days

8. A, B and C can complete a work separately in 24, 36 and 48 days respectively. They started together but C left after 4 days of start and A left 3 days before the completion of work. In how many days will the work be completed?

A. 15 days B. 18 days
C. 10 days D. 12 days

9. A and B can do a piece of work in 12 days. B and C together can do it in 15 days. If A is twice as good a work man as C, find in what time B alone can do it?

A. 12 days B. 15 days
C. 20 days D. 24 days

10. If 4 men or 6 boys can finish a piece of work in 20 days, in how many days can 6 men and 11 boys finish it?

A. 20 days B. 12 days
C. 8 days D. 6 days

11. 5 men and 2 boys together in 1 hour do 4 times work as 1 man and 1 boy do in an hour. Determine the ratio of the specified times of works of a man and a boy.

A. 4 : 1 B. 1 : 4
C. 3 : 4 D. 2 : 1

12. 75 boys finish a work in 24 days. How many men will complete twice the work in 20 days while 2 men do in a day same work as 3 boys do in a day?

A. 50 B. 24
C. 60 D. 120

13. If 3 men and 4 boys together earn ₹ 264 in 8 days and 2 men and 3 boys together earn ₹ 184 in the same time, then in how many days will 6 men and 7 boys together earn ₹ 315?

A. 5 days B. 6 days
C. 9 days D. 10 days

14. Pipes A and B can fill a tank in 36 and 48 hours respectively. Both pipes are opened together to fill the tank. After some time pipe B is closed and the tank is full in 24 hours. For how much time was the pipe B opend?

A. 20 hrs. B. 16 hrs.
C. 48 hrs. D. 36 hrs.

15. A tank is normally filled in 9 hours. But because of a leak in its bottom it takes 3 hours more to fill. How much time will it take the leak to empty the full tank?
A. 12 hrs. B. 24 hrs.
C. 36 hrs. D. 48 hrs.

16. Three pipes A, B and C are connected to a tank. A and B can fill it in 20 and 30 minutes respectively, while C can empty it in 15 minutes. If A, B and C are kept open successively for 1 minute each, how soon will the tank be filled?
A. 176 min B. 167 min
C. 180 min D. 170 min

17. Two pipes A and B can separately fill a cistern in $7\frac{1}{2}$ minutes and 5 minutes respectively and a waste pipe C can carry off 14 litres per minute. If all the pipes are opened when the cistern is full, it is emptied in 1 hour. How many litres does it hold?
A. 40 litres B. 20 litres
C. 36 litres D. 30 litres

18. Two workers A and B are engaged to do a work. A working alone takes 8 hours more to complete the job than if both worked together. If B worked alone, he would need $4\frac{1}{2}$ hours more to complete the job than they both working together. What time would they take to do the work together?
A. 4 hours B. 5 hours
C. 6 hours D. 7 hours

19. Savita can do a piece of work in 20 days. Tripti is 25% more efficient than Savita. The number of days taken by Tripti to do the same piece of work is :
A. 15 B. 16
C. 18 D. 25

20. A machine P can print one lakh books in 8 hours, machine Q can print the same number of books in 10 hours while machine R can print them in 12 hours. All the machines are started at 9 a.m. while machine P is closed at 11 a.m. and the remaining two machines complete the work. Approximately at what time will the work be finished?
A. 11:30 a.m. B. 12 noon
C. 12:30 p.m. D. 1 p.m.

21. A and B can do a piece of work in 45 days and 40 days respectively. They began to do the work together but A leaves after some days and then B completed the remaining work in 23 days. The number of days after which A left the work was :
A. 6 B. 8
C. 9 D. 12

22. A, B and C are employed to do a piece of work for ₹ 529. A and B together are supposed to do $\frac{19}{23}$ of the work and B and C together $\frac{8}{23}$ of the work. What amount should A be paid?
A. ₹ 315 B. ₹ 345
C. ₹ 355 D. ₹ 375

23. A sum of money is sufficient to pay A's wages for 21 days and B's wages for 28 days. The same money is sufficient to pay the wages of both for :
A. 12 days B. $12\frac{1}{4}$ days
C. 14 days D. $24\frac{1}{2}$ days

24. 12 men complete a work in 9 days. After they have worked for 6 days, 6 more men join them. How many days will they take to complete the remaining work?
A. 2 days B. 3 days
C. 4 days D. 5 days

25. 10 men and 15 women together can complete a work in 6 days. It takes 100 days for one man alone to complete the same work. How many days will be required for one woman alone to complete the same work?
A. 90 B. 125
C. 145 D. none of these

26. 12 men can complete a piece of work in 4 days. While 15 women can compete the same work in 4 days. 6 men start working on the job and after working for 2 days, all of them stopped working. How many women should be put on the job to complete the remaining work, if it is to be completed in 3 days?
A. 15 B. 18
C. 22 D. Data inadequate

27. 10 women can complete a work in 7 days and 10 children take 14 days to complete the work. How many days will 5 women and 10 children take to complete the work?
A. 3 B. 5
C. 7 D. Cannot be determined

28. Twenty-four men can complete a work in sixteen days. Thirty-two women can complete the same work in twenty-four days. Sixteen men and sixteen women started working and worked for twelve days. How many more men are to be added to complete the remaining work in 2 days?
A. 16 B. 24
C. 36 D. 48

29. A pump can fill a tank with water in 2 hours. Because of a leak, it took $2\frac{1}{3}$ hours to fill the tank. The leak can drain all the water of the tank in:

A. $4\frac{1}{3}$ hrs B. 7 hrs
C. 8 hrs D. 14 hrs

30. Two pipes A and B together can fill a cistern in 4 hours. Had they been opened separately, then B would have taken 6 hours more than A to fill the cistern. How much time will be taken by A to fill the cistern separately?
A. 1 hr B. 2 hrs
C. 6 hrs D. 8 hrs

31. A tank is filled in 5 hours by three pipes A, B and C. The pipe C is twice as fast as B and B is twice as fast as A. How much time will pipe A alone take to fill the tank?
A. 20 hrs
B. 25 hrs
C. 35 hrs
D. Cannot be determined

32. Bucket P has thrice the capacity as bucket Q. It takes 60 turns for bucket P to fill the empty drum. How many turns it will take for both the buckets P and Q, having each turn together to fill the empty drum?
A. 30 B. 40
C. 45 D. 90

33. A large tanker can be filled by two pipes A and B in 60 and 40 minutes respectively. How many minutes will it take to fill the tanker from empty state if B is used for half the time and A and B fill it together for the other half?
A. 15 min B. 20 min
C. 27.5 min D. 30 min

34. A leak in the bottom of a tank can empty the full tank in 8 hours. An inlet pipe fills water at the rate of 6 litres a minute. When the tank is full, the inlet is opened and due to the leak, the tank is empty in 12 hours. How many litres does the cistern hold?
A. 7580 B. 7960
C. 8290 D. 8640

35. Two pipes can fill a tank in 20 and 24 minutes respectively and a waste pipe can empty 3 gallons per minute. All the three pipes working together can fill the tank in 15 minutes. The capacity of the tank is :
A. 60 gallons B. 100 gallons
C. 120 gallons D. 180 gallons

ANSWERS

1	2	3	4	5	6	7	8	9	10
D	B	B	C	B	D	B	A	C	D
11	**12**	**13**	**14**	**15**	**16**	**17**	**18**	**19**	**20**
D	D	A	B	C	B	A	C	B	D
21	**22**	**23**	**24**	**25**	**26**	**27**	**28**	**29**	**30**
C	B	A	A	D	A	C	B	D	C
31	**32**	**33**	**34**	**35**					
C	C	D	D	C					

SOME SELECTED EXPLANATORY ANSWERS

1. In 16 days the work is completed by 24 men.

∴ In 1 day the work is completed by

$= 24 \times 16 = 384$ men.

Similarly in 1 day the work is completed by

$= 72 \times 8 = 576$ women

and in 1 day the work is completed by

$= 24 \times 32 = 768$ children

∵ 384 men = 768 children

$\therefore \quad 10 \text{ men} = \frac{768 \times 10}{384} = 20$ children

and 576 women = 768 children

$\therefore \quad 15 \text{ women} = \frac{768 \times 15}{576} = 20$ children

∴ (20 + 20 + 24) = 64 children

∵ 24 children complete the work in 32 days.

∴ 64 children complete the work in $= \frac{32 \times 24}{64}$

$= 12$ days.

2. Work done by 3 man + 9 women + 6 children in 1 day

$$= \frac{3 \times 1}{24 \times 18} + \frac{9 \times 1}{12 \times 54} + \frac{6 \times 1}{54 \times 16}$$

$$= \frac{1}{144} + \frac{1}{72} + \frac{1}{144} = \frac{1}{36}$$

∵ $\frac{1}{36}$ work is done in 1 day.

∴ 1 work is done in $= \frac{1 \times 36}{1} = 36$ days.

3. 80 persons can do the job in 16 days by working = 6 hrs daily

1 person can do the job in 16 days by working $= 6 \times 80$ hrs daily

1 person can do the job in 1 day by working $= 6 \times 80 \times 16$ hrs daily

64 persons can do the job in 15 days by working

$$= \frac{6\times80\times16}{64\times15} \text{ hrs} = 8 \text{ hr.}$$

5. 5 labourers in 5 hrs. can make = 5 mats

1 labourer in 5 hrs. can make $= \frac{5}{5}$ mats

1 labourer in 1 hrs. can make $= \frac{5}{5\times5}$ mats

10 labourers in 10 hrs. can make

$$= \frac{5\times10\times10}{5\times5} = 20 \text{ mats.}$$

6. A's 9 day's work $= \frac{1}{2}$

$\therefore$ A's 1 day's work $= \frac{1}{2\times9} = \frac{1}{18}$

B's 6 day's work $= \frac{1}{3}$

$\therefore$ B's 1 day's work $= \frac{1}{3\times6} = \frac{1}{18}$

$\therefore$ (A + B)'s 1 day's work $= \frac{1}{18} + \frac{1}{18}$

$$= \frac{2}{18} = \frac{1}{9}$$

$\therefore$ A and B both together will complete the work in 9 days.

8. Let the work be completed in x days. Therefore, A worked for $x - 3$ days, B for x days and C for 4 days.

A's 1 day's work $= \frac{1}{24}$

B's 1 day's work $= \frac{1}{36}$

and, C's 1 day's work $= \frac{1}{48}$

$$\therefore (x-3)\times\frac{1}{24} + x\times\frac{1}{36} + 4\times\frac{1}{48} = 1$$

$$\Rightarrow \frac{x-3}{24} + \frac{x}{36} + \frac{1}{12} = 1$$

$$\Rightarrow \frac{3x-9+2x+6}{72} = 1$$

$$\Rightarrow 5x - 3 = 72$$

$$\Rightarrow 5x = 75$$

$$\Rightarrow x = \frac{75}{5} = 15$$

Hence, the work was completed in 15 days.

11. Work of (5 men + 2 boys) = 4 (work of 1 man + 1 boy)

= Work of 4 men + 4 boys

$\Rightarrow$ Work of (5 men – 4 men) = Work of (4 boys – 2 boys)

$\Rightarrow$ Work of 1 man = Work of 2 boys

$\therefore$ Required ratio = 2 : 1

12. Work of 3 boys = Work of 2 men

$\therefore$ Work of 1 boy = Work of $\frac{2}{3}$ man

$\therefore$ Work of 75 boys = Work of $\frac{2}{3}\times75$

= 50 men

$\because$ 1 work is completed in 24 days by 50 men

$\therefore$ 1 work is completed in 1 day by 50×24 men

$\therefore$ 2 works are completed in 20 days by

$$\frac{50\times24\times2}{20} = 120 \text{ men.}$$

13. Let men be denoted by m and boys by b.

$3m + 4b$ earn in 8 days = ₹ 264

$\therefore$ $3m + 4b$ earn in 1 day $= \frac{264}{8} =$ ₹ 33 ...(*i*)

And, $2m + 3b$ earn in 8 days= ₹ 184

$\therefore$ $2m + 3b$ earn in 1 day $= \frac{184}{8} =$ ₹ 23 ...(*ii*)

By (*ii*) × 3 – (*i*) × 2, we have

$6m + 9b - 6m - 8b = 69 - 66$

$\Rightarrow$ $1b$ = ₹ 3

$\therefore$ $3m + 4b = 33$

$3m = 33 - 4 \times 3 = 21$

$1m = 7$

$\therefore$ Wages for 1 day of 6 men and 7 boys

$= 6 \times 7 + 7 \times 3$

= 42 + 21 = ₹ 63

$\therefore$ ₹ 63 are wages for 1 day

$\therefore$ ₹ 315 are wages for $= \frac{315}{63} = 5$ days.

15. Without the leak, in 1 hour $\frac{1}{9}$ of the tank is filled.

$\therefore$ In extra 3 hrs, the pipe fills $\frac{3}{9} = \frac{1}{3}$ of the tank more.

In (9 + 3) = 12 hours, the leak empties this extra $\frac{1}{3}$ of the tank.

$\therefore$ The leak will empty the full tank in $12\times\frac{3}{1} = 36$ hrs.

16. $\left(\frac{1}{20}+\frac{1}{30}-\frac{1}{15}\right) = \frac{1}{60}$ of the tank is filled in first 3 minutes

$$\frac{1}{20}+\frac{1}{30} = \frac{5}{60}$$

$$1-\frac{5}{60} = \frac{55}{60}$$

This $\frac{55}{60}$ part of the tank will be filled in $3 \times 55 = 165$ minutes.

The last $\frac{5}{60}$ part will be filled in 2 minutes, successively by A and B.

Total time = 165 + 2 = 167 minutes.

17. Cistern filled by (A + B) in 1 minute

$$= \left(\frac{2}{15}+\frac{1}{5}\right) = \frac{1}{3}$$

Net emptying work done by (A + B + C) in 1 minute

$$= \frac{1}{60}$$

Work done by C in 1 minute = $\left(\frac{1}{60}+\frac{1}{3}\right)=\frac{7}{20}$

$\therefore$ C alone can empty the cistern in $\left(\frac{20}{7}\right)$ minutes

Water thrown out by C in 1 minute = 14 litres

Water thrown out by C in $\left(\frac{20}{7}\right)$ minutes =

$$\left(14\times\frac{20}{7}\right) = 40 \text{ litres}$$

$\therefore$ Capacity of the cistern = 40 litres.

18. Let A and B together take x hours to complete the work. Then,

A alone takes $(x + 8)$ hrs and B alone takes $\left(x+\frac{9}{2}\right)$ hrs to complete the work. Then,

$$\frac{1}{(x+8)}+\frac{1}{\left(x+\frac{9}{2}\right)} = \frac{1}{x}$$

$$\Rightarrow \quad \frac{1}{(x+8)}+\frac{2}{(2x+9)} = \frac{1}{x}$$

$$\Rightarrow \quad x(4x + 25) = (x + 8)(2x + 9)$$

$$\Rightarrow \quad 2x^2 = 72$$

$$\Rightarrow \quad x^2 = 36$$

$$\Rightarrow \quad x = 6.$$

19. Ratio of times taken by Savita and Tripti

= 125 : 100

= 5 : 4.

Suppose Tripti takes x days to do the work.

$5 : 4 :: 20 : x \Rightarrow x = \left(\frac{4\times20}{5}\right)$

$\Rightarrow x = 16$ days.

Hence, Tripti takes 16 days to complete the work.

21. (A + B)'s 1 day's work = $\left(\frac{1}{45}+\frac{1}{40}\right) = \frac{17}{360}$.

Work done by B in 23 days = $\left(\frac{1}{40}\times23\right)$

$= \frac{23}{40}$.

Remaining work = $\left(1-\frac{23}{40}\right) = \frac{17}{40}$.

Now, $\frac{17}{360}$ work was done by (A + B) in 1 day.

$\frac{17}{40}$ work was done by (A + B) in $\left(1\times\frac{360}{17}\times\frac{17}{40}\right)$

= 9 days.

$\therefore$ A left after 9 days.

22. Work done by A = $\left(1-\frac{8}{23}\right) = \frac{15}{23}$.

$\therefore$ A : (B + C) = $\frac{15}{23}:\frac{8}{23}$ = 15 : 8.

So, A's share = ₹$\left(\frac{15}{23}\times529\right)$ = ₹ 345.

23. Let total money be ₹ x.

A's 1 day's wages = ₹$\frac{x}{21}$, B's 1 day's wages

$= ₹\frac{x}{28}$.

$\therefore$ (A + B)'s 1 day's wages = ₹$\left(\frac{x}{21}+\frac{x}{28}\right)$ = ₹$\frac{x}{12}$.

$\therefore$ Money is sufficient to pay the wages of both for 12 days.

24. 1 man's 1 day's work = $\frac{1}{108}$.

12 men's 6 day's work = $\left(\frac{1}{9}\times6\right) = \frac{2}{3}$.

Remaining work = $\left(1-\frac{2}{3}\right) = \frac{1}{3}$.

18 men's 1 day's work = $\left(\frac{1}{108}\times18\right) = \frac{1}{6}$.

$\frac{1}{6}$ work is done by them in 1 day.

$\therefore$ $\frac{1}{3}$ work is done by them in $\left(6\times\frac{1}{3}\right)$= 2 days.

25. 1 man's 1 day's work = $\frac{1}{100}$. (10 men + 15 women)'s

1 day's work = $\frac{1}{6}$.

15 women's 1 day's work = $\left(\frac{1}{6}-\frac{10}{100}\right) = \frac{1}{15}$.

1 woman's 1 day's work = $\frac{1}{225}$

$\therefore$ 1 woman alone can complete the work in 225 days.

26. 1 man's 1 day's work = $\frac{1}{48}$; 1 woman's 1 day's work

= $\frac{1}{60}$.

6 men's 2 day's work = $\left(\frac{6}{48}\times 2\right) = \frac{1}{4}$.

Remaining work = $\left(1-\frac{1}{4}\right) = \frac{3}{4}$.

Now, $\frac{1}{60}$ work is done in 1 day by 1 woman.

So, $\frac{3}{4}$ work will be done in 3 days by $\left(60\times\frac{3}{4}\times\frac{1}{3}\right)$

= 15 women.

27. 1 woman's 1 day's work = $\frac{1}{70}$; 1 child's 1 day's work

= $\frac{1}{140}$.

(5 women + 10 children)'s 1 day's work

$$= \left(\frac{5}{70}+\frac{10}{140}\right) = \frac{1}{7}.$$

$\therefore$ 5 women and 10 children will complete the work in 7 days.

29. Tank emptied by the leak in 1 hour = $\left(\frac{1}{2}-\frac{3}{7}\right) = \frac{1}{14}$.

$\therefore$ Leak will empty the tank in 14 hrs.

30. Let the cistern be filled by pipe A alone in x hours.

Then, pipe B will fill it in $(x + 6)$ hours.

$$\therefore \quad \frac{1}{x}+\frac{1}{(x+6)} = \frac{1}{4}$$

$$\Rightarrow \quad \frac{x+6+x}{x(x+6)} = \frac{1}{4}$$

$$\Rightarrow \quad x^2 - 2x - 24 = 0$$

$$\Rightarrow \quad (x-6)(x+4) = 0$$

$$\Rightarrow \quad x = 6.$$

32. Let capacity of P be x litres. Then, capacity of Q = $\frac{x}{3}$ litres.

Capacity of the drum = $60x$ litres.

Required number of turns = $\dfrac{60x}{\left(x+\frac{x}{3}\right)} = 45.$

33. Part filled by (A + B) in 1 minute

$$= \left(\frac{1}{60}+\frac{1}{40}\right) = \frac{1}{24}.$$

Suppose the tank is filled in x minutes.

Then $\quad \frac{x}{2}\left(\frac{1}{24}+\frac{1}{40}\right) = 1$

$$\Rightarrow \quad \frac{x}{2}\times\frac{1}{15} = 1$$

$$\Rightarrow \quad x = 30 \text{ min.}$$

35. Work done by the waste pipe in 1 minute

$$= \frac{1}{15}-\left(\frac{1}{20}+\frac{1}{24}\right) = \left(\frac{1}{15}-\frac{11}{120}\right) = -\frac{1}{40}.$$

[–ve sign means emptying]

$\therefore$ Volume of $\frac{1}{40}$ part = 3 gallons.

Volume of whole = (3 × 40) gallons
= 120 gallons.

10

Time & Distance

IMPORTANT FORMULAE

1. Speed = Distance ÷ Time
2. Distance = Time × Speed
3. Time = Distance ÷ Speed
4. x km/hr = $\left(x \times \frac{5}{18}\right)$ m/sec
5. x m/sec = $\left(x \times \frac{18}{5}\right)$ km/hr.
6. If the speed of a body is changed in the ratio $m : n$, then the ratio of the time taken changes in the ratio $n : m$.
7. When a man covers a certain distance with a speed of x km/h and another equal distance at the rate of y km/h, then for the whole journey, the average speed is given by

   Average speed = $\frac{2xy}{x+y}$ km/h.
8. The time taken by a train in passing a signal post or a telegraph pole or a man standing near a railway line

   $= \frac{\text{Length of the train}}{\text{Speed of the train}}$
9. The time taken by a train of length x passing a railway bridge or a platform or a tunnel or a train of length y at rest = $\frac{x+y}{\text{Speed}}$
10. (*a*) Time taken by faster train of length x and speed u to pass the slower train of length y and speed v in the same direction = $\frac{x+y}{u-v}$

    (*b*) Time taken by the trains in passing each other while moving in opposite direction = $\frac{x+y}{u+v}$
11. (*a*) Time taken by the train of length x and speed u to cross a man moving with speed v in same direction = $\frac{x}{u-v}$

    (*b*) Time taken by the train to cross a man moving in the opposite direction = $\frac{x}{u+v}$
12. If two trains start at the same time from two points A and B towards each other and after crossing, they take a and b hours in reaching B and A respectively. Then, A's speed : B's speed = $(\sqrt{b} : \sqrt{a})$

SOLVED EXAMPLES

Example 1: Distance between two railway stations A and B is 1536 kms. A train covers a journey between A to B at the uniform speed of 60 km/hr and returns from B to A at the uniform speed of 40 km/hr. What is the average speed of the train during the whole journey?

Solution: Average speed of train = $\frac{2 \times 40 \times 60}{40+60}$

$= \frac{2 \times 40 \times 60}{100}$ = 48 km/hr.

Example 2: A car covers the first 39 km of its journey in 45 minutes and covers the remaining 25 km in 35 minutes. What is the average speed of the car?

Solution: Total time = $\frac{45}{60} + \frac{35}{60} = \frac{4}{3}$ hrs.

and total distance = 39 + 25 = 64 km

∴ Average speed = $\frac{64 \times 3}{4}$ = 48 km/hr.

Example 3: A train covered a distance of 1235 km in 19 hours. Also, the average speed of a car is four-fifth the average speed of the train. How much distance will the car cover in 22 hours?

Solution: Speed of the train = $\frac{1235}{19}$ = 65 km/hr

∴ Speed of the car = $65 \times \frac{4}{5}$ = 52 km/hr

∴ Reqd. distance = 52 × 22 = 1144 km.

Example 4: A 320 metre long train moving with an average speed of 120 km/hr crosses a platform in 24 seconds. A man crosses the same platform in 4 minutes. What is the speed of man in metre/second?

Solution: Let the length of the platform be x metre. Then,

∵ Speed of the train = 120 km/hr = $120 \times \frac{5}{18}$

$= \frac{100}{3}$ m/sec.

From question,

$\because$ $320\text{m} + x = \dfrac{100}{3} \times 24$

$\therefore$ $x = 800 - 320$

$= 480$ metre.

$\therefore$ Reqd. speed of man $= \dfrac{480 \text{ m}}{4 \times 60 \text{ sec}} = 2$ m/sec.

Example 5: A 260 metre long train crosses a platform thrice its length in 80 seconds. What is the speed of the train in km/hour?

Solution: Speed of the train

$= \dfrac{(260+780)}{80} \times \dfrac{18}{5}$ km/hr.

$= 46.8$ km/hr.

MULTIPLE CHOICE QUESTIONS

1. A boat while travelling upstream covers a distance of 18 km at the speed of 3 km/hr, whereas while travelling downstream it covers the same distance at a speed of 9 km/hr. What is the speed of the boat in still waters?

A. 3 km/hr B. 5 km/hr
C. 7 km/hr D. None of these

2. The speed of a car is 1.5 times the speed of a bus. If the car travels at the speed of 60 km/hr what will be the difference in the time taken by the bus and the time taken by the car to cover 720 km?

A. 5 hours B. 6 hours
C. 4 hours D. 8 hours

3. A car covers a distance of 540 km in 9 hours. Speed of a train is double the speed of the car. Two-third the speed of the train is equal to the speed of a bike. How much distance will the bike cover in 5 hours?

A. 450 km B. 360 km
C. 400 km D. 500 km

4. A train travels at a speed of 30 km per hour for 12 minutes and then for the next 8 minutes at a speed of 45 km per hour. What is the average speed for the journey?

A. 40 km/hr B. 36 km/hr
C. 43 km/hr D. 46 km/hr

5. The ratio of the speeds of two trains is 7 : 8. If the second train runs 400 km in 5 hrs., find the speed of the first train.

A. 60 km/hr B. 55 km/hr
C. 70 km/hr D. 75 km/hr

6. A train travels from Dehradun to Delhi at a speed of 40 km/hr and returns at 60 km/hr. Find the average speed for the entire journey—

A. 50 km/hr B. 45 km/hr
C. 48 km/hr D. 52 km/hr

7. A car runs at 60 km per hr. A man runs at one-third the speed of the car and reaches office from his house in 15 minutes. How far is his office from his house?

A. 20 km B. 10 km
C. 5 km D. 4 km

8. Two men A and B start walking simultaneously from P to Q, a distance of 21 kms, at the speeds of 3 km and 4 km an hour respectively. B reaches Q, returns immediately and meets A at R. Then the distance from P to R is:

A. 6 km B. 12 km
C. 18 km D. 24 km

9. A monkey climbing up a greased pole ascends 12 metres and slips down 5 metres in alternate minutes. If the pole is 63 metres high, how long will it take him to reach the top?

A. $12\frac{7}{16}$ B. $7\frac{12}{16}$
C. $16\frac{7}{12}$ D. $10\frac{7}{12}$

10. A carriage driving in a fog passed a man who was walking at the speed of 6 km/hr. in the same direction. He could see the carriage for 4 minutes and it was visible to him up to a distance of 200 metres. The speed of the carriage is:

A. 9 km/hr B. 4 km/hr
C. 6 km/hr D. 8 km/hr

11. A train met with an accident 3 hours after starting, which detains it for one hour, after which it proceeds at 75% of its original speed. It arrives at the destination 4 hours late. Had the accident taken place 150 km further along the railway line, the train would have arrived only $3\frac{1}{2}$ hours late. Find the length of the trip:

A. 1200 km B. 1000 km
C. 2100 km D. 1050 km

12. Two places A and B are 162 kms apart. A train leaves A for B and at the same time another train leaves B for A. The two trains meet at the end of 6 hours. If the train travelling from A to B travels 8 km per hr. faster than the other. Then the speed of the faster train is:

A. $9\frac{1}{2}$ km/h B. 27 km/hr
C. $17\frac{1}{2}$ km/hr D. 19 km/hr

13. A train takes 18 seconds to pass completely through a station 162 metres long and 15 seconds to pass completely through another station 120 metres long. The length of the train is:

A. 60 m B. 75 m
C. 90 m D. 100 m

14. A man covered a distance of 3990 km partly by air, partly by sea and remaining by land. The time spent in air, on sea and on land is in the ratio 1 : 16 : 2 and the ratio of average speeds is 20 : 1 : 3 respectively. If total average speed is 42 km per hr, Then the distance covered by sea is :

A. 1502 kms B. 2520 kms
C. 1520 kms D. 1500 kms

15. A man is walking at a speed of 10 km/hr. After every kilometre, he takes rest for 5 minutes. How much time will he take to cover a distance of 5 km?

A. 1 hr B. 50 min
C. 20 min D. 30 min

16. Walking $\frac{5}{6}$ of his usual speed, a man reaches the station 10 min late. His usual time to reach the station is :

A. 30 min B. 40 min
C. 50 min D. 60 min

17. A is four times as fast as B and B is six times as fast as C. A certain distance is covered by B in 28 minutes. Then the time taken by C to cover the same distance is :

A. 2 : 48 hrs B. 2 : 40 hrs
C. 2 : 28 hrs D. 3 hrs

18. It takes eight hours for a 600 km journey, if 120 km is done by train and the rest by car. It takes 20 minutes more, if 200 km is done by train and the rest by car. The ratio of the speed of the train to the speed of the car is :

A. 2 : 3 B. 6 : 9
C. 3 : 4 D. 4 : 3

19. A bus takes two hours less for a journey of 300 km if its speed is increased by 5 kmph from its normal speed. The normal speed is

A. 30 kmph B. 25 kmph
C. 35 kmph D. 50 kmph

20. A boy goes to school with a speed of 3 km/hr and returns to the village with a speed of 2 km/hr. If he takes 5 hours in all, the distance between the village and the school is :

A. 6 km B. 7 km
C. 8 km D. 9 km

21. Rahim covers a certain distance in 14 hrs 40 min. He covers one half of the distance by train at 60 km/hr and the rest half by road at 50 km/hr. The distance travelled by him is :

A. 960 km B. 720 km
C. 1000 km D. 800 km

22. Two men start together to walk to a certain destination, one at 3.75 km an hour and another at 3 km an hour. The former arrives half an hour before the latter. The distance is:

A. 9.5 km B. 8 km
C. 7.5 km D. 6 km

23. A thief steals a car at 1.30 p.m. and drives it at 40 km an hour. The theft is discovered at 2 p.m. and the owner sets off in another car at 50 km an hour. He will overtake the thief at:

A. 3.30 p.m. B. 4 p.m.
C. 4.30 p.m. D. 6 p.m.

24. A man travels 35 km partly at 4 km/hr and at 5 km/hr. If he covers former distance at 5 km/hr and later distance at 4 km/hr, he could cover 2 km more in the same time. The time taken to cover the whole distance at original rate is :

A. 9 hours B. 7 hours
C. $4\frac{1}{2}$ hours D. 8 hours

25. A certain distance is covered at a certain speed. If half of this distance is covered in double the time, the ratio of the two speeds is :

A. 4 : 1 B. 1 : 4
C. 2 : 1 D. 1 : 2

26. A person sees a train passing over 1 km long bridge. The length of the train is half that of bridge. If the train clears the bridge in 2 minutes, the speed of the train is :

A. 50 km/hr B. 45 km/hr
C. 60 km/hr D. 30 km/hr

27. Two trains are running on parallel lines in the same direction at a speed of 50 km and 30 km per hour respectively. The faster train crosses a man in slower train in 18 seconds. The length of the faster train is :

A. 170 metres B. 100 metres
C. 98 metres D. 85 metres

28. A train speeds past a pole in 15 seconds and speeds past a platform 100 metres long in 25 seconds. Its length in metres is :

A. 200 B. 150
C. 50 D. Data inadequate

29. A train overtakes two persons who are walking in the same direction in which the train is goning, at the rate of 2 kmph and 4 kmph and passes them completely in 9 and 10 seconds respectively. The length of the train is :

A. 72 metres B. 54 metres
C. 50 metres D. 45 metres

30. A boat travels upstream from B to A and downstream from A to B in 3 hours. If the speed of the boat in still water is 9 km/hr and the speed of the current is 3 km/hr, the distance between A and B is :

A. 4 km B. 6 km
C. 8 km D. 12 km

31. Speed of a boat in standing water is 6 km/hr and the speed of the stream is 1.5 km/hr. A man rows to a place at a distance of 22.5 km and comes back to the starting point. The total time taken by him, is :

A. 6 hrs 30 min. B. 8 hrs 24 min.
C. 8 hrs D. 4 hrs 12 min.

32. A man can row $9\frac{1}{3}$ km/hr in still water and he finds that it takes him thrice as much time to row up than as to row down the same distance in river. The speed of the current is :

A. $3\frac{1}{3}$ km/hr B. $3\frac{1}{9}$ km/hr
C. $1\frac{1}{4}$ km/hr D. $4\frac{2}{3}$ km/hr

33. Sound is said to travel in air at about 1100 feet per second. A man hears the axe striking the tree, $\frac{11}{5}$ seconds after he sees it strike the tree. How far is the man from the wood chopper?

A. 2197 ft B. 2420 ft
C. 2500 ft D. 2629 ft

34. Asha left for city A from city B at 5.20 a.m. She travelled at the speed of 80 km/hr for 2 hours 15 minutes. After that the speed was reduced to 60 km/hr. If the distance between two cities is 350 kms, at what time did Asha reach city A?

A. 9.20 a.m. B. 9.25 a.m.
C. 9.35 a.m. D. None of these

35. A motorist covers a distance of 39 km in 45 minutes by moving at a speed of x kmph for the first 15 minutes, then moving at double the speed for the next 20 minutes and then again moving at his original speed for the rest of the journey. Then, x is equal to :

A. 31.2 B. 36
C. 40 D. 52

36. A train when moves at an average speed of 40 kmph, reaches its destination on time. When its average speed becomes 35 kmph, then it reaches its destination 15 minutes late. Find the length of journey.

A. 30 km B. 40 km
C. 70 km D. 80 km

37. With a uniform speed a car covers the distance in 8 hours. Had the speed been increased by 4 km/hr, the same distance could have been covered in $7\frac{1}{2}$ hours. What is the distance covered?

A. 420 km
B. 480 km
C. 640 km
D. Cannot be determined

38. A 270 metres long train running at the speed of 120 kmph crosses another train running in opposite direction at the speed of 80 kmph in 9 seconds. What is the length of the other train?

A. 230 m B. 240 m
C. 260 m D. 320 m

39. Two trains running in opposite directions cross a man standing on the platform in 27 seconds and 17 seconds respecively and they cross each other in 23 seconds. The ratio of their speeds is :

A. 1 : 3 B. 3 : 2
C. 3 : 4 D. 2 : 3

40. A boat covers a certain distance downstream in 1 hour, while it comes back in $1\frac{1}{2}$ hours. If the speed of the stream be 3 kmph, what is the speed of the boat in still water?

A. 12 kmph B. 13 kmph
C. 14 kmph D. 15 kmph

ANSWERS

1	2	3	4	5	6	7	8	9	10
D	B	C	B	C	C	C	C	C	A
11	12	13	14	15	16	17	18	19	20
A	C	C	C	B	C	A	C	B	A
21	22	23	24	25	26	27	28	29	30
D	C	B	D	A	B	B	B	C	D
31	32	33	34	35	36	37	38	39	40
C	D	B	D	B	C	B	A	B	D

SOME SELECTED EXPLANATORY ANSWERS

1. Speed of boat in still water $= \frac{9+3}{2} = 6$ km/hr.

2. Speed of bus $= 60 \times \frac{1}{1.5} = 40$ km/hr.

Hence, required difference $= \frac{720}{40} - \frac{720}{60}$

$= 18 - 12 = 6$ hrs.

3. Speed of the car $= \frac{540}{9} = 60$ km/hr.

$\therefore$ Speed of train $= 60 \times 2 = 120$ km/hr.

and speed of bike $= 120 \times \frac{2}{3} = 80$ km/hr.

$\therefore$ Reqd. distance $= 80 \times 5 = 400$ km.

4. In 12 minutes, the train with 30 km per hour speed goes

$= 30 \times \frac{5}{18} \times 60 \times 12$ m

$= \frac{5 \times 5 \times 20 \times 12}{1000}$ km $= 6$ km

In 8 minutes, the train with 45 km per hr. speed goes

$= 45 \times \frac{5}{18} \times 60 \times \frac{8}{1000}$ km $= 6$ km

Average speed $= \frac{12 \text{ km}}{20 \text{ min}} = \frac{12}{20} \times 60$ km/hr

$= 36$ km/hr.

5. $x : \frac{400}{5} = 7 : 8$

$\Rightarrow x \times \frac{5}{400} = \frac{7}{8} \quad \therefore \quad x = 70$ km/hr.

6. Reqd. average speed $= \frac{2 \times 40 \times 60}{(40+60)}$

$= \frac{2 \times 2400}{100} = 48$ km/hr.

8. Let A and B meet after time t hours.

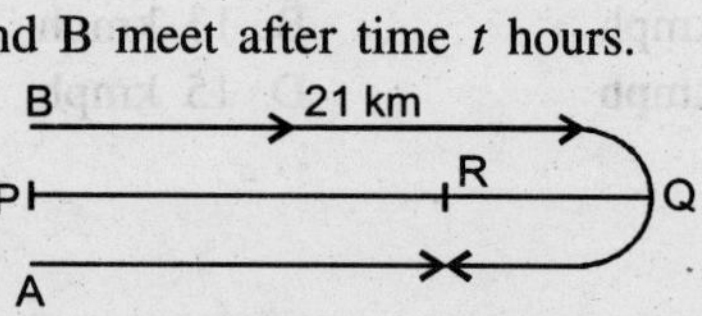

Distance covered by A in t hours $= 3t$ km.
Distance covered by B in t hours $= 4t$ km.
Total distance covered by A and B $= (3t + 4t)$ km

$= 7t$ km.

From the diagram we can see that the total distance covered by A and B is equal to twice the distance between P and Q.

$\therefore \quad 7t = 2 \times 21$

$t = \frac{2 \times 21}{7} = 6$ hours

Distance (PR) $= 6 \times 3 = 18$ kms.

9. From question,
In 2 minutes the net ascending of the monkey is $12 - 5 = 7$ metres. So, to cover 63 metres to above process is repeated $\frac{63}{7} = 9$ times. Since on 9th time, it will climb to the top.

Hence, Time taken to cover 56 metres $= \frac{56 \times 2}{7}$

$= 16$ minutes

Remaining distance $= 63 - 56 = 7$ metres

Time taken to ascend 7 metres $= \frac{7}{12}$ minutes

$\therefore$ Total time taken $= 16 + \frac{7}{12} = 16\frac{7}{12}$ minutes.

10. The distance covered by man in 4 minutes

$= \frac{6 \times 1000 \times 4}{60} = 400$ metres

The distance coverd by carriage in 4 minutes $= 200 + 400 = 600$ metres

$\therefore$ Sepeed of carriage $= \frac{600}{4} \times \frac{60}{1000}$ km/hr.

$= 9$ km/hr.

12. Both trains meet after 6 hours.

$\therefore$ The relative speed of two trains $= \frac{162}{6} = 27$ km/hr.

The speed of the slower train starting from B $= \frac{27-8}{2}$

$= \frac{19}{2} = 9\frac{1}{2}$ km/hr.

$\therefore$ The speed of the faster train $= 9\frac{1}{2} + 8 = 17\frac{1}{2}$ km/hr.

13. Let the length of the train $= x$ metres
Then, in 18 sec. the train travels $(x + 162)$ metres ...(*i*)

and in 15 sec. the train travels $(x + 120)$ metres

$\therefore$ In $(18 - 15) = 3$ sec. the train travels $(x + 162) - (x + 120) = 42$ m

$\therefore$ In 1 sec the train travels $= \frac{42}{3} = 14$ metes ...(*ii*)

$\therefore$ In 18 sec. the train travels $= 14 \times 18$

$= 252$ metres ...(*iii*)

From equations (*i*) and (*iii*)

$\therefore \quad x + 162 = 252$

$\Rightarrow \quad x = 252 - 162 = 90$

$\therefore$ Length of the train $= 90$ metres.

14. Total distance travelled = 3990 km

Ratio of time spent = 1 : 16 : 2

$\therefore$ Ratio of time × speeds = 20 × 1 : 16 × 1 : 2 × 3

= 20 : 16 : 6

Sum of the ratios = 20 + 16 + 6

= 42

$$\therefore \text{ Distance covered by sea} = \frac{3990}{42} \times 16$$

= 1520 kms.

15. Total time = walking time + rest time

$$\text{Walking time} = \frac{5}{10} \text{ hour} = \frac{1}{2} \text{ hour} = 30 \text{ min}$$

Rest time = 4 × 5 min = 20 min

$\therefore$ Required time taken = 30 min + 20 min

= 50 min.

17. Let the speed of C be x kmph

$\therefore$ B's speed = $6x$ kmph

and A's speed = 4 × $6x$ kmph = $24x$ kmph

The distance covered by B in 28 minutes

$$= 6x \times \frac{28}{60} \text{ km}$$

$$= \frac{28x}{10} \text{ km} = \frac{14x}{5} \text{ km}$$

The time taken by C to cover the same distance

$$= \frac{\frac{14x}{5} \text{ km}}{x \text{ kmph}} = \frac{14}{5} \text{ hrs}$$

$$= 2\frac{4}{5} \text{ hours} = 2 \text{ hrs } 48 \text{ minutes}$$

19. Let s be the normal speed and t be the normal time taken by the bus

$\therefore$ $s \times t = 300$ ...(1)

new speed = $s + 5$

new time = $t - 2$

$\therefore$ $(s + 5)(t - 2) = 300$

$$(s+5)\left(\frac{300}{s} - 2\right) = 300 \text{ (from (1))}$$

$(s + 5)(300 - 2s) = 300s$

$300s - 2s^2 + 1500 - 10s = 300s$

$\Rightarrow$ $2s^2 + 10s - 1500 = 0$

$\Rightarrow$ $2s^2 + 60s - 50s - 1500 = 0$

$\Rightarrow$ $2s(s + 30) - 50(s + 30) = 0$

$\Rightarrow$ $(s + 30)(2s - 50) = 0$

$\therefore$ $s = 25$

$\therefore$ The normal speed of the bus is 25 kmph.

20. Let the required distance be x km. Then,

$$\frac{x}{3} + \frac{x}{2} = 5$$

$\Rightarrow$ $2x + 3x = 30$

$\Rightarrow$ $x = 6.$

21. Let the total distance be x km. Then,

$$\frac{x}{2} \times \frac{1}{60} + \frac{x}{2} \times \frac{1}{50} = \frac{44}{3}$$

$$\text{or} \quad \frac{x}{120} + \frac{x}{100} = \frac{44}{3}$$

or $5x + 6x = 8800$

or $x = 800.$

$\therefore$ Required distance = 800 km.

22. Let the distance be x km. Then,

$$\frac{x}{3} - \frac{x}{3.75} = \frac{1}{2}$$

$$\text{or} \quad \frac{3.75x - 3x}{3 \times 3.75} = \frac{1}{2}$$

or $1.5x = 3 \times 3.75$

$$\text{or} \quad x = \frac{3 \times 3.75}{1.5} = 7.5 \text{ km}$$

23. Distance covered by thief in (1/2) hour = 20 km.

Now, 20 km is compensated by the owner at a relative speed of 10 km/hr in 2 hours.

So, he overtakes the thief at 4 p.m.

24. Suppose the man covers first distance in x hrs and second distance in y hrs. Then,

$4x + 5y = 35$

and $5x + 4y = 37.$

Solving these equations, we get $x = 5$ and $y = 3$.

$\therefore$ Total time taken = (5 + 3) hrs = 8 hrs.

25. Let x km be covered in y hrs. Then,

$$\text{1st speed} = \left(\frac{x}{y}\right) \text{km/hr.}$$

$$\text{2nd speed} = \left(\frac{x}{2} \div 2y\right) \text{km/hr} = \left(\frac{x}{4y}\right) \text{km/hr.}$$

$$\therefore \text{ Ratio of speed} = \frac{x}{y} : \frac{x}{4y} = 1 : \frac{1}{4} = 4 : 1.$$

26.
$$\text{Distance covered in } \frac{2}{60} \text{ hr} = \left(1 + \frac{1}{2}\right) \text{km}$$

$$= \frac{3}{2} \text{ km.}$$

$$\text{Distance covered in 1 hr} = \left(\frac{3}{2} \times \frac{60}{2}\right) \text{km} = 45 \text{ km.}$$

Hence, speed of the train = 45 km/hr.

27. Relative speed = (50 – 30) km/hr = 20 km/hr.

$$= \left(20 \times \frac{5}{18}\right) \text{m/sec} = \left(\frac{50}{9}\right) \text{m/sec.}$$

Let the length of faster train be x

Then, $x \times \frac{9}{50} = 18 \Rightarrow x = \frac{18 \times 50}{9} = 100$ m.

28. Let the length of the train be x metres and its speed be y metres/sec.

Then, $\frac{x}{y} = 15 \Rightarrow y = \frac{x}{15}$.

Now, $\frac{x+100}{25} = \frac{x}{15} \Rightarrow x = 150$ m.

30. Speed in downstream = (9 + 3) km/hr = 12 km/hr.
Speed in upstream = (9 – 3) km/hr = 6 km/hr.
Let the distance AB = x km.

Then, $\frac{x}{6} + \frac{x}{12} = 3 \Rightarrow 2x + x = 36$

$\Rightarrow x = 12$.

$\therefore$ Distance AB = 12 km.

31. Speed in upstream = (6 – 1.5) km/hr = 4.5 km/hr.
Speed in downstream = (6 + 1.5) km/hr
= 7.5 km/hr.

Total time taken = $\left(\frac{22.5}{4.5} + \frac{22.5}{7.5}\right)$ hrs. = 8 hrs.

32. Let speed in upstream = x km/hr
Then, speed in downstream = $3x$ km/hr

$\therefore$ Speed in still water = $\frac{1}{2}(x + 3x)$ km/hr
= $2x$ km/hr

Speed of the current = $\frac{1}{2}(3x - x)$ km/hr
= x km/hr

$\therefore 2x = \frac{28}{3}$ or $x = \frac{14}{3} = 4\frac{2}{3}$ km/hr.

33. Distance = $\left(1100 \times \frac{11}{5}\right)$ feet = 2420 feet.

34. Distance covered in 2 hrs 15 min

i.e., $2\frac{1}{4}$ hrs = $\left(80 \times \frac{9}{4}\right)$ km = 180 km.

Time taken to cover remaining distance

$= \left(\frac{350-180}{60}\right)$ hrs $= \frac{17}{6}$ hrs

$= 2\frac{5}{6}$ hrs = 2 hrs 50 min.

Total time taken = (2 hrs 15 min + 2 hrs 50 min)
= 5 hrs 5 min.

So, Asha reached city A at 10.25 a.m.

35. $x \times \frac{15}{60} + 2x \times \frac{20}{60} + x \times \frac{10}{60} = 39$

$\Rightarrow \frac{x}{4} + \frac{2x}{3} + \frac{x}{6} = 39$

$\Rightarrow 3x + 8x + 2x = 468$

$\Rightarrow x = 36$.

36. Difference between timings = 15 min = $\frac{1}{4}$ hr.

Let the length of journey be x km.

Then, $\frac{x}{35} - \frac{x}{40} = \frac{1}{4}$

$\Rightarrow 8x - 7x = 70$

$\Rightarrow x = 70$ km.

37. Let the distance be x km. Then,

$\frac{x}{7\frac{1}{2}} - \frac{x}{8} = 4 \Rightarrow \frac{2x}{15} - \frac{x}{8} = 4$

$\Rightarrow x = 480$ km.

38. Relative speed = (120 + 80) km/hr

$= \left(200 \times \frac{5}{18}\right)$ m/sec $= \left(\frac{500}{9}\right)$ m/sec.

Let the length of the other train be x metres.

Then, $\frac{x+270}{9} = \frac{500}{9}$

$\Rightarrow x + 270 = 500$

$\Rightarrow x = 230$.

40. Let the speed of the boat in still water be x kmph. Then,
Speed in downstream = $(x + 3)$ kmph,
Speed in upstream = $(x - 3)$ kmph.

$\therefore (x + 3) \times 1 = (x - 3) \times \frac{3}{2}$

$\Rightarrow 2x + 6 = 3x - 9$

$\Rightarrow x = 15$ kmph.

11 Mensuration

Triangle

1. Perimeter = sum of three sides
2. Area = $\frac{1}{2} \times \text{base} \times \text{height}$, or

   Area = $\sqrt{s(s-a)(s-b)(s-c)}$

   where *a, b, c*, are the lengths of the sides of triangle and $s = \frac{a+b+c}{2}$

Right Angled Triangle : It is one whose one of the angles is right angle, *i.e.*, 90°.

1. $(\text{Hypotenuse})^2 = (\text{Perpendicular})^2 + (\text{Base})^2$
2. Area = $\frac{1}{2} \times \text{Base} \times \text{Perpendicular}$

Equilateral Triangle : All three sides are equal in length and all three angles are equal to 60°.

1. Area = $\frac{\sqrt{3}}{4} \times (\text{Side})^2$
2. Area = $\frac{(\text{Height})^2}{\sqrt{3}}$
3. Height = $\frac{\sqrt{3}}{2} \times \text{side}$
4. Perimeter = 3 × side

Isosceles Triangle : Two sides are equal in lengths.

1. Area = $\frac{b}{4}\sqrt{4a^2 - b^2}$

   where a = lengths of equal sides

   b = length of unequal side
2. In an isosceles right triangle,

   (*a*) Hypotenuse = $\sqrt{2} \times \text{congruent side } (a)$

   (*b*) Area = $\frac{1}{2} \times a^2$

   (*c*) Perimeter = $\sqrt{2} \times a\left(\sqrt{2}+1\right)$

Rectangle :

1. Area = length(l) × breadth(b)
2. Perimeter = $2(l + b)$
3. Diagonal = $\sqrt{l^2 + b^2}$

Square :

1. Area = $(\text{Side})^2$
2. Perimeter = 4 × side
3. Diagonal = $\text{side} \times \sqrt{2}$

Parallelogram: Area = Base × Height.

Trapezium : Area = $\frac{1}{2} \times \text{Height} \times$ (Sum of parallel sides).

Here, height is the distance between the two parallel sides.

Rhombus :

1. Area = $\frac{1}{2} \times \text{Product of diagonals}$
2. Side = $\sqrt{\left(\frac{d_1}{2}\right)^2 + \left(\frac{d_2}{2}\right)^2}$, where d_1 and d_2 are diagonals
3. Perimeter = 4 × side

Quadrilateral : Area = $\frac{1}{2} \times$ One diagonal × (Sum of perpendicular to it from the opposite vertices)

$$= \frac{1}{2} \times d \times (a+b)$$

Circle :

1. Diameter = 2 × Radius
2. Area = $\pi r^2 = \frac{\pi}{4} d^2$; where d = diameter = $\sqrt{\frac{4A}{\pi}}$
3. Circumference = $2\pi r = \pi d$
4. Radius = $\frac{\text{Circumference}}{2\pi} = \frac{\sqrt{\text{Area}}}{\pi}$
5. Length of an Arc = $\frac{\theta}{360°} \times 2\pi r$
6. Area of sector = $\frac{\theta}{360°} \times \pi r^2 = \frac{1}{2} \times \text{Arc} \times r$

Polygon :

1. Interior angle + Exterior angle = 180°
2. Each interior angle = $\left(\frac{2n-4}{n}\right) \times 90°$

   where n = number of sides

3. Sum of Exterior angles = 360°
4. Perimeter = Number of sides × Length of side.
5. For an equilateral triangle of side 'a'

(a) radius of inscribed circle $= \frac{a}{2\sqrt{3}}$

and side of the triangle $= 2\sqrt{3}r$,

(b) radius of circumcircle $= \frac{a}{\sqrt{3}}$

6. Area of regular polygon $= \frac{1}{2}$(No. of sides) (Radius of the inscribed circle)
7. Area of regular hexagon $= \frac{3\sqrt{3}}{2}(\text{side})^2$

$= 2.598\ (\text{side})^2$

8. Area of a regular octagon

$= 2(\sqrt{2}+1)(\text{side})^2 = 4.828\ (\text{side})^2$

9. Area of quadrilateral, A

$= \sqrt{s(s-a)(s-b)(s-c)(s-d)}$

where, $s = \frac{a+b+c+d}{2}$

VOLUME AND SURFACE AREA OF SOLIDS

Cuboid : A cuboid has six faces, each one a ractangle. It has 12 edges. For example, a rectangular brick.

Let Length = l, Breadth = b and Height = h, then,

1. Volume = (Length × Breadth × Height)
2. Whole Surface Area = $2(lb + bh + lh)$
3. Diagonal $= \sqrt{l^2+b^2+h^2}$
4. Area of 4 walls of a room $= 2 \times h\ (l + b)$

Cube : In a cube, Length = Breadth = Height

1. Volume $= (l)^3$
2. Length $= \sqrt[3]{\text{Volume}}$
3. Whole Surface Area $= 6\ l^2$
4. Diagonal $= l \times \sqrt{3}$
5. Lateral Surface Area $= 4\ l^2$

Cylinder :

1. Volume $= \pi r^2 h$
2. Curved Surface Area $= 2\pi rh$
3. Total Surface Area $= 2\pi r(r + h)$

where r = radius, h = height

Spherical Cell :

1. Volume $= \frac{4}{3}\pi(R^3 - r^3)$
2. Total Surface Area= $4\pi(R^2 - r^2)$

where R = Outer radius

r = Inner radius

Sphere :

1. Volume $= \frac{4}{3}\pi r^3$
2. Surface Area $= 4\pi r^2$

Semi-sphere :

1. Volume $= \frac{2}{3}\pi r^3$
2. Curved surface area $= 2\pi r^2$
3. Total surface area $= 3\pi r^2$

Cone :

1. Slant height $(l) = \sqrt{r^2+h^2}$
2. Volume $= \frac{1}{3}\pi r^2 h$
3. Curved surface area $= \pi rl$
4. Total surface area $= \pi r\ (l + r)$
5. If the depth of the frustum of a cone be k and the radii of its ends are r_1 and r_2, then

(i) Slant height of the frustum of a cone

$= \sqrt{k^2+(r_1-r_2)^2}$

(ii) Curved surface of the frustum

$= \pi(r_1 + r_2)\ l.$

(iii) Volume $= \frac{\pi k}{3}\left(r_1^2 + r_1 r_2 + r_2^2\right)$

SOLVED EXAMPLES

Example 1: What is the longest rod that can be placed in a room which is 7 m long, 9 m broad and $\sqrt{39}$ m high?

Solution: Required length of rod

$= \sqrt{7^2+9^2+(\sqrt{39})^2} = \sqrt{49+81+39}$

$= \sqrt{169} = 13$ m.

Example 2: The total area of a circle and a square is equal to 5450 sq. cm. The diameter of the circle is 70 cms. What is the sum of the circumference of the circle and the perimeter of the square?

Solution: Let side of the square = x cm

Also $\pi \times 35 \times 35 + x^2 = 5450$

$\Rightarrow \frac{22}{7} \times 35 \times 35 + x^2 = 5450$

$\Rightarrow x^2 = 5450 - 3850 = 1600$

$\therefore x = 40$ cm

$\therefore$ Required sum $= \pi \times d + 4x$

$= \left(\frac{22}{7} \times 70 + 4 \times 40\right) = 380$ cm.

Example 3: Perimeter of a circle is equal to the prmimeter of a square whose area is 484 cm². What is the radius of the circle?

Solution: Side of a square $= \sqrt{\text{Area}} = \sqrt{484} = 22$ cm.

Perimeter of square = 4 × side = 4 × 22 = 88 cm.

$\therefore \quad 2\pi r = 88$

$\Rightarrow \quad 2\times\frac{22}{7}\times r = 88$

$\Rightarrow \quad r = \frac{88\times7}{2\times22} = 14$ cm.

Example 4: The perimeter of a square is equal to twice the perimeter of a rectangle of length 8 cm and breadth 7 cm. What is the circumference of a semicircle whose diameter is equal to the side of the square? (Rounded off to the two decimal place)

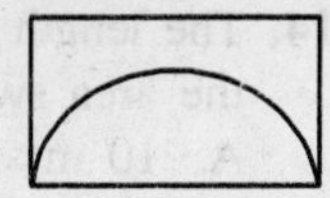

Solution: $\because$ Perimeter of the square

= 2 × Perimeter of rectangle

= 2(8 + 7) × 2 = 60 cm

One side of the square = 15 cm

Radius of the semicircle (R) = $\frac{15}{2}$ cm

$\therefore$ Perimeter of the semicircle = πR + 2R

$= \frac{22}{7}\times\frac{15}{2}+15 = 23.57 + 15$

= 38.57 cm

MULTIPLE CHOICE QUESTIONS

1. What is the area of a circle whose radius is equal to the side of a square whose perimeter is 112 metres?

A. 176 sq m B. 2504 sq m
C. 284 sq m D. None of these

2. The sum of the circumference of a circle and the perimeter of a rectangle is 132 cm. The area of the rectangle is 112 sq cm and breadth of the rectangle is 8 cms. What is the area of the circle?

A. 616 sq cm B. 540 sq cm
C. 396 sq cm D. Cannot be determined

3. The total area of a circle and a rectangle is equal to 1166 sq cm. The diameter of the circle is 28 cm. What is the sum of the circumference of the circle and the perimeter of the rectangle if the length of the rectangle is 25 cm?

A. 186 cm B. 182 cm
C. 184 cm D. Cannot be determined

4. What would be the cost of laying a carpet on a floor which has its length and breadth in the respective ratio of 32 : 21 and where its perimeter is 212 feet, if the cost per square foot of laying the carpet is ₹ 2.5?

A. ₹ 6,720 B. ₹ 5,420
C. ₹ 7,390 D. Cannot be determined

5. A triangle's perimeter is 25 cms. Which of the following may be true or is a possibility?

(*a*) The sides are 7 cms., 7 cms. and 11 cms.
(*b*) It is an equilateral triangle.
(*c*) The value of sides can be in integer only.

A. Only (*a*) B. Only (*a*) and (*b*)
C. Only (*c*) D. Only (*b*) and (*c*)

6. What will be the cost of building a fence around a circular field with area equal to 18,634 sq. metres; if the cost of building the fence per metre is ₹ 365?

A. ₹ 1,76,660 B. ₹ 68,01,410
C. ₹ 2,43,250 D. ₹ 56,60,220

7. The area of a square is 196 sq cms whose side is half the radius of a circle. The circumference of the circle is equal to breadth of a rectangle. If perimeter of the rectangle is 712 cm. What is the length of the rectangle?

A. 196 cm B. 186 cm
C. 180 cm D. 190 cm

8. The circumference of a circular plot is 484 metres. Find out the area of that circular plot—

A. 15246 m^2 B. 18634 m^2
C. 20328 m^2 D. 13552 m^2

9. A room measures 22 dm by 16 dm. I wish to buy a carpet for the floor leaving an uncarpeted margin 2 dm wide along each of the shorter sides of the room and a margin 0.5 dm wide along each of the longer sides of the room. If the price of the carpet is ₹ 500 per sq. metre, what is the cost of the whole carpet required?

A. ₹ 500 B. ₹ 2700
C. ₹ 2000 D. ₹ 1350

10. A paper is in the form of a rectangle ABCD where AB = 22 cm and BC = 14 cm. A semi-circular portion with segment BC as a diameter is cut off. Find the area of remaining paper.

A. 231 cm^2 B. 213 cm^2
C. 321 cm^2 D. 200 cm^2

11. How many plants can be put in a circular flower bed whose circumference is 880 dm allowing 35 dm^2 for each plant?

A. 880 B. 1760
C. 1000 D. 1500

12. An athletic track 14 m wide consists of two straight sections 120 m long joining semi-circular ends whose inner radius is 35 m. Calculate the area of the track :

A. 5670 m^2 B. 7065 m^2
C. 5607 m^2 D. 7056 m^2

13. A square park has each side of 100 m. At each corner of the park, there is a flower bed in the form of a quadrant of radius 14 m. Then the area of the remaining part of the park is :

A. 9384 m^2 B. 9834 m^2
C. 9000 m^2 D. 8900 m^2

14. The length of minute hand of a clock is 14 cm. Then the area swept by the minute hand in one minute is:

A. 10 m^2 B. 12.26 m^2
C. 20.26 m^2 D. 10.26 m^2

15. Find the area of ring between two concentric circles whose circumference are 77 cm and 55 cm.

A. 770 cm^2 B. 321 cm^2
C. 231 cm^2 D. 230 cm^2

16. A rectangle water reservoir is 10.8 metres long and 3.75 metres wide at base. Water flows into it at the rate of 18 m per sec. through the pipe having the cross section 7.5 cm × 4.5 cm. Then the height to which the water will rise in the reservoir in 30 minutes is :

A. 7.2 m B. 2.7 m
C. 3.7 m D. 7.3 m

17. A rectangular sheet of 44 cm × 18 cm is rolled along its length and a cylinder is formed. Then the volume of cylinder is :

A. 7227 cm^3 B. 7272 cm^3
C. 2727 cm^3 D. 2772 cm^3

18. How many metres of cloth 5 metre wide will be required to make a conical tent, the radius of whose base is 7 metre and height is 24 metre?

A. 100 m B. 110 m
C. 550 m D. 55 m

19. The surface area of a sphere whose volume is 4851 cubic metres is :

A. 1386 m^2 B. 1380 m^2
C. 1286 m^2 D. 3186 m^2

20. A hollow sphere of external and internal diameter 4 cm and 2 cm respectively, is melted into a cone of base diameter 8 cm. Then the height of the cone is :

A. 12 cm B. 14 cm
C. 20 cm D. 24 cm

21. The diamensions of a metallic rod are 19 cm × 4 cm × 2 cm and each side of a metallic cube is 4 cm. Both are melted and recast into a new cube. Find the length of edge of cube so formed.

A. 4 cm B. 5 cm
C. 6 cm D. 7 cm

22. An agricultural field is in the form of a rectangle of length 35 metres and width 15.4 metres. A pit 5.5 metres long, 4 metres wide and 2.5 metres deep is dug in the corner of the field and the earth taken out of the pit is spread uniformly over the remaining area of the field. The extent to which the level of the field has been raised is :

A. 16.6 cm B. 10.6 cm
C. 16.1 cm D. 6.10 cm

23. The side of a square exceeds the side of the another square by 4 cm and the sum of areas of two squares is 400 sq. cm. Find the dimensions of the square.

A. 8 cm, 12 cm B. 10 cm, 14 cm
C. 12 cm, 16 cm D. 14 cm, 18 cm

24. The cost of levelling a rectangular field at the rate of 85 paise per square metre is ₹ 624.75. Then the perimeter of the field if its sides are in the ratio of 5 : 3 is :

A. 35 m B. 21 m
C. 112 m D. 49 m

25. The diagonal of a rectangular field is 15 m and its area is 108 sq. m. What will be the total expenditure in fencing the field at the rate of ₹ 5 per metre?

A. ₹ 441 B. ₹ 420
C. ₹ 210 D. ₹ 120

26. The minute hand of a clock is 10 cm long. The area of the face of the clock described by the minute hand between 9 AM and 9.35 AM is :

A. 140 cm^2 B. 183.3 cm^2
C. 180 cm^2 D. 175.3 cm^2

27. The length of a rectangular plot is 60% more than its breadth. If the difference between the length and the breadth of that rectangle is 24 cm, what is the area of that rectangle?

A. 2400 sq. cm B. 2480 sq. cm
C. 2560 sq. cm D. Data inadequate

28. The area of a rectangle is 460 square metres. If the length is 15% more than the breadth, what is the breadth of the rectangular field?

A. 15 metres B. 26 metres
C. 34.5 metres D. None of these

29. The ratio between the length and the breadth of a rectangular field is 3 : 2. If only the length is increased by 5 metres, the new area of the field will be 2600 sq. metres. What is the breadth of the rectangular field?

A. 40 metres B. 60 metres
C. 65 metres D. Cannot be determined

30. If the length and breadth of a rectangular plot be increased by 50% and 20% respectively, then how many times will its area be increased?

A. $1\frac{1}{3}$ B. 2
C. $3\frac{2}{5}$ D. None of these

31. The length of a rectangle is decreased by *r*%, and the breadth is increased by (*r* + 5)%. Find *r*, if the area of the rectangle is unaltered.

A. 5 B. 8
C. 10 D. 20

32. The length and breadth of the floor of the room are 20 feet and 10 feet respectively. Square tiles of 2 feet length of different colours are to be laid on the floor. Black tiles are laid in the first row on all sides. If white

tiles are laid in the one-third of the remaining and blue tiles in the rest, how many blue tiles will be there?
A. 16 B. 24
C. 32 D. 48

33. A park square in shape has a 3 metre wide road inside it running along its sides. The area occupied by the road is 1764 square metres. What is the perimeter along the outer edge of the road?
A. 576 metres B. 600 metres
C. 640 metres D. Data inadequate

34. What will be the length of the diagonal of that square plot whose area is equal to the area of a rectangular plot of length 45 metres and breadth 40 metres?
A. 42.5 metres B. 60 metres
C. 75 metres D. Data inadequate

35. The length of one pair of opposite sides of a square is increased by 5 cm on each side: the ratio of the length and the breadth of the newly formed rectangle becomes 3 : 2. What is the area of the original square?
A. 25 sq. cm B. 81 sq. cm
C. 100 sq. cm D. 225 sq. cm

36. What will be the ratio between the area of a rectangle and the area of a triangle with one of the sides of the rectangle as base and a vertex on the opposite side of the rectangle?
A. 1 : 2
B. 2 : 1
C. 3 : 1
D. Data inadequate

37. A cow is tethered in the middle of a field with a 14 feet long rope. If the cow grazes 100 sq. ft. per day, then approximately what time will be taken by the cow to graze the whole field?
A. 2 days B. 6 days
C. 18 days D. 24 days

38. A circular ground whose diameter is 35 metres, has a 1.4 m broad garden around it. What is the area of the garden in square metres?
A. 160.16 B. 176.16
C. 196.16 D. None of these

39. The cost of the paint is ₹ 36.50 per kg. If 1 kg of paint covers 16 square feet, how much will it cost to paint outside of a cube having 8 feet each side?
A. ₹ 692 B. ₹ 768
C. ₹ 876 D. ₹ 972

40. The capacity of a cylindrical tank is 246.4 litres. If the height is 4 metres, what is the diameter of the base?
A. 1.4 m B. 2.8 m
C. 14 m D. 28 m

ANSWERS

1	2	3	4	5	6	7	8	9	10
D	A	B	A	A	A	C	B	D	A
11	**12**	**13**	**14**	**15**	**16**	**17**	**18**	**19**	**20**
B	D	A	D	C	B	D	B	A	B
21	**22**	**23**	**24**	**25**	**26**	**27**	**28**	**29**	**30**
C	B	C	C	C	B	C	D	A	D
31	**32**	**33**	**34**	**35**	**36**	**37**	**38**	**39**	**40**
D	A	B	B	C	B	B	D	C	D

SOME SELECTED EXPLANATORY ANSWERS

1. ∵ Radius of the circle $= \frac{112}{4} = 28$ m

∴ Area of the circle $= \frac{22}{7}\times 28\times 28 = 2464\ m^2$.

2. Length of the rectangle $= \frac{112}{8} = 14$ cm.

∴ Perimeter of the rectangle = 2(14 + 8) = 44 cm.

∴ Circumference of the circle = 132 – 44 = 88 cm.

∴ $r = \frac{88\times 7}{2\times 22} = 14$ cm

∴ Area of the circle

$= \frac{22}{7}\times 14\times 14 = 616$ sq cm.

3. From question—

∵ Diameter of the circle = 28 cm.

⇒ Area of the circle $= \frac{22}{7}\times(14)^2 = 616\ cm^2$

∴ Area of the rectangle = 1166 – 616 $= 550\ cm^2$

⇒ Breadth of the rectangle $= \frac{550}{25} = 22$ cm.

$\therefore$ Required sum = $2\times\frac{22}{7}\times(14)+2(25+22)$

= 88 + 94 = 182 cm.

4. Length of the floor = $\frac{212\times32}{2\times(32+21)}$ = 64 ft.

and breadth the floor = $\frac{64\times21}{32}$ = 42 ft.

$\therefore$ Area of the floor = 64 × 42 = 2688 sq. ft.

$\therefore$ Reqd. cost = ₹ 2688 × 2.5 = ₹ 6720.

5. Only statement A may be true or possibility.

6. Area of the circular field = 18634 m^2

$\therefore$ $\pi r^2 = 18634$

$\Rightarrow$ $r^2 = 18634\times\frac{7}{22} = 5929$

$\Rightarrow$ $r = 77$ m

$\therefore$ Length of the fence = $2\times\frac{22}{7}\times77$

= 484 m

$\therefore$ Total cost = 484 × 365

= ₹ 176660.

7. One side of the square = $\sqrt{196}$ = 14 cm

$\therefore$ Radius of the circle = 28 cm

$\therefore$ Circumference of the circle = $2\times\frac{22}{7}\times28$

= 176 cm

$\therefore$ Breadth of the rectangle = 176 cm

$\therefore$ Length of the rectangle = $\frac{712}{2}-176$

= 356 – 176

= 180 cm.

8. Let r be the radius of the circular plot then from problem,

$\because$ The circumference of the circular plot = 484 m

$\Rightarrow$ $2\pi r = 484$

$r = 484\times\frac{7}{22}\times\frac{1}{2}$ = 77 metres

$\therefore$ Area of the circular plot = $\pi r^2 = \frac{22}{7}\times(77)^2$

= 18634 m^2.

10. Area of the whole paper ABCD = 22 × 14

= 308 cm^2

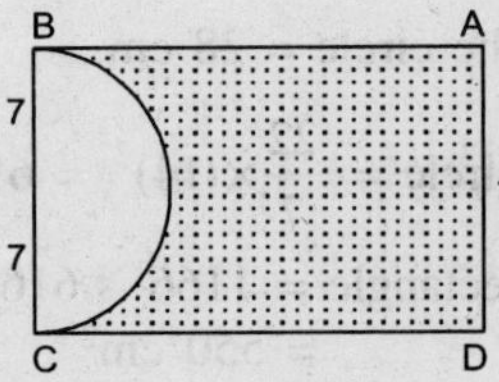

Radius of the semi-circle = $\frac{1}{2}\times14$ cm = 7 cm

$\therefore$ Area of the semi-circle = $\frac{1}{2}\pi r^2$

$= \frac{1}{2}\times\frac{22}{7}\times7\times7$

= 77 cm^2

$\therefore$ Area of the remaining part of the paper

= 308 cm^2 – 77 cm^2 = 231 cm^2.

11. Circumference of flower bed = 880 dm

$\Rightarrow$ $2\pi r = 880$ dm

$\Rightarrow$ $r = \frac{880}{2\pi}$ dm $= \frac{880\times7}{2\times22}$ = 140 dm

Area of flower bed = $\pi r^2 = \frac{22}{7}\times(140)^2$ dm^2

$\therefore$ Required number of plants

$= \frac{\frac{22}{7}\times140\times140}{35}$ dm^2

$= \frac{22}{7}\times\frac{140\times140}{35}$ = 1760.

13. Area of each quadrant = $\frac{\theta}{360}\times\pi r^2$

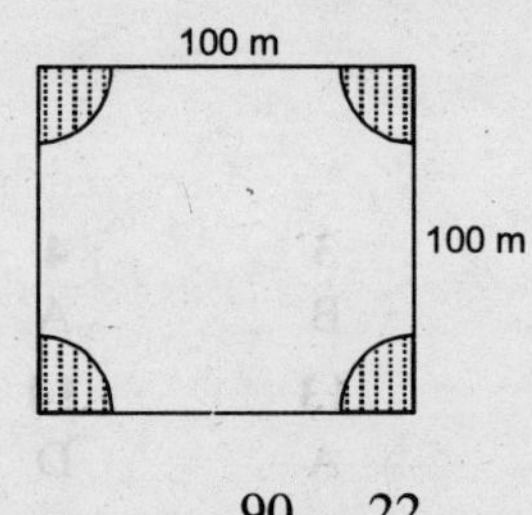

$= \frac{90}{360}\times\frac{22}{7}\times14\times14$ m^2 = 154 m^2

$\therefore$ Area of 4 quadrants = 154 × 4 m^2 = 616 m^2

Area of park = (100)^2 = 10,000 m^2

Hence area of the remaining part of the park

= 10,000 – 616 = 9,384 m^2.

14. Length of minute hand of clock = r = 14 cm

Central angle covered in 1 minute = $\frac{360}{60} = 6°$

$\therefore$ Area swept in one minute = Area of sector with central angle 6°

$= \frac{\theta}{360}\times\pi r^2 = \frac{6}{360}\times\frac{22}{7}\times14\times14$

$= \frac{154}{15}$ = 10.26 cm^2.

15. Let R be the radius of outer circle and r the radius of inner circle.

Circumference of outer circle = 77 cm

$\therefore \quad 2\pi R = 77$

$\Rightarrow R = \frac{77}{2\pi} = \frac{77\times7}{2\times22} = \frac{49}{4}$ cm

Circumference of inner circle = 55 cm

$\therefore 2\pi r = 55 \quad \Rightarrow \quad r = \frac{55}{2\pi} = \frac{55\times7}{2\times22} = \frac{35}{4}$ cm

$\therefore$ Area of ring $= \pi(R^2 - r^2)$

$= \frac{22}{7}\times\left(\frac{49}{4}+\frac{35}{4}\right)\left(\frac{49}{4}-\frac{35}{4}\right) = 231 \text{ cm}^2$

16. Volume of water that flows in one second

$= (18 \times 0.075 \times 0.045) \text{ m}^3$

Volume of water that flows in 30 minutes

$= 18 \times 0.075 \times 0.045 \times 30 \times 60$

$= 109.35 \text{ m}^3.$

Area of the base of the reservoir $= 10.8 \times 3.75 \text{ m}^2$

$\therefore$ Height of water $= \frac{109.35}{10.8\times3.75} = 2.7$ metres.

17. Let r cm be the radius of the base and h cm be the height. Then, $h = 18$ cm.

Now, circumference of the base = Length of the Sheet

$\Rightarrow$ Circumference = 44 cm

$\Rightarrow \quad 2\pi r = 44$

$\Rightarrow \quad 2\times\frac{22}{7}\times r = 44$

$\Rightarrow \quad r = 7$ cm

Volume of the cylinder $= \pi r^2 h \text{ cm}^3$

$= \frac{22}{7}\times(7)^2\times18 \text{ cm}^3$

$= 2772 \text{ cm}^3$

18. Slant height, $l = \sqrt{r^2+h^2}$

$= \sqrt{7^2+24^2} = 25$ metres

Curved Surface area of cone $= \pi rl$

$= \frac{22}{7}\times7\times25 = 550 \text{ m}^2$

$\therefore$ Area of the cloth required $= 550 \text{ m}^2$

$\therefore$ Length of the cloth $= \frac{550}{5} = 110$ metres.

21. Volume of the metallic rod $= 19 \times 4 \times 2$

$= 152$ cu.cm.

Volume of the cube $= 4^3 = 64$ cu. cm.

Total volume of both after melting them

$= 152 + 64 = 216$ cu.cm.

$\therefore$ Volume of new cube = 216 cu. cm.

$\Rightarrow$ Side = 6 cm.

23. Let S_1 and S_2 be the two squares. Let the side of the square S_1 be x cm. Then the side of the square $S_2 = (x + 4)$ cm.

$\therefore$ Area of square $S_1 = x^2$

and, Area of square $S_2 = (x + 4)^2$

$\therefore$ According to the question,

$x^2 + (x + 4)^2 = 400$

Solving the equation, weget

$x = -16$ and 12

But x cannot be negative

$\therefore$ Side of the square $S_1 = 12$ cm

and that of square $S_2 = x + 4 = 12 + 4$

$= 16$ cm.

26. Angle described by minute hand in 60 minutes = 360°

Angle described by minute hand in 35 minutes

$= \frac{360}{60}\times35 = 210°$

$\therefore$ The required area swept by the minute hand = Area of sector $r = 10$ cm

and, $\theta = 210°$

$= \frac{22}{7}\times10\times10\times\frac{210}{360}$

$= 183.3 \text{ cm}^2.$

27. Let breadth = x cm. Then, length $= \left(\frac{160}{100}x\right)$ cm

$= \frac{8}{5}x$ cm.

So, $\frac{8}{5}x - x = 24 \quad \Rightarrow \quad \frac{3}{5}x = 24$

$\Rightarrow \quad x = \left(\frac{24\times5}{3}\right) = 40.$

$\therefore$ Length = 64 cm, Breadth = 40 cm.

Area $= (64 \times 40) \text{ cm}^2 = 2560 \text{ cm}^2.$

28. Let breadth = x metres. Then, length

$= \left(\frac{115x}{100}\right)$ metres.

$\therefore \quad x\times\frac{115x}{100} = 460$

$\Rightarrow \quad x^2 = \left(\frac{460\times100}{115}\right) = 400$

$\Rightarrow \quad x = 20.$

29. Let length = $(3x)$ metres and breadth $= (2x)$ metres.

Then, $(3x + 5) \times 2x = 2600$

$\Rightarrow \quad 6x^2 + 10x - 2600 = 0$

$\Rightarrow \quad 3x^2 + 5x - 1300 = 0$

$\Rightarrow \quad (3x + 65)(x - 20) = 0$

$\Rightarrow \quad x = 20.$

$\therefore$ Breadth $= 2x = 40$ m.

31. Let original length = x and original breadth = y.
Then, original area = xy.

$$\text{New area} = \left[\frac{(100-r)}{100}\times x\right]\left[\frac{(105+r)}{100}\times y\right]$$

$$= \left[\left(\frac{10500-5r-r^2}{10000}\right)xy\right]$$

$$\therefore \left(\frac{10500-5r-r^2}{10000}\right)xy = xy$$

$\Rightarrow r^2 + 5r - 500 = 0$

$\Rightarrow (r + 25)(r - 20) = 0 \Rightarrow r = 20.$

32. Area left after laying black tiles
= [(20 – 4) × (10 – 4)] sq. ft = 96 sq. ft.

Area under white tiles = $\left(\frac{1}{3}\times 96\right)$ sq. ft
= 32 sq. ft

Area under blue tiles = (96 – 32) sq. ft
= 64 sq. ft

Number of blue tiles = $\frac{64}{(2\times 2)}$ = 16.

33. Let the length of the outer edge be x metres. Then, length of the inner edge = $(x - 6)$ m.

$\therefore \quad x^2 - (x - 6)^2 = 1764$

$\Rightarrow x^2 - (x^2 - 12x + 36) = 1764$

$\Rightarrow \quad 12x = 1800$

$\therefore \quad x = 150.$

$\therefore$ Required perimeter = $(4x)$ m
= (4 × 150) m
= 600 m.

34. Area = (45 × 40) m^2

$\Rightarrow \frac{1}{2}\times (\text{diagonal})^2 = 1800$

$\Rightarrow$ diagonal = 60 m.

35. Let original length of each side = x cm.
Then, its area = (x^2) cm^2.
Length of rectangle formed = $(x + 5)$ cm and its breadth = x cm.

$\therefore \frac{x+5}{x} = \frac{3}{2} \Rightarrow 2x + 10 = 3x \Rightarrow x = 10.$

$\therefore$ Original length of each side = 10 cm and its area = 100 cm^2.

36. Area of rectangle = lb sq. units.

Area of the triangle = $\frac{1}{2}lb$ sq. units.

$\therefore$ Required ratio = $lb : \frac{1}{2}lb = 2 : 1.$

37. Area of the field grazed = $\left(\frac{22}{7}\times 14\times 14\right)$ sq. ft
= 616 sq. ft.

Number of days taken to graze the field = $\frac{616}{100}$ days
= 6 days (approx.).

38. Radius of the ground = 17.5 m.
Radius of inner circle = (17.5 – 1.4) m = 16.1 m.
Area of the garden = $\pi \times [(17.5)^2 - 16.1)^2]$ m^2

$$= \left[\frac{22}{7}\times(17.5+16.1)(17.5-16.1)\right] m^2$$

$$= \left(\frac{22}{7}\times 33.6\times 1.4\right) m^2 = 147.84\ m^2.$$

39. Surface area of the cube = (6×8^2) sq. ft
= 384 sq. ft.

Quantity of paint required = $\left(\frac{384}{16}\right)$ kg = 24 kg.

$\therefore$ Cost of painting = ₹ (36.50 × 24) = ₹ 876.

12 Permutation & Combination

IMPORTANT RESULTS

(*i*) Number of permutations of n distinct things taken r at a time, $0 \le r \le n$
$= n(n-1)(n-2)\dots(n-r+1)$
$= n!/(n-r)! = {}^nP_r$.
This is equivalent to filling r places by r objects taking from n distinct objects.

(*ii*) The number of permutations of n distinct objects taken all at a time $= n!$

(*iii*) The number of combinations of n objects taken r at a time, $0 \le r \le n$. $= {}^nC_r$
$= \dfrac{n!}{(n-r)!.r!}$.

(*iv*) The number of permutations of n dissimilar things taken r at a time when each thing can be repeated any number of times $= n^r$.

(*v*) The number of combinations of n distinct objects taken r at a time when any object may be repeated any number of times
$=$ coefficient of x^r in $(1 + x + x^2 + \dots + x^r)^n$
$=$ coefficient of x^r in $(1-x)^n = {}^{n+r-1}C_r$.

(*vi*) Number of combinations of n distinct things taken r at a time when p particular things always occur $= {}^{(n-p)}C_{(r-p)}$.

(*vii*) The number of permutations of n distinct things taken r at a time when p particular things always occur $= {}^{(n-p)}C_{(r-p)}.r!$.

(*viii*) Number of combinations of n distinct things taken r at a time when p particular things never occur $= {}^{(n-p)}C_r$.

(*ix*) Number of permutations (arrangements) of n distinct things taken r at a time when p particular things never occur $= {}^{(n-p)}C_r.r!$.

(*x*) Number of permutations of n things, taken all at a time when p_1 are alike of one kind, p_2 are alike of second kind, ..., p_r of them are alike of the r-th kind $p_1 + p_2 + \dots + p_r \le n$, and remaining things are all different $= n!/\{p_1!p_2! \dots p_r!\}$.

(*xi*) If $3n$ things are to be divided into three equal groups, then the number of ways $= \dfrac{(3n)!}{n!.n!.n!.3!}$.

(*xii*) If $3n$ things are to be divided equally between 3 persons (*i.e.,* division of $3n$ things into 3 equal groups with permutation of groups) then the number of ways $= \dfrac{(3n)!}{(n!)^3}$.

(*xiii*) Greatest value of ${}^nC_r$
${}^nC_r$ is greatest when $r = n/2$ if n is even
$r = (n-1)/2$ or $(n+1)/2$ if n is odd.

(*xiv*) ${}^nC_r = {}^nC_{n-r}$.

(*xv*) ${}^nC_r + {}^nC_{r+1} = {}^{n+1}C_{r+1}$.

(*xvi*) ${}^nC_r = {}^nC_s \Rightarrow r = s$ or $r + s = n$.

(*xvii*) The number of circular permutations of n different things taken all at a time $= (n-1)!$

(*xviii*) The number of arrangements of n persons on a round table $= (n-1)!$.

(*xix*) The number of arrangements of n flowers to make a garland $= \dfrac{1}{2}(n-1)!$.

SOLVED EXAMPLES

Example 1: In how many different ways can the letters of the word 'PRIDE' be arranged ?

Solution: Required number of arrangements
$= {}^5P_5 = 5! = 120$

Example 2: In how many different ways can the letters of the word 'PRAISE' be arranged ?

Solution: Required number of ways
$= {}^6P_6 = 6! = 720$

Example 3: On a shelf there are 4 books on Economics, 3 books on Management and 4 books on Statistics. In how many different ways can the books be arranged so that the books on Economics are kept together ?

Solution: Required number of different ways

$= {}^8P_8 \times {}^4P_4 = \lfloor 8 \times \lfloor 4$

$= 40320 \times 24 = 967680$

Example 4: In how many different ways can the letters of the word 'PRETTY' be arranged?

Solution: Required ways $= \frac{\lfloor 6}{\lfloor 2} = 360$

Example 5: In how many different ways can the letters of the word 'PEANUT' be arranged ?

Solution: Required ways $= {}^6P_6 = \lfloor 6$

$= 6 \times 5 \times 4 \times 3 \times 2 \times 1 = 720.$

MULTIPLE CHOICE QUESTIONS

1. How many seven digit numbers can be made from the number 3428651, using each digit only once ?

A. 49 B. 180
C. 5040 D. 2520

2. How many words can be formed from the letters of the word 'DIRECTOR' so that the vowels are always together?

A. 2106 B. 1206
C. 360 D. 2160

3. In how many ways a committee of 5 members can be selected from 6 men and 5 ladies, consisting of 3 men and 2 ladies?

A. 120 B. 220
C. 200 D. 320

4. In how many ways can the letters of the word 'APPLE' be arranged?

A. 720 B. 120
C. 60 D. 180

5. How many words can be formed by using all the letters of the word. 'ALLAHABAD'?

A. 3780 B. 1890
C. 7560 D. 2520

6. How many words can be formed from the letters of the word 'SIGNATURE' so that the vowels always come together?

A. 720 B. 1440
C. 2880 D. 17280

7. In how many different ways can the letters of the word 'BANKING' be arranged so that the vowels always come together?

A. 120 B. 240
C. 360 D. 720

8. In how many ways can a group of 5 men and 2 women be made out of a total of 7 men and 3 women?

A. 63 B. 90
C. 126 D. 45

9. In a group of 6 boys and 4 girls, four children are to be selected. In how many different ways can they be selected such that at least one boy should be there?

A. 159 B. 194
C. 205 D. 209

10. A box contains 2 white balls, 3 black balls and 4 red balls. In how many ways can 3 balls be drawn from the box, if at least one black ball is to be included in the draw?

A. 32 B. 48
C. 64 D. 96

11. In how many ways can a team of 3 boys and 3 girls be selected from 5 boys and 4 girls?

A. 14 B. 40
C. 18 D. 140

12. The number of ways of selecting 9 balls from 6 red, 5 white and 5 blue if each selection consists of 3 balls of each colour is :

A. 1200 B. 20000
C. 2200 D. 2000

13. In how many ways can we select 4 red balls and 3 blue balls from 8 red balls and 9 blue balls?

A. 8550 B. 5850
C. 5880 D. 8580

14. Find the number of diagonals of a decagon

A. 45 B. 35
C. 55 D. 25

15. In how many ways can a student choose a programme of 5 courses if 9 courses are available and 2 specific courses are compulsary for every student?

A. 53 B. 14
C. 35 D. 28

16. Seven different lecturers are to deliver lectures in seven periods of a class on a particular day. A, B and C are three of the lecturers. The number of ways in which a routine for the day can be made such that A delivers his lecuture before B, and B before C, is

A. 420 B. 120
C. 210 D. 840

17. The number of words that can be made by rearranging the letters of the word APURBA so that vowels and consonants alternate is

A. 18 B. 35
C. 36 D. 24

18. The number of ways in which the letters of the word ARTICLE can be rearranged so that the even places are always occupied by consonants is

A. 576 B. ${}^4C_3 \times (4!)$
C. 2(4!) D. 288

19. The number of ways in which a couple can sit around a table with 6 guests if the couple take consecutive seats is

A. 1440 B. 720
C. 5040 D. 2880

20. The number of ways in which 20 different pearls of two colours can be set alternately on a necklace, there being 10 pearls of each colour, is

A. $9! \times 10!$ B. $5(9!)^2$
C. $(9!)^2$ D. $(18!)^2$

21. There are 4 mangoes, 3 apples, 2 oranges and 1 each of 3 other varieties of fruits. The number of ways of selecting at least one fruit of each kind is

A. 10! B. 9!
C. 4! D. 5!

22. The number of ways in which 6 different balls can be put in two boxes of different sizes so that no box remains empty is

A. 62 B. 64
C. 36 D. 60

23. The number of arrangements of the letters of the word BHARAT taking 3 at a time is

A. 72 B. 120
C. 14 D. 48

24. Find the number of different permutations of the letters of the word BANANA.

A. 16 B. 60
C. 61 D. 36

25. A box contains two white balls, three black balls and four red balls. In how many ways can three balls be drawn from the box if at least one black ball is to be included in the draw?

A. 16 B. 32
C. 64 D. 128

26. All the letters of the word EAMCET are arranged in possible ways. The number of such arrangement in which not two vowels are adjacent to each other is:

A. 360 B. 144
C. 72 D. 54

27. The number of arrangements which can be made by using all the letters of the word LAUGH, if the vowels are adjacent, is:

A. 10 B. 24
C. 48 D. 120

28. There are 10 lamps in a hall. Each one of them can be switched on independently. The number of ways in which the hall can be illuminated is:

A. 10^2 B. 1023
C. 210 D. 10!

29. Everybody in a room shakes hands with everybody else. The total number of hand shakes is 66. The total number of persons in the room is:

A. 11 B. 12
C. 8 D. 14

30. 20 persons are invited for a party. The number of ways in which they and the host can be seated at a circular table, if two particular persons be seated on either side of the host, is equal to:

A. 18! . 2!
B. 18! . 3!
C. 19! . 2!
D. 19! . 3!

31. The number of ways in which 5 prizes be distributed among 4 boys, while each boy is capable of having any number of prizes is:

A. 6^4 B. 4^5
C. $4! \cdot 2^4$ D. 6 . (4!)

32. In an examination there are three multiple choice questions and each question has 4 choices. Number of ways in which a student can fail to get all answers correct, is:

A. 11 B. 12
C. 27 D. 63

33. 7 men and 7 women are to sit round a table so that there is a man on either side of a woman. The number of seating arrangement is:

A. $(7!)^2$ B. $(6!)^2$
C. $6! \times 7!$ D. 7!

34. Three men have 4 coats, 5 waist coats and 6 caps. In how many ways can they wear them?

A. 178200
B. 172000
C. 172800
D. 1720800

35. To fill 12 vacancies there are 25 candidates of which 5 are from scheduled castes. If 3 of the vacancies are reserved for scheduled caste candidates while the rest are open to all, find the number of ways in which the selection can be made.

A. 4974200
B. 4947200
C. 4972400
D. 4927400

ANSWERS

1	2	3	4	5	6	7	8	9	10
C	D	C	C	C	D	D	A	D	C
11	**12**	**13**	**14**	**15**	**16**	**17**	**18**	**19**	**20**
B	D	C	C	C	D	C	A	A	B
21	**22**	**23**	**24**	**25**	**26**	**27**	**28**	**29**	**30**
C	A	A	B	C	C	C	B	B	A
31	**32**	**33**	**34**	**35**					
B	D	C	C	A					

SOME SELECTED EXPLANATORY ANSWERS

1. Required number of ways = $^7P_7 = \underline{|7} = 5040$

2. In the given word, we treat the vowels IEO as one letter.

Thus, we have DRCTR (IEO).

This group has 6 letters of which R occurs 2 times and others are different.

Number of ways arranging these letters $= \frac{6!}{2!} = 360.$

Now 3 vowels can be arranged among themselves in 3! = 6 ways.

∴ Required number of ways = (360 × 6) = 2160.

3. (3 men out of 6) and (2 ladies out of 5) are to be chosen.

∴ Required number of ways

$= (^6C_3 \times {^5C_2})$

$= \left(\frac{6\times5\times4}{3\times2\times1}\times\frac{5\times4}{2\times1}\right) = 200.$

4. The word 'APPLE' contains 5 letters, 1A, 2P, 1L and 1E.

∴ Required number of ways

$= \frac{5!}{(1!)(2!)(1!)(1!)} = 60.$

5. The word 'ALLAHABAD' contains 9 letters, namely 4A, 2L, 1H, 1B and 1D.

∴ Required number of words

$= \frac{9!}{(4!)(2!)(1!)(1!)(1!)} = 7560.$

6. The word 'SIGNATURE' contains 9 different letters. When the vowels IAUE are taken together, they can be supposed to form an entity, treated as one letter.

Then, the letters to be arranged are SGNTR (IAUE).

These 6 letters can be arranged in $^6P_6 = 6! = 720$ ways.

The vowels in the group (IAUE) can be arranged amongst themselves in $^4P_4 = 4! = 24$ ways.

∴ Required number of words = (720 × 24) = 17280.

7. In the word 'BANKING', we treat the two vowels AI as one letter. Thus, we have BNKNG (AI).

This has 6 letters of which N occurs 2 times and the rest are different.

Number of ways of arranging these letters

$= \frac{6!}{(2!)(1!)(1!)(1!)(1!)} = 360.$

Now, 2 vowels AI can be arranged in 2! = 2 ways.

∴ Required number of ways = (360 × 2) = 720.

8. Required number of ways

$= (^7C_5 \times {^3C_2})$

$= (^7C_2 \times {^3C_1}) = \left(\frac{7\times6}{2\times1}\times3\right) = 63.$

9. We may have (1 boy and 3 girls) or (2 boys and 2 girls) or (3 boys and 1 girl) or (4 boys).

∴ Required number of ways

$= (^6C_1 \times {^4C_3}) + (^6C_2 \times {^4C_2}) + (^6C_3 \times {^4C_1}) + (^6C_4)$

$= (^6C_1 \times {^4C_1}) + (^6C_2 \times {^4C_2}) + (^6C_3 \times {^4C_1}) + (^6C_2)$

$= (6\times4)+\left(\frac{6\times5}{2\times1}\times\frac{4\times3}{2\times1}\right)+\left(\frac{6\times5\times4}{3\times2\times1}\times4\right)+\left(\frac{6\times5}{2\times1}\right)$

= (24 + 90 + 80 + 15) = 209.

10. We may have (1 black and 2 non-black) or (2 black and 1 non-black) or (3 black)

∴ Required number of ways

$= (^3C_1 \times {^6C_2}) + (^3C_2 \times {^6C_1}) + (^3C_3)$

$= \left(3\times\frac{6\times5}{2\times1}\right)+\left(\frac{3\times2}{2\times1}\times6\right)+1 = (45 + 18 + 1) = 64.$

13. 4 red balls from 8 red balls can be selected in $^8C_4$ ways 3 blue balls from 9 balls can be selected in $^9C_3$ ways

∴ Total number of selection = $^8C_4 \times {^9C_3}$

$= \frac{8\times7\times6\times5}{1\times2\times3\times4}\times\frac{9\times8\times7}{1\times2\times3}$

$= 70 \times 84 = 5880.$

14. A decagon has 10 vertices

Join any two vertices in $^{10}C_2$ lines 10 lines will be sides.

$\therefore$ Number of diagonals $= {}^{10}C_2 - 10$

$= \frac{10\times9}{1\times2} - 10$

$= 45 - 10 = 35$

16. As the order of A, B and C is not to change they are to be treated identical in arrangement. So, the required

Number of ways $= \frac{7!}{3!}$.

$= \frac{7\times6\times5\times4\times3\times2\times1}{3\times2\times1} = 840.$

18. The number of ways to fill the three even places by 4 consonants $= {}^4P_3$.

After filling the even places, remaining places can be filled in $^4P_4$ ways.

So, the required number of words

$= {}^4P_3 \times {}^4P_4$

$= 4! \times 4! = 24 \times 24 = 576.$

19. A couple and 6 guests can be arranged in $(7 - 1)!$ ways. But the two people forming the couple can be arranged among themselves in 2! ways.

$\therefore$ The required number of ways $= 6! \times 2!$

$= 720 \times 2 = 1440.$

20. Ten pearls of one colour can be arranged in 1/2 (10 – 1)! ways. The number of arrangements of 10 pearls of the other colour in 10 places between the pearls of the first colour = 10!.

$\therefore$ The required number of ways $= \frac{1}{2}\times9!\times10!$

$= 5 \times 9! \times 9! = 5(9!)^2.$

22. Each ball can be put in 2 ways (either in one box or the other).

$\therefore$ 6 balls can be put in $2 \times 2 \times ...$ to six times, i.e., 2^6 ways. But in two of the ways one box is empty. So, the required number of ways $= 2^6 - 2. = 62.$

23.

Possibilities	*Selections*	*Arrangements*
One pair, one different	$^1C_1 \times {}^4C_1$	$^1C_1\times{}^4C_1\times\frac{3!}{2!} = 12$
Three different	$^5C_3$	$^5C_3 \times 3! = 60$

$\therefore$ The required number of ways = 72

24. BANANA

3A's, 2N's, B, i.e., 6 letters, 3 alike of one type and 2 of another type. Number of words taken all at a time is $\frac{6!}{3!\,2!} = \frac{6\times5\times4}{2} = 60.$

25. The required number of ways

$= {}^3C_1 \times {}^6C_2 + {}^3C_2 \times {}^6C_1 + {}^3C_3 = 64.$

26. First we place 3 consonant in 3! ways and then at four places (2 between them and 2 on sides) 3 vowels in which one vowels is repeated can be placed in $^4P_3/2!$ ways.

Hence, required number $= 3!\ {}^4P_3/2! = 72.$

27. Considering two vowels together as a letter, there are 4 letters in all which can be arranged in 4! ways while 2 vowels can also be arranged in 2! ways.

$\therefore$ Total number of arrangements $= 4!\ .\ 2! = 48.$

28. Each bulb has two choices, either switched on or off.

$\therefore$ Required number $= 2^{10} - 1 = 1023$

(Since in one way, when all switches are off, the hall will not be illuminated.)

29. Let there be n persons in a room.

$\therefore$ Total no. of shakehands $= {}^nC_2 = 66$

$\Rightarrow \quad \frac{1}{2}n(n-1) = 66$

$\Rightarrow \quad n^2 - n - 132 = 0$

$\Rightarrow \quad (n + 11)(n - 12) = 0$

$\Rightarrow \quad n = 12 \qquad (\because n \neq -11)$

30. Host can sit on any seat. Two particular guests can sit on either sides of host in 2! = 2 ways and remaining 18 guests in (18)! ways.

Hence, total ways = 2.(18)!.

31. Since every prize can be given by any of the four boys, so number of ways $= 4 \times 4 \times 4 \times 4 \times 4 = 4^5$.

32. Each question can be answered in 4 ways and all questions can be answered correctly in only one way, so the required number of ways $= 4^3 - 1 = 63.$

33. First the seven women sit round the table in 6! ways and then seven men will sit in the spaces between every pair of two women in 7! ways.

$\therefore$ Required seating arrangements $= 6! \times 7!$.

34. $^4P_3 \times {}^5P_3 \times {}^6P_3 = 172800.$

35. $5C_3 \times {}^{22}C_9 = 4974200.$

13

Probability

1. **Random Experiment:** If the result of an experiment is not certain and is any one of the several possible outcomes, the experiment is called a trial or a random experiment.
2. **Sample Space:** The set of all possible outcome of an experiment is called the sample space provided no two or more of these outcomes can occur simultaneously and exactly one of these outcomes must occur whenever the experiment is conducted.
3. **Events:** The outcomes of an experiment, *i.e.*, sample points of the sample space are usually known as simple events and any subset of the sample space 'S' is called an event.

   Thus throwing of a dice is an experiment, S = {1, 2, 3, 4, 5, 6} is the sample space, {1}, ... {6} are simple events and {1, 2}, etc., are events. The empty set ϕ is also an event as $\phi \subset S$ and it is called an impossible event. The sample space S is also a subset of S and so it is also an event. S represents the sure event, *i.e.*, certainty.
4. **Equally Likely Events:** A set of events is said to be equally likely if taking into consideration all the relevant factors there is no reason to expect one of them in preference to others.

   For example, when a fair coin is tossed, the occurrence of a tail or a head are equally likely.
5. **Exhaustive Events:** A set of events is said to be exhaustive if the performance of the experiment always results in the occurrence of atleast one of them.

   For example, if we throw a dice, then the events A_1 = {1, 2}, A_2 = {2, 3, 3} are not exhaustive as we can get 5 as outcome of the experiment which is not the member of any of the events A_1 and A_2. If we consider the events E_1 = {1, 2, 3} and E_2 = {2, 4, 5, 6}, then the set E_1, E_2, is exhaustive.
6. **Mutually Exclusive Events:** A set of events is said to be mutually exclusive if they have no point in common, *i.e.*, happening of one of them eliminates the happening of any of the remaining events. Thus E_1, E_2, E_3, ... are mutually exclusive iff $E_i \cap E_j = \phi$ for $i \neq$ E_1 and E_2 are mutually exclusive.
7. **Complement of An Events:** The complement of an event A, denoted by $\overline{A}$, A' or A^c, is the set of all sample points of the space other than the sample points in A.

   e.g., In the experiment of throwing a fair dice, S = {1, 2, 3, 4, 5, 6}. If A = {1, 3, 5, 6}, then $\overline{A}$ = {2, 4}

   Note that $A \cap \overline{A} = S$.
8. **Classical Definition of Probability:** If there are *n* exhaustive mutually exclusive and equally likely outcomes of an experiment and *m* of them are favourable to an event A, the probability of the happening of A is defined as the ratio *m*/*n*.

   Thus, denoting the probability of the happening of an event A by *p*, we have $p = m/n$.

   Clearly *p* is a positive number not greater than unity, so that $0 \leq p \leq 1$.

   Since the number of cases in which the event A will not happen is $n - m$, the probability *q* that the event will not happen is given by $q = \frac{(n-m)}{n} = 1 - \frac{m}{n}$ $= 1 - p. \therefore p + q = 1$

   If probability of happening of an event A is 1, then A is certain event and if probability of happening of an event A is 0, then A is impossible event.
9. **Odds in Favour and Odds against an Event:** As a result of an experiment if *a* of the outcomes are favourable to an event E and *b* of the outcomes are against it, then we say that odds are *a* to *b* in favour of E or odds are *b* to *a* against E.

   Thus, odds in favour of an event E

   $$= \frac{\text{number of favourable cases}}{\text{number of unfavourable cases}}.$$

   Similarly, odds against an event E

   $$= \frac{\text{number of unfavourable cases}}{\text{number of favourable cases}}.$$

   If odds in favour of an event are $a : b$ then the probability of the occurrence of that event is $\frac{a}{a+b}$

and the probability of the non-occurrence of that event is $\frac{b}{(a+b)}$.

Addition Theorem:

(*a*) If 'A' and 'B' are any two events in a sample space S, then $P(A \cup B) = P(A) + P(B) - P(A \cap B)$.

(*b*) If 'A' and 'B' are mutually exclusive then $P(A \cap B) = 0$ so that $P(A \cup B) = P(A) + P(B)$.

(*c*) If A is any event in S, then $P(A') = 1 - P(A)$.

Total Probability Theorem:

The probability that one of several mutually exclusive events $A_1, A_2, ..., A_n$ will happen, is the sum of the probabilities of the separate events. In symbol,

$$P(A_1 + A_2 + ... + A_n) = P(A_1) + P(A_2) + ... + P(A_n).$$

Conditional Probability:

The probability of B under the assumption that A has occurred is called the conditional probability of B under the condition that the event A has taken place and is denoted by P(B/A). P(B/A) is read as "probability at the event B given A^B."

Conditional Probability Theorem:

If A and B are any two events in the sample space S, the conditional probability of B relative to A is given by

$$P(B/A) = \frac{P(B \cap A)}{P(A)} = \frac{n(B \cap A)}{n(A)}, A \neq \phi.$$

Some Important Remarks about Coins, Dice and Playing Cards:

(*a*) **Coins:** A coin has a head side and a tail side. If an experiment consist of more than a coin, coins are considered to be distinct if not otherwise stated.

(*b*) **Dice:** A die (cubical) has six faces marked 1, 2, 3, 4, 5, 6. We may have tetrahedral (having four faces 1, 2, 3, 4) or pentagonal (having five faces 1, 2, 3, 4, 5) die. As in the case of dice, if we have more than one die, all dice are considered to be distinct if not otherwise stated.

(*c*) **Playing Cards:** A pack of playing cards usually contain 52 cards.

There are 4 suits (spade, heart, diamond and club) each having 13 cards. There are two colours — red (heart and diamond) and black (spade and club) each having 26 cards.

In thirteen cards of each suit, there are 3 face cards namely king, queen and jack, so there are in all 12 face cards (4 kings, 4 queens and 4 jacks).

Also there are 16 honours cards, 4 of each suit namely ace, king, queen and jack.

SOLVED EXAMPLES

Example 1: A bag contains 7 Red balls, 4 Green balls and 5 Yellow balls. What is the probability that 3 balls drawn at random are either Green or Yellow?

Solution: Total ways $= {}^{16}C_3 = \frac{\underline{|16}}{\underline{|13}\ \underline{|3}} = 560$

Favourable ways $= {}^4C_3 \times {}^5C_3 = \frac{\underline{|4}}{\underline{|3}\ \underline{|1}} \times \frac{\underline{|5}}{\underline{|3}\ \underline{|2}}$

$= 4 \times 10 = 40.$

$\therefore$ Reqd. Probability $= \frac{40}{560} = \frac{1}{14}$.

Example 2: A basket contains three Red and four Green balls. If four balls are drawn at random from the basket, what is the probability that two are Red and two are Green?

Solution: Reqd. Probability $= \frac{{}^3C_2 \times {}^4C_2}{{}^7C_4} = \frac{3 \times 6}{35}$

$= \frac{18}{35}$.

Example 3: A box contains 8 Red, 16 Blue, 4 Yellow and 12 Black bolls. Two balls are picked up randomly. What is the chance that both are Red?

Solution: Total results $= {}^{40}C_2 = \frac{40 \times 39}{2} = 780$

and favourable results $= {}^8C_2 = \frac{\underline{|8}}{\underline{|2}\ \underline{|6}} = \frac{8 \times 7}{2}$

$= 28.$

$\therefore$ Required probability $= \frac{28}{780} = \frac{7}{195}$.

Example 4: A bag contains 8 Red, 16 Blue, 12 Black and 4 Yellow marbles. One marbles is picked up randomly. What is the probability that it is not Blue?

Solution: Total results $= {}^{40}C_1 = 40$

Favourable results when it is Blue $= {}^{16}C_1 = 16$.

$\therefore$ Favourable results when it is not Blue

$= 40 - 16 = 24$

$\therefore$ Reqd. probability $= \frac{24}{40} = 0.6$.

MULTIPLE CHOICE QUESTIONS

Directions (Qs. 1 to 3) : *Study the information carefully to answer the questions that follow.*

A bucket contains 8 Red, 3 Blue and 5 Green marbles.

1. If 4 marbles are drawn at random, what is the probability that 2 are Red and 2 are Blue?

A. $\frac{11}{16}$ B. $\frac{3}{16}$

C. $\frac{11}{72}$ D. $\frac{3}{65}$

2. If 2 marbles are drawn at random, what is the probability that both are Green?

A. $\frac{1}{12}$ B. $\frac{5}{16}$

C. $\frac{2}{7}$ D. None of these

3. If 3 marbles are drawn at random, what is the probability that none is Red?

A. $\frac{3}{8}$ B. $\frac{1}{16}$

C. $\frac{1}{10}$ D. None of these

Directions (Qs. 4 to 5) : *Study the information and answer the questions that follow.*

A basket contains 3 Red balls, 5 Blue balls and 2 Green balls.

4. If three balls are drawn at random, what is the probability that none is Blue?

A. $\frac{1}{6}$ B. $\frac{1}{12}$

C. $\frac{1}{5}$ D. $\frac{3}{10}$

5. If four balls are drawn at random, what is the probability that at least one is Green?

A. $\frac{2}{3}$ B. $\frac{1}{2}$

C. $\frac{1}{3}$ D. $\frac{2}{5}$

6. If two balls are drawn at random, what is the probability that both are Red?

A. $\frac{1}{5}$ B. $\frac{3}{10}$

C. $\frac{3}{5}$ D. $\frac{1}{15}$

7. A dice is thrown once. Find the probability of getting a number greater than 3.

A. $\frac{2}{3}$ B. $\frac{1}{3}$

C. $\frac{1}{2}$ D. $\frac{3}{2}$

8. A bag contains 5 Red balls, 8 White balls, 4 Green balls and 7 Black balls. If one ball is drawn at random, find the probability that it is Black.

A. $\frac{7}{24}$ B. $\frac{5}{24}$

C. $\frac{5}{6}$ D. $\frac{1}{4}$

9. Find the probability that a number selected from the numbers 1 to 25 is not a prime number when each of the given number is equally to be selected.

A. $\frac{9}{25}$ B. $\frac{16}{25}$

C. $\frac{21}{25}$ D. $\frac{14}{25}$

10. A bag contains 3 Red balls and 5 Black balls. A ball is drawn at random from the bag. What is the probability that the ball drawn is not Red?

A. $\frac{3}{8}$ B. $\frac{5}{8}$

C. $\frac{1}{8}$ D. $\frac{7}{8}$

11. A die is thrown once. Find the probability of getting an odd number.

A. $\frac{1}{2}$ B. $\frac{3}{2}$

C. $\frac{2}{3}$ D. $\frac{1}{3}$

12. A letter is chosen at random from the word REED the probability of choosing an E is –

A. $\frac{3}{4}$ B. $\frac{1}{2}$

C. $\frac{1}{4}$ D. $\frac{2}{3}$

13. A card is drawn at random from a well-shuffled deck of playing cards. Find the probability that the card drawn is a card of spade or an ace.

A. $\frac{11}{13}$ B. $\frac{1}{26}$

C. $\frac{4}{13}$ D. $\frac{9}{13}$

14. If a letter is chosen at random from the word TRIANGLE.
What is the probability that it a vowel?

A. $\frac{3}{8}$ B. $\frac{5}{8}$

C. $\frac{1}{4}$ D. $\frac{3}{4}$

15. What is the chance that a leap year, selected at random, will contain 53 Sundays?

A. $\frac{52}{365}$ B. $\frac{52}{366}$

C. $\frac{2}{7}$ D. $\frac{5}{7}$

16. In a book of 300 pages, if a page is opened at random; find the probability that the sum of the digits in its number is 10.

A. $\frac{8}{15}$ B. $\frac{9}{15}$

C. $\frac{7}{75}$ D. $\frac{14}{15}$

17. A die is thrown once. Find the probability of getting an even number.

A. $\frac{1}{3}$ B. $\frac{1}{2}$

C. $\frac{1}{4}$ D. $\frac{1}{6}$

18. Two friends were born in the year 2000. What is the probability that they have the same birthday?

A. $\frac{1}{500}$ B. $\frac{1}{730}$

C. $\frac{1}{365}$ D. $\frac{1}{366}$

19. A number is chosen at random from among the first 30 natural numbers. The probability of the number chosen being prime is :

A. $\frac{1}{3}$ B. $\frac{3}{10}$

C. $\frac{1}{30}$ D. $\frac{11}{30}$

20. The probability that a number selected at random from the set of numbers {1, 2, 3, 4, ..., 100} is a cube is:

A. $\frac{1}{25}$ B. $\frac{2}{25}$

C. $\frac{3}{25}$ D. $\frac{4}{25}$

21. A single letter is selected at random from the word 'COLLEGE'. Find the probability of getting a vowel.

A. $\frac{3}{7}$ B. $\frac{4}{7}$

C. $\frac{5}{7}$ D. $\frac{2}{7}$

22. A box contains 5 Red balls, 4 Green balls and 7 White balls. A ball is drawn at random from the box. Find the probability that the ball drawn is neither Red nor White.

A. $\frac{2}{7}$ B. $\frac{3}{4}$

C. $\frac{1}{4}$ D. $\frac{1}{2}$

23. In a cricket match, a batsman hits boundary 8 times out of 40 balls he plays. Find the probability that he didn't hit boundary.

A. 0.8 B. 0.6

C. 0.5 D. 0.4

24. Two coins are tossed simultaneously. The probability of getting atmost one head is :

A. $\frac{1}{4}$ B. $\frac{3}{4}$

C. $\frac{1}{2}$ D. $\frac{1}{4}$

25. A bag contains 50 coins and each coin is marked from 51 to 100. One coin is picked at random. The probability that the number on the coin is not a prime number is :

A. $\frac{1}{5}$ B. $\frac{3}{5}$

C. $\frac{2}{5}$ D. $\frac{4}{5}$

26. In a football match, a player makes 4 goals from 10 penalty kicks. The probability of converting a penalty kick into a goal by the player, is :

A. $\frac{1}{4}$ B. $\frac{1}{6}$

C. $\frac{1}{3}$ D. $\frac{2}{5}$

27. Three unbiased coins are tossed. What is the probability of getting at least 2 heads?

A. $\frac{1}{4}$ B. $\frac{1}{2}$

C. $\frac{1}{3}$ D. $\frac{1}{8}$

28. In a simultaneous throw of two dice, what is the probability of getting a total of 7?

A. $\frac{1}{6}$ B. $\frac{1}{4}$

C. $\frac{2}{3}$ D. $\frac{3}{4}$

29. What is the probability of getting a sum 9 from two throws of a dice?

A. $\frac{1}{6}$ B. $\frac{1}{8}$

C. $\frac{1}{9}$ D. $\frac{1}{12}$

30. Tickets numbered 1 to 20 are mixed up and then a ticket is drawn at random. What is the probability that the ticket drawn bears a number which is a multiple of 3?

A. $\frac{3}{10}$ B. $\frac{3}{20}$

C. $\frac{2}{5}$ D. $\frac{1}{2}$

31. A card is drawn from a pack of 52 cards. The probability of getting a queen of club or a king of heart is :

A. $\frac{1}{13}$ B. $\frac{2}{13}$

C. $\frac{1}{26}$ D. $\frac{1}{52}$

32. Two cards are drawn together from a pack of 52 cards. The probability that one is a spade and one is a heart, is :

A. $\frac{3}{20}$ B. $\frac{29}{34}$

C. $\frac{47}{100}$ D. $\frac{13}{102}$

33. A bag contains 6 white and 4 red balls. Three balls are drawn at random. What is the probability that one ball is red and the other two are white?

A. $\frac{1}{2}$ B. $\frac{1}{12}$

C. $\frac{3}{10}$ D. $\frac{7}{12}$

34. A box contains 20 electric bulbs, out of which 4 are defective. Two bulbs are chosen at random from this box. The probability that at least one of these is defective, is :

A. $\frac{4}{19}$ B. $\frac{7}{19}$

C. $\frac{12}{19}$ D. $\frac{21}{95}$

35. Two dice are tossed. The probability that the total score is a prime number is :

A. $\frac{1}{6}$ B. $\frac{5}{12}$

C. $\frac{1}{2}$ D. $\frac{7}{9}$

ANSWERS

1	2	3	4	5	6	7	8	9	10
D	A	A	B	A	D	C	A	B	B
11	12	13	14	15	16	17	18	19	20
A	B	C	A	C	C	B	D	A	A
21	22	23	24	25	26	27	28	29	30
A	C	A	B	D	D	B	A	C	A
31	32	33	34	35					
C	D	A	B	B					

SOME SELECTED EXPLANATORY ANSWERS

1. Total number of ways = $^{16}C_4 = \frac{\lfloor 16}{\lfloor 12 \lfloor 4}$

$= 1820$

And favourable number of ways = $^{8}C_2 \times {^3C_2}$

$= \frac{\lfloor 8}{\lfloor 6 \lfloor 2} \times \frac{\lfloor 3}{\lfloor 1 \lfloor 2} = 84$

$\therefore$ Reqd. probability = $\frac{84}{1820} = \frac{3}{65}$.

2. Total number of ways = $^{16}C_2 = \frac{\lfloor 16}{\lfloor 14 \lfloor 2} = 120$

And favourable number of ways = $^5C_2$

$= \frac{\lfloor 5}{\lfloor 3 \lfloor 2} = 10$

$\therefore$ Reqd. probability = $\frac{10}{120} = \frac{1}{12}$.

3. Total number of ways = $^{16}C_3 = \frac{\lfloor 16}{\lfloor 13 \lfloor 3} = 560$

Total number of ways when all the marbles are Red

$= {}^8C_3 = \frac{\lfloor 8}{\lfloor 5 \lfloor 3} = 56$

∴ Probability when all the three are Red

$= \frac{56}{560} = \frac{1}{10}$

∴ And probability when none is Red $= 1 - \frac{1}{10}$

$= \frac{9}{10}$.

4. Probability that none is blue

$= \frac{^5C_3}{^{10}C_3} = \frac{10}{120} = \frac{1}{12}$

5. Probability that none is green

$= \frac{^8C_4}{^{10}C_4} = \frac{70}{210} = \frac{1}{3}$

∴ Probability that at least one is green

$= 1 - \frac{1}{3} = \frac{2}{3}$

6. Probability that both are red

$= \frac{^3C_2}{^{10}C_2} = \frac{3}{45} = \frac{1}{15}$

7. Sample space S = {1, 2, 3, 4, 5, 6} ; $n(S) = 6$
Let A = "getting a number grater than 3" then
A = {4, 5, 6} ; $n(A) = 3$

$P(A) = \frac{n(A)}{n(S)} = \frac{3}{6} = \frac{1}{2}$.

9. We are given numbers from 1 to 25.
$n(S) = 25$.
Prime numbers from 1 to 25 are 2, 3, 5, 7, 11, 13, 17, 19, 23.
These are 9 prime numbers.
Let E denote the event 'Number drawn is not prime'.

∴ $n(E) = 25 - 9 = 16$

$p(E) = \frac{n(E)}{n(S)} = \frac{16}{25}$.

10. Total number of balls in the bag = 3 + 5 = 8
Let E denote the event 'ball drawn is red'.

∴ $n(E) = 3$ and $n(S) = 8$.

(*i*) $p(E) = \frac{n(E)}{n(S)} = \frac{3}{8}$

(*ii*) $p(\text{not red}) = p(\bar{E})$

$= 1 - p(E) = 1 - \frac{3}{8} = \frac{5}{8}$.

11. Sample S = {1, 2, 3, 4, 5, 6} ; $n(S) = 6$
Let F denote the event 'getting an odd number'.
F = {1, 3, 5} ; $n(S) = 3$

∴ $P(F) = \frac{n(F)}{n(S)} = \frac{3}{6} = \frac{1}{2}$.

12. The sample space is $\{R, E_1, E_2, D\}$
There are 2 ways to choose the letter E from the four letters

∴ $P(E) = \frac{2}{4} = \frac{1}{2}$.

13. There are 52 cards in a deck of 52 playing cards. There are 13 spades and 4 aces in 52 playing cards. Out of 13 spades there is one ace.
∴ Favourable no ways for the card drawn to be spade or an ace = 13 spades + 3 aces = 16.

∴ P(a card of spade or an ace) $= \frac{16}{52} = \frac{4}{13}$.

14. Since there are 3 vowels out of a total of 8 letters, the probability of choosing a vowel is 3/8.

15. A Leap year has 366 days. In a leap year, here are 52 complete weeks (52 Sundays) and 2 days over these two days can be as follows.
Sunday and Monday, Monday and Tuesday, Tuesday and Wednesday, Wednesday and Thursday, Thursday and Friday, Friday and Saturday, Saturday and Sunday.
Out of these 7 possibilities, only two (Saturday and Sunday; Sunday and Monday) are favourable for having 53 Sundays)

Required chance $= \frac{2}{7}$.

17. A die is thrown once. Its sample space is
S = {1, 2, 3, 4, 5, 6}, $n(S) = 6$
Let E : an even number, then E = {2, 4, 6} Number of favourable cases = $n(E) = 3$

∴ $P(E) = \frac{n(E)}{n(S)} = \frac{3}{6} = \frac{1}{2}$.

18. Year 2000 is a leap year. There are 366 days in the year 2000. Total number of cases in which two friends can be born on the same day are 366.
Out of 366 days, number of favourable ways in which two friends have the same birthday = 1

∴ Required probability $= \frac{1}{366}$.

19. The prime numbers from first 30 natural numbers are 2, 3, 5, 7, 11, 13, 17, 19, 23, 29.
We find that there are 10 prime numbers from 1 to 30.
∴ Probability that the number chosen being prime is

$= \frac{10}{30} = \frac{1}{3}$.

20. From the natural numbers 1 to 100 the four number 1, 8, 27 and 64 are cubes

$\therefore$ Required Probability $= \frac{4}{100} = \frac{1}{25}$.

23. Let A denote the event that the batsman did not hit a boundary.

We have, Total number of trials = 40

Number of trials in which the event A happened

$= 40 - 8 = 32$

$\therefore \quad P(A) = \frac{32}{40} = \frac{4}{5} = 0.8.$

27. Here S = {HH, HT, TH, TT}.

Let E = Event of getting at least one head = {HT, TH, HH}.

$\therefore \ P(E) = \frac{n(E)}{n(S)} = \frac{4}{8} = \frac{1}{2}.$

28. We know that in a simultaneous throw of two dice, $n(S) = 6 \times 6 = 36$.

Let E = Event of getting a total of 7 = {(1, 6), (2, 5), (3, 4), (4, 3), (5, 2), (6, 1)}.

$\therefore \ P(E) = \frac{n(E)}{n(S)} = \frac{6}{36} = \frac{1}{6}.$

29. In two throws of a die, $n(S) = (6 \times 6) = 36$.

Let E = Event of getting a sum 9 = {(3, 6), (4, 5), (5, 4), (6, 3)}.

$\therefore \ P(E) = \frac{n(E)}{n(S)} = \frac{4}{36} = \frac{1}{9}.$

30. Here, S = {1, 2, 3, 4, ..., 19, 20}.

Let E = Event of getting a multiple of 3 = {3, 6, 9, 12, 15, 18}.

$\therefore \ P(E) = \frac{n(E)}{n(S)} = \frac{6}{20} = \frac{3}{10}.$

31. Here $n(S) = 52$.

Let E = Event of getting a queen of club or a king of heart.

Then, $n(E) = 2$.

$\therefore \ P(E) = \frac{n(E)}{n(S)} = \frac{2}{52} = \frac{1}{26}.$

32. Let S be the sample space. Then,

$n(S) = {}^{52}C_2 = \frac{(52 \times 51)}{(2 \times 1)} = 1326.$

Let E = Event of getting 1 spade and 1 heart.

$\therefore$ $n(E)$ = Number of ways of choosing 1 spade out of 13 and 1 heart out of 13

$= ({}^{13}C_1 \times {}^{13}C_1) = (13 \times 13) = 169.$

$\therefore \ P(E) = \frac{n(E)}{n(S)} = \frac{169}{1326} = \frac{13}{102}.$

33. Let S be the sample space. Then,

$n(S)$ = Number of ways of drawing 3 balls out of 10

$= {}^{10}C_3 = \frac{(10 \times 9 \times 8)}{(3 \times 2 \times 1)} = 120.$

Let E = Event of drawing 1 red and 2 white balls

$\therefore$ $n(E)$ = Number of ways of drawing 1 red ball out of 4 and 2 white balls out of 6

$= ({}^4C_1 \times {}^6C_2) = \left(4 \times \frac{6 \times 5}{2 \times 1}\right) = 60.$

$\therefore \ P(E) = \frac{n(E)}{n(S)} = \frac{60}{120} = \frac{1}{2}.$

34. P (None is defective)

$= \frac{{}^{16}C_2}{{}^{20}C_2} = \left(\frac{16 \times 15}{2 \times 1} \times \frac{2 \times 1}{20 \times 19}\right) = \frac{12}{19}.$

P (at least one is defective) $= \left(1 - \frac{12}{19}\right) = \frac{7}{19}.$

35. Clearly, $n(S) = (6 \times 6) = 36$.

Let E = Event that the sum is a prime number.

Then, E = {(1, 1), (1, 2), (1, 4), (1, 6), (2, 1), (2, 3), (2, 5), (3, 2), (3, 4), (4, 1), (4, 3), (5, 2), (5, 6), (6, 1), (6, 5)}

$\therefore \quad n(E) = 15$

$\therefore \quad P(E) = \frac{n(E)}{n(S)} = \frac{15}{36} = \frac{5}{12}.$

14

Data Interpretation

TABLES

Tables are often used in reports, magazines and newspaper to present as set of numerical data. It is one of the easiest and most accurate ways of presenting data. One of the main purposes of tables is to make complicated information easier to understand. Hence, the advantage of presenting data in a table is that one can see the information at a glance. We present below an example of tabular presentation of annual expenditure of 5 families for last 4 years.

Annual Expenditure of 5 families (in ₹ Thousands)

Years → Families ↓	2005	2006	2007	2008
A	35	50	55	60
B	40	60	65	75
C	45	50	70	80
D	30	40	45	50
E	50	55	60	70

Essentials of a Tables:

(i) **Title:** Heading of the table
(ii) **Stub:** The section of the table containing row headings in called Stub.
(iii) **Column Captions :** The heading of each column is designated as column caption.
(iv) **Body**
(v) **Footnotes**
(vi) **Source**

Example 1. Directions: *Study the folloiwng table and arswer the questions given below:*

Categories of students according to their enrolment in various courses in a vocational college.

Courses	*Categories (Faculty)*			
	Arts		*Science*	
	Boys	*Girls*	*Boys*	*Girls*
1. Business Management	45	25	65	25
2. Typewriting	186	23	32	20
3. Stenography	120	25	58	12
4. Typewriting and Stenography	100	12	5	3

1. If 60% boys and 70% girls pass in their respective courses, then total pass percentage is:
A. 58% B. 66% C. 54% D. 62%

2. The total number of students who have opted the course of Business Management and the course of Stenography exceeds the number of students studying in typewriting course by:
A. 44% B. 66% C. 33% D. 55%

3. In which of the following courses, percentage of girls is the highest?
A. Stenography B. Business Management
C. Typewriting D. Typewriting and Stenography

4. In the art faculty number of students who have opted the course of stenography is how much per cent of the total student in the faculty?

A. 27% B. 10% C. 18% D. 15%

5. Total number of boys studying in the college exceeds the total number of girls in the college by:

A. 321% B. 280% C. 308% D. 250%

Solution 1. $\because$ Total boys = 45 + 186 + 120 + 100 + 65 + 32 + 58 + 5 = 611

$\therefore$ Number of boys who pass in their respective courses = 60% of 611 = $611 \times \frac{60}{100} = 366.60$

Total number of girls = 25 + 23 + 25 + 12 + 25 + 20 + 12 + 3 = 145

Number of girls who pass in their respective courses = 70% of 145 = $\frac{145 \times 70}{100} = 101.50$

$\therefore$ Combined pass per cent = $\frac{(366.6 + 101.5) \times 100}{611 + 145} = \frac{46810}{756} = 61.92 \simeq 62\%$

Solution 2. $\because$ Number of students in Business Management course = 160

Number of students in typewriting course = 261

Number of students in stenography corse = 215

$\therefore$ Required per cent = $\left(\frac{160 + 215 - 261}{261}\right) \times 100 \approx 44\%$

Solution 3. Percentage of girls in Business Management course = $\frac{(25 + 25) \times 100}{160} = 31.25$

Percentage of girls in typewriting course = $\frac{(23 + 20) \times 100}{261} = 16.47$

Percentage of girls in stenography course = $\frac{(25 + 12) \times 100}{215} = 12.56$

And percentage of girls in typewriting and stenography course = $\frac{(12 + 3) \times 100}{120} = 12.50$

Hence, it is clear from the above that percentage of girls in Business Management course is the highest.

Solution 4. Required percentage = $\frac{(25 + 120) \times 100}{536} \times 100 \simeq 27$

Solution 5. Total number of boys = 45 + 65 + 186 + 32 + 120 + 58 + 100 + 5 = 611

Total number of girls = 25 + 23 + 25 + 12 + 25 + 20 + 12 + 3 = 145

$\therefore$ Required percentage = $\left(\frac{611 - 145}{145}\right) \times 100 \simeq 321\%$

Example 2. Directions: *Figures of biscuit exports from a country during various years are given in the following table. Study the table carefully and answer the questions that follows:*

VOLUME OF EXPORT OF BISCUITS FROM A COUNTRY

Year	*Quantity (in lakh tins)*	*Price (in crores ₹)*
2004	100	150
2005	75	150
2006	150	330
2007	160	400
2008	200	500

1. During which year per tin export price remained at lowest level?
A. 2007 B. 2006 C. 2005 D. 2004

2. During which years price per tin remained at the same level?
A. 2006 and 2008 B. 2006 and 2007 C. 2007 and 2008 D. 2006 and 2004

3. The export price of biscuit increased from 2004 to 2008 by how much per cent?
A. 100 B. $116\frac{2}{3}$ C. $233\frac{1}{3}$ D. 350

4. The difference between the number of biscuit tins exported in 2008 and that in 2007 is:
A. 40 B. 40,000 C. 4,00,000 D. 40,00,000

Solution 1. Export price per tin in 2004 = $\frac{₹\ 150 \text{ crores}}{₹\ 100 \text{ lakhs}} = ₹\ 150$

Export price per tin in 2005 = $\frac{₹\ 150 \text{ crores}}{₹\ 75 \text{ lakhs}} = ₹\ 200$

Export price per tin in 2006 = $\frac{₹\ 330 \text{ crores}}{₹\ 150 \text{ lakhs}} = ₹\ 220$

Export price per tin in 2007 = $\frac{₹\ 400 \text{ crores}}{₹\ 160 \text{ lakhs}} = ₹\ 250$

Export price per tin in 2008 = $\frac{₹\ 500 \text{ crores}}{₹\ 200 \text{ lakhs}} = ₹\ 250$

Hence, it is clear that per tin export price was at lowest level in 2004.

Solution 2. From the explanation given in the answer of Q.No. 1, it is clear that per tin export price was same in 2007 and 2008.

Solution 3. From the level of 2004 percentage increase in export price in 2008 = $\left(\frac{500-150}{150}\right)\times 100 = 233\frac{1}{3}\%$

Solution 4. Required difference = 200 – 160 = 40 lakh tins = 40,00,000 tins.

Example. 3. Directions: *Study the following table carefully and answer the questions given below:*

Number of Employees Working in Various departments of a Factory

Year	*Production Deptt.*	*Sale Deptt.*	*Procurement Deptt.*	*Administration & Accounts Deptt.*	*Research & Development Deptt.*
2003	150	25	50	45	75
2004	225	40	45	62	70
2005	450	65	30	90	73
2006	470	73	32	105	70
2007	500	80	35	132	74
2008	505	75	36	130	75

1. During which of the following years number of employees working in production deptt. of the factory was less than 50% of the total employees of the factory?
A. 2003 B. 2005 C. 2006 D. 2007

2. During which year the number of employees working in each department was more than those in the previous year and nearly equal to next year?
A. 2007 B. 2006 C. 2005 D. 2004

3. In which department of the factory number of employees remained nearly same during the years from 2003 to 2008?
A. Production B. Sale
C. Research and Development D. Administration and Accounts
E. Procurement

4. In which department of the factory number of employees are maximum from 2003 to 2008?

A. Production B. Sale
C. Research and Development D. Administration and Accounts
E. Sale and Procurement

5. During which year number of employees in the factory became nearly double the total number of employees working in the factory during 2003.

A. 2008 B. 2007 C. 2006 D. 2005

Solution 1. Number of total employees in 2003 = 150 + 25 + 50 + 45 + 75 = 345

∴ Percentage number of employees working in producton deptt. = $\frac{150 \times 100}{345}$ = 43.48%

Number of total employees in 2004 = 225 + 40 + 45 + 62 + 70 = 442

∴ Percentage number of employees working in production deptt. = $\frac{225 \times 100}{442}$ = 50.9%

Number of total employees in 2005 = 450 + 65 + 30 + 90 + 73 = 708

∴ Percentage number of employees working in proudction deptt. = $\frac{450 \times 100}{708}$ = 63.56%

Total number of employees in 2006 = 470 + 73 + 32 + 105 + 70 = 750

∴ Percentage number of employees working in production deptt. = $\frac{470 \times 100}{750}$ = 62.67%

Total number of employees in 2007 = 500 + 80 + 35 + 132 + 74 = 821

∴ Percentage number of employees working in production deptt. = $\frac{500 \times 100}{821}$ = 60.9%

Total number of employees in 2007 = 505 + 75 + 36 + 130 + 75 = 821

∴ Percentage number of employees working in production in deptt. = $\frac{505 \times 100}{821}$ = 61.51%

Hence, it is clear that percentage number of employees working in production deptt. was less than 50% in 2003.

Solution 2. It is clear from the table that it was the year 2007 when the number of employees working in each department was more than that in the previous year and nearly equal to next year.

Solution 3. According to the table in the research and development department number of employees remained nearly same during the year from 2003 to 2008.

Solution 4. Production.

Solution 5. Total number of employes in 2003 = 345
Total number of employees in 2005 = 708
Hence, it was the year 2005 when total number of employees in the factory became nearly double the total number of emplyees working in the factory during 2003.

Example. 4. *Study the table carefully to answer the questions that follow.*

Number of students studying in six different classes of six different schools.

Class → School ↓	V	VI	VII	VIII	IX	X
P	152	160	145	156	147	144
Q	148	166	150	155	157	143
R	161	152	140	145	143	165
S	159	142	149	140	142	168
T	147	144	158	163	154	150
U	150	160	162	160	161	140

1. What is the respective ratio of students studying in class IX of schools Q and R together to those studying in class VI of schools S and T together?
A. 143 : 150 B. 150 : 143 C. 127 : 181 D. 181 : 127

2. Number of students studying in class X from school P forms approximately what per cent of the total number of students studying in class X from all schools together?
A. 9 B. 12 C. 16 D. 24

3. Total number of students studying in school T from all classes together forms what per cent of total number of students studying in school S from all classes together?
A. 100.26 B. 101.78 C. 102.64 D. 103.52

4. The number of students studying in class VII from school U forms what per cent of the total number of students from all the classes together from that school?
A. 15.93 B. 16.14 C. 17.36 D. 18.28

5. Which class has the maximum number of students from all schools together?
A. V B. VI C. VII D. X

Solution 1. Number of students of class IX of schools Q and R = 157 + 143 = 300
Number of students of class VI of schools S and T = 142 + 144 = 286
Hence, required ratio = 300 : 286 = 150 : 143

Solution 2. Total number of students of class X from all schools = 144 + 143 + 165 + 168 + 150 + 140 = 910

Required percentage = $\frac{144}{910} \times 100 = 15.82 \simeq 16\%$

Solution 3. Total number of students of school T from all classes = 147 + 144 + 158 + 163 + 154 + 150 = 916
Total number of students of school S from all classes = 159 + 142 + 149 + 140 + 142 + 168 = 900

Required percentage = $\frac{916}{900} \times 100 = 101.78\%$

Solution 4. Total number of students of school U from all the classes = 150 + 160 + 162 + 160 + 161 + 140 = 933

Required percentage = $\frac{162}{933} \times 100 = 17.36\%$

Solution 5. Total number of students of class V from all the schools = 152 + 148 + 161 + 159 + 147 + 150 = 917
Total number of students of class VI from all the schools = 160 + 166 + 152 + 142 + 144 + 160 = 924
Total number of students of class VII from all the schools = 145 + 150 + 140 + 149 + 158 + 162 = 904
Total number of students of class VIII from all the schools = 156 + 155 + 145 + 140 + 163 + 160 = 919
Total number of students of class IX from all the schools = 147 + 157 + 143 + 142 + 154 + 161 = 904
Total number of students of class X from all the schools = 144 + 143 + 165 + 168 + 150 + 140 = 910
Hence, the class VI has maximum number of students.

BAR GRAPHS

Given quantity of a bar graph can be compared by the height or length. A bar graph can have either vertical or horizontal bars. You can compare different quantities or the same quantity at the different times. The bars may be placed adjacent to each other or may be separated from each other by spaces depending upon the problem.

For Example: **Registration of New Vehicles in Delhi (in thousands)**

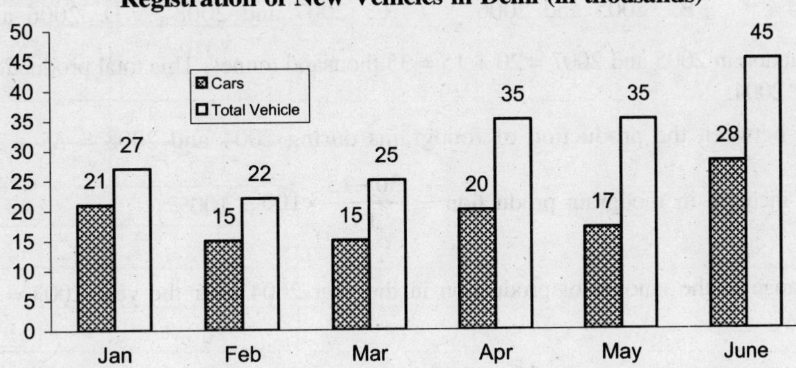

CUMULATIVE GRAPH

These are usually bar or line graphs where the height of the bar or line is divided up proportionally among various quantities presented in the graph. The representation of quantities may be done in terms of either percentage of the total or in absolute figures. These are also called sub-divided graphs. Thus, cumulative graph may be conveniently used for making comparisons.

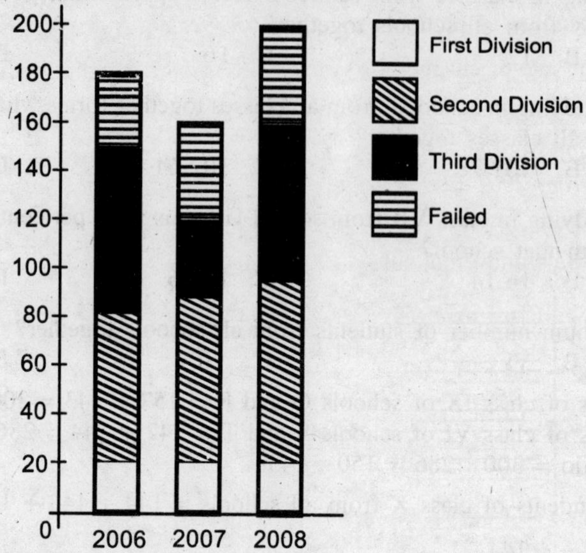

Example 1: Directions: *The following bar graph shows the production of food grains in India during certain year. Study the graph carefully and answer the questions given below:*

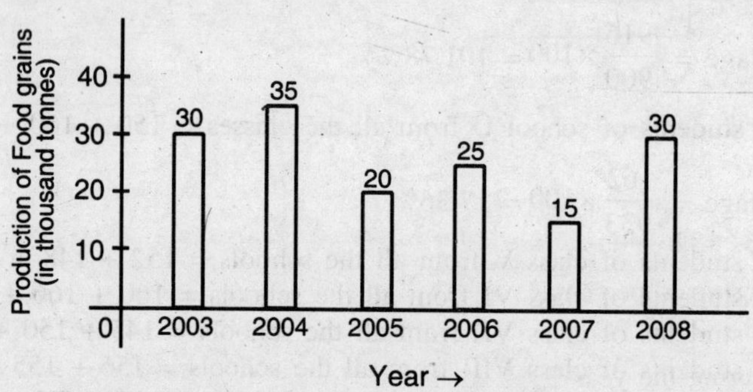

1. Total food grain production in 2005 and 2007 is equal to the production during which of the following years?

A. 2003 B. 2004 C. 2006 D. 2008

2. The difference between the foodgrain productions of 2004 and 2008 is:

A. 500 tonnes B. 1000 tonnes C. 5000 tonnes D. 10000 tonnes

3. Percentage increase in the production of foodgrains in the year 2008 over the year 2007 is:

A. 15% B. 30% C. 50% D. 100%

4. Rate of change in quantity of production in which of the following two consecutive years is at the minimum level?

A. 2003 and 2004 B. 2005 and 2006 C. 2007 and 2008 D. 2006 and 2007

Solution 1. Total production in 2005 and 2007 = 20 + 15 = 35 thousand tonnes. This total production is equal to the production in the year 2004.

Solution 2. Difference between the production of foodgrains during 2004 and 2008 = 35 – 30 = 5 thousand tonnes.

Solution 3. Percentage increase in foodgrain production = $\frac{30-15}{15} \times 100 = 100\%$

Solution 4. Rate of change in the amount of production in the year 2004 over the year 2003 = $\frac{35-30}{30} \times 100$

$= 16.67\%$

Rate of change in the amount of production in the year 2006 over the year 2005 = $\frac{(25-20)}{20} \times 100 = 25\%$

Rate of change in the amount of production in the year 2008 over the year 2007 = $\frac{(30-15)}{15} \times 100 = 100\%$

Rate of change in the amount of production in the year 2007 over the year 2006 = $\frac{(25-15)}{15} \times 100 = 66.67\%$

Hence, it is clear that the rate of change in the amount of production during the consecutive years 2007 and 2008 is at the highest level.

Example 2: Directions: *The bar graph as shown below gives information about the sale and profit details of a departmental store during the years from 2001–2008. Study the graph carefully and answer the questions asked here under.*

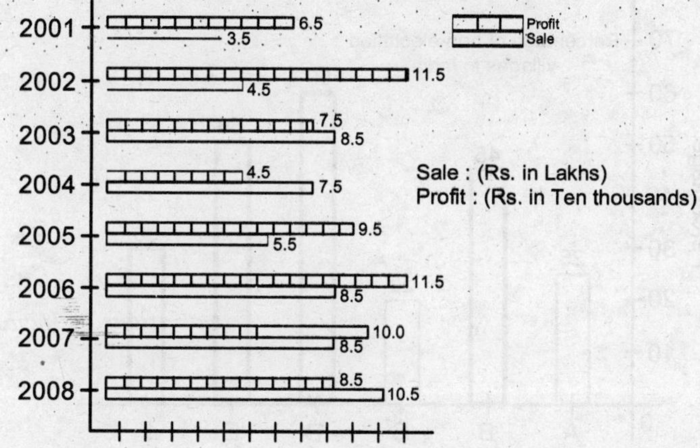

1. Mean of annual increase in sale from 2003 to 2008 (rupees in lakhs) is:
A. 0.1 B. 0.2 C. 0.3 D. 0.4

2. Annual mean profit of the store (rupees in ten thousands) is approximately:
A. 8.5 B. 8.6 C. 8.7 D. 9.0

3. During which of the following years percentage of profit earned by the store on the total sale was at the highest level?
A. 2001 B. 2002 C. 2003 D. 2005

4. Assuming the profit earned during the year 2001 as base (100), the profit made by the store during the year 2008 was:
A. 76 B. 105 C. 121 D. 131

5. During which year between 2001 to 2006 profit made by the store as compared to the previous year was more than 100%?
A. 2008 B. 2007 C. 2005 D. 2003

Solution 1. Mean of annual increase in sale from 2003 to 2008 = $\frac{10.5-8.5}{5} = .4$

Solution 2. Annual mean profit of the store (rupees in ten thousands)

$$= \frac{(6.5+11.5+7.5+4.5+9.5+11.5+10.0+8.5)}{8} = \frac{69.5}{8} = 8.7 \text{ (approx.)}$$

Solution 3. Percentage of profit on the total sale in 2001 = $\frac{6.5 \times 10000 \times 100}{3.5 \times 100000} = 18.57$

Percentage of profit on the total sale in 2002 = $\frac{11.5 \times 10000 \times 100}{4.5 \times 100000} = 25.56$

Percentage of profit on the total sale in 2003 = $\frac{7.5 \times 10000 \times 100}{8.5 \times 100000} = 8.82$

And percentage of profit on the total sale in 2005 = $\frac{9.5 \times 10000 \times 100}{5.5 \times 100000} = 17.27$

∴ Percentage of profit on the total sale in 2002 was the highest level.

Solution 4. Required profit = $\frac{100 \times 8.5}{6.5} = 131$

Solution 5. Profit per cent during the year 2005 as compared to the previous year = $\frac{(9.5 - 4.5) \times 100}{4.5} = 111.11\%$

Hence, profit in the year 2005 was more than 100% as compared to the previous year.

Example 3: Directions: *Study the graph given below and answer the questions asked there under.*

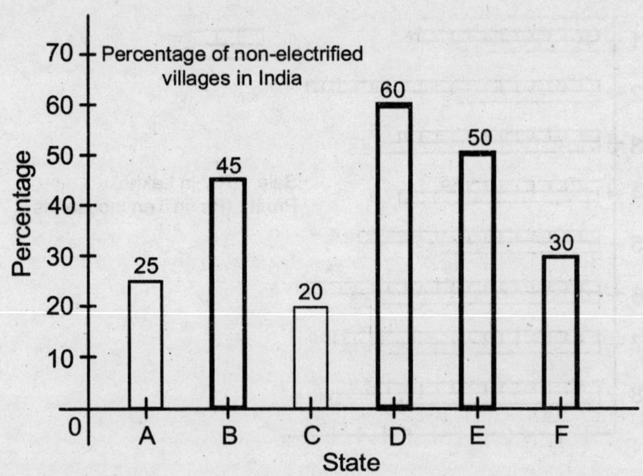

1. According to the table given below which state of India has the highest percentage of electrified villages?
   A. A B. B C. C D. D E. E
2. If central govt. decides to extend financial assistance for prompt electrification and begins with the state having the least electrified villages, then in the order of preference which of the following states will be fourth?
   A. C B. B C. A D. E E. F
3. How many states have at least 60% or more electrified villages?
   A. Five B. Three C. Four D. Two E. One
4. In which of the following states percentage of electrified villages is double the percentage of electrified villages in state (D)?
   A. C B. F C. A D. B E. E
5. What is the percentage of electrified villages in the state (A)?
   A. 65% B. 25% C. 35% D. 75% E. None of these

Solution 1. In the state (C) percentage of non-electrified villages is the least. It means the percentage of electrified villages for the state (C) is the highest.

Solution 2. The order of preference of the least electrified villages is given as: D, E, B, F, A, C.

Solution 3. In A, C and F states, percentage of non-electrified villages are 25, 20 and 30 per cent respectively. Hence, percentages of electrified villages for these three states are 75, 80 and 70 per cent respectively.

Solution 4. Percentage of electrified villages in state D = 100 – 60 = 40
Percentage of electrified villages in state C = 100 – 20 = 80
Hence, the percentage of electrified villages in the state C is double the percentage of electrified villages in state D.

Solution 5. Percentage of electrified villages in the state A = 100 – 25 = 75%

Example 4: Directions: *The following bar chart shows the amount accrued from additional taxation by the government during certain successive years. Study the chart carefully and answer the questions that follow:*

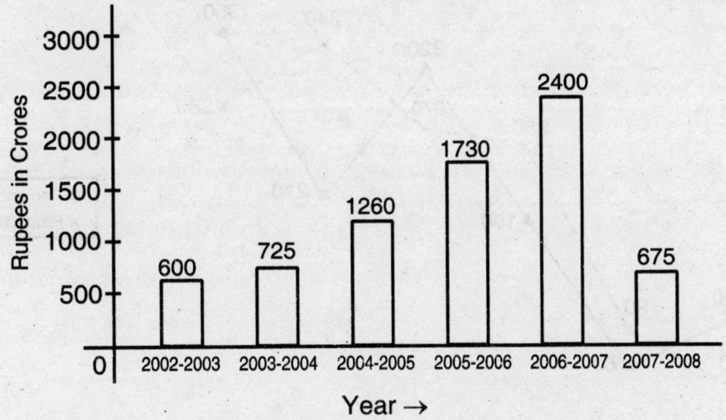

1. In between which of the following two successive financial years increase in taxation was at the highest level?
   A. 2002-2003 and 2003-04 B. 2005-06 and 2006-07
   C. 2006-07 and 2007-08 D. 2004-05 and 2005-06
2. Total increase in taxation was how much per cent between the years 2004-05 and 2005-06?
   A. 37.3% B. 35.4% C. 38.5% D. 36.6%
3. Percentage decrease in taxation in between the years 2006-07 and 2007-08 was:
   A. 71.8% B. 68.8% C. 72.6% D. 75%
4. Between the years 2005-06 and 2006-07 the taxation was increased by:
   A. ₹ 670 crores B. ₹ 650 crores C. ₹ 700 crores D. ₹ 715 crores
5. What is the average amount of taxation during all these years?
   A. ₹ 1233.7 crores B. ₹ 1433.5 crores C. ₹ 1231.7 crores D. ₹ 1400 crores

Solution 1. Between the financial years 2005-06 and 2006-07, maximum increase of ₹ 670 crores had been registered in taxation.

Solution 2. Percentage of total increase $= \frac{(1730-1260)}{1260} \times 100 = \frac{470 \times 100}{1260} = 37.3\%$

Solution 3. Percentage of total decrease in taxation between the years 2006-07 and 2007-08

$$= \frac{(2400-675)}{2400} \times 100 = \frac{1725 \times 100}{2400} = 71.8\%$$

Solution 4. Increase in taxation in between the financial years 2005-06 and 2006-07

$$= ₹\ (2400 - 1730) = ₹\ 670 \text{ crores}$$

Solution 5. Average of the total taxation during the years $= \frac{600+725+1260+1730+2400+625}{6}$

$$= \frac{7390}{6} = ₹\ 1231.7 \text{ crores}$$

LINE GRAPHS

Line graphs are used to show how a quantity changes continuously. If the line goes up, the quantity is increasing; if the line goes down, the quantity is decreasing; if the line is horizontal, the quantity is not changing.

For Example:

Demand and Production of rubber (in thousand tons) in various years

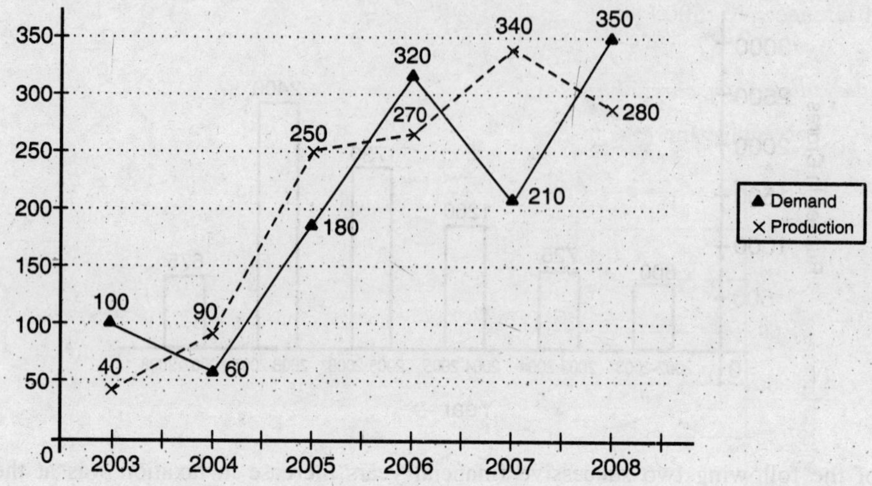

Example 1: Directions: *Study carefully the graph given below and answer the questions that follow:*

1. Price of a machine during 2005 was:

   A. ₹ $5\frac{1}{3}$ thousand B. ₹ 50 thousand C. ₹ 5103 D. ₹ 3 thousand

2. The percentage decrease in the production in 2007 over the previous year 2006 was:

   A. 5% B. 20% C. 25% D. 30%

3. The difference in revenue collection from the sale of machines during 2007 and 2008 was:

   A. ₹ 10 lakhs B. ₹ 1 crore C. ₹ 4 crores D. ₹ 6 crores

4. Had the company sold its machines in 2008 at 25% inhanced price of 2007, what would have been the total selling price of machines produced during the year?

   A. ₹ 35 crores B. ₹ 20 crores C. ₹ 12.5 crores D. ₹ 15.5 crores

5. Difference between the price of a machine in the year 2006 and that in the eyar 2007 was:

   A. ₹ 1500 B. ₹ 2500 C. ₹ 1800 D. ₹ 3200

Solution 1. No. of machines produced during the year 2005 = 15000

And price of the machines = ₹ 8,0000000

∴ Price of 1 machine in 2005 = ₹ $\frac{80000000}{15000}$ = ₹ $5\frac{1}{3}$ thousand

Solution 2. Percentage decrease in production of machines during 2007 over the previous year 2006

$$= \frac{(25-20)\times 100}{25} = 20\%$$

Solution 3. Difference in revenue collected from sale of machines during 2007 and 2008

= 16 – 10 = ₹ 6 crores.

Solution 4. On enhancing price by 25% in 2008, total price of the machines produced during the years 2008

$$= \frac{10\times 125}{100} = ₹\ 12.5 \text{ crores}$$

Solution 5. Price of 1 machine in 2006 = $\frac{80000000}{25000}$ = ₹ 3200

Price of 1 machine in 2007 = $\frac{100000000}{20000}$ = ₹ 5000

∴ Required difference = ₹ (5000 – 3200) = ₹ 1800

Example 2: Directions: *Study the graph carefully to answer the questions that follow:*

Percentage Increase in profit of two companies over the years

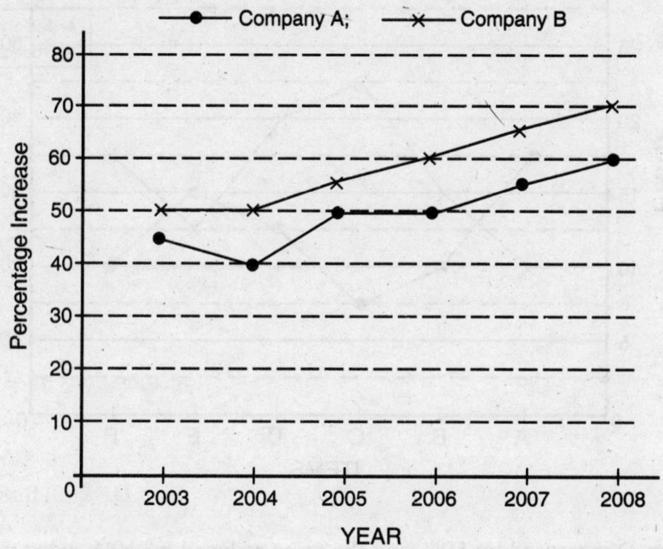

1. What is the per cent increase in profit of company A in the year 2007 from the previous year?
A. 5 B. 10 C. 45 D. 60

2. What is the per cent increase in profit of company A in the year 2008 from the previous year?
A. 7.64 B. 9.09 C. 8.12 D. 10.11

3. What is the per cent increase in profit of company A in the year 2005 from the previous year?
A. 10 B. 15 C. 20 D. 25

4. If the profit of company B in the year 2003 was ₹ 6,79,995, what would its profit have been in the year 2002?
A. ₹ 4,24,530 B. ₹ 4,53,330 C. ₹ 5,01,500 D. ₹ 5,53,330

5. Based on the graph, which of the following statements is True?
A. Company B has made the highest profit in the year 2008
B. Company A has made the lowest profit in the year 2004
C. Company B has made more amount of the profit than company A over the years
D. There is no increase in the profit of company B in the year 2005 from the previous year.

Solution 1. Required percentage = $\frac{55-50}{50}\times 100 = 10\%$

Solution 2. Required percentage = $\frac{60-55}{55}\times 100 = 9.09\%$

Solution 3. Required percentage = $\frac{50-40}{40}\times 100 = 25\%$

Solution 4. Required profit = $\frac{100}{150}\times 6{,}79{,}995 =$ ₹ 4,53,330

Solution 5. It is clear from the graph that company B has made the highest profit in the year 2008, i.e., 70%

Example 3: Directions: *Study the following graph carefully to answer these questions:*

Quantity of Various items sold and price per kg

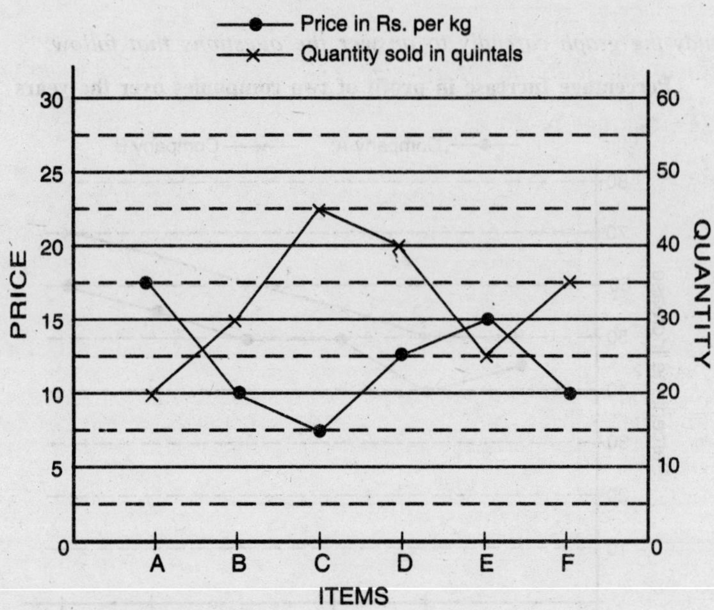

1. If the quantity sold of item D increased by 50% and the price reduced by 10%, what was the total value of the quantity sold for Item D?
   A. ₹ 675 B. ₹ 6750 C. ₹ 67500 D. ₹ 67550
2. What is the ratio between the total values of quantity sold for items E and F respectively?
   A. 3 : 2 B. 5 : 7 C. 7 : 5 D. 15 : 14
3. If the price as well as the quantity sold is increased by 20% for item A, what is the total value of quantity sold for item A?
   A. ₹ 42000 B. ₹ 48500 C. ₹ 49000 D. ₹ 50400
4. What is the average price per kg. of items A, B and C?
   A. ₹ 7.50 B. ₹ 9 C. ₹ 9.50 D. ₹ 11.67
5. Total value of quantity sold for item C is what per cent of the total value of the quantity sold for item E?
   A. 85 B. 87.5 C. 90 D. 111

Solution 1. New quantity of item D = $\frac{150}{100} \times 40 = 60$ quintal

New price per kg of item D = $\frac{90}{100} \times 12.50 =$ ₹ 11.25

Since, total value = ₹ (60 × 100 × 11.25) = ₹ 67500

Solution 2. Required ratio = (25 × 100 × 15) : (35 × 100 × 10) = 15 : 14

Solution 3. New price of item A per kg = $\frac{120}{100} \times 17.50 =$ ₹ 21

New quantity of item A = $\frac{120}{100} \times 20 = 24$ quintal

Since, total value = ₹ (24 × 100 × 21) = ₹ 50400

Solution 4. Average price per kg of items A, B and C= $\dfrac{17.50+10+7.50}{3}$ = ₹ 11.67

Solution 5. Required percentage = $\dfrac{45\times100\times750}{25\times100\times15}\times100$ = 90%

Example 4: Directions: *Study the following graph carefully and answer the questions that follow:*

Percentage profit earned by two companies over the given years

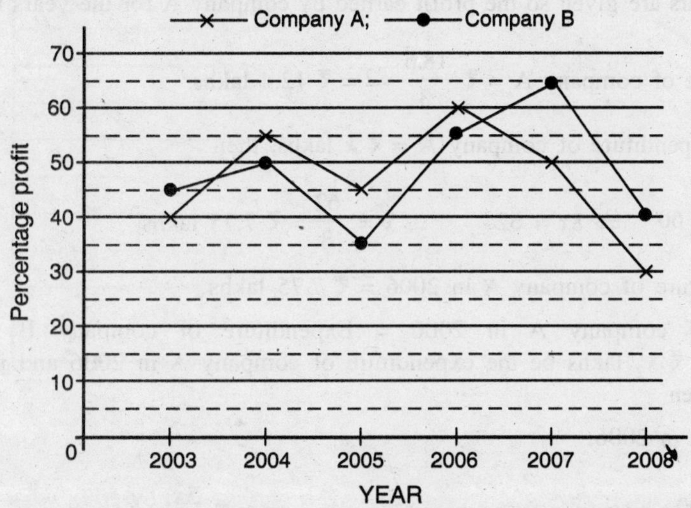

1. If the income of company A in 2005 was equal to its expenditue in 2007, what was the ratio between company's expenditure in the years 2005 and 2007 respectively?

A. 19 : 20 B. 20 : 19 C. 20 : 29 D. 29 : 20

2. If the total expenditure of the two companies in 2008 was ₹ 18 lakhs and expenditure of companies A and B in that year in the ratio of 4 : 5 respectively, then what was the income of company B in that year (in lakh ₹)?

A. 8 B. 10 C. 10.4 D. 14

3. If the total income of company A in all the years together was equal to the total expenditure of company B in all the years together, which was ₹ 265 lakhs, then what was the total percentage profit earned by Company A for all the years together?

A. 37 B. 45 C. 52 D. Cannot be determined

4. If the income of company B in 2006 was ₹ 18.6 lakhs and ratio of incomes of company A and B in 2006 was 2 : 3, what was the expenditure of company A in 2006 (in lakhs)?

A. 7.75 B. 9.75 C. 12 D. 12.4

5. If the income of company A in 2006 was equal to the total expenditure of company B in 2007, then what was the ratio of expenditure of company A in 2006 to the income of company B in 2007?

A. 10 : 13 B. 13 : 10 C. 25 : 66 D. 66 : 25

Solution 1. Let income of company A in 2005 = Expenditure of Company A in 2007 = ₹ x and also expenditure of company A in 2005 = ₹ x_1; then

For company A in 2005,

$$\frac{x-x_1}{x_1}\times100 = 45 \qquad \Rightarrow 29x_1 = 20x \qquad \therefore\ x_1 = \frac{20x}{29}$$

Since, required ratio = $\dfrac{20x}{29} : x$ = 20 : 29

Solution 2. In 2008, expenditure of company B = $\frac{5}{4+5} \times 18$ lakhs = ₹ 10 lakhs.

Let, income of company B in 2008 = ₹ x lakhs; then

$$\frac{x-10}{10} \times 100 = 40 \qquad \Rightarrow x - 10 = 4 \qquad \therefore x = ₹\ 14 \text{ lakhs}$$

Hence, income of company B in 2008 = ₹ 14 lakhs

Solution 3. Here, total income of company A in all the years together = ₹ 265 lakhs, while individual profit percentages for different years are given so the profit earned by company A for the years together can not be determined.

Solution 4. In 2006, income of company A = ₹ $\frac{18.6}{3} \times 2$ = ₹ 12.4 lakhs

Let in 2006, expenditure of company A = ₹ x lakhs, then

$$\frac{12.4 - x}{x} \times 100 = 60 \quad \Rightarrow 8x = 62 \qquad \therefore x = \frac{62}{8} = ₹\ 7.75 \text{ lakhs}$$

Hence, expenditure of company A in 2006 = ₹ 7.75 lakhs

Solution 5. Let income of company A in 2006 = Expenditure of company B in 2007 = ₹ x lakhs also ₹ x_1 lakhs, and ₹ x_2 lakhs be the expenditure of company A in 2006 and income of company B in 2007 respectively; then

For company A in 2006,

$$\frac{x - x_1}{x_1} \times 100 = 60 \qquad \Rightarrow 8x_1 = 5x \qquad \therefore x_1 = ₹\ \frac{5x}{8} \text{ lakhs}$$

For company B in 2007,

$$\frac{x_2 - x}{x} \times 100 = 65 \qquad \Rightarrow 20x_2 = 33x \qquad \therefore x_2 = ₹\ \frac{33x}{20} \text{ lakhs}$$

Since, required ratio = $\frac{5x}{8} : \frac{33x}{20} = \frac{5x}{8} \times 40 : \frac{33x}{20} \times 40$ = 25 : 66

PIE CHARTS

These are used to show the share of various sectors in the total. They usually show the percentage share of each sector in the whole (taken as 100%). In such representation the total quantity in question is distributed over a total angle of 360°. The area of each sector is proportional to the relative frequency of the class represented by the sector.

$$\text{Sector angle} = \frac{\text{Class frequency}}{\text{Total frequency}} \times 360°$$

For Example: ***Distribution of Expenditure of a family***

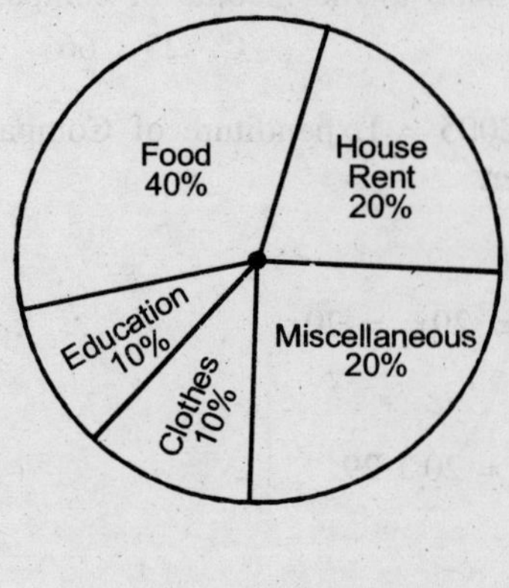

Example 1: Directions: *The pie-chart given below shows the household expenditure of a family on different items. Study the chart carefully and answer the questions that follows:*

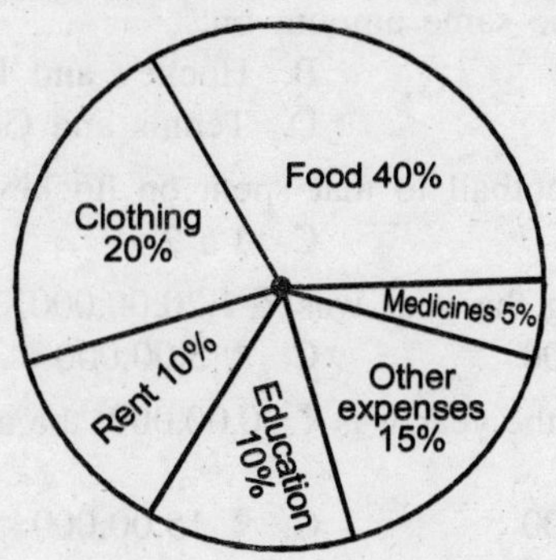

1. If the family's expenditure on education is ₹ 375 then the same amount is being expent by this family on which of the following items?
A. Medicines B. Other expenses C. Rent D. Clothing

2. If the family expends ₹ 750 per month on food, what is the annual expenditure of this family on education?
A. ₹ 2150 B. ₹ 1022.50 C. ₹ 2250 D. ₹ 1400

3. If total expenditure of this family is ₹ 4500, find the expenditure of the family on clothing?
A. ₹ 800 B. ₹ 900 C. ₹ 840 D. ₹ 950

4. What is the angle subtended on the centre by the segment representing other expenses?
A. 64° B. 54° C. 36° D. 15°

5. What is the ratio of total expenses on rent and clothing to the total expenses on medicines and other expenses of this family?
A. 2 : 3 B. 3 : 4 C. 3 : 5 D. 3 : 2

Solution 1. Both segments representing education and rent respectively are each 10%. Hence, family's expenditure on these two items will be equal, *i.e.*, ₹ 375 will be expent on rent.

Solution 2. Here, 40% = ₹ 750

$\Rightarrow 10\% = \frac{750}{40} \times 10 = ₹\ 187.50$

∴ Annual expenditure of the family on education = 187.50 × 12 = ₹ 2250

Solution 3. Expenditure on clothing = $\frac{20}{100} \times 4500 = ₹\ 900$

Solution 4. Central angle for 15% = $\frac{15}{100} \times 360° = 54°$

Solution 5. Expenditure on rent and clothing = 10% + 20% = 30%
Expenditure on medicine and other expenses = 5% + 15% = 20%
∴ Ratio of the two = 30% : 20% = 3 : 2

Example 2: Directions: *The pie chart, drawn here, shows the spending of a country on various sports during a particular year. Study the graph carefully and answer the questions that follow:*

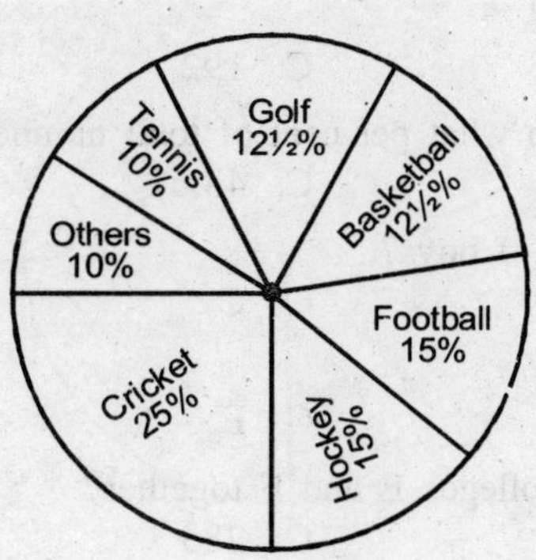

1. Graph shows that the most popular game of the country is:
A. Football B. Hockey C. Cricket D. Tennis

2. Out of the following the country spent the same amount on:
A. Hockey and Cricket B. Hockey and Football
C. Hockey and Golf D. Tennis and Golf

3. The ratio of the total amount spent on football to that spent on hockey is:
A. 2 : 1 B. 1 : 1 C. 1 : 2 D. 3 : 2

4. If the total amount spent on sports during the year was ₹ 1,20,00,000, how much was spent on basketball?
A. ₹ 16,00,000 B. ₹ 18,00,000 C. ₹ 3,00,000 D. ₹ 15,00,000

5. If the total amount spent on sports during the year was ₹ 30,00,000, the amount spent on cricket and hockey together was:
A. ₹ 18,00,000 B. ₹ 12,00,000 C. ₹ 15,00,000 D. ₹ 20,00,000

Solution 1. According to the graph, expenditure on cricket is the maximum. Therefore cricket is the most popular game.

Solution 2. According to the graph, equal amount has been spent on Hockey and Football.

Solution 3. Expenditure on the game of Football = 15% of total amount spent
Expenditure on the game of Hockey = 15% of total amount spent.
Therefore, the ratio of the expenditure on the two games = 1 : 1

Solution 4. 100% = ₹ 1,20,00,000
∴ Expenditure on the game of Basketball = 12½% of 1,20,00,000

$$= \frac{25 \times 1{,}20{,}00{,}000}{100} = ₹\ 15{,}00{,}000$$

Solution 5. 100% = ₹ 30,00,000
∴ Expenditure on the games of Cricket and Hockey = 25% + 15% = 40%
∴ Total expenditure on the games of Cricket and Hockey = 20% of 30,00,000

$$= \frac{40 \times 30{,}00{,}000}{100} = ₹\ 12{,}00{,}000.$$

Example 3: Directions: *Study the pie-charts carefully to answer the questions that follow.*

Percentage of Students in Six different Colleges
Total number of Students = 3500

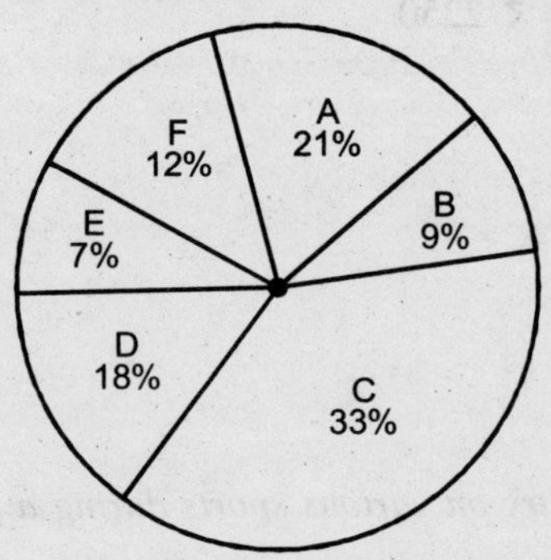

Percentage of Girls in each of the Colleges
Total number of Girls = 1800

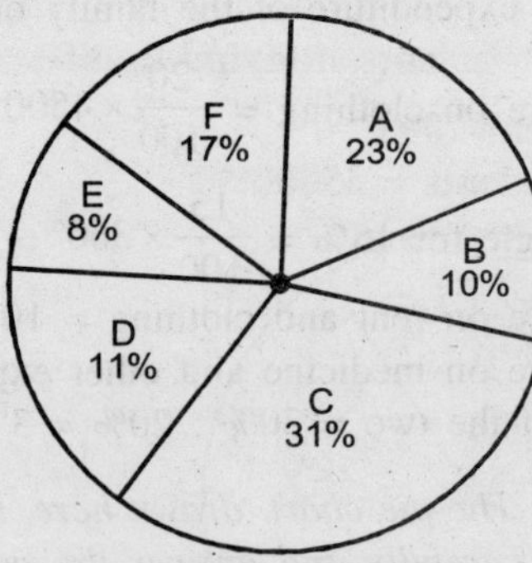

1. What is the number of girls in college D?
A. 176 B. 188 C. 192 D. 198

2. The number of boys from College A form what per cent of total number of students from that college?
A. 22.83 B. 38.41 C. 43.67 D. 56.29

3. Which college has the maximum number of boys?
A. A B. B C. C D. D

4. Which college has lowest number of girls?
A. B B. D C. E D. F

5. What is the total number of boys from Colleges E and F together?
A. 215 B. 251 C. 283 D. 310

Solution 1. Number of girls in College D = $\frac{11}{100} \times 1800 = 198$

Solution 2. Number of students from College A = $\frac{21}{100} \times 3500 = 735$

Number of girls from College A = $\frac{23}{100} \times 1800 = 414$

Since, number of boys from college A = 725 – 414 = 321

Required percentage = $\frac{321}{735} \times 100 = \frac{2140}{49} = 43.67\%$

Solution 3. Number of boys in college A = $\frac{21}{100} \times 3500 - \frac{23}{100} \times 1800 = 735 - 414 = 321$

Number of boys in college B = $\frac{9}{100} \times 3500 - \frac{10}{100} \times 1800 = 315 - 180 = 135$

Number of boys in college C = $\frac{33}{100} \times 3500 - \frac{31}{100} \times 1800 = 1155 - 558 = 597$

Number of boys in college D = $\frac{18}{100} \times 3500 - \frac{11}{100} \times 1800 = 630 - 198 = 432$

Hence, number of boys in college C has maximum.

Solution 4. It is clear from the pie-chart that percentage of girls in college E is minimum, hence, college E has lowest number of girls.

Solution 5. Number of boys from college E = $\frac{7}{100} \times 3500 - \frac{8}{100} \times 1800 = 245 - 144 = 101$

Number of boys from college F = $\frac{12}{100} \times 3500 - \frac{17}{100} \times 1800 = 420 - 306 = 114$

Hence, their total = 101 + 114 = 215

Example 4: Directions: *Study the following pie-diagrams carefully to answer these questions.*

Number of students studying in different faculties in the years 2008 and 2009 from State X

Year-2008
Total students = 35000

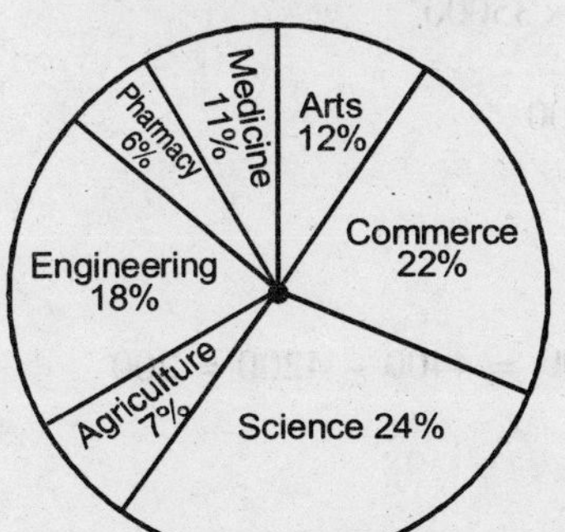

Year-2009
Total students = 40000

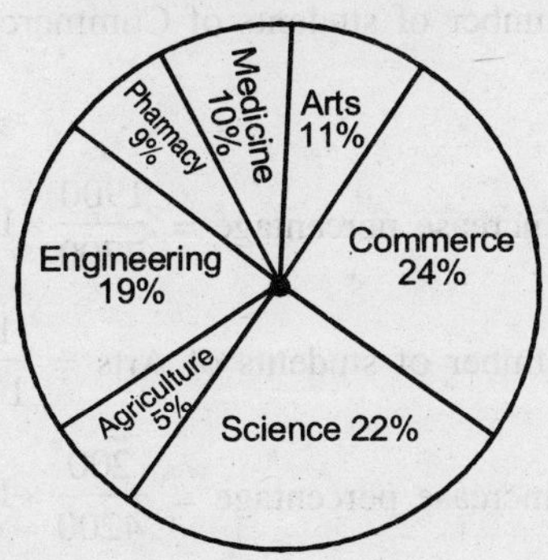

1. In which faculty there was decrease in the number of students from 2008 to 2009?
A. Arts B. Agriculture C. Pharmacy D. None of these

2. What was the approximate percentage increase in the number of students of Engineering from year 2008 to 2009?
A. 15 B. 17 C. 20 D. 25

3. In which of the following faculties the percentage increase in the number of students was minimum from 2008 to 2009?
A. Science B. Commerce C. Arts D. Medicine

4. What is the ratio between the number of students studying Pharmacy in the years 2008 and 2009 respectively?
A. 4 : 3 B. 5 : 7 C. 7 : 12 D. 8 : 13

5. In the year 2008, the number of students studying Arts and Commerce together is what per cent of the number of students studying these subjects together in 2009?
A. 76 B. 79 C. 82 D. 85

Solution 1. In 2008, number of students in Arts = $\frac{12}{100} \times 35000 = 4200$

In 2009, number of students in Arts = $\frac{11}{100} \times 40000 = 4400$

In 2008, number of students in Agriculture = $\frac{7}{100} \times 35000 = 2450$

In 2009, number of students in Agriculture = $\frac{5}{100} \times 40000 = 2000$

In 2008, number of students in Pharmacy = $\frac{6}{100} \times 35000 = 2100$

In 2009, number of students in Pharmacy = $\frac{9}{100} \times 40000 = 3600$

Hence, in Agriculture there was decrease in the number of students from 2008 to 2009.

Solution 2. In 2008, the number of students in Engineering = $\frac{18}{100} \times 35000 = 6300$

In 2009, the number of students in Engineering = $\frac{19}{100} \times 40000 = 7600$

Required percentage = $\frac{7600 - 6300}{6300} \times 100 = \frac{1300}{63} = 20.6 \approx 20\%$

Solution 3. Increase in number of students of Science = $\frac{22}{100} \times 40000 - \frac{24}{100} \times 35000$

$= 8800 - 8400 = 400$

Hence, Increase percentage = $\frac{400}{8400} \times 100 = 4\frac{16}{21}\%$

Increase in number of students of Commerce = $\frac{24}{100} \times 40000 - \frac{22}{100} \times 35000$.

$= 9600 - 7700 = 1900$

Hence, their increase percentage = $\frac{1900}{7700} \times 100 = 24\frac{52}{77}\%$

Increase in number of students of Arts = $\frac{11}{100} \times 40000 - \frac{12}{100} \times 35000 = 4400 - 4200 = 200$

Hence, their increase percentage = $\frac{200}{4200} \times 100 = 4\frac{16}{21}\%$

Increase in number of students of Medicine = $\frac{10}{100} \times 40000 - \frac{11}{100} \times 35000 = 4000 - 3850 = 150$

Hence, their increase percentage = $\frac{150}{3850} \times 100 = 3\frac{69}{77}\%$

So, in Medicine the percentage increase in the number of students was minimum from 2008 to 2009.

Solution 4. In 2008, the number of students studying Pharmacy = $\frac{6}{100} \times 35000 = 2100$

In 2009, the number of students studying Pharmacy = $\frac{9}{100} \times 40000 = 3600$

Hence, their ratio = 2100 : 3600 = 7 : 12

Solution 5. In 2008, the number of students studying Arts and Commerce together = $\frac{(12+22)}{100} \times 35000 = 34 \times 350 = 11900$

In 2009, the number of students studying Arts and Commerce together = $\frac{(11+24)}{100} \times 40000 = 35 \times 400 = 14000$

Hence, required percentage = $\frac{11900}{14000} \times 100 = 85\%$

Example 5: *Study the following Graph carefully and answer the questions given below:*

Percentage of students in various courses (A, B, C, D, E, F) and Percentage of girls out of these.
Total students = 1200 (800 girls + 400 boys)

(Total girls = 800)

Percentage in Various Courses

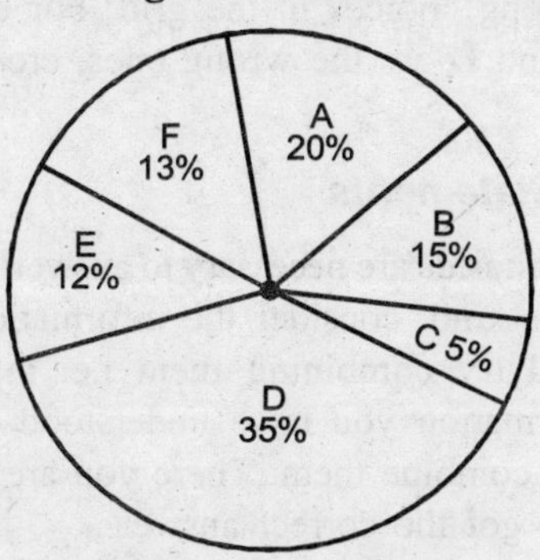

Percentage of Girls in Various courses

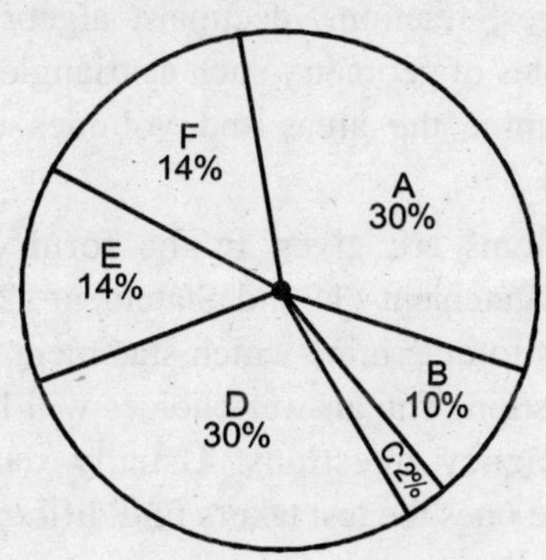

1. For which course is the number of boys the minimum?
(a) A *(b)* C *(c)* E *(d)* F

2. For course D what is the respective ratio of boys and girls?
(a) 3 : 4 *(b)* 3 : 5 *(c)* 4 : 5 *(d)* 5 : 6

Solution 1. Number of boys in course 'A' = $\frac{20}{100} \times 1200 - \frac{30}{100} \times 800 = 240 - 240 = 0$

Number of boys in course 'C' = $\frac{5}{100} \times 1200 - \frac{2}{100} \times 800 = 60 - 16 = 44$

Number of boys in course 'E' = $\frac{12}{100} \times 1200 - \frac{14}{100} \times 800 = 144 - 112 = 32$

Number of boys in course 'F' = $\frac{13}{100} \times 1200 - \frac{14}{100} \times 800 = 156 - 112 = 44$

Hence, in course 'A' number of boys is minimum *i.e.*, zero.

Solution 2. For course D,

Number of boys = $\frac{35}{100} \times 1200 - \frac{30}{100} \times 800 = 420 - 240 = 180$

Number of girls = $\frac{30}{100} \times 800 = 240$

Since, required ratio = 180 : 240 = 3 : 4

15

Data Sufficiency

Data sufficiency questions primary concern is to test how well you can reason quantitatively. This differs widely from quantitative section which tests your ability of manipulating numbers whereas **data sufficiency** tests your reasoning ability.

Concepts necessary to solve Data Sufficiency Questions

The data sufficiency questions covers maths and algebra concepts that are taught in High schools and colleges. In addition to basic arithmetic, you can expect questions testing your knowledge of averages, fractions, decimals, algebra, factoring, and basic principles of geometry such as triangles, circles, and how to determine the areas and volumes of simple geometric shapes.

Data sufficiency problems are given in the form of numbered statements like Statement (1) and Statement (2). The task of the test taker is to determine which statement is required to answer the question. The answer choices will be the same for data sufficiency questions. Usually data sufficiency problems are the ones the test takers find difficult to answer.

Rather than relying on your own assumptions, find out the information contained in the statement. Separate assumptions from the meaning hidden in the statement. By doing this it becomes easier to identify the correct answer.

Elimination of Answer Choices

There is one more tactic to separate the wrong answer choices from the right ones. It is nothing but simple memorization of answer choices. Practice memorizing the answer choices while preparing with practice questions.This will in turn helps a test taker to memorize the answers in the test section. Then rephrase them and arrange them. This should be done to eliminate the wrong answer choices.

Avoid Lengthy Calculations

This is the best tactic to avoid ending up with wrong answers. Calculating by depending only upon the formulae is not sufficient. This will result in long calculation. The trick for short calculations is to calculate the problem according to the information given in the statement.

Diagrams

Some of the data sufficiency problems contain figures or diagrams along with the statement. These diagrams contain information relevant to the answers. Identify the information in those diagrams and try fitting them with answer choice. This will help you in short calculations and in saving time.

Drawing Grids

To separate the assumptions, draw a grid on a scratch paper and write the alphabetical order of the choices. Now start eliminating the wrong choices in the grid. For example if the choices B, C and D are the wrong ones, cross them in the grid.

Combining the Statements

In some cases both choices are necessary to answer a question. In situations of this kind, consider the information in both the statements and try combining them i.e. information. Rephrase the information you have understood from both the statements and combine them. There you are! You will find that you have got the correct answer.

Evaluating Statements on their Own

There is a technique which is applicable for quantitative section, "Do not rely on what your mind conceives". Some statements seem to have the same idea. But the statements are designed in such a way so as to trick you. Evaluating statements is a common area where test takers tend to make errors.

Evaluating each statement on its own is the only way to avoid such errors. For example, assume that statement A is X and it may be either an odd number or even number. And also assume that statement B is Y, which might be an odd, even or prime number.

More over for real numbers used in data sufficiency, this method will end up in short and systematic calculation. As a result you will find the answer choice which is relevant to answer.

When an individual follows these strategies for data sufficiency, he/she will find it easier to answer the questions within the given time. Above all nothing can beat constant practice to be more efficient in data sufficiency section.

MULTIPLE CHOICE QUESTIONS

Directions (Qs. 1 to 7): *Each question is followed by two statements, I and II. Indicate your responses based on the following directives:*

Mark (A) if the question can be answered using I alone but not using II alone.

Mark (B) if the question can be answered using II alone but not using I alone.

Mark (C) if the question can be answered using either statement alone.

Mark (D) if the question can be answered using I and II together but not using I or II alone.

1. A man's annual salary for three consecutive years is in the ratio 4 : 5 : 6. What is the total savings of the man at the end of third year?

I. In the first two years, the man saved 30% of his salary which is ₹ 30,000 less than the third year's savings.

II. The ratio of income and expenditure of the man for the three years is 30 : 11.

2. The numerator and denominator of a fraction are positive integers. Is the numerator less than the denominator?

I. If both the numerator and the denominator are increased by 1, the fraction becomes 3/4

II. If both the numerator and the denominator are decreased by 1, the fraction becomes 5/7

3. In an exam, each correct answer gets 1 mark, a wrong answer –1/3 mark and no mark for the unanswered questions. How many marks did Kiran score in that exam, if the mark he scored is divisible by 3?

I. Out of a total of 50 questions in the paper, Kiran attempted 39.

II. 18 questions attempted by Kiran were wrong.

4. Raja and his wife have a son. The sum of the present ages of Raja and his son is 70 years. Find the present age of Raja's wife.

I. When Raja's wife would attain Raja's age, the sum of their ages would be 140 years.

II. When Raja's son would attain Raja's wife's age the sum of the ages of Raja and his son would be 160 years.

5. What is the value of the positive integer x?

I. $3x + 4y + 5z = 12$; y and z are positive integers.

II. $3x + 4y + 5z = 0$; y and z are integers.

6. Amit says, "After x years, my age will be 2 times the age of Bimal". Bimal says, "After $2x$ years my age will be same as the current age of Amit". What is the value of x?

I. The age difference between Amit and Bimal is 5 years.

II. After $2x$ years Amit is five times as old as Bimal is now.

7. A maid is appointed for 30 days with the condition that for every day that she worked, she would get ₹ 50 and for every day she did not work, ₹ 10 would be deducted from her earnings. At the end of 30 days can she take home more than ₹ 500?

I. She worked for at least 16 days.

II. She worked for at most 20 days.

Directions (Qs. 8 to 14): *Each question is followed by two statements, I and II Indicate your responses based on the following directives:*

Mark (A) if the question can be answered using one of the statements alone, but cannot be answered using the other statement alone.

Mark (B) if the question can be answered using either statement alone.

Mark (C) if the question can be answered using I and II together but not using I or II alone

Mark (D) if the question cannot be answered even using I and II together.

8. A number was increased by $x\%$ and then further increased by $y\%$. Was it effectively increased by more than 50%?

I. $x + y = 50$.

II. $x = 20$ and $y = 30$.

9. What was the cost price of an article which was sold for ₹ 144?

I. The percentage profit made was numerically equal to the cost price in rupees.

II. Profit made on the article was 45% of selling price.

10. Praveen and Rakesh started a business with investments in the ratio 2 : 3. The profit made by them at the end of a year was ₹ 60000. Find Praveen's share of profit

I. Praveen received ₹ 12000 less thatn Rakesh as his share.

II. Praveen and Rakesh stayed in the business till the end of the year.

11. A sum of ₹ P was lent for 2 years at R% p.a. at simple interest. Find R.

I. If ₹ (P + 1000) was lent for 2 years at R% p.a. at simple interest, the interest fetched would be ₹ 200 more than the interest received.

II. If ₹ P was lent for 2 years at (R + 10)% p.a. at simple interest, the interest fetched would be ₹ 200 more than the interest received.

12. A person bought an article and sold it at a loss of 10%. What is the cost price of the article?

I. If he had bought it for 20% less and sold it for ₹ 55 more, he would have made a profit of 40%.

II. If he had bought it for 20% more and sold it for ₹ 135 more, he would have made a profit of 20%.

13. Find the profit made by Somu on selling his watch.

I. If the cost price of Somu was 20% less, his profit would have been ₹ 60 more.

II. If the selling price of Somu was 20% more, his profit would have been ₹ 80 more.

14. A sum of ₹ P was lent at R% p.a. simple interest for 2 years. Find the simple interest.

I. The difference in the simple interest and the compound interest, compound interest being compounded annually on ₹ P at R% p.a. for 2 years is ₹ 10.

II. The interest earned on ₹ P in the first year in ₹ 10.

Directions (Qs. 15 to 17): *Each question is followed by two statements, I and II. Answer each question using the following instructions:*

Mark (A) if the question can be answered using statement I alone but not by using statement II alone.

Mark (B) if the question can be answered using statement II alone but not by using statement I alone.

Mark (C) if the question can be answered by using both the statements together but not by either of the statements alone.

Mark (D) if the question cannot be answered by using either of the statements alone.

Mark (E) if the question cannot be answered on the basis of the two statements.

15. When a number is divided by 18, the quotient and the remainder are 3 and x respectively. What is the value of x?

I. When the number is divided by 15, the remainder is 1.

II. When the number is divided by 8, the remainder is 2.

16. S represents the sum of the cubes of three consecutive integrers. Is S divisible by 18?

I. The middle number among the integers is divisible by 3

II. the middle number among the integers is divisible by 2.

17. A, B and C arē natural numbers. Is (A + 2B + 8C) divisible by 4?

I. B and C are even numbers.

II. A is an odd number.

Directions (Qs. 18 to 19): *Each question is followed by two statements, I and II. Indicate your responses based on the following directives:*

Mark (A) if the question can be answered using I alone but not using II alone.

Mark (B) if the question can be answered using II alone but not using I alone.

Mark (C) if the question can be answered using I and II together but not using I or II alone.

Mark (D) if the question can not be answered using I and II together.

18. Is $x > y$?

I. $\log_2 x = \log_4 y$

II. $\log x - \log 4 = \log y - \log 16$

19. A is a positive integer. Is A divisible by 3?

I. $10^A - 1$ is divisible by 37.

II. $10^A + 1$ is divisible by 11.

Directions (Qs. 20 to 24): *Each question is followed by two statements, I and II. Answer each question using the following instructions:*

Mark (A) if the question can be answered by using statement I alone but not by using statement II alone.

Mark (B) if the question can be answered using statement II alone but not by using statement I alone.

Mark (C) if the question can be answered by using both the statements together but not by either of the statements alone.

Mark (D) if the question can be answered by using either of the statements alone.

Mark (E) if the question cannot be answered on the basis of the two statements.

20. Are the roots of the equation $x^2 + ax + k^2 = 0$ real where a and k are real?

I. k is a root of the equation.

II. The square of the sum of the roots is greater than the product of the roots.

21. The series S is in geometric progression. Find the common ratio.

I. The sum of the first three terms of S is 26 and all terms of S are integers.

II. If the first term is multiplied by 4 and the second term is doubled, then the first three terms of S are in arithmetic progression.

22. The ratio of the sums of the first 13 terms of two APs with positive common differences is equal to the ratio of the sums of the first 23 terms. What is the ratio of the third terms of the two APs?

I. The 4th terms of the two APs are 1 and 2 respectively.

II. The 6th term of the second AP is 0.

23. Find the sum of the first twenty five terms of an arithmetic progression.

I. Four times the twelfth term of the arithmetic progression is equal to thrice the sixteenth term of the progression.

II. Thirteenth term of the progression is 4.

24. How many terms of a A.P., whose first term is 10, must be taken to make a sum of 24?
I. Common difference d is –2.
II. 4th term is 4.

Directions (Qs. 25 to 27): *Each question is followed by two statements, I and II. Indicate your responses based on the following directives:*

Mark (A) if the question can be answered using I alone but not using II alone.

Mark (B) if the question can be answered using II alone but not using I alone.

Mark (C) if the question can be answered using I and II together but not using I or II alone.

Mark (D) if the question cannot be answered even using I and II together.

25. Is the value of x non-negative?
I. $|3x - 2| = 10$.
II. $|x - 5| = 5$.

26. A and B are real numbers. Is $A = B$?
I. $A^2 = B^3$
II. $A^3 = B^2$

27. Is X > Y?
I. (X + Y) > X.
II. (X – Y) > 2Y.

Directions (Qs. 28 to 36): *Each question is followed by two statements, I and II. Answer each question using the following instructions:*

Mark (A) if the question can be answered by using statement I alone but not by using statement II alone.

Mark (B) if the question can be answered by using statement II alone but not by using statement I alone.

Mark (C) if the question can be answered by using both the statements together but not by either of the statements alone.

Mark (D) if the question can be answered by using either of the statements alone.

Mark (E) if the question cannot be answered on the basis of the two statements.

28. The average ages of the boys and the girls in a group are 20 years and 18 years respectively. Is the average age of the group more than 19 years?
I. There are at least as many boys as there are girls in the group.
II. There are at most as many boys as there are girls in the group.

29. Find the average of N natural numbers.
I. If 3/4th of the numbers are increased by 1 and the remaining are decreased by 3, the average of the numbers would be 2.5
II. If half of the N numbers are doubled and the remaining are tripled, the average of the numbers would be 6.75.

30. A vessel had X litres of milk. Y litres was withdrawn from it and then replaced with an equal amount of water. Find the present ratio of milk and water in it.
I. X = 10 Y
II. X – Y = 90

31. A boy has a total of 45 one-rupee and fifty paise coins. Find the total value of the coins with the boy.
I. The average value per coin is $75\frac{5}{9}$ paise.
II. The value of the rupee coins is more than that of the fifty paise coins by 1350 paise.

32. A dishonest milkman professes to sell his milk at cost price but he mixes it with water. Find the percentage of water in the mixture.
I. By adding water the milkman gains 25%.
II. Milk in the new mixture is 80%.

33. The average weight of boys in a class is 28 kg and that of girls is 21 kg. Find the average weight of the class.
I. The ratio of boys to girls is 5 : 2.
II. Number of boys in the class is 30.

34. The ratio of milk to water in a mixute is 5 : 1. What quantity of water should be added to make it 5 : 3?
I. The amount of water to be added is 1/3rd of that the original mixute.
II. The ratio of half the amount of milk to water in the original mixture is 5 : 2.

35. The number of students in section A is 20 and that of section B is 25. If 5 students from section A go to section B does the average weight of class B increase or decrease?
I. The weight of the heaviest student in section A is 50 kg and the weight of the lightest student in section B is 35 kg.
II. The weight of the lightest student in section A is 45 kg and the weight of the heaviest student in section B is 40 kg.

36. Which is the smallest of the three numbers, whose average is 14?
I. One of the numbers is equal to the average of the three numbers.
II. The largest and the smallest of the numbers are equally spaced from the middle number.

Directions (Qs. 37 to 43): *Each question is followed by two statements, I and II. Indicate your responses based on the following directives:*

Mark (A) if the question can be answered using I alone but not using II alone.

Mark (B) if the question can be answered using II alone but not using I alone.

Mark (C) if the question can be answered using either statement alone.

Mark (D) if the question can be answered using I and II together but not using I or II alone.

37. Satish, Rakesh and Prakash completed a job for a total of ₹ 1000. Find the share of Satish in this amount.
I. The ratio of the times for which Satish, Rakesh and Prakash worked is 3 : 2 : 1.
II. The ratio of the parts of the job completed by Satish, Rakesh and Prakash is 3 : 2 : 1.

38. P, Q and R can complete a job in 20 days, x days and y days respectively. All the three worked together for 6 days and then P left. Q and R completed the remaining part of the job. Who completed the maximum part of the job?
I. $4x = 3y$
II. $x = 30$

39. B can do a piece of work in 12 days. In how many days can A alone do the work?
I. A and B together completed the work and distributed the total earnings in the ratio of 3 : 2.
II. Two days work of A is equal to three days work of B.

40. A and C take 20 days to do a piece of work. B and C take 15 days to do the same work. How many days does C alone take to complete the work?
I. A and B together take 12 days to complete the work.
II. A, B and C together take 10 days to complete the work.

41. A can work twice as fast as B. C can work thrice as fast as B. How many days does A alone take to complete a piece of work?
I. B and C take 15 days to complete the work.
II. A and C take 12 days to complete the work.

42. Two taps P and Q together take 3 minutes 36 seconds to fill an empty tank. How many minutes does a drain pipe R take to empty the tank when it is full?
I. P and R can fill the tank in 9 minutes.
II. Q alone takes 9 minutes to fill the tank.

43. A certain number of people can complete a work in 18 days. How many more men are required to complete the work 3 days early?
I. 270 men can complete the work in a day.
II. If the number of people is half the original number, they complete half the work in 18 days.

Directions (Qs. 44 to 62): *Each question is followed by two statements, I and II. Indicate your responses based on the following directives:*

Mark (A) if the question can be answered using one of the statements alone, but cannot be answered using the other statement alone.
Mark (B) if the question can be answered using either statement alone.
Mark (C) if the question can be answered using I and II together but not using I or II alone.
Mark (D) if the question cannot be answered even using I and II together.

44. Praveen travelled from A to B at a certain speed and returned at a lesser speed. Was his forward travel time more than 4 hours?
I. The sum of his forward and return travel times was 8 hours.
II. His forward travel time was 1 hour less than his return travel time.

45. Rajesh started from his house for his office at a kmph and reached his office b minutes early. If he started from his house at b kmph, he would have reached his office 'a' minutes late. Find the usual time he would take to reach his office.
I. $a + b = 10$
II. $\dfrac{ab}{a-b} = 12$

46. If Satish travelled from his home to school at a speed a % higher than his usual speed, he would reach school x minutes early. If he instead travelled at b% lower than his usual speed, he would reach school y minutes late. Is $x > y$?
I. $a = 10$
II. $b = 10$

47. Car A started from P towards Q at x kmph at 8:00 a.m. Car B started from P towards Q at y kmph at 10:00 a.m. At what time B overtook A?
I. $x : y = 2 : 3$
II. PQ = 360 km

48. Car X started from A towards B. Car Y started simultaneously from B towards A. X and Y take a hours and b hours respectively to reach their destinations from their meeting point. Find the time they would take to meet.
I. $a = \dfrac{2500}{b}$
II. AB = 5000 km.

49. A train crosses a bridge in 1minute at a speed of 20 kmph. What is the length of the bridge?
I. The length of the trian is 200 m.
II. A man can cross the same bridge in 5 minutes.

50. Is the radius of a cylinder C equal to its height?
I. When the radius and the height of C are interchanged, its volume remains unchanged.
II. When the radius and the height of C are interchanged, its total surface area remains unchanged.

51. S is a square and R is a rectangle but not a square. Is the side of S less than the length of R?
I. The areas of S and R are equal.
II. The reciprocal of the side of S is equal to the average of the reciprocals of the length and the breadth of R.

52. R is a rectangle. C_1 and C_2 are cylinders. Is the radius of the base of C_2 more than that of C_1?

I. C_1 was formed by folding R along its length.

II. C_2 was formed by folding R along its breadth.

53. A sphere of radius 128 cm is melted and "*n*" small spheres are formed. Find '*n*'.

I. Radius of the each small sphere is 4 cm.

II. Radius of the large sphere is 32 times the radius of each small sphere.

54. What is the ratio of the areas of the squares ABCD and PQRS?

I. The ratio of the perimeters of ABCD and PQRS is 3 : 2.

II. The ratio of the lengths of the diagonals of ABCD and PQRS is 3 : 2.

55. Find the angle of elevation of the top of the tower at a point on the ground when the height of the flag staff standing on the tower is 7 metres.

I. Height of the tower is 30 meters.

II. The shadow of the tower is $\sqrt{3}$ times of the height of the tower.

56. A bag, has a total of 12 white and black balls. Threre are more white balls than black balls in it. Find the number of black balls in it.

I. The probability of drawing a white ball is at most 2/3

II. The probability of drawing a black ball is at most 1/3

57. Ashok purchased some apples, mangoes and oranges, each of which cost him ₹ 9, ₹ 8 and ₹ 3 respectively. Find the number of apples purchased.

I. He spent a total of ₹ 67 in purchasing them.

II. The number of apples purchased is a multiple of the number of mangoes purchased but not equal in number.

58. At a party, every person shook hands with every other person exactly once. Find the number of persons who attended the party.

I. The number of handshakes at the party is 300.

II. If instead of attending the party, each person sent a greeting card to every other person then the number of cards required is 90.

59. What is the area of the triangle AOB?

I. AOB is a right angled isosceles triangle right angled at O the origin.

II. The midpoint of AB is (3, 4)

60. Find the distance between the towers.

I. The heights of the towers are 80 metres and 62.5 metres respectively.

II. The line joining the tops of the towers makes an angle 45° with the horizontal line.

61. If A and B are two independent events, then find $P(A \cap B)$.

I. P (A) = P(B) = 0.4

II. $P(A \cup B) = 0.64$

62. Is $a = b$?

I. ${}^nP_a = {}^nP_b$

II. ${}^nC_a = {}^nC_b$

Directions (Qs. 63 to 73): *Each question is followed by two statements, I and II. Answer each question using the following instructions:*

Mark (A) if the question can be answered by using statement I alone but not by using statement II alone.

Mark (B) if the question can be answered using statement II alone but not by using statement I alone.

Mark (C) if the question can be answered by using both the statements together but not by either of the statements alone.

Mark (D) if the question can be answered by using either of the statements alone.

Mark (E) if the question cannot be answered on the basis of the two statements.

63. Each face of a cube is painted with exactly one of the colours—Red, Blue and Green. It is then cut into 125 smaller and identical cubes. How many of the smaller cubes have all the three colours on them?

I. No two adjacent faces of the large cube were painted with the same colour.

II. Two faces of the large cube were painted in red, two faces in blue and the remaining two faces in green colour.

64. A team of three members is to be formed from a group of two managers A and B, two assistants C and D, and two accountants E and F. Who is definitely selected?

I. A is selected only if D is selected. C and F are not selected together.

II. Only one among B and E is selected.

65. All apples are not bananas. All babanas are oranges. All oranges are tasty. Is A tasty?

I. A is an apple.

II. A is not a apple.

66. All human beings are monkeys. All monkeys are vegetarian. No vegetarian is a flesh eater. Is Ravi a flesh eater?

I. Ravi is a human being.

II. Ravi is not a human being.

67. All cats are tigers. All rats are dogs. No tiger is a dog. Is X a rat?

I. X is a tiger.

II. X is a cat.

68. In a group of people, each one watches at least one of the two games—Football and Cricket. Find the ratio of the number of people who watch Football to the number of people who watch Cricket.

I. 50% of the people who watch Football also watch Cricket.
II. 40% of the people who watch Cricket also watch Football.

69. In a colony, 50 persons read "*The Hindu*". Thirty among those who read "*The Hindu*" also read "*The Telegraph*". How many persons read '*The Telegraph*' but not "*The Hindu*"?
I. 20 persons read neither "*The Hindu*" nor "*The Telegraph*".
II. 60 persons do not read "*The Hindu*".

70. In a class 50% of the students passed in English. What percentage of students failed in both English and Maths?
I. 50% of the students passed in Maths.
II. 25% of the students passed in English and Maths.

71. In a class of 100 students, if 40 students failed in Maths, 40 students filed in Science and 30 students failed in Geography, how many students were passed only in Science?
I. 20 students were failed in all the three subjefts and 30 students were failed in none of the three subjects.
II. 15 students were failed only in Science.

72. If Kishore wakes up in the morning, he goes to the office. Only if Kishore does not go to the office, then he meets his boss. Only if Kishore meets his boss, he watches a movie in the evening. Did Kishore watch a movie on Monday evening?
I. Kishore went to the office on Monday.
II. Kishoere did not go to the office on Monday.

73. Each one of Rohan, Sanjay and Sunil plays a game among Cricket, Football and Hockey. Does Sanjay play Cricket?
I. Rohan either plays Cricket or Football. If Sanjay plays Cricket then Rohan plays Hockey.
II. Sunil plays either Football or Hockey, Sanjay plays either Cricket or Football and Rohan neither plays Cricket nor Football.

Directions (Qs. 74 to 81): *Each question is followed by two statements, I and II. Answer each question using the following instructions.*

Mark (A) if the question can be answered using statement I alone.

Mark (B) if the question can be answered using statement II alone.

Mark (C) if the question can be answered using both the statements together.

Mark (D) if the question cannot be answered using both the statements and some additional information is required to answer the question.

74. Six persons—Saroj, Sathya, Sumit, Sujal, Sambhabi and Sukla are sitting in a row facing same direction. Who is sitting at the extreme left of the row?
I. Sukla is sitting to the right of Sumit but to the immediate left of Saroj and none of Saroj, Sumit and Sukla is sitting at the extreme ends.
II. Sumit and Sujal, neither of whom is sitting at the extreme right, are sitting to the right of Sambhabi, who is two places away tc the right of Saroj.

75. Abhishek, Kishore, Pratap, Ramkumar and Vijay are sitting around a circular table not necessarily in the same order. Is Pratap sitting next to Vijay?
I. Pratap and Ramkumar sit next to each other.
II. Abhishek and Kishore do not sit in adjacent chairs.

76. Six persons—Mita, Mini, Monk, Motu, Mithu and Mala are standing in a queue, may not be in the same order. Is Motu the third person from the front end of the queue?
I. Motu is standing behind Mini and is standing ahead of Mala.
II. Mithu is three places ahead of Mita.

77. Five friends Anand, Bharat, Cindra, Dravid and Estlla sit in a row facing a particular direction, not necessarily in the same order. Who is sitting exactly at the middle of the row?
I. If Bharat and Dravid interchange their positions, then Dravid is to the immediate right of Estlla.
II. If Anand interchanges his position with Cindra, then Estlla is the only person sitting adjacent to him.

78. Six persons—Ashmit, Bikash, Charan, Dinesh, Ekta and Firoj are sitting around a circular table. Who is sitting opposite Ashmit?
I. Neither Charan nor Dinesh is adjacent to or opposite Bikash.
II. Charan is not sitting adjacent to Ekta and Firoj is not sitting adjacent to Dinesh.

79. Seven books—physics, chemistry, sociology, maths, biology, history and geography are kept in a stack, may not be in the same order. Is the history book kept at the middle?
I. The physics book is kept at the bottom, but not immediately below the history book.
II. The number of books above the chemistry book is the same as the number of books below the biology book. Exactly three books are there between the history book and the geography book.

80. In a queue of 36 persons, what is the position of Manoj from the front end?
I. Manoj is equidistant from Dinesh and Komal and Komal is the 13th person from the front end of the queue.
II. Suresh is the 15th person from the rear end of the queue and there are 15 persons in between Suresh and Manoj.

81. Ravi wanted to go from his house to his office. He started from his house and travelled 15 km. He then took a right turn and travelled 12 km before taking a left turn and travelling further 6 km to reach junction B. How far and in which direction is his office from his home?

I. After reaching junction B, he travelled 8 km to reach his office

II. On reaching junction B, he took a right turn to travel towards east another 8 km to reach his office.

Directions (Qs. 82 to 96): *Each question is followed by two statements, I and II. Indicate your response based on the following directives:*

Mark (A) if the question can be answered using one of the statements alone, but cannot be answered using the other statement alone.

Mark (B) if the question can be answered using either statement alone.

Mark (C) if the question can be answered using I and II together but not using I or II alone.

Mark (D) if the question cannot be answered even using I and II together.

82. If each alphabet has a specific code and the word APPEAL is coded as 122344 and the word PILLAR is coded as 1124. , then find the code for the word CATERPILLAR.

I. The word REPEAT is coded as 233467.

II. The word PLACATE is coded as 1223479.

83. Is X the cousin of Y?

I. X is the son of the father of Y.

II. Y is the daughter of the mother of X.

84. Six persons P,Q, R, S, T and U attended a party. Among them there are two couples. Is Q the husband of S?

I. P's father's mother's only child's spouse is Q.

II. R is the father of Q and they belong to different gender.

85. Is P the youngest person among P, Q, R, S and T?

I. At most one person is yonger than P.

II. Exactly three persons are elder than P.

86. Avinash, Adarsh and Shashank are the top rankers in three different subjects Maths, Physics and Chemistry not necessarily in the same order. Among these no two students got the same rank in any two subjects. Who cannot be the top aanker in Chemistry?

I. Avinash got better rank than Shashank in two subjects.

II. Shashank is the 2nd ranker is Maths.

87. Is A the mother of B?

I. B is the son of A?

II. A is the sister-in-law of C whose husband is D.

88. In a code language if "man is great" is coded as "# ¥@" then what is the code for "cow is god"?

I. In the code language, "man are superman" is coded as "¥αδ" and "pinky is god" is coded as "w#λ".

II. In the code language, "cow is great" is coded as "β#@" and "man is god" is coded as "#λ¥".

89. A is the father-in-law of D whose only son is E.Is E the father of G?

I. D is the grand father of G.

II. G is the grand son of C whose husband is D.

90. If "God is Great" is coded as "Do Re Mi" and as "Humans are Good" is coded as "So La Si", then find the Code for "God".

I. "God is Good" is coded as "Re Si Do".

II. "Good are Great" is coded as "Mi Si So".

91. Five persons—P, Q, R, S, and T belong to five different cities—Mumbai, Chennai, Hyderabad, Bangalore and Delhi. Which city does Q belong to?

I. Either P or Q belongs to Bangalore. Either R or S belongs to Chennai.

II. T belongs to either Delhi or Mumbai. R belongs to either Hyderabad or Delhi.

92. A, B, C, D and E are five consecutive years of which only one is a leap year. Find which one among them is the leap year.

I. The total number of days, in A, B, C together is more than the number of days in C, D, E.

II. The total number of days in A, B, C together is equal to the number of days in C, D, E.

93. Is X a leap year?

I. In the year X, July has the same calendar as that of January.

II. In the year X October is not having the same calendar as that of January.

94. Which day is 1 March this year?

I. 1 January this year was a Sunday.

II. There was only one day of the week that occurred 53 times last year.

95. Three boxes are labelled as X, Y and Z. One box contains only red balls, another box contains only green balls while the third box contains both red and green balls. Which box contains balls of only green colour?

I. A ball is picked up from the box labelled Z and it is found to be green in colour

II. Two balls are picked each from the boxes labelled X and Y, it is found that they are in red colour.

96. A, B, C and D are four members of a family. Among them, there is a Doctor, a Lawyer, a Professor and an Engineer. Is B the Professor?

I. Among them there is a couple who have two children. The Doctor is not the husband of Professor.

II. A, the Engineer is the brother of Lawyer C. B and C belong to the same gender while A and B belong to different genders.

ANSWERS

1	2	3	4	5	6	7	8	9	10
D	C	D	D	A	A	A	B	B	B
11	**12**	**13**	**14**	**15**	**16**	**17**	**18**	**19**	**20**
A	B	C	A	A	B	B	B	D	A
21	**22**	**23**	**24**	**25**	**26**	**27**	**28**	**29**	**30**
B	C	B	E	B	C	C	B	A	A
31	**32**	**33**	**34**	**35**	**36**	**37**	**38**	**39**	**40**
D	D	A	E	B	E	B	A	C	C
41	**42**	**43**	**44**	**45**	**46**	**47**	**48**	**49**	**50**
C	D	A	A	A	C	A	A	A	B
51	**52**	**53**	**54**	**55**	**56**	**57**	**58**	**59**	**60**
B	C	B	B	D	D	C	B	C	C
61	**62**	**63**	**64**	**65**	**66**	**67**	**68**	**69**	**70**
A	C	A	C	E	A	D	C	C	C
71	**72**	**73**	**74**	**75**	**76**	**77**	**78**	**79**	**80**
A	A	D	B	C	D	C	C	B	B
81	**82**	**83**	**84**	**85**	**86**	**87**	**88**	**89**	**90**
B	C	B	B	D	C	C	A	D	D
91	**92**	**93**	**94**	**95**	**96**				
D	A	B	D	A	C				

SOME SELECTED EXPLANATORY ANSWERS

1. Neither of the statements is independently sufficient, as we do not know the savings of the third year.

Using both the statements, we have 30% of $(4k + 5k) + 30{,}000$ = Savings of the third year.

As the ratio of income and expenditure = 30 : 11, the ratio of income and savings becomes 30 : 19.

Combining the two statements, total income of 3 years $= 4k + 5k + 6k = 15k$

Total savings of 3 years = 30% of $4k$ + 30% of $5k$ + (30% of $4k$ + 30% of $5k + 30{,}000$) $= 5.4k + 30{,}000$

Ratio of 3 year's income to year's savings $= (15k) : (5.4k + 30{,}000) = 30:19$.

This equation can be solved for k; and savings can be calculated.

2. Let the fraction be $\frac{n}{d}$

Using statement I,

$$\frac{n+1}{d+1} = \frac{3}{4}$$

$$4n + 4 = 3d + 3$$

$$4n = 3d - 1$$

$$\therefore \quad n = \frac{3}{4}d - \frac{1}{4} < \frac{3}{4}d. \quad \therefore \frac{n}{d} < \frac{3}{4} < 1$$

I is sufficient.

Using statement II,

$$\frac{n-1}{d-1} = \frac{5}{7}$$

Let $n - 1 = 5k$ and $d - 1 = 7k$ where k is a positive integer.

$n = 5k + 1$ and $d = 7k + 1$

As k is a positive integer, $5k < 7k$. $\therefore n < d$.

II is sufficient.

3. From statement I alone, we cannot find the score as we do not know the number of mistakes he committed. His mark can be 3 or 15 or 27 or 39.

From statement II alone, we cannot find the number of marks he got as we do not know the number of questions he attempted. Each statement is not sufficient.

From statements I and II, we can find his marks as 39 $- 18 - \frac{18}{3} = 15$.

4. Let the present ages of Raja, his wife and his son be x years, y years and z years respectively.

$$x + z = 70 \qquad ...(1)$$

Using statement I,

When Raja's wife would attain Raja's age, she would be x years old.

She would attain Raja's age $(x - y)$ years from now.

Raja's age then $= x + x - y = (2x - y)$ years.

$$x + 2x - y = 140$$

$$3x - y = 140 \quad ...(2)$$

we have two equations (1) and (2) with three unknowns x, y and z.

$\therefore$ y cannot be found,

I is not sufficient.

Using statement II, when Raja's son would attain Raja's wife's age, he would be y years old.

He would attain Raja's wife's age $(y - z)$ years from now.

Raja's age then $= (x + y - z)$ years.

$$x + y - z + y = 150$$

$$x + 2y - z = 150 \quad ...(3)$$

We have two equations (1) and (2) with three unknowns x, y and z.

$\therefore$ y cannot be found.

II is not sufficient.

Using both statements,

From (1) $\Rightarrow$ $z = 70 - x$

From (2) $\Rightarrow$ $x = \dfrac{140+y}{3}$

$$\therefore \quad \left(\frac{140+y}{3}\right)+2y-\left(70-\left(\frac{140+y}{3}\right)\right)=160$$

y can be found.

Both statements taken together are sufficient to answer the question.

5. From statement I,

$3x + 4y + 5z = 12$

Since $x, y, z \in z^+$ (a set of positive integers)

The value of each of x, y, z should be 1

$(3)(1) + (4)(1) + 5(1) = 12$. So $x = 1$

Statement I alone is sufficient.

From statement II

$3x + 4y + 5z = 0$

y and z are integers.

x is positive so y and z are negative.

If $y = z = -1$; then $3x = 9 \Rightarrow x = 3$

If $y = z = -2$; then $3x = 18 \Rightarrow x = 6$.

So unique values of x is not possible.

So statement II alone is not sufficient.

6. Let the present age of Amit be a years.

Let the present age of Bimal be b years.

Given

$$a + x = 2(b + x) \rightarrow \text{(I)}$$

Also, $\quad b + 2x = a \rightarrow$ (II)

From (I) and (II), we have

$$b + 3x = 2b + 2x$$

$$b = x$$

$$= x + 2x = a \text{ so } a = 3x$$

From statement I

$$a - b = 5$$

thus, $3x - x = 5$

Hence x can be found.

So statement I alone is sufficient.

From statement II,

$a + 2x = 5b$

As $a = 3x$ and $b = x$,

we get, $5x = 5x$

From this, x can't be found out, so

Statement II alone is not sufficient.

7. Let the number of days the maid was present be x.

$\therefore$ The number of days she was absent be $(30 - x)$

$$50x - (30 - x)10 > 500$$

$$5x - (30 - x) > 50$$

$$5x - 30 + x > 50$$

$$6x > 80$$

$$x > 13.\overline{3}$$

Statement I

Number of days the maid was present is greater than $13.\overline{3}$,

$\therefore$ She will get more than ₹ 500

Statement I is sufficient.

Statement II

She was present at most for 20 days.

Statement II is not sufficient to answer the question.

8. Let the number be N initially

Its value after the x% increase $= \text{N}+\dfrac{x}{100}\text{N}$

$$= \text{N}\left(1+\frac{x}{100}\right)$$

Its value after y% increase

$$= \text{N}\left(1+\frac{x}{100}\right)+\frac{y}{100}\left[\text{N}\left[\left(1+\frac{x}{100}\right)\right]\right]$$

$$= \text{N}\left(1+\frac{x}{100}\right)\left(1+\frac{y}{100}\right) = \text{N}\left(1+\frac{x+y}{100}+\frac{xy}{100^2}\right)$$

Precentage increase in the number

$$= \frac{\text{N}\left(1+\dfrac{x+y}{100}+\dfrac{xy}{100^2}\right)-\text{N}}{\text{N}}(100)\%$$

$$= \frac{x+y}{100} + \frac{xy}{100^2}(100)\% = \left(x+y+\frac{xy}{100}\right)\%$$

Using statement I, $x + y + \frac{xy}{100} > x + y = 50$

I is sufficient.

Using statement II, x and y are known.

∴ effective percentage increase in the number can be found.

II is sufficient.

Either of the statements is sufficient.

9. Given SP = ₹ 144, if c is the cost price.

From statement I, $c(1 + c/100) = 144$

$$100c + c^2 = 14{,}400$$

$c^2 + 100c - 14{,}400 = 0$ or $c = 80$ or -180.

As c cannot be negative c is 80.

From statement II,

$$\frac{SP-CP}{SP} \times 100 = 45$$

$$\frac{144-CP}{144} \times 100 = 45$$

So, CP can be calculated.

Either of the statements is sufficient.

10. Let the share of Praveen be ₹ p.

Share of Rakesh = ₹ $(60000 - p)$.

Using statement I, $p = 60000 - p - 12000$

$$p = 24000$$

I is sufficient.

Using statement II, ratio of profits = ratio of investments.

$$\therefore \quad p = \frac{2}{2+3}(60000) = 24000$$

II is sufficient. Either of the statements is sufficient.

11. Using statement I, extra interest $= (1000)\left(\frac{R}{100}\right)(2)$

$$= 200$$
$$R = 10$$

I is sufficient.

Using statement II, extra interest $= (P)\left(\frac{10}{100}\right)(2)$

$$= 200$$
$$P = 2000$$

II is not sufficient.

12. Given, SP = 0.9 (CP)

From statement I, 0.8 (CP) + 55 = (1.4)(0.9)(CP). Hence, CP can be found by solving it with the given equation.

From statement II, 1.2 CP + 135 = 1.2(1.2) (CP). Hence, CP can be found by solving it together with the given equation. Hence, each statement by itself is sufficient.

13. Let the cost price of Somu and his selling price be ₹ x and y respectively.

Using statement I, extra profit = decrease in the cost price $= \frac{20}{100}x = ₹\ 60$

$$x = ₹\ 300$$

y is unknown.

∴ profit cannot be found. I is not sufficient.

Using statement II, extra profit = increase in the selling price $= \frac{20}{100}y = 80$

$$y = 400$$

x is unknown. ∴ profit cannot be found.

II is not sufficient.

Using both statements, x and y are both known.

∴ profit can be found.

Both statements taken together are required to answer the question.

14. Using statement I, different between CI and SI for 2 years $= P\left(\frac{R}{100}\right)^2 = 10$

P and R are both unknown. I is not sufficient.

Using statement II, $\left(\frac{PR}{100}\right) = 10$

Simple interest $= P\left(\frac{R}{100}\right)2 = 20$

Since the interest for one year is known, interest for two years can be found.

∴ Statement II is sufficient.

15. When the number is divided by 18 the quotient is 3, so the number is greater than or equal to 54 and less than 71.

From statement I, the value of the number is only 61 as from the number 54 to the number less than 74 there is only one number (61) which leaves remainder 1 when divided by 15.

So when 61 is divided by 18 the remainder is 7. hence $x = 7$.

Statement I alone is sufficient.

From Statement II, the possible values of the number are 58 and 66.

When 58 is divided by 18 the remainder is 4.

When 66 is divided by 18 the remainder is 12. Hence, x can be 4 or 12. As x is not unique, statement II alone is not sufficient.

16. Using Statement I,

If the numbers are 2, 3 and 4S = 99.

In his case, S is not divisible by 18.

If the numbers are –1, 0, 1, S = 0.

In this case, S is divisible by 18.

I is not sufficient.

Using Statement II,

If the numbers are –1, 0 and 1, S = 0.

If the numbers are 1, 2 and 3, S = 36.

If the numbers are 3, 4 and 5, S = 216.

It can be seen that in all the cases considered, S is divisible by 18.

∴ S must always be divisible by 18.

II is sufficient.

17. From statement I, B and C are even numbers.

So 2B + 8C is divisible by 4, but we do not know whether A is divisible by 4 or not.

So statement I alone is not sufficient.

From statement II

A is an odd numbers so (A) + (2B + 8C)

= (odd) + (even) = odd

(A + 2B + 8C) is an odd number

Odd number are not divisible by even numbers.

So A + 2B + 8C is not divisible by 4.

Statement II alone is sufficient.

18. Using statement I, $\log_2 x = \dfrac{\log_2 y}{\log_2 2^2} = \dfrac{\log_2 y}{2} = \log_2 y^{\frac{1}{2}}$

$\therefore\ x = y^{\frac{1}{2}}$

If $y = \dfrac{1}{4}$ then $x = \dfrac{1}{2}$. In this case, $x > y$. If $y = 4$, $x = 2$. In this case, $x < y$.

Statement I is not sufficient.

Using statement II log x – log 4 = log y – log 16

$$\log\left(\frac{x}{4}\right) = \log\left(\frac{y}{16}\right)$$

$$\frac{x}{4} = \frac{y}{16}$$

$$y = 4x. \quad \therefore \quad y > x$$

Using statement II we can answer the question.

19. From statement I, if A = 1, $10^A - 1 = 0$.

If A = 3, $10^A - 1 = 999$.

In both cases $10^A - 1$ is divisible by 37.

Only in the second case, A is divisible by 3.

I is not sufficient.

From statement II, if A = 1, $10^A + 1 = 11$

If A = 3, $10^A + 1 = 1001$

In both cases, $10^A + 1$ is divisible by 11.

Only in the second case, is divisible by 3.

II is not sufficient.

Even both statements taken together are not sufficient to answer the question.

20. Using statement I, if one root is real, the other is also real, as all the coefficients are real.

Using statement II, $a^2 > k^2$

Let $a^2 = ck^2$ where $c > 1$

The discrimination $= \sqrt{a^2 - 4k^2} = \sqrt{(c-4)k^2}$

If $c \geq 4$ roots are real otherwise they are complex.

We can answer the question from statement I but not from statement II.

21. Let the first terms be a and the common ratio be r.

Using statement I, $a + ar + ar^2 = 26$

$\Rightarrow\ a(1 + r + r^2) = 26$

As all the terms are integers, a and r are integers.

Hence a can be 1 or 2 or 13 or 26

When $a = 1$, $1 + r + r^2 = 26$,

$r^2 + r - 25 = 0$

$$r = \frac{-1 \pm \sqrt{(-1)^2 - 4(1)(25)}}{2} = \frac{-1 \pm \sqrt{101}}{2}$$

Similarly, we can find the value of r for other values of a.

Only when $a = 2$ or 26, we get the corresponding integer value of r.

From statement II, if the first three terms of S are a, ar and ar^2, then we have $4a + ar^2 = 2(2ar)$

$\Rightarrow\ r^2 + 4 = 4r$

$\Rightarrow\ r^2 - 4r + 4 = 0$

$\Rightarrow\ (r - 2)^2 = 0$

$\Rightarrow\ r = 2$

∴ Statement II alone is sufficient.

22. Let the two APs be a, $a + 2d$. .... and b, $b + e$, $b + 2e$, ...

We have $\dfrac{a+6d}{b+6e} = \dfrac{a+11d}{a+11e} = \dfrac{d}{e}$

$\left(\text{If } \frac{N_1}{D_1} = \frac{N_2}{D_2},\right.$ both expression are equal to $\left.\frac{N_2 - N_1}{D_2 - D_1}\right)$

Therefore $\frac{a+6d}{b+6e} = \frac{d}{e} = \frac{-6d}{-6e} = \frac{a}{b}$

The ratio of any two corresponding terms is the same. From I, it follows that the ratio of any two corresponding terms is 1/2 provided, both the terms are not 0.

From II, it follows that the third terms are not 0 (as $d > 0$ and $e > 0$ and $a + 5d = b + 5e = 0$)

Therefore, even statement II is needed to make sure that the third terms are not both 0. We need both the statements.

23. From statement I, if a is the first term and d is the common ratio, then we have $4(a + 11d) = 3(a + 15d)$. But with this, we cannot find any value.

From statement II, we know the thirteenth term. Sum of the first twenty-five terms

$= 25/2[a + (a + 24d)]$

$= 25/2[2(a + 12d)] = 25/2[2 \times t_{13}]$

As we know t_{13}, we can find the sum from statement II alone.

24. $a = 10$, $s = 24$

$24 = n/2[2a + (n - 1)d]$ ..... (1)

From statement I, $d = -2$

On solving $n = 3$ or 8, so not sufficient.

From statement II, $10 + 3d = 4 \Rightarrow d = -2$

Substituting $d = -2$ in (1) $n = 3$ or 8, so not sufficient.

25. Statement I

$|3x - 2| = 10$

$3x - 2 = 10$

$3x = 12$	$-3x + 2 = 10$
$x = 4$	$-8 = 3x$

$x = \frac{-8}{3}$

Statement I is not sufficient.

Statement II; $|x - 5| = 5$.

$x - 5 = 5$ or $-x + 5 = 5$

$x = 10$ or $x = 0$

Statement II is sufficient.

26. Statement I,

A = 8, B = 4, In this case, A > B. Say A = 0, B = 0. In this case, A = B.

I is not sufficient.

Statement II,

Say A = 4, B = 8. In this case, A < B.

Say A = 0, B = 0. In this case, A = B. II is not sufficient.

Using both statements,

$A^2 = B^3$ and $A^3 = B^2$

$\therefore A = B^{\frac{3}{2}}$ and $A^3 = \left(B^{\frac{3}{2}}\right)^3 = B^2$

$B^2\left(B^{\frac{5}{2}} - 1\right) = 0$

$B^2 = 0$ or $B^{\frac{5}{2}} = 1$

$B = 0$ or $B^5 = 1^2 = 1$

$B = 0$ or 1.

If $B = 0, A = 0$

If $B = 1, A = 1$

In either case, A = B.

Both statements taken together are required to answer the question.

27. From statement I alone, as $(X + Y) > X \Rightarrow Y > 0$. So, we cannot conclude whether X > Y or X < Y. Hence I alone is not sufficient. From statement II alone, as $(X - Y) > 2Y \Rightarrow X > 3Y$

For X = –3, Y = –2

X > 3Y but X < Y

For X = –1, Y = –2

X > 3Y and X > Y

Hence II alone is not sufficient.

From statements I and II together, we have Y > 0 and X > 3Y.

For Y > 0, 3Y > Y and X > 3Y (given) hence X > Y.

So, I and II together are sufficient to answer the question.

28. The average age of the group will be more than 19 years if there are more boys than girls.

Using statement I, if there are more boys than girls, average age will be more than 19 years. If there are equal number of boys and girls it will be 19 years. I is not sufficient.

Using statement II, if there are equal number of boys and girls the average age will be 19 years. If there are more girls than boys, the average age will be less than 19 years. In any case, the average will never be more than 19 years. II is sufficient.

29. Using statement I, new sum = 3/4(N) (+1) + 1/4(N) (–3) initial sum

∴ old average (–1) new average = 2.5. I is sufficient.

Using statement II, let the averages of the half which

is doubled and that which is tripled be a and b respectively. Initial sum = $\frac{N}{2}a+\frac{N}{2}b$.

Old average = $\frac{a+b}{2}$.

New sum = $\frac{N}{2}(2a)+\frac{N}{2}(3b)$ = 6.75 N

i.e., $a + b = 6.75 - b/2$.

b is unknown. $\therefore$ $\frac{a+b}{2}$ can not be found.

II is not sufficient.

30. After Y litres was withdrawn, the vessel would have (X – Y) litres of milk.

$\therefore$ the vessel would finally have X – Y litres of milk and Y litres of water

$\therefore$ required ratio = $\frac{X-Y}{Y}=\frac{X}{Y}-1$

Using statement I,

$\frac{X}{Y}$ = 10. I is sufficient.

Using statement II, neither both X and Y known nor is $\frac{X}{Y}$ known.

$\therefore$ the question cannot be answered with the information in statement II. Ii is not sufficient.

31. Using statement I, total value of the coins = $75\frac{5}{9}(45)$ paise.

I is sufficient.

Let the number of one rupee coins and fifty paise coins be a and b respectively.

$a + b = 45$ and $100a - 50b = 1350$

solving these, $a = 24$ and $b = 21$.

The question cannot be answered. II is sufficient.

32. From statement I, we can know that the milkman added 20% water. From statement II, we can know that the milkman added 20% water.

33. Statement I, average weight = 5/7(28) + 2/7(21)

= 26kg.

Statement II is not sufficient as the number of girls are unknown.

34. From statement I, we do not get an idea of either the total volume of the initial mixture or of the volume of either of the components. Hence, statement I alone is not sufficient.

From statement II also, we do not get any information regarding the volumes of the total mixture or of the components and hence, statement II alone is not sufficient. Even when we take both the statements together, we have the same problem.

35. From statement I, if the total weight of 5 students who are transferred from section A to section B is more than 35(5) = 175 kg, then the average weight of the class B will increase. Otherwise, we can't say anything as we don't know the average weight of the section A. So first statement alone is not sufficient.

From statement II, if the total weight of the five students who are transferred from section A to section B is more than 40(5) = 200 kg, the average weight of class B will increase. The total weight of the five students is definitely more than 45(5) = 225 kg. Therefore, the average weight of section B will increase. We can answer the question from II alone.

36. Statement I says that one number is 14 and does not give any information about the remaining numbers. Statement II tells us that the numbers are of the form $14 - d$, 14, $14 + d$, but do not give any information to determine d uniquely. Even if we take both the statements together, we cannot determined uniquely and hence data is still insufficient.

37. As we do not have any information about the efficiencies. I alone is not sufficient.

Ratio of wages = Ratio of parts of the job done. We obtain the ratio of the parts of the job done only from statement II. II is sufficient.

38. Part of the job completed by P = $\frac{6}{20}=\frac{3}{10}$

Remaining part = $\frac{7}{10}$

This will be completed by Q and R. Parts of the job completed by Q and R will be in the ratio $\frac{1}{x}:\frac{1}{y}=y:x$

Using statement I, $\frac{y}{x}=\frac{4}{3}$

Parts of the remining part done by Q and R can be found. I is sufficient.

Using statement II, y is unknown.

$\therefore$ Parts of the remaining part done by Q and R cannot be found. II is not sufficient.

39. From statement I,

The earnings are distributed in the ratio of work done, as one days work ratio of A and B is 3 : 2, so the time taken by A and B is in the ratio 2 : 3.

Hence we can find the time taken by A to complete the work.

Statement I alone is sufficient.
From statement II,

A's 2 day's work = $3\times\frac{1}{12}$ = 1/4

A's 1 day's work = 1/8

So time taken by A to complete the work is 8 days

So statement II alone is sufficient. Hence either statement alone is sufficient.

40. Question statement gives, A + C do 1/20 ...(1) of the work in 1 day B + C do 1/15 ...(2) of the work in 1 day.

Statement I gives A + B do 1/12 in 1 day ...(3) we can get C from the three equations and hence, statement I alone is sufficient.

Statement II gives A + B + C do 1/10 in 1 day ...(4). By solving (1), (2) and (4), we can get C and hence, statement II alone is also sufficient.

41. Already in the question statement, relationship between efficiencies of A, B and C is given.

From statements I, B and C 1 day's work = 1/15

As B and C can be expressed in terms of A, time taken by A can be found.

∴ Statement I alone is sufficient.

From statement II, A and C 1 day's work 1/12

As C can be expressed in terms of A, time taken by A can be found.

∴ Statement II alone is sufficient.

42. Given P + Q = 1/216.

Statement I gives P + R = 1/540. It is clear that statement I alone is not sufficient as it gives another equation, but not the values of P and R. Statement II gives Q = 1/540. This statement also gives one more equation and hence is not sufficient by itself to find R. When both the statements are taken together, we get three equations and by solving we get P, Q and R and hence can answer the question.

43. Statement I gives that the work requires 270 man days and hence the number of men required to complete in 15 days is 18. Hence, statement I alone is sufficient. Statement II does not mention anything about the total quantity of work and hence is insufficient.

44. Let the forward and return travel times of Praveen be f hours and r hours respectively.

Using statement I, $f + r = 8$ ...(1)

Forward speed > Return speed

$\therefore f < r$

$\therefore f < 4$ as $f + r = 8$

I is sufficient.

Using statement II, $f = r - 1$

r is unknown

∴ the question cannot be answered. II is not sufficient.

45. Let the usual time be t hours.

$$\text{Distance} = a\left(t-\frac{b}{60}\right)\text{km} = b\left(t+\frac{a}{60}\right)\text{km}$$

$$\Rightarrow t = \frac{ab}{30(a-b)}$$

Using statement I,

$\frac{ab}{a-b}$ is not know.

t cannot be found

I is not sufficient.

Using statement II,

$\frac{ab}{a-b}$ is known.

t can be found

II is sufficient.

46. Let the usual time of travel of Satish be t minutes. If his speed was a % higher, it would be $\frac{100+a}{100}$ times his usual speed. His time would then be $\left(\frac{100}{100+a}\right)t$ minutes. If his speed was b% lower, it would be $\frac{100-b}{100}$ times his usual speed. His time would then be $\left(\frac{100}{100-b}\right)t$ minutes.

If $b = a$, $x = t - \frac{100t}{100+a}$ and $y = \frac{100t}{100-a} - t$

$$x - y = 2t - \left(\frac{(100t)(200)}{100^2 - a^2}\right)$$

$\therefore y > x$ whenever $a = b$

Both statements taken together are required to answer the question.

47. Distance travelled by A by 10 : 00 a.m. = $2x$ km.

Time of overtaking = $\frac{2x}{y-x}$ hours after 10:00 a.m.

$$\frac{2x}{y-x} = \frac{2}{\frac{y}{x}-1}$$

If $\frac{y}{x}$ is known, or both y and x known, the required time can be found. Using statement I,

$$\frac{y}{x} = \frac{3}{2}$$

I is sufficient.

Using statement II,

neither $\frac{y}{x}$ nor both y and x are known. II is not sufficient.

48. Let the meeting point be M.

Let the time taken to meet be t hours.

X would take t hours to travel AM for which Y would take b hours. Y would take t hours to travel BM for which X would take a hours.

Let the speeds of X and Y be x kmph and y kmph respectively.

$$AM = xt = yb \quad ...(1)$$

$$BM = yt = xa \quad ...(2)$$

Multiplying (1) and (2),

$$(xt)(yt) = (yb)(xa)$$

$$t > 0. \quad \therefore t = \sqrt{ab}$$

If ab or both a and b are known, then t can be found.

Using statement I, $ab = 2500$

I is sufficient.

Using statement II, neither ab nor both a and b are known.

II is not sufficient.

49. Time taken to cross the bridge

$$= \frac{\text{length of the bridge + length of the train}}{\text{speed of the train}}$$

Since we know, the time taken by the train to cross the bridge, the train's speed and the length of the train, so we can uniquely determine the length of the bridge.

Statement I alone is sufficient.

Statement II alone is not sufficient as we do not have any information about the trains speed.

50. Let the radius and the height of C be r and h respectively.

Using statement I, $\pi r^2h = \pi h^2 r$.

$$\pi rh\,(r - h) = 0$$

$$rh > 0$$

$$\therefore \quad r - h = 0 \quad i.e., \quad r = h$$

I is sufficient.

Using statement II, $2\pi\, r(r + h) = 2\pi h(h + r)$

$$2\pi(r^2 - h^2) = 0 \text{ i.e., } 2\pi(r + h)(r - h) = 0$$

$$2\pi(r + h) \neq 0$$

$$\therefore \quad r - h = 0 \quad i.e., \ r = h$$

II is sufficient.

Either of the statements is sufficient.

51. R is not a square.

$\therefore$ Its lenght is not equal to its breadth. Let the side of S be a. Let the lenght and the breadth of R be ℓ and b respectively

$$\ell \neq b.$$

$$\therefore \quad \ell > b$$

Using I, $a^2 = \ell b < \ell^2$

$$\therefore \quad a < \ell$$

I is sufficient.

Using statement II, $\frac{1}{a} = \frac{\frac{1}{\ell}+\frac{1}{b}}{2} > \frac{\frac{2}{\ell}}{2} = \frac{1}{\ell}$

$$\therefore \quad a < \ell$$

II is sufficient.

$\therefore$ Either of the statements is sufficient.

52. Let the length and the breadth of R be ℓ and b respectively.

As the information about C_1 and C_2 are given in different statements, the question cannot be answered by using either of the statements. Using both statements, the length of R would form the circumference of C_1 and its breadth would form the height of C_1.

$$\therefore \quad \text{Radius of } C_1 = \frac{\text{lenght of R}}{2\pi}$$

$$\text{Similarly, radius of } C_2 = \frac{\text{breadth of R}}{2\pi}$$

$\therefore$ Radius of $C_2 \leq$ radius of C_1.

$\therefore$ Radius of C_2 can never exceed the radius of C_1.

Both statements taken together are required to answer the question.

53. Volume of the sphere $= \frac{4}{3}\pi r^3$

Where r is radius of the sphere.

from statement I,

$$n = \frac{\frac{4}{3}\pi(128)^3}{\frac{4}{3}\pi(4)^3}$$

So statement I alone is sufficient.

From statement II, $\frac{R}{r} = 32$

$$n = \frac{\frac{4}{3}\pi R^3}{\frac{4}{3}\pi r^3} = \left(\frac{R}{r}\right)^3 = (32)^3$$

So statement II alone is also sufficient.

54. From statement I, the ratio of the sides will also be 3 : 2 or the ratio of the areas will be 3 : 2 i.e., 9 : 4. From statement II, the area of the square = (diagonal)/2 we have the ratio of two areas of squares will be $\frac{3^2}{2}:\frac{2^2}{2}$ = 9 : 4. Hence each statement alone is sufficient.

55. Statement I

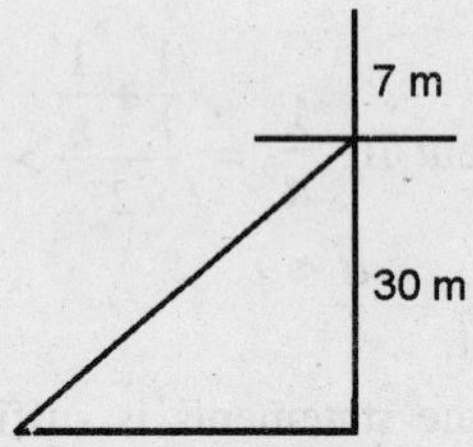

Data is not sufficient to compare the sides and get the angle of the right angled triangle.

∴ Statement I alone is not sufficient.

Statement II

Shadow of the tower is $\sqrt{3}$ times of the tower. But, the point at which the angle is measured is not known.

∴ Statement II alone is not sufficient.

Both the statements even combined the question cannot be answered.

56. There are more white balls than black balls in a total of 12 balls.

∴ There are more than 6 white balls and less than 6 black balls. Let us say there are w white balls and b black balls.

Using statement I, $\frac{w}{12} \le \frac{2}{3}$

$w \le 8$

∴ $w = 7$ or 8

∴ $b = 5$ or 4

Statement I alone is not sufficient

Using statement II, $\frac{b}{12} \ge \frac{1}{3}$

∴ $b \ge 4$

∴ $b = 4$ or 5

Statement II alone is not sufficient

Using both the statements we get,

$b = 4$ or 5, *i.e.*, when both the statements combined together also we cannot answer the question.

57. Given the cost of each of apple, mango and orange respectively is ₹ 9, ₹ 8 and ₹ 3

From statement I Ashok spent an amount of ₹ 67. The following table gives the possible number of fruits of each variety

Apples	Mangoes	Oranges	Total cost
1	2	14	67
2	2	11	67
3	2	8	67
4	2	5	67

We cannot uniquely determine the number of apples purchased.

∴ Statement I alone is not sufficient.

From statement II, we do not know the total amount spent. We cannot find the number of apples.

But combining the statements I and II only the 4th row of table i.e., 4 apples, 2 mangoes and 5 oranges satisfies the given conditions.

∴ Number of apples purchased = 4.

58. From statement I, the number of hand shakes = 300 *i.e.*, $^nC_2 = 300$

$\Rightarrow \frac{n(n-1)}{2} = 300$ or $n(n-1) = 600 = 25\ (24)$

$\Rightarrow n = 25$

∴ Statement I is sufficient.

From statement II: The number of cards required = 90

$\Rightarrow {}^nP_2 = 90 \Rightarrow n(n-1) = 10\ (9) \Rightarrow n = 10$

∴ Statement II is also sufficient.

We can answer the question, either of the statement.

59. From statement I, we get the location of one vertex. This is not enough to find the area of the triangle

∴ Statement I alone is not sufficient.

From statement II: We know only the mid-point of AB. We cannot find the area of the triangle.

But combining the two statements, the triangle is right angled and isosceles. The length of the median to the hypotenuse is 5. Each leg in $5\sqrt{2}$ and the are is 25. We can get the area by combining the two statements.

60. Since angle of elevation is not given in statement I.

∴ I alone is not sufficient.

Since heights of the towers are not given

Statement II alone is also not sufficient.

When I and II are combined

We get AB = 80 m; CD = 62.5 m and ∠EDB = 45°.

In Δ DEB tan 45 = BE/DE

DE = BE = AB – AE = 80 – 62.5 = 17.5

∴ The distance between two towers is 17.5 m.

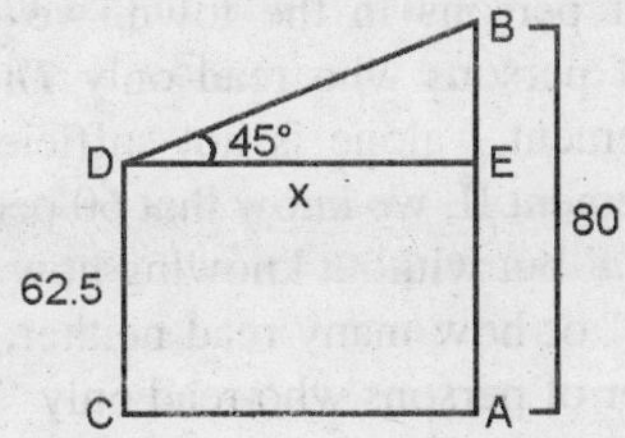

Combining both the statements, we can answer the question.

61. Since A and B are independent

$P(A \cap B) = P(A)\,P(B)$ ______(2)

Using statement I, $P(P \cap B) = (0.4)^2 = 0.16$.

I is sufficient to answer the question.

Using statement II, $P(A \cup B) = P(A) + P \cup B - P(A \cap B)$

$0.64 + P(A).\ P(B) = P(A) + P(B)$

We have only one equation with two unknowns P(A) and P(B)

So P(A) and P(B) cannot be found.

II alone is not sufficient to answer the question.

62. Statement I

$\therefore\ {}^nP_a = {}^nP_b \Rightarrow a \neq b\ (\because\ {}^6P_5 = {}^6P_6)$

$a = b + 1 = n.$

Statement I is not sufficient.

Using statement II. ${}^nC_a = {}^nC_b \Rightarrow a \neq b$

$\because {}^5C_3 = {}^5C_2$ or $a = n - b.$

$\therefore$ Combining both the statements, we get $a = b$.

63. From statement I, we can find the number of cubes which will have all three colours on them.

Since no two adjacent faces are painted with the same colour, opposite faces are painted with the same colour. Cubes which are at the corners of the original cube will have three colours on them. There will be eight such cubes.

Statement I alone is sufficient.

From statement II, we cannot find the number of cubes which will have all the 3 colurs on them.

If adjacent cubes are painted in the same colour, then 2 cubes will have all three colours on them.

Whereas if opposite faces are painted in the same colour, then 8 cubes will have three colour on them.

Thus statement B alone is not sufficient.

64. From I either ADC,ADF, ADE, ADB, BCD, DCE, DBF, DEF, BCE can be selected. So, I alone is not sufficient.

From II as we do not have any information about the remaining people we cannot determine who is definitely selected.

So, II alone is not sufficient.

Combining I and II, among four people B, C, E and F only two can be selected, as we need three people D must be selected. (A need not be selected every time as his selection depends on D).

65. The statements gives are as follows.

	Subject	Predicate
(*i*) particular negative	Apples	Bananas
(*ii*) universal affirmative	Bananas	Oranges
(*iii*) universal affirmative	Oranges	Tasty

From statement I,

A is an apple ⇒ A may or may not be a banana. (from (*i*))

Hence, I alone is not sufficient.

From statement II,

A is not an apple ⇒ A may or may not be a banana.

Hence, II alone is not sufficient.

Even by using both the statements we cannot answer the question.

66. The statements given are as follows.

	Subject	Predicate
(*i*) Universal affirmative	Humans	Monkeys
(*ii*) Universal affirmative	Monkey	Vegetarian
(*iii*) Universal negative	Vegetarian	East flesh

From statement I,

Ravi is human ⇒ Ravi is monkey (from (*i*))

⇒ Ravi is vegetarian (from (*ii*)) ⇒ Ravi is not a flesh eater (from (*iii*)).

Hence, statement I alone is sufficient.

From statement II, Ravi is not a human. So we cannot say if Ravi would be a flesh eater or not.

67. The statements can be classified as follows.

	Subject	Predicate
(*i*) Universal affirmative	Cats	Tiger
(*ii*) Particular affirmative	Cats	Dogs
(*iii*) Universal affirmative	Rats	Dogs
(*iv*) Universal negative	Tiger	Dogs

From statement I, X is a tiger ⇒ X is not dog (from (*iv*)) ⇒ X is not rat (from (*iii*)).

Hence, I alone is sufficient.

From statement II,

X is cat ⇒ X is tiger (from (*i*)) ⇒ X is not dog (from (*iv*))

⇒ X is not rat (from (*iii*)).

Hence, II alone is sufficient.

Thus each statement alone is sufficient.

68. From statement I, the ratio of the number of persons who love watching both the games to those who love watching football = 1/2

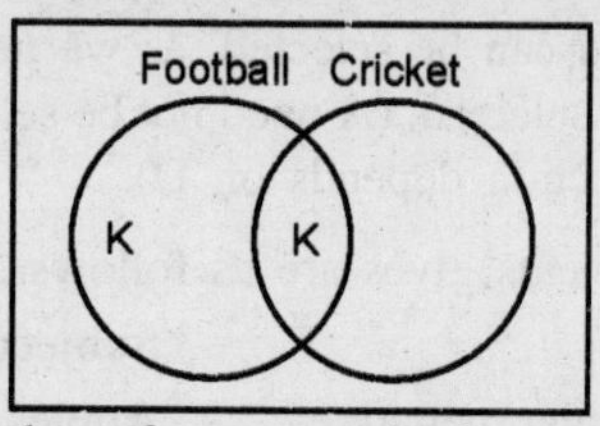

Le the number of persons who love watching Football be 2K.

∴ The number of persons who watch both is K.

But we do not know as to how many love to watch Cricket.

So statement I alone is not sufficient.

From statement II, the ratio of the no. of persons who love watching both the games to those who love watching Cricket = 2/5.

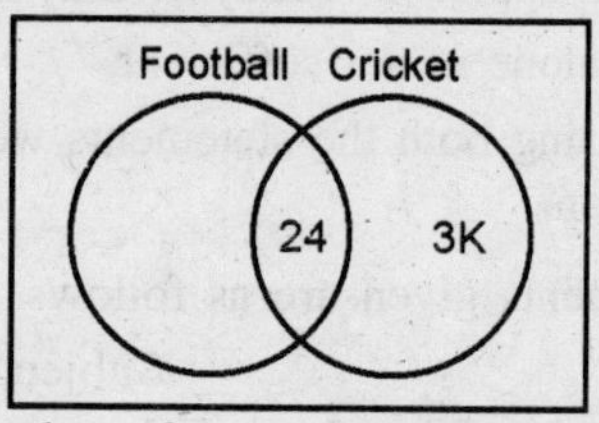

Let the number of persons who love watching cricket be $5K_1$ $\frac{40}{100}(5K_1) = 2K_1$

But we do not know the number of persons who love watching Football.

So statement II alone is not sufficient.

Using both the statements together those who love watching both the games $K = 2K_1$

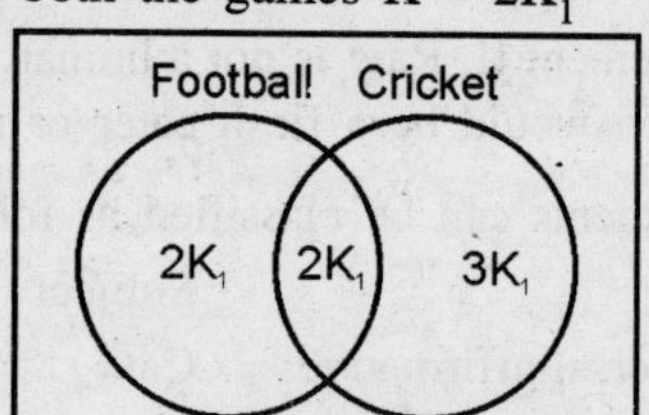

$n(F) = 4K_1$ $\quad n(C) = 5K_1$

The required ratio = $\frac{n(F)}{n(C)} = \frac{4}{5}$

We can answer the question using both the statements together.

69.

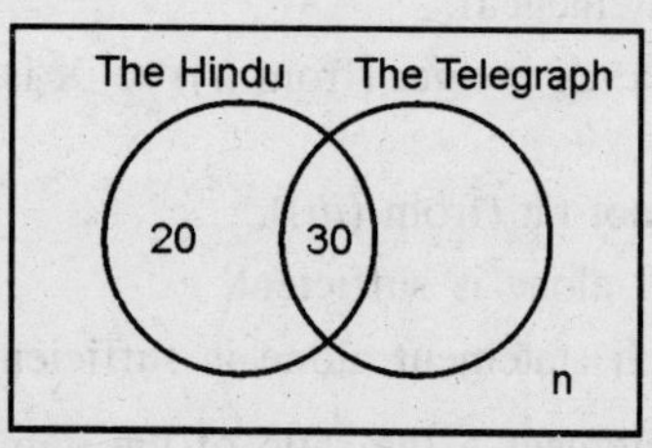

From statement I, we know that n = 20, but without knowing how many read '*The Telegraph*' or the total number of persons in the town, we cannot find the number of persons who read only *The Telegraph.* Thus statement I alone is not sufficient.

From statement II, we know that 60 people do not read '*The Hindu*' but without knowing how many read '*The Telegraph*' or how many read neither, we cannot find the number of persons who read only '*The Telegraph*'. Hence statement II alone is not sufficient.

Using both the statements together, we can find the number of persons reading only '*The Telegraph*'.

The number of persons not reading '*The Hindu*' = 60

The number of person who read '*The Telegraph*' only = 60 – 20 = 40.

70. From statement I alone we do not know the percentage of students who passed in only Maths or in only English or both the subjects. Hence statement I alone is not sufficient.

From statement II alone, we do not know the percentage of students who pass only in Maths or Maths. Statement B alone is not sufficient.

Combining both the statements, those who passed only in English = 50% – 25% = 25%

Those who passed only in Maths = 50% – 25%

Those who have passed either in English or Maths = 25% + 25% + 25% = 75%

Those who failed in both English and Maths = 100% – 75% = 25%.

71. The given information can be represented in the following venn-diagram, where the venn-diagram represents the number of students who passed in Maths, Science and Geography.

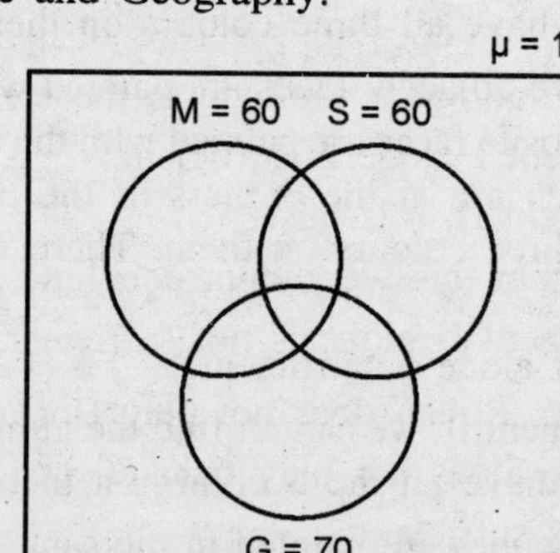

Now from I alone, the venn-diagram is as follows:

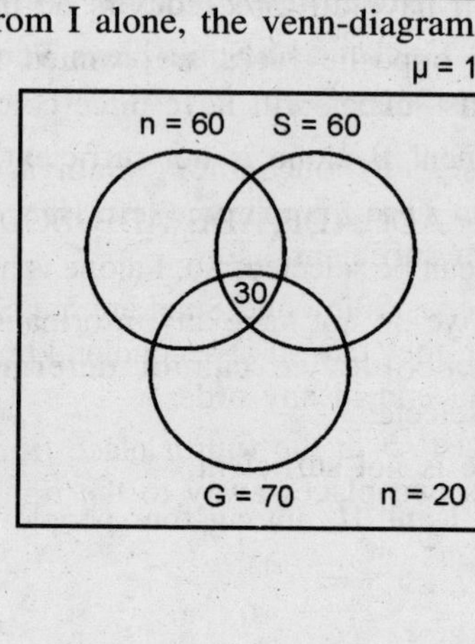

From the given information we can find out the following:

Exactly three = 30, none = 20.

Exactly 1 + Exactly 2 + Exactly 3 = 80 ...(1)

Also, exactly 1 + 2 × exactly 2 + 3 × exactly 3 = 190 ...(2)

By subtracting (1) from (2), we get

Exactly 2 + 2 × exactly 3 = 190 – 80

Exactly 2 = 110 – 2 × 30 = 50

From (1),

Exactly 1 = 0.

The number of students who were passed only in science is 0.

I alone is sufficient.

From II alone, there is no information regarding exactly 3.

Hence, II alone is not sufficient.

72. Let the statements be denoted by p, q, r and s respectively, where

p = Kishore wakes up in the morning.

q = he goes to office.

r = he meets his boss.

s = he watches a movie in the evening.

The problem statements and their implications are as follows.

If p, then q; $p \Rightarrow q$ and $\sim q \Rightarrow \sim p$

Only if $\sim p$, then r; $r \Rightarrow \sim q$ and $\sim q \Rightarrow \sim r$.

Only if r, then s; $s \Rightarrow r$ and $\sim r \Rightarrow \sim s$

From statement (I) q is true.

$\sim q \Rightarrow \sim r$, *i.e.*, he does not meet his boss on Monday.

$\Rightarrow$ He does not watch a movie in the evening.

Thus statement I alone is sufficient.

From statement II, q is false.

$\sim q \Rightarrow \sim p$, from this we cannot conclude anything.

Hence statement II alone is not sufficient.

73. From I alone, Rohan does not play Hockey.

Hence Sanjay does not play Cricket.

So I alone is sufficient.

From II alone, Rohan plays Hockey,

Sanjay plays Cricket and Sunil plays Football.

So, II alone is sufficient.

74. From I alone, any one among Sathya, Sujal and Sambhabi can sit at the extreme left. Hence statement I alone is not sufficient.

From II alone, as Sumit and Sujal are not sitting at the extreme right, they can sit at second and third position from the right end in any order.

Sambhabi must sit at the fourth place from the right end as she is two places away to the right of Saroj.

75. From statement I, we cannot conclude if Pratap is sitting next to Vijay for we only know that Pratap and Ramkumar are sitting next to each other.

So statement I alone is not sufficient.

From statement II, we only know that Abishek and Kishore fo not sit next to each other. From this we cannot conclude if Pratap is sitting next to Vijay or not. So statement II alone is not sufficient.

Using both the statements together. We can conclude the Pratap is not sitting next to Vijay.

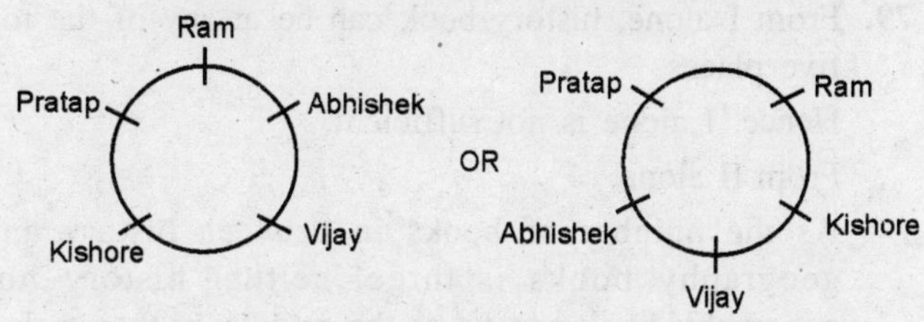

The above two possiblilities show that Vijay must be sitting between Kishore and Abishek, so he cannot be next to Pratap.

76. From I alone, we cannot get the definite position of any of the persons.

Hence I alone is not sufficient.

In II alone, only partial information is given.

∴ Hence, II alone is also not sufficient.

Combining I and II also, we cannot get that who is standing at the third position from the front end of the queue.

77. From statement I, we know that B and E are adjacent to each other and B is to the immediate right of E, but as we do not know the positions of the other people, we cannot determine who is sitting in the middle of the row.

So, statement I alone is nor sufficient.

From statement II as E is the only person sitting adjacent to C, C is at either of the ends of the row.

C E A/B/D A/B/D A/B/D

A/B/D A/B/D A/B/D E C

So, statement II alone is not sufficient.

From I and II, we get,

C E B D A or

C E B A D

Hence, we can find that Bharat is sitting at the middle of the row.

78. From I alone,

We get the positions of Bikash, Charan and Dinesh as follows:

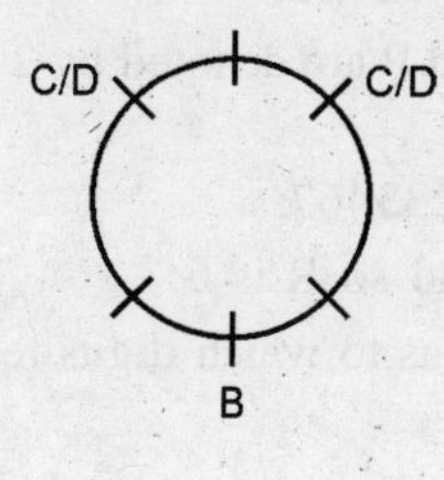

From II alone,

As Charan and Ekta are not adjacent to each other. Ekta can sit any of the other three places. Therefore, we cannot determine the position of Ashmit.

Combining I and II, we get Charan and Ekta are sitting opposite each other. Similarly, Firoj is sitting opposite Dinesh.

∴ Bikash must sit opposite Ashmit.

∴ Both the statements together are sufficient.

79. From I alone, history book can be at any of the top five places.

Hence, I alone is not sufficient.

From II alone,

As the number of books in between history and geography books is three, neither history nor geography book can be at the middle of the stack.

II alone is sufficient.

80. From I alone, there is no information regarding the number of people between Manoj and Komal.

Hence, I alone is not sufficient.

From II alone, as Suresh is the 15th person from the rear end, Manoj cannot be behind Suresh as there are 15 persons in between them.

Manoj mus be ahead of Suresh.

There are $(15 + 15) = 30$ people behind Manoj in the queue.

Position of Manoj is $(36 - 30) = 6$th from the front end.

∴ II alone is sufficient.

81. From the given information, we do not know in which direction he started his journey. With the information given in the question and that given in statement I, we can find the distance but not the direction.

Hence, I alone is not sufficient.

Statement II, we know the direction in which he is finally travelling. Hence, we can find the initial direction in which he started the journey.

Hence, using II alone we can find both distance dnd direction.

82. From statement I, we can find the code for the letters L, E, R and T.

APPEAL 122344

PILLAR 112456

So L is 1. A and P are denoted by 2 and 4 or by 4 and 2. Thus E is 3.

REPEAT 233467

T must be 7 and so R is 6.

But cannot say as to which digits represents C, A and P respectively.

Thus statement I alone is not sufficient.

From statement II,

APPEAL 122344

PILLAR 112456

PLACATE 1223479

We cannot say the digits representing C, T, I and R.

We acan say that A is represented by 2 and P by 4.

So statement II alone is not sufficient.

Now, by using both the statements together, we can conclude the alphabets and the corresponding digits will be as follows.

C	A	T	E	R	P	I	L
9	2	7	3	6	4	5	1

So the code for the word CATERPILLAR is 11223456679

Thus we can answer the question by using both the statements together.

83. From statement I,

Z_1 is the father of Y and X. X is the brother of Y since X is the son. Hence we can say that X is not the cousin of Y.

First statement alone is sufficient.

From statement II, Y is the daughter of the mother of X. Therefore Y is the sister of X.

Hence we can say that X is not the cousin of Y.

Second statement alone is also sufficient.

Thus, either of the statements is also sufficient.

84. From statement I, P's father's mother's only child is none other than P's father. As his spouse is Q, Q is the mother of P. As Q is the female she cannot be the husband.

So, statement I alone is sufficient.

From statement II, R is the father of Q and they belong to different gender which implies that Q is a female. So Q cannot be the husband.

So statement II alone is sufficient.

85. From statement I,

P may or may not be the youngest.

From statement II,

since exactly three persons are elder to P, the fourth person may be of equal age as of P or may be younger than P. Hence second statement alone is not sufficient.

Using both the statements also we have the possibility that the fourth person may be younger than P, as at the most one person is younger than P or of same age as P. Hence P may or may not be the youngest.

86. From statement I alone as we do not have any information regarding the remaining two students, we cannot determine who is not the top ranker in chemistry. From statement II, alone also we cannot determine who is not the topper in Chemistry.

Combining the statements I and II, we know that Shashank got 2nd rank in Maths. As Avinash got a better rank than Shashank in two subjects, one of them should be Maths. [As Shashank got 2nd rank in Maths, he has to get 1st rank in either Physics or Chemistry.] So, to get better rank than Shashank, Avinash has to get the first rank in Maths. So, he cannot get the first rank in Chemistry.

87. From statement I, we cannot say if A is the mother of B, or A can be the father of B. So statement I alone is not sufficient.

From statement II, we cannot say whether A is the mother of B or not as nothing is mentioned about B. So statement II alone is not sufficient.

Using both the statements together we can say that A is the mother of B for in statement I, it is given that B is A's son. Where as in statement. II, it is given that A is a female. So by using both we can conclude that A is B's mother.

88. From I alone, we can find the code for "man" as ¥ and 'is' as #.

I alone is not sufficient.

From II alone, we can find out the code for "is" as '#'.

'great' as @.

'cow' as 'β'.

'god' as 'λ'.

∴ II alone is sufficient.

89. From statement I, we cannot say if E is the father of G. D being the grandfather of G, does not make E as the father of G, or it is possible that D had a daughter whose child is G.

So statement I, alone is not sufficient.

From statement II, we cannot say regarding E being G's father or not as we do not know if D had any other child other than E.

So statement II, alone is not sufficient.

Even by using both the statements together, we cannot conclude whether E is the father of G or not.

Thus both the statements together are not sufficient to answer.

90. From statement I, we can find the code for the words Great and Good.

"God is Great" coded as "Do Re Mi"

"God is Good" codes as "Re Si Do"

We see that "God" and "is" is common among the two statements given. So the code for "God is" is "Do Re" in any order.

Code	Word
Si	Good
Mi	Great

But we cannot say what is the code for the word "God".

So statement I alone is not sufficiet.

From statement II, we can find the code for the words:

Words	Code
Humans	La
Great	Mi

But we cannot find the code for the word God.

Now, by using both the statements together we can conclude that the code for the word "God is" is "Do Re"

Words	Code
Great	Mi
Good	Si
Are	So
Humans	La

But individually, we cannot say what will be the code for the word God. So even by using both the statements together we cannot answer regarding the code for the word God.

91. From I alone,

Bangalore – P or Q

Chennai – R or S

Q may belong to any city other than Chennai.

∴ I alone is not sufficient.

From II alone,

T – Delhi/Mumbai

R – Hyderabad/Delhi

Q can be from any city.

∴ II alone is also not sufficient.

Combining I and II we get

S is from Chennai.

Q can be from any city other than Chennai.

∴ Neither of the statements is sufficient.

92. Let us consider the years such that only one among them is a leap year.

A	2001	2002	2003
B	2002	2003	2004
C	2003	2004	2005
D	2004	2005	2006
E	2005	2006	2007

But we take years where the centre year is not a leap year, we get any one of the five years as leap year.

A	1896	1895	1894	1893	1900
B	1897	1896	1895	1894	1901
C	1898	1897	1896	1895	1902
D	1899	1898	1897	1896	1903
E	1900	1899	1898	1897	1904

From statement I, we cannot say which are among them is the leap year.

Since A + B + C > C + D + E

A + B > D + E

It is possible that one of A or B is the leap year, but we cannot say which one.

Statement I alone is not sufficient.

From statement II, we can conclude that year C is the leap year.

Since (A + B + C) = (C + D + E)

A + B = D + E

Since there is only one leap year among the given 5 years, it has to be year C. So statement II alone is sufficient.

93. In a leap year, calendar of January matches with that of July. So, the year X is a leap year.

I alone is sufficient.

In any non-leap year, the calendar of January matches with that of October. As October is not matching with January the year X is a leap year. So, II alone is sufficient.

94. Statement I does not specify whether the year is a leap year or not. Statement II also is not sufficient as this suggests that last year was not a leap year.

Using both the statements also the question cannot be answered as we do not know whether the given year is leap or not.

95. Using statement I alone the box labelled Z may contain only green or both green and red.

Using statement II alone, we can conclude that Z has only green colour.

Hence only statement II alone is sufficient.

96. From I, as there is no information regarding A, B C and D we cannot determine whether B is the professor or not. So, I alone is not sufficient.

From II, we do not have the relation between B and D to A and C. So II alone is not sufficient.

Combining I and II, we get A and C are the children of B and D. As A is a male B is a female and D is a male. Hence, B is the Doctor and D is the professor as doctor is not the husband, doctor is female.

Analytical & Logical Reasoning

1 Letter Series

In letter series, the letters follow a definite order. The given series of letters can be in natural order or in reverse order or combination of both. The letters may be skipped or repeated or consecutive. The given series may be single or may even comprise of two different series merged at alternate positions. While attempting questions on letter series one should note the pattern of alphabet series.

Alphabets in natural series are :

A B C D E F G H I J K L M N O P Q R S T U V W X Y Z

A ↓ 1st, E ↓ 5th, J ↓ 10th, O ↓ 15th, T ↓ 20th, Y ↓ 25th

Alphabets in reverse series are :

Z Y X W V U T S R Q P O N M L K J I H G F E D C B A

Z ↓ 1st, V ↓ 5th, Q ↓ 10th, L ↓ 15th, G ↓ 20th, B ↓ 25th

Note : On reaching Z, the series restarts from A and on reaching A, it restarts from Z.

MULTIPLE CHOICE QUESTIONS

Directions : *In each of the following series determine the order of the letters. Then from the given options select the one which will complete the given series.*

1. B Y C X D W E ?
(*a*) S (*b*) T
(*c*) U (*d*) V

2. A D C G E ?
(*a*) G (*b*) J
(*c*) I (*d*) L

3. L N P R T ?
(*a*) U (*b*) V
(*c*) W (*d*) Y

4. X O I F ?
(*a*) D (*b*) F
(*c*) B (*d*) E

5. D Q G O J M ?
(*a*) K (*b*) P
(*c*) M (*d*) L

6. B A F E J I P O ? U
(*a*) V (*b*) T
(*c*) S (*d*) Q

7. Z A A Y B B X C ?
(*a*) W (*b*) C
(*c*) V (*d*) D

8. B D G K ? V
(*a*) N (*b*) P
(*c*) Q (*d*) M

9. V R O K ? D
(*a*) L (*b*) I
(*c*) H (*d*) J

10. Z X V T ?
(*a*) Q (*b*) S
(*c*) R (*d*) P

11. Z A Y B X C W D V E ?
(*a*) U D (*b*) U C
(*c*) T R (*d*) U F

12. B C D B D D B E D B F D B G ?
(*a*) B D (*b*) B F
(*c*) H B (*d*) D B

13. Z Y X C B A W V U F E D ?
(*a*) S T R (*b*) T S R
(*c*) R Q P (*d*) I H G

14. B D A C F H E G ?
(*a*) J L (*b*) I K
(*c*) J K (*d*) K L

15. A Z Y B X W C V U D T S E ?
(*a*) R S (*b*) S T
(*c*) R Q (*d*) Q R

16. bs, dt, fu, ?
(*a*) xz (*b*) ya
(*c*) hv (*d*) wc

17. cmw, hrb, ?, rbl, wgq, blv
(*a*) mwg (*b*) lvf
(*c*) lwg (*d*) mxg

18. dfe, jih, mln, ?, vut
(*a*) oqp (*b*) psr
(*c*) prq (*d*) rsp

19. ?, ayw, gec, mki, sqo
(*a*) zxw (*b*) bzw
(*c*) usq (*d*) may

20. eac, gce, ieg, ?
(*a*) jhi (*b*) jgi
(*c*) kgi (*d*) khi

21. bl, fp, jt, ?, rb, vf
(*a*) mw (*b*) nx
(*c*) oy (*d*) mx

22. ?, siy, oeu, kaq, gwm, cri
(*a*) wnc (*b*) wnb
(*c*) vnc (*d*) wmc

23. AZXBVTCR (..?..) (...?...)
(*a*) P, D (*b*) E, O
(*c*) Q, E (*d*) O, Q

24. QPO, SRQ, UTS, WVU, (?)
(*a*) XVZ (*b*) ZYA
(*c*) YXW (*d*) VWX

25. HS, JQ, LO, NM, (?)
(*a*) PK (*b*) RH
(*c*) PL (*d*) TG

26. AZ, BY, CX, ?
(*a*) EF (*b*) GH
(*c*) IJ (*d*) DW

27. DCXW, FEVU, HGTS, ?
(*a*) LKPO (*b*) ABYZ
(*c*) JIRQ (*d*) LMRS

28. AB, DEF, HIJK, ?, STUVWX
(*a*) MNOPQ (*b*) LMNOP
(*c*) LMNO (*d*) QRSTU

29. G, H, J, M, ?, V
(*a*) T (*b*) S
(*c*) R (*d*) Q

30. Y, W, U, S, Q, (?), (?)
(*a*) N, J (*b*) M, L
(*c*) J, R (*d*) O, M

EXPLANATORY ANSWERS

1. (*d*) : There are two alternate series.

+1 +1 +1
B Y C X D W E V
−1 −1 −1

Series I : BCDE (natural order)
Series II : YXWV (reverse order)

2. (*b*) : There are two alternate series.

+2 +2
A D C G E J
+3 +3

Series I : ACE (follows +2 pattern)
Series II : DGJ (follows +3 pattern)

3. (*b*) : The series follows the +2 pattern. *i.e.*

L N P R T V
+2 +2 +2 +2 +2

4. (*b*) : The series follows the sequence of the difference decreasing by 3 *i.e.*

X O I F F
−9 −6 −3 −0

5. (*c*) : There are two alternate series.

+3 +3 +3
D Q G O J M M
−2 −2

Series I : DGJM (follows +3 pattern)
Series II : QOM (follows −2 pattern)

6. (*a*) : Each vowel (AEIOU) is preceded by the letter that comes next to it in the natural alphabetical series.

B A F E J I P O V U

7. (*b*) : There are three alternate series. The second and the third set are the same, *i.e.*,

−1 −1 −1
Z AA Y BB X CC
+1 +1

Series I : ZYX (in reverse order)
Series II and *III* : ABC (in natural order)
The letter 'C' needs to be repeated.

8. (*b*) : The difference between the letters increases at each step after beginning with two.

B D G K P V
+2 +3 +4 +5 +6

9. (*c*) : The letters are in reverse series and the difference is four and three alternately.

V R O K H D
−4 −3 −4 −3 −4

10. (*c*) : The letters in this series are moved two steps backward.

Z X V T R

−2 −2 −2 −2

11. (*d*) : There are two alternate series

−1 −1 −1 −1 −1

Z A Y B X C W D V E U F

+1 +1 +1 +1 +1

Series I : ZYXWVU (in reverse order)

Series II : ABCDEF (in natural order)

12. (*d*) : The actual letters which are moving one step forward in the series are packed between B and D.

B C D B D D B E D B F D B G D

+1 +1 +1 +1

13. (*b*) : The letters in the series are in group of three. The letters in the groups are in consecutive reverse series and are in two alternate sections.

ZYX CBA WVU FED TSR

14. (*a*) : The letters in the series are arranged in following order (Four letters are mixed)

A B C D is written as BDAC; and

E F G H as FHEG. Similarly,

I J K L will be JLIK.

The answer option is JL, *i.e.*, (a)

15. (*c*) : There are two alternate series.

+1 +1 +1 +1

A Z Y B X W C V U D T S E R Q

+1 +1 +1 +1

Series I : ABCDE (in natural series)

Series II : ZY XW VU TS RQ (double letter in reverse series)

16. (*c*) : The first letters of the terms are alternate while the second letters are in alphabetical order.

17. (*a*) : All the letters of each term are moved five steps forward to obtain the corresponding letters of the next term.

18. (*d*) : There is a gap of three letters between the first letter of each term and the last letter of the next term.

19. (*c*) : Each term in the series consists of alternate letters in reverse order. The first letter of each term and the last letter of the next term are also alternate.

20. (*c*) : The first letters of the terms are alternate and similarly the second and the third letters.

21. (*b*) : Each letters of a term of the series is four steps ahead of the corresponding letter of the preceding term.

22. (*d*) : The letters in each term are moved four steps backward to obtain the letters of the next term.

23. (*a*) : The first, fourth and seventh letters are in alphabetical order. So, tenth letter would be the letter after C i.e., D. Also, the second and third letters are alternate and in reverse order and so are the fifth and sixth letters.

24. (*c*) : Each term in the series consists of three consecutive letters in reverse order. The first letter of each term and the last letter of the next term are the same.

25. (*a*) : The first letter of each term is moved two steps forward and the second letter is moved two steps backward to obtain the corresponding letters of the next term.

26. (*d*) : The first letter of each term is moved one step forward and the second letter is moved one step backward to obtain the corresponding letters of the next term.

27. (*c*) : First two letters of each term are in reverse order. Similarly third and fourth letters are also in reverse order. Besides, the second letter of each term is the letter next to the first letter of the preceding term.

28. (*a*) : The number of letters in the terms goes on increasing by 1 at each step. Each term consists of letters in alphabetical order. The last letter of each term and the first letter of the next term are alternate.

29. (*d*) : The first, second, third, fourth and fifth terms are moved one, two, three, four and five steps respectively forward to obtain the successive terms.

30. (*d*) : The series consists of alternate letters in reverse order.

2 Wrong Letter Series

In this type of series, the candidates are not required to find the letter or group of letters which will complete the given series but, they have to identify the letter or number which is wrong or misfit in the given series.

MULTIPLE CHOICE QUESTIONS

Directions : *Which letter(s) in each of the following series is wrong or is misfit in the series?*

1. A E H O U
(*a*) U (*b*) O
(*c*) H (*d*) E

2. C H M S W B
(*a*) C (*b*) S
(*c*) B (*d*) W

3. X S N I C Y
(*a*) Y (*b*) C
(*c*) S (*d*) I

4. Z A W B X C
(*a*) D (*b*) C
(*c*) X (*d*) W

5. M L O N Q P R
(*a*) R (*b*) O
(*c*) Q (*d*) L

6. D K R Y F L
(*a*) L (*b*) D
(*c*) R (*d*) Y

7. L N Q T W Z C F
(*a*) C (*b*) Q
(*c*) L (*d*) F

8. XW, DC, CB, NM, PQ
(*a*) NM (*b*) CB
(*c*) PQ (*d*) XW

9. B E I N S A I
(*a*) A (*b*) E
(*c*) S (*d*) I

10. Z T P K H F
(*a*) Z (*b*) P
(*c*) T (*d*) F

11. DFH, KMO, RTU, YAC, FHJ
(*a*) KMO (*b*) YAC
(*c*) RTU (*d*) DFH

12. UWT, PRO, KMJ, EGD, ACZ
(*a*) UWT (*b*) KMJ
(*c*) PRO (*d*) EGD

13. E, H, L, P, W, D, L
(*a*) L (*b*) W
(*c*) P (*d*) H

14. W, T, Y, V, A, X, D, Z, E
(*a*) D (*b*) Y
(*c*) T (*d*) W

15. BX, EV, HT, KR, MP
(*a*) EV (*b*) MP
(*c*) KR (*d*) HT

16. ZBX, WEU, THS, QKO
(*a*) ZBX (*b*) QKO
(*c*) WEU (*d*) THS

17. MAN, TEN, PIN, GBN, SUN
(*a*) TEN (*b*) MAN
(*c*) GBN (*d*) SUN

18. AM, CP, ER, GT, IV, KX
(*a*) AM (*b*) ER
(*c*) IV (*d*) KX

19. A, P, C, M, F, J, J, H, O, D
(*a*) P (*b*) M
(*c*) F (*d*) H

20. US, RV, OY, MB, IE, FH
(*a*) US (*b*) MB
(*c*) IE (*d*) FH

EXPLANATORY ANSWERS

1. (*c*) : The series is made with vowels only. AEIOU. I should be in place of H.

2. (*b*) : The letters in the series are moved five steps forward.

C H M R W B

+5 +5 +5 +5 +5

R should be in place of S

(The series restarts from A on reaching Z)

3. (*b*) : The pattern in the series is –5, *i.e.,*

X S N I D Y

–5 –5 –5 –5 –5

D should be in place of C.

4. (*d*) : There are two alternate series

+1 +1

Z A Y B X C

–1 –1

Series I : ZYX (reverse series)

Series II : ABC (natural series)

Y should be in place of W.

5. (*a*) : Two consecutive letters are written backwards.

ML ON QP SR

S should be in place of R.

6. (*a*) : The pattern in this series in moving the letters seven steps forward.

D K R Y F M

+7 +7 +7 +7 +7

M should be in place of L.

7. (*c*) : The pattern in the series is +3

K N Q T W Z C F

+3 +3 +3 +3 +3 +3 +3

K should be in place of L.

8. (*c*) : The series is made with any two consecutive letters written backwards.

XW DC CB NM QP

← ← ← ← ←

Q should come before P in the series.

9. (*c*) : The difference between the letters is increased by one at each step.

B E I N T A I

+3 +4 +5 +6 +7 +8

T should be in place of S.

10. (*b*) : The difference between the letters is decreased by one at each step.

Z T O K H F

–6 –5 –4 –3 –2

O should be in place of P.

11. (*c*) : In place of RTU, it should be RTV.

12. (*d*) : In place of EGD, it should be FHE.

13. (*e*) : Q should comes at place of P, then series will be E, E + 3, E + 7, E + 12, ...

14. (*a*) :

+2 +2 +2 +2

W T Y V A X C Z E

–3 –3 –3 –3

15. (*b*) :

–2 –2 –2 –2

BX EV HT KR NP

+3 +3 +3 +3

16. (*d*) :

+3 +3 +3

Z B X W E U T H R Q K O

–2 –2 –2 –2

17. (*c*) : Middle term is occupied by next vowel term, so GON should come in place of GBN.

18. (*a*) :

AN CP ER GT IV KX

+2 +2 +2 +2 +2

+2 +2 +2 +2 +2

19. (*d*) :

A P C M F J J G O D

+2 +3 +4 +5

–3 –3 –3 –3

20. (*b*) :

US RV OY LB IE FH

–3 –3 –3 –3 –3

+3 +3 +3 +3 +3

3 Repeated Series

In this type of series small letters of the alphabet are used to make a set of letters which are repeated. The candidate has to find the set of letters which will fit the blanks left in the given series in such a manner that one section of the series is further repeated in the same manner.

MULTIPLE CHOICE QUESTIONS

Directions : *Which of the following groups of letters will complete the given series?*

1. ab_ _ _b_bbaa_
(*a*) babba (*b*) abaab
(*c*) abbab (*d*) baaab

2. aa_ab_ _aaa_a
(*a*) baaa (*b*) abab
(*c*) aaab (*d*) aabb

3. _baa_aab_a_a
(*a*) baab (*b*) abab
(*c*) aaba (*d*) aabb

4. _a cca_ccca_acccc_aaa
(*a*) ccaa (*b*) acca
(*c*) caac (*d*) caaa

5. c_bbb_ _abbbb_abbb_
(*a*) abccb (*b*) bacbb
(*c*) aabcb (*d*) abacb

6. ac_cab_baca_aba_acac
(*a*) bcbb (*b*) aacb
(*c*) babb (*d*) acbc

7. _ _aba_ _ba_ab
(*a*) abbab (*b*) bbaba
(*c*) baabb (*d*) abbba

8. _ _babbba_a_ _
(*a*) bbaba (*b*) babbb
(*c*) baaab (*d*) ababb

9. k_mk_lmkkl_kk_mk
(*a*) lklm (*b*) lkmk
(*c*) lkmm (*d*) lkml

10. abc_d_bc_d_b_dd
(*a*) decdb (*b*) dadac
(*c*) cdabe (*d*) bacde

11. b_abbc_bbca_bcabb_ab
(*a*) acba (*b*) acaa
(*c*) cacc (*d*) cabc

12. aca_ac_ _a_ac
(*a*) babc (*b*) aaac
(*c*) cacc (*d*) caca

13. ba_cb_b_bab_?
(*a*) acbb (*b*) bcaa
(*c*) cabb (*d*) bacc

14. ab_aa_caab_aab_a
(*a*) bcbc (*b*) bbca
(*c*) cbcc (*d*) caba

15. _bbcaa_bcaa_bc_a_bca
(*a*) bacab (*b*) abbab
(*c*) abcba (*d*) bcaab

16. a_bccb_ca_cca_baab_c
(*a*) accab (*b*) abcaa
(*c*) bacaa (*d*) ababc

17. b_dabbcd_b_c_ab
(*a*) cabd (*b*) bcad
(*c*) dcba (*d*) acbd

18. a_ba_caacb_bc_
(*a*) acbb (*b*) cbab
(*c*) cbaa (*d*) cbba

19. c_baa_aca_cacab_acac_bca
(*a*) acbaa (*b*) cbaac
(*c*) bccab (*d*) bbcaa

20. _bcc_ac_aabb_ab_cc
(*a*) bacab (*b*) abaca
(*c*) aabca (*d*) bcaca

21. ab_ccca_bccc_bbcc_
(*a*) abbc (*b*) bbac
(*c*) bbca (*d*) cabc

22. _cbc_a_bcaac_ca
(*a*) aaba (*b*) caab
(*c*) bcab (*d*) aacb

23. ab_ba_ _ba_
(*a*) abba (*b*) baab
(*c*) baba (*d*) abab

24. yx_yx_yxz_xzy_zyxz
(*a*) zzyx (*b*) xxzy
(*c*) yyzx (*d*) yzxz

25. xxxy_y_xxy_yxx_
(*a*) xyxy (*b*) yxyx
(*c*) yyxx (*d*) xxyy

26. aa_aabb_b_aa_aabb_bb
(*a*) abbab (*b*) bbbaa
(*c*) babba (*d*) aabbb

27. ab_aabc_abc_
(*a*) aab (*b*) caa
(*c*) bca (*d*) cba

28. c_abca_b_aab
(*a*) bbc (*b*) bca
(*c*) aac (*d*) cab

29. cc_bac_bbacc_baccb_a
(*a*) cbca (*b*) bcbc
(*c*) bcbb (*d*) bbbc

30. aaab_aaaaba_a_abaa
(*a*) baa (*b*) aaa
(*c*) aba (*d*) bba

31. nl_n_mn_m_lm
(*a*) mlnn (*b*) lnnm
(*c*) mlnl (*d*) mlln

32. lm_nnll_mn_llmm_nl
(*a*) lmmn (*b*) lnnm
(*c*) mmnn (*d*) mnll

33. aba_caab_cca_bac_a
(*a*) cbac (*b*) caac
(*c*) bccb (*d*) abcb

34. _cd_bc_ab_dabc_a
(*a*) badcd (*b*) ccddb
(*c*) abccd (*d*) ccbab

35. _cabcc_bc_abcca_
(*a*) babc (*b*) acac
(*c*) cacb (*d*) bacc

36. mn_om_oom_oo_noo
(*a*) onnm (*b*) nmmo
(*c*) mnoo (*d*) nmon

37. _lmn_ll_nol_mno
(*a*) nlmo (*b*) mnoo
(*c*) omlm (*d*) loml

38. xx_zzx_yzz_xyz_
(*a*) xyzx (*b*) zyxx
(*c*) yxxy (*d*) yxxz

39. y_xx_yxx_yyx_xy
(*a*) xyxx (*b*) xxxy
(*c*) yxxy (*d*) xxyx

40. a_bab_ _b_abb
(*a*) abab (*b*) bbab
(*c*) aabb (*d*) babb

EXPLANATORY ANSWERS

1. (*d*) : The series is abbaab, abbaab
2. (*c*) : The series is aaaaba, aaaaba
3. (*b*) : The series is aba, aba, aba, aba
4. (*d*) : The series is c,a,cc,aa, ccc, aaa, cccc, aaaa
5. (*a*) : The series is cabbbb, cabbbb, cabbbb
6. (*b*) : The series is acac, abab, acac, abab, acac
7. (*a*) : The series is ab, ab, ab, ab, ab, ab
8. (*b*) : The series is bababb, bababb
9. (*d*) : The series is klmk, klmk, klmk, klmk
10. (*b*) : The series is abcdd, abcdd, abcdd
11. (*d*) : The series is bcab, bcab, bcab, bcab, bcab
12. (*c*) : The series is ac, ac, ac, ac, ac, ac
13. (*d*) : The series is babc, babc, babc
14. (*c*) : The series is abca, abca, abca, abca
15. (*b*) : The series is abbca, abbca, abbca, abbca
16. (*d*) : The series is aabcc, bbcaa, ccabb, aabcc
17. (*a*) : The series is bcdab, bcdab, bcdab
18. (*c*) : The series is acbabca, acbabca
19. (*a*) : The series is cabaac, acabca, cabaac, acabca
20. (*a*) : The series is bbccaa, ccaabb, aabbcc.
21. (*b*) : The series is abbccc, abbccc.
22. (*d*) : The series is acbca, acbca, acbca.
23. (*d*) : The series is ab, ab, ab, ab, ab.
24. (*a*) : The series is yxz, yxz, yxz, yxz, yxz, yxz.
25. (*b*) : The series is xxx, yyy, xxx, yyy, xxx.
26. (*a*) : The series is aaaaa, bbbbb, aaaaa, bbbbb.
27. (*b*) : The series is abca, abca, abca.
28. (*c*) : The series is caab, caab, caab.
29. (*c*) : The series is ccbba, ccbba, ccbba, ccbba.
30. (*b*) : The series is aaabaa, aaabaa, aaabaa.
31. (*d*) : The series is nlm, nlm, nlm, nlm.
32. (*c*) : The series is lmmnnl, lmmnnl, lmmnnl.
33. (*b*) : The series is abacca, abacca, abacca.
34. (*a*) : The series is bcda, bcda, bcda, bcda.
35. (*c*) : The series is ccab, ccab, ccab, ccab.
36. (*a*) : The series is mnoo, mnoo, mnoo, mnoo.
37. (*d*) : The series is llmno, llmno, llmno.
38. (*d*) : The series is xxyzz, xxyzz, xxyzz.
39. (*a*) : The series is yxxxy, yxxxy, yxxxy.
40. (*b*) : The series is abb, abb, abb, abb.

4 Number Series

In this type of series, the set of given numbers in a series are related to one another in a particular pattern or manner. The relationship between the numbers may be (*i*) consecutive odd/even numbers; (*ii*) consecutive prime numbers; (*iii*) squares/cubes of some numbers with/without variation of addition or subtraction of some number; (*iv*) sum/product/difference of preceding numbers; (*v*) addition/subtraction/multiplication/division by some number; and (*vi*) many more combinations of the relationships given above.

MULTIPLE CHOICE QUESTIONS

Directions : *In the following questions, select the number(s) from the given options for completing the given series.*

1. 3, 9, 27, 81, 243, ?
(*a*) 486 (*b*) 729
(*c*) 972 (*d*) 359

2. 1, 6, 12, 19, 27, ?
(*a*) 38 (*b*) 35
(*c*) 36 (*d*) 54

3. 2, 6, 14, 30, 62, ?
(*a*) 126 (*b*) 128
(*c*) 120 (*d*) 130

4. 8, 48, 16, 96, 32, ?
(*a*) 192 (*b*) 150
(*c*) 64 (*d*) 288

5. 2, 8, 14, 24, 34, 48, ?
(*a*) 66 (*b*) 62
(*c*) 58 (*d*) 64

6. 4, 9, 19, 34, 54, ?, 109
(*a*) 89 (*b*) 84
(*c*) 74 (*d*) 79

7. 3, 15, 35, 63, 99, ?
(*a*) 144 (*b*) 143
(*c*) 121 (*d*) 169

8. 2, 3, 6, 18, 108, ?
(*a*) 1944 (*b*) 1658
(*c*) 648 (*d*) 1008

9. 1, 2, 3, 2, 3, 5, 4, 5, ?
(*a*) 9 (*b*) 6
(*c*) 10 (*d*) 7

10. 7776, 1296, 216, 36, 6, ?
(*a*) 6 (*b*) 0
(*c*) 3 (*d*) 1

11. 1, 8, 27, 64, 125, ?
(*a*) 172 (*b*) 176
(*c*) 216 (*d*) 189

12. 2, 3, 5, 7, 11, 13, ?
(*a*) 19 (*b*) 57
(*c*) 31 (*d*) 17

13. 80, 64, 48, 32, 16, ?
(*a*) 4 (*b*) 0
(*c*) 8 (*d*) 1

14. 3, 8, 13, 24, 41, ?
(*a*) 65 (*b*) 75
(*c*) 70 (*d*) 80

15. 9, 81, 90, 810, 819, ?
(*a*) 7371 (*b*) 900
(*c*) 8100 (*d*) 1638

16. 0, 8, 24, 48, 80, ?
(*a*) 110 (*b*) 96
(*c*) 120 (*d*) 140

17. 2, 4, 8, 3, 9, 27, 4, 16, ?
(*a*) 64 (*b*) 32
(*c*) 48 (*d*) 24

18. 27, 28, 25, 25, 23, 22, 21, ?
(*a*) 20 (*b*) 21
(*c*) 19 (*d*) 18

19. 80, 63, 72, 72, 64, 81, 56, ?
(*a*) 96 (*b*) 98
(*c*) 89 (*d*) 90

20. 0, 5, 22, 57, ?, 205
(*a*) 198 (*b*) 116
(*c*) 172 (*d*) 92

21. 19200, 1920, 240, 40, 10, ?
(*a*) 1 (*b*) 0
(*c*) 5 (*d*) 10

22. 1, 2, 1, 3, 2, 7, 13, ?
(*a*) 39 (*b*) 26
(*c*) 92 (*d*) 108

23. 2, 4, 4, 8, 16, 16, 256, ?
(*a*) 64 (*b*) 36
(*c*) 180 (*d*) 32

24. 3, 6, 9, 36, 27, 216, 81, ?
(*a*) 1296 (*b*) 648
(*c*) 243 (*d*) 1944

25. 19, 4, 14, 7, 10, 11, 7, ?
(*a*) 16 (*b*) 15
(*c*) 17 (*d*) 23

26. 1, 2, 3, 5, 8, ?
(*a*) 11 (*b*) 13
(*c*) 15 (*d*) 9

27. 64, 16, 32, 8, ?
(*a*) 24 (*b*) 32
(*c*) 4 (*d*) 16

28. 2, 3, 5, 10, 20, 40, ?
(*a*) 65 (*b*) 75
(*c*) 60 (*d*) 80

29. 3, 15, 90, 630, 5040, ?
(*a*) 35280 (*b*) 40320
(*c*) 45360 (*d*) 10080

30. 1890, 1890, 630, 126, 18, ?
(*a*) 6 (*b*) 9
(*c*) 2 (*d*) 3

EXPLANATORY ANSWERS

1. (*b*) : The numbers in the series are multiplied by 3 to get the next numbers.

2. (*c*) : The difference between the numbers in the series increases by 1, after beginning from 5, *i.e.,*

1 6 12 19 27 36
+5 +6 +7 +8 +9

3. (*a*) : The difference between the numbers in the series doubles each time, after beginning from 4, *i.e.,*

2 6 14 30 62 126
+4 +8 +16 +32 +64

4. (*a*) : There are two alternate series and the numbers are multiplied by 2.

×2 ×2
8 48 16 96 32 192
×2 ×2

Series I : 8, 16, 32 *Series II* : 48, 96, 192

5. (*b*) : The sequence in the series is :

+2 +3 +4 +5
–3 –3 –3 –3

The difference increases by 4 at alternate step.

6. (*d*) : The difference between the numbers increases by 5, after beginning from 5, *i.e.,*

4 9 19 34 54 79 109
+5 +10 +15 +20 +25 +30

7. (*b*) : The series follows the sequence of the squares of even numbers beginning from 2, minus 1.

8. (*a*) : Every third number in the series is the product of previous two numbers.

2 3 6 18 108 1944
2 × 3
3 × 6
6 × 18
18 × 108

9. (*a*) : In this series three numbers from a set. The first two numbers of each set are in natural order and the third number is the sum of first and second numbers. The first number of the next set begins with double the first number of the previous set.

×2 ×2
2 3 2 3 5 4 5 9
1 + 2 2 + 3 4 + 5

10. (*d*) : The numbers in this sequence are divided by 6 each time.

11. (*c*) : The sequence in the series is cube of numbers in their natural order.

12. (*d*) : The series consists of prime numbers in increasing order.

13. (*b*) : The numbers in the series are decreasing with the difference of 16 at each step.

14. (*c*) : The sequence in the series is (number + next number) + addition of natural number increasing by 1 at each step, after beginning from 2.

3 8 13 24 41 70
(3+8) + 2
(8 + 13) + 3
(13 + 24) + 4
(24 + 41) + 5

15. (a) : The sequence in the series is × 9, + 9 which is repeated.

16. (c) : The sequence of the series consists of the product of 2 even numbers in natural order, *i.e.,*

0 8 24 48 80 120

↓ ↓ ↓ ↓ ↓ ↓

(0 × 2) (2 × 4) (4 × 6) (6 × 8) (8 × 10) (10 × 12)

17. (a) : In this series 3 numbers form a set. The first number is in natural order. The second number is the square of above number. The third number is the cube of first number.

18. (c) : There are two alternate series :

−2 −2 −2

27 28 25 25 23 22 21 19

−3 −3 −3

Series I : 27, 25, 23, 21 (following −2 pattern)

Series II : 28, 25, 22, 19 (following −3 pattern)

19. (d) : There are two alternate series :

−8 −8 −8

80 63 72 72 64 81 56 90

+9 +9 +9

Series I : 80, 72, 64, 56 (following −8 pattern)

Series II : 63, 72, 81, 90 (following +9 pattern)

20. (b) : The series follows this sequence : cube of natural numbers starting from 1 minus odd numbers starting from 1.

0 5 22 57 116 205

↓ ↓ ↓ ↓ ↓ ↓

1^3-1 2^3-3 3^3-5 4^3-7 5^3-9 6^3-11

21. (c) : The sequence in the series is divided by even numbers in descending order, starting from 10, *i.e.,*

19200 1920 240 40 10 5

÷10 ÷8 ÷6 ÷4 ÷2

22. (c) : The sequence in the series is the product of previous two numbers −1, +1 alternately, *i.e.,*

1 2 1 3 2 7 13 92

(1 x 2) −1

(2 x 1) +1

(1 x 3) −1

(3 x 2) +1

(2 x 7) −1

(7 x 13) +1

23. (d) : There are two alternate series :

×2 ×4 ×16

2 4 4 8 16 16 256 32

×2 × 2 × 2

Series I : 2, 4, 16, 256 (next number is the square of previous number)

Series II: 4, 8, 16, 32 (number is multiplied by 2 to get the next number)

24. (a) : There are two alternate series :

×3 ×3 ×3

3 6 9 36 27 216 81 1296

×6 ×6 ×6

Series I : 3, 9, 27, 81 (number is multiplied by 3 to get the next number)

Series II : 6, 36, 216, 1296 (number is multiplied by 6 to get the next number)

25. (a) : There are two alternate series :

−5 −4 −3

19 4 14 7 10 11 7 16

+3 +4 +5

Series I : 19, 14, 10, 7 (The sequence followed is −5, −4, −3)

Series II : 4, 7, 11, 16 (The sequence followed is +3, +4, +5)

26. (b) : The numbers in the series are sum of two numbers preceding them.

1 2 3 5 8 13

(1+2)

(2+3)

(3+5)

(5+8)

27. (d) : There are two alternate series and the numbers are divided by 2.

÷ 2 ÷ 2

64 16 32 8 16

÷ 2

28. (d) : The sequence in the series is that a number is the sum of all the numbers preceding it.

2 3 5 10 20 40 80

(2+3)

(2+3+5)

(2+3+5+10)

(2+3+5+10+20)

(2+3+5+10+20+40)

29. (c) : The sequence in the series is ×5, ×6, ×7, ×8, ×9.

3 15 90 630 5040 45360

×5 ×6 ×7 ×8 ×9

30. (c) : The sequence of numbers is divided by odd numbers in increasing order starting from 1.

1890 1890 630 126 18 2

÷ 1 ÷ 3 ÷ 5 ÷ 7 ÷ 9

5

Wrong Number Series

In this type of series, the given series may be complete but what needs to be identified is the number in the given completed series which is disturbing the sequential pattern of that series and does not fit in with the relationship shared between the other numbers.

MULTIPLE CHOICE QUESTIONS

Directions : *In the given series find the number which is wrong.*

1. 5, 25, 120, 625, 3125, 15625
(*a*) 15625 (*b*) 625
(*c*) 120 (*d*) 5

2. 4, 8, 11, 22, 18, 36, 24, 50
(*a*) 8 (*b*) 22
(*c*) 36 (*d*) 24

3. 2, 4, 12, 24, 72, 142, 432
(*a*) 432 (*b*) 12
(*c*) 142 (*d*) 72

4. 2, 3, 4, 4, 6, 8, 9, 12, 16
(*a*) 3 (*b*) 9
(*c*) 6 (*d*) 12

5. 97, 91, 86, 83, 79, 77, 76, 76
(*a*) 86 (*b*) 76
(*c*) 91 (*d*) 83

6. 7, 11, 11, 9, 15, 7, 19, 5, 23, 1
(*a*) 5 (*b*) 7
(*c*) 1 (*d*) 9

7. 3, 7, 12, 28, 48, 118, 192, 448
(*a*) 12 (*b*) 118
(*c*) 28 (*d*) 7

8. 10, 100, 1100, 11000, 111000, 1210000
(*a*) 1210000 (*b*) 11000
(*c*) 100 (*d*) 111000

9. 24576, 6144, 1536, 386, 96, 24
(*a*) 386 (*b*) 6144
(*c*) 96 (*d*) 1536

10. 11, 13, 15, 17, 19, 23, 29, 31, 37
(*a*) 13 (*b*) 23
(*c*) 15 (*d*) 31

11. 36, 43, 49, 54, 60, 61, 63, 64
(*a*) 60 (*b*) 63
(*c*) 54 (*d*) 43

12. 3, 10, 41, 206, 1236, 8660
(*a*) 10 (*b*) 41
(*c*) 206 (*d*) 1236

13. 17, 25, 37, 50, 65, 82, 101
(*a*) 25 (*b*) 17
(*c*) 101 (*d*) 65

14. 12, 20, 38, 42, 56, 72
(*a*) 20 (*b*) 38
(*c*) 56 (*d*) 72

15. 5, 6, 11, 22, 42, 88, 176
(*a*) 5 (*b*) 42
(*c*) 176 (*d*) 11

16. 905, 180, 175, 35, 30, 6, 1
(*a*) 6 (*b*) 1
(*c*) 175 (*d*) 905

17. 5040, 840, 160, 42, 14, 7
(*a*) 7 (*b*) 160
(*c*) 14 (*d*) 840

18. 3, 32, 363, 3993, 43923
(*a*) 363 (*b*) 3
(*c*) 32 (*d*) 43923

19. 3, 18, 38, 78, 123, 178, 243
(*a*) 78 (*b*) 3
(*c*) 123 (*d*) 18

20. 0, 1, 2, 2, 4, 4, 6, 8, 8, 10
(*a*) 6 (*b*) 0
(*c*) 1 (*d*) 10

21. 5, 12, 16, 24, 27
(*a*) 12 (*b*) 16
(*c*) 24 (*d*) 27

22. 27, 36, 45, 60, 71, 99
(*a*) 27 (*b*) 45
(*c*) 60 (*d*) 71

23. 81, 64, 49, 35, 25, 16
(*a*) 64 (*b*) 35
(*c*) 25 (*d*) 16

24. 625, 125, 25, 5, 0
(*a*) 0 (*b*) 125
(*c*) 5 (*d*) 625

25. 25, 19, 17, 13, 11, 7
(*a*) 25 (*b*) 17
(*c*) 13 (*d*) 7

EXPLANATORY ANSWERS

1. (*c*) : The numbers in the series are multiplied by 5 to get the next number.
∴ 125 should be in place of 120.

2. (*d*) : Two numbers form a pair. The first number increases by 7 for the next pair and the second number is the double of first number.

4 8 11 22 18 36 25 50
×2 ×2 ×2 ×2
+7 +7 +7

∴ 25 should be in place of 24.

3. (*c*) : There are two alternate series and in each series, the numbers are multiplied by 6 to get the next number.

×6 ×6 ×6
2 4 12 24 72 144 432
×6 ×6

∴ 144 should be in place of 142.

4. (*b*) : There are three alternate series and in each series, the numbers are multiplied by 2 to get the next number.

×2 ×2
2 3 4 4 6 8 8 12 16
×2 ×2
×2 ×2

Series I : 2, 4, 8
Series II : 3, 6, 12
Series III : 4, 8, 16
∴ 8 should be in place of 9.

5. (*d*) : The difference between the consecutive numbers in the series decreases by 1 at each step.

97 91 86 82 79 77 76 76
−6 −5 −4 −3 −2 −1 −0

∴ 82 should be in place of 83.

6. (*c*) : There are two alternate series :

+4 +4 +4 +4
7 11 11 9 15 7 19 5 21 3
−2 −2 −2 −2

Series I : 7, 11, 15, 19, 23 (addition of 4)
Series II : 11, 9, 7, 5, 3 (subtraction of 2)
∴ 3 should be in place of 1.

7. (*b*) : There are two alternate series and in each series, the numbers are multiplied by 4 to get the next number.

×4 ×4 ×4
3 7 12 28 48 112 192 448
×4 ×4 ×4

∴ 112 should be in place of 118.

8. (*d*) : The numbers in this series are multiplied by 10 and 11 alternately is, *i.e.,* × 10, × 11.

10 100 1100 11000 121000 1210000
×10 ×11 ×10 ×11 ×10

∴ 121000 should be in place of 111000.

9. (*a*) : The numbers in this series are divided by 4 to get the next number.

24576 6144 1536 384 96 24
÷4 ÷4 ÷4 ÷4 ÷4

∴ 384 should be in place of 386.

10. (*c*) : The series comprises of prime numbers in increasing order. Only 15 is an exception.

11. (*a*) : The difference between two consecutive numbers is decreasing by 1 at each step.

36 43 49 54 58 61 63 64
+7 +6 +5 +4 +3 +2 +1

∴ 58 should be in place of 60.

12. (*d*) : The sequence followed in this series is :

$3 \times 3 + 1 = 10$
$10 \times 4 + 1 = 41$
$41 \times 5 + 1 = 206$
$206 \times 6 + 1 = 1237$; and
$1237 \times 7 + 1 = 8660$

∴ 1237 should be in place of 1236.

13. (*a*) : The numbers in the series are 1 plus the squares of numbers in natural order starting from 4.

17 26 37 50 65 82 101
↓ ↓ ↓ ↓ ↓ ↓ ↓
4^2+1 5^2+1 6^2+1 7^2+1 8^2+1 9^2+1 10^2+1

∴ 26 should be in place of 25.

14. (*b*) : The sequence followed in this series is :
$3 \times 4 = 12$
$4 \times 5 = 20$
$5 \times 6 = 30$
$6 \times 7 = 42$
$7 \times 8 = 56$
$8 \times 9 = 72$
$\therefore$ 30 should be in place of 38.

15. (*b*) : The numbers in the series are the sum of all the numbers preceding them.

5 6 11 22 44 88 176

(5 + 6)
(5 + 6 + 11)
(5 + 6 + 11 + 22)
(5 + 6 + 11 + 22 + 44)
(5 + 6 + 11 + 22 + 44 + 88)

$\therefore$ 44 should be in place of 42.

16. (*d*) : The sequence followed in the series is $\div$ 5, − 5 which is repeated.

900 180 175 35 30 6 1
$\div 5$ -5 $\div 5$ -5 $\div 5$ -5

$\therefore$ 900 should be in place of 905.

17. (*b*) : The numbers in the series are divided by numbers in reverse order starting from 6 to get the next number.

5040 840 168 42 14 7
$\div 6$ $\div 5$ $\div 4$ $\div 3$ $\div 2$

$\therefore$ 168 should be in place of 160.

18. (*c*) : The numbers in this series are multiplied by 11 to get the next number.

3 33 363 3993 43923
$\times 11$ $\times 11$ $\times 11$ $\times 11$

$\therefore$ 33 should be in place of 32.

19. (*b*) : 3 is the only prime number in the series.

20. (*d*) : There are two alternate series :

+2 +2 +2 +2
0 1 2 2 4 4 6 8 8 16
$\times 2$ $\times 2$ $\times 2$ $\times 2$

Series I : 0, 2, 4, 6, 8, (addition of 2)
Series II : 1, 2, 4, 8, 16 (multiplication by 2)
$\therefore$ 16 should be in place of 10.

21. (*c*) : The sequence in the series is +7, +4 which is repeated, *i.e.,*

5 12 16 23 27
+7 +4 +7 +4

$\therefore$ 23 should be in place of 24.

22. (*d*) : All other numbers are divisible by 3.

23. (*b*) : The sequence is the square of the number in decreasing order starting from 9, *i.e.,*
$9^2 = 9 \times 9 = 81$
$8^2 = 8 \times 8 = 64$
$7^2 = 7 \times 7 = 49$
$6^2 = 6 \times 6 = 36$
$5^2 = 5 \times 5 = 25$
$4^2 = 4 \times 4 = 16$
$\therefore$ 36 should be in place of 35.

24. (*a*) : The number in the series are divided by 5 to get the next number

625 125 25 5 1
$\div 5$ $\div 5$ $\div 5$ $\div 5$

$\therefore$ 1 should be in place of 0 digit.

25. (*a*) : Other numbers are prime numbers.

6 Word Analogy

In Analogy Tests, the relationship between two given words is established and then applied to the other words. The type of relationship may vary, so while attempting such questions the first step is to identify the type of relationship.

MULTIPLE CHOICE QUESTIONS

Directions : *In the questions given below one term is missing. Based on the relationship of the two given words find the missing term from the given options.*

1. HUNGER : FOOD : : THIRST : ?
(*a*) Water (*b*) Drink
(*c*) Tea (*d*) Coffee

2. HUNTER : GUN : : WRITER : ?
(*a*) Book (*b*) Pen
(*c*) Poem (*d*) Page

3. WOOL : SHEEP : : SILK : ?
(*a*) Saree (*b*) String
(*c*) Silkworm (*d*) Moth

4. FOOD : STOMACH : : FUEL : ?
(*a*) Engine (*b*) Automobile
(*c*) Rail (*d*) Aeroplane

5. WATER : SAND : : OCEAN : ?
(*a*) Island (*b*) River
(*c*) Desert (*d*) Waves

6. ADULT : BABY : : FLOWER : ?
(*a*) Seed (*b*) Bud
(*c*) Fruit (*d*) Butterfly

7. WRITER : READER : : PRODUCER : ?
(*a*) Creator (*b*) Contractor
(*c*) Creature (*d*) Consumer

8. ENTRANCE : EXIT : : LOYALTY : ?
(*a*) Treachery (*b*) Patriotism
(*c*) Fidelity (*d*) Reward

9. MOTHER : MATERNAL : : FATHER : ?
(*a*) Eternal (*b*) Detrimental
(*c*) Paternal (*d*) Formidable

10. PEARL : NECKLACE : : FLOWER : ?
(*a*) Plant (*b*) Garden
(*c*) Petal (*d*) Bouquet

11. ALPHABET : WORD : : WORD : ?
(*a*) Sound (*b*) Music
(*c*) Sentence (*d*) Dictionary

12. LIFE : DEATH : : HOPE : ?
(*a*) Cry (*b*) Pain
(*c*) Despair (*d*) Sad

13. GOOD : BAD : : VIRTUE : ?
(*a*) Blame (*b*) Sin
(*c*) Despair (*d*) Vice

14. BIRD : FLY : : SNAKE : ?
(*a*) Timid (*b*) Clatter
(*c*) Crawl (*d*) Hole

15. CAT : MOUSE : : BIRD : ?
(*a*) Cage (*b*) Trap
(*c*) Eagle (*d*) Worm

16. STATE : EXILE
(*a*) Police : Arrest
(*b*) Judge : Convict
(*c*) Constitution : Amendment
(*d*) Church : Excommunicate

17. CAPRICIOUSNESS : RELIABILITY
(*a*) Extemporaneous : Predictability
(*b*) Unreliable : Inhuman
(*c*) Tenacious : Practicality
(*d*) Arbitrary : Whimsical

18. LOATH : COERCION
(*a*) Detest : Caressing
(*b*) Irritate : Caressing
(*c*) Irate : Antagonism
(*d*) Reluctant : Persuasion

19. SCALES : FISH
(*a*) Lady : Dress (*b*) Tree : Leaves
(*c*) Bird : Feather (*d*) Skin : Man

20. TREE : SAPLING
(*a*) Hut : Mansion (*b*) Giant : Dwarf
(*c*) Horse : Foal (*d*) Ant : Elephant

21. CHALK : BLACKBOARD
(*a*) Door : Handle (*b*) Table : Chair
(*c*) Ink : Paper (*d*) Type : Paint

22. PRIMEVAL : MEDIEVAL
(*a*) Dinosaur : Dragon
(*b*) Gorilla : Soldier
(*c*) Evolution : Revelation
(*d*) Thorn : Rose

23. TRILOGY : NOVEL
(*a*) Rice : Husk
(*b*) Milk : Cream
(*c*) Fabric : Weaving
(*d*) Serial : Episode

24. PEDANT : ERUDITION
(*a*) Prude : Modesty
(*b*) Blunt : Politician
(*c*) Diplomats : Tactless
(*d*) Enemy : Friendly

25. FURY : IRE
(*a*) Convulsion : Spasm
(*b*) Amusement : Happiness
(*c*) Joke : Laugh
(*d*) Dispassion : Emotion

26. BaBy : TaTa : : LiLy : ?
(*a*) PooL (*b*) ROse
(*c*) HaNd (*d*) DoWN

27. BCDA : STUR : : KLMJ : ?
(*a*) VWXU (*b*) EFHG
(*c*) SRTU (*d*) QSRP

28. AEI : LPT : : CGK : ?
(*a*) OSV (*b*) RUY
(*c*) TXC (*d*) FJN

29. RUX : TRP : : BEH : ?
(*a*) SQN (*b*) QON
(*c*) QOM (*d*) QNL

30. CART : ART : : FOUR : ?
(*a*) RUN (*b*) TWO
(*c*) QUE (*d*) OUR

31. FIK : JGO : : DFR : ?
(*a*) BIO (*b*) HDV
(*c*) GCU (*d*) FLP

32. LJH : KKI : : CIA : ?
(*a*) BJB (*b*) DHB
(*c*) BJC (*d*) BBJ

33. ACE : HIL : : MOQ : ?
(*a*) SVW (*b*) TUX
(*c*) RTW (*d*) WUS

34. BCDE : WVUT : : QRST : ?
(*a*) EFHG (*b*) JIHG
(*c*) POML (*d*) GEDC

35. PNLJ : IGEC : : USQO : ?
(*a*) HJLN (*b*) LNJH
(*c*) NLJH (*d*) JHNL

36. DIMO : DMIO : : JUVR : ?
(*a*) JVRU (*b*) JRVU
(*c*) JVUR (*d*) JUVR

EXPLANATORY ANSWERS

1. (*a*) : Hunger is satiated by food, thirst by water.

2. (*b*) : Weapon of a hunter is a gun, weapon of a writer is a pen.

3. (*c*) : Wool is obtained from sheep, silk is obtained from silkworm.

4. (*a*) : Food is consumed in stomach, fuel is consumed in engine.

5. (*c*) : The related words are near opposites.

6. (*b*) : The youngone of an adult is a baby and that of a flower is a bud.

7. (*d*) : A writer aims to please the readers by his writings, a producer aims to please the consumers by his products.

8. (*a*) : The related words are opposites.

9. (*c*) : Relations on the mother's side are maternal and on the father's side paternal.

10. (*d*) : Many pearls make a necklace, many flowers make a bouquet.

11. (*c*) : More than one alphabet make a word, more than one word make a sentence.

12. (*c*) : The related words are opposites.

13. (*d*) : The related words are opposites.

14. (*c*) : Birds fly, snakes crawl.

15. (*d*) : Cat chases the mouse, bird chases the worm.

16. (*d*) : Punishment of leaving the State is exile, punishment of leaving the Church is excommunication.

17. (*c*) : The related words are antonyms.

18. (*d*) : Loathing is the result of constant coersion, reluctance is the result of constant persuasion.

19. (*d*) : Covering of the fish is scales, covering of man is skin.

20. (*c*) : Young tree is a sapling, young horse is a foal.

21. (*c*) : Chalk is used to write on blackboard, ink is used to write on paper.

22. (*a*) : The related words represent almost the same time period and answer option represents almost the same sized creatures.

23. (*d*) : Trilogy is section of a novel, episode is a section of a serial.

24. (*a*) : The related words are synonyms.

25. (*b*) : Fury causes ire, amusement causes happiness.

26. (*c*) : In each group the alternate letters are capitals.

27. (*a*) : In each group the first three letters are consecutive and they follow the fourth letter.

ABCD : RSTU : : JKLM : UVWX

28. (*d*) : In each group the letters jump three letters between them, *i.e.,* they are moving to the fourth letter.

AEI : LPT : : CGK : FJN

+4 +4   +4 +4   +4 +4   +4 +4

29. (*c*) : The letters in first set are jumping two letters, *i.e.,* moving three steps forward and in the second they are jumping one letter, *i.e.* moving two steps backward.

RUX : TRP : : BEH : QOM

+3 +3   –2 –2   +3+3   –2 –2

30. (*d*) : The first set of letters drop the first letter to get the second set.

CART : ART : : FOUR : OUR

31. (*b*) : The three letters in first set are moved +4, –2, and +4 steps respectively.

FIK : JGO : : DFR : HDV

+4, –2, +4   +4, –2, +4

32. (*a*) : The letters in the first set are moved –1, +1 and +1 steps respectively.

LJH : KKI : : CIA : BJB

–1, +1, +1   –1, +1, +1

33. (*b*) : The three letters are moved +7, +6 and +7 steps forward respectively.

ACE : HIL : : MOQ : TUX

+7, +6, +7   +7, +6, +7

34. (*b*) : The consecutive letters in the first set are in natural order and in the second set, they are in reverse order.

BCDE : WVUT : : QRST : JIHG

35. (*c*) : The letters are moved seven steps backwards.

PNLJ : IGEC : : USQO : NLJH

–7, –7, –7, –7   –7, –7, –7, –7

36. (*c*) : Only the middle letters are reversed to obtain the second set of letters.

DIMO : DMIO : : JUVR : JVUR

7 Number Analogy

In number analogy also, the relationship between the given numbers is detected and then applied to the second part to find the missing numbers. This relationship between the numbers can be based on any of the following patterns: *(i)* numbers can be odd/even/prime numbers; *(ii)* numbers can be multiples of one number; *(iii)* numbers can be squares/cubes of different numbers; *(iv)* some numbers can be added to/subtracted from/multiplied to/divided into the first number to get the second number; *(v)* the second number can be the sum/product/difference of the digits of first number; and *(vi)* combinations of any mathematical calculations given above can apply to the relationship between the two given numbers.

MULTIPLE CHOICE QUESTIONS

Directions : *In the following questions, select the number from the given options which follows the same relationship as shared between the first two numbers.*

1. 1 : 11 : : 2 : ?
(*a*) 20 (*b*) 22
(*c*) 24 (*d*) 44

2. 18 : 27 : : 22 : ?
(*a*) 42 (*b*) 39
(*c*) 33 (*d*) 54

3. 14 : 20 : : 16 : ?
(*a*) 23 (*b*) 10
(*c*) 48 (*d*) 32

4. 8 : 27 : : 64 : ?
(*a*) 277 (*b*) 125
(*c*) 250 (*d*) 99

5. $\frac{1}{7} : \frac{1}{14} : : \frac{1}{9} : ?$
(*a*) $\frac{1}{88}$ (*b*) $\frac{1}{80}$
(*c*) $\frac{1}{81}$ (*d*) $\frac{1}{18}$

6. 0.16 : 0.0016 : : 1.02 : ?
(*a*) 10.20 (*b*) 0.102
(*c*) 0.0102 (*d*) 1.020

7. 5 : 24 : : 8 : ?
(*a*) 65 (*b*) 63
(*c*) 62 (*d*) 64

8. 23 : 53 : : 8 : ?
(*a*) 66 (*b*) 57
(*c*) 27 (*d*) 19

9. 6 : 9 : : 7 : ?
(*a*) 4 (*b*) 14
(*c*) 10 (*d*) 28

10. 7 : 28 : : 2 : ?
(*a*) 8 (*b*) 16
(*c*) 24 (*d*) 12

11. 65 : 30 : : 44 : ?
(*a*) 79 (*b*) 62
(*c*) 28 (*d*) 16

12. 99 : 76 : : 24 : ?
(*a*) 1 (*b*) 13
(*c*) 9 (*d*) 7

13. 11 : 35 : : 17 : ?
(*a*) 3 (*b*) 22
(*c*) 58 (*d*) 10

14. 663 : 884 : : 221 : ?
(*a*) 332 (*b*) 554
(*c*) 773 (*d*) 442

15. 16 : 0.16 : : ?
(*a*) 2 : 0.02 (*b*) 7 : 0.007
(*c*) 1.3 : 0.13 (*d*) 0.01 : 0.001

16. $3 : \frac{1}{3} : : ?$
(*a*) 6 : 12 (*b*) 5 : 2/15
(*c*) 8 : 1/8 (*d*) 9 : 27

17. 43 : 34 : : 52 : ?
(*a*) 49 (*b*) 25
(*c*) 36 (*d*) 64

18. 65 : 13 : : 180 : ?
(*a*) 93 (*b*) 36
(*c*) 133 (*d*) 102

19. 882 : 447 : : 881 : ?
(*a*) 444 (*b*) 445
(*c*) 446 (*d*) 447

20. 125 : 27 : : 343 : ?
(*a*) 729 (*b*) 64
(*c*) 216 (*d*) 512

21. 30 : 42 : : 56 : ?
(*a*) 92 (*b*) 21
(*c*) 38 (*d*) 72

22. 190 : 10 : : 102 : ?
(*a*) 4 (*b*) 7
(*c*) 3 (*d*) 5

23. 95 : 38 : : 167 : ?
(*a*) 110 (*b*) 120
(*c*) 113 (*d*) 134

24. 4 : 36 : : 6 : ?
(*a*) 63 (*b*) 54
(*c*) 35 (*d*) 30

25. 6 : 12 : : 20 : ?
(*a*) 50 (*b*) 30
(*c*) 42 (*d*) 38

26. 6 : 18 : : 4 : ?
(*a*) 2 (*b*) 6
(*c*) 8 (*d*) 16

27. 10 : 20 : : 30 : ?
(*a*) 45 (*b*) 60
(*c*) 50 (*d*) 70

28. 92 : 69 : : 46 : ?
(*a*) 57 (*b*) 23
(*c*) 31 (*d*) 19

29. 63 : 9 : : 49 : ?
(*a*) 12 (*b*) 3
(*c*) 36 (*d*) 7

30. 2 : 11 : : ?
(*a*) 6 : 17 (*b*) 8 : 43
(*c*) 5 : 41 (*d*) 7 : 35

EXPLANATORY ANSWERS

1. (*b*) : The first number is repeated to obtain the second number.

2. (*c*) : In the given set, the numbers are multiples of 9 and in the second set, multiples of 11.

18 : 27 : : 22 : 33
↓ ↓ ↓ ↓
9 × 2 9 × 3 11 × 2 11 × 3

3. (*a*) : The relationship between the numbers is :

14 : 20 : : 16 : 23
↓ ↓ ↓ ↓
7 × 2 (7 × 3) –1 8 × 2 (8 × 3) –1

4. (*b*) : The numbers are cubes of different numbers.

8 : 27 : : 64 : 125
↓ ↓ ↓ ↓
2^3 3^3 4^3 5^3

5. (*d*) : The first fraction is multiplied by half to obtain the second fraction.

$\frac{1}{7} : \frac{1}{14} :: \frac{1}{9} : \frac{1}{18}$ (each $\times\frac{1}{2}$)

6. (*c*) : The decimals are divided by 100.

0.16 : 0.0016 : : 1.02 : 0.0102 (each ÷ 100)

7. (*b*) : The second number is square of first number minus 1.

5 : 24 : : 8 : 63
$(5^2) - 1$ $(8^2) - 1$

8. (*d*) : All the numbers are prime numbers.

9. (*c*) : The second number is three more than the first number.

6 : 9 : : 7 : 10 (each +3)

10. (*a*) : The second number is four times the first.

7 : 28 : : 2 : 8 (each × 4)

11. (*d*) : The second number is the product of digits of first number.

65 : 30 → (6 × 5) : : 44 : 16 → (4 × 4)

12. (*a*) : The second number is 23 less than the first number.

99 : 76 : : 24 : 1 (each –23)

13. (*a*) : All the numbers are odd numbers.

14. (*d*) : The digits at tens and hundreds place is same but the digit at the units place is half the other identical digits.

66 3 : 88 4 : : 22 1 : 44 2

15. (*a*) : Of the two related numbers, the second number is the result of first number divided by 100.

16 : 0.16 :: 2 : 0.02

÷ 100 ÷ 100

16. (*c*) : Of the two related numbers, the second number is the part fraction of the first number, *i.e.* 3 is related to one-third $\left(\frac{1}{3}\right)$. Similarly, 8 will be related to one-eighth $\left(\frac{1}{8}\right)$.

17. (*b*) : The digits are written in reverse order in the second number.

43 : 34 :: 52 : 25

18. (*b*) : The first number is divided by 5 to get the second number.

65 : 13 :: 180 : 36

÷ 5 ÷ 5

19. (*c*) : The relationship between two numbers is :

88 2 : 44 7 :: 88 1 : 44 6

÷ 2 ÷ 2

+5 +5

20. (*a*) : The numbers are cubes of different odd numbers.

125 : 27 :: 343 : 729

↓ ↓ ↓ ↓

5^3 3^3 7^3 9^3

21. (*d*) : The numbers are squares of different numbers plus the number itself.

30 : 42 :: 56 : 72

↓ ↓ ↓ ↓

$5^2 + 5$ $6^2 + 6$ $7^2 + 7$ $8^2 + 8$

22. (*c*) : The second number is the sum of digits of first number.

190 : 10 :: 102 : 3

$(1+9+0)$ $(1+0+2)$

23. (*a*) : The second number is 57 less than the first number.

95 : 38 :: 167 : 110

–57 –57

24. (*b*) : The second number is 9 times the first number.

4 : 36 :: 6 : 54

× 9 × 9

25. (*b*) : The numbers are squares of different numbers minus the number itself.

6 : 12 :: 20 : 30

↓ ↓ ↓ ↓

$3^2 - 3$ $4^2 - 4$ $5^2 - 5$ $6^2 - 6$

26. (*c*) : Half the square of first number is the second number.

6 : 18 :: 4 : 8

$6^2 \div 2$ $4^2 \div 2$

27. (*b*) : The second number is double the first number.

10 : 20 :: 30 : 60

× 2 × 2

28. (*b*) : The difference between the two numbers is 23.

92 : 69 :: 46 : 23

–23 –23

29. (*d*) : The first number is divided by 7 to get the second number.

63 : 9 :: 49 : 7

÷ 7 ÷ 7

30. (*c*) : The sum of the digits of the second number is equal to the first number.

2 : 11 :: 5 : 41

$(1+1)$ $(4+1)$

8

Odd One Out

In this type of classification, four words are given out of which three are almost same in matter or meaning and only one word is different from the common three. One has to find out the word which is different from the rest.

MULTIPLE CHOICE QUESTIONS

Directions : *In each of the following questions, three words are alike in some manner. Spot the odd one out.*

1. (*a*) Green (*b*) Red (*c*) Colour (*d*) Orange

2. (*a*) Stable (*b*) Hole (*c*) Canoe (*d*) Sty

3. (*a*) Nose (*b*) Eyes (*c*) Skin (*d*) Teeth

4. (*a*) Venus (*b*) Moon (*c*) Pluto (*d*) Mars

5. (*a*) Happy (*b*) Gloomy (*c*) Lively (*d*) Cheerful

6. (*a*) Cone (*b*) Circle (*c*) Triangle (*d*) Rectangle

7. (*a*) Lead (*b*) Mercury (*c*) Copper (*d*) Iron

8. (*a*) Kite (*b*) Bird (*c*) Radar (*d*) Jet

9. (*a*) Knee (*b*) Shoulder (*c*) Ankle (*d*) Palm

10. (*a*) Deluge (*b*) Calamity (*c*) Catastrophe (*d*) War

11. (*a*) Cub (*b*) Chicken (*c*) Pig (*d*) Pup

12. (*a*) Rabbit (*b*) Crocodile (*c*) Earthworm (*d*) Snail

13. (*a*) Tree (*b*) Leaf (*c*) Bush (*d*) Herb

14. (*a*) Doctor (*b*) Teacher (*c*) Engineer (*d*) Diver

15. (*a*) Trot (*b*) Equestrian (*c*) Derby (*d*) Grunt

16. (*a*) Ornate (*b*) Pleasant (*c*) Decorate (*d*) Beautify

17. (*a*) Polo (*b*) Chess (*c*) Ludo (*d*) Squash

18. (*a*) Tutor (*b*) Principal (*c*) Pupil (*d*) Professor

19. (*a*) Pond (*b*) River (*c*) Stream (*d*) Brook

20. (*a*) Quotation (*b*) Duty (*c*) Tax (*d*) Octroi

21. (*a*) Root (*b*) Tree (*c*) Branch (*d*) Flower

22. (*a*) Mumbai (*b*) Chandigarh (*c*) Lucknow (*d*) Hyderabad

23. (*a*) Immortal (*b*) Eminence (*c*) Perpetual (*d*) Everlasting

24. (*a*) Spinach (*b*) Potato (*c*) Carrot (*d*) Ginger

25. (*a*) Van (*b*) Aeroplane (*c*) Helicopter (*d*) Transport

Directions : *Three of the following four in each question are alike in a certain way and so form a group. Select the group of letters that does not belong to that group.*

26. (*a*) ACE (*b*) LOR (*c*) GIK (*d*) VXZ

27. (*a*) TSR (*b*) LKJ (*c*) PQO (*d*) HGF

28. (*a*) EF LM (*b*) KJ SR (*c*) XW HG (*d*) ED YX

29. (*a*) JOPK (*b*) BOPC (*c*) QOPR (*d*) TOPS

30. (*a*) DfH (*b*) MoQ (*c*) UwY (*d*) lnO

31. (*a*) JKkL (*b*) OPpQ (*c*) DEEf (*d*) VWwX

32. (*a*) BdfH (*b*) FHJL (*c*) RTvX (*d*) uVwX

33. (*a*) DFHEG (*b*) TWXUV (*c*) OQSPR (*d*) JLNKM

34. (*a*) FEUV (*b*) DCXW (*c*) BAZY (*d*) HGTS

35. (*a*) UTSR (*b*) XYZW (*c*) ONML (*d*) IHGF

Directions : *In each of the following questions, there are four options. Three numbers, in these options, are alike in certain manner. Only one number does not fit in. Choose the one which is different from the rest.*

36. (*a*) 1948 (*b*) 2401 (*c*) 966 (*d*) 1449

37. (*a*) 182 (*b*) 169 (*c*) 130 (*d*) 158

38. (*a*) 129 (*b*) 130 (*c*) 131 (*d*) 132

39. (*a*) 3215 (*b*) 9309 (*c*) 4721 (*d*) 2850

40. (*a*) 1776 (*b*) 2364 (*c*) 1976 (*d*) 3776

41. (*a*) 64 (*b*) 84 (*c*) 16 (*d*) 36

42. (*a*) 24 (*b*) 90 (*c*) 54 (*d*) 36

43. (*a*) 7658 (*b*) 1234 (*c*) 9876 (*d*) 6543

44. (*a*) 3 (*b*) 9 (*c*) 5 (*d*) 7

45. (*a*) 6450 (*b*) 1776 (*c*) 2392 (*d*) 3815

Directions : *In the following questions select the pair which is different from the other three.*

46. (*a*) Chair - Furniture
(*b*) Shirt - Garment
(*c*) Necklace - Jewellery
(*d*) Bogie - Engine

47. (*a*) Crayon - Paper (*b*) Pencil - Lead (*c*) Pen - Ink (*d*) Brush - Paint

48. (*a*) War - Peace (*b*) Real - Natural (*c*) Premiere - First (*d*) Wrath - Anger

49. (*a*) Finger - Thimble (*b*) Head - Cap (*c*) Waist - Tiara (*d*) Foot - Shoe

50. (*a*) Day - Night (*b*) Clever - Foolish (*c*) Clear - Blurred (*d*) Arrive - Come

Directions : *In the following questions, which of the following pair of letters is different from the other three?*

51. (*a*) FGH - HIJ (*b*) PQR - RST (*c*) MNO - OPQ (*d*) CDE - DEF

52. (*a*) JuM - jUm (*b*) iLo - Ilo (*c*) PSa - psA (*d*) ZeX - zEx

53. (*a*) NQT - JMP (*b*) CFI - RUX (*c*) ADG - FGH (*d*) SVY - ORU

54. (*a*) DXD - XDX (*b*) KUK - UKU (*c*) FHF - EHE (*d*) RSR - SRS

55. (*a*) AYT - BZU (*b*) FNG - EMF (*c*) RWO - QVN (*d*) HJD - GIC

56. (*a*) POT - TOP (*b*) TAN - ANT (*c*) BIN - NIB (*d*) DUB - BUD

57. (*a*) LT - HF (*b*) IN - VA (*c*) ZE - XM (*d*) RB - SC

58. (*a*) MQU - CGK (*b*) BGL - TYD (*c*) RWB - FKP (*d*) JOT - DIN

59. (*a*) JLN - KMO (*b*) GIK - HJL (*c*) DFJ - EGI (*d*) OQS - PRT

60. (*a*) URS - UrS (*b*) JPL - JpL (*c*) PQM - PqM (*d*) MoN - MON

Directions : *Which of the following pair of numbers is different from the other three pairs?*

61. (*a*) 28, 4 (*b*) 63, 7 (*c*) 56, 8 (*d*) 35, 5

62. (*a*) 3, 11 (*b*) 8, 16 (*c*) 5, 13 (*d*) 14, 24

63. (*a*) 13, 156 (*b*) 12, 144 (*c*) 15, 180 (*d*) 16, 176

64. (*a*) 32, 13 (*b*) 46, 20 (*c*) 51, 24 (*d*) 72, 45

65. (*a*) 5, 25 (*b*) 9, 81 (*c*) 4, 64 (*d*) 6, 36

66. (*a*) 91, 10 (*b*) 57, 12 (*c*) 69, 15 (*d*) 72, 13

67. (*a*) 20, 10 (*b*) 45, 27 (*c*) 15, 12 (*d*) 30, 18

68. (*a*) 7, 3 (*b*) 13, 9 (*c*) 11, 7 (*d*) 17, 8

69. (*a*) 10, 20 (*b*) 40, 50 (*c*) 30, 40 (*d*) 50, 60

70. (*a*) 16, 26 (*b*) 3, 4 (*c*) 26, 24 (*d*) 27, 25

EXPLANATORY ANSWERS

1. (c) : All others are types of colour.

2. (c) : Canoe is a boat. Others are resting places of birds/ animals.

3. (d) : All others are sense organs.

4. (b) : All others are planets.

5. (b) : All others are expressions of joy.

6. (a) : All others are geometrical figures.

7. (b) : All others are solid metals.

8. (c) : All others are flying objects. Radar traces the objects in sky.

9. (d) : All others are joints in human body.

10. (d) : All others are nature-bound. Only war is man-made.

11. (c) : All others are young ones of animals.

12. (a) : All others are crawling animals.

13. (b) : All others are types of vegetation.

14. (d) : All others are professions, but only Diver's profession comprises of diving under water.

15. (d) : All other terms are related with activities of a horse.

16. (b) : All other words have the same meaning.

17. (a) : All others are indoor games.

18. (c) : All others are instructors. Pupil learns from the instructor.

19. (a) : All others are running forms of water.

20. (a) : All others are forms of taxes.

21. (a) : All others are parts of tree which are above the ground.

22. (b) : All are capitals but Chandigarh is the capital of two states.

23. (b) : All others have the same meaning.

24. (a) : All others are underground vegetables.

25. (d) : All others are means of transport.

26. (b) : The sequence in each group is +2. Only option (*b*) has sequence in +3, *i.e.*,

ABCDE   LMNOPQR
+2 +2   +3 +3

GHIJK   VWXYZ
+2 +2   +2 +2

27. (c) : The sequence of alphabet in each group is in reverse order. Only option (*c*) has sequence in disturbed order.

28. (a) : Two consecutive alphabet in each group are in reverse sequence (–1), *i.e.*,

KJ SR; XW HG; ED YX
–1 –1 –1 –1 –1 –1

Only in option (*a*) the sequence is in natural order (+1), *i.e.*,

E F   L M
+1   +1

29. (d) : In each group, letters 'OP' are common. The two corner alphabet are in natural order (+1); *i.e.*,

JOPK ; BOPC ; QOPR
+1 +1 +1

Only in option (*d*) they are in reverse order (–1); *i.e.*,

TOPS
–1

30. (d) : In other groups, only the alphabet in the centre is of lower case. In this option letter 'L' on the left is also in lower case.

31. (c) : In other groups, the third letter which is a repeat of the second alphabet is in lower case.

32. (d) : In each group, the sequence of the alphabet, irrespective of the case, is +2; *i.e.*,

B D f H ; F H J L ; R T v X
+2 +2 +2   +2 +2+2   +2 +2 +2

Only in option (*d*) the sequence is in natural order (+1), *i.e.*,

u V w X
+1 +1 +1

33. (b) : In each group; the alphabet at positions-first, fourth, second, fifth and third, form a natural sequence.

In option (*b*), the alphabet at positions first, fourth, fifth, second and third, form the natural sequence.

34. (a) : In each group, two alphabets in the corner and two alphabets in the centre correspond to their reverse order positioned alphabet. *i.e.*,

natural order → A B C D E F G H I J K L M
reverse order → Z Y X W V U T S R Q P O N
natural order → N O P Q R S T U V W X Y Z
reverse order → M L K J I H G F E D C B A

As such—
D corresponds with W and
C corresponds with X.
B corresponds with Y and
A corresponds with Z.
H corresponds with S and
G corresponds with T.

Similarly,
F should correspond with U and
E should correspond with V;
i.e. letters 'UV' should be written as 'VU'

35. (b) : In each group, the alphabet are in reverse order.

36. (a) : Other numbers are divisible by 7.

37. (d) : Other numbers are multiples of 13.

38. (c) : 131 is a prime number.
39. (b) : In other numbers, no digit is repeated.
40. (b) : In other numbers, the last two digits are same.
41. (b) : All other numbers are perfect squares.
42. (a) : In other numbers, the sum of both the digits is 9.
43. (a) : In others, the digits are consecutive in natural or reverse series.
44. (b) : All others are prime numbers.
45. (d) : All other numbers are divisible by 2.
46. (d) : Bogie is a part of train which is a type of conveyance. Chair, Shirt and Necklace are types of furniture, garment and jewellery respectively.
47. (a) : The medium used for writing with pencil is lead, with pen it is ink and with brush, it is paint. With crayon it should be wax.
48. (a) : The pair of words are opposite. Other pairs are synonyms.
49. (c) : Tiara is worn on the head.
50. (d) : Other words are opposite to each other.
51. (d) : In all other groups, the letters are in natural series and the last letter of first part is the first letter of the second part.
52. (b) : In all other groups, only the vowel is in lower case in the first part and in second part the case is reversed.
53. (c) : In all other groups, the letters jump two letters in between them.
54. (c) : In all other groups, the single letter in first part is repeated in the second and vice versa.
55. (a) : In all other groups, the letters in the first part are one step forward than the corresponding letters in the second part.
56. (b) : In all other groups, the letters in the first part are written backwards in the second part.
57. (d) : All other groups have letters made by straight lines only.
58. (a) : In all other groups, the letters jump four letters in between then.
59. (c) : In all others groups, the letters jump one letter between them.
60. (d) : In all other groups, the second part is same as first except that the middle letter is changed to small letter.
61. (b) : In all other pairs 7 times the second number is the first number. In this option 9 times 7 is the first number.
62. (d) : In all other groups, the difference between the two numbers is 8.
63. (d) : In other groups 12 times, the first number is the second number. In this option 11 times the first number is the second number.
64. (a) : 13 is a prime number.
65. (c) : In all other pairs, the second number is a perfect square of the first.
66. (d) : In all other pairs, the second number is the sum of the digits of the first number.
67. (a) : In all other pairs, the numbers are divisible by 3.
68. (d) : In other groups, the difference between the numbers is 4.
69. (a) : In this option, the second number is double the first number. In other groups the difference between the two numbers is 10.
70. (a) : In other groups, the first digits of the two numbers are same (consider 3 and 4 as 03 and 04).

9

Coding and Decoding

Coding is a secretive language which is used to change the representation of the actual term/word/value. This coded language can be framed by (*i*) moving the letters one or more steps forward or backward; (*ii*) substituting numbers for letters and vice–versa; (*iii*) writing the letters of the given word in reverse order in part or in whole; and (*iv*) replacing the letters in their natural series by the same positioned letters in their reverse series.

There is variety in ways of coding. Coding language is not only for words and numbers but also for hiding a group of words, statements or even sentences. This form of coding pattern may appear to be confusing but after solving only a few questions it is very easy to understand. Questions based on this coding pattern require no moving of steps or straining efforts of calculations, but only quick tallying or comparing ability. The codes can be letters or numbers.

MULTIPLE CHOICE QUESTIONS

Directions : *In the following questions select the right option which indicates the correct code for the word or letter given in the question.*

1. If MUSK is coded as 146816, then ZERO will be coded as :

(*a*) 113811 (*b*) 122912
(*c*) 15915 (*d*) 2651815

2. If BAD is coded as 7, HIS as 9, LOW will be coded as:

(*a*) 50 (*b*) 8
(*c*) 23 (*d*) 5

3. In a certain code LIBERATE is written as 56403170, TRIBAL will be written in the same code as :

(*a*) 734615 (*b*) 736415
(*c*) 136475 (*d*) 034615

4. In certain military code, SYSTEM is written as SYSMET, and NEARER as AENRER, what will be the code for FRACTION?

(*a*) CRAFNOIT (*b*) FRCAITNO
(*c*) CARFNOIT (*d*) FRACNOIT

5. If CRUDE is written as BSTED, then MOIST will be coded as :

(*a*) NNJRU (*b*) LNHRS
(*c*) NPJTU (*d*) LPHTS

6. In a certain code ALPACA is written as ACAPLA. How will ANIMAL be written in that code?

(*a*) LAMINA (*b*) ALAMIN
(*c*) LAMNIA (*d*) AAMLIN

7. In a certain code FINGER is written as DGLECP. What will be the code for KIDNEY?

(*a*) IGBLCW (*b*) IGCLBW
(*c*) IBCGLE (*d*) IGBKCV

8. In a certain code QUESTION is written as NXBVQLLQ. How will REPLY be coded?

(*a*) YHMOV (*b*) OBMVI
(*c*) VHSOB (*d*) OHMOV

9. If in a certain code SKEW is coded as PNCY, then what will JXQV will stand for?

(*a*) MUTS (*b*) MUST
(*c*) MTSU (*d*) STUM

10. In a certain code LONDON is written as MPOEPO. What will IVOHSZ mean in the same code?

(*a*) HUNGRY (*b*) HUNDRY
(*c*) GRUNHY (*d*) HONDUS

11. In a certain code language 8514 is a code for HEAD, 3945 for RIDE and 057 for BEG. What will be the code for GRADE?

(*a*) 71345 (*b*) 73415
(*c*) 74135 (*d*) 73145

12. If MOTHERLAND is coded as 9501623748, how will DREAM be coded?

(*a*) 82697 (*b*) 86297
(*c*) 82769 (*d*) 82679

13. If OATH is coded in a certain language as TEYL, then how will WORD be coded?

(*a*) BWRH (*b*) HRWB
(*c*) BSWH (*d*) CSXI

14. If FINANCE is coded as GKQESIL, then how will BANK be coded in the same manner?
(*a*) CBOL (*b*) CDRP
(*c*) CCQO (*d*) CCPN

15. PLANNING is coded in a certain language as UFFHSCSA. How will AUTHORITY be coded in the same language?
(*a*) FOYBTLNND (*b*) FYOTBNNLT
(*c*) FBOYTLNTN (*d*) FBOYTNLTN

Directions : *In the following questions study the coded patterns and then select the right option from the given alternatives.*

16. In a certain language, (a) 'go ju mi' stands for 'plenty of money'; (b) pao ju go nei vu' for 'money creates lots of problems'; (c) 'kol vu nei' for 'problems create tension'; and (d) 'sol tun ju haw' for 'still money is needed'. Which of the following words stand for 'money'?
(*a*) nei (*b*) ju
(*c*) haw (*d*) go

17. In a certain language, (a) 'FOR' stands for 'old is gold'; (b) 'ROT' stands for 'gold is pure'; (c) 'ROM' stands for 'gold is costly'. How will 'pure old gold is costly' be written?
(*a*) TFROM (*b*) FOTRM
(*c*) FTORM (*d*) TOMRF

18. In a certain code '415' means 'milk is hot'; '18' means 'hot soup'; and '895' means 'soup is tasty'. What number will indicate the word 'tasty'?
(*a*) 9 (*b*) 8
(*c*) 5 (*d*) 4

19. In a certain code '643' means 'she is beautiful', '593' means 'he is handsome', and '567' means 'handsome meets beautiful'. What number will indicate the word 'meets'?
(*a*) 5 (*b*) 3
(*c*) 7 (*d*) 6

20. In a certain code language, (a) 'dugo hui mul zo' stands for 'work is very hard'; (b) 'hui dugo ba ki' for 'Bingo is very smart'; (c) 'nano mul dugo' for 'cake is hard', and (d) 'mul ki qu' for 'smart and hard'. Which of the following words stand for 'Bingo'?
(*a*) jalu (*b*) dugo
(*c*) ki (*d*) ba

Directions : *Read the given coded information and choose the correct answer from the given options.*

21. If 'water' is called 'blue', 'blue' is called 'red', 'red' is called 'white', 'white' is called 'sky', 'sky' is called 'rain', 'rain' is called 'green', 'green' is called 'air' and 'air' is called 'table', which of the following is the colour of milk?
(*a*) white (*b*) rain
(*c*) sky (*d*) green

22. If 'light' is called 'dark', 'dark' is called 'green', 'green' is called 'blue', 'blue' is called 'red', 'red' is called 'white' and 'white' is called 'yellow', what is the colour of blood?
(*a*) red (*b*) dark
(*c*) white (*d*) yellow

23. If 'sky' is called 'sea', 'sea' is called 'water', 'water' is called 'air', 'air' is called 'cloud' and 'cloud' is called 'river', then what do we drink when thirsty?
(*a*) sky (*b*) air
(*c*) water (*d*) sea

24. If 'yellow' means 'red', 'white' means 'green', 'red' means 'orange', 'blue' means 'white' and 'green' means 'blue', then the colour of sky is :
(*a*) white (*b*) green
(*c*) blue (*d*) yellow

25. If 'land' is called 'sky', 'sky' is called 'air', 'air' is called 'water', 'water' is called 'sand' and 'sand' is called 'solid', where do fishes swim?
(*a*) air (*b*) sky
(*c*) water (*d*) sand

EXPLANATORY ANSWERS

1. (*b*) : The coded number signifies the position of the alphabet in its reverse order of the alphabetical series (ZYXW...)

M U S K → MUSK
↓ ↓ ↓ ↓
14th 6th 8th 16th → 146816

Similarly,

Z E R O → ZERO
↓ ↓ ↓ ↓
1st 22nd 9th 12th → 122912

2. (*d*) : The coded number is the sum of number digits signifying the position of the alphabet in the natural order.

B A D
↓ ↓ ↓
2nd 1st 4th *i.e.*, 2 + 1 + 4 = 7

Similarly,

H I S
↓ ↓ ↓
8th 9th 19th *i.e.*, 8 + 9 + 19 = 36
further, 3 + 6 = 9

Also,

L O W
↓ ↓ ↓
12th 15th 23rd *i.e.*, 12 + 15 + 23 = 50
further, 5 + 0 = 0

3. (*b*) : The letters of the word TRIBAL are picked from LIBERATE. So will be the coded numbers.

L I B E R A T E → given word
5 6 4 0 3 1 7 0 → codes

Similarly,

T R I B A L → word to be coded
7 3 6 4 1 5 → answer codes

4. (*c*) : The word is divided into two equal parts and the letters of each part are written in reverse order.

SYS TEM SYS MET NEA RER AEN RER

Similarly,

FRAC TION CARF NOIT

5. (*d*) : The code is formed by moving the letters one step backwards and one step forward alternately.

CRUDE BSTED
−1
+1
−1
+1
−1

Similarly,

MOIST LPHTS
−1
+1
−1
+1
−1

6. (*a*) : The letters of the word are written backwards.

ALPACA ACAPLA

Similarly,

ANIMAL LAMINA

7. (*a*) : The word is coded by moving the letters two steps backwards.

FINGER DGLECP
−2
−2
−2
−2
−2
−2

Similarly,

KIDNEY IGBLCW
−2
−2
−2
−2
−2
−2

8. (*d*) : The letters of the word are coded by moving three steps backward and three steps forward alternately.

QUESTION NXBVQLLQ
−3
+3
−3
+3
−3
+3
−3
+3

Similarly,

REPLY OHMOV
−3
+3
−3
+3
−3

9. (*b*) : The letters are decoded by moving the letters +3, −3, +2 and −2 steps respectively.

PNCY SKEW
+3
3
+2
−2

Similarly,

JXQV MUST
+3
−3
+2
−2

10. (*a*) : The letters of the coded word are moved one step backward.

MPOEPO LONDON
−1
−1
−1
−1
−1
−1

Similarly,

I V O H S Z H U N G R Y
−1
−1
−1
−1
−1
−1

11. (*d*) : The word GRADE is framed by letters in the given words. So, in order to find the code for GRADE select the respective number codes.

H E A D R I D E B E G → letters
8 5 1 4 3 9 4 5 0 5 7 → codes

So,
G R A D E →letters
7 3 1 4 5 →answer codes

12. (*d*) : The letters of the word are coded by numbers. So to find the code for DREAM select the respective numbers.

M O T H E R L A N D → letters
9 5 0 1 6 2 3 7 4 8 → codes

So, D R E A M → letters
8 2 6 7 9 → answer codes

13. (*c*) : The letters are coded by moving five and four steps forward alternately.

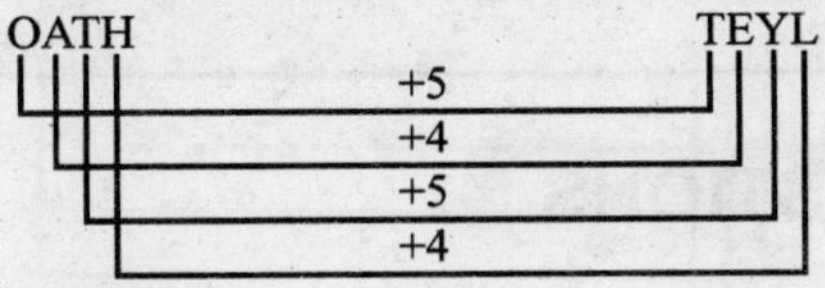

Similarly,

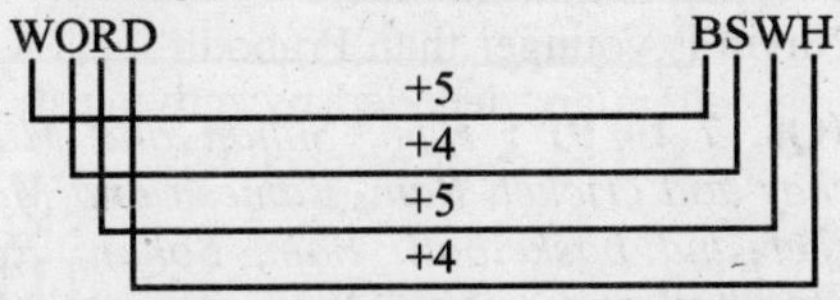

14. (*c*) : The letters of the word are coded by moving one step ahead and increasing the difference by one.

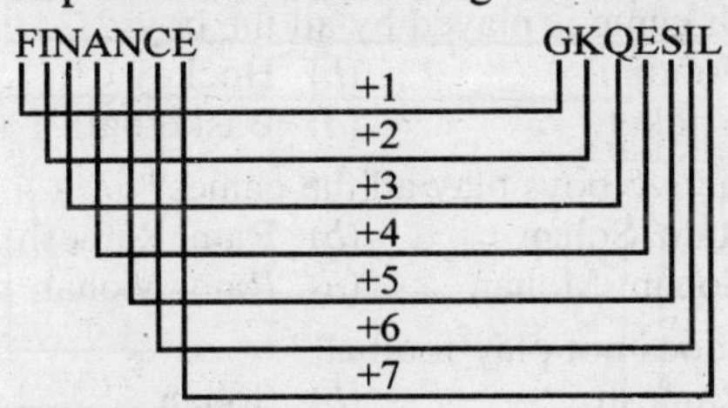

Similarly,

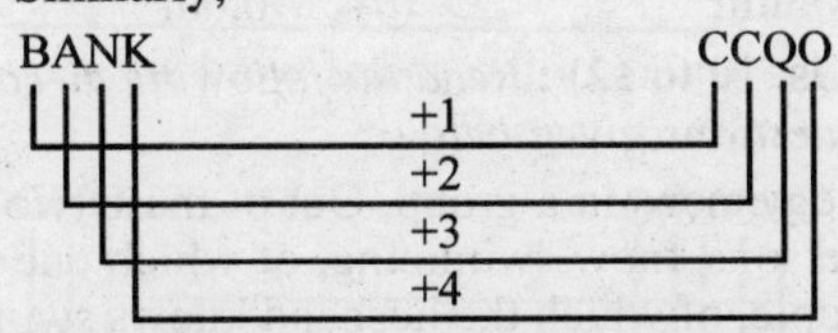

15. (*a*) : The letters of the word are coded by moving five steps forward and six steps backward alternately.

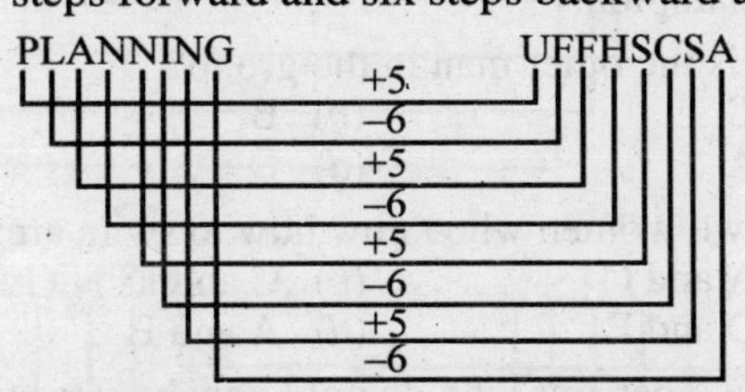

Similarly,

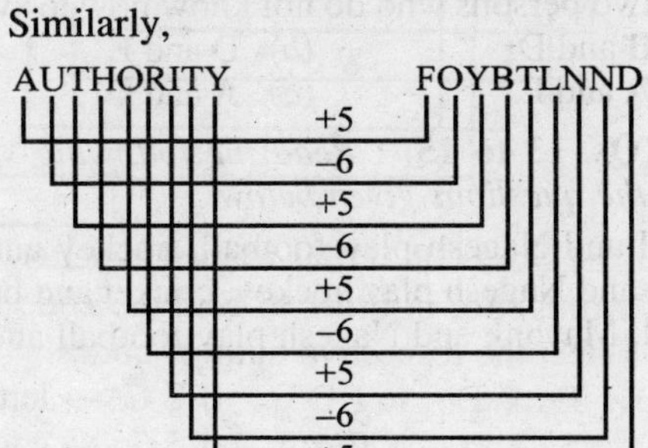

16. (*b*) :

	Code	*Sentence*
1.	go *ju* mi	plenty of *money*
2.	pao *ju* go nei vu	*money* creates lots of problems
3.	kol vu nei	problems create tension
4.	sol tun *ju* haw	still *money* is needed

In 1st, 2nd and 4th codes and their sentences the word 'ju' is repeated and so is 'money'.

17. (*a*) :

Code *Sentence*

1. FOR old is gold

2. ROT gold is pure

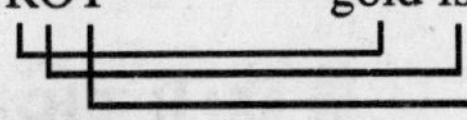

3. ROM gold is costly

Therefore,

F stands for old
O stands for is
R stands for gold
T stands for pure
M stands for costly

So, 'pure old gold is costly' will be written as 'TFROM'.

18. (*a*) :

	Code	*Sentence*
1.	415	milk is hot
2.	18	hot soup
3.	895	soup is *tasty*

From 3rd code and its sentence neither number '9' is repeated nor the word 'tasty'.

19. (*c*) :

	Code	*Sentence*
1.	643	she is beautiful
2.	593	he is handsome
3.	567	handsome *meets* beautiful

From 3rd code and its sentence, neither number '7' nor the word 'meets' is repeated.

20. (*d*) :

	Code	*Sentence*
1.	*dugo hui* mul zo	work *is very* hard
2.	*hui dugo* **ba** *ki*	**Bingo** *is very smart*
3.	nano mul *dugo*	cake is *hard*
4.	mul *ki* qu	*smart* and hard

From 2nd code and its sentence, neither 'ba' nor 'Bingo' is repeated.
(Words repeated are in italics)

21. (*c*) : Colour of milk is 'white' and 'white' is called 'sky'.

22. (*c*) : Colour of blood is 'red' and 'red' is called 'white'.

23. (*b*) : We drink 'water' when we are thirsty and 'water' is called 'air'.

24. (*a*) : Colour of sky is 'blue' and 'blue' means 'white'.

25. (*d*) : Fishes swim in 'water' and 'water' is called 'sand'.

10

Statement Analysis

In these type of questions, a few statements are given. Certain facts are broken up and mentioned in these statements. What is required is to analyse the statements, arrange and sort out the given facts and then answer the questions related to the given statements.

MULTIPLE CHOICE QUESTIONS

1. Among five friends, A is heavier than B; C is lighter than D; B is lighter than D but heavier than E. Who among them is the heaviest?

(*a*) B (*b*) C
(*c*) A (*d*) Can't say

2. Pune is bigger than Jhansi, Sitapur is bigger than Chittor. Raigarh is not as big as Jhansi, but is bigger than Sitapur. Chittor is not as big as Sitapur. Which is the smallest?

(*a*) Jhansi (*b*) Pune
(*c*) Chittor (*d*) Sitapur

3. Ajay works more than Ram. Alok works as much as Raju. Pankaj works less than Alok. Ram works more than Alok. Who works the most of all?

(*a*) Ajay (*b*) Ram
(*c*) Alok (*d*) Raju

4. Among five friends P, Q, R, S and T, who is the youngest? To arrive at the answer which of the following information given in the statements (A) and (B) is sufficient?

(A) R is younger than P and T.
(B) S is younger than Q.

(*a*) Only A alone is sufficient
(*b*) Either A or B is sufficient
(*c*) Both A and B together are needed
(*d*) Both A and B together are not sufficient

5. A is elder to B while C and D are elder to E who lies between A and B. If C be elder to B, which one of the following statements is necessarily true?

(*a*) E is elder to B (*b*) A is elder to C
(*c*) C is elder to D (*d*) D is elder to C

6. Suresh is as much older than Kamal as he is younger than Prabodh. Navin is as old as Kamal. Which of the following statements is wrong?

(*a*) Suresh is older than Navin
(*b*) Kamal is younger than Suresh
(*c*) Prabodh is not the oldest
(*d*) Navin is younger than Prabodh

Directions (Qs. 7 to 9) : *Ram, Sohan and Mohan play football, hockey and cricket. Ram, Ramesh and Mohan play hockey, cricket and basketball. Ram, Sohan, Rahim and Mohan play football and cricket. Now answer the following questions based on the above statements :*

7. Which game is played by all the boys?

(*a*) Football (*b*) Hockey
(*c*) Cricket (*d*) Basketball

8. Which two boys play all the games?

(*a*) Ram, Sohan (*b*) Ram, Ramesh
(*c*) Sohan, Mohan (*d*) Ram, Mohan

9. Who does not play football?

(*a*) Ramesh (*b*) Sohan
(*c*) Rahim (*d*) Mohan

Directions (Qs. 10 to 12) : *Read the following directions and answer the questions given below :*

There are five persons in a group. Out of these two are men. Only three persons know swimming, of which one is a man. There is a couple, of which the husband knows swimming. A is the younger sister of D and B is the husband of E. C is a swimming champion.

10. Who is the other man in the group?

(*a*) C (*b*) B
(*c*) A (*d*) D

11. The two women who know how to swim are:

(*a*) A and C (*b*) C and D
(*c*) D and E (*d*) A and E

12. The two persons who do not know how to swim are :

(*a*) B and D (*b*) D and E
(*c*) A and E (*d*) A and D

Directions (Qs. 13 to 15) : *Read the following statements and answer the questions given below :*

Rajat, Sushil and Nagesh play football, hockey and cricket. Rajat, Ramu and Nagesh play hockey, cricket and basketball. Rajat, Sushil, Mayank and Nagesh play football and cricket.

13. Which game is played by all the boys?
(*a*) Hockey (*b*) Basketball
(*c*) Football (*d*) Cricket

14. Who does not play football?
(*a*) Rajat (*b*) Nagesh
(*c*) Sushil (*d*) Ramu

15. Which of the following two boys play all the games?
(*a*) Nagesh, Rajat (*b*) Mayank, Ramu
(*c*) Ramu, Nagesh (*d*) Sushil, Mayank

Directions (Qs. 16 to 20) : *Read the following information and answer the questions given below:*

(*i*) P, Q, R, S, T and U are six members in a family in which there are two married couples.
(*ii*) T a teacher is married to the doctor who is mother of R and U.
(*iii*) Q the lawyer is married to P.
(*iv*) P has one son and one grandson.
(*v*) Of the two married ladies one is a housewife.
(*vi*) There is also one student and one male engineer in the family.

16. Who among the following is the housewife?
(*a*) Q (*b*) P
(*c*) S (*d*) T

17. Which of the following represents the group of females in the family?
(*a*) QTR (*b*) PSR
(*c*) PSU (*d*) Data inadequate

18. Which of the following is true about the granddaughter in the family?
(*a*) She is a student (*b*) She is an engineer
(*c*) She is a lawyer (*d*) Data Inadequate

19. How R is related to U?
(*a*) Brother (*b*) Sister
(*c*) Brother or Sister (*d*) Data inadequate

20. How is P related to R?
(*a*) Grandfather (*b*) Mother
(*c*) Sister (*d*) Grandmother

EXPLANATORY ANSWERS

1. (*d*) : The five friends in descending order of weight are: A/D, B/C, E or A/D, B, C/E. Either A or D is the heaviest.

2. (*c*) : The order of cities in descending order of size is : Pune, Jhansi, Raigarh, Sitapur, Chittor.

3. (*a*) : On the basis of doing work, the descending order will be : Ajay, Ram, Alok/Raju, Pankaj.

4. (*d*) : Statements are not inter-related.

5. (*a*) : The order in descending seniority will be : A/C/D, E, B.

6. (*c*) : On the basis of age the descending order will be : Prabodh, Suresh, Kamal/Navin.

Chart for Answers 7 to 9

Boy	*Games Played*
Ram	Football, Hocky, Cricket, Basketball
Sohan	Football, Hockey, Cricket
Mohan	Football, Hockey, Cricket, Basketball
Ramesh	Hockey, Cricket, Basketball
Rahim	Football, Cricket.

7. (*c*) **8. (*d*)** **9. (*a*)**

For Answers 10 to 12 the information chart will be :

A. Woman ((younger sister of D); knows swimming.
B. Man (husband of E); knows swimming
C. Woman; Swimming champion
D. Man (brother of A)
E. Woman (wife of B)

From the couple B & E, B (husband) knows swimming. C is a swimming champion. Three persons know how to swim of which only B is a man. So, the two women who know how to swim are C and A (the younger sister of D). There are two men in the group of five. One is B and the other will be D.

10. (*d*) **11. (*a*)** **12. (*b*)**

Chart for Answers 13 to 15

Boy	*Games Played*
Rajat	— Football, Hockey, Cricket, Basketball
Sushil	— Football, Hockey, Cricket, Basketball
Nagesh	— Football, Hockey, Cricket, Basketball
Ramu	— Hockey, Cricket, Basketball
Mayank	— Football, Cricket

13. (*d*) **14. (*d*)** **15. (*a*)**

Chart for Answers 16 to 20

P (female/housewife) — Q (male/lawyer) *Grandparents*

T (male/teacher) — S (female/doctor) *Parents*

R, U (of R and U, one is a male engineer and one is a student) Children

16. (*b*) : One married lady S who is mother of R and U is a doctor. Q the lawyer is married to P. Of the two married ladies one is a house-wife. As Q is a lawyer the other married lady who is a housewife should be P.

17. (*d*) : Sex of R and U is not given.

18. (*a*) : P has one son T and one grandson either R or U one of whom is an engineer. So, the granddaughter is a student.

19. (*c*) : Sex of R and U is not given, but as both are children of S and one is a male engineer, the relationship between the two is of brother and sister.

20. (*d*)

11

Place Arrangement

Place arrangement generally refers to the positioning of persons or objects in a manner indicated by set of information given. One has to understand the order of placement and then attempt questions following the given information.

MULTIPLE CHOICE QUESTIONS

Directions: *In the following questions, understand the arrangement pattern and then select the right answer from the given options :*

1. Five boys are sitting in a row. Raghu is not adjacent to Shyam or Amit. Ajay is not adjacent to Shyam. Raghu is adjacent to Mayank. If Mayank is at the middle in the row, then Ajay is adjacent to whom out of the following?

(*a*) Amit (*b*) Raghu
(*c*) Mayank (*d*) Shyam

2. Mini is to the right of Rajni but to the left of Ananta. Saya is to the right of Mini but to the left of Jaya. Who is on the extreme left if all the girls are facing North?

(*a*) Jaya (*b*) Mini
(*c*) Rajni (*d*) Saya

3. O, P, Q, R, S and T are standing on a bench according to their height. P is taller than O but shorter than S. Only S is taller than T. R is shorter than P but taller than Q. Who is the shortest?

(*a*) O (*b*) Q
(*c*) P (*d*) Cannot be said

4. Five personalities are living in a multistoried building. Mr. Effortless lives in a flat above Mr. Active, Mr. Charge lives in a flat below Mr. Diligent, Mr. Active lives in a flat above Mr. Diligent and Mr. Behaved lives in a flat below Mr. Charge. Who lives in the topmost flat?

(*a*) Mr. Charge (*b*) Mr. Diligent
(*c*) Mr. Effortless (*d*) Mr. Behaved

5. Six friends are sitting in a circle and playing cards. Kenny is to the left of Danny. Michael is in-between Bob and John. Roger is in between Kenny and Bob. Who is sitting to the right of Michael?

(*a*) Danny (*b*) John
(*c*) Kenny (*d*) Bob

6. Four girls A, B, C and D are sitting in a circle. B and C are facing each other. Which of the following is definitely true?

(*a*) A is to the left of C
(*b*) D is to the left of C
(*c*) A and D are facing each other
(*d*) A is not between B and C

7. Brijesh, Jayesh, Amar and Praveer are playing a game of cards. Amar is to the right of Jayesh who is to the right of Brijesh. Who is to the right of Amar?

(*a*) Brijesh (*b*) Praveer
(*c*) Brijesh or Praveer (*d*) Jayesh

8. In a pile of 10 books there are 3 of History, 3 of Hindi, 2 of Maths, and 2 of English. Taking from above there is an English book between a History and Maths book, a History book between a Maths and an English book, a Hindi book between an English and a Maths book, a Maths book between two Hindi books, and two Hindi books between a Maths and a History book. Book of which subject is at the sixth position from the top?

(*a*) English (*b*) Hindi
(*c*) History (*d*) Maths

9. Five persons A, B, C, D and E are sitting in a row facing you such that D is on the left of C and B is on the right of E. A is on the right of C and B is on the left of D. If E occupies a corner position, then who is sitting in the centre?

(*a*) A (*b*) B
(*c*) C (*d*) D

10. Six friends A, B, C, D, E and F are standing in a circle. B is between F and C; A is between E and D; F is to the left of D. Who is between A and F?

(*a*) C (*b*) B
(*c*) D (*d*) E

EXPLANATORY ANSWERS

1. (*b*) : The order of sitting is :
Amit, Shyam, Mayank, Ajay, Raghu
or
Ajay, Raghu, Mayank, Amit, Shyam

2. (*c*) : The order in which the girls are positioned is :
Rajni, Mini, Ananta, Saya, Jaya
or
Saya, Jaya, Ananta
or
Saya, Ananta, Jaya

3. (*d*) : In descending order of height, the standing positions are :

S		S
T		T
P	*or*	P
R		R
O		Q
Q		O

Either O or Q is the shortest. The informa-tion given is not enough to clarify the answer.

4. (*c*) : The personalities living in flats in multi-storied building are in order given below :
Mr. Effortless
Mr. Active
Mr. Diligent
Mr. Charge
Mr. Behaved

5. (*d*) : The order in which the friends are sitting is :

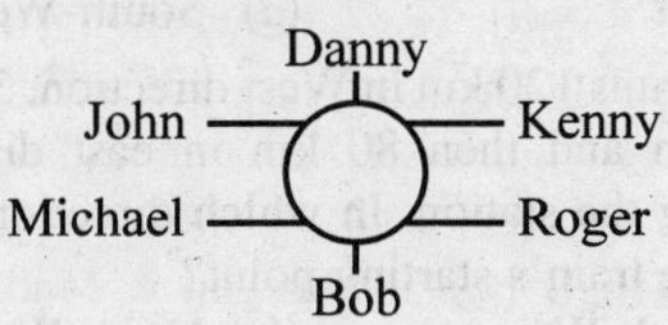

6. (*c*) : The sitting positions are :

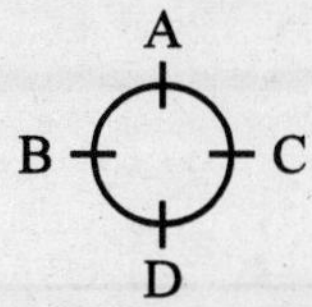

or

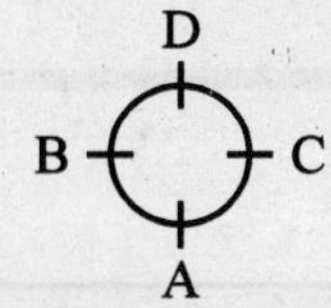

7. (*b*) : The order from left to right is :
Brijesh, Jayesh, Amar, Praveer

8. (*b*) : The pile of books is in the order :
1st — History
English
Maths
History
English
6th — Hindi
Maths
Hindi
Hindi
10th — History

9. (*d*) : Sitting order while facing us is :
A, C, D, B, E

10. (*c*) : The pattern of standing is :

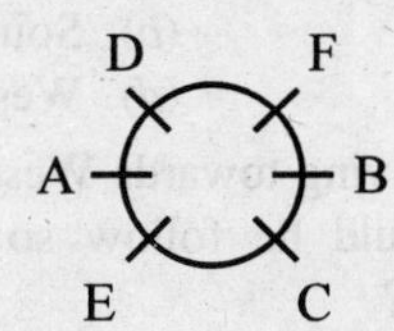

12

Direction Sense

In these type of tests, the directions in questions needs to be perceived. Such questions are based on the direction chart.

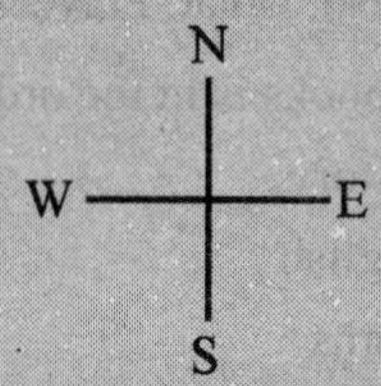

N = North S = South E = East W = West

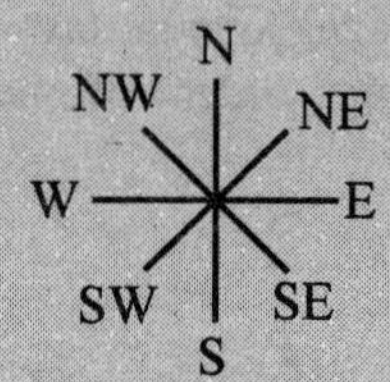

The sense of the different directions are guided by the left and right turns or angular turns.

MULTIPLE CHOICE QUESTIONS

Directions : *In the following questions, select the right answer from the given options to depict the correct direction/distance.*

1. Kittu walks towards East and then towards South. After walking some distance he turns towards West and then turns to his left. In which direction is he walking now?

(*a*) North (*b*) South
(*c*) East (*d*) West

2. A person is driving towards West. What sequence of directions should he follow so that he is driving towards South?

(*a*) left, right, right (*b*) right, right, left
(*c*) left, left, left (*d*) right, right, right

3. Richa drives 8 km to the South, turns left and drives 5 km. Again, she turns left and drives 8 km. How far is she from her starting point?

(*a*) 3 km (*b*) 5 km
(*c*) 8 km (*d*) 13 km

4. Dingi runs 40 km towards North then turns right and runs 50 km. He turns right and runs 30 km, and once again turns right and runs 50 km. How far is he from his starting point?

(*a*) 90 km (*b*) 50 km
(*c*) 10 km (*d*) 5 km

5. If North is called North-West, North-West is called West, West is called South-West and so on. What will South-East be called?

(*a*) East (*b*) West
(*c*) North-East (*d*) South-East

6. A man travels 100 km towards South. From there he turns right and travels 100 km and again turns right to travel 50 km. Which direction is he in from his starting point?

(*a*) North (*b*) North-East
(*c*) East (*d*) South-West

7. A train runs 120 km in West direction, 30 km in South direction and then 80 km in east direction before reaching the station. In which direction is the station from the train's starting point?

(*a*) South-West (*b*) North-West
(*c*) South-East (*d*) South

8. Facing the West direction, Priya jogs for 20 m, turns left and goes further 40 m. She turns left again and jogs for 20 m. Then she turns right to go 20 m to reach the park. How far is the park from her starting point and in which direction?

(*a*) 20 m South (*b*) 40 m West
(*c*) 60 m South (*d*) 100 m East

9. If all the directions are rotated, *i.e.,* if North is changed to West and East to North and so on, then what will come in place of North-West?

(*a*) South-West (*b*) North-East
(*c*) East-North (*d*) East-West

10. A and B start together from one point. They walk 10 km towards North. A turns left and covers 5 km whereas B turns right and covers 3 km. A turns left again and covers 15 km whereas B turns right and covers his 15 km. How far is A from B?

(*a*) 18 km (*b*) 10 km
(*c*) 5 km (*d*) 8 km

11. Tarun is walking towards East. What direction he should not follow if he should walk towards North?

(*a*) right, right, left, right, right
(*b*) right, right, left, left, left
(*c*) right, right, right
(*d*) right, left, right, left

12. Sony and Moni start walking from a point. Sony walks in West direction and Moni in South direction. After covering 20 km, Soni turns left and walks 15 km. Moni walks 10 km, turns left and walks 5 km. Soni, then turns left again and walks 25 km, whereas Moni turns right and walks 5 km. How far are Sony and Moni from each other?

(*a*) 5 km
(*b*) They are back at the starting point
(*c*) They are at same place at the finishing point
(*d*) Data is insufficient

13. I was walking in South-East direction. After a while I turned 90° to the right and walked ahead. Later I turned 45° to the right. In which direction am I walking now?

(*a*) North (*b*) North-East
(*c*) South-West (*d*) East

14. Two friends Jack and Bunny start a race, and together they run for 50 mts. Jack turns right and runs 60 mts while Bunny turns left and runs 40 mts. Then Jack turns left and runs 50 mts while Bunny turns right and runs 50 mts. How far are the two friends now from each other?

(*a*) 60 mts (*b*) 100 mts
(*c*) 20 mts (*d*) 150 mts

15. A policeman left his police post and proceeded south 4 km, after hearing a loud sound from point A. On reaching the place he heard another sound and proceeded 4 km to his left to the point B, only to find that the sound was coming from left of B. From B he proceeded left to reach that place 4 km away. In which direction he has to go to reach his police post?

(*a*) North (*b*) South
(*c*) East (*d*) West

EXPLANATORY ANSWERS

1. (*b*) :
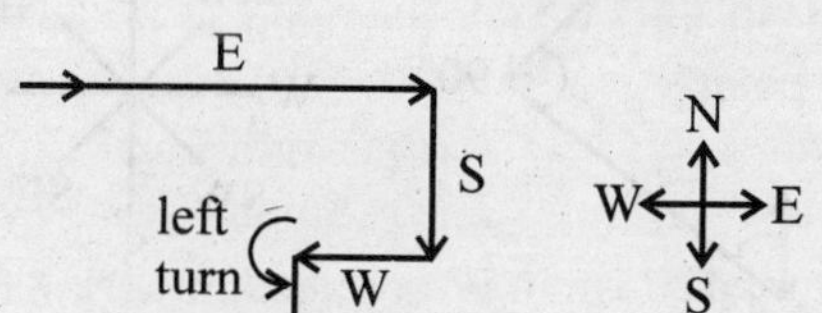

2. (*d*) :
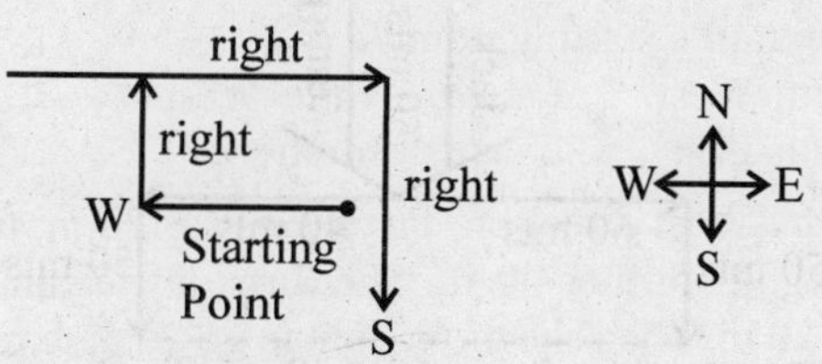

3. (*b*) :
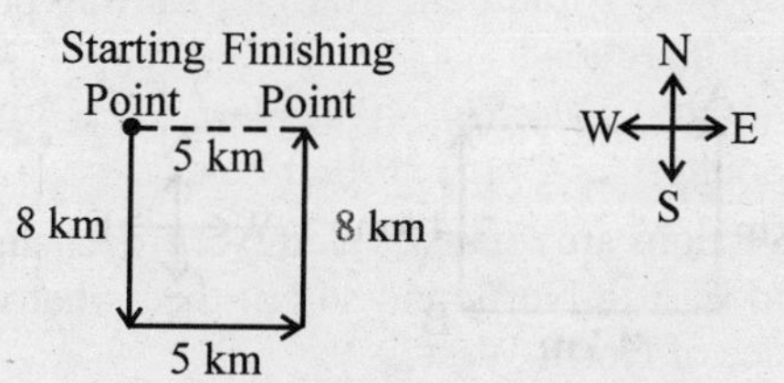

4. (*c*) :
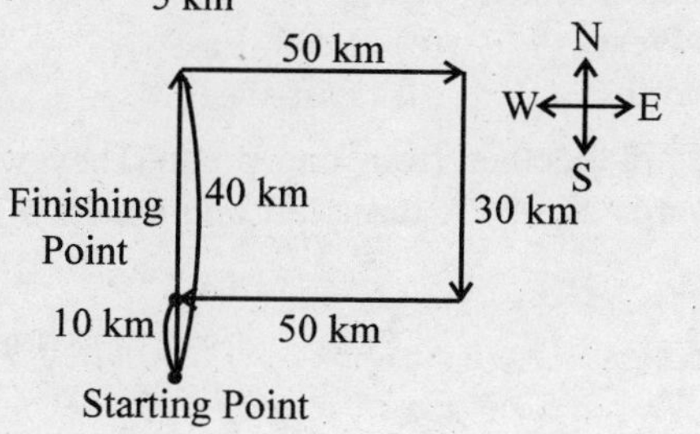

5. (*a*) :
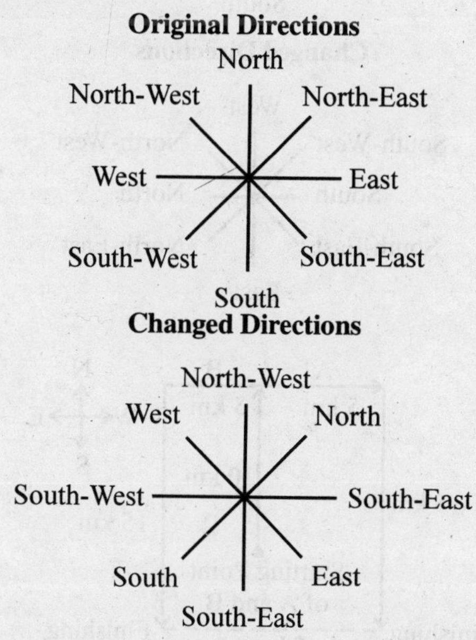

6. (*d*) :
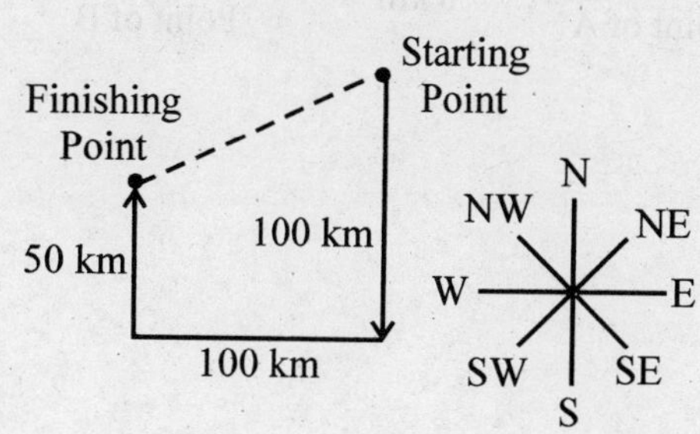

7. (a) :

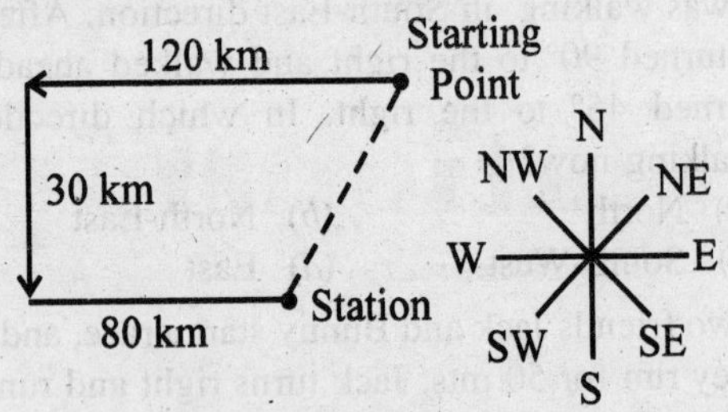

8. (c) : (40 + 20) = 60 metres South

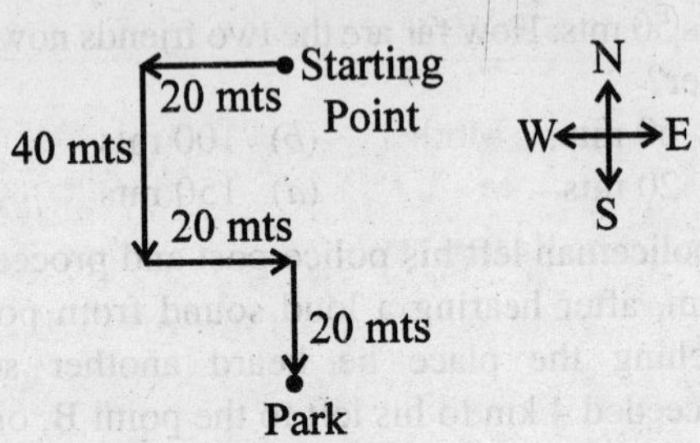

9. (a) : Original Directions

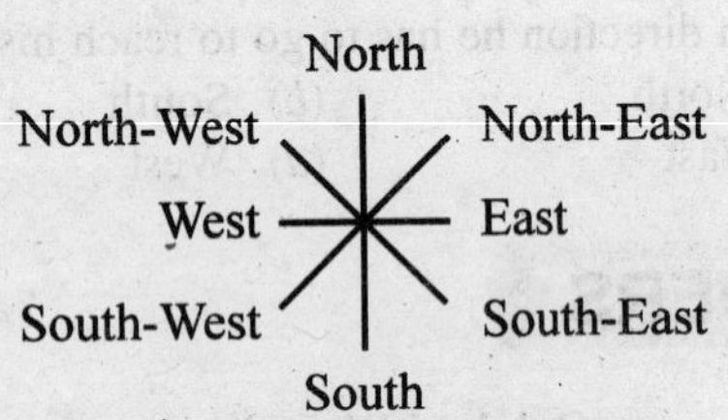

Changed Directions

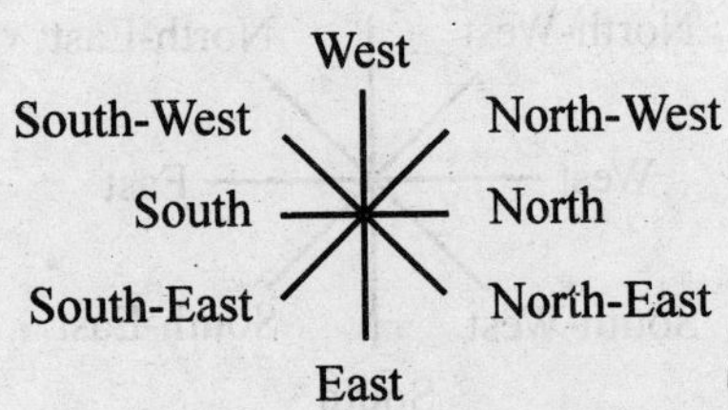

10. (d) :

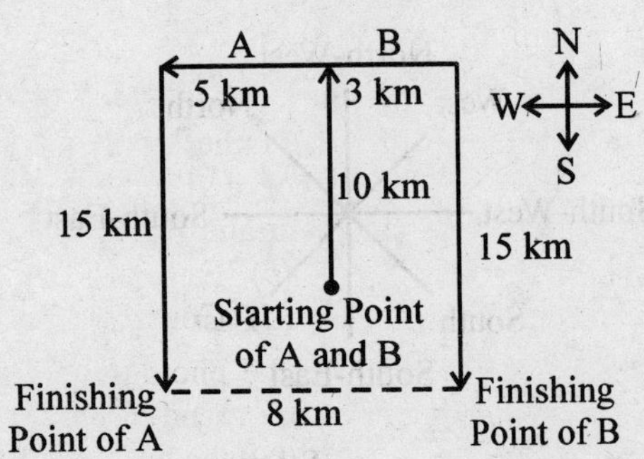

11. (d) :

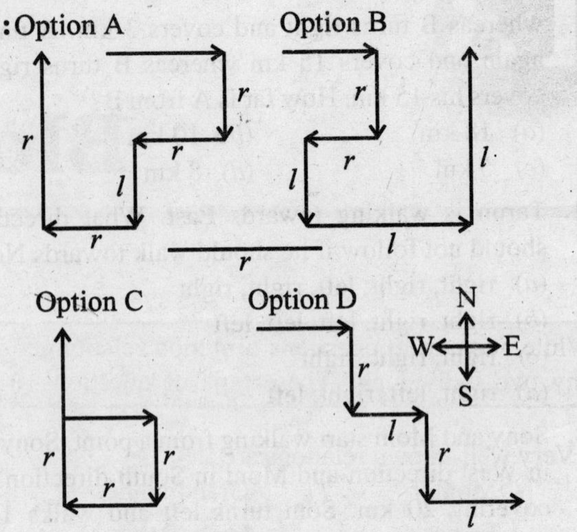

12. (c) :

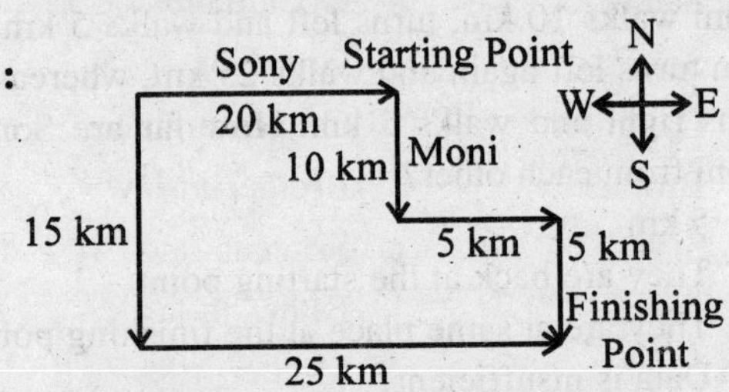

13. (a) : The direction of the movement is :

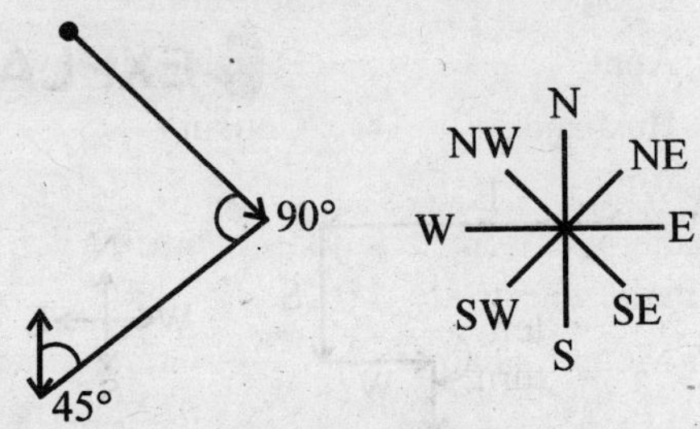

14. (b) : The track of both the friends is :

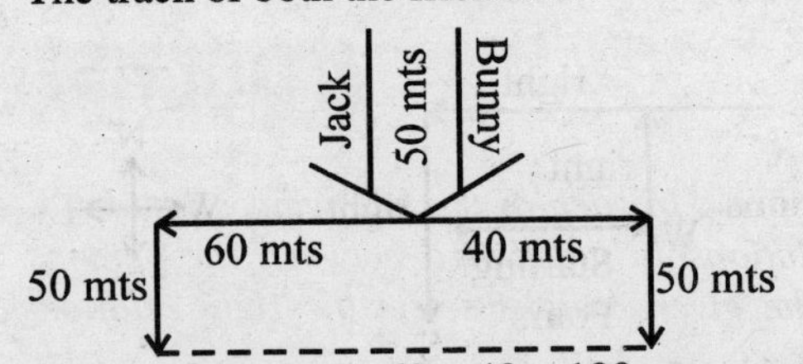

The two friends are 60 + 40 = 100 mts apart

15. (d) :

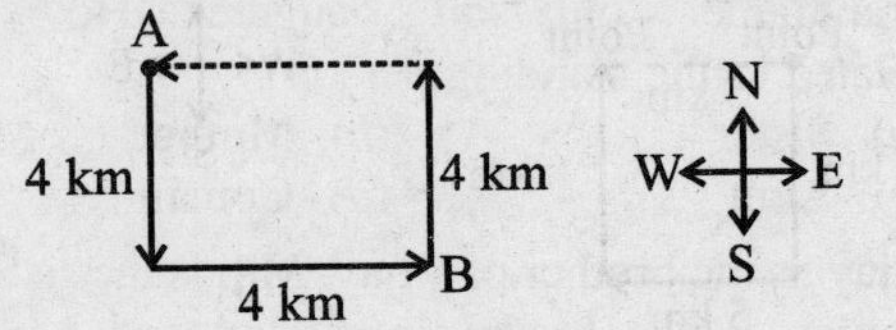

13 Blood Relationships

While attempting questions on blood relations, one should be clear of all the relation patterns that can exist between any two individuals. These type of questions are given mainly to test one's relationship ability.

Very well-known relations are :

Mother	Grandmother
Father	Grandfather
Son	Grandson
Daughter	Granddaughter
Brother	Brother-in-law
Sister	Sister-in-law
Niece	Father-in-law
Nephew	Mother-in-law
Uncle	Son-in-law
Aunt	Daughter-in-law
Husband	Cousin
Wife	

The patterns of some relationships which help in solving questions in these tests are :

Father's *or* Mother's Father	—	Grandfather (Paternal *or* Maternal)
Father's *or* Mother's Mother	—	Grandmother (Paternal *or* Maternal)
Father's *or* Mother's Son	—	Brother
Father's *or* Mother's Daughter	—	Sister
Father's Brother	—	Paternal Uncle
Father's Sister	—	Paternal Aunt
Mother's Brother	—	Maternal Uncle
Mother's Sister	—	Maternal Aunt
Uncle *or* Aunt's Son *or* Daughter	—	Cousin
Son's Wife	—	Daughter-in-law
Daughter's Husband	—	Son-in-law
Husband's *or* Wife's Brother	—	Brother-in-law
Husband's *or* Wife's Sister	—	Sister-in-law
Brother's Wife	—	Sister-in-law
Sister's Husband	—	Brother-in-law
Brother's Son	—	Nephew
Brother's Daughter	—	Niece

MULTIPLE CHOICE QUESTIONS

Directions : *In each of the following questions keenly study the relationship mentioned between the persons, and then from the given options select the right relationship as the answer.*

1. A lady said, "The person standing there is my grandfather's only son's daughter". How is the lady related to the standing person?
(*a*) Sister (*b*) Mother
(*c*) Aunt (*d*) Cousin

2. Ajay is the brother of Vijay. Mili is the sister of Ajay. Sanjay is the brother of Rahul and Mehul is the daughter of Vijay. Who is Sanjay's Uncle?
(*a*) Rahul (*b*) Ajay
(*c*) Mehul (*d*) Data inadequate

3. A man introduced the boy coming with him as "He is son of the father of my wife's daughter". What relation did the boy bear to the man?
(*a*) Son-in-law (*b*) Son
(*c*) Brother (*d*) Father

4. If Amit's father is Billoo's father's only son and Billoo has neither a brother nor a daughter, what is the relationship between Amit and Billoo?
(*a*) Uncle — Nephew (*b*) Father — Daughter
(*c*) Father — Son (*d*) Cousins

5. Pointing to a woman in the photograph a man said, "She is the daughter of my grand-mother's only son. How is the woman related to the man?
(*a*) Mother (*b*) Daughter
(*c*) Sister-in-law (*d*) Sister

6. Pointing to a photograph, a woman said, "She is the only daughter of my mother's father." How is the woman related to the person in the photograph?

(*a*) Mother (*b*) Grandmother
(*c*) Daughter (*d*) Cannot be determined

7. Ram is the brother of Shyam and Mahesh is the father of Ram. Jagat is the brother of Priya and Priya is daughter of Shyam. Who is the uncle of Jagat ?
(*a*) Shyam (*b*) Mahesh
(*c*) Ram (*d*) Data insufficient

8. Introducing a man, a woman said, "His wife is the only daughter of my father". How is the man related to the woman?
(*a*) Husband (*b*) Father
(*c*) Father-in-law (*d*) Brother

9. If Maya is the only daughter of Richa's grandmother's brother, how is Maya's daughter related to Richa?
(*a*) Niece (*b*) Cousin
(*c*) Aunt (*d*) Mother

10. Pointing to a woman, a man said, "Her husband's mother is the wife of my father's only son". How is the man related to the woman?
(*a*) Son (*b*) Brother-in-law
(*c*) Uncle (*d*) Father-in-law

EXPLANATORY ANSWERS

1. (*a*) :

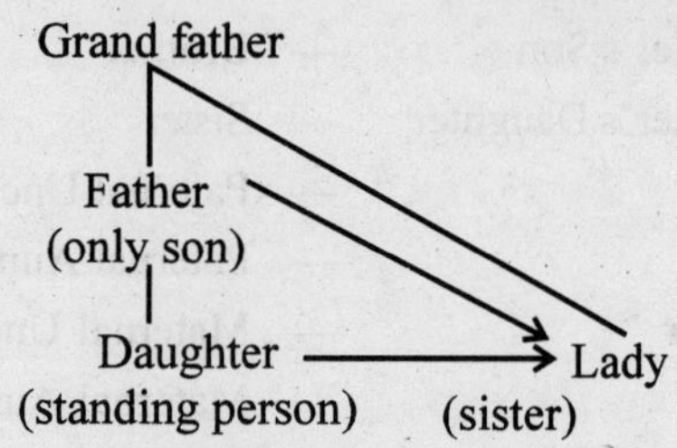

Lady's grandfather's son is lady's father and father's daughter will only be lady's sister.

2. (*d*) :

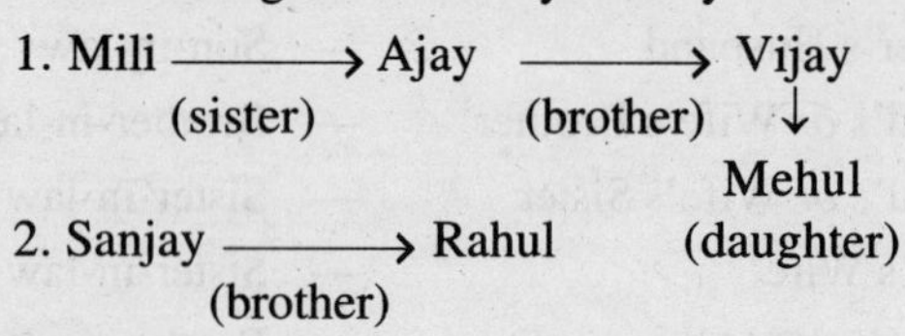

There are two sets of relationship. Information given is incomplete and no relation can be established between the two sets.

3. (*b*) : The relationship chart based on problem is :

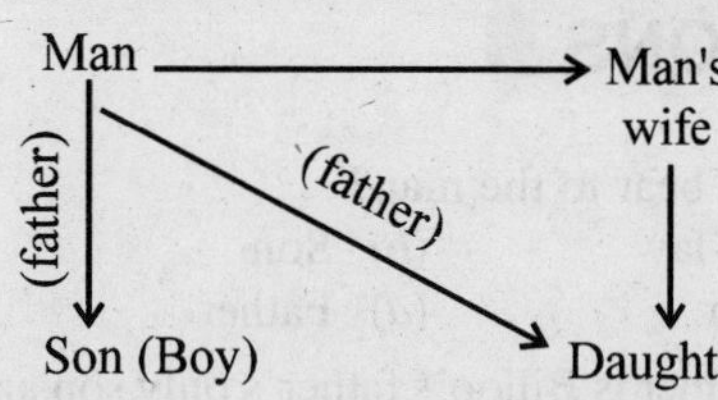

'Father of the man's wife's daughter' is the man himself and the boy in question is the man's son.

4. (*c*) : The relationship chart based on problem is :

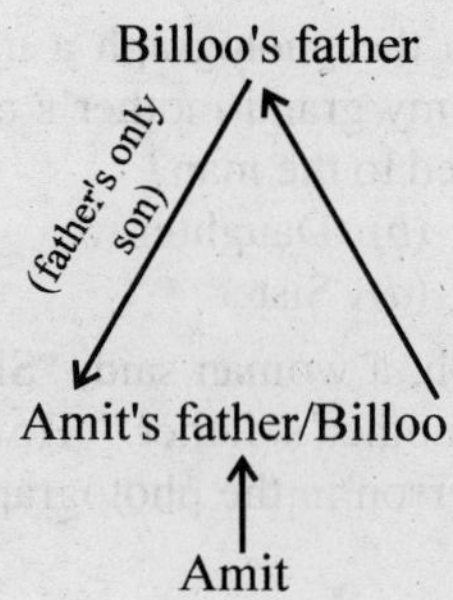

Amit's father is Billoo's father's only son means Billoo is the only son in question also, he is the father of Amit. It must be noted that Billoo has no brother which means he is single and also, when he has no daughter, Amit is his only son.

5. (*d*) :

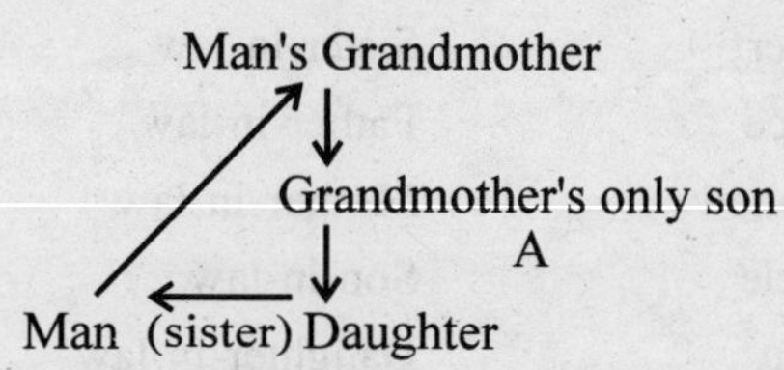

'My grandmother's only son' is the father of the man, and 'daughter of my grand-mother's only son' is the sister of the man.

6. (*c*) :

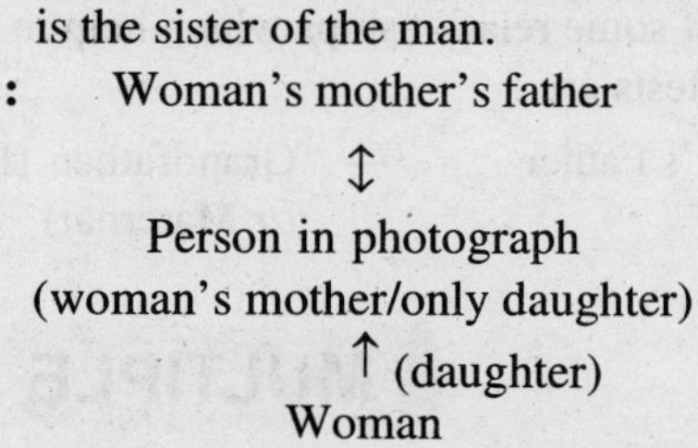

'Only daughter of my mother's father' is the person in the photograph and she is also the mother of the woman. So, the woman is the daughter of the person in the photo-graph.

7. (*c*) :

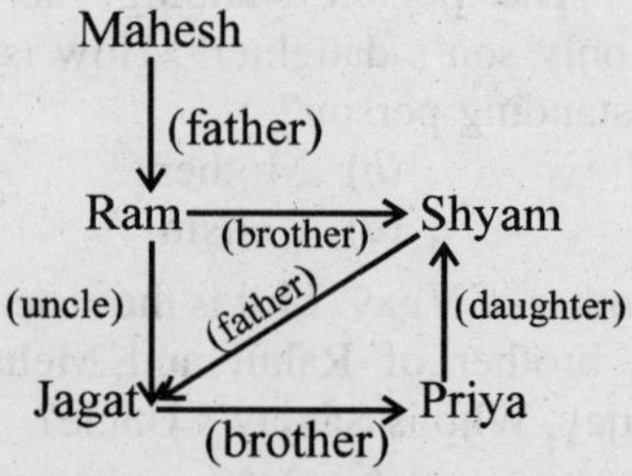

Jagat is brother of Priya and Priya is daughter of Shyam. So Shyam is also the father of Jagat. Ram is the brother of Shyam. So, Jagat's father's brother Ram is the uncle of Jagat.

8. **(*a*) :**

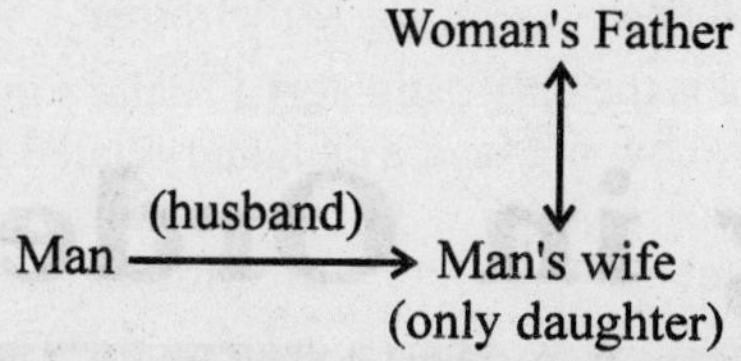

'Only daughter of my father' is the woman herself and the man is her husband.

9. **(*b*) :**

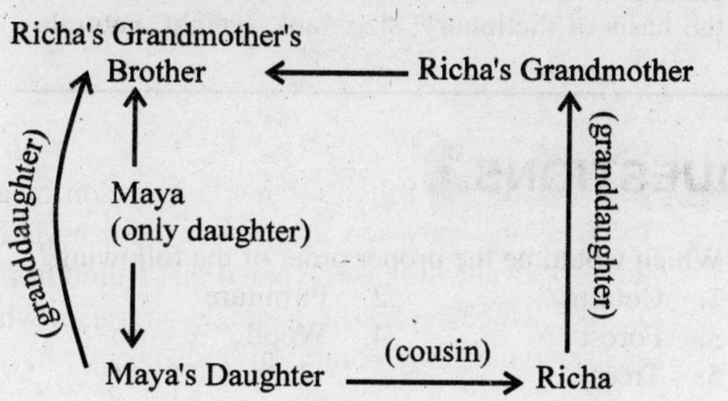

Both Maya's Daughter and Richa are granddaughters of a brother and a sister respectively. So Maya's daughter is the cousin of Richa.

10. **(*d*) :**

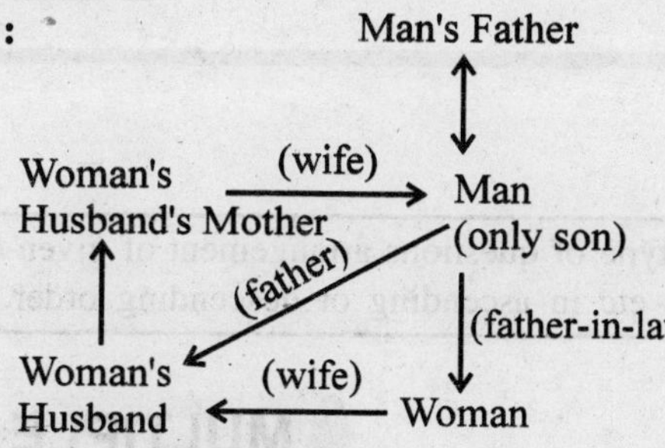

'My father's only son' is the man himself. 'Her husband's mother' is the wife of the man and so the man is the father of the woman's husband. As the woman is the wife of man's son, the man is the father-in-law of the woman.

14 Arranging in Order

In these type of questions arrangement of given options is done on the basis of dictionary, size, rank, weight, natural sequence etc in ascending or descending order.

MULTIPLE CHOICE QUESTIONS

1. Which one would be a meaningful order of the following?

1. Key 2. Door
3. Lock 4. Room
5. Switch on

(a) 5 1 2 4 3 (b) 4 2 1 5 3
(c) 1 3 2 4 5 (d) 1 2 3 5 4

2. If the following words are rearranged in the alphabetical order as in a dictionary, which will be the third word?

(a) Amphibian (b) Amorphous
(c) Ambiguous (d) Amphidextrous

3. In a telephone directory, which of the follow-ing names will appear in the third?

(a) Randhir (b) Randesh
(c) Rama (d) Ramesh

4. Which one would be a meaningful order of the following?

1. Mother 2. Child
3. Milk 4. Cry
5. Smile

(a) 3 2 1 5 4 (b) 2 4 1 3 5
(c) 1 5 2 4 3 (d) 2 4 3 1 5

5. Which would be the proper order of the following?

1. Rainbow 2. Rain
3. Sun 4. Happy
5. Child

(a) 4 2 3 5 1 (b) 4 5 1 2 3
(c) 2 1 4 3 5 (d) 2 3 1 5 4

6. Which would be the proper order of the following? (in ascending order)

1. Trillion 2. Thousand
3. Billion 4. Hundred
5. Million

(a) 4 2 5 3 1 (b) 1 5 3 2 4
(c) 4 2 3 5 1 (d) 1 2 3 4 5

7. Which would be the proper order of the following?

1. Country 2. Furniture
3. Forest 4. Wood
5. Trees

(a) 4 3 2 1 5 (b) 2 4 1 5 3
(c) 1 3 5 4 2 (d) 1 2 3 4 5

8. If the following words are arranged in natural order, what will come in the last place in ascending order?

1. Captain 2. Brigadier
3. Major 4. Lieutenant-General
5. Lieutenant

(a) Lientenant-General (b) Brigadier
(c) Captain (d) Major

9. If these words are arranged according to dictionary then what word will come in the third place?

1. Originality 2. Originally
3. Originator 4. Original
5. Originate

(a) Originality (b) Originator
(c) Originate (d) Originally

10. What will be the natural order of the following?

1. Independence Day 2. Christmas
3. Diwali 4. Holi
5. Republic Day

(a) 5 4 1 3 2 (b) 5 1 4 3 2
(c) 4 1 5 3 2 (d) 2 3 4 1 5

11. If the following words are arranged in ascending order what word will be in the fourth place?

1. Air commodore
2. Wing Commander
3. Pilot Officer
4. Squadron Leader
5. Air Marshal

(a) Air Marshal
(b) Air Commodore
(c) Wing Commander
(d) Pilot Officer

12. What will be the natural order of the following words according to size?

1.	Dog	2.	Horse
3.	Ant	4.	Giraffe
5.	Mouse		

(*a*) 3 5 1 2 4 (*b*) 3 1 5 2 4
(*c*) 3 5 1 4 2 (*d*) 3 5 2 1 4

13. Which would be the proper order of the following? (in ascending order)

1.	Trillion	2.	Thousand
3.	Billion	4.	Hundred
5.	Million		

(*a*) 4 2 5 3 1 (*b*) 1 5 3 2 4
(*c*) 4 2 3 5 1 (*d*) 1 2 3 4 5

14. What would be the meaningful order of the following?

1.	Table	2.	Tree
3.	Wood	4.	Seed
5.	Plant		

(*a*) 1 3 2 4 5 (*b*) 4 5 3 2 1
(*c*) 4 5 2 3 1 (*d*) 1 2 3 4 5

15. Which one would be a meaningful order of the following?

1.	Snake	2.	Grass
3.	Eagle	4.	Frog
5.	Insect		

(*a*) 3 2 1 4 5 (*b*) 5 2 1 4 3
(*c*) 2 5 4 1 3 (*d*) 2 4 5 3 1

16. Which one would be a meaningful order of the following?

1.	Windows	2.	Walls
3.	Floor	4.	Foundation
5.	Roof	6.	Room

(*a*) 4 2 1 5 3 6 (*b*) 4 3 5 6 2 1
(*c*) 4 5 3 2 1 6 (*d*) 4 1 5 6 2 3

17. Which would be a meaningful order of the following?

1.	Index	2.	Contents
3.	Title	4.	Chapters
5.	Introduction		

(*a*) 2 3 4 5 1 (*b*) 3 2 5 4 1
(*c*) 5 1 4 2 3 (*d*) 3 2 5 1 4

18. What would be the proper order of the following :

1.	Decameter	2.	Meter
3.	Kilometer	4.	Centimeter
5.	Milimeter		

(*a*) 1 4 3 2 5 (*b*) 5 4 1 2 3
(*c*) 5 4 3 2 1 (*d*) 5 4 2 1 3

EXPLANATORY ANSWERS

1. (*c*) : With the Key you open a Lock, then open the Door to enter the Room and Switch on the light.

2. (*a*) : The dictionary order of the words is— **Ambiguous,** Amorphous, Amphibian, Amphidextrous.

3. (*b*) : In telephone directory the order of names is— Ramesh, **Rama,** Randesh, Randhir.

4. (*b*) : When a Child Cries the Mother gives him Milk and then he Smiles.

5. (*d*) : When it rains in the Sun a Rainbow is formed and the Child is Happy to see it.

6. (*a*) : The words are arranged in order of increasing amount.

7. (*c*) : The proper order of the words is—Country, Forest, Trees, Wood, Furniture. In a Country there are Forests comprising Trees the Wood of which are used to make Furniture.

8. (*a*) : The arrangement of ranks in ascending order is— Lieutenant, Captain, Major, Brigadier, **Lieutenant-General.**

9. (*d*) : The dictionary order of words is—Original, Originality, **Originally,** Originate, Originator.

10. (*a*) : The natural order of National Holidays in Calendar is—Republic Day (January), Holi (March), Independence Day (August), Diwali (October/November), Christmas (December).

11. (*b*) : The arrangement of ranks in ascending order is— Pilot Officer, Squadron Leader, Wing Commander, **Air Commodore,** Air Marshal.

12. (*a*) : The natural order of living beings according to increasing size is—Ant, Mouse, Dog, Horse, Giraffe.

13. (*a*) : The words are arranged in order of increasing amount.

14. (*c*) : From a Seed a Plant is grown into a Tree whose Wood is used to make a Table.

15. (*c*) : In the Grass there are many Insects. Frog who eats them is a prey of Snake who in turn is the prey of an Eagle.

16. (*a*) : The order of words depict the right order of constructing a building/room.

17. (*b*) : A Title comprises Contents of Introduction for Chapter Index.

18. (*d*) : The proper order of measurement in increasing order is—Milimeter, Centimeter, Meter, Decameter, Kilometer.

15

Rows and Ranks

These type of problems need easy calculations to find out the number of objects in a row, lane or queue or to find a person's rank in a class of certain number of students; or to find the total number of students.

MULTIPLE CHOICE QUESTIONS

1. In a row of trees, one tree is fifth from either end of the row. How many trees are in the row?

(*a*) 11 (*b*) 8
(*c*) 10 (*d*) 9

2. Jaya ranks 5th in a class of 53. What is her rank from the bottom in the class?

(*a*) 49th (*b*) 48th
(*c*) 47th (*d*) 50th

3. Mohan ranks twenty-first in a class of sixty-five students. What will be his (Mohan's) rank if the lowest candidate is assigned rank 1?

(*a*) 44th (*b*) 45th
(*c*) 46th (*d*) Data inadequate

4. If Rahul finds that he is 12th from the right in a line of boys and 4th from the left, how many boys should be added to the line such that there are 28 boys in the line?

(*a*) 12 (*b*) 14
(*c*) 20 (*d*) 13

5. In a row of boys, Rajan is tenth from the right and Suraj is tenth from the left. When Rajan and Suraj interchange their positions, Suraj will be twenty-seventh from the left. Which of the following will be Rajan's position from the right?

(*a*) Tenth
(*b*) Twenty-sixth
(*c*) Twenty-ninth
(*d*) None of these

6. Mahesh and Suresh are ranked 11th and 12th respectively from the top in a class of 41 students. What will be their respective ranks from the bottom?

(*a*) 32nd and 33rd
(*b*) 29th and 30th
(*c*) 30th and 31st
(*d*) 31st and 30th

7. Uma ranked 8th from the top and 37th from bottom in a class. How many students are there in the class?

(*a*) 47 (*b*) 46
(*c*) 45 (*d*) None of these

8. In a queue, Sadiq is 14th from the front and Joseph is 17th from the end, while Jane is in between Sadiq and Joseph. If Sadiq be ahead of Joseph and there be 48 persons in the queue, how many persons are there between Sadiq and Jane?

(*a*) 5 (*b*) 6
(*c*) 7 (*d*) 8

9. Rohan ranked eleventh from the top and twenty-seventh from the bottom among the students who passed the annual examination in a class. If the number of students who failed in the examination was 12, how many students appeared for the examination?

(*a*) 48
(*b*) 49
(*c*) 50
(*d*) Cannot be determined

10. Some boys are sitting in a row. P is sitting fourteenth from the left and Q is seventh from the right. If there are four boys between P and Q, how many boys are there in the row?

(*a*) 19 (*b*) 21
(*c*) 25 (*d*) 23

EXPLANATORY ANSWERS

1. (*d*):

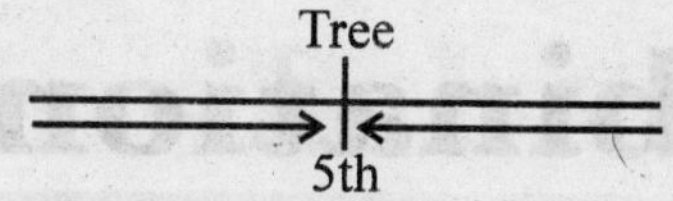

Total number of trees in the row are :
(5 + 5) –1 = 9.

2. (*a*) :

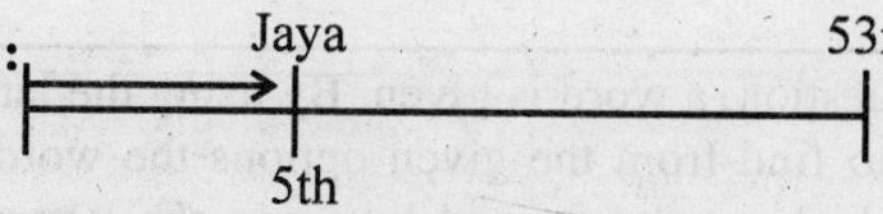

Jaya's rank from the bottom is :
(53 – 5) +1 = 49th.

3. (*b*) :

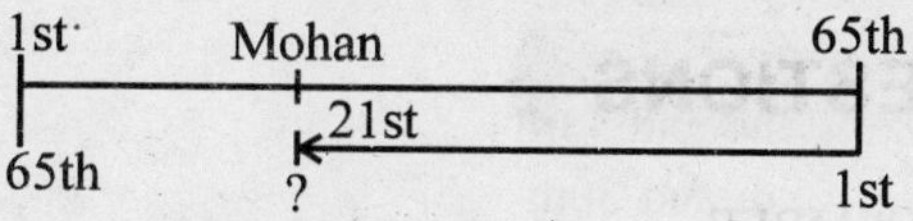

Note : Mohan's rank from the last or the question asked means the same.

Mohan's rank is (65 – 21) +1 = 45th.

4. (*d*) :

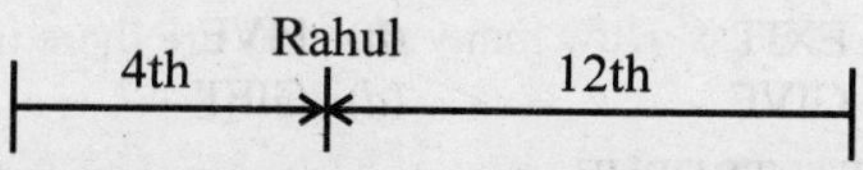

The number of boys in the line are :
(4 + 12) – 1 = 15
To make a line of 28 boys, (28 –15) *i.e.* 13 more boys are needed.

5. (*d*) :

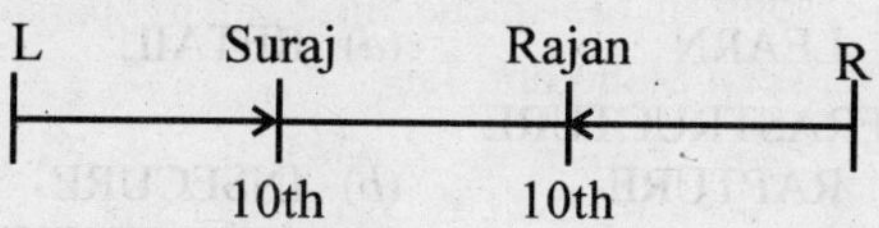

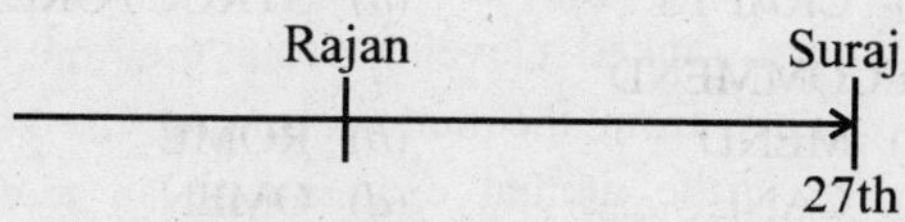

As the position of boys is equal from both ends, Rajan will also be 27th from the right after changing positions.

6. (*d*) :

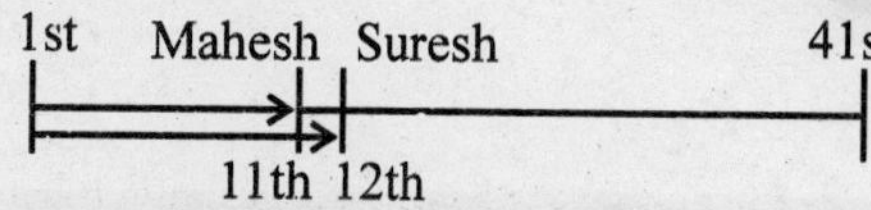

Mahesh's position from bottom is :
(41 – 11) + 1 = 31st
Suresh's position from bottom is :
(41 – 12) +1 = 30th.

7. (*d*) :

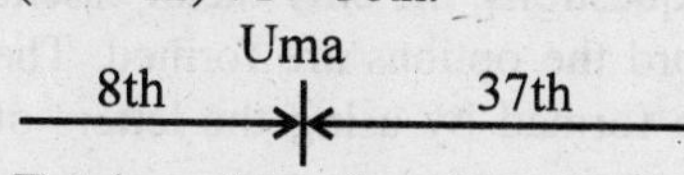

Total number of students in the class are :
(8 + 37) – 1 = 44.

8. (*d*) :

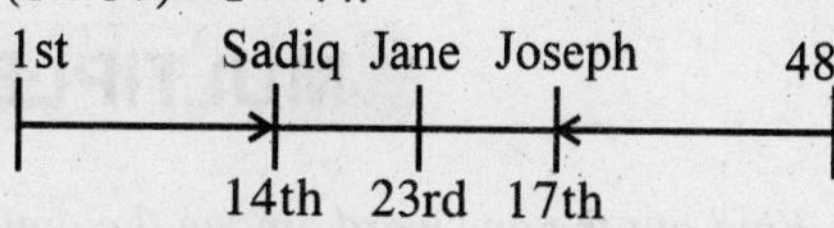

Sadiq's position from front : 14th
Joseph's position from last :
17th or 48 – 17 + 1 = 32nd from front.
Middle postion between Sadiq & Joseph

$$= \frac{(32-14)}{2} + 14 = 23 \text{ rd.}$$

Hence, Jane position is 23rd from front
Person between Sadiq & Jane
= 23 – 14 – 1 = 8.

9. (*b*) :

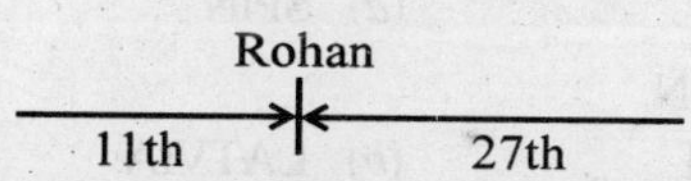

Number of students who passed the examination
(11+ 27) – 1 = 37
Those who failed = 12
Total number of students who appeared in the examination = 37 + 12 = 49.

10. (*c*) :

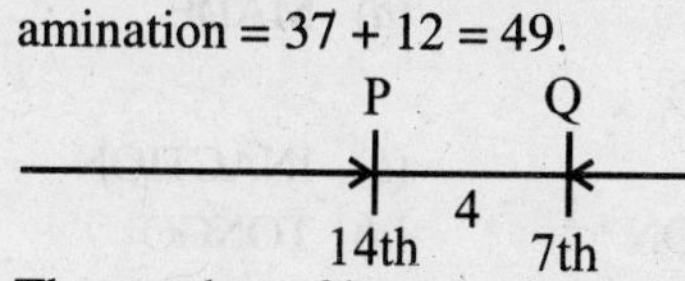

The number of boys in the row are :
(14 + 4 + 7) = 25.

16 Permutations and Combinations

In these type of questions, the only factor essential is alertness. In every question, a word is given. By using the letters of this given word the options are formed. The candidates are required to find from the given options the word (i) which cannot be formed by using the letters of the given word or (ii) which can be formed by using the letters of the given word.

MULTIPLE CHOICE QUESTIONS

Directions : *Find out the one word among the options which cannot be formed by using the letters of the word as given in each question.*

1. ROTATION
(*a*) TORN (*b*) NOTE
(*c*) TART (*d*) RAIN

2. PHILOSOPHY
(*a*) SOIL (*b*) SHIP
(*c*) SOLO (*d*) SPIN

3. SLAVATION
(*a*) SNORT (*b*) LATVIA
(*c*) SALIVA (*d*) AVAIL

4. ACADEMY
(*a*) DEMY (*b*) MACE
(*c*) DIRE (*d*) MADE

5. INCOGNITO
(*a*) GOING (*b*) INACTION
(*c*) IGNITION (*d*) TONGO

6. JUDICIARY
(*a*) INJURY (*b*) CADDY
(*c*) DICY (*d*) ACRID

7. DOCTRINE
(*a*) CRUST (*b*) DOCTOR
(*c*) TIRED (*d*) CREED

8. EDUCATED
(*a*) DATE (*b*) CUTE
(*c*) EAST (*d*) DUCT

9. INSUFFICIENT
(*a*) ENTICE (*b*) SCENT
(*c*) SUFFICE (*d*) THENCE

10. DECEMBER
(*a*) REDEEM (*b*) DECREE
(*c*) BRACED (*d*) MEMBER

11. FUGITIVE
(*a*) EXIT (*b*) FIVE
(*c*) GIVE (*d*) GIFT

12. CATASTROPHE
(*a*) TASTE (*b*) CHEAP
(*c*) POUCH (*d*) STARE

13. TORRENTIAL
(*a*) TRAIL (*b*) MENTAL
(*c*) LEARN (*d*) RETAIL

14. INFRASTRUCTURE
(*a*) RAPTURE (*b*) INSECURE
(*c*) CRAFTS (*d*) STRUCTURE

15. RECOMMEND
(*a*) MEND (*b*) ROME
(*c*) CANE (*d*) OMEN

Directions : *Find out the one word among the options which can be formed by using the letters of the word as given in each question.*

16. INVESTIGATE
(*a*) INVERT (*b*) GLIDE
(*c*) STING (*d*) ACTED

17. MAJORITY
(*a*) MORE (*b*) JURY
(*c*) READ (*d*) TRAY

18. WATERMELON
(*a*) MAKER (*b*) WRITE
(*c*) TOWER (*d*) NOVEL

19. PREDICTION
(*a*) DESIRE (*b*) CREDIT
(*c*) ACTION (*d*) PICKED

20. BARGAIN
(*a*) GRAIN (*b*) BARGE
(*c*) ANGRY (*d*) TRING

21. THERMOSTAT
(*a*) MOTHER (*b*) STAMEN
(*c*) THRUST (*d*) HOIST

22. LEARNED
(*a*) DREAM (*b*) CLEAR
(*c*) ELDER (*d*) DRAPE

23. ADVENTURE
(*a*) AWARE (*b*) EVENT
(*c*) TRUCE (*d*) DRIED

24. THANKSGIVING
(*a*) AVENGE (*b*) HAUNTS
(*c*) GRAINS (*d*) SAVING

25. NOCTURNAL
(*a*) CRUST (*b*) TRAIL
(*c*) CORAL (*d*) OCEAN

26. PUNCTUATED
(*a*) DEATH (*b*) PANEEC
(*c*) CUANEP (*d*) TAULP

27. CHARMED
(*a*) HEAD (*b*) CHUM
(*c*) DECK (*d*) MORE

28. ULTIMATUM
(*a*) AUTUMN (*b*) LIMPID
(*c*) UTMOST (*d*) TUMULT

29. HARBINGER
(*a*) BIRTH (*b*) RANCH
(*c*) ABODE (*d*) GRAIN

30. OTHERWISE
(*a*) SHORT (*b*) WASTE
(*c*) HOVER (*d*) ITEMS

31. DERELICT
(*a*) TRACE (*b*) CREED
(*c*) LINER (*d*) ERODE

32. SINISTER
(*a*) TIMER (*b*) SISTER
(*c*) TEASE (*d*) STRAIN

33. WORKSHOP
(*a*) SHOCK (*b*) ROWER
(*c*) PORCH (*d*) WHOOP

34. DEVELOPING
(*a*) PLAIN (*b*) GREED
(*c*) DEVIL (*d*) VOICE

35. RESIDENTIAL
(*a*) GRIND (*b*) TRADE
(*c*) LOVER (*d*) STRAP

36. BLANDISHMENT
(*a*) BOARD (*b*) METAL
(*c*) SHAPE (*d*) CRASH

37. PAINSTAKER
(*a*) POKER (*b*) STALK
(*c*) TAKEN (*d*) PRIDE

38. BELLIGERENT
(*a*) GREEN (*b*) LEGAL
(*c*) BLOAT (*d*) INFER

39. UNDISCHARGED
(*a*) CHANGED
(*b*) DISARMED
(*c*) GROUNDED
(*d*) SHARPEN

40. LEVIGATE
(*a*) VEIN (*b*) GLAD
(*c*) EAST (*d*) GIVE

ANSWERS

1	2	3	4	5	6	7	8	9	10
(*b*)	(*d*)	(*a*)	(*c*)	(*b*)	(*a*)	(*a*)	(*c*)	(*d*)	(*c*)
11	**12**	**13**	**14**	**15**	**16**	**17**	**18**	**19**	**20**
(*a*)	(*c*)	(*b*)	(*a*)	(*c*)	(*c*)	(*d*)	(*c*)	(*b*)	(*a*)
21	**22**	**23**	**24**	**25**	**26**	**27**	**28**	**29**	**30**
(*a*)	(*c*)	(*b*)	(*d*)	(*c*)	(*c*)	(*a*)	(*d*)	(*d*)	(*a*)
31	**32**	**33**	**34**	**35**	**36**	**37**	**38**	**39**	**40**
(*b*)	(*b*)	(*d*)	(*c*)	(*b*)	(*b*)	(*c*)	(*a*)	(*a*)	(*d*)

17

Symbol Substitution

Questions in these category are easy to attempt. Candidates must be quick in substituting symbols and calculations.

MULTIPLE CHOICE QUESTIONS

1. If "+" means "–"; "–" means "×"; "×"means "÷" and "÷" means "+", then
$15 \times 5 \div 10 + 5 - 3 = ?$
(*a*) 9.5 (*b*) 0
(*c*) – 2 (*d*) 24

2. If "+" means "×"; "–" means "÷"; "÷"means "+" and "×" means "–", then what will be the value of 20 ÷ 40 – 4 × 5 + 6 = ?
(*a*) 60 (*b*) 1.67
(*c*) 150 (*d*) 0

3. If × stands for addition, < for subtraction, + stands for division, > for multiplication, – stands for equal to, ÷ for greater than, and = stands for less than, state which of the following is true?
(*a*) $5 \times 3 < 7 \div 8 + 4 \times 1$
(*b*) $3 \times 4 > 2 - 9 + 3 < 3$
(*c*) $5 > 2 + 2 = 10 < 4 \times 8$
(*d*) $3 \times 2 < 4 \div 16 > 2 + 4$

4. If → stands for subract, ← stands for add, ↓↑↑ stands for multiply, ↓↓ stands for divide, ↔ for greater than, ⟷ stands for equal to, then which of the following alternatives is true?
(*a*) 4 ← 6 ↑↑ 2 ⟷ 3 → 12 ← 12
(*b*) 10 ↓↓ 5 ↑↑ 5 ⟷ 9 → 3 ← 4
(*c*) 15 ↑↑ 2 → 5 ⟷ 12 ↓↓ 4 ← 3
(*d*) 13 ↓↓ 13 ← 1 ↔ 20 → 5 ↑↑ 2

5. If Δ denotes =; + denotes >, – denotes <, □ denotes ≠, × denotes > and ÷ denotes < then a + b – c denotes
(*a*) b Δ c □ a (*b*) b □ a ÷ c
(*c*) a ÷ b × c (*d*) b – a + c

6. If '✱' denotes '×', 'Δ' denotes '÷', '□' denotes '–', '●' denotes '+', 'α' denotes '=' and 'β' denotes ≠, then which of the following euations is correct?
(*a*) 2 □ 10 ✱ 4 Δ 5 α 5 ● 12 Δ 6
(*b*) 27 Δ 9 ● 6 β 3 ✱ 6 □ 9
(*c*) 4 Δ 2 ✱ 0 α 7 Δ 1 ✱ 0
(*d*) 5 ● 6 Δ 3 □ 2 α 8 Δ 4 ✱ 3

7. If ↓ stands for '÷', ↑ stands for '×', → stands for '+' and ← stands for '–', then
25 ↓ 5 → 3 ↑ 6 ← 8 = ?
(*a*) 9 (*b*) 12
(*c*) 16 (*d*) 15

8. If the + and × signs of the following equations are interchanged, which will be the correct equation?
(*a*) $7 \times 5 + 3 = 20$ (*b*) $4 + 9 \times 1 = 42$
(*c*) $6 \times 5 + 8 = 46$ (*d*) $2 + 11 \times 4 = 28$

9. If '+' stands for multiplication, '×' stands for addition, '÷' stands for subtraction and '–' stands for division, then what will be the result of the following equation?
$7 \times 4 \div 10 \times 2 + 5 = ?$
(*a*) 7 (*b*) 0
(*c*) 11 (*d*) 15

10. If 'A' means '÷', 'B' means '+', 'C' means '×' and 'D' means '–', then
12 C 4 A 24 D 10 B 1 = ?
(*a*) $11\frac{1}{3}$ (*b*) 23
(*c*) – 7 (*d*) $16\frac{4}{5}$

EXPLANATORY ANSWERS

1. (c) : $15 \div 5 + 10 - 5 \times 3$
$3 + 10 - 15 = -2$

2. (d) : $20 + 40 \div 4 - 5 \times 6$
$20 + 10 - 30 = 0$

3. (c) : (a) $5 + 3 - 7 > 8 \div 4 + 1$
$1 > 3$
(b) $3 + 4 \times 2 = 9 \div 3 - 3$
$11 = 0$
(c) $5 \times 2 \div 2 < 10 - 4 + 8$
$5 < 14$
(d) $3 + 2 - 4 > 16 \times 2 \div 4$
$1 > 8$

4. (b) : (a) $4 + 6 \times 2 = 3 - 12 + 12$
$16 = 3$
(b) $10 \div 5 \times 5 = 9 - 3 + 4$
$10 = 10$
(c) $15 \times 2 - 5 = 12 \div 4 + 3$
$25 = 6$
(d) $13 \div 13 + 1 > 20 - 5 \times 2$
$2 > 10$

5. (d) : What is given is $a > b < c$
The equations are :
(a) $b = c \neq a$ which is wrong
(b) $b \neq a < c$ which is wrong
(c) $a < b > c$ which is wrong
(d) $b < a > c$ which is correct
Therefore, (d) is the answer.

6. (c) : The solved equations will be :
(a) $2 - 10 \times 4 \div 5 = 5 + 12 \div 6$
$2 - 8 = 5 + 2$
$-6 = 7$ which is wrong
(b) $27 \div 9 + 6 \neq 3 \times 6 - 9$
$3 + 6 \neq 18 - 9$
$9 \neq 9$ which is wrong
(c) $4 \div 2 \times 0 = 7 \div 1 \times 0$
$2 \times 0 = 7 \times 0$
$0 = 0$ which is correct
(d) $5 + 6 \div 3 - 2 = 8 \div 4 \times 3$
$5 = 6$ which is wrong

7. (d) : $25 \div 5 + 3 \times 6 - 8$
$5 + 18 - 8 = 15$

8. (c) : After interchanging the signs the equations are :
(a) $7 + 5 \times 3 = 22$ which is wrong
(b) $4 \times 9 + 1 = 37$ which is wrong
(c) $6 + 5 \times 8 = 46$ which is correct
(d) $2 \times 11 + 4 = 26$ which is wrong

9. (c) : $7 + 4 - 10 + 2 \times 5$
$7 + 4 - 10 + 10 = 11$

10. (c) : $12 \times 4 \div 24 - 10 + 1$
$2 - 10 + 1 = -7$

18 Missing Numbers

Playing with numbers and mathematical skills are needed to attempt these type of tests. The candidates have to work out the right combination of arithmetical symbols to arrive at the answer options which will take the place of the interrogation sign in the given questions.

MULTIPLE CHOICE QUESTIONS

Directions : *In each question given below which one number can be placed at the sign of interrogation?*

1.

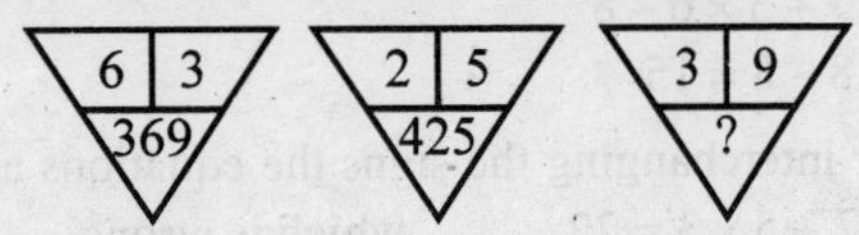

(*a*) 693 (*b*) 939
(*c*) 981 (*d*) 993

2.

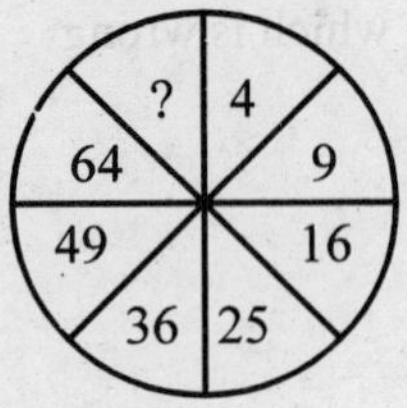

(*a*) 68 (*b*) 100
(*c*) 72 (*d*) 81

3.

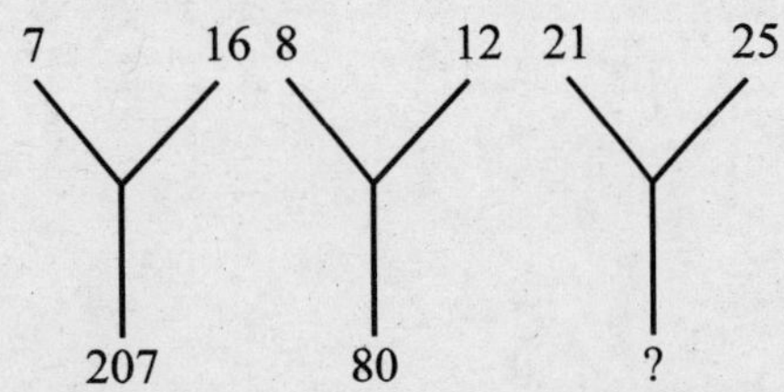

(*a*) 425 (*b*) 184
(*c*) 241 (*d*) 210

4.

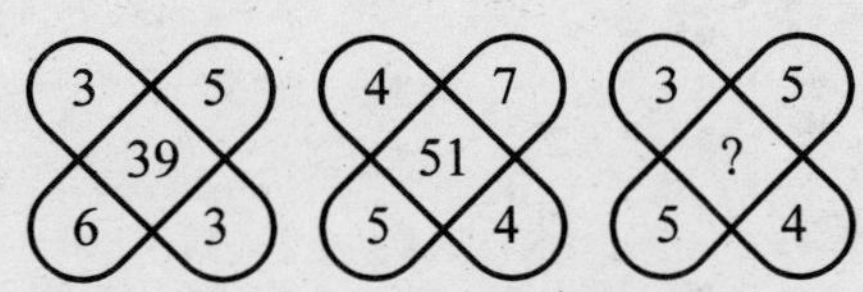

(*a*) 35 (*b*) 37
(*c*) 45 (*d*) 48

5.

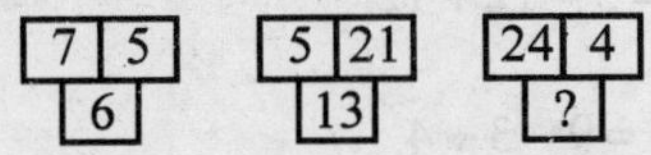

(*a*) 4 (*b*) 8
(*c*) 20 (*d*) 14

6.

14	9	4
12	7	2
10	5	0
16	11	?

(*a*) 9 (*b*) 6
(*c*) 3 (*d*) 7

7.

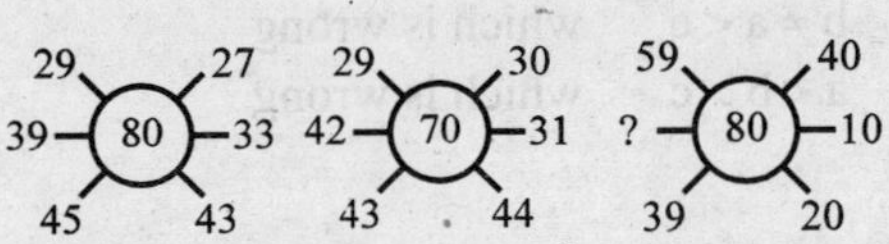

(*a*) 69 (*b*) 49
(*c*) 50 (*d*) 60

8.

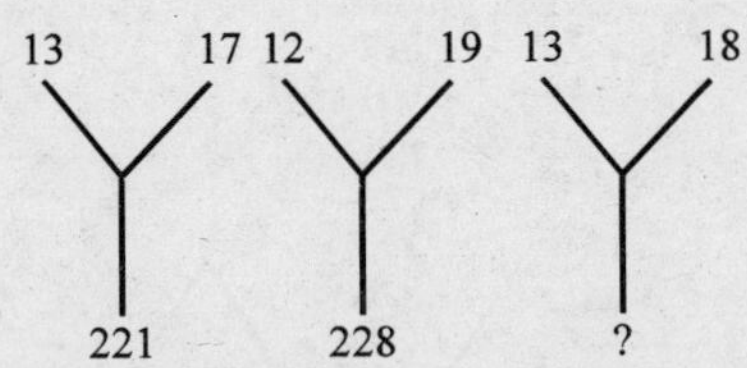

(*a*) 31 (*b*) 229
(*c*) 234 (*d*) 312

9.

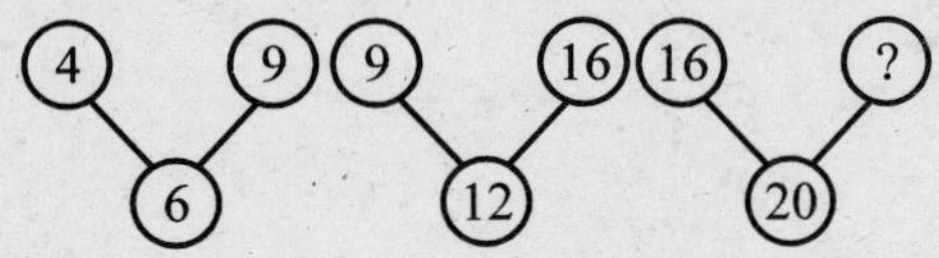

(*a*) 21 (*b*) 25
(*c*) 50 (*d*) 60

10.

51	(11)	61
64	(30)	32
35	(?)	43

(*a*) 25 (*b*) 27
(*c*) 32 (*d*) 37

11.

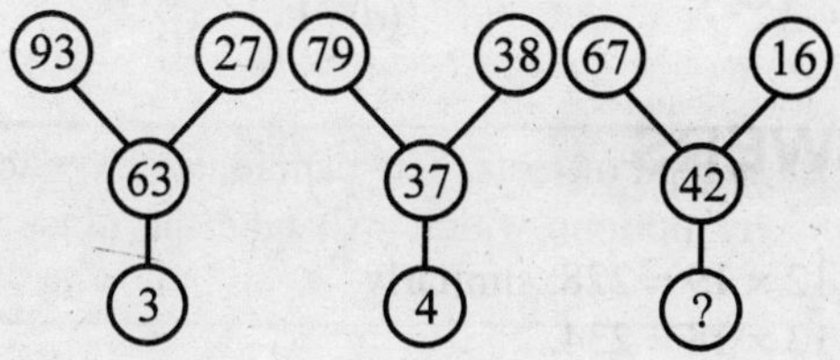

(*a*) 5 (*b*) 6
(*c*) 8 (*d*) 9

12.

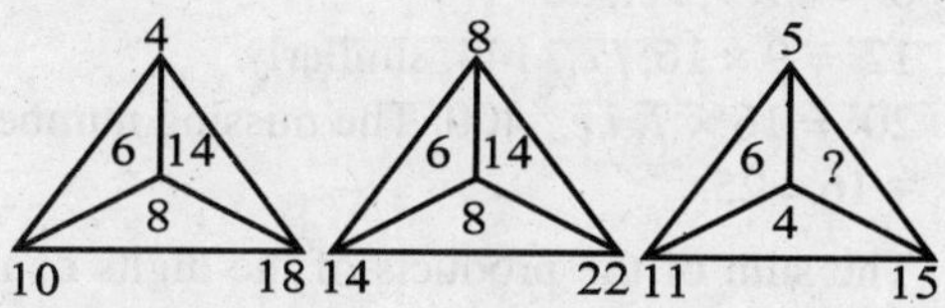

(*a*) 8 (*b*) 14
(*c*) 10 (*d*) 6

13.

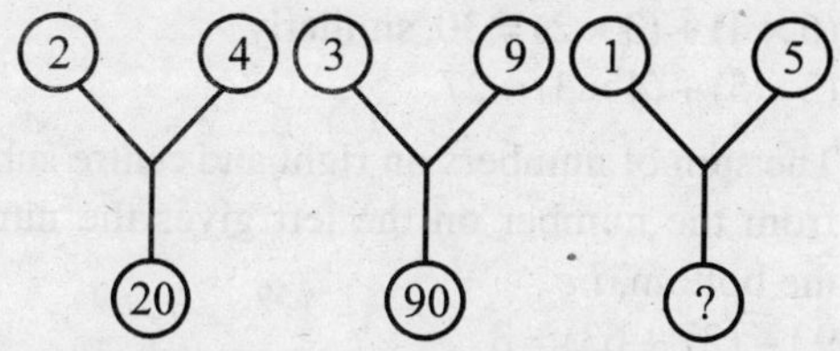

(*a*) 20 (*b*) 25
(*c*) 26 (*d*) 75

14.

27	22	50
13	12	26
9	2	?

(*a*) 12 (*b*) 39
(*c*) 18 (*d*) 24

15.

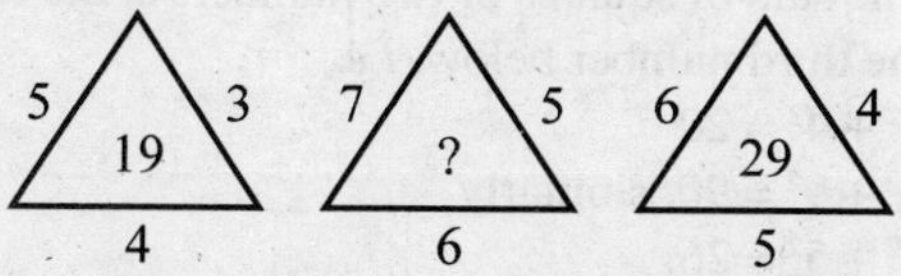

(*a*) 25 (*b*) 47
(*c*) 37 (*d*) 41

16.

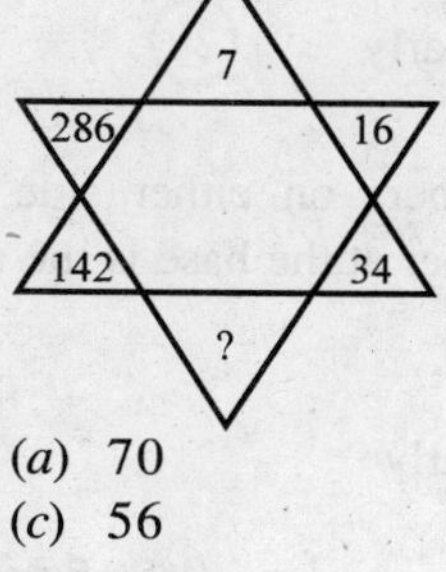

(*a*) 70 (*b*) 68
(*c*) 56 (*d*) 92

17.

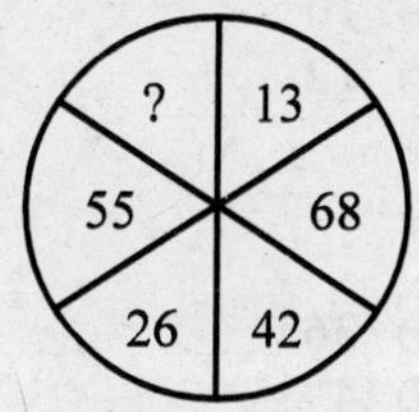

(*a*) 41 (*b*) 37
(*c*) 29 (*d*) 25

18.

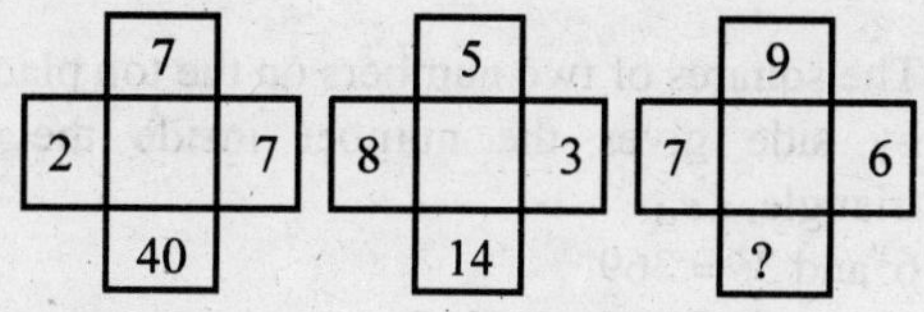

(*a*) 72 (*b*) 68
(*c*) 82 (*d*) 96

19.

42	(21)	22
78	(?)	84
162	(18)	99

(*a*) 12 (*b*) 13
(*c*) 60 (*d*) 72

20.

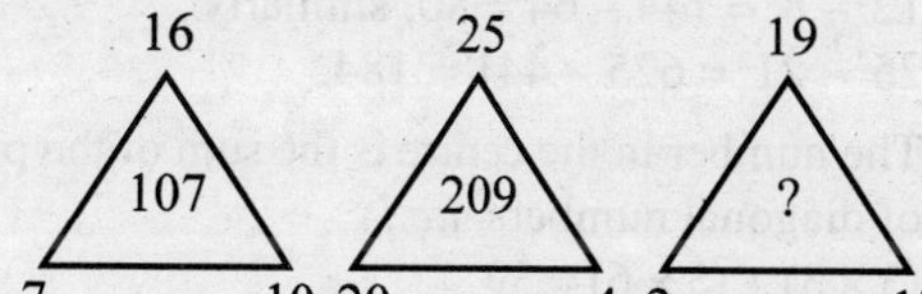

(*a*) 68 (*b*) 93
(*c*) 175 (*d*) 217

21.

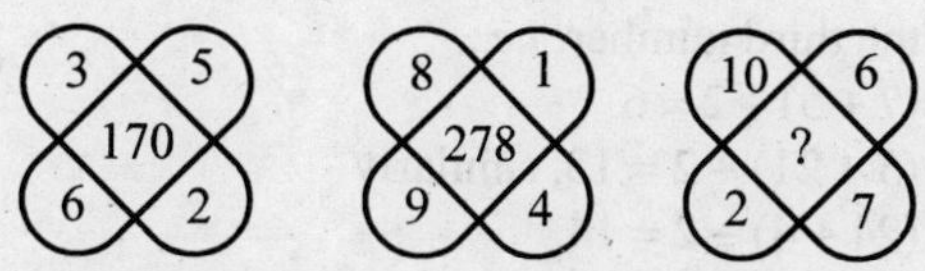

(*a*) 960 (*b*) 628
(*c*) 830 (*d*) 492

22.

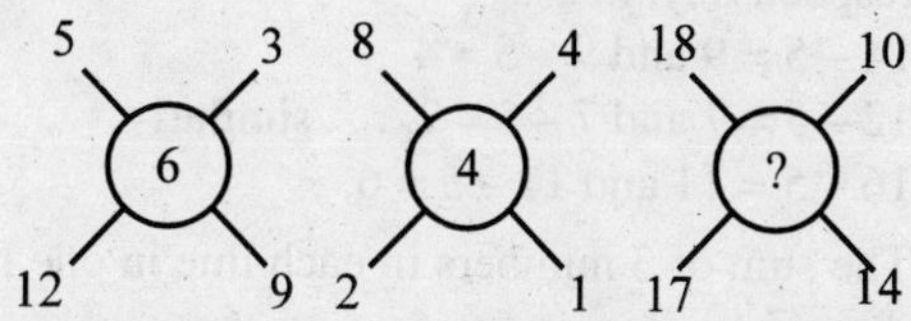

(*a*) 18 (*b*) 10
(*c*) 36 (*d*) 24

23.

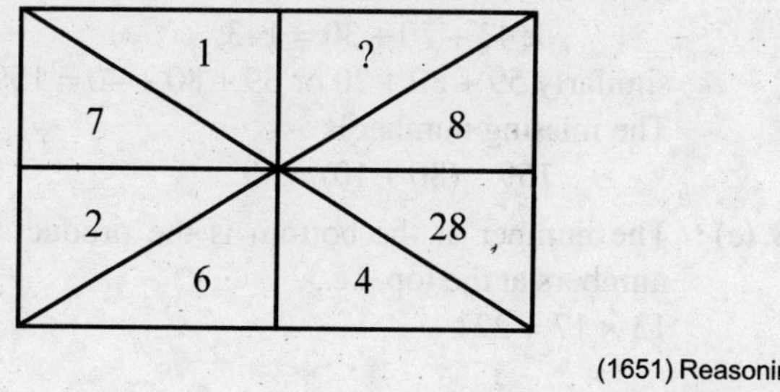

(*a*) 24 (*b*) 10
(*c*) 32 (*d*) 12

24.

6	(40)	4
3	(12)	3
7	(?)	2

(*a*) 51 (*b*) 36
(*c*) 22 (*d*) 4

25.

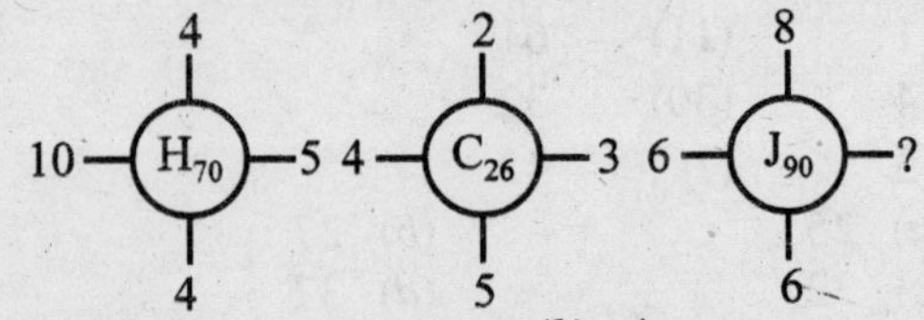

(*a*) 5 (*b*) 4
(*c*) 2 (*d*) 1

EXPLANATORY ANSWERS

1. (*c*) : The squares of two numbers on the top placed side by side gives the number inside the bottom triangle, *i.e.*,
6^2 and $3^2 = 369$
2^2 and $5^2 = 425$, similarly
3^2 and $9^2 = 981$.

2. (*d*) : Starting from number 4 the numbers are the squares of numbers in natural order *i.e.*, $2^2 = 4$, $3^2 = 9$, $4^2 = 16 \ldots\ldots 9^2 = 81$.

3. (*b*) : The number at the bottom is the difference of the squares of two numbers at the top, *i.e.*,
$16^2 - 7^2 = 256 - 49 = 207$
$12^2 - 8^2 = 144 - 64 = 80$, similarly
$25^2 - 21^2 = 625 - 441 = 184$.

4. (*b*) : The number in the centre is the sum of the products of diagonal numbers, *i.e.*,
$(3 \times 3) + (5 \times 6) = 39$
$(4 \times 4) + (7 \times 5) = 51$, similarly
$(3 \times 4) + (5 \times 5) = 37$.

5. (*d*) : Sum of two numbers on the top divided by 2 gives the third number, *i.e.*,
$(7 + 5) \div 2 = 6$
$(5 + 21) \div 2 = 13$, similarly
$(24 + 4) \div 2 = 14$.

6. (*b*) : The numbers in 2nd and 3rd columns are 5 less than the nunbers in 1st and 2nd columns respectively, *i.e.*,
$14 - 5 = 9$ and $9 - 5 = 4$
$12 - 5 = 7$ and $7 - 5 = 2$, . . . similarly
$16 - 5 = 11$ and $11 - 5 = 6$.

7. (*a*) : The sum of 3 numbers in each line in one figure is same, *i.e.*,
$29 + 80 + 43$ or $39 + 80 + 33$
or $45 + 80 + 27 = 152$
$29 + 70 + 44$ or $42 + 70 + 31$
or $43 + 70 + 30 = 143$,
similarly $59 + 80 + 20$ or $39 + 80 + 40 = 159$.
The missing number is :
$159 - (80 + 10) = 69$.

8. (*c*) : The number at the bottom is the product of two numbers at the top, *i.e.*,
$13 \times 17 = 221$
$12 \times 19 = 228$, similarly
$13 \times 18 = 234$.

9. (*b*) : Square of number at the bottom is equal to the product of two numbers at the top, *i.e.*,
$6^2 = 4 \times 9$, *i.e.*, 36
$12^2 = 9 \times 16$, *i.e.*, 144, similarly
$20^2 = 16 \times ?$, *i.e.*, 400. The missing number is $400 \div 16 = 25$.

10. (*b*) : The sum of the products of the digits of numbers in 1st and 3rd columns is the number in the 2nd column, *i.e.*,
$(5 \times 1) + (6 \times 1) = 11$
$(6 \times 4) + (3 \times 2) = 30$, similarly
$(3 \times 5) + (4 \times 3) = 27$.

11. (*d*) : The sum of numbers on right and centre subtracted from the number on the left gives the number at the bottom, *i.e.*,
$93 - (27 + 63) = 3$
$79 - (38 + 37) = 4$, similarly
$67 - (16 + 42) = 9$.

12. (*c*) : The number inside each triangle is the difference of the numbers at its base *i.e.*
$10 - 4 = 6$, $18 - 4 = 14$ and $18 - 10 = 8$
$14 - 8 = 6$, $22 - 8 = 14$ and $22 - 14 = 8$, similarly
$11 - 5 = 6$, $15 - 5 = 10$ and $15 - 11 = 4$.

13. (*c*) : The sum of squares of two numbers at the top gives the third number below, *i.e.*,
$2^2 + 4^2 = 20$
$3^2 + 9^2 = 90$, similarly
$1^2 + 5^2 = 26$.

14. (*a*) : The sum of numbers in 1st and 2nd column plus 1 is the number in the 3rd column, *i.e.*,
$27 + 22 + 1 = 50$
$13 + 12 + 1 = 26$, similarly
$9 + 2 + 1 = 12$.

15. (*d*) : The product of numbers on either side of the triangle plus the number at the base is the number inside the triangle, *i.e.*,
$(5 \times 3) + 4 = 19$
$(6 \times 4) + 5 = 29$, similarly
$(7 \times 5) + 6 = 41$.

16. (*a*) : Clockwise starting from number 7, the next number is obtained by doubling the number and adding 2, *i.e.*,
$(7 \times 2) + 2 = 16$
$(16 \times 2) + 2 = 34$ . . . , similarly
$(34 \times 2) + 2 = 70$
$(70 \times 2) + 2 = 142$
$(142 \times 2) + 2 = 286$.

17. (*c*) : The difference between the numbers in opposite sectors is 13, *i.e.*,
$26 - 13 = 13$
$68 - 55 = 13$, similarly
The missing number is $42 - 13 = 29$
($42 + 13 = 55$ is not given as option).

18. (*b*) : The number at the bottom is obtained by subtracting the sum of two numbers in the centre grid line from the square of the number at the top, *i.e.*,
$7^2 - (2 + 7) = 40$
$5^2 - (8 + 3) = 14$, similarly
$9^2 - (7 + 6) = 68$.

19. (*b*) : The number inside the brackets is obtained by multiplying the number on the left by 2 and then dividing the product by the sum of digits of number on the right, *i.e.*,
$(42 \times 2) \div (2 + 2) = 21$
$(162 \times 2) \div (9 + 9) = 18$, similarly
$(78 \times 2) \div (8 + 4) = 13$.

20. (*a*) : Subtracting the sum of squares of two numbers at the base from the square of number at the apex gives the number inside the triangle, *i.e.*,
$16^2 - (7^2 + 10^2) = 107$
$25^2 - (20^2 + 4^2) = 209$, similarly
$19^2 - (2^2 + 17^2) = 68$.

21. (*c*) : The number in the centre is the product of all the 4 numbers minus 10, *i.e.*,
$(3 \times 5 \times 2 \times 6) - 10 = 170$
$(8 \times 1 \times 4 \times 9) - 10 = 278$, similarly
$(10 \times 6 \times 7 \times 2) - 10 = 830$.

22. (*d*) : The number inside the circle is the product of difference of two numbers above and difference of two numbers below, *i.e.*,
$(5 - 3)\ (12 - 9) = 6$
$(8 - 4)\ (2 - 1) = 4$, similarly
$(18 - 10)\ (17 - 14) = 24$.

23. (*a*) : Starting from number 1 anticlockwise the number in the diagonally opposite section is its multiplication by 4, *i.e.*,
$1 \times 4 = 4$, $7 \times 4 = 28$, $2 \times 4 = 8$, similarly
$6 \times 4 = 24$.

24. (*a*) : Square of the number on the left plus the number on the right is the number within brackets, *i.e.*,
$6^2 + 4 = 40$
$3^2 + 3 = 12$, similarly
$7^2 + 2 = 51$.

25. (*b*) : Letter H is 8th in order of alphabetical series. Taking the sum of numbers placed vertically outside the circle + 8; multiplying it by the number on the right; then subtracting from the product the number on the left, gives the number inside the circle, *i.e.*,
Step I → $4 + 8 + 4 = 16$
Step II → $16 \times 5 = 80$
Step III → $80 - 10 = 70$
Letter C is 3rd in order, so
Step I → $2 + 3 + 5 = 10$
Step II → $10 \times 3 = 30$
Step III → $30 - 4 = 26$
Similarly, J is 10th in order, so
Step I → $8 + 10 + 6 = 24$
Step II → $24 \times ?$
Step III → $(24 \times ?) - 6 = 90$
Simplifying the above equation :
$24 \times ? = 90 + 6$, *i.e.*, 96
$? = 96 \div 24 = 4$.

19

Alphabet Problems

Alphabet problems are fun to attempt. They are based on alphabetical series in natural as well as reverse order.

Natural Order

A B C D E F G H I J K L M N O P Q R S T U V W X Y Z

Reverse Order

Z Y X W V U T S R Q P O N M L K J I H G F E D C B A

Note : The series starts from A on reaching Z and from Z on reaching A. Of these A E I O U are vowels and the rest are consonants.

MULTIPLE CHOICE QUESTIONS

1. Which letter should be ninth letter to the left of ninth letter from the right if the first half of the alphabet is reversed?

(*a*) I (*b*) D
(*c*) F (*d*) E

2. Starting from the fifth letter from the left, if twelve letters are written in reverse order, then which letter will be the seventh to the left of the fourteenth letter from the right?

(*a*) N (*b*) H
(*c*) L (*d*) O

3. What letter will come in the centre of sixth letter from the right and thirteenth letter from the left?

(*a*) Q (*b*) R
(*c*) P (*d*) S

4. If in the word "DISTURBANCE", the first letter is interchanged with the last letter, the second letter is interchanged with the tenth letter and so on, which letter would come after the letter "T" in the newly formed word?

(*a*) I (*b*) U
(*c*) N (*d*) S

5. If it is possible to make a meaningful word with the third, the fifth, the seventh and the tenth letters of the word PROJECTION. If no such word can be made, give 'O' as the answer. If more than one such word can be made, give 'M' as the answer.

(*a*) T (*b*) N
(*c*) O (*d*) M

6. A meaningful word is made if we take the first, fourth, fifth, seventh, tenth, eleventh and the twelfth letters of the word "FELICITATIONS". Which of the following will be the fifth letter of that word from the right end of that word?

(*a*) T (*b*) C
(*c*) N (*d*) I

7. On rearranging the jumbled spelling of the word SKARTINS, a language is obtained. What is the fifth letter from the right of the rearranged word?

(*a*) N (*b*) K
(*c*) R (*d*) S

8. If the 1st, 3rd, 5th, 7th, 10th and 13th letters of the word "ENTERTAINMENT" are used to make a meaningful word, then what two letters will come in the centre?

(*a*) AM (*b*) TT
(*c*) RE (*d*) NA

9. On rearranging the jumbled spelling of the word GRUBY a sport is obtained. What is the letter in the centre?

(*a*) R (*b*) B
(*c*) G (*d*) U

10. If the letters 'ERVSECI' can be rearranged to form a meaningful word what will be the fifth letter from the right?

(*a*) R (*b*) V
(*c*) E (*d*) None of these

11. From the word HASTEN how many independent meaningful English words can be made without changing the order of the letters and using each letter only once?
(*a*) 1 (*b*) 2
(*c*) 3 (*d*) 4

12. What letter comes in the middle of sixteenth letter from the right and seventeenth letter from the left?
(*a*) L (*b*) N
(*c*) P (*d*) M

13. If the alphabetical series is written in reverse order, what letters will come on either side of the seventh letter to the left of seventeenth letter from the right?
(*a*) X, Z (*b*) W, Y
(*c*) Y, W (*d*) Z, X

14. How many pairs of letters are there in the word BUCKET which have as many letters between them in the word as in the alphabet?
(*a*) One (*b*) Two
(*c*) Three (*d*) Four

15. If the position of the fifth and twelfth letters of the word GLORIFICATIONS is interchanged and likewise the position of the fourth and fourteenth letters, the third and tenth letters, the second and eleventh letters and the first and thirteenth letters are interchanged, which of the following will be twelfth letter from the right end?
(*a*) T (*b*) O
(*c*) R (*d*) I

EXPLANATORY ANSWERS

1. (*d*) : MLKJIGHFEDCBANOPQRSTUVWXYZ
9th 9th

2. (*d*) : ABCDPONMLKJIHGFEQRSTUVWXYZ
7th 14th

3. (*a*) :

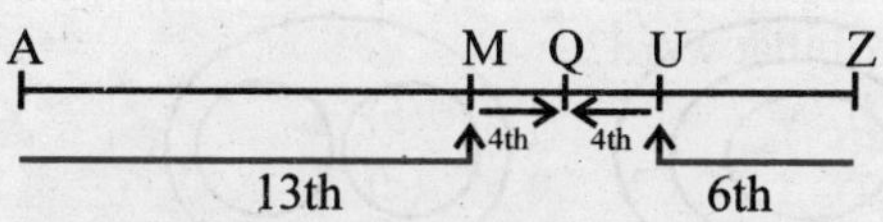

13th letter from left is 'M' and 6th letter from right is 'U'. The letter in the centre of 'M' and 'U' is 'Q'.

4. (*d*) : E C N A B R U T S I D

5. (*d*) : P R O J E C T I O N
The 3rd, 5th, 7th and 10th letters are OETN.
The words formed are TONE and NOTE.

6. (*b*) : F E L I C I T A T I O N S
The 1st, 4th, 5th, 7th, 10th, 11th and 12th lettes are FICTION.
The word is FICTION and 5th letter from right is 'C'.

7. (*d*) : The word is SANSKRIT and 5th letter from right is 'S'.

8. (*b*) : E N T E R T A I N M E N T
The 1st, 3rd, 5th 7th, 10th and 13th letters are ETRAMT.
The word formed is MATTER.

9. (*c*) : The sport is RUGBY. The letter in the centre is 'G'.

10. (*a*) : The word is SERVICE and fifth letter from right is 'R'.

11. (*b*) : The words formed are : HAS, TEN.

12. (*b*) :

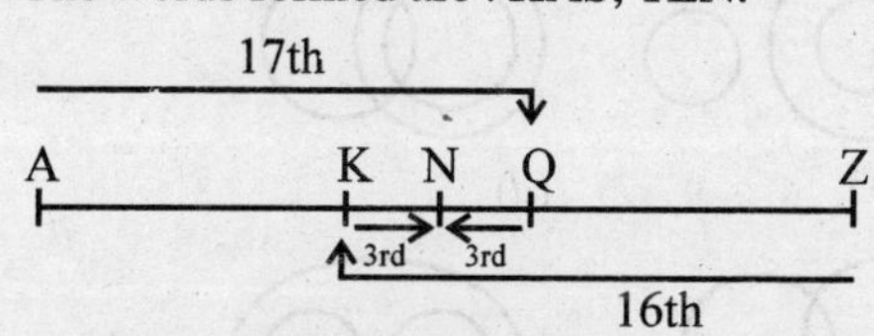

16th letter from right is 'K' and 17th letter from left is 'Q'. The letter in the middle of 'K' and 'Q' is 'N'.

13. (*c*) :

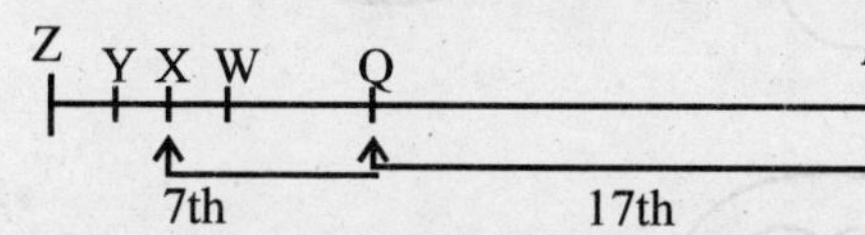

17th letter from right is 'Q', 7th letter to the left of 'Q' is 'X' and letter on either side of 'X' in this alphabetical series are 'Y' and 'W'.

14. (*a*) : BUCKET : C and E have K in the word and D in the alphabetical series.
D

15. (*a*) : N I T S O F I C A O L I G R
12th
12th letter from right is T'.

20

Logical Diagram

In these type of questions, a set of five different figures is given as options. Each figure represents a logical pattern of certain groups of related words wherein each word represents a class. One has to identify the most appropriate logical figure for the set of words given.

MULTIPLE CHOICE QUESTIONS

Directions : *From the five logical Diagrams, select one which best illustrates the relationship among three given classes in the questions 1 to 10.*

(*a*)

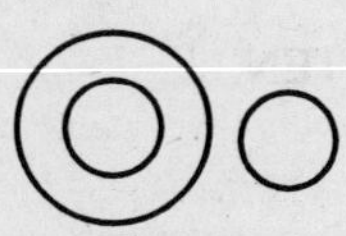

(*b*)

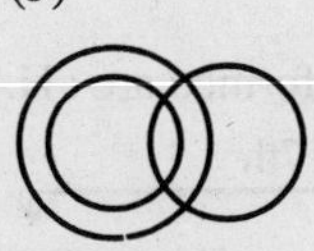

(*c*)

(*d*)

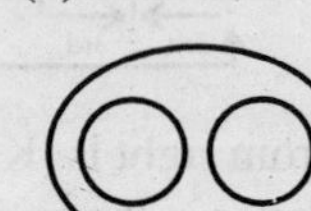

(*e*)

1. Birds, fruits, mangoes
2. Criminals, lawyers, bandits
3. Swimmers, bachelors, men
4. Smart, engineers, women
5. Vegetables, potatoes, brinjals
6. Grapes, sweet, fruit
7. Doctors, architects, humans
8. Scholars, people, Indians
9. Children, naughty, studious
10. Pens, pencils, stationery

Directions : *From the five logical diagrams select one which best illustrates the relationship among three given classes in questions 11 to 20.*

(*a*)

(*b*)

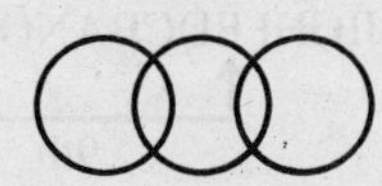

(*c*)

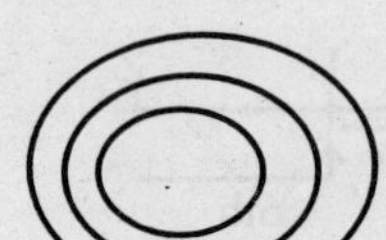

(*d*)

(*e*)

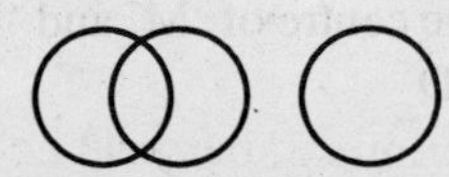

11. Ink, crayons, red
12. Canines, dogs, pups
13. Table, chair, stool
14. Ripe, mangoes, basket
15. Scholars, studious, illiterates
16. Cars, ships, means of conveyance
17. Age, number, thirteen
18. Country, state, continent
19. Father, parent, mother
20. Iron, metal, mercury

Directions : *From the five logical diagrams select one which best illustrates the relationship among three given classes in the questions 21 to 30.*

(*a*)

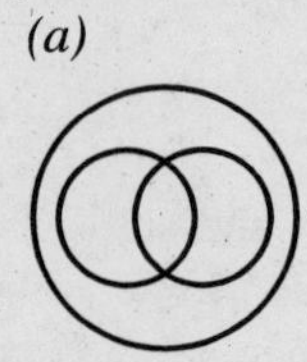

(*b*)

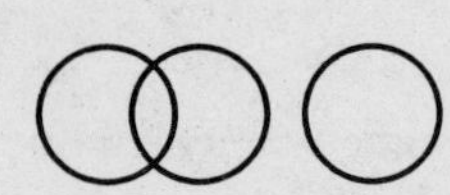

(c)

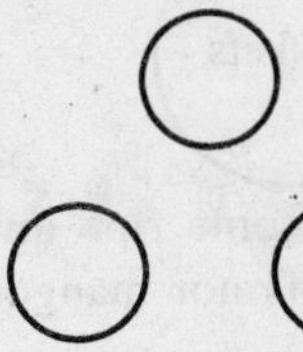

(d)

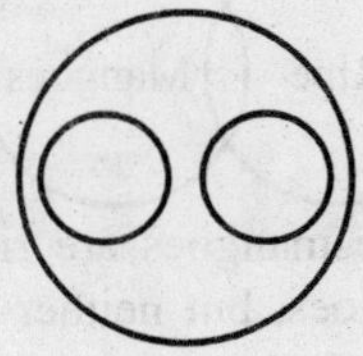

(e)

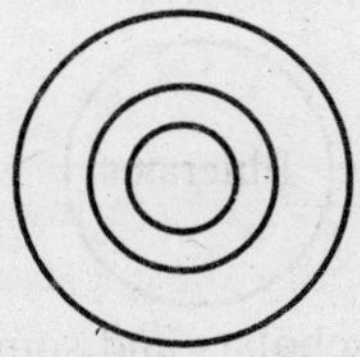

21. Lizards, Reptiles, Crocodiles
22. Whales, Tortoise, Fishes
23. Birds, Crows, Parrots
24. Sisters, Women, Mothers
25. Plants, Scientists, Men
26. Novels, Album, English
27. Sentences, Words, Paragraph
28. Beef, Mutton, Flesh
29. Principal, Student, Parent
30. Lily, Rose, Flower.

EXPLANATORY ANSWERS

1. (*a*) :

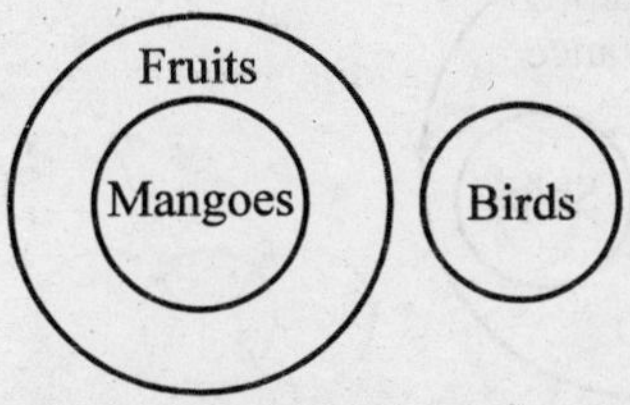

All mangoes are fruits, but neither fruits, nor mangoes can be birds.

2. (*a*) :

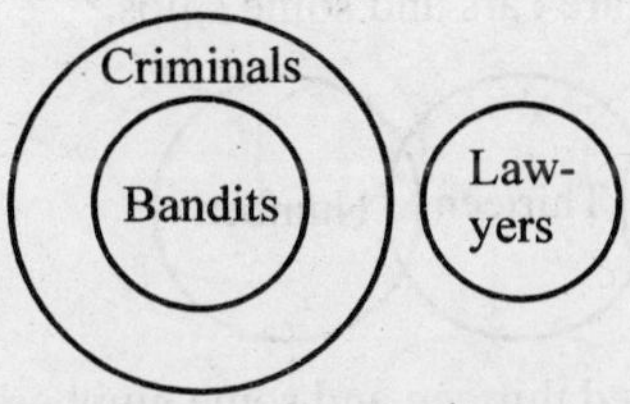

All bandits are criminal, but neither criminals nor bandits can be lawyers.

3. (*b*) :

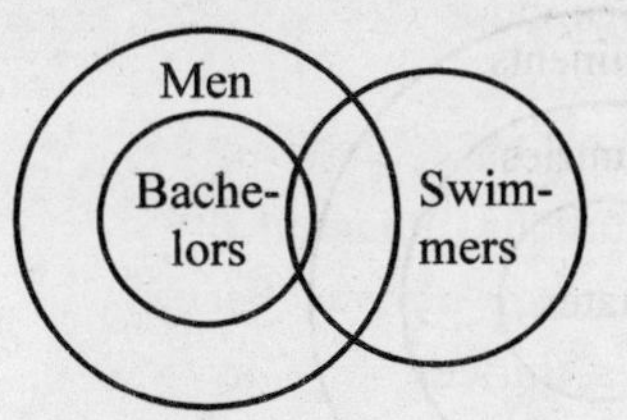

All bachelors are men and some men and bachelors can be swimmers.

4. (*c*) :

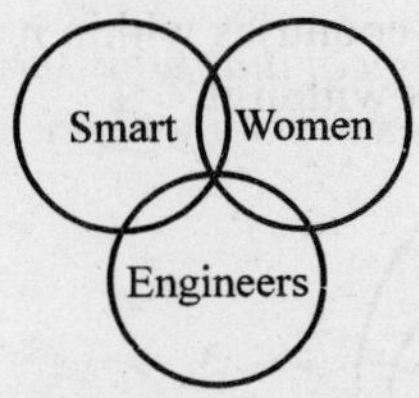

Some women can be smart and some women can be engineers and vice–versa. Some engineers can be women and some engineers can be smart and vice–versa.

5. (*d*) :

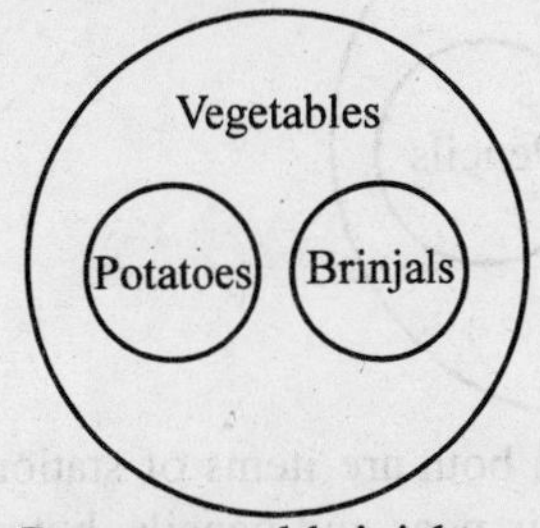

Potatoes and brinjals are vegetables but they have nothing in common. Some vegetables are potatoes and some are brinjals.

6. (*b*) :

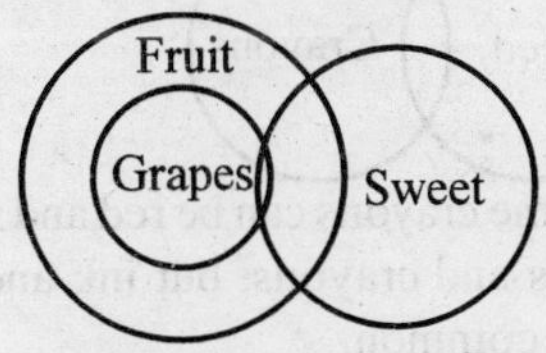

Some grapes are sweet and all grapes are fruit. But not all that is sweet is fruit.

7. (*d*) :

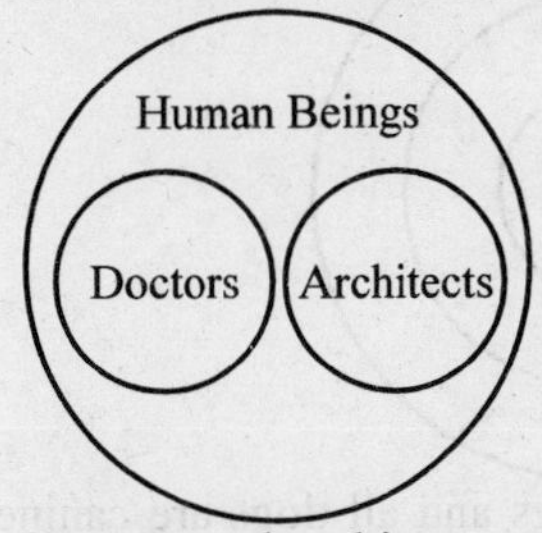

Doctors and architects are two separate classes, but all doctors and architects are humans and some humans are either doctors or architects.

8. (*e*) :

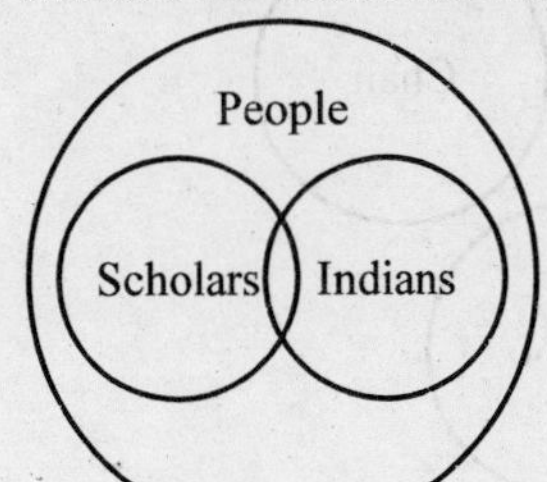

Some Indians can be scholars and some scholars can be Indians. All scholars and Indians are people.

9. (*c*) :

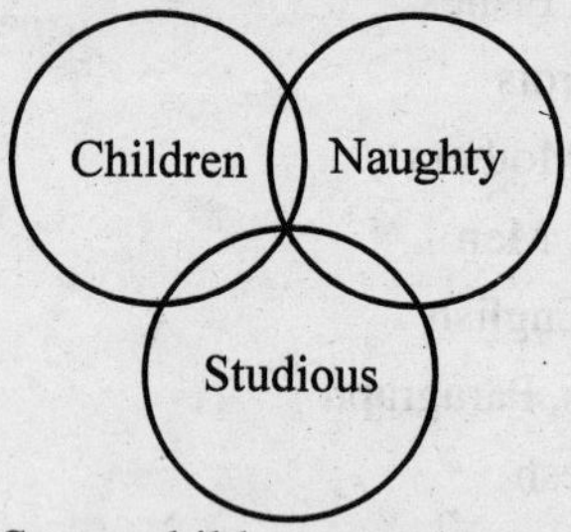

Some children can be naughty and some can be studious, some studious can be children and some naughty. Some naughty can be studious and some children.

10. (*d*) :

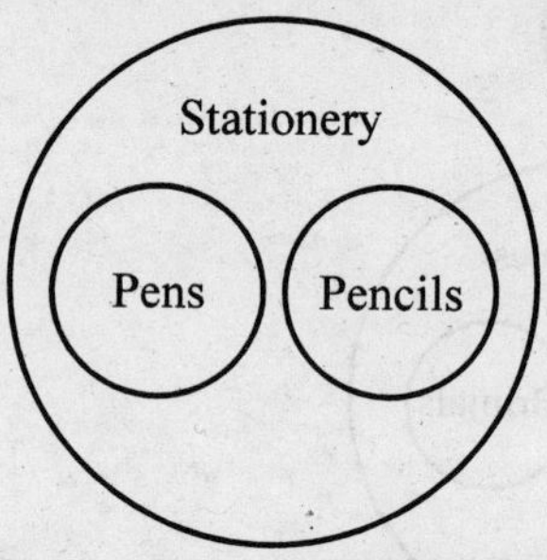

Pens and pencils both are items of stationery and some stationery is pens and pencils, but pens and pencils are two separate classes.

11. (*b*) :

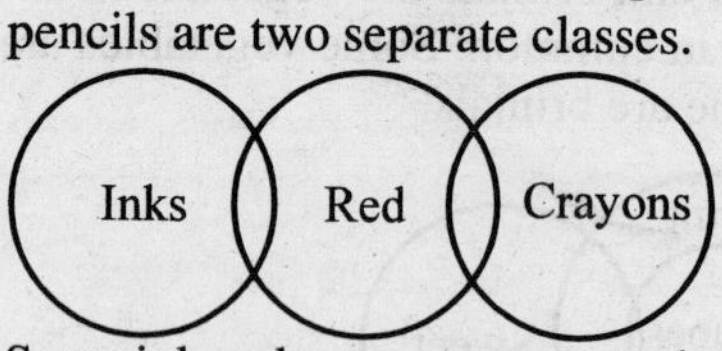

Some ink and some crayons can be red and some red units can be inks and crayons, but ink and crayon have nothing in common.

12. (*c*) :

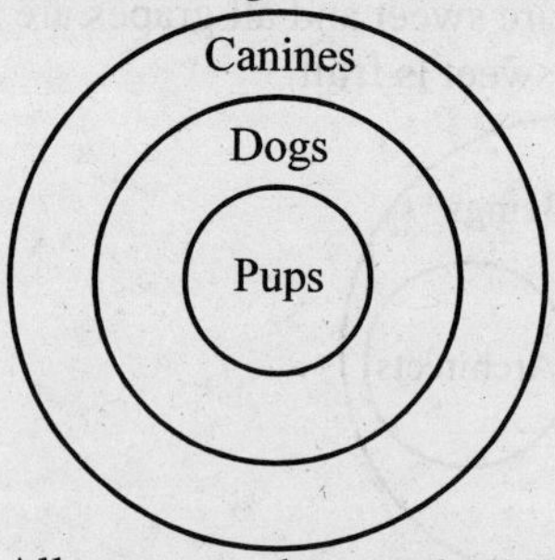

All pups are dogs and all dogs are canines. Some canines are dogs, of which some are pups.

13. (*a*) :

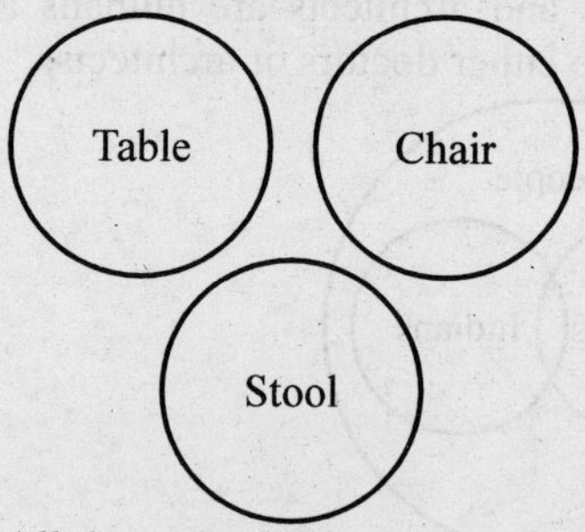

All three are different classes.

14. (*e*) :

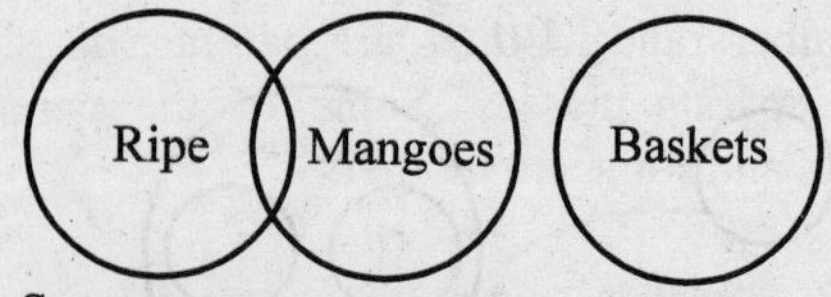

Some mangoes are ripe and some ripe units are mangoes, but neither ripe units nor mangoes can be baskets.

15. (*e*) :

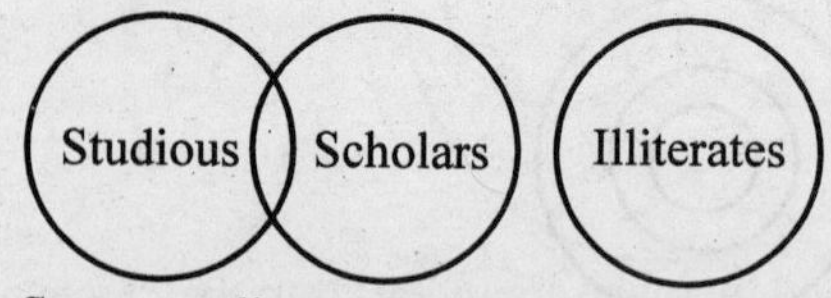

Some studious can be scholars and some scholars can be studious, but neither studious nor scholars can be illiterates.

16. (*d*) :

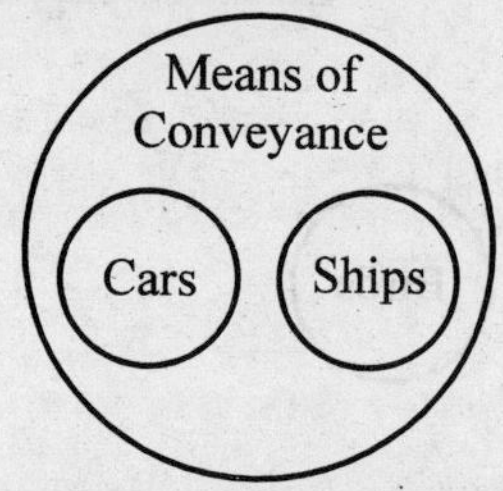

All cars and all ships are means of conveyance, but neither is contained in the other. Some means of conveyance are cars and some ships.

17. (*b*) :

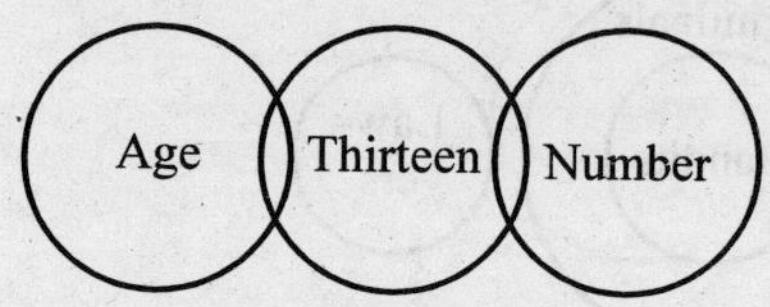

Some are aged thirteen and some numbers are thirteen. Some thirteen units are numbers and some ages. Age and number have nothing in common.

18. (*c*) :

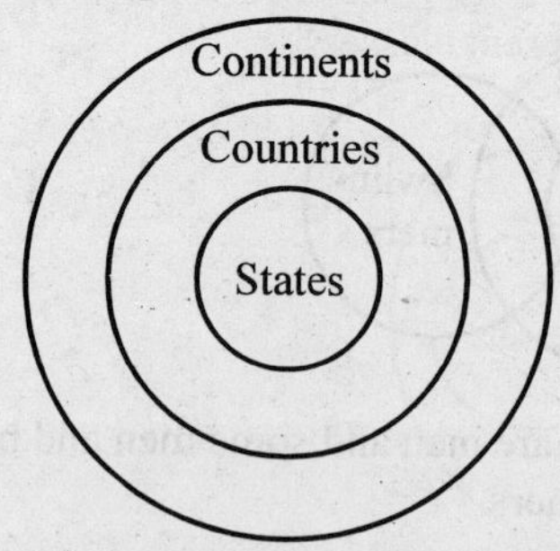

All continents contain countries within it and each country contains states within it.

19. (*d*) :

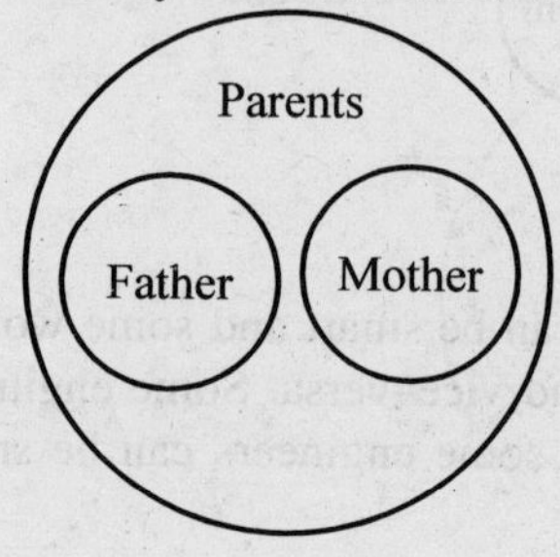

All fathers and mothers are parents, but they are two separate classes. Some of the parents are fathers and some are mothers.

20. (*d*) :

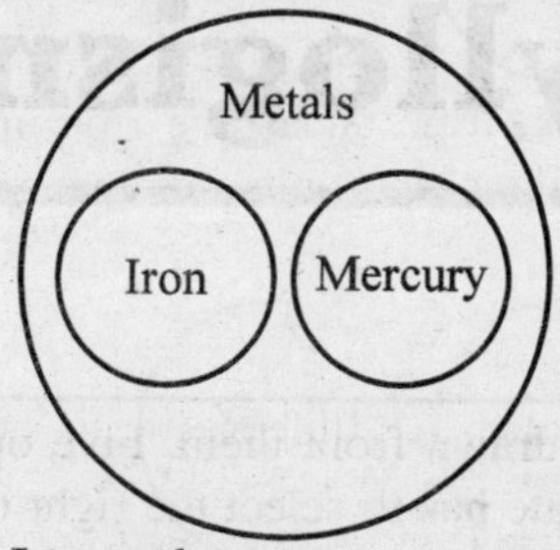

Iron and mercury are metals, but they have nothing in common. Some metals are iron and some mercury.

21. (*d*) :

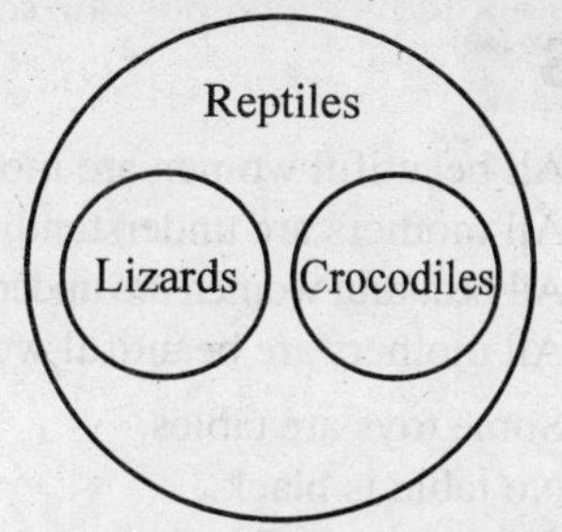

All lizards and crocodiles are reptiles, but neither is contained in the other. Some reptiles are lizards and some crocodiles.

22. (*c*) :

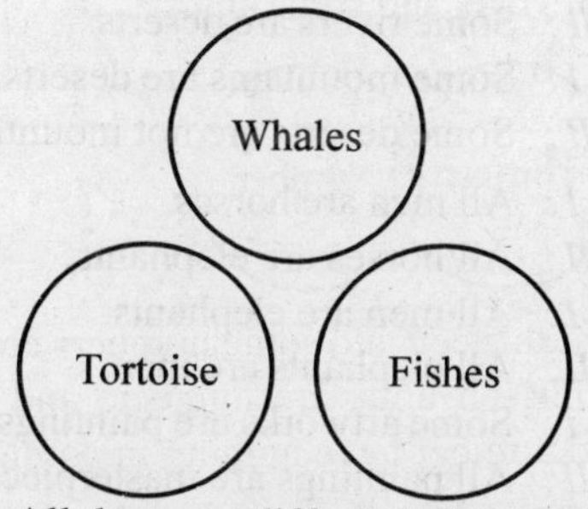

All three are different classes.

23. (*d*) :

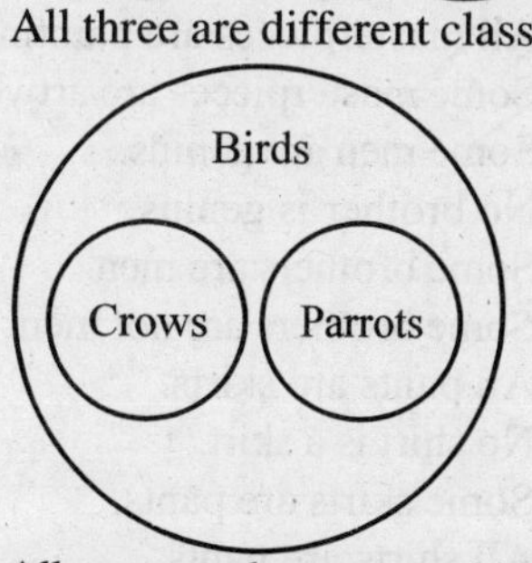

All crows and parrots are birds but neither is contained in the other. Some birds are crows and some parrots.

24. (*a*) :

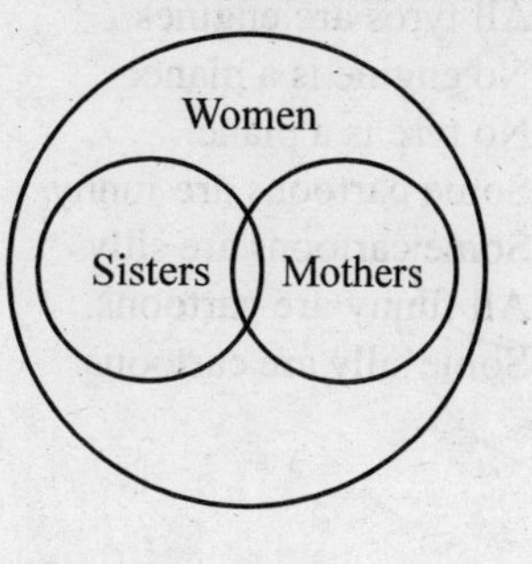

Some sisters can be mothers and some mothers can be sisters. All sisters and mothers are women.

25. (*b*) :

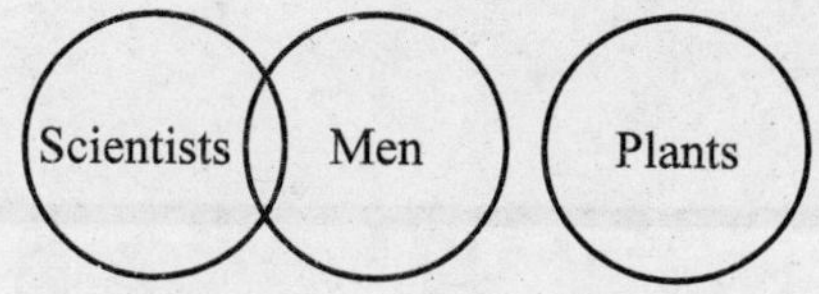

Some scientists can be men and some men can be scientists, but neither scientists nor men can be plants.

26. (*b*) :

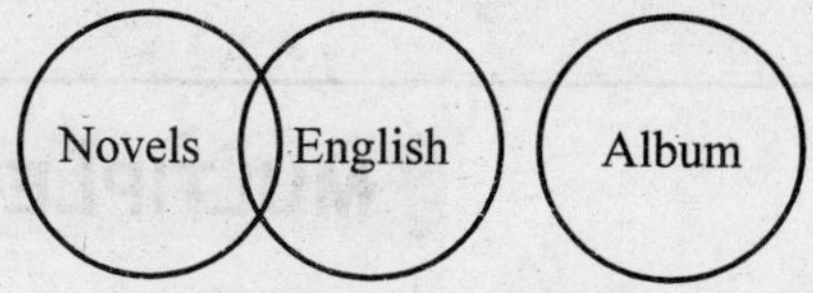

Some novels can be in English and some written material in English can be novels, but neither novels nor English can be album.

27. (*e*) :

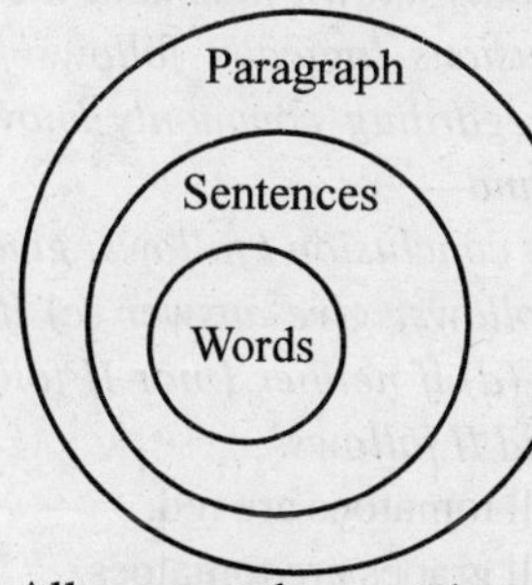

All paragraphs contain sentences within it and all sentences contain words within it.

28. (*d*) :

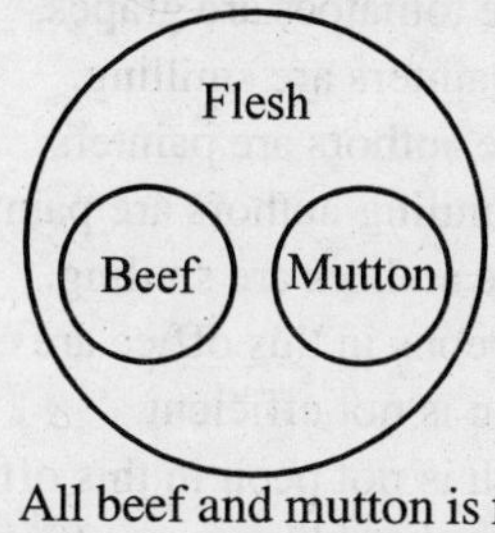

All beef and mutton is flesh of some animals. Some flesh is beef and some mutton.

29. (*b*) :

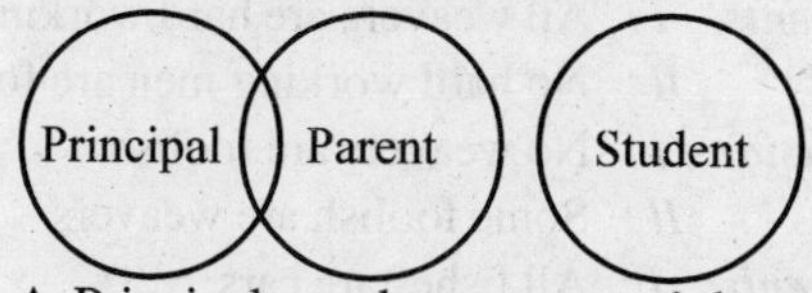

A Principal can be a parent and parent can be a principal. A student is a different class.

30. (*d*) :

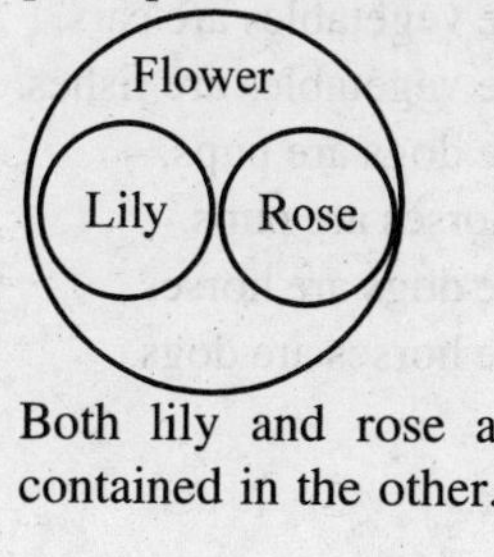

Both lily and rose are flowers. But neither is contained in the other.

21 Syllogism

In this reasoning pattern, the two premises are followed by two or four conclusions drawn from them. Five options (*a*), (*b*), (*c*), (*d*) and (*e*) are given as answers. Based on the two statements the candidate has to select the right option as answer.

MULTIPLE CHOICE QUESTIONS

Directions : *In each question below are given two statements followed by two conclusions numbered I and II. You have to take the two given statements to be true even if they seem to be at variance from commonly known facts and then decide which of the given conclusions logically follows from the two given statements, disregarding commonly known facts. Read both the statements and—*

Give answer (a) if only conclusion I follows; give answer (b) if only conclusion II follows; give answer (c) if either I or II follows; give answer (d) if neither I nor II follows and give answer (e) if both I and II follows.

1. *Statements* I: All tomatoes are red.
II: All grapes are tomatoes.
Conclusions I: All grapes are red.
II: Some tomatoes are grapes.

2. *Statements* I: All painters are smilling.
II: Some authors are painters.
Conclusions I: All smiling authors are painters.
II: Some authors are smiling.

3. *Statements* I: All peons in this office are efficient.
II: Ramu is not efficient.
Conclusions I: Ramu is not peon in this office.
II: Ramu should be more efficient.

4. *Statements* I: All weavers are hard working.
II: No hard working men are foolish.
Conclusions I: No weavers are foolish.
II: Some foolish are weavers.

5. *Statements* I: All fishes are cars.
II: All cars are vegetables.
Conclusions I: Some vegetables are cars.
II: Some vegetables are fishes.

6. *Statements* I: Some dogs are pups.
II: All horses are pups.
Conclusions I: Some dogs are horses.
II: Some horses are dogs.

7. *Statements* I: All beautiful women are mothers.
II: All mothers are understanding.
Conclusions I: All beautiful women are understanding.
II: All mothers are beautiful women.

8. *Statements* I: Some toys are tables.
II: No table is black.
Conclusions I: Some toys are black.
II: Some toys are not black.

9. *Statements* I: All rivers are mountains.
II: Some rivers are deserts.
Conclusions I: Some mountains are deserts.
II: Some deserts are not mountains.

10. *Statements* I: All men are horses.
II: All horses are elephants.
Conclusions I: All men are elephants.
II: All elephants are men.

11. *Statements* I: Some artworks are paintings.
II: All paintings are masterpieces.
Conclusions I: All masterpieces are paintings.
II: Some masterpieces are artworks.

12. *Statements* I: Some men are genius.
II: No brother is genius.
Conclusions I: Some brothers are men.
II: Some brothers are not men.

13. *Statements* I: All pants are skirts.
II: No shirt is a skirt.
Conclusions I: Some skirts are pants.
II: All shirts are pants.

14. *Statements* I: All planes are tyres.
II: All tyres are engines.
Conclusions I: No engine is a plane.
II: No tyre is a plane.

15. *Statements* I: Some cartoons are funny.
II: Some cartoons are silly.
Conclusions I: All funny are cartoons.
II: Some silly are cartoons.

Directions : *In the following questions, select the set of conclusion which logically follows from the given statements.*

16. *Statements* *I* : All foxes are plates.
II : All plates are trees.
Conclusions *I* : All foxes are trees.
II : All trees are foxes.
III : Some trees are foxes.
IV : Some trees are plates.
(*a*) All conclusions are correct.
(*b*) Only conclusions I, III, and IV follow.
(*c*) Only conclusions II, III, and IV follow.
(*d*) Only conclusions I and IV follow.
(*e*) None of the above.

17. *Statements* *I* : Some cubes are squares.
II : All squares are circles.
Conclusions *I* : All cubes are circles.
II : Some circles are cubes.
III : Some circles are squares.
IV : All squares are cubes.
(*a*) Only conclusion I follows.
(*b*) Only conclusion I, II and III follow.
(*c*) All conclusions are correct.
(*d*) Only conclusions II and III follow.
(*e*) None of the above.

18. *Statements* *I* : All cups are goats.
II : All goats are tins.
Conclusions *I* : All goats are cups.
II : All tins are goats.
III : No cups are tins.
IV : No tins are cups.
(*a*) Only conclusions III and IV follow.
(*b*) Only conclusios I and II follow.
(*c*) Only conclusions I, II and III follow.
(*d*) All conclusions are correct.
(*e*) None of the above.

19. *Statements* *I* : All bombs are bags.
II : Some bags are jets.
Conclusions *I* : All bombs are jets.
II : All jets are bombs.
III : Some jets are bombs.
IV : Some bombs are jets.
(*a*) Only conclusion III follows.
(*b*) Only conclusions I and II follow.
(*c*) All conclusions are correct.
(*d*) Only conclusions III and IV follow.
(*e*) None of these.

20. *Statements* *I* : Some thorns are jackets.
II : Some jackets are boats.
Conclusions *I* : No thorns are boats.
II : All jackets are boats.
III : Some boats are thorns.
IV : No jackets are thorns.
(*a*) Either conclusion I or IV follows.
(*b*) Either conclusion I or II follows.
(*c*) Either conclusion I or III follows.
(*d*) No conclusion is correct.
(*e*) All conclusions are correct.

EXPLANATORY ANSWERS

1. (*e*) : When all tomatoes are red and all grapes are tomatoes, then all grapes are also red. When all grapes are tomatoes, then some tomatoes must be grapes. Therefore, both conclusions I and II are correct.

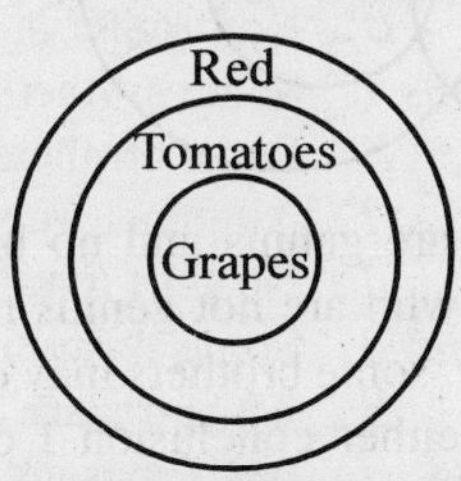

2. (*b*) : When all painters are smiling and some authors are painters, then some authors are smiling. Therefore, only conclusion II is correct.

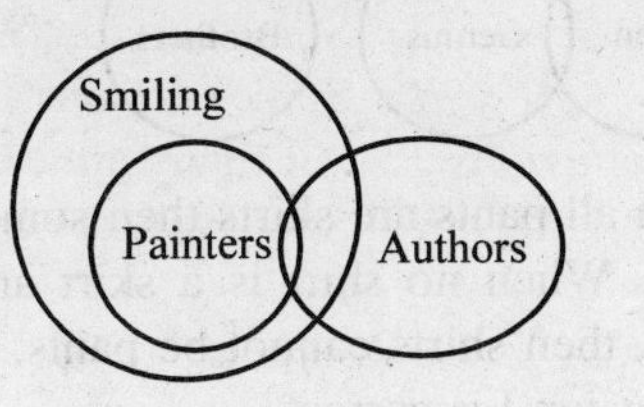

3. (*a*) : When all the peons of the office are efficient, then Ramu cannot be a peon in this office. Therefore, only conclusion I is correct.

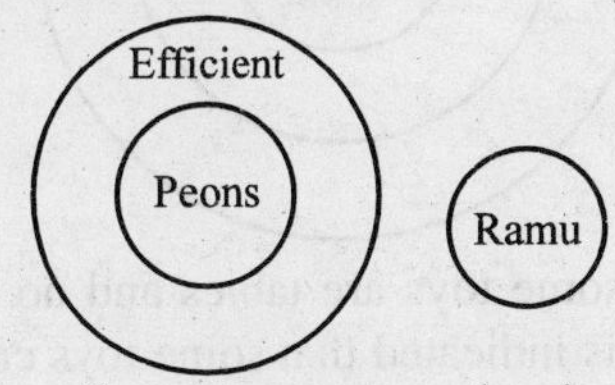

4. (*a*) : When all weavers are hardworking and no hardworking men are foolish, then no weavers are foolish. Therefore, only conclusion I is correct.

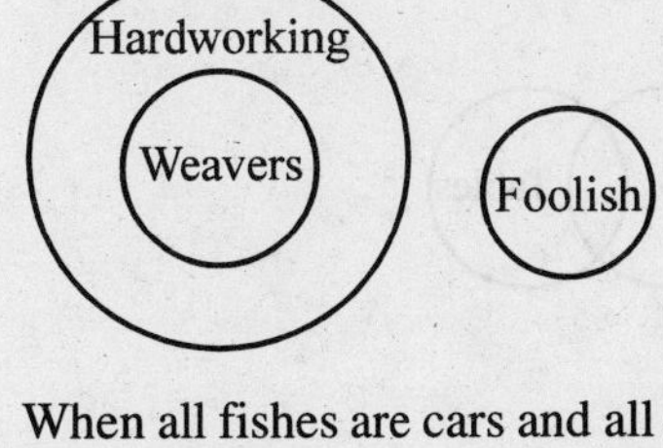

5. (*e*) : When all fishes are cars and all cars are vegetables, then all fishes will naturally be vegetables. This means that some vegetables are fishes. And when all cars are vegetables, then some vegetables will be cars naturally. Therefore, both the conclusions I and II are correct.

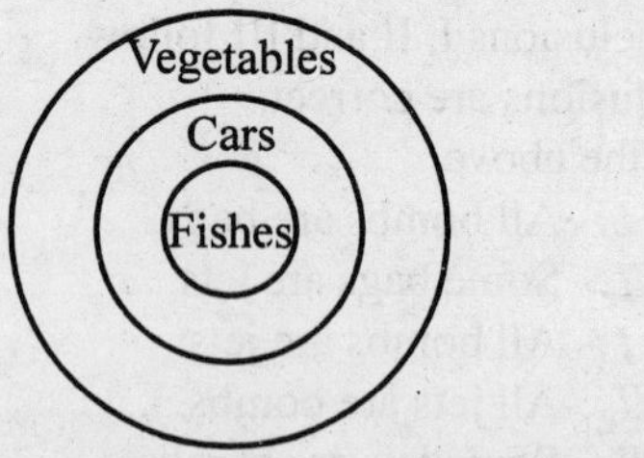

6. (*d*) : No relationship can be established between the two statements. Therefore, neither conclusion I nor conclusion II is correct.

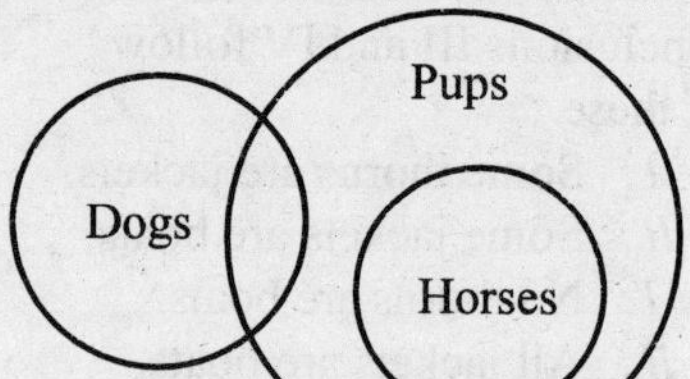

7. (*a*) : When all beautiful women are mothers and all mothers are understanding, then naturally all beautiful women are understanding. All mothers need not be beautiful women. Therefore, only conclusion I is correct.

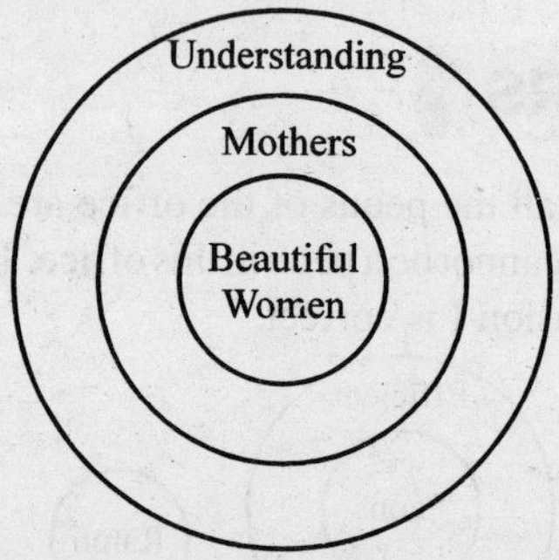

8. (*c*) : When some toys are tables and no table is black, then it is indicated that some toys can be black, as all toys are not tables. On the other hand, some toys may not be black. Therefore, there is a possibility that some toys may or may not be black. As such, either conclusion I or conclusion II can be correct.

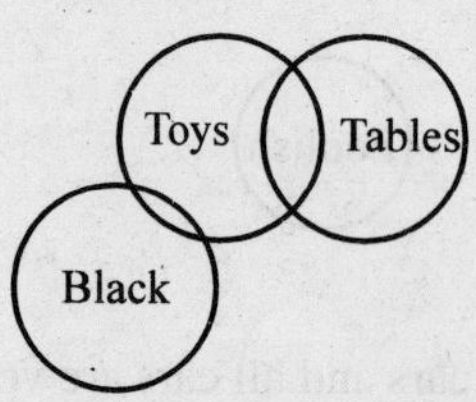

9. (*e*) : When all rivers are mountains and some rivers are deserts, then some deserts cannot be mountains and also, some mountains need not be deserts. Therefore, both conclusion I and conclusion II are correct.

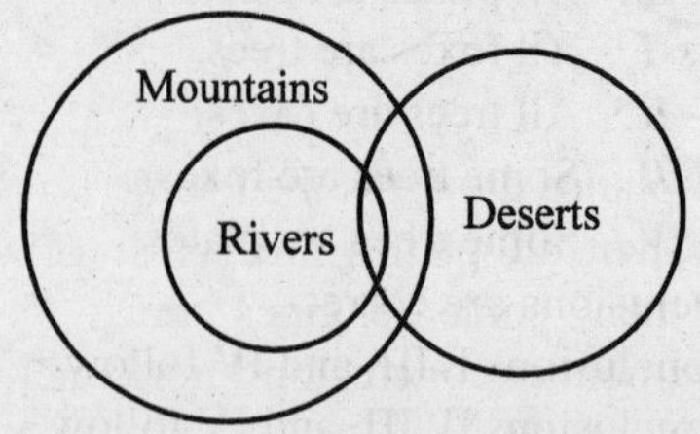

10. (*a*) : When all men are horses and all horses are elephants then, naturally all men are elephants, but all elephants need not be men. Therefore, only conclusion I is correct.

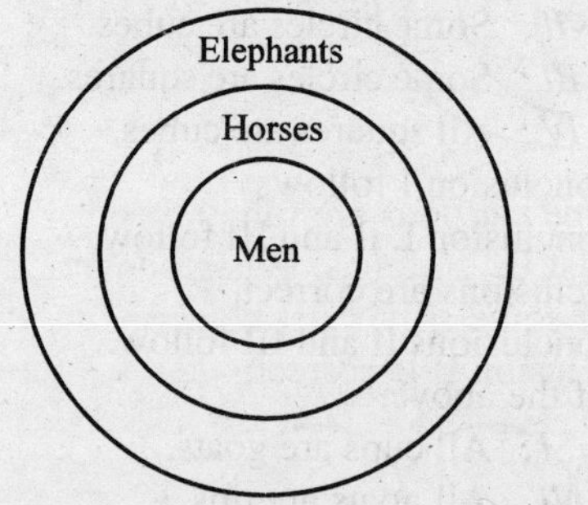

11. (*b*) : When some artworks are paintings which are all masterpieces, then some masterpieces must be artwork. When all paintings are masterpieces then some (not all) masterpieces must be paintings. Therefore, only conclusion II is correct.

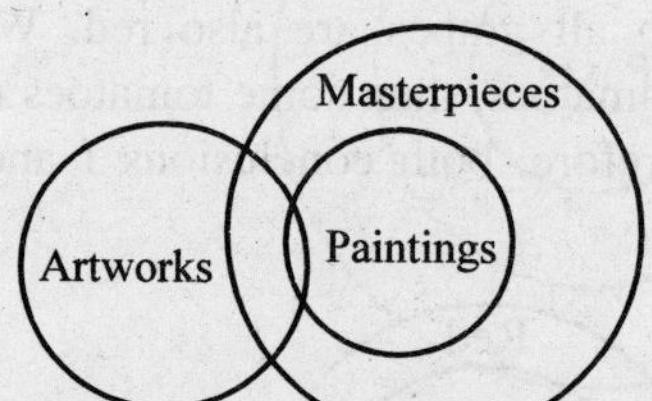

12. (*c*) : Some men are genius and no brother is a genius. Some men who are not genius may or may not be brothers, *i.e.* some brothers may or may not be men. Therefore, either conclusion I or conclusion II is correct.

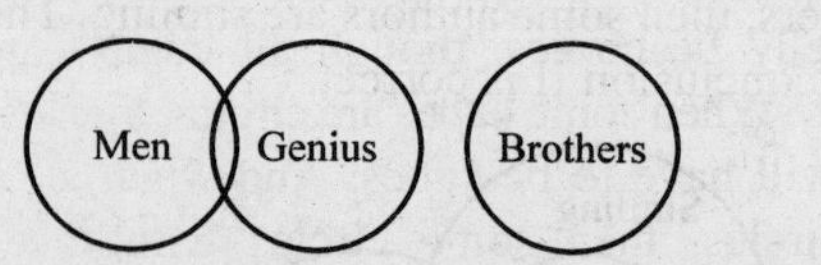

13. (*a*) : When all pants are skirts then some skirts must be pants. When no shirt is a skirt and all pants are skirts, then shirts cannot be pants. Therefore, only conclusion I is correct.

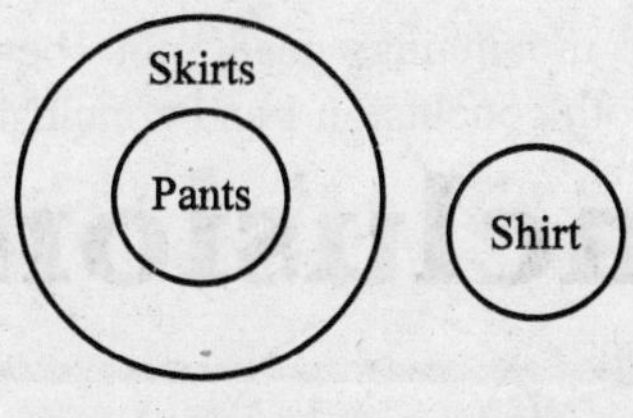

14. (*d*) : When all planes are tyres then some tyres must be planes. When all planes are tyres which are all engines then some engines must be planes. Therefore, neither conclusion I nor conclusion II is correct.

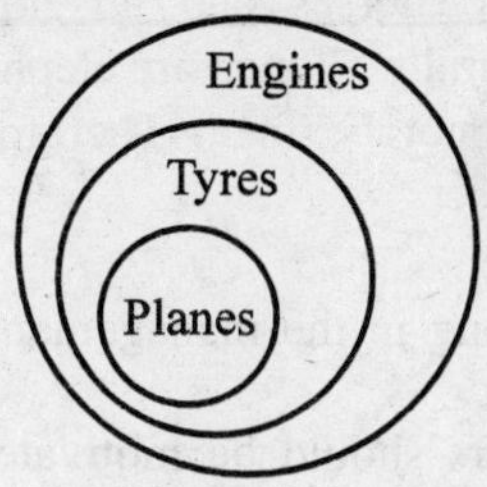

15. (*b*) : When some cartoons are funny then some (not all) funny must be cartoons. When some cartoons are silly, then some silly must be cartoons. Therefore, only conclusion II is correct.

16. (*b*) :

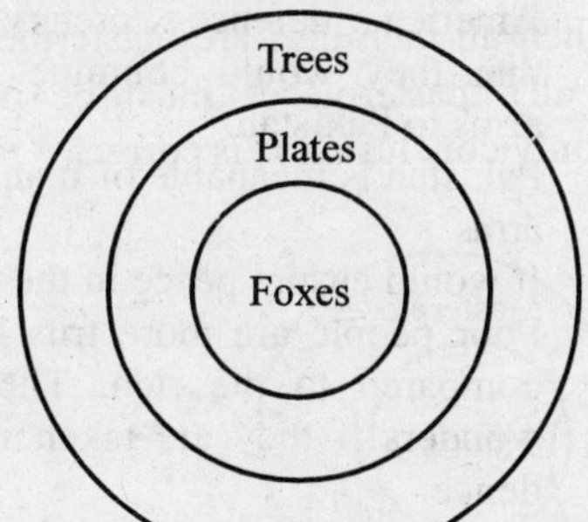

When all foxes are plates and all plates are trees, then naturally all foxes will be trees and some trees will then have to be foxes. And when all plates are trees then some trees will have to be plates. Therefore, only conclusions I, III and IV are correct.

17. (*d*) : When it is given that some cubes are squares and all squares are circles, then some cubes will naturally be circles, though all cubes cannot be circles. When some cubes are circles, then some circles will have to be cubes. And when all squares are circles, then some circles will have to be squares. As per given statement II, all squares are circles. Therefore all squares cannot be cubes. As such only conclu-sions II and III are correct.

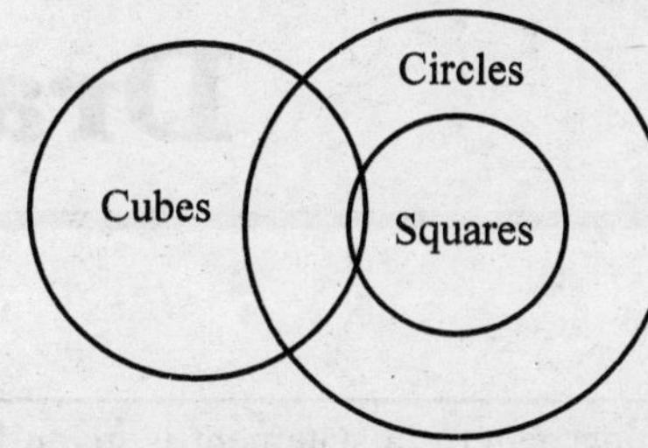

18. (*e*) : When all cups are goats, then only some goats can be cups. When all goats are tins, then only some tins can be goats. When all cups are goats and all goats are tins, then naturally all cups are tins and some tins must be cups. Therefore, all the conclusions are incorrect.

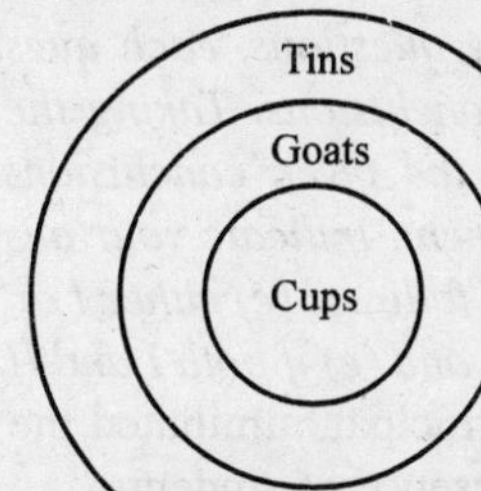

19. (*e*) : When all bombs are bags and some bags are jets then all bombs cannot be jets. 'Some bags' indicate that there is no chance of some bombs being jets, or some jets being bombs. Further all jets cannot be bombs. Therefore, all conclusions are incorrect.

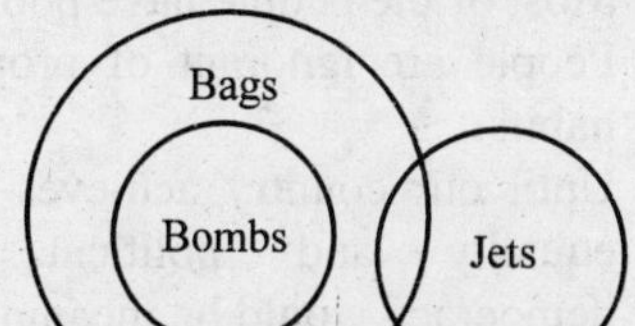

20. (*d*) : When some thorns are jackets and some jackets are boats, then some thorns may be boats but some boats cannot be thorns. As per Statement I, some thorns are jackets, some jackets need to be thorns. When Statement II conveys that some jackets are boats, then it is clear that all jackets cannot be boats. Therefore, no particular given conclusion can be drawn from the given statements.

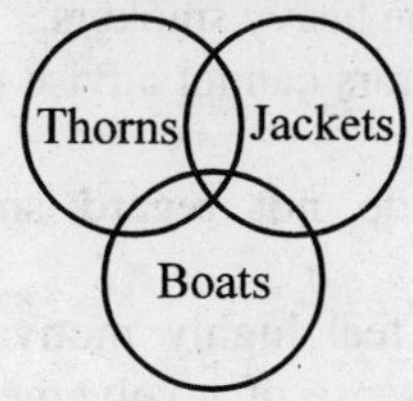

22 Drawing Conclusions

In this reasoning pattern, a statement is given followed by two conclusions. The statement is to be taken as the fact. Then based on it one has to decide which of the conclusion(s) definitely follows or does not follow from the given statement.

MULTIPLE CHOICE QUESTIONS

Directions : *In the following questions, each question has a statement followed by two conclusions. Taking the statement to be true, decide which of the given conclusions definitely follows from the given statement. Indicate your answer as (a) if only I follows; (b) if only II follows; (c) either I or II follows; (d) if neither I nor II follows; and (e) if both I and II follow.*

1. *Statement* : The Principal humiliated the teacher in the presence of students.
Conclusions I : The Principal did not like the teacher.
II : The teacher was not popular amongst his students.

2. *Statement* : Good health is dependent on right eating habits. Most of the people do not follow any rule regarding eating.
Conclusions I : Most of the people have poor health.
II : People are ignorant of proper eating habits.

3. *Statement* : Until our country achieves economic equality and political freedom, democracy would be meaningless.
Conclusions I : Political freedom and democracy go hand in hand.
II : Economic equality leads to real political freedom and democracy.

4. *Statement* : Black cloud follows thunder. Rains follow thunder.
Conclusions I : Thunder is the cause of rain.
II : Black cloud is the cause of thunder.

5. *Statement* : Practice is better than precept. Most doctors are heavy smokers.
Conclusions I : Most doctors cannot advise people not to smoke.
II : Doctors do not regard smoking as injurious.

6. *Statement* : Workers feel highly motivated when they get sense of involvement by participating in the management of companies.
Conclusions I : Workers should be motivated to produce more.
II : Workers should be allowed to participate in the management of companies.

7. *Statement* : Industrial revolution which first of all started in Europe has brought about modern age.
Conclusions I : Disparity between rich and poor results in revolution.
II : Revolution overhauls society.

8. *Statement* : America's defence secretary reiterated that they would continue to supply arms to Pakistan.
Conclusions I : Pakistan is incapable of manufacturing arms.
II : It would ensure peace in the region.

9. *Statement* : Poor people are more trust-worthy as compared to the rich. They can do wonders if they are taken into confidence.
Conclusions I : Poor people are wonderful people.
II : By and large rich people are not trust-worthy.

10. *Statement* : Government has spoiled many top ranking financial institutions by appointing bureaucrats as Directors of these institutions.
Conclusions I : Government should appoint Directors of the financial institutes taking into consi-deration the expertise of the person in the area of finance.
II : The Director of the financial institute should have exper-tise commensurate with the financial work carried out by the institute.

11. *Statement* : The Constitutional amendment carried out just last month prohibits the employment of child labour in any organisation.

Conclusions I: The employers must now abide by this amendment to the Constitution.

II: Children below 14 years will now be engaged in acquiring education.

12. *Statement* : People who speak too much against the dowry are those who had taken it themselves.

Conclusions I: It is easier said than done.

II: People have double standards.

13. *Statement* : Vegetable prices are soaring in the market.

Conclusions I: Vegetables are becoming a rare commodity.

II: People cannot eat vegetables.

14. *Statement* : As far as the rate of literacy is concerned, there is hardly any difference between the States of Kerala and West Bengal, but one is ahead of the other in respect of the percentage of population unemployed.

Conclusions I: Unemployment is more in Kerala than in West Bengal.

II: West Bengal has higher unemployment than Kerala.

15. *Statement* : He tirelessly argues for the replacement of the present system of allocating work to the teaching staff in the department by another system which will give the teachers higher job satisfaction.

Conclusions I: No system is free from problems and therefore there should be no system.

II: Any system you may evolve but it should give higher job satisfaction to the teachers.

16. *Statement* : The unequal allocation of funds by the Government for literacy and employment generation programmes stands out clearly when we analyse the attitude of government towards these two areas.

Conclusions I: Government has more favourable attitude towards literacy programmes than employment generation programmes.

II: Government has more favourable attitude towards employment generation programmes than literacy programmes.

17. *Statement* : Amul chocolates sold enough in comparison to all other chocolate brands in India.

Conclusions I: No other chocolate is sold in India.

II: The selling data of every chocolate brand is available in India.

18. *Statement* : Wearing Khadi is not only good for our health but also for our nation.

Conclusions I: Economy of India also depends upon Khadi sector.

II: It is difficult to maintain Khadi clothes.

19. *Statement* : Ideas given by our ancestors that were once discarded as uneconomical and unviable, turn out to be as functional and inevitable in present cicumstances.

Conclusions I: In ancient period ideas were considered either completely functional or totally infeasible.

II: Ideas cannot change from time to time.

20. *Statement* : The Supreme Court gave a judgement that the maintenance of old age parents is the responsibility of the married girls, if they do not have brothers.

Conclusions I: Constitution is always interpreted to help oppressed people out.

II: Before the Supreme Court gave its verdict, a married girl must have denied to pay for the maintenance of her parents.

21. *Statement* : Rapid growth presupposes efficiency which in turn requires a meritocracy, which runs counter to the principle of representation.

Conclusions I: Meritorious persons yield better results.

II: Rapid growth and the demand for representation are not complimentary.

22. *Statement* : The average number of students per teacher is 50 in urban areas, whereas, it is 60 in rural areas. The national average is 55.

Conclusions I: The student–teacher ratio in the rural areas is higher than in the urban areas.

II: More students study with the same teacher in the rural areas as compared to those in the urban areas.

23. *Statement* : The rise of the Japanese economy has triggered a revolution in the recent past as a result of their dedication and commitment.

Conclusions I: Revolution leads to economic prosperity.

II: Hard work and sincerity result in economic prosperity.

24. *Statement* : Power consumption in every family has been doubled during the last five years.

Conclusions I: There is a lot of development in the society.

II: Power rates have become cheaper.

25. *Statement* : The knowledge of content does not matter much, what matters is whether you can orally express it or not.

Conclusions I: Art of delivery and knowledge of content are not necessarily interdependent.

II: Oral expression is an acquired skill.

26. *Statement* : Skyscrapers are coming up with a high speed in cosmopolitan cities.

Conclusions I: There is a vertical growth in cosmopolitan cities.

II: There is a lot of speed pressure in cosmopolitan cities.

27. *Statement* : Now more and more mothers have developed positive attitude towards breast feeding.

Conclusions I: Mothers had wrong notions about breast feeding.

II: Breast feeding is proved good for mother and child both.

28. *Statement* : Of the ten fishermen caught in the seastorm, nine managed to return to the shore. Parveen has not yet returned after four days.

Conclusions I: Parveen got killed in the storm.

II: Parveen has survived the storm.

29. *Statement* : Now you don't need an import licence to own a VCR.

Conclusions I: VCRs are now manufactured indigenously.

II: VCRs are now freely permitted to be imported.

30. *Statement* : Use of cosmetics is hazardous.

Conclusions I: Cosmetics have got side effects.

II: Cosmetics are expensive.

EXPLANATORY ANSWERS

1. (*a*): Only conclusion I follows. Conclusion II is not confirmed by the given statement.

2. (*a*): Conclusion I definitely follows. Conclusion II is not confirmed by the given statement.

3. (*b*): Only conclusion II follows. Conclusion I cannot be drawn from the given statement.

4. (*b*): Only conclusion II follows. Conclusion I is wrong as rains are the cause of thunder.

5. (*e*): Both conclusion I and conclusion II follow from the statements as the doctors who are heavy smokers indicate that they do not regard smoking as injurious and thus, cannot advise people not to smoke.

6. (*e*): Both conclusions I and II follow from the statement as involvement of workers boosts production and benefits the companies.

7. (*b*): Only conclusion II follows. Conclusion I is not related the statement.

8. (*d*): Neither conclusion I nor conclusion II follows. Both the conclusions are unrelated to the statement.

9. (*b*): Only conclusion II follows. Conclusion I is not related to the statement.

10. (*e*): Both conclusion I and conclusion II follow.

11. (*a*): Only conclusion I follows. Conclusion II is not related to the statement.

12. (*e*): Both conclusion I and conclusion II follow as people who talk too much often do not practise what they say.

13. (*d*): Neither conclusion I nor conclusion II is confirmed by the statement.

14. (*d*): Neither conclusion I nor conclusion II follows as the precentage of unemployment in each state is not mentioned.

15. (*b*): Only conclusion II follows. Conclusion I runs counter to the statement as a system is necessary.

16. (*d*): Neither conclusion I nor conclusion II is confirmed by the statement.

17. (*b*): Only conclusion II follows as the data of every chocolate brand shows the comparision of Amul chocolates with other brands. Coclusion I is not confirmed by the statement.

18. (*a*): Only conclusion I follows. Conclusion II is not related to the statement.

19. (*d*): Neither conclusion I, nor conclusion II follows; both the conclusions are unrelated.

20. (*b*): Only conclusion II follows, because the Supreme Court judgement must have come in such a case and not in vaccum.

21. (*e*): Both conclusion I and conclusion II follow as the statement presupposes better results with merit and contradiction between merit and representation.

22. (*b*): The conclusion I runs counter to the facts mentioned in the statement. Only conclusion II follows.

23. (*b*): The conclusion I is not confirmed by the statement. Only conclusion II follows.

24. (*d*): Neither conclusion I nor conclusion II definitely follows from the statement.

25. (*a*): Only conclusion I follows. Conclusion II is not related to the statement.

26. (*e*): Appearance of skyscrapers in cosmopolitan cities is the manifestation of space pressure in vertical growth. Thus both conclusion I and II follow.

27. (*e*): Both conclusion I and conclusion II follow.

28. (*c*): It is not certian if Parveen is killed in the storm or has survived the storm. Thus, either conclusion I or conclusion II follows from the statement.

29. (*b*): Conclusion II alone follows from the statement. Conclusion I relates the indigenous production of VCRs. It is not related to the statement.

30. (*a*): Only conclusion I follows definitely from the statement. Conclusion II is not related to the statement.

23

Drawing Assumptions

In this type of reasoning, a statement is given and assumptions are drawn from it. An assumption is something assumed, supposed and taken for granted. The implicity of the assumptions in question. The implication means the hidden meaning, something which is derived from and as such based upon the statement.

MULTIPLE CHOICE QUESTIONS

Directions : *In each question below is given a statement followed by two assumptions numbered I and II. An assumption is something supposed or taken for granted. You have to consider the statement and the following assumptions and decide which of the assumptions is implicit in the statement. Give answer—*

(*a*) If assumption I is implicit.
(*b*) If assumption II is implicit.
(*c*) If either I or II is implicit.
(*d*) If neither I nor II is implicit.
(*e*) If both I and II are implicit.

1. *Statement* : Apart from the entertainment value of television, its educational value cannot be ignored.
Assumptions I: People take the television to be means of entertainment only.
II: The educational value of television is not realised properly.

2. *Statement* : Ten candidates who were on the waiting list could finally be admitted to the course.
Assumptions I: Wait-listed candidates get admission with difficulty.
II: A large number of candidates were on the waiting list.

3. *Statement* : In Mumbai, railway trains are indispensable for people in the suburbs to reach their places of work on time.
Assumptions I: Railway trains are the only mode of transport available in the suburbs of Mumbai.
II: Only railway trains run punctually.

4. *Statement* : The mangoes are too cheap to be good.
Assumptions I: When the mango crop is abundant, the prices go down.
II: The lower the selling price, the inferior is the quality of the commodity.

5. *Statement* : Of all the newspapers published in Delhi, 'The Time and Space' has the largest number of readers.
Assumptions I: Volume of readership of all the newspapers in Delhi is known.
II: No newspaper in Delhi other than 'The Time and Space' has large readership.

6. *Statement* : 'Smoking Prohibited'. A notice in a railway compartment.
Assumptions I: Some persons like to smoke in the compartment.
II: Some persons smoke in the compartment.

7. *Statement* : A friend in need is a friend indeed.
Assumptions I: Relatives are of no help in difficult circumstances.
II: Taking help from a friend is desirable.

8. *Statement* : When in Chennai, stay at Hotel 'X' for comfort and good service, because my best friend has recommended it.
Assumptions I: My best friend is trustworthy.
II: Comfort and good service cannot be expected from all hotels.

9. *Statement* : A friend in need is a friend indeed.
Assumptions I: A person needs help from others on certain occasions.
II: A friend is expected to help when needed.

10. *Statement* : 'Smoking Prohibited.' A notice in a railway compartment.

Assumptions I : Generally, people expect a notice.

II : Smoking is injurious to health.

Directions : *In each question below is given a statement followed by three assumptions number I, II and III. An assumption is something supposed or taken for granted. You have to consider the statement along with the assumptions and decide which of the assumptions is implicit in the statement. Then decide which of the alternatives is the correct answer.*

11. *Statement :*

"Use Riya Cold Cream for fair complexion"—an advertisement.

Assumptions :

I. People like to use cream for fair complexion.

II. People are easily fooled.

III. People respond to advertisements.

(*a*) Only I is implicit.
(*b*) Only II is implicit.
(*c*) Only I and III are implicit.
(*d*) Only I and II are implicit.
(*e*) All are implicit.

12. *Statement :*

"Slogans against smoking in office should be put on the notice board"—an employee in an office suggests.

Assumptions :

I. The employee felt that his suggestion will be considered.

II. People smoke in the office.

III. Some people will stop smoking after reading the slogans.

(*a*) Only I is implicit.
(*b*) Only I and II are implicit.
(*c*) Only II and III are implicit.
(*d*) None is implicit.
(*e*) All are implicit.

13. *Statement :*

"I want to present him a book on techniques of yoga on his birthday"—A tells B.

Assumptions :

I. A and B will be invited for the birthday celebrations.

II. The person to whom the book is to be presented, is not keeping good health.

III. Book is the best gift for birthdays.

(*a*) Only I and II are implicit.
(*b*) Only II and III are implicit.
(*c*) Only I and III are implicit.
(*d*) None is implicit.
(*e*) All are implicit.

14. *Statement :*

Keeping in view the financial constraint, the management institution has decided to charge at the time of providing employment in various organisations, a placement fee of Rs. 25000 from the organisation in which the student will be provided the employment.

Assumptions :

I. It will help in increasing the demand of the students belonging to the management institution.

II. The amount collected in this way will be purposeful.

III. It may be possible that the organisation providing employment may select less number of students in future.

(*a*) None follows.
(*b*) Only II and III follow.
(*c*) Only I follows.
(*d*) Only I and II follow.
(*e*) None of these.

15. *Statement :*

In order to reduce the gap between income and expenditure the company has decided to increase the price of its product from next month.

Assumptions :

I. The rate will remain more or less same after the increase.

II. The expenditure will more or less remain same in near future.

III. The rival companies will also increase the price of the similar product.

(*a*) All are implicit.
(*b*) Only II and III are implicit.
(*c*) Only I and II are implicit.
(*d*) Only III is implicit.
(*e*) None of these.

16. *Statement :*

The professor announced in the class that the next periodical examination will be held on 15th of the next month.

Assumptions :

I. All the students may appear in the examination.

II. The college will remain open on 15th of the next month.

III. The students can study till 15th of the next month to pass the examination.

(*a*) Only II and III are implicit.
(*b*) Only II is implicit.
(*c*) Only I and II are implicit.
(*d*) Only III is implicit.
(*e*) None of these.

17. *Statement :*

"We must introduce objective type tests to improve our examinations for admission to MBA"–The Chairman of the Admission Committe tells the Committee.

Assumptions :

I. The admission at present is directly through the interviews.

II. The Admission Committee is desirous of improving the admission examinations.
III. The Chairman himself is an MBA.
(*a*) Only I is implicit.
(*b*) Only II is implicit.
(*c*) Only I and II are implicit.
(*d*) Only I and III are implicit.
(*e*) None is implicit.

18. *Statement :*

"Since no university has responded to our proposal, we shall not depute any faculty member for the course this year"–A notice in a department.

Assumptions :
I. The proposal has been received by the universities.
II. The proposal had also indicated financial provisions.
III. The proposal was not received by the universities.
(*a*) Only I is implicit.
(*b*) Only III is implicit.
(*c*) Only either I or III is implicit.
(*d*) None is implicit.
(*e*) All are implicit.

19. *Statement :*

"Wanted a two-bedroom flat in the court area for immediate possession"—advertisement.

Assumptions :
I. Flats are available in court area.
II. Some people will respond to the advertisement.
III. It is a practice to give such an advertisement.
(*a*) None is implicit.
(*b*) Only I and II are implicit.
(*c*) Only II is implicit.
(*d*) All are implicit.
(*e*) None of these.

20. *Statement :*

A group of friends decided to go for a picnic to Avon during the next holiday season to avoid crowd of people.

Assumptions:
I. Generally many people do not go to Avon.
II. People prefer other spots to Avon.
III. Many people do not know about Avon.
(*a*) All are implicit.
(*b*) Only II is implicit.
(*c*) Only I and II are implict.
(*d*) None is implicit.
(*e*) None of these.

21. *Statement :*

"Buy 'Y' T.V. for better sound quality"–an advertisement.

Assumptions:
I. 'Y' T.V. is the only T.V. in the market.
II. 'Y' T.V. is the costliest.
III. People generaly ignore such advertisements.
(*a*) Only I is implicit. (*b*) Only II is implicit.
(*c*) Only III implicit. (*d*) All are implicit.
(*e*) None is implicit.

22. *Statement :*

"Look at her audacity. Madhu has not replied to my letter"—A tells B.

Assumptions :
I. Madhu received his letter.
II. Madhu did not receive his letter.
III. The letter was sent by post.
(*a*) Only I is implicit. (*b*) Only II is implicit.
(*c*) Only III is implicit. (*d*) All are implicit.
(*e*) None is implicit.

23. *Statement :*

"Put a notice on the Board that all the employees should come on time to office."—An officer tells his assistant.

Assumptions :
I. All the employees come late.
II. Employees read such notice on the Board.
III. Employees will follow the instructions.
(*a*) Only I and II are implicit.
(*b*) Only III is implicit.
(*c*) Only I and III are implicit.
(*d*) Only II and III are implicit.
(*e*) All are implicit.

24. *Statement :*

The telephone company informed the subs-cribers through a notification that those who do not pay their bills by the due date will be charged penalty for every defaulting day.

Assumptions :
I. Majority of the people may pay their bills by the due date to avoid penalty.
II. The money collected as penalty may set off the losses due to delayed payment.
III. People generally pay heed to such notices.
(*a*) None is implicit.
(*b*) Only II and III are implicit.
(*c*) Only I and II are implicit.
(*d*) All are implicit.
(*e*) None of these.

25. *Statement :*

Considering the tickets sold during the last seven days, the circus authorities decided to continue the show for another fortnight which includes two weekends.

Assumptions :
I. People may not turn up on weekdays.
II. The average number of people who will be visiting circus will more or less be same as that of the last seven days.

III. There may not be enough response at other places.
(*a*) None is implicit.
(*b*) Only II is implicit.
(*c*) Both I and II are implicit.
(*d*) All are implicit.
(*e*) None of these.

26. *Statement :*

"If you want best training in computers, join our classes"–an advertisement.

Assumptions:

I. People want training related with computers.
II. People read advertisements.
III. People want quality training.
(*a*) Only I and II are implicit.
(*b*) Only III is implicit.
(*c*) Only I and III are implicit.
(*d*) Only II and III are implicit.
(*e*) All are implicit.

27. *Statement :*

The national air carrier has decided to start a weekly air service from town 'A' to town 'B'.

Assumptions :

I. There will be enough passengers to make the operation economically viable.
II. Other carriers may not start such service.
III. The people staying around these towns can afford the cost of air travel.
(*a*) All are implicit.
(*b*) Only I is implicit.
(*c*) Both II and III are implicit.
(*d*) Both I and II are implicit.
(*e*) None of these.

28. *Statement :*

Prabodh wrote a second letter to his mother after a month as he did not receive any reply to the first letter.

Assumptions :

I. Prabhodh's mother did not receive the letter.
II. The letter generally reaches within a fortnight.
III. His mother promptly sends reply to his letters.
(*a*) Only II and III are implicit.
(*b*) Only III is implicit.
(*c*) None is implicit.
(*d*) Only I and II are implicit.
(*e*) None of these.

29. *Statement :*

The school authority decided to open a summer school this year in the school compound for the students in the age range of 7–14 years.

Assumptions :

I. All the students will attend the summer school.
II. All the parents will prefer to remain in city than going out of town for enabling their children to attend the summer school.
III. Those who cannot afford to go out of station will send their children to summer school.
(*a*) Only II implicit.
(*b*) Only II and III are implicit.
(*c*) None is implicit.
(*d*) All are implicit.
(*e*) Only III is implicit.

30. *Statement :*

In view of the recent spurt in sugar prices in the open market the government has asked the dealers to release vast quantity of imported sugar in the open market.

Assumptions :

I. The dealers will follow the government directive.
II. The sugar prices will come down.
III. The price of indigenous sugar will remain unchanged.
(*a*) None is implicit.
(*b*) Only I and II are implicit.
(*c*) Only I and III are implicit.
(*d*) All are implicit.
(*e*) Only II and III are implicit.

EXPLANATORY ANSWERS

1. (*e*) : Both assumptions I and II are implicit.

2. (*a*) : Assumption II is not implicit in the statement. Only assumption I is implicit.

3. (*d*) : Neither assumption I nor assumption II is implicit in the statement.

4. (*b*) : Assumption I is not implicit in the statement, because it states the principle of supply and prices. Only assumption II is implicit which states the principle of price-quality parity.

5. (*a*) : Assumption I is implicit in the statement because the largest readership may be established only if the volume of readership of all the newspapers is known.

6. (*b*) : Assumption II suggests that some persons smoke in the compartment. 'Smoking Prohibited' is the notice for those persons.

7. (*d*) : The statement highlights the criterion of a true friend.

8. (*e*) : Both assumptions I and II are implicit because *(i)* the advice is from a trustworthy best friend, and *(ii)* all hotels cannot provide comfort and good service.

9. (*b*) : Only assumption II is implicit because only a true friend helps in need.

10. (*d*) : None of the assumptions is implicit. The statement does not imply that people expect a notice or smoking is injurious to health.

11. (*c*) : The advertisement itself denotes that people like to use cream for fair complexion and they do respond to advertisements. Advertisements are for awareness of the product not for fooling people.

12. (*e*) : Suggestion for putting up the notice is made because some people must be smoking in office. Notice board is read by people in office and some may stop smoking after reading the slogans.

13. (*c*) : According to A presents are best gifts for birthdays and when A has bought the book he expects to be invited for the birthday celebration. Nothing is mentioned about health of the birthday person.

14. (*b*) : The amount collected will lessen the financial constraint and there is every possibility for selecting less number of students because of the fee charged.

15. (*e*) : The increased amount is not mentioned and nothing is said about the rival companies. As the rates are increased to meet the expenditure, the latter may remain same. Only assumption II is implicit.

16. (*d*) : The students have the time to study till 15th of the next month. All students may or may not appear in the examination and the announcement does not inform anything about the college being open till the mentioned date.

17. (*b*) : Assumption II is obvious from the statement. Nothing is mentioned about the interviews or the qualification of the Chairman of the Admission Committee.

18. (*d*) : The universities have or have not received the proposal is a matter of doubt and the matter of the proposal is not mentioned.

19. (*c*) : It is a fact that people do read advertisement and some of them will respond to them. Availability of the flat in court area is doubtful and it is a matter of need, not practice, to give such an advertisement.

20. (*a*) : The friends decided to go to Avon in order to avoid crowd. This means that many people do not know about Avon (as known places attract crowds) and that is why they do not go there. People prefer known places instead of Avon.

21. (*e*) : The degree of comparison 'better sound' indicates there are other T.V.s also. Nothing is mentioned about the price of the T.V. in advertisement or its rivals. Advertisements enhancing the quality of any product are not ignored by the people.

22. (*c*) : Madhu received the letter or not is doubtful. As A is expecting a reply, he had certainly sent the letter.

23. (*d*) : Notice boards are read by the employees and the instructions in the notice are meant to be followed. However, all the employees cannot come late to office.

24. (*e*) : After receiving the notification people will pay heed to the notice and make payments on time to avoid penalty. Nothing is said about the utility of money collected as penalty. Therefore, assumptions I and III are implicit.

25. (*b*) : The circus show is continued for another fortnight because the people visiting the circus are expected to be more or less same as in the earlier days. People turn up every day and the response at other places cannot be assumed.

26. (*e*) : Advertisement is given because people who want quality training related with computers will read such advertisements.

27. (*a*) : The reason for starting the service is that there are enough passengers. Other carriers may not start the service as national air carrier has taken the decision to do it. Now people staying in towns A and B can afford the air travel as weekly services may have some benefits for the passengers.

28. (*a*) : Prabodh's mother received the letter or not is doubtful. Before writing the second letter Prabodh had waited for 15 days to receive the reply knowing that his mother gives a prompt reply.

29. (*c*) : The assumptions are not related to the statements.

30. (*b*) : The dealers have to follow the government directives and when vast quantity of sugar is released in the market, its prices will come down.

24

Course of Action

In these type of questions, a situation is presented and some courses of action are suggested in the context of that situation. The candidates are required to determine which course of action follows immediately from the given statement without letting their own perception influence the decision.

MULTIPLE CHOICE QUESTIONS

Directions (Qs. 1 to 20) : *In each question below is given a statement followed by two courses of action numbered I and II. A course of action is a step or administrative decision to be taken for improvement, follow-up or further action in regard to the problems, policy, etc. on the basis of the information given in the statement. You have to assume everything in the statement to be true, then decide which of the given suggested courses of action logically follows for pursuing.*

Give answer (a) if only I follows; give answer (b) if only II follows; give answer (c) if either I or II follows; give answer (d) if neither I nor II follows and give answer (e) if both I and II follow.

1. *Statement :*
Although the Indian economy is still heavily dependent on agriculture, its share in global agricultural trade is less than the share of agricultural exports to total exports.
Courses of Action :
(i) Efforts should be made to increase our agricultural production.
(ii) The exports of non-agricultural commodities should be reduced.

2. *Statement :*
About 30 to 40 per cent of children who are enrolled, do not attend school on any given day.
Courses of Action :
(i) More schools should be started.
(ii) Reasons for this absenteeism should be found out.

3. *Statement :*
Certain mining industries in Gujarat may come to a standstill because of the notification issued by the Ministry of Environment and Forest banning mining operations and industries alike within 25 kms of National Park, the bird sanctuary and reserve forest areas.
Courses of Action :
(i) The Ministry should be asked to immediately withdraw the notification.
(ii) The Government should make efforts to shift the parks, sanctuaries and reserve forests to other non-mining areas.

4. *Statement :*
The Indian electronic component industry venturing into the West European markets faces tough competition from the Japanese.
Courses of Action :
(i) India should search for other international markets for its products.
(ii) India should improve the quality of the electronic components to compete with the Japanese in capturing these markets.

5. *Statement :*
Air export volumes have increased substantially over the past decade causing backlogs and difficulties for air cargo agents because of increased demand for space and service.
Courses of Action :
(i) Airlines and air cargo agents should jointly work out a solution to combat the problem.
(ii) The reasons for the increase in the volume of air export should be found out.

6. *Statement :*
The world conference on 'Education of All' took place in Thailand in 1990. Widely attended conference endorsed the Framework for Action for Meeting the Basic Learning Needs of all Children.
Courses of Action :
(i) India should suitably implement the Action points of this conference.
(ii) India should also immediately organise this type of conference.

7. *Statement :*

Huge amount of resources are required to develop tourist places in a country like India which is endowed with vast coastal lines, rivers, forests, temples, etc.

Courses of Action :

(i) More tourist-resorts along the coastal line only should be started.

(ii) The tourist-potential of India should be exploited.

8. *Statement :*

Financial stringency prevented the Manipur Government from paying salaries to its employees since April this year.... a report in December.

Courses of Action :

(i) The Manipur Government should immediately curtail the staff strength at least by 30 per cent.

(ii) The Manipur Government should reduce wasteful expenditure and arrange to pay the salaries of its employees.

9. *Statement :*

Reports of steep and continued decline in the inflows into the Gobindsagar reservoir of the Bhakra Dam, coupled with a depleted stock of steam coal with the thermal power plants in the north, may lead to a serious power crisis in the region.

Courses of Action :

(i) The supply of steam coal to the thermal power plants needs to be immediately stepped up by the Government.

(ii) The Government should set up hydraulic power plants on other rivers in the region.

10. *Statement :*

The police department has come under a cloud with recent revelations that at least two senior police officials are suspected to have been involved in the illegal sale of a large quantity of weapons from the State police armoury.

Courses of Action :

(i) A thorough investigation should be ordered by the State Government to bring out all those who are involved into the illegal sale of arms.

(ii) State police armoury should be kept under Central Government's control.

11. *Statement :*

Assembly and Parliamentary elections should be held simultaneously.

Courses of Action :

(i) The Constitution should be amended for this purpose.

(ii) The President should be given responsibility for deciding it.

12. *Statement :*

Express Highways are essential for rapid transportation via roads.

Courses of Action :

(i) Tenders for Express Highways should be issued soon..

(ii) Only Express Highways should be built from now on for the rapid development of the country.

13. *Statement :*

The production of silk in State 'X' has come down drastically due to loopholes in implementation of sericulture development programme despite availability of funds.

Courses of Action :

(i) Government of State 'X' should immediately study the loopholes and take corrective action to plug the loopholes.

(ii) The silk producers of State 'X' must be given technical help for this year.

14. *Statement :*

In order to speed up the work of assessment of papers of H.S.C. Examination, the examination board in State 'X' has decided to appoint temporary and inexperienced teachers of Junior Colleges as examiners.

Courses of Action :

(i) Many of the teachers should be given higher compensation as they are putting hard efforts in spite of lack of knowledge and expertise.

(ii) These teachers need to be given proper orientation by experienced examiners and centralised assessment may be made under supervision of moderators.

15. *Statement :*

If the retired Professors of the same Institutes are also invited to deliberate on re-structuring of the organisation, their contribution may be beneficial to the Institute.

Courses of Action :

(i) Management may seek opinion of the employees before calling retired Professors.

(ii) Management should involve experienced people for the systematic re-structuring of the organisation.

16. *Statement :*

Experts Committee on Industrial Relations has revealed in their recent report that not only the internal problems of the organisations were successfully managed through quality circles but also employees' motivation level was increased significantly.

Courses of Action :

(i) Organisations should organise orientation programmes for their employees on quality circles.

(ii) Employees should be encouraged to join quality circle programmes.

17. *Statement :*
Monsoons may fail this year as its parameters are not giving consistent results.
Courses of Action :
(*i*) A more detailed study regarding the parameters of efficiency should be taken up.
(*ii*) Water conservation methods should be strictly implemented from now onwards.

18. *Statement :*
Researchers are feeling agitated as libraries are not equipped to provide the right information to the right users at the right time in the required format. Even the users are not aware about the various services available for them.
Courses of Action :
(*i*) All the information available in the libraries should be computerised to provide faster services to the users.
(*ii*) Library staff should be trained in computer operations.

19. *Statement :*
One of the problems facing the food processing industry is the irregular supply of raw material. The producers of raw materials are not getting a reasonable price.
Courses of Action :
(*i*) The government should regulate the supply of raw material to other industries also.
(*ii*) The government should announce an attractive package to ensure regular supply of raw material for food processing industry.

20. *Statement :*
India's performance in the last Olympic Games was very poor. Not even a single medal could be bagged by the players. Government had spent Rs. 5 crores in training and deputing a team of players to participate in the Olympic Games.
Courses of Action :
(*i*) India should stop sending players to the future Olympic games.
(*ii*) Government should immediately set up an inquiry commission to find out the reasons for India's dismal performance.

EXPLANATORY ANSWERS

1. (*d*) : India's share in global agricultural trade is less than the share of agricultural exports to total exports. Therefore, there is no necessity of increasing the agricultural production. Reducing the exports of non-agricultural commodity will not affect the share of agricultural exports. So, neither course of action follows.

2. (*d*) : About 30 to 40 per cent of enrolled children do not attend schools. Therefore, the decision to start more schools is pointless. The statement does not mention that a particular group of children do not attend schools. This means the percentage of absent children comprises of various children who are absent at different times. There cannot be specific reasons for this absenteeism. So, neither course of action follows.

3. (*d*) : The department cannot withdraw the notification because environment pollution must be checked around the game sanctuary and reserve forest areas. Further, it is coherent to shift the mining industries. So, neither course of action follows.

4. (*e*) : In the face of tough competition from the Japanese, India should either improve the quality of electronic components to capture the market or search for other international markets where Japanese have not captured the market. So, both the course of action follow.

5. (*a*) : The reasons for the increase in the volume of air exports according to the statement is, increase in demand for space and service. Therefore, the airlines and air cargo agents should jointly work out a solution to combat the problem. So, only course of action I follows.

6. (*e*) : In a country like India where illiteracy rate is still high, such conferences should be organised and India must suitably implement the action points of the "Education for All" conference. So, both the courses of action follow.

7. (*b*) : Considering the amount of resources required to develop tourist resorts in India it is necessary to exploit the tourist potential. So, only course of action II follows.

8. (*b*) : Curtailing the staff strength would only hinder the work. Instead, the government can reduce wasteful expenditure and arrange to pay the salaries of its employees. So, only course of action II follows.

9. (*a*) : To check the spurt of serious power crisis in the region, the government should immediately step up the supply of steam coal to the thermal power plant. Setting up of hydraulic power plants is an entirely new proposal. So, only course of action I follows.

10. (*a*) : It is unlikely that the Central Government should keep a check an every state police armoury. But, a thorough investigation can bring out the culprits and the police department can continue to function with the force of efficient police officers. So, only course of action I follows.

11. (c) : If the Assembly and Parliamentary elections are held simultaneously, it will save a lot of time and finances. But this is possible only if the Constitution is amended for this purpose. If the Constitution cannot be amended, the President should be given the responsibility for deciding it. So, either courses of action follows.

12. (d) : Express highways are essential for rapid transportation via roads but other ways of rapid transportation are also available. These highways require exorbitant expenditure and it is not possible to build only such highways for a country's rapid development. So, neither course of action follows.

13. (*a*) : Course of action I is an instant response to the given statement. Any technical help is useless in presence of loopholes. So, only course of action I follows.

14. (*b*) : Inexperienced teachers must be given proper orientation so that they assess the papers rightly. Compensation is a reward for efficient work and not just hard work. So, only course of action II follows.

15. (b) : Retired professors are expected to be more experienced than the present employees. The management should consider their opinion for the re-structuring of the organisation. So, only course of action II follows.

16. (*e*) : Orientation programmes bring forth the internal problems of an organisation which can then be solved and also, the employees get motivated to come forward. So, both the courses of action follow.

17. (*e*) : If the parameters are faulty, they should be corrected and conservation is necessary for the prevention of resources for future. So, both the courses of action follow.

18. (*e*) : In order to maintain a systematic function libraries should be computerised and library staff should be trained in computer operations. So, both the courses of action follow.

19. (*b*) : At present the food processing industry needs the attention of the Government and not other industries. A regular supply of raw material can resolve the problem. So, only course of action II follows.

20. (*b*) : Poor performance can be improved if the reasons are apparent. Stepping out of such games is an immature decision. So, only course of action II follows.

25

Venn Diagrams

In these type of questions, diagrammatic representation presents a logical illustration of particular class or statements based on which the questions are asked. A clear view of the diagram makes the concepts clear for attempting such questions.

MULTIPLE CHOICE QUESTIONS

1. What is the number which is common to only two geometrical figures?

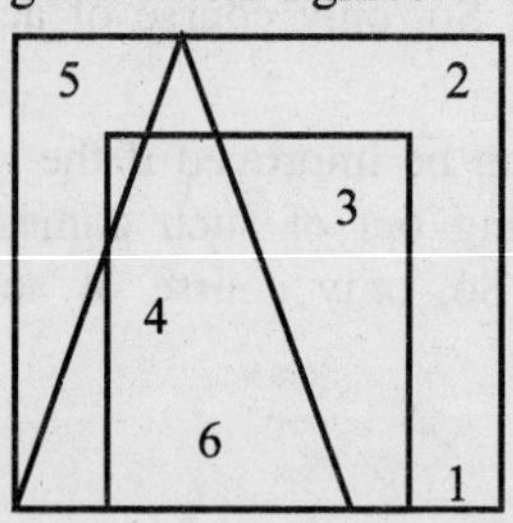

(*a*) 4 (*b*) 3
(*c*) 5 (*d*) 2

Directions (Qs. 2 and 3) : *In the following diagram, rectangle represents Hindi Announcers, circle represents English Announcers, square represents French Announcers, and triangle represents German Announcers.*

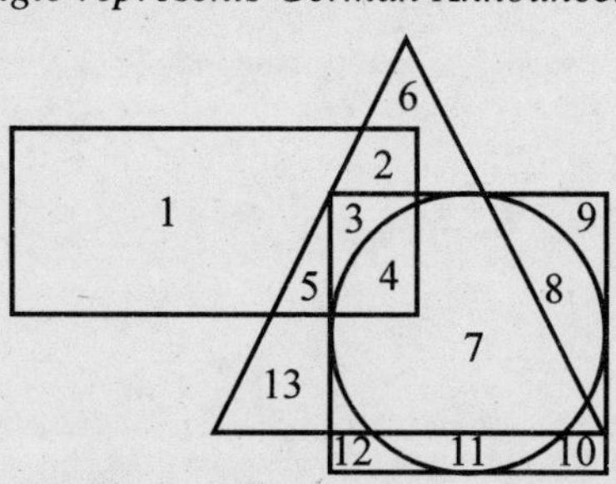

2. Which area represents those announcers who can present programmes in Hindi, French and German only?
(*a*) 1 (*b*) 2
(*c*) 3 (*d*) 4

3 Which area represents those announcers who can present programmes in French and English only?
(*a*) 7 (*b*) 9
(*c*) 11 (*d*) 13

Directions (Qs. 4 and 5) : *Study the diagram to answer these questions.*

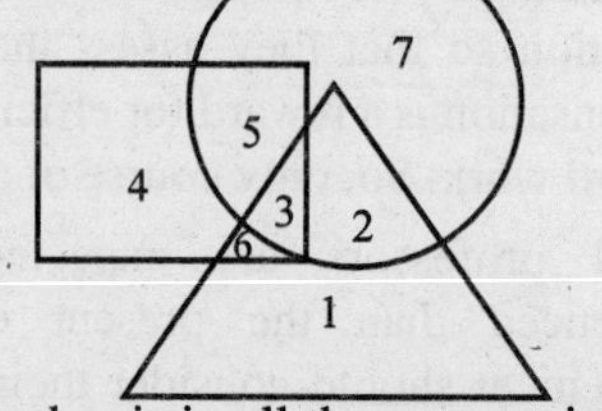

4. Which number is in all the geometrical figures?
(*a*) 5 (*b*) 6
(*c*) 2 (*d*) 3

5. Number 6 is in :
(*a*) Rectangle and triangle
(*b*) Circle and traingle
(*c*) Rectangle and circle
(*d*) Rectangle only

Directions (Qs. 6 to 9) : *In the following diagram*

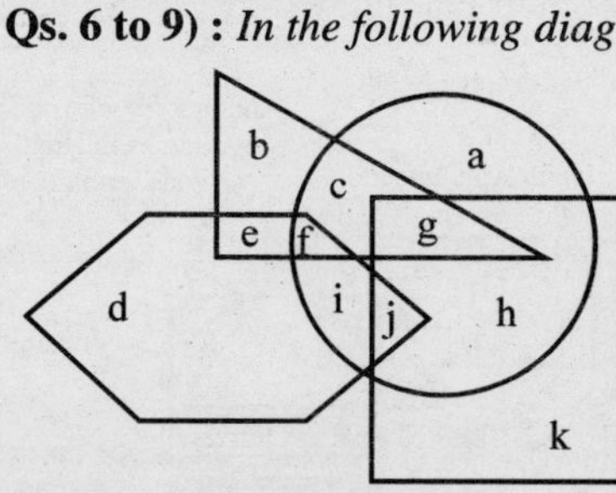

The Circle represents players
The Triangle represents outdoor games
The Hexagon represents indoor games and
The Square represents national level players
Study the diagram and answer the questions given below:

6. The letter in the section representing the players who play indoor games at national level is :
(*a*) f (*b*) i
(*c*) j (*d*) g

7. The letter representing the section of outdoor as well as indoor game players who do not play at the national level is :

(*a*) c (*b*) f
(*c*) e (*d*) i

8. The section representing national level players who do not play either outdoor or indoor games but still come under the category of players is :
(*a*) k (*b*) g
(*c*) c (*d*) h

9. Persons who play outdoor games but do not come under the category of players are represented in the section marked :
(*a*) b (*b*) c
(*c*) a (*d*) d

Directions (Qs. 10 to 13) : *Study the diagram given below.*

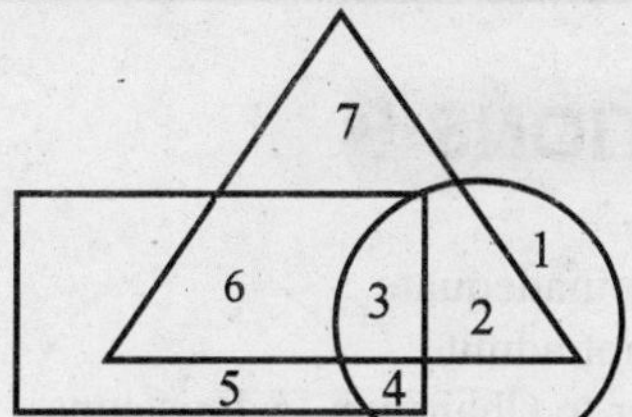

A college provides three different activities, students union represented by triangle, literary society represented by rectangle and social service league represented by circle.

10. Those who take part in both literary society and social service league but not in students union are represented by :
(*a*) 3 & 4 (*b*) 5 & 6
(*c*) 5 & 1 (*d*) 4

11. Those who take part in students union but not in social service league are represented by :
(*a*) 2 & 7 (*b*) 6 & 7
(*c*) 6 (*d*) 7

12. Those students who are members of literary society only and not of any other activity are represented by:
(*a*) 2 (*b*) 5
(*c*) 3 & 4 (*d*) 3

13. Those students who are members of all three groups are represented by :
(*a*) 2 (*b*) 3
(*c*) 4 (*d*) 6

EXPLANATORY ANSWERS

1. *(b)* :

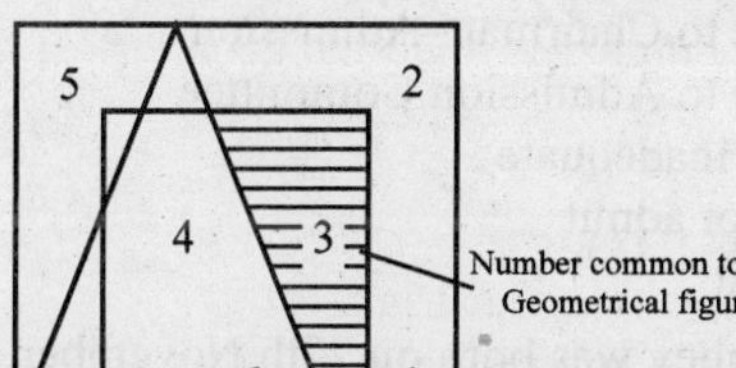

Note : Numbers 4 and 6 are common to all three geometrical figures.

2. *(c)*, **3.** *(c)* :

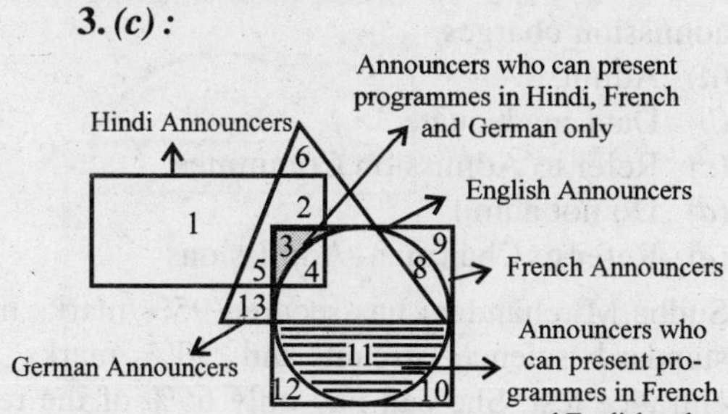

4. *(d)*, **5.** *(a)* :

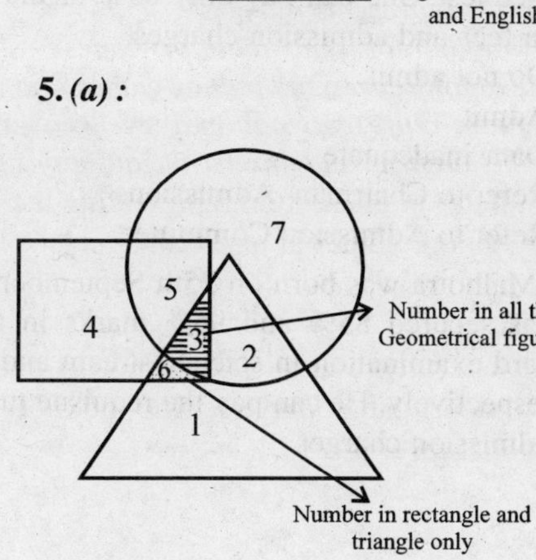

6. *(c)*, **7.** *(b)*, **8.** *(d)*, **9.** *(a)* :

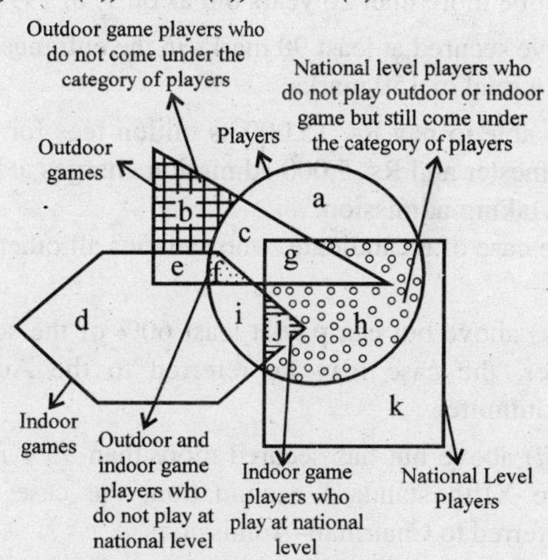

10. *(d)*, **11.** *(b)*, **12.** *(b)*, **13.** *(b)* :

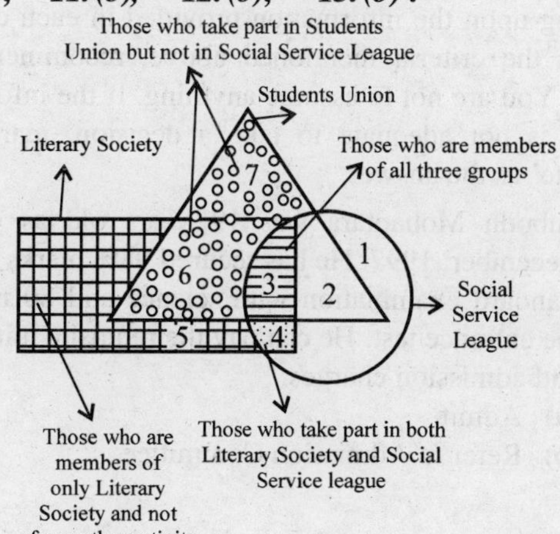

26

Decision Making

In this type of tests certain categories of information are given followed by certain criteria. Further, the candidates are given a few cause that needs to be analysed. They are required to take the right decision, from the given options, after comparing the information in each case with the given information and criteria.

MULTIPLE CHOICE QUESTIONS

Directions (Qs. 1 to 5) : *Read the following information carefully and answer the questions given below it.*

Following are the criteria for admitting a student in the first year engineering course in a college.

The student must—

(*i*) have passed XII standard examination in science with at least 80% marks.

(*ii*) not be more than 20 years old as on 1. 8. 1998.

(*iii*) have secured at least 90 marks in the entrance test out of a total of 150 marks.

(*iv*) be able to pay Rs. 15,000 as tuition fees for the first semester and Rs. 5,000 admission charges at the time of taking admission.

In the case of a candidate, who satisfies all other criteria except at—

(*a*) (*iv*) above but can pay at least 60% of the stipulated fees, the case may be referred to the Admission Committee.

(*b*) (*iii*) above but has secured more than 95% marks in the XIIth standard examination, the case may be referred to Chairman–Admissions.

You are given the following cases as on 1. 8. 1998. Depending upon the information provided in each case and based on the criteria mentioned above, recommend your decision. You are not to assume anything. If the information provided is not adequate to take a decision, mark 'Data inadequate' as the answer.

1. Subodh Mohaptara was 19 years old as on 20th December, 1997. He has secured 98% marks in XIIth standard examination with science and 80 marks in the entrance test. He can pay the requisite tuition fees and admission charges.
(*a*) Admit
(*b*) Refer to Admission Committee
(*c*) Data inadequate
(*d*) Do not admit
(*e*) Refer to Chairman–Admissions

2. Reema Jaiswal was born on 20th July, 1978. She has secured 85% and 75% marks in XIIth standard with science and entrance test respectively. She can pay the requisite tuition fees and admission charges.
(*a*) Refer to Chairman–Admission
(*b*) Refer to Admission Committee
(*c*) Data inadequate
(*d*) Do not admit
(*e*) Admit

3. Ashok Dubey was born on 27th November, 1978. He has secured 90% marks in the XIIth standard examination with science and 95 marks in the entrance test. He can pay Rs. 10,000 tuition fees and Rs. 3,500 admission charges.
(*a*) Admit
(*b*) Data inadequate
(*c*) Refer to Admission Committee
(*d*) Do not admit
(*e*) Refer to Chairman–Admissions

4. Sudha Mirchandani has secured 95% marks in XIIth standard science stream and 70% marks in the entrance test. She can pay only 65% of the requisite tuition fees and admission charges.
(*a*) Do not admit
(*b*) Admit
(*c*) Data inadequate
(*d*) Refer to Chairman–Admissions
(*e*) Refer to Admission Committee

5. Salil Malhotra was born on 25th September in 1979. He has secured 85% and 95% marks in the XIIth standard examination in science stream and entrance test respectively. He can pay the requisite tuition fees and admission charges.

(a) Admit
(b) Do not admit
(c) Refer to Chairman–Admissions
(d) Refer to Admission Committee
(e) Data inadequate

Directions (Qs. 6 to 12) : *Read the following information to answer the given questions :*

Following are the criteria for selecting candidates for Research Fellowship :

The candidate must—

(i) be a post–graduate with minimum of 65% marks.
(ii) not be more than 30 years as on 15. 10. 1993.
(iii) have at least 3 years research experience.
(iv) have diploma in Statistics.
(v) have secured at least 50% marks in the entrance test.
(vi) have finalised the topic for research.

However, in case a candidate who fulfills all other criteria except

(a) (iii) above, but has M. Phil degree, should be referred to Dean.
(b) (iv) above, should be referred to Chairman.
(c) (i) above, but has at least 55% marks in post–graduation and 65% in the entrance test, should be given fellowship.
(d) (v) above, but has at lest 45% marks, should be wait–listed.

Based on these criteria and information provided below, decide the course of action in each case. You are not to assume anything. If the data provided is not adequate to decide the given course of action, your answer will be "data inadequate". These cases are given to you as on 15. 10. 1993.

6. 26 years old, Mamta Deo is a post graduate with 58%. She has got four years' research experience and has finalised topic for research. She got 70% marks in diploma in Statistics and 54% marks in the entrance examination.
(a) Fellowship not to be granted
(b) Refer to Dean
(c) Refer to Chairman
(d) Grant Fellowship
(e) Data inadequate

7. Shushila Nair has 65% marks in her post–graduation and is M. Phil. She has 5 years research experience. She has secured 65% in the entrance test and has finalised the research topic. Her date of birth is 11. 8. 65.
(a) Grant fellowship (b) Wait–list
(c) Refer to Dean (d) Refer to Chairman
(e) Data inadequate

8. Ameer Khan has got 58% in the entrance test. He has 68% in his post-graduation and 55% in M. Phil. He has got diploma in Statistics and has also finalised the topic for research.
(a) Grant Fellowship (b) Refer to Dean
(c) Refer to Chairman (d) Wait-list
(e) Data inadequate

9. 28 years old Neetu Singh is M. Phil with 60% marks. She has secured 70% marks in entrance examination and has finalised the topic for research. She has also got diploma in Statistics.
(a) Grant Fellowship
(b) Fellowship not to be granted
(c) Refer to Chairman
(d) Refer to Dean
(e) Data inadequate

10. Sadhana Suman has got diploma in Statistics with 60% marks and post-graduation with 56% marks. Her date of birth is 12. 12. 1968. She has got one year research experience and is still doing her M. Phil. She has secured 65% marks in entrance exam and has also finalised the research topic.
(a) Grant Fellowship
(b) Fellowship not to be granted
(c) Refer to Dean
(d) Data inadequate
(e) Wait-list

11. Amar Sikka is 28 years old and has got 65% marks in post-graduation and 60% marks in M. Phil. He has got 65% marks in entrance examination and has finalised the topic for research. He also has a diploma in Statistics.
(a) Grant Fellowship
(b) Fellowship not to be granted
(c) Refer to Dean
(d) Refer to Chairman
(e) Data inadequate

12. Madan Paul has got 5 years' research experience and has finalised the research topic. He has got 56% marks in post-graduation. His date of birth is 15. 2. 1966. He has also got diploma in Statistics. He got 70% marks in the entrance test.
(a) Grant Fellowship
(b) Fellowship not to be granted
(c) Refer to Dean
(d) Data inadequate
(e) None of these

Directions (Qs. 13 to 19) : *Read the following information to answer the given questions :*

Following are the conditions for selecting candidates for Research Fellowship :

The candidate must—

(i) be a post-graduate with minimum of 60% marks.
(ii) not be more than 30 years as on 1. 9. 1993.
(iii) have at least 3 years' research experience.
(iv) have diploma in statistics.
(v) have secured at least 55% marks in the entrance test.
(iv) have finalised the topic for research.

However, in the case of a candidate who fulfills all other criteria except—

(*a*) (*iii*) above but has M. Phil degree, should be given fellowship.
(*b*) (*iv*) above should be referred to Dean.
(*c*) (*i*) above but has atleast 55% marks in post–graduation, should be wait-listed.
(*d*) (*v*) above but has atleast 50% marks, should be referred to Chairman.

Based on these criteria and information provided below, decide the course of action in each case. You are not to assume anything. If the data provided is not adequate to decide the given course of action, your answer will be "data inadequate". These cases are given to you as on 1. 9. 1993.

13. Vijay Gupta is 24 years old and has got 58% marks in the entrance examination. He has secured 63% marks in his post-graduation and 55% marks in M. Phil. He has got diploma in Statistics and has also finalised the topic for research.
(*a*) Grant Fellowship (*b*) Refer to Dean
(*c*) Refer to Chairman (*d*) Wait-list
(*e*) Data inadequate

14. Ajoy Dwivedi has secured 65% marks in the post-graduation and has secured 5 years research experience. He has secured 65% marks in the entrance test and has finalised the research topic. His date of birth is 11. 8. 1965.
(*a*) Grant Fellowship (*b*) Wait-list
(*c*) Refer to Dean (*d*) Data inadequate
(*e*) None of these

15. Madan Soren has got diploma in Statistics with 60% marks and post-graduation with 56% marks. His date of birth is 12. 12. 1968. He has got 1 year research experience and is still doing his M. Phil. He has secured 60% marks in the entrance exam and has also finalised the research topic.
(*a*) Grant fellowship
(*b*) Fellowship not to be granted
(*c*) Refer to Dean
(*d*) Data inadequate
(*e*) Wait-list

16. 26 years old Mamta Kulkarni is M. Phil with 60% marks. She has secured 70% marks in the entrance examination and has finali-sed the topic for research. She has also got diploma in Statistics.
(*a*) Grant Fellowship
(*b*) Fellowship not to be granted
(*c*) Refer to Chairman
(*d*) Refer to Dean
(*e*) Data inadequate

17. Chander Manoj has got 5 years' research experience and has finalised the research topic. He has got 56% marks in post-graduation and 60% marks in the entrance test. His date of birth is 15. 2. 1966. He has also got diploma in Statistics.
(*a*) Grant Fellowship
(*b*) Fellowship not to be granted
(*c*) Refer to Dean
(*d*) Data inadequate
(*e*) None of these

18. Raman Hooda is 28 years old and has got 65% marks in Post-graduation and 60% marks in M. Phil. He has got 65% marks in the entrance examination and has finalised the topic for research. He also has a diploma in Statistics.
(*a*) Grant Fellowship
(*b*) Fellowship not to be granted
(*c*) Refer to Dean
(*d*) Refer to Chairman
(*e*) Data inadequate

19. 26 years old Janaki Deo is a post-graduate with 58% marks. She has got four years research experience and has finalised the topic for research. She got 70% marks in the diploma in Statistics and 54% marks in the entrance examination.
(*a*) Fellowship not to be granted
(*b*) Refer to Dean
(*c*) Refer to Chairman
(*d*) Grant Fellowship
(*e*) Data inadequate

Directions (Qs. 20 to 27) : *Read the following information carefully and answer the questions given below :*

Following are the criteria of promotion from Junior Officer's Cadre to Senior Officer's Cadre in an organisation :

The candidate must—

(*a*) have completed atleast 5 years in the organisation.
(*b*) have secured 65% marks in the written test for promotion.
(*c*) have secured 60% marks in the Group Discussion.
(*d*) have secured 70% marks in the interview.
(*e*) have good record of his work performance.
(*f*) have good communication skill and get along well with his colleagues.
(*g*) not be more than 40 years and less than 30 years as on 1. 9. 93.
(*h*) have good academic record with an average of at least 65% marks.

However, in the case of a candidate who—

(*i*) satisfies all other conditions except (*d*) above but has secured 75% marks in the written test and 65% marks in he Group Discussion, the case is to be referred to the General Manager (Personnel)–GM(P) for the decision.

(*j*) satisfies all other criteria except (*h*) above but has secured an average of more than 60% marks, the case is to be referred to the Managing Director (MD) of the organisation.

Now read the information provided in the case of each candidate in each of the questions given below and decide on the basis of the information provided and based on the above conditions, which of the courses of action you would suggest. These cases are given as on 5. 9. 1993. (Remember you are not to assume anything which is not provided in the question). If complete information is not provided, the answer would be Data inadequate.

20. 38 years old Rajesh has secured 65% marks in the written test for promotion, 73% marks in interview and 62% marks in Group Discussion. He has good record of his work performance, good communication skills and gets along well with his colleagues. He has good academic record with an average of 61% marks and has completed 7 years in the organisation.

(*a*) Refer to GM(P) (*b*) Refer to MD
(*c*) Promote (*d*) Do not promote
(*e*) Data inadequate

21. 34 years old Sudha has secured 60% marks in the written test for promotion, 72% marks in inteview and 69% marks in Group discussion. She has good communication skill and gets along well with her colleagues. Her record of work performance is good and she has completed 6½ years in the organisation.

(*a*) Refer to MD (*b*) Do not promote
(*c*) Refer to GM(P) (*d*) Promote
(*e*) Data inadequate

22. Surekha has secured 70% marks in written test for promotion, 69% marks in Group Discussion and 72% marks in the interview. She has a good academic record with an average of 67% marks. She has good record of work performance and gets along well with her colleagues. She has good communication skill and has completed 7 years in the organisation.

(*a*) Do not promote (*b*) Promote
(*c*) Refer to GM(P) (*d*) Refer to MD
(*e*) Data inadequate

23. 39 years old Pawan has secured 66% marks in written promotion test and has a good academic record with an average of 62% marks. He has secured 65% marks in Group Discussion and 72% marks in interview. He has good record of his work performance, communication skill and gets along well with his colleagues. He has completed 6 years in the organisation.

(*a*) Do not promote (*b*) Refer to MD
(*c*) Refer to GM(P) (*d*) Promote
(*e*) Data inadequate

24. 34 years old Sudhir has secured 76% marks in the written test for promotion, 66% marks in Group Discussion and 67% marks in interview. He has good academic record with an average of 68% marks, good communication skill and gets along well with his colleagues. His work performance is good and he has completed 6 years in the organisation.

(*a*) Promote (*b*) Refer to MD
(*c*) Refer to GM(P) (*d*) Do not promote
(*e*) Data anadequate

25. 32 years old Manjula has good academic record with an average of 66% marks. She has secured 67% marks in Group Discussion, 75% marks in interview and 60% marks in written test for promotion. She has good communication skill and gets along well with her colleagues. She has good record of her work performance and has completed 7 years in the organisation.

(*a*) Do not promote (*b*) Refer to MD
(*c*) Promote (*d*) Refer to GM(P)
(*e*) Data inadequate

26. 31 years old Krishna secured 65% marks in written test for promotion, 72% marks in interview and 62% marks in Group Discussion. He has good academic record with an average of 67% marks and good communication skill. He has completed 9 years in the organisation. He gets easily annoyed and irritated with his colleagues and his record of work performance since the last two years is just average.

(*a*) Promote (*b*) Refer to GM(P)
(*c*) Refer to MD (*d*) Do not promote
(*e*) Data inadequate

27. 33 years old Neelima has secured 63% marks in Group Discussion, 71% marks in interview and 66% marks in written test for promotion. She hase good academic record with an average of 68% marks and has good communication skill. She gets along well with her colleagues and has good record of her work performance. She has completed 6 years in the organisation.

(*a*) Refer to GM(P) (*b*) Do not promote
(*c*) Promote (*d*) Refer to MD
(*e*) Data inadequate

Directions (Qs. 28 to 35) : *Read the following information carefully and answer the questions based on the given information. Following are the criteria for admitting a student in Medical course.*

The student must—

(*a*) have passed XIIth Std. Science examination with Biology amd have secured at least 60% marks.
(*b*) be of 18 years of age as on September 1, 1994.
(*c*) have obtained 70% marks in the entrance test.
(*d*) be able to pay Rs. 20,000 at the time of admission.

In the case of a candidate, who satisfies all other criteria except at—

I. (*c*) above, but has obtained 90% marks in the XIIth Std. Science examination, should be referred to the Principle.

II. (*d*) above, but can pay Rs. 10,000 at the time of admission, can be provisionally admitted.

You are given the following cases as on September 1, 1994. Depending upon the information provided in each case and based on the criteria mentioned above, recommend your decision. You are not to assume anything. Give answer

(*a*) if the students is to be admitted;
(*b*) if the student is not to be admitted;
(*c*) if the student is to be referred to the principle;
(*d*) if the student is to be admitted provisionally, and give answer;
(*e*) if the data are inadequate.

28. Mahesh Sagar has secured 75% marks in the entrance test and can pay admission fee of Rs. 15,000 at the time of admission. He has secured 60% marks in the XIIth Std. Science examination with Biology.

29. Sunil Agarwal has passed XIIth Std. Science examination with Biology and secured 95% marks. His date of birth is July 15, 1976. He has obtained 60% marks in the entrance test. He can pay the fee of Rs. 20,000 at the time of admission.

30. Jayaesh Ozha was 17 years old as on 11th September, 1993. He has secured 90% marks in his XIIth Std. Science examination with Biology. He has secured 75% marks in the entrance test and can pay the admission fee of Rs. 20,000.

31. Pramod Lohia was born on 5th April, 1973. He has secured 80% marks in the XIIth Std. Science examination with Biology and can pay admission fee of Rs. 20,000. He has secured 85% marks in the entrance test.

32. Usha Jethmalani has secured 68% marks in her XIIth Std. Examination in Science with Biology and has secured 75% marks in the entrance test. She was born on 20th October, 1975. She can pay the admission fee of Rs. 20,000.

33. Sudha Malhotra has secured 70% marks in the entrance test and 60% marks in her XIIth Std. Science Examinaion with Biology. She can pay the admission fee of Rs. 20,000 and was born on 7th November, 1978.

34. Rajesh Gupta has secured 90% marks in the XIIth Std. Science Examination with Biology and 60% marks in the entrance test. He was born on 3rd Obtober, 1973. He can pay the admission fee of Rs. 20,000.

35. Prasad Mehra was born on 4th July, 1976. He has secured 80% marks in the entrance test. He has secured 85% marks in the XIIth Std. Science Examination with Biology. He can pay the admission fee of Rs. 15,000.

Directions (Qs. 36 to 45) : *Study the following information to answer the questions.*

The following are the criteria for organising the Training Programme of an Institute, in different Hotels.

To organise the programme in Hotel Taj, the following criteria must be fulfilled :

(*i*) The Programme Coordinator should be of the rank of Deputy Director or Joint Director.
(*ii*) The Programme should be in one of these areas—HRD, Advertising, Computers or Statistics.
(*iii*) The duration of the Programme should not be more than seven days.
(*iv*) The fee per participant should not be less than Rs. 5,000/-
(*v*) The number of participants should be at least 50.

If all the other criteria are fulfilled except--

(*a*) the duration of the course is more than seven days—the programme is to be organised in Hotel Ritz.
(*b*) the Programme Coordinator is of the rank of Assistant Director, but the fee per participant is more than Rs. 7,000, the Programme will be organised in Taj.
(*c*) the number of participants being less than 50 but more than 30, the Programme will be organised in Hotel Sideways.
(*d*) the fee, per participant is less than Rs. 5000 but more than Rs. 3,500, the Programme should be organised in Hotel Sansy.
(*e*) the Programme is in other than the areas mentioned in (*ii*) above, but the Programme Coordinator is of Joint Director level, the Programme should be organised in Hotel Sideways.

Based on the above criteria and the information provided in each question, decide about the appropriate course of action. You are not to assume anything. If some information is not available to decide the course of action, the answer would be "data inadequate".

Give answer

(*a*) if the Programme is to be organised in Hotel Taj;
(*b*) if the Programme is to be organised in Hotel Sideways;
(*c*) if the Programme is to be organised in Hotel Ritz;
(*d*) if the Programme is to be organised in Hotel Sansy; and
(*e*) if the data are inadequate.

36. Prof. V. D. Dixit, Assistant Director, has offered four days' programme for 60 participants in the area of HRD. The fee per participant is Rs. 8,000.

37. Dr. Ashwani Dolke, Joint Director is organising five days' programme on Research Methodology. The total number of participants are 60 and the fee per participant is Rs. 7,000.

38. Dr. (Mrs.) Saroj Rao, Deputy Director of the institute wants to organise a programme for 55 candidates with a fee of Rs. 6,000 per participant. The duration of the course is four days and it is in the area of computers.

39. A Training Programme on Advertising is to be organised for 45 participants.The fee per participant is Rs. 6,000 and the duration of the course is six days.

40. A Training Programme on Statistics is proposed by the Deputy Director with Rs. 3,30,000 fee for 50 participants. The duration will be 8 days.

41. A five days' Training Programme by Joint Director is to be organised in the field of Statistics. The total fee for the Programme is Rs. 2,10,000 for 50 participants.

42. Nikhil Mukesh, Joint Director, is an expert on Computers. He is offering five days' programme on Mathematics for 60 participants. The fee per participant is Rs. 5,300.

43. A five days' Training Programme for 55 participants is to be organised by Shri Jacob Singh, Assistant Director. The fee per participant is Rs. 8,000.

44. Dr. Narayan Shiva, Deputy Director is offering a Programme on HRD with a total fee of Rs. 1,92,500 at the rate of Rs. 5,500 per participant. The duration of the course is five days.

45. Mrs. Shiela D'souza, Deputy Director has submitted a proposal to organise four days' Programme on Computers. The fee per participant would be Rs. 4,000 for 60 participants.

Directions (Qs. 46 to 55) : *Study the following information carefully and answer the questions given below it :*

Following are the criteria for allotment of residential accomodation by an organisation to its employees. The employee must—

(*a*) have completed at least 10 years in this unit of the organisation but of which at least 4 years in the supervisory cadre.

(*b*) not have more than five members in the family.

(*c*) have at least 5 years of service remaining, the retirement age being 58 years.

(*d*) not have his/her own house.

(*e*) not be staying in a house owned by his/her spouse.

In the case of an employee who—

I. satisfies all other criteria except at (*a*) above and joined the organisation as a manager, should be referred to the Director–Finance.

II. satisfies all other criteria except at (*c*) above and at present working as a Senior Manager, is to be referred to the Managing–Director.

III. has been transferred from another city, the condition (*a*) may be waived.

Based on the above criteria and on the basis of the information provided in case of each employee in each of the following questions, you have to decide whether or not, accommodation is to be provided or the case is to be referred to higher authority. Please note that that you are not to assume anything in the case of any of the employees. All the cases are presented before you on 31st July, 1992.

Mark answer (*a*) if the employee is to be provided with accommodation.

Mark answer (*b*) if the employee is not to be provided with accommodation.

Mark answer (*c*) if the complete information as regards all the above criteria are not provided in the question.

Mark answer (*d*) if the case is to be referred to the Director–Finance.

Mark answer (*e*) if the case is to be referred to the Managing Director.

46. Hemant Adhikari has joined the organisation four years back as an Officer. He stays with his wife and two children. Neither he nor his wife owns a house.

47. Medha Pradhan has joined the organisation in 1975 when she was twenty-eight years old as an Officer. She is unmarried and stays with her mother in a rented house. She does not own a house.

48. Geeta Gokhale has been working in the organisation for the last ten years, out of which five years as an Officer. Her date of birth is 8th June 1940. She stays with her husband and three children. Neither she nor her husband owns a house.

49. Vipin Ghade has recently been transferred from another office and was 53 years old as on 8th February, 1992. He has been working in the organisation for the last twenty years out of which ten years as a Manager. There are four members in his family. Neither he nor his wife owns a house.

50. Ashok Patkar has joined the organisation as a Manager in 1986 and was 30 years old as on 15th August, 1989. He has four members in his family. Neither he nor his wife owns a house.

51. Vikas Kulkarni was 52 years old as on 5th May, 1990. He has been working in the organisation for the last eight years out of which five years as an officer. He stays with his wife and one son. Neither he nor his wife owns a house.

52. Shirish Apte is a Senior Manager for the last four years and has been working in the organisation for the last twelve years. He has three members including himself in the family. Neither he nor his wife owns a house. He will be retiring from the organisation in 1998.

53. Sudha Karkare was promoted as a Supervisor five years back after joining the organisation in 1979 as a clerk. She stays with her husband and two children in a rented house and she does not own any house. She was 42 years old as on 23rd April, 1991.

54. Kubal Das stays with his wife and three children. He has been working in the organisation for the last twelve years out of which five years as a Manager. His date of birth is 2nd July 1945. He stays in a rented house.

55. Satish Aggrawal has been working in this unit for the last 15 years and has one son and two daughters along with his wife in the family. Neither he nor his wife owns a house. His date of birth is 20th January, 1950.

Directions (Qs. 56 to 63) : *Read the following information carefully and answer the questions given below :*

Following are the conditions for drawing a list of suitable candidates to be called for interview after a written test for recruitment is conducted for supervisory level persons for a reputed company providing software and services.

The candidate must—

(*a*) be holding a degree in basic science with 60% or above or engineering degree with 45% and above marks.

(*b*) have passed the written examination with 65% or above marks.

(*c*) be in the age group of 24 years to 30 years as on 1. 4. 1997.

(*d*) have experience in a computer company for a minimum period of 3 years after having obtained diploma in computer with 60% or above marks.

(*e*) be presently drawing a monthly salary of Rs. 8,000 and above.

In case of the applicant, who satisfies all other criteria except—

(I) at (*a*) above, be referred to the Manager (Recruitment).

(II) at (*b*) above but has obtained more than 75% marks in M.Sc. or Engineering degree, be referred to the Assistant General Manager (A.G.M.).

(III) at (*c*) above, be referred to the General Manager (G.M.).

Based on these criteria and information provided below, decide the course of action in each case. You are not to assume anything. If the data provided is not adequate to decide the given course of action, your answer will be "Data inadequate". The cases are given to you as on 1. 4. 1997.

56. Bimbadevi, aged 27 years, is a graduate in Science with diploma in computers and has been working for the last three and half years in a computer firm. Her present monthly salary is Rs. 10,500. She has passed her graduation, diploma and witten examination with 55%, 70% and 68% marks respectively.

(*a*) Do not call for interview
(*b*) Refer to G.M.
(*c*) Refer to Manager (Recruitment)
(*d*) Data inadequate
(*e*) Call for interview

57. Kasam Bhure did his B. Sc. with 70% marks followed by diploma in computers with 69% marks. He is employed from June 1993 in a computer firm with a salary of Rs. 9,700. He was born in Octorber 1973. He has passed the written examination with 72% marks.

(*a*) Call for interview
(*b*) Refer to Manager (Recruitment)
(*c*) Data inadequate
(*d*) Refer to G.M.
(*e*) Do not call for interview

58. Kamalnath has been working in computer company for the last 6 years after completing his diploma in computers with 62% marks. He has passed his Engineering degree and written examination with 55% and 68% marks respectively. He has completed 28 years of age in September 1995.

(*a*) Data inadequate (*b*) Do not call for interview
(*c*) Call for interview (*d*) Refer to G.M.
(*e*) Refer to A.G.M.

59. Niranjan Singh did his diploma in Computers at the age of 22 years after completing his B.Sc. Immediately after completing his dimploma, he got job in a computer company and is employed for the last four years with a salary of Rs. 10,500. He has got more than 65% marks in all the examinations including the written examination.

(*a*) Data inadequate
(*b*) Refer to Manager (Recruitment)
(*c*) Do not call for interview
(*d*) Refer to A.G.M.
(*e*) Call for interview

60. Deepa Sahi is working in supervisory capacity for the last 4 years in a computer company after having completed her Engineering degree with 55% and diploma in computers with 70% marks. She has secured 72% marks in the written examintion. Her date of birth is 15 January, 1967 and her present salary is Rs. 10,000.

(*a*) Data inadequate
(*b*) Call for interview
(*c*) Do not call for interview
(*d*) Refer to Manager (Recruitment)
(*e*) Refer to G.M.

61. Mina Khan completed her B.Sc. at the age of 21 years, completed her one year's diploma in computers immediately and got a job from June 1994. She draws a monthly salary of Rs. 11,000. She has cleared all her examinations including written examination with a minimum of 68% marks.

(*a*) Refer to G.M.
(*b*) Refer to A.G.M.
(*c*) Refer to Manager (Recruitment)
(*d*) Do not call for interview
(*e*) Data inadequate

62. Jemi Desouza, born on 12. 2. 1968, is employed in a computer company for the last five years and is presently drawing a monthly salary of Rs. 9,500. He has successfully completed his degree in Engineering with 64% marks, diploma in computers and written examination with 65% and 60% marks respectively.

(*a*) Do not call for interview
(*b*) Call for interview
(*c*) Data inadequate
(*d*) Refer to G.M.
(*e*) Refer to Manager (Recruitment)

63. Anand Gupta has done his Master's degree in Science and then completed his diploma in Computers with 64% marks from a reputed institute. For the last 4 years, he has been working in a computer company. His age is 27 years and his present salary is Rs. 8,500. He obtained 68% marks in the written examination.
(*a*) Refer to Manager (*b*) Data inadequate
(*c*) Call for interview (*d*) Do not call for interview
(*e*) Refer to A.G.M.

Directions (Qs. 64 to 70) : *Trinity Health Club gives 40 per cent concession in monthly fees to those who satisfy anyone of the following criteria :*

(*i*) All children in the age group of 5 to 12 years who come from low income group and produce income and birth certificates.
(*ii*) All girls/women up to 25 years of age who represent their school or college in sports and games.
(*iii*) Unemployed married women in the age group of 35 to 50 years.
(*iv*) Senior citizens of age 50 years and above.
(*v*) All physically handicapped persons who produce relevent medical certificates.
(*vi*) All people who donate blood once a year or social workers/volunteers of registered social/cultural organisations.
(*vii*) Servicemen/Ex-Servicemen of Police/Defence Forces.
(*viii*) Sons and daughters of donors/founder members of the Trinity Health Club.

Decide in each of the following whether the person is 'eligible', find out the category/ies which is/are applicable to him/her for taking this decision and indicate your answer accordingly. The number indicated in the brackets in the choice indicates the category/ies mentioned above. Do not assume any information which is not given.

64. Bela, a 9 year old school girl and daughter of a merchant, has participated in dance and music competitions. Her father, who is 55 years old, has given large donations for organising blood camps. Her uncle is one of the founders of Trinity Club.
(*a*) Eligible under (*i*) only
(*b*) Eligible under (*i*), (*ii*), (*iv*) and (*viii*)
(*c*) Eligible under (*ii*) and (*viii*) only
(*d*) Eligible under (*i*), (*ii*) and (*vi*) only
(*e*) Not eligible

65. Nina Gupta, a college-going married unemployed woman of 22 years, arranges donations for physically handicapped students. She has represented her school at the State–level elocution competition.
(*a*) Eligible under (*viii*) only
(*b*) Not Eligible
(*c*) Eligible under (*ii*) only
(*d*) Eligible under (*ii*), (*iii*) only
(*e*) Eligible under (*ii*), (*iii*), (*vii*) and (*viii*)

66. Mridula is a young married and employed lady police officer. Her husband is a captain in the army. She is a sportswoman and represents police forces in national level competitions. She occasionally donates blood for her sick mother who is 55 years old.
(*a*) Eligible under (*vii*) only
(*b*) Eligible under (*ii*) and (*iii*) only
(*c*) Not eligible
(*d*) Eligible under (*ii*), (*iii*) and (*iv*) only
(*e*) Eligible under (*vi*) and (*vii*) only

67. Mrs. Ratna, wife of Major Ravi Singh, is active in evening parties and clubs. She encourages people to donate to the Trinity club. She is 32 years old and unemployed.
(*a*) Not eligible
(*b*) Eligible under (*iii*) only
(*c*) Eligible under (*viii*) only
(*d*) Eligible under (*vii*) only
(*e*) Eligible under (*iii*), (*vi*), (*vii*) and (*viii*)

68. Mr. Rane established a registered trust for social work after his retirement from army at the age of 48 years. Since then he has been actively carrying out social work for the last eight years. His son is the founder member of "Trinty".
(*a*) Eligible under (*iv*) only
(*b*) Not eligible
(*c*) Eligible under (*iv*) and (*vi*) only
(*d*) Eligible under (*iv*), (*vi*) and (*vii*)
(*e*) Eligible under (*iv*), (*vi*), (*vii*) and (*viii*)

69. Rita is an 11 years old school girl who can produce her birth certificate. Her parents are senior citizens and regular blood donors. Though a medically–certified physically handicapped person, she represents her school in national competition for the physically handicapped.
(*a*) Eligible under (*i*), (*ii*) and (*iv*)
(*b*) Eligible under (*ii*) and (*v*) only
(*c*) Not eligible
(*d*) Eligible under (*iv*) and (*v*) only
(*e*) Eligible under (*i*), (*ii*) and (*vi*) only

70. Dipti who represents her school in Badminton is the 14–year old daughter of physically handicapped parents having very low income. Her mother aged 40 years is unemployed. She has birth certificate and low income group certificate.
(*a*) Eligible under (*i*) only
(*b*) Eligible under (*ii*) only
(*c*) Eligible under (*i*), (*ii*) and (*v*)
(*d*) Eligible under (*i*), (*ii*), (*iii*), (*v*) only
(*e*) Not eligible

Directions (Qs. 71 to 80) : *A reputed educational foundation is to select meritorious graduate students for the award of scholarships for higher studies abroad who fulfill the following criteria :*

(*i*) Age : More than 20 years but less than 25 years, as on 1. 4. 1997.
(*ii*) Minimum Percentage of marks : B. Sc.–60% or BA–50% or B. Com–55%.
(*iii*) Prizes/awards : Should have won at least one either in Essay, Debate or Sports.
(*iv*) Annual Family Income : Not to be more than Rs. 1,00,000.
(*v*) Domicile : Must be staying in India for the last 15 years or above, as on 1. 4. 1997.

Concessions : One has to fulfill the above conditions. However, certain concessions are available to the following categories :

(*a*) Wards of Farmers—relaxation in the maximum age limit upto five years.
(*b*) SC/ST applicants—relaxation in marks upto 5%.

Consideration :

(*c*) Not fulfilling criterion (*iii*) above—the case may be referred to the Secretary of the Foundation.
(*d*) Not fulfilling criterion (*iv*) above—the case may be referred to the Chairman of the Foundation.

You have to study the information provided for a candidate in each of the following questions and decide the course of action.

Give answer—

(*a*) If the candidate is to be selected.
(*b*) If the candidate is to be selected under any concession.
(*c*) If the case is to be referred to the Secretary.
(*d*) If the case is to be referred to the Chairman, and
(*e*) If the candidate is not to be selected.

Note : The candidate is to be considered "Not to be selected", if any required information (creteria *i* to *v*) is not provided.

71. Hailing from a poor scheduled tribe family living near Rourkela since 1949, 21 year old Ratan got his B. A. degree with 46% marks. His father, a school teacher and sole earner in the family has a monthly income of Rs. 4,200. Ratan won the first prize in a District Level Essay competition.

72. Family of 22 year old Antony is settled at Cochin since 1969. He has 59% marks in B. A. and had won a gold medal in the All India Lions Club Essay Comeption. His father—a farmer— has an annual income of Rs. 79,000.

73. Yogesh, who will be 23 years old on 1. 1. 1999, passed B. Com. with 51 per cent marks. His father, a Government employee, has been working at Srinagar since 1979 whose present monthly salary is Rs. 8,200. Yogesh got the 3rd prize in the State Essay Championship.

74. Twenty-three years old Mandakini has 58% marks in B.Com. and had won the 1st prize in her final year Debate Competition. Her annual family income is about Rs. 1.25 lakh. Her father is settled in Shillong for the last 21 years.

75. Mahasweta was born on 4. 3. 1979 and recently passed B. Sc. with 63% marks. Her father, a lecturer at a Junior College at Ranchi since 1978, has an annual family income of Rs. 82,000. She won a third prize in the State Athletic Championship.

76. Irfan, son of a Foreman in Indonesia, came to India with his family in 1970 and since then settled at Aligarh. He got a first prize in an International Debate Competition. He is a graduate in science with 69% marks. His annual family income is Rs. 96,000.

77. 24 years old Jagdeep has 57% marks in B. Com. and has bagged the first prize in College Music Competition. His father, an industrial workman and sole earner in the family, has a monthly income of Rs. 7,800 and is settled in Nasik since 1958.

78. 26 years old Rukmini has been staying with her father at Chandbali village in Tripura since her birth. Her father, a farmer, has an annual income of Rs. 78,000. She obtained 59% marks in B. Com. and had won a prize in College Sports Competition.

79. Born on 21. 2. 1976, Rachna completed her B. Sc. in March 1996. Her father is a farmer settled in Madhya Pradesh for the last 32 years. She is a State Champion in Badminton. Her annual family income is Rs. 91,000.

80. Chelaram, son of a farmer who has been staying in Haryana for the last five decades, passed B.A. with 51% marks. He has got a second prize in the District Level Weightlifting Championship. His date of Birth is 3. 9. 1974 and his father's annual income is Rs. 83,000.

EXPLANATORY ANSWERS

1. (*e*) : Criterion *(iii)* is not satisfied but Criterion *(b)* gives the decision.

2. (*d*) : Both Criterion *(iii)* and Criterion *(b)* are not satisfied.

3. (*c*) : Criterion *(iv)* is not satisfied but Criterion *(a)* gives the decision.

4. (*c*) : Criterion *(ii)* is not given.

5. (*a*)

6. (*a*) : Both Criterion *(i)* and Criterion *(c)* are not satisfied.

7. (*d*) : Criterion *(iv)* is not given but Criterion *(b)* gives the decision.

8. (*e*) : Both Criteria *(ii)* and *(iii)* are not given and Criterion *(a)* alone cannot give the decision.

9. (*e*) : Both Criterion *(i)* and Criterion *(iii)* are not given and criterion *(a)* alone cannot give the decision.

10. (*b*) : Both Criterion *(iii)* and Criterion *(a)* are not satisfied.

11. (*c*) : Criterion *(iii)* is not satisfied but Criterion *(a)* gives the decision.

12. (*a*) : Criterion *(i)* is not satisfied but Criterion *(c)* gives the decision.

13. (*a*) : Criterion *(iii)* is not satisfied but Criterion *(a)* gives the decision.

14. (*c*) : Criterion *(iv)* is not satisfied but Criterion *(b)* gives the decision.

15. (*b*) : Both Criterion *(iii)* and Criterion *(a)* are not satisfied.

16. (*e*) : Criterion *(i)* is not given and Criterion *(c)* is not satisfied.

17. (*e*) : Criterion *(i)* is not satisfied but Criterion *(c)* gives the decision of "wait–list." This option is not given.

18. (*a*) : Criterion *(iii)* is not satisfied but Criterion *(a)* gives the decision.

19. (*a*) : Both Criterion *(i)* and Criterion *(iii)* are not satisfied, Criterion *(c)* and Criterion *(d)* gives different decisions.

20. (*b*) : Criterion *(h)* is not satisfied but Criterion *(j)* gives the decision.

21. (*b*) : Criterion *(b)* is not satisfied.

22. (*e*) : Criterion *(g)* is not given.

23. (*b*) : Criterion *(h)* is not satisfied but Criterion *(j)* gives the desicion.

24. (*c*) : Criterion *(d)* is not satisfied but Criterion *(i)* gives the decision.

25. (*a*) : Criterion *(b)* is not satisfied.

26. (*d*) : Both Criterion *(e)* and Criterion *(f)* are not satisfied.

27. (*c*)

28. (*e*) : Criterion *(a)* is not given.

29. (*c*) : Criterion *(c)* is not satisfied but Criterion *(I)* gives the decision.

30. (*b*) : Criterion *(b)* is not satisfied.

31. (*a*) **32. (*a*)**

33. (*b*) : Criterion *(b)* is not satisfied.

34. (*c*) : Criterion *(c)* is not satisfied but Criterion *(I)* gives the decision.

35. (*d*) : Criterion *(d)* is not satisfied but Criterion *(II)* gives the decision.

36. (*a*) : Criterion *(i)* is not satisfied but Criterion *(b)* gives the decision.

37. (*b*) : Criterion *(ii)* is not satisfied but Criterion *(e)* gives the decision.

38. (*a*)

39. (*e*) : Criterion *(i)* is not given.

40. (*c*) : Criterion *(iii)* is not satisfied but Criterion *(a)* gives the decision.

41. (*d*) : Criterion *(iv)* is not satisfied but Criterion *(d)* gives the decision.

42. (*b*) : Criterion *(ii)* is not satisfied but Criterion *(e)* gives the decision.

43. (*e*) : Criterion *(ii)* is not given.

44. (*b*) : Criterion *(v)* is not satisfied but Criterion *(c)* gives the decision (No. of participants is 1, 92, 500 ÷ 5,500, *i.e.,* 35)

45. (*d*) : Category *(iv)* is not satisfied but category *(d)* gives the decision.

46. (*c*) : Criterion *(c)* is not given and Criterion *(a)* is not satisfied.

47. (*a*) **48. (*a*)**

49. (*b*) : Both Criterion *(c)* and Criterion *(II)* are not satisfied.

50. (*a*)

51. (*b*) : Both Criterion *(a)* and Criterion *(c)* are not satisfied.

52. (*e*) : Criterion *(c)* is not satisfied but Criterion *(II)* gives the decision.

53. (*a*)

54. (*c*) : Criterion *(d)* is not given.

55. (*c*) : Criterion *(a)* is not given.

56. (*c*) : Criterion *(a)* is not satisfied but Criterion *(I)* gives the decision.

57. (*d*) : Criterion *(c)* is not satisfied but Criterion *(III)* gives the decision.

58. (*a*) : Criterion *(e)* is not given.

59. (*e*)

60. (*e*) : Criterion *(c)* is not satisfied but Criterion *(III)* gives the decision.

61. (*d*) : Both Criterion *(a)* and Criterion *(d)* are not satisfied and Criterion *(I)* alone cannot give the decision.

62. (*a*) : Both Criterion *(b)* and Criterion *(II)* are not satisfied.

63. (*b*) : Neither Criterion *(a)* nor Criterion *(II)* is given.

64. (*c*) : Both Criterion *(ii)* and Criterion *(viii)* are satisfied.

65. (*c*) : Only Criterion *(ii)* is satisfied.

66. (*e*) : Both Criteria *(vi)* and *(vii)* are satisfied.

67. (*a*) : None of the Criterion is satisfied.

68. (*d*) : Criteria *(iv), (vi)* and *(vii)* are satisfied.

69. (*b*) : Both Criterion *(ii)* and Criterion *(v)* are satisfied.

70. (*b*) : Only Criterion *(ii)* is satisfied.

71. (*b*) : Criterion *(ii)* is not satisfied but Criterion *(b)* gives the decision.

72. (*a*)

73. (*e*) : Criterion *(ii)* is not satisfied and Criterion *(b)* is not given.

74. (*d*) : Criterion *(iv)* is not satisfied but Criterion *(d)* gives the decision.

75. (*e*) : Criterion *(i)* is not satisfied.

76. (*e*) : Criterion *(i)* is not given.

77. (*a*)

78. (*b*) : Criterion *(i)* is not satsified but Criterion *(a)* gives the decision.

79. (*e*) : Criterion *(ii)* is not given and so, remain unsatisfied.

80. (*a*)

27 Distinguishing Arguments

In this type of reasoning one has to decide upon the forcefulness of an argument. One should keep in mind that the argument should not be (*i*) an assumption; (*ii*) generalised (it should be specific); (*iii*) ambiguous; (*iv*) hidden as an implication; (*v*) an example or quotation. Also, the argument should be factually correct and in conformity with the prevailing ideas of truth.

In making decisions about important questions, it is desirable to be able to distinguish between 'strong' arguments and 'weak' arguments so far as they relate to the questions. 'Strong' arguments must be both important and directly related to the question. 'Weak' arguments may not be directly related to the question and may be of minor importance or may be related to the trivial aspect of the question.

MULTIPLE CHOICE QUESTIONS

Directions : *Each question below is followed by two arguments numbered I and II. You have to decide which of the given arguments is a 'strong' argument and which is a 'weak' argument. Give answer :*

(*a*) If only argument I is strong.
(*b*) If only argument II is strong.
(*c*) If either I or II is strong.
(*d*) If neither I nor II is strong.
(*e*) If both I and II are strong.

1. Should higher education be restricted only to the most deserving ones?
Arguments :
I. No, it is against the principle of democracy.
II. Yes, only deserving people are entitled in this country.

2. Should strikes in the field of education be banned?
Arguments :
I. Yes, it is against professional ethics.
II. Yes, it affects the students adversely.

3. Should young entrepreneurs be employed?
Arguments :
I. Yes, that will help in industrial development of the country.
II. Yes, that will reduce the burden on the employment market.

4. Guides published by private agencies should be banned.
Arguments :
I. Yes, because only then the students will start reading textbooks.
II. No, because these are of immense help to the weak students in the eleventh hour.

5. Should India manufacture atom bombs?
Arguments :
I. Yes, it is imperative to protect the sove-reignty and integrity of the country.
II. No, this will create imbalance in the power of nations in this region.

6. Should open book systems be introduced in examinations?
Arguments :
I. Yes, because it will avoid mass copying.
II. No, because then all students will get 100% marks.

7. Should smoking be prohibited?
Arguments :
I. Yes, it is wrong to smoke away millions of good money.
II. No, it will throw thousands of workers in the tobacco industry out of employment.

8. Should there be a uniform syllabus for all universities in India?
Arguments :
I. Yes, why not?
II. No, in a vast country like India, variation is necessary to suit local needs.

9. Should religion be banned?
Arguments :
I. Yes, it develops fanaticism in people.
II. No, religion binds people together.

10. Should educated people work in villages?

Arguments :

I. Yes, because they can revolutionize agri-culture and revamp rural atmosphere.

II. No, the educated should be employed in cities.

11. Should there be complete ban on strike by Government employees in India?

Arguments :

I. Yes, this is the only way to teach discipline to the employees.

II. No, this deprives the citizens of their democratic rights.

12. Should India have no military force at all?

Arguments :

I. No, other countries in the world do not believe in non-violence.

II. Yes, many Indians believe in non-violence.

13. Should India encourage exports when most things are insufficient for internal use itself?

Arguments :

I. Yes, we have to earn foreign exchange to pay for our imports.

II. No, even selective encouragement would lead to shortages.

14. Should total ban be put on trapping wild animals?

Arguments :

I. Yes, trappers are making a lot of money.

II. No, ban on hunting and trapping are not effective.

15. Should fashionable dresses be banned?

Arguments :

I. Yes, fashions keep changing and hence consumption of cloth increases.

II. No, fashionable clothes are a person's self-expression and therefore his/her fundamental right.

16. Should luxury hotels be banned in India?

Arguments :

I. Yes, they are places from where international criminals operate.

II. No, affluent foreign tourists will have no place to stay.

17. Should cutting of trees be banned altogether?

Arguments :

I. Yes, it is very necessary to do so to restore ecological balance.

II. No, a total ban would harm timber based industries.

18. Should education upto graduation be made compulsory for all?

Arguments :

I. Yes, it is very desirable.

II. No, what is the use?

19. Should telecasting feature films be stopped?

Arguments :

I. Yes, young children are misguided by the feature films.

II. No, this is the only way to educate the masses.

20. Should taxes on colour television be further increased?

Arguments :

I. Yes, colour television is a luxury item and only rich people buy them.

II. No, television are bought by the poor too.

21. Should voting age in India be lowered to 16?

Arguments :

I. Yes, why should it not be so in India when in some other countries the voting age is 16 ?

II. No, it will be foolish to do so when even the marriage age for boys is 21.

22. Should religion be taught in our schools?

Arguments :

I. Yes, do the parents not wish to develop their wards into mature individuals?

II. No, how can one dream of such a step when we want the young generation to fulfill its role.

23. Should family planning be made compulsory in India?

Arguments :

I. Yes, looking at the miserable conditions in India, there is no other go.

II. No, in India there are people of various religions and family planning is against the tenets of some of the religions.

24. Should correspondence courses at university level in India be discontinued?

Arguments :

I. Yes, only interactoin between teachers and students and among students while learning helps total development of personality at the University level.

II. No, such a step of discontinuation is the craze of the elite, who have no knowledge of what is meant by poverty.

25. Should hunting be banned?

Arguments :

I. Yes, it has been proved to be a definite environmental hazard.

II. No, what will hunters do?

26. Should those who receive dowry, despite the law prohibiting it, be punished?

Arguments :

I. Yes, because those who violate the law must be punished.

II. No, dowry system is firmly rooted in the society since time immemorial.

27. Should religion be taught in our schools?
Arguments :
I. No, ours is a Secular State.
II. Yes, teaching religion helps inculcate moral values among children.

28. Should octroi be abolished?
Arguments :
I. Yes, it will eliminate an important source of corruption.
II. No, it would adversely affect government revenue.

29. Should promotion of personnel in an organisa-tion be made only on the basis of seniority?
Arguments :
I. Yes, otherwise senior persons will become disinterested in their work.
II. No, it does injustice to the really deserving persons.

30. Should colleges be given the status of a University?
Arguments :
I. Yes, managements of colleges will appreciate it.
II. No, colleges even in Western countries have not yet been given such a status.

31. Should workers be allowed to participate in the management of factories?
Arguments :
I. Yes, it is one of the demands of workers.
II. No, the management will oppose this idea.

32. .Should competitive examinations for electing candidates for jobs consist of objective tests only?
Arguments :
I. Yes, the assessment of objective tests is reliable.
II. No, the number of questions to be answered in the objective tests is always very large.

33. Should military training be made compulsory for all?
Arguments :
I. Yes, it is very desirable.
II. No, what is the use?

34. Should India stop nuclear programme?
Arguments :
I. Yes, China desires so.
II. No, the nation must always remain up-to-date in its defence preparedness.

35. Should killing of wild animals be banned altogether?
Arguments :
I. Yes, it is very necessary to restore ecological balance.
II. No, a total ban would harm the leather industry.

36. Are modern girls really against dowry?
Arguments :
I. Yes, they want to end the age-old atrocities caused by dowry.
II. No, they think dowry is their lawful right and a means to their status in society.

37. Can all scientists be good singers?
Arguments :
I. Yes, because musical notes are scientifically based.
II. No, because good singers are gifted with an inborn talent for singing.

38. Inter-caste marriages will lead to National Integration.
Arguments :
I. Yes, this will lead to intermingling of various castes and thus casteism will disappear.
II. This will lead to a lot of male-adjustment of married couples, especially when the rituals and dogmas of different castes differ a good deal.

39. Simple living and high thinking is a good principle of life.
Arguments :
I. This makes one a great person.
II. All great men follow this principle.

40. Should number of public holidays be reduced?
Arguments :
I. No, there is going to be hue and cry.
II. Yes, it must be done as early as possible.

41. Should all the schools be made co-educational?
Arguments :
I. Yes, otherwise how else do we promote education of girl child. More girls should be enrolled in the co-educational schools.
II. No, in a conservative society like ours, parents have many reservations against co-education and it may hamper education of girls.

42. Should tabloids be banned?
Arguments :
I. No, how can such a stupid thing be ever even thought of.
II. Yes, no problem. We have many good newspapers.

43. Should students have their union?
Arguments :
I. Yes, because they shall fight for the cause of the students.
II. No, because it gives a chance to the politicians to play their dirty game through the unions.

44. Should wild life be protected?
Arguments :
I. Yes, otherwise some species could disappear.
II. No, the wild beasts are a menace to the neighbouring villages, so they should be-destroyed.

45. Should Government stop spending huge amounts of money on international sports?

Arguments :

I. Yes, this money can be utilised for the upliftment of the poor.

II. No, sportspersons will be frustrated and will not get international exposure.

46. Should all the transport corporations be handed over to the private organisations?

Arguments :

I. Yes, there will be significant change in the quality and punctuality of services.

II. No, there would not be job secuity for the employees at all the levels.

47. Sanskrit should become the national language of the country.

Arguments :

I. It is our cultural heritage.

II. It is understood by all the Indians.

48. Economics has been called the 'dismal science', because :

Arguments :

I. Economics is not a science.

II. Political economists have been predic-ting that ever increasing population will ultimately lead to catastrophic conditions.

49. Should violence be depicted in films?

Arguments :

I. Yes, because it makes the films very interesting and useful.

II. No, because this spreads violence.

50. India is a poor country and therefore should not try to spend too much moeny on launching of rockets and satellites.

Arguments :

I. Yes, launching them successfully does not in any way help the country.

II. It drains away the meagre resources of the country.

EXPLANATORY ANSWERS

1. (*a*) : In a democracy, every person has equal rights, so argument I is strong.

2. (*b*) : Argument II is directly related to the statement. Argument I is weak because strike is a big power used by employees for their defence.

3. (*e*) : Both argument I and argument II are strong and directly related to the statement.

4. (*b*) : Argument I is based on assumption but argument II follows the statement because guides are helpful for short-time preparation.

5. (*a*) : Argument I is strong because it is directly related to the statement and argument II is weak as it is based on assumption.

6. (*d*) : Both argument I and argment II are based on assumption.

7. (*d*) : Argument I is ambiguous and argument II is based on assumption.

8. (*b*) : Argument I is the implication of the statement but considering the diversity factor in India, assumption II is strong.

9. (*b*) : Only argument II is strong. Religion does not develop fanaticism, but it will be mis-used by the people.

10. (*a*) : Argument I is strong as education would not be wasted by doing working in village, but it will be used for all round development.

11. (*b*) : Argument I is based on assumption and argument II is important and directly related to the statement.

12. (*d*) : Neither argument I nor argument II is strong as both are based on assumptions.

13. (*a*) : Argument II is based on assumption but argument I is strong.

14. (*d*) : Neither argument I nor argument II is strong.

15. (*b*) : Nothing is important about the 'consumption of clothes' in the statement. However, argument II is important and strong because the manner of dressing is a persons's own choice.

16. (*b*) : Argument I is not strong. Argument II is related to the statement.

17. (*e*) : Both argument I and argument II are strong and directly related to the statement.

18. (*d*) : Argument I means that all students want to be graduates, which is wrong, and argument II is pessimistic approach. As such, neither argument I nor argument II is strong.

19. (*d*) : Neither argument I nor argument II is strong. Argument I is not strong because films also educate masses. Argument II is weak because films are not the only way to educate the masses.

20. (*d*) : Neither argument I nor argument II is strong. Television which is a means of entertainment is bought by those who can afford it, and not only by the rich people. A really poor person cannot afford to buy a television not even at the present price.

21. (*d*) : Neither the argument I nor argument II is strong. It is no sound argument that other countries have done so. Similarly, marriage age is not directly related to age of voting.

22. (*d*) : Neither of the arguments is strong because maturity and playing a successful role has no connection with teaching of religion in our schools.

23. (*a*): Only argument I is strong. Planning is necessary to avoid chaos.

24. (*a*) : If the objective of education is total development of personality, then argument I is sound.

25. (*a*) : Only argument I is strong. Hunting has only been a pleasurable sport. Hunters can do something better.

26. (*a*) : Only argument I is strong. Social evils must be uprooted and not carried on.

27. (*a*) : Only argument I is strong. Teaching religion may lead to communal differences and disparities.

28. (*e*) : Both argument I and argument II are strong. On one hand the illegal method of collecting octroi will be checked, but on the other the Government will suffer revenue losses.

29. (*e*) : Both argument I and argument II are equally forceful. Senior persons look forward to promotions. Also, a not so good senior may be an obstacle in the path of prosperity of a really deserving person.

30. (*d*) : Neither argument I nor argument II is strong. The management may not be capable to keeping up the status. Rules in western countries do not apply in Indian education system as compulsory.

31. (*d*) : Neither argument I nor argument II is strong. Management is the sole decider of the actions to be taken.

32. (*e*) : Both argument I and argument II are strong. The number of questions in such tests is always large but it also shows the candidate's indepth knowledge of the subject.

33. (*d*) : Neither argument I nor argument II is strong.

34. (*b*) : Argument II is strong, because national defence preparedness is vital for our country.

35. (*e*) : Both the arguments I and II are strong. Such a step would restore ecological balance but it would harm the leather industry.

36. (*e*) : In our society today though there are girls who want to end the system, they even consider it as their lawful right. So, both the arguments are strong.

37. (*b*) : Scientifically based musical notes does not mean that scientists can be good singers. Singing is a talent. Only argument II is strong.

38. (*c*) : The two arguments are contrary to each other. So, either of the two arguments can be strong.

39. (*e*) : Both argument I and argument II are strong.

40. (*d*) : Neither the issue of holidays raise a hue and cry, nor the necessity of urgency makes sense. So, neither argument I nor argument II is strong.

41. (*b*) : Making schools co-educational is not the solution for promoting education of girl child. Argument II is strong as what is said is very likely to happen.

42. (*d*) : Neither argument I nor argument II is strong.

43. (*e*) : Both argument I and argument II are strong.

44. (*a*) : What is once destroyed cannot be recovered and wild life species must be protected. Only argument I is strong. Wild life can be protected in zoos or safaris. So, argument II is weak.

45. (*b*) : Upliftment of the poor has its own government quota. On the other hand facilities must be provided to able sportsmen. So, only argument II is strong.

46. (*e*) : Competition is more among private organisations as they seek efficiency. So, both argument I and argument II are strong.

47. (*e*) : Both argument I and argument II are strong.

48. (*b*) : Argument I is wrong. Argument II is strong.

49. (*c*) : Violence is the base of action films but on the other hand this also harms immature minds. So, either argument I or argument II is strong.

50. (*e*) : For a country like India rocket launching costs a lot as it drains away the resources. So, both argument I and argument II are strong.

28 Inferal Conclusions

In this type of reasoning a statement is followed by inferences drawn from it. From these inferences only one definitely follows which is the hidden proposition of the sentence and it is the right answer to the question.

MULTIPLE CHOICE QUESTIONS

Directions : *In each question one or more statements are given followed by various inferential conclusions. The inferential conclusion which can be drawn, without assuming anything in addition to the information given, is the correct answer.*

1. Unless you have atleast two years work experience, you stand limited chance to be selected for the job. This is reflected in the selections of about last eight years.

(*a*) Minimum stipulation of two year's work experience is made very clear to everybody aspiring for the job.

(*b*) Most applicants have much more than two year's work experience.

(*c*) Prior work experience is considered an important advantage in performing the job.

(*d*) If you have more than two year's work experience, you are bound to be selected.

2. Some students, who get only average marks in school, become highly successful later in their profession though these are only exceptions.

(*a*) There is no link whatsoever between academic performance and success in the profession.

(*b*) There are other exceptions also— like good school performance but utter failure otherwise.

(*c*) Only those who perform well in school are given good jobs.

(*d*) Some students under-perform during school years.

3. Various offices of an organisation are usually connected by an internal telephone system called INTERCOM.

(*a*) PBX is manually operated and controlled by an operator.

(*b*) A good internal communication system is essential for dealing with inter-departmental business.

(*c*) Communication made over the intercom can be kept secret.

(*d*) A maximum of 10 pins can be connected together in the intercom between a distance of 3000 meters.

4. High rate of population growth is linked closely with the fact of dwindling area under forest cover.

(*a*) Forests are inimical to human growth.

(*b*) There is less danger to human population from wild animals.

(*c*) Decreasing space of forest cover acts as incentive for humans to reproduce at a faster rate.

(*d*) Population growth leads to demand for more land.

5. Word Processors are screen-based systems in which the typed matter is displayed on a screen.

(*a*) The text can be stored on hard disc, floppy disc for subsequent amendment or printing.

(*b*) It is manually and electrically operated.

(*c*) Editing is possible by inserting new material or deleting unwanted material.

(*d*) It holds data in memory for future use.

6. The least technological development in the field of copying is photo-copying system.

(*a*) It does not have capacity to reduce or enlarge the script of photograph.

(*b*) It is an electrically-operated machine which produces an image of the original copy.

(*c*) Every copy produced looks as original, impressive and attractive.

(*d*) It is fast, easy to operate and cheaper and elegant in producing copies.

7. The population of big cities is increasing day by day but the civic amenities and support services are the same and not expanding.

(*a*) Birth rate is higher in cities as compared to villages.

(*b*) The cities are dying.

(*c*) If civic services are improved, the rate of population growth would further increase.
(*d*) More and more people are using services which have limited capacity.

8. Women being poorly educated, 33 per cent reservation for women in legislative bodies wiil mean rule by the poorly educated over the educated people.
(*a*) Uneducated women are elected by uneducated women.
(*b*) 33 per cent of the elected women will be uneducated.
(*c*) The number of uneducated persons in the legislative bodies will rise significantly.
(*d*) Educated persons will loose voice in the legislature.

9. Fax machine uses telephone to transmit any form of written or printed matter, diagrams, photographs from one location to another.
(*a*) It is a method of editing, amending and printing of any document.
(*b*) Equipments have to be on both receiving and sending ends.
(*c*) This machine is known as Fascimile Telegraphy.
(*d*) It can send copy of document to long distances by combining the speed of telephone with the reproduction facility of the office copier.

10. Many types of calculators are used in offices for calculations, day to day billing, accounting etc.
(*a*) Calculator can be used for arithmetical processes.
(*b*) Electronic calculator can be used for multiple functions.
(*c*) It can be used for making payments to a large number of people.
(*d*) It enables you to add, subtract, multiply, divide, calculate percentage and square roots.

11. Most of the people convey that they get bright and excellent ideas in the early morning hours.
(*a*) Ideas come only in the early morning hours.
(*b*) Chances of getting bright ideas at night are less.
(*c*) People are not able to express clearly how they get ideas.
(*d*) After sound sleep of two hours mind becomes fresh.

12. The average age of both men and women in India has gone up considerably over the last 70 years.
(*a*) Rekha would probably live longer than her grandmother.
(*b*) There is a wider difference between the average age of men and women.
(*c*) There seems to be a small difference between the average age of men and women.
(*d*) Drugs to increase longevity have been found.

13. Population explosion in India is due to high rate of illiteracy in the rural areas.
(*a*) Illiterates in urban areas practice family planning.
(*b*) Urban areas have very small percentage of the total population.
(*c*) Rural areas contribute significantly to population growth.
(*d*) Urban people are all literates.

14. Computer is a device which automatically accepts data, processes in a sequence and supply the results.
(*a*) A Computer has five main parts—input, output, storage, calculating and control.
(*b*) The exact photographs and pictures can be drawn with the help of this device.
(*c*) The physical parts such as the keyboard, the central processing unit and the printer are known as HARDWARE.
(*d*) The programmes and data are called SOFTWARE.

15. E-mail is the quickest means of communication in printed form.
(*a*) The message is produced at both the sending and receiving installations.
(*b*) In big cities telegram, connected with teleprinter service are transmitted.
(*c*) Incoming messages are typed automatically at the receiving end.
(*d*) Messages are sent from one computer to another with the help of modem, over the telephone lines.

ANSWERS

1	2	3	4	5	6	7	8	9	10
(*c*)	(*d*)	(*c*)	(*d*)	(*b*)	(*a*)	(*d*)	(*d*)	(*a*)	(*c*)
11	**12**	**13**	**14**	**15**					
(*b*)	(*a*)	(*c*)	(*b*)	(*d*)					

29

Matrices

In these type of questions, a set of numbers or letters are given in a figure which is divided into different cells. Some logic is applied to form a pattern in rows or columns. One of the cells is left empty. The candidates are required to analyse the logic and from the given options select one which will complete the given incomplete matrix.

MULTIPLE CHOICE QUESTIONS

Directions (Qs. 1 to 5) : *In each question given below which one number when placed at the sign of interrogation shall complete the number matrix.*

1.

4	5	6
2	3	7
1	8	3
21	98	?

(*a*) 16 (*b*) 73
(*c*) 76 (*d*) 94

2.

3	6	8
5	8	4
4	7	?

(*a*) 9 (*b*) 8
(*c*) 7 (*d*) 6

3.

1	4	9	?
1	2	3	4
2	4	6	?

(*a*) 16 and 8 (*b*) 25 and 5
(*c*) 36 and 4 (*d*) 49 and 7

4.

7	16	9
5	21	16
9	?	4

(*a*) 29 (*b*) 21
(*c*) 13 (*d*) 42

5.

6	6	8
5	7	5
4	3	?
120	126	320

(*a*) 12 (*b*) 16
(*c*) 8 (*d*) 4

Directions (Qs. 6 to 10) : *In questions given below, a word is represented by only one set of numbers as given in any one of the alternatives. The sets of numbers given in the alternatives are represented by two classes of alphabets as in the 2 matrices given below. The columns and rows of matrix I are numbered from 0 to 4 and that of matrix II from 5 to 9. A letter from these matrices can be represented first by its row and next by column number. Identify the set for the word given in each question.*

6.

	0	1	2	3	4
0	E	A	T	S	H
1	A	H	T	A	S
2	E	H	A	S	T
3	H	E	A	T	S
4	S	H	T	A	E

	5	6	7	8	9
5	O	R	K	L	P
6	L	P	O	R	K
7	O	K	R	P	L
8	P	R	K	L	O
9	R	L	K	O	P

REAP

(*a*) 56, 00, 22, 59 (*b*) 68, 21, 22, 86
(*c*) 96, 00, 01, 99 (*d*) 86, 34, 24, 69

7.

	0	1	2	3	4
0	C	A	F	E	D
1	C	D	A	E	D
2	D	E	C	F	E
3	A	D	D	D	C
4	D	C	A	C	A

	5	6	7	8	9
5	T	R	R	S	M
6	S	T	M	R	S
7	R	S	P	T	P
8	P	S	T	R	T
9	P	M	P	M	S

FARM

(*a*) 02, 30, 85, 65 (*b*) 20, 31, 76, 68
(*c*) 23, 12, 68, 96 (*d*) 44, 43, 87, 57

8.

	0	1	2	3	4
0	C	H	H	N	N
1	I	N	C	B	B
2	N	E	C	H	I
3	I	C	N	E	I
4	B	B	E	C	I

	5	6	7	8	9
5	Y	P	S	P	K
6	K	T	S	R	R
7	K	Y	R	Y	Y
8	P	R	K	P	R
9	S	Y	S	T	S

PREY

(*a*) 82, 76, 34, 77 (*b*) 80, 70, 43, 12
(*c*) 94, 87, 43, 56 (*d*) 88, 77, 33, 79

9.

	0	1	2	3	4
0	D	K	A	E	C
1	C	D	K	A	E
2	K	C	E	A	D
3	K	C	D	E	A
4	E	D	A	K	C

	5	6	7	8	9
5	P	L	O	T	N
6	T	P	N	L	O
7	P	N	T	O	L
8	O	N	T	P	L
9	L	O	P	N	T

COLD

(*a*) 10, 85, 79, 24 (*b*) 31, 99, 77, 22
(*c*) 30, 66, 86, 43 (*d*) 44, 96, 95, 22

10.

	0	1	2	3	4
0	M	M	R	M	R
1	O	F	R	F	N
2	R	N	O	R	O
3	F	O	F	M	O
4	F	M	R	R	F

	5	6	7	8	9
5	T	G	J	T	G
6	P	J	G	G	U
7	U	T	P	T	J
8	U	T	T	J	P
9	G	U	G	T	P

TURF

(*a*) 57,86, 44, 11 (*b*) 86, 75, 20, 32
(*c*) 77, 68, 41, 44 (*d*) 89, 97, 12, 30

EXPLANATORY ANSWERS

1. (*d*) : $4^2 + 2^2 + 1^2 = 21$, *i.e.*, $16 + 4 + 1 = 21$
$5^2 + 3^2 + 8^2 = 98$, *i.e.*, $25 + 9 + 64 = 98$
$6^2 + 7^2 + 3^2 = 36 + 49 + 9 = 94$.

2. (*d*) : Sum of numbers in each row is 17, *i.e.*,
$3 + 6 + 8 = 17$
$5 + 8 + 4 = 17$
So, $17 - (4 + 7) = 6$.

3. (*a*) : The root of number in first row is the number in the second row and this number multiplied by 2 is the number in the third row, *i.e.*,
$\sqrt{1} = 1$ and $1 \times 2 = 2$
$\sqrt{4} = 2$ and $2 \times 2 = 4$
$\sqrt{9} = 3$ and $3 \times 2 = 6$
So, square of 4, *i.e.*, 16 in the first row and 4 × 2, *i.e.*, 8 in the third row completes the matrix.

4. (*c*) : $7 + 9 = 16$
$5 + 16 = 21$
$9 + 4 = 13$.

5. (*c*) : $6 \times 5 \times 4 = 120$
$6 \times 7 \times 3 = 126$
So, 320 ÷ (8 × 5), *i.e.*, 8 in the third row completes the matrix.

6. (*a*) : R is represented by row 5, column 6
E by row 0, column 0
A by row 2, column 2
P by row 5, column 9.

7. (*c*) : F is represented by row 2, column 3
A by row 1, column 2
R by row 6, column 8
M by row 9, column 6.

8. (*d*) : P is represented by row 8, column 8
R by row 7, column 7
E by row 3, column 3
Y by row 7, column 9.

9. (*a*) : C is represented by row 1, column 0
O by row 8, column 5
L by row 7, column 9
D by row 2, column 4.

10. (*b*) : T is represented by row 8, column 6
U by row 7, column 5
R by row 2, column 0
F by row 3, column 2.

30

Next Figure in Series

In this form of non-verbal series, which are the most common, four or five consecutive problem figures form a definite sequence and one is required to select the one figure from the given set of Answer Figures that will continue the same sequence.

One has to try different set of moves, changes, replacements, rotations, repetitions and a lot more variations to arrive at the logical pattern making the series. Practising alone will sharpen one's skill of solving such sequences.

MULTIPLE CHOICE QUESTIONS

Directions (Q. 1–20) : *In each of the following questions which one of the five answer figures given below should come after the problem figures if the sequence are continued?*

1. Problem Figures

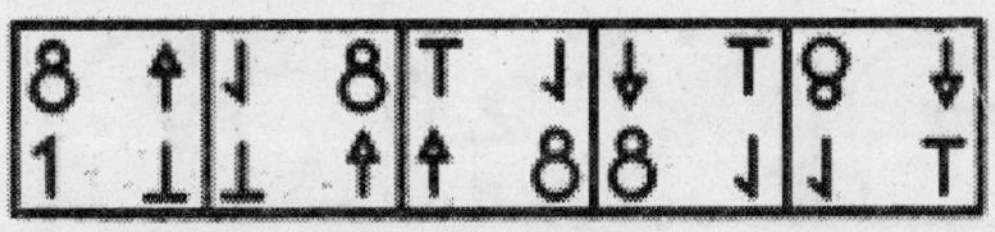

Answer Figures

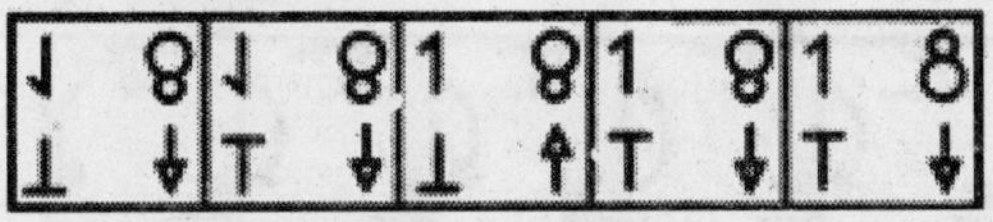

(*a*) (*b*) (*c*) (*d*) (*e*)

2. Problem Figures

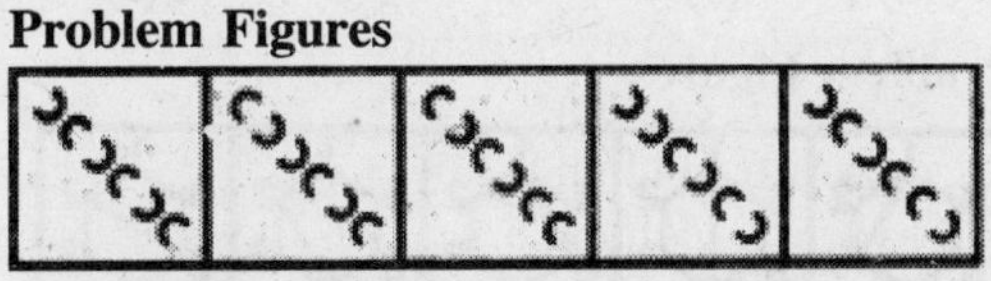

Answer Figures

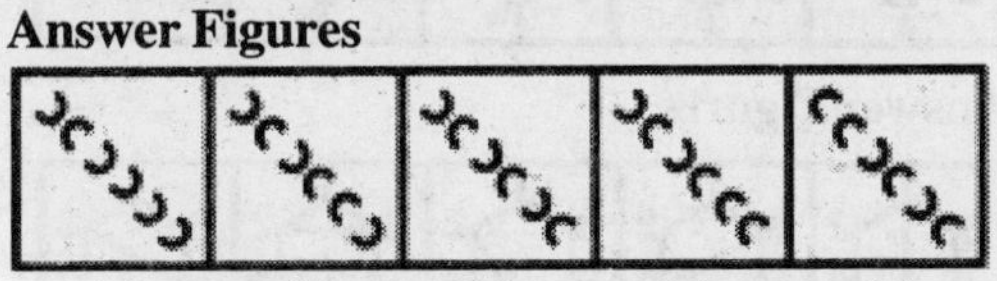

(*a*) (*b*) (*c*) (*d*) (*e*)

3. Problem Figures

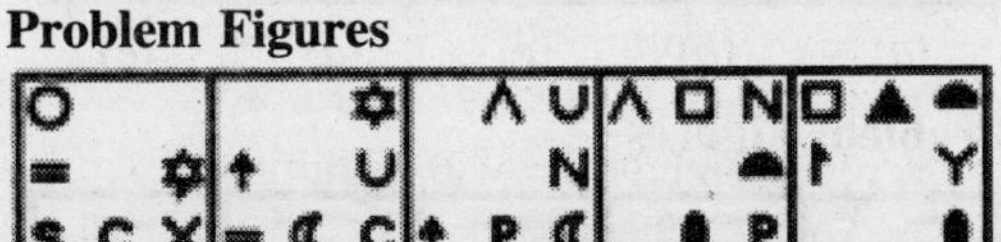

Answer Figures

(*a*) (*b*) (*c*) (*d*) (*e*)

4. Problem Figures

Answer Figures

(*a*) (*b*) (*c*) (*d*) (*e*)

5. Problem Figures

Answer Figures

(*a*) (*b*) (*c*) (*d*) (*e*)

6. Problem Figures

Answer Figures

(*a*) (*b*) (*c*) (*d*) (*e*)

7. Problem Figures

Answer Figures

(a) (b) (c) (d) (e)

8. Problem Figures

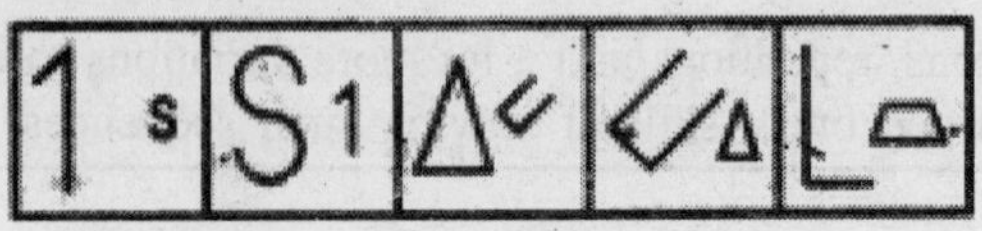

Answer Figures

(a) (b) (c) (d) (e)

9. Problem Figures

Answer Figures

(a) (b) (c) (d) (e)

10. Problem Figures

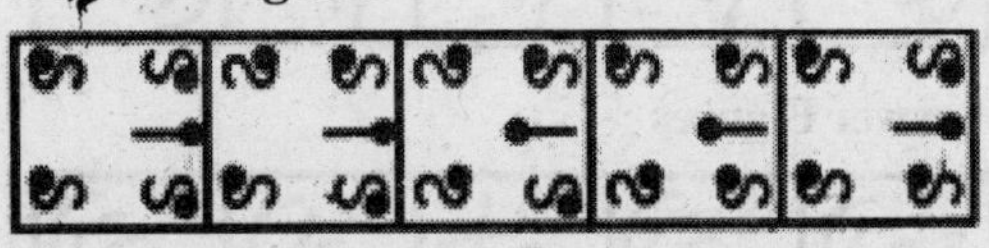

Answer Figures

(a) (b) (c) (d) (e)

11. Problem Figures

Answer Figures

(a) (b) (c) (d) (e)

12. Problem Figures

Answer Figures

(a) (b) (c) (d) (e)

13. Problem Figures

Answer Figures

(a) (b) (c) (d) (e)

14. Problem Figures

Answer Figures

(a) (b) (c) (d) (e)

15. Problem Figures

15.

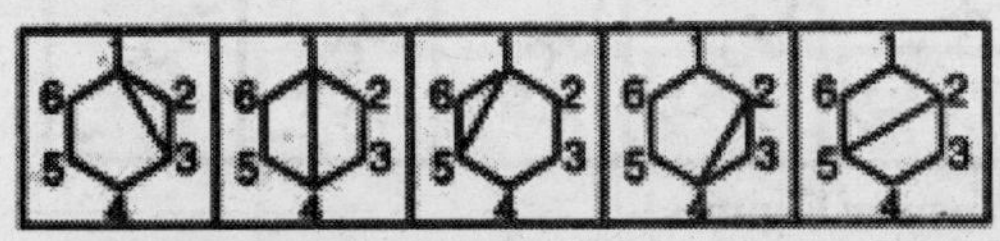

Answer Figures

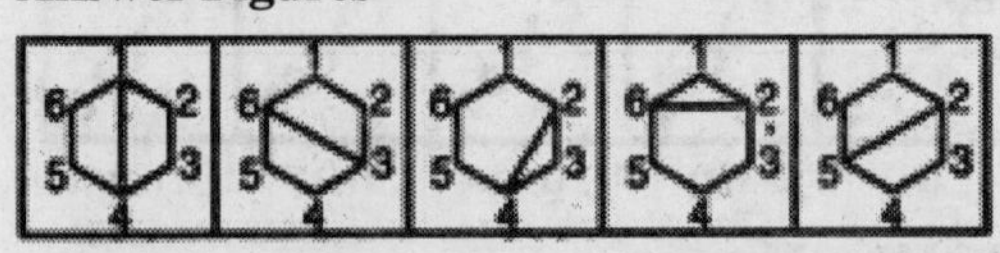

(a) (b) (c) (d) (e)

16. Problem Figures

Answer Figures

(a) (b) (c) (d) (e)

17. Problem Figures

Answer Figures

(a) (b) (c) (d) (e)

18. Problem Figures

Answer Figures

(a) (b) (c) (d) (e)

19. Problem Figures

Answer Figures

(a) (b) (c) (d) (e)

20. Problem Figures

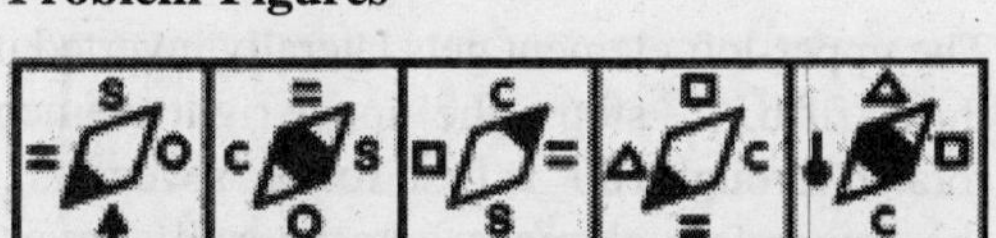

Answer Figures

(a) (b) (c) (d) (e)

21. Problem Figures

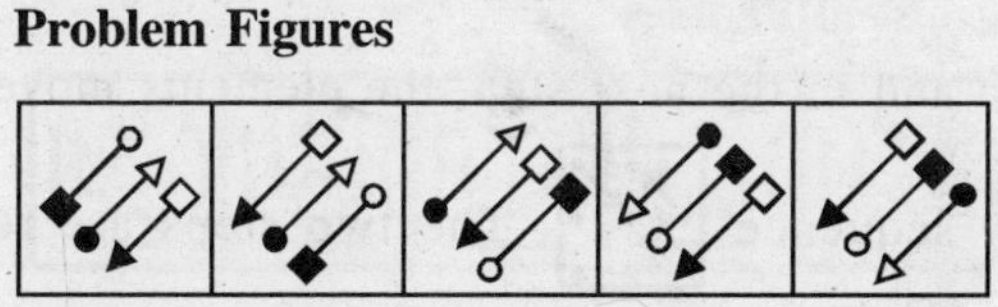

Answer Figures

(a) (b) (c) (d) (e)

22. Problem Figures

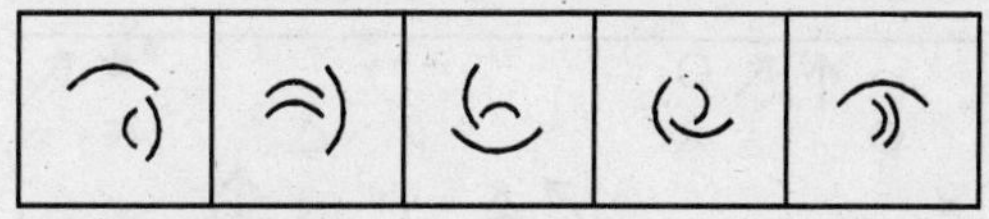

Answer Figures

(a) (b) (c) (d) (e)

23. Problem Figures

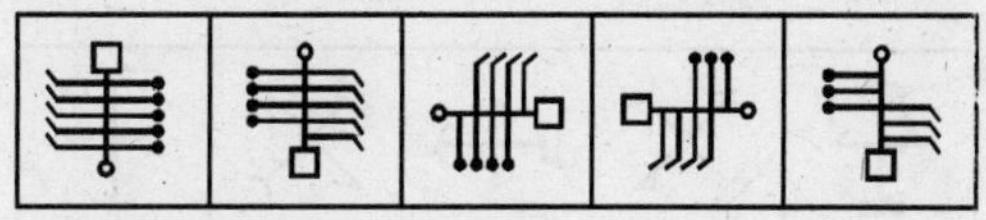

Answer Figures

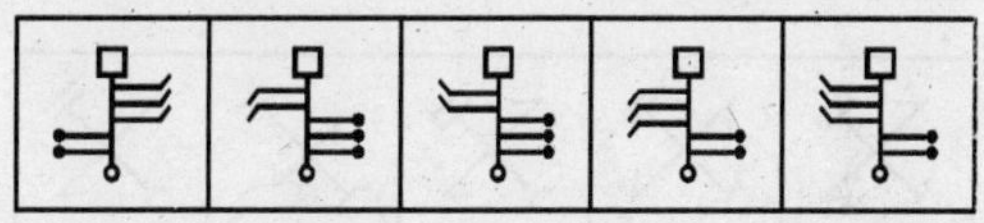

(a) (b) (c) (d) (e)

24. Problem Figures

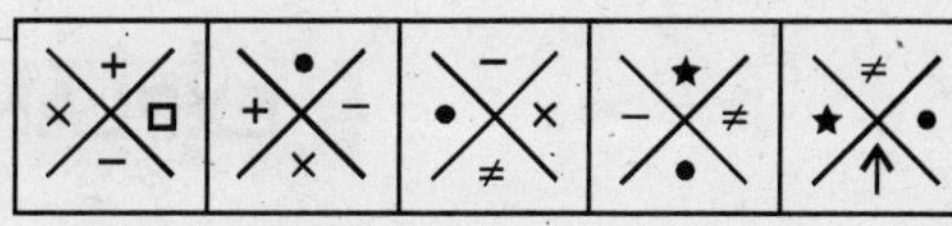

Answer Figures

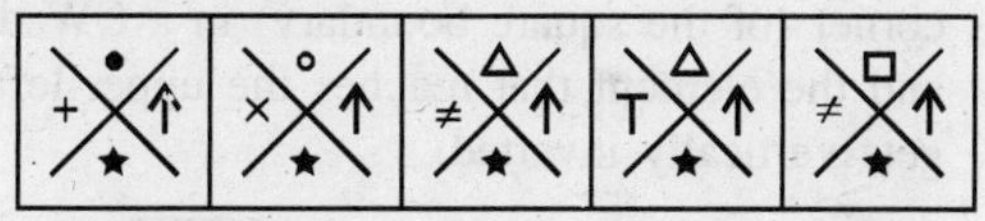

(a) (b) (c) (d) (e)

25. Problem Figures

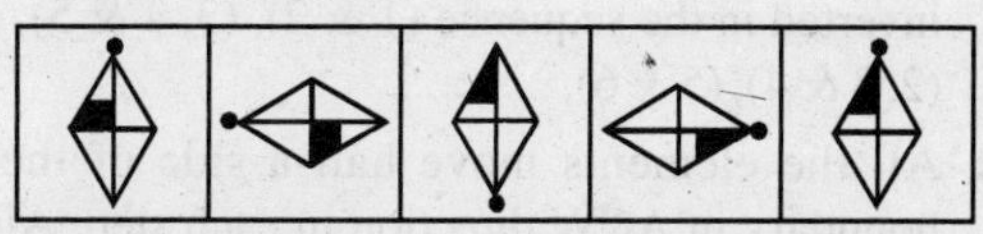

Answer Figures

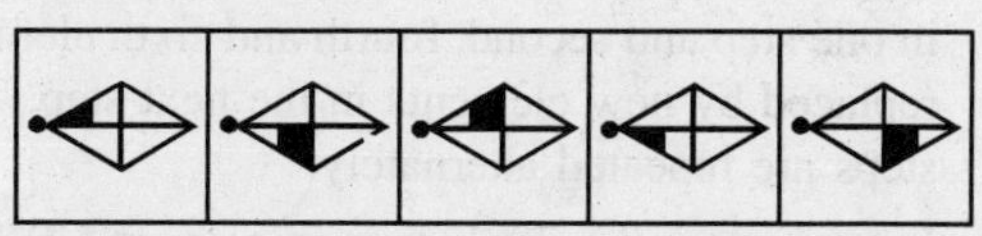

(a) (b) (c) (d) (e)

26. Problem Figures

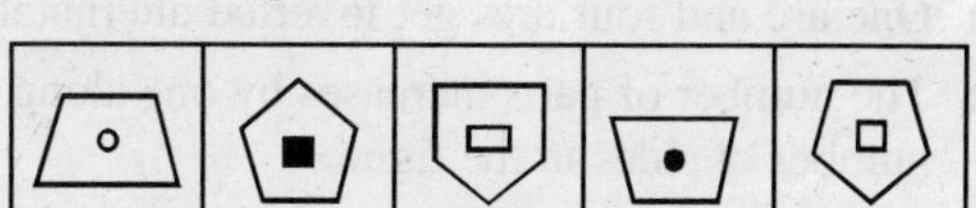

Answer Figures

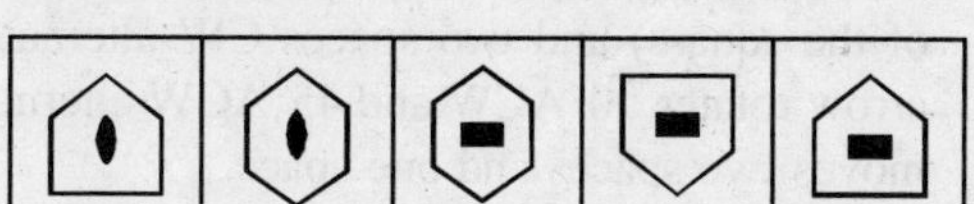

(a) (b) (c) (d) (e)

27. Problem Figures

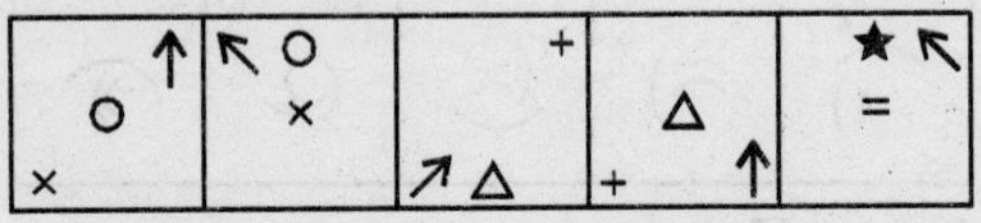

Answer Figures

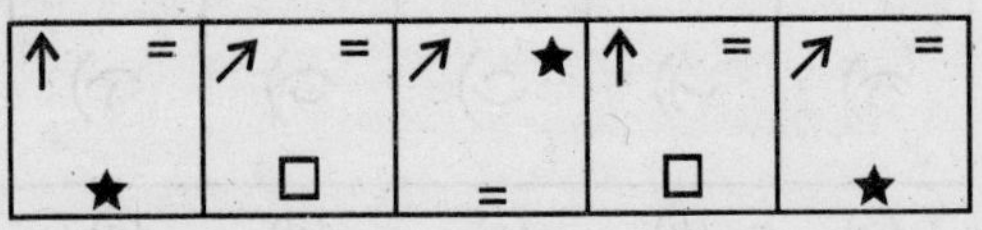

(a) (b) (c) (d) (e)

28. Problem Figures

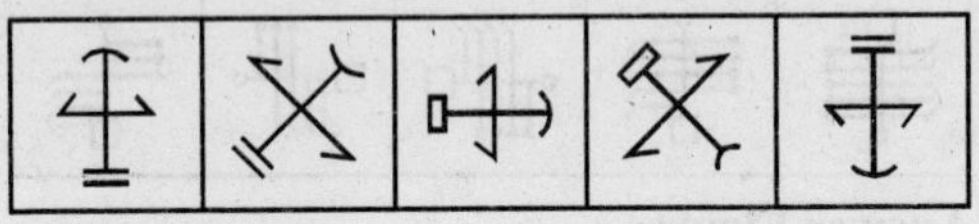

Answer Figures

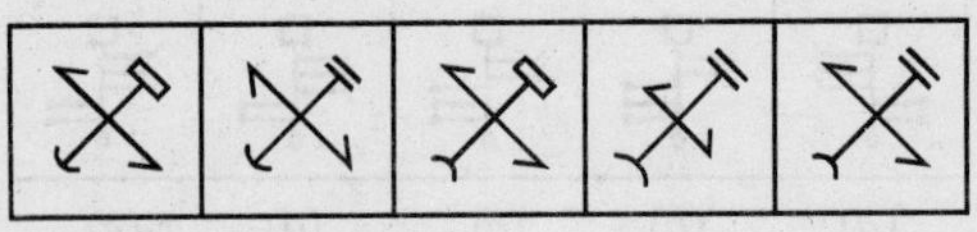

(a) (b) (c) (d) (e)

29. Problem Figures

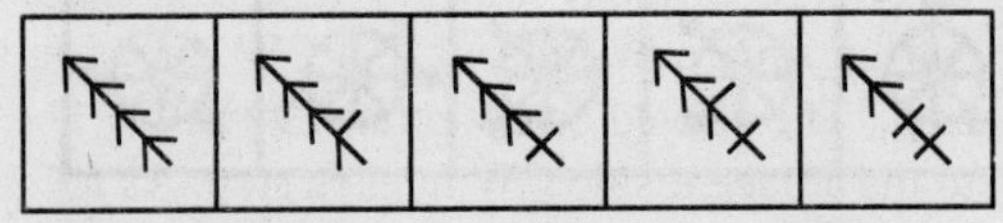

Answer Figures

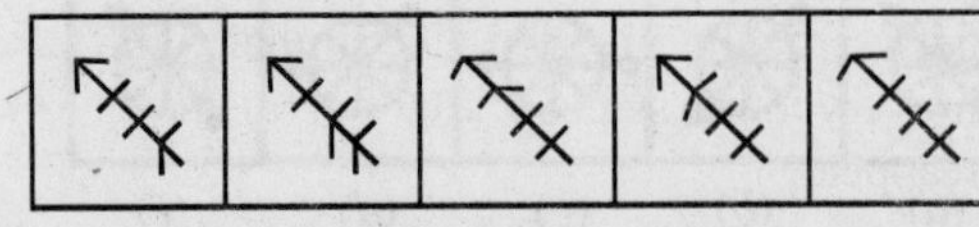

(a) (b) (c) (d) (e)

30. Problem Figures

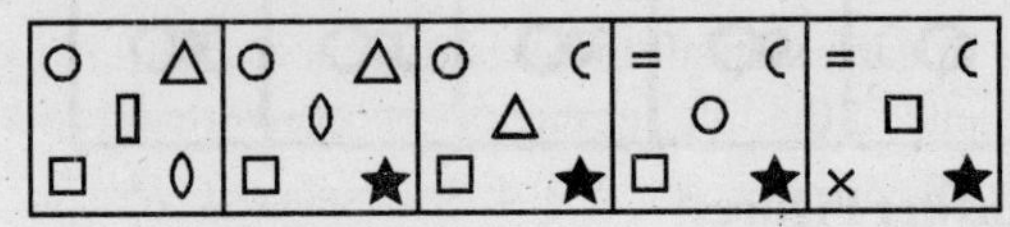

Answer Figures

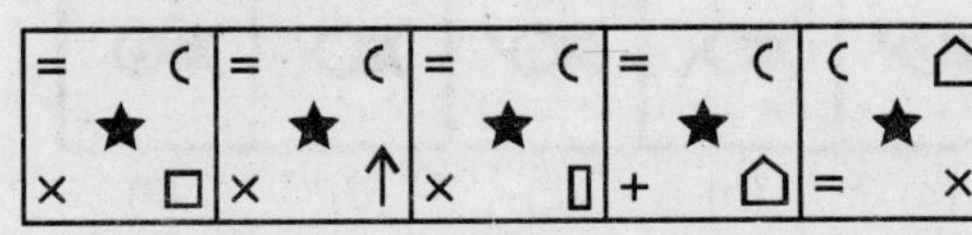

(a) (b) (c) (d) (e)

EXPLANATORY ANSWERS

1. (*d*) : In each step, all the elements move to the adjacent corner (of the square boundary) in a CW direction and the element that reaches the upper-left corner gets vertically inverted.

2. (*c*) : We can label the arcs as shown . The arcs get inverted in the sequence (1 & 2), (3, 4 & 5), (6 & 1), (2, 3 & 4), (5 & 6), ...

3. (*d*) : All the elements move half-a-side of the square boundary in ACW direction in each step. Also, first, third and fifth elements are replaced by new elements in one step and second, fourth and sixth elements are replaced by new elements in the next step. The two steps are repeated alternately.

4. (*a*) : In each step, the dot moves one space CW and the arrow moves two spaces CW.

5. (*c*) : One arc and four arcs get inverted alternately.

6. (*e*) : The number of parts increases by one along with the number of sides in the figure.

7. (*c*) : The pin rotates 45°CW and 90°CW alternately and moves one space (each space is equal to half-a-side of the square) and two spaces CW alternately. The arrow rotates 90°ACW and 45°ACW alternately and moves two spaces and one space.

8. (*e*) : In one step, the two elements interchange positions and the smaller element gets enlarged while the larger element gets reduced in size. In the next step, the smaller element is replaced by a new small element and the larger element is replaced by a new large element.

9. (*c*) : In each step, the elements move in the order .

10. (*b*) : The upper-left element gets laterally inverted in first, third, fifth..... steps; the upper-right element gets rotated through 180° is first, fourth, seventh,.... steps; the lower-left element gets laterally inverted in second, fourth, sixth, ... steps; the lower-right element gets rotated through 180° in third, sixth,... steps and the pin at the middle-right position gets laterally inverted in every second step.

11. (*e*) : In one step, the elements move in the sequence  and in the next step, the elements move in the sequence 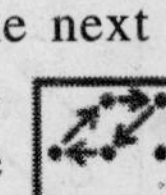. The two steps are repeated alternately.

12. (*e*) : One of the pins gets inverted in each step. The pins get inverted sequentially from right to left.

13. (*c*) : The number of symbols added sequentially is 3, 2, 5, 2, 7, 2, ... These symbols are added to form a sequence of 1, 2, 3, 4, 5, 6 identical symbols.

14. (*d*) : The bent pin gets laterally inverted and vertically inverted alternately and moves to the adjacent side (of the square boundary) in a CW direction in each step,. The curved arrow rotates 90°ACW and moves to the adjacent side CW in each step. As for the third element — in one step the outer 'L' shaped symbol gets laterally inverted and in the next step the element gets vertically inverted. Or, similar figure reappears in every fourth step. So, after fig. (*e*), the fig. (*b*) should reappear.

15. (*d*) : The diagonals of the hexagon are drawn in a sequence— (1, 3), (1, 4), (1, 5), (2, 4), (2, 5), (2, 6), (3, 5), (3, 6), (4, 6).

16. (*e*) : In each step all the existing curves get inverted and one of the line segments gets converted into a curve.

17. (*b*) : The pin moves 1, 2, 3, 4, 5, ... spaces CW sequentially. The arrow moves two steps ACW each time.

18. (*b*) : A new feature is added at each step (in a set order).

19. (*a*) : One circle is replaced by a new symbol in each step and this replacement takes place on either ends alternately. The complete figure rotates 45°ACW in each step.

20. (*e*) : The shading moves upwards sequentially from lower-left corner to the upper-right corner of the rhombus and once it reaches the upper-right corner, it moves to the lower-left corner in the next step. Also, in each step, all the symbols move to the adjacent side (of the rhombus) in a CW direction and the symbol that reaches the LHS position gets replaced by a new symbol.

21. (*e*) : First the places of two corner elements are interchanged; secondly, the places of two elements from the top are interchanged while the bottom element is inverted; finally the places of two elements from the bottom are interchanged while the top element is inverted. The above steps are repeated.

22. (*d*) : The outermost arc is turned 90° clockwise at each step; the middle arc is turned 90° anticlockwise at each step; and the smallest arc is turned 90° clockwise in alternate figures.

23. (*e*) : The figure is rotated by 180° degrees and one dot bar is removed, then the figure is rotated 90° anticlockwise and one bent line is removed. The above rotations are repeated and the dot bar and bent line are removed in a set pattern.

24. (*c*) : All the four elements are moved one step anticlockwise and starting from the square, anticlockwise one by one the elements are replaced by entirely new elements.

25. (*e*) : The rhombus shape is turned 90° anticlockwise at each step, and the shade is moved one segment anticlockwise.

26. (*e*) : Figure one is inverted in figure four and the unshaded circle is shaded. Figure two is inverted in figure five and the shaded square is unshaded. Similarly, figure three should be inverted in answer figure and the unshaded triangle should be shaded.

27. (*e*) : The arrow is moved anticlockwise in a set order and its three directions are in pendulum order. The other two elements are same in first and second figures, same in third and fourth figures and similarly, should be same in fifth and answer figures. One of them is moved vertically and the other is moved diagonally.

28. (*e*) : Observe the series. Figure five is an inverted image of figure one. To continue, the answer figure should be an inverted image of figure two.

29. (*d*) : The bent line segments are straightened one by one from the bottom in a set pattern.

30. (*b*) : Starting from the bottom right element, anticlockwise, one by one the positions of the corner element and the element in the centre are interchanged and then the element in the corner is replaced by a new element.

31

Non-Verbal Analogy

Analogy is a process of reasoning between two parallel cases. It relates to agreement or correspon-dence in certain respects between two things. It is a process whereby the underlying relationship that exists between two figures, designs or patterns is determined. Under the process, one has to discover the features common to the two figures or designs. This common feature is a model or base. The question seeks solution on the basis of this model or base. In non-verbal analogy questions, there are two sets of figures which are known as *(i)* Problem Figures, and *(ii)* Answer Figures. The Problem Figures are divided into two parts. The Answer Figures are given below the Problem Figures.

MULTIPLE CHOICE QUESTIONS

Directions : *The second figure in the first unit of the Problem Figures bears a certain relationship to the first figure. Similarly, one of the figures in the Answer Figures bears the same relationship to the first figure in the second unit of the Problem Figures. Locate the figure which would fit the question mark.*

1. Problem Figures

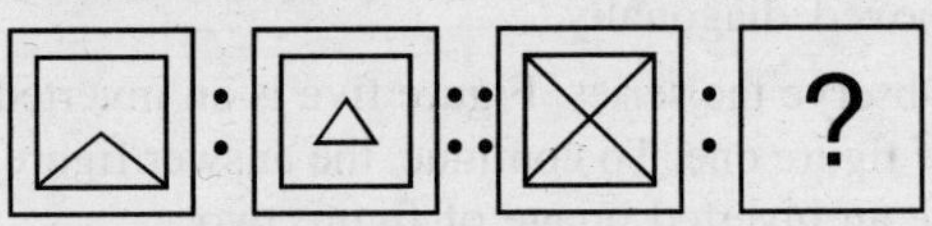

Answers Figures

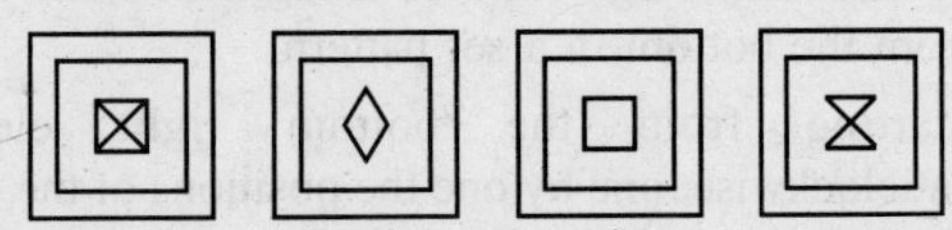

(a) *(b)* *(c)* *(d)*

2. Problem Figures

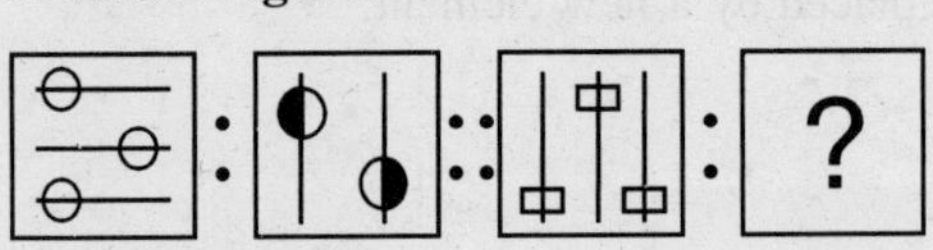

Answer Figures

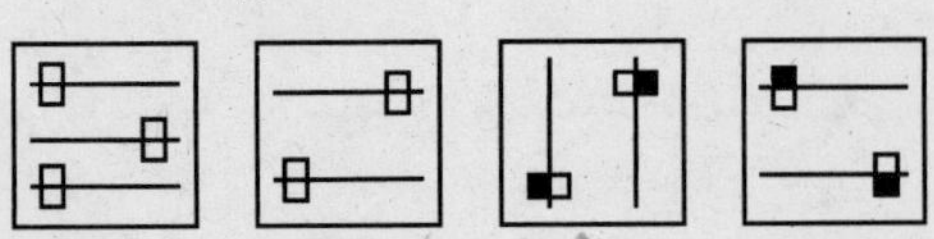

(a) *(b)* *(c)* *(d)*

3. Problem Figures

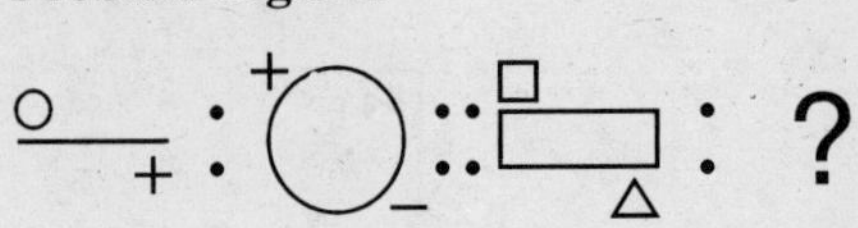

Answer Figures

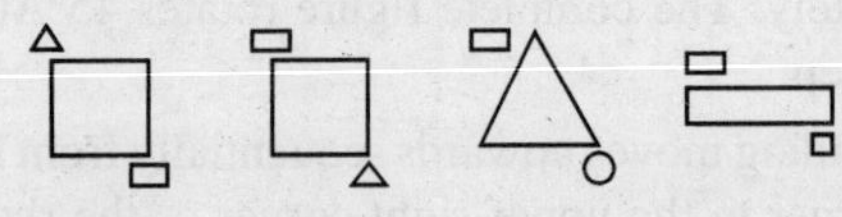

(a) *(b)* *(c)* *(d)*

4. Problem Figures

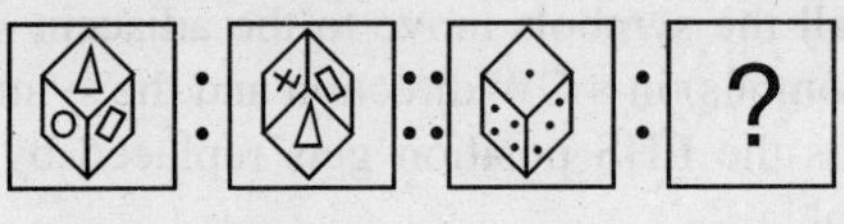

Answer Figures

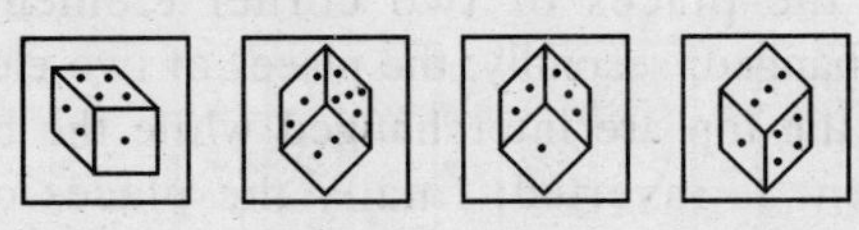

(a) *(b)* *(c)* *(d)*

5. Problem Figures

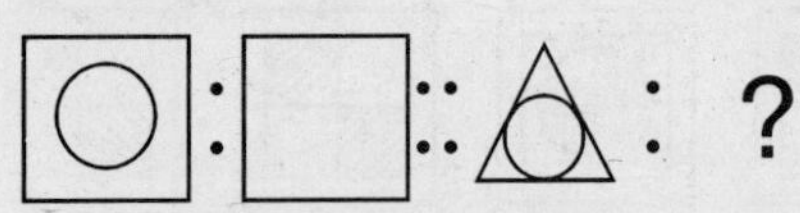

Answer Figures

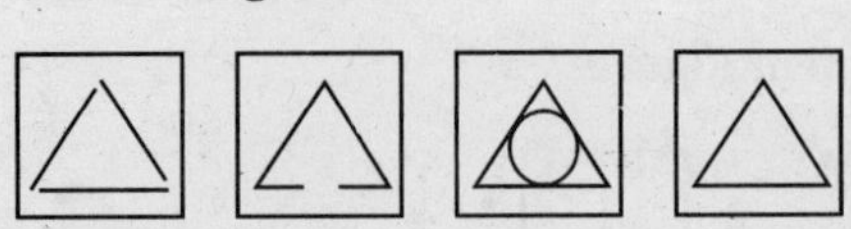

(a) *(b)* *(c)* *(d)*

6. Problem Figures

Answer Figures

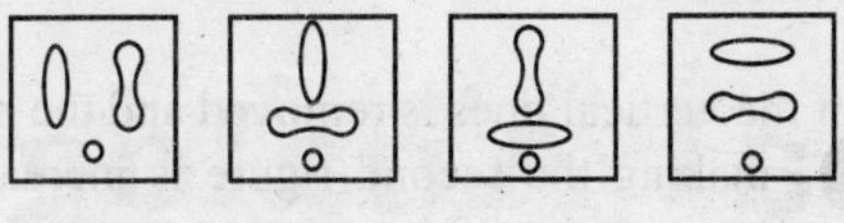

(*a*) (*b*) (*c*) (*d*)

7. Problem Figures

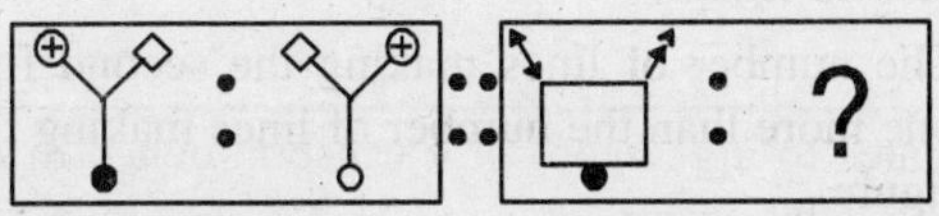

Answer Figures

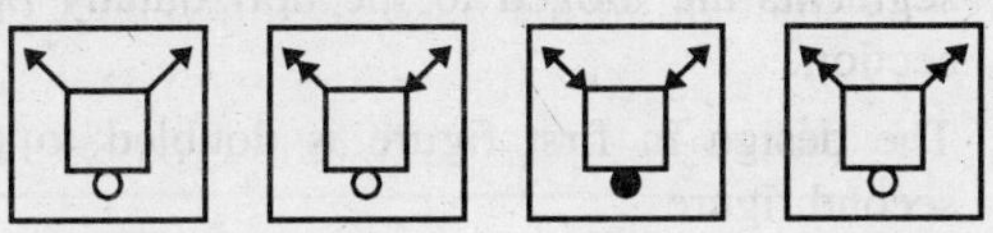

(*a*) (*b*) (*c*) (*d*)

8. Problem Figures

Answer Figures

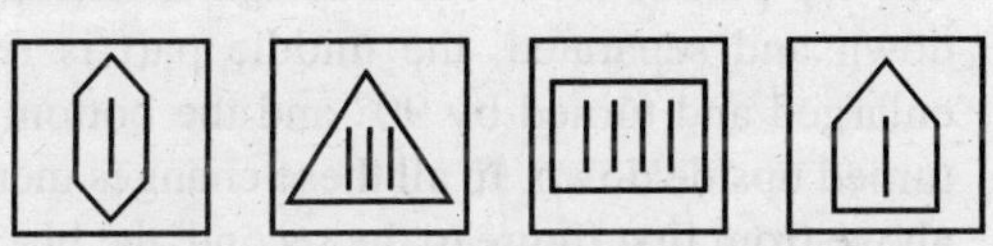

(*a*) (*b*) (*c*) (*d*)

9. Problem Figures

Answer Figures

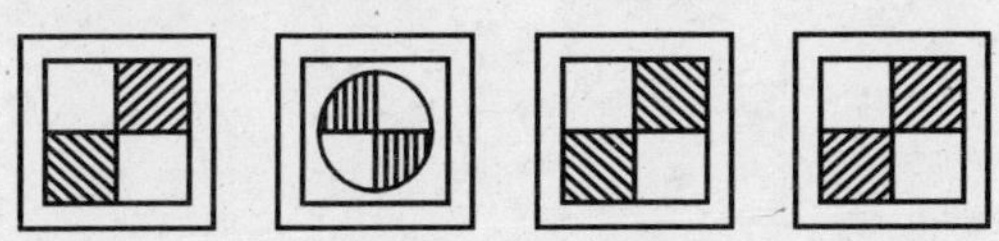

(*a*) (*b*) (*c*) (*d*)

10. Problem Figures

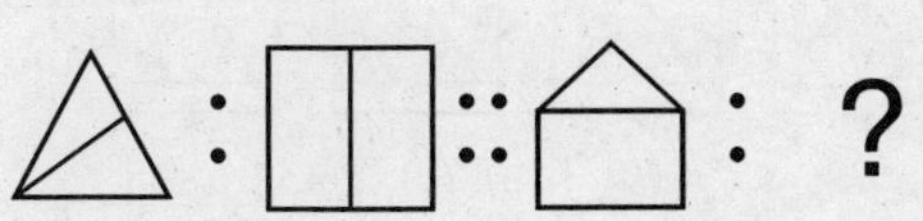

Answer Figures

(*a*) (*b*) (*c*) (*d*)

11. Problem Figures

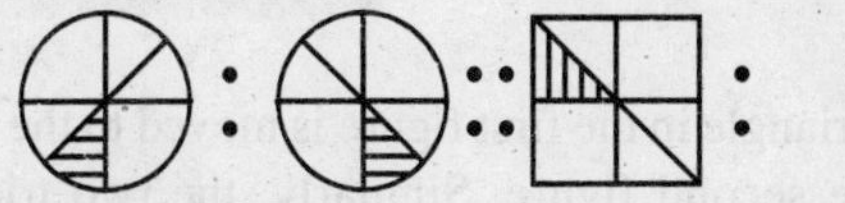

Answer Figures

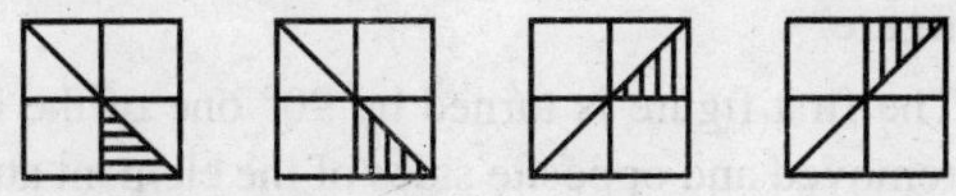

(*a*) (*b*) (*c*) (*d*)

12. Problem Figures

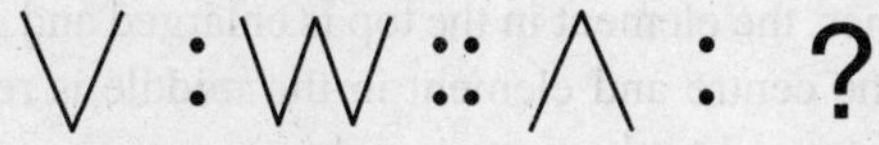

Answer Figures

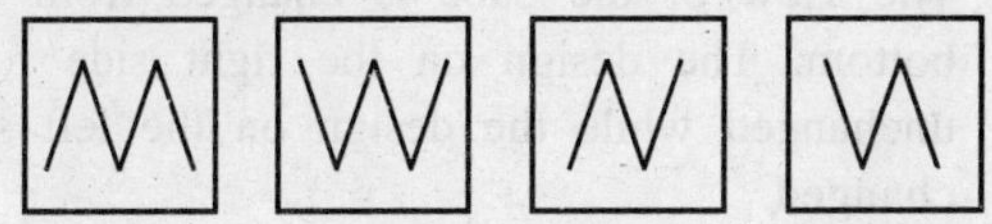

(*a*) (*b*) (*c*) (*d*)

13. Problem Figures

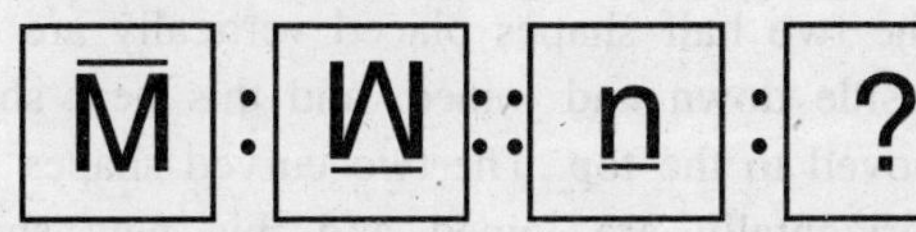

Answer Figures

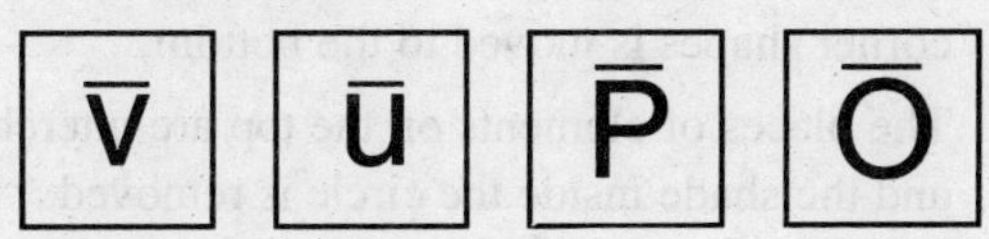

(*a*) (*b*) (*c*) (*d*)

14. Problem Figures

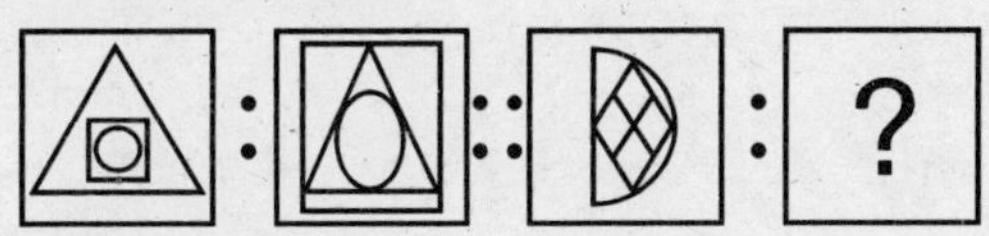

Answer Figures

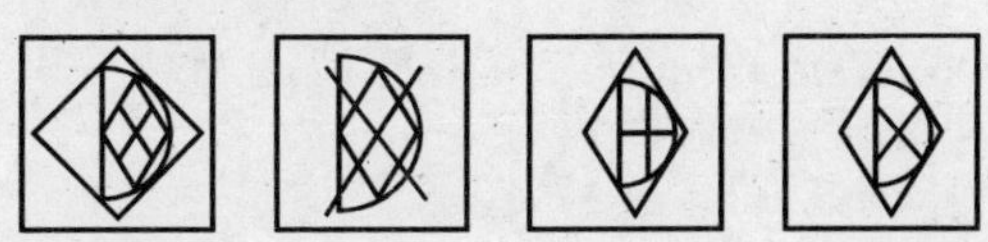

(*a*) (*b*) (*c*) (*d*)

15. Problem Figures

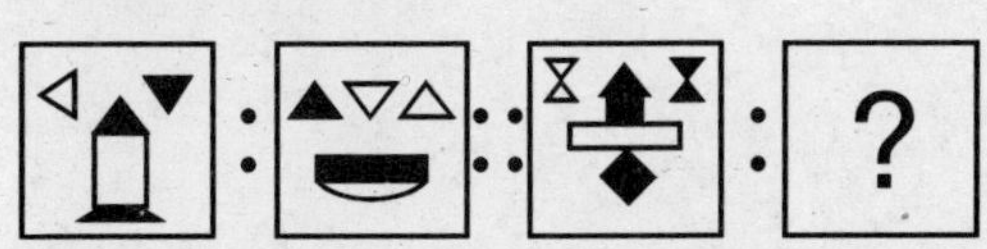

Answer Figures

(*a*) (*b*) (*c*) (*d*)

EXPLANATORY ANSWERS

1. (*d*) : The triangle in the first figure is moved to the centre of the second figure. Similarly, the two triangles joined at the apex are moved to the centre in answer figure.

2. (*d*) : The first figure is turned by 90° one of the bars is removed and opposite sides of the element attached to the bar are shaded to get the second figure.

3. (*a*) : The element at the bottom is moved to the diagonal corner, the element in the top is enlarged and moved to the centre and element in the middle is reduced and moved to the bottom right corner.

4. (*c*) : The view of the cube is changed from top to bottom. The design on the right side remains unchanged while the design on the left side is changed.

5. (*d*) : The inner shape in the first figure is removed to get the second figure.

6. (*b*) : The two half shapes placed vertically are turned upside down and joined, and this new shape is moved to the top. The two curved shapes placed horizontally are joined and this new shape is moved to the centre. One of the two remaining corner shapes is moved to the bottom.

7. (*b*) : The places of elements on the top are interchanged and the shade inside the circle is removed.

8. (*a*) : One of the vertical lines is removed and the number of lines making the second figure is increased by one.

9. (*a*) : First figure is rotated 90° anticlockwise to get the second figure.

10. (*a*) : The number of lines making the second figure is one more than the number of lines making the first figure.

11. (*c*) : The diagonal line is turned by 90° and the line segments are moved to the horizontally opposite section.

12. (*a*) : The design in first figure is doubled to get the second figure.

13. (*b*) : The first figure is turned upside down to get the second figure.

14. (*d*) : The middle and the inner most shapes are enlarged and the outermost shape is reduced and placed in the middle of the two enlarged shapes.

15. (*c*) : The shape on the top left side is turned by 90° and the shape on the top right is turned upside down. The top part of the vertical design is turned upside down and separated, the middle part is reduced/ enlarged and turned by 90° and the bottom part is turned upside down. In all these changes mentioned above from first figure to the second, the black areas are made white and vice versa.

32 Non-Verbal Odd One Out

Classification means arranging the given content in groups or classes having qualities of same kind. In classification type questions, the figures or items are sorted out in groups on the basis of their similarities in qualities in shapes, size, pattern, structure, genus, order, species, grade, style, constituents and other specifications, and thus the answer is found out.

MULTIPLE CHOICE QUESTIONS

Directions : *In each of the following questions one of the figures is different from the rest. Spot the figure.*

1. 

(a) (b) (c) (d) (e)

2. 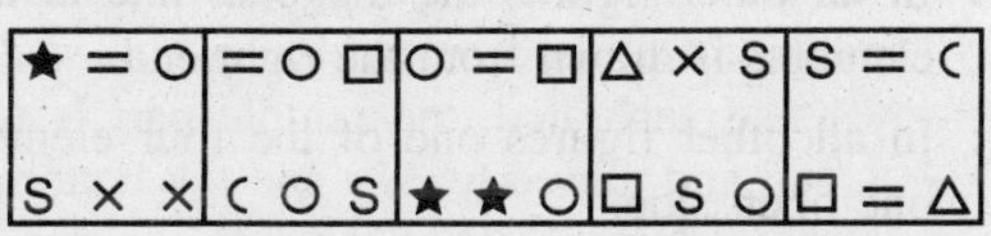

(a) (b) (c) (d) (e)

3. 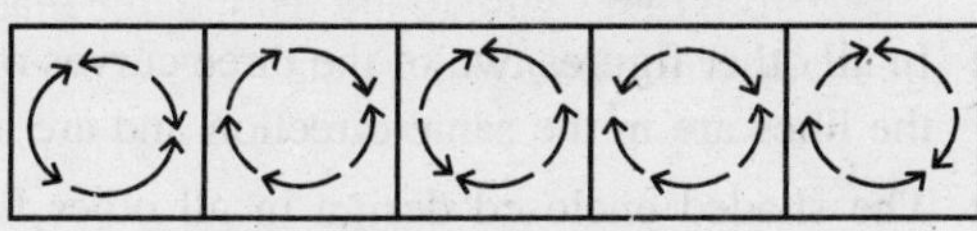

(a) (b) (c) (d) (e)

4. 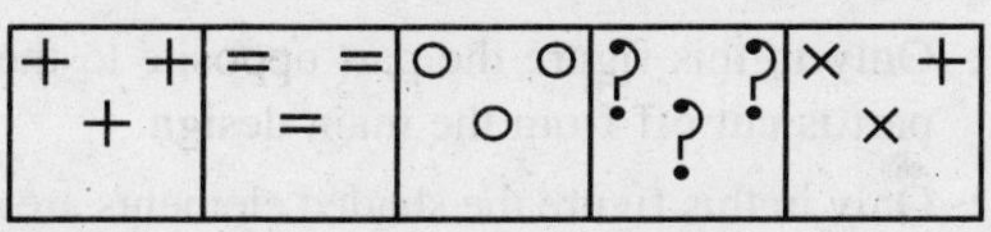

(a) (b) (c) (d) (e)

5. 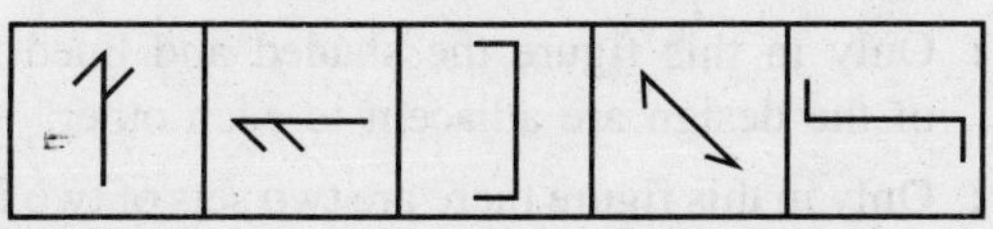

(a) (b) (c) (d) (e)

6. 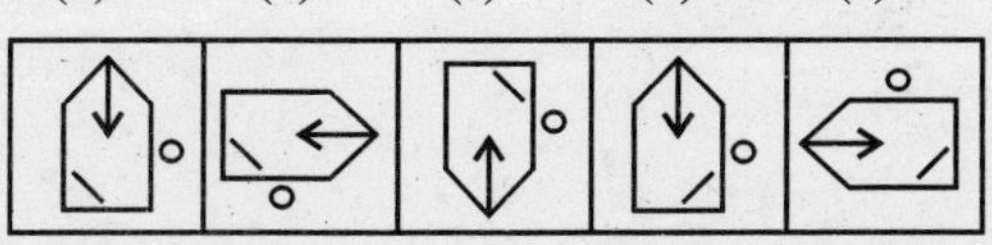

(a) (b) (c) (d) (e)

7. 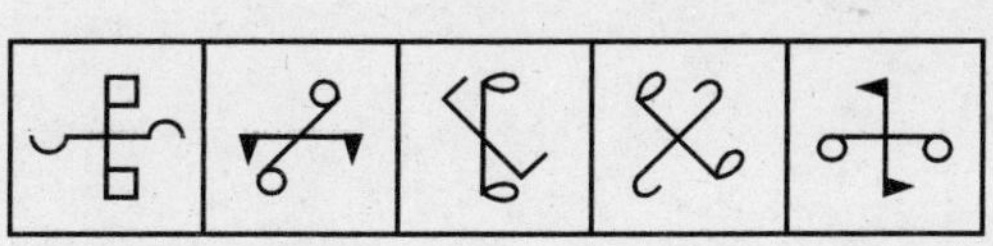

(a) (b) (c) (d) (e)

8. 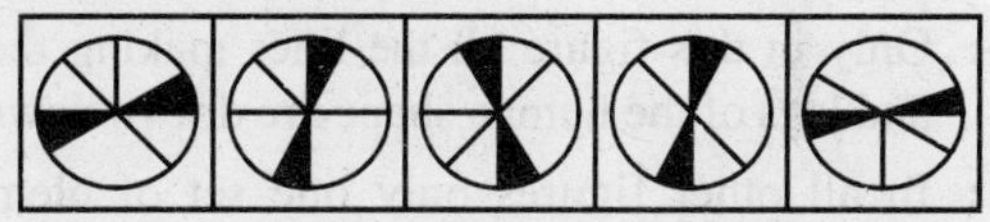

(a) (b) (c) (d) (e)

9. 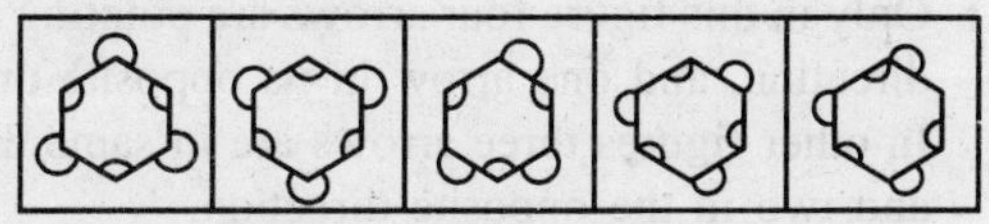

(a) (b) (c) (d) (e)

10. 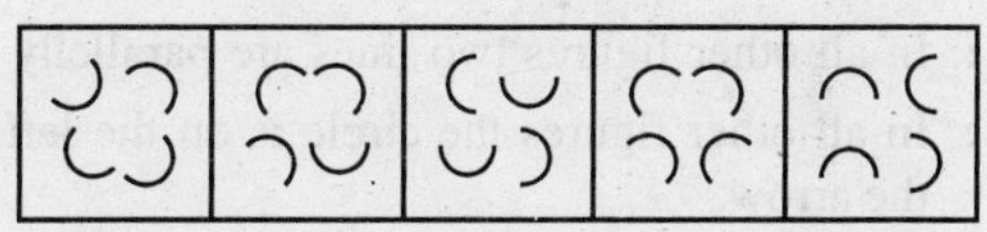

(a) (b) (c) (d) (e)

11. 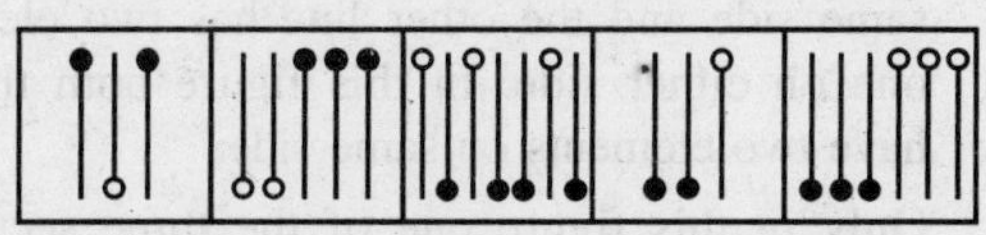

(a) (b) (c) (d) (e)

12. 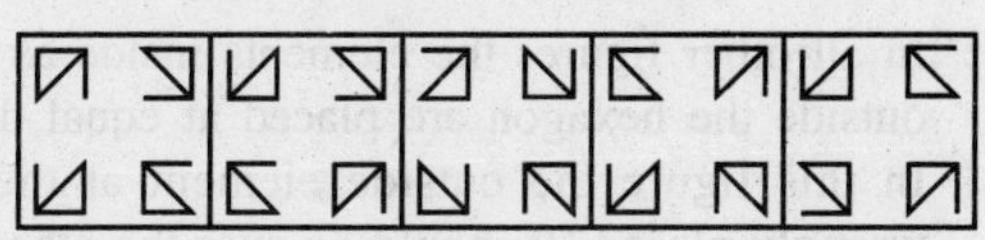

(a) (b) (c) (d) (e)

13. 

(a) (b) (c) (d) (e)

14. 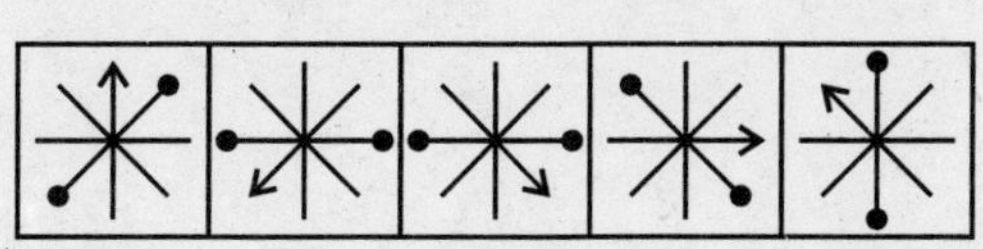

(a) (b) (c) (d) (e)

15. 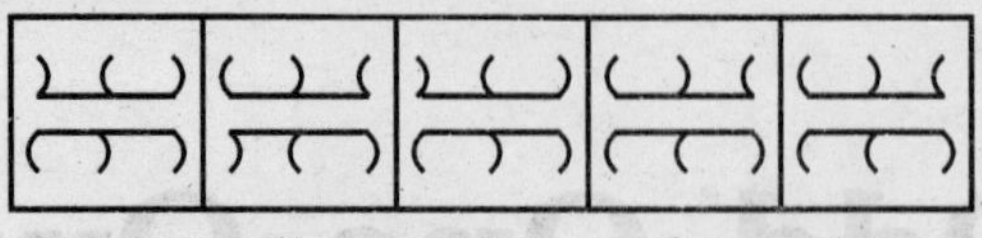

(a) (b) (c) (d) (e)

16. 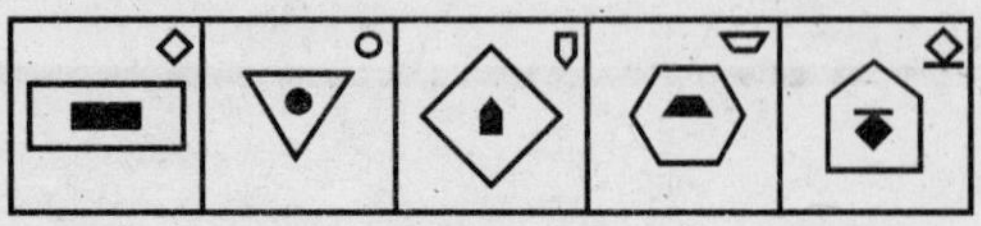

(a) (b) (c) (d) (e)

17. 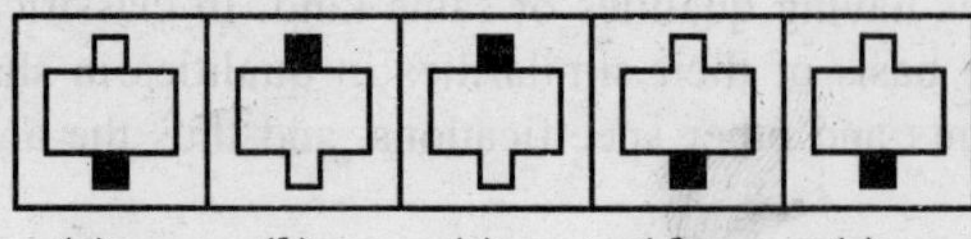

(a) (b) (c) (d) (e)

18. 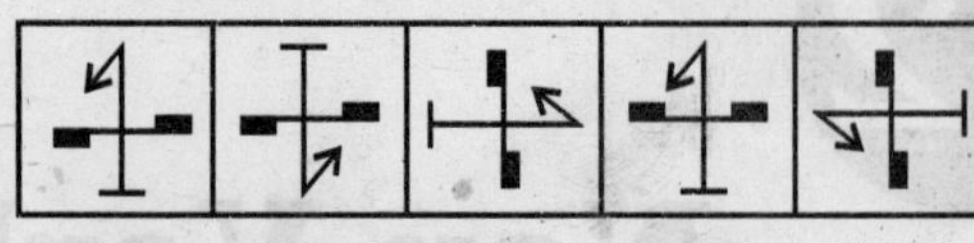

(a) (b) (c) (d) (e)

19. 

(a) (b) (c) (d) (e)

20. 

(a) (b) (c) (d) (e)

EXPLANATORY ANSWERS

1. (*d*) : Only in this figure all the lines making the hands and legs of the human shape are drawn downwards.

2. (*c*) : In all other figures only one set of elements is identical.

3. (*b*) : Only in this figure four arrows are pointing in same direction, and one arrow in its opposite direction. In other figures three arrows are in same direction and two in the opposite direction.

4. (*e*) : All other figures have three identical elements.

5. (*d*) : In all other figures two lines are parallelly drawn.

6. (*c*) : In all other figures the circle is on the left side of the arrow.

7. (*c*) : In all other figures one line has two elements on same side and the other line has two ele-ments, one on either side. In this figure both the lines have two elements on same side.

8. (*d*) : Only in this figure one of the three sections in between the shaded sections is not marked.

9. (*e*) : In all other figures the elements inside as well as outside the hexagon are placed at equal distance. In this figure the outside element at the top is wrongly placed (It should be near the other corner of the same line).

10. (*d*) : In all other figures two elements are in same direction and the other two elements are opposite to each other.

11. (*e*) : Only in this figure the two types of elements are equal in number.

12. (*c*) : In all other figures the diagonal line in all other elements is drawn from the right end.

13. (*a*) : In all other figures one of the four elements has one open side.

14. (*c*) : In all other figures the arrow is to the left of the dot.

15. (*b*) : In all other figures two of the three curves on one of the lines are in the same direction and are adjacent.

16. (*a*) : The shaded enclosed design in all other figures is same as the small design at the top.

17. (*a*) : Only in this figure the part opposite to the shaded part is cut off from the main design.

18. (*d*) : Only in this figure the shaded elements are on same side of the line.

19. (*b*) : Only in this figure the shaded and lined sections of the design are adjacent to each other.

20. (*e*) : Only in this figure there are two sets of two identical elements.